ANNUAL REVIEW OF BIOCHEMISTRY

ANNUAL REVIEW
OF BIOCHEMISTRY

VOLUME 73, 2004

CHARLES C. RICHARDSON, *Editor*
Harvard Medical School

ROGER D. KORNBERG, *Associate Editor*
Stanford University School of Medicine

CHRISTIAN R.H. RAETZ, *Associate Editor*
Duke University Medical Center

JEREMY W. THORNER, *Associate Editor*
University of California, Berkeley

www.annualreviews.org   science@annualreviews.org   650-493-4400

ANNUAL REVIEWS
4139 El Camino Way • P.O. Box 10139 • Palo Alto, California 94303-0139

ANNUAL REVIEWS
Palo Alto, California, USA

International Standard Serial Number: 0066-4154
International Standard Book Number: 0-8243-0873-5
Library of Congress Catalog Card Number: 32-25093

TYPESET BY CADMUS PROFESSIONAL COMMUNICATIONS, LINTHICUM, MD
PRINTED AND BOUND BY QUEBECOR WORLD PRINTING, KINGSPORT, TN

Annual Review of Biochemistry
Volume 73, 2004

CONTENTS

ERRATA
    An online log of corrections to *Annual Review of Biochemistry* chapters
    may be found at http://biochem.annualreviews.org/errata.shtml

RELATED ARTICLES

The Bacterial RecA Protein as a Motor Protein, Michael M. Cox

DNA Mismatch Repair: Molecular Mechanisms and Biological Function,
Mark J. Schofield and Peggy Hsieh

From the *Annual Review of Nutrition,* Volume 24 (2004)

Extracellular Thiols and Thiol/Disulfide Redox in Metabolism,
Siobhan E. Moriarty-Craige and Dean P. Jones

Sulfur Amino Acid Metabolism: Pathways for Production and Removal of Homocysteine and Cysteine, Martha H. Stipanuk

Mammalian Zinc Transporters, Juan P. Liuzzi and Robert J. Cousins

Retinoic Acid Receptors and Cancers, Dianne Robert Soprano, Pu Qin, and Kenneth J. Soprano

From the *Annual Review of Pharmacology and Toxicology,*
Volume 44 (2004)

Biochemical Mechanism of Nitroglycerin Action and Tolerance: Is This Old Mystery Solved? Ho-Leung Fung

Protein Sulfenic Acids in Redox Signaling, Leslie B. Poole,
P. Andrew Karplus, and Al Claiborne

Membrane Trafficking of G Protein–Coupled Receptors,
Christopher M. Tan, Ashley E. Brady, Hilary Highfield Nickols,
Qin Wang, and Lee E. Limbird

From the *Annual Review of Plant Biology,* Volume 55 (2004)

Biosynthesis and Accumulation of Sterols, Pierre Benveniste

The Ubiquitin 26S Proteasome Proteolytic Pathway, Jan Smalle and
Richard D. Vierstra

DNA Methylation and Epigenetics, Judith Bender

Alexander Rich

Annu. Rev. Biochem. 2004. 73:1–37
doi: 10.1146/annurev.biochem.73.011303.073945
Copyright © 2004 by Annual Reviews. All rights reserved
First published online as a Review in Advance on February 10, 2004

THE EXCITEMENT OF DISCOVERY

Alexander Rich

*Department of Biology, Massachusetts Institute of Technology, Cambridge,
Massachusetts 02139-4307*

■ **Abstract**   I had the good luck to start research at the dawn of molecular biology
when it was possible to ask fundamental questions about the nature of the nucleic
acids and how information is transferred in living systems. The search for answers led
me into many different areas, often with the question of how molecular structure
leads to biological function. Early work in this period provided some of the roots
supporting the current explosive developments in life sciences. Here I give a brief
account of my development, describe some contributions, and provide a hint of the
exhilaration in discovering new things. Most of all, I had the good fortune to have
inspiring teachers, stimulating colleagues, and excellent students.

CONTENTS

EARLY FAMILY LIFE

In the years just before and immediately after World War I, my parents immigrated separately as young adults to the United States from Russia. My father arrived in 1913 and lived first in New York and Connecticut. The chaos surrounding immigration and the necessity to support himself and his family made it impossible to go to high school. Instead, he began to learn the dry cleaning trade. His parents had a farm in the Berkshire Mountains in western Massachusetts, and it was there that he met my mother who had immigrated in 1920 and lived with her family in Hartford, Connecticut. They married and settled in Hartford. A first son David was born in 1922, and I followed in 1924. In early 1925, the family moved to Springfield, Massachusetts, where my father had set up a dry cleaning business. Most of my childhood was spent in Springfield, which was at the time a heavily industrialized city on the banks of the Connecticut River. It had a number of industries, including gun manufacturers, especially in the U.S. Armory, which had produced guns since 1794.

Even though neither of our parents had a high school education, they valued education enormously. My father had a difficult time with his business and once the Depression set in, he, like many others, struggled just to find employment. However, he admonished my brother and me to get an education because, as he stressed, you can never lose the education, even though you may lose jobs or your possessions. Although our financial position was precarious, my brother and I never felt deprivation; after all, everyone else we knew was in a similar situation.

David was precocious, and even though we were only two years apart in age, during our childhood the psychological distance was much greater. He read widely, seemed to understand everything and usually interacted with older people. All of this, I suspect, provided a spur for me to catch up. The Springfield public schools that we attended provided a sound education. The school system was populated by a large number of very capable female teachers—in that era women did not have access to many other professions. I attended a technical high school that provided a good background in mathematics but largely taught its students how to operate machine lathes and other industrial equipment. It was taken for granted that we would work to supplement the family income whenever we could.

As a young boy, I seemed to have a great thirst for information and knowledge in many different fields. I was especially intrigued by the wonders of Nature and can clearly recall meeting regularly with two or three neighborhood boys at about age nine. I described to them what I knew about the organization of the Universe and pondered about what really was beyond the stars that we could see. Such questions left me with a strong sense of wonder and awe. In high school, I was befriended by an English teacher who was very interested in science. He sponsored an after-school group to discuss relativity theory. Certainly, we did not go very far into the theory, but this reinforced my enthusiasm for learning new things. In the summer of 1941, after my junior year at high school, I applied for and obtained a position working in the U.S. Armory in Springfield. My job was that of a barrel rifler. Thus, I ran the machines that put the helical grooves in the inside of rifle barrels. This was a highly skilled operation and required a knowledge of many machine manipulations. I worked the shift from 11:00 P.M. to 7:00 A.M. and kept the job during my entire senior year at high school. At age 17, it did not seem too difficult to keep to this schedule, and I found that I could do some of my homework while running the machines, especially memorizing Milton's poems.

The high school English teacher urged me to apply to college, and I was awarded a tuition scholarship to Harvard College. In August 1942, I resigned from my position as a Master Rifler in the U.S. Armory and moved 90 miles east to Cambridge, Massachusetts, to begin a college education.

HARVARD YEARS

Entering college in the fall of 1942 had a heightened sense of tension and excitement. Most of us knew we would be going off to the military soon. My academic interests were not fixed, but on an intuitive level I knew that I wanted to study chemistry, physics, and mathematics, as well as philosophy. However, the real fascination came from learning about atoms and molecules. After some uncertainty, I finally decided to major in biochemical sciences. This allowed me to take a large number of physical chemistry courses.

After a short time I enlisted in the Navy V12 Program, which was designed to train Naval officers. The program kept us at Harvard but had interesting features, such as a 6:00 A.M. wake-up time, followed by a three- to four-mile run along the Charles River before breakfast. Many of the instructors complained that the Navy students seemed to fall asleep in their 9:00 A.M. classes. Because there was a shortage of graduate teaching assistants, I volunteered to grade papers in physics courses, and that brought me in contact with several outstanding professors. By June 1944, I had completed the accelerated Navy program and was then sent to work in a hospital at a submarine base for six months. After that, I was sent to Syracuse Medical School. Upon being discharged from the Navy in January 1946, I returned to Harvard College to complete the requirements for my undergraduate degree and also to start a research program.

John T. Edsall, a prominent protein biochemist, was my tutor. Working with him was a great stimulus. Edsall had been studying amino acids by Raman spectroscopy and had demonstrated the zwitterion nature of amino acids. For my undergraduate thesis I extended these studies to a number of rarer amino acids, as well as dipeptides, which allowed us to identify the Raman bands associated with stretching vibrations of the peptide bond (1). In addition, I was able to carry out an infrared spectral analysis of urea, working with E. Bright Wilson in the Chemistry Department. This involved a point-by-point measurement using a homemade IR machine, slowly ratcheting through the spectrum. It took me 30 h of continuous operation to obtain the complete spectrum of urea. In two years, that machine was replaced by an automated system that shortened the time for a similar spectral analysis by almost three orders of magnitude.

At that point I faced a career dilemma. I had a well-developed interest in pursuing the physical chemistry of biological materials, but at the same time, medical school was interesting, giving me a detailed view of the biology of one organism. After much internal debate and discussions with John Edsall, I applied to complete my last two medical school years at Harvard Medical School. This decision had the added advantage that I could continue to work with John Edsall during that time since his office and laboratory were then in the medical school. It also gave me an excellent opportunity to learn how to fractionate and characterize proteins.

Many people visited John Edsall during this period, and one such visitor was J.D. Bernal from Cambridge, England. I was impressed by Bernal's intensity and the breadth of his knowledge. I came to know Bernal several years later when I was in Cambridge, England, where he was appropriately known as Sage.

During my senior year at medical school, I had discussions with John Edsall about my future. John Edsall had never taken an internship after medical school, but rather had gone on directly to do biophysical work in muscle proteins. Having decided to follow a similar path, I wrote to a number of scientists whom Edsall recommended, including Linus Pauling. I was pleased to receive an acceptance from Linus Pauling and was delighted when my application for a National Research Council Fellowship was accepted. This was before the NIH had a postdoctoral fellowship program, and the NSF was yet to be invented. Thus, I graduated in high spirits, spent a few months in Europe, and then showed up in Pasadena, California, on October 1, 1949.

CALTECH: 1949–1954

When I was ushered into Linus Pauling's office shortly after my arrival at Caltech, I was dressed in a formal suit and tie appropriate for an East Coast medical school graduate. Sitting behind a desk in a flamboyant Hawaiian shirt was Linus Pauling, who greeted me with a warm smile. It was clear that this was a different world from the one I had left in Boston. At the time of my arrival, Pauling had already published the important paper dealing with sickle cell anemia and its characterization as a molecular disease. This was a revolutionary description of disease, and it changed the way research was interpreted. Pauling's accomplishments at this time were already legendary. He had applied quantum mechanics to chemistry and written his monumental work, *The Nature of the Chemical Bond*, the first full treatment of the relationship between molecular structure and the properties of substances. The predicted quantum mechanical orbitals made possible a rational understanding of the way molecules were built up, both organic and inorganic. My initial interests were in pursuing chemical theory, and Pauling suggested that I do some valence-state calculations for carbon atoms. At the same time, he suggested that I use my medical background to look for other cases of blood diseases that might also be associated with molecular defects in hemoglobin. In surveying a number of patients at the Los Angeles Children's Hospital, I found a young patient who had Cooley's anemia, now known as thalassemia, whose blood contained almost 100% fetal hemoglobin (2). I developed a method for analyzing the blood; this was one of my first publications while at Caltech. However, the fetal hemoglobin was normal, so it was not useful for studying molecular defects.

The organization of the Pauling lab was quite different from that which I had seen before. Pauling had an office in one corner of the Crellin Laboratory and in the other corner was a large room housing ~15 postdoctoral fellows, each with a desk in the open room. This was a very stimulating environment where we had intense discussions. There were a large number of talented postdoctoral fellows at Caltech. During that period I became especially close to Jack Dunitz, David Davies, and Robert Shulman, all of whom became highly productive scientists. Every few days Pauling would walk into the large room, and we would have some animated discussions, usually related to thoughts that Pauling had about different subjects. He would often make suggestions about topics that were worth pursuing.

I volunteered to work as a teaching assistant in the general chemistry course that Pauling taught. In the laboratory for the course, I joined Norman Davidson, who had recently joined the faculty and was working on shock-wave kinetics. We became close friends and talked often about biological and chemical phenomena. It may be that these discussions helped him move in later years into biological studies. During the same period, I became very friendly with Max Delbrück, who was in the Biology Department in an adjoining building. Jim Watson appeared briefly in 1949 and then left to go to Copenhagen to work with Herman Kalckar. One of the stimulating features of being a postdoctoral fellow at Caltech was the complete access that it gave to members of the faculty. Caltech was a small school at the time, and there was a great deal of interaction between young and old members of the scientific community. In the Chemistry Department, I became close to Verner Schomaker, who had an extensive knowledge of chemistry, and Jerry Vinograd, who arrived later. Among the many talented graduate students were Martin Karplus, Gary Felsenfeld, Matt Meselson, and Hardin McConnell.

One evening I was visiting Pauling at his home, where he was staying because he had a cold. I was there to discuss research, and he showed me a large book that had just arrived containing the results of a Faraday Society meeting in theoretical chemistry. He tossed the book down and said, "This is like whipping a dead horse." He explained that in the 1930s he had tried to obtain closed solutions for the equations describing the orbitals in molecules leading to the arrangement of bonding sites. However, he was unable to get exact solutions, and since then, people were using approximations piled upon other approximations, a process that Pauling thought was not very productive. This was, of course, before computers were developed, and Pauling's assessment of the futility was appropriate at that time. That evening had a great impact on me; I thought, "Here is an area where Linus Pauling was unable to solve the problems. Why am I fussing around doing valence-state calculations by approximations that are unlikely to be much better than those of many other people." After reflecting on this for some time, I decided to switch my research into X-ray structural analysis. Stimulated by Pauling's statement that knowing the structure of molecules led to their properties and function, I started by working with Dick Marsh, a crystallographer who was analyzing the structure of silk fibroins. At that same time, I started a crystal X-ray analysis with Jack Dunitz and Leslie Orgel, a theoretical chemistry postdoctoral fellow recently arrived from Chicago. We solved the three-dimensional structure of ferrocene (3), which had already been solved in projection. Today, this seems like a simple molecule, but at that time in the absence of computers or diffractometers, it required a great deal of work to obtain the data, measure the intensities visually, then carry out calculations with the aid of primitive punched-card systems that did not help very much.

During this period, I was taking a large number of courses and was very impressed by a thermodynamics course taught by Jack Kirkwood. Working with Kirkwood's graduate student Irwin Oppenheim, and Martin Karplus, we decided to take very good notes and then write Kirkwood's lectures in book form. This turned out to be a great method for learning thermodynamics, and the thick book that we produced was distributed in mimeograph form to many people. Several years later, Oppenheim expanded that book with Kirkwood to bring out his thermodynamics text.

I usually worked late into the evening in the laboratory and met Richard Feynman, who had recently arrived at Caltech. He was unmarried then, and it was also his habit to work late in the evening. We frequently got together and went to a local bar at the end of the evening where we would sit and talk about many things. He had a fund of stories that I listened to with great eagerness, many of which were eventually taped and produced as a collection. In 1952, I married my wife Jane, and we moved into a house in Pasadena near Caltech. Her interests were in childhood development. However, she did participate somewhat in my work by measuring the visual intensities of a large number of reflections in our three-dimensional X-ray analysis of the structure of ferrocene.

Quite often I would visit Pauling in his office. Invariably the secretary waved me in, and he was willing to discuss questions that I posed. On one occasion, I came into his office, and there he sat with his large feet resting on the desk. His chair was tipped back, and he was speaking into a dictating machine. I asked what he was doing. He said, "I'm writing a book on general chemistry." "Are you able to just dictate it?" I asked. "Oh, yes," he said, "I dictate the book and then get a copy back from the secretary and correct her mistakes." I considered this quite remarkable. However, over the years I adopted this same mode of writing, organizing the material for a paper in my head and then dictating it.

In December 1952, Linus Pauling suggested that I try to obtain X-ray diffraction patterns of DNA. This was after he had already generated a three-stranded model for DNA with bases on the outside. Shortly after publication, he realized it was incorrect. Pauling had been denied a passport to attend a meeting in London at which time Maurice Wilkins had displayed some diffraction photographs of DNA. Pauling wrote to Wilkins to ask if he could see these diffraction patterns. Wilkins declined because, he explained, they were still under analysis by his colleagues and himself. Thus, Pauling asked me to try to collect diffraction data.

The X-ray facilities at Caltech were not set up for fiber diffraction; I had a far less than optimal camera with a fairly weak, conventional X-ray tube. Thus, exposures took over 24 h, and it took some time to obtain oriented photographs. I began to get well-oriented photographs a few months later. Shortly after that, however, Pauling informed me that his son Peter had written to him that two "English" fellows at the Cavendish Laboratory had made a model of DNA with two strands in which the bases were on the inside. A short time later, their April publication in *Nature* was available, and it was clear that their structure both fit the data and solved an important biological problem in that it implied an obvious method for DNA replication.

In September 1953, Pauling organized a conference on the structure of proteins that included nucleic acids as well. A British group from the Cavendish Laboratory was there, including Bragg, Kendrew, Perutz, and Crick. By then, Jim Watson had returned to do postdoctoral work with Max Delbrück. Other attendees included Maurice Wilkins and John Edsall. Max Perutz presented his first evidence showing that a heavy atom could be introduced into the hemoglobin crystal, making it possible to solve the phase problem. Watson and Crick presented their work on DNA, and Wilkins showed the results of his X-ray studies. I gave a brief paper on how DNA and the alpha helix might interact, but the ideas were not notable. In particular, I failed to recognize that it would be possible for an alpha helix to enter into the major groove of the DNA molecule. During that period, my interest had changed to focus on RNA structure, to find whether RNA could form a double helix. Jim Watson, who was not fully comfortable in confining himself to work in the Delbrück lab, was interested in the same question. This made it natural to join forces and work together on this problem.

I became acquainted with and charmed by Francis Crick, who had such a bubbling enthusiasm. Crick invited me to come to Cambridge, England to use their more intense rotating anode machine, if I ever got results with the RNA project that would benefit by using this advanced facility. This turned out to be a significant invitation that I acted upon two years in the future.

One possibility for separating the DNA chains for replication was that the molecule might also turn left-handed. Working with wooden ball-and-stick models that the Pauling Laboratory had in abundance, I tried to build a left-handed version of the right-handed B-DNA model of Watson and Crick. I could show that left-handed B-DNA would not be stable; however, I had occasion to revisit the problem some 25 years later. At one point, Pauling poked his head in the door of the model room and said, "Alex, work hard on this problem. I like most important discoveries to be made in Pasadena." This comment was typical of Pauling, who had a great competitive sense, especially regarding the Bragg Laboratory in Cambridge, England. Because he had already discovered the alpha helix and the beta sheet as well as the sickle cell discovery of molecular disease, I could see why he would make such a statement, even though he had clearly missed out on the double helix.

Jim Watson and I collected samples of RNA from many people. By then, I had developed reasonable skill in making oriented fibers, formed by slowly withdrawing a glass rod from a sticky nucleic acid droplet using a microscope for the slow movement. The X-ray diffraction patterns were not encouraging. They all looked similar, somewhat suggestive of a double helix but with a different distribution of intensity. Leslie Orgel was interested in the work on RNA and, together with Jim Watson, the three of us discussed these problems extensively. It was thought that RNA was important in directing the synthesis of proteins. Just how that occurred was mysterious and unknown. George Gamow, a distinguished theoretical physicist and a friend of Max Delbrück, arrived at Caltech, very much taken by the problem of how proteins are made under the direction of the nucleic acids. He proposed the first code for determining the sequence of amino acids from the DNA double helix (4). His code was incorrect, but it stimulated a great deal of work in an effort to develop the actual code used in nature. Later, we wrote a review considering various codes (5). Gamow was a lively personality and a great practical joker. He had the idea of setting up a club of people interested in the coding problem, and he decided that the group would be defined by a tie. It became known as the RNA Tie Club and consisted of some 20 members, one for each amino acid. Many of the members were friends of Gamow, but it also included many of the principal workers in the then-small field of nucleic acid research as well as a number of physicists such as Dick Feynman, Edward Teller, and others. The main purpose of the club was to circulate manuscripts dealing with the code. It was in one of these RNA Tie Club manuscripts that Francis Crick first developed the adapter hypothesis suggesting that a small RNA molecule might act as an intermediate between the amino acids and the nucleic acid molecule that codes for protein. Later, it was realized that these adapters had been discovered by Mahlon Hoagland and Paul Zamecnik (6) and eventually opened the door to the role of transfer RNA in protein synthesis.

Our thoughts in 1953 and 1954 were much more primitive. We wondered whether it was possible for the RNA molecule to have a form that would bind specific amino acids in some kind of a linear array. Such models seem quite naïve today, but that is where thinking about this problem first began. In the work on RNA fibers, Jim Watson and I used RNA preparations in which the base ratios varied widely. Some were nearly complementary, as found in DNA, and others were far removed. Nonetheless, the patterns remained about the same. This continued to be a puzzle. Of course, at the time, we did not understand the variety of different types of RNA in the preparations. And there was some question about whether RNA was a linear polymer, as was the case for DNA. In the literature, there was discussion of the fact that chains could also be attached to the $2'$ hydroxyl group of the ribose, thereby giving rise to a complicated net, instead of a linear polymer. Watson and I published two papers describing these results, one in *Nature* (7) and another sent to *PNAS* (8) by Linus Pauling. We included a diffraction pattern but had no clear interpretation of it.

The National Research Council Fellowship was awarded for a three-year period. At the end of that time in 1952, I enlisted in the U.S. Public Health Service so that I could continue my research. Having finished my medical training after World War II, I was subjected to the doctor draft for the Korean War. However, working as a "yellow beret" at NIH offered an alternative. I was hired by Seymour Kety, who was then organizing the research groups for the newly formed National Institute of Mental Health. He appointed me Chief of the Section on Physical Chemistry, but because the large research Building 10 was being built in Bethesda, I remained at Caltech for another two years. As a Section Chief, I was required to hire coworkers. Jack Dunitz was interested in continuing work in the general biological area, and he agreed to join me at NIH. By June 1954, the building had been completed. Jane and I packed our belongings and started out on the long trip from Pasadena to Bethesda, Maryland. As I reflected on my five years of work at Caltech, I concluded that I learned quite a bit but actually accomplished very little. The publications I had were what I considered routine, and I had serious doubts about whether I could, in fact, make important discoveries.

SCIENCE AT NIH

The contrast between Caltech and NIH was very great. Whereas Caltech was small in size and had people from all branches of science, NIH was extremely large and was staffed predominantly by experts in biomedical research. As I moved into the new laboratory in Building 10, I was faced with the job of assembling the right equipment to carry out research. In a short time Jack Dunitz arrived from Caltech. By then, I had written a letter to David Davies, who had left Caltech to return to England,where he was working in a chemical company. I urged him to come and join the research at NIH, since I thought the phosphates that we would be looking at (the nucleic acids) would be more interesting than the inorganic chemicals that he would be working with. A letter was also sent to former Pauling graduate student Gary Felsenfeld, urging him to join the lab.

I determined to continue working on the question of RNA structure, and the question of whether it could form a double helix. With somewhat better X-ray cameras than those available at Caltech, I continued to study the diffraction patterns from oriented RNA preparations. However, they did not provide insight. This research effort took an abrupt change early in 1955 when Severo Ochoa came to give a seminar at NIH, describing the enzyme polynucleotide phosphorylase that had just been isolated by Marianne Grunberg-Manago in his laboratory (9). It converted nucleoside diphosphates into long RNA chains. He agreed to make some of these polymers available to me, and within a short time with some help from Leon Heppel at NIH, we were able to make our own enzyme and our own polymers. I found that the random copolymer containing adenine and uracil residues produced a diffraction pattern very similar to that found in naturally occurring RNA. These synthetic chains were linear and that made me feel that the likelihood of a significant amount of branched RNA molecules was not very great. The preparation of polyadenylic acid began to produce an interesting pattern highly suggestive of a helical complex. After working on that for some time, it became apparent that more data could be obtained with a more powerful X-ray source and better cameras. Accordingly, I wrote to Francis Crick, and he invited me to Cambridge, England, where I could use the improved equipment that had been developed at the Medical Research Council unit in the Cavendish Laboratory.

CAMBRIDGE, ENGLAND, AND THE STRUCTURE OF COLLAGEN

In July 1955, I flew to Cambridge, England, for a short visit. At roughly the same time, Jim Watson arrived in Cambridge, as well as Leslie Orgel. We were thus able to resume intensive discussions on RNA structure that we had begun at Caltech, with the significant addition of Francis Crick. Improved diffraction patterns of polyadenylic acid were obtained by myself and by Jim Watson. Eventually, we solved its structure working with Francis Crick and David Davies (10). However, the direction of my research would soon change rather abruptly.

Francis Crick had generously offered to have me stay at his house in Portugal Place, together with his wife Odile and their two young daughters. One Saturday morning the new issue of *Nature* arrived, and Francis looked through it. Over breakfast I asked him if there was anything interesting in the journal. He replied that there was an article by Bamford and colleagues at the Courtauld's Laboratory, describing a new form of polyglycine called polyglycine II. They did not know the structure but presented IR spectra and a powder X-ray diagram. As we continued eating breakfast, the idea developed that perhaps we could solve the structure by working with skeletal brass molecular models that were available in the laboratory. The usual form of polyglycine has an extended polypeptide chain with each residue related to the previous one by a translation and a rotation of 180°. Once we were in the laboratory, we realized that, if we changed the rotation from 180° to 120°, we could build a trigonal lattice in which each polyglycine chain formed hydrogen bonds with six other chains around it. The lattice so constructed clearly predicted the position and intensities (11) observed in the published powder pattern. We realized that we had solved its structure by early afternoon of the same day. This was a striking example of the power of using a molecular model to interpret the structure of regular macromolecules.

After we had written the paper and sent it in, Francis and I were having afternoon tea outside the laboratory. I mentioned to him that the C-H bond of the glycine residues where longer side chains would be attached for other amino acids was in the same plane as the N-H bond in the adjoining peptide linkage. To me this meant it would be easy to form a pyrollidine ring. We wondered if we could take three chains from the polyglycine lattice and make a repeating sequence of glycine-proline-proline as a model for the unsolved structure of collagen. It was known at the time that collagen had a large amount of glycine and dipeptide sequences of Gly-Pro and Hydroxypro-Gly. Thus, a repeat of this type seemed reasonable. We returned to the laboratory and found that by building such a model, then twisting it so that we had a coiled coil, we could make a model that predicted some of the known reflections of the highly characteristic collagen fiber diffraction pattern. In our haste we almost overlooked the fact that, because of the symmetry, there were two such models that could be generated from the same lattice. After building six-foot-high models, we placed them at one end of a long basement corridor in the Cavendish with a powerful point light source at the other end. On a sheet of paper we plotted the projection of the model and its approximate atomic coordinates. A reducing pantograph was used to drill small holes at atomic centers in a metallic strip about 12 cm in length. With this "mask," we made a trip to another laboratory that had a large optical diffractometer. By looking at the diffraction patterns, we could see that one of the two similar models seemed a better fit to the collagen diffraction pattern. We published a short note in *Nature* (12) describing the structure and later a longer, detailed article comparing the diffraction pattern of this model with the observed collagen pattern (13).

The effect of this work was that my stay in the Crick household, originally planned to be a few weeks, was extended to somewhat over six months, during which time my wife came and joined us. I felt bad about imposing on the Cricks for such a long period, but my wife and I had the opportunity to reciprocate several years later when the Cricks spent a term staying with us in our house in Cambridge, Massachusetts.

Although the collagen work had a different direction from the focus on the nucleic acids, it had a strong positive effect on my psyche. For one thing, I began to develop some self-assurance in my ability to carry out research and make discoveries. I believe that a form of "scientific maturation" is an important component in developing a confident thrust into research work. Some time later, Linus Pauling was invited to attend a meeting held in England on collagen, and I presented the results of our work in this area. At an intermission, a number of the English scientists asked if it was true that I had trained to do research in the Pauling laboratory. Pauling replied that I had spent a long time in the laboratory without accomplishing very much, but I must have learned quite a bit. He smiled at me as he made this comment. I felt that it was probably a fair assessment.

RNA STRUCTURE AND THE DISCOVERY OF HYBRIDIZATION

On returning to NIH from England, work on RNA structure continued with David Davies. In 1956, we mixed solutions of polyriboadenylic acid (poly rA) and polyribouridylic acid (poly rU) and discovered a remarkable transformation revealed by fiber X-ray diffraction. These two molecules actually reacted with each other to produce a double helix! A brief note was sent to the *Journal of the American Chemical Society* in June of that year (14), describing the work with a preliminary interpretation of the diffraction pattern. At the same time, Bob Warner published a short note, describing hypochromism when the two polymers are mixed (15). Our X-ray patterns clearly indicated the formation of a helical complex that was not present in either of the two individual polymers. Furthermore, the pattern had significant differences compared with those produced from DNA.

This result, which seems so obvious today, generated a great deal of skepticism at the time. While walking down a long corridor at NIH, I met Herman Kalckar, an eminent Danish biochemist. I mentioned that we discovered that poly rA and poly rU formed a double helix. Kalckar was incredulous. "You mean without an enzyme?" he asked. His attitude was justified, since the only double helix known at that time was one made with the DNA polymerase enzyme that Arthur Kornberg had purified (16). Other critics thought it was highly unlikely that polymers containing over 1000 nucleotides would be able to disentangle themselves and form a regular structure. They believed it would be hopelessly entangled. Somewhat later a distinguished polymer chemist told me that my interpretation was wrong since both polymer chains were highly charged and that they would never combine. These were examples of the thinking at that time.

Two weeks after sending off the 1956 *JACS* note, I wrote a letter to Linus Pauling, describing these results. The letter reveals a sense of incredulity on my part this reaction could happen and that it was "completely reproducible." This was the first demonstration that RNA molecules could form a double helix. It was also the first hybridization reaction and, as we pointed out in the *JACS* letter, "this method of forming a two-stranded helical molecule by simply mixing two substances can be used for a variety of studies" (14).

An important method for studying the nucleic acids was measuring their absorbance in the ultraviolet. For some time, it had been known that polymerization of nucleotides resulted in a decrease in absorbance at 260 nm. Bob Warner had described the hypochromism on mixing (15). Although the mechanism of hypochromicity was not understood at the time, it was a useful tool for analysis. Insight into the reaction of poly rA and poly rU was obtained by Gary Felsenfeld, who carefully measured hypochromicity in mixtures of varying composition and showed that it fell to a very sharp minimum at a 1:1 mol ratio (17). We also discovered that addition of divalent cations such as magnesium would change the picture dramatically (18), leading to the formation of a three-stranded molecule. We concluded that a second strand of polyuridylic acid bound in the major groove of the poly rA–poly rU duplex. Addition of this strand did not increase the diameter of the molecule and neatly accounted for the 50% increase in sedimentation constant. It was proposed that the uracil in the second poly rU strand bound N3 and O4 to the N7 and N6 of adenine. This proposal was considerably strengthened two years later by the X-ray analysis of the co-crystal containing 9-methyl adenine and 1-methyl thymine by Karst Hoogsteen (19) which had the same hydrogen bonding. The 1957 discovery of the RNA triplex pointed out the structural complexity inherent in RNA molecules. Over the next several years, a variety of polynucleotide interactions were studied (20–23), leading to the formation of other two- and three-stranded molecules.

MOLECULAR BIOLOGY AT MIT

In the 1950s when biomedical research was expanding rapidly, the NIH was regarded as a prime recruiting ground for universities. After receiving several offers for faculty positions in university biology or chemistry departments, I decided to move so that I could work with students. I was particularly attracted by an offer from the then rather small Biology Department at MIT because it could attract excellent students, and several people at MIT, such as Jerry Wiesner and Victor Weisskopf, felt strongly that molecular biology would grow at the Institute. Accordingly, in 1958, I left NIH and moved to the Biology Department at MIT. Once again, I found myself moving into empty rooms, but the generous program of NIH grant support soon allowed me to equip the rooms and resume research.

Shortly after I arrived, I was asked if I would be willing to have Antioch Work Study students come to work in my laboratory for a three- or six-month period. I readily accepted, as these students were known to be highly motivated and well qualified. This program continued at MIT for several years during which time I had as undergraduates people such as Joan Steitz and Mario Capecchi, among others, who made valuable contributions to the laboratory's research program. At the same time a number of talented people came to the MIT laboratories for postdoctoral work. David Blow, originally from Cambridge, England, began working with me at NIH and moved to MIT where he continued working for another year. Among others from England were researchers such as Tony North and David Green, both of whom were trained crystallographers. Ken-ichi Tomita from Japan, Paul Knopf, and Hank Sobell were also among the early postdoctoral fellows.

INFORMATION TRANSFER AND DNA-RNA HYBRIDIZATION

In the late 1950s, a key question was, How does DNA "make" RNA? Several investigators at that time had shown that crude preparations of an RNA polymerase activity would incorporate ribonucleotides into RNA using a DNA template, but the mechanism was not at all clear. It was widely believed that information transfer went from DNA to RNA, but how did that happen? When I was at NIH, I met Gobind Khorana, who was then visiting the laboratories of Arthur Kornberg and Leon Heppel. We became friends, and when I later learned that he had been able to synthesize oligomers of poly deoxythymidylic acid (poly dT) (24), I asked if he would be willing to make a sample available. I told him I wanted to study the possibility that this could make a double helix with an RNA molecule, even though it was known that the RNA backbone was significantly different from the DNA backbone due to the 2' hydroxyl group in RNA. At the time it was not obvious that they could combine.

Nonetheless, in 1960 I showed that these two molecules could accommodate each other to form a hybrid helix containing one strand of poly dT and one strand of poly rA (25), as seen from hypochromism and other studies. This was the first experimental demonstration of a hybrid helix, and the discovery of messenger RNA was still one year in the future. It immediately provided experimental support for a model of how DNA could "make" RNA, using complementary base-pairing, as in DNA replication. A year later in 1961, experiments by J. Hurwitz with a purified RNA polymerase preparation demonstrated that this was the mechanism underlying information transfer from DNA to RNA (26). The reaction between poly dT and poly rA was the first experimental demonstration that the two different backbones could adapt to each other in this method of information transfer. The reaction was also the first hybridization of a DNA molecule with an RNA molecule. The same hybridization reaction is widely used today in the purification of eukaryotic mRNA by hybridizing poly dT to their poly rA tails.

Later in 1960, Marmur, Doty, and their colleagues demonstrated that it was possible to renature naturally occurring denatured DNA duplexes by incubating them at an intermediate temperature that would allow the single strands to anneal together with the correct sequence (27, 28). A year later, this annealing method was also adopted to form DNA–RNA hybrids in viral systems (29). Together with my graduate student Howard Goodman, we established, using hybridization experiments, that tRNA was encoded by DNA (30). Our knowledge of RNA at the time was so primitive that we had to rule out RNA replication for tRNA synthesis.

EARLY THOUGHTS ON THE ORIGIN OF LIFE

In 1961, I was asked to contribute an article to a volume dedicated to Albert Szent-Gyorgyi, and I decided to write about the evolution and origin of life. This was prompted by the current ideas largely espoused by the Russian biochemist Oparin, who believed that life began through the creation of primitive protein molecules that provided the necessary environment for developing cells and cell replication. However, the greatest weakness in a theory of this type was it did not really explain the evolution of nucleic acid–mediated protein synthesis. It seemed more reasonable to suggest that life began with nucleic acids.

I suggested that primitive polynucleotide chains could act as follows: "We postulate that the primitive polynucleotide chains are able to act as a template or a somewhat inefficient catalyst for promoting the polymerization of the complementary nucleotide residues to build up an initial two-stranded molecule. . . . It may be reasonable to speculate that the hypothetical stem or parent polynucleotide molecule was initially an RNA-like polymer. . . " (31). It was apparent by that time that RNA molecules could contain genetic information, as in the RNA viruses. They also were known to play a central role in the synthesis of proteins. DNA was thus regarded as a specialized derivative molecule that evolved later in a form that contained specific information and carried out the molecular replication inherent in propagating living systems.

At the time it was not known how mRNA was made in vivo. Thus, I suggested, "mRNA may be made in vivo as complementary copies of one or both strands of DNA. If both strands are active, then the DNA would produce two RNA strands that are complementary to each other. Only one of these might be active in protein synthesis, and the other strand might be a component of the control or regulatory system" (31). This is perhaps the first statement of what we now know as "antisense" RNA. It also suggested that RNA molecules might have a control or regulatory function, which now appears to be the case with the newly emerging information about the role of micro-RNAs or RNAi.

This article, published in 1962, is probably the oldest statement regarding the role of RNA in the origin of life. Only after many years did the concept of "The RNA World" develop together with the enormous growth and recognition of its ribozyme activities. It is interesting that in the early days of the development of molecular biology, very large features of biological systems were unknown. It was possible to speculate then and anticipate features that could be investigated only several years in the future.

DNA IS FOUND IN ORGANELLES

In 1962, a student, Mike Vaughn, came into my office, saying that electron micrographs of organelles showed string-like features that might be DNA. Together with a postdoctoral fellow, Ed Chun, we decided to explore the possibility, using the newly developed method of density gradient centrifugation. Chloroplasts were isolated from several organisms and all had a DNA species distinct from the nucleus. Another band appeared in some pelleted preparations that we suggested came from contaminating mitochondria (32). This approach was widely adopted, opening a vast field of research. This is an example of how new areas of biology opened up during that period.

SINGLE-CRYSTAL STRUCTURES

X-ray diffraction studies of nucleic acid fibers were carried out extensively by M. Wilkins, R. Franklin, and colleagues in the 1950s and 1960s, but it was realized that the limitations of such studies were enormous. In fiber X-ray diffraction, a rather small number of reflections are registered. However, the number of variables needed to define the structure (at least $3N$, where N is the number of atoms) is so great that it was clear that fiber diffraction could not "prove" a structure. It could only say that a particular conformation was compatible with the limited diffraction data from fibers. Starting in the early 1960s, there were many single-crystal X-ray diffraction studies involving co-crystals of purines and pyrimidine derivatives such as those initiated by Karst Hoogsteen (19). These studies were useful in obtaining information about components of nucleic acid structure. For example, together with Hank Sobell and Ken Tomita, we solved a co-crystal of cytosine and guanine derivatives showing that they were held together by three hydrogen bonds (33), not two, as initially suggested by Watson and Crick. Linus Pauling had already emphasized this point based on general structural considerations (34). Several co-crystals were solved of derivatives of adenine and uracil or adenine and thymine during this period in our MIT laboratory (35–37), as well as elsewhere (38). At the same time selective base-pairing could be demonstrated in solution. The center of the double helix is hydrophobic owing to base stacking. Together with Yoshimasa Kyogoku and Dick Lord, we studied the hydrogen bonding of purine and pyrimidine derivatives by their IR spectra in chloroform solution. It showed that adenine selectively paired with thymine or uracil derivatives, while guanine paired with cytosine derivatives (39).

But a disturbing trend emerged from the crystallographic studies. Only Hoogsteen base-pairing was found between adenine and thymine/uracil derivatives. This led some investigators to suggest that the double helix might be held together by Hoogsteen pairing. The calculated diffraction pattern of such a helix had many similarities to that predicted by a double helix held together by Watson-Crick base pairs, even though the fit was not good (40). Thus, the question remained: What is the real structure of the double helix?

The Double Helix at Atomic Resolution

The first single-crystal structures of a double helix were solved in 1973 by my students and postdoctoral fellows, including John Rosenberg and Ned Seeman. This was before it was possible to synthesize and obtain oligonucleotides in significant quantities suitable for crystallographic experiments. However, we succeeded in crystallizing two dinucleoside phosphates, the RNA oligomers GpC (41) and ApU (42). The significant point in this analysis was that the resolution of the diffraction pattern was 0.8 Å. Atomic resolution allowed us to visualize not only the sugar phosphate backbone in the form of a double helix, but also the positions of ions and water molecules. It could be shown that extending the structure using the symmetry of the two base pairs made it possible to generate RNA double helices that were quite similar to the structures that had been deduced from studies of double-helical fibers of RNA. The bond angles and distances from these structures provided the library of acceptable angles and distances and, in addition, gave rise to the nomenclature for identifying torsion angles in the sugar phosphate backbone.

The GpC structure had the anticipated base pairs connected by three hydrogen bonds. However, the ApU structure showed for the first time that Watson-Crick base pairs formed when the molecule was constrained in a double helix (Figure 1), as opposed to the Hoogsteen base pairs that were favored in the single-crystal complexes of adenine with uracil derivatives. I mailed preprints of these to several people, including Jim Watson. He phoned me, saying that after having read the ApU manuscript, he had his first good night's sleep in 20 years! This indicated to him that the uncertainty about the organization of the double helix was resolved. The significance of the double helix at atomic resolution was recognized by the editors of *Nature* who, in their "News and Views" commentary, called it the "missing link" and recognized that "the many pearls offered" helped resolve one of the big uncertainties in nucleic acid structure ["News and Views" (1973) *Nature*, 243:114].

These structures capped the effort that I had started some 20 years earlier, which started to make progress in 1956 with the recognition that poly rA and poly rU would form a double helix. Here, at last, was the demonstration at atomic resolution of the details of that structure. High-resolution crystallographic analysis of larger fragments of the double helix (DNA or RNA) did not emerge until almost a decade later with the availability of chemically synthesized and purified oligonucleotides in large enough quantities to permit single-crystal diffraction analysis.

A single-crystal X-ray structure of a hybrid helix did not appear until 1982 when, together with Andy Wang and colleagues, we solved the structure of a DNA–RNA hybrid linked to double-helical DNA (43). This was 22 years after the hybrid helix was first observed (25). It showed that the dilemma of two different backbone conformations was resolved by having the DNA strand adopt the RNA duplex conformation. This had been inferred from fiber diffraction

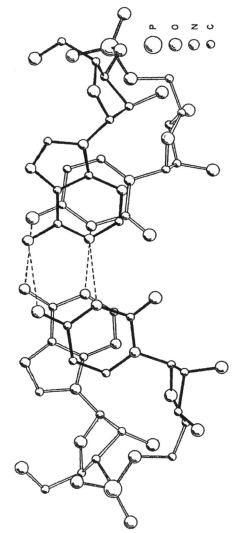

Figure 1 The 0.8 Å resolution double helical structure of the dinucleoside phosphate ApU, as displayed on the cover of *Nature* magazine (42).

studies and has remained a constant feature, reflecting the relative conformational flexibility of the DNA backbone compared with the less flexible RNA strand.

POLYSOME DISCOVERIES AND PROTEIN SYNTHESIS

Experiments in 1961 revealed the presence of a rapidly metabolizing fraction of RNA called mRNA. Evidence supporting the triplet code (44) suggested that very large mRNA molecules could be formed. While attending a Gordon Conference in 1961, I began to think about how ribosomes would be organized on mRNA. The ~150 amino acids in hemoglobin chains, for example, required 450 nucleotides or an mRNA over 1500 Å long if the bases were stacked. This was much larger than the ~200 Å-sized ribosomes in which protein synthesis occurred.

On returning to MIT, I talked with my first graduate student Jon Warner and suggested that he try to see if there were several ribosomes associated with individual mRNA strands. His first experiments using *Escherichia coli* were unsuccessful. He then joined Paul Knopf to work with rabbit reticulocytes, which have the advantages of containing no nucleus; they can be lysed by gentle methods and they synthesize mostly hemaglobin. Their experiments were enormously successful. Sedimenting the lysate on a sucrose density gradient revealed that polymerized radioactive amino acids were found in a rapidly sedimenting fraction but were not found with individual ribosomes. In a shallow sucrose gradient, a series of peaks of both optical density and radioactivity were found. By simple counting it was possible to separate fractions containing two, three, four, or five ribosomes. Although electron microscopy studies were not very common in those days, we brought these samples to Cecil Hall, an electron microscopist in the department, and the organization of what we called polysomes was clearly visible. A cluster of five ribosomes that synthesize hemaglobin was found acting on the same mRNA strand. A thin strand could be observed running between the ribosomes, both in shadowed electron microscope preparations and in preparations negatively stained with uranyl acetate (Figure 2). They could be readily dissociated by a brief exposure to ribonuclease.

Two papers were sent in for publication: one dealing with the electron microscopic characterization (45) and the other with biochemical experiments (46). *Science* put the electron microscopic field of polysomes synthesizing hemoglobin on the cover of the magazine. The article described our interpretation with ribosomes moving along the mRNA strand at the same time as the polypeptide chains were elongating. This was a radically different view of protein synthesis, and many *Science* readers objected vigorously to this paper. Editor Phil Abelson sent me a sheaf of these letters. Most of the comments required no answer since the second paper with the biochemistry came out the following month. However, some of the readers' comments afforded a certain

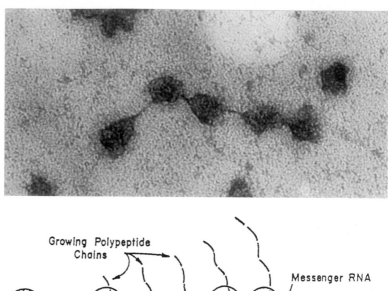

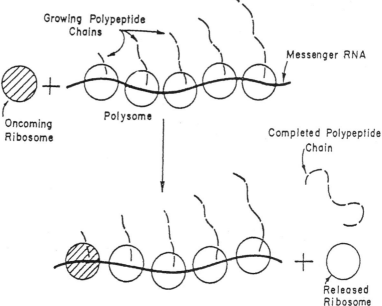

Figure 2 (*Top*) Negatively stained electron micrograph of rabbit reticulocyte polysomes synthesizing hemoglobin, showing the thin mRNA strand (43). (*Bottom*) A schematic model of polysome function is shown, as demonstrated experimentally (47).

level of amusement. For example, several stated that these were "artifacts," and that they had been seen in electron microscopic sections for years.

The model suggested that longer mRNA strands coding for larger polypeptide chains would have larger polysomes. At about that time Jacques Monod visited Salvador Luria at MIT and came to my laboratory where I showed him these recent results. He expressed great delight because it solved a problem that he had been pondering, namely, how could you have the coordinate synthesis of several different peptide chains in unison. It was apparent from the model that the mRNA could be polycistronic, and the ribosomes moving along the chain would synthesize one polypeptide chain after the other, insuring a stoichiometric synthesis.

The discovery of polysomes changed thinking in the field of protein biosynthesis in several ways. First, the electron microscopic pictures offered a graphic visualization of the assembly mechanism, implying that ribosomes move along the message and the polypeptide chain elongates. Although there had been limited previous discussion of ribosomal movement along the message, images of the polysomes clearly indicated a dynamic assembly structure, strongly reinforcing the movement of ribosomes. Further experiments were done with polysomes to investigate this mechanism. Experiments in vitro with labeled single ribosomes by my student Howard Goodman (47) showed that they could attach to the end of polysomes, while further incubation led to the release of individual ribosomes and completed polypeptide chains (Figure 2). Although movement of ribosomes over mRNA seems obvious now, this notion required a reordering of the thinking of many people in the field. There was some resistance to these ideas for the first year or so, but subsequent experiments in my laboratory and several others showed the universality of the mechanism.

An example of the extent to which the polysome discovery jolted the thinking of scientists in the field was graphically illustrated to me at a meeting on hemoglobin at Columbia University's Arden House in November 1962, shortly after submitting the two polysome papers. After I presented the polysome data, including my interpretation, Fritz Lipmann took strong exception. He told me that I had misinterpreted the data. I was surprised at the vigor of his criticism and a little puzzled at the same time. Several weeks later Lipmann came to Boston and visited me at MIT. He apologized and told me that he had been mistaken when he objected to my interpretation. I was quite impressed that such a distinguished scientist would go out of his way to tell a younger colleague that he had been mistaken.

The Mechanism of Protein Synthesis

The isolation of polysomes on a sucrose gradient meant that the ribosomes actually involved in protein synthesis could be separated from the large number of inactive monomer ribosomes. This made it possible to uncover further aspects of the translation machinery. For example, it was generally assumed that there was one polypeptide chain per ribosome. Isolating active ribosomes made it possible to demonstrate experimentally that there is one growing polypeptide chain per ribosome (48). Of far greater interest, however, was our ability to analyze the number of transfer RNA (tRNA) molecules per active ribosome, as this had a direct bearing on the mechanism of protein synthesis.

Rabbit reticulocyte cells have already lost their nucleus and no longer synthesize RNA. This made it possible for Jon Warner to do a rather clean analysis of the system by incubating the reticulocytes with radioactive adenine or cytosine (49). These molecules could penetrate into the interior, where they were incorporated into ribonucleoside triphosphates. Although RNA was not synthesized, the 5' CCA ends of tRNA molecules were continually cleaved off and then enzymatically readded to tRNA. The radioactive nucleotides thus labeled the ends of tRNA molecules and nothing else. This showed that two tRNA molecules were bound to active ribosomes, but only one was bound to inactive ribosomes at the top of the gradient. Several investigators had shown previously that ribosomes would bind one tRNA in an exchangeable fashion, and attempts were made to formulate a mechanism whereby a single tRNA would be involved in protein synthesis; however, the mechanism seemed implausible. With the presence of two tRNA molecules, it was possible to make a different interpretation. We postulated that the two tRNA-binding sites occupy adjacent codons. We called one Site A, which bound aminoacyl tRNA, and the other Site P, which bound peptidyl tRNA. We envisioned these two sites acting in a coordinated manner to transfer the growing polypeptide chain and to move the mRNA codon from Site A to Site P. We suggested that it constituted the basis of ribosomal movement relative to the messenger strand. This statement, formulated in 1964, rapidly became the standard description for interpreting ribosomal movement and tRNA activity in protein synthesis. The recent determination of the three-dimensional structure of the ribosome has confirmed the interpretation and provided a structural basis for understanding how this mechanism works.

The statement describing the movement of tRNA molecules in the ribosome postulated engagement of the transfer RNA molecule to the mRNA, and at the same time, it is the substrate for the ribosomal peptidyl transferase enzyme that transfers the peptide chain from the tRNA in the P site onto the amino acid of the aminoacyl tRNA in the A site. The structural and geometric background for understanding where these two events occur had to await determination of the three-dimensional folding of transfer RNA molecules.

In 1967, Leonard Malkin and I asked to what extent the growing polypeptide chain in rabbit reticulocyte polysomes was accessible to externally added proteolytic enzymes (50). These experiments demonstrated that a resistant fragment shielded by the ribosome was the most recently synthesized segment, and it was attached to the tRNA molecule. It was shown that the shielded segment contained 30–35 amino acids. An extended conformation at 3.6 Å/amino acid implied a protected segment of about 100 Å in length. The structure protecting the chain came to be referred to as the polypeptide tunnel. The recent three-dimensional structure determination of the 50S subunit at high resolution has revealed a tunnel of that approximate size (51). The large shielded section of polypeptide probably stabilized its attachment to the ribosome. For very large polypeptide chains, it was possible that the chains could start to fold in a native conformation while they were still attached to the ribosome. This had been observed at an earlier date as active β-galactosidase could be detected on bacterial polysomes synthesizing that protein (52).

Polypeptides and Polyesters

The first indication that the ribosomal peptidyl transferase could catalyze the formation of esters as well as peptides was obtained in experiments with analogs of puromycin in which the α-amino group was replaced by a hydroxyl group (53). This study revealed that many of the characteristics of ester formation were similar to those found in the peptide formation of the puromycin reaction itself. These experiments encouraged my student Steve Fahnestock to incorporate an ester linkage in the viral coat protein of the R17 bacteriophage by providing all of the amino acids needed in a segment of the viral coat protein, except for one provided by chemically modified phenylalanyl tRNA^{phe}. Treatment with HONO yielded an α-hydroxy phenyllactyl tRNA (54). The coat protein was synthesized with an ester linkage specifically introduced at one position in the chain, identified explicitly by hydrolyzing the synthesized chain in the presence of alkali and characterizing the fragments produced by the reaction.

This 1971 experiment was the first example of a nonnatural residue incorporated into a protein. In recent years, a large number of nonnatural residues have been incorporated into proteins, and this field is now burgeoning to produce many proteins with specialized features. Finally, experiments were carried out incorporating α-hydroxyl phenyllactyl tRNA^{phe} in an in vitro synthesis directed by polyuridylic acid (55). This yielded ribosome-catalyzed polyester formation in which every residue contained ester linkages rather than peptide linkages. It was the first completely nonnatural polymer made by ribosomes directed by mRNA.

CRYSTAL STRUCTURE OF tRNA

Methods for purifying tRNA improved during the 1960s, and there was an increase in attempts to form single crystals. This effort was very frustrating because it was easy to fail, and most people involved in the effort failed repeatedly. In 1968, working with postdoctoral fellow Sung Hou Kim, we were able to obtain single crystals of *E. coli* tRNA^{phe} (56). Three other groups also obtained single crystals of various tRNAs in that same year, and all of these crystals were poor in that they were somewhat disordered and the resolution was limited. Our earliest crystals diffracted to ~20 Å. By the next year, we were able to get crystals that diffracted to 6 or 7 Å resolution, and a study of the three-dimensional Patterson function using the 12 Å data from the crystals of *E. coli* tRNA^{Fmet} yielded approximate molecular dimensions of 80 × 25 × 35 Å (57). These crystals represented progress of a sort, but the frustration was great because they were not suitable for solving the structure.

Aided by Gary Quigley and others in the lab, we spent the next two years looking at many different purified tRNA preparations and explored many different crystallization procedures. By 1971, we reached an exciting turning point: Yeast tRNA^{phe} could be crystallized in a simple orthorhombic unit cell with a resolution of 2.3 Å (58)! These were the first crystals of tRNA suitable for analysis. The key event in making these crystals was the incorporation of spermine, a naturally occurring polyamine. The spermine bound specifically to yeast tRNA^{phe} and stabilized it so that it made a high-resolution crystal. This was an important discovery at the time. The stabilization effect of spermine on yeast tRNA^{phe} made it possible to form good crystals in other lattices as well. Analysis of the crystal diffraction pattern showed that it had a characteristic helical distribution of diffracting intensities when viewed in one direction but did not show a helical distribution when viewed at right angles. This was taken as evidence that short helical segments containing 4–7 base pairs were found in the molecule, a result entirely consistent with the cloverleaf folding of tRNA molecules suggested by Holley and colleagues after sequencing the first tRNA molecule (59). This discovery opened the door to the ultimate solution of the structure of yeast tRNA^{phe}.

Tracing the Backbone of Yeast tRNA^{phe} and Solving the Structure

Myoglobin was the first protein whose three-dimensional structure was solved. The structure was revealed at various levels of resolution. An important milestone was the tracing of the polypeptide chain that showed how the myoglobin molecule is organized as a series of α-helical and single-stranded regions folded together. This was our first glimpse of how a protein molecule is folded, even though the high-resolution structure had not been completed.

The structure of yeast tRNA^{phe} was revealed in a similar, gradual way. Crystallographic research moved more slowly in the early 1970s than today. Computers were primitive; advanced area detectors, cryo-crystallography, and synchrotron beams were things in the future. In the lab, Bud Suddath built a cold room around our primitive diffractometer to stabilize the crystals. However, before work could continue, heavy-atom derivatives had to be discovered that would be useful for phasing the diffraction pattern of a crystalline nucleic acid molecule. This had never been done before, and it took considerable time to discover appropriate derivatives. Three different types of heavy atoms were developed containing platinum, osmium, or samarium ions. The osmium residue was very important because it was known to form complexes with ribonucleotides involving both the 2′ and 3′ hydroxyl groups. Only one pair of *cis* hydroxyl groups was found at the 3′ CCA end of the tRNA molecule. In order to gain some appreciation of the geometry, the structure of an osmium–adenosine complex was solved, which enabled us to visualize the interaction (60). The single osmium derivative in the tRNA crystal made it possible to identify the 3′ end of the tRNA chain (61). The samarium ions were very useful, and they occupied more than one site. The platinum residue was somewhat less valuable since it was only useful for 5.5 Å data. An interim electron density map at 5.5 Å made it possible to uncover the external shape of portions of the molecule. However, the true shape and fold of the molecule was not revealed until a map at 4 Å was produced in 1973 (61).

At 4 Å resolution, peaks were seen throughout the electron density map that were due to the electron-dense phosphate groups. We knew a great deal about the distance constraints between adjacent phosphate groups in a polynucleotide chain, and this made it possible to look for peaks between 5 and 7 Å apart. Tracing the chain led to the discovery that the tRNA molecule had an unusual L-shape. The CCA acceptor helix was collinear with the T pseudo-U helix, and it is almost at right angles to the anticodon stem that is collinear with the dihydro U stem. The molecule had the shape of an L, with the amino acid acceptor 3′ hydroxyl group at one end of the L and the anticodon loop at the other over 70 Å away. At the corner of the L, there was a complex folding of the T pseudo-U and dihydro U loops. A perspective diagram of the L-shaped fold was published (Figure 3), as well as a sample of the electron density map showing double-helical regions in which the intense peaks associated with phosphate groups are separated from each other by a region of lower electron density due to the base pairs.

The L-shaped folding of the tRNA polynucleotide chain was a dramatic and surprising discovery, especially the separation between the acceptor site and the anti-codon. The backbone tracing (Figure 3) was published on the front page of *The New York Times* on January 13, 1973, together with a discussion of its role in protein synthesis. No one had anticipated that the molecule would organize in this fashion. Even at 4 Å resolution, this folding was compatible with much experimental data concerning tRNA molecules. For example, it was known that photo activation of *E. coli* tRNA^{val} resulted in the formation of a photo dimer involving the 4 thio-U residue in position 8 and the cytosine in position 13. In the 4 Å folding of the polynucleotide chain, these two bases were in close proximity, and the distance between the phosphate groups of these two residues was short enough to allow formation of the photo dimer (61).

The L-shaped tRNA molecule is now a standard feature of molecular biology, having been found in virtually all tRNA molecules, even when they are complexed to aminoacyl synthetase enzymes. The significance of the folding is twofold. First, it revealed that the 3′ acceptor end is over 70 Å away from the anticodon loop, which has implications for understanding the interaction between tRNA molecules and tRNA aminoacyl synthetases. Second and most important, it suggested that the interaction of the tRNA molecules with mRNA occurs at one end of the L, whereas the segment responsible for forming the peptide bond is considerably removed from the site. This makes it possible to have great specificity with many interactions at either end of the molecule due to this separation.

At that time I made a proposal regarding the movement of tRNA molecules in the ribosome (62). A codon occupies ~10 Å (3 × 3.4Å) when the bases are stacked. Thus, the center-to-center distance between the A-site and P-site codon is 10 Å. The chain tracing showed that there was an anticodon stem, and it was ~20 Å in diameter. Thus the paradox: How can two tRNAs simultaneously occupy the A and P sites? The answer, I suggested, was that the mRNA "bends or turns a corner" between the A and P site, allowing both tRNAs to make contact. In the recent paper tracing the path of mRNA in the ribosome (63), a kink of ~45° is seen between the A-site and P-site codons. Thus, the tRNA chain tracing was valuable in anticipating some features of the system.

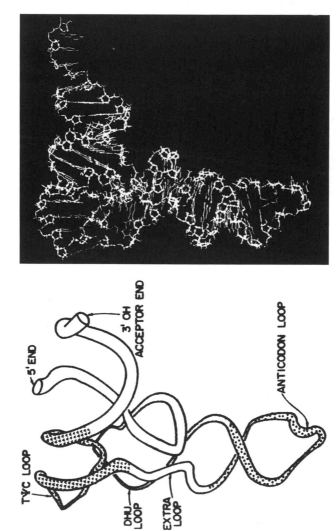

Figure 3 (*Left*) At 4 Å resolution the fold of the tRNA[phe] chain could be visualized, as shown in this perspective diagram (61). (*Right*) The 3 Å tRNA[phe] structure reveals the complete interactions of the L-shaped molecule, as shown on the cover of *Science* (64).

Today, we are accustomed to seeing a variety of complex ribonucleotide molecules in which double-helical segments and single-chain segments are juxtaposed to make complex structures with a variety of functions, especially in ribozymes. The beginning of our understanding of the manner in which complex polynucleotide chains can fold started with this first tracing of yeast tRNA^{phe} visualized at 4 Å resolution. This tracing was seen in more detail a year later in our 3 Å analysis of the folding of yeast tRNA^{phe} in the orthorhombic lattice (64). Simultaneously, Aaron Klug and colleagues published the 3 Å structure of the same spermine-stabilized yeast tRNA^{phe} in the monoclinic lattice (65). Both papers confirmed the L-shaped folding of the polynucleotide chain, even though the lattice was different. These 3 Å structures were very similar and revealed in great detail the manner in which base-pairing of nucleotides, both in the double-helical regions and in the single-stranded regions, stabilizes the molecule. The folding was held together by a variety of hydrogen-bonding interactions, including many in the nonhelical regions of the molecule. These hydrogen-bonding interactions included the formation of triplexes and other interactions. Recognition of the importance of these alternative types of hydrogen bonds explained why the model builders of that period, trying to anticipate the structure of tRNA, were all incorrect. They relied excessively on Watson-Crick base-pair interactions and did not recognize the stabilizing effect of many other types of hydrogen bonds. The L-shaped folding was predicted to be a general conformation found in all tRNA molecules (66). Subsequent work amply verified the relative constancy of the hydrogen-bonding networks (67, 68).

We have known of the L-shaped folding of tRNA molecules since 1973. The full understanding of why this particular folding is robust, and of the manner in which modified nucleotides are important to this folding, is still a work in progress. However, for understanding the interaction of tRNA molecules with the ribosome during protein synthesis, the L-shaped folding provides the central information regarding its interactions and movements in protein synthesis.

I have focused on the work of my colleagues and myself in uncovering some basic aspects of RNA structure and protein biosynthesis over the 20-year period 1954–1974. However, the detailed mechanism could not be understood until the more recent developments elucidating the three-dimensional structure of the ribosome. The earlier work in our lab and others set the stage for understanding the mechanisms that we now are beginning to envision in more detail in the machinery of protein synthesis.

SEARCH FOR LIFE ON MARS

My earlier interest in problems related to the origin of life led me to participate in the biological experiments carried out on the Mars Landers in the Viking Mission. Starting in 1969, a team was assembled that included Joshua Lederberg, Norman Horowitz, Chuck Klein, and others. We met periodically in California to design the experiments in the search for life on Mars. The discussions leading to the selection of experiments often had a philosophic character since we asked what were the fundamentals for life and how might they be transformed in the harsh environment of the Martian surface. For example, would a Martian "tree" largely grow underground to shield itself from hard UV radiation, much in the manner of the endolithic organisms that are found in the Antarctic? For this reason, I visited the Antarctic when I was a member of the National Science Board and was able to retrieve rock samples with photosynthetic organisms embedded in the rock.

In July 1976, I was living in Pasadena as a Visiting Professor at Caltech and was present at the Jet Propulsion Lab with other Viking Mission scientists as the first pictures came back from the surface of Mars. It was an extremely tense gathering because we did not know whether the Lander would actually land safely and be functional. It was electrifying to watch the slow, line-by-line creation of the first photograph taken on the surface of Mars.

In August 1976, I was scheduled to present a paper at the International Biochemistry Congress in Amsterdam, and I brought to the Congress a report of experimental results from the first month or so of the Mission. A great deal of excitement was generated at the time since our first findings looked as if they might be positive. Only later did we begin to understand the complexities of what we observed. Although our first publication from the Mission (69) suggested both positive and negative results, in the end we concluded that much of what we were looking at had to do with the chemistry of the Martian surface due to either the addition of water or heat. Overall, this experience helped to maintain my enthusiasm for research on extraterrestrial life.

INTERNATIONAL SCIENCE AND PUGWASH MEETINGS

In the late 1950s, together with many others, I was strongly concerned about the threat of nuclear war. When I was invited to attend a Pugwash meeting in 1959, I readily assented. These were meetings of scientists from the East and the West. Linus Pauling and John Edsall both attended the meeting in Pugwash, Nova Scotia. Up to 1972, I attended 13 of these meetings in various countries, attended by scientists from the Soviet Union, the People's Republic of China, and many other countries. These served as useful vehicles for discussing new ideas and helped to minimize the tension then growing between East and West. At a meeting in Moscow in 1960, I met a number of scientists from the Soviet Union and adopted a mode of operation. I usually wrote joint papers with the Soviet scientists, presenting suggestions on alleviating tensions in the nuclear arms race. At a meeting in London in 1962, a Russian colleague and I developed the idea that we might use automated seismographs to monitor nuclear testing, the so-called "black boxes." The proposal that we drafted was then signed by several other American and Soviet scientists and became a document of the conference.

It was also sent to world leaders. At one stage, the Soviet Union contemplated using such automated seismographs, and it was one of the elements that facilitated the eventual signing of the Limited Test Ban Agreement in 1963. Extending over many years, I developed warm friendships with a number of Russian scientists, including V.A. Engelhardt and A.A. Bayev, who was active in the field of transfer RNA sequencing. The meetings contained elements of both politics and science. Although these meetings took considerable time, both in preparation and execution, I felt the effort was worthwhile since it addressed a major problem facing our society—the prevention of nuclear war.

DNA TURNS LEFT-HANDED

It was not until the late 1970s that the development of DNA synthesis made it possible to carry out single crystal X-ray diffraction studies that could "prove" the structure. In 1978 I met a Dutch organic chemist, Jacques Van Boom, who could synthesize DNA oligomers. He made d(CG)$_3$, and Andy Wang crystallized it and discovered it diffracted to 0.9 Å resolution. Heavy atoms were used to solve the structure, which revealed, remarkably, a left-handed double helix with two antiparallel chains held together by Watson-Crick base pairs (70). Every other base had rotated around the glycosyl bonds so that the bases alternated in *anti* and *syn* conformations along the chain. The zigzag arrangement of the backbone (hence, Z-DNA) was different from the smooth, continuous coil seen in B-DNA (Figure 4). The general response to this unusual structure was amazement, coupled with skepticism.

The relationship between Z-DNA and the more familiar right-handed B-DNA began to be apparent from the earlier work of Pohl & Jovin (71), who showed that the UV circular dichroism of poly (dG-dC) nearly inverted in 4 M sodium chloride solution. The suspicion that this was due to a conversion from B-DNA to Z-DNA was confirmed by examining the Raman spectra of these solutions and the Z-DNA crystals (72). The conversion to left-handed Z-DNA was associated with a "flipping over" of the base pairs so that they were upside down in their orientation relative to what would be found in B-DNA. Sequences that most readily converted had alternations of purines and pyrimidines, especially alternations of C and G (73). Alfred Nordheim showed that it also occurred quite easily with alternations of CA on one strand and TG on the other strand (74), and many other sequences were shown to be capable of forming Z-DNA (75).

This discovery stimulated a burst of research from a large number of chemists who were very interested in studying DNA conformational changes. It tended to leave most biologists rather puzzled, since the ionic conditions suitable for stabilizing Z-DNA were very far from those present in a cell. This view changed somewhat with the discovery that negative supercoiling would also stabilize Z-DNA (76). Supercoiling was known to be a part of biological systems, and it

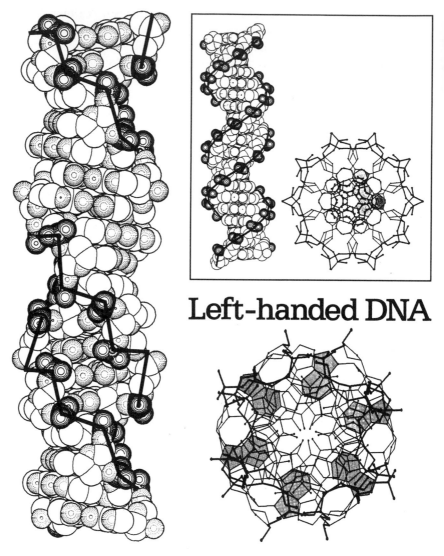

Figure 4   A diagram showing a comparison between B-DNA and Z-DNA with solid lines running from phosphate to phosphate, as shown on the cover of *Nature* (70).

suggested a connection between this alternative conformation and biological phenomena.

Does Z-DNA Have a Biology?

Research work on the biology of Z-DNA progressed very slowly. By the mid-1980s after several years of research in which nothing definitive emerged about Z-DNA, most biologists were very skeptical about its role. Many felt that Z-DNA was a nonfunctional conformational phenomenon. My conviction was very simple. Here was an alternative DNA conformation, which could, in principle, form under in vivo conditions, and I felt it likely that it would be used because evolution is opportunistic. The challenge was to find out how it was used.

The first indications of a biological role for Z-DNA came from immunological work. In collaboration with David Stollar, we found that, unlike B-DNA, Z-DNA is highly antigenic, yielding polyclonal (77) and monoclonal (78) antibodies. Characterization of these antibodies by Eileen Lafer led to the discovery that Z-DNA-specific antibodies are found in human autoimmune diseases, especially systemic lupus erythematosus (79). Antibodies to Z-DNA also provided a useful tool for characterizing chromosome organization. They bound specifically to the interband regions of the *Drosophila* polytene chromosomes; the binding was particularly strong in the puff regions, the sites of enhanced transcriptional activity (80). Ciliated protozoa have two nuclei: the macronucleus, which is the site of transcription, and the micronucleus, which contains DNA involved in sexual reproduction. Anti-Z-DNA antibodies stained the macronucleus of the ciliated protozoan *Stylonychia*, but not its micronucleus (81). These were the first data to suggest a connection between Z-DNA and transcriptional activity.

An important advance came with the work of Liu & Wang (82) in 1987 on the interaction of RNA polymerase with DNA during transcription. They pointed out that the moving complex does not rotate around helical DNA, but instead plows straight through. Because the ends of the DNA molecule are fixed, the DNA behind the moving polymerase was unwound and subjected to negative torsional strain, while positive torsional strain developed in front. Further evidence came later from the work of P. Shing Ho and colleagues, who found a high concentration of sequences favoring Z-DNA formation near the transcription start site (83). To study the association with transcription more directly, I collaborated with Burkhardt Wittig and colleagues, using a technique developed by Peter Cook at Oxford. Mammalian cells were encapsulated in agarose microbeads; mild detergent treatment lysed the cytoplasmic membrane, permeabilizing the nuclear membrane but leaving the nucleus otherwise intact. The resulting "entrapped" nuclei replicated DNA at nearly the in vivo rate, and they were able to carry out transcription (84). Using biotinylated monoclonal antibodies against Z-DNA, the level of Z-DNA was shown to be regulated by torsional strain (85). An increase in transcriptional activity of the embedded nuclei resulted in a parallel increase in the amount of Z-DNA (86). Using a UV laser pulse for protein-DNA cross linking, the biotinylated anti-Z-DNA antibodies were linked to DNA. This made it possible to isolate DNA restriction fragments bound to the antibody. With cultured human cells, three regions upstream of the *c-myc* gene formed Z-DNA when *c-myc* was expressed. However, these regions quickly reverted to B-DNA upon switching off *c-myc* transcription (87). Nonetheless, the actin gene control retained its Z-DNA at all times.

The picture that then emerged was that the negative torsional strain induced by the movement of RNA polymerase stabilized Z-DNA formation near the transcription start site. Even though topoisomerases tried to relax the DNA, the continued movement of RNA polymerases generated more negative torsional strain than the topoisomerases could relax. However, upon cessation of transcription, topoisomerases rapidly converted it back to the right-handed B conformation. Thus, Z-DNA was seen as a metastable conformation, forming and disappearing depending upon physiological activities.

Z-DNA Binding Proteins

If Z-DNA were to have biological functions, it seemed highly likely that a class of proteins should bind to it specifically. The challenge was to isolate such proteins that bound selectively to Z-DNA with high affinity. The first successful method was developed in the lab by Alan Herbert (88). A gel shift assay was used with radioactive-labeled, chemically stabilized Z-DNA in the presence of a ~20,000-fold excess of B-DNA and single-stranded DNA. A Z-DNA-binding protein was found to be a nuclear RNA editing enzyme (89), called doublestranded RNA adenosine deaminase (or ADAR1). This enzyme acts on doublestranded segments formed in pre-mRNA, binding to the duplex and selectively deaminating adenosine, yielding inosine. Ribosomes interpret inosine as guanine. Thus, ADAR1 can alter the amino acid sequence of a DNA-encoded protein. The functional properties of the edited protein (with the amino acid alteration) are often different from those of the unedited protein. The editing enzyme is found in all metazoa; it acts to increase the functional diversity of proteins transcribed from a given locus (90).

Proteolytic dissection of the editing enzyme ADAR1 led to a domain from the N terminus called $Z\alpha_{ADAR1}$ (91). $Z\alpha_{ADAR1}$ was found to contain all of the Z-DNA binding properties associated with the editing enzyme, and it bound Z-DNA tightly with a low nanomolar K_d. This domain was used to create a conformationally specific restriction endonuclease that would only cut Z-DNA. The naturally occurring restriction enzyme Fok1 has a DNA recognition domain and a separate nuclease domain attached to it. A chimeric molecule was made by Yang Kim in which the $Z\alpha$ domain of the editing enzyme ADAR1 was covalently linked with the nuclease domain of the Fok1 restriction enzyme (92). This restriction enzyme bound and cleaved plasmids only when they were supercoiled and Z-DNA was present.

In 1999, Thomas Schwartz, a graduate student, discovered that the purified $Z\alpha_{ADAR1}$ domain could be co-crystallized with $d(CG)_3$. The crystal structure solved at 2.1 Å resolution (93) revealed that the DNA was in a form virtually identical to that seen in the first Z-DNA crystal (70). The 70-amino acid binding domain was found to adopt a helix-turn-helix β-sheet motif (winged helix) in which the recognition helix and the β-sheet bound to five successive phosphate groups in the zigzag backbone of Z-DNA, and it recognized the *syn* conformation of guanine.

It is possible that the Z-DNA binding domain of ADAR1 targets Z-DNA forming regions of some transcriptionally active genes, as only they have Z-DNA. $Z\alpha_{ADAR1}$ appears to be active in vivo in the editing of certain transcripts where it may target the gene (94); however, its role in RNA editing is not resolved. Bernie Brown and Ky Lowenhaupt showed that $Z\alpha_{ADAR1}$ also binds tightly to left-handed Z-RNA (95). This may be related to the role of the editing enzyme in modifying measles and other negative-strand RNA viruses, which are extensively hyperedited.

The co-crystal structure of $Z\alpha_{ADAR1}$ and Z-DNA made it possible to identify those amino acids important for Z-DNA recognition. A computer search rapidly revealed other proteins with similar sequence motifs. One is the protein DLM1 which is up-regulated in tissues in contact with tumors and is also interferon-induced. The co-crystal structure of a domain of DLM1 ($Z\alpha_{DLM1}$) and $d(CG)_3$ was solved at a resolution of 1.85 Å, and it showed that this second protein domain recognizes Z-DNA in a manner very similar to that found with $Z\alpha_{ADAR1}$, but with a few variations (96). This second structure clearly indicated that a family of such proteins exists.

Viruses Use Z-DNA Binding Proteins

Another member of this family of proteins is E3L, found in poxviruses such as vaccinia. These are large DNA viruses that reside in the cytoplasm of cells and produce a number of proteins that help to abort the interferon response of the host cell. E3L is a small 25-Kd protein necessary for pathogenicity. When vaccinia virus is given to a mouse, the mouse dies in about a week. However, in a virus that has a mutated or missing E3L, it is no longer pathogenic for the mouse, even though the virus can still reproduce in cell culture (97). To demonstrate the pathogenicity of the vaccinia virus in the mouse and its relationship to possible Z-DNA binding activities of E3L, a collaboration was set up with Bertram Jacobs. Chimeric viruses were created in which the N-terminal domain of vaccinia E3L (Z_{E3L}) was removed, and either the domains $Z\alpha_{ADAR1}$ or $Z\alpha_{DLM1}$ were inserted. In carrying out these domain swaps, a little more than a dozen amino acids in the domain remained unchanged, but over 50 other amino acids were changed. Nonetheless, the chimeric viruses were as pathogenic for mice as the wild type (98). Other experiments were carried out in which mutations in the chimeric virus that weakened Z-DNA binding were also shown to weaken pathogenicity. Similar mutations in the wild type weakened mortality. Loss of Z-DNA binding led to loss of pathogenicity. It is likely that the Z_{E3L} domain binds to Z-DNA formed near the transcription start site of certain genes, thereby impairing the antiviral response of the host cell (98). This is a new example of the way in which viruses seek to exploit features of the host cell in order to overcome the host defense mechanisms.

A small molecule or drug can probably be made that will bind to the Z-DNA binding pocket of the E3L molecule. This drug should prevent mice from dying when infected with vaccinia virus. It may also be active in humans to prevent untoward effects due to vaccination. More significant is the fact that the E3L protein of the closely related variola virus, the agent of smallpox, is virtually identical to the vaccinia E3L (98). Hence, such a drug binding to E3L may develop into a treatment of smallpox.

To my great surprise, work on Z-DNA and its binding proteins has led us to the possibility of developing a therapy for certain viral diseases, including smallpox. This is a striking example of serendipity in scientific research. And, of course, it helps to maintain the excitement that is an inherent part of the scientific enterprise.

In this outline, I have described the gradual unfolding of a research trail. My driving force has always been curiosity, and one of the intrinsic rewards of a life in science is the excitement of uncovering some aspect of nature.

ACKNOWLEDGMENTS

Here I acknowledge the contributions of my colleagues who played an essential role in this work. The students, postdoctoral fellows, and coworkers are too numerous to mention in this brief account, but the pleasure of working with them has been very great. Together we shared speculations and many disappointments, but also many successes. Finally, I am thankful for the enlightened attitude in this country and elsewhere about the importance of research, as this has provided the material support that made it possible to pursue a life in science.

The *Annual Review of Biochemistry* is online at http://biochem.annualreviews.org

LITERATURE CITED

1. Edsall JT, Otvos JW, Rich A. 1950. *J. Am. Chem. Soc.* 72:474–77
2. Rich A. 1952. *Proc. Natl. Acad. Sci. USA* 38:187–96
3. Dunitz JD, Orgel LE, Rich A. 1956. *Acta Crystallogr.* 9:373–75
4. Gamow G. 1954. *Nature* 173:318–20
5. Gamow G, Rich A, Yas M. 1956. In *Advances in Biological and Medical Physics,* 4:23–68. New York: Academic
6. Hoagland MB, Zamecnik PC, Stephenson ML. 1957. *Biochim. Biophys. Acta* 24:215–16
7. Rich A, Watson JD. 1954. *Nature* 173: 995–96
8. Rich A, Watson JD. 1954. *Proc. Natl. Acad. Sci. USA* 40:759–64
9. Grunberg-Manago M, Ortiz PJ, Ochoa S. 1955. *Science* 122:907–10
10. Rich A, Davies DR, Crick FHC, Watson JD. 1961. *J. Mol. Biol.* 3:71–86
11. Crick FHC, Rich A. 1955. *Nature* 176: 780–81
12. Rich A, Crick FHC. 1955. *Nature* 176: 915–16
13. Rich A, Crick FHC. 1961. *J. Mol. Biol.* 3:483–506
14. Rich A, Davies DR. 1956. *J. Am. Chem. Soc.* 78:3548
15. Warner RC. 1956. *Fed. Proc.* 15:379
16. Kornberg A, Lehman IR, Bessman MJ, Simms ES. 1956. *Biochim. Biophys. Acta* 21:197–98

17. Felsenfeld G, Rich A. 1957. *Biochim. Biophys. Acta* 26:457–68
18. Felsenfeld G, Davies DR, Rich A. 1957. *J. Am. Chem. Soc.* 79:2023–24
19. Hoogsteen K. 1959. *Acta Crystallogr.* 12:822–23
20. Rich A. 1957. In *The Chemical Basis of Heredity*, ed. WD McElroy, B Glass, pp. 557–62. Baltimore: Johns Hopkins Univ. Press
21. Rich A. 1958. *Nature* 181:521–25
22. Rich A. 1958. *Biochim. Biophys. Acta* 29:502–9
23. Davies DR, Rich A. 1958. *J. Am. Chem. Soc.* 80:1003
24. Tener GM, Khorana HG, Markham R, Pol EH. 1958. *J. Am. Chem. Soc.* 80:6223–30
25. Rich A. 1960. *Proc. Natl. Acad. Sci. USA* 46:1044–53
26. Furth JJ, Hurwitz J, Goldmann M. 1961. *Biochem. Biophys. Res. Commun.* 4:362–67
27. Doty P, Marmur J, Eigner J, Schildkraut C. 1960. *Proc. Natl. Acad. Sci. USA* 46:461–76
28. Marmur J, Lane D. 1960. *Proc. Natl. Acad. Sci. USA* 46:453–61
29. Hall BD, Spiegelman S. 1961. *Proc. Natl. Acad. Sci. USA* 47:137–46
30. Goodman HM, Rich A. 1962. *Proc. Natl. Acad. Sci. USA* 48:2101–9
31. Rich A. 1962. In *Horizons in Biochemistry*, ed. M Kasha, B Pullman, pp. 103–26. New York: Academic
32. Chun EHL, Vaughn MH, Rich A. 1963. *J. Mol. Biol.* 7:130–41
33. Sobell HM, Tomita K, Rich A. 1963. *Proc. Natl. Acad. Sci. USA* 49:885–92
34. Pauling L, Corey RB. 1956. *Arch. Biochem. Biophys.* 65:164–68
35. Mathews FS, Rich A. 1964. *J. Mol. Biol.* 8:89–95
36. Katz L, Tomita K, Rich A. 1965. *J. Mol. Biol.* 13:340–50
37. Tomita K, Katz L, Rich A. 1967. *J. Mol. Biol.* 30:545–49
38. Voet D, Rich A. 1970. *Prog. Nucleic Acid Res. Mol. Biol.* 10:183–265
39. Kyogoku Y, Lord R, Rich A. 1966. *Science* 154:518–20
40. Arnott S, Wilkins MHF, Hamilton LD, Langridge R. 1965. *J. Mol. Biol.* 11:391–402
41. Day RO, Seeman NC, Rosenberg JM, Rich A. 1973. *Proc. Natl. Acad. Sci. USA* 70:849–53
42. Rosenberg JM, Seeman NC, Kim JJP, Suddath FL, Nicholas HB, Rich A. 1973. *Nature* 243:150–54
43. Wang AH-J, Fujii S, van Boom JH, van der Marel GA, van Boeckel SAA, Rich A. 1982. *Nature* 299:601–4
44. Crick FHC, Barnett L, Brenner S, Watts-Tobin RJ. 1961. *Nature* 192:1227–32
45. Warner JR, Rich A, Hall CE. 1962. *Science* 138:1399–403
46. Warner JR, Knopf PM, Rich A. 1963. *Proc. Natl. Acad. Sci. USA* 149:122–29
47. Goodman HM, Rich A. 1963. *Nature* 199: 318–22
48. Warner JR, Rich A. 1964. *J. Mol. Biol.* 10:202–11
49. Warner JR, Rich A. 1964. *Proc. Natl. Acad. Sci. USA* 51:1134–41
50. Malkin LI, Rich A. 1967. *J. Mol. Biol.* 26:329–46
51. Ban N, Nissen P, Hansen J, Moore PB, Steitz TA. 2000. *Science* 289:905–20
52. Kiho Y, Rich A. 1964. *Proc. Natl. Acad. Sci. USA* 51:111–18
53. Fahnestock S, Neumann H, Shashoua V, Rich A. 1970. *Biochemistry* 9:2477–83
54. Fahnestock S, Rich A. 1971. *Nat. New Biol.* 229:8–10
55. Fahnestock S, Rich A. 1971. *Science* 173: 340–43
56. Kim S-H, Rich A. 1968. *Science* 162: 1381–84
57. Kim S-H, Rich A. 1969. *Science* 166: 1621–24
58. Kim S-H, Quigley G, Suddath FL, Rich A. 1971. *Proc. Natl. Acad. Sci. USA* 68:841–45
59. Holley RW, Apgar J, Everett GA, Madi-

son JT, Marguisse M, et al. 1965. *Science* 147:1462–65

60. Conn JF, Kim JJ, Suddath FL, Blattman P, Rich A. 1974. *J. Am. Chem. Soc.* 96:7152–53

61. Kim S-H, Quigley GJ, Suddath FL, McPherson A, Sneden D, et al. 1973. *Science* 179:285–88

62. Rich A. 1974. In *Ribosomes*, ed. M Nomura, A Tissiere, P Lengyel, pp. 871–84. Cold Spring Harbor, NY: Cold Spring Harbor Lab. Press

63. Yusupova GZ, Yusupova MM, Cate JHD, Noller HF. 2001. *Cell* 106:233–41

64. Kim S-H, Suddath FL, Quigley GJ, McPherson A, Sussman JL, et al. 1974. *Science* 185:435–39

65. Robertus JD, Ladner JE, Finch JT, Rhodes D, Brown RS, et al. 1974. *Nature* 250:546–51

66. Kim S-H, Sussman JL, Suddath FL, Quigley GJ, McPherson A, et al. 1974. *Proc. Natl. Acad. Sci. USA* 71:4970–74

67. Quigley GJ, Wang AH-J, Seeman NC, Suddath FL, Rich A, et al. 1975. *Proc. Natl. Acad. Sci. USA* 72:4866–70

68. Quigley GJ, Rich A. 1976. *Science* 194:796–806

69. Klein HP, Horowitz NH, Levin GV, Oyama VI, Lederberg J, et al. 1976. *Science* 194:99–105

70. Wang AH-J, Quigley GJ, Kolpak FJ, Crawford JL, van Boom JH, et al. 1979. *Nature* 282:680–86

71. Pohl FM, Jovin TM. 1972. *J. Mol. Biol.* 67:375–96

72. Thamann TJ, Lord RC, Wang AH-J, Rich A. 1981. *Nucleic Acids Res.* 9:5443–57

73. Rich A, Nordheim A, Wang AH-J. 1984. *Rev. Biochem.* 53:791–846

74. Nordheim A, Rich A. 1983. *Proc. Natl. Acad. Sci. USA* 80:1821–25

75. Feigon J, Wang AH-J, van der Marel GA, van Boom JH, Rich A. 1985. *Science* 230:82–84

76. Peck LJ, Nordheim A, Rich A, Wang JC.

1982. *Proc. Natl. Acad. Sci. USA* 79:4560–64

77. Lafer EM, Moller A, Nordheim A, Stollar BD, Rich A. 1981. *Proc. Nat. Acad. Sci. USA* 78:3546–50

78. Moller A, Gabriels JE, Lafer EM, Nordheim A, Rich A, Stollar BD. 1982. *J. Biol. Chem.* 257:12081–85

79. Lafer EM, Valle RPC, Moller A, Nordheim A, Schur P, et al. 1983. *J. Clin. Invest.* 71:314–21

80. Nordheim A, Pardue ML, Lafer EM, Moller A, Stollar BD, Rich A. 1981. *Nature* 294:417–22

81. Lipps HJ, Nordheim A, Lafer EM, Ammermann D, Stollar BD, Rich A. 1983. *Cell* 32:435–41

82. Liu LF, Wang JC. 1987. *Proc. Natl. Acad. Sci. USA* 84:7024–27

83. Schroth GP, Chou P-J, Ho PS. 1992. *J. Biol. Chem.* 267:11846–55

84. Jackson DA, Yuan J, Cook PR. 1988. *J. Cell. Sci.* 90:365–78

85. Wittig B, Dorbic T, Rich A. 1989. *J. Cell. Biol.* 108:755–64

86. Wittig B, Dorbic T, Rich A. 1991. *Proc. Natl. Acad. Sci. USA* 88:2259–63

87. Wittig B, Wolfl T, Dorbic T, Vahrson W, Rich A. 1992. *EMBO J.* 11:4653–63

88. Herbert AG, Rich A. 1993. *Nucleic Acids Res.* 21:2669–72

89. Herbert A, Lowenhaupt K, Spitzner J, Rich A. 1995. *Proc. Natl. Acad. Sci. USA* 92:7550–54

90. Bass BL. 2002. *Annu. Rev. Biochem.* 71:817–46

91. Herbert A, Alfken J, Kim Y-G, Mian IS, Nishikura K, Rich A. 1997. *Proc. Natl. Acad. Sci. USA* 94:8421–26

92. Kim Y-G, Kim PS, Herbert A, Rich A. 1997. *Proc. Natl. Acad. Sci. USA* 94:12875–79

93. Schwartz T, Rould MA, Lowenhaupt K, Herbert A, Rich A. 1999. *Science* 284:1841–45

94. Herbert A, Rich A. 2001. *Proc. Natl. Acad. Sci. USA* 98:12132–37

95. Brown BA, Lowenhaupt K, Wilbert CM,

Hanlon EB, Rich A. 2000. *Proc. Natl. Acad. Sci. USA* 97:13532–36

96. Schwartz T, Behlke J, Lowenhaupt K, Heinemann U, Rich A. 2001. *Nat. Struct. Biol.* 8:761–65

97. Brandt TA, Jacobs BL. 2001. *J. Virol.* 75:850–56

98. Kim Y-G, Muralinath M, Brandt T, Pearcy M, Hauns K, et al. 2003. *Proc. Natl. Acad. Sci. USA* 100:6974–79

Annu. Rev. Biochem. 2004. 73:39–85
doi: 10.1146/annurev.biochem.73.011303.073723
Copyright © 2004 by Annual Reviews. All rights reserved
First published online as a Review in Advance on January 20, 2004

MOLECULAR MECHANISMS OF MAMMALIAN DNA REPAIR AND THE DNA DAMAGE CHECKPOINTS

Aziz Sancar,[1] Laura A. Lindsey-Boltz,[1]
Keziban Ünsal-Kaçmaz,[1] and Stuart Linn[2]

[1]Department of Biochemistry and Biophysics, University of North Carolina School of
Medicine, Chapel Hill, North Carolina 27599-7260; email: aziz_sancar@med.unc.edu,
llindsey@med.unc.edu, kezi_unsal@med.unc.edu
[2]Division of Biochemistry and Molecular Biology, University of California, Berkeley,
California 94720-3202; email: slinn@socrates.berkeley.edu

Key Words damage recognition, excision repair, PIKK family, Rad17-RFC/9-1-1 complex, signal transduction

■ **Abstract** DNA damage is a relatively common event in the life of a cell and may lead to mutation, cancer, and cellular or organismic death. Damage to DNA induces several cellular responses that enable the cell either to eliminate or cope with the damage or to activate a programmed cell death process, presumably to eliminate cells with potentially catastrophic mutations. These DNA damage response reactions include: (*a*) removal of DNA damage and restoration of the continuity of the DNA duplex; (*b*) activation of a DNA damage checkpoint, which arrests cell cycle progression so as to allow for repair and prevention of the transmission of damaged or incompletely replicated chromosomes; (*c*) transcriptional response, which causes changes in the transcription profile that may be beneficial to the cell; and (*d*) apoptosis, which eliminates heavily damaged or seriously deregulated cells. DNA repair mechanisms include direct repair, base excision repair, nucleotide excision repair, double-strand break repair, and cross-link repair. The DNA damage checkpoints employ damage sensor proteins, such as ATM, ATR, the Rad17-RFC complex, and the 9-1-1 complex, to detect DNA damage and to initiate signal transduction cascades that employ Chk1 and Chk2 Ser/Thr kinases and Cdc25 phosphatases. The signal transducers activate p53 and inactivate cyclin-dependent kinases to inhibit cell cycle progression from G1 to S (the G1/S checkpoint), DNA replication (the intra-S checkpoint), or G2 to mitosis (the G2/M checkpoint). In this review the molecular mechanisms of DNA repair and the DNA damage checkpoints in mammalian cells are analyzed.

0066-4154/04/0707-0039$14.00

CONTENTS

INTRODUCTION

The primary structure of DNA is constantly subjected to alteration by cellular metabolites and exogenous DNA-damaging agents. These alterations may cause simple base changes or they may cause more complex changes including deletions, fusions, translocations, or aneuploidy. Such alterations may ultimately lead to cellular death of unicellular organisms or degenerative changes and aging of multicellular organisms.

DNA damages can perturb the cellular steady-state quasi-equilibrium and activate or amplify certain biochemical pathways that regulate cell growth and division and pathways that help to coordinate DNA replication with damage removal. The four types of pathways elicited by DNA damage known, or presumed, to ameliorate harmful damage effects are DNA repair, DNA damage checkpoints, transcriptional response, and apoptosis (1) (Figure 1). Defects in any of these pathways may cause genomic instability (2). This review focuses upon the mechanisms of DNA repair and the DNA damage checkpoints.

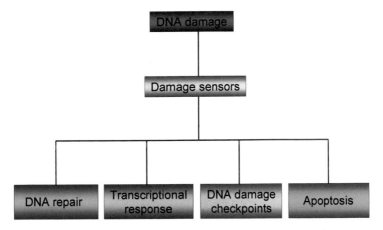

Figure 1 DNA damage response reactions in mammalian cells. The four responses (DNA repair, transcriptional response, DNA damage checkpoints, and apoptosis) may function independently, but frequently a protein primarily involved in one response may participate in other responses.

SUBSTRATES AND SIGNALS FOR DNA REPAIR AND DNA DAMAGE CHECKPOINTS

Proteins that bind to a specific sequence or structure in DNA, in contrast to enzymes with simple substrates, must recognize their target in the vast excess of related structures. Damage sensors not only bind undamaged DNA in search of damage, but they also contact undamaged DNA during specific binding. Therefore, they generally have nonnegligible affinity for undamaged DNA. Since the amount of undamaged DNA vastly exceeds that of damaged DNA, DNA damage sensors spend far more time associated with undamaged DNA than with damaged DNA. Yet, these sensors carry out their specific functions in the presence of high concentrations of nonspecific DNA, because damage recognition is usually a multistep reaction. There is thus a low probability that all of the steps will occur subsequent to the initial contact with undamaged DNA. However, even with multistep recognition, the discrimination between undamaged and damaged DNA is not absolute, so that neither DNA repair nor the DNA damage checkpoints should be envisioned as operated by molecular switches. Rather, both processes are operative at all times, but the magnitudes of the repair or checkpoint reactions are amplified by the presence of DNA damage.

DNA damages include covalent changes in DNA structure and noncovalent anomalous structures, including base-pair mismatches, loops, and bubbles arising from a string of mismatches. The first are processed by DNA repair and recombination pathways, whereas the latter are processed by mismatch repair pathways (3, 4). This review focuses mostly on cellular responses to covalent alterations in DNA structure. Examples of abnormal DNA structures that induce such responses are listed below (Figure 2).

Replication, Recombination, and Repair Intermediates

During these processes, fork structures, bubbles, Holliday structures, and other nonduplex DNA forms are generated, possibly providing high-affinity binding sites for repair/checkpoint proteins. In particular, stalled and collapsed replication forks elicit DNA damage response reactions (5, 6).

DNA Base Damages

Base damages include O^6-methylguanine, thymine glycols, and other reduced, oxidized, or fragmented bases in DNA that are produced by reactive oxygen species or by ionizing radiation. Ultraviolet (UV) radiation also gives rise to these species indirectly by generating reactive oxygen species, as well as producing specific products such as cyclobutane pyrimidine dimers and (6-4) photoproducts. Chemicals form various base adducts, either bulky adducts by large polycyclic hydrocarbons or simple alkyl adducts by alkylating agents. Nearly half of chemotherapeutic drugs, including cisplatin, mitomycin C, psoralen, nitrogen mustard, and adriamycin, make base adducts, and a major challenge in chemotherapy is to find drugs that damage DNA without invoking DNA repair or checkpoint responses in cancer cells.

DNA Backbone Damages

Backbone damages include abasic sites and single- and double-strand DNA breaks. Abasic sites are generated spontaneously, by the formation of unstable base adducts or by base excision repair. Single-strand breaks are produced directly by damaging agents, or single-strand breaks and gaps in the range of 1–30 nucleotide (nt) are produced as intermediates of base and nucleotide excision repair. Breaks induced by oxidative damage often retain a residue of the sugar at the site of the break. Double-strand breaks are formed by ionizing radiation and other DNA-damaging agents. In addition, double-strand breaks are essential intermediates in recombination.

Cross-links

Bifunctional agents such as cisplatin, nitrogen mustard, mitomycin D, and psoralen form interstrand cross-links and DNA-protein cross-links. DNA-protein cross-links may also be produced by reaction of the aldehyde form of abasic sites with proteins (7). Although noncovalent, but tight, DNA-protein complexes are not considered damage in the strict sense, some extremely tight complexes may elicit the same cellular responses as covalent lesions by constituting roadblocks to transcription or replication.

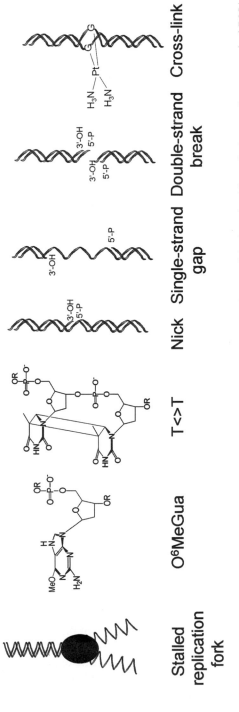

Stalled replication fork   O⁶MeGua   T<>T   Nick   Single-strand gap   Double-strand break   Cross-link

Figure 2 DNA lesions and structures that elicit DNA response reactions. Some of the base backbone lesions and noncanonical DNA structures that elicit DNA response reactions are shown. O⁶ MeGua indicates O⁶-methyldeoxyguanosine, T<>T indicates a cyclobutane thymine dimer, and the cross-link shown is cisplatin G-G interstrand cross-link.

DNA DAMAGE RECOGNITION

Several strategies are utilized for damage recognition to initiate DNA damage responses: direct recognition (bimolecular), multistep recognition (molecular matchmakers and combinatorial), recognition by proxy, and recognition of DNA repair intermediates.

Direct Damage Recognition

In its simplest form, direct damage recognition consists of complementarity between a particular DNA damage and cognate protein, usually an enzyme. Photolyase and DNA glycosylases employ this recognition mechanism. In cases where the enzyme- or enzyme-substrate crystal structures are available, recognition is not by a simple binary reaction involving surface complementarity, but is rather by an induced fit mechanism, as illustrated by *Escherichia coli* photolyase (8–11).

E. coli photolyase repairs cyclobutane pyrimidine dimers by a photoinduced electron transfer from the flavin cofactor, located deep within the core of the protein, to the pyrimidine dimer, which lies within the DNA duplex (Figure 3). For efficient electron transfer, the flavin and the pyrimidine dimer need to be in direct contact. To achieve such contact, the enzyme first interacts with the backbone of the DNA duplex and then, upon encountering the region containing the pyrimidine dimer, makes specific contacts between the distorted duplex backbone (12) and ionic- or H-bond-forming residues on the enzyme surface. This interaction results in a low-affinity and low-specificity enzyme-substrate (E-S) complex. At the dimer site, the H-bonds between the two pyrimidines of the dimer and the complementary purines are either weakened or broken, allowing the pyrimidine dimer to rotate around the axis of the phosphodiester backbone. This rotation permits the dimer to enter a cleft in the enzyme that leads to the flavin in the core. Thus, from the perspective of substrate recognition, this binding mechanism provides two independent proofreading steps (backbone recognition and dinucleotide flipping). Photolyase also has a third proofreading opportunity at the chemical step: Because electron transfer cannot occur to normal bases even if an undamaged dinucleotide were to fit into the active site cavity, it would be rejected without alteration.

Multistep Damage Recognition

MOLECULAR MATCHMAKERS   A molecular matchmaker is a protein that, through a process utilizing ATP hydrolysis, brings two compatible, but otherwise solitary macromolecules together, promotes their association by a conformational change and then dissociates from the complex (13, 14). Examples of molecular matchmakers include Replication Factor C (RFC) in eukaryotic DNA replication and UvrA and XPC in bacterial and mammalian nucleotide excision repair, respec-

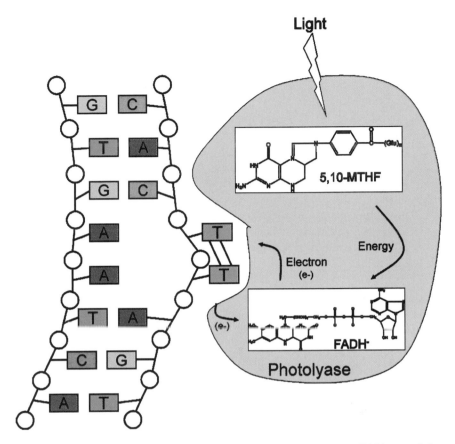

Figure 3 Direct repair by photoreactivation. Photolyase binds to DNA containing a pyrimidine dimer in a light-independent reaction and flips the dimer out into the active site pocket. Catalysis is initiated by light. The photoantenna cofactor, methenyltetrahydrofolate (5,10-MTHF), absorbs a photon and transfers the excitation energy to the catalytic cofactor, FADH⁻. Then, the excited state FADH⁻ transfers an electron to the pyrimidine dimer, splitting the dimer into two pyrimidines. The electron returns to the flavin radical to regenerate FADH⁻, and the enzyme dissociates from the repaired DNA.

tively. RFC loads PCNA onto DNA and then dissociates, allowing PCNA to act as a DNA polymerase clamp and confer high processivity upon the polymerase. UvrA recognizes DNA damage and loads UvrB onto the damage site, and in the process, a conformational change in both UvrB and DNA takes place such that a very stable UvrB-DNA complex is formed. UvrA must then dissociate from this complex to allow the formation of the UvrB-UvrC-DNA complex in which the dual incisions are made.

COMBINATORIAL RECOGNITION   This recognition mechanism is common for transcriptional regulation (15–17), but relatively rare in DNA repair. Genes generally contain multiple regulatory elements and, depending upon which of these sites are occupied, transcription might be turned on or off or the level may be fine-tuned. In DNA repair, there is only one recognition DNA element to which the repair factors may bind, but nevertheless, combinatorial recognition is utilized in a broad sense for several DNA repair functions. For example, photolyase can bind to and repair pyrimidine dimers without the aid of another protein. In the absence of light, however, photolyase binds to the dimer and instead facilitates the binding of UvrA and the subsequent nucleotide excision processes (18). Since UvrA and photolyase can each bind to the damage independently but simultaneously, the synergistic action of the two proteins for promoting excision repair is combinatorial. Similarly, in human nucleotide excision repair, three damage recognition proteins, RPA, XPA, and XPC, act in a cooperative, but not combinatorial, manner to achieve high-specificity recognition.

Recognition by Proxy

Proteins that carry out functions not directly related to DNA repair can become part of the recognition process, as exemplified by transcription-coupled repair in *E. coli*. When RNA polymerase encounters a lesion in the transcribed strand, it arrests at the damage site and, in so doing, creates a target for proteins with affinity for RNA polymerase in the elongation mode (19, 20). One such protein is the bacterial transcription-repair coupling factor (TRCF) that binds to the stalled RNA polymerase and recruits the damage-recognition complex, $UvrA_2UvrB_1$, to the damage site (21). TRCF also facilitates the dissociation of $UvrA_2$ from the $UvrA_2UvrB_1$-DNA complex, accelerating the rate of formation of the rate-limiting $UvrB_1$-DNA complex. Consequently, lesions in the transcribed strand are marked by RNA polymerase and are excised more rapidly than lesions in the nontranscribed strand (or nontranscribed regions of the genome) (22, 23). Because the RNA polymerase does not actively recognize damage but simply arrests at the damage site and thus helps recruit the repair machinery, it is not a matchmaker that actively searches for the target; instead it provides a proxy mechanism of recognition. Transcription-coupled repair occurs in mammalian cells as well (22). However, the molecular details of the mechanism are not known.

Recognition of DNA Repair Intermediates

DNA excision repair pathways involve removal of damages and possibly adjacent DNA nucleotides and replacement with newly synthesized DNA or, in a recombination mechanism, with DNA from the homologous duplex. The repair reactions therefore produce as intermediates single-strand nicks, gaps, flap structures, double-strand breaks, or joint molecules (such as Holliday structures).

Such structures might be recognized by other damage-sensing systems, which may in turn initiate an alternative set of reactions that are not part of the direct repair pathway. For example, a gap generated by nucleotide excision repair might be utilized by the recombination system to initiate homologous recombination (24) or by a DNA damage-checkpoint pathway to arrest cell cycle progression (25).

DNA REPAIR MECHANISMS

DNA repair pathways can be divided into five categories: direct repair, base excision repair, nucleotide excision repair, double-strand break repair, and repair of interstrand cross-links. These are described individually below.

Direct Repair

There are two direct repair mechanisms in the majority of organisms: the photoreversal of UV-induced pyrimidine dimers by DNA photolyase and the removal of the O^6-methyl group from O^6-methylguanine (O^6MeGua) in DNA by methylguanine DNA methyltransferase. Photolyase is not present in many species, including humans, whereas methylguanine DNA methyltransferase has nearly universal distribution in nature.

PHOTOLYASE   Many species from each of the three kingdoms of life and even some viruses have photolyase, whereas many species in all three kingdoms do not (11). Photolyase repairs UV-induced cyclobutane pyrimidine dimers and (6-4) photoproducts using blue-light photons as an energy source (Figure 3). Photolyase is a monomeric protein of 55–65 kDa with two chromophore cofactors, a pterin in the form of methenyltetrapydrofolate and a flavin in the form of $FADH^-$. There are two types of photolyases, one that repairs cyclobutane pyrimidine dimers (photolyase) and the other that repairs (6-4) photoproducts (6-4 photolyase) (8, 10, 11, 26). The structures and reaction mechanisms of the two types are similar, and the reaction mechanism of E. coli photolyase will be used for illustration (Figure 3). The folate, which is located at the surface of the enzyme and functions as a photo-antenna, absorbs a violet/blue-light photon (350–450 nm) and transfers the excitation energy to the flavin cofactor. The excited $FADH^-$ transfers an electron to the cyclobutane pyrimidine dimer to generate a dimer radical anion that splits into two canonical pyrimidines concomitant with back electron transfer to restore the flavin to its catalytically competent form. Placental mammals lack photolyase but contain two proteins with high sequence and structural similarities to photolyase, but no repair function. These proteins, named cryptochromes, function as photoreceptors for setting the circadian clock (11, 27).

METHYLGUANINE DNA METHYLTRANSFERASE   This is a small protein of 20 kDa that does not contain a cofactor and is ubiquitous in nature. Like photolyase, it is presumed to recognize damage by three-dimensional diffusion, and after forming a low-stability complex with the DNA backbone at the damage site, it is thought to flip-out the O^6MeGua base into the active site cavity (28), wherein the methyl group is transferred to an active site cysteine (29). The protein then dissociates from the repaired DNA, but the C-S bond of methylcysteine is stable, and therefore, after one catalytic event the enzyme becomes inactivated and is accordingly referred to as a suicide enzyme. Mice lacking O^6MeGua methyl-transferase are highly susceptible to tumorigenesis by alkylating agents (30). Some human tumors or cell lines resistant to drugs that alkylate DNA are defective in mismatch repair. O^6MeGua has a high frequency of pairing to T, and the O^6MeGua•T base pair activates the mismatch repair system, which leads to futile excision and resynthesis of the T residue, eventually initiating an apoptotic response. Inactivation of the mismatch repair system by either mutation or promoter silencing prevents the futile cycle and the ensuing apoptosis, resulting in resistance to alkylating drugs (31).

In addition to these relatively ubiquitous direct repair enzymes, there are two other direct repair enzymes with more limited phylogenetic distribution: spore photoproduct lyase of spore-forming bacteria, which repairs UV-induced spore photoproducts that form in DNA-A (32), and oxidative methyl transferase in bacteria (AlkB) and humans (hABH1–3), which repairs 1-methyladenine and 3-methylcytosine (33–36). This latter enzyme uses DNA or RNA as a substrate, requires Fe^{2+} and 2-oxoglutarate, and releases the hydroxylated methyl group as formaldehyde.

Base Excision Repair

Base excision repair is initiated by a DNA glycosylase that releases the target base to form an abasic (AP) site in the DNA (Figure 4). AP sites can also be a direct damage product (37–40). There are DNA glycosylases that recognize

---

Figure 4 Base excision repair mechanisms in mammalian cells. A damaged base is removed by a DNA glycosylase to generate an AP site. Depending on the initial events in base removal, the repair patch may be a single nucleotide (short patch) or 2–10 nucleotides (long patch). When the base damage is removed by a glycosylase/AP lyase that cleaves the phosphodiester bond 3′ to the AP site, APE1 endonuclease cleaves the 5′ bond to the site and recruits Pol β to fill in a 1-nt gap that is ligated by Lig3/XRCC1 complex. When the AP site is generated by hydrolytic glycosylases or by spontaneous hydrolysis, repair usually proceeds through the long-patch pathway. APE1 cleaves the 5′ phosphodiester bond, and the RFC/PCNA-Polδ/ε complex carries out repair synthesis and nick translation, displacing several nucleotides. The flap structure is cleaved off by FEN1 endonuclease and the long-repair patch is ligated by Ligase 1.

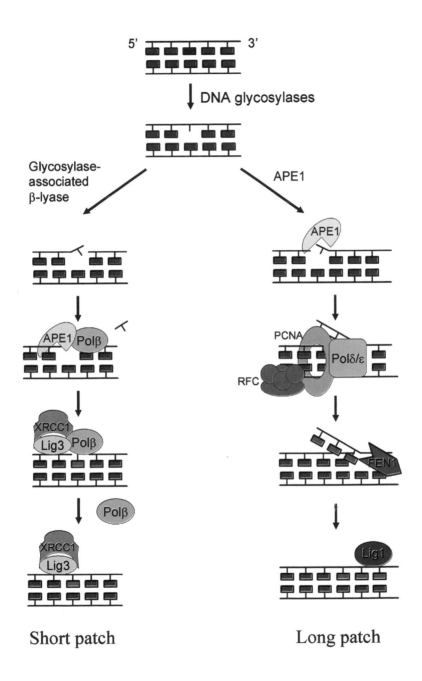

Short patch   Long patch

oxidized/reduced bases, alkylated (usually methylated) bases, deaminated bases (e.g., uracil, xanthine), or base mismatches. Organisms generally contain enzymes recognizing uracil (uracil-DNA glycosylase), alkylated purines (methyl-purine glycosylase), oxidized/reduced pyrimidines (homologues of *E. coli* endonuclease III), or oxidized purines (homologs of *E. coli* Fapy glycosylase or 8-oxoguanine glycosylase). Some DNA glycosylases are simple glycosylases, catalyzing only the hydrolytic removal of the base so as to form an apurinic/apyrimidinic (AP) site, whereas others cleave off the base by a lyase mechanism and catalyze a subsequent AP lyase reaction. In the AP lyase reaction, a Schiff's base is formed with the 1'-aldehyde of the AP deoxyribose, labilizing the 2'-hydrogen to give a β-elimination reaction, leaving a 5'-phosphomonoester and a 3'-unsaturated sugar phosphate residue (39). Generally, lyase reactions are associated with glycosylases that remove oxidized bases, but not with those that remove normal or alkylated bases.

After the lyase reaction, the 3'-sugar residue is generally removed by an AP endonuclease incising 5' to the abasic sugar to form a gap that is filled by DNA polymerase, and the resulting nick is ligated. In cases where the glycosylase lacks lyase activity, the 5' incision is first made by APE1 in mammalian cells, and the abasic sugar can be removed by the dRP lyase activity of DNA polymerase β (Polβ) (41, 42), which concurrently fills in the 1-nucleotide gap. The 1-nucleotide replacement pathway is called the short-patch base excision repair. In mammalian cells, DNA Polβ, APE1, and DNA ligase III-XRCC1 are utilized for short-patch base excision repair. In an alternative mechanism, long-patch base excision repair, APE1 makes the 5' incision to the AP site, and then the combination of DNA Polδ/ε, PCNA, and FEN1 displaces the strand 3' to the nick to produce a flap of 2–10 nucleotides, which is cut at the junction of the single- to double-strand transition by FEN1 endonuclease. A patch of the same size is then synthesized by Polδ/ε with the aid of PCNA and is ligated by DNA ligase 1 (43, 44). Polβ can also take part in a long-patch base excision repair mechanism (45). Depending upon the tissue type, one or the other type of base excision repair may predominate. However, it is thought that in general, base excision repair initiated by glycosylases is short patch and that initiated by AP sites resulting from "spontaneous hydrolysis" or oxidative base loss is long patch.

The damage recognition mechanisms of DNA glycosylases are similar to that of DNA photolyase. The initial recognition is by diffusion. Minor backbone distortions and the H-bond acceptor and donor changes that occur as a result of base damage are recognized at low specificity and affinity, which is followed by base flipping to generate a high-specificity and -affinity complex (46). Enzyme-substrate co-crystals of uracil DNA glycosylase and methyl-purine DNA glycosylase indicate that these enzymes flip-out the base from the duplex by a so-called pinch-push-pull mechanism (40). Binding of the enzyme to the damage site compresses the duplex on either side of the damaged base, thereby enabling the enzyme to insert amino acid side chains into the helix in the immediate vicinity of the damaged base. This facilitates the rotational movement of the base around the phosphodiester bond axis, flipping the base out into the active site cavity of the enzyme, wherein the glycosylic bond is finally attacked by the active site residues. Hence, glycosylases also employ multiple "proofreading" mechanisms to achieve specificity. Despite these safeguards, however, some glycosylases such as methylpurine glycosylases from *E. coli*, yeast, and humans release guanine in addition to N^3-methyladenine and, in fact, *E. coli* 3-mAde DNA glycosylase II releases all four normal bases at low, but physiologically significant levels (47). Thus attacking normal DNA is a price that has to be paid by glycosylases with relatively wide substrate ranges.

Nucleotide Excision Repair

Nucleotide excision repair (excision repair) is the major repair system for removing bulky DNA lesions formed by exposure to radiation or chemicals, or by protein addition to DNA. The damaged bases are removed by the "excision nuclease," a multisubunit enzyme system that makes dual incisions bracketing the lesion in the damaged strand (48–51) (Figure 5). The excision nuclease can also remove all simple single-base lesions. Because of the wide substrate range, excision repair cannot recognize the specific chemical groups that make up the lesion, but is thought to recognize the phosphodiester backbone conformations created by the damage. Recognition of abnormal backbone conformations is sometimes referred to as indirect readout, in contrast to direct readout, which refers to the recognition of a particular DNA sequence through reading out the H-bond donor and acceptor code in the major and minor grooves of the duplex (52). H-bonding, or formation of salt bridges with the backbone, has a considerable degree of flexibility because of the lack of a requirement for strict directionality for forming H-bonds or ionic bonds with negatively charged residues. However, indirect readout is insufficient to explain the ability of the excision nuclease to recognize lesions which in some cases unwind and kink the duplex whereas in others, such as psoralen monoadducts, stabilize the helix (48).

The basic steps of nucleotide excision repair are (*a*) damage recognition, (*b*) dual incisions bracketing the lesion to form a 12–13-nt oligomer in prokaryotes or a 24–32-nt oligomer in eukaryotes, (*c*) release of the excised oligomer, (*d*) repair synthesis to fill in the resulting gap, and (*e*) ligation. As noted above in *E. coli* and other prokaryotes, excision repair is carried out by three proteins, UvrA, UvrB, and UvrC. In humans, excision repair is carried out by 6 repair factors (RPA, XPA, XPC, TFIIH, XPG, and XPF•ERCC1) composed of 15 polypeptides (53–55), and defects in excision repair cause a photosensitivity syndrome called xeroderma pigmentosum (XP), which is characterized by a very high incidence of light-induced skin cancer (56) (Table 1). In both prokaryotes and eukaryotes, the excision follows a similar path (49): an ATP-independent, low-specificity recognition, followed by ATP-dependent DNA unwinding and formation of a long-lived DNA-protein complex with some members of the excision nuclease system, and, finally, dual incisions by two nucleases. Despite these mechanistic similarities, there is no evolutionary relationship between the

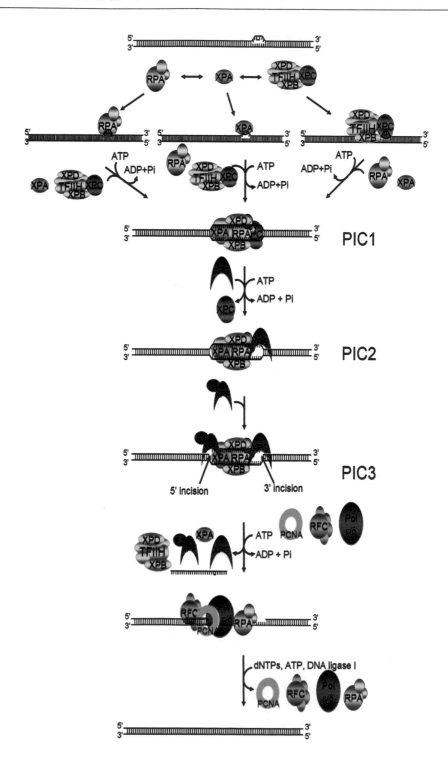

TABLE 1 Six factors of the human excision nuclease

Factor	Proteins (yeast homolog)	Activity	Role in repair
XPA	XPA/p31 (Rad14)	DNA binding	Damage recognition
RPA	p70	DNA binding	Damage recognition
	p32	Replication factor	
	p11		
XPC	XPC/p106 (Rad4)	DNA binding	Damage recognition
	HR23B/p58 (Rad23)		Molecular matchmaker
TFIIH	XPB/ERCC3/p89 (Rad25)	DNA-dependent ATPase	Unwinding the duplex
	XPD/ERCC2/p80 (Rad3)	Helicase	Kinetic proofreading
	p62 (Tfb1)	General transcription factor	
	p52 (Tfb2)		
	p44 (Ssl1)		
	p34 (Tfb4)		
XPG	XPG/ERCC5/p135 (Rad2)	Nuclease	3'-incision
XPF-ERCC1	XPF/ERCC4/p112 (Rad1)	Nuclease	5'-incision
	ERCC1/p33 (Rad10)		

excision nuclease proteins of prokaryotes and eukaryotes. This is surprising as the proteins involved in direct repair, base excision repair, mismatch repair, and recombination are evolutionarily conserved among the eukaryotic and prokaryotic kingdoms. Another significant difference between the prokaryotic and eukaryotic excision repair systems is that the prokaryotic proteins, UvrA, UvrB, and UvrC, do not perform any significant function outside nucleotide excision repair, whereas three of the six eukaryotic excision repair factors play essential roles in replication (RPA), transcription (TFIIH), or recombination (XPF·ERCC1).

Figure 5 Nucleotide excision repair in human cells. The DNA damage is recognized by the cooperative binding of RPA, XPA, and XPC-TFIIH, which assemble at the damage site in a random order. The four repair factors form a complex at the binding site, and if the binding site is damage-free, ATP hydrolysis by the XPB and XPD helicases dissociates the complex (kinetic proofreading). If the site contains a lesion, ATP hydrolysis unwinds the duplex by about 25 bp around the lesion, making a stable preincision complex 1 (PIC1) at the damage site. XPG then replaces XPC in the complex to form a more stable preincision complex 2 (PIC2). Finally, XPF·ERCC1 is recruited to the damage site to form preincision complex 3 (PIC3). The damaged strand is incised at the 6th $\pm$ 3 phosphodiester bond, 3' to the damage by XPG, and the 20th $\pm$ 5 phosphodiester bond 5' to the damage by XPF·ERCC1. The resulting 24–32 oligomer is released, and the gap is filled by Polδ/ϵ with the aid of replication accessory proteins PCNA and RFC.

The damage recognition factors for the human excision nuclease are RPA, XPA, and XPC (57–59). Each of these is a DNA-binding protein with some preference for damaged DNA (59–63). However, the discrimination between undamaged and damaged nucleotides afforded by the individual factors is of insufficient magnitude to confer the necessary in vivo specificity (57, 64). The specificity problem in mammalian nucleotide excision repair has been solved by combining two fundamental mechanisms utilized in high-specificity macromolecular interactions: (*a*) cooperative binding and (*b*) kinetic proofreading. In cooperative binding, specificity is achieved by binding of two or more interacting proteins to adjacent or overlapping sites on DNA (or any macromolecular lattice) such that occupancy of one site by a protein facilitates the occupancy of the adjacent sites by the other proteins by increasing the local concentration of the other proteins by protein-protein interaction (65). Thus, considering the relative abundance of the repair factors (RPA is the most abundant), their DNA affinities (XPC has the highest specific and nonspecific binding constants), and their affinities for each other [RPA to XPA (60) and XPA and XPC to TFIIH (66, 67)], the unique properties of each of the three damage recognition factors are likely utilized in a cooperative manner to assemble the excision nuclease regardless of the order of binding (recognition by random rather than by ordered assembly). The moderate specificity provided by cooperative binding is amplified by the kinetic proofreading function of TFIIH. This is a six-subunit transcription/repair factor (68) with 3′ to 5′ and 5′ to 3′ helicase activities imparted by the XPB and XPD subunits, respectively. In kinetic proofreading, one or more high-energy intermediates are placed in the reaction pathway such that nonspecific (incorrect) products can be aborted at any step along the reaction pathway in a manner that slows down the reaction rate somewhat without seriously compromising the rate of the specific reaction to a physiologically unacceptable level (69). In excision repair, once TFIIH is recruited by the three damage recognition factors to form PIC1 (preincision complex 1), the DNA is unwound by about 20 bp at the assembly site. If the assembly is at a nondamage site, ATP hydrolysis by TFIIH leads to the disassembly of the complex (kinetic proofreading). In contrast, PIC1 formed at a damage site is more stable, and the unwound DNA constitutes a high-affinity binding site for XPG. The XPG protein has higher affinity to unwound DNA than a duplex (70), and therefore binding of XPG, concurrent with dissociation of XPC from the complex, leads to formation of PIC2, which, by virtue of XPG entry and XPC exit, provides an additional layer of specificity. At this step the 3′ incision may take place. However, as a nick can be easily ligated, PIC2 formation is not necessarily irreversible and kinetic proofreading may take place at this step as well. Finally, the entry of XPF•ERCC1 into the complex to form PIC3 results in the irreversible dual incisions and release of the excised oligomer.

To summarize, damage recognition by human excision nuclease utilizes cooperative binding and kinetic proofreading to provide a physiologically relevant degree of specificity at a biologically acceptable rate. However, it must be noted that despite all of these thermodynamic and kinetic safeguards aimed at ensuring excision of only damaged nucleotides, the discrimination of human excision nuclease between damaged and undamaged DNA is not absolute, and the enzyme excises oligomers from undamaged DNA (gratuitous repair) at a significant and potentially mutagenic rate (71).

As noted above, in both prokaryotes and eukaryotes, transcribing DNA is repaired at a faster rate than nontranscribing DNA (22, 23, 72). This transcription-coupled repair requires the CSA and CSB gene products (mutated in Cockayne's syndrome) in humans (20, 73, 74) and the Mfd protein in *E. coli* (75). In vivo data indicate that in humans, but not in yeast, XPC is dispensable for transcription-coupled repair (76). This repair mode is well understood in *E. coli* but has not been reconstituted in vitro with eukaryotic proteins, and therefore its precise mechanism in humans remains to be elucidated. In addition to transcription, excision repair is also affected by chromatin structure, which exerts a strong inhibitory effect (77–80). Chromatin remodeling factors (81–83) and histone modification (84) partially alleviate the inhibitory effect of nucleosomes.

Finally, the role of an enigmatic protein called DDB (*D*amaged *D*NA *B*inding protein) in nucleotide excision repair deserves some comment. Of all the human damaged DNA-binding proteins identified to date, DDB has the highest affinity for damaged DNA (85, 86) as exemplified by a $K_D = 5 \times 10^{-10}$ for (6-4) photoproducts (87). Ironically, DDB is not required for repair (48–51), and it is uncertain whether DDB plays a direct role in excision repair or in any other repair reaction. DDB is a heterodimer of p127, the product of the *DDB1* gene, and p48, the product of the *DDB2* gene (88). Mutations in *DDB2* give rise to xeroderma pigmentosum group E (XP-E) (89), a mild form of the disease. There are contradictory findings on the direct involvement of DDB in excision repair (74, 90–92). However, it appears that, both in vivo (93) and in vitro (64, 94), DDB has no or only a minor role in the excision nuclease function. DDB interacts with several cellular and viral proteins involved in transcriptional regulation and other cellular processes (95). Current evidence suggests that DDB might be involved in a general cellular response to DNA damage rather than a specific repair pathway. Two recent studies support such a model. In one study, it was found that nontransformed fibroblasts from XP-E patients had greatly reduced basal and damage-inducible p53 levels and the levels of p53-regulated proteins (96). These cells are deficient in caspase 3 activation and apoptosis, and as a consequence, more resistant than normal cells to killing by UV irradiation. Another study found that a complex consisting of DDB1, DDB2, Cul4A, and Roc1 exhibited a robust ubiquitin ligase activity and that this core complex was associated with the 8-subunit COP9 signalosome (CSN) in which the ubiquitin ligase activity of the core complex was inhibited (97). Interestingly, this study also identified an essentially identical complex in which the DDB2 subunit was replaced with the CSA protein. An interesting model was proposed for how these two complexes may regulate global repair and transcription-coupled repair by stimulating ubiquitin ligase activity or suppressing it in the DDB2/CSN and CSA/CSN complexes, respectively.

Finally, in addition to the well-characterized nucleotide excision and base excision repair pathways, in a number of organisms, including *Schizosaccharomyces pombe*, there exists an alternative excision repair pathway in which an enzyme called UV dimer endonuclease incises DNA immediately 5' to the damage, and then the damage is removed by a 5' to 3' exonuclease (98, 99). Similarly, AP endonucleases such as the Nfo of *E. coli* and the APE1 in humans can incise immediately 5' to many nonbulky oxidatively damaged bases to initiate their removal by 5' to 3' exonucleases (100).

Double-Strand Break Repair and Recombinational Repair

Double-strand breaks are produced by reactive oxygen species, ionizing radiation, and chemicals that generate reactive oxygen species. Double-strand breaks are also a normal result of V(D)J recombination and immunoglobulin class-switching processes, but occur unnaturally during replication as a consequence of replication fork arrest and collapse. Double-strand breaks are repaired either by homologous recombination (HR) or nonhomologous end-joining (NHEJ) mechanisms (Figure 6) (101–105).

HOMOLOGOUS RECOMBINATION (HR)   This process has three steps: strand invasion, branch migration, and Holliday junction formation. The Holliday junction is then resolved into two duplexes by the structure-specific endonucleases, resolvases. Strand invasion and branch migration is initiated by Rad51 in eukaryotes (106), or RecA in prokaryotes (6, 107). In eukaryotes, Rad52, Rad54, Rad55, Rad57, BRCA1, and BRCA2 are also involved in homologous recombination, but the precise roles of these proteins are unclear. The Mre11/Rad50/NBS1 (M/R/N) complex may process the termini of the double-strand break before initiation of strand invasion by Rad51. The MUS81·MMS4 heterodimer resolves the Holliday junctions or the topologically equivalent four-strand intermediates arising from replication fork regression ("chicken foot") (108–111). The distinguishing property of homologous recombination is that the information lost from the broken duplex is retrieved from a homologous duplex. In cases where the two duplexes are not exactly homologous, gene conversion may take place.

SINGLE-STRAND ANNEALING   A double-strand break repair mechanism that may be considered a transitional pathway between HR and NHEJ is the so-called single-strand annealing (SSA) repair mechanism. In this case, the ends of the duplex are digested by an exonuclease, possibly the M/R/N complex (112–114), until regions of some homology on the two sides of the break are exposed. These regions are then paired, and the nonhomologous tails are trimmed off, so that the two duplex ends can be ligated. SSA is invariably associated with information

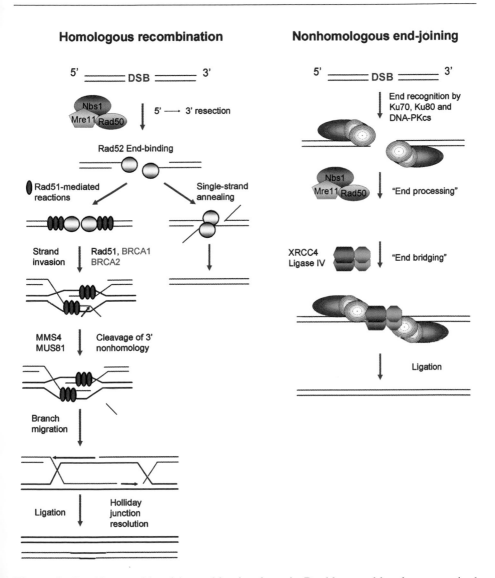

Figure 6   Double-strand break/recombinational repair. Double-strand breaks are repaired by either homologous recombination, which in eukaryotes depends on the Rad51-family proteins—orthologs of the bacterial RecA recombinase, or by nonhomologous end-joining mediated by the DNA-PK complex. A key intermediate in homologous recombination is the Holliday intermediate, in which the two recombining duplexes are joined covalently by single-strand crossovers. Resolvases such as MUS81·MMS4 cleave the Holliday junctions to separate the two duplexes. In the single-strand annealing (SSA) mechanism, the duplex is digested by a 5′ to 3′ exonuclease to uncover microhomology regions that promote pairing, trimming, and ligation. BRCA1 and BRCA2 are also involved in homologous recombination, but their precise roles are unclear.

loss due to the exonuclease action. In addition to the M/R/N complex, Rad52 and RPA are also likely to participate in the SSA pathway.

NONHOMOLOGOUS END-JOINING (NHEJ)   In this form of repair in eukaryotes, the Ku heterodimer binds to the two ends of a double-strand break and recruits DNA-PKcs (115) and the ligase4-XRCC4 heterodimer, which then ligates the two duplex termini regardless of whether the two ends come from the same chromosome (116, 117). The M/R/N complex may also participate in NHEJ, particularly when this pathway is utilized for V(D)J recombination.

Genetic data indicate that HR is important for the recovery of collapsed replication forks. In contrast, NHEJ is essential for V(D)J recombination and is thought to be the major pathway for repair of double-strand breaks induced by ionizing radiation and radiomimetic agents. The relative contribution of SSA to the repair of double-strand breaks is unknown.

Cross-link Repair

Many chemotherapeutic drugs and bifunctional DNA-damaging agents induce interstrand DNA cross-links. In *E. coli* and in *Saccharomyces cerevisiae*, nucleotide excision repair and homologous recombination systems work coordinately or in tandem to remove the cross-link (118, 119). In humans, it is unknown whether nucleotide excision repair plays a significant role in repairing cross-links. In CHO cells, mutations in XPF (ERCC4) and ERCC1 render cells extremely sensitive to cross-linking agents, whereas mutations in other excision repair genes have a less drastic effect (120). This has been taken as evidence that the structure-specific XPF·ERCC1 nuclease plays a special role in cross-link repair aside from its function in nucleotide excision repair.

When cross-linked DNA is used as a substrate for the human excision nuclease in vitro, two activities are detected. One requires all six excision nuclease factors and results in dual incisions, both of which occur 5′ to the lesion and excise a fragment free of damage that is 22–28 nt in length from either strand of the duplex (121, 122). This gap might initiate a recombination reaction that may ultimately remove the cross-link or might initiate a futile repair synthesis reaction in which the cell excises an oligonucleotide, fills in the gap, and then re-excises the oligonucleotide. The second activity involves XPF·ERCC1, which, in an RPA-dependent reaction, digests the DNA in a 3′ to 5′ direction past the cross-link and thus converts the interstrand cross-link to a single-strand diadduct (122).

Cross-links induce double-strand breaks during replication both in vivo and in vitro (123, 124), presumably due to replication fork collapse and consequent nuclease attack. The following is a working model for cross-link repair in mammalian cells (Figure 7). Starting at the double-strand break induced by replication, the XPF·ERCC1 nuclease degrades one of the cross-linked strands in a 3′ to 5′ direction past the cross-link and thus converts the interstrand cross-link into an intrastrand dinucleotide adduct adjacent to a double-strand break (122).

Recombination proteins, including Rad51, Rad52, XRCC2, XRCC3, and RPA, would then act in concert to generate the classic Holliday junction intermediate that would be resolved by the MUS81·MMS4 complex. In addition to this presumably major pathway, there is evidence both in yeast and in humans for a minor cross-link repair pathway involving the entire nucleotide excision repair system and the error-prone DNA polymerases (125, 126). Finally, cells from patients with Fanconi's anemia (FA) are extremely sensitive to cross-linking agents (127). Recent findings indicate that the FA complex (composed of FANC-A, -C, -E, and -F proteins), together with the M/R/N complex and BRCA1/BRCA2 (identical to FA-B and FA-D1), participate in both cross-link repair and double-strand break repair by homologous recombination (127).

Various repair pathways may share enzymes and reaction intermediates. Conversely, particular lesions might be repaired by more than one pathway. In the latter case, various pathways might compete for the same substrate, interfering with one another's function, or might cooperate in removing the lesion. Likewise, it is unclear whether a particular damage-specific binding protein can act as a nucleation site for more than one repair pathway. For example, O^6-methylguanine can be repaired by direct repair, by base excision repair, or by nucleotide excision repair, and it is unknown whether the binding of RPA to the lesion to initiate nucleotide excision repair interferes with direct and base excision repair pathways or instead stimulates them. Similarly, it is not yet clear whether the DNA repair proteins participate directly in the DNA damage checkpoint responses.

DNA DAMAGE CHECKPOINTS

What Are the DNA Damage Checkpoints?

DNA damage checkpoints are biochemical pathways that delay or arrest cell cycle progression in response to DNA damage (128). Originally a checkpoint was defined as a specific point in the cell cycle when the integrity of DNA was examined ("checked") before allowing progression through the cell cycle (129, 130). The meaning of the term checkpoint has recently become more ambiguous, as the term has been applied to the entire ensemble of cellular responses to DNA damage, including the arrest of cell cycle progression, induction of DNA repair genes, and apoptosis. There is some justification for this expansion of the definition, because activation of proteins involved in cell-cycle arrest also leads to induction of genes that participate in DNA repair and apoptosis. However, DNA repair pathways are functional in the absence of damage-induced cell-cycle arrest, and apoptosis can occur independently of the cell-cycle arrest machinery. Accordingly, the term "DNA damage checkpoint" should be reserved for the events that slow or arrest cell-cycle progression in response to DNA damage.

A further question is whether checkpoints are specific points during the cell cycle at which genomic integrity is assessed (131). As the biochemical basis of

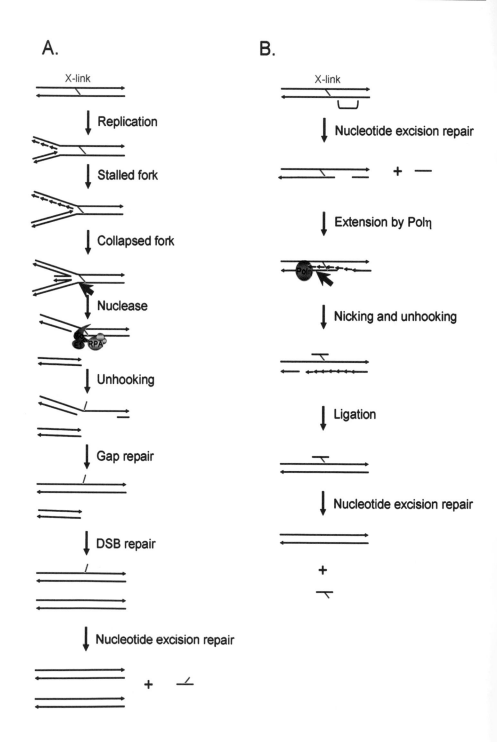

checkpoint responses becomes better defined, it also becomes clear that they are not cellular counterparts of international border checkpoints, but instead analogous to constant surveillance and response systems in that they continuously monitor the integrity of the genome and control cell-cycle progression accordingly (131).

All eukaryotic cells have four phases within the cell cycle, G1, S, G2, and M, and one outside, G_0. In some unicellular organisms, G1 or G2 are extremely short and difficult to observe, but in mammalian somatic cells, the phases are well-defined and represent stages in the life of a cell in which distinct biochemical reactions take place. In an unperturbed cell cycle, the transition points G1/S and G2/M, as well as S-phase progression, are tightly controlled, and available evidence indicates that the same proteins involved in regulating the orderly progression through the cell cycle are also involved in the checkpoint responses. Thus, the DNA damage checkpoints are not unique pathways activated by DNA damage, but rather are biochemical pathways operative under normal growth conditions that are amplified upon an increase in DNA damage. Whether there are thresholds for the level of damage necessary to elicit the checkpoint responses is not known.

Although the checkpoint pathways are operational during the entire cell cycle and hence may slow down the cell cycle at any point during the four phases, the slowing or arrest of progression from one phase to another represents qualitative changes in the cellular state as opposed to quantitative changes that may occur within states. The term checkpoint is defined based upon the interstate transition that is being inhibited by DNA damage as the G1/S, intra-S, and G2/M checkpoints.

Figure 7   Cross-link (X-link) repair. (A) In the error-free mechanism, a replication fork induces a double-strand break upon encountering a cross-link. XPF·ERCC1, aided by RPA, then digests the duplex 3′ to 5′ past the cross-link. The resulting intermediate is annealed to a homologous duplex and is converted to a duplex with a dinucleotide cross-link by a recombination reaction. Prior to or following the resolution of the Holliday intermediate, the dinucleotide cross-link is eventually eliminated by nucleotide excision repair. Note that in this mode of repair, only one of the six excision repair factors, XPF·ERCC1, is required for the initial conversion of the interstrand cross-link into a single-strand lesion. (B) In the error-prone pathway, the 6-factor excision nuclease makes dual incisions 5′ to the cross-link in one or the other strand to produce a 26-nucleotide gap 5′ to the cross-link. An error-prone polymerase (such as Polη) fills in the gap and may synthesize past the cross-link to generate a triple-stranded intermediate that may constitute a substrate for the excision nuclease system. Following damage removal by excision repair, the gap is filled in and ligated.

TABLE 2   DNA damage checkpoint proteins

Protein function	Mammals	*S. pombe*	*S. cerevisiae*
Sensors			
RFC-like	Rad17	Rad17	Rad24
PCNA-like	Rad9	Rad9	Ddc1
	Rad1	Rad1	Rad17
	Hus1	Hus1	Mec3
PI3-Kinases (PIKK)	ATM	Tel1	Tel1
	ATR	Rad3	Mec1
PIKK binding partner	ATRIP	Rad26	Ddc2/Lcd1/Pie1
Mediators			
	MDC1		
	53BP1		
	TopBP1	Cut5	Dpb11
	Claspin	Mrc1	Mrc1
	BRCA1	Crb2/Rph9	Rad9
Transducers			
Kinase	Chk1	Chk1	Chk1
	Chk2	Cds1	Rad53

Molecular Components of the DNA Damage Checkpoints

The DNA damage checkpoint, like other signal transduction pathways, conceptually has three components: sensors, signal transducers, and effectors (Table 2). Extensive genetic analysis in yeast and more recently in human cells, combined with rather limited biochemical studies, have helped to identify proteins involved in damage sensing, signal transduction, and effector steps of the DNA damage checkpoints (Figure 8). However, there is not an absolute demarcation between the various components of the checkpoint. For example, the damage sensor, ATM, also functions as a signal transducer. Moreover, a fourth class of checkpoint proteins, called mediators, has been identified. This class includes BRCA1, Claspin, 53BP1, and MDC1 and is conceptually placed between sensors and signal transducers. However, as in the case of ATM functioning as both a sensor and a transducer, these mediator proteins also appear to participate in more than one step of the checkpoint response.

. Although the G1/S, intra-S, and the G2/M checkpoints are distinct, the damage sensor molecules that activate the various checkpoints appear to either be shared by all three pathways or to play a primary sensor role in one pathway and a back-up role in the others. Similarly, the signal-transducing molecules, which are protein kinases and phosphatases, are shared by the different checkpoints to

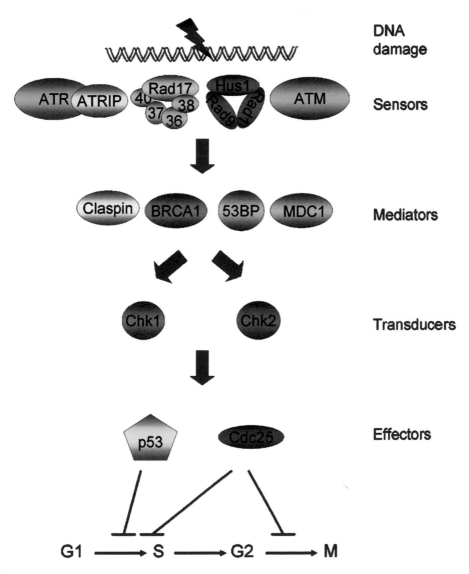

Figure 8 Components of the DNA damage checkpoints in human cells. The damage is detected by sensors that, with the aid of mediators, transduce the signal to transducers. The transducers, in turn, activate or inactivate other proteins (effectors) that directly participate in inhibiting the G1/S transition, S-phase progression, or the G2/M transition.

varying degrees (1, 132). The effector components (proteins that inhibit phase transition) of the checkpoints are, naturally, what gives the checkpoints their unique identities. However, various sensors, mediators, and signal transducers may play more prominent roles in one checkpoint than in others (133, 134), as is apparent when the checkpoints are analyzed separately, below.

SENSORS   As in DNA repair, DNA damage checkpoints require the recognition of DNA damage to initiate subsequent events. Although it is commonly believed that the sensors for repair and for the checkpoint response are distinct, future research will likely reveal considerable overlap at the damage detection step. Clearly, RPA is an important damage sensor for both nucleotide excision repair (48, 50) and the DNA damage checkpoints (135). Nevertheless, two groups of proteins have been identified as checkpoint-specific damage sensors: the two phosphoinositide 3-kinase-like kinase (PIKK) family members, ATM and ATR (136), and the RFC/PCNA (clamp loader/polymerase clamp)-related Rad17-RFC/9-1-1 complex (137).

ATM and ATR proteins   Mutations in ATM (*a*taxia *t*elangiectesia *m*utated) cause ataxia-telangiectasia (A-T) in humans, a condition primarily characterized by cerebellar degeneration, immunodeficiency, genome instability, clinical radiosensitivity, and cancer predisposition (138). ATM is a 350-kDa oligomeric protein containing many HEAT motifs (139, 140). It exhibits significant sequence homology to the phosphoinositide 3-kinases, but lacks lipid kinase activity (138, 141). It does have protein kinase activity, and this activity is stimulated in vivo by agents that induce double-strand breaks (142) and reportedly by linear DNA in vitro (143, 144). Atomic force electron microscopy shows that ATM binds preferentially to DNA termini, apparently in monomeric form (145). Upon exposure of cells to ionizing radiation, ATM phosphorylates many proteins, including Chk2, p53 (142, 146), NBS1 (147), BRCA1 (148), and itself (140) at serines and threonines in the sequence context of SQ or TQ. However, neighboring sequences also confer specificity—for example, $S_{15}Q$ of p53 is phosphorylated by ATM but not $S_{37}Q$. Indeed, a comprehensive study with synthetic peptides has uncovered a hierarchy of substrate preferences for ATM (149). Autophosphorylation of ATM converts the oligomer into monomers, which appear to be the active form of the enzyme for the checkpoint response (140).

ATR was discovered in the human genome database as a gene with sequence homology to ATM and SpRad3, hence the name ATR (*A*TM and *R*ad3 related) (150). The gene encodes a protein of 303 kDa with a C-terminal kinase domain and regions of homology to other PIKK family members. Knockout of ATR in mice results in embryonic lethality (151, 152), and mutations causing partial loss of ATR activity in humans have been associated with the human autosomal recessive disorder Seckel syndrome, which shares features in common with A-T (153). ATR, like ATM, is a protein kinase with specificity for S and T residues in SQ/TQ sequences, and it phosphorylates essentially all of the proteins that are phosphorylated by ATM. In contrast to ATM, ATR is activated in vivo by UV light rather than by ionizing radiation, and it is the main PIKK family member that initiates signal transduction following UV irradiation (132).

The relative specificity of ATR for UV irradiation raised the possibility that the enzyme might directly recognize specific damages formed by this agent. Assays for electrophoretic mobility shift, DNA pulldown with immobilized ATR, ATR pulldown with immobilized DNA, and UV-induced DNA-protein cross-linking showed that ATR binds directly to DNA (154). The enzyme binds to a 50-bp duplex containing a (6-4) photoproduct with about twofold higher affinity than undamaged duplex. Analysis of ATR-DNA interactions by electron microscopy further showed that binding to DNA increases with increasing UV irradiation. Furthermore, when linear DNA was used, ATR was rarely seen at the DNA termini, indicating that, in contrast to ATM, ATR does not recognize double-strand breaks. When phosphorylation of p53 by ATR was measured, it was found that undamaged DNA stimulates the kinase activity (155, 156) and that UV-irradiated DNA increased the rate of phosphorylation moderately and in a dose-dependent manner (154). Thus, at present it appears that ATM is a sensor and transducer responding to double-strand breaks, and ATR serves an analogous role for base damages, at least from UV irradiation.

The prototype of the PIKK family, DNA-PK, is a heterotrimer of a 450-kDa catalytic subunit (DNA-PKcs) and a dimer of Ku70 and Ku80. The Ku70/Ku80 dimer binds to DNA ends and recruits DNA-PKcs, which then becomes activated as a DNA-dependent protein kinase (115). It has been suggested that ATM and ATR are recruited to DNA through the intermediacy of a similar DNA-binding partner. Indeed, in both *S. cerevisiae* and *S. pombe*, the ATR homologs (scMec1 and spRad3, respectively) were shown to be in a complex with scDdc2 and spRad3. Although the existence of an ATM-associated partner has not yet been reported in mammalian cells, ATR interacts with an 86-kDa protein, ATRIP (*ATR Interacting Protein*), that has very limited homology to scDdc2 and spRad26 (157). Although DNA recruitment studies in *S. cerevisiae* showed that the scMec1-scDdc2 complex functions in a manner analogous to DNA-PKcs/Ku70/Ku80 (i.e., scMec1 requires scDdc2 for DNA binding), this is not the case for ATR (135, 154). ATRIP is not required for the binding of ATR to naked or RPA-covered single-strand DNA or double-strand DNA (135, 154). However, ATRIP may confer some specificity for the binding of the ATR-ATRIP complex to RPA coated DNA rather than naked DNA (135).

Rad17-RFC and the 9-1-1 complex   The Rad17-RFC complex is a checkpoint-specific structural homolog of the replication factor, RFC. The replicative form of RFC is a heteropentamer composed of p140, p40, p38, p37, and p36. In Rad17-RFC, the p140 subunit is replaced by the 75-kDa Rad17 protein (158–161). Electron microscopy has revealed that the two complexes have similar structures: a globular shape with a deep groove running down the length of the complex (162, 163).

The 9-1-1 (Rad9-Rad1-Hus1) complex is the checkpoint counterpart of PCNA, a homotrimer with a ring-like structure. Although the Rad9, Rad1, and Hus1 proteins have little sequence homology to PCNA, or to one another, molecular modeling suggested that they may form a PCNA-like structure (164) and indeed the three proteins were found to interact by a variety of methods (161, 165–168). Electron microscopic analysis showed that they form a heterotrimeric ring of 5-nm diameter, with a 2-nm hole that is very similar to that of PCNA (162, 163). During replication, RFC binds to the primed template junction, displaces DNA Polα, and recruits PCNA, which it clamps around the DNA duplex in an ATP-dependent reaction. PCNA recruits Polδ to replace RFC in the replicative complex, and then PCNA-Polδ carries out highly processive DNA synthesis (169). Rad17-RFC and the 9-1-1 complex are thought to perform functions analogous to RFC and PCNA during the checkpoint response, but details of the process are unknown. Genetic data in fission yeast indicate that Rad17-RFC and the 9-1-1 complex are necessary for all checkpoint responses (137, 170, 171), but that the spRad3-spRad26 complex responds to DNA-damage independently of other checkpoint proteins (172). In fact, in vivo biochemical experiments show that in budding yeast the 9-1-1 complex equivalent (scDdc1-scRad17-scMec3) is recruited to double-strand breaks introduced by HO endonuclease independently of recruitment of scMec1 (ATR) (173, 174). In human cells, chromatin binding experiments have yielded similar results (175): When cells are exposed to either ionizing radiation or UV light, ATR and the 9-1-1 complex become associated with chromatin independently of one another. Rad17-RFC was found to be bound to chromatin at all times regardless of DNA damage.

In vitro studies have shown that Rad17-RFC is a DNA-stimulated ATPase that binds both to DNA and the 9-1-1 complex (161). Binding to DNA is ATP-independent, but the formation of the Rad17-RFC/9-1-1 complex requires ATP. Furthermore, Rad17-RFC recruits the 9-1-1 complex to a nicked or gapped circle (but not to a covalently closed duplex) or to an artificial template/primer structure in an ATP-dependent manner. RFC is a molecular matchmaker that loads PCNA onto DNA in a catalytic manner fueled by ATP hydrolysis. Rad17-RFC appears to load the 9-1-1 complex onto DNA in a similar manner in both humans and *S. cerevisiae* (176, 177). Thus, Rad17-RFC/9-1-1 clearly performs a function similar to that of RFC/PCNA pair. What confers specificity to the checkpoint clamp loader/checkpoint clamp is unknown. However, phosphorylation of hRad17 by ATR upon DNA damage is necessary for the DNA damage checkpoint response (178, 179).

MEDIATORS   These proteins simultaneously associate with damage sensors and signal transducers at certain phases of the cell cycle and as a consequence help provide signal transduction specificity. The prototype mediator is the scRad9 protein (180, 181), which functions along the signal transduction pathway from scMec1 (ATR) to scRad53 (Chk2). Another mediator, Mrc1 (*m*ediator of *r*eplication checkpoint), found in both *S. cerevisiae* and *S. pombe* (133, 134), is expressed only during S phase and is necessary for S-phase checkpoint signaling from scMec1/spRad3 to scRad53/spCds1. In humans, three proteins that contain the BRCT protein-protein interaction module fit into the mediator category: the p53 binding protein, 53BP1 (182, 183); the topoisomerase binding protein, TopBP1 (184); and the mediator of DNA damage checkpoint 1, MDC1 (185–187). These proteins interact with damage sensors such as ATM, repair proteins such as BRCA1 and the M/R/N complex, signal transducers such as Chk2, and even effector molecules such as p53. The DNA damage checkpoint response is abrogated in cells that have decreased levels of, or lack, these proteins.

In addition to these bona fide mediators, other proteins such as H2AX, BRCA1, the M/R/N complex, and SMC1 (structural maintenance of chromatin 1) play essential roles in the activation of checkpoint kinases. As these proteins also play direct roles in DNA repair, sister chromatid pairing, and segregation, they cannot simply be considered to be mediators. Similarly, the human Claspin, which has marginal sequence homology to yeast Mrc1 and was originally thought to be a mediator based on studies conducted with *Xenopus* Claspin in egg extracts (188, 189), appears to function more like a sensor because, even though it is required for xChk1 phosphorylation by xATR, xClaspin is recruited to stalled replication forks independently of xATR and xRad17 (190).

SIGNAL TRANSDUCERS   In humans, there are two kinases, Chk1 and Chk2, with a strictly signal transduction function in cell cycle regulation and checkpoint responses (137, 191, 192). These kinases were identified based on homology with yeast scChk1 (193) and scRad53/spCds1 (194–197), respectively. Both Chk1 and Chk2 are S/T kinases with moderate substrate specificities. In mammalian cells, the double-strand break signal sensed by ATM is transduced by Chk2 (197, 198), and the UV-damage signal sensed by ATR is transduced by Chk1 (132, 199). However, there is some overlap between the functions of the two proteins, as is apparent in discussing the individual checkpoints, below. Of special significance, Chk1 (-/-) causes embryonic lethality in mice (200, 201), whereas Chk2 (-/-) mice are viable and appear to exhibit near-normal checkpoint responses (202). Mutations in hChk2 cause a Li-Fraumeni-like cancer-prone syndrome in humans (203).

EFFECTORS   In humans, three phosphotyrosine phosphatases, Cdc25A, -B, and -C, dephosphorylate the cyclin-dependent kinases that act on proteins directly involved in cell-cycle transitions. Phosphorylation of these Cdc25 proteins by the checkpoint kinases creates binding sites for the 14-3-3 adaptor proteins, of which there are 8 isoforms. Phosphorylation inactivates the Cdc25 proteins by excluding them from the nucleus, by causing proteolytic degradation, or both. Unphosphorylated Cdc25 proteins promote the G1/S transition by dephosphorylating Cdk2 and promote the G2/M transition by dephosphorylating Cdc2 phosphotyrosine (204).

The G1/S Checkpoint

The G1/S checkpoint prevents cells from entering the S phase in the presence of DNA damage by inhibiting the initiation of replication (Figure 9). Under suitable conditions, cells in the G1 phase of the cell cycle become committed to enter the S phase at a stage called the restriction point in mammalian cells and start in budding yeast (205). The restriction point precedes the actual start of DNA synthesis by about 2 h in human cells. If there is DNA damage, however, entry into S phase is prevented regardless of whether the cells have passed the restriction point. Current evidence suggests the following sequence of events for the G1/S checkpoint in human cells. If the DNA damage is double-strand breaks caused by ionizing radiation or radiomimetic agents, ATM is activated and phosphorylates many target molecules, notably p53 and Chk2. These phosphorylations result in the activation of two signal transduction pathways, one to initiate and one to maintain the G1/S arrest (204). The reaction that initiates the G1/S arrest is phosphorylation of Chk2, which in turn phosphorylates Cdc25A phosphatase, causing its inactivation by nuclear exclusion and ubiquitin-mediated proteolytic degradation (206, 207). Lack of active Cdc25A results in the accumulation of the phosphorylated (inactive) form of Cdk2, which is incapable of phosphorylating Cdc45 to initiate replication at the sites of preformed ORC-ORI complexes. If the DNA damage is by UV light or UV-mimetic agents, the signal is sensed by ATR, Rad17-RFC, and the 9-1-1 complex, leading to phosphorylation of Chk1 by ATR. The activated Chk1 then phosphorylates Cdc25A, leading to G1 arrest.

Whether the initial arrest is caused by the ATM-Chk2-Cdc25A pathway or the ATR-Chk1-Cdc25A pathway, this rapid response is followed by the p53-mediated maintenance of G1/S arrest, which becomes fully operational several hours after the detection of DNA damage (204). In the maintenance stage, ATM or ATR phosphorylates Ser15 of p53 directly and Ser20 through activation of Chk2 or Chk1, respectively (142, 146, 208–210). The phosphorylation of p53

→

Figure 9   The G1/S checkpoint. DNA damage is sensed by ATM after double-strand breaks or by ATR, Rad17-RFC, and the 9-1-1 complex after UV-damage. ATM/ATR phosphorylates Rad17, Rad9, p53, and Chk1/Chk2 that in turn phosphorylates Cdc25A, causing its inactivation by nuclear exclusion and ubiquitin-mediated degradation. Phosphorylated and inactivated Cdk2 accumulates and cannot phosphorylate Cdc45 to initiate replication. Maintenance of the G1/S arrest is achieved by p53, which is phosphorylated on Ser15 by ATM/ATR and on Ser20 by Chk1/Chk2. Phosphorylated p53 induces p21[WAF-1/Cip1] transcription, and p21[WAF-1/Cip1] binds to the Cdk4/CycD complex, thus preventing it from phosphorylating Rb, which is necessary for the release of the E2F transcription factor and subsequent transcription of S-phase genes. p21[WAF-1/Cip1] also binds to and inactivates the Cdk2/CycE complex, thus securing the maintenance of the G1/S checkpoint.

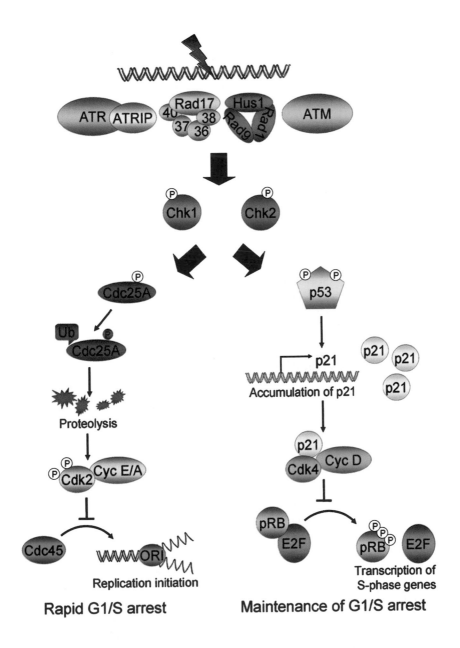

inhibits its nuclear export and degradation, thus resulting in increased levels of p53 (211). p53 activates its target genes, including p21$^{WAF-1/Cip1}$, which binds to and inhibits the S-phase-promoting Cdk2-CyclinE complex, thereby maintaining the G1/S arrest. p21$^{WAF-1/Cip1}$ also binds to the Cdk4-CyclinD complex and prevents it from phosphorylating Rb (212). The phosphorylation of Rb results in its release of the E2F transcription factor, which is required for the transcription of S-phase genes in order for S phase to proceed (204, 213).

An important issue in studies on the G1/S checkpoint is the identity of the initiating signal. Several studies in budding yeast suggest that processing of the damage by excision repair may be necessary to initiate the checkpoint response (214–217). When G1-arrested yeast cells are UV-irradiated, entry into S phase is delayed when the cells are released from the arrest. This checkpoint response is dependent on a functional excision repair system and on the yeast homologs of ATR, Rad17-RFC, and the 9-1-1 complex. When excision repair–deficient cells are subjected to the same treatment, they progress into S phase with no delay. Once they enter the S phase, however, the checkpoint is activated, presumably via replication blocks at UV photoproducts. These data have been interpreted to mean that single-strand gaps generated by excision repair were the necessary intermediates for activating the G1/S checkpoint by UV irradiation. Experiments with *Xenopus* egg extracts have generally supported this model (190, 218–220). There is less evidence for a role of excision repair in checkpoint activation in mammalian cells. At low doses of UV irradiation, however, p53 accumulation during G1 is defective in *XPA* (-/-) cells, which cannot make UV-induced nicks or gaps (221). At high doses of UV, by contrast, p53 does accumulate during G1 even in excision repair-defective cells. Indeed, a recent report shows that in budding yeast, activation of the G1/S checkpoint at moderately high UV doses is independent of excision repair (222). It is possible that cyclobutane pyrimidine dimers, which predominate at low doses of UV, do not constitute a significant signal for checkpoint activation, but that (6-4) photoproducts, which become a substantial fraction of the damage at high UV doses, can be recognized by the damage sensors for checkpoint activation.

The Intra-S-Phase Checkpoint

The intra-S-phase checkpoint is activated by damage encountered during the S phase or by unrepaired damage that escapes the G1/S checkpoint and leads to a block in replication (215). Although there is some evidence for the slowing down of replication fork progression by an active mechanism such as the sequestration of PCNA by p21$^{WAF-1/Cip1}$ (212) or the ubiquitination of PCNA by scRad6 (223), the predominant mechanism of S-phase arrest is the inhibition of firing of late origins of replication (224–229). The original definition of a checkpoint implied a biochemical pathway that ensures the completion of a reaction in a prior phase of the cycle before transition to the next phase, so that the intra-S-phase checkpoint does not strictly meet that definition of a checkpoint (110). However, as the molecular bases of checkpoints have become clearer, it has become evident that the intra-S checkpoint meets the current definition of a checkpoint as "a biochemical regulatory pathway which dictates the progression of cell cycle events (rather than phase transitions) in an orderly manner and that prevents the initiation of certain biochemical reactions before completion of the others within the cell" (131).

The damage sensors for the intra-S checkpoint encompass a large set of checkpoint and repair proteins. When the damage is a frank double-strand break or a double-strand break resulting from replication of a nicked or gapped DNA, ATM, the M/R/N complex, and BRCA1 are all required for the activation of the checkpoint (230) (Figure 10). It is presumed that all of these repair proteins function as sensors because biochemical data show that all bind to either double-strand breaks (ATM) or to special branched DNA structures (the M/R/N complex, BRCA1, BRCA2) (104, 127, 231–234). In addition to actively participating in DNA repair, these proteins also activate the intra-S checkpoint by a kinase signaling cascade. It appears that the S-phase checkpoint initiated by double-strand breaks proceeds through two pathways. The well-understood ATM-Chk2-Cdc25A-Cdk2 pathway is strictly a checkpoint response. A second pathway, which depends on phosphorylation of SMC1 by ATM with the aid of BRCA1, FANCD2, and NBS1, likely plays a role in cell-cycle arrest, but more importantly, activates the recovery process mediated by these numerous recombination proteins, including the cohesins SMC1 and SMC3 (235, 236). The two branches of the S-phase checkpoint are outlined in Figure 10.

In vivo biochemical analysis shows that H2AXγ (phosphorylated H2AX), 53BP1, BRCA1, Mre11, FANCD2, and SMC1 colocalize to foci containing the double-strand breaks aided by the MDC1 mediator (185–187). The details of formation of these so-called ionizing radiation induced foci (IRIF), presumed to be checkpoint/repair factories, are not known. In fact, in contrast to ATR, ATM fails to exhibit clear focal localization after DNA damage (132). Functional evidence indicates, however, that at some point ATM must localize to the IRIF because it phosphorylates H2AX, 53BP1, NBS1, BRCA1, FANCD2, and SMC1, which do localize to these sites. A plausible scenario is that the double-strand break repair proteins bind to the break site where they are then phosphorylated and activated by ATM, initiating the ATM-NBS1/FANCD2/BRCA1-SMC1 pathway. The latter plays an active role in repair of double-strand break or recovery of collapsed replication forks, and also likely plays a role in S-phase checkpoint signaling.

A most striking feature of the ionizing radiation-induced S-phase checkpoint is a phenomenon called *radioresistant DNA synthesis* (RDS), which illustrates the intimate relationship between this checkpoint and double-strand break repair. In wild-type cells, ionizing radiation causes what appears to be an immediate cessation of ongoing DNA synthesis. In contrast, in cells from patients with A-T, a disease characterized by extreme sensitivity to ionizing radiation, DNA synthesis is not inhibited by radiation (RDS), and the cells appear to go through S phase without any delay (224). RDS is also observed in cells with mutations

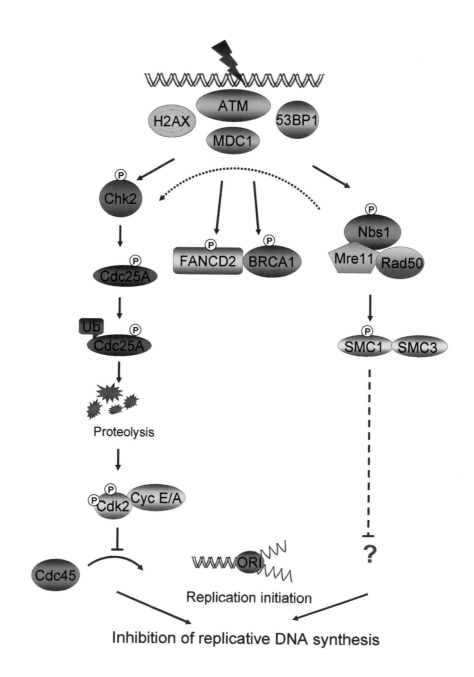

in subunits of the M/R/N complex (232, 237, 238) and in FANCD2 (239). However, RDS is not necessarily associated with increased sensitivity to ionizing radiation. Thus, although BRCA1 mutants exhibit RDS, they are only mildly sensitive to ionizing radiation; and NBS1 and FANCD2 mutants, which also have the RDS phenotype, are only marginally sensitive to ionizing radiation, but very sensitive to cross-linking agents. Perhaps ATM, the loss of which results in RDS and extreme sensitivity to ionizing radiation, possesses a nonredundant function in the S-phase checkpoint, which encompasses both checkpoint and repair/recovery reactions, so that its loss results in a more severe radiation-sensitivity phenotype.

In contrast to ATM as a sensor for double-strand breaks, when DNA is damaged by UV or chemicals that make bulky base lesions, the main PIKK family damage sensor is the ATR protein, or more precisely, the ATR-ATRIP heterodimer (132, 157). ATR binds to chromatin (240) and can bind directly to UV-induced lesions (154), or to RPA-coated single-stranded DNA (135) generated from the repair or replication of these lesions, and becomes activated. Activated ATR phosphorylates Chk1, which in turn phosphorylates and downregulates Cdc25A, and thus inhibits firing of replication origins (229) (Figure 11). Recent evidence from *Xenopus* egg extracts suggests that the presence of single-stranded DNA during S phase induces an ATR-dependent checkpoint that inhibits origin firing by downregulating Cdc7-Dbf4 protein kinase activity, which is required for Cdc45 binding to chromatin (241). Therefore, the downregulation of either Cdk2/Cyclin E (CDK complex) or Cdc7/Dbf4 (DDK complex) may result in the inhibition of late replication origins during the intra-S checkpoint. As for the ATM-initiated checkpoint, ATR-initiated signaling also results in the phosphorylation of BRCA1 (242, 243), NBS1, and other targets, promoting the recovery of stalled/regressed/collapsed replication forks and thereby coordinating the inhibition of replication initiation with the recovery of active replication forks by homologous recombination and related processes.

Genetic analyses of yeast have demonstrated that DNA Polymerase ε (Polε) may play a role in sensing DNA damage during S phase (244, 245). Polε is required for coordinated and efficient chromosomal DNA replication in eukaryotes (246–249) and has also been implicated in nucleotide excision (250, 251), base excision (252), and recombinational repair (253). The catalytic subunit

---

Figure 10 The ATM-regulated intra-S-phase checkpoint. In response to double-strand breaks induced by ionizing radiation, ATM triggers two cooperating parallel cascades to inhibit replicative DNA synthesis. ATM, through the intermediacy of MDC1, H2AX, and 53BP1, phosphorylates Chk2 on Thr68 to induce ubiquitin-mediated degradation of Cdc25A phosphatase. The degradation locks the S phase–promoting Cyclin E/Cdk2 in its inactive, phosphorylated form and prevents the loading of Cdc45 on the replication origin. ATM also initiates a second pathway by phosphorylating NBS1 of the M/R/N complex, as well as SMC1, BRCA1, and FANCD2.

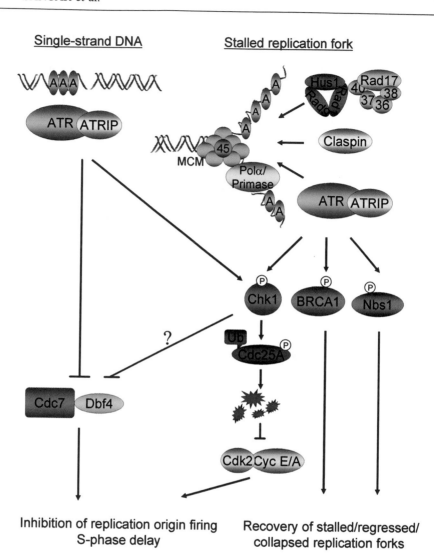

Figure 11 The ATR-mediated intra-S checkpoint. The ATR-ATRIP complex, Rad17-RFC, the 9-1-1 complex, and Claspin are independently recruited to RPA(A)-coated single-stranded regions of the stalled replication fork. ATR then phosphorylates Chk1 and other substrates, and activated Chk1 phosphorylates Cdc25A, which leads to inactivation of Cdk2/Cyclin E complex. Single-strand DNA gaps can also be sensed by ATR and following ATR activation, origin firing and consequently DNA replication is inhibited through downregulation of Cdc7/Dbf4 protein kinase activity.

of Polε (261 kDa) is composed of two domains: an N-terminal domain containing polymerase and exonuclease motifs and a 120-kDa C-terminal domain (254). Interestingly, in budding and fission yeasts, the C-terminal domain of Polε, but not the N-terminal polymerase domain, is essential for viability (245, 255, 256), indicating that the checkpoint sensor function of Polε is more important for cellular physiology than its polymerase functions. Mutations in the essential C-terminal domain abolish the inducibility of the RNR3 subunit of ribonucleotide reductase in response to DNA damage or hydroxyurea (244). In mammalian cells, the C terminus of Polε interacts with Mdm2, perhaps as part of a DNA damage response (257, 258).

Studies with *S. cerevisiae* and *S. pombe* have shown that lesions in replicating DNA cause replication fork arrest, and that recombination proteins, such as Rad51, the M/R/N complex, and the MUS81·MMS4 resolvase (259), and the replication proteins, such as the error-prone Polζ (260), aid in the recovery of the stalled replication fork (231). S-phase checkpoint mutants lack a coordinated replication fork arrest and retrieval/recombinational process. As a consequence, intermediates with long single-stranded regions accumulate following DNA damage in these mutants (261) and eventually lead to potentially lethal double-strand breaks (226, 262). Although these double-strand breaks are expected to occur in a stochastic manner, evidence indicates that in both *S. cerevisiae* Mec1 mutants (263) and in human ATR mutants (264), the breaks preferentially occur during G2 in specific regions of chromosomes that under physiological conditions replicate slower than the rest of the genome [called *replication slow zones* (RSZ) in yeast and *Fragile* sites (FRA) in humans]. These breaks are thought to occur in unreplicated chromosomal regions from stalled forks that escape the scMec1/ATR-mediated intra-S checkpoint and are, in part, the source of chromosome rearrangements often seen in tumor cells (264).

The G2/M Checkpoint

The G2/M checkpoint prevents cells from undergoing mitosis in the presence of DNA damage. Depending on the type of DNA damage, the ATM-Chk2-Cdc25 signal transduction pathway and/or the ATR-Chk1-Cdc25 pathway is activated to arrest the cell cycle following DNA damage in G2 (199, 265, 266) (Figure 12). As in other checkpoints, with certain types of DNA lesions, such as those created by UV light, ATR-Chk1 signaling initiates cell-cycle arrest, but the maintenance of the arrest is then performed by ATM-Chk2 signaling (132). With other types of lesions, such as ionizing radiation-induced double-strand breaks, the order of action is reversed (265, 266). In any event, checkpoint kinases inhibit the entry into mitosis by downregulating Cdc25 and upregulating Wee1, which together control Cdc2/CyclinB activity (267). Initially, Cdc25C was thought to be the effector of the G2/M checkpoint (268). Subsequently, it was found that mouse Cdc25C (-/-) cells have a normal G2/M checkpoint (269). In contrast, disruption of the Chk1-Cdc25A pathway abrogates ionizing radiation-induced S and G2 checkpoints (270), indicating that the Cdc25A phosphatase is the main effector of the G2/M checkpoint. Upon phosphorylation, the Cdc25 phosphatase binds to the 14-3-3 proteins, becomes sequestered in the cytoplasm, and is degraded by the ubiquitin-proteasome pathway (195, 268, 271). Inactivation of Cdc25 leads to

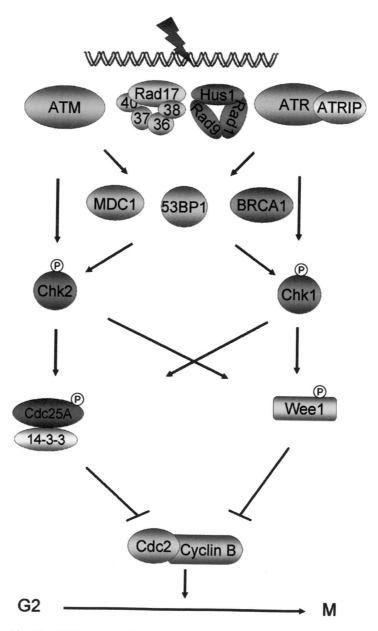

Figure 12 The G2/M checkpoint. The ionizing radiation- (ATM) and UV damage-responsive sensor proteins (ATR-ATRIP, Rad17-RFC, and 9-1-1) are recruited to the damage site. The mediator proteins such as MDC1, BRCA1 and/or 53BP1 communicate the DNA damage signal to Chk1 and/or Chk2, thereby regulating the Cdc2/CyclinB, Wee1, and Cdc25A proteins that are crucial for the G2/M transition by changing their expression, phosphorylation, and cellular localization.

accumulation of Y15-phosphorylated Cdc2 and mitotic arrest. Initiation of the G2/M checkpoint may not be as simple as presented above. MAP kinases p38γ (272) and p38α (273) have been implicated in the G2/M checkpoint responses to ionizing radiation and UV, respectively. Clearly, the interfacing of two major cellular signal transduction pathways, the MAP kinase- and the checkpoint response pathways, deserves more study.

The Replication Checkpoint (S/M Checkpoint)

The replication checkpoint (also referred to as the S/M checkpoint) is the process by which mitosis is inhibited while DNA replication is ongoing or blocked (128). In both the G2/M and replication checkpoints, the ATR-Chk1-Cdc25 signal transduction pathway is utilized to inhibit mitosis, although the initiating signals for the two checkpoints are different. Ongoing replication or replication forks blocked by DNA damage or nucleotide starvation initiate the replication checkpoint. Evidence from in vitro studies with *Xenopus* egg extracts indicates that the initiating signal is a component of the replication fork, and experiments with various replication inhibitors suggested that the signal might actually be the RNA primer of Okazaki fragments (274). However, more recent experiments have raised questions about the true identity of the replication checkpoint signal, as the inhibitors used to assign it to RNA primers inhibited the entire replisome assembly (275).

The replication checkpoint has been observed in all model systems, including yeast (110) and mammalian cells (266). Surprisingly, however, recent evidence indicates that in mice the replication checkpoint activated by hydroxyurea and aphidicolin is independent of ATM and ATR, but the UV- and ionizing radiation-induced replication checkpoint is dependent on these damage sensor/signal transducer kinases (266), raising the possibility that replication forks and DNA damage during S phase may inhibit mitosis by different signaling mechanisms.

CONCLUSIONS

DNA damage activates several distinct biochemical pathways. First, DNA repair enzymes of varying complexities recognize and eliminate the damage. Second, DNA damage activates DNA damage checkpoints, which arrest cell cycle progression. Under most circumstances, checkpoints aid in cellular survival. Third, DNAα damage activates transcription of certain genes (transcriptional response). The role of the transcriptional response in cell survival is unknown. Finally, apoptosis in metazoans is the programmed cell death that is activated by either cell death ligands or DNA damage, and serves to eliminate superfluous or deregulated and dangerous cells.

The four DNA damage response pathways described above can and do function independently under certain circumstances. However, under most conditions there is extensive interaction between these response reactions. Clearly, the individual players in the checkpoint responses serve not only to delay the cell cycle, but also to mediate DNA repair, directly and indirectly. Exactly how this occurs and, in the greater sense, how these agents assess DNA damage both quantitatively and qualitatively so as to choose between mediating DNA repair or apoptosis are outstanding areas for study. These aspects of the DNA damage response pathways must be considered in future research for a better understanding of the cell's response to genotoxicants and for the development of strategies aimed at targeting these pathways for cancer prevention and chemotherapy.

ACKNOWLEDGMENTS

This work was supported by National Institutes of Health grants GM32833 (to A.S.), GM20830 (to L.A.L.-B.), and GM59424 (to S.L.), and by the Miller Institute for Basic Research in Science (A.S.).

The *Annual Review of Biochemistry* is online at http://biochem.annualreviews.org

LITERATURE CITED

1. Zhou BB, Elledge SJ. 2000. *Nature* 408: 433–39
2. Kolodner RD, Putnam CD, Myung K. 2002. *Science* 297:552–57
3. Modrich P, Lahue R. 1996. *Annu. Rev. Biochem.* 65:101–33
4. Kolodner RD, Marsischky GT. 1999. *Curr. Opin. Genet. Dev.* 9:89–96
5. Cox MM, Goodman MF, Kreuzer KN, Sherratt DJ, Sandler SJ, Marians KJ. 2000. *Nature* 404:37–41
6. Cox MM. 2002. *Mutat. Res.* 510:107–20
7. Minko IG, Zou Y, Lloyd RS. 2002. *Proc. Natl. Acad. Sci. USA* 99:1905–9
8. Sancar A. 1994. *Biochemistry* 33:2–9
9. Deisenhofer J. 2000. *Mutat. Res.* 460: 143–49
10. Todo T. 1999. *Mutat. Res.* 434:89–97
11. Sancar A. 2003. *Chem. Rev.* 103:2203–38
12. Park H, Zhang KJ, Ren YJ, Nadji S, Sinha N, et al. 2002. *Proc. Natl. Acad. Sci. USA* 99:15965–70
13. Orren DK, Selby CP, Hearst JE, Sancar A. 1992. *J. Biol. Chem.* 267:780–88
14. Sancar A, Hearst JE. 1993. *Science* 259: 1415–20
15. Johnson PF, McKnight SL. 1989. *Annu. Rev. Biochem.* 58:799–839
16. Chen L. 1999. *Curr. Opin. Struct. Biol.* 9:48–55
17. Naar AM, Lemon BD, Tjian R. 2001. *Annu. Rev. Biochem.* 70:475–501
18. Sancar A, Franklin KA, Sancar GB. 1984. *Proc. Natl. Acad. Sci. USA* 81:7397–401
19. Tornaletti S, Hanawalt PC. 1999. *Biochimie* 81:139–46
20. Friedberg EC. 1996. *Annu. Rev. Biochem.* 65:15–42
21. Selby CP, Sancar A. 1993. *Science* 260: 53–58
22. Mellon I, Spivak G, Hanawalt PC. 1987. *Cell* 51:241–49
23. Mellon I, Hanawalt PC. 1989. *Nature* 342:95–98
24. Cox MM. 2001. *Annu. Rev. Genet.* 35: 53–82
25. Zhou J, Ahn J, Wilson SH, Prives C. 2001. *EMBO J.* 20:914–23
26. Todo T, Takemori H, Ryo H, Ihara M, Matsunaga T, et al. 1993. *Nature* 361: 371–74

27. Sancar A. 2000. *Annu. Rev. Biochem.* 69:31–67

28. Daniels DS, Tainer JA. 2000. *Mutat. Res.* 460:151–63

29. Lindahl T, Sedgwick B, Sekiguchi M, Nakabeppu Y. 1988. *Annu. Rev. Biochem.* 57:133–57

30. Kawate H, Sakumi K, Tsuzuki T, Nakatsuru Y, Ishikawa T, et al. 1998. *Proc. Natl. Acad. Sci. USA* 95:5116–20

31. Modrich P. 1997. *J. Biol. Chem.* 272:24727–30

32. Fajardo-Cavazos P, Salazar C, Nicholson WL. 1993. *J. Bacteriol.* 175:1735–44

33. Koonin EV, Aravind L. 2001. *J. Mol. Biol.* 307:1271–92

34. Falnes PO, Johansen RF, Seeberg E. 2002. *Nature* 419:178–82

35. Trewick SC, Henshaw TF, Hausinger RP, Lindahl T, Sedgwick B. 2002. *Nature* 419:174–78

36. Aas PA, Otterlei M, Falnes PO, Vagbo CB, Skorpen F, et al. 2003. *Nature* 421:859–63

37. Wilson SH. 1998. *Mutat. Res.* 407:203–15

38. Memisoglu A, Samson L. 2000. *Mutat. Res.* 451:39–51

39. McCullough AK, Dodson ML, Lloyd RS. 1999. *Annu. Rev. Biochem.* 68:255–85

40. Mol CD, Parikh SS, Putnam CD, Lo TP, Tainer JA. 1999. *Annu. Rev. Biophys. Biomol. Struct.* 28:101–28

41. Matsumoto Y, Kim K. 1995. *Science* 269:699–702

42. Beard WA, Wilson SH. 2000. *Mutat. Res.* 460:231–44

43. Frosina G, Fortini P, Rossi O, Carrozzino F, Raspaglio G, et al. 1996. *J. Biol. Chem.* 271:9573–78

44. Klungland A, Lindahl T. 1997. *EMBO J.* 16:3341–48

45. Prasad R, Dianov GL, Bohr VA, Wilson SH. 2000. *J. Biol. Chem.* 275:4460–66

46. Roberts RJ, Cheng X. 1998. *Annu. Rev. Biochem.* 67:181–98

47. Berdal KG, Johansen RF, Seeberg E. 1998. *EMBO J.* 17:363–67

48. Sancar A. 1996. *Annu. Rev. Biochem.* 65:43–81

49. Petit C, Sancar A. 1999. *Biochimie* 81:15–25

50. Wood RD. 1997. *J. Biol. Chem.* 272:23465–68

51. Wood RD. 1999. *Biochimie* 81:39–44

52. Travers AA. 1989. *Annu. Rev. Biochem.* 58:427–52

53. Mu D, Park CH, Matsunaga T, Hsu DS, Reardon JT, Sancar A. 1995. *J. Biol. Chem.* 270:2415–18

54. Mu D, Hsu DS, Sancar A. 1996. *J. Biol. Chem.* 271:8285–94

55. Evans E, Moggs JG, Hwang JR, Egly JM, Wood RD. 1997. *EMBO J.* 16:6559–73

56. Cleaver JE. 1968. *Nature* 218:652–56

57. Mu D, Wakasugi M, Hsu DS, Sancar A. 1997. *J. Biol. Chem.* 272:28971–79

58. Wakasugi M, Sancar A. 1998. *Proc. Natl. Acad. Sci. USA* 95:6669–74

59. Wakasugi M, Sancar A. 1999. *J. Biol. Chem.* 274:18759–68

60. He Z, Henricksen LA, Wold MS, Ingles CJ. 1995. *Nature* 374:566–69

61. Sugasawa K, Ng JM, Masutani C, Iwai S, van der Spek PJ, et al. 1998. *Mol. Cell.* 2:223–32

62. Sugasawa K, Okamoto T, Shimizu Y, Masutani C, Iwai S, Hanaoka F. 2001. *Genes Dev.* 15:507–21

63. Missura M, Buterin T, Hindges R, Hubscher U, Kasparkova J, et al. 2001. *EMBO J.* 20:3554–64

64. Reardon JT, Sancar A. 2003. *Genes Dev.* 17:2539–51

65. Ptashne M, Gann A. 2002. *Genes and Signals.* Cold Spring Harbor, NY: Cold Spring Harbor Lab. Press

66. Park CH, Mu D, Reardon JT, Sancar A. 1995. *J. Biol. Chem.* 270:4896–902

67. Drapkin R, Reardon JT, Ansari A, Huang JC, Zawel L, et al. 1994. *Nature* 368:769–72

68. Egly JM. 2001. *FEBS Lett.* 498:124–28

69. Hopfield JJ. 1974. *Proc. Natl. Acad. Sci. USA* 71:4135–39

70. Hohl M, Thorel F, Clarkson SG, Scharer OD. 2003. *J. Biol. Chem.* 278:19500–8

71. Branum ME, Reardon JT, Sancar A. 2001. *J. Biol. Chem.* 276:25421–26

72. Bohr VA, Smith CA, Okumoto DS, Hanawalt PC. 1985. *Cell* 40:359–69

73. Venema J, Mullenders LH, Natarajan AT, van Zeeland AA, Mayne LV. 1990. *Proc. Natl. Acad. Sci. USA* 87:4707–11

74. Hanawalt PC. 2002. *Oncogene* 21: 8949–56

75. Selby CP, Witkin EM, Sancar A. 1991. *Proc. Natl. Acad. Sci. USA* 88:11574–78

76. Venema J, van Hoffen A, Natarajan AT, van Zeeland AA, Mullenders LH. 1990. *Nucleic Acids Res.* 18:443–48

77. Smerdon MJ, Conconi A. 1999. *Prog. Nucleic Acid Res. Mol. Biol.* 62:227–55

78. Thoma F. 1999. *EMBO J.* 18:6585–98

79. Hara R, Mo J, Sancar A. 2000. *Mol. Cell. Biol.* 20:9173–81

80. Green CM, Almouzni G. 2002. *EMBO Rep.* 3:28–33

81. Ura K, Araki M, Saeki H, Masutani C, Ito T, et al. 2001. *EMBO J.* 20:2004–14

82. Hara R, Sancar A. 2002. *Mol. Cell. Biol.* 22:6779–87

83. Hara R, Sancar A. 2003. *Mol. Cell. Biol.* 23:4121–25

84. Wang D, Hara R, Singh G, Sancar A, Lippard SJ. 2003. *Biochemistry* 42: 6747–53

85. Chu G, Chang E. 1988. *Science* 242: 564–67

86. Tang J, Chu G. 2002. *DNA Repair* 1:601–16

87. Reardon JT, Nichols AF, Keeney S, Smith CA, Taylor JS, et al. 1993. *J. Biol. Chem.* 268:21301–8

88. Keeney S, Chang GJ, Linn S. 1993. *J. Biol. Chem.* 268:21293–300

89. Nichols AF, Ong P, Linn S. 1996. *J. Biol. Chem.* 271:24317–20

90. Wakasugi M, Shimizu M, Morioka H, Linn S, Nikaido O, Matsunaga T. 2001. *J. Biol. Chem.* 276:15434–40

91. Hwang BJ, Toering S, Francke U, Chu G. 1998. *Mol. Cell. Biol.* 18:4391–99

92. Hwang BJ, Ford JM, Hanawalt PC, Chu G. 1999. *Proc. Natl. Acad. Sci. USA* 96:424–28

93. Itoh T, Nichols A, Linn S. 2001. *Oncogene* 20:7041–50

94. Kazantsev A, Mu D, Nichols AF, Zhao X, Linn S, Sancar A. 1996. *Proc. Natl. Acad. Sci. USA* 93:5014–18

95. Shiyanov P, Nag A, Raychaudhuri P. 1999. *J. Biol. Chem.* 274:35309–12

96. Itoh T, O'Shea CO, Linn S. 2003. *Mol. Cell. Biol.* 23:7540–53

97. Groisman R, Polanowska J, Kuraoka I, Sawada J, Saijo M, et al. 2003. *Cell* 113: 357–67

98. Alleva JL, Zuo S, Hurwitz J, Doetsch PW. 2000. *Biochemistry* 39:2659–66

99. McCready SJ, Osman F, Yasui A. 2000. *Mutat. Res.* 451:197–210

100. Ischenko AA, Saparbaev MK. 2002. *Nature* 415:183–87

101. Petrini JH. 1999. *Am. J. Hum. Genet.* 64:1264–69

102. Ferguson DO, Alt FW. 2001. *Oncogene* 20:5572–79

103. Khanna KK, Jackson SP. 2001. *Nat. Genet.* 27:247–54

104. D'Amours D, Jackson SP. 2002. *Nat. Rev. Mol. Cell. Biol.* 3:317–27

105. Rouse J, Jackson SP. 2002. *Science* 297: 547–51

106. Sung P. 1994. *Science* 265:1241–43

107. Kowalczykowski SC. 2000. *Trends Biochem. Sci.* 25:156–65

108. Chen XB, Melchionna R, Denis CM, Gaillard PH, Blasina A, et al. 2001. *Mol. Cell.* 8:1117–27

109. Kaliraman V, Mullen JR, Fricke WM, Bastin-Shanower SA, Brill SJ. 2001. *Genes Dev.* 15:2730–40

110. Boddy MN, Russell P. 2001. *Curr. Biol.* 11:R953–56

111. Ogrunc M, Sancar A. 2003. *J. Biol. Chem.* 278:21715–20

112. Paull TT, Gellert M. 1998. *Mol. Cell.* 1:969–79

113. Trujillo KM, Yuan SSF, Lee EYHP, Sung P. 1998. *J. Biol. Chem.* 273: 21447–50

114. Carney JP, Maser RS, Olivares H, Davis EM, Le Beau M, et al. 1998. *Cell* 93:477–86

115. Gottlieb TM, Jackson SP. 1993. *Cell* 72:131–42

116. Ramsden DA, Gellert M. 1998. *EMBO J.* 17:609–14

117. McElhinny SAN, Snowden CM, McCarville J, Ramsden DA. 2000. *Mol. Cell. Biol.* 20:2996–3003

118. Van Houten B, Gamper H, Holbrook SR, Hearst JE, Sancar A. 1986. *Proc. Natl. Acad. Sci. USA* 83:8077–81

119. Jachymczyk WJ, von Borstel RC, Mowat MR, Hastings PJ. 1981. *Mol. Gen. Genet.* 182:196–205

120. Hoy CA, Thompson LH, Mooney CL, Salazar EP. 1985. *Cancer Res.* 45:1737–43

121. Bessho T, Mu D, Sancar A. 1997. *Mol. Cell. Biol.* 17:6822–30

122. Mu D, Bessho T, Nechev LV, Chen DJ, Harris TM, et al. 2000. *Mol. Cell. Biol.* 20:2446–54

123. McHugh PJ, Sones WR, Hartley JA. 2000. *Mol. Cell. Biol.* 20:3425–33

124. Bessho T. 2003. *J. Biol. Chem.* 278: 5250–54

125. Greenberg RB, Alberti M, Hearst JE, Chua MA, Saffran WA. 2001. *J. Biol. Chem.* 276:31551–60

126. Zheng H, Wang X, Warren AJ, Legerski RJ, Nairn RS, et al. 2003. *Mol. Cell. Biol.* 23:754–61

127. D'Andrea AD, Grompe M. 2003. *Nat. Rev. Cancer* 3:23–34

128. Nyberg KA, Michelson RJ, Putnam CW, Weinert TA. 2002. *Annu. Rev. Genet.* 36:617–56

129. Weinert TA, Hartwell LH. 1988. *Science* 241:317–22

130. Hartwell LH, Weinert TA. 1989. *Science* 246:629–34

131. Nasmyth K. 1996. *Science* 274:1643–45

132. Abraham RT. 2001. *Genes Dev.* 15:2177–96

133. Alcasabas AA, Osborn AJ, Bachant J, Hu F, Werler PJ, et al. 2001. *Nat. Cell Biol.* 3:958–65

134. Tanaka K, Russell P. 2001. *Nat. Cell Biol.* 3:966–72

135. Zou LZ, Elledge SJ. 2003. *Science* 300: 1542–48

136. Durocher D, Jackson SP. 2001. *Curr. Opin. Cell Biol.* 13:225–31

137. Melo J, Toczyski D. 2002. *Curr. Opin. Cell Biol.* 14:237–45

138. Shiloh Y. 1997. *Annu. Rev. Genet.* 31:635–62

139. Perry J, Kleckner N. 2003. *Cell* 112: 151–55

140. Bakkenist CJ, Kastan MB. 2003. *Nature* 421:499–506

141. Savitsky K, Bar-Shira A, Gilad S, Rotman G, Ziv Y, et al. 1995. *Science* 268: 1749–53

142. Banin S, Moyal L, Shieh S, Taya Y, Anderson CW, et al. 1998. *Science* 281: 1674–77

143. Gately DP, Hittle JC, Chan GK, Yen TJ. 1998. *Mol. Biol. Cell* 9:2361–74

144. Chan DW, Son SC, Block W, Ye R, Khanna KK, et al. 2000. *J. Biol. Chem.* 275:7803–10

145. Smith GC, Cary RB, Lakin ND, Hann BC, Teo SH, et al. 1999. *Proc. Natl. Acad. Sci. USA* 96:11134–39

146. Canman CE, Lim DS, Cimprich KA, Taya Y, Tamai K, et al. 1998. *Science* 281:1677–79

147. Lim DS, Kim ST, Xu B, Maser RS, Lin J, et al. 2000. *Nature* 404:613–17

148. Cortez D, Wang Y, Qin J, Elledge SJ. 1999. *Science* 286:1162–66

149. Kim ST, Lim DS, Canman CE, Kastan MB. 1999. *J. Biol. Chem.* 274:37538–43

150. Cimprich KA, Shin TB, Keith CT, Schreiber SL. 1996. *Proc. Natl. Acad. Sci. USA* 93:2850–55

151. Brown EJ, Baltimore D. 2000. *Genes Dev.* 14:397–402

152. de Klein A, Muijtjens M, van Os R, Ver-

hoeven Y, Smit B, et al. 2000. *Curr. Biol.* 10:479–82

153. O'Driscoll M, Ruiz-Perez VL, Woods CG, Jeggo PA, Goodship JA. 2003. *Nat. Genet.* 33:497–501

154. Ünsal-Kaçmaz K, Makhov AM, Griffith JD, Sancar A. 2002. *Proc. Natl. Acad. Sci. USA* 99:6673–78

155. Hall-Jackson CA, Cross DA, Morrice N, Smythe C. 1999. *Oncogene* 18:6707–13

156. Lakin ND, Hann BC, Jackson SP. 1999. *Oncogene* 18:3989–95

157. Cortez D, Guntuku S, Qin J, Elledge SJ. 2001. *Science* 294:1713–16

158. Griffiths DJ, Barbet NC, McCready S, Lehmann AR, Carr AM. 1995. *EMBO J.* 14:5812–23

159. Kondo T, Matsumoto K, Sugimoto K. 1999. *Mol. Cell. Biol.* 19:1136–43

160. Green CM, Erdjument-Bromage H, Tempst P, Lowndes NF. 2000. *Curr. Biol.* 10:39–42

161. Lindsey-Boltz LA, Bermudez VP, Hurwitz J, Sancar A. 2001. *Proc. Natl. Acad. Sci. USA* 98:11236–41

162. Griffith JD, Lindsey-Boltz LA, Sancar A. 2002. *J. Biol. Chem.* 277:15233–36

163. Shiomi Y, Shinozaki A, Nakada D, Sugimoto K, Usukura J, et al. 2002. *Genes Cells* 7:861–68

164. Venclovas C, Thelen MP. 2000. *Nucleic Acids Res.* 28:2481–93

165. St Onge RP, Udell CM, Casselman R, Davey S. 1999. *Mol. Biol. Cell* 10:1985–95

166. Volkmer E, Karnitz LM. 1999. *J. Biol. Chem.* 274:567–70

167. Burtelow MA, Kaufmann SH, Karnitz LM. 2000. *J. Biol. Chem.* 275:26343–48

168. Burtelow MA, Roos-Mattjus PM, Rauen M, Babendure JR, Karnitz LM. 2001. *J. Biol. Chem.* 276:25903–9

169. Yuzhakov A, Kelman Z, Hurwitz J, O'Donnell M. 1999. *EMBO J.* 18:6189–99

170. Paulovich AG, Armour CD, Hartwell LH. 1998. *Genetics* 150:75–93

171. Longhese MP, Foiani M, Muzi-Falconi M, Lucchini G, Plevani P. 1998. *EMBO J.* 17:5525–28

172. Edwards RJ, Bentley NJ, Carr AM. 1999. *Nat. Cell Biol.* 1:393–98

173. Kondo T, Wakayama T, Naiki T, Matsumoto K, Sugimoto K. 2001. *Science* 294:867–70

174. Melo JA, Cohen J, Toczyski DP. 2001. *Genes Dev.* 15:2809–21

175. Zou L, Cortez D, Elledge SJ. 2002. *Genes Dev.* 16:198–208

176. Bermudez VP, Lindsey-Boltz LA, Cesare AJ, Maniwa Y, Griffith JD, et al. 2003. *Proc. Natl. Acad. Sci. USA* 100:1633–38

177. Majka J, Burgers PM. 2003. *Proc. Natl. Acad. Sci. USA* 100:2249–54

178. Bao S, Tibbetts RS, Brumbaugh KM, Fang Y, Richardson DA, et al. 2001. *Nature* 411:969–74

179. Post S, Weng YC, Cimprich K, Chen LB, Xu Y, Lee EYHP. 2001. *Proc. Natl. Acad. Sci. USA* 98:13102–7

180. Vialard JE, Gilbert CS, Green CM, Lowndes NF. 1998. *EMBO J.* 17:5679–88

181. Gilbert CS, Green CM, Lowndes NF. 2001. *Mol. Cell* 8:129–36

182. Schultz LB, Chehab NH, Malikzay A, Halazonetis TD. 2000. *J. Cell Biol.* 151:1381–90

183. Wang B, Matsuoka S, Carpenter PB, Elledge SJ. 2002. *Science* 298:1435–38

184. Yamane K, Wu XL, Chen JJ. 2002. *Mol. Cell. Biol.* 22:555–66

185. Goldberg M, Stucki M, Falck J, D'Amours D, Rahman D, et al. 2003. *Nature* 421:952–56

186. Lou ZK, Minter-Dykhouse K, Wu XL, Chen JJ. 2003. *Nature* 421:957–61

187. Stewart GS, Wang B, Bignell CR, Taylor AM, Elledge SJ. 2003. *Nature* 421:961–66

188. Kumagai A, Dunphy WG. 2000. *Mol. Cell.* 6:839–49

189. Kumagai A, Dunphy WG. 2003. *Nat. Cell Biol.* 5:161–65

190. Lee J, Kumagai A, Dunphy WG. 2003. *Mol. Cell.* 11:329–40

191. Rhind N, Russell P. 2000. *J. Cell. Sci.* 113:3889–96

192. McGowan CH. 2002. *BioEssays* 24:502–11

193. Walworth N, Davey S, Beach D. 1993. *Nature* 363:368–71

194. Rhind N, Russell P. 1998. *Curr. Opin. Cell Biol.* 10:749–58

195. Sanchez Y, Wong C, Thoma RS, Richman R, Wu RQ, et al. 1997. *Science* 277:1497–501

196. Brown AL, Lee CH, Schwarz JK, Mitiku N, Piwnica-Worms H, Chung JH. 1999. *Proc. Natl. Acad. Sci. USA* 96:3745–50

197. Hirao A, Kong YY, Matsuoka S, Wakeham A, Ruland J, et al. 2000. *Science* 287:1824–27

198. Matsuoka S, Rotman G, Ogawa A, Shiloh Y, Tamai K, Elledge SJ. 2000. *Proc. Natl. Acad. Sci. USA* 97:10389–94

199. Zhao H, Piwnica-Worms H. 2001. *Mol. Cell. Biol.* 21:4129–39

200. Takai H, Tominaga K, Motoyama N, Minamishima YA, Nagahama H, et al. 2000. *Genes Dev.* 14:1439–47

201. Liu QH, Guntuku S, Cui XS, Matsuoka S, Cortez D, et al. 2000. *Genes Dev.* 14:1448–59

202. Jack MT, Woo RA, Hirao A, Cheung A, Mak TW, Lee PW. 2002. *Proc. Natl. Acad. Sci. USA* 99:9825–29

203. Bell DW, Varley JM, Szydlo TE, Kang DH, Wahrer DCR, et al. 1999. *Science* 286:2528–31

204. Bartek J, Lukas J. 2001. *Curr. Opin. Cell Biol.* 13:738–47

205. Pardee AB. 2002. *J. Biol. Chem.* 277:26709–16

206. Molinari M, Mercurio C, Dominguez J, Goubin F, Draetta GF. 2000. *EMBO Rep.* 1:71–79

207. Falck J, Mailand N, Syljuasen RG, Bartek J, Lukas J. 2001. *Nature* 410:842–47

208. Kastan MB, Lim DS. 2000. *Nat. Rev. Mol. Cell. Biol.* 1:179–86

209. Ryan KM, Phillips AC, Vousden KH. 2001. *Curr. Opin. Cell Biol.* 13:332–37

210. Chehab NH, Malikzay A, Stavridi ES, Halazonetis TD. 1999. *Proc. Natl. Acad. Sci. USA* 96:13777–82

211. Zhang YP, Xiong Y. 2001. *Science* 292:1910–15

212. Harper JW, Adami GR, Wei N, Keyomarsi K, Elledge SJ. 1993. *Cell* 75:805–16

213. Lin WC, Lin FT, Nevins JR. 2001. *Genes Dev.* 15:1833–44

214. Siede W, Friedberg AS, Dianova I, Friedberg EC. 1994. *Genetics* 138:271–81

215. Paulovich AG, Hartwell LH. 1995. *Cell* 82:841–47

216. Neecke H, Lucchini G, Longhese MP. 1999. *EMBO J.* 18:4485–97

217. Gerald JN, Benjamin JM, Kron SJ. 2002. *J. Cell. Sci.* 115:1749–57

218. Guo Z, Kumagai A, Wang SX, Dunphy WG. 2000. *Genes Dev.* 14:2745–56

219. Stokes MP, Van Hatten R, Lindsay HD, Michael WM. 2002. *J. Cell Biol.* 158:863–72

220. Lupardus PJ, Byun T, Yee MC, Hekmat-Nejad M, Cimprich KA. 2002. *Genes Dev.* 16:2327–32

221. Nelson WG, Kastan MB. 1994. *Mol. Cell. Biol.* 14:1815–23

222. Zhang H, Taylor J, Siede W. 2003. *J. Biol. Chem.* 278:9382–87

223. Hoege C, Pfander B, Moldovan GL, Pyrowolakis G, Jentsch S. 2002. *Nature* 419:135–41

224. Painter RB, Young BR. 1980. *Proc. Natl. Acad. Sci. USA* 77:7315–17

225. Santocanale C, Diffley JF. 1998. *Nature* 395:615–18

226. Tercero JA, Diffley JF. 2001. *Nature* 412:553–57

227. Costanzo V, Robertson K, Ying CY, Kim E, Avvedimento E, et al. 2000. *Mol. Cell.* 6:649–59

228. Kastan MB, Zhan Q, el-Deiry WS, Carrier F, Jacks T, et al. 1992. *Cell* 71:587–97

229. Heffernan TP, Simpson DA, Frank AR, Heinloth AN, Paules RS, et al. 2002. *Mol. Cell. Biol.* 22:8552–61

230. Howlett NG, Taniguchi T, Olson S, Cox B, Waisfisz Q, et al. 2002. *Science* 297: 606–9

231. Osborn AJ, Elledge SJ, Zou L. 2002. *Trends Cell Biol.* 12:509–16

232. Connelly JC, Leach DR. 2002. *Trends Biochem. Sci.* 27:410–18

233. Venkitaraman AR. 2002. *Cell* 108: 171–82

234. Yang HJ, Jeffrey PD, Miller J, Kinnucan E, Sun YT, et al. 2002. *Science* 297: 1837–48

235. Kim ST, Xu B, Kastan MB. 2002. *Genes Dev.* 16:560–70

236. Yazdi PT, Wang Y, Zhao S, Patel N, Lee EYHP, Qin J. 2002. *Genes Dev.* 16: 571–82

237. Stewart GS, Maser RS, Stankovic T, Bressan DA, Kaplan MI, et al. 1999. *Cell* 99:577–87

238. Varon R, Vissinga C, Platzer M, Cerosaletti KM, Chrzanowska KH, et al. 1998. *Cell* 93:467–76

239. Taniguchi T, Garcia-Higuera I, Xu B, Andreassen PR, Gregory RC, et al. 2002. *Cell* 109:459–72

240. Hekmat-Nejad M, You ZS, Yee MC, Newport JW, Cimprich KA. 2000. *Curr. Biol.* 10:1565–73

241. Costanzo V, Shechter D, Lupardus PJ, Cimprich KA, Gottesman M, Gautier J. 2003. *Mol. Cell.* 11:203–13

242. Tibbetts RS, Cortez D, Brumbaugh KM, Scully R, Livingston D, et al. 2000. *Genes Dev.* 14:2989–3002

243. Lee JS, Collins KM, Brown AL, Lee CH, Chung JH. 2000. *Nature* 404:201–4

244. Navas TA, Zhou Z, Elledge SJ. 1995. *Cell* 80:29–39

245. Feng W, D'Urso G. 2001. *Mol. Cell. Biol.* 21:4495–504

246. Waga S, Masuda T, Takisawa H, Sugino A. 2001. *Proc. Natl. Acad. Sci. USA* 98:4978–83

247. Zlotkin T, Kaufmann G, Jiang Y, Lee MY, Uitto L, et al. 1996. *EMBO J.* 15:2298–305

248. Pospiech H, Kursula I, Abdel-Aziz W, Malkas L, Uitto L, et al. 1999. *Nucleic Acids Res.* 27:3799–804

249. D'Urso G, Nurse P. 1997. *Proc. Natl. Acad. Sci. USA* 94:12491–96

250. Nichols AF, Sancar A. 1992. *Nucleic Acids Res.* 20:2441–46

251. Shivji MK, Podust VN, Hübscher U, Wood RD. 1995. *Biochemistry* 34: 5011–17

252. Wang Z, Wu X, Friedberg EC. 1993. *Mol. Cell. Biol.* 13:1051–58

253. Jessberger R, Podust V, Hubscher U, Berg P. 1993. *J. Biol. Chem.* 268: 15070–79

254. Li Y, Pursell ZF, Linn S. 2000. *J. Biol. Chem.* 275:23247–52

255. Dua R, Levy DL, Campbell JL. 1999. *J. Biol. Chem.* 274:22283–88

256. Kesti T, Flick K, Keranen S, Syvaoja JE, Wittenberg C. 1999. *Mol. Cell.* 3: 679–85

257. Vlatkovic N, Guerrera S, Li Y, Linn S, Haines DS, Boyd MT. 2000. *Nucleic Acids Res.* 28:3581–86

258. Asahara H, Li Y, Fuss J, Haines DS, Vlatkovic N, et al. 2003. *Nucleic Acids Res.* 31:2451–59

259. Haber JE, Heyer WD. 2001. *Cell* 107: 551–54

260. Kai M, Wang TS. 2003. *Genes Dev.* 17:64–76

261. Sogo JM, Lopes M, Foiani M. 2002. *Science* 297:599–602

262. Lopes M, Cotta-Ramusino C, Pellicioli A, Liberi G, Plevani P, et al. 2001. *Nature* 412:557–61

263. Cha RS, Kleckner N. 2002. *Science* 297: 602–6

264. Casper AM, Nghiem P, Arlt MF, Glover TW. 2002. *Cell* 111:779–89

265. Xu B, Kim ST, Lim DS, Kastan MB. 2002. *Mol. Cell. Biol.* 22:1049–59

266. Brown EJ, Baltimore D. 2003. *Genes Dev.* 17:615–28

267. Yarden RI, Pardo-Reoyo S, Sgagias M, Cowan KH, Brody LC. 2002. *Nat. Genet.* 30:285–89

268. Peng CY, Graves PR, Thoma RS, Wu ZQ, Shaw AS, Piwnica-Worms H. 1997. *Science* 277:1501–5

269. Chen MS, Hurov J, White LS, Woodford-Thomas T, Piwnica-Worms H. 2001. *Mol. Cell. Biol.* 21:3853–61

270. Zhao H, Watkins JL, Piwnica-Worms H. 2002. *Proc. Natl. Acad. Sci. USA* 99:14795–800

271. Furnari B, Rhind N, Russell P. 1997. *Science* 277:1495–97

272. Wang XF, McGowan CH, Zhao M, He LS, Downey JS, et al. 2000. *Mol. Cell. Biol.* 20:4543–52

273. Bulavin DV, Higashimoto Y, Popoff IJ, Gaarde WA, Basrur V, et al. 2001. *Nature* 411:102–07

274. Michael WM, Ott R, Fanning E, Newport J. 2000. *Science* 289:2133–37

275. You ZZ, Kong L, Newport J. 2002. *J. Biol. Chem.* 277:27088–93

Annu. Rev. Biochem. 2004. 73:87–106
doi: 10.1146/annurev.biochem.73.011303.073706
Copyright © 2004 by Annual Reviews. All rights reserved
First published online as a Review in Advance on March 11, 2004

Cytochrome C-Mediated Apoptosis

Xuejun Jiang[1] and Xiaodong Wang[2]

[1]*Cell Biology Program, Memorial Sloan-Kettering Cancer Center, New York, New York 10021; email: jiangx@mskcc.org*
[2]*Howard Hughes Medical Institute and Department of Biochemistry, University of Texas Southwestern Medical Center, Dallas, Texas 75390-9050; email: xwang@biochem.swmed.edu*

Key Words   caspase, mitochondria, Bcl-2, apoptosome

■ **Abstract**   Apoptosis, or programmed cell death, is involved in development, elimination of damaged cells, and maintenance of cell homeostasis. Deregulation of apoptosis may cause diseases, such as cancers, immune diseases, and neurodegenerative disorders. Apoptosis is executed by a subfamily of cysteine proteases known as caspases. In mammalian cells, a major caspase activation pathway is the cytochrome c-initiated pathway. In this pathway, a variety of apoptotic stimuli cause cytochrome c release from mitochondria, which in turn induces a series of biochemical reactions that result in caspase activation and subsequent cell death. In this review, we focus on the recent progress in understanding the biochemical mechanisms and regulation of the pathway, the roles of the pathway in physiology and disease, and their potential therapeutic values.

CONTENTS

INTRODUCTION

In the middle of the last century, the concept of apoptosis, or programmed cell death, emerged with its unique and dynamic morphological features that are distinguishable from senescence or necrosis, such as cell shrinkage, plasma membrane blebbing, chromatin condensation, nuclear membrane breakdown, and formation of small vesicles from the cell surface also known as apoptotic bodies (1). After apoptosis, the apoptotic bodies are rapidly engulfed by phagocytes, and thus a potential inflammatory response is avoided (1). This deliberate physiological cell suicide concept was proved molecularly in the 1990s by Horvitz and colleagues by showing that in *Caenorhabditis elegans*, there is an intrinsic signaling pathway controlling the cell death of a group of specific neuronal cells during development (2).

In the *C. elegans* apoptosis pathway, there are both positive and negative regulators of cell death (3–5). The pathway is initiated by Egl-1, which functions to antagonize the negative regulator Ced-9. Egl-1 causes Ced-9 to release its inhibition on Ced-4, which in turn recruits and activates Ced-3, a cysteine protease and the executioner of apoptosis. Interestingly, Ced-9 is homologous to human Bcl-2 (6), which was originally identified as an oncogene product, because translocation and subsequent overexpression of the gene causes B-cell lymphoma (7–11). Later, it was found that the oncogenic property of Bcl-2 is due to its activity to protect cells from death (12, 13). Therefore apoptosis and oncogenesis are linked together: Cancer can develop as a result of not only overproliferation of cells but also inhibition of normal physiological cell death. In addition to Ced-9, the executing molecule of the *C. elegans* apoptosis pathway, Ced-3 also has many homologs in mammals. The homology of Ced-9 and Ced-3 with mammalian apoptotic proteins indicates a highly conserved cell death mechanism utilized by worms and mammals.

Ced-3 and its mammalian homologs are cysteine proteases called caspases (14, 15). Caspases are normally inactive in their zymogen form or proform. During apoptosis, a procaspase is proteolytically cleaved to generate a small subunit and a large subunit, and two cleaved caspase molecules form a heterotetramer, which is the active form of the enzyme. On the basis of structural studies of many caspases associated with specific peptide inhibitors (16–20), and a more recent study on the structure of free caspase-7, in both proform and activated form (21), a general mechanism for caspase activation becomes clear. The proteolytic cleavage of a caspase can induce a dramatic conformational change that exposes the catalytic pocket of the enzyme, and therefore results in its activation. The proteolytic activation of caspases can be achieved either by autocatalysis or by an upstream protease. A caspase that cleaves and activates itself is called an initiator caspase. Once an initiator caspase is activated, it can trigger a cascade to activate downstream executioner caspases. Subsequently, the activated executioner caspases cleave numerous cellular targets to destroy normal cellular functions, activate other apoptotic factors, inactivate antiapoptotic proteins, and eventually lead to apoptotic cell death (14, 15). The central role of caspase activity in apoptosis is further underscored by the observation that inhibition of caspase activity can block apoptosis and all classical morphological changes associated with the process (14, 15). Therefore, understanding caspase activation is essential for apoptosis research and development of therapies for apoptosis-related diseases.

THE DEATH RECEPTOR-MEDIATED CASPASE ACTIVATION PATHWAY

One of the caspase activation pathways being characterized in mammals is the cell surface death receptor-mediated pathway (22). This pathway is initiated by extracellular hormones or agonists that belong to the tumor necrosis factor (TNF) superfamily, including TNFα, Fas/CD95 ligand, and Apo2 ligand/TRAIL. These agonists recognize and activate their corresponding receptors, members of TNF/NGF receptor family, such as TNFR1, Fas/CD95, and Apo2. Then, via a series of protein-protein interactions involving domains, which include the death domain and the death effector domain, the receptors will recruit specific adaptor proteins to form a complex called the death-inducing signaling complex (DISC). DISC recruits and activates the initiator caspases, caspase-8 or caspase-10, probably by bringing the procaspases close enough in proximity so that they can cleave each other. These activated initiator caspases trigger a caspase cascade and subsequent cell death by activating downstream executioner caspases, such as caspase-3 and caspase-7.

Genetic evidence indicates that the cell surface death receptor-mediated apoptosis is critical for normal immune system function. For example, mutations on Fas and Fas ligand in humans can lead to a complicated immune disorder known as the autoimmune lymphoproliferative syndrome (ALPS) (23, 24), a resemblance of murine lymphoproliferation (lpr) and generalized lymphoproliferative disorder (gld) caused by Fas and Fas ligand mutations, respectively (25–27).

Identification of caspase-8/caspase-10 activation by the TNF pathway revealed an important apoptosis mechanism. However, this pathway could not answer many outstanding questions in the field. First, this receptor-mediated pathway does not explain the involvement of the Bcl-2 family members in apoptosis, whose worm counterpart Ced-9 negatively regulates the worm caspase, Ced-3. Second, there are numerous cases showing non-receptor-mediated caspase activation. Third, molecular cloning has identified many putative initiator caspases in addition to caspase-8 and caspase-10. All these questions set the stage for the cytochrome *c*-initiated caspase activation pathway, a pathway with multiple components homologous to the players in the *C. elegans* apoptosis pathway.

THE CYTOCHROME *C*-INITIATED CASPASE ACTIVATION PATHWAY

In 1995, the laboratory of Xiaodong Wang set out to study the mechanisms of caspase activation using an in vitro biochemical approach. Initially, using a cell-free system to study apoptosis seemed unfeasible because the programmed

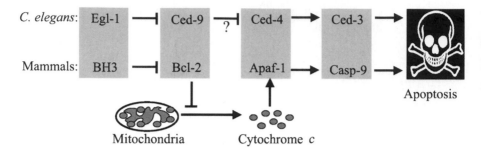

Figure 1 A comparison of the *C. elegans* programmed cell death pathway and the mammalian cytochrome *c*-mediated death pathway. The homologous molecules of the two pathways are labeled with shaded boxes. The question mark indicates that the mechanism by which Ced-9 inhibits Ced-4, whether via a direct interaction or not, has not yet been directly demonstrated.

cell death requires intact cellular architecture, and processing of cells to a cell-free state virtually kills them. However, apoptosis is unique. Caspase activation is one of its hallmarks, and preparation of a cell-free system from naïve, healthy cells does not activate caspases even though it kills all the cells (28). Therefore, it is theoretically possible to study caspase activation in vitro.

Then the question became how to initiate the caspase cascade in the cell-free system. The laboratory found that addition of the nucleotide dATP, or the less potent ATP, induced activation of caspase-3, a major executioner caspase in HeLa cell cytosolic extracts. This assay led to purification of the first protein required for dATP-triggered caspase-3 activation, which turned out to be cytochrome *c* (28). Subsequently, the other two components required for dATP-initiated caspase-3 activation were identified as Apaf-1, which is the binding partner of dATP and homologous to the *C. elegans* protein Ced-4 (29), and caspase-9, an initiator caspase homologous to *C. elegans* caspase Ced-3 (30).

The role of cytochrome *c* in activating apoptosis seemed puzzling at first glance because it is an essential protein in energy production and is located inside the mitochondria. But the pivotal role of cytochrome *c* in apoptosis was quickly confirmed in large by two results. The first one was the identification of its downstream binding partner, Apaf-1, a homolog of *C. elegans* Ced-4 (29). The second was the demonstration that Bcl-2 inhibits cell death by preventing cytochrome *c* release from mitochondria (31). Therefore, the discovery of a cytochrome *c*-mediated caspase activation pathway delineated a mammalian caspase activation pathway that is the counterpart of the *C. elegans* cell death pathway (Figure 1), and it led to identification of the mechanism by which the oncogene product Bcl-2 prevents apoptosis. As illustrated in Figure 1, in the mammalian pathway, the Egl-1 homologs are BH3-only proteins, such as Bim, Bid, Bad, Box, Noxa, and Puma. The Ced-9 homologs are the antideath members of the Bcl-2 family, such as Bcl-2, Bcl-XL, and Mcl-1. The Ced-4 homolog is

Apaf-1, and the Ced-3 homolog is caspase-9. An obvious difference between these two pathways is the function of Bcl-2 in mammalian cells and Ced-9 in the worms. Although these proteins are homologous to each other, Bcl-2 functions to inhibit cytochrome *c* release from mitochondria and thereby prevent downstream caspase activation, whereas Ced-9 is believed to directly inhibit the activity of Ced-4 to recruit and activate the worm caspase, Ced-3 (3–5).

The physiologic roles of the cytochrome *c*-mediated caspase activation pathway are intriguing. Much knowledge was gained from targeted gene disruption studies in mice, and the gene for every component of the pathway downstream of mitochondria, including cytochrome *c*, has been knocked out (32–37). In vitro studies showed that the embryonic fibroblast cells with Apaf-1, caspase-9, caspase-3, or cytochrome *c* knocked-out are resistant to various apoptotic stimuli. At the whole-animal level, the predominant phenotype shared by the knockout of Apaf-1, caspase-9, and caspase-3 genes is a severe developmental defect in the central nervous system (CNS) that results in the protrusion of brain tissue from the forehead and perinatal lethality. These results indicate an essential role of this apoptotic pathway in brain development. However, it is peculiar to see that this is the only predominant phenotype because apoptosis has been demonstrated to be involved in development of other body processes, such as the immune system. Therefore, there must be other tissue-specific pathways involved in the development of these organs, for example, the cell surface receptor-mediated death pathway.

On the other hand, the cytochrome *c*-mediated pathway is still likely to be involved in other biological events, including immune system homeostasis and elimination of damaged or harmful cells, during the normal life span after birth. It is unfortunate that conventional gene disruption of Apaf-1, caspase-9, and caspase-3 all result in lethality upon birth, raising the necessity to engineer more sophisticated, tissue and time-specific conditioning-knockout or transgenic animal models. Until now, most supporting evidence for the critical roles of the pathway in after-birth life is from tissue culture studies. For example, numerous experiments revealed that the pathway is essential for stress-induced and genotoxic-induced cell death, indicating the pathway plays a pivotal role to protect the organism from deadly diseases such as cancer (33, 34, 38, 39). This notion is supported by the finding that many malignant human melanoma cells, from both cancer patients and established cell lines, lose expression of Apaf-1 protein and are resistant to chemotherapy or P53-induced apoptosis (40).

Besides its essential role in CNS development and mediating stress-induced apoptosis, another function of the cytochrome *c*-mediated caspase activation pathway is that it can always serve as an amplifying/accelerating route for other apoptotic pathways, such as the death receptor pathway (15, 41, 42) and the cytotoxic T lymphocyte-mediated pathway (43, 44). In these pathways, although downstream executioner caspases can be directly activated by upstream proteases, i.e., caspase-8/10 for the death receptor pathway (41, 42) and Granzyme B for cytotoxic T lymphocyte pathway (43), these upstream proteases can also cleave and activate Bid, a prodeath member of the Bcl-2 family. The truncated Bid (t-Bid) is targeted to mitochondria to induce cytochrome c release and the subsequent cytochrome c-mediated caspase activation pathway (41). This amplification process is particularly important in certain cells known as type II cells. In these cells, the death receptor-mediated apoptosis has to be mediated by the mitochondria pathway, and as a result it can be blocked by antiapoptotic members of the Bcl-2 family (45). Mechanistically, the death receptor-mediated caspase cascade in type II cells is inhibited by inhibitor of apoptosis (IAP) proteins, which are inhibitors of caspases, and the inhibitory activity of IAPs needs to be antagonized by a protein released from mitochondria, SMAC/DIABLO, to ensure progression of programmed cell death (46, 47). (IAP and SMAC/DIABLO will be discussed in detail below.) The amplification role of the cytochrome c pathway for other apoptosis pathways has profound therapeutic implications: When drugs targeted to both the cytochrome c pathway and other pathways are combined, a synergistic effect could be achieved, even though separate usages only have poor effect. This cocktail therapy might have huge benefits, especially for cancers with type II cell origin.

MITOCHONDRIA, AN ORGANELLE FOR LIFE AND DEATH

Many cellular structures and organelles are damaged or destroyed as a consequence of apoptosis (1). However, mitochondria, the organelle essential for life, is not only affected passively but is also actively involved in promoting apoptosis, as revealed by the cytochrome c-mediated caspase activation pathway. More mitochondrial proapoptotic proteins have been discovered. As was the case in the discovery of the apoptotic function of cytochrome c, the other proteins were not necessarily found from studies designed to target a potential apoptotic function of mitochondria. For example, another proapoptotic activity, SMAC/DIABLO, was not realized to be a mitochondrial protein until its identification and characterization (48, 49).

The IAP family of proteins (50) preluded discovery of SMAC/DIABLO. IAP proteins inhibit caspase activity by directly binding to the active enzymes (51, 52). These proteins contain single or multiple baculovirus IAP repeat (BIR) domains, which are responsible for the caspase inhibitory activity (50). It is likely that IAP proteins serve to inhibit residual or unwanted caspase activity in healthy cells. But then the question is, When cells are committed to apoptosis, is there a specific mechanism to antagonize the inhibitory role of IAP and thereby render more potent caspase activity? Such a mechanism was identified by two independent groups that applied distinct approaches. Du et al. (48) observed that dATP-initiated caspase-3 activation in HeLa cell cytosolic extracts could be greatly enhanced by a HeLa cell membrane fraction solubilized with detergent. The protein responsible for the enhancement was purified and found to be, again, a mitochondrial protein. This novel mitochondrial protein was called the second mitochondria-derived activator of caspase (SMAC), and it was found that SMAC enhances caspase activation via antagonizing IAP function. At about the same time, Vaux and colleagues (49) made the same discovery by directly searching for IAP antagonists using a coimmunoprecipitation approach. They named the protein DIABLO (direct IAP binding protein with low pI).

There is documentation that expression of various IAP proteins is aberrantly upregulated in certain cancer tissues. For example, Survivin, a member of the IAP family, is overexpressed in most cancer cells (53); another member, ML-IAP/Livin, was originally identified because of its overexpression in human melanoma (54); cIAP1 is overexpressed in esophageal squamous cell sarcoma (55); and the cIAP2 locus is translocated and results in a fusion in mucosa-associate lymphoid lymphoma (56). Therefore, upregulation of IAPs might contribute to oncogenesis, and they can be categorized as oncogenes just like antiapoptotic members of the Bcl-2 family. However, it should be noted that the oncogenecity of Survivin might be due to its function in cytokinesis rather than that in apoptosis (57).

The oncogenic nature of IAP proteins makes them potential drug targets in IAP-overexpressing cancers. A mechanism-based drug design strategy is to develop drugs mimicking SMAC/DIABLO function. Structural studies of IAP-caspase complexes (58–60), SMAC (61), and SMAC-IAP complexes (62, 63) provide valuable insight for this purpose. It was found that the first four amino acids of mature SMAC, AVPI interact tightly with the BIR3 domain of xIAP, and the first residue alanine fits perfectly into a groove of the BIR3 domain. Consistently, a single mutation of this N-terminal alanine of SMAC to any other tested residue completely abolishes the ability of SMAC to interact with and suppress IAP activity. Similarly, addition of a single residue in front of this critical alanine does the same thing. More strikingly, the small peptide AVPI and other synthetic peptides with conserved alterations of the three later residues can also interact and suppress IAP, though with less potency than the SMAC protein (61). This result provides a promising approach for designing drug leads to attack IAP proteins. Practically, to make a feasible AVPI-like drug lead, cell permeability, peptide stability, and many other parameters should all be taken into account.

Discoveries of the mitochondrial protein SMAC/DIABLO in addition to cytochrome *c* as a proapoptotic player further suggest a central role of mito-chondria in programmed cell death. And the story does not end here. Another mitochondrial protein, Omi/HtrA2, can also function as SMAC/DIABLO to antagonize the caspase inhibitory activity of IAP (64–68). A difference between SMAC and Omi/HtrA2 is that the latter is also a serine protease that can proteolytically cleave and inactivate IAP proteins, and therefore it is presumably a more efficient IAP suppressor than SMAC (69, 70). In addition to proteins that can trigger or enhance caspase activation, mitochondria also release proapoptotic proteins with functions unrelated to caspase activation. Such proteins include apoptosis inducing factor (AIF) (71) and endonuclease G (Endo G) (72). These proteins are involved in DNA fragmentation and subsequent chromosomal condensation, a hallmark morphological feature of programmed cell death.

Why do cells utilize so many mitochondrial proteins as apoptotic factors? Strategically, it is an efficient and safe mechanism. In normal cells, these proteins are all in mitochondria, but their targets are in either cytoplasm (e.g., Apaf-1 for cytochrome c and IAP for SMAC/DIABLO) or the nucleus (e.g., genomic DNA for Endo G). This spatial segregation ensures that the proteins perform their killing functions only when they are deliberately released from the organelle during apoptosis. Although these proteins are transiently exposed to the cytoplasm when newly synthesized, their apoptotic functions require them to be processed inside of mitochondria. For example, cytochrome c has to be folded into the mature, heme-bound form to activate Apaf-1 (28), and the mitochondria target sequence of SMAC/DIABLO has to be removed in order to antagonize IAP (48, 49). Another advantage of spatial separation is that these proteins can be multifunctional, i.e., they can have other functions inside mitochondria in normal cells. For example, cytochrome c is an essential component of the mitochondrial electron transfer chain, Omi/HtrA2 can function as molecular chaperone and degrade denatured proteins (73–75), and AIF, with a pyridine nucleotide-disulphide oxidoreductase domain, can protect cells from oxidative stress (76). Whether SMAC/DIABLO and Endo G also have nonapoptotic functions is not clear.

In summary, a central role of mitochondria in mammalian apoptosis has been firmly established, though more detailed work is needed for better understanding. The organelle is involved in both caspase-dependent (cytochrome c, SMAC/ DIABLO, and Omi/HtrA2) and caspase-independent (Endo G and AIF) cell death mechanisms, and release of the mitochondrial death proteins is closely regulated by the Bcl-2 family of proteins. The apoptotic function of mitochondria is probably not limited to vertebrates. Originally, it was believed that mitochondria did not have apoptotic function in *C. elegans* because the Bcl-2 homolog Ced-9 was thought to directly inhibit Ced-4 rather than inhibit mitochondria-releasing events (2). However, this proposed mechanism for Ced-9 still lacks direct biochemical demonstration, and the mitochondrial localization of Ced-4 and Ced-9 (77) suggests a connection of the worm death pathway with the organelle. Furthermore, recent studies strongly support an apoptotic role for mitochondria in *C. elegans*, carried out by the worm counterparts of Endo G and AIF, two mitochondrial proteins (78–80).

REGULATION OF MITOCHONDRIAL FUNCTIONS BY THE Bcl-2 FAMILY

Retrospectively, it is now clear why protein release from the mitochondria has to be precisely regulated by a big family of proteins, the Bcl-2 family members, and why malfunction of these proteins has severe consequences, such as B-cell lymphoma caused by translocation of Bcl-2 locus. Mechanistically, although a complete picture has been elusive, considerable progress has been made on regulation of mitochondrial protein release by Bcl-2 proteins, summarized in Figure 2.

The Bcl-2 family can be divided into two subgroups, prodeath members and antideath members (81, 82). Within the same subgroup, the exact function of individual members can still be quite different. Among the prodeath members, Bak and Bax have been categorized as the last gateway of cytochrome *c* release, and their homooligomerization on the mitochondrial membrane is essential for release (83, 84). Bak/Bax function appears to be regulated by mitochondria-specific lipids and proteins, such as cardiolipin (85, 86) and VDAC2 (87), but whether any mitochondrial protein, such as VDAC (88–90), adenine nucleotide translocase (ANT) (91, 92), or components of permeability transition pore (91, 93, 94) are indispensable for Bak/Bax-induced protein release is still under debate. Other prodeath members, mainly BH3-only proteins, are thought to directly induce Bak/Bax oligomerization or to antagonize the antideath Bcl-2 members, and their regulation is very different. For example, Bid can be activated by caspase cleavage as discussed earlier. In response to growth hormones, Bad is phosphorylated by the PI3 K-Akt survival pathway and is thus inactive (95–97). Puma (98) and Noxa (99) are transcriptionally upregulated by p53 after DNA damage, and Bim is regulated both by phosphorylation (100–102) and by transcription (103).

The antiapoptotic Bcl-2 members prevent mitochondrial protein release by interacting with and inhibiting both Bak/Bax and BH3-only proteins. Interestingly, recent evidence from the laboratory of Xiaodong Wang indicates an apical function for the antideath member Mcl-1 that distinguishes it from Bcl-2 or Bcl-XL (104). Mcl-1 is a quick-turnover protein, and it can be degraded by the ubiquitination-proteasome pathway. When cells are treated with various apoptotic signals, Mcl-1 protein level decreases dramatically, due to a blockage of its synthesis as well as a possible acceleration of its degradation. And Mcl-1 disappearance is a prerequisite for downstream apoptotic events, such as Bcl-XL inactivation, Bim dephosphorylation, Bax translocation, Bax/Bak oligomerization, and subsequent cell death (Figure 2).

The function of the Bcl-2 family may be more than just regulation of mitochondria. New observations emerge suggesting that the Bcl-2 family can also regulate endoplasmic reticulum integrity and that this regulation is also important for apoptosis (105, 106). Moreover, a model suggesting an inhibitory role of Bcl-2 on caspase(s) upstream of cytochrome *c* release, via interaction with a yet-to-be-identified mammalian Ced-4 homolog other than Apaf-1, is still viable (107). This model of Bcl-2 function, rooted from its *C. elegans* homolog Ced-9, has, however, no direct experimental support. Yet evidence has been presented that in human fibroblasts transformed with the adenoviral oncogene E1A, cytotoxic stress can induce caspase-2 activation, which is upstream of and is required for cytochrome *c* release in this specific context (108).

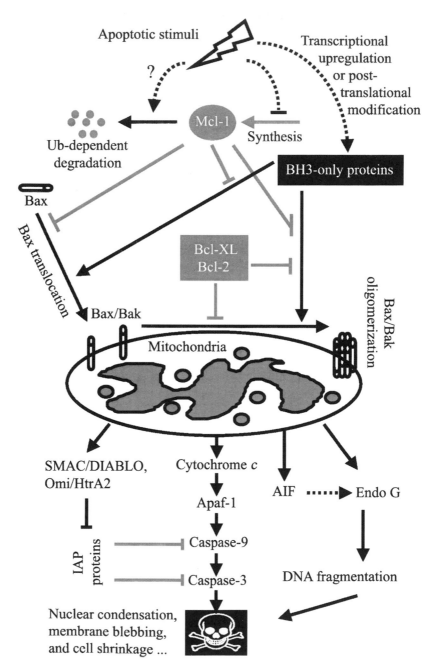

Figure 2 Regulation of the mitochondria apoptotic pathway by the Bcl-2 family members. Regulation by the Bcl-2 family members and the events downstream of mitochondria are shown.

THE APOPTOSOME, A CASPASE ACTIVATION MACHINERY

After release from mitochondria, the biochemistry of how cytochrome c triggers caspase activation is very complex. It was found that cytochrome c can interact with the C-terminal WD40 repeats of Apaf-1 and that this interaction is required for activation of the pathway (109). Further, a de novo reconstitution was achieved by using purified recombinant Apaf-1, procaspase-9, procaspase-3, and highly purified horse cytochrome c (110). When all the proteins are incubated together in the presence of nucleotide dATP/ATP, caspase-3 is activated. A striking phenomenon observed in this in vitro system is that Apaf-1 and cytochrome c are induced into a huge complex in a dATP/ATP-dependent manner. This complex was termed the apoptosome (110).

Binding of nucleotide to Apaf-1 is critical for apoptosome formation and is stimulated by cytochrome c (111). In the Ced-4 homologous domain of Apaf-1, there are classical Walker's A and B boxes, and they are believed to be the dATP/ATP binding and hydrolysis motifs (29). Using a rapid filtration assay, it was shown that cytochrome c stimulated dATP binding to Apaf-1 (111). This finding provides a biochemical explanation of how this essential protein for life can function as a death inducer. A surprising finding from this study is that nucleotide hydrolysis by Apaf-1 is not required for caspase activation because dATP, but not dADP, is associated with the mature apoptosome. There is no correlation between dATP hydrolysis and caspase activation, and a nonhydrolyzable ATP analog, AMPPCP, can also trigger apoptosome formation and caspase-3 activation (111).

Walker's boxes are also present in the Apaf-1 homologs, Ced-4 in *C. elegans*, and DARK in *Drosophila* (112). Presumably, nucleotide binding is also required for the ability of these two proteins to activate caspase, though formal experimental tests are needed for a final conclusion. Interestingly, sequence alignment of DARK predicts that its Walker's boxes have only nucleotide binding activity but not hydrolysis activity (112). Again, this prediction lacks experimental demonstration. But if this is true, it confirms the result that nucleotide hydrolysis by Apaf-1 is not required for caspase activation. Yet it also raises the question— why does mammalian Apaf-1 but not its *Drosophila* homolog possess nucleotide hydrolysis activity? Speculatively, the nucleotide hydrolysis activity of mammalian Apaf-1 can provide a safeguard mechanism because dATP or ATP, but not their hydrolyzed products, can drive the formation of a functional apoptosome complex. Thus, in a healthy cell, low levels of unwanted apoptosome complexes can be discharged by Apaf-1 nucleotide hydrolysis activity before they can reach their targets, such as initiator caspases, to cause irreversible damages.

The apoptosome machinery also provides a unique mechanism for caspase-9 activation. Unlike the conventional caspase activation mechanism in which a proper proteolytic cleavage of caspase is both necessary and sufficient for its activation, activation of caspase-9 by the apoptosome requires a constant association of the enzyme with the oligomeric death machinery (111, 113). As a matter of fact, as long as the caspase is in this million-dalton complex, proteolytic processing is not really required for its activity (114). This observation raised the question—why is caspase-9 processed during apoptosis? A potential answer came from a biochemical-structural study (114). It was found that the caspase inhibitory protein XIAP can only interact with processed caspases (51, 52). For caspase-9, autocleavage of human caspase-9 at D315 exposes a new N terminus that starts with ATPF, similar to the N terminus of SMAC, AVPI. It was then confirmed by mutagenesis that this newly exposed sequence is required for interaction of the processed caspase-9 with XIAP (114). On the basis of this work, it is reasonable to assume the purpose of autocleavage of caspase-9 is to ensure that leaky, unwanted apoptosome-caspase-9 activity can be blocked by IAP proteins. In addition to the D315 autocleavage site, human caspase-9 also has a D330 site that can be recognized and cleaved by caspase-3, an executioner caspase in the downstream of caspase-9 (115). Kinetic studies showed that once this site is cleaved by caspase-3, the apoptosome-caspase-9 holoenzyme is eightfold more active than the D330A caspase-9 mutant associated with the apoptosome (autocleavage at D315 does not have this effect), suggesting this cleavage functions as a positive feedback loop (116). Although this is a reasonable hypotheses, physiological relevance of these two cleavage events needs to be addressed experimentally. The only apparent approach, although difficult, might be gene knock-in experiments.

Recently, a 27-Å three-dimensional structure of the apoptosome complex has been solved using cryo-electron microscopy (EM) technology (117). The structure gives insight on how the apoptosome assembles, how it activates caspase-9, and why activation of the caspase is distinct from the conventional caspase activation mechanism. The structure revealed that the apoptosome is composed of seven molecules of Apaf-1, and they form a symmetrical wheel-like structure. In the apoptosome complex, Apaf-1 interacts with the adjacent Apaf-1 molecules via their N-terminal CARD domains to form a central hub region, and the C-terminal WD40 repeats are extended to form the outside ring. On the basis of electron density, it was proposed that there was only one cytochrome c associated with each Apaf-1, though an early kinetic study suggested that there were two (118, 119). The central hub region is also the location for caspase-9 recruitment based on a cryo-EM study of the apoptosome when complexed with procaspase-9. Because there are seven Apaf-1 CARD domains in each apoptosome hub, caspase-9 can be highly enriched locally. Furthermore, it is likely that the CARD domain interaction between Apaf-1 and caspase-9 induces the enzyme to a fully extended, active conformation that cannot be achieved by proteolytic cleavage alone. It has been suggested that like other caspases, caspase-9 also needs to form a heterotetramer in the apoptosome to be active (117, 120, 121). However, direct evidence is required to confirm this assumption.

Incorporating all the biochemical and structural studies, a model illustrating the detailed biochemical mechanism of cytochrome c-induced caspase activation is presented in Figure 3. Upon sensing a variety of apoptotic stimuli, cytochrome c is released from mitochondria and associates with the apoptotic mediator Apaf-1 in its

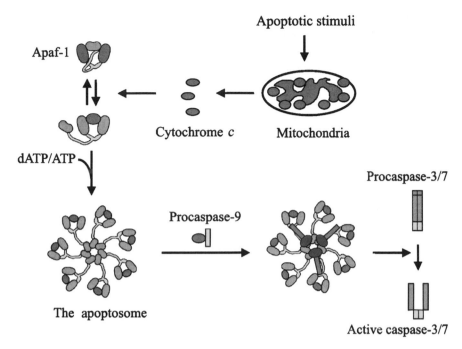

Figure 3   The mechanisms of apoptosome formation and caspase activation initiated by cytochrome *c* release.

C-terminal WD40 repeats. After association with cytochrome *c*, Apaf-1 switches from a rigid conformation to a more flexible one such that the nucleotide dATP/ATP binding activity of its Walker's motif is greatly facilitated. Binding of dATP/ATP in turn triggers formation of the active, seven-span symmetrical complex, the apoptosome, via interaction among the N-terminal CARD domains of the individual Apaf-1 molecules. The apoptosome subsequently recruits procaspase-9 into its central hub region through CARD domain interaction, and a conformational change of the enzyme is induced. Therefore caspase-9 and the apoptosome form an active holoenzyme to activate downstream executioner caspases, such as caspase-3 and caspase-7, which eventually lead to programmed cell death. In this pathway, IAP proteins function to inhibit caspase activity, and they can be overcome by mitochondrial proteins SMAC/DIABLO and Omi/HtrA2, as discussed above.

ADDITIONAL REGULATIONS OF THE CYTOCHROME *C* PATHWAY

It was long suspected that cytochrome *c*-mediated caspase activation has additional regulatory mechanisms based upon certain observations. For example, in the initial dATP-triggered caspase-3 activation assay, up to 1-mM dATP is required to activate apoptosome formation and subsequent caspase activation in HeLa cell extracts, whereas in the final reconstitution system, micromolar levels of dATP are sufficient to activate these events (28, 110, 111). More importantly, the cellular concentration of dATP is ~10 micromoles, and it does not increase during apoptosis (122). This discrepancy of nucleotide concentration points to a potential regulation at the level of dATP binding to Apaf-1. Furthermore, the fact that the nucleotide binding motif is within the Ced-4 domains of Apaf-1 and its *C. elegans* and *Drosophila* counterparts also indicates the conservation and significance of nucleotide binding in this pathway. Another observation is from studies on an ovarian cancer cell line SKOV-3. Cytochrome *c* and dATP failed to activate caspase-3 in SKOV-3 cell extracts, even though the expression levels of Apaf-1, caspase-9, caspase-3, and IAP were all normal (123). It was found that apoptosome formation in the cell extracts could still be induced, but subsequent recruitment and activation of caspase-9 by the complex was defective (123). These results suggest a novel regulation at the level of caspase-9 activation by the apoptosome complex, and this regulation is repressed in the ovarian cancer cells. In addition, it was reported that multiple heat shock proteins, such as Hsp90, Hsp70, and Hsp27, could inhibit cytochrome *c*-initiated caspase activation by directly interacting with Apaf-1 or other players in the pathway (124–128). However, the physiological relevance of these in vitro experiments is not yet clear. Also, mounting evidence indicates a regulatory role of the signaling lipid ceramide in apoptosis, and under specific conditions, ceramide might function in the upstream of mitochondria via Bax activation, although the detailed molecular mechanism is still obscure (128a, 128b).

A recent chemical-biological study, initiated by a high-throughput screen to search for chemicals that can activate caspase-3 in HeLa cell extracts, sheds light on understanding the additional regulation of the cytochrome *c* pathway (129). An interesting analogy is that the high-throughput screen performed in this study is very similar to the original dATP assay (28), which can be viewed as the result of a primitive, infant form of chemical screen. These two screens share a common feature strategically: unlike many screens with defined targets, these screens were designed to study a biochemical event without knowing the direct targets and were therefore likely to identify new pathways. As a result, the earlier screen led to the discovery of the cytochrome *c*-initiated caspase activation pathway (28). The later screen fished out PETCM (α-(trichloromethyl)-4-pyridineethanol) as a caspase-3 activator (Figure 4), and further studies using PETCM revealed a regulatory pathway controlling the apoptosome machinery (129).

Biochemical studies of PETCM-triggered caspase-3 activation showed that the compound can activate apoptosome formation by suppressing the inhibitory effect of prothymosin-α (ProT), an oncogene product. After apoptosome formation, caspase-9 recruitment and activation can be enhanced by PHAP proteins, which are putative tumor suppressors. Previously, although these two proteins

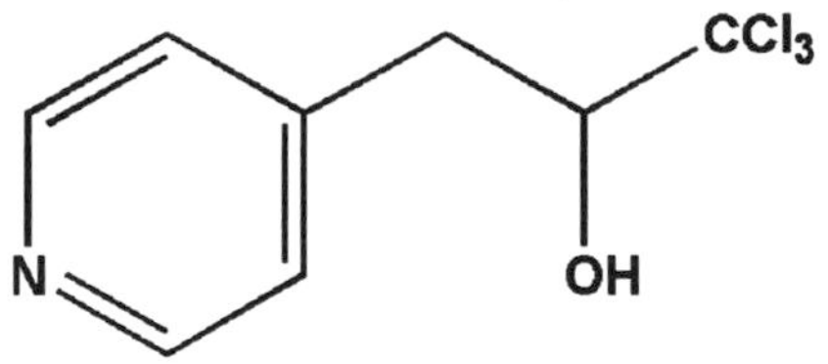

Figure 4   Covalent structure of PETCM (α-(trichloromethyl)-4-pyridineethanol).

were characterized as an oncoprotein and tumor suppressor, the biochemical activities responsible for these properties were not known (130–133). Therefore, the activities of these proteins in apoptosis might contribute to their functions in oncogenesis. Also, this provides an explanation of why a high amount of nucleotide is required for caspase activation in the crude system. It is likely that in cells, a PETCM-like signal is required to suppress inhibition of dATP binding to Apaf-1 by ProT during apoptosis, whereas in the cell extracts prepared from naïve HeLa cells, high amounts of dATP have to be used to overcome this suppression.

The oncogenic and tumor suppressive property of ProT and PHAP suggest that malfunction of this death regulatory pathway might result in tumorigenesis. This hypothesis is consistent with many reports showing overexpression of ProT in tumors, which include breast cancer, colon cancer, lung cancer, and liver cancer (134–137). Therefore, this new death regulatory pathway presents a potential target for cancer chemotherapy. In fact, development of cancer therapy was the original purpose of the chemical screen that uncovered PETCM. However, to make PETCM a realistic drug lead from the current form with an apparent EC50 of 0.05 mM (129), the compound has to be modified to get better potency as well as cell permeability. Additionally, its direct target needs to be identified in order to perform structure-based modification.

It is also possible that this pathway is involved in brain development and the neurodegenerative disorder spinocerebella ataxia type-1 (SCA1) because PHAP is upregulated in the early brain developmental stage characterized by massive apoptosis (138–140), and the protein has been reported to interact with Ataxin-1 protein, whose mutation by insertion of polyglutamine tract is the genetic cause of SCA1 (141).

Further studies of this pathway will focus on (*a*) identification of new components of the pathway (which include the direct PETCM target and the physiological signals whose function is mimicked by PETCM), (*b*) investigation of the physiological roles of the pathway, and (*c*) clinical applications. Overall, this new pathway revealed how complicated regulation of the cytochrome *c*-mediated apoptosis pathway can be and how it functionally interacts with multiple proteins involved in oncogenesis.

PERSPECTIVES

To date, the cytochrome *c*-mediated caspase activation pathway is well established, and its physiological and pathological significance is overwhelmingly supported by studies at biochemical, genetic, and cellular levels. The pathway is under precise regulation in a time- and spatial-specific manner, both before and after cytochrome *c* release. However, there are still many important questions. A few examples include, Mechanistically, how do the Bcl-2 family members, Bax and Bak in particular, interact with mitochondria to control release of cytochrome *c* and other proteins? Developmentally, how is this pathway regulated in a tissue-specific manner, especially in the nervous system? In terms of regulation, interaction of this pathway with other signaling networks, especially those functioning in oncogenesis, needs to be extensively investigated. And clinically, although this pathway is a promising cancer therapy target, the theoretical value is yet to be translated into medicine. Taking all of these into account, this complex and important pathway still presents an enormous challenge for both basic biological research and therapeutic exploration in the future.

ACKNOWLEDGMENTS

We thank members of Xiaodong Wang's laboratory for helpful comments and suggestions and Elie Traer for critical reading. The work in our laboratory is supported by grants from the Howard Hughes Medical Institute, National Institutes of Health, and Welch Foundation.

The *Annual Review of Biochemistry* is online at http://biochem.annualreviews.org

LITERATURE CITED

1. Kerr JF, Wyllie AH, Currie AR. 1972. *Br. J. Cancer* 26:239–57
2. Horvitz HR, Shaham S, Hengartner MO. 1994. *Cold Spring Harbor Symp. Quant. Biol.* 59:377–85
3. Hengartner MO. 1999. *Recent Prog. Horm. Res.* 54:213–24
4. Liu QA, Hengartner MO. 1999. *Ann. NY Acad. Sci.* 887:92–104
5. Horvitz HR. 1999. *Cancer Res.* 59: S1701 6
6. Hengartner MO, Horvitz HR. 1994. *Cell* 76:665–76
7. Tsujimoto Y, Jaffe E, Cossman J, Gorham J, Nowell PC, Croce CM. 1985. *Nature* 315:340–43
8. Tsujimoto Y, Cossman J, Jaffe E Croce CM. 1985. *Science* 228 1440–43
9. Cleary ML, Smith SD, Sklar J. 1986. *Cell* 47:19–28
10. Graninger WB, Seto M, Boutain B Goldman P, Korsmeyer SJ. 1987 *J. Clin. Invest.* 80:1512–15
11. Raffeld M, Wright JJ, Lipford E, Coss man J, Longo DL, et al. 1987. *Cance Res.* 47:2537–42
12. Vaux DL, Cory S, Adams JM. 1988 *Nature* 335:440–42
13. Hockenbery D, Nunez G, Milliman C Schreiber RD, Korsmeyer SJ. 199C *Nature* 348:334–36

14. Thornberry NA, Lazebnik Y. 1998. *Science* 281:1312–16
15. Budihardjo I, Oliver H, Lutter M, Luo X, Wang XD. 1999. *Annu Rev. Cell Dev. Biol.* 15:269–90
16. Walker NP, Talanian RV, Brady KD, Dang LC, Bump NJ, et al. 1994. *Cell* 78:343–52
17. Wilson KP, Black JA, Thomson JA, Kim EE, Griffith JP, et al. 1994. *Nature* 370:270–75
18. Rotonda J, Nicholson DW, Fazil KM, Gallant M, Gareau Y, et al. 1996. *Nat. Struct. Biol.* 3:619–25
19. Blanchard H, Kodandapani L, Mittl PR, Marco SD, Krebs JF, et al. 1999. *Struct. Fold. Des.* 7:1125–33
20. Watt W, Koeplinger KA, Mildner AM, Heinrikson RL, Tomasselli AG, Watenpaugh KD. 1999. *Struct. Fold. Des.* 7:1135–43
21. Chai JJ, Wu Q, Shiozaki E, Srinivasula SM, Alnemri ES, Shi YG. 2001. *Cell* 107:399–407
22. Ashkenazi A, Dixit VM. 1998. *Science* 281:1305–8
23. Jackson CE, Fischer RE, Hsu AP, Anderson SM, Choi YN, et al. 1999. *Am. J. Hum. Genet.* 64:1002–14
24. Jackson CE, Puck JM. 1999. *Curr. Opin. Pediatr.* 11:521–27
25. Adachi M, Watanabe-Fukunaga R, Nagata S. 1993. *Proc. Natl. Acad. Sci. USA* 90:1756–60
26. Takahashi T, Tanaka M, Brannan CI, Jenkins NA, Copeland NG, et al. 1994. *Cell* 76:969–76
27. Lynch DH, Watson ML, Alderson MR, Baum PR, Miller RE, et al. 1994. *Immunity* 1:131–36
28. Liu XS, Kim CN, Yang J, Jemmerson R, Wang XD. 1996. *Cell* 86:147–57
29. Zou H, Henzel WJ, Liu XS, Lutschg A, Wang XD. 1997. *Cell* 90:405–13
30. Li P, Nijhawan D, Budihardjo I, Srinivasula SM, Ahmad M, et al. 1997. *Cell* 91:479–89
31. Yang J, Liu XS, Bhalla K, Kim CN, Ibrado AM, et al. 1997. *Science* 275: 1129–32
32. Kuida K, Zheng TS, Na SQ, Kuan CY, Yang D, et al. 1996. *Nature* 384: 368–72
33. Yoshida H, Kong YY, Yoshida R, Elia AJ, Hakem A, et al. 1998. *Cell* 94:739–50
34. Cecconi F, Alvarez-Bolado G, Meyer BI, Roth KA, Gruss P. 1998. *Cell* 94:727–37
35. Hakem R, Hakem A, Duncan GS, Henderson JT, Woo M, et al. 1998. *Cell* 94:339–52
36. Kuida K, Haydar TF, Kuan CY, Gu Y, Taya C, et al. 1998. *Cell* 94: 325–37
37. Li K, Li YC, Shelton JM, Richardson JA, Spencer E, et al. 2000. *Cell* 101: 389–99
38. Soengas MS, Alarcon RM, Yoshida H, Giaccia AJ, Hakem R, et al. 1999. *Science* 284:156–59
39. Wang XD. 2001. *Genes Dev.* 15: 2922–33
40. Soengas MS, Capodieci P, Polsky D, Mora J, Esteller M, et al. 2001. *Nature* 409:207–11
41. Luo X, Budihardjo I, Zou H, Slaughter C, Wang XD. 1998. *Cell* 94: 481–90
42. Li HL, Zhu H, Xu CJ, Yuan JY. 1998. *Cell* 94:491–501
43. Barry M, Heibein JA, Pinkoski MJ, Lee SF, Moyer RW, et al. 2000. *Mol. Cell. Biol.* 20:3781–94
44. Lord SJ, Rajotte RV, Korbutt GS, Bleackley RC. 2003. *Immunol. Rev.* 193:31–38
45. Scaffidi C, Fulda S, Srinivasan A, Friesen C, Li F, et al. 1998. *EMBO J.* 17:1675–87
46. Srinivasula SM, Datta P, Fan XJ, Fernandes-Alnemri T, Huang ZW, Alnemri ES. 2000. *J. Biol. Chem.* 275:36152–57
47. Sun XM, Bratton SB, Butterworth M,

MacFarlane M, Cohen GM. 2002. *J. Biol. Chem.* 277:11345–51

48. Du CY, Fang M, Li YC, Li L, Wang XD. 2000. *Cell* 102:33–42
49. Verhagen AM, Ekert PG, Pakusch M, Silke J, Connolly LM, et al. 2000. *Cell* 102:43–53
50. Deveraux QL, Reed JC. 1999. *Genes Dev.* 13:239–52
51. Ekert PG, Silke J, Hawkins CJ, Verhagen AM, Vaux DL. 2001. *J. Cell Biol.* 152:483–90
52. Bratton SB, Walker G, Srinivasula SM, Sun XM, Butterworth M, et al. 2001. *EMBO J.* 20:998–1009
53. Ambrosini G, Adida C, Altieri DC. 1997. *Nat. Med.* 3:917–21
54. Vucic D, Stennicke HR, Pisabarro MT, Salvesen GS, Dixit VM. 2000. *Curr. Biol.* 10:1359–66
55. Imoto I, Yang ZQ, Pimkhaokham A, Tsuda H, Shimada Y, et al. 2001. *Cancer Res.* 61:6629–34
56. Dierlamm J, Baens M, Wlodarska I, Stefanova-Ouzounova M, Hernandez JM, et al. 1999. *Blood* 93:3601–9
57. Uren AG, Wong L, Pakusch M, Fowler KJ, Burrows FJ, et al. 2000. *Curr. Biol.* 10:1319–28
58. Riedl SJ, Renatus M, Schwarzenbacher R, Zhou Q, Sun CH, et al. 2001. *Cell* 104:791–800
59. Huang YH, Park YC, Rich RL, Segal D, Myszka DG, Wu H. 2001. *Cell* 104:781–90
60. Chai JJ, Shiozaki E, Srinivasula SM, Wu Q, Datta P, et al. 2001. *Cell* 104:769–80
61. Chai JJ, Du CY, Wu JW, Kyin S, Wang XD, Shi YG. 2000. *Nature* 406:855–62
62. Liu ZH, Sun CH, Olejniczak ET, Meadows RP, Betz SF, et al. 2000. *Nature* 408:1004–8
63. Wu G, Chai JJ, Suber TL, Wu JW, Du CY, et al. 2000. *Nature* 408:1008–12
64. Suzuki Y, Imai Y, Nakayama H, Takahashi K, Takio K, Takahashi R. 2001. *Mol. Cell* 8:613–21
65. Martins LM, Iaccarino I, Tenev T, Gschmeissner S, Totty NF, et al. 2002. *J. Biol. Chem.* 277:439–44
66. Hegde R, Srinivasula SM, Zhang Z, Wassell R, Mukattash R, et al. 2002. *J. Biol. Chem.* 277:432–38
67. van Loo G, van Gurp M, Depuydt B, Srinivasula SM, Rodriguez I, et al. 2002. *Cell Death Differ.* 9:20–26
68. Verhagen AM, Silke J, Ekert PG, Pakusch M, Kaufmann H, et al. 2002. *J. Biol. Chem.* 277:445–54
69. Yang QH, Church-Hajduk R, Ren JY, Newton ML, Du CY. 2003. *Genes Dev.* 17:1487–96
70. Jin S, Kalkum M, Overholtzer M, Stoffel A, Chait BT, Levine AJ. 2003. *Genes Dev.* 17:359–67
71. Susin SA, Lorenzo HK, Zamzami N, Marzo I, Snow BE, et al. 1999. *Nature* 397:441–46
72. Li LY, Luo L, Wang XD. 2001. *Nature* 412:95–99
73. Spiess C, Beil A, Ehrmann M. 1999. *Cell* 97:339–47
74. Faccio L, Fusco C, Chen A, Martinotti S, Bonventre JV, Zervos AS. 2000. *J. Biol. Chem.* 275:2581–88
75. Gray CW, Ward RV, Karran E, Turconi S, Rowles A, et al. 2000. *Eur. J. Biochem.* 267:5699–710
76. Klein JA, Longo-Guess CM, Rossmann MP, Seburn KL, Hurd RE, et al. 2002. *Nature* 419:367–74
77. Chen FL, Hersh BM, Conradt B, Zhou Z, Riemer D, et al. 2000. *Science* 287:1485–89
78. Parrish J, Li LL, Klotz K, Ledwich D, Wang XD, Xue D. 2001. *Nature* 412:90–94
79. Wang XC, Yang CL, Chai JJ, Shi YG, Xue D. 2002. *Science* 298:1587–92
80. Parrish JZ, Yang CL, Shen BH, Xue D. 2003. *EMBO J.* 22:3451–60
81. Gross A, McDonnell JM, Korsmeyer SJ. 1999. *Genes Dev.* 13:1899–911

82. Adams JM, Cory S. 2001. *Trends Biochem. Sci.* 26:61–66
83. Wei MC, Zong WX, Cheng EH, Lindsten T, Panoutsakopoulou V, et al. 2001. *Science* 292:727–30
84. Zong WX, Lindsten T, Ross AJ, MacGregor GR, Thompson CB. 2001. *Genes Dev.* 15:1481–86
85. Lutter M, Fang M, Luo X, Nishijima M, Xie XS, Wang XD. 2000. *Nat. Cell Biol.* 2:754–61
86. Kuwana T, Mackey MR, Perkins G, Ellisman MH, Latterich M, et al. 2002. *Cell* 111:331–42
87. Cheng EH, Sheiko TV, Fisher JK, Craigen WJ, Korsmeyer SJ. 2003. *Science* 301:513–17
88. Priault M, Chaudhuri B, Clow A, Camougrand N, Manon S. 1999. *Eur. J. Biochem.* 260:684–91
89. Shimizu S, Narita M, Tsujimoto Y. 1999. *Nature* 399:483–87
90. Shimizu S, Shinohara Y, Tsujimoto Y. 2000. *Oncogene* 19:4309–18
91. Marzo I, Brenner C, Zamzami N, Jurgensmeier JM, Susin SA, et al. 1998. *Science* 281:2027–31
92. Bauer MK, Schubert A, Rocks O, Grimm S. 1999. *J. Cell Biol.* 147:1493–502
93. Eskes R, Antonsson B, Osen-Sand A, Montessuit S, Richter C, et al. 1998. *J. Cell Biol.* 143:217–24
94. Pastorino JG, Tafani M, Rothman RJ, Marcineviciute A, Hoek JB, et al. 1999. *J. Biol. Chem.* 274:31734–39
95. Zha JP, Harada H, Yang E, Jockel J, Korsmeyer SJ. 1996. *Cell* 87:619–28
96. Datta SR, Dudek H, Tao X, Masters S, Fu HA, et al. 1997. *Cell* 91:231–41
97. del Peso L, Gonzalez-Garcia M, Page C, Herrera R, Nunez G. 1997. *Science* 278:687–89
98. Nakano K, Vousden KH. 2001. *Mol. Cell* 7:683–94
99. Oda E, Ohki R, Murasawa H, Nemoto J, Shibue T, et al. 2000. *Science* 288:1053–58
100. Biswas SC, Greene LA. 2002. *J. Biol. Chem.* 277:49511–16
101. Lei K, Davis RJ. 2003. *Proc. Natl. Acad. Sci. USA* 100:2432–37
102. Putcha GV, Le SY, Frank S, Besirli CG, Clark K, et al. 2003. *Neuron* 38:899–914
103. Dijkers PF, Medema RH, Lammers JW, Koenderman L, Coffer PJ. 2000. *Curr. Biol.* 10:1201–4
104. Nijhawan D, Fang M, Traer E, Zhong Q, Gao WH, et al. 2003. *Genes Dev.* 17:1475–86
105. Scorrano L, Oakes SA, Opferman JT, Cheng EH, Sorcinelli MD, et al. 2003. *Science* 300:135–39
106. Zong WX, Li C, Hatzivassiliou G, Lindsten T, Yu QC, et al. 2003. *J. Cell Biol.* 162:59–69
107. Marsden VS, O'Connor L, O'Reilly LA, Silke J, Metcalf D, et al. 2002. *Nature* 419:634–37
108. Lassus P, Opitz-Araya X, Lazebnik Y. 2002. *Science* 297:1352–54
109. Hu YM, Ding LY, Spencer DM, Nunez G. 1998. *J. Biol. Chem.* 273:33489–94
110. Zou H, Li YC, Liu HS, Wang XD. 1999. *J. Biol. Chem.* 274:11549–56
111. Jiang XJ, Wang XD. 2000. *J. Biol. Chem.* 275:31199–203
112. Rodriguez A, Oliver H, Zou H, Chen P, Wang XD, Abrams JM. 1999. *Nat. Cell Biol.* 1:272–79
113. Rodriguez J, Lazebnik Y. 1999. *Genes Dev.* 13:3179–84
114. Srinivasula SM, Hegde R, Saleh A, Datta P, Shiozaki E, et al. 2001. *Nature* 410:112–16
115. Srinivasula SM, Fernandes-Alnemri T, Zangrilli J, Robertson N, Armstrong RC, et al. 1996. *J. Biol. Chem.* 271:27099–106
116. Zou H, Yang RM, Hao JS, Wang J, Sun CH, et al. 2003. *J. Biol. Chem.* 278:8091–98
117. Acehan D, Jiang XJ, Morgan DG,

Heuser JE, Wang XD, Akey CW. 2002. *Mol. Cell* 9:423–32

118. Purring C, Zou H, Wang XD, McLendon G. 1999. *J. Am. Chem. Soc.* 121: 7435–36

119. Purring-Koch C, McLendon G. 2000. *Proc. Natl. Acad. Sci. USA* 97: 11928–31

120. Renatus M, Stennicke HR, Scott FL, Liddington RC, Salvesen GS. 2001. *Proc. Natl. Acad. Sci. USA* 98:14250–55

121. Shiozaki EN, Chai J, Rigotti DJ, Riedl SJ, Li P, et al. 2003. *Mol. Cell* 11:519–27

122. Mesner PW Jr, Bible KC, Martins LM, Kottke TJ, Srinivasula SM, et al. 1999. *J. Biol. Chem.* 274:22635–45

123. Liu JR, Opipari AW, Tan LJ, Jiang YB, Zhang YJ, et al. 2002. *Cancer Res.* 62:924–31

124. Pandey P, Saleh A, Nakazawa A, Kumar S, Srinivasula SM, et al. 2000. *EMBO J.* 19:4310–22

125. Saleh A, Srinivasula SM, Balkir L, Robbins PD, Alnemri ES. 2000. *Nat. Cell Biol.* 2:476–83

126. Beere HM, Wolf BB, Cain K, Mosser DD, Mahboubi A, et al. 2000. *Nat. Cell Biol.* 2:469–75

127. Bruey JM, Ducasse C, Bonniaud P, Ravagnan L, Susin SA, et al. 2000. *Nat. Cell Biol.* 2:645–52

128. Pandey P, Farber R, Nakazawa A, Kumar S, Bharti A, et al. 2000. *Oncogene* 19:1975–81

128a. Kolesnick RN, Kronke M. 1998. *Annu. Rev. Physiol.* 60:643–65

128b. Kolesnick R, Fuks Z. 2003. *Oncogene* 22:5897–906

129. Jiang XJ, Kim HE, Shu HJ, Zhao YM, Zhang HC, et al. 2003. *Science* 299: 223–26

130. Pineiro A, Cordero OJ, Nogueira M. 2000. *Peptides* 21:1433–46

131. Chen TH, Brody JR, Romantsev FE, Yu JG, Kayler AE, et al. 1996. *Mol. Biol. Cell* 7:2045–56

132. Brody JR, Kadkol SS, Mahmoud MA, Rebel JM, Pasternack GR. 1999. *J. Biol. Chem.* 274:20053–55

133. Bai JN, Brody JR, Kadkol SHS, Pasternack GR. 2001. *Oncogene* 20:2153–60

134. Magdalena C, Dominguez F, Loidi L, Puente JL. 2000. *Br. J. Cancer* 82:584–90

135. Mori M, Barnard GF, Staniunas RJ, Jessup JM, Steele GD Jr, Chen LB. 1993. *Oncogene* 8:2821–26

136. Sasaki H, Nonaka M, Fujii Y, Yamakawa Y, Fukai I, et al. 2001. *Surg. Today* 31:936–38

137. Wu CG, Habib NA, Mitry RR, Reitsma PH, van Deventer SJ, Chamuleau RA. 1997. *Br. J. Cancer* 76:1199–204

138. Matsuoka K, Taoka M, Satozawa N, Nakayama H, Ichimura T, et al. 1994. *Proc. Natl. Acad. Sci. USA* 91:9670–74

139. Mutai H, Toyoshima Y, Sun W, Hattori N, Tanaka S, Shiota K. 2000. *Biochem. Biophys. Res. Commun.* 274: 427–33

140. Radrizzani M, Vila-Ortiz G, Cafferata EG, Di Tella MC, Gonzalez-Guerrico A, et al. 2001. *Brain Res.* 907:162–74

141. Matilla A, Koshy BT, Cummings CJ, Isobe T, Orr HT, Zoghbi HY. 1997. *Nature* 389:974–78

Annu. Rev. Biochem. 2004. 73:107–146
doi: 10.1146/annurev.biochem.73.011303.074004
Copyright © 2004 by Annual Reviews. All rights reserved
First published online as a Review in Advance on April 2, 2004

Nuclear Magnetic Resonance Spectroscopy of High-Molecular-Weight Proteins

Vitali Tugarinov,[1,2,3] Peter M. Hwang,[2] and
Lewis E. Kay[1,2,3]

*Departments of Medical Genetics,[1] Biochemistry,[2] and Chemistry,[3] University of
Toronto, Ontario, Canada M5S 1A8; email: vitali@pound.med.utoronto.ca,
peter@pound.med.utoronto.ca, kay@pound.med.utoronto.ca*

Key Words   protein NMR, resonance assignments, methyl TROSY, deuteration

■ **Abstract**   Recent developments in NMR spectroscopy, which include new experiments that increase the lifetimes of NMR signals or that precisely define the orientation of internuclear bond vectors with respect to a common molecular frame, have significantly increased the size of proteins for which quantitative structural and dynamic information can be obtained. These experiments have, in turn, benefited from new labeling strategies that continue to drive the field. The utility of the new methodology is illustrated by considering applications to malate synthase G, a 723 residue enzyme, which is the largest single polypeptide chain for which chemical shift assignments have been obtained to date. New experiments developed specifically to address the complexity and low sensitivity of spectra recorded on this protein are presented. A discussion of the chemical information that is readily available from studies of systems in the 100 kDa mol wt range is included. Prospects for membrane protein structure determination are discussed briefly in the context of an application to an *Escherichia coli* enzyme, PagP, localized to the outer membrane of gram-negative bacteria.

CONTENTS

0066-4154/04/0707-0107$14.00   **107**

INTRODUCTION

Overall Goals

In the past decade, solution NMR spectroscopy has continued to evolve as a powerful tool for the study of biomolecular structure and dynamics. The development of multidimensional, multinuclear NMR spectroscopy (1) in parallel with new labeling methodologies (2) has significantly increased the size of molecules that are now amenable to NMR studies. The introduction of media for obtaining weak alignment of solute molecules (3–7), along with new experiments for measuring the resultant residual couplings (8), and chemical shift changes that accompany alignment (9) have facilitated the measurement of powerful orientational restraints. This, in turn, has led to substantial improvements in the accuracy of NMR-derived structures, especially in cases where only low densities of other restraints are available (10). The development of transverse relaxation-optimized spectroscopy (TROSY) experiments (11, 12) in which the lifetimes of NMR signals are substantially increased has significantly impacted the size limitations that have plagued NMR studies of macromolecules in the past. These advances in concert hold the promise for many interesting applications involving a wide range of biochemical systems.

The new methodologies have been described in detail in a series of reviews, and the interested reader is referred to the literature (1, 12–20). In this review, we choose to not be comprehensive and to not focus on what has been reviewed previously. Instead, we concentrate primarily on one protein that our laboratory has studied over the past several years, malate synthase G, MSG, and illustrate concepts and approaches using this system. MSG is a monomeric enzyme, composed of 723 residues (82 kDa) (21), and at present is close to a factor of two larger than other proteins that have been assigned by multidimensional NMR methods. Experiments that work well for proteins of 40 kDa fail, in some cases, for this large system, and in this review we describe new experimental approaches that circumvent such problems. As described below, MSG is a four-domain enzyme that binds a number of ligands. We present results from studies that measure the relative orientation of domains as a function of ligand, the thermodynamics and kinetics of ligand binding, as well as hydrogen exchange and site-specific dynamics. An application to the study of structure and dynamics of a membrane enzyme, PagP, (22) is also discussed; it illustrates how the new methodologies have also impacted the study of this important group of molecules. These examples provide the reader with a feel for the information that is available from NMR studies of large proteins and the strong complimentarity between results from NMR and X-ray crystallography. Building on the labeling strategies described herein, a methyl-TROSY experiment is presented that compliments TROSY spectroscopy of backbone amide correlations and promises to extend NMR studies further to even higher mol wt systems.

A Brief Introduction to MSG

MSG catalyzes the Claisen condensation of glyoxylate with an acetyl group of acetyl-coenzyme A (acetyl-CoA), producing malate, an intermediate in the citric-acid-cycle (Figure 1a). The abstraction of a proton from the methyl group of the acetyl-CoA thioester presents both a kinetic and a thermodynamic challenge for the weak bases that are typically available in proteins, and it has led to an interest in structural studies of this enzyme. The crystal structure of MSG complexed with magnesium and glyoxylate has been solved at a resolution of 2.0 Å (21). As illustrated in Figure 1b, MSG is composed of four main domains: (a) a centrally located core domain of the molecule based on a highly stable $\beta 8/\alpha 8$ barrel fold, (b) an N-terminal α-helical clasp linked to the first strand of the barrel by a long extended loop, (c) an α/β domain appended to the molecular core, and (d) the C-terminal end of the enzyme consisting of a five-helix plug connected to the barrel by an extended loop. The active site of the enzyme is located in a cleft at the interface between the C-terminal plug and loops at the C-terminal ends of several of the β-strands of the core barrel (21). In order to complement the existing crystallographic structural studies on the glyoxylate-bound state of the enzyme, our laboratory has initiated studies on the apo form of the protein, as described in detail below.

BACKBONE AND SIDE CHAIN ASSIGNMENTS IN LARGE PROTEINS

Backbone Resonance Assignments

A first step in any detailed study of protein structure and/or dynamics involves backbone resonance assignments. That is, the frequency of absorption of each NMR active spin along the backbone must be determined to be used subsequently as an "atomic signature," which identifies each site in the protein. A number of important developments in the past decade have led to a significant increase in the size of macromolecules for which backbone resonances can be obtained. A major advance in this regard has been the use of highly deuterated

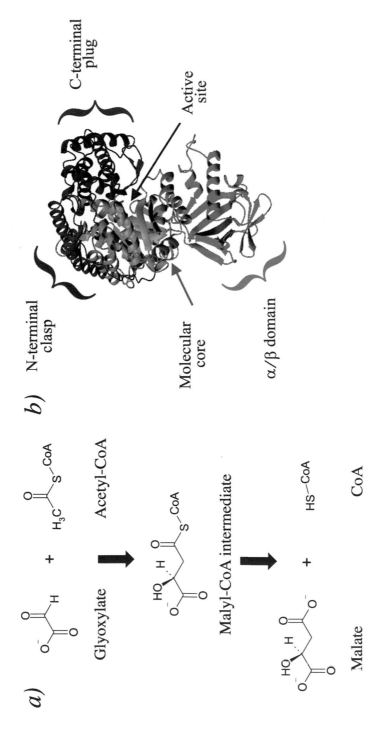

Figure 1 (*a*) Reaction catalyzed by MSG. (*b*) Ribbon diagram of MSG illustrating the four domains of the molecule. MSG coordinates were obtained from PDB accession code (1d8c) (21). Reproduced with permission from Tugarinov & Kay (48).

protein samples, leading to attenuation of the relaxation rates of NMR active nuclei and concomitant sensitivity and resolution gains in spectra (15, 16, 23, 24). Another important contribution has been the development of ^{1}HN-^{15}N TROSY (11) in which the components of ^{1}HN and ^{15}N signals that relax slowly in macromolecules at high magnetic fields are selected. TROSY experiments are also best conducted with samples that are highly deuterated. Not surprisingly, sample preparation is one of the key (and most time consuming) steps in any NMR study of a large protein. Many of the issues have been reviewed previously (15); aspects of sample preparation that required new approaches in the case of MSG are discussed below.

IN VITRO REFOLDING OF LARGE PROTEINS OVEREXPRESSED IN D$_2$O-BASED MEDIA
Overexpression of proteins in pure D$_2$O using [^{2}H,^{13}C]-D-glucose as the main carbon source leads to the production of perdeuterated proteins, with deuterium incorporation also at exchangeable backbone and side chain positions. Although subsequent protein purification and sample preparation steps are normally conducted in ^{1}H$_2$O over the course of several days, ^{2}H $\rightarrow$ ^{1}H exchange may be incomplete for amides in very stable structural motifs. Because state-of-the-art NMR methodologies for protein backbone assignments rely on a high occupancy of protons at the amide sites, incomplete ^{2}H $\rightarrow$ ^{1}H exchange can lead to severe losses (or total absence) of NMR signals (25, 26). This problem necessitates the development of efficient protocols for unfolding/refolding of proteins in ^{1}H$_2$O in order to achieve complete exchange in a reasonable time frame.

In order to demonstrate the importance of unfolding/refolding in the case of MSG, a fully protonated sample of the protein was dissolved in D$_2$O at 25°C and ^{1}HN-^{15}N TROSY-HSQC spectra were recorded over a period of 2 months. Approximately 120 backbone amides remained protonated after 7 days, and roughly 70 were not completely exchanged after a period of 6 weeks. Although near complete backbone assignments were reported for a number of large systems without the need for refolding (27–29), these studies involved proteins composed of several noncovalently linked identical units. The molecular core of each of these relatively small units may be less stable and/or more accessible to solvent than that of a large monomeric molecule.

In vitro refolding of large multidomain proteins is problematic due to complex and poorly understood thermodynamics and kinetics of the folding process (30–32). Several methods are commonly in use; these include (a) unfolding the protein with denaturant (6 M GuHCl or urea) followed by slow refolding using dialysis, (b) refolding His-tagged proteins on a metal-affinity column using a gradient with decreasing denaturant concentrations, or (c) partial protein denaturation (~2 M GuHCl) followed by fast dilution refolding. The latter approach was used previously for NMR studies of perdeuterated MBP in our laboratory (25). However, in the case of MSG, these methods failed to provide sufficiently high yields of refolded protein (26). In contrast, complete denaturation of MSG with 6 M GuHCl and subsequent fast dilution refolding into a denaturant-free buffer gave refolding yields of 60 ± 15% (26). Rapid dilution from a fully unfolded state was necessary to avoid the aggregation and ensuing precipitation that occur if a partially denatured state is allowed to accumulate. For multidomain proteins with less favorable refolding properties, it may be necessary to resort to refolding chromatography with immobilized minichaperones (33) or to a fast dilution method with the addition of nondetergent sulfobetaines (chemical chaperones) (34). In some cases the addition of binding substrates (small molecules or metal ions) and/or protein stabilizing agents (e.g., sucrose and trehalose) to the refolding medium can help to stabilize the folded state, thereby minimizing aggregation of partially folded intermediates (35). Once refolding is complete, the structural integrity of the protein can be established by comparing peak positions in 2D ^{1}HN-^{15}N correlation spectra with the subset of peaks appearing in spectra of protein that has not been refolded.

BACKBONE ASSIGNMENTS OF LARGE PROTEINS USING 4D TROSY   The complete backbone assignments of a number of systems of high-mol wt but in which the spectral complexity is moderate have been reported in the past several years. Included in this list is the 110 kDa homo-octameric protein, 7,8-dihydroneopterin aldolase (27), as well as a number of 8-stranded β-barrel membrane proteins, OmpX (36, 37), OmpA (38), and PagP (22) dissolved in lipid detergents, with effective mol wt in the 50–60 kDa range. In the case of MSG, both the high-mol wt and the large number of residues (723, correlation time of 37 ns at 37°C) necessitated the use of four-dimensional (4D) TROSY (26). In the design of these experiments, special care has been taken to optimize sensitivity. When one considers that a 1 mM solution of a 100 kDa monomeric protein consists of ~10% by weight protein, it is clear that for many systems it will be necessary to record spectra at concentrations significantly below 1 mM. Concentrations ranging from 0.5–0.9 mM were commonly used for MSG. Low concentrations, fast relaxing signals, and peak overlap contribute to the challenges associated with studies of large monomeric proteins.

A suite of 4D assignment experiments has been developed over the past several years to address the issues mentioned above. Initially a pair of 4D TROSY experiments were described; these included the 4D TROSY-HNCACO (providing correlations of the form $[\omega_{C\alpha}(i),\omega_{CO}(i),\omega_N(i),\omega_{HN}(i)]$ and in some cases $[\omega_{C\alpha}(i\text{-}1),\omega_{CO}(i\text{-}1),\omega_N(i),\omega_{HN}(i)]$) and the 4D TROSY-HNCOCA (providing correlations of the form $[\omega_{C\alpha}(i\text{-}1),\omega_{CO}(i\text{-}1),\omega_N(i),\omega_{HN}(i)]$) (39). Backbone chemical shifts are obtained by matching either ^{13}C$^\alpha$ -^{13}CO or ^{1}HN-^{15}N chemical shift pairs for different residues from these 4D data sets, as described in some detail below. Approximately 95% of the expected intra- and inter-residue correlations were obtained in HNCACO and HNCOCA data sets, respectively, of maltose binding protein at 5°C, where the overall tumbling time of the protein is 46 ns (39). Later, these experiments were supplemented by the introduction of the 4D TROSY-HNCO$_{i\text{-}1}$CA$_i$ (providing correlations $[\omega_{C\alpha}(i),\omega_{CO}(i\text{-}1),\omega_N(i),\omega_{HN}(i)]$ and in some cases $[\omega_{C\alpha}(i\text{-}1),\omega_{CO}(i\text{-}1),\omega_N(i),\omega_{HN}(i)]$) (40) and by the use of the 4D ^{15}N,^{15}N-edited NOESY (41, 42), so that ambiguities arising in the course of the assignments could be resolved (see below). Backbone assignments of a dimeric form of the human tumor suppressor protein p53 (67 kDa homodimer, 279 residues) have been reported (28), utilizing this array of four 4D data sets and a pair of 3D TROSY data sets that correlate backbone amide and side chain ^{13}C$^\beta$ chemical shifts (39, 43). These early studies demonstrated the utility of 4D-TROSY for the investigation of proteins with moderate numbers of residues. However, the full potential of this methodology is realized only in applications to systems of increasing complexity, such as MSG. Below we explain in some detail the process of resonance assignments in MSG using 4D TROSY NMR experiments (26).

Figure 2 illustrates the procedure for backbone resonance assignments for the segment of polypeptide chain in MSG extending from the amide of Ala 33 to the amide of Phe 35 (Figure 2a). Prior to the start of the backbone assignment process, it is useful to simplify the ^{1}HN-^{15}N correlation map by recording a modified 2D HN(CACB) experiment in which only amide correlations of Ala and those residues immediately following Ala are selected (26). The distinct Ala ^{13}C$^\beta$ chemical shift makes this selection particularly straightforward. Alanine is one of the most abundant amino acids in protein sequences (the most abundant in MSG, 10.1%) and serves as a good starting point for resonance assignments. Figure 2b shows a region of such a spectrum in which correlations from only Ala residues (or, rarely, residues following Ala) are observed.

Figure 2c–i illustrates how the family of 4D data sets are used for sequential assignment. Starting from Ala 33 (shown in red in Figure 2b), a ^{13}C$^\alpha$ – ^{13}CO slice of the 4D HNCACO at the amide ^{1}HN and ^{15}N chemical shifts of Ala 33 (Figure 2c) gives the assignments of the ^{13}C$^\alpha$ (55.0 ppm) and ^{13}CO(181.4 ppm) shifts of this residue because the (^{1}HN,^{15}N) correlation of Ala 33 is well resolved in the ^{1}HN-^{15}N HSQC spectrum. A weaker crosspeak in this plane (Figure 2c) appears at ^{13}C$^\alpha$/^{13}CO chemical shifts of Ala 32. Next, the (^{1}HN,^{15}N) slice of the 4D HNCOCA at ^{13}C$^\alpha$(55.0 ppm)/^{13}CO(181.4 ppm) corresponding to the shifts of Ala 33 (Figure 2d) allows the immediate assignment of the ^{1}HN/^{15}N chemical shifts of the subsequent residue—Ala 34 (shown in blue)—because the corresponding crosspeak is unique in this slice of the 4D HNCOCA. Figure 2e–g shows slices from 4D data sets at the ^{1}HN(8.07 ppm)/^{15}N(121.8 ppm) shifts of Ala 34. In this case, the 4D HNCACO spectrum alone is not sufficient for the assignment of ^{13}C$^\alpha$/^{13}CO shifts of Ala 34 as there are five residues in MSG with approximately the same pairs of backbone amide shifts (five peaks in Figure 2g). To choose the right correlation out of the five possibilities, 4D HNCOCA (Figure 2e) and 4D HNCO$_{i-1}$CA$_i$ (Figure 2f) data sets are used. The ^{13}C$^\alpha$/^{13}CO chemical shifts of Ala 33 are known, so the ^{13}C$^\alpha$ shift of Ala 34 can be obtained by identifying a peak in the 4D HNCO$_{i-1}$CA$_i$ spectrum at the same ^{13}CO position as that of Ala 33 and a ^{13}C$^\alpha$ position different from that of Ala 33 (such a crosspeak is shown in blue in Figure 2f). The assignment of the ^{13}CO of Ala 34 is then immediately obtained from the HNCACO slice (Figure 2g), thereby resolving the

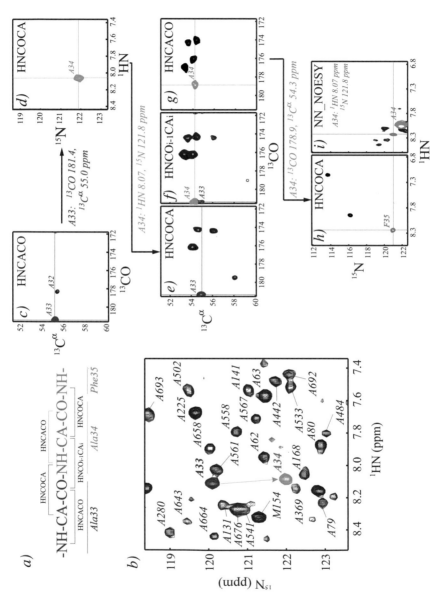

Figure 2 Sequential backbone assignments of the Ala 33–Phe 35 segment of MSG using 4D TROSY. Adapted from Tugarinov et al. (26).

fivefold degeneracy described above. In the absence of chemical shift degeneracies, the $^{13}C^{\alpha}$(54.3 ppm)/^{13}CO(178.9 ppm) chemical shifts of Ala 34 are sufficient to assign the amide shifts of Phe 35 from the 4D HNCOCA, Figure 2*h*. However, because of the threefold degeneracy of this pair of carbon shifts (three crosspeaks in Figure 2*h*), the correct assignment can only be made with additional information. In this case, the ambiguity can be resolved using ^{1}HN-^{1}HN NOEs from a 4D ^{15}N,^{15}N-edited NOESY data set (Figure 2*i*). The correct ^{1}HN,^{15}N assignment for Phe 35 is shown in green in Figure 2*h*, which coincides with the position of the NOE crosspeak in Figure 2*i* marked with green crosshairs. Generally, a high degree of degeneracy among $^{13}C^{\alpha}$/^{13}CO pairs of chemical shifts is observed, and the 4D NOESY data set is very helpful in these cases. In fact, a recent study indicates that NOE information may be essential for assignment of certain residues in MSG (44). Typically, in β-sheets and loops, sequential ^{1}HN-^{1}HN NOE crosspeaks cannot be detected, and in cases of chemical shift degeneracy, assignments are made based on the connectivities from 3D TROSY experiments that correlate amide shifts with inter- or intraresidue $^{13}C^{\beta}$ shifts. The residues of MSG were connected sequentially, as described above, starting from Ala or Gly until another residue of an unambiguous type (one of Ala, Gly, Ser, or Thr) was reached. The connected stretches, e.g., Ala(Gly)-(X)$_n$-Ala, were positioned in the MSG sequence taking into account residue-type information of all the intervening residues X based on $^{13}C^{\beta}$ chemical shifts.

In favorable circumstances, a single sample of the refolded [U-^{15}N,^{13}C,^{2}H]-labeled protein is sufficient to record all NMR experiments mentioned in this section. The redundancy of information available from the combination of 4D and 3D data sets ensures a high degree of reliability of the resulting assignments.

COVALENT MODIFICATIONS AND SECONDARY STRUCTURE FROM HETERONUCLEAR NMR DATA   Deviations of $^{13}C^{\alpha}$, $^{13}C^{\beta}$, and ^{13}CO chemical shifts from mean random coil values corrected for deuterium isotope shifts can be used to predict the secondary structure of a protein on the basis of the chemical shift index of Wishart & Sykes (45) or using the database approach of Bax and coworkers (46). In the case of MSG, the NMR-based secondary structure prediction was in good agreement with that derived from the X-ray structure (26). In addition to secondary structure, which emerges from the initial assignments, it is also possible in some cases to observe rare covalent modifications of the polypeptide chain. Figure 3*a* illustrates an isoaspartyl linkage resulting from the deamidation of the side chain of an Asn residue via a cyclic imide intermediate (30). In Figure 3*b*, strips from 3D TROSY-HNCACB (correlations primarily of the form $[\omega_{C\alpha}(i)/\omega_{C\beta}(i),\omega_N(i),\omega_{HN}(i)]$) and 3D TROSY-HN(CO)CACB ($[\omega_{C\alpha}(i\text{-}1)/\omega_{C\beta}(i\text{-}1),\omega_N(i),\omega_{HN}(i)]$) data sets have been used to distinguish between the expected -Asn(305)-Gly(306)- connectivity and the observed -IsoAsp(305)-Gly(306)-connection. Crosspeaks involving $^{13}C^{\alpha}$ and $^{13}C^{\beta}$ nuclei are of opposite signs in these spectra, illustrated in the HNCACB strip plot at the amide shifts of Asn 305, with the crosspeak of the carbon directly attached to the carbonyl carbon

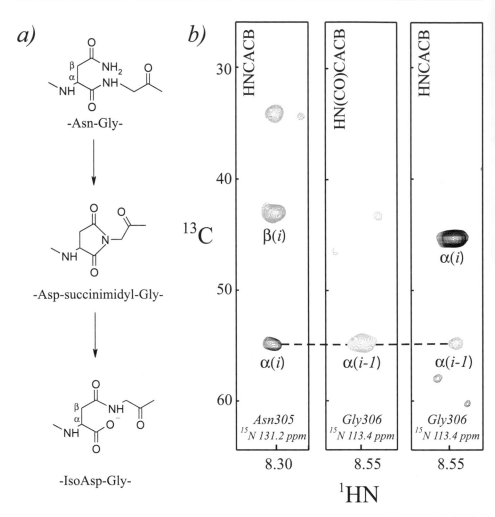

Figure 3 (*a*) The deamidation pathway of the Asn 305 side chain in MSG resulting in the isoaspartyl linkage between Asn 305 and Gly 306. (*b*) Slices from 3D HNCACB and HN(CO)CACB spectra extracted at ^{1}HN and ^{15}N chemical shifts of Asn 305 and Gly 306. Negative peaks are shown with gray contours. Adapted from Tugarinov et al. (26).

(^{13}C$^{\alpha}$, for a typical peptide bond) plotted positive (in all of the spectra in the Figure). The compelling evidence that the regular -Asn-Gly- linkage does not exist comes from the negative sign of the crosspeak in the HN(CO)CACB spectrum at the ^{1}HN,^{15}N shifts of Gly306 and the ^{13}C$^{\alpha}$ chemical shift of Asn 305. Conversely, the data of Figure 3*b* are consistent with the -IsoAsp-Gly- structure. Iso-linkages occur predominantly in Asn-Gly pairs (30). Earlier, Chazin et al. (47) used ^{1}H NMR spectroscopy to confirm one such linkage in calbindin D$_{9k}$. Of special interest is the fact that the isoaspartyl linkage in MSG was identified exactly in the middle of a flexible stretch (residues 300–310) for which coordinates could not be determined by X-ray crystallography (21). Two β-strands are predicted in this region from the chemical shift index, with the Asn 305-Gly 306 pair possibly forming a flexible β-hairpin connection between them. The unusual flexibility of this segment is supported by ^{15}N relaxation data (26, 48) and is likely to be a prerequisite for the formation of the cyclic imide intermediate.

Assignments of Side Chains in Large Proteins

The success of NMR methodology for assignment of backbone spins in proteins in the 100 kDa range has encouraged the development of new methods for assignment of side chains (49–51). A main strategy in our laboratory has been to focus on assignments of protonated methyl groups, specifically Ile(δ1), Leu, and Val methyls, in otherwise perdeuterated proteins. This labeling pattern preserves many of the important features of perdeuteration with respect to sensitivity gains in spectra, while maintaining a critical number of protons at important side chain positions for further studies of protein structure (52–54) and dynamics (55, 56). ^{1}H-^{13}C correlation spectra of methyls are, in general, of high quality even in applications involving high-mol-wt systems (57), because many of the transitions that contribute to methyl crosspeaks are long lived (58) and because the intensity of each correlation is derived from three degenerate methyl protons. The strategies for biosynthetic incorporation of protons into Ile(δ1), Leu, and Val methyl groups of perdeuterated proteins are well established (52, 53). A concise overview of these biosynthetic methods is given below.

THE IMPORTANCE OF α-KETOACIDS AS BIOSYNTHETIC PRECURSORS FOR METHYL-CONTAINING RESIDUES IN PROTEINS   It has been long known that certain α-ketoacids can serve as biosynthetic precursors of a number of methyl-containing amino acids (Ala, Ile, Leu, and Val) in proteins overexpressed in minimal media (59). The list of α-ketoacids that are routinely used in our laboratory for a variety of applications is shown in Figure 4. Addition of α-ketoisovalerate (compound I) to [D$_2$O, ^{15}N, U-^{13}C,^{2}H-glucose]-based minimal media for protein expression in *Escherichia coli* leads to the production of uniformly ^{15}N,^{13}C, highly deuterated, Leu, Val methyl-protonated proteins (53), whereas the addition of α-ketobutyrate (compound II) generates proteins with methyl protonation restricted to the Ile δ1 position (52). Typically, both α-ketobutyrate and α-ketoisovalerate are added to growth media to generate highly deuterated proteins with protonation at methyl positions of Ile (δ1), Leu, and Val.

In addition to selective methyl protonation in deuterated, uniformly ^{15}N,^{13}C-labeled proteins, different labeling patterns involving side chains of Ile, Leu, and Val are possible by using some of the other precursors shown in Figure 4. In an effort to produce proteins with ^{13}C only at methyl sites in a cost-effective manner for high throughput screening of ligands, the Abbott group (60) has developed synthetic methods for production of ^{13}C methyl-labeled α-ketobutyric and

I

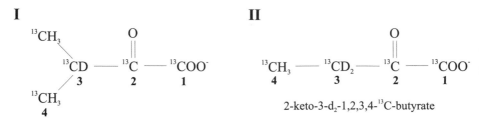

2-keto-3-methyl-^{13}C-3-d$_1$-1,2,3,4-^{13}C-butyrate

II

2-keto-3-d$_2$-1,2,3,4-^{13}C-butyrate

III

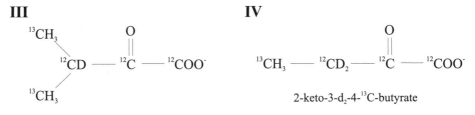

2-keto-3-methyl-^{13}C-3-d$_1$-4-^{13}C-butyrate

IV

2-keto-3-d$_2$-4-^{13}C-butyrate

V

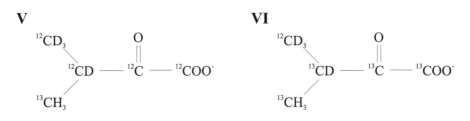

2-keto-3-methyl-d$_3$-3-d$_1$-4-^{13}C-butyrate

VI

2-keto-3-methyl-d$_3$-3-d$_1$-1,2,3,4-^{13}C-butyrate

Figure 4   Chemical formulae of the set of isotopically labeled α-ketoacids used as biosynthetic precursors for methyl-containing residues in proteins. Sodium salts of α-ketobutyric and α-ketoisovaleric acids protonated at the methyl and the 3-positions are commercially available and can be quantitatively exchanged to 3-^{2}H at high pH in D$_2$O (53). These precursors are added in the amounts of ~50 mg (butyric acid) and ~80 mg (valeric acid) per liter of [D$_2$O; U-^{13}C,^{2}H-glucose]-based growth media ~1 h prior to induction of protein overexpression to produce Ile (δ1 only), Leu, and Val methyl protonated-[^{13}C,^{2}H] labeled proteins (53).

α-ketoisovaleric acids, compounds III and IV in Figure 4. An alternative and apparently even more cost-effective synthetic strategy for production of α-ketoacids with this labeling pattern using Grignard chemistry has been proposed recently by Wagner and coworkers (61). Our laboratory has demonstrated that for assignments of methyls in MSG a labeling strategy that makes use of compound VI in Figure 4 is particularly advantageous. Below we describe in detail optimal labeling strategies and recently developed NMR experiments for assignments of methyl groups in MSG, because standard approaches that have worked so well for smaller systems failed in this case.

ASSIGNMENTS OF Ile, Leu, AND Val METHYL GROUPS USING COSY-BASED HN-DETECTED NMR EXPERIMENTS   The commonly used NMR approach for assignment of Ile, Leu, and Val methyl ^{13}C and ^{1}H resonances involves a transfer of magnetization originating on methyl groups to the backbone $^{13}C^\alpha$ carbon using homonuclear ^{13}C-^{13}C Hartmann-Hahn mixing schemes (62). Subsequently the magnetization is relayed through carbonyl spins to the amide nitrogens and protons for detection (63–67). Because the backbone spins are already assigned, correlation of side chain and backbone chemical shifts immediately leads to the assignment of the side chain ^{13}C and ^{1}H spins. These experiments have been used for assignment of methyl groups in a highly deuterated maltose binding protein (MBP, 42 kDa) sample at 600 MHz and 37°C (25) and for assignment of methyls in a large number of smaller systems. Recently, ^{1}HN-^{15}N TROSY versions of these experiments were used by Wüthrich and coworkers (68) to assign methyl groups of the membrane protein OmpX in micelles (effective mol wt ~60 kDa). For larger systems and at higher magnetic field strengths, experiments that avoid the transfer through fast-relaxing carbonyl nuclei are attractive alternatives. However, the branch points in each of the side chain carbon scaffolds of Ile, Leu, and Val degrade the efficiency of $^{13}C^m \rightarrow {}^{13}C^\alpha$ magnetization transfer. For example, consider Ile. Signal originating on $^{13}C^m$ is transferred up the side chain. At the $^{13}C^\beta$ carbon, magnetization can go in one of two ways, $^{13}C^\beta \rightarrow {}^{13}C^\alpha$ or $^{13}C^\beta \rightarrow {}^{13}C^{\gamma 2}$. The former transfer is productive in the sense that only the signal transferred to $^{13}C^\alpha$ will ultimately be detected, whereas the later transfer siphons off the signal. This bifurcation leads to a reduction in sensitivity of almost a factor of 2.5, which is unacceptable for applications to high-mol-wt proteins where signal to noise is limited in the first place. With this in mind, our laboratory has designed COSY-based experiments that transfer magnetization down the side chain assignment of Ile methyl groups in large proteins (49). Each transfer step is sequential in this approach, affording control over the flow of magnetization, unlike the case in TOCSY-transfer. This is possible for Ile because the chemical shifts of all ^{13}C spins at the branch point are well separated so that magnetization transfer from $^{13}C^\beta$ to $^{13}C^{\gamma 2}$ (unproductive) can be prevented by application of $^{13}C^{\gamma 2}$-selective 180° pulses (49). The efficacy of this approach was demonstrated on a sample of MBP at 5°C (correlation time of 46 ns, equivalent to ~100 kDa protein at 37°C) for which both the resolution and sensitivity were considerably improved over the more traditional TOCSY-based experiment (49).

Unfortunately, the approach described for Ile is not possible for Val and Leu side chains in which pro-R and pro-S methyls (and Leu $^{13}C^\gamma$ spins) all resonate at similar frequencies. We have, therefore, modified the labeling protocol so that proteins are produced with Val and Leu residues ^{13}C-enriched at only a single methyl site. This labeling strategy involves the use of α-ketoisovalerate ^{13}C-labeled (nonstereospecifically) at one methyl, while the other is of the ^{12}CD$_3$ variety (compound VI in Figure 4) to generate a linear ^{13}C spin system for these residues, effectively avoiding the problem associated with the branch point, described above. The gain in sensitivity that results outweighs the losses associated with the effective twofold dilution of the methyl groups (50). A methyl protonated {I(δ1 only), L(^{13}CH$_3$,^{12}CD$_3$), V(^{13}CH$_3$,^{12}CD$_3$)} U-[^{15}N,^{13}C,^{2}H] sample of MSG was produced using U-[^{13}C]-4-[^{1}H] α-ketobutyrate (compound II) and [^{13}CH$_3$/^{12}CD$_3$]-α-ketoisovalerate (compound VI) (50).

Figure 5a–c illustrates the magnetization flow in the array of COSY-based HN-detected NMR experiments that have been recorded on MSG at 37°C (276 protonated methyl groups). Three-dimensional data sets containing correlations of the form [ω$_{Cm}$(i),ω$_N$(i),ω$_{HN}$(i)] and [ω$_{Hm}$(i),ω$_N$(i),ω$_{HN}$(i)] are generated (50). Figure 5d,e shows planes from these 3D data sets, extracted at the amide ^{15}N shift of Leu 202 and used to assign the ^{13}C and ^{1}H methyl chemical shifts of this residue. In order to ascertain which ^{1}H shift is to be paired with which ^{13}C, a high-resolution ^{1}H-^{13}C 2D correlation map is employed, Figure 5f,g, showing clearly that only a single ^{1}H/^{13}C combination is possible. Using this set of experiments 90%, 55%, and 88% of the Ile, Leu, and Val methyls could be assigned in MSG (50).

METHYL-DETECTED OUT-AND-BACK EXPERIMENTS FOR ASSIGNMENT OF Ile, Leu, AND Val METHYLS IN LARGE PROTEINS   A drawback of the HN-detected experiments described above is that the net magnetization that can be transferred from the methyl to the amide site is effectively that associated with a single proton (rather than the reservoir from all 3 of the methyl protons) (50). As an alternative, "out-and-back" experiments can be employed in which signal both originates and is detected on the methyl groups, circumventing losses associated with net transfer to amides (50). The magnetization transfer steps in this class of

---

→

Figure 5   (a-c) Schematic diagrams of the magnetization flow in the HN-detected experiments for assignments of Ile(δ1), Leu and Val methyls (note that the methyls in Leu and Val are of the ^{13}CH$_3$/^{12}CD$_3$ variety). (d,e) ^{13}C^m-HN (correlations of the form [ω$_{Cm}$(i),ω$_N$(i),ω$_{HN}$(i)]) and ^{1}H^m-HN (correlations of the form [ω$_{Hm}$(i),ω$_N$(i),ω$_{HN}$(i)]) slices at the ^{15}N chemical shift of Leu 202 from HN-COSY experiments recorded on U-[^{2}H,^{15}N,^{13}C], [^{13}CH$_3$]-Ile(δ1) [^{13}CH$_3$/^{12}CD$_3$]-Leu,Val MSG, 800 MHz, 37°C. (f,g) regions of a 2D ^{1}H-^{13}C constant time (CT) HMQC spectrum showing the corresponding crosspeaks of Leu 202. ^{13}C^m and ^{1}H^m chemical shifts from 3D spectra in panels (d,e) indicated with solid and dashed lines correspond to peak positions in the CT-HMQC shown with solid and dashed lines, respectively, illustrating how proton-carbon connectivities can be established from a 2D ^{1}H-^{13}C correlation map. Adapted from Tugarinov & Kay (50).

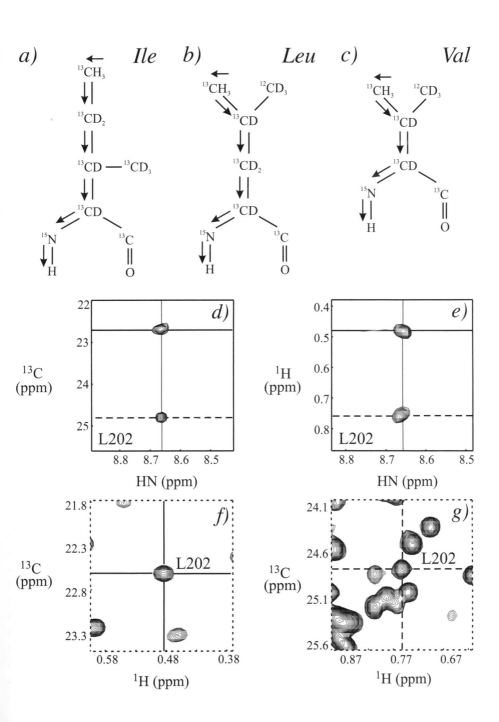

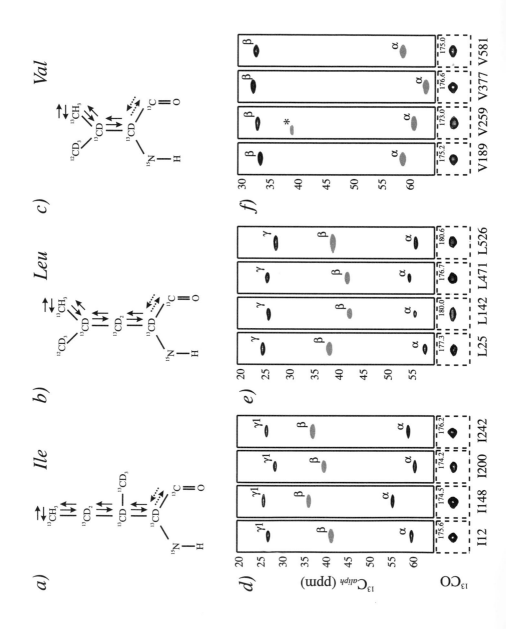

experiment are illustrated schematically in Figure 6a–c. A single 3D data set with correlations of the form $[\omega_{C\gamma,C\beta,C\alpha}(i), \omega_{Cm}(i), \omega_{Hm}(i)]$ for Ile and Leu and $[\omega_{C\beta,C\alpha}(i), \omega_{Cm}(i), \omega_{Hm}(i)]$ for Val is obtained. Additional experiments can be performed to relay signal to the ^{13}CO spin and back, providing correlations of the form $[\omega_{CO}(i), \omega_{Cm}(i), \omega_{Hm}(i)]$. As with the HN-detected experiments, these schemes benefit from Leu and Val-$[^{13}CH_3/^{12}CD_3]$ isotopic labeling and from using selective pulses to prevent the transfer of magnetization from $^{13}C^\beta$ to $^{13}C^{\gamma 2}$ in Ile (50). On average, sensitivity gains ranging from factors of 5 to 10 relative to the HN-detected data sets were observed for MSG (50).

Figure 6d–f shows ^1H_m-$^{13}C_{aliph}$ and ^1H_m-^{13}CO strips from the data sets described above for selected residues of MSG. The sequence-specific assignments of Ile, Leu, and Val methyls can be obtained by matching three ^{13}C frequencies ($^{13}C^\alpha$, $^{13}C^\beta$, and ^{13}CO) to those available from the compiled list of backbone and $^{13}C^\beta$ assignments. Note that, unlike HN-detected data sets, the methyl-detected experiments inherently contain information on ^1H-^{13}C connectivities within a given methyl group, simplifying the assignment process.

The high sensitivity of the methyl-detected experiments significantly reduces the requirements for both NMR instrument time and long-term protein stability. Less than a week of machine time was required to record all of these experiments on a 0.9 mM sample of MSG, whereas close to three weeks of NMR time was employed for the HN-detected experiments. On the basis of data from the methyl-detected data sets exclusively, 95% of Ile δ1, 64% of Leu, and 93% of Val methyls could be assigned. Degeneracy in Leu $^{13}C^\alpha/^{13}C^\beta$ chemical shift pairs, lower sensitivity of the CO experiment, and strong scalar coupling between $^{13}C^\delta$ and $^{13}C^\gamma$ nuclei in 15 Leu spin-systems precluded full assignment of Leu methyls. Combining the connectivities obtained in both methyl- and HN-data sets gave assignments for 95%, 88%, and 99% of Ile, Leu, and Val residues with a high degree of confidence. Finally, a 3D $[^{13}C$-F_1, ^{13}C-$F_2]$-edited NOESY spectrum of a $\{L(^{13}CH_3,^{13}CH_3), V(^{13}CH_3,^{13}CH_3)\}$ U-$[^{15}N,^2H]$ MSG sample (mixing time of 30 ms) was used to obtain intramethyl NOE correlations. In addition to confirming our assignments, this data set was used to assign Leu methyls in cases for which assignment of only one of the two methyls was available via the through-bond experiments described above, owing to strong coupling effects for example. In this way, the number of Leu methyls that was assigned was increased to 91%.

Figure 6   (a-c) Schematic diagrams of the magnetization transfer steps employed in the methyl-detected experiments recorded on U-$[^2H,^{15}N,^{13}C]$, $[^{13}CH_3]$-Ile(δ1) $[^{13}CH_3/^{12}CD_3]$-Leu,Val MSG. (d-f) Selected $^1H^m$-$^{13}C^{aliph}$ strips, with correlations at $[\omega_{C\gamma,C\beta,C\alpha}(i), \omega_{Cm}(i), \omega_{Hm}(i)]$ (upper row, solid frame) and $^1H^m$-^{13}CO strips, with correlations at $[\omega_{CO}(i), \omega_{Cm}(i), \omega_{Hm}(i)]$ (dashed-lined frame) for Ile (d), Leu (e) and Val (f). Negative peaks ($^{13}C^\beta$ for Ile and Leu; $^{13}C^\alpha$ for Val) are shown in gray. The peak labeled * derives from another spin system and is more intense on another slice. Adapted from Tugarinov & Kay (50).

STRUCTURAL AND DYNAMIC INFORMATION FROM NMR STUDIES OF LARGE PROTEINS

The chemical shifts assigned from the experiments described in the previous sections form the basis for NMR studies of biomolecular structure and dynamics. MSG is a particularly rich system for such studies. A high-resolution X-ray structure of the glyoxylate bound state has been published (21), and thus a structural framework is available, on the one hand, to interpret dynamics and binding data that can be readily obtained from NMR. On the other hand, structures of other liganded states or of the apo state of the enzyme have not been determined, and NMR studies can provide insight into how the domains of the protein change in response to ligand binding. In this context, NMR can be used in concert with the X-ray structure to rapidly generate high-quality models of the protein in its different states. Below, we illustrate a number of applications involving MSG and establish that the same sort of quantitative studies that have to date been reserved for applications involving small proteins can also be performed on larger systems, albeit with some modifications to the existing methodology. Finally, a structural and dynamic study of the membrane enzyme PagP is presented, illustrating that in favorable cases the methodology can also be applied successfully to small membrane proteins.

Domain Orientation in MSG from Residual Dipolar Couplings and Chemical Shift Changes Upon Alignment

Many spin interactions that contain useful structural (70, 71) and potentially dynamic information (72) average to zero for molecules that tumble in isotropic solution. The introduction of a small amount of molecular alignment (typically on the order of 0.1%) restores these interactions, at least partially, so that they can be easily measured (3). In this case, the dipolar coupling between pairs of proximal NMR active spins is nonzero, and the magnitude of this coupling is related to the orientation of the vector connecting the spins in a molecular alignment frame (3, 71). This dipolar coupling leads to peak splittings in spectra in a manner analogous to scalar couplings. A large number of experiments, including schemes optimized for applications involving high-mol-wt proteins, such as MSG (73–75), have recently been developed to facilitate the measurement of such couplings. Chemical shift changes upon alignment also provide structural information that is complementary to that derived from dipolar couplings (76–79). Because in both cases information is obtained that relates orientations of bond vectors or axis systems (chemical shift tensors) to a single molecular frame, the couplings or shift changes measured provide particularly powerful structural restraints.

The ^1HN-^{15}N dipolar coupling ($^1D_{H-N}$) is by far the largest of those accessible in uniformly deuterated, $^{15}N,^{13}C$-labeled proteins, and a total of 415 such couplings ranging from -40 to $+35$ Hz have been measured for the apo form of MSG (48). These couplings were supplemented by 320 ^{13}CO chemical shift changes that occur upon alignment. A number of important technical issues emerge in these measurements concerning the application to large proteins, and the interested reader is referred to the literature (48, 73, 74).

In the case of MSG, our goal was to use these restraints to understand how the domains in the enzyme change orientation in response to ligand binding. In particular, it was believed that the position of the C-terminal domain of the protein might differ in apo (open) and substrate-bound (closed) states of the molecule (21). Earlier circular dichroism studies of both yeast and maize malate synthases indicated that significant conformational changes likely occur in the enzyme upon ligand binding (80, 81). Domain reorientations that occur upon ligand binding in pyruvate kinase (82), a close structural analogue of MSG, have been observed from X-ray studies (83). In addition, dimeric citrate synthase (84) that possesses a totally different fold but catalyzes a similar reaction, undergoes a large conformational change in which its smaller domain rotates from the main body of the dimer by $\sim$18° (21, 84).

It is straightforward to use the dipolar couplings and chemical shifts recorded on the unligated protein to reorient the domains from the X-ray ligated state to the solution apo conformation. Using the X-ray structures of the domains on an individual basis, molecular alignment frames were determined for each domain by minimizing the difference between measured couplings and those calculated on the basis of the structure of the domain. Figure 7 compares calculated couplings obtained using the glyoxylate-bound X-ray structure versus experimental couplings from the apo solution state for each of the four domains in the protein (see Figure 1). The excellent correlation in all cases indicates that the intradomain structures of the glyoxylate-bound and apo forms of the protein are the same. Also included in the Figure are the Euler angles, {α, β, and γ} that effectively describe the transformation from the domain orientation described by the X-ray structure to the solution orientation. Of note, Euler angles from the four domains are equivalent to within experimental error. Therefore, contrary to predictions based on earlier studies with *E. coli* MSG and related proteins, the domain orientations in the apo form are not significantly different from those in the glyoxylate-bound structure determined by X-ray crystallography (21). The correlation between experimentally measured carbonyl chemical shift changes and those calculated from X-ray coordinates of the MSG-glyoxylate complex for the whole molecule (Figure 7*e*) and for the C-terminal plug (Figure 7*f*) confirm the conclusions of the dipolar coupling analysis. Thus, the gross features of the enzyme do not change upon ligand binding; much more subtle structural changes are involved. In an attempt to characterize these changes as well as the kinetics and thermodynamics of ligand binding, we

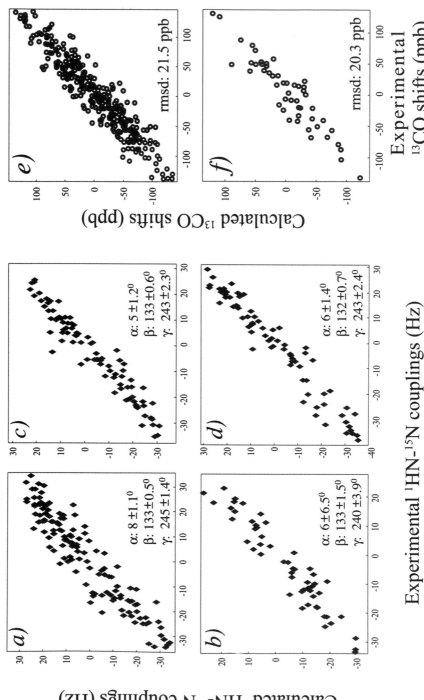

have measured a variety of NMR parameters that are sensitive to motions spanning a wide range of timescales.

Quantitative NMR Studies of Ligand Binding to MSG

NMR is a particularly powerful probe of ligand-induced structural and dynamical changes in macromolecules, and not surprisingly, large numbers of binding studies have been reported over a period of many years. NMR is also a powerful tool for determining the kinetic parameters of ligand binding not easily accessible by other methods (85). Below we summarize the sort of information that can be obtained from studies of systems as large as MSG.

[15]N RELAXATION AND HYDROGEN EXCHANGE IN Apo AND LIGAND BOUND FORMS OF MSG ^{15}N spin relaxation measurements of apo MSG establish that the protein tumbles isotropically in solution, with correlation times for each of the domains ranging from 35 to 37 ns (37°C) (26). The X-ray structure shows that all the domains in the glyoxylate-bound form make extensive surface contacts with the core of the molecule, with contact surface areas of $\sim$2300, $\sim$1330, and $\sim$2000 Å^2 between the core and the N-terminal, α/β, and the C-terminal domains, respectively. These close contacts contribute to the overall globular shape of the enzyme and ensure that the molecule tumbles as a single entity, despite the fact that the domains are connected to the core of the protein through flexible linkers. [15]N $T_{1\rho}$ values (related inversely to line width) have been measured for well-separated peaks of the protein in the apo, pyruvate-bound (pyruvate serves as a glyoxylate-mimicking inhibitor of the enzymatic reaction), and pyruvate/acetyl-CoA-bound states, and these values are plotted as a function of residue number in Figure 8a–c. The relaxation times are homogeneous throughout the sequence for all forms of the protein, with the exception of several residues in loops and linkers (48). Interestingly, the regions that are flexible on the ps-ns timescale in solution are also disordered in the crystal state (Figure 8d). In order to establish whether slower processes might be affected by binding, hydrogen exchange with solvent was measured using 3D TROSY-HNCO type (86) exchange experiments (87). The exchange rates of 28 amides could be quantified accurately (rates greater than $\sim$3 s^{-1}) and are shown in Figure 8e for the apo state of MSG. In general, regions with high exchange rates correlate well with those having elevated ^{15}N transverse relaxation times. Similar exchange profiles

---

Figure 7 Experimental versus calculated $^1\text{D}_{HN}$ values for the (a) core domain, (b) N-terminal α-clasp, (c) α/β domain, and (d) C-terminal plug of MSG. Experimental versus calculated carbonyl shifts upon alignment of MSG for (e) the whole molecule and (f) residues of the C-terminal plug. The alignment tensor obtained from the full set of $^1\text{D}_{HN}$ data was used for the calculation of carbonyl shifts. MSG was aligned using Pf1 phage, $\sim$12 mg/ml (4). Reproduced with permission from Tugarinov & Kay (48).

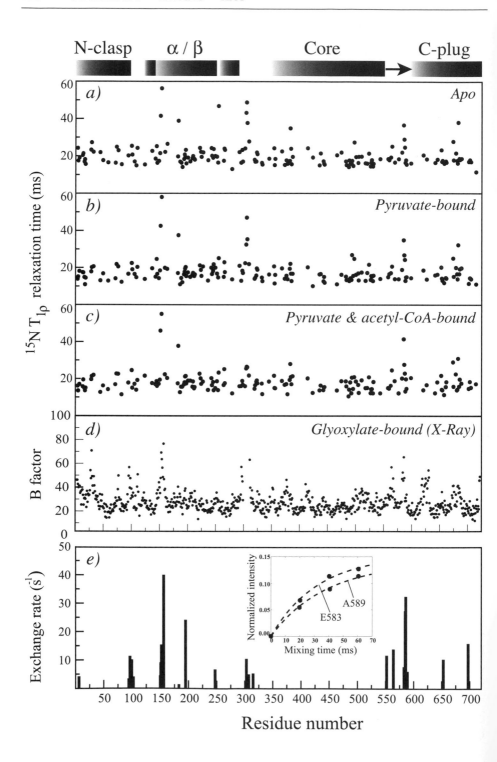

are obtained for the various bound states of the protein. The absence of significant changes in hydrogen exchange and ^{15}N relaxation profiles suggests that the changes in dynamics that do accompany binding, like the changes in structure, are minor. Notably, there is not rigidification of the linker between the C-terminal domain and the molecular core, in contrast to predictions in the literature.

KINETICS AND THERMODYNAMICS OF LIGAND BINDING TO MSG   Figure 9a illustrates a well-resolved section from the ^{1}HN-^{15}N TROSY-HSQC spectrum of apo MSG, with trajectories of peak displacements resulting from the titration of apo MSG with pyruvate, indicated with arrows. The equilibrium dissociation constants (K_d) of various MSG complexes can be determined from the shift changes that accompany ligand binding (48, 88). The best-fit curves in both ^{15}N and ^{1}HN dimensions for Val 620 titrated with pyruvate and glyoxylate are shown in Figure 9b,c. Estimated K_d values averaged over 20 well-resolved peaks with significant shifts in either ^{1}HN or ^{15}N dimensions are 1.02 $\pm$ 0.15 mM and 600 $\pm$ 70 μM for pyruvate and glyoxylate, respectively. As expected, glyoxylate, the physiological substrate of MSG, binds stronger than pyruvate.

The binding of glyoxylate or pyruvate occurs in the intermediate exchange regime on the ^{15}N and ^{1}HN chemical shift timescales for many resonances in MSG, as evidenced by considerable line-broadening at intermediate ligand concentrations (Figure 9a). It is therefore straightforward to obtain quantitative estimates of the kinetic parameters of ligand binding by fitting peak line shapes as a function of added ligand. The exchange kinetics of binding were determined for both pyruvate and glyoxylate assuming that the binding process obeys a simple second-order reaction,

$$P + L \underset{k_{off}}{\overset{k_{on}}{\longleftrightarrow}} PL,$$

in which P, L, and PL denote free protein, free ligand, and complex, respectively. Figure 9d,e shows two examples of line-shape simulations from the titration data for glyoxylate (^{1}H dimension of Ile 482) and pyruvate (^{15}N dimension of Val 620). The on rates for pyruvate derived from line-shape analysis vary between

---

Figure 8   (a-c) ^{15}N $T_{1\rho}$ relaxation times versus residue number for a subset of 165 residues of MSG as a function of different binding states of the molecule; (d) crystallographic B factors versus residue number of the glyoxylate-bound enzyme; (e) solvent exchange rates for the amides of apo MSG versus residue number. The inset shows the build-up of peak intensities of Glu 583 and Ala 589 as a function of mixing time. Adapted from Tugarinov & Kay (48).

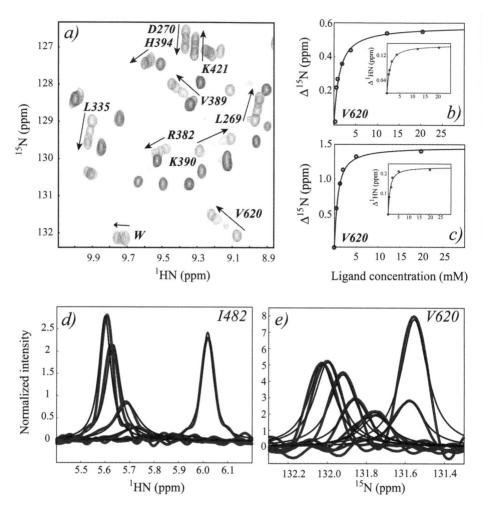

Figure 9 (*a*) Titration of [^{15}N,^{2}H]-labeled MSG with pyruvate. A selected region of the ^{1}HN-^{15}N TROSY-HSQC is shown. (*b,c*) K_d determination using chemical shift changes of Val 620 upon titration with (*b*) pyruvate and (*c*) glyoxylate. (*d,e*) Examples of line-shape simulations for (*d*) Ile 482 (^{1}HN dimension) in titration with glyoxylate and (*e*) Val 620 (^{15}N dimension) in titration with pyruvate. Experimental line shapes are shown in red, and the results of the simulations are shown in blue. Modified from Tugarinov & Kay (48).

0.8×10^6 and 1.5×10^6 M^{-1}s^{-1} and are slightly higher for glyoxylate, 1.5×10^6-3.8×10^6 M^{-1}s^{-1}. These on rates are about two orders of magnitude lower than biomolecular diffusion-controlled rates ($\sim 10^9$ M^{-1}s^{-1}) and are similar to values reported for substrates of many enzymes (10^6-10^8 M^{-1}s^{-1}) (30, 32). The entry of ligands into the binding crevice of apo MSG might require subtle structural rearrangements consistent with a large number of small chemical shift changes that were quantified upon ligand binding (48), and this may account for the decreased on rates.

FUTURE PROSPECTS—DE NOVO STRUCTURAL STUDIES OF MSG   At this point we are well poised to attempt de novo structural studies of MSG. To this end NOESY data sets quantifying methyl-methyl, methyl-HN, and HN-HN distances are being recorded. These distance restraints will be used in concert with the ^{1}HN-^{15}N dipolar couplings and ^{13}CO shift changes upon alignment, which have been measured previously, to attempt to generate global folds of the enzyme. Figure 10 illustrates the quality of structures obtained using a similar approach for β-cyclodextrin-loaded maltose binding protein. In Figure 10a, the 10 lowest energy structures calculated on the basis of distance restraints of the type indicated above in combination with hydrogen bond and dihedral angle data are shown (54), and Figure 10b shows the structures obtained when dipolar coupling data are included (77). Of note, the relative orientation of the two domains in this protein differs by 10° between solution and crystal conformations (77, 89), but in the apo- and maltotriose-loaded forms of the protein, the interdomain structures obtained from NMR and X-ray analysis are identical (90). This emphasizes the importance of a method alternative to X-ray crystallography for structural studies of multidomain proteins (89, 91).

Structure and Dynamics of Membrane Proteins by Solution NMR Spectroscopy—A Case Study of PagP

PagP is an outer membrane enzyme found in gram-negative bacteria that catalyzes the transfer of a palmitate chain from the *sn*-1 position of a phospholipid molecule in the inner leaflet of the outer membrane to lipid A, in the outer leaflet, forming hepta-acylated lipid A (92), Figure 11a. This modification confers resistance to certain antimicrobial peptides that are part of the host innate immune system (93). In an effort to understand the catalytic mechanism of the enzyme as well as the potential role of dynamics in its function, NMR studies on this protein have been initiated.

EXPRESSION AND RECONSTITUTION OF PagP   Membrane proteins are notoriously difficult to overexpress, presenting a significant obstacle to structural studies. In some favorable cases, the membrane protein of interest can be made to accumulate in a misfolded form. For example, gram-negative bacterial outer membrane proteins require a signal sequence to direct them to the outer membrane. When the signal sequence is removed, the protein forms dense inactive aggregates within the cytoplasm (94). Although nonfunctional expression is generally avoided due to problems associated with reconstitution, this is not a deterrent to solution NMR studies because unfolding and refolding are often required steps for working with large perdeuterated systems (see In Vitro Refolding of Proteins, above). In fact, this approach is the one that we used to obtain sufficient quantities of protein for the analyses described below.

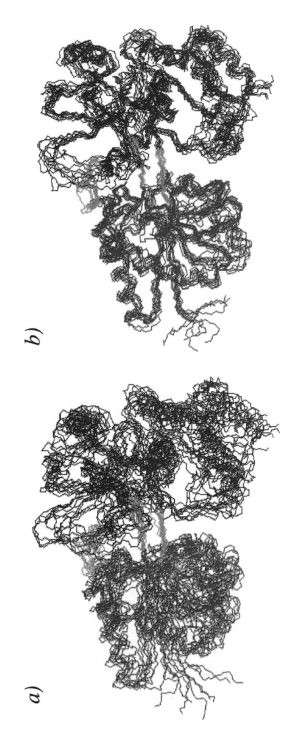

Figure 10 The 10 lowest-energy NMR structures of maltose binding protein (MBP), complexed to β-cyclodextrin (not shown) calculated from (*a*) NOE (826 HN-HN, 769 HN-CH₃, 348 CH₃-CH₃), hydrogen bond (48) and dihedral angle (464) restraints, and (*b*) NOE, hydrogen bond, dihedral angle, carbonyl chemical shift changes upon alignment (278) and residual dipolar coupling (279 ^{1}H-^{15}N, 275 ^{13}C$^\alpha$-^{13}CO, 261 ^{15}N-^{13}CO) restraints. Domains are in red and blue, and the linkers between them are in green. Modified from (77).

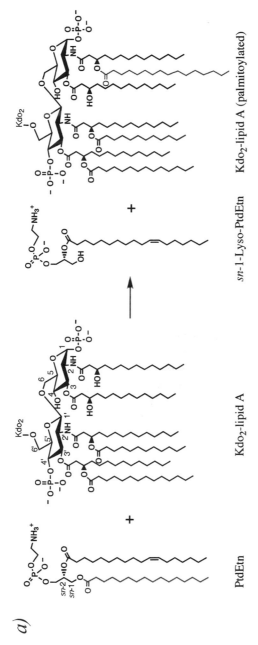

PtdEtn   Kdo₂-lipid A   *sn*-1-Lyso-PtdEtn   Kdo₂-lipid A (palmitoylated)

Figure 11   (*a*) PagP-catalyzed palmitoylation of lipid A involving acyl transfer from PtdEtn (phosphatidylethanolamine) to lipid A. (*b*) Ribbon diagram of PagP based on the structure determined in DPC. The protein is embedded in a membrane with phospholipids on the inner leaflet and lipopolysaccharide (lipid A) on the outside. The orientation of the protein relative to the bilayer is not available from the NMR experiments. Residues colored in red have backbone ^{15}N T_1s that are less than 80% of the value predicted for a 20 nanosecond overall correlation time [see Hwang et al. (22) for details]. Yellow residues have NMR signals too faint to be detected in ^{1}HN–^{15}N HSQC spectra, and green residues have weak signal due to conformational exchange. All other residues are shown in blue. The key catalytic residues are indicated. Adapted from Hwang et al. (22).

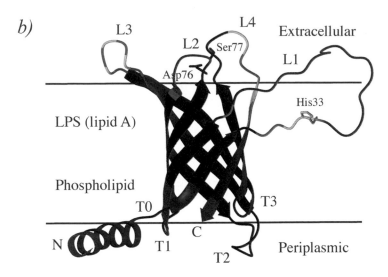

Figure 11   Continued

A number of β-barrel membrane proteins have been successfully overexpressed and reconstituted (94), making this class of membrane protein particularly attractive for structural work. Denatured PagP in 6 M guanidine hydrochloride could be refolded completely by rapid dilution into dodecylphosphocholine containing buffer, as monitored by CD spectroscopy. Samples of PagP in *n*-octyl-β-D-glucoside were also prepared and studied, as described in Hwang et al. (22). Recently developed lipopeptide detergents, amphipathic helical peptides fused to a long-chain fatty acid, are particularly promising for the study of membrane proteins (95). Excellent spectra were obtained of PagP in one such detergent, LPD-14.

STRUCTURE DETERMINATION OF PagP   Once a well-folded and stable sample has been produced, there is still no guarantee that it will be amenable to NMR analysis. Fortunately, an ^{1}HN-^{15}N HSQC spectrum run on an ^{15}N-labeled sample is sufficient for determining whether it is worthwhile to continue with structural studies at this stage.

There are currently three solution NMR structures of multi-span integral membrane proteins, including OmpX (37), OmpA (38), and PagP (22). For all three, nonfunctional expression in highly deuterated, ^{15}N, ^{13}C media followed by reconstitution produced homogeneous samples of perdeuterated ^{15}N,^{13}C-labeled protein, protonated at all exchangeable sites. In the case of PagP, backbone chemical shift assignments were carried out using conventional TROSY-based 3D NMR experiments (86, 97). The ^{13}C$^\alpha$, ^{13}C$^\beta$, and ^{13}CO chemical shifts obtained during the course of the assignment are very reliable indicators of secondary structure (45), and empirically based programs, such as TALOS, can be used to extract backbone dihedral angle restraints for most residues found within secondary structure elements (46).

Structural calculations for PagP were carried out using HN-HN NOEs, dihedral angle restraints, and hydrogen bond restraints, the latter established on the basis of hydrogen exchange experiments (22). For many protein topologies this information is insufficient to accurately define global folds. However, in the case of β-barrels, cross-strand NOEs are extremely effective at constraining the structure. Figure 11*b* shows a ribbon diagram of PagP, with the residues important for catalytic activity highlighted. The active site is thus located at the extracellular membrane interface in an optimal position for interacting with the polar groups of lipid A. This also implies that phospholipids, which are located exclusively in the inner leaflet, must somehow migrate to the outer leaflet for the acylation of lipid A to occur, Figure 11*a*.

Although the barrel of PagP is well defined in the solution NMR structure, the N-terminal α-helix could not be properly positioned using only HN-HN NOEs and dihedral angle restraints. In this regard an approach based on selective methyl labeling (see above) might be useful, as described recently in the context of structural studies of OmpX (68). Dipolar couplings (see above) would be of utility in reorienting the helical "domain" relative to the barrel. However, because the linker region to is ill-defined, restraints measured in a pair of alignment media would be necessary to uniquely define the helix orientation (99).

DYNAMICS OF PagP BY NMR   A recent crystal structure of PagP in lauryldimeth-ylamine oxide has shown that a detergent molecule is bound inside the barrel (G.G. Privé, personal commnication), very likely occupying the binding site for the phospholipid *sn*-1 acyl chain (92). Moreover, because the active site resides in the center of the barrel, the β-strands of the enzyme must come apart somewhat to allow substrate entry. Studying the conformational fluctuations in PagP is key, therefore, to understanding its function.

The main advantage of high-resolution NMR over other structural techniques lies in its ability to measure how a structure changes as a function of time. Dynamic information on the ps-ns timescale can be obtained from ^{15}N spin relaxation experiments of the sort described above for MSG. The results of such a study are color coded on the structure in Figure 11*b*. There are three highly mobile clusters in the protein with ps-ns timescale motions (indicated by red). Correlations from amino acids in yellow could not be observed in HSQC spectra, suggesting that amide positions of these residues are dynamic on a μs-ms timescale, while those residues colored in green have intensities that are attenuated, again pointing to dynamics in this time regime. Interestingly, regions with slow motions tend to localize to the extracellular loops, indicating that significant structural rearrangements are occurring here. Such conformational fluctuations may be important to allow substrate access to the active site. Still slower timescale motions can be studied from hydrogen exchange experiments. The results from all of the dynamics studies together suggest that entry of substrate into the barrel may occur by breaking contacts between strands A and B or between strands F and G. In this regard, it is of interest that both pairs of strands have a number of proline residues and therefore contain less than the maximal number of bridging hydrogen bonds.

TROSY IN METHYL GROUPS—NEW PERSPECTIVES FOR ^{1}H-^{13}C SPECTROSCOPY OF LARGE PROTEINS AND SUPRAMOLECULAR COMPLEXES

Brief Introduction

New technologies have recently emerged that hold promise for the qualitative study of structure and dynamics in supramolecular complexes. Particularly exciting in this regard are the experiments of Wüthrich and coworkers (100, 101), based on the original work of Dalvit (102), in which polarization is transferred between ^{1}HN and ^{15}N coupled spin pairs due to interference between ^{1}H chemical shift anisotropy and ^{1}HN-^{15}N dipolar interactions. These so called CRIPT and CRINEPT pulse schemes have been used to record ^{1}HN-^{15}N correlation maps of ^{15}N-labeled GroES in a GroES-GroEL complex (900 kDa) and to obtain information about the structure and the dynamics of the assembly (29). The experiments described above focus on backbone positions of a protein. It is of considerable interest to develop complementary approaches for studying side chains, and for reasons that should come as no surprise based on the discussion in the Section on Backbone and Side Chain Assignments above, we have targeted methyl groups as probes of structure and dynamics in this context.

The Methyl-TROSY Effect—A Qualitative Explanation

A detailed analysis of the eight transitions that contribute to ^{13}C magnetization in a methyl group and the 10 ^{1}H transitions from which the ^{1}H signal is derived shows that the relaxation properties are not uniform. In the high-mol-wt limit and considering only the dominant dipolar contributions to relaxation, it can be shown that six of the eight ^{13}C lines are long lived and two relax rapidly (103, 104). Conversely, 50% of the ^{1}H signal decays with a large time constant, with the other half rapidly dephasing (105–107). The construction of a TROSY experiment for methyl groups is predicated, therefore, on designing a pulse scheme in which the fast relaxing transitions are isolated from those that relax slowly. This, of course, is the idea behind any of the amide-based TROSY pulse schemes as well. However, methyl groups are much more complex than ^{1}HN-^{15}N spin pairs, and initial attempts to design TROSY-based experiments for methyls in our group failed.

In a recent series of papers, it has been shown that the simplest of ^{1}H-^{13}C correlation experiments, the HMQC pulse sequence (108, 109), is an optimized TROSY experiment for methyl groups, with 1/2 of the resulting signal derived

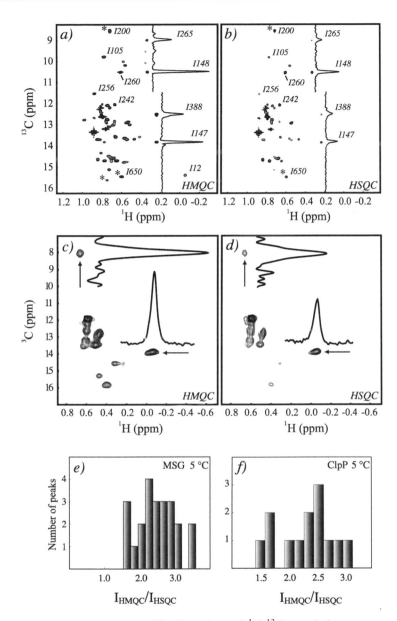

Figure 12 (*a,b*) A comparison of Ile-δ1 regions of ^{1}H-^{13}C correlation maps recorded at 5°C (800 MHz) on a U-[^{15}N,^{2}H], Ileδ1-[^{13}C,^{1}H] sample of MSG using (*a*) HMQC and (*b*) HSQC pulse sequences (0.8 mM, 173 min/spectrum). Peaks marked with * are aliased. (*c,d*) The same comparison for a U-[^{15}N,^{2}H], Ileδ1-[^{13}C,^{1}H] sample of ClpP protease at 5°C, 800 MHz (1 mM in monomer, spectra recorded in 1.5 h). (*e,f*) Histograms of intensity ratios in the HMQC and HSQC spectra (I_{HMQC}/I_{HSQC}) obtained for MSG (*e*) and ClpP (*f*). Adapted from Tugarinov et al. (58).

exclusively from the slowly relaxing transitions of both nuclei (58, 110). That is, the slow (50%) and fast relaxing (50%) pathways are isolated (neglecting relaxation contributions from external ^1H spins). Conversely, the much more popular HSQC experiment (111) intermixes fast and slowly relaxing transitions so that only 3/16 of the net signal derives from coherences that relax slowly throughout the course of the experiment. More detailed explanations are given in the literature (58, 110) and are beyond the scope of the present discussion. Suffice it to say that the TROSY effect observed in methyl groups using the HMQC experiment results from the cancellation of dipolar (both ^1H-^1H and ^1H-^{13}C) fields in a manner that is completely magnetic field independent, so long as the system tumbles slowly ($\omega_C \tau_M \gg 1$, in which ω_C is the ^{13}C Larmor frequency and τ_M is the tumbling time). Below, we compare spectra of regular HMQC and HSQC experiments recorded on two samples of large perdeuterated methyl-protonated proteins at 5°C and illustrate some of the benefits described above.

EXPERIMENTAL VERIFICATION WITH HIGH-MOLECULAR-WEIGHT PROTEINS   As with ^1HN-^{15}N TROSY, the intensity of correlations in methyl-TROSY spectra are also sensitive to contributions to relaxation from external spins (58). The experiments are thus best performed using highly deuterated samples. As an initial test of the methodology, perdeuterated protein samples (dissolved in D_2O) were prepared with [^{13}C,^1H]-labeled Ile δ1 methyl groups. Proteins with this labeling pattern can be obtained using α-ketobutyrate [^{13}C,^1H]-labeled at the methyl position (compound IV in Figure 4) and [^2H,^{12}C]-D-glucose as carbon sources in D_2O-based minimal media growths. Restricting protonation to Ile δ1 sites ensures that the only external source of relaxation for a given methyl derives from other Ile δ1 methyls in the protein. In the case of MSG, the average sum of effective distances from a given methyl to all external protons is ~5.5 Å using this labeling scheme, so that the majority of Ile methyls are well isolated.

Figure 12a,b shows ^1H-^{13}C methyl correlation maps recorded on U-[^{15}N,^2H], Ileδ1-[^{13}C,^1H] MSG in D_2O at 5°C, 800 MHz, using HMQC (panel a) and HSQC (panel b) pulse schemes. The correlation time of MSG at 5°C in D_2O is ~120 ns (equivalent to a molecule in the 200–250 kDa range at 37°C), estimated as described in Tugarinov et al. (58). For comparison, both spectra are plotted at the same level, and it is clear that the HMQC data set has significantly higher sensitivity. A histogram of the ratios of signal to noise in HMQC and HSQC

Figure 13   A comparison of a selected region of HMQC and HSQC spectra of U-[^{15}N,^2H], MSG (samples are 0.71 mM), 37°C, 800 MHz, with Leu,Val methyl groups labeled as indicated. The average, normalized signal to noise ratio, $<SN>_N$, for each spectrum is indicated. All spectra are recorded and processed identically and are shown with the same contour levels. Reproduced with permission from (113).

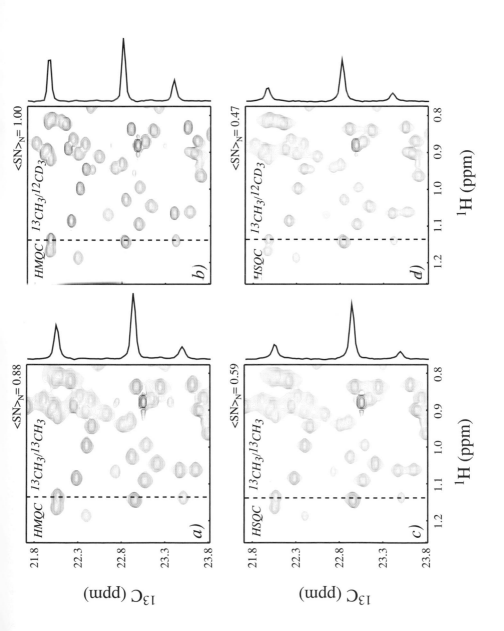

spectra recorded at 5°C, (I_{HMQC}/I_{HSQC}) is shown in Figure 12*e*. On average, a gain in signal of a factor of 2.6 is noted in the HMQC data set.

Figure 12*c,d* demonstrates the improvements in sensitivity of HMQC versus HSQC methyl correlation maps that have been recorded on the U-[^{15}N,^{2}H], Ileδ1-[^{13}C,^{1}H]-labeled protease ClpP in D_2O at 5°C. This complex consists of 14 identical subunits, each of 193 residues (total mol wt 305 kDa), with 16 isoleucines/subunit (112). At 5°C, the rotational correlation time of ClpP in D_2O is estimated in the 400–450 ns range. At least 14 of the expected 16 correlations are counted in the HMQC spectrum of Figure 12*c*. As shown in the histogram of Figure 12*f*, sensitivity gains of correlations in HMQC relative to HSQC maps vary between a factor of 1.5 to 3.1 depending on the residue. Sensitivity gains lower than those theoretically predicted (58) may result from conformational heterogeneity (chemical exchange), which affects HMQC spectra to a greater extent than HSQC data sets. Indeed, significant line broadening was observed for several peaks in spectra of ClpP recorded at higher temperatures.

The high quality of HMQC spectra recorded at 5°C on the pair of systems described above suggests that methyl-TROSY will prove useful in the study of very large proteins and supramolecular complexes.

Optimal Isotopic Labeling Methods for TROSY in Methyl Groups

The labeling scheme involving selective protonation of Ile δ1 methyl groups is of somewhat limited utility because only a single type of probe is available for structural and dynamical studies, and it would clearly be advantageous to include protonation at Leu and Val methyl sites. As described above, however, the introduction of protons comes at the expense of degrading the TROSY effect because the added protons are effective sources of external relaxation. In Leu and Val residues in which two methyl groups are side by side, intraresidue ^{1}H-^{1}H dipolar interactions are especially detrimental to the TROSY effect. With this in mind, a labeling scheme has been proposed in which each Leu and Val is labeled with one ^{13}CH$_3$ and one ^{12}CD$_3$ methyl (using compound V of Figure 4) (113). As we show below, the sensitivity loss associated with the twofold dilution of methyl protons in Leu and Val residues is significantly more than compensated for (both in terms of increased sensitivity and resolution) by the improved relaxation properties of the remaining NMR active methyls.

Figure 13 shows selected regions of HMQC and HSQC data sets recorded on a pair of MSG samples (37°C, τ_C = 45 ns) with equal protein concentrations (0.71 ± 0.05 mM). These include (*a*) a U-[^{2}H,^{15}N] Leu,Val-[^{13}CH$_3$/^{13}CH$_3$] sample and (*b*) a U-[^{2}H,^{15}N] Leu,Val-[^{13}CH$_3$/^{12}CD$_3$] sample—both prepared using the appropriately labeled α-ketoisovalerate (compounds III and V in Figure 4). The average normalized signal-to-noise ratios ($<SN>_N$) for all well-separated correlations in the full spectra are indicated above each plot. A comparison of the panels *b* and *d* in Figure 13 establishes that the same benefits observed in HMQC data sets relative to the corresponding HSQC maps in the case of

Ile-protonated samples (Figure 12) are also seen for samples with Leu and Val methyls labeled with $[^{13}CH_3/^{12}CD_3]$. Namely, the HMQC spectrum is of significantly better resolution with more than a factor of two improvement in $<SN>_N$ relative to its HSQC counterpart.

The HMQC data sets in panels *a* and *b* of Figure 13 show that despite the loss of a factor of two in concentration of NMR active methyls in the case of $[^{13}CH_3/^{12}CD_3]$ Leu/Val methyl labeling, spectra are nonetheless of improved sensitivity and resolution relative to those recorded on Leu,Val-$[^{13}CH_3/^{13}CH_3]$ MSG samples. The differences between HMQC spectra recorded on the two sets of labeled samples are even more pronounced for MSG at 5°C ($\tau_C \approx 120$ ns) with an average $<SN>$ improvement of close to a factor of two in favor of the Leu,Val-$[^{13}CH_3/^{12}CD_3]$-labeled sample. The proposed Leu,Val-$[^{13}CH_3/^{12}CD_3]$ labeling scheme is, therefore, critical for the optimization of the methyl TROSY effect.

Simultaneous protonation of Ile $\delta 1$ and $[^{13}CH_3/^{12}CD_3]$-labeling of Leu and Val methyls in deuterated protein samples is desirable to maximize the number of available methyl sites for NMR studies. Figure 14 shows a comparison of HMQC and HSQC correlation maps recorded on a U-$[^2H,^{15}N]$ Ile$\delta 1$-$[^{13}CH_3]$ Leu,Val-$[^{13}CH_3/^{12}CD_3]$ MSG sample at 5°C. A gain of 2.6 in $<SN>$ and significant improvements in resolution are noted for the HMQC data set, with essentially no differences between Leu, Val correlations in spectra recorded on U-$[^2H,^{15}N]$ Ile$\delta 1$-$[^{13}CH_3]$ Leu,Val-$[^{13}CH_3/^{12}CD_3]$ and U-$[^2H,^{15}N]$ Leu,Val-$[^{13}CH_3/^{12}CD_3]$ MSG samples.

CONCLUDING REMARKS

Recent developments in NMR spectroscopy have been highlighted through a discussion of their application to studies of a high-mol-wt single polypeptide enzyme, malate synthase G (723 residues, 82 kDa). Backbone ^1HN, ^{15}N, $^{13}C^\alpha$, ^{13}CO, and side chain $^{13}C^\beta$ assignments of the protein have been obtained from a series of 4D TROSY-based triple resonance experiments—the first step in any detailed analysis of protein structure and dynamics. These assignments have facilitated a quantitative study of domain orientation and dynamics as a function of a variety of ligands that bind to the protein. Near complete side chain methyl assignments have been made using new labeling schemes and pulse sequences with improved sensitivity and resolution. Studies of the dynamics of side chains and how such dynamics change in response to ligand binding are now also possible. Advances in solution spectroscopy of water-soluble proteins can be put to good use in the study of membrane proteins, illustrated here in the context of the determination of the global fold of PagP, an eight-stranded β-barrel membrane protein. Finally, TROSY, originally developed for amide ^1HN, ^{15}N spin pairs has been extended to $^{13}CH_3$ methyl groups, so that both backbone and now side chain structure and dynamics can be probed in very high-mol-wt systems.

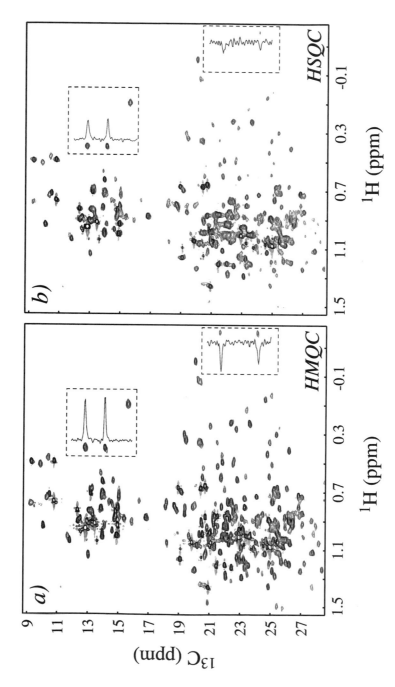

Figure 14 A comparison of HMQC (*a*) and HSQC (*b*) ^{1}H-^{13}C correlation maps recorded on a U-[^{2}H,^{15}N], Ileδ1-[^{13}CH$_3$] Leu,Val-[^{13}CH$_3$/^{12}CD$_3$] sample of MSG at 5°C ($\tau_c \approx$ 118ns), 800 MHz (0.63 mM, net acquisition time 2.5 h/spectrum). Adapted from (113).

ACKNOWLEDGMENTS

This work was supported by a grant from the Canadian Institutes of Health Research (CIHR) to L.E.K. We thank Dr. W.-Y. Choy (University of Toronto) for providing Figure 10 of this review. V.T. acknowledges financial support from the Human Frontiers Science Program. L.E.K. holds a Canada Research Chair in Biochemistry and is a member of the Protein Engineering Network Centres of Excellence.

The *Annual Review of Biochemistry* is online at http://biochem.annualreviews.org

LITERATURE CITED

1. Bax A, Grzesiek S. 1993. *Acc. Chem. Res.* 26:131–38
2. Goto NK, Kay LE. 2000. *Curr. Opin. Struct. Biol.* 10:585–92
3. Tjandra N, Bax A. 1997. *Science* 278: 1111–14
4. Hansen MR, Mueller L, Pardi A. 1998. *Nat. Struct. Biol.* 5:1065–74
5. Clore GM, Starich MR, Gronenborn AM. 1998. *J. Am. Chem. Soc.* 120: 10571–72
6. Ruckert M, Otting G. 2000. *J. Am. Chem. Soc.* 122:7793–97
7. Sass J, Cordier F, Hoffman A, Cousin A, Omichinski JG, et al. 1999. *J. Am. Chem. Soc.* 121:2047–55
8. Bax A, Kontaxis G, Tjandra N. 2001. *Methods Enzymol.* 339:127–74
9. Cornilescu G, Marquardt J, Ottiger M, Bax A. 1998. *J. Am. Chem. Soc.* 120: 6836–37
10. Clore GM, Starich MR, Bewley CA, Cai M, Kuszewski J. 1999. *J. Am. Chem. Soc.* 121:6513–14
11. Pervushin K, Riek R, Wider G, Wüthrich K. 1997. *Proc. Natl. Acad. Sci. USA* 94:12366–71
12. Pervushin K. 2001. *J. Biomol. NMR* 20:275–85
13. LeMaster DM. 1990. *Q. Rev. Biophys.* 23:133–74
14. Clore GM, Gronenborn AM. 1991. *Science* 252:1309–99
15. Gardner KH, Kay LE. 1998. *Annu. Rev. Biophys. Biomol. Struct.* 27:357–406
16. Farmer BT, Venters RA. 1998. In *Biological Magnetic Resonance*, ed. NR Krishna, LJ Berliner, pp. 75–120. New York: Kluwer Acad./Plenum
17. Prestegard JH. 1998. *Nat. Struct. Biol. NMR Suppl.* 5:517–22
18. Wider G, Wüthrich K. 1999. *Curr. Opin. Struct. Biol.* 9:594–601
19. Kanelis V, Forman-Kay JD, Kay LE. 2001. *IUBMB Life* 52:291–302
20. Bax A. 2003. *Protein Sci.* 12:1–16
21. Howard BR, Endrizzi JA, Remington SJ. 2000. *Biochemistry* 39:3156–68
22. Hwang PM, Choy WY, Lo EI, Chen L, Forman-Kay JD, et al. 2002. *Proc. Natl. Acad. Sci. USA* 99:13560–65
23. Grzesiek S, Anglister J, Ren H, Bax A. 1993. *J. Am. Chem. Soc.* 115:4369–70
24. Yamazaki T, Lee W, Arrowsmith CH, Muhandiram DR, Kay LE. 1994. *J. Am. Chem. Soc.* 116:11655–66
25. Gardner KH, Zhang X, Gehring K, Kay LE. 1998. *J. Am. Chem. Soc.* 120: 11738–48
26. Tugarinov V, Muhandiram R, Ayed A, Kay LE. 2002. *J. Am. Chem. Soc.* 124: 10025–35
27. Salzmann M, Pervushin K, Wider G, Senn H, Wüthrich K. 2000. *J. Am. Chem. Soc.* 122:7543–48
28. Mulder FAA, Ayed A, Yang D, Arrow-

smith CH, Kay LE. 2000. *J. Biomol. NMR* 18:173–76

29. Fiaux J, Bertelsen EB, Horwich AL, Wüthrich K. 2002. *Nature* 418:207–21
30. Creighton TE. 1993. *Proteins: Structures and Molecular Properties.* New York: Freeman
31. Creighton TE, ed. 1992. *Protein Folding.* New York: Freeman
32. Fersht A. 1999. *Structure and Mechanism in Protein Science.* New York: Freeman
33. Altamirano MM, Golbik R, Zahn R, Buckle AM, Fersht AR. 1997. *Proc. Natl. Acad. Sci. USA* 94:3576–78
34. Goldberg ME, Expert-Bezançon N, Vuillard L, Rabilloud T. 1995. *Fold. Des.* 1:21–27
35. Lilie H, Schwarz E, Rudolph R. 1998. *Curr. Opin. Biotechnol.* 9:497–501
36. Fernandez C, Hilty C, Bonjour S, Adeishvili K, Pervushin K, Wüthrich K. 2001. *FEBS Lett.* 504:173–78
37. Fernandez C, Adeishvili K, Wüthrich K. 2001. *Proc. Natl. Acad. Sci. USA* 98: 2358–63
38. Arora A, Abildgaard F, Bushweller JH, Tamm LK. 2001. *Nat. Struct. Biol.* 8:334–38
39. Yang D, Kay LE. 1999. *J. Am. Chem. Soc.* 121:2571–75
40. Konrat R, Yang D, Kay LE. 1999. *J. Biomol. NMR* 15:309–13
41. Venters RA, Metzler WJ, Spicer LD, Mueller L, Farmer BT. 1995. *J. Am. Chem. Soc.* 117:9592–93
42. Grzesiek S, Wingfield P, Stahl S, Kaufman J, Bax A. 1995. *J. Am. Chem. Soc.* 117:9594–95
43. Salzmann M, Wider G, Pervushin K, Senn H, Wüthrich K. 1999. *J. Am. Chem. Soc.* 121:844–48
44. Coggins BE, Zhou P. 2003. *J. Biomol. NMR* 26:93–111
45. Wishart DS, Sykes BD. 1994. *J. Biomol. NMR* 4:171–80
46. Cornilescu G, Delaglio F, Bax A. 1999. *J. Biomol. NMR* 13:289–302

47. Chazin WJ, Kordel J, Thulin E, Hofmann T, Drakenberg T, Forsen S. 1989. *Biochemistry* 28:8646–53
48. Tugarinov V, Kay LE. 2003. *J. Mol. Biol.* 327:1121–33
49. Tugarinov V, Kay LE. 2003. *J. Am. Chem. Soc.* 125:5701–6
50. Tugarinov V, Kay LE. 2003. *J. Am. Chem. Soc.* 125:13868–78
51. Eletsky A, Moreira O, Kovacs H, Pervushin K. 2003. *J. Biomol. NMR* 26:167–79
52. Gardner KH, Kay LE. 1997. *J. Am. Chem. Soc.* 119:7599–600
53. Goto NK, Gardner KH, Mueller GA, Willis RC, Kay LE. 1999. *J. Biomol. NMR* 13:369–74
54. Mueller GA, Choy WY, Yang D, Forman-Kay JD, Venters RA, Kay LE. 2000. *J. Mol. Biol.* 300:197–212
55. Nicholson LK, Kay LE, Baldisseri DM, Arango J, Young PE, et al. 1992. *Biochemistry* 31:5253–63
56. Mulder FAA, Mittermaier A, Hon B, Dahlquist FW, Kay LE. 2001. *Nat. Struct. Biol.* 8:932–35
57. Gardner KH, Rosen MK, Kay LE. 1997. *Biochemistry* 36:1389–401
58. Tugarinov V, Ollerenshaw J, Hwang P, Kay LE. 2003. *J. Am. Chem. Soc.* 125: 10420–28
59. Gottschalk G. 1986. *Bacterial Metabolism.* Berlin: Springer Verlag
60. Hajduk PJ, Augeri DJ, Mack J, Mendoza R, Yang JG, et al. 2000. *J. Am. Chem. Soc.* 122:7898–904
61. Gross JD, Gelev VM, Wagner G. 2003. *J. Biomol. NMR* 25:235–42
62. Fesik SW, Eaton HL, Olejniczak ET, Zuiderweg ERP, McIntosh LP, Dahlquist FW. 1990. *J. Am. Chem. Soc.* 112: 886–88
63. Montelione GT, Lyons BA, Emerson SD, Tashiro M. 1992. *J. Am. Chem. Soc.* 114:10974–75
64. Logan TM, Olejniczak ET, Xu RX, Fesik SW. 1993. *J. Biomol. NMR* 3:225–31

65. Grzesiek S, Anglister J, Bax A. 1993. *J. Magn. Reson. B* 101:114–19
66. Gardner KH, Konrat R, Rosen MK, Kay LE. 1996. *J. Biomol. NMR* 8:351–56
67. Lin Y, Wagner G. 1999. *J. Biomol. NMR* 15:227–39
68. Hilty C, Fernandez C, Wider G, Wüthrich K. 2002. *J. Biomol. NMR* 23:289–301
69. Deleted in proof
70. Bastiaan EW, MacLean C, van Zijl PCM, Bothner-By AA. 1987. *Annu. Rep. NMR Spectrosc.* 9:35–77
71. Tolman JR, Flanagan JM, Kennedy MA, Prestegard JH. 1995. *Proc. Natl. Acad. Sci. USA* 92:9279–83
72. Tolman JR, Flanagan JM, Kennedy MA, Prestegard JH. 1997. *Nat. Struct. Biol.* 4:292–97
73. Yang D, Venters RA, Mueller GA, Choy WY, Kay LE. 1999. *J. Biomol. NMR* 14:333–43
74. Kontaxis G, Clore GM, Bax A. 2000. *J. Magn. Reson.* 143:184–96
75. Permi P. 2000. *J. Biomol. NMR* 17:43–54
76. Ottiger M, Tjandra N, Bax A. 1997. *J. Am. Chem. Soc.* 119:9825–30
77. Choy WY, Tollinger M, Mueller GA, Kay LE. 2001. *J. Biomol. NMR.* 21:31–40
78. Lipsitz RS, Tjandra N. 2001. *J. Am. Chem. Soc.* 123:11065–66
79. Wu Z, Tjandra N, Bax A. 2001. *J. Am. Chem. Soc.* 123:3617–18
80. Beeckmans S, Khan AS, Kanarek L, Van Driessche E. 1994. *Biochem. J.* 303:413–21
81. Schmid G, Durchschlag H, Biedermann G, Eggerer H, Jaenicke R. 1974. *Biochem. Biophys. Res. Commun.* 58:419–26
82. Larsen TM, Laughlin LT, Holden HM, Rayment I, Reed GH. 1994. *Biochemistry* 33:6301–9
83. Larsen TM, Laughlin LT, Holden HM, Rayment I, Reed GH. 1997. *Arch. Biochem. Biophys.* 345:199–206
84. Remington SJ, Wiegand G, Huber R. 1982. *J. Mol. Biol.* 158:111–52
85. Sandstrom J. 1982. *Dynamic NMR Spectroscopy.* New York: Academic
86. Yang D, Kay LE. 1999. *J. Biomol. NMR* 13:3–10
87. Hwang TL, van Zijl PC, Mori S. 1998. *J. Biomol. NMR.* 11:221–26
88. Johnson PE, Brun E, MacKenzie LF, Withers SG, McIntosh LP. 1999. *J. Mol. Biol.* 287:609–25
89. Skrynnikov NR, Goto NK, Yang D, Choy WY, Tolman JR, et al. 2000. *J. Mol. Biol.* 295:1265–73
90. Evenas J, Tugarinov V, Skrynnikov NR, Goto NK, Muhandiram R, Kay LE. 2001. *J. Mol. Biol.* 309:961–74
91. Fischer MW, Losonczi JA, Weaver JL, Prestegard JH. 1999. *Biochemistry* 38:9013–22
92. Bishop RE, Gibbons HS, Guina T, Trent MS, Miller SI, Raetz CR. 2000. *EMBO J.* 19:5071–80
93. Guo L, Lim KB, Poduje CM, Daniel M, Gunn JS, et al. 1998. *Cell* 95:189–98
94. Buchanan SK. 1999. *Curr. Opin. Struct. Biol.* 9:455–61
95. McGregor CL, Chen L, Pomroy NC, Hwang P, Go S, et al. 2003. *Nat. Biotechnol.* 21:171–76
96. Deleted in proof
97. Salzmann M, Pervushin K, Wider G, Senn H, Wüthrich K. 1998. *Proc. Natl. Acad. Sci. USA* 95:13585–90
98. Deleted in proof
99. Al-Hashimi HM, Valafar H, Terrell M, Zartler ER, Eidsness MK, Prestegard JH. 2000. *J. Magn. Reson.* 143:402–6
100. Riek R, Fiaux J, Bertelsen EB, Horwich AL, Wüthrich K. 2002. *J. Am. Chem. Soc.* 124:12144–53
101. Riek R, Pervushin K, Wüthrich K. 2000. *Trends Biochem. Sci.* 25:462–68
102. Dalvit C. 1992. *J. Magn. Reson.* 97:645–50
103. Kay LE, Bull TE. 1992. *J. Magn. Reson.* 99:615–22

104. Kay LE, Torchia DA. 1991. *J. Magn. Reson.* 95:536–47

105. Werbelow LG, Marshall AG. 1973. *J. Magn. Reson.* 11:299–313

106. Müller N, Bodenhausen G, Ernst RR. 1987. *J. Magn. Reson.* 75:297–334

107. Kay LE, Prestegard JH. 1987. *J. Am. Chem. Soc.* 109:3829–35

108. Bax A, Griffey RH, Hawkings BL. 1983. *J. Magn. Reson.* 55:301–15

109. Mueller L. 1979. *J. Am. Chem. Soc.* 101: 4481–84

110. Ollerenshaw JE, Tugarinov V, Kay LE. 2003. *Magn. Reson. Chem.* 41:843–52

111. Bodenhausen G, Rubin DJ. 1980. *Chem. Phys. Lett.* 69:185–89

112. Wang J, Harting JA, Flanagan JM. 1997. *Cell* 91:447–56

113. Tugarinov V, Kay LE. 2004. *J. Biomol. NMR.* 28:165–72

Annu. Rev. Biochem. 2004. 73:147–76
doi: 10.1146/annurev.biochem.73.012803.092429
Copyright © 2004 by Annual Reviews. All rights reserved
First published online as a Review in Advance on March 26, 2004

Incorporation of Nonnatural Amino Acids Into Proteins

Tamara L. Hendrickson[1], Valérie de Crécy-Lagard[2], and Paul Schimmel[2]

[1]*Johns Hopkins University, Department of Chemistry, 3400 N. Charles Street, Baltimore, Maryland 21218; email: tamara.hendrickson@jhu.edu*
[2]*The Scripps Research Institute, 10550 North Torrey Pines Road, La Jolla, California 92037; email: vcrecy@scripps.edu, schimmel@scripps.edu*

Key Words aminoacyl-tRNA synthetase, suppression, orthogonal, protein biosynthesis

■ **Abstract** The genetic code is established by the aminoacylation of transfer RNA, reactions in which each amino acid is linked to its cognate tRNA that, in turn, harbors the nucleotide triplet (anticodon) specific to the amino acid. The accuracy of aminoacylation is essential for building and maintaining the universal tree of life. The ability to manipulate and expand the code holds promise for the development of new methods to create novel proteins and to understand the origins of life. Recent efforts to manipulate the genetic code have fulfilled much of this potential. These efforts have led to incorporation of nonnatural amino acids into proteins for a variety of applications and have demonstrated the plausibility of specific proposals for early evolution of the code.

CONTENTS

0066-4154/04/0707-0147$14.00
147

INTRODUCTION

The basic mechanics of protein translation wherein a nucleic acid sequence is converted into proteins was defined during the 1950s and 1960s. The key advance was delineation of the relationship between triplet nucleotide codons and amino acids (1). The seminal work of Nirenberg, Khorana, Crick, Brenner, and coworkers defined codons as trinucleotides that specified individual amino acids, identified the need for starting and terminating a polypeptide, predicted that the code was degenerate because there were 64 codons but only 20 amino acids, and hypothesized that an adaptor was needed to make the connection between nucleic acid and polypeptide sequences (1).

During this same era, the hypothesis of an adaptor was followed by the discovery of transfer RNAs (the adaptors), aminoacyl tRNA synthetases (AARSs), and messenger RNAs. The aminoacylation of tRNAs by synthetases was established as the biochemical basis for the genetic code. While the details of protein translation were being worked out, the possibility that this machinery could be manipulated to introduce nonnatural amino acids into proteins was recognized. Such manipulations were realized in a variety of ways. Summarized below are historical and recent investigations of protein modification through incorporation of virtually any natural or nonnatural amino acids at one or more specified site in a protein.

EARLY EFFORTS TO UNDERSTAND AND BREAK BARRIERS WITHIN THE CODE

Hydrolysis of a Misactivated Amino Acid

The aminoacyl-tRNAs (AA-tRNAs) are the focal point for most efforts to incorporate nonnatural amino acids into proteins. Most typically the 20 amino

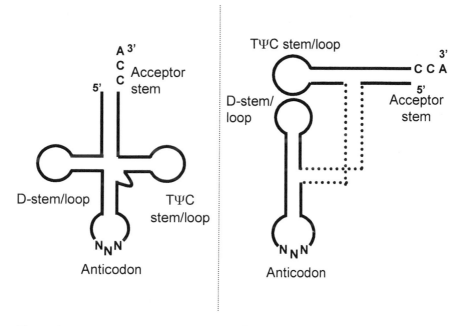

Figure 1   Secondary (*left*) and tertiary (*right*) structures of tRNA. The trinucleotide anticodon is shown as NNN and the acceptor stem ends as CCA. The 3'-terminal adenosine is aminoacylated by the AARSs.

acids are enzymatically joined to the 3'-ends of their cognate tRNAs in a two-step aminoacylation reaction.

$$E + ATP + AA \rightarrow E(AA\text{-}AMP) + PPi \qquad\qquad 1.$$

$$E(AA\text{-}AMP) + tRNA \rightarrow AA\text{-}tRNA + AMP + E \qquad\qquad 2.$$

In the first reaction, the amino acid AA is condensed with ATP to form the highly reactive aminoacyl adenylate. This step is commonly referred to as amino acid activation (for protein synthesis). In the second reaction—known as the transfer reaction, the activated amino acid is transferred to the 3' (or 2')-hydroxyl of the tRNA cognate to the amino acid. The tRNA's anticodon bears the trinucleotide that codes for the amino acid (Figure 1). Because the code is degenerate, there is (in most instances) more than one tRNA for each amino acid. The tRNAs that accept the same amino acid are known as isoacceptors. Thus, each AARS is highly specific for a single encoded amino acid and for an isoacceptor set of tRNAs. [Some glutamyl- and aspartyl-tRNA synthetases are exceptions that can aminoacylate two different sets of tRNAs (2).]

　　Pauling argued that particularly for an enzyme like isoleucyl-tRNA synthetase (IleRS) the differentiation of cognate isoleucine from valine was inherently difficult, because only hydrophobic forces could be used to distinguish between the two side chains as they differed by one methylene group (3). Consistent with this prediction, early work demonstrated activation of valine (to form Val-AMP) by IleRS. Remarkably, addition of tRNA^{Ile} to the IleRS·(Val-AMP) complex resulted in hydrolysis of the adenylate, so that valine did not end up stably attached to tRNA^{Ile} (4). Subsequently, a novel editing activity for these enzymes was discovered (5–7). This activity cleared mischarged amino acids from a specific tRNA [for example, the deacylation of Val-tRNA^{Ile} by IleRS (6)] and appeared to be general to many of the AARSs (5). Indeed, editing of misactivated amino acids is now recognized as a widespread phenomenon used by AARSs to ensure accuracy in tRNA aminoacylation (reviewed in 8). By probing the active site for aminoacylation, the early studies also suggested a distinct site for editing (5). Subsequent genetic, biochemical, and structural work elaborated in detail the nature of this second site.

Changing the Relationship Between an Amino Acid and a tRNA

Work on the concept of tRNA identity (reviewed in 9) provided many examples of differential recognition of cognate tRNAs between species. This early work laid the foundation for expression of heterologous AARSs in vivo as a means to incorporate the wrong amino acid into a protein at a specific position. Eventually, the practical concept of orthogonal synthetase-tRNA pairs was put forth, that is, the idea that a heterologous synthetase-tRNA pair could be introduced into an organism and operate as an extra pair that was "orthogonal" to the existing homologous set of 20 AARSs (with their cognate tRNA isoacceptors). Orthogonality means that the new AARS does not mischarge any of the tRNAs from the host organism and the new tRNA is not a substrate for any host AARS. Thus, an organism engineered in this way has 21 noncross-reacting AARSs that can, in principle, be further manipulated so that the new twenty-first pair brings in a novel amino acid (see below).

Missense, Nonsense, and Frameshift Suppression

Early examples showed strain-specific misincorporation of amino acids into proteins in response to a mutant-sense (missense) codon, read-through of a stop (nonsense) codon, or a frameshift of the gene's open reading frame. These phenomena were strain-specific, that is, they occurred only in strains carrying specific "suppressor" phenotypes (Su$^+$) (reviewed in 10). Suppressor phenotypes were soon traced to mutated tRNA genes (11, 12). Similarly, frameshift mutations, where the trinucleotide open reading frame was shifted by one base, were identified and traced to tRNAs with four-base anticodons [reviewed in (13)]. The first reported example was a mutant tRNA^{Gly} with a four-base anticodon (14, 15).

These variations in the code further demonstrated that the relationship between mRNAs and the resulting sequence of a coded protein was more variable than expected. The observations also suggested that organisms could tolerate some variability. Suppression of UAG amber stop codons was used for rational misincorporation of natural and nonnatural amino acids into proteins (discussed below). More recently, methods to exploit frameshift suppression as a mechanism to insert nonnatural amino acids into proteins have also been developed (discussed below).

BUILDING ON HISTORICAL METHODS

Use of Amino Acid Auxotrophs

THE EARLY YEARS   That noncanonical amino acids could be incorporated into proteins was discovered from the study of amino acid analogs synthesized by plants [up to 800 analogs have been identified to date (16)]. A subset of these amino acids (listed in Table 1) are toxic to microorganisms because they are misincorporated into proteins in place of a related canonical amino. In strains auxotrophic for an encoded amino acid, high levels of substitution by an analog could be achieved. The tolerance to the analog varied from case to case. For example, misincorporation of norleucine or pentafluorophenylalanine inhibited growth. However, these analogs are only bacteriostatic and not bactericidal. On the other hand, misincorporation of canavanine leads to cell death. In contrast, selenomethionine or trifluoroleucine are well tolerated. The main mechanism that keeps analogs from being incorporated into proteins is provided by the aminoacyl-tRNA synthetases (17). If a synthetase's specificity is bypassed, then mischarging of an analog onto a tRNA results in insertion of the analog into a growing polypeptide chain. This realization provided an intellectual framework for strategies to incorporate analogs into proteins.

SELECTION FOR TOTAL REPLACEMENT OF AN AMINO ACID BY AN ANALOG   An unnatural amino acid can replace its natural counterpart in only a few cases. The replacement of methionine by selenomethionine (18) has been extensively used for phase determination in protein structure studies (19). Also, with epigenetic adaptation that is not well understood, cells can replace leucine with trifluoroleucine [Orgel proposed that specific enzymes could be overexpressed in the adaptive phase (20)]. However, the possibility cannot be ruled out that a low percentage ($< 5\%$) of the natural amino acid (e.g., methionine versus selenomethionine, leucine versus trifluoroleucine) was still present.

Selection methods have also been used in attempts to replace a canonical amino acid with an analog. Replacement of tryptophan with 4-fluorotryptophan (in *Bacillus subtilis* and, separately, in *Escherichia coli*) was attempted with experimental evolution techniques (21, 22). These experiments demonstrated that total replacement of a coded amino acid could be achieved with a small number of genome-wide mutations. In *B. subtilis*, complete replacement of the amino acid with the analog was achieved. In contrast, exposure of *E. coli* to the analog resulted in selection for more efficient use of the canonical amino acid.

TABLE 1 Multi-site incorporation of analogs into *Escherichia coli* proteins

Analog	Target AARS	Whole cells	Purified proteins	Applications
Azetidine-2-carboxylic acid	ProRS	(144)		
3,4-Dehydroproline	ProRS	60% (145)		
Perthiaproline	ProRS		(146)	Drug carrier
Canavanine	ArgRS	(147)		Measure of stress resistance
Ethionine	MetRS	(148)	(149)	
Norleucine	MetRS	38% (150)		Increased enzyme activity (26)
	LeuRST252Y		(151)	
	IleRS$_{Ala}$*			
Selenomethionine	MetRS	100% (18)	(19)	Crystallography
Aminohexanoic acid	MetRS		(149)	
Telluromethionine				Crystallography
Homoallylglycine	MetRS		(24)	Alkene function-ality
Homopropargylglycine	LeuRST252Y[a]		(151)	Staudinger ligation (134)
2-Butynylglycine			(134)	
Azidohomolanine				
Transcrotylglycine	MetRS[a]		(25)	
Allyglycine	LeuRST252Y[a]			
7-Azatryptophan	TrpRS	50% (152)	(153)	
4-Fluorotryptophan	TrpRS	100%[b] (21) 99.97%[c] (22)	(154)	NMR
5-Fluorotryptophan	TrpRS		(154)	NMR
7-Fluorotryptophan				
β-(Thienopyrrolyl)alanines	TrpRS		(155)	Chromophore
β-Selenolo(3,3-β)pyrrolyl-alanine				
Aminotryptophans	TrpRS		(156)	PH sensors
Trifluoroleucine	IleRS		(157)	Enhanced hydrophobicity
Norvaline	LeuRST252Y[a]	20%	(151)	
	IleRS$_{Ala}$*			

TABLE 1   (*Continued*)

Analog	Target AARS	Whole cells	Purified proteins	Applications
4-Azaleucine	LeuRS	(158)		
Trifluoroleucine	LeuRS	100%^{d} (159)	(160)	
Hexafluoroleucine	LeuRS^a		(161)	Stability
Furanomycin	IleRS	(162)		
O-Methylthreonine	IleRS	(163)		
	IleRS$_{Ala}$*			
S-Methylcysteine	IleRS$_{Ala}$*			
O-Methylserine				
Selenocysteine		(164)		
Azaserine		(17)		
p-Fluorophenylalanine	PheRS	75% (150)	(165)	Tracer
o-Fluorophenylalanine				
m-Fluorophenylalanine				
β-Thienylalanine	PheRS	50% (166)		
p-Chlorophenylalanine	PheRSA294G		(167)	
p-Bromophenylalanine	PheRSA294G^a		(168)	Crystallography
p-Iodophenylalanine			(130)	
p-Ethynylphenylalanine				
p-Cyanophenylalanine				
p-Azidophenylalanine				
p-Acetylphenylalanine	PheRSA294GT251G^a		(169)	Ketone functionality
1,2,4-Triazole-3-alanine	HisRS		(170, 171)	
2-Methylhistidine	HisRS		(172)	
3-Fluorotyrosine	TyrRS	(173)	(165)	
3-Chlorotyrosine				
Azatyrosine	TyrRSF130S		(174)	

*V. Pezo, D. Metzgar, T.L.H. Hendrickson, W.F. Waas, S. Hazebrouck, V. Döring, P. Marlière, P. Schimmel, and V. de Crécy-Lagard, manuscript submitted).

^{a}Overexpressed.

^{b}In *B. subtilis*.

^{c}*In E. coli.*

^{d}With adaptation.

MULTISITE MISINCORPORATION IN OVEREXPRESSED PROTEINS   Most amino acid analogs are too toxic to promote sustained exponential growth. However, modified proteins can be overexpressed in nondividing cells if enough biomass has been generated prior to induction. Several groups used auxotrophic strains to express proteins of interest by transcription of the robust T7 or P_{Lac} promoters (23, 24). By washing cells and replacing the exogenously added canonical amino acid with its analog just before inducing gene expression (or, alternatively, by balancing initial amino acid concentrations so that the natural amino acid is consumed when induction starts), high levels of misincorporation (80% to 99%) of nonnatural analogs into target proteins can be achieved, with good yields of the purified proteins (10–100 mg/L). For analogs that are activated poorly by the relevant AARS, overexpression of the AARS is sometimes needed to allow efficient misincorporation (25). To diversify further the chemistry of misincorporated analogs, mutations in the amino acid activation pocket of the AARS can be introduced to broaden side-chain specificity. The repertoire of amino acids that have been introduced via modification of the active sites of AARSs is presently limited, but can in principle be expanded greatly (Table 1). A recent application of this method by Arnold and collaborators has shown that the peroxygenase activity of the cytochrome P450 BM-3 domain can be increased twofold by replacing all methionines with norleucines (26).

MISSENSE AND NONSENSE SUPPRESSION   The phenomenon of informational suppression was pivotal for understanding the genetic code (27, 28) and ultimately led to widely used tools (29). Mutations of tRNA genes led to suppressors of nonsense (30), missense, and frameshift mutations (31, 32). When harboring suppressor tRNAs, cells could be grown with up to 70% readthrough of stop codons (33). Although suppressors were not originally used to incorporate unnatural amino acids into proteins, they showed that the wrong "natural" amino acid could be inserted in vivo into specific locations of proteins. This observation further demonstrated that the ribosome does not proofread amino acids during translation.

Posttranslational Modification of Cysteine

The unique chemistry of the sulfhydryl side chain of cysteine allows its modification by thiol-specific reagents containing a variety of substituents, thus effectively expanding the genetic code of a single protein after it has been purified (34). These methods were initially limited to proteins containing a single cysteine but site-directed mutagenesis has enlarged the possibilities (35). By removal of unwanted cysteines, and placement of a cysteine at any desired site whatsoever, proteins containing a single cysteine can be chemically changed by thiol-modifying reagents—and then screened for desired activities. These methods allow the introduction of new chemistries (36).

Using methanosulfate reagents, Jones and coworkers produced subtilisin variants in which the ratio of esterase to amidase activity was increased up to 50-fold (37). In another vein, Distefano and coworkers linked a pyrodoxamine analogue to Cys117 of an adipocyte lipid binding protein and thereby produced a semisynthetic biocatalyst that reductively aminated a variety of α-keto acids (36).

In contrast, Silverman & Harbury (38) randomly substituted cysteine in place of valine in overexpressed proteins. These substitutions were accomplished by concomitantly expressing *E. coli* cysteinyl-tRNA synthetase and tRNA^{Cys} that was modified to decode GUC codons for valine (39). In this particular study, the resulting statistical ensemble of substituted proteins was used to footprint the structure at single amino acid resolution (38).

IN VITRO AND SEMI-IN-VITRO METHODS

In Vitro Translation with Chemically Altered Aminoacylated tRNAs

CHEMICAL ALTERATION AND AMINOACYLATION OF tRNAs   In vitro translation systems have been isolated from many organisms and have been used to manipulate translation. In early studies, Johnson and coworkers demonstrated the utility of these systems for nonnatural amino acid incorporation (40): LysRS was used to synthesize Lys-tRNA^{Lys}, and the ϵ-amino group was then acetylated with N acetoxysuccinimide. The N-ϵ-Ac-Lys was ribosomally incorporated into hemoglobin using a rabbit reticulocyte cell-free translation system. These efforts were expanded by Johnson & Cantor (41) to modify lysine with fluorescent probes (42) and with cross-linking reagents for affinity and photo-affinity labeling (43). In these instances, the nonnatural lysine analog was incorporated into proteins at all lysine codons, rather than at one specific site.

Hecht and coworkers expanded the utility of misacylated tRNAs by developing chemical methods to covalently attach a nonnatural amino acid to the 3'-end of tRNA^{Phe} (44, 45). The gene for tRNA^{Phe} was truncated so as to lack the 3'-terminal adenosine and then transcribed in vitro. The dinucleotide AppA was chemically aminoacylated with an N-terminally protected amino acid, which had been activated by carbonyldiimidazole. RNA ligase could insert an aminoacyl adenylate from this modified dinucleotide into the truncated form of tRNA^{Phe} (lacking the 3'-terminal A) to generate a chemically modified aminoacyl-tRNA^{Phe} (44). The attached amino acid could be varied as desired. The synthesis of chemically aminoacylated tRNAs was later improved by synthesizing the dinucleotide pCpA, aminoacylated with a protected amino acid of choice. This dinucleotide was more efficiently ligated, in this case to a tRNA^{Phe} truncated to lack the terminal pCpA dinucleotide (45). In both cases, the amino acid in the resultant aminoacyl-tRNA was N-terminally protected so that these tRNAs could not be readily used for ribosome-based translation systems. The work showed, however, that tRNAs could be chemically manipulated and consequently laid the foundation for further development of chemically aminoacylated tRNAs.

Later, Hecht and coworkers synthesized several pyroglutamylaminoacyl-tRNAs, effectively using the cyclic amino acid pyroglutamate as an N-terminal protecting group during synthetic manipulations. The enzyme pyroglutamate aminopeptidase was subsequently used to remove the pyroglutamate, to generate a fully deprotected and functional aminoacyl-tRNA. These modified tRNAs were competent substrates for in vitro translation using *E. coli* 70S ribosomes (46).

Schultz and coworkers (47) expanded the utility of chemical tRNA amino-acylation by introducing two key variations into the aminoacyl dinucleotide synthesis originally developed by Hecht. First, they protected the amino acid amine with a variety of cleavable protecting group such as *o*-nitrophenylsulfenyl chloride (48–50). Second, the synthetic route to aminoacylated pCpA was simplified by replacing the cytidine with a deoxycytidine, to generate pdCpA (47). This protected aminoacyl-pdCpA was still a substrate for RNA ligase and could be appended onto truncated tRNAs. Importantly, deprotection of the amino acid could be readily performed, either chemically or photolytically, to generate a translationally competent aa-tRNA (49, 50). Additional methods for chemical aminoacylation have subsequently been developed, including the use of micelles during the aminoacylation of pdCpA (51) and the evolution of ribozymes capable of aminoacylating different tRNAs (52) (see below).

NONSENSE AND SENSE SUPPRESSION IN VITRO   The utility of chemically amino-acylated tRNAs for in vitro translation of full-length proteins was first demon-strated by Schultz and coworkers (48, 53). The flexibility of synthetic tRNA aminoacylation reactions was combined with the specificity of amber suppres-sion. The amber-codon reading tRNA was chemically aminoacylated (with a series of natural and nonnatural amino acids) and used for in vitro translation. Unlike earlier attempts that were not position-specific, several phenylalanine analogs (*p*-fluorophenylalanine, *p*-nitrophenylalanine, and homophenylalanine) were incorporated into a β-lactamase test protein at an internal position modified to a UAG amber triplet (replacing the codon for Phe66). Yields of β-lactamase, using this in vitro technique, were sufficient for enzyme purification and characterization and ranged from 1 to 3 μg/ml. With these methods, a variety of nonnatural amino acids and probes have been introduced into specific proteins (49, 50).

Similarly, Chamberlin and coworkers treated an amber suppressor Tyr-tRNA$^{Tyr}_{CUA}$ with Na^{125}I, to generate radiolabeled ^{125}I-Tyr-tRNA^{Tyr}. This tRNA was also a competent suppressor tRNA in a rabbit reticulocyte in vitro translation system. ^{125}I-Tyr was incorporated specifically into position 9 of a 16 residue peptide (54).

The yields achievable with the in vitro translation systems have been low (range 1–10 μg/ml) owing mainly to both the difficulty of obtaining high amounts of misacylated tRNAs and the low ratio of suppression/termination at amber codons due to the presence of the release factor (RF1) in the cell extracts. The recent development of a reconstituted in vitro translation system, in which RF1 can be omitted (55), allows both high yields of protein expression (160 μg/ml/h in batch mode) and efficient suppression.

Recently, Roberts and colleagues have begun to investigate the combination of in vitro amber and sense suppression with mRNA display (see 56). Messenger RNA display facilitates peptide selection experiments by generating libraries of covalent mRNA-peptide complexes. Selection experiments can target the peptide (or protein); following selection, the covalently bound mRNA can be read by RT-PCR to determine the sequence of the selected peptide. mRNA display was used for selection of peptides containing biocytin at positions denoted by amber codons and, separately, by sense codons. In each case, the suppressor tRNA was chemically aminoacylated with biocytin. Multiple rounds of selection led to an amber suppression efficiency of >85% and sense suppression of the GUA (valine) codon at levels ranging from 40% to 50%.

SELEX of Ribozymes

A drawback to chemical aminoacylation of tRNA is that synthetic techniques required to make aminoacylated pdCpA are not convenient for many biochemists. In contrast, Suga and coworkers used in vitro selections to generate RNA ribozymes that aminoacylate tRNAs with phenylalanine and phenylalanine analogs; these catalysts offer an alternative to chemical aminoacylation. In one study, a precursor tRNA comprised of an amber suppressor and a varied 5'-leader sequence was used to select for self-aminoacylation with phenylalanine (57). The selected aminoacyl precursor was treated with RNase P to remove the leader, thus yielding full-length Phe-tRNA. The leader sequence was also capable of intermolecular aminoacylation and, further, was functional when covalently bound to resin. This latter feature simplified subsequent purification of charged tRNA products.

In vitro ribozyme selection has also been used to generate catalysts for tRNA aminoacylation that have broad specificity. In this case, Suga and coworkers selected a ribozyme that can transfer any activated amino acid from the 5'-terminus of the ribozyme to the 3' end of a tRNA (52). This ribozyme is nonspecific for the amino acid, but selectively aminoacylates tRNA^{fMet} via recognition of the tRNA acceptor stem and anticodon loop. The dual recognition of the acceptor stem and anticodon parallels that of many aminoacyl tRNA synthetases. Thus, tRNA^{fMet} is the "cognate" substrate for the ribozyme, designated BC28. Mutations in BC28 can be introduced to program other tRNA specificities, based on the formation of new base pairs with the anticodon loop of a different tRNA. This ribozyme thus offers a promising tool for the aminoacylation of any tRNA with any amino acid.

More recently, Suga and coworkers (58) developed a resin-immobilized ribozyme capable of aminoacylating a wide variety of tRNAs with various analogs of phenylalanine. This "Flexiresin" lacks the tRNA specificity of BC28 and its mutants because it only interacts with the tRNA acceptor stem. However, Flexiresin provides, in principle, a rapid and economical route to aminoacylation with variants of phenylalanine (59).

Chemical Synthesis of Functional Enzymes

Solid phase peptide synthesis (SPPS) (see 61), has been used to manipulate many small proteins. Commercial peptide synthesizers led to reports of the synthesis of small polypeptides, including among others insulin-like growth factor (62), the chemokine IL-8 (63), rubredoxin (as both D- and L-enantiomers, which were mixed together to make the racemic form) (64, 65), and human immunodeficiency virus 1 (HIV-1) protease (66). Kent and colleagues (66) synthesized the 99 amino acid HIV-1 protease and showed it to be kinetically competent. A variant HIV-1 protease, containing two N-α-aminobutyric acid (Abu) replacements for the enzyme's two cysteine residues, was also constructed. The crystal structure of this variant HIV-1 protease was solved, providing structural insight into the dimer interface of the enzyme (67). The enantiomer of wild-type HIV-1 protease was also chemically synthesized with all D-amino acids and, as expected, this enzyme's chiral substrate specificities were reversed from those of the wild-type enzyme (68).

An important application of SPPS with D-amino acids is for the selection of RNA aptamers that bind tightly to the active site of a target protein (69). These aptamers are selected in the conventional way, using RNA molecules comprised of natural bases with their L-ribose sugar, but the selection is conducted using an all-D-isomer polypeptide. Once the sequence of the desired aptamer is determined, it can then be synthesized as the enantiomeric RNA, known as the spiegelmer. The spiegelmer binds to the natural protein target made up of L-amino acids. The advantage of the RNA spiegelmer is its resistance to nuclease degradation.

The major limitation of SPPS is that synthetic peptides are typically feasible up to about 100 amino acids in length (but see below). Misfolding of synthetic peptides is common, and a requirement for the formation of specific disulfide bonds can also limit the usefulness of the synthetic methodology. Special techniques have been developed to overcome some of these problems. For example, sequential disulfide bond formation has been used to improve the yield of the native conformation in polypeptides with multiple disulfide bonds (70).

Peptide Ligation

CYANOGEN BROMIDE CLEAVAGE AND RELIGATION   Cyanogen bromide cleaves proteins after methionine to generate a C-terminal homoserine lactone (Figure 2). This lactone is highly reactive to nucleophiles. Furthermore, in certain cases, peptides containing this lactone can be condensed with a second peptide to regenerate a full-length protein (71). Religation introduces a homoserine for methionine substitution into the full-length protein. The religation technique has been used to graft natural and nonnatural amino acid-containing synthetic peptides into regions of cytochrome c (full-length cytochrome c is 104 amino acids) (72–74). For example, Gray, Imperiali, and colleagues replaced His72 with a bipyridyl-alanine analog [(S)-2-amino-3-(2,2'-bipyrid-4-yl)propanoic acid or 4Bpa]. In this case, residues 1–65 of horse heart cytochrome c were generated by

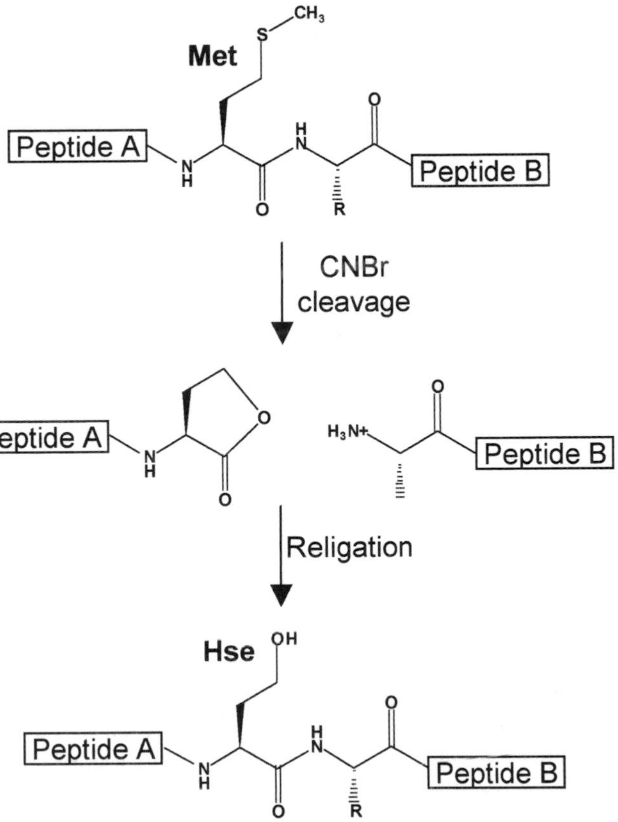

Figure 2 Cyanogen bromide (CNBr) cleavage at methionine and religation. Peptide B can be replaced with a synthetic peptide to introduce nonnatural amino acids into the polypeptide product.

cyanogen bromide cleavage of wild-type protein. A synthetic peptide corresponding to residues 66–104 was prepared by SPPS (with 4Bpa replacing His72). The two fragments were ligated to generate full-length 4Bpa-containing cytochrome c. The bipyridyl ligand was then used to specifically introduce a $Ru(bpy)_3^{2+}$ redox center onto the surface of the protein, thus enabling experiments to test electron transfer between the ruthenium and the protein-bound heme iron (74).

PROTEIN LIGATION   Cyanogen bromide cleavage and religation is not applicable to all proteins. In contrast, in vitro protein ligation, based on a thioester exchange reaction (Figure 3), is more broadly useful for incorporation of nonnatural amino acids into proteins having a wide range of sizes (for recent reviews, see 75, 76). As shown in Figure 3, peptide ligation arises from the incubation of two peptides:

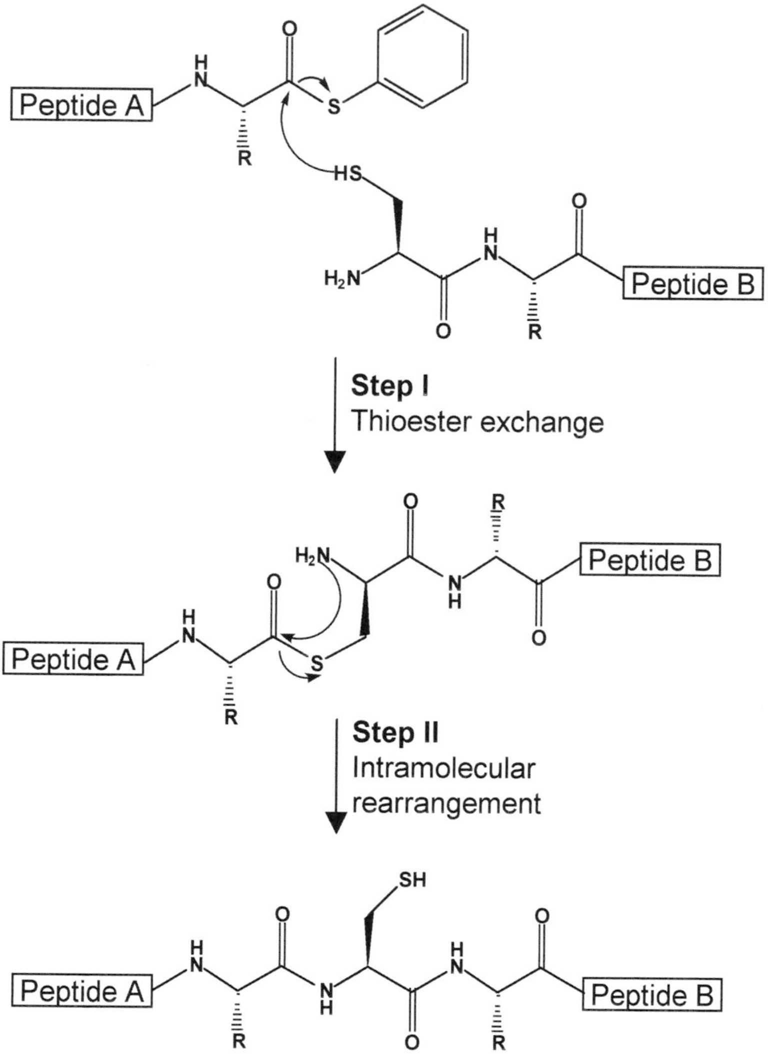

Figure 3 Peptide ligation. This chemical method condenses a thioester modified peptide (*peptide A*) with a peptide containing an N-terminal cysteine (*peptide B*). Each fragment can be synthesized by SPPS or expressed in vivo.

an N-terminal fragment with an activated C-terminal thioester (peptide A) and a C-terminal fragment with an N-terminal cysteine residue (peptide B). The cysteine undergoes thioester exchange and then rearranges to generate the more stable amide bond. Cole, Muir, and colleagues expanded the applicability of peptide ligation to include proteins expressed in vivo. In this work, addition of a self-cleaving intein and a chitin-binding domain to the C terminus of a protein was used to convert the C-terminal carboxylate to a thioester that could then be used in a condensation reaction (77, 78) With the recent developments using proteins synthesized in vivo, the size of the ligated protein product seems essentially unlimited. Muir and coworkers synthesized an active, tetraphosphorylated analog of the Type I TGFβ receptor, containing a single methionine to norleucine substitution (79). The catalytic activity of this 50-kD protein was similar to that of its phosphorylated wild-type counterpart and was more than an order of magnitude more active than the dephosphorylated version. Peptide ligation has also been used to synthesize glycopeptides (80), nonhydrolyzable phosphorylated proteins (78, 81–84), and a membrane-bound potassium channel (85), as well as for isotope labeling of specific proteins (86).

Import of AA-tRNAs into Eukaryotic Cells

By coinjecting *Xenopus* oocytes with a mischarged amber-reading tRNA and mRNA containing an amber codon at a given position, site-specific insertion of any desired amino acid was achieved in eukaryotes (87). For the amber-reading tRNA to be effective in this system, a high suppression efficiency and lack of recognition by the host aminoacyl tRNA synthetases were critical. For this purpose, a mutant form of *Saccharomyces cerevisiae* tRNA^{Phe} (88) and a variant of *Tetrahymena* tRNA^{Gln} (87) were developed. These particular tRNAs were used by Dougherty (89) to study the mechanism of ligand-gated ion channels, by inserting novel amino acid probes into nicotinic acetylcholine receptor.

Coinjection of mRNA and misacylated tRNA allows a wide range of analogs to be incorporated. Because the amino acid is not introduced into the media except as attached to tRNA, toxicity is bypassed. A major limitation is that stoichiometric quantities of mischarged tRNA and mRNA are needed. Thus, this technique is useful for applications where sensitive assays requiring limited quantities of modified protein are available. Additional methods for introducing tRNAs and mRNA into cells have been developed by RajBhandary and coworkers. *E. coli* amber and ochre (reading UAA) suppressor tRNAs were imported into mammalian COS1 cells, leading to specific suppression (90). With two distinct suppressor tRNAs, two different unnatural amino acids can be introduced into the same protein. In other work, Vogel and coworkers microinjected CHO cells with suppressor tRNAs charged with leucine (91), and Monahan et al. introduced amber suppressor tRNAs misacylated with different amino acid analogs into several mammalian cells lines by electroporation (92).

ORTHOGONAL AND NEW AARS/tRNA PAIRS

Rational Design of Orthogonal AARS/tRNA Pairs in Bacteria and Eukaryotes

The desire to incorporate an unnatural amino acid at a specific site of a protein expressed in vivo gave rise to the concept of orthogonality (see above). Here

TABLE 2   Orthogonal AARS/tRNA pairs

Host organism	AARS	tRNA	Reference
Mammalian cells	*E. coli* GlnRS	*E. coli* tRNA$^{Gln}_{CUA}$ derivative	(96)
E. coli[a]	*S. cerevisiae* PheRS	*S. cerevisiae* tRNA$^{Phe}_{CUA}$	(175)
E. coli	*S. cerevisiae* GlnRS	*S. cerevisiae* tRNA$_2$ Gln derivative	(176)
E. coli	*M. janaschii* TyrRS	*M. janaschii* tRNA$^{Tyr}_{CUA}$ derivative	(94)
E. coli	*S. cerevisiae* TyrRS derivative	tRNA$_2$^{fmet} derivative	(95)
S. cerevisiae	*E. coli* GlnRS derivative	*Human initiator* tRNA	(95)
Mammalian cells	*E. coli* TyrRS mutant	*Bacillus stearothermophilus* tRNA$^{Tyr}_{CUA}$ derivative	(177)
S. cerevisiae	*E. coli* TyrRS	*E. coli* tRNA$^{Tyr}_{CUA}$	(100)
E. coli	*Methanobacterium thermoautotrophicum* LeuRS	*Halobacterium* sp. NRC-1 tRNA$^{Leu}_{CUA, UCA, UCCU}$ variants	(99)

[a]*p*-F-Phe resistant strain, PheRSA294S.

again, as with the microinjection experiments, the idea is to use tRNA constructs that are not charged by the host organism. The suppressor tRNA must, therefore, be cointroduced with an AARS that charges that specific tRNA but none of the host tRNAs. This AARS-tRNA pair is orthogonal to the host pairs.

Truly orthogonal AARS-tRNA pairs for prokaryotes have now been developed (93). In general, two approaches have been used. In one case, a distant AARS (e.g., an *S. cerevisiae* AARS) was expressed in *E. coli*. In other instances, host AARS-tRNA pairs were mutated so that they fulfilled the orthogonality requirements. (Of course, in the latter situation, the wild-type pair was also retained.) Optimization of each pair was typically required in order to completely achieve orthogonality (93). Examples of orthogonal pairs expressed in *E. coli* include the *Methanococcus janaschii* TyrRS/tRNA$^{Tyr}_{CUA}$ pair (94) and *S. cerevisiae* TyrRS combined with an *E. coli* tRNA$^{fMet}_2$ variant that is charged only by the yeast synthetase (95) (see Table 2). The first orthogonal pair to be developed in eukaryotes was *E. coli* GlnRS/tRNA^{Gln} that was imported into COS cells (96).

Manipulation of AARS Specificity

Efforts to isolate mutant tRNAs that activate a noncognate amino acid analog started with the identification of variants that no longer activate the cognate amino acid. To this end, Schultz and coworkers combined a positive selection step, which yielded AARS variants that activated the desired nonnatural analog,

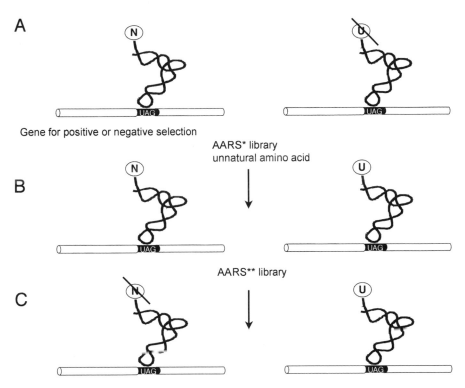

Figure 4   Strategy to select for an orthogonal pair specific for a given unnatural amino acid. (*A*) The starting point is an organism containing an orthogonal AARS/tRNA$_{CUA}$ pair charging a natural amino acid N but not the unnatural amino acid U. At the same time, an amber codon (UAG) is inserted into a gene that codes for a selectable marker such as antibiotic resistance, GFP, or the Gal4 repressor. (*B*) Positive selection for mutant AARS* variants that allow suppression in the presence of the analog. (*C*) Negative screen or selection for AARS** variant that rejects the natural amino acid.

with a negative selection screen to eliminate those that still activated the natural amino acid (Figure 4). This *tour de force* allowed specific misincorporation of *O*-methyltyrosine at a position specified by a stop codon (94). This technique was expanded to include incorporation of a wide range of tyrosine analogs, including fluorescent and photoactive probes, and ketone "handles" (Table 3). At present, the principal limitation is the spectrum of possible analogs that can be activated by a given AARS scaffold [for example, *M. jannaschii* TyrRS (94)]. However, the numbers of amino acid variations that can be incorporated into proteins are continuing to increase, particularly as methods allowing multiple and rapid selection and counter-selection steps are improved (97). In addition, several orthogonal AARS-tRNA pairs are being developed to accommodate many different amino acid types (98, 99). Although this technology was first optimized

TABLE 3   Site-specific in vivo misincorporation

Amino acid	Application	Reference
O-Methyl-L-tyrosine		(94)
L-2-Naphthylalanine		(178)
p-Azido-L-phenylalanine		(101, 133, 179)
p-Benzoyl-L-phenylalanine	Photocrosslinking	(101, 132, 133)
O-Allyl-L-tyrosine (*OAY*)	Alkene functional group	
p-Isopropyl-L-phenylalanine		(97, 180)
p-Amino-L-phenylalanine		
p-Carboxyl-L-phenylalanine		
p-Methoxy-L-phenylalanine		(101, 130)
m-Methoxy-L-phenylalanine		
p-Iodo-L-phenylalanine		
p-Bromo-L-phenylalanine		
p-Acetyl-L-phenylalanine	Keto functional group	(101, 135)
m-Acetyl-L-phenylalanine	In vivo labeling	(181)
m-Acetyl-L-phenylalanine	Glycoprotein	(136)

in bacteria, a positive/negative selection system was subsequently developed in *S. cerevisiae* (100, 101). Other strategies to achieve site-specific incorporation have been based on phage display (102).

In the examples described above the unnatural amino acid had to be exogenously added to the medium. Thus, a transport system for the unnatural amino acid was required. To bypass this requirement and create an organism with a true 21 amino acid genetic code, genes for the biosynthetic pathway for *p*-aminophenylalanine (*p*AF), as well as a unique *p*AF-AARS and cognate tRNA^{pAF}, were introduced into *E. coli* (103). The *p*AF biosynthesis genes came from the chloramphenicol-producing *Streptomyces venezualeae*. By combining the biosynthetic pathway for the analog with the analog-specific AARS, Schultz and coworkers created an organism that, under induction with IPTG, can autonomously introduce *p*AF at all amber codons. This approach cannot at present be used for sustained growth of the organism.

NOVEL CODONS

Four- and Five-Base Codons

The modern genetic code is composed entirely of trinucleotide codons, each of which has been assigned. Thus, the three-base code makes it difficult to introduce new amino acids with their own specific codons. The discovery of +1 frameshift suppressor tRNAs (containing four-base anticodons complementary to mutant four-base codons) made clear that the ribosomal machinery could handle codon/anticodon pairs that were greater than three nucleotides in length. Building from this discovery, designed four-base and five-base codon/anticodon pairs became useful for creating new codon/anticodon pairs for the incorporation of nonnatural amino acids (104).

Using an *E. coli* in vitro translation system, Hohsaka and colleagues evaluated a series of four-base codons (e.g., AGGU, CGGU, CCCU, CUCU, CUAU, GGGU, CGGG and GGGC) for incorporation of *p*-nitrophenylalanine (ntrPhe) alone and *p*-nitrophenylalanine together with β-anthranilolyl-L-α,β-diaminopropionic acid (atnDap) into streptavidin. A series of four-base-reading yeast tRNA^{Phe}s was designed. Each tRNA was transcribed in vitro (missing the 3'-terminal CA) and chemically aminoacylated by ligating synthetic *p*-ntrPhe-pdCpA or atnDAP-pdCpA to the truncated tRNA using T4 RNA ligase (44, 45, 47). Of 24 pairs, 17 novel four-base-containing tRNAs were competent in translation. (105, 106) The use of five-base codons in in vitro translation systems has also been reported (107).

In contrast to the in vitro work, Magliery et al. developed four-base codons for translation in *E. coli* (13). Mutagenesis of the β-lactamase gene was followed by a selection strategy. Two specific serine codons were independently changed [Ser70 (essential) and Ser124 (nonessential)] to four randomized bases. A library of tRNA^{Ser} variants was also constructed. In this library, the anticodon loop of tRNA^{Ser} was also randomized. The loop was expanded from the normal 7 to either 8 or 9 bases. Colonies were assessed for growth in the presence of ampicillin, to test for synthesis of β-lactamase. A series of selections led to the identification of 14 different four-base codons that, when incorporated in place of the active site Ser70 codon in the β-lactamase gene, resulted in moderate ampicillin resistance. A variant of the selection scheme devised by Schultz and colleagues was used to determine the optimal length for unnatural codon/anticodon pairs (108). Two-, three-, four-, five-, and six-base codons were examined in a selection with tRNAs containing anticodon loops ranging from 6 to 10 nucleotides. *E. coli* tolerated three-, four-, and five-base codons, with a preference for n+4 length anticodon loops (n being the number of bases in the corresponding codon). A number of efficient four- and five-base anticodon-containing tRNAs were generated. Four- and five-base codons have not yet been utilized for the incorporation of nonnatural amino acids in vivo (99).

New Base Pairs in Transcription and Translation

Unnatural base pairs that are competent substrates for transcription and translation offer an alternative approach to codon expansion. These bases can be used to increase the number of trinucleotide codons accessible to the translation apparatus. This scheme avoids the challenges faced by frameshifting with expanded codons. Benner and coworkers demonstrated that isoguanosine and

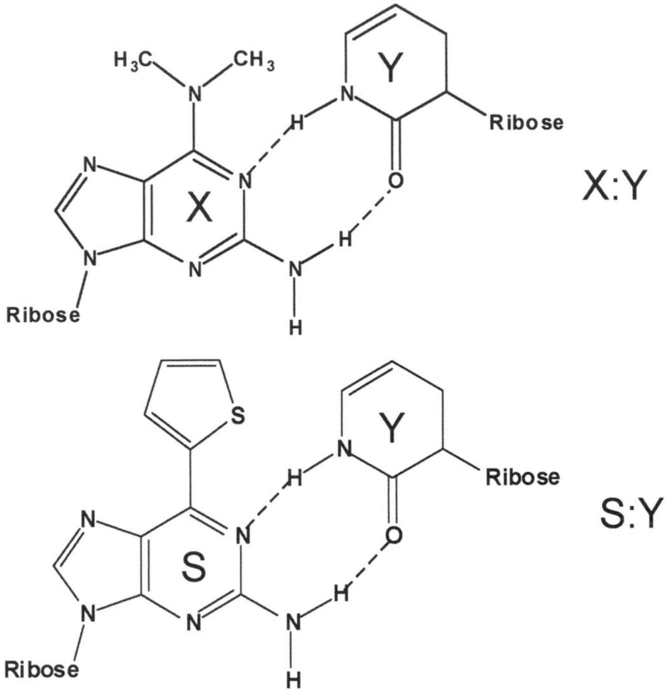

Figure 5 Nonnatural base pairs can be used to expand the number of available codons in mRNA. X, 2-amino-6-dimethylaminopurine; Y, pyridin-2-one; S, 2-amino-6-(thienyl)purine.

isocytidine could be incorporated into DNA and RNA by appropriate polymerases in a template-directed manner (109, 110). This isoC:isoG base pair presented a different hydrogen bonding arrangement than either A:T or G:C base pairs.

Ohtsuki et al. (111) and Hirao et al. (112) expanded the number of synthetic base pairs that can be incorporated into RNA and, in turn, demonstrated their utility for the incorporation of nonnatural amino acids into proteins. The new bases were 2-amino-6-(N,N-dimethylamino)purine (X), pyridin-2-one (Y), and 2-amino-6-(2-thienyl)purine (S) (Figure 5). Y pairs with either X or S, but not with any of the canonical bases (A, G, C, U). The three new bases are substrates for template-dependent RNA polymerase, when the synthetic template contains the respective complementary bases. Tyrosine 32 of the human C-Ha-Ras protein was chosen as the target site for modification. mRNA encoding for Ras with a YAG codon at position 32 was synthesized by transcription of a special DNA template. [The DNA template was generated by inserting (by ligation) a synthetic oligonucleotide containing S (at codon 32) between 5′ and 3′ *ras* gene fragments.] A combination of chemical synthesis and RNA ligation was used to convert yeast tRNA^{Tyr} into a new tRNA with a modified CUS anticodon. The tRNA$^{Tyr}_{CUS}$ was aminoacylated with 3-chlorotyrosine by wild-type *S. cerevisiae* tyrosyl-tRNA synthetase. The combination of a modified Ras mRNA and Tyr-tRNA$^{Tyr}_{CUS}$ was functional in vitro and resulted in production of full-length Ras with 3-chlorotyrosine at position 32. The activity of this modified Ras was not determined.

INSIGHTS INTO DEVELOPMENT OF THE GENETIC CODE

Strategies for the in vivo incorporation of unnatural amino acids typically rely on avoiding ambiguity at all costs. Other strategies can capitalize upon ambiguity and rely either on inactivating proofreading or on imposing a positive selection for ambiguity. These strategies can give insights into the possible mechanism of evolution of the genetic code.

Infiltration of the Code via Ambiguous Intermediate Steps

As mentioned above, AARSs establish an accurate genetic code through the aminoacylation of tRNAs. In some cases, structural similarities between amino acids lead to misactivations. AARSs execute proofreading to eliminate misactivated amino acids before the errors are translated into proteins. Nine of the 20 synthetases have been demonstrated to have editing activities (see 8). Elimination of the editing activity restores ambiguity to tRNA aminoacylation and is therefore a way to facilitate misincorporation of noncognate amino acids into proteins. Inactivation of the editing activities of IleRS, ValRS, LeuRS, and AlaRS allowed misincorporation of many analogs into proteins (Table 1). Surprisingly, up to 20% misincorporation of Abu at Val codons or of norvaline at Ile codons was tolerated in whole cells simply by disrupting the editing activities of either ValRS or IleRS (113; V. Pezo, D. Metzgar, T.L.H. Hendrickson, W.F. Waas, S. Hazebrouck, V. Döring, P. Marlière, P. Schimmel, and V. de Crécy-Lagard, manuscript submitted). A scheme for the positive selection of ambiguous strains resulted in isolation of mutants with editing deficiencies that maintain ambiguous tRNAs (39). Cells carrying *thyA* alleles with a Cys146 → Ile codon substitution were auxotrophic for dT. (This auxotrophy occurs because Cys146 is essential for catalysis by thymidylate synthase.) But, under specific circumstances, cysteine can replace dT as a growth factor, if alleles of *ileS* that encode editing-deficient AARSs are present. In these cases, cysteine is misacylated onto tRNA^{Ile} and inserted at codon 146 of the mutant *thyA* mRNA. These selections allow the evolution of strains under ambiguous conditions for extended generations.

A Statistical Code as an Evolutionary Intermediate

All theories of the origin of the genetic code envision a tentative, primitive code that was gradually refined. In this regard, an early codon is thought to have specified more than one amino acid, where any of several amino acids that are chemically similar could be inserted at a given position in a growing polypeptide chain. Examples, among others, of related amino acids are aspartate and glutamate; or isoleucine, leucine, and valine; or alanine, serine, and threonine. Thus, an early form of a specific protein, such as lactate dehydrogenase, would be comprised of a statistical mixture of related sequences. Within this mixture, certain microspecies would be more active than others. Selective pressure favored the microspecies that were most fit. The emergence of these microspecies was achieved by limiting (eventually to one) the number of amino acids that were specified by a single codon. The eventual one codon—one amino acid relationship of the modern code was achieved in significant part by the acquisition of editing activities by tRNA synthetases.

The 20 AARSs are organized into two structurally distinct classes of 10 enzymes each (115–117). These classes are referred to as class I and class II. Each class can be divided into 3 subclasses (referred to as subclass a, b, or c) that group together the most closely related synthetases. Significantly, AARSs in the same subclass correspond to those with related amino acid substrates. For example, glutamyl- and glutaminyl-tRNA synthetases are in subclass Ib, whereas aspartyl- and asparaginyl-tRNA synthetases are in subclass IIb. Likewise, phenylalanyl-tRNA synthetase is in subclass IIc, whereas tyrosinyl- and tryptophanyl-tRNA synthetases are in subclass Ic. Thus, in the early development of the code, a single enzyme could have activated either tyrosine or tryptophan. As the code developed and became more refined, the gene for this enzyme duplicated and the encoded proteins developed into separate enzymes for each of the two amino acids. Indeed, TyrRS and TrpRS are close homologs, and simple variations at the active site give rise to their distinct specificities (118).

To test these ideas, the editing activity of ValRS and, separately, of IleRS has been disrupted (113, 119, V. Pezo, D. Metzgar, T.L.H. Hendrickson, W.F. Waas, S. Hazebrouck, V. Döring, P. Marlière, P. Schimmel, and V. de Crécy-Lagard, manuscript submitted). Bacterial cells harboring either of these editing-deficient AARSs produce statistical proteins, as stated above. Conditions were found where a strain harboring an editing-deficient IleRS grew to a higher yield than that achieved by the wild-type isogenic strain. These conditions are those where the cognate amino acid—isoleucine—is limiting, but where nonnatural alternatives such as norvaline or valine are available. Thus, under isoleucine-starvation conditions, the editing-deficient strain can supply an amino acid other than Ile at isoleucine codons, whereas the wild-type strain may jump these hungry codons (120). These experiments demonstrate that cells harboring statistical proteins could, in principle, play an important role in the development of early organisms, particularly when some amino acids became limiting and alternatives were available.

APPLICATIONS AND IMPORTANCE

Pharmaceutical Applications of Nonnatural Amino Acids

Modified peptides are key pharmaceuticals for the treatment of a wide variety of diseases. A prominent class of compounds in this category are the protease inhibitors. HIV protease inhibitors (121) represent a now-classic example of peptides containing nonnatural amino acids. For example, dipeptide HIV protease inhibitors, containing a nonhydrolyzable peptide backbone at the site of proteolysis, are transition-state inhibitors of HIV protease (122). New efforts in HIV protease inhibitor design are also targeting the challenge of drug resistance by introducing monocycles into peptides (123). Other examples whereby proteases are therapeutically targeted with nonnatural peptides or peptide mimetics include angiotensin-converting enzyme inhibitors (e.g., ramipril as a treatment for hypertension), the secretases (key proteases thought to be relevant to the emergence of Alzheimer's disease) (124), and the essential secreted aspartyl proteases from *Candida albicans*, as new antifungal targets (125, 126).

Nonnatural Amino Acids as Biophysical Probes

SPIN LABELS   A spin-labeled nonnatural amino acid, containing nitroxyl 1-oxyl-2,2,5,5-tetramethylpyrroline, was first site-specifically incorporated into a T4 lysozyme using in vitro translation and suppression of an amber stop codon, to generate a protein that had an electron paramagnetic resonance (EPR) signal that could be quantified (127). A second spin-labeled amino acid (2,2,6,6-tetramethylpiperidine-1-oxyl-4-amino- 4-carboxylic acid or TOAC) was chemically incorporated into peptides by SPPS, as a tool to investigate folding transitions in peptide helices (128).

FLUORESCENT AMINO ACIDS   Amber suppression and in vitro translation has also been used to site-specifically incorporate the fluorescent 7-azatryptophan into T4 lysozyme (127). The emission spectrum of this modified T4 lysozyme had a 10-nm red-shift, compared with wild-type. Two nonnatural amino acids, a fluorophore and a quencher, have been site-specifically incorporated into streptavidin, using an in vitro translation system and two orthogonal four-base codons (106).

Additionally, nonnatural amino acid incorporation methods have been used to modify the fluorescent properties of green fluorescent protein (GFP). In one study, an *E. coli* strain that was auxotrophic for tryptophan was grown in the presence of 4-aminotryptophan. Incorporation of this nonnatural amino acid into GFP created a new "Gold" fluorescent protein with a $\lambda_{max} = 574$ nm (129). Site-specific replacement in vivo of GFP Tyr66 by an array of phenylalanine analogs using amber suppression and an orthogonal AARS/tRNA pair demonstrated that the fluorescence emission of GFP could be tuned over a range of more than 150 nm (130).

Finally, an alanine derivative of an environmentally sensitive fluorophore, 6-dimethylamino-2-acylnapthalene (Aladan), was incorporated into two different potassium channels by injection of synthetic aladanyl-tRNA into *Xenopus* oocytes and by SPPS into the IgG-binding domain GB1 (131). Several different substituted GB1 domains were evaluated, and the strength of the Aladan emission and its λ_{max} were shown to be highly dependent on the residue's local electrostatic environment.

CROSS-LINKING AND PROBES FOR CHEMICAL MODIFICATION   The photoaffinity label *p*-benzoyl-L-phenylalanine (*p*Bpa) has been incorporated into proteins via in vitro translation (127) and, more recently, in vivo via the generation of an orthogonal AARS/tRNA suppressor pair and an amber stop codon. In this latter work, a modified glutathione-S-transferase was overexpressed in *E. coli* with *p*Bpa in place of Phe52. Irradiation of pBpa in this mutant was used to covalently dimerize glutathione-*S*-transferase (132, 133). This method should prove generally applicable to a wide array of protein targets and may prove useful for identifying protein-protein interactions in vivo.

In vivo systems for introducing azide functionality in place of methionine (134) or a ketone in response to an amber codon (135) have also been developed. The advantage to these two chemical moieties is that they are not normally found in proteins and each is uniquely reactive: a protein containing a nonnatural azide can be chemically modified by triarylphosphine reagents (Staudinger ligation) (134) and a protein with a ketone can be converted into an oxime by treatment with a substituted hydrazide or hydroxylamine (135). Thus, the nature of the chemical modification can be controlled by the R groups on the reactants, perhaps to introduce a wide array of functionalities into proteins. For example, staphylococcal protein A was generated to contain the keto amino acid, *p*-acetyl-L-phenylalanine, at position 7. This keto amino acid mutation was exploited to covalently append aminooxy saccharide derivatives onto the protein (136).

FLUORINATED AMINO ACIDS AS SPECTROSCOPIC AND MECHANISTIC PROBES Because fluorine is similar in size to hydrogen (but much more electronegative), fluorinated amino acids can often be incorporated into proteins or peptides with minimal spatial impact (reviewed in 137). These modified peptides and proteins can have dramatically improved stabilities (137), have novel nonfluorescent spectroscopic properties (138), and can be used as mechanistic probes (139).

Nonnatural Amino Acids as Tools to Study Signal Transduction

The caged phosphoserine (CPS) analog (1-(2-nitrophenyl) ethyl phosphoserine) (140) contains a photo-labile-protecting group that, upon irradiation with UV light, is cleaved to generate phosphoserine. Imperiali and coworkers recently demonstrated the effectiveness of decaging of CPS for observing phosphoserine-dependent protein-protein interactions (141). CPS and an environmentally sensitive fluorophore, 6-(2-dimethylaminonaphthoyl)alanine (DANA), were incorporated by SPPS into a peptide that normally binds the cell cycle regulation intermediate protein 14–3–3. The resulting peptide (Ac-Arg-Leu-DANA-Arg-CPS-Leu-Pro-Ala-CONH$_2$) lost its binding properties; however, irradiation to release phosphoserine generated a peptide that bound to 14–3–3 (quantified by the change in DANA fluorescence).

CONCLUSIONS

The first attempts to introduce nonnatural amino acids into proteins resulted in modified proteins that were useful for research, such as proteins with internal probes placed at specific sites (40–47, 127). The recent development of new genetic and chemical methods has brought the field to the point where commercial—including medical—applications are plausible. For example, attachment of polyethylene glycol (sometimes called PEGylation) to therapeutic proteins (biologics) such as α-interferon can improve in vivo activity, extend circulating lifetime, increase solubility, and reduce immunogenicity (142). Current methods for PEGylation typically give heterogenous products. For example, N-hydroxysuccinimide PEG can be used to attach PEG to the ϵ-amino groups of lysine side chains (143). The result is species with one, two, three, or more modifications, all in admixture. For therapeutic applications in humans, biologics must be manufactured in a way that gives a consistent product. Heterogenous products are difficult to produce in a way that is compliant with the requirement for consistency. Introduction of a nonnatural amino acid at a single site, combined with use of a nonnatural amino-acid-reactive PEG, would result in a homogenous PEGylated protein that is far easier to produce with consistency. Many other applications for human therapeutics are now also conceivable.

From the standpoint of research into the origins of life, manipulations of aminoacyl tRNA synthetases are inevitably the main focus. Manipulation of synthetases to incorporate nonnatural amino acids has illuminated the central role of the editing reactions (113, 119). Without editing, the genetic code is ambiguous and statistical proteins are generated freely. The viability of organisms that harbor statistical proteins, and the selective pressures that introduced editing reactions, are of great interest. These questions also raise the possibility that early proteins in living organisms contained less than 20 amino acids. Long-term selections in cells bearing editing defects could be designed to eliminate at least 1 of the 20 canonical amino acids, and not replace the one eliminated with an alternative. Experiments of this sort explore the range of possibilities for the kinds of primitive living systems that could exist, at least in principle.

ACKNOWLEDGMENTS

This work was supported by grants GM23562 and GM15539 from the National Institutes of Health, by grant CA92577 from the National Cancer Institute, and by Grant MCB-0128901 from the National Science Foundation.

The *Annual Review of Biochemistry* is online at http://biochem.annualreviews.org

LITERATURE CITED

1. Crick FHC, Barnett L, Brenner S, Watts-Tobin RJ. 1961. *Nature* 192:1227–32
2. Ibba M, Söll D. 2000. *Annu. Rev. Biochem.* 69:617–50
3. Pauling L. 1958. In *Festschrift fuer Prof. Dr. Arthur Stoll Siebzigsten*, pp. 597–602. Basel, Switz: Birkhauser-Verlag
4. Baldwin AN, Berg P. 1966. *J. Biol. Chem.* 241:839–45
5. Schreier AA, Schimmel PR. 1972. *Biochemistry* 11:1582–89
6. Eldred EW, Schimmel PR. 1972. *J. Biol. Chem.* 247:2961–64
7. Yarus M. 1972. *Proc. Natl. Acad. Sci. USA* 69:1915–19
8. Hendrickson TL, Schimmel P. 2003. See Ref. 182, pp. 34–64
9. Giegé R, Sissler M, Florentz C. 1998. *Nucleic Acids Res.* 26:5017–35
10. Garen A. 1968. *Science* 160:149–59
11. Goodman HM, Abelson J, Landy A, Brenner S, Smith JD. 1968. *Nature* 217:1019–24
12. Andoh T, Ozeki H. 1968. *Proc. Natl. Acad. Sci. USA* 59:792–99
13. Magliery TJ, Anderson JC, Schultz PG. 2001. *J. Mol. Biol.* 307:755–69
14. Yourno J. 1972. *Nat. New Biol.* 239:219–21
15. Riddle DL, Carbon J. 1973. *Nat. New Biol.* 242:230–34
16. Fowden L. 2001. *Amino Acids* 20:217–24
17. Fowden L, Lewis D, Tristram H. 1967. *Adv. Enzymol. Relat. Areas Mol. Biol.* 29:89–163
18. Cohen G, Bowie CD. 1957. *C. R. Acad. Sci. Ser. B* 244:680–83
19. Hendrickson WA, Horton JR, LeMaster DM. 1990. *EMBO J.* 9:1665–72
20. Orgel LE. 1964. *J. Mol. Biol.* 9:208
21. Wong JT-F. 1983. *Proc. Natl. Acad. Sci. USA* 80:6303–6
22. Bacher JM, Ellington AD. 2001. *J. Bacteriol.* 183:5414–25
23. Budisa N, Minks C, Alefelder S, Wenger W, Dong F, et al. 1999. *FASEB J.* 13:41–51
24. van Hest JC, Tirrell DA. 1998. *FEBS Lett.* 428:68–70
25. Kiick KL, van Hest JC, Tirrell DA. 2000. *Angew. Chem. Int. Ed. Engl.* 39:2148–52
26. Cirino PC, Tang Y, Takahashi K, Tirrell DA, Arnold FH. 2003. *Biotechnol. Bioeng.* 83:729–34
27. Berger H, Yanofsky C. 1967. *Science* 156:394–97
28. Brenner S, Barnett L, Katz ER, Crick FH. 1967. *Nature* 213:449–50
29. Miller JH. 1991. *Methods Enzymol.* 208:543–63
30. Eggertsson G, Soll D. 1988. *Microbiol. Rev.* 52:354–74
31. Murgola EJ. 1985. *Annu. Rev. Genet.* 19:57–80
32. Bruce AG, Atkins JF, Wills N, Uhlenbeck O, Gesteland RF. 1982. *Proc. Natl. Acad. Sci. USA* 79:7127–31
33. Kleina LG, Masson JM, Normanly J, Abelson J, Miller JH. 1990. *J. Mol. Biol.* 213:705–17
34. Kenyon GL, Bruice TW. 1977. *Methods Enzymol.* 47:407–30

35. Wynn R, Richards FM. 1993. *Protein Sci.* 2:395–403
36. Qi D, Tann CM, Haring D, Distefano MD. 2001. *Chem. Rev.* 101:3081–111
37. Plettner E, Khumtaveeporn K, Shang X, Jones JB. 1998. *Bioorg. Med. Chem. Lett.* 8:2291–96
38. Silverman JA, Harbury PB. 2002. *J. Biol. Chem.* 277:30968–75
39. Döring V, Marlière P. 1998. *Genetics* 150:543–51
40. Johnson AE, Woodward WR, Herbert E, Menninger JR. 1976. *Biochemistry* 15:569–75
41. Johnson AE, Cantor CR. 1977. *Methods Enzymol.* 46:180–94
42. Crowley KS, Liao S, Worrell VE, Reinhart GD, Johnson AE. 1994. *Cell* 78:461–71
43. Krieg UC, Walter P, Johnson AE. 1986. *Proc. Natl. Acad. Sci. USA* 83:8604–08
44. Hecht SM, Alford BL, Kuroda Y, Kitano S. 1978. *J. Biol. Chem.* 253:4517–20
45. Heckler TG, Chang LH, Zama Y, Naka T, Chorghade MS, Hecht SM. 1984. *Biochemistry* 23:1468–73
46. Roesser JR, Xu C, Payne RC, Surratt CK, Hecht SM. 1989. *Biochemistry* 28:5185–95
47. Robertson SA, Noren CJ, Anthony-Cahill SJ, Griffith MC, Schultz PG. 1989. *Nucleic Acids Res.* 17:9649–60
48. Noren CJ, Anthony-Cahill SJ, Griffith MC, Schultz PG. 1989. *Science* 244:182–88
49. Ellman J, Mendel D, Anthony-Cahill S, Noren CJ, Schultz PG. 1991. *Methods Enzymol.* 202:301–36
50. Mendel D, Cornish VW, Schultz PG. 1995. *Annu. Rev. Biophys. Biomol. Struct.* 24:435–62
51. Nenomiya K, Kurita T, Hohsaka T, Sisido M. 2003. *Chem. Commun.* 17:2242–43
52. Bessho Y, Hodgson DR, Suga H. 2002. *Nat. Biotechnol.* 20:723–28
53. Anthony-Cahill SJ, Griffith MC, Noren CJ, Suich DJ, Schultz PG. 1989. *Trends Biochem. Sci.* 14:400–3
54. Bain JD, Diala ES, Glabe CG, Dix TA, Chamberlin AR. 1989. *J. Am. Chem. Soc.* 111:8013–14
55. Shimizu Y, Inoue A, Tomari Y, Suzuki T, Yokogawa T, et al. 2001. *Nat. Biotechnol.* 19:751–55
56. Frankel A, Li S, Starck SR, Roberts RW. 2003. *Curr. Opin. Struct. Biol.* 13:506–12
57. Saito H, Kourouklis D, Suga H. 2001. *EMBO J.* 20:1797–806
58. Murakami H, Bonzagni NJ, Suga H. 2002. *J. Am. Chem. Soc.* 124:6834–35
59. Murakami H, Kourouklis D, Suga H. 2003. *Chem. Biol.* 10:1077–84
60. Deleted in proof
61. Merrifield B. 1996. *Protein Sci.* 5:1947–51
62. Li CH, Yamashiro D, Gospodarowicz D, Kaplan SL, Vliet GV. 1983. *Proc. Natl. Acad. Sci. USA* 80:2216–20
63. Clark-Lewis I, Moser B, Walz A, Baggiolini M, Scott GJ, Aebersold R. 1991. *Biochemistry* 30:3128–35
64. Zawadzke LE, Berg JM. 1992. *J. Am. Chem. Soc.* 114:4002–3
65. Zawadzke LE, Berg JM. 1993. *Proteins* 16:301–5
66. Schneider J, Kent SB. 1988. *Cell* 54:363–68
67. Wlodawer A, Miller M, Jaskolski M, Sathyanarayana BK, Baldwin E, et al. 1989. *Science* 245:616–21
68. Milton RC, Milton SC, Kent SB. 1992. *Science* 256:1445–48
69. Nolte A, Klussmann S, Bald R, Erdmann VA, Furste JP. 1996. *Nat. Biotechnol.* 14:1116–19
70. Kellenberger C, Hietter H, Luu B. 1995. *Pept. Res.* 8:321–27
71. Offord RE. 1972. *Biochem. J.* 129:499–501
72. Wallace CJ, Offord RE. 1979. *Biochem. J.* 179:169–82
73. Wallace CJ, Clark-Lewis I. 1992. *J. Biol. Chem.* 267:3852–61

74. Wuttke DS, Gray HB, Fisher SL, Imperiali B. 1993. *J. Am. Chem. Soc.* 115: 8455–56

75. Dawson PE, Kent SB. 2000. *Annu. Rev. Biochem.* 69:923–60

76. Muir TW. 2003. *Annu. Rev. Biochem.* 72:249–89

77. Muir TW, Sondhi D, Cole PA. 1998. *Proc. Natl. Acad. Sci. USA* 95:6705–10

78. Cole PA, Courtney AD, Shen K, Zhang ZS, Qiao YF, et al. 2003. *Acc. Chem. Res.* 36:444–52

79. Flavell RR, Huse M, Goger M, Trester-Zedlitz M, Kuriyan J, Muir TW. 2002. *Org. Lett.* 4:165–68

80. Miller JS, Dudkin VY, Lyon GJ, Muir TW, Danishefsky SJ. 2003. *Angew. Chem. Int. Ed. Engl.* 42:431–34

81. Lu W, Gong DQ, Bar-Sagi D, Cole PA. 2001. *Mol. Cell.* 8:759–69

82. Wang DX, Cole PA. 2001. *J. Am. Chem. Soc.* 123:8883–86

83. Lu W, Shen K, Cole PA. 2003. *Biochemistry* 42:5461–68

84. Zhang Z, Shen K, Lu W, Cole PA. 2003. *J. Biol. Chem.* 278:4668–74

85. Valiyaveetil FI, MacKinnon R, Muir TW. 2002. *J Am. Chem. Soc.* 124: 9113–20

86. Cowburn D, Muir TW. 2001. *Methods Enzymol.* 339:41–54

87. Saks ME, Sampson JR, Nowak MW, Kearney PC, Du F, et al. 1996. *J. Biol. Chem.* 271:23169–75

88. Nowak MW, Kearney PC, Sampson JR, Saks ME, Labarca CG, et al. 1995. *Science* 268:439–42

89. Dougherty DA. 2000. *Curr. Opin. Chem. Biol.* 4:645–52

90. Kohrer C, Xie L, Kellerer S, Varshney U, RajBhandary UL. 2001. *Proc. Natl. Acad. Sci. USA* 98:14310–15

91. Ilegems E, Pick HM, Vogel H. 2002. *Nucleic Acids Res.* 30:e128

92. Monahan SL, Lester HA, Dougherty DA. 2003. *Chem. Biol.* 10:573–80

93. Magliery TJ, Pastrnak M, Anderson JC, Santoro SW, Herberich B, et al. 2003. See Ref. 182, pp. 95–114

94. Wang L, Brock A, Herberich B, Schultz PG. 2001. *Science* 292:498–500

95. Kowal AK, Kohrer C, RajBhandary UL. 2001. *Proc. Natl. Acad. Sci. USA* 98: 2268–73

96. Drabkin HJ, Park HJ, RajBhandary UL. 1996. *Mol. Cell Biol.* 16:907–13

97. Santoro SW, Wang L, Herberich B, King DS, Schultz PG. 2002. *Nat. Biotechnol.* 20:1044–48

98. Wang L, Schultz PG. 2001. *Chem. Biol.* 8:883–90

99. Anderson JC, Schultz PG. 2003. *Biochemistry* 42:9598–608

100. Chin JW, Cropp TA, Chu S, Meggers E, Schultz PG. 2003. *Chem. Biol.* 10: 511–19

101. Chin JW, Cropp TA, Anderson JC, Mukherji M, Zhang Z, Schultz PG. 2003. *Science* 301:964–67

102. Pastrnak M, Schultz PG. 2001. *Bioorg. Med. Chem.* 9:2373–79

103. Mehl RA, Anderson JC, Santoro SW, Wang L, Martin AB, et al. 2003. *J. Am. Chem. Soc.* 125:935–39

104. Hohsaka T, Sisido M. 2002. *Curr. Opin. Chem. Biol.* 6:809–15

105. Hohsaka T, Ashizuka Y, Taira H, Murakami H, Sisido M. 2001. *Biochemistry* 40:11060–64

106. Taki M, Hohsaka T, Murakami H, Taira K, Sisido M. 2002. *J. Am. Chem. Soc.* 124:14586–90

107. Hohsaka T, Ashizuka Y, Murakami H, Sisido M. 2001. *Nucleic Acids Res.* 29:3646–51

108. Anderson JC, Magliery TJ, Schultz PG. 2002. *Chem. Biol.* 9:237–44

109. Switzer C, Moroney SE, Benner SA. 1989. *J. Am. Chem. Soc.* 111:8322–23

110. Piccirilli JA, Krauch T, Moroney SE, Benner SA. 1990. *Nature* 343:33–37

111. Ohtsuki T, Kimoto M, Ishikawa M, Mitsui T, Hirao I, Yokoyama S. 2001. *Proc. Natl. Acad. Sci. USA* 98:4922–25

112. Hirao I, Ohtsuki T, Fujiwara T, Mitsui T, Yokogawa T, et al. 2002. *Nat. Biotechnol.* 20:177–82

113. Döring V, Mootz HD, Nangle LA, Hendrickson TL, de Crécy-Lagard V, et al. 2001. *Science* 292:501–4

114. Deleted in proof

115. Eriani G, Delarue M, Poch O, Gangloff J, Moras D. 1990. *Nature* 347:203–6

116. Cusack S, Berthet-Colominas C, Hartlein M, Nassar N, Leberman R. 1990. *Nature* 347:249–55

117. Webster T, Tsai H, Kula M, Mackie GA, Schimmel P. 1984. *Science* 226:1315–17

118. Yang X-L, Schimmel P, Ribas de Pouplana L. 2003. *Proc. Natl. Acad. Sci. USA* 100:15376–80

119. Nangle LA, De Crécy-Lagard V, Döring V, Schimmel P. 2002. *J. Biol. Chem.* 277:45729–33

120. Gallant JA, Lindsley D. 1998. *Proc. Natl. Acad. Sci. USA* 95:13771–76

121. Abdel-Rahman HM, Al-Karamany GS, El-Koussi NA, Youssef AF, Kiso Y. 2002. *Chem. Biol.* 9:1905–22

122. Kiso Y, Matsumoto H, Mizumoto S, Kimura T, Fujiwara Y, Akaji K. 1999. *Biopolymers* 51:59–68

123. Mak CC, Brik A, Lerner DL, Elder JH, Morris GM, et al. 2003. *Bioorg. Med. Chem.* 11:2025–40

124. Dewachter I, Van Leuven F. 2002. *Lancet Neurol.* 1:409–16

125. Bein M, Schaller M, Korting HC. 2002. *Curr. Drug Targets* 3:351–57

126. Hruby VJ. 2002. *Nat. Rev. Drug Discov.* 1:847–58

127. Cornish VW, Benson DR, Altenbach CA, Hideg K, Hubbell WL, Schultz PG. 1994. *Proc. Natl. Acad. Sci. USA* 91:2910

128. McNulty JC, Silapie JL, Carnevali M, Farrar CT, Griffin RG, et al. 2000. *Biopolymers* 55:479–85

129. Bae JH, Rubini M, Jung G, Wiegand G, Seifert MHJ, et al. 2003. *J. Mol. Biol.* 328:1071–81

130. Wang L, Xie J, Deniz AA, Schultz PG. 2003. *J. Org. Chem.* 68:174–76

131. Cohen BE, McAnaney TB, Park ES, Jan YN, Boxer SG, Jan LY. 2002. *Science* 296:1700–3

132. Chin JW, Schultz PG. 2002. *Chembiochem* 3:1135–37

133. Chin JW, Martin AB, King DS, Wang L, Schultz PG. 2002. *Proc. Natl. Acad. Sci. USA* 99:11020–24

134. Kiick KL, Saxon E, Tirrell DA, Bertozzi CR. 2002. *Proc. Natl. Acad. Sci. USA* 99:19–24

135. Wang L, Zhang Z, Brock A, Schultz PG. 2003. *Proc. Natl. Acad. Sci. USA* 100: 56–61

136. Liu H, Wang L, Brock A, Wong CH, Schultz PG. 2003. *J. Am. Chem. Soc.* 125:1702–3

137. Yoder NC, Kumar K. 2002. *Chem. Soc. Rev.* 31:335–41

138. Reid PJ, Loftus C, Beeson CC. 2003. *Biochemistry* 42:2441–48

139. Yee CS, Chang MCY, Ge J, Nocera DG, Stubbe J. 2003. *J. Am. Chem. Soc.* 125: 10506–7

140. Rothman DM, Vazquez ME, Vogel EM, Imperiali B. 2002. *Org. Lett.* 4:2865–68

141. Vazquez ME, Nitz M, Stehn J, Yaffe MB, Imperiali B. 2003. *J. Am. Chem. Soc.* 125:10150–51

142. Molineux G. 2003. *Pharmacotherapy* 23:S3–8

143. Bory C, Boulieu R, Souillet G, Chantin C, Guibaud P, Hershfield MS. 1991. *Adv. Exp. Med. Biol.* 309A:173–76

144. Fowden L, Neale S, Tristram H. 1963. *Nature* 35:35–38

145. Fowden L, Richmond MH. 1963. *Biochem. Biophys. Acta.* 71:459

146. Budisa N, Minks C, Medrano FJ, Lutz J, Huber R, Moroder L. 1998. *Proc. Natl. Acad. Sci. USA* 95:455–59

147. Schwartz JH, Maas WK. 1960. *J. Bacteriol.* 49:794–99

148. Smith RC, Salmon WD. 1965. *J. Bacteriol.* 89:686

149. Budisa N, Steipe B, Demange P, Eckerskorn C, Kellermann J, Huber R. 1995. *Eur. J. Biochem.* 230:788–96

150. Cowie DB, Cohen GD, Bolton ET, Robichon-Szulmajster H. 1959. *Biochem. Biophys. Acta* 34:39–46

151. Tang Y, Tirrell DA. 2002. *Biochemistry* 41:10635–45

152. Brawerman G, Ycas M. 1957. *Arch. Biochem. Biophys.* 68:112–17

153. Soumillion P, Jespers L, Vervoort J, Fastrez J. 1995. *Protein Eng.* 8:451–56

154. Ross JB, Szabo AG, Hogue CW. 1997. *Methods Enzymol.* 278:151–90

155. Budisa N, Alefelder S, Bae JH, Golbik R, Minks C, et al. 2001. *Protein Sci.* 10:1281–92

156. Budisa N, Rubini M, Bae JH, Weyher E, Wenger W, et al. 2002. *Angew. Chem. Int. Ed. Engl.* 41:4066–69

157. Wang P, Tang Y, Tirrell DA. 2003. *J. Am. Chem. Soc.* 125:6900–6

158. Stieglitz B, Calvo JM. 1971. *J. Bacteriol.* 108:95–104

159. Rennert OM, Anker HS. 1963. *Biochemistry* 2:471–76

160. Tang Y, Ghirlanda G, Vaidehi N, Kua J, Mainz DT, et al. 2001. *Biochemistry* 40:2790–96

161. Tang Y, Tirrell DA. 2001. *J. Am. Chem. Soc.* 123:11089–90

162. Kohno T, Kohda D, Haruki M, Yokoyama S, Miyazawa T. 1990. *J. Biol. Chem.* 265:6931–35

163. Smulson ME, Rabinovitz M, Breitman TR. 1967. *J. Bacteriol.* 94:1890–95

164. Muller S, Heider J, Bock A. 1997. *Arch. Microbiol.* 168:421–27

165. Minks C, Huber R, Moroder L, Budisa N. 2000. *Anal. Biochem.* 284:29–34

166. Cohen GN, Munier R. 1959. *Biochem. Biophys. Acta* 31:347–56

167. Ibba M, Hennecke H. 1995. *FEBS Lett.* 364:272–75

168. Sharma N, Furter R, Kast P, Tirrell DA. 2000. *FEBS Lett.* 467:37–40

169. Datta D, Wang P, Carrico IS, Mayo SL, Tirrell DA. 2002. *J. Am. Chem. Soc.* 124:5652–53

170. Schlesinger S, Schlesinger MJ. 1967. *J. Biol. Chem.* 242:3369–72

171. Soumillion P, Fastrez J. 1998. *Protein Eng.* 11:213–17

172. Schlesinger S, Schlesinger MJ. 1969. *J. Biol. Chem.* 244:3803–9

173. Ravel JM, White MN, Shive W. 1965. *Biochem. Biophys. Res. Commun.* 20:352–59

174. Hamano-Takaku F, Iwama T, Saito-Yano S, Takaku K, Monden Y, et al. 2000. *J. Biol. Chem.* 275:40324–28

175. Furter R. 1998. *Protein Sci.* 7:419–26

176. Liu DR, Schultz PG. 1999. *Proc. Natl. Acad. Sci. USA* 96:4780–85

177. Sakamoto K, Hayashi A, Sakamoto A, Kiga D, Nakayama H, et al. 2002. *Nucleic Acids Res.* 30:4692–99

178. Wang L, Brock A, Schultz PG. 2002. *J. Am. Chem. Soc.* 124:1836–37

179. Chin JW, Santoro SW, Martin AB, King DS, Wang L, Schultz PG. 2002. *J. Am. Chem. Soc.* 124:9026–27

180. Zhang Z, Wang L, Brock A, Schultz PG. 2002. *Angew. Chem. Int. Ed. Engl.* 41:2840–42

181. Zhang Z, Smith BA, Wang L, Brock A, Cho C, Schultz PG. 2003. *Biochemistry* 42:6735–46

182. Lapointe J, Briakier-Gingras L, eds. 2003. *Translation Mechanism.* New York: Eurekah.com/Kluwer Acad.

Annu. Rev. Biochem. 2004. 73:177–208
doi: 10.1146/annurev.biochem.73.071403.160049
Copyright © 2004 by Annual Reviews. All rights reserved
First published online as a Review in Advance on April 2, 2004

REGULATION OF TELOMERASE BY TELOMERIC PROTEINS

Agata Smogorzewska and Titia de Lange

*The Rockefeller University, New York, New York 10021; email:
delange@mail.rockefeller.edu, asmogorzewska@partners.org*

Key Words   aging, cancer, human, telomere, yeast

■ **Abstract**   Telomeres are essential for genome stability in all eukaryotes. Changes in telomere functions and the associated chromosomal abnormalities have been implicated in human aging and cancer. Telomeres are composed of repetitive sequences that can be maintained by telomerase, a complex containing a reverse transcriptase (hTERT in humans and Est2 in budding yeast), a template RNA (hTERC in humans and Tlc1 in yeast), and accessory factors (the Est1 proteins and dyskerin in humans and Est1, Est3, and Sm proteins in budding yeast). Telomerase is regulated in *cis* by proteins that bind to telomeric DNA. This regulation can take place at the telomere terminus, involving single-stranded DNA-binding proteins (POT1 in humans and Cdc13 in budding yeast), which have been proposed to contribute to the recruitment of telomerase and may also regulate the extent or frequency of elongation. In addition, proteins that bind along the length of the telomere (TRF1/TIN2/tankyrase in humans and Rap1/Rif1/Rif2 in budding yeast) are part of a negative feedback loop that regulates telomere length. Here we discuss the details of telomerase and its regulation by the telomere.

CONTENTS

THE END REPLICATION PROBLEM

The advent of linear chromosomes created a significant challenge for DNA replication. The problem, referred to as the end replication problem (1, 2), originates from the use of short RNAs to prime DNA synthesis. Removal of these primers results in 8–12 nucleotide (nt) gaps that do not impede the duplication of circular genomes because each gap can be closed by extending a preceding Okazaki fragment. However, on a linear template, the last RNA that primed lagging-strand synthesis will leave a gap that can not be filled. In the absence of a telomere maintenance system, many eukaryotes (fungi, trypanosomes, flies, mosquitos) lose terminal sequences at ~3–5 bp/end/division, a modest rate predicted by the end replication problem (3–6; O. Dreesen and G. Cross, personal communication; J. Cooper, personal communication). Human and mouse telomeres shorten much faster (50–150 bp/end/cell division) (7–9); this suggests that chromosome ends, in these organisms, might be actively degraded. If telomere erosion is not balanced by elongation, telomeres will progressively shorten, eventually leading to chromosome instability and cell death. Therefore, the long-term proliferation of all eukaryotic cells, including cells giving rise to the germline, requires a mechanism to counteract telomere attrition. Here we review the mechanisms by which telomeric DNA is maintained and discuss how telomere-associated proteins regulate this process.

THE CONSEQUENCES OF TELOMERE DYSFUNCTION

The telomeric nucleoprotein complex allows cells to distinguish natural chromosome ends from DNA breaks [reviewed in (10, 11)]. Without telomere protection, chromosome ends activate DNA damage response pathways that signal cell cycle arrest, senescence, or apoptosis. Telomeres also prevent inappropriate DNA repair reactions, such as exonucleolytic degradation and ligation of one end to another. When telomere function is impaired, fusion of unprotected chromosome ends can generate dicentric chromosomes, which are unstable in mitosis and wreck havoc in the genome.

Telomeres have received considerable attention since the realization that changes in their structure and function occur during cancer development and aging. Many human cell types display telomere erosion, a process that is though to limit the proliferative capacity of transformed cells and has the hallmark of a tumor suppressor system. In most human cancer, the telomere barrier has been bypassed through the activation of a telomere maintenance system, making telomere replication an attractive target for therapeutic intervention. Although the programmed shortening of human telomeres may be effective in limiting the cancer burden early in life, the same program may have detrimental consequences late in life. In the aged, short telomeres are predictive of diminished health and longevity, and at least one human premature aging syndrome is associated with compromised telomere function (12, 13). Diminishing telomere function late in life may even promote genome instability and therefore contribute to the higher incidence of cancer in the aged. The role of telomeres in cancer and aging has been reviewed extensively elsewhere (14, 15).

TELOMERE MAINTENANCE BY TELOMERASE

The most versatile and widely used method of telomere maintenance is based on telomerase (Figure 1) (16, 17). A two-component ribonucleoprotein enzyme, telomerase contains a highly conserved reverse transcriptase [telomerase reverse transcriptase, TERT (18–20)] and an associated template RNA [telomerase RNA component, TERC, also referred to as TR or TER (21–24)]. TERT is most closely related to the reverse transcriptases of non-LTR retroposons and group II introns (23), and like these RTs, it extends the 3′ end of a DNA rather than an RNA primer (25). The primer for telomerase is the chromosome terminus, which can be positioned on an alignment site in TERC such that the 3′ end of the telomere is adjacent to the short (often 6 nt) template sequence (Figure 1A and B). Extension of the telomere terminus results in the addition of one telomeric repeat, and repeated alignment and extension steps can endow chromosome ends with the direct repeat arrays typical of telomeres. Although the sequence and size of telomerase RNAs are highly variable, they share structural motifs (Figure 1C) (26, 27), which may mediate the interaction with TERT, or control of the alignment, extension, and translocation steps. After elongation of the 3′ end, C-strand synthesis is presumably required to create double-stranded telomeric DNA, but the details of this step have only been examined in ciliates [(28–30), reviewed in (31)]. In addition, Tetrahymena telomeres have a precisely defined terminal structure that is generated by nucleolytic processing (32, 33), and it will be interesting to learn whether similar terminus transactions are required in other organisms.

In most unicellular organisms, telomerase has a housekeeping function, and its core components are always expressed. In contrast, telomerase is strongly suppressed in the human soma, a phenotype also observed in old world monkeys and new world primates (but not in prosimians, such as lemurs) [(18, 20, 34, 35); reviewed in (36)]. Robust telomerase activity is restricted to ovaries, testes, and highly proliferative tissues. This regulation place exists primarily at the level of

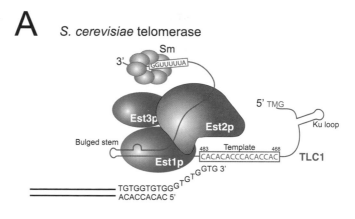

A   *S. cerevisiae* telomerase

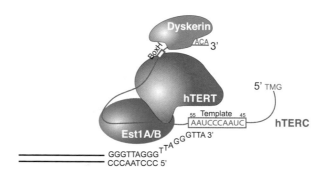

B   Human telomerase

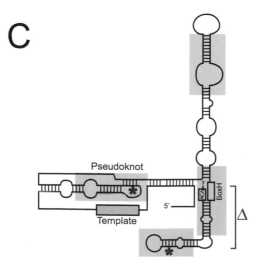

C

transcription of the *hTERT* gene; hTERC is virtually ubiquitous (24). The repression of *hTERT* transcription involves multiple genes previously implicated in tumorigenesis, which include Menin, the Mad/Myc pathway, and the TGFβ target Sip1 [(37); reviewed in (38)].

Exogenous expression of hTERT in primary human fibroblasts is sufficient to reconstitute telomerase activity and to counteract telomere erosion. The resulting telomere maintenance immortalizes most human cell types (39–41). Like primary cells, tumor cells require a telomere maintenance system for long-term proliferation, and in the majority of cases, this is provided by upregulation of hTERT [reviewed in (36)]. Telomerase activity per se does not induce transformation (42), and although telomerase is necessary for immortalization, *hTERT* is not an oncogene (43, 44). Conversely, oncogenic transformation does not require telomerase activity, and cells with very long telomeres can be fully transformed into a tumorigenic phenotype in vitro without a telomere maintenance system (45). Similarly, certain childhood tumors that originate in young cells with long telomeres can be cancerous and metastatic even though they lack telomerase. However, the extensive proliferation of cells during the prolonged multistep tumorigenesis pathway that leads to most adult human cancers is predicted to exhaust the telomere reserve, necessitating telomerase activation (46).

Once hTERT and hTERC are expressed, they have to be properly assembled and targeted to chromosome ends. Some of the biogenesis of telomerase is likely to take place in the nucleolus because GFP tagged hTERT is localized to nucleoli in G1, from which it moves to the nucleoplasm in S/G2 (47). Similarly, yeast Est2p, the TERT component of telomerase in *Saccharomyces cerevisiae*, is enriched in the nucleolus upon overexpression (48). Attempts to visualize telomerase at telomeres have failed. The only physical evidence for the association of telomerase with chromosome ends comes from chromatin precipitation studies in yeast showing that Est2p is present on telomeres during G1 and S phase (49, 50).

The presence of Est2p at telomeres in S phase is expected based on the finding that telomerase can extend chromosome ends during and immediately after DNA replication (51). Telomerase can even extend a telomere-like substrate in cells arrested in mitosis, and in this setting, its action is dependent on components of lagging-strand synthesis (52). It is not known whether telomerase can act on both newly replicated (sister) telomeres, and it remains to be determined whether telomerase can also act before DNA replication.

Figure 1   Telomerase holoenzyme in yeast and man. *A*. Budding yeast telomerase docked at a telomere 3' end. *B*. Human telomerase docked at a telomere 3' end. *C*. Conserved structural motifs in vertebrate telomerase RNAs [after (27)]. The positions of *DKC* mutations in the *hTERC* gene are indicated in red.

TELOMERASE ACCESSORY FACTORS

The telomerase holoenzyme often contains additional proteins that are not required for catalysis per se. In *S. cerevisiae*, telomerase is composed of the usual reverse transcriptase and RNA core components (Est2 and TLC1, respectively) and two accessory factors, Est1, which binds to a bulged stem in TLC1, and Est3 (Figure 1*A*) (3, 53–55). Although Est1 and Est3 are not required for in vitro telomerase activity (56, 57), mutations in these genes lead to progressive telomere shortening, the so-called ever shorter telomeres (est) phenotype (3, 54). This est phenotype is also observed for strains lacking the core components of telomerase and points to a complete failure in telomere maintenance (21, 54). In addition, TLC1 RNA, which is generated by RNA polymerase II and contains a trimethylguanosine cap, has an association with Sm proteins (Figure 1*A*), previously implicated in snRNP biogenesis (58).

Accessory factors have also been found for human telomerase. The human genome contains at least three *EST1* orthologs, two of which (*EST1A* and *B*) were recently shown to encode telomerase associated proteins, suggesting a conserved role for Est1 in telomerase regulation [(59, 60); reviewed in (61)]. A confounding issue in the analysis of the *EST1A* gene is its role in nonsense-mediated decay (62). Mammalian Est3p orthologs have not been identified to date, and there is no indication that the mammalian telomerase RNA interacts with Sm proteins. Instead, human telomerase has an important interaction with another RNA binding protein, dyskerin (63). Dyskerin is a putative pseudouridine synthase that has been proposed to play a role in ribosomal processing because it binds to many small nucleolar RNAs (snoRNA) (64). Like the snoRNAs, hTERC contains a H/ACA motif that constitutes the dyskerin binding site (63) (Figure 1*B* and *C*). The H/ACA motif is conserved among vertebrate telomerase RNAs (27), but it is absent from yeast and ciliate telomerase RNAs.

Evidence in favor of the functional significance of the binding of dyskerin to hTERC comes from the genetics of a rare human disease, dyskeratosis congenita (DKC) (65). The X-linked form of DKC is due to a mutation in dyskerin, whereas the autosomal dominant form is due to mutations in the hTERC gene (63). DKC is classically described as a triad of muco-cutaneous changes that include abnormal skin pigmentation, nail dystrophy, and mucosal leukoplakia (65). The most profound defect in DKC and the leading cause of death is bone marrow failure. Additional symptoms include developmental delay, short stature, extensive dental caries/loss, hair loss/gray hair, pulmonary disease, and increased incidence of cancer. Patients with dyskerin mutations have fivefold less hTERC than unaffected siblings, implicating dyskerin in processing or stability of the telomerase RNA (63). Their telomerase activity is diminished, and these defects correlate with shorter telomeres and chromosome end fusions, which are pathognomonic for telomere dysfunction (63, 66).

Because dyskerin deficiency affects both telomerase RNA and ribosomal RNA, it is difficult to establish the contribution of telomere dysfunction to the

DKC (67). However, the fact that some DKC patients have a mutation in the *hTERC* gene (12) (Figure 1*C*), shows conclusively that this disease can be induced by a telomere defect. In each case of autosomal DKC, the expression of *hTERC* is diminished, and affected individuals have very short telomeres. The phenotype of these heterozygous patients is probably due to haploinsufficiency of the human telomerase RNA; a similar situation is seen in mice lacking mTerc (68). In addition to DKC, mutations in the *hTERC* gene can cause aplastic anemia, further strengthening the link between telomerase function and bone marrow maintenance (69, 70).

Two- and three-hybrid screens as well as coimmunoprecipitation experiments have suggested that human telomerase has potential interactions with a large number of additional factors. Although some of these interacting proteins may play a role in biogenesis, stability, and localization of telomerase, the functional significance of most of these interactions has not been established [reviewed in (36, 71, 72)].

TELOMERASE-INDEPENDENT TELOMERE LENGTH CHANGES

Telomerase is not the only activity that affects the length of telomeres. In human cells and in fungi, telomeres can be maintained by a recombination-based mechanisms, referred to as ALT in human cells and as the survivor pathway in yeast [reviewed in (73, 74)]. Furthermore, telomeres can be shortened by exonucleolytic attack, and they can undergo large sudden deletions. The latter, termed telomere rapid deletion (TRD), has been proposed to constitute a second sizing mechanism for telomeres in *S. cerevisiae* [reviewed in (75)], and it is anticipated that similar deletions could affect telomere length in mammals. Although this review is focused on the telomerase pathway, it is possible that some of the regulatory events discussed below do not act directly on telomerase but affect one or more of these other telomere lengthening or shortening events.

REGULATION OF TELOMERASE AT THE TELOMERE TERMINUS: THE ROLE OF CDC13

A priori, the telomere terminus is expected to be a prime site for telomerase regulation. By analogy to the control of RNA polymerases, regulation of telomerase could take place at the level of the recruitment to the telomere terminus, at the initiation of elongation, or at the rate and processivity of the elongation cycles. Indeed, telomere maintenance in *S. cerevisiae* is primarily regulated by a telomere terminus specific factor, Cdc13 (Figure 2) [reviewed by (76)]. Its initial identification as a cell division cycle mutant reflects the essential role of *CDC13* in the protection of telomeres (77). Cells lacking *CDC13* function accumulate single-stranded DNA at

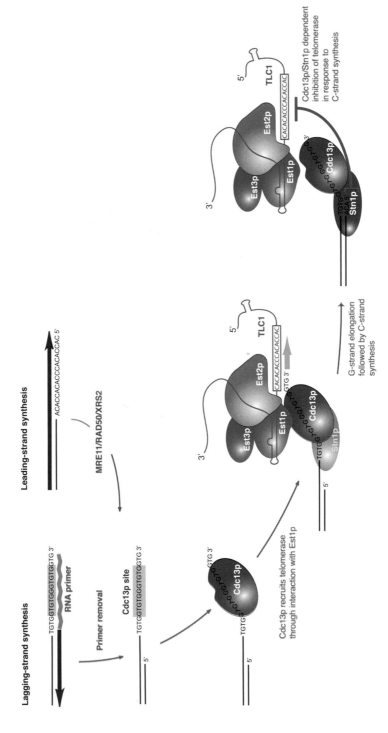

Figure 2 Recruitment of yeast telomerase. Speculative model of how the interaction of Cdc13 with the telomerase holoenzyme regulates recruitment of telomerase to the telomere.

chromosome ends, which induces a *RAD9*-dependent cell cycle arrest (77). However, *CDC13* was independently identified in a screen for *est* genes as EST4 [a mutation referred to as *cdc13–2^{est}* (54, 78)], pointing to a crucial role for Cdc13 in telomerase-mediated telomere maintenance.

Cdc13 is a single-stranded DNA-binding protein with a preference for the G-rich strand of yeast telomeric DNA (78, 79) (Figure 2). The current model for its function proposes that Cdc13 interacts with Est1, thereby recruiting telomerase to the telomere terminus. In support of this model, the telomere maintenance defect of the *cdc13–2^{est}* mutant can be suppressed by a specific mutation in EST1 (*est1–60*), which has an est phenotype on its own. Such allele-specific suppression is most easily interpreted as a restoration of a physical interaction. In agreement, the mutations represent a charge swap in which the phenotype of a Glu->Lys mutation in *cdc13–2^{est}* is suppressed by the reverse (Lys->Glu) change in *est1–60* (80). Thus, the est phenotype of *cdc13* mutants could be explained if the Cdc13-Est1 interaction is necessary to recruit telomerase to the telomere terminus.

The Cdc13-Est1 telomerase recruitment model is consistent with a number of gene fusion experiments in which Cdc13 or its DNA-binding domain (DBD) were fused to protein components of the telomerase complex, i.e., Est2, Est1, and Est3. The resulting fusions rescue the telomere maintenance defects of *cdc13–2^{est}* and *est1Δ* strains (80, 81). For example, a fusion of the Cdc13 DBD to Est2 suppresses the requirement for Est1 in telomere maintenance. Collectively, these experiments suggest that Cdc13 interacts with Est1 to recruit telomerase to the very end of the telomere and that this recruitment step is essential for telomere maintenance (Figure 2).

Est1 may have a second role in addition to bridging the interaction between telomerase and Cdc13. In cells that express a Cdc13-Est2 fusion, the presence of Est1 results in much longer telomeres, suggesting a positive regulatory role that is independent of recruitment (80, 81). Furthermore, certain mutant Est1 alleles lack this positive regulatory function, whereas others are specifically defective in recruitment but still can stimulate telomere elongation in the Cdc13-Est2 fusion context (82). These separation-of-function mutations argue that Est1 plays multiple roles in telomere maintenance. Because Est1 does not affect the catalytic activity of telomerase as measured in cell lysates (56, 57), new assays may be required to reveal how Est1 affects telomerase in vivo.

Indirect evidence suggests that recruitment of telomerase involves multiple steps. ChIP experiments have shown that Est2 can bind to telomeres in G1 and that this association is not dependent on Cdc13 (49). One possibility is that Est2 first binds to telomeric chromatin in G1 and subsequently becomes positioned at the telomere terminus by Cdc13. Indeed, Cdc13 deficiency has a substantial effect on the presence of Est2 at telomeres in S phase (49). Potentially, the G1 recruitment of Est2 could be mediated by an interaction between TLC1 and one or more proteins in the telomeric complex. Evidence indicating such an interaction came from overexpression of TLC1, which was found to interfere with telomeric silencing (21). Later studies indicated that this attribute of TLC1 is dependent on a genetic interaction between a stem-loop structure in TLC1 and the NHEJ protein, Ku (83). Ku is a component of the telomeric chromatin in yeast, it may therefore facilitate recruitment of the Est2/TLC1 complex to telomeres.

Because Cdc13 binds to single-stranded DNA, it is pertinent to ask when its binding site is available at telomeres and how single-stranded telomeric DNA is generated (Figure 2). Although long (>50 nt), single-stranded 3′ tails are only observed in late S phase (84); G1 telomeres have shorter 3′ overhangs that are still sufficient to recruit Cdc13 (R.J. Wellinger, personal communication). How are these overhangs created? One candidate is the Mre11/Rad50/Xrs2 complex, which is known to act as a nuclease in certain settings [reviewed in (85, 86)]. An indirect assay performed on nocodazole blocked (G2/M) cells implicated the Mre11 complex in the loading of Cdc13 (87). However, the in vitro nuclease activity of Mre11 complex has the wrong (3′->5′) polarity (88), and mutations in the nuclease domain of *MRE11* do not have a telomere maintenance defect (89, 90). Furthermore, most *rad50Δ* strains do not have an est phenotype, and their telomere shortening rates are moderate compared to est strains (91), indicating that other pathways for Cdc13 loading must be available. Cdc13 binding sites could simply be created passively by DNA replication (Figure 2) when the last RNA primer of lagging-strand DNA synthesis is removed. Extension of the lagging end should be sufficient to counteract all telomere attrition. In addition, genetic and physical assays suggest an interaction of Cdc13 or one of its protein partners with the machinery executing lagging-strand DNA synthesis (52, 92–94). Perhaps this provides Cdc13 with an alternative way to arrive at telomeres while they are in the process of DNA replication.

In addition to its main role as a positive regulator of telomere maintenance, Cdc13 also limits telomere elongation. This is deduced from the telomere elongation phenotype of certain mutations in *CDC13* or the gene for its interacting partner Stn1 (92, 93). For instance, in strains carrying the *cdc13–5* mutation, telomerase elongates telomeres to four times their usual length. The telomeres also have excessive G overhangs in late S phase that become duplex with delayed kinetics, suggesting a defect in the coordination of lagging-strand synthesis with telomere elongation. Overexpression of Stn1 suppresses both the telomere elongation and G-strand overhang phenotypes, pointing to Stn1 as a critical factor in this aspect of telomere replication (Figure 2). Consistent with the idea that Stn1 controls C-strand synthesis, overexpression of Stn1 also suppressed the inappropriate telomere elongation in DNA polymerase α mutants (92, 93). Lundblad and colleagues (92) proposed a two-step model in which Cdc13 would first recruit telomerase to the telomere, allowing extension of the G strand. Subsequently, Cdc13 together with Stn1 would promote C-strand synthesis with this event and limit further elongation of the telomere by telomerase (Figure 2).

TELOMERE LENGTH HOMEOSTASIS: *CIS*-ACTING CONTROL BY FACTORS BINDING TO DUPLEX TELOMERIC REPEATS

When cells use telomerase-independent methods of telomere maintenance, the length of the individual telomeres is highly variable from chromosome end to chromosome end. By contrast, when telomerase is available, for instance in wild-type yeast or in mammalian tumor cells, telomeres are stably maintained within a relatively narrow size distribution. In these settings, there is a balance between the replicative attrition of telomeres and their elongation by telomerase (95–97). Crosses between closely related species of mice showed that this stable length setting is under genetic control (98, 99).

Telomere length is influenced by the level of telomerase expression but also depends on a control pathway that acts in cis at each individual telomere (Figure 3). The earliest observations on *cis*-acting telomere length control were made by Blackburn and colleagues (96, 100), who introduced an exogenous linear plasmid into budding yeast and found that cells added new telomeres with the same length as the endogenous telomeres. Similarly, a new telomere can be formed after transfection of a telomere seed into mammalian cells; in this process the new telomere undergoes gradual lengthening until it matches the host cell telomeres (101–103). Similar growth of newly formed telomeres was noted in ES cells that had healed a *I-SceI*-cut chromosome with the addition of a new telomere (104). During these telomere healing events, the other telomeres in the cell remain stable, indicating that telomere length control acts in *cis* at each individual chromosome end. To achieve such control, the length of each individual telomere has to be monitored and regulated independently. Obviously, *cis*-acting length control cannot be exerted through changes in the expression of telomerase. Rather, telomeres engage factors that modulate how telomerase acts at the telomere terminus. Thus, like other chromosomal elements, such as enhancers and replication start sites, telomeres recruit a polymerase and locally control its action.

Negative Feedback Control by the Yeast RAP1/RIF1/RIF2 Complex

As telomeres become longer, their further extension by telomerase is progressively inhibited (105). This is an imprecise and stochastic process that keeps telomeres within a broad size range. Telomere length control involves a negative feedback loop in which the addition of new telomeric repeats by telomerase creates binding sites for a telomerase regulator (Figure 3A). In budding yeasts, the main *cis*-acting regulator of telomere length is the repressor/activator protein 1, Rap1 [(106), reviewed in (107)]. *S. cerevisiae* Rap1 has a central Myb-type DNA-binding domain, which binds to a loosely defined recognition site present ~20 times within the heterogeneous TG1–3 tract of yeast telomeres (108–110).

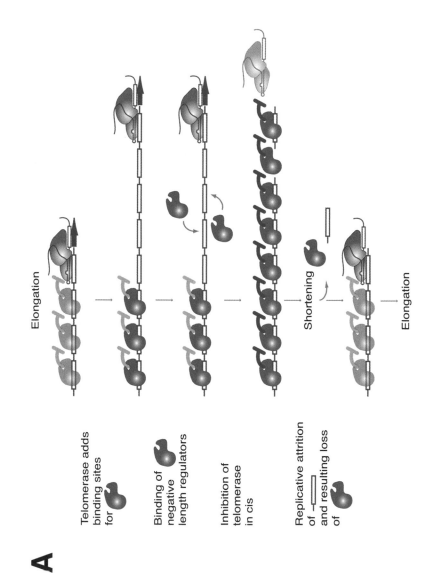

A

Elongation

Telomerase adds
binding sites
for

Binding of
negative
length regulators

Inhibition of
telomerase
in cis

Replicative attrition
of
and resulting loss
of

Shortening

Elongation

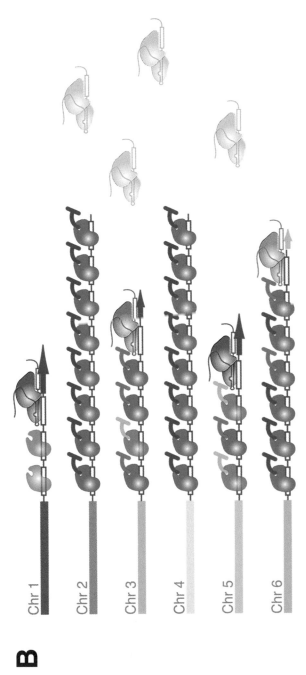

Figure 3 Telomere length homeostasis. The figure displays the principles of *cis*-acting telomere length regulation by telomere binding factors. *A*. Depiction of the negative feedback regulation of telomerase at individual telomeres. Examples of the *cis*-acting negative regulators are TRF1, Taz1, and Rap1. *B*. Schematic representation of different telomere lengthening states within one nucleus. Short telomeres (chromosomes 1 and 5, in this example) are elongated by telomerase, whereas within the same nucleus, long telomeres (chromosomes 2 and 4) are not elongated.

Its C terminus is a protein interaction domain that is crucial for telomere length regulation and gene silencing (111–122). Two telomere length regulators, the Rap1 interacting factors Rif1 and Rif2, bind to the Rap1 C terminus, and the same domain recruits the silencing proteins Sir3 and Sir4. Rap1 also contains a BRCT protein interaction domain in its N terminus and a *trans*-activation domain that is important for transcriptional regulation (107, 123).

Although *S. cerevisiae RAP1* is essential, barring assessment of its null phenotype, overexpression studies and several *rap1* temperature-sensitive mutants reveal phenotypes consistent with its role as a negative regulator of telomere length (111, 112, 114). Furthermore, deletion of the nonessential *RIF1* and *RIF2* genes results in extensive telomere elongation (113, 121). Tethering experiments showed that the number of Rap1 molecules bound at each individual chromosome end serves as a gauge for the length of the telomeric repeat array (120). In these studies, the C terminus of Rap1, which includes the region where Rif1 and Rif2 interact, was fused to the DNA-binding domain of Gal4 and tethered to an engineered telomere with subtelomeric Gal4-binding sites. A negative correlation was seen between the number of Gal4 sites and the stable length setting of that telomere, suggesting that as more Rap1 C termini were tethered to the telomere, the final telomere length was shorter. Experiments of this type established the *cis*-acting nature of the feedback control and demonstrated that the number of telomeric repeats at individual chromosome ends is sensed through the number of bound Rap1 molecules (105, 120, 124, 125). Elegant studies in *Kluyveromyces lactis* further confirmed this model and also illuminated the particular importance of the most terminal telomeric repeats in the Rap1p counting mechanism (126–129).

The current challenge is to determine how Rap1 exerts its control. The Rap1 pathway may be connected to a second pathway for telomere length control (discussed below) that involves the DNA damage response kinases Tel1 and Mec1. The effects of Rap1, Rif1, and Rif2 on telomere length are greatly decreased in cells lacking the DNA damage response kinase Tel1 (130, 131). For instance, in *tel1*Δ cells, the counting of Rap1 at telomeres is diminished, and the *rap1–17* mutation, which normally generates extremely long telomeres, no longer has this effect on telomere length (130, 131). One interpretation is that Rap1/Rif1/Rif2 act on Tel1 and Mec1, but the epistasis relationships are not completely straightforward, and other interpretations have been offered (132).

Negative Feedback Control by the Mammalian TRF1 Complex

Telomere length control in human cells has been studied in tissue culture systems using immortalized cell lines, which usually maintain their telomeres at a stable length setting. The design of these experiments has to take into account the extensive variation in telomere length. Although tumor telomeres are usually stably maintained, their length can range from ~2 to greater than 20 kb probably due to genetic changes incurred during tumorigenesis [reviewed in (133)]. This variability bars direct comparison between different tumor cell lines. Furthermore, within each tumor cell line, there can be extensive variation in telomere length between subclones, probably due to epigenetic changes. For instance, subclones of workhorse tumor cell lines, such as HeLa, 293, and HT1080 cells, can vary widely with regard to telomere length setting, telomerase levels, and telomere dynamics (initial growth or shortening after subcloning) (134; B. van Steensel and T. de Lange, unpublished information). Because of this variation, the effect of exogenously expressed genes can not be evaluated in a small number of transfected clones. This problem can be circumvented by studying a large number of individually altered cells simultaneously [for instance, by retroviral infection (135)] or by using cell lines in which inducible gene expression is used to control for clonal variation (97).

Using such inducible gene expression systems, a feedback loop that controls human telomere length was identified (97, 136). The main control is exerted by the TTAGGG repeat binding factor 1 (TRF1), a small dimeric protein with a C-terminal Myb-type DNA-binding domain that has exquisite specificity for the sequence TTAGGGTTAG (137–141) (Figure 4). TRF1 binds to the duplex telomeric TTAGGG repeat array, and the total number of TRF1 molecules per chromosome end is correlated with the length of the telomeric tract. ChIP experiments on different cell lines showed that TRF1 immunoprecipitates 20% to 30% of the telomeric DNA, regardless of whether the telomeres were 4 kb or 25 kb, indicating that longer telomeres contain much more TRF1 (142). Furthermore, immunofluorescence studies showed that the TRF1 signal increases with telomere length (136). Thus, TRF1 behaves analogous to Rap1 in that the amount of TRF1 present at telomeres reflects their lengths.

The role of TRF1 in telomere length control was revealed by changing its expression level in the Tet-inducible HTC75 line, a subclone of the human fibrosarcoma cell line HT1080 (97, 136). Overexpression of TRF1 caused telomeres to gradually shorten until a new length setting was achieved. Inversely, partial inhibition of TRF1 through expression of a dominant negative allele resulted in progressive elongation of the telomeres to a new equilibrium length. These telomere length changes occurred even though the telomerase activity was not altered, consistent with a *cis*-acting regulatory pathway. Furthermore, TRF1 levels do not affect the rate of telomere shortening in telomerase negative cells, indicating that TRF1 alters telomere length through an effect on telomere elongation (143). Hence, it was proposed that TRF1 controls the action of telomerase at each individual telomere (97). By tethering a lacI-TRF1 fusion to a subtelomeric array of lacO sites, Gilson and colleagues were able to provide direct proof for the idea that TRF1 can limit telomere elongation in cis (144).

According to the model (Figure 3*B*), a long telomere recruits a large number of TRF1 molecules, which block telomerase from adding more repeats. Conversely, a telomere that is short contains less TRF1 and has a greater chance of being elongated. As a consequence, all telomeres in a given cell line will eventually converge to a similar median telomere length setting. At this equilib-

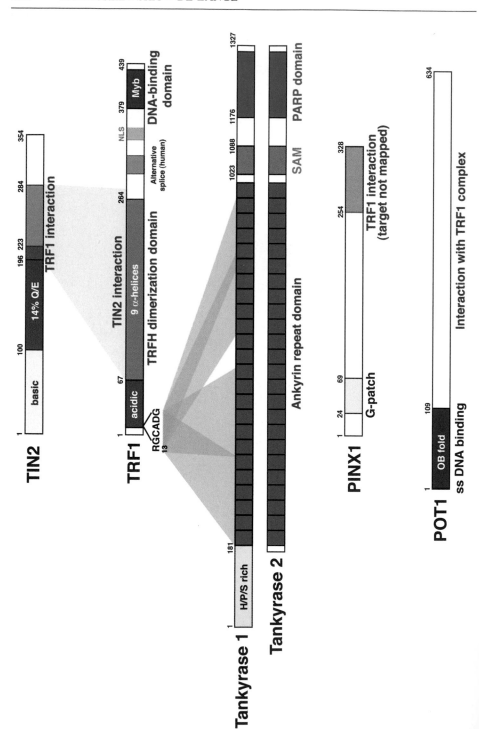

rium length, the amount of telomere-bound TRF1 is sufficient to prevent inappropriate elongation by telomerase but low enough to allow the enzyme to counteract telomere shortening. The final stable length setting in a cell population is determined by the telomerase activity, the rate of telomere shortening, and the levels of telomere length control factors such as TRF1. Changes in each of these parameters can reset telomere length to a new equilibrium. Under extreme circumstances, for instance when one of these factors is absent, no new equilibrium is reached, and telomeres are ultimately lost (e.g., in cells lacking telomerase) or display a runaway elongation phenotype in which telomere continue to lengthen [so far only observed in *K. lactis* (128)].

TRF1 Partners: Tankyrase 1 and 2, TIN2, and PINX1

The binding of TRF1 to telomeres can be inhibited by two related enzymes, tankyrase 1 and 2 (145–151) (Figure 4). The tankyrases are poly(ADP-ribose) polymerases (PARPs) that were originally identified as TRF1 interacting proteins (Figure 4). The two enzymes are nearly identical in amino acid sequence and form both homo- and heterodimers, suggesting that they are functionally similar (or identical) (148, 149, 152). It is likely that the tankyrases also have a multitude of nontelomeric functions, because they are present in the Golgi, nuclear pore complexes, and centrosomes, where they have additional interacting partners (147, 149–151, 153, 154). Tankyrases can ADP-ribosylate TRF1 in vitro, and this modification diminishes the ability of TRF1 to bind to telomeric DNA (146). Forced overexpression of tankyrase 1 in the nucleus results in removal of TRF1 from telomeres in vivo as determined by IF and ChIP (142, 145, 148). Consistent with these findings, overexpression of tankyrase 1 leads to telomere elongation (145), the phenotype seen upon TRF1 inhibition (97).

A second interacting partner of TRF1, TIN2 (Figure 4), can also affect telomere length (135). TIN2 is a small protein with no known domains apart from its C-terminal TRF1 binding domain. It can form a ternary protein complex with both TRF1 and tankyrase, and the presence of TIN2 appears to stabilize the TRF1-tankyrase interaction (J. Ye & T. de Lange, submitted). Conversely, tankyrase promotes the interaction between TRF1 and TIN2. The formation of this ternary complex may be important for the nuclear import of tankyrase. Tankyrase lacks a nuclear localization signal and is predominantly cytoplasmic. But it can be brought into the nucleus through interaction with TRF1 (154), and TIN2 may facilitate this process. In vitro, TIN2 protects TRF1 from being modified by the tankyrases, explaining the stabilizing effect of TIN2 on the TRF1-tankyrase interaction (J. Ye & T. de Lange, submitted). Furthermore, RNAi mediated inhibition of TIN2 results in loss of TRF1 from telomeres, and

Figure 4 The TRF1 telomere length regulation complex. Domain structure and features of TRF1 and its partners are shown.

this effect is reversed by 3AB, a tankyrase inhibitor. Thus, TIN2 appears to protect TRF1 from being modified by tankyrase. The modulation of tankyrase by TIN2 can explain how tankyrase can accumulate on telomeres even though the enzyme has the ability to dislodge its telomere tethering partner, TRF1.

A fourth TRF1 interacting protein, PINX1 has been proposed to affect telomere length control (155). PINX1 can inhibit telomerase in vitro, and it has been suggested that PINX1 affects telomere length by altering the telomerase activity throughout the nucleus. Such *trans*-acting control of telomerase is not consistent with the proposed role for other components of the TRF1 complex, which are thought to act in cis at individual chromosome ends. It will be interesting to see how these various mechanisms of length control are integrated. Strikingly, PINX1 is the only component of the TRF1 complex that is conserved in budding yeast. Deletion of the budding yeast ortholog of PINX1 (Gno1p) affects rRNA maturation but has no effect on telomere length (156). Human PINX1 is concentrated in the nucleolus (155), the site of both rRNA and, possibly, telomerase maturation.

So far, there is no model that integrates the effects of TRF1, TIN2, tankyrase, and PINX1 on telomere length control. To a great extent, this is due to the fact that these studies have involved overexpression strategies with the associated concern of whether certain phenotypes reflect the real function of the protein or an effect of overexpression (e.g., through titration of other factors). Only in the case of TRF1 has its role as a negative regulator of telomere length been confirmed by the opposing phenotypes of overexpression of the full length protein and a dominant negative allele. Therefore, it will be important to analyze the inhibition phenotype of the TRF1 interacting factors with alternative strategies. The fact that deletion of Trf1 from the mouse genome leads to early embryonic death (157) does not bode well for using mouse genetics to address these questions. Instead, RNAi approaches may be a better alternative.

Telomere Length Control by POT1: Connecting the TRF1 Complex to the Telomere Terminus

One of the main challenges in the dissection of telomere length control is to determine how proteins bound to the duplex telomeric DNA regulate telomerase. The dilemma is that telomerase acts at the 3′ overhang at a considerable distance from most of the regulatory factors, such as TRF1 or Rap1. A recent analysis of human POT1 has shed light on this question.

POT1 was identified based on its sequence similarity to proteins that bind to single-stranded telomeric DNA in ciliates (158). The human version of POT1 has a single-stranded DNA-binding domain in its N terminus, which allows the proteins to bind to arrays of the sequence TAGGGTTAG with great sequence specificity [158; D. Loayza, H. Parsons, K. Hoke, J. Donigian, and T. de Lange, (213)] (Figure 4). In vivo, POT1 associates with telomeres, and this binding is diminished when TRF2 is inhibited, a situation that leads to degradation of the telomeric overhang (142). These findings show that POT1 is a single-strand telomeric DNA-binding factor.

However, an N-terminal truncation form of POT1 (POT1$^{\Delta OB}$), which lacks the DNA-binding domain, can still associate with telomeres, indicating that binding to single-stranded DNA is not necessary for the association of POT1 with telomeres (142). This second mechanism for telomere association depends on an interaction of POT1 with the TRF1 complex and is proposed to be crucial for telomere length control (142). Endogenous POT1 can be removed from telomeres through inhibition of TRF1, and as was shown for the TRF1 complex, longer telomeres contain more POT1. These data are consistent with POT1 being recruited to the telomeric chromatin by the TRF1 complex and indicate that POT1, like the TRF1 complex, could function as a protein-counting device to measure telomere length.

The role of POT1 in telomere length homeostasis is apparent from the telomere elongation phenotype of POT1$^{\Delta OB}$. When this mutant is expressed, the endogenous POT1 is repressed (through an unknown mechanism) so that the only version of POT1 at telomeres is the POT1$^{\Delta OB}$ protein. Telomerase positive cells expressing POT1$^{\Delta OB}$ show immediate and extensive telomere elongation. Their telomeres grow from a median of ~6 kb to 20 kb in the course of 40 PD, which is an unusually high rate of telomere elongation and suggests a complete lack of telomerase inhibition. This elongation occurs even though the telomeres contain large amounts of TRF1 and its interacting proteins. Apparently, the displacement of full-length POT1 by POT1$^{\Delta OB}$ has abrogated the ability of the TRF1 complex to control telomerase. It was therefore proposed that POT1 functions downstream of the TRF1 complex to relay the negative regulation to the telomere terminus (142).

The model for POT1-mediated telomere length control proposes that the loading of POT1 on the single-stranded telomeric DNA inhibits telomerase from elongating the telomere (Figure 5A). As telomeres get longer, more TRF1 complex is present at the chromosome end, increasing the chance of POT1 being present on the single-stranded telomeric DNA where it would block telomerase. The model is based on the finding that the binding of POT1 to telomeres is greatly improved by its association with the TRF1 complex present on the double-stranded telomeric repeat array. For instance, using ChIP, it was found that removal of the TRF1 complex also diminished the association of POT1 with telomeres, even though the length of the single-stranded telomeric DNA was unaffected (142). Thus, through TRF1-mediated loading, POT1 could function to transduce information about the length of the telomere to the telomere terminus.

How does POT1 inhibit telomerase? It could be as simple as blocking access to the 3′ end (Figure 5A). POT1 has some preference to bind to its recognition site at a 3′ end [158, 159; D. Loayza, H. Parsons, K. Hoke, J. Donigian, and T. de Lange, (213)], and its physical presence there may simply preclude telomerase from accessing the end. A second model is based on the unusual architecture of telomeres (Figure 5B). Mammalian telomeres have been observed in an altered

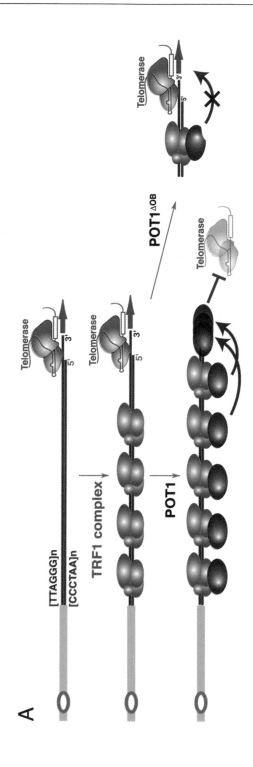

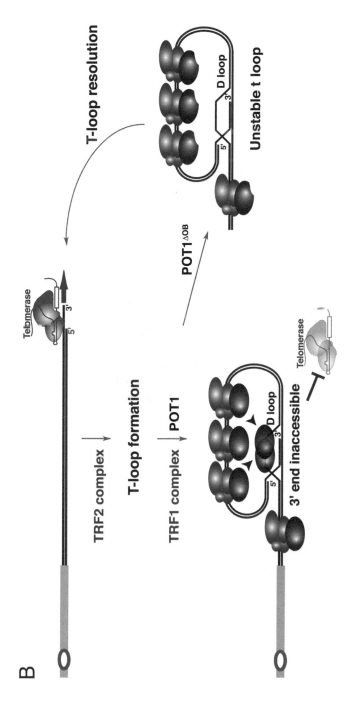

Figure 5 Proposed role for POT1 in telomere length regulation by the TRF1 complex.

conformation, called t-loops (160), which are large duplex loops formed through the strand-invasion of the G-strand overhang into the duplex part of the telomere. How t loops are created in vivo is not yet known, but in vitro, TRF2 can remodel telomeric DNA into t-loop like structures (161). In the t-loop configuration, the single-stranded 3′ overhang of the telomere terminus is thought to be base-paired to the C-strand sequences. Because telomerase requires an unpaired 3′ end (17, 162), the telomere terminus is unlikely to be accessible to telomerase when telomeres are in t loops. Based on its biochemical features, POT1 should have the ability to bind to the displaced TTAGGG repeats at the base of the t loop (the D loop) (213). Potentially the binding of POT1 to the D loop could stabilize t loops (e.g., by preventing branch-migration) and thereby block telomerase from gaining access to the 3′ telomere terminus. Both models explain why POT1$^{\Delta OB}$ abrogates the ability of the TRF1 complex to control telomere length. Although the TRF1 complex can still recruit POT1$^{\Delta OB}$ to the telomeric chromatin, this form of POT1 can not inhibit telomerase because it lacks single-stranded DNA-binding activity (Figure 5A and B).

POT1 is similar to Cdc13 in *S. cerevisiae* in that they both use an OB-fold to bind to single-stranded telomeric DNA (158, 163). This is a protein motif that is used to recognize single-stranded nucleic acids in numerous settings, which include DNA replication (e.g., replication protein A). As discussed above, the main Cdc13 functions are to protect chromosome ends from degradation and to recruit telomerase. POT1 may be functionally similar because deficiency in pot1 in *Schizosaccharomyces pombe* results in rapid telomere loss (158). Furthermore, human POT1 has been proposed to play a role in telomerase recruitment on the basis of studies in which transfected POT1 induced telomere elongation in a subset of clones (164). For both the protective role of human POT1 and to establish whether POT1 is necessary to recruit telomerase, it will be imperative to execute POT1 inhibition studies, for instance using RNAi. Conversely, it will be of interest to establish whether Cdc13 can act as a transducer for the Rap1-dependent telomere length control pathway in *S. cerevisiae*. Given that the *cdc13–5* mutant results in telomere elongation (92), this is a possibility worth pursuing.

FROM YEAST TO MAN: DRASTIC CHANGES IN THE TELOMERE LENGTH CONTROL COMPLEX

Although mammalian TRF1 and Rap1 of *S. cerevisiae* both bind to duplex telomeric DNA and function to control telomere length, these proteins are not orthologs. In fact, the budding yeast protein most closely related to TRF1, Tbf1 (165–167), has not been implicated in telomere biology, and genes for TIN2 and tankyrase are absent from the budding yeast genome. By contrast, mammalian cells do contain a Rap1 ortholog, hRap1, which binds to telomeres and affects their length (168, 169). Overexpression of hRap1 can result in telomere short ening, suggesting that hRap1, like scRap1, is a negative regulator of telomere maintenance. Several hRap1 truncation mutants have a telomere elongation phenotype, perhaps because they act as dominant negative alleles (169). Remarkably, hRap1 is not a DNA-binding protein, and its association with telomeres depends on interaction with TRF2, a TRF1 paralog (168) (Figure 6). TRF2 is essential for the protection of chromosome ends [reviewed in (10)] and also contributes to the length regulation of telomeres (136), probably in part through its interaction with hRap1. For human Rap1, the C terminus functions to recruit the protein to telomeres, and telomere length regulation is dependent on the Myb domain and the N-terminal BRCT domain (169). By contrast, in *S. cerevisiae* Rap1, the Myb domain tethers Rap1 to telomeres, and the C terminus recruits the telomere length regulators Rif1 and Rif2. A human ortholog of the yeast Rap1 interacting factor, Rif1, was recently identified, but there is as yet no indication that this protein is a Rap1 interacting factor or associates with telomeres. Rather, Rif1 plays an important role in the ATM-dependent DNA damage response (J. Silverman, H. Takai, S. Buonomo, and T. de Lange, submitted).

TRF1 does have an ortholog in *S. pombe*, the telomere binding protein Taz1 (170) (Figure 6). Like TRF1 and scRap1, Taz1p is a negative regulator of telomere length with taz1-strains showing dramatic telomere elongation (170). Interestingly, Taz1 interacts with the fission yeast ortholog of Rap1 as well as with an ortholog of Rif1p, which is not a Rap1 interacting factor in this organism (171, 172). Both spRap1 and spRif1 behave as negative regulators of telomere length, indicating functional conservation.

Thus, both in mammals and in *S. pombe*, the telomeric complex is built upon a TRF-like factor that interacts with Rap1, whereas in *S. cerevisiae*, Rap1 binds directly to telomeric DNA, and there is no TRF-like protein at telomeres (Figure 6). *S. cerevisiae* Rap1 must have evolved the ability to function as a protein-counting device for telomere length measurements, perhaps co-opting Rif1 and Rif2 in the task of controlling telomerase. With regard to this remarkable rearrangement in the telomeric complex, it will be of interest to learn about the fate (or origin) of Rif2, but so far no orthologs have been found outside the budding yeasts.

This drastic change in architecture of the telomeric complex may relate to the altered sequence of *S. cerevisiae* telomeres (Figure 6). Most eukaryotes have TTAGGG repeats as telomeric DNA or carry a closely related sequence (for instance, TTAGGC in worms, TTTAGGG in plants, and TTACAGG in *S. pombe*), indicating a high degree of sequence conservation in the telomerase template region and a coupled conservation of the telomeric DNA recognition factors. However, the budding yeasts have noncanonical telomeric repeats that are often highly irregular in sequence and diverge very rapidly. We have proposed that the budding yeasts experienced a telomerase catastrophy in which the template region was altered leading to telomeres with a different sequence that could not bind the cognate TRF-like factor (168). The model suggests that these altered telomeric sequences bound budding yeast Rap1 and other essential

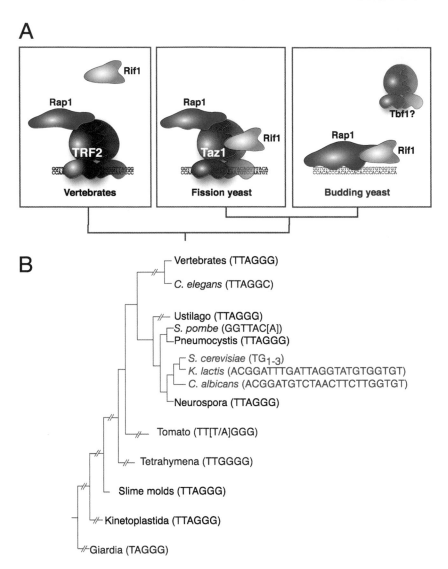

Figure 6 Evolution of the telomeric complex [modified from (168)]. *A*. Relationship between telomeric proteins in vertebrates, fission yeast, and budding yeast. TRF2, Taz1, and Tbf1 are structurally related. In budding yeast, Rap1 binds telomeric DNA, whereas human and fission yeast Rap1 bind to telomeres via TRF2 and Taz1, respectively. Budding yeast Rif1 binds to Rap1, whereas fission yeast Rif1 binds to Taz1. The role of human Rif1 at telomeres has not been established. *B*. Exceptional telomeric DNA in the budding yeasts. A selection of eukaryotes, their approximate evolutionary relationship, and their telomeric sequences are shown. Most eukaryotes have telomeres with TTAGGG repeats or closely related sequences. The budding yeasts (*blue*) stand out as having noncanonical telomeres that rapidly diverge.

factors mediating telomere function. The now dispensable TRF-like binding module was eventually lost or altered. Budding yeast Tbf1 may well be a vestigial TRF that has lost its place in the telomeric complex. In agreement with this scenario, present day budding yeast telomeres can be converted to TTAGGG repeats without loss of viability or length regulation (173, 174).

DNA DAMAGE RESPONSE PATHWAYS AND THE CONTROL OF TELOMERE MAINTENANCE

Several lines of evidence indicate that telomere maintenance is influenced by components of the DNA damage response pathway. The simplest interpretation is that telomeres resemble damaged DNA (perhaps during or immediately after their replication) and that the associated activation of the DNA damage response pathway regulates telomere maintenance by telomerase. In budding and fission yeast, telomere maintenance is strictly dependent on the presence of one of the two DNA damage response kinases, *TEL1* and *MEC1* in *S. cerevisiae* (175) and *tel1+* and *rad3+* in *S. pombe* (176). When one of the two kinases is missing, telomeres are short (more so for *tel1Δ*) but stable (177, 178). When both are absent, an est phenotype is seen (175, 176). Epistasis analysis showed that *TEL1* acts together with *MRE11/RAD50/XRS2* in telomere maintenance (91, 179), a relationship that is reminiscent of the TM (Tel1/Mre11 complex) checkpoint for double-stranded breaks (180, 181). Perhaps the TM checkpoint needs to detect unprocessed telomere ends in order for telomerase to become activated. If this checkpoint is not available, the Mec1p checkpoint can compensate.

What are the targets of Tel1 and Mec1 in the telomere maintenance pathway? It is unlikely that telomerase itself is regulated because *tel1Δ mec1Δ* cells have wild-type levels of the enzymatic activity (132). The DNA repair function of the Mre11 complex can be enhanced by a Tel1-dependent DSB signal (180), so a more likely candidate is the Mre11 complex. Perhaps phosphorylation stimulates the ability of the Mre11 complex to facilitate Cdc13p loading. As discussed above, genetic evidence indicates that Tel1 and Mec1 are affected by (or have an effect on) the Rap1 complex, but the details of this interaction are yet to be resolved.

It is also possible that Rif1 and Rif2 are targets of Tel1 and Mec1 signaling. If Rif1 and Rif2 are absent, yeast can maintain telomeric DNA without the help of Tel1 and Mec1 (132), raising the possibility that these kinases counteract inhibitory effects of Rif1 and Rif2. The recent finding that human Rif1 is regulated by the Tel1 ortholog *ATM* makes this scenario particularly attractive (J. Silverman, H. Takai, S. Buonomo, and T. de Lange, submitted). However, other studies place Rif1 and Rif2 upstream of Tel1 in the telomere length control pathway (131).

The connection between telomeres and the DNA damage response is not limited to *TEL1/MEC1* signaling. Telomere maintenance and length control in budding yeast is also influenced by a number of other DNA damage response factors, which include Rad17, Rad53, Mec3, and Ddc1, and a similar set of genes affects telomere length in fission yeast (182–184). Furthermore, both in budding and fission yeast, deletion of Ku leads to very short but stable telomeres (91, 185–189). This phenotype is not related to the deficiency in NHEJ, because loss of DNA ligase IV does not affect telomere length (185, 187).

As many aspects of the DNA damage response pathways are highly conserved, and much of what is learned in *S. cerevisiae* and *S. pombe* will be a guide for studies of other eukaryotes, which include mammals. However, some critical aspects of both the DNA response pathway and the telomeric complex have changed over the course of evolution, so it is important to verify all regulatory pathways in each species. Several lines of evidence implicate the DNA damage response pathway in telomere length control in mammals. Peripheral blood lymphocytes from ataxia telangiectasia (A-T) patients show significant telomere shortening compared to age-matched normal donors (190) and *ATM* deficient mice have slightly shortened telomeres as well as extrachromosomal telomeric DNA (191). However, because no telomere maintenance defect was found in an extensive survey of telomerase-positive A-T cell lines (192), the ATM kinase may not play a role in the telomerase pathway per se but may affect other parameters that result in telomere length changes (e.g., telomere shortening rates).

The case is also not clear for the role of Ku70/80, DNA-PKcs, and PARP-1 in telomere length control. Although *DNA-PKcs* null mice have normal telomere length (193, 194), mice with the *DNA-PKcs* SCID mutation have been reported to have longer telomeres (193, 195). Perhaps the SCID mutation, though deficient in NHEJ, may be a gain of function mutation for the role of DNA-PKcs in telomere length control. It is much harder to explain discrepancies in reports on Ku80. Even though two groups measured telomere lengths in the same *Ku86*[-/-] mouse strain, one report shows shorter telomeres (196), and the other found no change (197). Finally, mice lacking *PARP-1* have shorter telomeres, but this phenotype is only seen when *p53* is also absent (198, 199).

Clearly, dissection of these pathways in mammals is at an early stage. Similarly, the role of DNA damage response genes in telomere length maintenance in organisms, as diverged as trypanosomes (Ku) (200), worms (Mrt2 and Hus1) (201–203), and plants (Ku and Mre11) (204–206), is now beginning to be addressed.

A final gene that merits discussion in the context of the DNA damage response is *TEL2*. This gene was among the first telomere length regulators identified in *S. cerevisiae* (178), yet its role in telomere maintenance has remained somewhat of a mystery. Budding yeast *TEL2* is an essential gene that encodes a protein that can bind double-stranded and single-stranded TG1–3 sequences (207–209). A mutant allele of *TEL2* (*tel2–1*) gives rise to moderately shortened telomeres (178). Tel2 appears to act in the Tel1 pathway because *tel2–1 tel1–1* double mutant cells have a telomere length defect typical of *TEL1* deficiency (178). Furthermore, consistent with Tel2 functioning in the Tel1 pathway, perturbations in the Rap1p telomere length control do not reset telomere length in *tel2–1* cells (207).

The *Caenorhabditis elegans* ortholog of *TEL2* is *rad-5*, a gene required for the DNA damage checkpoints in this organism (202). Like other checkpoint mutants, *rad-5* mutants do not undergo apoptosis after irradiation, and their germline cells fail to arrest in response to DNA damage (202). This phenotype is also seen with *mrt-2* worms, mutant for the *C. elegans* ortholog of *RAD17*, and with worms lacking normal *hus-1* function (op241) (201–203, 210). *Rad-5* is allelic with *clk-2*, a gene that affects biological rhythms and life span (202, 211, 212). By contrast to what is observed in the *S. cerevisiae tel2–1* mutant, worms with either the *rad-5* or *clk-2* mutation have normal telomere length (202). Conversely, *tel2–1* has no checkpoint defect (202). Because each of these mutations map to different parts of the Tel2 protein, further mutational dissection will be required to determine whether *TEL2/rad-5* is a conserved telomere length regulator and DNA damage checkpoint gene. The link between DNA damage response and telomere maintenance does exist in *C. elegans*; both *hus-1* and *mrt-2* have an est-like progressive telomere loss phenotype (201, 203). A human ortholog of *TEL2* has been identified, but its function at telomeres has not been established.

ACKNOWLEDGMENTS

We apologize to our colleagues whose work was not cited here due to space limitations. We are indebted to Kristina Hoke, Diego Loayza, Richard Wang, Joshua Silverman, Jeffrey Ye, and other members of the de Lange lab for comments on this manuscript. TdL is grateful to Vicki Lundblad, David Shore, Alessandro Bianchi, Roger Reddel, Kurt Runge, Jack Griffith, Ginger Zakian, Eric Gilson, Joachim Lingner, Julie Cooper, Mundy Wellinger, Tom Petes, Carol Greider, and Lea Harrington for discussion and communication of unpublished data. Work on telomere length regulation in the de Lange laboratory is supported by a grant from the NCI (CA76027) and a Burroughs Wellcome Toxicology Scholar Award. AS was supported by NIH MSTP grant to the Weill Medical College of Cornell University/RU/MSKCC Tri-Institutional MD/PhD program and by a training grant to RU.

The *Annual Review of Biochemistry* is online at http://biochem.annualreviews.org

LITERATURE CITED

1. Watson JD. 1972. *Nat. New Biol.* 239: 197–201
2. Olovnikov AM. 1973. *J. Theor. Biol.* 41: 181–90
3. Lundblad V, Szostak JW. 1989. *Cell* 57:633–43
4. Levis RW. 1989. *Cell* 58:791–801
5. Johnson FB, Marciniak RA, McVey M, Stewart SA, Hahn WC, Guarente L.

2001. *EMBO J.* 20:905–13
6. Walter MF, Bozorgnia L, Maheshwari A, Biessmann H. 2001. *Insect. Mol. Biol.* 10:105–10
7. Niida H, Matsumoto T, Satoh H, Shiwa M, Tokutake Y, et al. 1998. *Nat. Genet.* 19:203–6
8. Blasco MA, Lee HW, Hande MP, Samper E, Lansdorp PM, et al. 1997.

Cell 91:25–34

9. Harley CB, Futcher AB, Greider CW. 1990. *Nature* 345:458–60
10. de Lange T. 2002. *Oncogene* 21:532–40
11. Blackburn EH. 2001. *Cell* 106:661–73
12. Vulliamy T, Marrone A, Goldman F, Dearlove A, Bessler M, et al. 2001. *Nature* 413:432–35
13. Cawthon RM, Smith KR, O'Brien E, Sivatchenko A, Kerber RA. 2003. *Lancet* 361:393–95
14. Maser RS, DePinho RA. 2002. *Science* 297:565–69
15. Wright WE, Shay JW. 2000. *Nat. Med.* 6:849–51
16. Greider CW, Blackburn EH. 1985. *Cell* 43:405–13
17. Greider CW, Blackburn EH. 1987. *Cell* 51:887–98
18. Nakamura TM, Morin GB, Chapman KB, Weinrich SL, Andrews WH, et al. 1997. *Science* 277:955–59
19. Lingner J, Hughes TR, Shevchenko A, Mann M, Lundblad V, Cech TR. 1997. *Science* 276:561–67
20. Meyerson M, Counter CM, Eaton EN, Ellisen LW, Steiner P, et al. 1997. *Cell* 90:785–95
21. Singer MS, Gottschling DE. 1994. *Science* 266:404–9
22. Shippen-Lentz D, Blackburn EH. 1990. *Science* 247:546–52
23. Greider CW, Blackburn EH. 1989. *Nature* 337:331–37
24. Feng J, Funk WD, Wang SS, Weinrich SL, Avilion AA, et al. 1995. *Science* 269:1236–41
25. Nakamura TM, Cech TR. 1998. *Cell* 92:587–90
26. Romero DP, Blackburn EH. 1991. *Cell* 67:343–53
27. Chen JL, Blasco MA, Greider CW. 2000. *Cell* 100:503–14
28. Vermeesch JR, Williams D, Price CM. 1993. *Nucleic Acids Res.* 21:5366–71
29. Vermeesch JR, Price CM. 1994. *Mol. Cell. Biol.* 14:554–66
30. Ray S, Karamysheva Z, Wang LB, Ship-

pen DE, Price CM. 2002. *Mol. Cell. Biol.* 22:5859–68
31. Price CM. 1997. *Biochemistry-Moscow* 62:1216–23
32. Jacob NK, Kirk KE, Price CM. 2003. *Mol. Cell* 11:1021–32
33. Jacob NK, Skopp R, Price CM. 2001. *EMBO J.* 20:4299–308
34. Kim NW, Piatyszek MA, Prowse KR, Harley CB, West MD, et al. 1994. *Science* 266:2011–15
35. Steinert S, White DM, Zou Y, Shay JW, Wright WE. 2002. *Exp. Cell Res.* 272: 146–52
36. Cong YS, Wright WE, Shay JW. 2002. *Microbiol. Mol. Biol. Rev.* 66:407–25
37. Lin SY, Elledge SJ. 2003. *Cell* 113: 881–89
38. Ducrest AL, Szutorisz H, Lingner J, Nabholz M. 2002. *Oncogene* 21:541–52
39. Ramirez RD, Morales CP, Herbert BS, Rohde JM, Passons C, et al. 2001. *Genes Dev.* 15:398–403
40. Vaziri H, Benchimol S. 1998. *Curr. Biol.* 8:279–82
41. Bodnar AG, Ouellette M, Frolkis M, Holt SE, Chiu CP, et al. 1998. *Science* 279:349–52
42. Morales CP, Holt SE, Ouellette M, Kaur KJ, Yan Y, et al. 1999. *Nat. Genet.* 21:115–18
43. Hahn WC, Dessain SK, Brooks MW, King JE, Elenbaas B, et al. 2002. *Mol. Cell. Biol.* 22:2111–23
44. Hahn WC, Counter CM, Lundberg AS, Beijersbergen RL, Brooks MW, Weinberg RA. 1999. *Nature* 400:464–68
45. Seger YR, Garcia-Cao M, Piccinin S, Cunsolo CL, Doglioni C, et al. 2002. *Cancer Cell* 2:401–13
46. Hiyama E, Hiyama K, Yokoyama T, Matsuura Y, Piatyszek MA, Shay JW. 1995. *Nat. Med.* 1:249–55
47. Wong JMY, Kusdra L, Collins K. 2002. *Nat. Cell Biol.* 4:731–36
48. Teixeira MT, Forstemann K, Gasser SM, Lingner J. 2002. *EMBO Rep.* 3:652–59

49. Taggart AK, Teng SC, Zakian VA. 2002. *Science* 297:1023–26
50. Smith CD, Smith DL, DeRisi JL, Blackburn EH. 2003. *Mol. Biol. Cell* 14:556–70
51. Marcand S, Brevet V, Mann C, Gilson E. 2000. *Curr. Biol.* 10:487–90
52. Diede SJ, Gottschling DE. 1999. *Cell* 99:723–33
53. Seto AG, Livengood AJ, Tzfati Y, Blackburn EH, Cech TR. 2002. *Genes Dev.* 16:2800–12
54. Lendvay TS, Morris DK, Sah J, Balasubramanian B, Lundblad V. 1996. *Genetics* 144:1399–412
55. Hughes TR, Evans SK, Weilbaecher RG, Lundblad V. 2000. *Curr. Biol.* 10:809–12
56. Cohn M, Blackburn EH. 1995. *Science* 269:396–400
57. Lingner J, Cech TR, Hughes TR, Lundblad V. 1997. *Proc. Natl. Acad. Sci. USA* 94:11190–95
58. Seto AG, Zaug AJ, Sobel SG, Wolin SL, Cech TR. 1999. *Nature* 401:177–80
59. Snow BE, Erdmann N, Cruickshank J, Goldman H, Gill RM, et al. 2003. *Curr. Biol.* 13:698–704
60. Reichenbach P, Hoss M, Azzalin CM, Nabholz M, Bucher P, Lingner J. 2003. *Curr. Biol.* 13:568–74
61. Lundblad V. 2003. *Curr. Biol.* 13: R439–41
62. Chiu SY, Serin G, Ohara O, Maquat LE. 2003. *RNA* 9:77–87
63. Mitchell JR, Wood E, Collins K. 1999. *Nature* 402:551–55
64. Luzzatto L, Karadimitris A. 1998. *Nat. Genet.* 19:6–7
65. Dokal I. 2000. *Br. J. Haematol.* 110: 768–79
66. Dokal I, Bungey J, Williamson P, Oscier D, Hows J, Luzzatto L. 1992. *Blood* 80:3090–96
67. Ruggero D, Grisendi S, Piazza F, Rego E, Mari F, et al. 2003. *Science* 299: 259–62
68. Hathcock KS, Hemann MT, Opperman KK, Strong MA, Greider CW, Hodes RJ.

2002. *Proc. Natl. Acad. Sci. USA* 99: 3591–96
69. Yamaguchi H, Baerlocher GM, Lansdorp PM, Chanock SJ, Nunez O, et al. 2003. *Blood* 102(3):916–18
70. Vulliamy T, Marrone A, Dokal I, Mason PJ. 2002. *Lancet* 359:2168–70
71. Ford LP, Wright WE, Shay JW. 2002. *Oncogene* 21:580–83
72. Collins K, Mitchell JR. 2002. *Oncogene* 21:564–79
73. Lundblad V. 2002. *Oncogene* 21:522–31
74. Reddel RR. 2003. *Cancer Lett.* 194: 155–62
75. Lustig AJ. 2003. *Nat. Rev. Genet.* 4: 916–23
76. Evans SK, Lundblad V. 2000. *J. Cell Sci.* 113:3357–64
77. Garvik B, Carson M, Hartwell L. 1995. *Mol. Cell. Biol.* 15:6128–38
78. Nugent CI, Hughes TR, Lue NF, Lundblad V. 1996. *Science* 274:249–52
79. Lin JJ, Zakian VA. 1996. *Proc. Natl. Acad. Sci. USA* 93:13760–65
80. Pennock E, Buckley K, Lundblad V. 2001. *Cell* 104:387–96
81. Evans SK, Lundblad V. 1999. *Science* 286:117–20
82. Evans SK, Lundblad V. 2002. *Genetics* 162:1101–15
83. Peterson SE, Stellwagen AE, Diede SJ, Singer MS, Haimberger ZW, et al. 2001. *Nat. Genet.* 27:64–67
84. Wellinger RJ, Wolf AJ, Zakian VA. 1993. *Cell* 72:51–60
85. Maser RS, Bressan DA, and Petrini JHJ. 2001. In *DNA Damage and Repair*, ed. MF Hoekstra, JA Nickoloff, pp. 147–72. Totowa, NJ: Humana
86. Haber JE. 1998. *Cell* 95:583–86
87. Diede SJ, Gottschling DE. 2001. *Curr. Biol.* 11:1336–40
88. Trujillo KM, Yuan SS, Lee EY, Sung P. 1998. *J. Biol. Chem.* 273:21447–50
89. Moreau S, Ferguson JR, Symington LS. 1999. *Mol. Cell. Biol.* 19:556–66
90. Tsukamoto Y, Taggart AK, Zakian VA. 2001. *Curr. Biol.* 11:1328–35

91. Nugent CI, Bosco G, Ross LO, Evans SK, Salinger AP, et al. 1998. *Curr. Biol.* 8:657–60

92. Chandra A, Hughes TR, Nugent CI, Lundblad V. 2001. *Genes Dev.* 15:404–14

93. Grandin N, Reed SI, Charbonneau M. 1997. *Genes Dev.* 11:512–27

94. Qi H, Zakian VA. 2000. *Genes Dev.* 14:1777–88

95. Counter CM, Avilion AA, LeFeuvre CE, Stewart NG, Greider CW, et al. 1992. *EMBO J.* 11:1921–29

96. Shampay J, Szostak JW, Blackburn EH. 1984. *Nature* 310:154–57

97. van Steensel B, de Lange T. 1997. *Nature* 385:740–43

98. Zhu L, Hathcock KS, Hande P, Lansdorp PM, Seldin MF, Hodes RJ. 1998. *Proc. Natl. Acad. Sci. USA* 95:8648–53

99. Starling JA, Maule J, Hastie ND, Allshire RC. 1990. *Nucleic Acids Res.* 18:6881–88

100. Shampay J, Blackburn EH. 1988. *Proc. Natl. Acad. Sci. USA* 85:534–38

101. Barnett MA, Buckle VJ, Evans EP, Porter AC, Rout D, et al. 1993. *Nucleic Acids Res.* 21:27–36

102. Hanish JP, Yanowitz JL, de Lange T. 1994. *Proc. Natl. Acad. Sci. USA* 91:8861–65

103. Sprung CN, Afshar G, Chavez EA, Lansdorp P, Sabatier L, Murnane JP. 1999. *Mutat. Res.* 429:209–23

104. Sprung CN, Reynolds GE, Jasin M, Murnane JP. 1999. *Proc. Natl. Acad. Sci. USA* 96:6781–86

105. Marcand S, Brevet V, Gilson E. 1999. *EMBO J.* 18:3509–19

106. Shore D, Nasmyth K. 1987. *Cell* 51:721–32

107. Shore D. 1994. *Trends Genet.* 10:408–12

108. Konig P, Giraldo R, Chapman L, Rhodes D. 1996. *Cell* 85:125–36

109. Gilson E, Roberge M, Giraldo R, Rhodes D, Gasser SM. 1993. *J. Mol. Biol.* 231:293–310

110. Longtine MS, Wilson NM, Petracek ME, Berman J. 1989. *Curr. Genet.* 16:225–39

111. Conrad MN, Wright JH, Wolf AJ, Zakian VA. 1990. *Cell* 63:739–50

112. Lustig AJ, Kurtz S, Shore D. 1990. *Science* 250:549–53

113. Hardy CF, Sussel L, Shore D. 1992. *Genes Dev.* 6:801–14

114. Kyrion G, Boakye KA, Lustig AJ. 1992. *Mol. Cell. Biol.* 12:5159–73

115. Kyrion G, Liu K, Liu C, Lustig AJ. 1993. *Genes Dev.* 7:1146–59

116. Liu C, Mao X, Lustig AJ. 1994. *Genetics* 138:1025–40

117. Moretti P, Freeman K, Coodly L, Shore D. 1994. *Genes Dev.* 8:2257–69

118. Buck SW, Shore D. 1995. *Genes Dev.* 9:370–84

119. Liu C, Lustig AJ. 1996. *Genetics* 143:81–93

120. Marcand S, Gilson E, Shore D. 1997. *Science* 275:986–90

121. Wotton D, Shore D. 1997. *Genes Dev.* 11:748–60

122. Ray A, Runge KW. 1998. *Mol. Cell. Biol.* 18:1284–95

123. Callebaut I, Mornon JP. 1997. *FEBS Lett.* 400:25–30

124. Grossi S, Bianchi A, Damay P, Shore D. 2001. *Mol. Cell. Biol.* 21:8117–28

125. Ray A, Runge KW. 1999. *Mol. Cell. Biol.* 19:31–45

126. Krauskopf A, Blackburn EH. 1996. *Nature* 383:354–57

127. Krauskopf A, Blackburn EH. 1998. *Proc. Natl. Acad. Sci. USA* 95:12486–91

128. McEachern MJ, Blackburn EH. 1995. *Nature* 376:403–9

129. McEachern MJ, Underwood DH, Blackburn EH. 2002. *Genetics* 160:63–73

130. Ray A, Runge KW. 1999. *Proc. Natl. Acad. Sci. USA* 96:15044–49

131. Craven RJ, Petes TD. 1999. *Genetics* 152:1531–41

132. Chan SW, Chang J, Prescott J, Blackburn EH. 2001. *Curr. Biol.* 11:1240–50

133. de Lange T. 1995. In *Telomeres*, ed. EH Blackburn, CW Greider, pp. 265–93. Cold Spring Harbor, NY: Cold Spring Harbor Lab. Press

134. Bryan TM, Englezou A, Dunham MA, Reddel RR. 1998. *Exp. Cell Res.* 239: 370–78

135. Kim SH, Kaminker P, Campisi J. 1999. *Nat. Genet.* 23:405–12

136. Smogorzewska A, van Steensel B, Bianchi A, Oelmann S, Schaefer MR, et al. 2000. *Mol. Cell. Biol.* 20:1659–68

137. Chong L, van Steensel B, Broccoli D, Erdjument-Bromage H, Hanish J, et al. 1995. *Science* 270:1663–67

138. Bianchi A, Smith S, Chong L, Elias P, de Lange T. 1997. *EMBO J.* 16:1785–94

139. Bianchi A, Stansel RM, Fairall L, Griffith JD, Rhodes D, de Lange T. 1999. *EMBO J.* 18:5735–44

140. Zhong Z, Shiue L, Kaplan S, de Lange T. 1992. *Mol. Cell. Biol.* 12:4834–43

141. Konig P, Fairall L, Rhodes D. 1998. *Nucleic Acids Res.* 26:1731–40

142. Loayza D, de Lange T. 2003. *Nature* 423:1013–18

143. Karlseder J, Smogorzewska A, de Lange T. 2002. *Science* 295:2446–49

144. Ancelin K, Brunori M, Bauwens S, Koering CE, Brun C, et al. 2002. *Mol. Cell. Biol.* 22:3474–87

145. Smith S, de Lange T. 2000. *Curr. Biol.* 10:1299–302

146. Smith S, Giriat I, Schmitt A, de Lange T. 1998. *Science* 282:1484–87

147. Kaminker PG, Kim SH, Taylor RD, Zebarjadian Y, Funk WD, et al. 2001. *J. Biol. Chem.* 276:35891–99

148. Cook BD, Dynek JN, Chang W, Shostak G, Smith S. 2002. *Mol. Cell. Biol.* 22: 332–42

149. Sbodio JI, Lodish HF, Chi NW. 2002. *Biochem. J.* 361:451–59

150. Sbodio JI, Chi NW. 2002. *J. Biol. Chem.* 277:31887–92

151. Seimiya H, Smith S. 2002. *J. Biol. Chem.* 277(16):14116–26

152. de Rycker M, Venkatesan RN, Wei C, Price CM. 2003. *Biochem. J.* 372(1): 87–96

153. Chi NW, Lodish HF. 2000. *J. Biol. Chem.* 275(49):38437–44

154. Smith S, de Lange T. 1999. *J. Cell Sci.* 112:3649–56

155. Zhou XZ, Lu KP. 2001. *Cell* 107:347–59

156. Guglielmi B, Werner M. 2002. *J. Biol. Chem.* 277(38):35712–19

157. Karlseder J, Kachatrian L, Takai H, Mercer K, Hingorani S, et al. 2003. *Mol. Cell. Biol.* 23:6533–41

158. Baumann P, Cech TR. 2001. *Science* 292:1171–75

159. Lei M, Baumann P, Cech TR. 2002. *Biochemistry* 41:14560–68

160. Griffith JD, Comeau L, Rosenfield S, Stansel RM, Bianchi A, et al. 1999. *Cell* 97:503–14

161. Stansel RM, de Lange T, Griffith JD. 2001. *EMBO J.* 20:E5532–40

162. Lingner J, Cech TR. 1996. *Proc. Natl. Acad. Sci. USA* 93:10712–17

163. Mitton-Fry RM, Anderson EM, Hughes TR, Lundblad V, Wuttke DS. 2002. *Science* 296:145–47

164. Colgin LM, Baran K, Baumann P, Cech TR, Reddel RR. 2003. *Curr. Biol.* 13: 942–46

165. Bilaud T, Koering CE, Binet-Brasselet E, Ancelin K, Pollice A, et al. 1996. *Nucleic Acids Res.* 24:1294–303

166. Brigati C, Kurtz S, Balderes D, Vidali G, Shore D. 1993. *Mol. Cell. Biol.* 13:1306–14

167. Koering CE, Fourel G, Binet-Brasselet E, Laroche T, Klein F, Gilson E. 2000. *Nucleic Acids Res.* 28:2519–26

168. Li B, Oestreich S, de Lange T. 2000. *Cell* 101:471–83

169. Li B, de Lange T. 2003. *Mol. Biol. Cell.* 12:5060–68

170. Cooper JP, Nimmo ER, Allshire RC, Cech TR. 1997. *Nature* 385:744–47

171. Chikashige Y, Hiraoka Y. 2001. *Curr. Biol.* 11:1618–23

172. Kanoh J, Ishikawa F. 2001. *Curr. Biol.* 11:1624–30

173. Alexander MK, Zakian VA. 2003. *EMBO J.* 22:1688–96

174. Brevet V, Berthiau AS, Civitelli L,

Donini P, Schramke V, et al. 2003. *EMBO J.* 22:1697–706

175. Ritchie KB, Mallory JC, Petes TD. 1999. *Mol. Cell. Biol.* 19:6065–75
176. Naito T, Matsuura A, Ishikawa F. 1998. *Nat. Genet.* 20:203–6
177. Craven RJ, Petes TD. 2000. *Mol. Cell. Biol.* 20:2378–84
178. Lustig AJ, Petes TD. 1986. *Proc. Natl. Acad. Sci. USA* 83:1398–402
179. Ritchie KB, Petes TD. 2000. *Genetics* 155:475–79
180. Usui T, Ogawa H, Petrini JH. 2001. *Mol. Cell* 7:1255–66
181. D'Amours D, Jackson SP. 2001. *Genes Dev.* 15:2238–49
182. Nakamura TM, Moser BA, Russell P. 2002. *Genetics* 161:1437–52
183. Longhese MP, Paciotti V, Neecke H, Lucchini G. 2000. *Genetics* 155:1577–91
184. Dahlen M, Olsson T, Kanter-Smoler G, Ramne A, Sunnerhagen P. 1998. *Mol. Biol. Cell* 9:611–21
185. Boulton SJ, Jackson SP. 1998. *EMBO J.* 17:1819–28
186. Gravel S, Larrivee M, Labrecque P, Wellinger RJ. 1998. *Science* 280:741–44
187. Baumann P, Cech TR. 2000. *Mol. Biol. Cell* 11:3265–75
188. Polotnianka RM, Li J, Lustig AJ. 1998. *Curr. Biol.* 8:831–34
189. Porter SE, Greenwell PW, Ritchie KB, Petes TD. 1996. *Nucleic Acids Res.* 24:582–85
190. Metcalfe JA, Parkhill J, Campbell L, Stacey M, Biggs P, et al. 1996. *Nat. Genet.* 13:350–53
191. Hande MP, Balajee AS, Tchirkov A, Wynshaw-Boris A, Lansdorp PM. 2001. *Hum. Mol. Genet.* 10:519–28
192. Sprung CN, Bryan TM, Reddel RR, Murnane JP. 1997. *Mutat. Res.* 379:177–84
193. Goytisolo FA, Samper E, Edmonson S, Taccioli GE, Blasco MA. 2001. *Mol. Cell. Biol.* 21:3642–51
194. Espejel S, Franco S, Sgura A, Gae D, Bailey SM, et al. 2002. *EMBO J.* 21:6275–87

195. Hande P, Slijepcevic P, Silver A, Bouffler S, van Buul P, et al. 1999. *Genomics* 56:221–23
196. di Fagagna FD, Hande MP, Tong WM, Roth D, Lansdorp PM, et al. 2001. *Curr. Biol.* 11:1192–96
197. Samper E, Goytisolo FA, Slijepcevic P, van Buul PP, Blasco MA. 2000. *EMBO Rep.* 1:244–52
198. di Fagagna FD, Hande MP, Tong WM, Lansdorp PM, Wang ZQ, Jackson SP. 1999. *Nat. Genet.* 23:76–80
199. Samper E, Goytisolo FA, Menissier-de Murcia J, Gonzalez-Suarez E, Cigudosa JC, et al. 2001. *J. Cell Biol.* 154:49–60
200. Conway C, McCulloch R, Ginger ML, Robinson NP, Browitt A, Barry JD. 2002. *J. Biol. Chem.* 277(24):21269–77
201. Ahmed S, Hodgkin J. 2000. *Nature* 403:159–64
202. Ahmed S, Alpi A, Hengartner MO, Gartner A. 2001. *Curr. Biol.* 11:1934–44
203. Hofmann ER, Milstein S, Boulton SJ, Ye M, Hofmann JJ, et al. 2002. *Curr. Biol.* 12:1908–18
204. Riha K, Watson JM, Parkey J, Shippen DE. 2002. *EMBO J.* 21:2819–26
205. Gallego ME, White CI. 2001. *Proc. Natl. Acad. Sci. USA* 98:1711–16
206. Bundock P, Hooykaas P. 2002. *Plant Cell* 14:2451–62
207. Runge KW, Zakian VA. 1996. *Mol. Cell. Biol.* 16:3094–105
208. Kota RS, Runge KW. 1999. *Chromosoma* 108:278–90
209. Kota RS, Runge KW. 1998. *Nucleic Acids Res.* 26:1528–35
210. Hartman PS, Herman RK. 1982. *Genetics* 102:159–78
211. Benard C, McCright B, Zhang Y, Felkai S, Lakowski B, Hekimi S. 2001. *Development* 128:4045–55
212. Lim CS, Mian IS, Dernburg AF, Campisi J. 2001. *Curr. Biol.* 11:1706–10
213. Loayza D, Parsons H, Donigian J, Hoke K, de Lange T. 2004. *J. Biol. Chem.* 279:13241–48

Annu. Rev. Biochem. 2004. 73:209–39
doi: 10.1146/annurev.biochem.73.011303.073844
First published online as a Review in Advance on February 26, 2004

CRAWLING TOWARD A UNIFIED MODEL OF CELL MOTILITY: Spatial and Temporal Regulation of Actin Dynamics

Susanne M. Rafelski[1] and Julie A. Theriot[2]

[1]Department of Biochemistry, Stanford University, Stanford, California 94305; email: susanner@stanford.edu
[2]Department of Biochemistry and Department of Microbiology and Immunology, Stanford University, Stanford, California 94305; email: theriot@stanford.edu

Key Words cytoplasmic structure, cytoskeleton, polymerization, self-organization, treadmilling

■ **Abstract** Crawling cells of various morphologies displace themselves in their biological environments by a similar overall mechanism of protrusion through actin assembly at the front coordinated with retraction at the rear. Different cell types organize very distinct protruding structures, yet they do so through conserved biochemical mechanisms to regulate actin polymerization dynamics and vary the mechanical properties of these structures. The moving cell must spatially and temporally regulate the biochemical interactions of its protein components to exert control over higher-order dynamic structures created by these proteins and global cellular responses four or more orders of magnitude larger in scale and longer in time than the individual protein-protein interactions that comprise them. To fulfill its biological role, a cell globally responds with high sensitivity to a local perturbation or signal and coordinates its many intracellular actin-based functional structures with the physical environment it experiences to produce directed movement. This review attempts to codify some unifying principles for cell motility that span organizational scales from single protein polymer filaments to whole crawling cells.

CONTENTS

INTRODUCTION

A crawling cell achieves its movement by protruding its front and retracting its rear to displace itself directionally. This type of amoeboid motility is the mechanism by which many different cell types, including slime molds, amoebas, leukocytes, fibroblasts, epithelial cells, and neuronal growth cones, fulfill their specific biological roles. Although these cells' overall migration occurs through related molecular and cellular mechanisms, their types of movement are very distinct and are optimized to their specific environments. Consider three examples of crawling cells moving in different environments with very different morphologies: the keratocyte, the neutrophil, and the neuronal growth cone [see the supplemental online movies 1, 2, and 3 (follow the Supplemental Material link from the Annual Reviews home page at http://www.annualreviews.org) and the section Coordination of Different Properties of a Cell to Produce Movement below]. Epithelial keratocytes function in wound healing, in which cells at the outer edge of an epithelial sheet move along the surface toward the wound as the cells deeper in the sheet follow (1). The keratocyte glides elegantly along its two-dimensional environment by protruding a single thin lamellipodium and retracting its rear in a highly coordinated manner. In contrast, a neutrophil intensely pushes its way through three-dimensional tissues to locate and consume invading microbes (2). It follows its target by extending thick pseudopods in many directions as it responds and adapts to a rapidly changing environment of external signals. Finally, the neuron extends a specialized structure, the growth cone, to find distant targets in the wiring of the nervous system. The growth cone explores its environment by extending dynamic filopodia and migrates along preestablished molecular paths laid out in the developing embryo (3). The protrusion-retraction mechanism in each of these cell types occurs through the assembly of complex dynamic organizations at the front and their coordinated disassembly at the rear of the cells. They each achieve their protrusive dynamics through the polymerization of actin and the regulation of higher-order multicomponent actin structures. Their molecular components have similar functionalities which include proteins stimulating actin nucleation, enhancing actin polymerization, accelerating actin depolymerization, and regulating the structural organization and dynamics of the complex actin networks in the protruding cellular regions. Yet the coordinated effort of all of these similar molecules, in the context of their specific cell type, produces very distinct morphologies during each cell's migration.

The tremendous advances in molecular biology and biochemistry that have occurred in the past 20 years have allowed researchers to begin identifying members of the long inventory list of individual proteins and structures contributing to crawling cellular motility. Their efforts have lead to the classification of specific protein interactions and their regulation within subcellular modules of cellular motility, such as the protrusion of lamellipodia and filopodia, adhesion to the substrate, and retraction of the rear. Now that classification and characterization of small subcellular modules of cell motility are well underway, a new challenge has come to the fore—understanding how the cell coordinates these modules to achieve robust whole-cell motility that is responsive to mechanical and chemical features of its environment. The cell must regulate its overall diverse biological functions within each hierarchical level beginning with the behavior of individual proteins at the nanometer size scale yet translating to behaviors, four orders of magnitude larger, over tens of microns at the level of the cell. Protein-protein interactions are dynamic over timescales of milliseconds to seconds, yet whole-cell motility can remain persistent for minutes to hours, four orders of magnitude longer. The cell must exert exact control over the nonadditive layers of complexity that such vast differences in size and timescales create. This review explores how the cell can regulate each of these levels, from individual molecules through the combinatorial nature of their interactions that create dynamic structures to the global behavioral changes mediated by these structures and needed for the cell's specific biological role. Several of the basic unifying principles necessary for cell motility are examined. One such principle is a "treadmilling" at all levels of the cell with assembly at the front and disassembly at the rear through which directed movement occurs. Additionally, the cell regulates transitions between dynamic subcellular self-organizing structures to be sensitive yet robust in its response to the environment. Finally, the cell coordinates these distinct structures and components in space and in time within its crowded, yet highly structured, cytoplasm, such that it can leverage their integrated functions to produce its movement.

EXPERIENCING THE CELLULAR ENVIRONMENT

One major challenge in taking information learned in simplified biochemical systems and applying it to the living cell is that the physical properties within the cell itself differ significantly from conditions in dilute aqueous solution. The cell's viscoelastic environment renders inertia insignificant (4), and the mechanical properties of cytoplasm are dependent on the speeds of physical deformations within the cell. The large number of subcellular structures, organelles and macromolecules creates a crowded environment where ability to move is strongly dependent on the size of the individual component. Movement over distances that are large relative to the size of the component cannot occur simply through diffusion (5). The cell therefore requires organized transport mechanisms to regulate the spatial distribution of its components in time. The spatial boundary created by the cell plasma membrane and typical cell volume of picoliters reduce the meaning of the molar concentration of a specific molecule; so it is important to consider the finiteness of its numbers, unlike in vitro solutions with a volume of microliters or milliliters. Fluctuations caused by thermal energy or inherent noise can be significant in effecting the outcome of a cellular process, for example by changing the immediate position of a molecule and, because of its finite numbers within the cell, considerably changing its distribution in the local environment. Appreciating the complexity and nonclassicality of the environment that any intracellular component experiences is fundamental to integrating information about these components and to understanding the global behavior of a cell as a whole.

Cells contain significant mass percentages of protein, which sometimes exceed one quarter of their weight. The large proportion of proteins in the cytoplasm implies that their intracellular distribution more closely resembles a protein crystal than a protein solution (6). Taking the sizes of the proteins and the other macromolecules in the cell into account, as well as the structures into which they can assemble, leads to a rather crowded and inhomogeneous view of the intracellular milieu. Goodsell's (7) correctly scaled depiction of the *Saccharomyces cerevisae* cytoplasm and its components, including the sizes and concentrations of soluble proteins, ribosomes, and cytoskeletal structures (Figure 1*a*), demonstrates the crowding well. Recently, advances in microscopy have allowed the three-dimensional visualization of *Dictyostelium discoideum* cytoplasm via cryoelectron tomography (Figure 1*b*) (8). The similarity in the spatial distribution of cytoplasmic components between these two images is striking and, although each includes several different components and spans different size scales, clearly demonstrates the degree of cytoplasmic crowding at all levels within the cell. Diffusion rates for small tracer molecules in fibroblast cytoplasm and 12% to 13% dextran or Ficoll solutions are similar; this observation leads to an estimate that only half of total cytoplasmic protein is in solution at any given moment (9). Interestingly, observations of the statistically lower than expected variability in isoelectric points and molecular weight distributions of the proteome dating back almost 20 years also led to suggestions that up to half of proteins may be confined to structures and not free in solution (6), a fact currently well accepted in light of more recent ultrastructural investigations of the cytoskeleton. An attempt to create interaction maps of the yeast proteome combined and analyzed extensive two-hybrid data and found only one large network of protein interactions linking over 1500 proteins in the proteome,

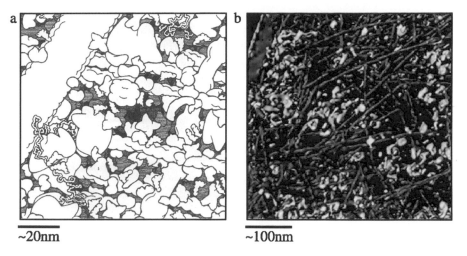

~20nm   **~100nm**

Figure 1 Visualization of the crowded cytoplasm. (*a*) A drawing of the *Saccharomyces* cytoplasm, which includes cytoskeletal filaments, ribosomes, mRNA, and soluble protein, created by considering their relative sizes and concentrations (scale bar ~20 nm). (*b*) A visualization of the three-dimensional *Dictyostelium* cytoplasm generated by cryoelectron tomography. Actin filaments are red, membrane is blue, and other macromolecular complexes, including ribosomes, are green (scale bar ~100 nm). These two depictions of the cellular cytoplasm demonstrate its severely crowded nature. (*a*) Reproduced and modified with permission from (7), Copyright (1993) Springer-Verlag. (*b*) Reproduced and modified with permission from (8), Copyright (2002) American Association for the Advancement of Science, http://www.sciencemag.org.

whereas all of the other networks contained fewer than 20 proteins (10). Individual proteins are thus highly interconnected in their interactions and are spatially regulated by their mutual interactions with the structural scaffold of the cell, demonstrating the highly ordered nature of the intracellular environment.

The crowded nature of the cytoplasm affects the properties of molecular interactions by favoring the folding of proteins, intramolecular associations, and the formation of oligomeric structures [reviewed in (11, 12)]. The size of a macromolecule and the presence of other macromolecules nearby reduce the volume and components from the local environment accessible to that macromolecule. The effective concentration of neighboring molecules is therefore increased and depends on their relative sizes. The rate-limiting step in a reaction can depend on the ability of two molecules to find each other in space or on the kinetics of their specific association, which determines the respective positive or negative effect of crowding on a cellular process (11, 12). Experiments have shown that crowding is the principal factor in decreasing diffusion of a small probe, while binding interactions and the microviscosity calculated in the absence of other interactions (fluid-phase viscosity) are not. In fact, the fluid-phase viscosity of small solutes is only 10% to 30% less in cytoplasm than in water [reviewed in (13)]. Although diffusion can be the principle mechanism of molecular movement for interactions on a ~ 20 nm scale, it cannot completely explain the mechanism by which molecules move over greater distances to perform their functions (5). It is therefore crucially important for the cell to regulate the spatial distribution of its proteins very tightly to control its functions and to foster mechanisms other than diffusion for long-distance transport of large components including protein oligomers.

One consequence of intracellular crowding is a size-dependent diffusion capability of tracers in cytoplasm indicative of a molecular sieving mechanism, probably owing to the actin cytoskeleton [reviewed in (9)]. The dense cytoplasmic meshwork of actin filaments excludes larger tracers as well as intracellular organelles from regions of the cell, including lamellipodia and pseudopodia (14), and accounts for most of the diffusional constraint within *Dictyostelium* (15). Specialized mechanisms to transport molecules to the cellular regions where they function are therefore necessary, especially when those regions are within a dense actin meshwork, such as at the leading edge of crawling cells. It has been shown in two enzymatic systems that crowding itself may increase sensitivity to the environment [reviewed in (16)]. In a region of high macromolecular or structural density, crowding decreases the absolute numbers of a specific molecule that are able to access that region and could make the region more sensitive to fluctuations in the spatial distribution of the molecule. Inherent noise and fluctuations within the cell provide opportunities for stochastic amplification of a process beyond an easily reversible state, leading to many possible outcomes in a signal-sensitive cell. Thus the physical properties of the cellular environment significantly contribute to the cell's overall spatial and temporal regulation of its components, especially in the dynamics of its coordinated movement.

TREADMILLING AT ALL LEVELS OF CELL MOTILITY

The basic unit of motility must be able to protrude in the direction of motion and retract in the opposite direction to displace itself from one point to the next. This basic principle applies to all levels of motility within the cell, from the treadmilling action of a single actin filament to the treadmilling of a large motile cell coordinating all of its components to protrude in the front and retract in the rear. To understand the mechanism by which a cell can displace itself in space, we must consider the mechanisms of displacement at all spatial and temporal scales therein.

The Inherent Asymmetry of Polymers

In considering basic principles of biological assembly, Crane (17) noticed that building a chain from a set of asymmetric objects in which each object can only interact with another in a specific spatial orientation created structures of a helical nature, regardless of how different the objects were in shape and size. Pauling (18) took this concept and applied it specifically to protein polymers. A single globular protein is always asymmetric. Therefore, if a protein is capable of interacting with an identical or distinct binding partner, the resulting dimer and subsequent multimers will preserve this asymmetry and form a helical polar polymer. The variety of possible polymer structures depend on the diameter of the protein, the pitch of the helix induced by polymerization, bond flexibility and geometry, and the energetic preferences of the protein subunits for interactions along different surfaces. Multiple individual polymer helices can interact with each other to stabilize their resulting structure (Figure 2a) (18). This basic concept explains the helical nature of various polymeric cellular structures as well as how the specific shapes of their subunits lead to different geometries, which includes the double protofilament helix of actin (Figure 2a), and the higher number of protofilaments in microtubules and flagella (19).

Combining Actin Filament Structural Asymmetry with Distance and Time

The globular asymmetric nature of the monomeric actin subunit itself thus creates the basis for the helical nature and structural polarity of a single polymeric actin filament. In this case, energetics would dictate that the ratio of the on and off rates between monomer and either end of the polymer must be equal. However, the structural asymmetry of the filament causes different specific interactions at each end that result in one (the barbed front end) having much faster kinetics than the other (the pointed back end). In order to transform the structural polarity of actin into a dynamic spatial structure, a temporal regulator is needed to translate the asymmetric kinetics inherent in the filament structure into asymmetric energetics (20). For actin, association with ATP and its subsequent hydrolysis provide this clock. The energy released by the nucleotide hydrolysis can induce a conformational change in the protein, which alters the associations between a subunit and the rest of the polymer based on the state of the bound nucleotide. A polymerization system that is differentially sensitive to the relative numbers of ATP- or ADP-bound subunits at each end is thus created. This coupling to a tiny clock regulates the temporal dynamic behavior with different energetics at each end [reviewed in (21)]. As a consequence, within a range of free monomer concentrations, the front end can polymerize while the back end depolymerizes causing a displacement in space, a process called treadmilling (Figure 2b) (22). This process has been observed for individual actin filaments in vitro (23) as well as for microtubules in vivo (24, 25).

Accelerating Treadmilling in Time and Expanding It in Space: The Actin Array

Actin filament dynamics are regulated spatially, through the intrinsic polarity of an asymmetric protein polymer, and temporally, through coupling this asymme-

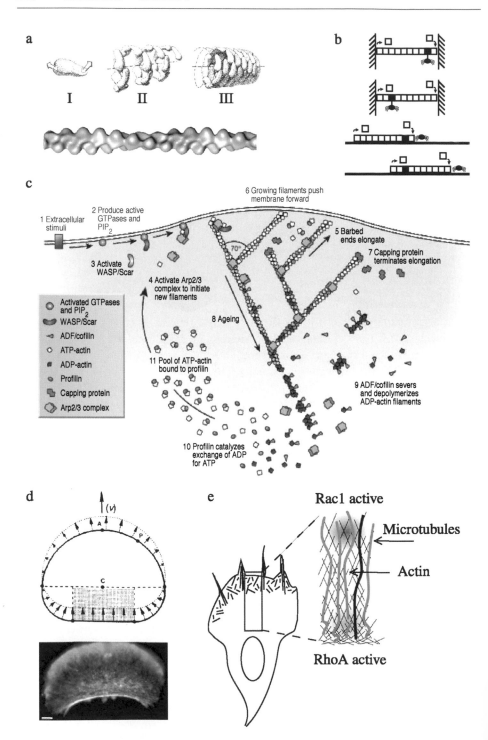

a

I   II   III

b

c

1 Extracellular
stimuli

2 Produce active
GTPases and
PIP₂

6 Growing filaments push
membrane forward

5 Barbed
ends elongate

7 Capping protein
terminates elongation

3 Activate
WASP/Scar

4 Activate Arp2/3
complex to initiate
new filaments

70°

8 Ageing

Activated GTPases
and PIP₂

WASP/Scar

ADF/cofilin

ATP-actin

ADP-actin

Profilin

Capping protein

Arp2/3 complex

11 Pool of ATP-actin
bound to profilin

9 ADF/cofilin severs
and depolymerizes
ADP-actin filaments

10 Profilin catalyzes
exchange of ADP
for ATP

d

(v)

c

e

Rac1 active

Microtubules

Actin

RhoA active

try to nucleotide hydrolysis, which together create the basic unit of motility, a treadmilling filament. At monomer concentrations similar to those inside cells, treadmilling is extremely slow (23, 26). The behavior of a single filament is therefore insufficient to span the large size of the cell or produce the rapid polymerization/depolymerization dynamics required for cellular movement. The

Figure 2   Treadmilling as considered at all levels of the cellular organizational hierarchy. (*a*) A protein polymer made of asymmetric globular subunits has an intrinsic helical nature. Two of these helical polymers can interact to form higher-level structures stabilized by their associations. The top shows an asymmetric globular protein subunit (I) that upon iterative self-association forms a helical polar polymer (II). Two of these helices may interact and form different structures depending on their specific intermolecular interactions, as seen in (III) and in the reconstruction of an actin filament on the bottom. (*b*) Leveraging the polar nature of the polymeric protein helices by adding a time-dependent variable to its association, the hydrolysis of nucleotide, provides the basis for the treadmilling of a single actin filament. This basic unit of motility can assemble at the front and disassemble at the rear, between two barriers (*top*) or along a surface (*bottom*), effectively moving an individual filament forward, and can produce force through its polymerization. (*c*) The regulation of actin treadmilling behavior in assembling dynamic protruding regions of the cell occurs by complex combinatorial control of its dynamics through the action of accessory proteins. This figure represents a recent model of the dendritic array, including many of the necessary components and their functions required to produce its treadmilling dynamics. (*d*) At the scale of the whole cell, as seen in this example of a keratocyte, the front must assemble structures to protrude as the rear disassembles and contracts at the same net rate, demonstrating a treadmilling of the whole cell as it moves forward (*top*). The arrows represent the extent of protrusion and retraction needed in a model of keratocyte motility. Actin (*blue*) localized at the leading edge and myosin II (*pink*) at the contracting rear of a keratocyte fragment provide the necessary protrusion-contraction dynamics for movement (*bottom*). (*e*) In larger cells, interactions between the actin and microtubule cytoskeletons allow for coordination of assembly and disassembly dynamics over greater distances and longer times. The association of microtubules to the actin network positively regulates their polymerization at the front and induces their breakage through contraction forces at the transition zone of the lamellipodium. These physical effects on microtubule dynamics can activate the global signalers, Rac1 and RhoA, in different regions of the cell, and their regulation further enhances the cell's treadmilling dynamics. (*a*) Reproduced and modified from (18) & #8211; by permission of The Royal Society of Chemistry and from (137), by copyright permission of The Rockefeller University Press. (*b*) Reproduced and modified with permission from (21), Copyright (2000) Blackwell Publishing. (*c*) Reproduced and modified with permission from (41), Copyright (2003) Nature Publishing Group, http://www.nature.com/. (*d*) Reproduced and modified with permission from (116), Copyright (1993) Nature Publishing Group, http://www.nature.com/ and from (61), Copyright (1999) with permission from Elsevier. (*e*) Reproduced and modified with permission from (62), Copyright (2003) Nature Publishing Group, http://www.nature.com/.

treadmilling actin filament requires further higher-order organization to leverage its basic unit of motility to function over greater distances and at greater speeds. Large complex actin structures have been observed in the protruding regions of cells (27, 28), and experiments show that treadmilling of this actin meshwork as a whole occurs during cellular motility (29–31). Ten years ago it was clear that molecular mechanisms must exist to allow for the regulated creation of nucleation sites at the cell periphery, desequestration of monomeric actin, and regulated depolymerization at the other end of the treadmilling meshwork (32). Tremendous progress has been made in identifying specific molecules involved in this process and the mechanisms by which they contribute to what is now known as "dendritic array treadmilling" (33). The coordinated effort of these proteins has been extensively reviewed (34–41).

The orientation of actin's assembling end, causing protrusion at the front, and its disassembling end at the back dictate the overall polarity of the dendritic array. The nucleotide hydrolysis and polymerization properties of an actin filament set up the spatial and temporal regulation of the array's behavior by modulating the association of various actin binding proteins that feedback on actin polymer dynamics. The speed at which a filament can polymerize or depolymerize, which is dependent on the surrounding components in its environment, and rates at which nucleotide hydrolysis occurs throughout the filament contribute to the overall dynamics of the array. The different conformations of ATP- and ADP-bound actin subunits within the filament control where accessory proteins bind preferentially and exert their function. The interactions between actin and its binding proteins leverage and amplify the temporal regulation intrinsically provided by the actin filament to provide a spatial framework for the dynamics of the dendritic array. The cell sets up the intrinsic outward polarity of the entire dendritic array by localizing necessary regulatory proteins to the spatial boundary of its membrane.

The assembly and protrusion of the front end of the array is enhanced by the Arp2/3 complex, which can nucleate branched actin filaments off the sides of other filaments, thereby increasing the availability of actin barbed ends for enhanced polymerization (42, 43). This activity is itself spatially controlled because there seems to be a preferential association of Arp2/3 with regions of actin filaments near the new barbed ends (44) and may be due to its preference for ATP-actin within the filament [reviewed in (37)]. Another spatial restriction to Arp2/3 complex activity is that the complex must be activated by an accessory protein, such as a member of the WASP family. These accessory proteins are in turn regulated by coactivators, such as phosphatidylinositide 4,5 bisphosphate (PIP$_2$) and CDC42, which are anchored to the membrane. In addition, actin branches can dissociate over time, upon which free actin filament pointed ends can anneal to other filaments deeper within the array. This debranching may be spatiotemporally regulated as well by phosphate release upon nucleotide hydrolysis on actin [reviewed in (38)]. A further layer of temporal regulation is possible owing to the capability of the Arp2/3 complex itself to hydrolyze ATP, although the role for this activity in regulating array dynamics is not yet clear (45, 46).

Some accessory proteins contribute to an increase in the number of free barbed ends through the severing of actin filaments, which include ADF/cofilin (34–41, 44) and gelsolin (34–41, 47). The number of available barbed ends is negatively regulated by barbed end-capping proteins, which can also prevent annealing of filaments but are in turn regulated by tropomyosin; this allows for filament annealing in their presence (38). In addition, tropomyosin protects filaments from severing, and it is spatially regulated, resulting in its absence from the leading edge and presence deeper in the lamellipodium (48). Such dual roles are also apparent for some other proteins. One is ADF/cofilin, which, in addition to severing filaments, increases the rate of depolymerization in order to speed the recycling of monomeric actin (49). Regulation of the degree of depolymerization can also occur through pointed end capping, via Tmod3, an isoform of tropomodulin (50). The overall depolymerization activity must be spatially regulated in order to ensure recycling of monomeric actin to the front end of the array. This regulation is demonstrated by the localization of activated cofilin to the lamellipodium in fibroblasts (51) and by the more localized absence of cofilin along the periphery of the leading edge in rapidly moving keratocytes (33). Recycling of ATP-actin to the leading edge occurs through sequestration of free monomers, via thymosin $\beta4$ and profilin, to control their concentration and promote nucleotide exchange (52). The preferential affinity of profilin for ATP-actin and spatially localized binding partners such as vasodilator-stimulated phosphoprotein (VASP) then allows profilin to act as a shuttle and increase the availability of polymerizable actin to the front of the array. Figure 2c shows a current model for many of the molecular components and their mechanisms.

Multicomponent combinatorial regulation of the actin polymerization dynamics creates a higher-level structure that self-organizes and continuously recycles the necessary molecular components to allow the array to treadmill forward. Polymerization against a load can itself produce a force [reviewed in (21)], and overall polymerization of the dendritic array is a mechanism by which protrusive force is generated at the leading edge of moving cells (31, 53). Although crawling cells have more complex requirements for their movement (see sections below), some intracellular pathogens, such as *Listeria monocytogenes* and *Shigella flexneri*, move solely by harnessing the dendritic array geometry to their surface through their own regulatory proteins, ActA and IcsA, respectively [reviewed in (54)]. Interestingly, even though *Listeria* and *Shigella* motility could be reconstituted with a subset of the proteins present in the array, a crowding agent was required (55), demonstrating an essential role for the crowded nature of the cell in regulating the dynamics of the array to produce movement. Persistent motion can be simulated by a simplified mathematical model through a section of a cell incorporating barbed end polymerization, branching near barbed ends, pointed end depolymerization, debranching, and de novo nucleation to ensure maintenance of free actin (56). Although the direction of this motion can switch through stochastic fluctuations in model behavior, it remains correlated with the direction in which the barbed ends point. In addition, the model creates an asymmetric filamentous actin distribution, as is seen in lamellipodia (57) [reviewed in (38)]. Long simulations of this model produce an imitation of the persistent random walk observed for crawling cells. Thus, the basic underlying spatial polarity the cell uses to create a treadmilling array for protrusion can in fact produce directed motion. While the dendritic array can span a large protruding region of the moving cell, some of its protein components must be rapidly turned over spatially to support its fast-paced dynamics, leading to a problem of delivery of components to the leading edge, which is considered later.

Coupling the Actin Array to Cell Contraction and the Adhesion/De-adhesion Cycle

When the dendritic array is assembled inside a membrane-bound cell, the protrusion it causes will not, by itself, allow the cell to crawl. A section of the cell will protrude because of polymerization dynamics but will not suffice for the whole cell to move unless the rear is able to retract, effectively allowing the cell to treadmill (Figure 2*d*). The mechanism of retraction is thought to be dependent on contraction of actin fibers via myosin II (58, 59). In keratocytes, myosin II thick filaments are colocalized to regions in the cell body and transition zone where actin filaments are in bundles or asters, increasing in concentration as the actin filament concentration decreases (57). This spatial regulation suggests a role for the contracting functions of mysosin II in the retracting rear. Fragments of keratocyte lamellipodia can polarize and move persistently (60), showing that the combination of the dendritic array treadmilling at the front and myosin II-mediated contraction at the rear is sufficient for their movement (Figure 2*d*) (61).

In cells much larger than the specialized, fast-moving keratocytes, these same principles hold, but additional levels of spatial and temporal regulation are required to coordinate the necessary protrusive and retractive elements of treadmilling over much larger distances. Well-regulated signaling cascades can allow for large regions of the cell to behave in a unified manner and transduce spatial information globally. The cell links the dynamics of the actin meshwork to that of the microtubule cytoskeleton, which due to its thicker and stiffer filaments can provide structural regulation over greater distances (Figure 2*e*). The interactions between the actin and microtubule cytoskeleton coupled to reinforcing signaling cascades set up a polymerization/contraction treadmill [reviewed in (62)]. In this treadmill, the polymerization activity of the microtubules at the protruding region activates the global signaler, Rac1, which in turn stimulates the polymerization of both actin and microtubules. As the microtubules become displaced toward the contracting region of the cell, they are broken through its compressive forces. The catastrophic shortening of microtubules in the contracting region activates another broad-acting signaler, RhoA, which stabilizes microtubules and stimulates the contractile actin structures (62). Thus the microtubule network is coupled to the actin-based protrusion/retraction organization, linking structural spatial information contained in the dynamic behavior to temporally regulated signaling molecules to coordinate polymerization and depolymerization over greater distances (Figure 2e).

The cell translocates across a solid substrate through a treadmilling process in which protrusion is coordinated with increased adhesion in the front and retraction with loss of adhesion in the rear, effectively requiring a treadmilling of adhesions. Many different cellular molecules interact with the different substrates on which the cell crawls, and they are regulated through signals and mechanosensory processes to allow for their necessary, rapid turnover [reviewed in (63)]. For efficient movement, the cell must have strong attachments to the surface in the protruding regions, as seen in keratocytes (64, 65). The action of myosin II in the rear pulls at the adhesions and weakens them, allowing for the cell body to be pulled toward the tightly adhering front. The localization of adhesions in these specialized cells is optimally coordinated with protrusion and retraction to allow for rapid movement. In larger cells, microtubules can negatively regulate focal adhesion when targeted to and closely associated with these sites (66). The cell thus engineers a direct link between protrusion and adhesion dynamics, regulated through the polarity of the cellular cytoskeleton, to provide the necessary coordination to move.

An impressive example of how the whole cell treadmills along a substrate is the nematode sperm cell [reviewed in (67)]. This cell contains very little actin, but it uses a completely different polymerizable protein to generate its movement. This major sperm protein (MSP) can polymerize into helical filaments and, because of its own specific shape and interactions, can further assemble into higher-order structures. MSP assembly is spatially regulated by a membrane-bound activator, MSP polymerization organizing protein (MPOP), which is only activated at sites where MSP polymerizes (68). The MSP polymerized assemblies consist of multiple branched and packed fiber complexes that span the protruding lamellipodium and assemble at the front end while disassembling at the rear. The protrusion and adhesion dynamics are both mediated in time and space by mechanisms that are pH-sensitive and can therefore be coordinated by the local environment. Two recent physical models have integrated information on the dynamics of the *Ascaris* sperm cell and the mechanics of the polymerized MSP gel (67, 69). These models can recapitulate the major features of crawling movement and have identified new possible experimental directions to further elucidate the mechanism by which cells treadmill forward by hierarchial temporal and spatial enhancement of the inherent asymmetry of protein self-association.

REGULATING THE HIERARCHY OF DYNAMIC ACTIN ORGANIZATIONAL STRUCTURES

The neutrophil, keratocyte, and growth cone all use similar molecular machinery to achieve net actin filament growth at the front and loss at the rear, but they manage to achieve such varied dynamic morphologies in their movement through the specific regulation of their functions and interactions in space and time. Any given cell must therefore be able to reorganize many of its different types of subcellular structures to provide the higher-level reorganization it needs to respond to a specific set of conditions, such as activation of signaling molecules upon exposure to a chemoattractant in a neutrophil. The intracellular local environment is full of random movement from thermal energy, and amplification of this movement results in constant fluctuations over variable size and time-scales. The cell must be able to create order out of these fluctuations and the transient interactions of its components and regulate their functions in space and time. The uniqueness of such dissipative self-organizing systems is that they must experience a constant flow of energy and matter to persist, and they have to minimize the rate of energy loss to reach their dynamic steady state (70, 71). These systems exploit the stochastic nature of the environment to create a dynamic, ordered organization by amplifying random fluctuations. If an environment is near a critical set of conditions that allow multiple outcomes, the stochastic dynamics of these conditions can determine the resulting behavior. Thus the ever-changing transient dynamics within the cell create the ability for a behavior or organized structure to form. Because of the finite number of molecules within the cell and the stochastic nature of their exact locations, even when targeted to a specific region of the cell, the relative ratios of necessary components can precisely regulate their ability to interact and self-organize into different higher-order functional structures. Once an organized steady state is reached, the dynamic functional unit has to persist until another is needed, meaning that the system must be less sensitive to certain components than to others that the cell perceives as important environmental signals, such as chemoattractant and repulsive agents. Good examples of self-organized units with differential sensitivities are the dynamic, self-organized structures created by microtubules and their motors in vitro (72). These structures were shown to be sensitive to the relative ratio of motors to microtubules, forming asters only after a motor concentration threshold was reached. A computational model of this process showed that the extent of aster formation was highly dependent on the speed of the motor along the microtubule, while changing the affinity of a fast-moving motor for the microtubule had less of an effect (72). Such results would ensure the formation and persistence of these robust dynamic structures in the presence of distinct motors with similar speeds.

Biochemical patterns have been shown to form if there is a local self-enhancing activity at balance with a more global inhibition [reviewed in (73)]. The amplification must be nonlinear in nature (73), for example by being autocatalytic through cooperativity or positive feedback, in order for the amplified behavior to dominate among surrounding fluctuations and persist in the presence of the global inhibitor. In the context of cell motility, patterns constitute dynamic, self-organized structures regulated in space and time, such as overall cell polarity. An example is when neutrophil-like HL60 cells respond to a chemoattractant by activating two distinct G proteins regulating two separate signaling pathways with different locations in the cell, one acting at the front and one at the rear, establishing cellular polarity (74). The pathway in the front activates positive feedback between Rac signaling and actin polymerization, amplifying protrusion, while the signal in the rear activates Rho signaling and promotion of actin-myosin contractile structures, inducing retraction. Polymerization and contraction are regulated in spatially distinct regions of the cell and inhibit each other, strengthening the amplification of the signal-induced process leading to large-scale morphological changes in the front and the rear (74). Here the self-organization of two functioning dynamic substructures spatially regulated within the cell requires amplification of a set of signals that feed back into reorganizing these structures to a point where they are at steady state and can exert their functions, allowing the cell to move.

The cell maintains its sensitivity by regulating the self-organization of intracellular structures through their own sensitivity to the ratios of their components. A molecule with a certain function can cause different outcomes based on its concentration relative to its binding partner (see α-actinin in a section below), or a multifunctional molecule can have different functions that become dominant based on the local ratios of the components (see VASP in a later section). The robustness of an outcome created by the sensitive self-organized structure can be regulated through relative insensitivity to concentrations of specific components (72) or through redundancy in molecules able to provide a specific function (see examples below). The next section explores specific examples of structures that are sensitive to the balance of their environment and examples of mechanisms to ensure robustness of necessary functions through redundancy in molecular components at all levels of cell motility. Their implications are considered within the context of subcellular self-organizing structures that are spatiotemporally regulated by the cell to produce motion.

Intrinsic Properties of Actin Contributing to Its Organization

There is more to the life of actin filaments than simply polymerization and depolymerization in coordination with nucleotide hydrolysis. Structures forming within filaments immediately upon polymerization can be different from those found later. This difference in structure could be related to nucleotide hydrolysis because filaments show different conformations of actin in the presence of ATP, ADP, or a transition state analog (75, 76). Additionally, various substructures are possible within a single actin filament. Actin subunits have been shown to be capable of a tilted conformation in some regions of a filament in the presence of unbound ADF/cofilin, thus changing the available interaction sites to binding partners along the filament under these conditions (77). Cross-linking studies have identified short-lived actin dimers that are thought to have various associative properties and eventually change conformation to that of the stable long-lived associations found within the filament [reviewed in (78)]. This special conformation can be positively regulated by high concentrations and specific types of divalent or multivalent cations. Filaments comprised of dimers interacting in this way are rough in appearance, and they may function as cross-linkers between antiparallel actin filaments or cause branching of filaments through their special geometry (78). The fact that this orientation of two actin monomers is short-lived and may induce an inherent branching capability illustrates a way that the associative interactions between actin itself influences the timing of the actin polymer structure and the subsequent spatial restrictions for interacting proteins. At higher divalent cation concentrations, other higher-order polymeric actin structures can be seen such as "two-layer rafts" in which shielding by the ions allows actin filaments to self-associate (79), and in the presence of other polycations, actin bundles are formed (80). In addition, experiments and models have shown that, within a crowded solution, actin filaments spontaneously separate into bundles (81), which has implications for the crowded environment of the cytoplasm. Specific differential conformations within actin filaments as well as higher-order associations in the absence of other binding proteins are dependent on the components in the environment and affect the actin structures that form and their mechanical properties.

In vitro purified actin polymerizes within minutes and forms a filamentous network, which over hours acquires the stiffer elastic properties of a gel (82). The rate at which gelation of the network occurs is not due to polymerization itself but rather homogenization of the network through redistribution of actin from regions of higher density to lower via its reptation (82). Recent microrheological experiments on living cells suggest that the mechanical properties of this gel may be those of a soft glassy material near its metastable glass transition, able to regulate its properties through changes in the dynamics of the cytoskeletal meshwork (83). Thus the global mechanical properties of actin meshworks are sensitive to local densities therein. The viscosity due to actin in the *Dictyostelium* cytoplasm can be regulated by the osmolarity of its environment (15). The relationship between this viscosity and cellular volume is nonlinear, indicating that the mechanical state of the actin meshwork within the cell is variable and sensitive to the volume. Given the nonlinear dependency of protein associations on the crowding of the environment (11), the mechanical properties of the actin meshwork may well be very sensitive to the volume of its environment. The examples in this section show that some specific higher-level organizations of actin networks are energetically preferred. In order to leverage these energetic advantages into rapid cell shape changes, the cell stabilizes and regulates the different structures found in organized actin networks through dynamic interactions with accessory proteins.

Sensitivity and Robustness in Multicomponent Subcellular Actin Structures

Although actin itself can spontaneously form higher-level assemblies, the cell regulates the spatial and temporal distribution of its different dynamic actin organizations through accessory actin binding proteins. These proteins can stabilize structurally and functionally distinct organizations such as actin gels or actin bundles (Figure 3a), which are sensitive to the relative concentrations of their protein components. In vitro the addition of α-actinin to actin induces a threshold concentration-dependent transition from a gel-like to a bundle-like organization that can be predicted by a simple model taking into account relative concentrations and affinities of different α-actinin proteins (84). Adding a spatial boundary condition, by introducing actin and different binding proteins into vesicles, induces phase transitions in the organizational structure of the multiprotein actin networks dependent on the size of the spatial constraint, relative concentrations of the components, and temperature (85). These transitions can be spatiotemporally regulated within a cell through the localizations of binding partners to specific structures, such as fimbrin to filopodia or α-actinin to stress fibers (86). Robustness of these structures is achieved through the high numbers of actin binding proteins in the cell, many with overlapping functions. For example, different isoforms of myosin in *Dictyostelium* can compensate for each other functionally when one or several are removed from the cell (87).

Whereas certain actin binding proteins cause specific structural reorganizations of the network based on their topology of interaction with actin, others can have multiple more complicated functions, which affect both spatial and temporal regulation of motility. Consider the Enabled/VASP (Ena/VASP) family of proteins. These proteins have been shown to interact with F-actin, have high affinities for profilin (thus recruit monomeric actin for polymerization), have actin nucleation capabilities independent of Arp2/3, have anticapping function, and can dramatically affect the organizational state of actin meshworks [reviewed in (88)]. They localize to tips of filopodia, focal contacts, and the leading edge of protruding lamellipodia, and their function is redundantly ensured because there are multiple members of this family that can compensate if one is deleted from a cell (88). In addition, the various functions of each of these members may have slightly different effects on the resulting dynamics these proteins control. Changing the concentrations of Ena/VASP relative to other proteins at the leading edge creates dramatic changes in the architecture of the filament network. An increase leads to elongated filaments, while a decrease results in a more highly branched organization (89), demonstrating the sensitivity of the environment to the balance of Ena/VASP's different functions.

Protrusion dynamics at the leading edge can be sensitive to other components as well. The Arp2/3 complex, in conjunction with its activating cofactors, creates a dendritic branched network. Its spatial organization, the angle between mother and daughter branches induced by the complex, may be optimized for cellular protrusion (90). Temporal regulation of Arp2/3 occurs through the combinatorial control of its carefully regulated activators and additional cofactors. An example of this is cortactin, which can stimulate Arp2/3 actin nucleation by itself as well as function synergistically with N-WASP (91) for further stimulation of nucleation. Cortactin interacts with its cofactor dynamin, a protein involved in endocytosis (92) that can effect Arp2/3-cortactin–dependent actin nucleation in a concentration-dependent manner. Dynamin stimulates nucleation at lower

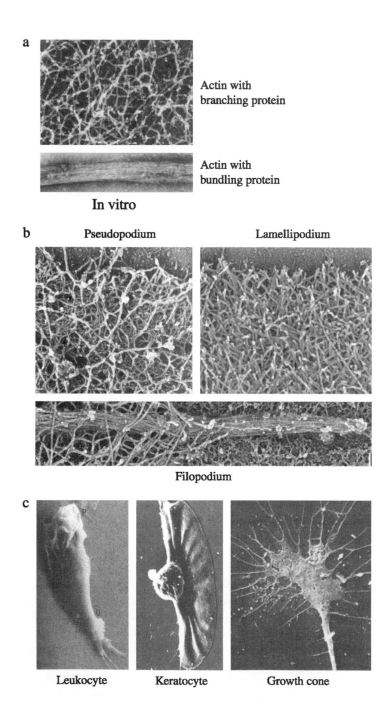

a

Actin with
branching protein

Actin with
bundling protein

In vitro

b   Pseudopodium   Lamellipodium

Filopodium

c

Leukocyte   Keratocyte   Growth cone

concentrations and inhibits at higher (93), thus regulating the effect of cortactin on actin nucleation by making nucleation sensitive to levels of other components in the local environment. The cell provides many such tuneable parameters, carefully and continuously regulating actin nucleation.

Just as the dynamics of actin nucleation in the dendritic network must be carefully regulated in space and time to control protrusion, nucleation must also be regulated in creating other types of actin organizational structures with nonprotrusive functions. Another family of proteins, the formins, has been identified with the ability to nucleate actin polymerization directly, and formin dynamics are sensitive to other proteins in the environment, such as profilin (94). These proteins are implicated in nucleating filamentous actin structures, such as stress fibers, yeast actin cables, and contractile actin bundles, needed for cyto-kinesis [reviewed in (94)]. In addition to providing a different nucleator for filamentous structures, the cell also inhibits Arp2/3 branched nucleation through proteins, such as eplin (95), that can stabilize bundled actin structures through spatial regulation of their activity. The complex multicomponent regulation of formin activation, its actin nucleation, and its stabilization of induced structures by

---

Figure 3 The various organizational structures produced by actin and its interacting accessory proteins can create distinct morphologies of movement. (*a*) In vitro experiments visualizing the structure of the actin filament meshwork in the presence of an actin cross-linking protein. Actin with ABP forms a gel-like organization (*top*), and actin with the Abl-related kinase, Arg, forms bundles (*bottom*). (*b*) Within a cell's protruding region, actin structures depend on the relative concentrations of the components in their environment. Ultrastructure of the tangled actin meshwork in a *D. discoideum* pseudopod (*left*), the branched dendritic array in a keratocyte lamellipodium (*middle*), and a bundled filopodium from the leading edge of a B16F1 mouse melanoma cell (*right*) are shown. The structural organization in the pseudopod is more disorganized and tangled, and it contains branches and long filaments throughout. The dendritic array in the lamellipodium is more ordered and displays more free ends. The filopodium emerges from a lamellar meshwork and is made of tightly bundled actin filaments, held together at the tip by a "filopodial tip complex." (*c*) SEM images of three cells with different morphologies in their protruding regions and movement are shown. The first is a leukocyte, ~15 μm long, with its pseudopod, labeled P (*left*). Next is a gliding keratocyte, ~28 μm in diameter, with its thin lamellipodium extending rightward in the direction of motion (*middle*). The last is the growth cone of a chick dorsal root ganglia (DRG) neuron with its many extending dynamic filopodia (*right*); the cell body is ~15 μm in diameter at its widest. (*a*) Reproduced and modified from (138) by copyright permission of The Rockefeller University Press and with permission from (139), Copyright (2001) National Academy of Sciences, U.S.A. (*b*) Reproduced and modified from (140), (57), and (97) by copyright permission of The Rockefeller University Press. (*c*) Reproduced and modified from (141) by permission of Oxford University Press, with permission from J. Lee, and from (142), Copyright (1985 *Journal of Neuroscience Research*), reprinted by permission of Wiley-Liss, Inc. a subsidiary of John Wiley & Sons, Inc.

inhibiting the other, branched, nucleation pathway indicate once again that activation of actin nucleation is carefully regulated and sensitive to the identity of proteins, their activity state, and their concentrations. These examples demonstrate the cell's combinatorial control over actin network structures, which are sensitive to the local environment. The complex spatiotemporal regulation of the effector components results in subcellular dynamic structures with distinct organizations.

Transitions Between Higher-Level Organizational Actin Structures

For dynamic cell motility, it is not sufficient that the cell be able to build various types of higher-order actin structures, it must also be able to switch among them. To extrapolate the biochemical observations described above to a cellular level, local spatial constraints and ratios of actin to its binding partners can sensitively effect the higher-level organizations of actin networks. The cell must be able to regulate the creation of branched and bundled actin networks, which it does through differential regulation of the nucleation of such structures and through the reorganization of one structure into the other. The transition between branched and bundled networks made of flexible polymers lies at a critical metastable boundary (96). Theoretical calculations show that small changes in the density of cross-linker can induce a phase change from one type of structure to the other. Thus the transition between actin bundles and branched structures occurs through changes in the relative levels of components in a specific region of the cell. Because actin filaments in a living cell turn over so rapidly, with average half-lives on the order of one minute, slight changes in component availability can cause large-scale transitions across phase boundaries even if the phase transition would normally be kinetically forbidden. This concept is beautifully demonstrated in recent work on the transition between lamellipodial and filopodial actin organization (Figure 3b) at the leading edge of cells and in an in vitro assay (97, 98). Filopodia can form through λ-precursors in which the dendritic organization of the leading edge allows for filaments to elongate, associate at their tips, and form bundles. Specific proteins, such as VASP, may regulate filopodial initiation by its anticapping function and allow bundles to form at the cell edge through organizing a tip complex. The filopodial bundles are subsequently stabilized by the presence of other actin bundling proteins, which include fascin throughout the whole filopodia and α-actinin in regions closer to the cell (97). Further, specific factors that may control the transition between the lamellipodial and filopodial organization were identified. In an in vitro assay, filopodial structures could be induced from actin clouds around beads coated with Arp2/3 activator through changing the concentrations of components in the environment. These structures were shown to be sensitive to the identity of the nucleator and to the concentration of capping protein (98). Thus the balance between the relative strengths of the nucleator and other necessary components, such as capping protein, and their resulting functions in the local environment could induce a large-scale reorganization of dynamic self-organizing actin structures. Such a global reorganization of intracellular structure is seen over brief time intervals in the activation of platelets during which an increase in intracellular calcium first activates gelsolin and then PIP_2, leading to a dramatic change of the actin meshwork, with the extension of lamellipodial and filopodial structures as the platelet flattens (99). Within a cell, such transitions are spatiotemporally regulated through sensitive responses to the local environment and more globally regulate the types of protrusive movement created.

To control the different actin organizations within the leading edge of the cell, global signaling molecules can induce one structure or another, as seen in the capability of Rac1 to induce lamellar ruffling and of RhoA to induce stress fibers in serum-starved 3T3 cells [reviewed in (100)]. In order to ensure that a cell can elicit its specific needed response, multiple protein interaction pathways can lead to the same effect and within their delicate mutual cross talk can add layers of regulation onto the final behavior. An example is seen in the regulation of filopodial and lamellipodial formation through interactions of various sets of proteins. CDC42 can directly interact with WASP and thereby regulate the nucleating activity of Arp2/3 [reviewed in (101)]. However, another interacting pathway among CDC42, IrsP53, and Ena/VASP proteins demonstrates an Arp2/3-independent mechanism of regulation [reviewed in (36)]. Recently the discovery of IrsP53 colocalizing with WAVE in lamellipodia and filopodia, independently of Ena/VASP proteins, has demonstrated an additional pathway through which the cell regulates the protrusion dynamics at its leading edge (102). Because the spatial and temporal regulation of protrusive actin structures can be a vital response by a cell to its environment, it provides multiple, redundant pathways to ensure the behavioral response required, such as the neutrophil's ability to continuously follow its microbial prey.

The fact that a cell has many different individual protein components that are able to provide similar functions is clear from considering intracellular pathogens. *Listeria* and *Shigella* both move intracellularly through the same mechanism, propelling themselves by usurping the cell's dendritic array organization. They do so by entering the cell's regulatory pathway at different places; *Listeria* provides its own Arp2/3 activator, and *Shigella* recruits N-WASP to activate Arp2/3. Recently, *Burkholderia pseudomallei* was shown to recruit Arp2/3 to its surface to create actin tails. Neither N-WASP nor Ena/VASP proteins were needed (103), indicating another mechanism by which an intracellular pathogen can hijack the cell's machinery. This demonstrates how different regulators can produce similar structures and how the same regulators can create different higher-order structures in varied environmental contexts.

COORDINATION OF DIFFERENT PROPERTIES OF A CELL TO PRODUCE MOVEMENT

For a cell to be motile, the balance between different protrusive organizations and their regulation must be well coordinated. Recall the example of the three different cells with distinct dynamic morphology moving through very dissimilar environments (Figure 3c). Despite their varied geometries, a lymphocyte crawls through three-dimensional tissues just as fast as a keratocyte on a two-dimensional surface, each moving in a way optimized for its environment (1, 60, 104). To achieve this rapid movement the cell must globally coordinate its behavior with the rest of its components in space and time as well as its environment to span the relatively large distances and long times required for whole-cell motility.

Coordination Within Dynamic Actin Structures for Cellular Motility

To consider how a cell coordinates its protrusive actin structures, it is useful to consider a simplified system in which a bead is uniformly coated with a nucleator (such as *Listeria* ActA, which activates Arp2/3) and is placed into cytoplasmic extract (104a). The nucleator recruits from its surroundings the necessary components that then self-organize into a dynamical structure that propels the bead using the same molecular mechanisms that allow a cell to extend a protrusive actin structure (see array treadmilling above). Stochastic fluctuations in the dynamics of the actin cloud that forms around an ActA-coated bead allow it to enhance and amplify local asymmetries that are then self-organized into an actin "comet tail" to move at steady state (105). The behavior of all the needed molecular components is inherently coordinated to form a dynamic protrusive structure similar to that seen inside cells. Cells must also coordinate signaling molecules to regulate these dynamics. Similar stochastic fluctuations in peripheral localization of activated CDC42 in yeast cells allows for amplification of actin nucleation and for the production of polarized CDC42 caps along a region of the cell periphery through a positive feedback loop (106).

While amplification of stochastic fluctuations can intrinsically induce a cell or cellular region to locally coordinate the interactions between its molecules to produce a protrusive behavior, this can also occur in response to an external signal. Neutrophils respond to the local presence of a chemoattractant by inducing polymerization toward this environmental signal and coordinating the formation of nucleation sites in order to amplify the protrusive behavior (107). The whole cell is then able to move in the direction of the chemoattractant. Likewise, stimulation of fibroblasts with platelet-derived growth factor induces global remodeling of the cytoskeleton. These dramatic changes may occur partly through the coordination of multiple proteins, which include actin, dynamin, cortactin, Arp2/3, gelsolin, and N-WASP, to form dynamic circular waves that move through the cell upon stimulation (108). These waves affect cortical actin structures, disassemble stress fibers, and are required for lamellipodial protrusion over large areas. Although they seem to occur only once upon stimulation (108), similar-looking waves of actin can occur in stationary, unstimulated B16F1 melanoma cells, MDCK cells, and in *Dictyostelium* [(109, 110) and S.M Rafelski, unpublished observation] and have been observed to create protrusive structures at cell boundaries (110). Thus large-scale dynamic organizations within the cell allow for coordination of a new state on a more global scale, for example the transition of a resting to a moving cell, through the coordination of many of the regulatory proteins to cover the large distances over which this remodeling must occur.

Coordination Between Physical and Chemical Processes

A cell experiences physical forces from its environment and from some of the biochemical processes occurring intracellularly, such as actin polymerization and myosin-dependent contraction. It must therefore coordinate the interactions between the biochemical molecular processes and the physical components in order to respond to its environment effectively. This section focuses on the interplay of interactions between physical and chemical processes at all levels within a cell. On a small scale, experiments have shown that the specific spatial distribution of an actin nucleator, ActA, on the surface of small (3 μm) vesicles not only allows for these vesicles to move by actin polymerization, but it also causes a partitioning of the forces generated during this process (111, 112). The combined forces compress the vesicle, and modeling of their distributions and magnitudes show that while there are many protrusive forces along the sides, there are strong retarding forces at the rear of the vesicle. ActA has nucleation activity via activation of Arp2/3 (113), causing protrusive force, and interacts with the comet tail via Ena/VASP proteins as well as Arp2/3 binding (114, 115), contributing a retarding force. These two types of forces partition into local subregions through the spatial polar distribution of ActA's biochemical activities along the vesicle surface, and they contribute to the physical forces and dynamics produced by the polymerizing actin comet tail. Taking a more global viewpoint, if one considers the general forces on the scale of a whole keratocyte that are produced in order for it to move (Figure 2d), they too are seen to be spatially partitioned. Protruding forces exist at the front, disappear at a transitional zone, and become retracting forces at the rear (116, 117). The cell as a whole, therefore, organizes its local behaviors to partition its forces globally for movement.

As is the case with many processes in the cell, causes and effects can become interchanged owing to complicated feedback loops. Physical changes in the environment can regulate biochemical pathways and the subsequent local and global reorganizations they produce. For example, locally produced forces may induce greater curvature at those regions of the cell membrane. Transient changes in membrane curvature induced by addition of phospholipid to platelets have been shown to cause filopodia to form (118). The change in curvature locally activates phosphoinositide 3-kinase (PI 3-kinase), which in turn activates multiple molecules, including PIP_2, known to be involved in modulating actin dynamics [reviewed in (119)]. This effect was also seen in fibroblasts in the absence of microtubules, giving a less well-defined cellular structure (118). In this example, a local change in the physical environment triggers signals implicated in regulating protrusive actin structures to reorganize the protrusion dynamics locally. To restrict the reorganization to a specific region, other structural organizations, such as the microtubule network, can regulate the extent of the effect these changes have on the cell.

As was discussed above, a cell needs to create and regulate traction to move across a substrate. The attachments to the surface need to be spatially well distributed to have a positive effect on the movement and must be short-lived not to arrest the cell completely. An example of precise regulation was directly shown through the transient interactions of Arp2/3 and vinculin at new sites of adhesion (120). Activators of Arp2/3 and vinculin stimulated their interaction, and cells expressing mutant vinculin unable to associate with the Arp2/3 complex showed decreased lamellipodial protrusion (120). These experiments demonstrate the direct link is transient in nature and allow polymerization and adhesion to be spatially coordinated between active Arp2/3 complex and active adhesion sites. In addition, the forces placed on initial sites of adhesion can regulate their development into more adhesive structures (121) and therefore their capability to coordinate with the rest of the cell for its migration. Locally exerted forces, such as pulling on a macrophage adhering to a bead, can regulate the polymerization dynamics, stimulating the formation of protrusive actin structures as seen when a macrophage engulfs a pathogen (122). Stretching forces occurring when a cell's protrusive rate is greater than its retractive rate can regulate calcium release in keratocytes through stretch-activated calcium channels (123). The resultant increase in intracellular calcium levels stimulates the retraction of the cell rear to restore the net steady-state balance between protrusion and retraction rates. Thus the physical forces exerted on the cell as it moves communicate with its intracellular molecule-based dynamic reorganizations to coordinate its movement.

Cellular coordination also occurs through global effects of volume changes and their mechanical consequences on the actin network. The spatial distribution of ion channels is regulated such that their effect at the leading edge of crawling cells induces water uptake and volume shrinkage at the contracting rear (124). Additionally, cells can exhibit dramatic actin flow, as seen in *Aplysia* growth cones, in the direction opposing motion (30). The hydrodynamic environment within the cell is therefore rapidly changing. Moving *Dictyostelium* cells display much greater diffusion of intracellular GFP than do stationary ones, making their effective viscosity 1.4 times lower when moving while not changing their volume significantly (15). In addition, diffusion is slightly enhanced in the protruding region of the cell, indicating local differences in the mechanical properties of the dynamically polymerizing actin structures. Experiments examining actin dynamics during cell protrusion show a rapid trafficking of monomeric actin to protrusive structures in fibroblasts faster than can be accounted for by diffusion (125). Other experiments investigating the relative distributions of 10 and 70 kDa dextrans in cytotoxic T-lymphocytes reveal a greater accessibility of the larger dextran to regions of cellular protrusion upon recognition of target cells (126), contrary to the predictions of simple diffusion or molecular sieving. Flow-like properties of the liquid phase of the intracellular environment may allow molecules moving through their crowded environment within actively polymerizing regions of a cell to exhibit dramatic mobility differences. In an extreme case, the sea cucumber *Thyone briareus* extends its long 90 μm acrosomal process fully within ten seconds through polymerization of actin, a process requiring an active flow-dependent transport of monomer to the tip of the extending process (127). Actin meshworks have gel-like mechanical properties, and it has been suggested that hydrodynamic pressure on gels may create channels (125). These hypothetical channels could provide a route for larger molecules or complexes to cover their required distances on shorter timescales simply through the inherent mechanical properties produced by dynamic polymerizing structures acting in coordination with cellular movement. This type of rapid transport directed to the leading edge of protruding regions has been previously considered in the context of myosin II-dependent contractions at the rear of the cell, providing the necessary intracellular flow (128). Further work is needed to understand this hydrodynamic flow mechanism and the positive effect its coordination throughout the cell could have on continuous protrusion dynamics.

Coordination Between Local Behaviors and Global Responses

For a cell to move and respond to its environment, a local behavior induced by an environmental signal clearly must be coordinated and coregulated with a global response to the extent required by the cell. Growth cones explore their environment and, upon discovering chemical signals, must change their direction of movement to follow their required path and stabilize microtubule bundles in the axons of their neurons (129, 130). Local changes in substrate adhesion on growth cones have been shown to interfere with their continuing retrograde flow and can induce large-scale cellular reorganization of the actin and microtubule network [reviewed in (131)]. Additionally, disruption of actin bundles at the periphery induces a collapse of a region of the protrusive structures nearby, leading to repulsive turning motion (132). Thus, a local cue can induce global reorganization within the cell, which results in a new large-scale behavior directly linked to the localized effect.

Now consider specific examples of regulated proteins affecting local protrusion dynamics and therefore the overall global response of the cell. The local activation of thymosin β4 in a keratocyte induces a local depolymerization of actin structures and a global response manifested in the whole cell, which pivots around that region through its coordinated protrusion-retraction dynamics (133). The multiple functions of the Ena/VASP proteins have been shown to have significant effects on the behavior of the dendritic array and of moving cells. The direct effect of Ena/VASP proteins on the dendritic array dynamics alone is exemplified in their strong positive effect on the rapid movement of pathogens, such as *Listeria*, for which dendritic array treadmilling is sufficient for movement. In *Dictyostelium* cells, DdVASP induces filopodial extensions and regulates actin polymerization (134). It also localizes to the regions of the cell that respond to chemoattractant. In the absence of DdVASP, cells still extend pseudopods but no longer adhere well. The combination of these effects causes DdVASP null mutants to respond poorly to chemoattractant, effectively decreasing their directional motility (134). In neutrophils, sequestration of VASP proteins from the cell periphery causes a concentration-dependent decrease in migration rate upon stimulation with fMLP as well as a decrease in polymerization capability (135). Additionally, Jurkat T cells require Ena/VASP proteins for the rapid polymerization of actin extensions toward their target (136). These observations demonstrate a link between local Ena/VASP-dependent enhancement of polymerization, possible effects on adhesion, and the global effect this protein family can have on efficient chemotaxis, migration, or active actin dynamics in these cells.

In contrast to the previously mentioned cell types in which Ena/VASP proteins generally enhance cell movement, in unstimulated fibroblasts it has been shown that although Ena/VASP proteins induce local rapid protrusive behavior, evident as ruffling (89), their absence induces protrusion of a larger, more persistent lamellipodium and more rapid overall cellular movement. These apparently contradictory results may be reconciled by considering the biological context of motility in these cell types. Neutrophils and *Dictyostelium* cells are fast moving, responding to their environment in an active and directed fashion. The coordination between their local actin dynamics and global cellular movement is optimized for a quick response, as are the actin rearrangements in activated T cells. Unstimulated fibroblasts, however, move at rates 20 times lower than *Listeria* or neutrophils and may respond to cues in their environment, but they are not inherently motile cells dependent on rapid movement for their biological function. Perhaps the coordination between local protrusion and global motility is more complex or less efficiently regulated in these cells, leading them to benefit from a slower, more persistent protruding structure. Fibroblasts are also five times larger in diameter and must therefore coordinate their efforts over a greater distance, which may affect the cell's capability to coordinate its behavior at a fast rate. Both a difference in the dynamics of the protrusive structures between these cell types and a possible difference in coordination efficiency caused by their varied sizes may explain the distinct effects of Ena/VASP proteins. These examples illustrate the complexity of cellular regulation of molecular, physical, and mechanical processes and the coordination between them to create motile behavior. They also demonstrate that the global outcome from the integration of local functions may be distinct in cells because of differences in their motile states, for example actively moving neutrophils in contrast to unstimulated fibroblasts. The effects that intracellular components can have on the dynamics of crawling cell motility should be considered within the context of the type and morphology of movements of various cell types as well as the functions that these cells perform in their supercellular environment.

STUDYING THE SCALES OF A CELL FROM MOLECULAR TO GLOBAL COORDINATION

The field is continuing to obtain specific detailed information on the identity, interactions, and dynamics of many molecules implicated in cellular motility and the ways in which a cell can integrate their functions, spanning the scales of proteins and molecular machines to the physical, mechanical, and dynamic properties of whole cells. In the continual attempt to understand the process by which a cell crawls, it is important to investigate the required interactions by perturbing cellular organizations at all of these levels. Biochemical experiments provide information on the identity of necessary components and on the dynamics of their interactions. Reconstituted systems of a limited number of proteins lead to an understanding of the specific capabilities of that subset of proteins in organizing structures and coregulating their functions. Experiments at the level of biochemical complexity of all molecular components of a cell, such as those using extracts, are needed to understand the basic principles of processes, without spatial and temporal cell boundary conditions, at an intermediate level of organization. Finally, direct investigation of the cell as a whole provides information on its overall temporal and spatial regulation as well as its physical and mechanical properties. Experiments at all of these levels are useful in providing frameworks for mathematical models of various cellular processes, which in turn can direct specific areas of further experiment and subsequently lead to more refined and accurate models. The integration of information learned from all levels of experiment must then be applied to the whole cell, to understand how it has been able to regulate the behavior of its components all along, from the individual protein through four or more orders of magnitude in space and time to the coordinated behavior of its movement. Likewise, additional investigation in the field can build on a mechanically sound, preexisting structure to coordinate biological, physical, and chemical perspectives to progress into unexplored territory.

ACKNOWLEDGMENTS

We thank all of the publishers and authors contributing to the composite figures in this review for their permission to include their work. We thank our colleagues in the field of cell motility and past and present members of the Theriot lab for valuable and stimulating discussions, especially Paula Giardini Soneral, Catherine Lacayo, and Cyrus Wilson. We also thank Anne Meyer and Catherine Lacayo for critical reading of the manuscript. J.A.T. is supported by grants from the National Institutes of Health, the American Heart Association, and the David and Lucile Packard Foundation. S.M.R. is supported by a National Science Foundation Predoctoral Fellowship.

The *Annual Review of Biochemistry* is online at http://biochem.annualreviews.org

LITERATURE CITED

1. Radice GP. 1980. *J. Cell Sci.* 44: 201–23
2. Zigmond SH, Lauffenburger DA. 1986. *Annu. Rev. Med.* 37:149–55
3. Tessier-Lavigne M, Goodman CS. 1996. *Science* 274:1123–33
4. Purcell E. 1977. *Am. J. Phys.* 45:3–11
5. Agutter PS, Wheatley DN. 2000. *BioEssays* 22:1018–23
6. Fulton AB. 1982. *Cell* 30:345–47
7. Goodsell D. 1993. *The Machinery of Life.* New York: Springer-Verlag. 68 pp.
8. Medalia O, Typke D, Hegerl R, Angenitzki M, Sperling J, Sperling R. 2002. *J. Struct. Biol.* 138:74–84
9. Luby-Phelps K. 2000. *Int. Rev. Cytol.* 192:189–221
10. Schwikowski B, Uetz P, Fields S. 2000. *Nat. Biotechnol.* 18:1257–61
11. Minton AP. 2001. *J. Biol. Chem.* 276: 10577–80
12. Ellis RJ. 2001. *Trends Biochem. Sci.* 26:597–604
13. Verkman AS. 2002. *Trends Biochem. Sci.* 27:27–33
14. Provance DW Jr, McDowall A, Marko M, Luby-Phelps K. 1993. *J. Cell Sci.* 106:565–77
15. Potma EO, de Boeij WP, Bosgraaf L, Roelofs J, van Haastert PJ, Wiersma DA. 2001. *Biophys. J.* 81:2010–19
16. Aon MA, Gomez-Casati DF, Iglesias AA, Cortassa S. 2001. *Cell Biol. Int.* 25:1091–99
17. Crane H. 1950. *Sci. Mon.* 70:376–89
18. Pauling L. 1953. *Discuss. Faraday Soc.* 13:170–76
19. Oosawa F, Asakura S. 1975. *Thermodynamics of the Polymerization of Protein.* London: Academic. 204 pp.
20. Hill TL, Kirschner MW. 1982. *Int. Rev. Cytol.* 78:1–125
21. Theriot JA. 2000. *Traffic* 1:19–28
22. Wegner A. 1976. *J. Mol. Biol.* 108: 139–50

23. Fujiwara I, Takahashi S, Tadakuma H, Funatsu T, Ishiwata S. 2002. *Nat. Cell Biol.* 4:666–73
24. Rodionov VI, Borisy GG. 1997. *Science* 275:215–18
25. Shaw SL, Kamyar R, Ehrhardt DW. 2003. *Science* 300:1715–18
26. Selve N, Wegner A. 1986. *J. Mol. Biol.* 187:627–31
27. Heath JP. 1983. *J. Cell Sci.* 60:331–54
28. Svitkina TM, Verkhovsky AB, Borisy GG. 1995. *J. Struct. Biol.* 115:290–303
29. Wang YL. 1985. *J. Cell Biol.* 101: 597–602
30. Forscher P, Smith SJ. 1988. *J. Cell Biol.* 107:1505–16
31. Theriot JA, Mitchison TJ. 1991. *Nature* 352:126–31
32. Theriot JA. 1994. *Semin. Cell Biol.* 5:193–99
33. Svitkina TM, Borisy GG. 1999. *J. Cell Biol.* 145:1009–26
34. Pollard TD, Blanchoin L, Mullins RD. 2000. *Annu. Rev. Biophys. Biomol. Struct.* 29:545–76
35. Pantaloni D, Le Clainche C, Carlier MF. 2001. *Science* 292:1502–6
36. Small JV, Stradal T, Vignal E, Rottner K. 2002. *Trends Cell. Biol.* 12:112–20
37. Welch MD, Mullins RD. 2002. *Annu. Rev. Cell Dev. Biol.* 18:247–88
38. Pollard TD, Borisy GG. 2003. *Cell* 112: 453–65
39. Carlier MF, Le Clainche C, Wiesner S, Pantaloni D. 2003. *BioEssays* 25: 336–45
40. dos Remedios CG, Chhabra D, Kekic M, Dedova IV, Tsubakihara M, et al. 2003. *Physiol. Rev.* 83:433–73
41. Pollard TD. 2003. *Nature* 422:741–45
42. Volkmann N, Amann KJ, Stoilova-McPhie S, Egile C, Winter DC, et al. 2001. *Science* 293:2456–59
43. Amann KJ, Pollard TD. 2001. *Nat. Cell Biol.* 3:306–10

44. Ichetovkin I, Grant W, Condeelis J. 2002. *Curr. Biol.* 12:79–84

45. Dayel MJ, Holleran EA, Mullins RD. 2001. *Proc. Natl. Acad. Sci. USA* 98:14871–76

46. Le Clainche C, Pantaloni D, Carlier MF. 2003. *Proc. Natl. Acad. Sci. USA* 100:6337–42

47. Falet H, Hoffmeister KM, Neujahr R, Italiano JE Jr, Stossel TP, et al. 2002. *Proc. Natl. Acad. Sci. USA* 99: 16782–87

48. DesMarais V, Ichetovkin I, Condeelis J, Hitchcock-DeGregori SE. 2002. *J. Cell Sci.* 115:4649–60

49. Carlier MF, Laurent V, Santolini J, Melki R, Didry D, et al. 1997. *J. Cell Biol.* 136:1307–22

50. Fischer RS, Fritz-Six KL, Fowler VM. 2003. *J. Cell Biol.* 161:371–80

51. Dawe HR, Minamide LS, Bamburg JR, Cramer LP. 2003. *Curr. Biol.* 13: 252–57

52. Goldschmidt-Clermont PJ, Furman MI, Wachsstock D, Safer D, Nachmias VT, Pollard TD. 1992. *Mol. Biol. Cell* 3:1015–24

53. Abraham VC, Krishnamurthi V, Taylor DL, Lanni F. 1999. *Biophys. J.* 77: 1721–32

54. Cameron LA, Giardini PA, Soo FS, Theriot JA. 2000. *Nat. Rev. Mol. Cell Biol.* 1:110–19

55. Loisel TP, Boujemaa R, Pantaloni D, Carlier MF. 1999. *Nature* 401:613–16

56. Sambeth R, Baumgaertner A. 2001. *Phys. Rev. Lett.* 86:5196–99

57. Svitkina TM, Verkhovsky AB, McQuade KM, Borisy GG. 1997. *J. Cell Biol.* 139:397–415

58. Wessels D, Soll DR, Knecht D, Loomis WF, De Lozanne A, Spudich J. 1988. *Dev. Biol.* 128:164–77

59. Jay PY, Pham PA, Wong SA, Elson EL. 1995. *J. Cell Sci.* 108:387–93

60. Euteneuer U, Schliwa M. 1984. *Nature* 310:58–61

61. Verkhovsky AB, Svitkina TM, Borisy GG. 1999. *Curr. Biol.* 9:11–20

62. Rodriguez OC, Schaefer AW, Mandato CA, Forscher P, Bement WM, Waterman-Storer CM. 2003. *Nat. Cell Biol.* 5:599–609

63. Zamir E, Geiger B. 2001. *J. Cell Sci.* 114:3583–90

64. Lee J, Jacobson K. 1997. *J. Cell Sci.* 110:2833–44

65. de Beus E, Jacobson K. 1998. *Cell Motil. Cytoskelet.* 41:126–37

66. Kaverina I, Krylyshkina O, Small JV. 1999. *J. Cell Biol.* 146:1033–44

67. Bottino D, Mogilner A, Roberts T, Stewart M, Oster G. 2002. *J. Cell Sci.* 115:367–84

68. LeClaire LL 3rd, Stewart M, Roberts TM. 2003. *J. Cell Sci.* 116:2655–63

69. Joanny JF, Julicher F, Prost J. 2003. *Phys. Rev. Lett.* 90:168102

70. Nicolis G, Prigogine I. 1977. *Self-Organization in Non-Equilibrium Systems.* New York: Wiley. 491 pp.

71. Kirschner M, Gerhart J, Mitchison T. 2000. *Cell* 100:79–88

72. Surrey T, Nedelec F, Leibler S, Karsenti E. 2001. *Science* 292:1167–71

73. Meinhardt H, Gierer A. 2000. *BioEssays* 22:753–60

74. Xu JS, Wang F, Van Keymeulen A, Herzmark P, Straight A, et al. 2003. *Cell* 114:201–14

75. Janmey PA, Hvidt S, Oster GF, Lamb J, Stossel TP, Hartwig JH. 1990. *Nature* 347:95–99

76. Orlova A, Egelman EH. 1992. *J. Mol. Biol.* 227:1043–53

77. Galkin VE, VanLoock MS, Orlova A, Egelman EH. 2002. *Curr. Biol.* 12: 570–75

78. Schoenenberger CA, Bischler N, Fahrenkrog B, Aebi U. 2002. *FEBS Lett.* 529:27–33

79. Wong GCL, Lin A, Tang JX, Li Y, Janmey PA, Safinya CR. 2003. *Phys. Rev. Lett.* 91:018103

80. Tang JX, Janmey PA. 1996. *J. Biol. Chem.* 271:8556–63

81. Kulp DT, Herzfeld J. 1995. *Biophys. Chem.* 57:93–102

82. Tseng Y, An KM, Wirtz D. 2002. *J. Biol. Chem.* 277:18143–50

83. Fabry B, Maksym GN, Butler JP, Glogauer M, Navajas D, Fredberg JJ. 2001. *Phys. Rev. Lett.* 87:148102

84. Wachsstock DH, Schwartz WH, Pollard TD. 1993. *Biophys. J.* 65:205–14

85. Limozin L, Sackmann E. 2002. *Phys. Rev. Lett.* 89:168103

86. Bretscher A, Weber K. 1980. *J. Cell Biol.* 86:335–40

87. Jung G, Wu X, Hammer JA 3rd. 1996. *J. Cell Biol.* 133:305–23

88. Kwiatkowski AV, Gertler FB, Loureiro JJ. 2003. *Trends Cell. Biol.* 13:386–92

89. Bear JE, Svitkina TM, Krause M, Schafer DA, Loureiro JJ, et al. 2002. *Cell* 109:509–21

90. Maly IV, Borisy GG. 2001. *Proc. Natl. Acad. Sci. USA* 98:11324–29

91. Weaver AM, Karginov AV, Kinley AW, Weed SA, Li Y, et al. 2001. *Curr. Biol.* 11:370–74

92. Sever S. 2002. *Curr. Opin. Cell Biol.* 14:463–67

93. Schafer DA, Weed SA, Binns D, Karginov AV, Parsons JT, Cooper JA. 2002. *Curr. Biol.* 12:1852–57

94. Evangelista M, Zigmond S, Boone C. 2003. *J. Cell Sci.* 116:2603–11

95. Maul RS, Song Y, Amann KJ, Gerbin SC, Pollard TD, Chang DD. 2003. *J. Cell Biol.* 160:399–407

96. Borukhov I, Bruinsma RF, Gelbart WM, Liu AJ. 2001. *Phys. Rev. Lett.* 86:2182–85

97. Svitkina TM, Bulanova EA, Chaga OY, Vignjevic DM, Kojima S, et al. 2003. *J. Cell Biol.* 160:409–21

98. Vignjevic D, Yarar D, Welch MD, Peloquin J, Svitkina T, Borisy GG. 2003. *J. Cell Biol.* 160:951–62

99. Hartwig JH, Barkalow K, Azim A, Italiano J. 1999. *Thromb. Haemost.* 82:392–98

100. Hall A. 1998. *Science* 279:509–14

101. Higgs HN, Pollard TD. 2001. *Annu. Rev. Biochem.* 70:649–76

102. Nakagawa H, Miki H, Nozumi M, Takenawa T, Miyamoto S, et al. 2003. *J. Cell Sci.* 116:2577–83

103. Breitbach K, Rottner K, Klocke S, Rohde M, Jenzora A, et al. 2003. *Cell. Microbiol.* 5:385–93

104. Miller MJ, Wei SH, Parker I, Cahalan MD. 2002. *Science* 296:1869–73

104a. Cameron LA, Footer MJ, van Oudenaarden A, Theriot JA. 1999. *Proc. Natl. Acad. Sci. USA* 96:4908–13

105. van Oudenaarden A, Theriot JA. 1999. *Nat. Cell Biol.* 1:493–99

106. Wedlich-Soldner R, Altschuler S, Wu LN, Li R. 2003. *Science* 299:1231–35

107. Weiner OD, Servant G, Welch MD, Mitchison TJ, Sedat JW, Bourne HR. 1999. *Nat. Cell Biol.* 1:75–81

108. Krueger EW, Orth JD, Cao H, McNiven MA. 2003. *Mol. Biol. Cell* 14:1085–96

109. Ballestrem C, Wehrle-Haller B, Imhof BA. 1998. *J. Cell Sci.* 111(12):1649–58

110. Vicker MG. 2002. *Exp. Cell Res.* 275:54–66

111. Giardini PA, Fletcher DA, Theriot JA. 2003. *Proc. Natl. Acad. Sci. USA* 100:6493–98

112. Upadhyaya A, Chabot JR, Andreeva A, Samadani A, van Oudenaarden A. 2003. *Proc. Natl. Acad. Sci. USA* 100:4521–26

113. Welch MD, Rosenblatt J, Skoble J, Portnoy DA, Mitchison TJ. 1998. *Science* 281:105–8

114. Gerbal F, Laurent V, Ott A, Carlier MF, Chaikin P, Prost J. 2000. *Eur. Biophys. J.* 29:134–40

115. Kuo SC, McGrath JL. 2000. *Nature* 407:1026–29

116. Lee J, Ishihara A, Theriot JA, Jacobson K. 1993. *Nature* 362:167–71

117. Oliver T, Dembo M, Jacobson K. 1999. *J. Cell Biol.* 145:589–604

118. Bettache N, Baisamy L, Baghdiguian S, Payrastre B, Mangeat P, Bienvenue A. 2003. *J. Cell Sci.* 116:2277–84

119. Yin HL, Janmey PA. 2003. *Annu. Rev. Physiol.* 65:761–89

120. DeMali KA, Barlow CA, Burridge K. 2002. *J. Cell Biol.* 159:881–91

121. Galbraith CG, Yamada KM, Sheetz MP. 2002. *J. Cell Biol.* 159:695–705

122. Vonna L, Wiedemann A, Aepfelbacher M, Sackmann E. 2003. *J. Cell Sci.* 116: 785–90

123. Lee J, Ishihara A, Oxford G, Johnson B, Jacobson K. 1999. *Nature* 400:382–86

124. Schwab A. 2001. *Am. J. Physiol. Renal Physiol.* 280:F739–47

125. Zicha D, Dobbie IM, Holt MR, Monypenny J, Soong DY, et al. 2003. *Science* 300:142–45

126. Waters JB, Oldstone MB, Hahn KM. 1996. *J. Immunol.* 157:3396–403

127. Olbris DJ, Herzfeld J. 1999. *Biophys. J.* 77:3407–23

128. Conrad PA, Giuliano KA, Fisher G, Collins K, Matsudaira PT, Taylor DL. 1993. *J. Cell Biol.* 120:1381–91

129. Morris JR, Lasek RJ. 1982. *J. Cell Biol.* 92:192–98

130. Tanaka EM, Kirschner MW. 1991. *J. Cell Biol.* 115:345–63

131. Suter DM, Forscher P. 2000. *J. Neurobiol.* 44:97–113

132. Zhou FQ, Waterman-Storer CM, Cohan CS. 2002. *J. Cell Biol.* 157:839–49

133. Roy P, Rajfur Z, Jones D, Marriott G, Loew L, Jacobson K. 2001. *J. Cell Biol.* 153:1035–48

134. Han YH, Chung CY, Wessels D, Stephens S, Titus MA, et al. 2002. *J. Biol. Chem.* 277:49877–87

135. Anderson SI, Behrendt B, Machesky LM, Insall RH, Nash GB. 2003. *Cell Motil. Cytoskelet.* 54:135–46

136. Krause M, Sechi AS, Konradt M, Monner D, Gertler FB, Wehland J. 2000. *J. Cell Biol.* 149:181–94

137. McGough A, Pope B, Chiu W, Weeds A. 1997. *J. Cell Biol.* 138:771–81

138. Niederman R, Amrein PC, Hartwig J. 1983. *J. Cell Biol.* 96:1400–13

139. Wang YX, Miller AL, Mooseker MS, Koleske AJ. 2001. *Proc. Natl. Acad. Sci. USA* 98:14865–70

140. Cox D, Ridsdale JA, Condeelis J, Hartwig J. 1995. *J. Cell Biol.* 128: 819–35

141. Kondo K, Yoshitake J. 1976. *J. Electron Microsc.* 25:99–102

142. Connolly JL, Seeley PJ, Greene LA. 1985. *J. Neurosci. Res.* 13:183–98

Annu. Rev. Biochem. 2004. 73:241–68
doi: 10.1146/annurev.biochem.73.011303.073626
First published online as a Review in Advance on April 8, 2004

ATP-Binding Cassette Transporters in Bacteria

Amy L. Davidson[1] and Jue Chen[2]

[1]Department of Molecular Virology and Microbiology, Baylor College of Medicine, Houston, Texas 77030; email: davidson@bcm.tmc.edu

[2]Department of Biological Sciences, Purdue University, West Lafayette, Indiana 47907; email: chenjue@purdue.edu

Key Words   structure, transport, coupling, conformational change, efflux

■ **Abstract**   ATP-binding cassette (ABC) transporters couple ATP hydrolysis to the uptake and efflux of solutes across the cell membrane in bacteria and eukaryotic cells. In bacteria, these transporters are important virulence factors because they play roles in nutrient uptake and in secretion of toxins and antimicrobial agents. In humans, many diseases, such as cystic fibrosis, hyperinsulinemia, and macular dystrophy, are traced to defects in ABC transporters. Recent advances in structural determination and functional analysis of bacterial ABC transporters, reviewed herein, have greatly increased our understanding of the molecular mechanism of transport in this transport superfamily.

CONTENTS

0066-4154/04/0707-0241$14.00

241

INTRODUCTION

Transporters belonging to the ATP-binding cassette (or ABC) superfamily couple the energy released from ATP hydrolysis to the translocation of a wide variety of substances into or out of cells and organelles. The establishment of the family traces its origin to the discovery that a protein responsible for multiple drug resistance in human cancers, alternately referred to as P-glycoprotein, MDR, or ABCB1 (1, 2), bore homology to the well-studied family of periplasmic binding protein-dependent transporters that mediate uptake of a large variety of solutes in gram-negative bacteria (3). P-glycoprotein was later found to reduce the effective concentration of chemotherapeutic drugs inside cells by pumping them out in an ATP-dependent fashion (4). Every ABC transporter appears to be made of four protein domains or subunits: two hydrophobic membrane-spanning domains (MSDs) that are presumed to constitute the translocation pathway or channel across the membrane and two hydrophilic nucleotide-binding domains (NBDs) that interact at the cytoplasmic surface to supply the energy for active transport. The term ATP-binding cassette, which refers to the NBD, was coined to describe the modular nature of these transporters (5); "traffic ATPase" has also been used to describe these transporters because they control the direction of movement of the transported compound (6) in and out of the cells (7).

ABC transporters now constitute one of the largest superfamilies of proteins known (8, 9): There are 48 ABC transporters in humans, and 80 in the gram-negative bacterium *Escherichia coli*. In the transport classification database (10) (http://www-biology.ucsd.edu/~msaier/transport/), the ABC family is currently subdivided into 22 subfamilies of prokaryotic importers, 24 subfamilies of prokaryotic exporters, and 10 subfamilies of eukaryotic proteins. Several other groups maintain online databases, with similar classifications, devoted exclusively to ABC transporters (11, 12). Transporters are assigned to subfamilies on the basis of function and phylogeny, and assignments generally correlate with substrate specificity, though there are exceptions (10, 13, 14). Even though there may be no homology between MSDs in different subfamilies, some degree of homology is maintained across the entire superfamily in the NBDs (25% to 30% identity), suggesting that a similar mechanism for coupling of transport to ATP hydrolysis is employed.

Bacterial Uptake Systems

All bacterial ABC transporters that mediate uptake utilize a high-affinity solute binding protein that is located in the periplasm of gram-negative bacteria and is either tethered to the cell surface (3) or fused to the transporter itself (15) in gram-positive bacteria. These binding protein-dependent transporters take up a wide variety of substrates, which include nutrients and osmoprotectants that range from small sugars, amino acids, and small peptides to metals, anions, iron chelators (siderophores), and vitamin B_{12}. In gram-negative organisms, small substrates gain access to the periplasm via diffusion through outer membrane pore-forming proteins known as porins (16). Larger compounds, such as vitamin

B_{12} and iron-siderophore complexes, are actively transported across the outer membrane through high-affinity transporters utilizing energy transduced from the electrochemical gradient across the cytoplasmic membrane (17).

The high-resolution structures of many periplasmic binding proteins have been determined (18). Three distinct folding patterns have been identified, all of which have two lobes with a substrate-binding site positioned in a cleft between them. The class I and class II binding proteins, represented by the galactose/ glucose binding protein and the maltose binding protein, respectively (19), differ primarily in the fold of the two domains. Both classes undergo a conformational change involving the bending of a hinge that joins the two lobes. The binding proteins take on a more open conformation in absence of the transported substrate, and substrate binding promotes bending at the hinge and closing the cleft around the substrate (18, 19). Bound substrates experience rapid exchange, indicating that the two lobes open and close readily even in the presence of substrate (20, 21). Substrate homologues that bind to the binding protein without inducing closure of the substrate-binding cleft may not be transported (21a). Class III binding proteins, represented by the Ferric siderophore binding protein (22) and vitamin B_{12} binding protein (23, 24), have a more rigid hinge structure and may not experience the same type of substrate-induced conformational changes that have been documented in the class I and class II binding proteins.

Binding proteins have two roles in transport; both of which were elucidated from the study of mutant transporters able to function in the absence of a binding protein (25, 26). Although transport still exhibits substrate specificity, the apparent K_m for substrate in the transport reaction is greatly increased in binding protein-independent mutants, indicating that binding proteins are responsible for the high-affinity transport that is characteristic of these transporters. The K_d for substrate-binding to the binding protein often dictates the K_m for substrate in the transport reaction (27). The reason why binding proteins are absolutely required in the wild type is that they stimulate the ATPase activity of the transporters (28, 29). The binding protein-independent mutants display substantially increased rates of ATPase activity as compared to the wild type in the absence of binding protein; this explains how they are able to transport maltose in the absence of binding protein (28, 30).

Further insight into the mechanism of translocation arose from the observation that the maltose binding protein (MBP), which normally displays a low affinity for the maltose transporter (MalFGK$_2$) as judged by the K_m for MBP in the transport reaction (25–100 μM) (31, 32), becomes tightly bound to the transporter in an intermediate conformation stabilized by the phosphate analogue vanadate (33). Following ATP hydrolysis, vanadate becomes trapped along with ADP in one of the two nucleotide binding sites in a conformation that is presumed to mimic the transition state of ATP hydrolysis (34, 36, 37). A trigonal bipyramidal coordination for vanadate, characteristic of the γ-phosphate in the transition state, is seen in the high-resolution structure of vanadate-ADP-inhibited myosin (35). The presence of high-affinity binding between MBP and

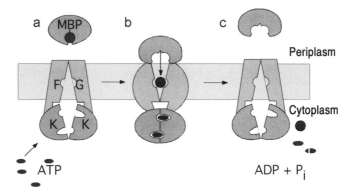

Figure 1 Model for maltose transport. (*a*) MBP, in a closed conformation with maltose bound, interacts with the transporter to initiate transport and hydrolysis. (*b*) In the presumed transition state for ATP hydrolysis, MBP is tightly bound to the transporter in an open conformation that has a lower affinity for maltose, and the transmembrane helices have reoriented to expose an internal sugar-binding site to the periplasm. (*c*) Following ATP hydrolysis, maltose is transported, and MBP is released as the transporter returns to its original conformation. MBP activates the ATPase activity of the transporter by bringing the two NBDs into close proximity, completing the nucleotide-binding sites at the dimer interface. Modified from (33, 90).

$MalFGK_2$ in the presence of vanadate suggests that MBP stimulates the ATPase activity of $MalFGK_2$ by stabilizing the transition state for ATP hydrolysis (33). MBP, trapped in a complex with $MalFGK_2$, no longer binds maltose with high affinity (33), suggesting that it may be open when tightly bound and promote release of maltose into the transmembrane translocation pathway (Figure 1).

Bacterial Efflux Systems

ABC transporters also function in efflux of substances from bacteria, which include surface components of the bacterial cell (such as capsular polysaccharides, lipopolysaccharides, and techoic acid), proteins involved in bacterial pathogenesis (such as hemolysin, heme-binding protein, and alkaline protease), peptide antibiotics, heme, drugs and siderophores (12). Often, the MSD will be fused to the NBD in the bacterial exporters, and two of these half-transporters will interact to complete the four-domain structure of the ABC transporter (38).

In gram-negative organisms, ABC transporters can mediate secretion of their protein substrates across both membranes simultaneously, bypassing the periplasmic space (39). This secretion pathway is referred to as type I secretion and involves two additional accessory proteins in addition to the ABC transporter (6, 39–41). For example, in the secretion of hemolysin (HlyA) from *E. coli*, the inner membrane ABC transporter HlyB interacts with an inner membrane fusion protein HlyD (42) and an outer membrane facilitator TolC (43). The structure of TolC reveals it is a channel that spans both the outer membrane and the periplasmic space of *E. coli* (44, 45). One domain forms a β-barrel typical of other outer membrane channels, and a second helical domain forms a long narrow cylinder that reaches 100 Å into the periplasm. Although the structure of a HlyD homologue has not been determined, these proteins clearly facilitate the contact between the inner membrane ABC transporter and the TolC channel (46, 47) and may even form part of the *trans*-periplasmic tunnel (48). Translocation through the type I system begins with recognition of the substrate via type I-specific signal sequences in the protein (49–53). Both the MSD and NBD of HlyB have been implicated in substrate recognition (53, 54) as well as in the cytoplasmic domain of HlyD (47). Following recognition, the ABC transporter HlyB presumably transports the unfolded protein across the inner membrane (55–58) and into the attached tunnel where folding might commence (48). The association of inner membrane transporter and outer membrane tunnel appears to be transient, triggered by the binding of substrate to the transport apparatus (46, 47, 59, 60). Transport through inner and outer membrane is tightly coupled; in the absence of TolC, the substrate remains in the cytoplasm rather than being released into the periplasm (61).

ABC proteins that are 35% identical to the mammalian multidrug efflux pump P-glycoprotein are present in bacteria (62, 63). The region of homology extends throughout both the MSD and NBD, suggesting that these ABC proteins may recognize and extrude similar substrates. One such protein, LmrA from *Lactococcus lactis*, has been categorized as a multidrug efflux pump. LmrA mimics the human multidrug resistance phenotype when expressed in lung fibroblasts (62), and expression of LmrA in *E. coli* cells lacking an endogenous multidrug efflux pump increases the resistance of the mutant cell to a large variety of antimicrobial agents (64). Another protein, MsbA from *E. coli*, essential for cell viability, has been categorized as a lipid exporter or lipid flippase (63, 65). At nonpermissive temperatures, cells with a temperature-sensitive mutation in *msbA* fail to translocate newly synthesized phospholipids and lipid A molecules from the inner leaflet of the inner membrane to either the outer leaflet or the outer membrane (66). Though it was originally thought that MsbA transported only lipids (67), purified MsbA can catalyze efflux of fluorescent drug analogues from reconstituted membrane vesicles (68) in keeping with its homology to LmrA and P-glycoprotein. Intriguingly, the *lmrA* gene is able to complement the temperature-sensitive allele of *msbA*, indicating that the physiologic function of LmrA, like MsbA, may be to flip lipids across the bilayer (68). Although it has been known for many years that drug-efflux pumps can contribute to multiple-antibiotic resistance of bacteria (69), most of these drug-efflux pumps are coupled to the proton motive force, and it is not clear whether any ABC transporters contribute to clinically significant multiple-antibiotic resistance in bacterial pathogens (70).

The drug-binding site or sites in P-glycoprotein have been localized to the MSD (71) and display a broad specificity for their mostly hydrophobic substrates. Insight into a mechanism by which these binding sites might accommodate different types of substrates has come from the structures of BmrR and QacR, transcriptional regulators involved in multidrug resistance (72, 73), and from AcrB, a non-ABC efflux pump (74, 75). Drugs with different structures interact with different residues within a large, flexible binding site. Experiments using fluorescent drug analogues in both LmrA and the mammalian P-glycoprotein suggest that these hydrophobic substrates are recruited to the drug-binding site from the inner leaflet of the membrane, where their concentration may be higher than in the cytosol, and released into the aqueous milieu (76, 76a). Alternatively, a lipid flippase would be expected to bind lipids from the inner leaflet and release into the outer leaflet of the membrane (77, 77a). In either case, cycles of ATP hydrolysis are proposed to be coupled to changes in the binding affinity and orientation (high-affinity-inward-facing versus low-affinity-outward-facing) of the drug-binding site(s) within the membrane (34, 78). The Lol transporter in *E. coli* represents an interesting variant of an efflux pump in which the substrates, lipoproteins destined for the outer membrane, are removed from the outer leaflet of the inner membrane and transferred to a periplasmic chaperone for translocation to the outer membrane (79).

In the past five years, several high-resolution structures of ABC transporters, primarily of bacterial or archael origin, have been determined. The focus of this review is to describe the recent insights into the mechanism of translocation and, in particular, ATP hydrolysis and energy coupling, which have been gained from these structures of isolated NBDs and the first intact transporters.

STRUCTURE OF THE ATP-BINDING CASSETTE

A Conserved Structure for Nucleotide-Binding Domains

ABC transporters have two NBDs or nucleotide-binding subunits, also known as ATP-binding cassettes, which power the transporter by binding and hydrolyzing ATP. Presumably, ATP binding and/or hydrolysis are coupled to conformational changes in the MSDs that mediate the unidirectional pumping of substrates across the membrane. Despite the large diversity of the transport substrates, the sequences of the ABC components are remarkably conserved among all ABC

Figure 2 Sequence alignment of all ABC ATPases whose structures have been determined. Sequences are ordered based on their homology to E.c.MalK. Critical conserved sequence motifs are highlighted. Secondary structure elements for E.c.MalK are shown above the sequence. The color schemes for the secondary structure as well as the conserved motifs are the same as in Figure 3.

```
                              s1              s2         s3   Walker A   h1

E.c.MalK    1   -----------MASVQLQNVTKAWGEVV----VSKDINLDIHEGEFVVFVGPSGCGKSTLLRMI
T.l.MalK    1   -----------MAGVRLVDVWKVFGEVT----AVREMSLEVKDGEFMILLGPSGCGKTTTLRMI
GlcV        1   -----------MVRIIVKNVSKVFKKGK--VVALDNVNINIENGERFGILGPSGAGKTTFMRII
MJ0796      1   ------------XIKLKNVTKTYKXGEEIIYALKNVNLNIKEGEFVSIXGPSGSGKSTXLNII
E.c.MsbA  351   --------------------YPGR--DVPALRNINLKIPAGKTVALVGRSGSGKSTIASLI
V.C.MsbA  351   ----------------------YQGK--EKPALSHVSFSIPQGKTVALVGRSGSGKSTIANLF
cTAP1     489   PPSGLLTPLHLEGLVQFQDVSFAYPNRP-DVLVLQGLTFTLRPGEVTALVGPNGSGKSTVAALL
HlyB        1   -------------DITFRNIRFRYKPD--SPVILDNINLSIKQGEVIGIVGRSGSGKSTLTKLI
HisP        1   --------MMSENKLHVIDLHKRYGGHE----VLKGVSLQARAGDVISIIGSSGSGKSTFLRCI
MJ1267      1   -------MRDTMEILRTENIVKYFGEFK----ALDGVSISVNKGDVTLIIGPNGSGKSTLINVI
BtuD        1   ------------MSIVMQLQDVAESTR-----LGPLSGEVRAGEILHLVGPNGAGKSTLLARM


                       s4       s5                           h2   s6   Q loop        h3

E.c.MalK   50   AGLETITSGDLFIGEKRMNDT--------------PPAERGVGMVFQSYALYPHLSVAENMSF
T.l.MalK   50   AGLEEPSRGQIYIGDKLVADPEKG---------IFVPPKDRDIAMVFQSYALYPHMTVYDNIAF
GlcV       52   AGLDVPSTGELYFDDRLVASNGKL---------IVPPEDR-KIGMVFQTWALYPNLTAFENIAF
MJ0796     52   GCLDKPTEGEVYIDNIKTNDLDDD---------ELTKIRRDKIGFVFQQFNLIPLLTALENVEL
E.c.MsbA  390   TRFYDIDEGEILMDGHDLREYT------------LASLRNQVALVSQNVHLFNDTVANNIAYA
V.C.MsbA  390   TRFYDVDSGSICLDGHDVRDYK------------LTNLRRHFALVSQNVHLFNDTIANNIAYA
TAP1      552   QNLYQPTGGQLLLDGKPLPQYE-------------HRYLHRQVAAVGQEPQVFGRSLQENIAYG
HlyB       50   QRFYIPENGQVLIDGHDLALAD------------PNWLRRQVGVVLQDNVLLNRSIIDNISLA
HisP       53   NFLEKPSEGAIIVNGQNINLVRDKDGQLKVADKNQLRLLRTRLTMVFQHFNLWSHMTVLENVME
MJ1267     54   TGFLKADEGRVYFENKDITNKEP-----------AELYHYGIVRTFQTPQPLKEMTVLENLLI
BtuD       47   AGMTS-GKGSIQFAGQPLEAWS------------ATKLALHRAYLSQQQTPPFATPVWHYLTL


                                          h4                          LSGGQ    h5

E.c.MalK   99   GLKLAGAKK------------EVINQRVNQVAEVLQLAH--LLDRKPKALSGGQRQRVAIGRT
T.l.MalK  105   PLKLRKVPR------------QEIDQRVREVAELLGLTE--LLNRKPRELSGGQRQRVAICDN
GlcV      106   PLTNMKMSK------------EFIRKRVEEVAKILDIHH--VINHFNELSGGQQQRVALARA
MJ0796    107   PLIFKYRGAX---------SGFERRKRAIKCLIDLGLEER-FANHKPNQLSGGQQQRVAIARA
E.c.MsbA  441   RTEQYSREQI----------ELAARMAYAMDFINKMDNGLDTVIGENGVLLSGGQRQRIAIARA
V.C.MsbA  441   AFGEVTREQI----------EQAARQAHAMEFIENMPQGLDTVIGENGTSLSGGQRQRIAIARA
TAP1      603   LTQKPTMEEI----------TAAAVKSGAHSFISGLPQGYDTEVDEAGSQLSGGQRQAVALARA
HlyB      101   NPG-MSVEKV----------IYAAKLAGAHDFISELREGYNTIVGEQGAGLSGGQRQRIAIARA
HisP      117   APIQVLGLS-----------KHDARERALKYLAKVGIDER-AQGKYPVHLSGGQQQRVSIARA
MJ1267    106   GEICPGESPLNSLFYKKWIPKEEEMVEKAFKILEFLKLSH--LYDRKAGELSGGQMKLVEIGRA
BtuD       95   H--QHDKTRT------------ELLNDVAGALALDDK--LGRSTNQLSGGEWQRVRLAAV


                     Walker B                                      H motif
                           s7                    h6              s8   h7        s9

E.c.MalK  148   LVAE-P------SVFLLDEPLSNLDAALRVQMRIEISRLHKRLGRTMIYVTHDQVEAMTLADKI
T.l.MalK  154   IVRK-P------QVFLLDEPLSNLDAKLRVRMRAELKKLQRQLGVTTIYVTHDVEAMTMGDRI
GlcV      155   LVKD-P------SLLLLDEPFSNLDARMRDSARALVKEVQSRLGVTLLVVSHDPADIFAIADRV
MJ0796    160   LANN-P------PIILADEPTGALDSKTGEKIXQLLKKLNEEDGKTVVVVTHDINVARFGE-RI
E.c.MsbA  495   LLRDSP-------ILILDEATSALDTESERAIQAALDELQK--NRTSLVIAHRLS-TIEKADEI
V.C.MsbA  495   LLRDAP-------ILILDEATSALDTESERAIQAALDELQK--NKTVLVIAHRLS-TIEQADEI
TAP1      657   LIRK-P------CVLILDDATSALDANSQLQVEQLLYESPERYSRSVLLITQHLS-LVEQADHI
HlyB      154   LVNN-P------KILIFDEATSALDYESEHVIMRNMHKICK--GRTVIIIAHRLS-TVKNADRI
HisP      168   LAME-P------DVLLFDEPTSALDPELVGEVLRIMQQLAEE-GKTMVVVTHEMGFARHVSSHV
MJ1267    168   LMTN-P------KMIVMDEPIAGVAPGLAHDIFNHVLEKAK--GITFLIIEHRLDIVLNYIDHL
BtuD      141   VLQITPQANPAGQLLLLLDEPMNSLDVAQQSALDKILSALCQQ-GLAIVMSSHDLNHTLRHAHRA


                   s9    s10     h8        h9

E.c.MalK  205   VVLDAGRVAQVGKPLELYHYPADRFVAGFIG-------------  235 (to 371)
T.l.MalK  211   AVMNRGVLQQVGSPDEVYDKPANTFVAGFIG-------------  241 (to 372)
GlcV      212   GVLVKGKLVQVGKPEDLYDNPVSIQVASLIG-------------  242 (to 353)
MJ0796    216   IYLKDGEVER-EKLRGFDDR------------------------  235 (end)
E.c.MsbA  550   VVVEDGVIVERGTHNDLLEH-RGVYAQLHKMQFGQ---------  582 (end)
V.C.MsbA  550   LVVDEGEIIERGRHADLLAQ-DGAYAQLHRIQFGE---------  582 (end)
TAP1      714   LFLEGGAIREGGTHQQLMEK-KGCYWAMVQAPADAPE-------  748 (end)
HlyB      209   IVMEKGKIVEQGKHKELLSEPESLYSYLYQLQSD----------  241 (end)
HisP      224   IFLHQGKIEEEGDPEQVFGNPQSPRLQQFLKGSLKKLEH-----  262 (end)
MJ1267    224   YVMFNGQIIAEGRGEEEIKNVLSDPKVVEIYIGE----------  257 (end)
BtuD      204   WLLKGGKMLASGRREEVLTPPNLAQAYGMNFRRLDIEGHRMLIS  247 (end)
```

transporters (Figure 2). Several conserved sequence motifs, such as the Walker A and Walker B motifs that are found in many ATPases (80), can be identified, and mutations in these regions often severely reduce or eliminate transport and ATPase activity (70, 81).

The structures of isolated NBDs from eight different transporters, which include both importers (HisP, GlcV, MJ1267, *E. coli* [E.c.MalK], *T. litoralis* [T.l.MalK]) and exporters (TAP, HlyB, MJ0796), have been reported (82–90). The structures, like the sequences, are very similar, and no clear distinction can be made between importer and exporter proteins. The structure of a NBD monomer can be divided into two subdomains (Figure 3a): a larger RecA-like (91) subdomain (colored green in Figures 2 and 3) consisting of two β-sheets and six α-helices and a smaller helical subdomain formed by three to four α-helices (colored cyan in Figures 2 and 3). The helical subdomain is specific to the ABC transporters and not seen in other ATPases. ATP is shown bound to the RecA-like subdomain, and the γ-phosphate is positioned close to the edge of one of the β-sheets where it interacts with several residues directly or via H_2O. The Walker A motif, also known as the P loop, follows β-strand 3 and forms a loop that binds to the phosphates of ATP or ADP (Figure 3 and Table 1). The Walker B motif forms β-strand 7, and the terminal aspartate coordinates the Mg^{2+} ion in the nucleotide-binding site through H_2O (85, 86, 89). A glutamate residue immediately following the Walker B binds to the attacking water and the Mg^{2+} ion (87, 92). This glutamate may be the catalytic base for hydrolysis because its mutation leads to complete inactivation of ATPase activity (87, 92). The Q loop, following β-strand 6, also known as the lid (83) or the γ-phosphate switch (85), contains a glutamine that binds to the Mg^{2+} ion and attacking water (87, 89). The structure of the Q loop, which joins the RecA-like subdomain to the helical subdomain, appears to be highly flexible, as reflected by high-B factors in most NBD structures. The flexibility of the Q loop may have an important implication in its function to couple hydrolysis to transport; this function will be discussed below. The H motif following β-strand 8, which has also been referred to as a switch (83), contains a highly conserved histidine residue that forms a hydrogen bond with the γ-phosphate of ATP (87, 90). The signature motif, also known as

Figure 3 Structure of the NBD (residues 1–235) of E.c.MalK with bound ATP (90). (*a*) Stereo view of the monomer. (*b*) The homodimer, consisting of molecules A and B, viewed down the local twofold axis. The RecA-like subdomain is green, and the helical subdomain is cyan. Different colors further distinguish the conserved segments: Walker A motif (*red*), LSGGQ motif (*magenta*), Walker B motif (*blue*), and the Q loop (*yellow*). The ATP is represented in ball-and-stick model [O atom (*red*), N atom (*blue*)]. The color schemes for the domains of the B molecule are similar to those of the A molecule, except that they are rendered in lighter hue. (*c*) Schematic diagram of the interaction between one of the two ATPs bound to the homodimer. Black lines represent van der Waals contacts, and blue lines correspond to hydrogen bonds and salt bridges. Panel *c* reproduced from (90).

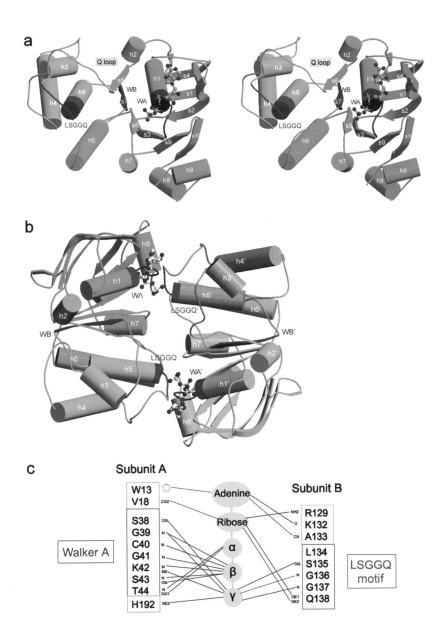

TABLE 1 Function of conserved motifs in the nucleotide-binding domain

Motif	Consensus sequence	Function	Supporting structures[a]
Walker A or P loop	GxxGxGKST[b]	ATP binding	HisP, MJ0796, MJ1267, Rad50, TAP1, GlcV, E.c.MalK
Q loop or lid	Q	a. TM subunit interaction	a. BtuCD
		b. Q H-bond to Mg	b. MJ0796 (E171Q), GlcV/ADP
		c. Binding to the attacking water	c. MJ0796 (E171Q)
LSGGQ or linker peptide or signature motif	LSGGQxQR[b]	ATP binding	Rad50, MJ0796 (E171Q), E.c.MalK
Walker B	hhhhD[b]	D makes a water-bridged contact with Mg^{2+}	GlcV (MgADP, MgAMP-PNP), MJ1267 (MgADP), MJ0796 (MgADP)
	E following Walker B	a. Binds to attacking water	a. MJ0796 (E171Q)
		b. Binds to Mg through a water	b. GlcV (MgADP, MgAMPPNP)
H motif or switch region	H	His H-bond to γ-phosphate	MJ0796 (E171Q), E.c.MalK

[a] References for structures are as follows: HisP (82), MJ0796 (85), MJ1267 (86), Rad50 (99), TAP1 (84), GlcV (89), E.c.MalK (90), BtuCD (105), MJ0796 (E171Q) (87).

[b] Where x represents any amino acid, and h represents a hydrophobic amino acid.

the LSGGQ motif, linker peptide, or C motif, has been used as the "signature" to identify ABC transporters and is the only major conserved motif that does not contact nucleotide in the monomer structure (82). The function of this motif was not clear until the correct dimeric arrangement of ABC proteins was established.

Dimeric Arrangement of Two Nucleotide-Binding Domains in the Transporter

All ABC transporters appear to have two NBDs, and ATP hydrolysis is highly cooperative. This is manifest either by the presence of positive cooperativity in ATP hydrolysis (29, 92, 93) or the loss of function of both nucleotide-binding sites following mutation or modification of a single site (94–97). Based on the structure of the first NBD monomer, HisP (82), and the sequences of atypical ABC transporters that appeared to have only one functional nucleotide-binding site (98), it was predicted that the NBDs would dimerize with ATP bound along the dimer interface, flanked by the Walker A motif of one subunit and the LSGGQ motif of the other (98). At first, it was not clear whether this was the true interface of the NBDs, because several other ABC crystallographic dimers had been reported (82, 83). Recently, three structures of isolated NBDs have been reported that form a "nucleotide-sandwich" dimer consistent with the predicted structure (98). These include Rad50 (99), an ABC-like ATPase, MJ0796, the NBD subunit of the LolD transporter from *Methanococcus jannaschii* (87), and E.c.MalK, the NBD component of the maltose transporter from *E. coli* (90). In the case of MJ0796, mutation of the catalytic glutamate immediately following the Walker B motif (E171Q) greatly enhanced the stability of the dimer and was necessary to crystallize the protein in this dimeric conformation. As shown for E.c.MalK (Figures 3*b* and 3*c*), residues in the Walker A motif of one subunit and the LSGGQ motif of the other subunit are engaged in extensive interactions with ATP.

Biochemical studies of the intact maltose transporter using vanadate have provided compelling evidence that the ATP-sandwich dimer represents the physiological conformation in the intact transporter. In the presence of UV light, vanadate, which is trapped in the position of the γ-phosphate of ATP (33), mediates highly specific photocleavage of the NBDs at the Walker A and LSGGQ motifs (100). Because the LSGGQ motif is distant from the Walker A motif in a NBD monomer (Figure 3*a*), the close approach of both motifs to the γ-phosphate could occur only across the dimer interface. Hence, the ATP-sandwich dimer appears to resemble the conformation of nucleotide-binding domains in the catalytically active state of the intact transporter when residues from both the Walker A and LSGGQ motifs contact the γ-phosphate.

Elucidation of the correct dimer interface for the NBDs in ABC transporters, together with the observation of nucleotide-dependent association of isolated NBDs (87, 92, 99), has provided answers to several key questions about ABC transporters. First, the functional role of the signature motif is now clearly established. In the dimeric NBD structures, the LSGGQ motif participates in ATP binding and hydrolysis (Figure 3). Second, the dimeric arrangement of the NBDs explains why all ABC transporters have two ABC components. Residues from both subunits are required to form the ATPase active sites, which are located right at the dimer interface. Third, the basis for cooperativity between ATP-binding sites can be explained if both sites must bind ATP before the NBDs can associate into a catalytically active conformation. The structures of MJ0796 (87), E.c.MalK (90), and Rad50 (99) all show two ATP molecules binding symmetrically at the dimer interface, suggesting that two ATPs must bind to form the closed dimer. In the alternating catalytic sites model for hydrolysis as originally proposed by Senior and colleagues (34) for P-glycoprotein, it was hypothesized that ATP binding at one site promoted ATP hydrolysis at the second site. Because only one ADP is trapped by vanadate (36, 37) and it can be trapped in either nucleotide-binding site (101, 101a), it is suggested that only one

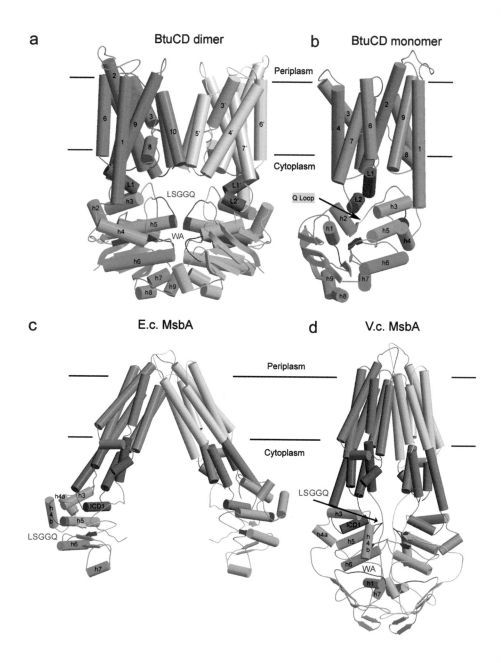

of the two bound ATPs is hydrolyzed per transport event and that the two sites alternate in catalysis (34). It has been difficult to determine whether one or both ATPs are hydrolyzed per transport event. In vivo measurements of growth yields in bacteria suggest that only one ATP is needed to transport one substrate into the cell (102), whereas a recent report using the purified and reconstituted OpuA transporter suggests that two ATPs are needed (103). Models incorporating either one (34) or two ATPs (104) have been proposed.

STRUCTURES OF INTACT TRANSPORTERS

Although the crystal structures of isolated NBDs provide a detailed picture of how these two ATP-binding components interact with each other, the high-resolution structure of an intact bacterial ABC importer should reveal how the NBDs interact with the MSDs and how the MSDs interact with each other. This type of information will form the basis for our understanding of the molecular mechanism by which ATP hydrolysis is coupled to transport. The high-resolution structures of two intact bacterial transporters have been determined: the vitamin B_{12} transporter, BtuCD (105), and the lipid A transporter, MsbA (106, 107).

The Vitamin B_{12} Importer: BtuCD

Vitamin B_{12} is transported into the cell via a periplasmic binding protein-dependent ABC transport system and the transporter itself consists of four subunits, arranged as two homodimers, a transmembrane BtuC dimer, and a nucleotide-binding dimer BtuD. The two BtuC subunits, each consisting of 10 transmembrane α-helices, form a homodimer with a potential translocation pathway located at the interface (Figure 4a). This pathway is at least partially accessible to the periplasm and appears closed at the cytoplasm. The two NBDs form a dimer similar to the ATP-sandwich dimer observed in MJ0796 (E171Q) and E.c.MalK (87, 90), except that ATP is absent from the crystal, and the Walker A sequence of one subunit is 4 A° further away from the LSGGQ motif of the opposite subunit. A large cavity is found at the center of the heterotetramer,

---

Figure 4   Structures of intact ABC transporters. (*a*) Structure of the dimeric vitamin B_{12} transporter BtuCD, front view with the molecular twofold axis running vertically. The MSDs (BtuC) are in gray, and the NBDs (BtuD) are the same colors as in Figure 3. The L loops (L1 and L2) are orange. (*b*) Side view of the BtuCD monomer, 75° from the view in panel *a*, shows docking of the L loop of BtuC into a cleft formed by helices 2, 3 and the Q loop of BtuD. (*c*) Structure of the E.c.MsbA dimer. The MSDs are gray, the ICDs are orange, and the NBDs are cyan and green. The location of the LSGGQ motif is labeled. Most of the RecA-like subdomain, which includes the Walker A motif, is disordered. (*d*) The structure of the V.c.MsbA dimer, colored the same as panel *c*.

below the predicted translocation pathway (Figure 4*a*). This water-filled channel may allow release of substrate into the cytoplasm without disturbing the BtuD dimer (105). The interface between the MSD subunits (BtuC) and the NBD subunits (BtuD) is provided mainly by the cytoplasmic L loop of BtuC and the Q loop on the surface of the BtuD subunit (Figure 4*b*). The L loop forms two short helices connected by a sharp turn (L1 and L2, colored orange in Figure 4*a* and *b*) and docks into a cleft on the surface of BtuD located between the RecA-like subdomain and the helical subdomain (Figure 4*b*). The L loop is likely to correspond to the cytoplasmic loop in other binding protein-dependent MSDs that contain a conserved EAA motif, which has been identified previously as a site of assembly for NBDs on MSDs (107b). Residues around the Q loop that line the cleft in BtuD make extensive contacts with the L1 and L2 helices. Two helices flanking the Q loop, helix 2 of the RecA-like subdomain, and helix 3 of the helical subdomain also make side chain contacts to the L loop of BtuC (Figure 4*b*). Not only is the Q loop a point of contact with the MSDs (105), it is also in contact with atoms in the ATP-binding site (85, 87). Thus, the presence or absence of ATP in the ATPase site has the potential to influence the conformation of the Q loop as well as its interaction with the MSD, thereby providing a pathway for coupling of ATP binding and hydrolysis to rearrangements in the TM region. The Q loop appears to be flexible in most of the ABC structures, perhaps due to the absence of the MSD domain and/or the MgATP. The cleft centered about the Q loop is found on the surface of all isolated NBDs whose structures are known. Hence, the general mechanism of attachment of NBDs to MSDs as well as the mechanism of coupling of transport to hydrolysis may well be conserved throughout the ABC transporter superfamily. In addition to the main contacts between the Q loop and L loop, helix 3 and the following loop also make multiple contacts with the cytoplasmic portion of TM1 in BtuC (Figure 4*a* and *b*). This loop, which connects helix 3 and helix 4, has high sequence and structural variability and may contribute to the specificity needed to prevent cross-reactivity of bacterial NBD subunits with noncognate MSDs (88, 108).

The Lipid A Flippase: MsbA

Two crystal structures of the efflux pump, MsbA, have also been determined, from *E. coli* (E.c.MsbA) (106) and *Vibrio cholera* (V.c.MsbA) (107). MsbA is a half-transporter, meaning that each MSD is fused to a NBD, and two subunits form the functional dimer. Several features of this efflux pump differ from the vitamin B_{12} importer. The six transmembrane helices of the MsbA monomer form a tight bundle (Figure 4*c* and 4*d*), as compared to the more complex architecture of BtuC (Figure 4*a* and *b*). The intracellular loops are generally larger than in BtuC and fold into an intracellular domain (ICD) that extends the transmembrane helical bundle into the cytoplasm (Figure 4*c*). Sequence alignments suggest that ICD may be a general feature for all exporters. In contrast to BtuCD in which the MSD subunits form a tight dimer interface that spans the membrane (Figure 4*a*), the MSDs of E.c.MsbA are tilted 40° relative to the normal of the membrane and contact each other only in the outer leaflet of the membrane with the effect of spreading the ICD and attached NBDs quite far apart in the cytoplasm in a very open V-shaped configuration (Figure 4*c*). The temperature-sensitive mutation that inactivates E.c.MsbA is located in the MSD dimer interface (66). In V.c.MsbA, which shares 68% sequence identity with E.c.MsbA, the tilt angle is reduced, which causes the transporter to take on a more closed conformation (Figure 4*d*). A patch of positive charge lines a hollow in the interior of the helical bundle in the inner leaflet of the membrane and is proposed as the transport substrate-binding site (106).

The conformation of the NBDs in the two MsbA structures also differs dramatically from that observed in BtuCD and from each other. In E.c.MsbA, the two NBDs are not in contact, and a substantial fraction of the RecA-like subdomain, including the Walker A motif, is disordered (106). The helical subdomain consists of four helices and mediates contact with the ICD. In contrast with BtuCD, where the LSGGQ motif faces into the dimer interface (Figure 4*a*), the LSGGQ motif of E.c.MsbA faces out, away from the twofold axis of symmetry (Figure 4*c*). In V.c.MsbA, the backbone of the entire NBD is resolved in the crystal structure, yet surprisingly, it takes on a conformation not observed in any other NBD to date (107), importer or exporter. In contrast to E.c.MsbA, the helical subdomains of the V.c.MsbA dimer are orientated similarly to those of BtuCD, with the LSGGQ motifs facing toward the dimer interface (Figure 4*d*). However, unlike all other RecA-like subdomains, where 10 β-strands and 6 α-helices pack tightly together, the V.c.MsbA RecA-like domain only consists of 6 β-strands and 3 α-helices. Moreover, there is an unusually large separation between the helical subdomain and the RecA-like subdomain. As a consequence, the dimer contacts observed in BtuCD and other Rad50-like dimers are absent in V.c.MsbA (99), and neither the Walker B motif nor the Q loop is similarly positioned to interact with MgATP as in other structures (Figure 4*d*). This conformation cannot exist in the glucose and maltose importers because the RecA-like subdomain of MsbA would collide with the C-terminal domain of the nucleotide-binding subunits MalK (90, 109) and GlcV (89). At the present time, it remains an open question as to whether the structural differences between BtuCD and MsbA imply different transport mechanisms for uptake and efflux systems, and this question is discussed further below.

CONFORMATIONAL CHANGES IN THE NUCLEOTIDE-BINDING DOMAINS DURING HYDROLYSIS

All of the current models for transport emphasize the key role of protein conformational change in the mechanism by which ATP hydrolysis is coupled to transport. In this section, we summarize the biochemical and structural evidence for conformational changes in the NBDs. Isolated NBDs undergo both an

ATP-dependent association (90, 92) and a γ-phosphate-induced rotation of the helical subdomain relative to the RecA-like subdomain (84, 85), both of which are likely to be an integral part of the mechanism of coupling of transport to ATP hydrolysis.

Evidence of Conformational Change in Nucleotide-Binding Domains

Several lines of evidence indicate that association and dissociation of the NBDs is a key feature of ABC transport. Although the Rad50 protein forms dimers in the presence of ATP (99), ATP-dependent dimerization of isolated transporter NBDs was not observed until the catalytic glutamate was replaced with glutamine (E171Q) in MJ0796 (92). This mutation eliminated hydrolysis and stabilized the ATP-bound dimer. Biochemical evidence in support of the association/dissociation hypothesis also comes from the intact maltose transporter system, in which a fluorescent probe, attached to cysteine 40 in the Walker A motif, becomes less accessible to solvent in the vanadate-trapped, transition state-like intermediate as compared to the ground state (110). Addition of ATP to the intact maltose transporter also enhances disulfide-crosslinking between cysteines introduced at position 85 in the Q loops of the MalK subunits (111). In the structure of E.c.MalK dimer, both of these residues lie along the dimer interface, and the changes are consistent with closure of the dimer interface during ATP hydrolysis (90).

Conformational Changes Revealed by *E. coli* MalK Structures

E.c.MalK is exceptional among NBDs whose structures have been determined because the wild-type protein crystallizes as a dimer in both nucleotide-free and ATP-bound forms (90), making it possible to visualize the MalK dimer in different conformational states. The reason for the increased stability of both the nucleotide-free and nucleotide-bound dimer became clear upon solving the structures. MalK belongs to a group of bacterial sugar transporters that contain an additional C-terminal domain of about 135 residues (3), and subunit-subunit interactions involving the C-terminal domain contribute substantially to the dimer interface (Figure 5). In MalK, this domain, called the regulatory domain, is also known to interact with regulatory proteins, such as enzyme IIA from the glucose-phosphotransferase system (112) and the transcriptional regulator MalT (113, 114). In contrast to the C-terminal domains, which maintain their inter-subunit contacts in all of the E.c.MalK homodimer structures, the N-terminal NBDs of E.c.MalK are in close contact only in the ATP-bound structure (Figure 3*b* and 5*a*). In two different nucleotide-free structures that were obtained, the NBDs separated to different degrees (Figure 5*a* and *b*). Hinge bending between the NBD and the regulatory domain is primarily responsible for the separation of NBDs (Figure 5*b*). The motion implied by these structures can be likened to that

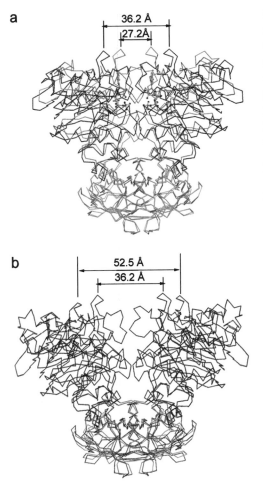

a

36.2 Å

27.2Å

b

52.5 Å

36.2 Å

Figure 5 The tweezer-like motion of E.c.MalK dimer. Closed, semiopen, and open structures of E.c.MalK homodimer with superimposed C-terminal regulatory domains. The distances between two H89 residues in a homodimer are indicated. (*a*) Closed form with bound ATP (*yellow*) and semiopen form without bound ATP (*blue*) are superimposed. The excellent overlap of the regulatory domains is evident by the green color, resulting from the combination of yellow and blue colors. (*b*) Superposition of the semiopen (*blue*) and the open (*red*) nucleotide-free structures. Figure is reproduced from (90).

of a pair of tweezers. The regulatory domains represent the handle that holds the two halves together, and the Q loops are positioned at the tips of the tweezers where they can move apart or together. The distance between the two histidines at position 89 in the Q loop increases by either 9 or 25 Å in moving from the closed form to the semiopen or open conformations, respectively (Figure 5*a* and 5*b*). Because the Q loop contacts the MSD of the intact transporter, the motions of the tweezers are likely to induce movements of the MSDs (90).

In addition to the movement of the entire NBDs, MalK also undergoes a hinge rotation of the helical subdomain relative to the RecA-like subdomain. In fact, the relative orientation of the helical subdomain to the RecA-like subdomain varies significantly among all known NBD structures (84, 85). The difference is observed not only in different proteins, but also for the same protein crystallized in different crystal lattices (89, 90), and even between molecules in the same

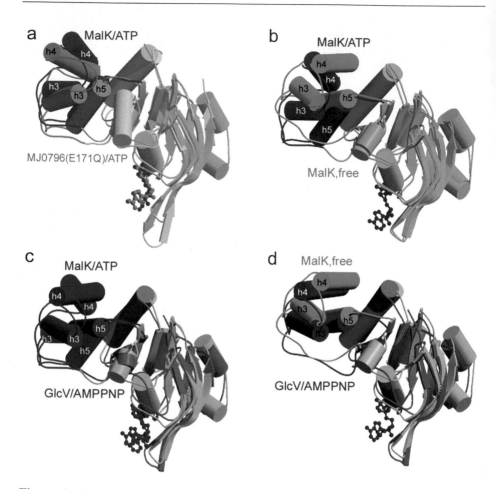

Figure 6   Helical subdomain rotation in NBDs. NBDs are superimposed based on their RecA-like subdomains. The RecA-like subdomains are rendered in lighter color and the helical subdomains in darker color. (*a*) Overlay of two ATP-bound structures: E.c.MalK (*blue*) and MJ0796(E171Q) (*yellow*). (*b*) Overlay of the ATP-bound (*blue*) and the open, nucleotide-free (*green*) E.c.MalK structures. (*c*) Overlay of the ATP-bound E.c.MalK (*blue*) and the AMPPNP-bound GlcV (*red*) structures. (*d*) Overlay of the open, ATP-free E.c.MalK (*green*) and the AMPPNP-bound GlcV (*red*) structures.

crystal lattice (83). The hinge rotation of the helical subdomain relative to the RecA-like subdomain ranges from 5° to 25°, depending on the structures that are compared. However, when the three available ATP-bound structures, HisP, E.c.MalK, and MJ0796(E171Q), are compared, the orientation of the helical subdomains are essentially the same (Figure 6*a* versus Figure 6*b*), indicating that in the presence of the γ-phosphate, the helical subdomain becomes locked into one orientation. Interestingly, the helical subdomain of GlcV in the presence of the nonhydrolyzable ATP analogue Mg/AMPPNP folds into an orientation more similar to the nucleotide-free than the ATP-bound form of E.c.MalK (Figure 6c versus Figure 6d), suggesting that AMPPNP is not a good mimic of ATP for ABC transporter proteins. AMPPNP and ATPγS, another nonhydrolyzable analogue, also failed to support dimerization of MJ0796(E171Q) (92). One could argue that the difference in the helical subdomain orientation is simply a crystallographic artifact, i.e., that protein-protein contacts inside a crystal lattice stabilize the helical subdomain in a particular orientation. However, the fact that the flexibility of the helical subdomain was observed only in structures lacking the γ-phosphate of ATP suggests that it has important mechanistic implications. In the context of the isolated NBD, it has been suggested that its function is to withdraw the signature motif, located N-terminal to helix 5, from the ATP binding site in order to facilitate release of ADP following hydrolysis (86). It is clear from structures of ATP-bound dimers that NBDs must dissociate to permit nucleotide binding and release. In the context of the intact transporter, it has been suggested that this hinge rotation is controlled by the MSDs to regulate the ATPase activity of the NBDs (87, 108). When the helical subdomain is rotated away from RecA subdomain, as in the nucleotide-free conformation (Figure 6), the two NBDs cannot approach closely enough to form the ATP-bound dimer. For example, this rotation could be tied to the mechanism by which drug binding to the MSD stimulates ATP hydrolysis by an efflux pump. The added flexibility of this hinge may also be important in adjusting the size of the opening of the translocation pathway to accommodate the size of the substrate.

Comparison of the BtuCD structure (105) to the E.c.MalK dimer structures (90) reveals that the BtuD dimer is very similar to the semiopen MalK configuration. Hence, there appears to be a role for both helical subdomain rotation and NBD closure in proceeding from the open conformation observed in the BtuCD structure to the closed-ATP-bound conformation. This observation also confirms that the conformational changes identified in the isolated MalK dimer are representative of what will occur in the intact transporter complex.

Two Physiological Conformational States for Nucleotide-Binding Domains of MsbA?

As we have discussed, nucleotide-free MsbA from *E. coli* and *V. cholera* crystallized in different conformations, neither of which contains an NBD dimer resembling either the BtuD dimer or the E.c.MalK dimers in their open and closed conformations. Given the good agreement between the BtuCD structure and the E.c.MalK dimer structures, and between these structures and biochemical characterizations of ABC transporters, serious questions have been raised about whether the MsbA structures should be viewed as conformational intermediates in the cycle of an efflux pump or as nonphysiologic conformations introduced during the purification or crystallization of the transporter (115–117). Assuming the two MsbA structures are different physiologic intermediates or snapshots of the catalytic cycle, then the conformational changes required to bring the open

E.c.MsbA conformation (Figure 4*c*) to the closed V.c.MsbA conformation (Figure 4*d*) are quite dramatic. Not only must each MsbA monomer rotate inward ~15° to bring the MSDs and NBDs close together, but the NBDs must also rotate around the ICDs by ~120°. The latter rotation would present the LSGGQ motif in the helical subdomain to the dimer interface as observed in the other NBD dimers. In addition, the author (107) indicates that the RecA-like subdomains of V.c. MsbA would have to swing ~180° toward the helical subdomains to convert the NBD dimer in V.c.MsbA to a dimer resembling BtuD that could bind ATP. Because the NBD of V.c.MsbA contains substantially fewer β-strands and α-helices than most NBDs, significant refolding of the structure would have to occur in order to bring this structure into compliance with the Rad50-like dimer interface.

Translation of this complex series of conformational changes into a plausible mechanism for translocation would, in our opinion, be extremely difficult, whereas the conformational changes demonstrated in the MalK dimer, as discussed in the next section, offer a simple explanation of how transport is coupled to hydrolysis in both importers and exporters in the ABC family.

MECHANISMS FOR COUPLING OF HYDROLYSIS TO TRANSPORT

Models for Coupling in Periplasmic Binding Protein-Dependent Transport

Substantial evidence has accumulated in support of a model in which ATP hydrolysis is coupled to conformational changes in the transporter that mediate the movement of substrate across the membrane. This concept is illustrated in the model for maltose transporter (Figure 1) in which an energetically favorable event (ATP hydrolysis) and an energetically unfavorable event (maltose transport against a concentration gradient) are linked via a common conformational intermediate. This conformation was trapped with vanadate acting as a transition state analogue (33). The concerted conformational changes in MBP and MalFGK$_2$ required to form this intermediate simultaneously promote both ATP hydrolysis and release of maltose from MBP directly into the mouth of the translocation pathway.

From the structure of the binding protein-dependent vitamin B$_{12}$ transporter in a nucleotide-free state, it is now possible to address what type of conformational changes need to occur in transmembrane regions to facilitate transport and how they might be coupled to conformational changes in NBDs. In the nucleotide-free structure of BtuCD, the proposed transmembrane translocation pathway appears to be at least partially open to the periplasm and closed at the cytoplasm (105), suggesting that a cytoplasmic gate must open at some point in the translocation cycle. The authors (105) propose a model in which the closure of the NBDs that

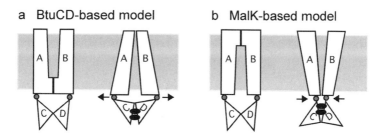

Figure 7 Two models for coupling of NBD closure to transmembrane movement. (*a*) BtuCD-based model. NBD closure coincides with opening of translocation pathway to cytoplasm. (*b*) E.c.MalK-based model. NBD closure coincides with closing of translocation pathway to cytoplasm and opening to periplasm.

accompanies binding and hydrolysis of ATP opens the gate by forcing the MSDs apart in a motion analogous to that of a toggle switch (Figure 7*a*). However, the E.c.MalK dimer structures (Figure 5) suggest that the residues in MalK and BtuD that make contact with the MSDs would be closer together in the ATP-bound form as compared to the free form, making it difficult to understand how closure of NBDs could mediate opening of a cytoplasmic gate in either system. Based on the E.c.MalK structures, a different model is proposed (90) that predicts movement of the MSDs opposite to the movement observed in the BtuCD-based model (Figure 7*b*). In the MalK-based model, the transmembrane pathway is open to the cytoplasmic side in the resting state, and ATP-induced closure of the NBDs mediates the closing of a gate at the cytoplasmic surface of the MSDs and the opening of a gate at the periplasmic surface, thereby alternating access to a central translocation pathway through the membrane. If the model in Figure 7*b* is correct, and it applies equally well to both maltose and B_{12} transport, then the conformation of BtuCD captured in the crystal may be an intermediate between the resting state and the transition state in which the cytoplasmic gate has already closed, but the periplasmic gate has not yet fully opened. The observed flexibility of the hinge between the RecA-like subdomain and the helical subdomain in the NBDs (Figure 6) could permit a more open conformation of BtuCD, which is observed in one of the MalK dimer structures (90). Clearly, more biochemical and structural analyses of intact ABC transporters are needed to clarify the mechanism of translocation.

Models for Coupling in Drug Efflux Systems

Just as maltose is transferred from a high-affinity binding site on one side of the membrane to a low-affinity binding site on the other side (Figure 1), drug transport is suggested to involve changes in drug-binding site affinity and orientation (inward versus outward facing) during a cycle of ATP hydrolysis (78). In the original model for alternating sites, Senior and colleagues (101) noted that, although ATP does not bind with high affinity to ABC transporters, ATP hydrolysis results in the formation of a high-energy state with ADP and P_i bound and suggests that relaxation of this state (i.e., the release of P_i) might be accompanied by the conformational changes that result in drug transport. They now postulate that both the formation and collapse of the transition state of catalysis are critical events for driving the conformational changes that mediate drug transport (78) as illustrated in our model for maltose transport (Figure 1). Binding of drug substrate, or periplasmic binding protein, promotes the formation of the catalytic transition state by promoting NBD closure, and subsequent ATP hydrolysis leads to NBD dimer reopening. In both drug efflux pumps and bacterial uptake systems, it appears that this closed state can be trapped either with a transition state nucleotide analogue (33, 37) or with ATP if hydrolysis is prevented through mutation of a key catalytic residue (92, 118–120).

Evidence for conformational coupling between nucleotide-binding sites and drug-binding sites in the MSD of drug efflux pumps has come from both vanadate-trapping experiments (37, 121, 122) and spectroscopic studies (123, 124) in prokaryotic and eukaryotic systems. In the bacterial LmrA protein, a combination of infrared spectroscopy and hydrogen/deuterium exchange kinetics was used to show that ATP hydrolysis triggered substantial changes in accessibility of the residues in the TM domain without accompanying changes in mean helix orientation (125), suggesting that helix translation or, in particular, helix rotation could alternately expose the drug-binding site to the lipid environment or to the aqueous extracellular environment (125). Binding of radiolabeled vinblastine to LmrA displays positive cooperativity, although both cooperativity and high-affinity binding are lost in the vanadate-trapped intermediate (122). These data are consistent with the interpretation that drug transport occurs coincidently with ATP hydrolysis, and the authors (122) suggest vanadate has stabilized an intermediate in which the drug-binding site is occluded during the switch from a high-affinity inward-facing state before hydrolysis to a low-affinity outward-facing state after hydrolysis. They propose a variation of the alternating catalytic sites model (34), likened to a two-cylinder engine, in which each half-transporter has its own drug carrier site, and while ATP hydrolysis occurs concomitantly with drug transport in one half of the transporter, the second half resets its drug-binding site from the outward-facing low-affinity conformation to the inward-facing high-affinity conformation (122). This model (Figure 8a) can be contrasted to models proposed for the mammalian P-glycoprotein (Figures 8b and 8c) in which one interchangeable drug-binding site is shared between the two halves of the pump (78) or in which both high- and low-affinity sites coexist, and loss of high-affinity binding in the transition state promotes transfer of drug from the high-affinity (on) site to the low-affinity (off) site (104). Although the first two models incorporate the concept that drug transport occurs each time ATP is hydrolyzed, the authors of the third model suggest that ATP hydrolysis at the second site is required to reset the system before transport can occur again (104). Both key concepts in this final model, the loss of high-affinity binding in the catalytic transition

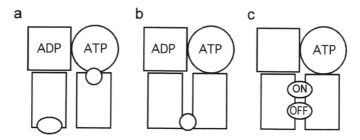

Figure 8   Three models for multidrug efflux pumps. (*a*) Model 1 (122). Each half-transporter has a drug-binding site, and during ATP hydrolysis, one site reorients from inward-facing to outward-facing while the other site reorients from outside to inside. (*b*) Model 2 (34). A single drug-binding site, shared between the MSDs, reorients from inward-facing to outward-facing following ATP hydrolysis. (*c*) Model 3 (104). Two binding sites coexist, an inward-facing (ON) site and an outward-facing (OFF) site. Loss of high-affinity binding to the ON site in the transition state promotes transfer of drug from the ON site to the OFF site.

state and the requirement for ATP hydrolysis in resetting the system, have recently been challenged on the grounds that they rely too heavily on the use of photoaffinity labeling by drugs to assess changes in drug-binding site affinity (126).

Other models for transport by P-glycoprotein have also been proposed. Higgins and colleagues report a lowered drug-binding affinity both in the presence of the nonhydrolyzable ATP analogue AMPPNP and in the vanadate-trapped transporter (127). Medium resolution structures of P-glycoprotein obtained by cryo-electron microscopy of two-dimensional crystals provide evidence for at least three distinct arrangements for the MSDs (128). Crystals of P-glycoprotein formed in the presence of AMPPNP differ from those obtained in the absence of nucleotide, and the vanadate-trapped species adopts a third conformation. On the basis of these observations, the authors suggest that nucleotide binding to P-glycoprotein rather than ATP hydrolysis drives the major conformational changes that reorient the drug-binding site from the inside to the outside of the cell (127). Consistent with this hypothesis, nonhydrolyzable nucleotide analogues or ATP in the absence of Mg will support opening of the mammalian cystic fibrosis transmembrane regulator channel (129).

A recent study based upon thermodynamic analysis of the intrinsic rate-limiting step of drug transport indicates that P-glycoprotein undergoes two distinct catalytic cycles in the presence and absence of drug with different rate-limiting transition states (130). In an extension of the alternating catalytic sites model (34), the authors propose (130) that drug is still bound at a high-affinity site in the drug-coupled transition state because the catalytic constant for ATP hydrolysis varies with different drugs. They also postulate that during ATP hydrolysis in the absence of drug, the drug-binding site is in the low-affinity form (130). This latter postulate clearly complicates interpretation of published vanadate-trapping experiments if, in fact, vanadate traps a conformation of the transporter seen only in the basal transition state and not in the drug-coupled transition state.

CONCLUDING REMARKS

Great advances have been made in the past several years in our understanding of the molecular mechanism of translocation in bacterial ABC transporters. Both biochemistry and structural biology have had equally important roles in these advances. As we have discussed, the progress in structural biology has been mired by seemingly contradictory results, and only by interpreting the structures based on the results of biochemical analyses and determining what is and is not plausible has it been possible to put together a reasonable model of the way transport occurs. Stabilization of conformational intermediates in the transport cycle, in particular through the use of vanadate, has been key to the advances made in biochemical characterization of ABC transporters, and it is hoped that crystal structures of these conformational intermediates alongside their ground state counterparts will further the field.

Although most of the recent progress in structural biology has been made with prokaryotic family members, it is likely that eukaryotic transporters will utilize the same mechanisms for translocation. The structure of TAP1, the NBD of the mammalian transporter involved in antigen processing, has been determined, and it is essentially the same as that seen in the prokaryotic NBD structures (84). Defects in many of the human ABC transporters have been linked to disease, and many of their physiologic functions have been deduced (9). It is hoped that, through the study of bacterial homologues, we will come to a better understanding of the structure, function, and mechanism of human protein action.

ACKNOWLEDGMENTS

Work in the Davidson lab was supported by grants from NIH (R01 GM49261) and the Welch Foundation (Q-1391). Work in the Chen lab was supported by the Pew Scholarship and an American Cancer Society Institutional Research Grant to the Purdue Cancer Center. We thank James Westbrooks, Kimberly Westbrooks, and Tasha Biesinger for their comments on the manuscript.

The *Annual Review of Biochemistry* is online at http://biochem.annualreviews.org

LITERATURE CITED

1. Chen C-J, Chin JE, Ueda K, Clark DP, Pastan I, et al. 1986. *Cell* 47:381–89
2. Gros P, Croop J, Housman D. 1986. *Cell* 47:371–80
3. Boos W, Lucht JM. 1996. In *Escherichia coli and Salmonella: Cellular and Molecular Biology*, ed. FC Neidhardt, R Curtiss III, JL Ingraham, ECC

Lin, KB Low, et al. pp. 1175–209. Washington, DC: ASM Press

4. Horio M, Gottesman MM, Pastan I. 1988. *Proc. Natl. Acad. Sci. USA* 85:3580–84

5. Hyde SC, Emsley P, Hartshorn MJ, Mimmack MM, Gileadi U, et al. 1990. *Nature* 346:362–65

6. Holland IB, Blight MA. 1999. *J. Mol. Biol.* 293:381–99

7. Ames GFL, Mimura CS, Shyamala V. 1990. *FEMS Microbiol. Rev.* 6:429–46

8. Dassa E, Hofnung M, Paulsen IT, Saier MH Jr. 1999. *Mol. Microbiol.* 32:887–89

9. Dean M, Hamon Y, Chimini G. 2001. *J. Lipid Res.* 42:1007–17

10. Saier MH Jr. 2000. *Microbiol. Mol. Biol. Rev.* 64:354–411

11. Quentin Y, Fichant G. 2000. *J. Mol. Microbiol. Biotechnol.* 2:501–4

12. Dassa E, Bouige P. 2001. *Res. Microbiol.* 152:211–29

13. Saurin W, Dassa E. 1994. *Protein Sci.* 3:325–44

14. Kuan G, Dassa E, Saurin W, Hofnung M, Saier MH. 1995. *Res. Microbiol.* 146:271–78

15. van der Heide T, Poolman B. 2002. *EMBO Rep.* 3:938–43

16. Delcour AH. 2003. *Front. Biosci.* 8:d1055–71

17. Postle K, Kadner RJ. 2003. *Mol. Microbiol.* 49:869–82

18. Quiocho FA, Ledvina PS. 1996. *Mol. Microbiol.* 20:17–25

19. Wilkinson AJ, Verschueren KHG. 2003. In *ABC Proteins: From Bacteria to Man*, ed. IB Holland, SPC Cole, K Kuchler, CF Higgins, pp. 187–207. London: Academic

20. Miller DM, Olson JS, Pflugrath JW, Quiocho FA. 1983. *J. Biol. Chem.* 258: 13665–72

21. Ledvina PS, Tsai AL, Wang Z, Koehl E, Quiocho FA. 1998. *Protein Sci.* 7:2550–59

21a. Hall JA, Thorgeirsson TE, Liu J, Shin YK, Nikaido H. 1997. *J. Biol. Chem.* 272:17610–14

22. Clarke TE, Ku SY, Dougan DR, Vogel HJ, Tari LW. 2000. *Nat. Struct. Biol.* 7:287–91

23. Borths EL, Locher KP, Lee AT, Rees DC. 2002. *Proc. Natl. Acad. Sci. USA* 99:16642–47

24. Karpowich NK, Huang HH, Smith PC, Hunt JF. 2003. *J. Biol. Chem.* 278: 8429–34

25. Treptow NA, Shuman HA. 1985. *J. Bacteriol.* 163:654–60

26. Ames GFL, Mimura CS, Holbrook SR, Shyamala V. 1992. *Adv. Enzymol. Relat. Areas Mol. Biol.* 65:1–47

27. Merino G, Boos W, Shuman HA, Bohl E. 1995. *J. Theor. Biol.* 177:171–79

28. Davidson AL, Shuman HA, Nikaido H. 1992. *Proc. Natl. Acad. Sci. USA* 89:2360–64

29. Liu CE, Liu PQ, Ames GFL. 1997. *J. Biol. Chem.* 272:21883–91

30. Petronilli V, Ames GFL. 1991. *J. Biol. Chem.* 266:16293–96

31. Manson MD, Boos W, Bassford PJ Jr, Rasmussen BA. 1985. *J. Biol. Chem.* 260:9727–33

32. Dean DA, Hor LI, Shuman HA, Nikaido H. 1992. *Mol. Microbiol.* 6:2033–40

33. Chen J, Sharma S, Quiocho FA, Davidson AL. 2001. *Proc. Natl. Acad. Sci. USA* 98:1525–30

34. Senior AE, Al-Shawi MK, Urbatsch IL. 1995. *FEBS Lett.* 377:285–89

35. Smith CA, Rayment I. 1996. *Biochemistry* 35:5404–17

36. Sharma S, Davidson AL. 2000. *J. Bacteriol.* 182:6570–76

37. Urbatsch IL, Sankaran B, Weber J, Senior AE. 1995. *J. Biol. Chem.* 270: 19383–90

38. Saurin W, Hofnung M, Dassa E. 1999. *J. Mol. Evol.* 48:22–41

39. Binet R, Letoffe S, Ghigo JM, Delepelaire P, Wandersman C. 1997. *Gene* 192:7–11

40. Salmond GP, Reeves PJ. 1993. *Trends Biochem. Sci.* 18:7–12
41. Koronakis V, Hughes C. 1993. *Semin. Cell Biol.* 4:7–15
42. Dinh T, Paulsen IT, Saier MH Jr. 1994. *J. Bacteriol.* 176:3825–31
43. Paulsen IT, Park JH, Choi PS, Saier MH Jr. 1997. *FEMS Microbiol. Lett.* 156:1–8
44. Koronakis V, Sharff A, Koronakis E, Luisi B, Hughes C. 2000. *Nature* 405: 914–19
45. Andersen C, Koronakis E, Bokma E, Eswaran J, Humphreys D, et al. 2002. *Proc. Natl. Acad. Sci. USA* 99:11103–8
46. Letoffe S, Delepelaire P, Wandersman C. 1996. *EMBO J.* 15:5804–11
47. Thanabalu T, Koronakis E, Hughes C, Koronakis V. 1998. *EMBO J.* 17: 6487–96
48. Holland IB, Benabdelhak H, Young J, Pimenta A, Schmitt L, Blight M. 2003. See Ref. 19, pp. 209–41
49. Havarstein LS, Holo H, Nes IF. 1994. *Microbiology* 140(Pt. 9):2383–89
50. Izadi-Pruneyre N, Wolff N, Redeker V, Wandersman C, Delepierre M, Lecroisey A. 1999. *Eur. J. Biochem.* 261: 562–68
51. Hui D, Morden C, Zhang F, Ling V. 2000. *J. Biol. Chem.* 275:2713–20
52. Michiels J, Dirix G, Vanderleyden J, Xi C. 2001. *Trends Microbiol.* 9:164–68
53. Benabdelhak H, Kiontke S, Horn C, Ernst R, Blight MA, et al. 2003. *J. Mol. Biol.* 327:1169–79
54. Zhang F, Sheps JA, Ling V. 1993. *J. Biol. Chem.* 268:19889–95
55. Delepelaire P, Wandersman C. 1998. *EMBO J.* 17:936–44
56. Palacios JL, Zaror I, Martinez P, Uribe F, Opazo P, et al. 2001. *J. Bacteriol.* 183:1346–58
57. Fernandez LA, de Lorenzo V. 2001. *Mol. Microbiol.* 40:332–46
58. Sapriel G, Wandersman C, Delepelaire P. 2002. *J. Biol. Chem.* 277:6726–32
59. Pimenta AL, Young J, Holland IB, Blight MA. 1999. *Mol. Gen. Genet.* 261:122–32
60. Balakrishnan L, Hughes C, Koronakis V. 2001. *J. Mol. Biol.* 313:501–10
61. Gray L, Mackman N, Nicaud JM, Holland IB. 1986. *Mol. Gen. Genet.* 205: 127–33
62. van Veen HW, Callaghan R, Soceneantu L, Sardini A, Konings WN, Higgins CF. 1998. *Nature* 391:291–95
63. Zhou Z, White KA, Polissi A, Georgopoulos C, Raetz CR. 1998. *J. Biol. Chem.* 273:12466–75
64. Putman M, van Veen HW, Degener JE, Konings WN. 2000. *Mol. Microbiol.* 36:772–73
65. Polissi A, Georgopoulos C. 1996. *Mol. Microbiol.* 20:1221–33
66. Doerrler WT, Reedy MC, Raetz CR. 2001. *J. Biol. Chem.* 276:11461–64
67. Doerrler WT, Raetz CR. 2002. *J. Biol. Chem.* 277:36697–705
68. Reuter G, Janvilisri T, Venter H, Shahi S, Balakrishnan L, van Veen HW. 2003. *J. Biol. Chem.* 278:35193–98
69. Levy SB. 1992. *Antimicrob. Agents Chemother.* 36:695–703
70. Poelarends GJ, Vigano C, Ruysschaert JM, Konings WN. 2003. See Ref. 19, pp. 243–62
71. Ambudkar SV, Dey S, Hrycyna CA, Ramachandra M, Pastan I, Gottesman MM. 1999. *Annu. Rev. Pharmacol. Toxicol.* 39:361–98
72. Zheleznova EE, Markham PN, Neyfakh AA, Brennan RG. 1999. *Cell* 96:353–62
73. Schumacher MA, Miller MC, Grkovic S, Brown MH, Skurray RA, Brennan RG. 2001. *Science* 294:2158–63
74. Yu EW, McDermott G, Zgurskaya HI, Nikaido H, Koshland DE Jr. 2003. *Science* 300:976–80
75. Yu EW, Aires JR, Nikaido H. 2003. *J. Bacteriol.* 185:5657–64
76. Bolhuis H, van Veen HW, Molenaar D, Poolman B, Driessen AJ, Konings WN. 1996. *EMBO J.* 15:4239–45

76a. Shapiro AB, Ling V. 1997. *Eur. J. Biochem.* 250:122–29

77. Margolles A, Putman M, van Veen HW, Konings WN. 1999. *Biochemistry* 38: 16298–306

77a. Ruetz S, Gros P. 1994. *Cell* 77:1071–81

78. Urbatsch IL, Tyndall GA, Tombline G, Senior AE. 2003. *J. Biol. Chem.* 278: 23171–79

79. Yakushi T, Masuda K, Narita S, Matsuyama S, Tokuda H. 2000. *Nat. Cell Biol.* 2:212–18

80. Walker JE, Saraste M, Runswick MJ, Gay NJ. 1982. *EMBO J.* 1:945–51

81. Shyamala V, Baichwal V, Beall E, Ames GFL. 1991. *J. Biol. Chem.* 266: 18714–19

82. Hung LW, Wang IX, Nikaido K, Liu PQ, Ames GFL, Kim SH. 1998. *Nature* 396:703–7

83. Diederichs K, Diez J, Greller G, Muller C, Breed J, et al. 2000. *EMBO J.* 19:5951–61

84. Gaudet R, Wiley DC. 2001. *EMBO J.* 20:4964–72

85. Yuan YR, Blecker S, Martsinkevich O, Millen L, Thomas PJ, Hunt JF. 2001. *J. Biol. Chem.* 276:32313–21

86. Karpowich N, Martsinkevich O, Millen L, Yuan Y, Dai PL, et al. 2001. *Structure* 9:571–86

87. Smith PC, Karpowich N, Millen L, Moody JE, Rosen J, et al. 2002. *Mol. Cell* 10:139–49

88. Schmitt L, Benabdelhak H, Blight MA, Holland IB, Stubbs MT. 2003. *J. Mol. Biol.* 330:333–42

89. Verdon G, Albers SV, Dijkstra BW, Driessen AJ, Thunnissen AM. 2003. *J. Mol. Biol.* 330:343–58

90. Chen J, Lu G, Lin J, Davidson AL, Quiocho FA. 2003. *Mol. Cell* 12:651–61

91. Story RM, Steitz TA. 1992. *Nature* 355:374–76

92. Moody JE, Millen L, Binns D, Hunt JF,

Thomas PJ. 2002. *J. Biol. Chem.* 277: 21111–14

93. Davidson AL, Laghaeian SS, Mannering DE. 1996. *J. Biol. Chem.* 271: 4858–63

94. Azzaria M, Schurr E, Gros P. 1989. *Mol. Cell. Biol.* 9:5289–97

95. Al-Shawi MK, Senior AE. 1993. *J. Biol. Chem.* 268:4197–206

96. Loo TW, Clarke DM. 1995. *J. Biol. Chem.* 270:22957–61

97. Davidson AL, Sharma S. 1997. *J. Bacteriol.* 179:5458–64

98. Jones PM, George AM. 1999. *FEMS Microbiol. Lett.* 179:187–202

99. Hopfner KP, Karcher A, Shin DS, Craig L, Arthur LM, et al. 2000. *Cell* 101:789–800

100. Fetsch EE, Davidson AL. 2002. *Proc. Natl. Acad. Sci. USA* 99:9685–90

101. Urbatsch IL, Sankaran B, Bhagat S, Senior AE. 1995. *J. Biol. Chem.* 270: 26956–62

101a. Hrycyna CA, Ramachandra M, Ambudkar SV, Ko YH, Pedersen PL, et al. 1998, *J. Biol. Chem.* 273:16631–34

102. Ferenci T, Boos W, Schwartz M, Szmelcman S. 1977. *Eur. J. Biochem.* 75:187–93

103. Patzlaff JS, van der Heide T, Poolman B. 2003. *J. Biol. Chem.* 278:29546–51

104. Sauna ZE, Ambudkar SV. 2000. *Proc. Natl. Acad. Sci. USA* 97:2515–20

105. Locher KP, Lee AT, Rees DC. 2002. *Science* 296:1091–98

106. Chang G, Roth CB. 2001. *Science* 293: 1793–800

107. Chang G. 2003. *J. Mol. Biol.* 330: 419–30

107b. Mourez M, Hofnung M, Dassa E. 1997. *EMBO J.* 16:3066–77

108. Jones PM, George AM. 2002. *Proc. Natl. Acad. Sci. USA* 99:12639–44

109. Samanta S, Ayvaz T, Reyes M, Shuman HA, Chen J, Davidson AL. 2003. *J. Biol. Chem.* 278:35265–71

110. Mannering DE, Sharma S, Davidson

AL. 2001. *J. Biol. Chem.* 376:
12362–68

111. Hunke S, Mourez M, Jehanno M, Dassa
     E, Schneider E. 2000. *J. Biol. Chem.*
     275:15526–34

112. Nelson SO, Postma PW. 1984. *Eur.
     J. Biochem.* 139:29–34

113. Reyes M, Shuman HA. 1988. *J. Bacteriol.* 170:4598–602

114. Kuhnau S, Reyes M, Sievertsen A, Shuman HA, Boos W. 1991. *J. Bacteriol.*
     173:2180–86

115. Thomas PJ, Hunt JF. 2001. *Nat. Struct.
     Biol.* 8:920–23

116. Davidson AL. 2002. *Science* 296:
     1038–40

117. Campbell JD, Biggin PC, Baaden M,
     Sansom MS. 2003. *Biochemistry*
     42:3666–73

118. Sauna ZE, Muller M, Peng XH,
     Ambudkar SV. 2002. *Biochemistry*
     41:13989–4000

119. Orelle C, Dalmas O, Gros P, Di Pietro
     A, Jault JM. 2003. *J. Biol. Chem.* 278:
     47002–8

120. Janas E, Hofacker M, Chen M, Gompf
     S, van der Does C, Tampe R. 2003.
     J. Biol. Chem. 278:26862–69

121. Ramachandra M, Ambudkar SV, Chen
     D, Hrycyna CA, Dey S, et al. 1998.
     Biochemistry 37:5010–19

122. van Veen HW, Margolles A, Muller M,
     Higgins CF, Konings WN. 2000.
     EMBO J. 19:2503–14

123. Liu R, Sharom FJ. 1996. *Biochemistry*
     35:11865–73

124. Grimard V, Vigano C, Margolles A,
     Wattiez R, van Veen HW, et al. 2001.
     Biochemistry 40:11876–86

125. Vigano C, Grimard V, Margolles A,
     Goormaghtigh E, van Veen HW, et al.
     2002. *FEBS Lett.* 530:197–203

126. Qu Q, Chu JWK, Sharom FJ. 2003.
     Biochemistry 42:1345–53

127. Martin C, Berridge G, Higgins CF,
     Mistry P, Charlton P, Callaghan R.
     2000. *Biochemistry* 39:11901–6

128. Rosenberg MF, Velarde G, Ford RC,
     Martin C, Berridge G. 2001. *EMBO J.*
     20:5615–25

129. Aleksandrov AA, Chang XB, Aleksandrov L, Riordan JR. 2000. *J. Physiol.*
     528:259–65

130. Al-Shawi MK, Polar MK, Omote H,
     Figler RA. 2003. *J. Biol. Chem.* 278:
     52629–40

Annu. Rev. Biochem. 2004. 73:269–92
doi: 10.1146/annurev.biochem.73.011303.073700

Structural Basis of Ion Pumping by Ca^{2+}-ATPase of the Sarcoplasmic Reticulum

Chikashi Toyoshima[1] and Giuseppe Inesi[2]

[1]Institute of Molecular and Cellular Biosciences, The University of Tokyo, Tokyo 113-0032, Japan; email:ct@iam.u-tokyo.ac.jp
[2]Department of Biochemsitry and Molecular Biology, University of Maryland Medical School, Baltimore, Maryland 21201-1503; email: ginesi@umaryland.edu

Key Words membrane protein, crystal structure, active transport, Ca^{2+} binding, P-type ATPase

■ **Abstract** The structures of the Ca^{2+}-ATPase (SERCA1a) have been deter mined for five different states by X ray crystallography. Detailed comparison of the structures in the Ca^{2+} bound form and unbound (but thapsigargin bound) form reveals that very large rearrangements of the transmembrane helices take place accompanying Ca^{2+} dissociation and binding and that they are mechanically linked with equally large movements of the cytoplasmic domains. The meanings of the rearrangements of the transmembrane helices and those of the cytoplasmic domains as well as the mechanistic roles of phosphorylation are now becoming clear. Furthermore, the roles of critical amino acid residues identified by extensive mutagenesis studies are becoming evident in terms of atomic structure.

CONTENTS

INTRODUCTION

Ca^{2+}-ATPase of skeletal muscle sarcoplasmic reticulum [sarco(endo)plasmic reticulum calcium ATPase 1, SERCA1] is an integral membrane protein of M_r 110 K consisting of a single polypeptide chain (1). The adult form in the fast-twitch skeletal muscle (SERCA1a) from rabbit contains 994 amino acid residues (2). It can transport 2 Ca^{2+} from the cytoplasm to the lumen of sarcoplasmic reticulum (SR) against a concentration gradient, in exchange for 2 or 3 H^+ per ATP hydrolyzed (3). In muscle cells, Ca^{2+} ions stored in SR are released through Ca^{2+} release channels for contraction. The Ca^{2+} ions released have to be pumped back into SR to cause relaxation. The Ca^{2+}-ATPase is responsible for this pumping process and maintains a 10^4-fold (0.1 μM versus 1.5 mM) concentration gradient across the membrane. Efficiency of the energy conversion is extremely high. Free energy required for transporting 2 mol of Ca^{2+} against such concentration gradient is ~12 kcal, neglecting the effect of membrane potential [for an example of such calculations see (3, 4)], whereas the free energy liberated by ATP hydrolysis is ~12 kcal/mol in typical circumstances (5). Accordingly, the conversion rate is close to 100%. However, the $2Ca^{2+}$/ATP ratio may not be realized due to slippage of the pump (6). SR Ca^{2+}-ATPase is also recognized as an important source of body heat (7).

The reaction mechanism is commonly interpreted according to an E1/E2 scheme (Figure 1). The classical model (8) postulates that, in the E1 state, the Ca^{2+}-binding sites have high affinity and are accessible from the cytoplasm, whereas in the E2 state, the Ca^{2+}-binding sites have low affinity and open to the lumen (extracellular side). Other members of E1/E2-type ATPases include Na^+K^+-ATPase and gastric H^+K^+-ATPase among others (9). They are commonly referred to as P-type ATPases, because the enzyme is autophosphorylated by ATP during the reaction cycle. This is a characteristic feature of the P-type ATPase, distinct from F_1F_0-type, for example. Transfer of bound Ca^{2+} is thought to take place between E1P and E2P. These two phosphorylated states are distinguished by their sensitivity to ADP (i.e., ATP is synthesized from ADP in E1P but not in E2P). Each principal state of the reaction cycle can also be characterized by a different susceptibility to trypsin and proteinase K (10).

The signature motif of the P-type ATPase superfamily is D-K-T-G-T-[LIVM]-[TIS], starting from the aspartate residue that is phosphorylated (Asp351

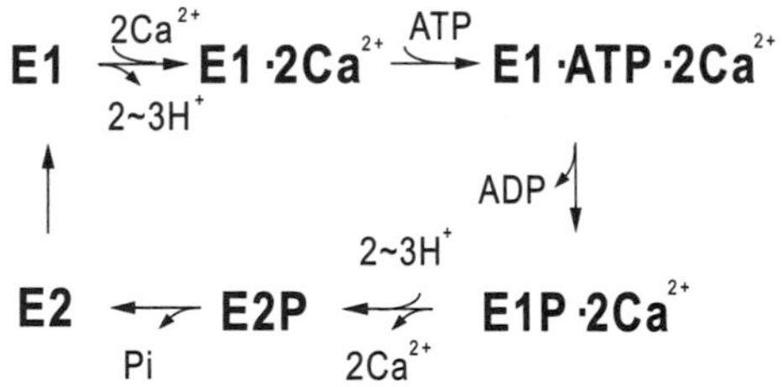

$$\text{E1} \underset{2\sim 3\text{H}^+}{\overset{2\text{Ca}^{2+}}{\rightleftharpoons}} \text{E1} \cdot 2\text{Ca}^{2+} \overset{\text{ATP}}{\longrightarrow} \text{E1} \cdot \text{ATP} \cdot 2\text{Ca}^{2+}$$

Figure 1 reaction scheme:

Figure 1 A simplified (forward direction only) reaction scheme according to the E1/E2 model. Although 2Ca^{2+} in the E1 species are explicitly shown in this diagram, they are often omitted in literature and also in this article. Mg^{2+} is also assumed to be bound for the ATP-bound or phosphorylated states. Thus, E1P, for example, should read E1P·Mg^{2+}·2Ca^{2+}.

in SERCA1a, Figure 2). This single motif uniquely identifies all P-type ATPases (11); there are, however, many other conserved motifs (Figure 2). Unlike other ATPases, P-type ATPases lack the P-loop (12). The organization of key residues around the phosphorylation site classifies P-type ATPases as members of the haloacid dehalogenase superfamily, which contains enzymes unrelated to nucleotides or phosphorylation (13). Within this superfamily, phosphoserine phosphatase (PSPase) is structurally the best characterized member (14). Furthermore, despite the different folding patterns, the key residues used for phosphoryl transfer and hydrolysis in P-type ATPases are identical to those in bacterial two-component regulator proteins (15). The latter uses phosphorylation for making a new molecular interface by exposing the hydrophobic surface. The mechanistic role of phosphorylation in ion pumping, however, has been unclear.

CRYSTALLOGRAPHY OF THE Ca^{2+}-ATPase

So far the structures of rabbit SERCA1a have been determined for five different states, although only two have been published. They are Ca^{2+}-bound E1·2Ca^{2+} state (at 2.6 Å resolution; PDB accession code 1EUL) (16) and Ca^{2+}-unbound E2 state stabilized with thapsigargin (TG) (17), a potent plant inhibitor [E2(TG) at 3.1 Å resolution; PDB accession code 1IWO] (18). In addition to these, the enzyme has been crystallized in the presence of Ca^{2+}, Mg^{2+}, and AMPPCP, a nonhydrolyzable analog of ATP (E1·AMPPCP, at 2.7 Å resolution), in the presence of Ca^{2+}, Mg^{2+}, AlF$_x$, and ADP (E1·AlF$_x$·ADP, at 2.8 Å resolution), and in the absence of Ca^{2+} but in the presence of Mg^{2+} and MgF$_x$ (E2·MgF$_4^{2-}$). (Here the dot refers to noncovalent binding.) E1·AMPPCP will be a close analog of E1·ATP·2Ca^{2+} in the reaction scheme in Figure 1. E1·AlF$_x$·ADP will represent the structure immediately after the hydrolysis of ATP (i.e., E1P·ADP·2Ca^{2+}) with AlF$_x$ as the stable phosphate analog (19), and

$E2 \cdot MgF_4^{2-}$ is an analog of E2P with MgF_4^{2-} as the phosphate analog (20). A comparison of these structures with those of PSPase in five different states (14) and biochemical evidence (21) suggest that $E2 \cdot MgF_4^{2-}$ corresponds to the "product state" with bound phosphate after hydrolysis of E2P. Thus, we do not yet have good analogs of E1P and E2P. Nevertheless, limited proteolysis experiments (10) showed that there are no large domain movements between $E1 \cdot P \cdot ADP$ and E1P and between E2P and $E2 \cdot MgF_4^{2-}$. Therefore, we have reasonably good atomic models to cover the whole reaction cycle of Ca^{2+}-ATPase. The $E1 \cdot 2Ca^{2+}$ structure was rerefined by one of the authors at 2.5Å resolution (C. Toyoshima, unpublished work) and is presented here.

ARCHITECTURE OF THE Ca^{2+}-ATPase

Ca^{2+}-ATPase is a tall molecule of about 150 Å high and 80 Å thick and comprises a large cytoplasmic headpiece, a transmembrane domain made of 10 (M1-M10) α-helices, and short lumenal loops (Figures 2 and 3). Four of the transmembrane helices, M2-M5, have long cytoplasmic extensions (Figure 2). The cytoplasmic headpiece consists of three domains, designated as A (actuator), N (nucleotide binding), and P (phosphorylation) domains. The A domain, whose functional role was not obvious at the time of the publication of the first crystal structure (16), functions as the "actuator" of the gating mechanism that regulates Ca^{2+} binding and release. The N domain, a long insertion between two parts forming the P domain, contains the adenosine binding site, whereas the γ-phosphate reacts with Asp351 in the P domain. The three cytoplasmic domains are widely split in $E1 \cdot 2Ca^{2+}$ but gather to form a compact headpiece in the other states (Figure 3). The A domain is directly connected to M1-M3, and the P domain to the M4 and M5 helices. The M5 helix is ~60 Å long, running from the lumenal surface to one end of the P domain, and it functions as the "spine" of the molecule. On the lumenal side, the loops are short, except for that of ~40 residues connecting the M7 and M8 helices (L78, Figures 2 and 3). The distance between the Ca^{2+}-binding sites and the phosphorylation site is longer than 50 Å.

Figure 2   Two-dimensional diagram of the structure, the amino acid sequence, and the results of mutation experiments of the skeletal muscle Ca^{2+}-ATPase (SERCA1a). Secondary structure elements (*boxes* for helices; *arrows* for β-strands), which are assigned with DSSP (60), follow those in the Ca^{2+}-bound form (16); the elements that change in E2(TG) (i.e., M1′, M2′, and A3) (18) are shown in dotted lines and are identified by italic letters. T1 and T2 are trypsin digestion sites and PrtK, major proteinase K digestion sites (61). Conserved residues of the P-type ATPases are shown in different fonts. Those conserved in 135 out of 159 sequences are shown in open italic letters; those conserved in only 90 sequences are in normal open letters (11). Mutation information was taken from (62).

In E1·2Ca^{2+}, both the A and N domains are likely to be mobile, and the crystal structure shown in Figure 3 may not necessarily show their most probable positions. In fact, their locations are substantially different in microcrystals of E1·2Ca^{2+} grown in the absence of sodium butyrate (22).

Ca^{2+}-ATPase as a Membrane Protein

The orientations and positions of atomic models with respect to the lipid bilayer are illustrated in Figure 3; the proteins are placed in the bilayer of dioleoylphosphatidylcholine (DOPC) generated by molecular dynamics calculation (Y. Sugita, M. Ikeguchi, and C. Toyoshima, unpublished results). Ca^{2+}-ATPase is fully functional when reconstituted in DOPC membrane (23). These orientations and positions with respect to the membrane were determined from crystallographic constraints and from comparison with different crystal forms. The orientations coincided with those in the original crystals. This figure alone shows many interesting features. For example, amphipathic segments of the models (i.e., short loops connecting transmembrane helices and a short amphipathic part of M1 (M1′ in Figure 3) are located within the interface region of the lipid bilayer, keeping hydrophobic residues (Phe57 is shown in Figure 3) within the hydrophobic region. This is an expected feature that supports the orientations of the models with respect to the bilayer. The positions of surface loops and those of interface-residing residues [Trp, Tyr, Lys, and Arg (24)] are also consistent with these orientations. The coordinates of the aligned models are available at the author's web site (http://www.iam.u-tokyo.ac.jp/StrBiol/models).

Organization of the Transmembrane Domain

As expected from the amino acid sequence, SERCA1 has 10 (M1-M10) transmembrane α-helices (Figure 2), two of which (M4 and M6) are partly unwound (Figures 4 and 5) through the whole reaction cycle. M6 and M7 are far apart and are connected by a long cytosolic loop that runs along the bottom of the P domain and is hydrogen bonded to M5 (Figure 3). M4-M6 and M8 contain the residues directly coordinating the two Ca^{2+} (Figure 4). The amino acid sequence is well conserved for M4 to M6 but not for M8 even within the members of closely related P-type ATPases, such as Na$^+$K$^+$- and H$^+$K$^+$-ATPases. As described later, all helices from M1 to M6 move considerably during the reaction cycle (Figures 4 and 5), whereas M7-M10 helices keep their positions (Figure 4), apparently acting as a membrane anchor. M7-M10 appear to have a specialized function in each subfamily and are indeed absent in bacterial type I P-type ATPases (9). Nevertheless, M7 contains a GxxG motif (Gly842 and Gly845) (26), often used for tight packing of transmembrane helices. In fact, M7 packs tightly with M5 at Gly770 (Figure 5), which works as a pivot of the bending of M5 (see below). Thus, M7 and M8 have important roles in SERCA1.

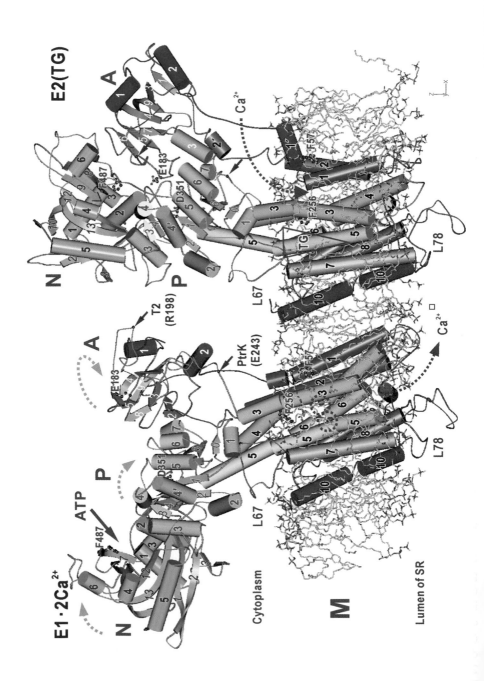

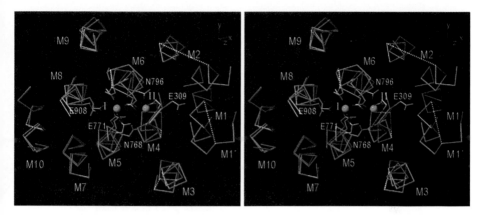

Figure 4 Stereo diagram showing the organization of the transmembrane helices, viewed from the cytoplasmic side approximately normal to the membrane. Only the cytoplasmic half (~10 residues) of each helix is shown as a Cα trace. E1·2Ca²⁺ is in violet, and E2(TG) is in light green. Two Ca²⁺ in the binding sites (I and II) are shown as spheres (*cyan*). Side chains of the residues directly coordinating Ca²⁺ are also shown [E1·2Ca²⁺ (atom color) and E2(TG) (*light green*)]. M1′ is the amphipathic helix lying on the membrane (Figures 2 and 3). Dotted lines connect corresponding residues in the two conformations.

Structure of the P Domain

The P domain contains the residue of phosphorylation, Asp351. There are three critical aspartate residues (Asp627, Asp703, and Asp707) clustered around the phosphorylation site (Figure 6), in addition to an absolutely conserved Lys residue, Lys684 (Figure 6). The P domain has a Rossmann fold, commonly found

---

Figure 3 E1·2Ca²⁺ and E2(TG) forms of Ca²⁺-ATPase in lipid bilayer, which was generated by molecular dynamics simulation of dioleoylyphosphatidylcholine (DOPC). Cylinders represent α-helices, arrows β-strands, which are numbered as in Figure 2. M3 and M5 helices in the unbound form are approximated with two and three cylinders, respectively. The color changes gradually from the N terminus (*blue*) to the C terminus (*red*). Two purple spheres (*circled*) in the transmembrane region of E1·2Ca²⁺ represent bound Ca²⁺ (sites I and II). TG binds to the space between M3 and M7 (*large triangle*). Orange arrows in E1·2Ca²⁺ indicate the direction of movement of the cytoplasmic domains during the change from E1·2Ca²⁺ to E2(TG). Magenta arrow in E2(TG) shows the proposed pathway (18) for Ca²⁺ to enter into the binding cavity. T2 trypsin digestion site and a proteinase K digestion site (PrtK) are also marked. F57 is at the boundary between M1 and M1′ (Figure 2). E183 is a critical residue (63) in the A domain. F256 is a key residue in TG binding (52). D351 is the residue of phosphorylation. F487 is a key residue in ATP binding (64). Prepared with Molscript (65).

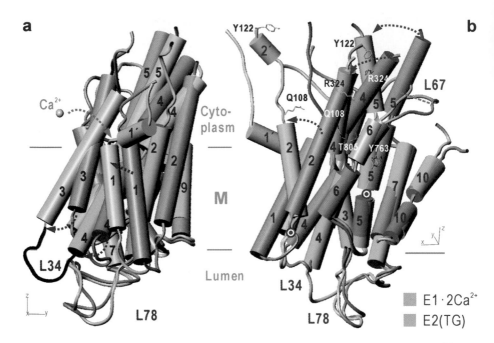

Figure 5 Rearrangement of transmembrane helices on the dissociation of Ca^{2+}. The models for $E1\cdot2Ca^{2+}$ (*violet*) and E2(TG) (*light green*) are superimposed and viewed from the right (*a*) and the rear (*b*) of that presented in Figure 3. The M5 helix lies along the plane of the paper, and M8 and M9 are removed in (*b*), so that "domino" movements of the M5, M4, and M2 helices can be seen clearly. Double circles (*red* and *white*) show pivot positions for M2 and M5. The orange broken line shows a critical hydrogen bond between the L67 loop and the M5 helix. Red arrows indicate the direction of movements during the change from $E1\cdot2Ca^{2+}$ to E2(TG). Cyan arrow shows the proposed pathway for the first Ca^{2+} ion (18).

in nucleotide binding proteins, consisting of a parallel β-sheet (seven strands in Ca^{2+}-ATPase) and associated short α-helices (Figure 6). The catalytic residue (i.e., Asp351 in Ca^{2+}-ATPase) is always at the C-terminal end of the first β-strand, which is connected to a long insertion, the N domain (Figure 2). The polypeptide chain returns to the P domain close to where it left (Figure 6). Thus, the P domain is formed by two regions far apart in the amino acid sequence. This is why the P-type ATPase was thought to be an orphan in evolution for a long time (13). This system requires Mg^{2+} for phosphorylation, although Ca^{2+} can substitute in ATP (and AMPPCP) binding (27, 28). Mg^{2+} is coordinated by carboxyl groups of Asp351 and Asp703, main chain carbonyl of Thr353, and two water molecules. This geometry is shared by all members of the haloacid dehalogenase superfamily. Lys684 appears to be particularly important for the binding of γ-phosphate of ATP. Although the position of the Cα of the

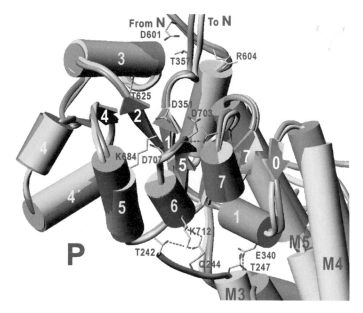

Figure 6 Organization of the P domain and the linkage with the transmembrane helices (18). Superimposition of the E1·2Ca^{2+} (*violet*) and E2(TG) (*light green*) crystal structures fitted with the P domain. The residues (in atom color) represent those in E2(TG). Orange broken lines show potential hydrogen bonds in E2(TG).

corresponding residue in phosphoglucomutase is one residue shifted (29), the position of the terminal nitrogen is precisely conserved.

Structure of the N Domain

N domain is the largest of the 3 cytoplasmic domains, consisting of residues approximately Asn359-Asp601 (Figure 2). Sequence similarity of this domain is low among the members of the P-type ATPase family, although the similarity in 3D structure is quite high, at least with Na$^+$K$^+$-ATPase (30). The N domain is connected to the P domain with two strands that have a β-sheet-like hydrogen bonding pattern. This will allow large domain movements with a precise orientation (31). Consecutive prolines (Pro602–603), which must be quite rigid, may serve as additional guides for orienting the domain movement. The N domain contains the binding site for the adenosine moiety of ATP (16). Phe487 (Figure 3) makes an aromatic adenine ring interaction, which is a common feature in many ATP binding sites. Lys515, a critical residue, is located at one end of the binding cavity. This residue can be labeled specifically with FITC at alkaline pH and has been used for many spectroscopic studies (32). The binding of the adenine ring is predominantly hydrophobic, devoid of hydrogen bonds (30). Lys515 and Glu442, another critical residue (33), appear to be important in lining the binding cavity.

Structure of the A Domain

The A domain is the smallest of the 3 cytoplasmic domains and consists of the N-terminal ~50 residues that form 2 short α-helices and ~110 residues between the M2 and M3 helices (Figure 2), which form a deformed jelly roll structure. The A domain contains a sequence motif [181]TGES, one of the signature sequences of the P-type ATPase family (Figure 2). This sequence makes a loop that comes very close to the phosphorylation site Asp351 in the E2 and E2P states (Figure 3). In the E2P state, this loop appears to be very important for shielding the aspartyl phosphate from bulk water. Because this domain is connected to the M1-M3 helices with loops of 10 residues or more (Figure 2), and connected to the P domain at only one place (around Gly156-Ala725) in E1·Ca^{2+} (Figure 7), the A domain is highly mobile. This feature allows the A domain to take different positions imposed by other domains and act as the actuator of the gates that regulate the binding and release of Ca^{2+} ions.

Structure of the Transmembrane Ca^{2+}-Binding Sites

It is well established that SR Ca^{2+}-ATPase has two high affinity transmembrane Ca^{2+}-binding sites (34), and the binding is cooperative (35). There was much confusion as to their locations, partly because these Ca^{2+}-binding sites do not appear to bind lanthanides (36). Debates still exist as to the presence of low affinity sites, in particular, in the lumenal region [e.g., (37)]. X-ray crystallography of the rabbit SERCA1a in 10 mM Ca^{2+} at pH 6.1 identified two binding sites in the transmembrane region but none outside the membrane (16). Valence search (38), which has proven to be powerful in identifying metal binding sites, showed no other peaks if the threshold was set higher than 1.6.

The two Ca^{2+}-binding sites (I and II) (39) are located side by side near the cytoplasmic surface of the lipid bilayer (Figure 3), even though the binding of two Ca^{2+} is sequential and cooperative (35). Site I, the binding site for the first Ca^{2+}, is located at the center of the transmembrane domain when viewed normal to the membrane (Figure 4) in a space surrounded by M5, M6, and M8 helices. Site I is formed by Asn768, Glu771 (M5), Thr799, Asp 800 (M6), Glu908 (M8), and two water molecules (Figure 8). Thus, no main chain oxygen atoms contribute to site I. Site II is nearly on the M4 helix when viewed normal to the membrane (Figure 4), and it is located slightly (~3 Å) closer to the cytoplasmic surface than site I (Figure 3). Site II is formed with the contribution of M4-M6 and without water molecules (Figures 4 and 8). The M4 helix is partly unwound (between Ile307-Gly310) and provides three main chain oxygen atoms to the coordination of Ca^{2+} (Figure 8). Asn796 and Asp800 (M6) provide one side chain oxygen atom, whereas Glu309 provides two to cap the bound Ca^{2+} (Figure 8). This arrangement of oxygen atoms is reminiscent of the EF-hand motif [e.g., (40)].

Thus, both site I and site II have seven coordinating oxygen atoms but of differing characteristics. Asp800, on the unwound part of M6, is the only residue

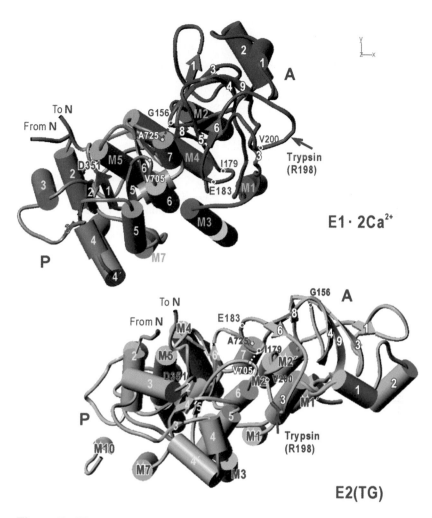

Figure 7 Views normal to the membrane of the cytoplasmic region of the Ca^{2+}-ATPase in E1·2Ca^{2+} and E2(TG) states. N domain is removed. All the helices (*cylinders*) and β-strands (*arrows*) are numbered according to Figure 2. Dotted lines show hydrogen bonds (using main chain carbonyl and amide groups) formed between the A and P domains. Rotation axis of the A domain comes near V200 and approximately perpendicular to the plane of the paper.

that contributes to both sites (Figures 8 and 9). Any substitutions to site I residues, except one for Glu908, totally abolish the binding of Ca^{2+} (41). In contrast, even double mutations of Glu309 and Asn796 leave 50% Ca^{2+} binding (41). Taken together, these mutagenesis experiments show that site II is the binding site for the second Ca^{2+} (41). Both Glu771 (M5) and Glu908 (M8) provide only one side chain oxygen to the coordination in site I; Gln can

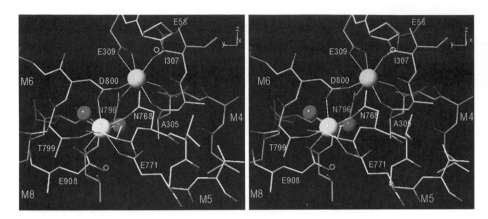

Figure 8   Stereo view of the transmembrane Ca^{2+}-binding sites. Viewed approximately parallel to the membrane plane. Two Ca^{2+} in the binding sites are shown as spheres (*cyan*) and water molecules are shown as smaller red spheres. White dotted circles indicate predicted protonation sites.

substitute for Glu908 to a large extent but not for Glu771 (41). Molecular dynamics simulation (Y. Sugita, M. Ikeguchi, and C. Toyoshima, unpublished results) and pKa calculations (N. Miyashita, Y. Sugita, and C. Toyoshima, unpublished work) indicate that Glu908 is protonated (Figure 8). The "unused" oxygen of Glu771 carboxyl is important for fixing Ala305, whose carbonyl oxygen contributes to site II, thereby establishing the cooperativity of Ca^{2+} binding.

STRUCTURAL CHANGES ACCOMPANYING THE DISSOCIATION OF Ca^{2+}

As illustrated in Figure 3, Ca^{2+}-ATPase undergoes large structural changes on the dissociation of bound Ca^{2+} to the cytoplasmic side (18). This is the reverse reaction of Ca^{2+} binding in Figure 1. The three cytoplasmic domains change their orientations and gather to form a compact headpiece. The P domain inclines 30° with respect to the membrane, and the N domain inclines 60° relative to the P domain, whereas the A domain rotates ~110° horizontally. The structure of each domain, however, is hardly altered (Figure 6). In contrast, some of the transmembrane helices are bent or curved (M1, M3, and M5) (Figure 3) or partially unwound (M2) (Figure 5); M1-M6 helices undergo drastic rearrangements (Figures 4 and 5) that involve shifts normal to the membrane (M1-M4). Thus, the structural changes are very large and global; they are mechanically linked and coordinated by the P domain.

P Domain as the Coordinator of the Transmembrane Helices

The M3-M5 helices are directly linked to the P domain by hydrogen bonds (Figure 6). M3 is connected to the P1 helix at the bottom of the P domain through a critical hydrogen bond involving Glu340 (42). The top part of M5 is integrated into the Rossmann fold (Figures 6 and 7) and moves together with the P domain as a single entity. M4 and M5 are "clamped" by forming a short antiparallel β-sheet (strands 7 and 0) (Figure 6). Thus, M4 is also linked to the P domain but much less rigidly compared to M5. M6 is also connected, though less directly, to the P domain through L67 (Figure 5), which is in turn linked to M5 through a critical (43, 44) hydrogen bond between Arg751 and the carbonyl of Arg819 (Figure 5). Thus, the M3-M6 helices are all linked to the P domain with flexible "joints" except for M5. This is an expected feature, because rather loose interfaces will be necessary for realizing multiple, largely different configurations in different functional states.

Accordingly, if the P domain inclines, for instance, due to the bending of M5, all these helices (M3-M6) will incline and generate movements that have components perpendicular to the bilayer (Figure 5). The distances of the movements depend on the distances from the pivoting point, located around Gly770 at the middle of the membrane (double circles in Figure 5). The lower part below Gly770 hardly moves. The shift is therefore small for M6 and large for M3 and M4; the entire M3 and M4 helices move downward during dissociation of Ca^{2+}, whereas M6 undergoes rather local changes around Asn796-Asp800. However, even this part appears to be controlled by the bending of M5 through a hydrogen bond between the main chain carbonyl of Thr799 and the hydroxyl of Tyr763 and more indirectly through a hydrogen bond between Thr805 and Gln108 (Figure 5).

Furthermore, when the P domain inclines, M4 impinges on the top part of M2 (Figure 5) and triggers the rotation of the A domain. The P domain itself pushes the A domain to rotate (Figure 7). This rotation eventually rearranges the M1-M3 helices (Figure 4). Thus, taken together, the movements of the M1-M6 helices are all coordinated by the P domain. It must be noted that the movements of the M1-M4 helices have components normal to the membrane, but they are generated by the change in inclination (Figure 5). The sole exception is the movement of the M1 helix, which is highly detached from the other helices. Also, the N-terminal part of M1 (M1′ in Figure 2 and 3) is amphipathic (Figure 2). Therefore, M1 can really be pulled out from the membrane into the cytoplasm by the A domain to which M1 is directly connected.

Ca^{2+}-Binding Sites in the E2 State and Proton Counter Transport

The most important movements of the transmembrane helices directly relevant to the dissociation of Ca^{2+} are (*a*) a shift of M4 toward the lumenal side by one turn of an α-helix (5.5 Å), (*b*) bending of the upper part of M5 (above Gly770) toward

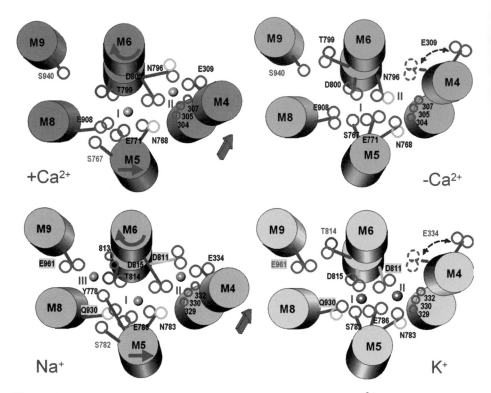

Figure 9 Schematic diagram of the cation binding sites in the Ca^{2+}-ATPase and in the Na^+K^+-ATPase in two conformations. Red, blue, and brown circles, respectively, represent oxygen, nitrogen, and carbon atoms. Small red circles indicate that main chain carbonyl oxygen atoms are used for coordination. Ca^{2+} ions are supposed to enter into the binding cavity by the conformation change of the side chain of Glu309 as indicated by dotted circles. Residue numbers of the Na^+K^+-ATPase refer to those of human $\alpha1$ isoform. Residue numbers in yellow rectangles indicate that they are different between the Ca^{2+}- and Na^+K^+-ATPases. The arrows show the movements of the helices in the transition from $E1\cdot2Ca^{2+} \rightarrow E2(TG)$ for the Ca^{2+}-ATPase and $E1\cdot3Na^+ \rightarrow E2\cdot2K^+$ for the Na^+K^+-ATPase.

M4, and (c) a nearly 90° rotation of the unwound part of M6 (18) (Figure 9). As a result, profound reorganization of the Ca^{2+}-binding residues takes place, and the number of coordinating oxygen atoms decreases (Figure 9). For site I, this is due to the movement of Asn768 toward M4 caused by the bending of M5; for site II, replacement of Asp800 by Asn796 (i.e., rotation of M6) is critical. Then, why is the large movement of M4 (and M3) toward the lumenal side necessary?

Homology modeling of the cation binding sites of Na^+K^+-ATPase (45) suggested that such movements are needed for counter transport. It was surprising to see that very regular coordination geometry could be made immediately for the K$^+$ binding sites of the Na$^+$K$^+$-ATPase, starting from the main chain conformation derived from the E2(TG) model of the Ca^{2+}-ATPase (18). The residues involved in the cation binding sites appear essentially the same between Ca^{2+}- and Na$^+$K$^+$-ATPases (Figure 9). The only critical difference is that Asn796 in Ca^{2+}-ATPase is substituted by Asp811 in Na$^+$K$^+$-ATPase. Asp811 is a key residue in K$^+$ binding, playing a similar role to Asp800 in Ca^{2+} binding of the Ca^{2+}-ATPase and Asp815 in Na$^+$ binding of the Na$^+$K$^+$-ATPase, by contributing to both sites I and II (Figure 9). The position of Asp811 (Asn796 in Ca^{2+}-ATPase) is one turn below Asp815 (Asp800 in Ca^{2+}-ATPase) toward the extracellular (lumenal) side. To form a cation binding site there, M4 must move toward the extracellular side so that carbonyl oxygens (Figures 8 and 9) can contribute to the binding. Hence, the movement of M4 ensures the release of one cation (Ca^{2+} or Na$^+$) and the binding of the other cation (H$^+$ or K$^+$) at the different level with respect to the membrane. This will certainly help in avoiding the competition between the binding cations.

This movement of M4 can be generated by the bending of the upper part of M5 (or by the inclination of the P domain) (Figure 5). This bending allows residues on different faces of the M5 helix to be used for cation binding (e.g., Ser782 can be used instead of Glu786 in K$^+$ binding at site I) (Figure 9). This will allow profound reorganization of the binding residues to accommodate a larger cation (K$^+$) with a higher affinity and also to ensure the release of Na$^+$.

It is established that Ca^{2+}-ATPase counter transports 2 (or 3) H$^+$ (3). However, it is still difficult to understand why such large movements are needed if the ions to be counter transported are H$^+$. If Ca^{2+}-ATPase emerged through Na$^+$K$^+$-ATPase, this would be understandable. However, this lineage is not supported from the phylogeny of P-type ATPases (11). It would therefore be very interesting to know what ions are counter transported by the pump protein ancestral to both Ca^{2+}-and Na$^+$K$^+$-ATPases.

It is questionable whether counter transport of H$^+$ by Ca^{2+}-ATPase has a relevant physiological role, because native SR membranes are leaky to H$^+$ (46).Yet, given the number of negatively charged residues clustered around the transmembrane Ca^{2+}-binding sites (Figures 8 and 9), there is no doubt that the structure is stable only if most of them are protonated or neutralized by other cations, such as Na$^+$ or K$^+$. For example, in E2(TG), Asn796, and Glu771 side chains appear to form two hydrogen bonds (Figure 4), which require the protonation of Glu771 carboxyl. Because Glu908 and Glu58 are expected to be protonated even in E1·2Ca^{2+} (Figure 8), there are only three more candidates for protonation, namely, the carboxyls of Glu771, Asp800, and Glu309. Mutagenesis studies (47, 48) and pKa calculation (N. Miyashita, Y. Sugita, and C. Toyoshima, unpublished results) suggest that Glu771 and Glu309 are protonated. If the transport of K$^+$ should be avoided, then proton binding would be the only possibility. Accordingly, we propose that proton counter transport is required for the stability of the ATPase in the absence of Ca^{2+}. Protons will be exchanged with Ca^{2+} on either side of the membrane.

Entry Pathway of Ca^{2+} to the Binding Sites

It is not obvious how Ca^{2+} reaches the transmembrane binding sites because of the lack of a vestibule similar to those found in ion channels (49). However, evidence indicating the involvement of Glu309, the residue capping the site II Ca^{2+}, is accumulating. In E2(TG) structure, Glu309 points outside the binding cavity and toward the cytoplasm (Figure 4). The amphipathic short helix lying on the cytoplasmic surface (M1 in Figure 3) and the loop connecting the A domain and M3 helix provide negatively charged residues around a path leading to Glu309 (18). This path is made by the large rotation of the A domain and is destroyed in E1·2Ca^{2+}. Thus, it is likely that the conformation change of the Glu309 side chain delivers Ca^{2+} to the binding cavity (Figure 9). The idea that Glu309 is the gating residue has also been proposed from mutagenesis studies (48) and isotope exchange experiments (G. Inesi, unpublished results).

This idea agrees well with the cooperative binding of Ca^{2+} (35). Site I is the binding site for the first Ca^{2+} and has a high affinity ($\sim$0.5 μM at pH 7.0) (50). Because Ca^{2+} reaches site I through site II, site II has to have substantially lower affinity than site I. After the binding of Ca^{2+} to site I, site II may become a binding site with even higher affinity (51). For this purpose, the region around Asp800 has to be flexible and fixed by the binding of the first Ca^{2+}.

Whether the first Ca^{2+} or the second Ca^{2+} releases the A domain to allow ATP to cross-link the N and P domains is an important question. Because the N796A mutant is not protected by AMPPCP against proteinase-K attack (27), it seems that the binding of the second Ca^{2+} is required to release the N domain. If this were not the case, strict stoichiometry (i.e., transport of 2 Ca^{2+} per ATP hydrolyzed) would not be realized. Our homology modeling shows that site I in the E2 state can accommodate K^+, which is substantially larger than Ca^{2+} (1.33 Å compared with 0.95 Å). Because site I entirely consists of side chain oxygen atoms, it is expected to have substantial flexibility. Therefore, the binding of the first Ca^{2+} may not require straightening of the M5 helix. However, it is expected that the binding of the first Ca^{2+} requires at least some rearrangement of certain transmembrane helices, because TG eliminates high affinity Ca^{2+} binding (17).

TG binds to Ca^{2+}-ATPase in a space between M3 and M7 (Figure 3) (18). TG has a complementary shape to this part of Ca^{2+}-ATPase in the E2 conformation. Binding of TG here will prevent movement of M3, the position of which is very different between E1Ca^{2+} and E2(TG) (Figure 5), and thereby locks the ATPase in the E2 conformation. The binding is hydrophobic and involves, if any, only one hydrogen bond. Extensive mutagenesis studies (52; G. Inesi, unpublished information) show that Phe256 (M3) (shown in Figure 3) and Ile765 (M5) are critical to TG binding. Thus, it is easy to understand that TG will impair the binding of the second Ca^{2+}; however, it would be more difficult to understand how TG interferes with the binding of the first Ca^{2+} unless it also requires rearrangements of the transmembrane helices.

Exit Pathway From the Binding Sites

The exit of Ca^{2+} on the lumenal side is not obvious either. In E1·2Ca^{2+} there are layers of hydrophobic residues below the binding cavity. Site I Ca^{2+} and the closest water molecule observed in the crystal structure (near the loop connecting M5 and M6) is >12 Å away from the membrane surface. In E2(TG), the loop connecting M3 and M4 comes close to the L78 loop (Figure 3), sealing the access pathway at the surface. Thus, in this aspect, the structure deviates from the classical E1/E2 models that postulate lumenally opened binding sites in E2 (8). However, in the normal reaction cycle in forward direction, the lumenal gate will open during the E1P $\rightarrow$ E2P transition and close after the binding of counter ions (H$^+$ in this case) (Figure 1). In the E2·MgF$_4$$^{2-}$ complex (20), which is an analog of E2·P state, the position of the A domain differs from E2(TG), apparently to open the lumenal gate by moving the M1-M4 helices (C. Toyoshima and H. Nomura, unpublished observation).

THE ROTATION OF THE A DOMAIN

As described above, cytoplasmic domains show drastically different arrangements between E1·Ca^{2+} and E2(TG). In particular, the A domain undergoes a 110° rotation, which appears to be another key event in active transport (10) (Figures 7 and 10). In fact, the A domain has different positions in all of the four states (Figure 10), and so does the M1 helix. This appears to be the way by which Ca^{2+}-ATPase regulates the cytoplasmic and lumenal gates of the ion pathway. To understand the rotation of the A domain, it is important to note that the A domain is directly linked to three (not just two) transmembrane helices (M1-M3) and that its position can be directly altered by the N and P domains and indirectly by the M4-M6 helices through the movements of M3 and the P domain.

Large bending of M5 is required for rotating the A domain horizontally, because the A domain is pushed by the P domain on the M2 side while loosely fixed on the M3 side (Figure 7). During E1·2Ca^{2+} $\rightarrow$ E2(TG), M2 changes its inclination by ~30°, apparently triggered by the movement of M4, which is in turn pushed by M5 (Figure 5). However, to achieve a ~110° rotation, the movement of M2 is clearly insufficient. It may work as a trigger of the A-domain movement, but during the transition, M2 detaches from M4 and may become free. Finally, M2 inclines more than M4 (Figure 5) until it is stabilized by the P7 helix in E2(TG) (Figure 7). If a single force is exerted onto the A domain, there will be a pivoting point, that is, conserved contact between two residues on different domains. In fact, rotation of the A domain appears to occur around an axis running close to Val200, which also moves ~4.5 Å in E1Ca^{2+} $\rightarrow$ E2(TG) (Figure 7). In E1·2Ca^{2+}, Gly156 (A), and Ala725 (P) make a hydrogen bond using main chain. This hydrogen bond breaks during E1Ca^{2+} $\rightarrow$ E2(TG) transition; instead, Leu179 (A) and Val705 (P) and also Leu180 (A) and Asn706

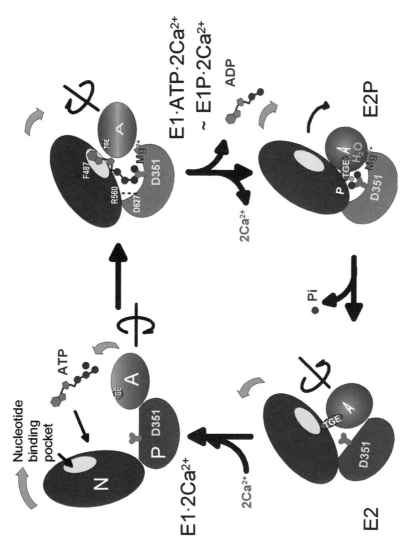

Figure 10 Schematic models for four distinct arrangements of the three cytoplasmic domains in the reaction cycle (Figure 1) of Ca^{2+}-ATPase. D351 is the residue of phosphorylation. Transfer of "occluded" Ca^{2+} to the lumenal side is thought to take place between E1P and E2P. The ^{183}TGE-loop is also shown.

(P) form hydrogen bonds in E2(TG) (Figure 7). Hence, the P domain alone is not sufficient to make the A-domain rotation. In reality, the movement of the A domain is more complicated, and its location in E2(TG) is stabilized by salt bridges (18) with the N domain (Figure 3). Thus, the rotation of the A domain is a result of coordinated movements of the N and P domains and transmembrane helices.

The A Domain as the Actuator Domain

In E1·2Ca^{2+}, all three cytoplasmic domains are well separated (Figures 3 and 10). The M1 helix is deeply embedded in the lipid bilayer (Figure 3). There is no vestibule that leads to the transmembrane Ca^{2+}-binding sites. The pathway from the lumenal side to the Ca^{2+}-binding sites is sealed by hydrophobic residues just below the Ca^{2+}-binding sites. In E2(TG), the three cytoplasmic domains gather to form a compact headpiece, stabilized by several hydrogen bonds formed between A-P and A-N domains. The M1 helix is pulled toward the cytoplasm (Figure 3); the N-terminal amphipathic part (M1′) (Figures 2, 3, and 5) is kinked and forms a short helix lying on the membrane surface. In E2·MgF$_4^{2-}$, the A domain is pulled toward the P domain (Figure 10) to form a tightly sealed hydrophobic environment around the phosphorylation residue; this allows a particular water molecule to attack the acylphosphate. The presence of Mg^{2+} in addition to phosphate (analog) in E2P appears critical for the positioning of the A domain. The kinked part of the M1 helix (M1′) is pulled further into the cytoplasm, compared to E2(TG). This moves the lumenal end of the M4 helix to open the lumenal gate. This movement is expected to be rather small, presumably just enough to release Ca^{2+} while preventing K$^+$ from entering. In any case, the length of the link between the A domain and the M1 helix must be critical but not the amino acid sequence itself, as demonstrated by a mutagenesis study (53).

In E1·AMPPCP and E1·AlF$_x$·ADP, which have virtually the same structures, the A domain is pushed into an extreme position by the N domain, which is crosslinked to the P domain by the nucleotide analogs (Figure 10). In this position, the link between the A domain and the M3 helix appears to be strained, whereas the M1 helix is pulled toward the cytoplasm but in an orientation different from that in E2(TG) or E2·MgF$_4^{2-}$. This appears to be the way of occluding the bound Ca^{2+} because it locks the side chain of Glu309. To make a transition from E1P to E2P, therefore, the link between the A domain and the M3 helix must be important in rotating the A domain. In fact, when this link is cut by proteinase K (Figure 3), the E1P → E2P transition is severely slowed (25). In any case, precise positioning of the A domain and the M1 helix must be crucial. A good example is provided by a mutagenesis study of Val200 (54) (Figure 7). Any modifications to this residue largely abolish ATPase activity. It is conceivable that this residue works as a spacer between the A and P domains by making a van der Waals contact with Pro681 (P), but no other roles could be identified from the crystal structures.

ROLE OF PHOSPHORYLATION IN ION PUMPING

Because the position of the A domain is affected by the N and P domains, the interfaces between them must be critical and should be altered during the reaction cycle. Binding of Ca^{2+} and phosphorylation (or ATP binding) will be the two most important factors. Binding of Ca^{2+} places strong constraints to the bending of the M5 helix, because the structure around the major pivoting point (Gly770) (Figure 5) becomes rigid. Limited bending of M5 is used in E1·AMPPCP and presumably in E1P also to make closed configurations different from that in E2P.

Phosphorylation or the binding of γ-phosphate makes the position of the hinge between the N and the P domains substantially higher and also changes the direction of the inclination of the N domain. The binding of γ-phosphate accompanies the dissociation of Mg^{2+} from ATP and the binding of Mg^{2+} to the P domain around Asp703 (Figure 6). This Mg^{2+} bound to the P domain directly alters the position of the A domain in E2P. Thus, one of the roles of phosphorylation is to change the interface between the cytoplasmic domains.

In this regard, it is interesting to note that the top part of the two halves (strands 1–4 and 5–7) of the central β-sheet in the P domain are staggered in unphosphorylated forms (Figure 6) but better aligned in phosphorylated forms, bringing the hinge region between the N and P domains closer to Asp351. This is a unique feature of the Ca^{2+}-ATPase among the members of the haloacid dehalogenase family. It helps to gain another ~30° inclination of the N domain as well as to change the direction of inclination. Thus, the domain movement of the N domain consists of two steps to achieve the ~90° inclination that is required for bridging the N and P domains by ATP.

ROLE OF THE CLOSED CONFIGURATION OF THE CYTOPLASMIC DOMAINS

Importance in the E2 State

As described, closed configurations of the cytoplasmic domains are important in regulating ion flow by locating the A domain in different positions. In the E2 state, a closed configuration of the headpiece appears to be important in two other aspects: (*a*) restriction of the delivery of ATP to the phosphorylation site and (*b*) restriction of the thermal movements of transmembrane helices (18).

The reaction cycle of Ca^{2+}-ATPase (Figure 1) is regulated essentially by Ca^{2+} alone. ATP can bind to the enzyme even when Ca^{2+} is absent, but without Ca^{2+}, the reaction cycle cannot proceed. With the arrangement of the cytoplasmic domains in E2(TG), the γ-phosphate of ATP comes close to but cannot reach the phosphorylation residue Asp351. It requires deeper inclination of the N domain or rotation of the A domain to release the N domain (Figure 10). This in turn requires the binding of Ca^{2+}, which will cause the shift of the transmembrane helices into different positions.

Because the large-scale conformation changes shown in Figure 3 occur without involvement of ATP or phosphate, such changes must be realized by thermal movements. In fact, in Na$^+$K$^+$- and H$^+$K$^+$-ATPases, transmembrane helices come out of the membrane in the absence of K$^+$ after proteolysis and mild heat treatment (55, 56). Also, SR vesicles with loaded Ca^{2+} show significant leakage (6), and thapsigargin can stop it. The closed configuration of the cytoplasmic domains in E2 conformation (Figure 3) will limit thermal movements of the transmembrane helices and therefore the leakage. Because the position of the A domain regulates the gates from and to the binding sites, it has to be fixed in the E2 conformation by the N and P domains.

ATP as a Cleavable Cross-Linker

What causes the E1P → E2P transition is an important unresolved problem. In view of the accelerating effect of ADP in making the complex of Ca^{2+}-ATPase with AlF$_x$ (19), the presence of ADP must be important for cross-linking the N and P domains. Then, it is likely that the dissociation of ADP is the trigger of the transition. In fact in E2·MgF$_4^{2-}$, the TGES loop in the A domain occupies the space where ADP was, stabilized by Mg^{2+} and MgF$_4^{2-}$, and makes the E2P state ADP insensitive. Does the TGES loop kick out ADP from the binding site? This seems unlikely because the TGES loop is located more than 25 Å away from ADP in the complex with ADP·AlF$_x$. A much more likely situation is that ADP leaves first, by thermal movement, and then the A domain rotates due to mechanical strain in the A-domain/M3 link. The hydrolysis of the aspartyl phosphate will destabilize the coordination of phosphate, Mg^{2+}, and the TGES loop, and eventually it triggers the E2P → E2 transition.

This consideration suggests that Ca^{2+}-ATPase uses ATP as a cleavable cross-link between the N and the P domain. It is known that acetylphosphate and carbamoylphosphate can function as substrates. They are obviously too small for crossbridging the N and the P domain. This means that neither the adenosine moiety nor triphosphate moiety is vital. A crucial feature appears to be the presence of an oxygen atom at the β-phosphate position. This oxygen is needed for positioning the Arg560 side chain to make a salt bridge with Asp627 (27) (Figure 10). Both Arg560 (27, 33, 57) and Asp627 (58) are critical residues.

CONCLUSION

As described above, we are beginning to understand that the Ca^{2+} pump uses thermal energy as its real fuel and is endowed with several devices to regulate thermal movements. It is established that ion pumping by these proteins does not involve a power stroke directly driven by ATP (59). The free energy liberated by ATP hydrolysis appears to be used in an implicit way to realize conformations that otherwise occur at negligible frequency. We regret that, at the time of writing, only two crystal structures of Ca^{2+}-ATPase were published, and we could not include in this review the figures showing details of the other states. Animations illustrating the domain movements in Ca^{2+}-ATPase should be very helpful, and they are available at the author's Web site (http://www.iam. u-tokyo.ac.jp/StrBiol/animations) in addition to the computer file of Figure 2.

ACKNOWLEDGMENTS

This work was supported in part by a Grant-in-Aid for Creative Scientific Research from the Ministry of Education, Culture, Sports, Science, and Technology 14GS0308 (CT), National Institutes of Health Program Project HL27867 (GI), and Human Frontier Science Program (CT and GI). We thank our colleagues in Tokyo and Baltimore, particularly Hiromi Nomura, who prepared nearly all the crystals, and Yuji Sugita, who gave us new insights by molecular dynamics simulation. We also thank Takeo Tsuda and Luc Raeymaekers for their help in making Figure 2. We are grateful to David B. McIntosh and Philippe Champeil for their help in improving the manuscript.

The *Annual Review of Biochemistry* is online at http://biochem.annualreviews.org

LITERATURE CITED

1. MacLennan DH, Brandl CJ, Korczak B, Green NM. 1985. *Nature* 316:696–700
2. Brandl CJ, deLeon S, Martin DR, MacLennan DH. 1987. *J. Biol. Chem.* 262:3768–74
3. Yu X, Carroll S, Rigaud JL, Inesi G. 1993. *Biophys. J.* 64:1232–42
4. Mathews CK, van Holde KE, Ahern KG. 2000. *Biochemistry*, pp. 341–42. San Francisco: Addison Wesley Longman. 3rd ed.
5. Berg JM, Tymoczko JL, Stryer L. 2002. *Biochemistry*. New York: Freeman. 894 pp. 5th ed.
6. Inesi G, de Meis L. 1989. *J. Biol. Chem.* 264:5929–36
7. de Meis L, Arruda AP, da Silva WS, Reis M, Carvalho DP. 2003. *Ann. NY Acad. Sci.* 986:481–88
8. de Meis L, Vianna AL. 1979. *Annu. Rev. Biochem.* 48:275–92
9. Møller JV, Juul B, le Maire M. 1996. *Biochim. Biophys. Acta* 1286:1–51
10. Danko S, Yamasaki K, Daiho T, Suzuki H, Toyoshima C. 2001. *FEBS Lett.* 505:129–35
11. Axelsen KB, Palmgren MG. 1998. *J. Mol. Evol.* 46:84–101
12. Saraste M, Sibbald PR, Wittinghofer A. 1990. *Trends Biochem. Sci.* 15:430–34
13. Aravind L, Galperin MY, Koonin EV. 1998. *Trends Biochem. Sci.* 23:127–29
14. Wang W, Cho HS, Kim R, Jancarik J, Yokota H, et al. 2002. *J. Mol. Biol.* 319:421–31
15. Johnson LN, Lewis RJ. 2001. *Chem. Rev.* 101:2209–42
16. Toyoshima C, Nakasako M, Nomura H, Ogawa H. 2000. *Nature* 405:647–55
17. Sagara Y, Inesi G. 1991. *J. Biol. Chem.* 266:13503–6

18. Toyoshima C, Nomura H. 2002. *Nature* 418:605–11
19. Troullier A, Girardet JL, Dupont Y. 1992. *J. Biol. Chem.* 267:22821–29
20. Murphy AJ, Coll RJ. 1992. *J. Biol. Chem.* 267:5229–35
21. Danko S, Daiho T, Yamasaki K, Kamidochi M, Suzuki H, Toyoshima C. 2001. *FEBS Lett.* 489:277–82
22. Ogawa H, Stokes DL, Sasabe H, Toyoshima C. 1998. *Biophys. J.* 75:41–52
23. Lee AG. 1998. *Biochim. Biophys. Acta* 1376:381–90
24. Chin CN, von Heijne G, de Gier JW. 2002. *Trends Biochem. Sci.* 27:231–34
25. Møller JV, Lenoir G, Marchand C, Montigny C, le Maire M, et al. 2002. *J. Biol. Chem.* 277:38647–59
26. Senes A, Ubarretxena-Belandia I, Engelman DM. 2001. *Proc. Natl. Acad. Sci. USA* 98:9056–61
27. Ma H, Inesi G, Toyoshima C. 2003. *J. Biol. Chem.* 278:28938–43
28. Shigekawa M, Wakabayashi S, Nakamura H. 1983. *J. Biol. Chem.* 258:14157–61
29. Lahiri SD, Zhang G, Dunaway-Mariano D, Allen KN. 2002. *Biochemistry* 41:8351–59
30. Hilge M, Siegal G, Vuister GW, Guntert P, Gloor SM, Abrahams JP. 2003. *Nat. Struct. Biol.* 10:468–74
31. Hayward S. 1999. *Proteins* 36:425–35
32. Bigelow DJ, Inesi G. 1992. *Biochim. Biophys. Acta* 1113:323–38
33. Clausen JD, McIntosh DB, Vilsen B, Woolley DG, Andersen JP. 2003. *J. Biol. Chem.* 278:20245–58
34. Clarke DM, Loo TW, Inesi G, MacLennan DH. 1989. *Nature* 339:476–78
35. Inesi G, Kurzmack M, Coan C, Lewis DE. 1980. *J. Biol. Chem.* 255:3025–31
36. Henao F, Orlowski S, Merah Z, Champeil P. 1992. *J. Biol. Chem.* 267:10302–12
37. Lee AG, East JM. 2001. *Biochem. J.* 356:665–83
38. Nayal M, Di Cera E. 1994. *Proc. Natl. Acad. Sci. USA* 91:817–21
39. Andersen JP, Vilsen B. 1995. *FEBS Lett.* 359:101–6
40. Glusker JP. 1991. *Adv. Protein Chem.* 42:1–76
41. Zhang Z, Lewis D, Strock C, Inesi G, Nakasako M, et al. 2000. *Biochemistry* 39:8758–67
42. Zhang Z, Sumbilla C, Lewis D, Summers S, Klein MG, Inesi G. 1995. *J. Biol. Chem.* 270:16283–90
43. Zhang Z, Lewis D, Sumbilla C, Inesi G, Toyoshima C. 2001. *J. Biol. Chem.* 276:15232–39
44. Sorensen TL, Andersen JP. 2000. *J. Biol. Chem.* 275:28954–61
45. Ogawa H, Toyoshima C. 2002. *Proc. Natl. Acad. Sci. USA* 99:15977–82
46. Meissner G, Young RC. 1980. *J. Biol. Chem.* 255:6814–19
47. Andersen JP. 1994. *FEBS Lett.* 354:93–96
48. Vilsen B, Andersen JP. 1998. *Biochemistry* 37:10961–71
49. Doyle DA, Morais CJ, Pfuetzner RA, Kuo A, Gulbis JM, et al. 1998. *Science* 280:69–77
50. Inesi G, Zhang Z, Lewis D. 2002. *Biophys. J.* 83:2327–32
51. Orlowski S, Champeil P. 1991. *Biochemistry* 30:352–61
52. Yu M, Lin J, Khadeer M, Yeh Y, Inesi G, Hussain A. 1999. *Arch. Biochem. Biophys.* 362:225–32
53. Daiho T, Yamasaki K, Wang G, Danko S, Iizuka H, Suzuki H. 2003. *J. Biol. Chem.* 278:39197–204
54. Kato S, Kamidochi M, Daiho T, Yamasaki K, Gouli W, Suzuki H. 2003. *J. Biol. Chem.* 278:9624–29
55. Gatto C, Lutsenko S, Shin JM, Sachs G, Kaplan JH. 1999. *J. Biol. Chem.* 274:13737–40
56. Lutsenko S, Anderko R, Kaplan JH. 1995. *Proc. Natl. Acad. Sci. USA* 92:7936–40
57. McIntosh DB, Clausen JD, Woolley DG, MacLennan DH, et al. 2003. *Ann. NY Acad. Sci.* 986:101–5
58. Maruyama K, Clarke DM, Fujii J, Inesi G,

Loo TW, MacLennan DH. 1989. *J. Biol. Chem.* 264:13038–42

59. Inesi G. 1985. *Annu. Rev. Physiol.* 47:573–601

60. Kabsch W, Sander C. 1983. *Biopolymers* 22:2577–637

61. Juul B, Turc H, Durand ML, Gomez de Gracia A, Denoroy L, et al. 1995. *J. Biol. Chem.* 270:20123–34

62. Wuytack F, Raeymaekers L, Missiaen L. 2002. *Cell Calcium* 32:279–305

63. Clarke DM, Loo TW, MacLennan DH. 1990. *J. Biol. Chem.* 265:14088–92

64. McIntosh DB, Woolley DG, Vilsen B, Andersen JP. 1996. *J. Biol. Chem.* 271:25778–89

65. Kraulis PJ. 1991. *J. Appl. Crystallogr.* 24:946–50

Annu. Rev. Biochem. 2004. 73:293–320
doi: 10.1146/annurev.biochem.72.121801.161455
Copyright © 2004 by Annual Reviews. All rights reserved
First published online as a Review in Advance on March 11, 2004

DNA Polymerase γ, The Mitochondrial Replicase[1]

Laurie S. Kaguni

*Department of Biochemistry and Molecular Biology, Michigan State University, East
Lansing, Michigan 48824-1319; email: lskaguni@msu.edu*

Key Words replication, fidelity, processivity, mutagenesis and disease,
mitochondrial toxicity, antiviral nucleoside inhibition

■ **Abstract** DNA polymerase (pol) γ is the sole DNA polymerase in animal
mitochondria. Biochemical and genetic evidence document a key role for pol γ in
mitochondrial DNA replication, and whereas DNA repair and recombination were
thought to be limited or absent in animal mitochondria, both have been demonstrated
in recent years. Thus, the mitochondrial replicase is also apparently responsible for
the relevant DNA synthetic reactions in these processes. Pol γ comprises a catalytic
core in a heterodimeric complex with an accessory subunit. The two-subunit holoenzyme is an efficient and processive polymerase, which exhibits high fidelity in
nucleotide selection and incorporation while proofreading errors with its intrinsic 3′
→ 5′ exonuclease. Incorporation of nucleotide analogs followed by proofreading
failure leads to mitochondrial toxicity in antiviral therapy, and misincorporation
during DNA replication leads to mitochondrial mutagenesis and dysfunction. This
review describes our current understanding of pol γ biochemistry and biology, and it
introduces other key proteins that function at the mitochondrial DNA replication fork.

CONTENTS

---

[1]The abbreviations used are: pol, DNA polymerase; dRP, deoxyribose phosphate; aa,
amino acid(s); PEO, progressive external ophthalmoplegia; exo, exonuclease; aaRS,
amino acyl-tRNA synthetase; nt, nucleotide(s); mtDNA, mitochondrial DNA; and mtSSB,
mitochondrial single-stranded DNA-binding protein.

INTRODUCTION

And Then There Were Many: A Short History of Animal Cell DNA Polymerases

In the mid-1970s, when this author entered the field of eukaryotic DNA polymerases (pol), there were three. Pol[1] α was thought to be the nuclear replicative polymerase; pol β, a versatile DNA repair enzyme; and pol γ, the mitochondrial DNA polymerase (1). Soon thereafter appeared pol δ, which exhibited mechanistic properties like pol γ but was found to be nuclear. And then a pol δ variant emerged and was later shown to be distinct; it was renamed pol ϵ (2). An intense period of mechanistic study of the five DNA polymerases ensued, and together with the development of an in vitro system for virus SV40 replication, the roles of α, δ, and ϵ in nuclear DNA replication grew more apparent, and a number of other key players at the replication fork were identified, which include the single-stranded DNA-binding protein RP-A, the polymerase clamp loader RF-C, and the processivity factor PCNA (2, 3).

It might be argued that there were two key features that allowed the discovery of the first five DNA pols, (*a*) relative abundance and (*b*) the development of differential assays based on substrate specificity and sensitivity to inhibitors. By comparison, the combination of bacterial precedent, biochemical genetics, and the emergent field of genomics contributed largely to the discovery of 10 novel DNA polymerases in only the last few years (4). These enzymes are involved in nuclear DNA repair and specialized DNA synthetic processes that contribute substantially to the maintenance of genetic integrity, yet they are largely dispensable for cell viability. Thus, in 2004, there remain only four replicative DNA pols in animal cells, and DNA polymerase γ remains the sole DNA polymerase found in animal mitochondria. As such, and in contrast to the functional specialization evident in nuclear DNA polymerases, pol γ appears to be uniquely responsible for all DNA synthetic reactions in replication, repair, and recombination in that critical organelle.

Structure-Function Relationships in DNA Pols

An extensive body of structural and functional data argue that all DNA pols share three common features. All appear to use an identical two-metal ion-catalyzed mechanism for phosphoryl transfer, and they share a common overall architecture of a right hand comprising palm, fingers, and thumb subdomains with distinct roles (5). Whereas the palm subdomain in which catalysis occurs is homologous among the four pol classes for which structures are available, the finger and thumb subdomains differ substantially, though they share some analogous structural features and function in similar ways. Furthermore, functional and structural data from studies of the DNA polymerases from several families argue that conformational transitions, which involve the fingers subdomain, occur during the catalytic cycle that affect both the fidelity and the efficiency of nucleotide polymerization. In composite, these allow detailed comparisons to be drawn with the relatively modest studies of mitochondrial DNA polymerase.

Finer levels of distinction have been used to classify all DNA polymerases into five families, largely according to amino acid sequence alignments, with the aid of extensive crystal structure analyses (6–8). Cellular replicative DNA pols derive from three of these families. The nuclear enzymes, pols α, δ, and ϵ, all belong to the family B group for which crystal structures are available of bacteriophage and archaeal representatives (9–11). The bacterial replicative enzyme, DNA polymerase III holoenzyme, belongs to the family C group that is currently not represented by a crystal structure. Mitochondrial DNA polymerase belongs to the family A group of which *Escherichia coli* DNA polymerase I is the prototype, and for which there exist a number of crystal structures. These include the Klenow fragments of *E. coli* and *Bacillus stearothermophilus* pol I, *Thermus aquaticus* DNA polymerase, and bacteriophage T7 DNA polymerase (12–15). T7 DNA polymerase is distinguished from the bacterial pols in the family A group by its heterodimeric structure in which the catalytic core is complexed with an accessory subunit. This association confers upon the holoenzyme specialized properties that serve to convert an otherwise pol I-like catalytic core best suited for a DNA repair function into an efficient, processive replicative form (16). Pol γ shares this distinction. Additional conserved sequence elements in the pol γ catalytic cores reinforce the designation of a γ-subclass within the family A group (17, 18). The critical, multipurpose cellular role of pol γ, its distinction as a highly faithful, catalytically efficient and processive replicative enzyme comprising only two subunits, and its relationship to human mitochondrial disease and evolution make it an interesting and worthy subject for an *Annual Review*.

DNA POLYMERASE γ, THE MITOCHONDRIAL REPLICASE

Discovery and Purification

Pol γ was identified in the early 1970s as an RNA-dependent DNA polymerase in human HeLa cells (19), and it was subsequently found to be present in animal mitochondria (20), representing less than 1% of the total cellular DNA polymerase activity. Though low relative abundance and uncontrolled proteolysis hindered progress in elucidating its physical properties and subunit structure, high-molecular-weight forms that set the stage for purification of mammalian forms were identified from chick (21) and *Drosophila* (22) embryos. Three key features enabled the purification of *Drosophila* pol γ (22). First, embryonic mitochondria were used as the starting material, increasing the specific activity and yield 8- and 20-fold, respectively, over that from adults. Second, mitochondria were purified by differential centrifugation and then extracted with both salt and detergent, which served to increase the yield ~ 15-fold as compared to the commonly used sucrose gradient sedimentation protocols. Third, a differential assay was developed using DNase I-activated calf thymus DNA in which activity is measured in the presence of 200 mM KCl, thereby reducing 10-fold the major contaminating activity of pol α that is at least 50-fold more abundant than pol γ. The native enzyme from *Drosophila* embryonic mitochondria was purified ~ 2500-fold, which produced a modest 10 μg of near-homogeneous enzyme from 200 g of embryos.

Physical Properties and Subunit Structure

The first high mol wt forms of pol γ from chick and *Drosophila* embryos had similar sedimentation coefficients and estimated molecular masses (21, 22). The *Drosophila* enzyme (7.6 S, 160,000 Da) was shown to be a heterodimer of 125 and 35 kDa subunits. The two-subunit composition of the *Drosophila* pol γ was documented further by physical and immunological approaches. An in situ activity gel analysis allowed the assignment of the DNA polymerase activity to the 125 kDa polypeptide in both crude and near-homogenous forms (22). Immunoprecipitation of a crude enzyme fraction showed it to have the same subunit composition as the near-homogenous form, suggesting its intact nature, and controlled proteolysis experiments showed that the 125 and 35 kDa subunits comprise unrelated polypeptides (23). Furthermore, a photochemical cross-linking analysis of enzyme-DNA complexes with UV light showed the larger, but not the smaller, subunit to make close contact with DNA, and limited trypsin digestion of the native enzyme showed that an ~ 65 kDa proteolytic fragment of the catalytic subunit retains DNA binding activity (23). Subsequent purifications of pol γ from frog (24), pig (25), and human cells (26) showed the vertebrate forms to contain a large catalytic subunit and several smaller polypeptides, some of which appeared to result from in vitro proteolysis. Our data taken together with the representative data from other systems led us to propose a consensus subunit structure for animal mitochondrial DNA polymerase in which a large polypeptide of 120–140 kDa containing the polymerase function is associated quantitatively with a smaller subunit of 35–50 kDa (23).

Sequence Organization and Activities of the Catalytic Core

While the subunit structure of animal pol γ was under investigation, it was established that the high fidelity of mitochondrial DNA polymerase was in part due to the presence of a second catalytic activity: $3' \rightarrow 5'$ exonuclease was demonstrated in the chick (27), pig (28), *Drosophila* (29), and frog (30) enzymes, although its subunit association remained unclear. A genomic clone of the *Saccharomyces cerevisiae* catalytic core was identified (31) that was found to have substantial amino acid sequence conservation with the three DNA polymerase and three $3' \rightarrow 5'$ exonuclease active site motifs in *E. coli* DNA pol I, providing strong suggestive evidence that the $3' \rightarrow 5'$ exonuclease is intrinsic to the catalytic core and placing the mitochondrial replicase in the family A group of DNA pols (32). Genetic and biochemical study of the yeast enzyme subsequently showed a contribution by its $3' \rightarrow 5'$ exonuclease of several 100-fold to error avoidance in vivo (33, 34). $3' \rightarrow 5'$ exonuclease activity in the catalytic core of animal pol γ was first demonstrated in a recombinant form of the *Drosophila* polypeptide (35), and it was subsequently studied both in the human catalytic core (36, 37) and in reconstituted recombinant forms of the *Drosophila* (38, 39) and human (40, 41) holoenzymes.

A third catalytic activity, 5'-deoxyribose phosphate (dRP) lyase, was recently discovered in association with frog pol γ (42, 43) and was demonstrated to reside in the human catalytic core (44). This enzyme activity functions in base excision repair to remove the 5'-terminal dRP sugar moiety that results from the upstream activities of glycosylases and AP endonucleases in the recognition and removal of damaged bases in DNA. The identification of an intrinsic dRP lyase activity in animal pol γ provides a mechanism for elimination of the dRP moiety necessary for single nucleotide gap filling in base excision repair in animal mitochondria. This repair-related functionality was subsequently extended to other family A DNA pols, which includes *E. coli* DNA pol I (43). Recent reviews provide further discussion of the role of pol γ in DNA repair (45, 46).

The isolation of yeast genomic clones from *Schizosaccharomyces pombe* (17, 47) and *Pichia pastoris* (17), and cDNA clones of the frog (17), human (48), and *Drosophila* (35, 48) catalytic cores identified six conserved sequence elements among the γ pols in addition to the exonuclease (exo) and pol motifs shared among all family A pols. Figure 1 presents a schematic alignment of the pol γ catalytic cores and shows that four of these lie between the exo and pol motifs, in the region that we have termed the spacer (49). In *E. coli* pol I this region contains 196 amino acid residues and comprises part of the exonuclease, which forms a cleft with the polymerase, that includes the thumb and part of its palm subdomains (12). In bacteriophage T7 pol, this region of 293 amino acid (aa) residues contains a 71 aa residue loop within the thumb subdomain that constitutes the interaction site with its accessory subunit, *E. coli* thioredoxin (14) (Figure 2). The spacer regions of the animal γ pols are much larger; they range from 437 amino acids in *Drosophila* to 482 amino acids in *Xenopus laevis*.

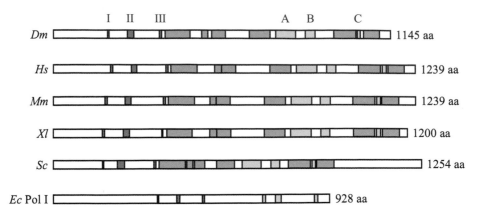

Figure 1 Schematic alignment of the catalytic cores of γ pols and *E. coli* DNA pol I. Pink and green boxes indicate the conserved Exo (I-III) and Pol (A-C) active site motifs, respectively, that are shared among family A pols, and blue boxes indicate γ-specific sequences.

Because they constitute up to 40% of the total length of the polypeptide, they might more properly be considered a domain. A number of biochemical and genetic studies of active site mutants have revealed strong conservation in the function of amino acid residues that are within the $3' \rightarrow 5'$ exo and DNA pol motifs (see below). However, no function of the spacer region of animal pol γs has yet been demonstrated.

Developmental lethality, characterized by disruption of adult visual system development and aberrant behavioral phenotypes, occurs in a *Drosophila* strain bearing a mutation of Glu[595] to Ala in the sequence just distal to the third conserved sequence block in the spacer (50). The developmental and behavioral phenotypes are concomitant with mitochondrial DNA (mtDNA) depletion in moribund third instar larvae (N. Luo and L.S. Kaguni, unpublished results). Similarly, several recent reports have linked the human mitochondrial disorder, progressive external ophthalmoplegia (PEO), to mutations in the pol γ catalytic-subunit gene (51, 52), implicating several amino acids that lie within the spacer region in addition to those that map near the region of the exonuclease and polymerase active site motifs. An ongoing mutational study in our laboratory suggests that conserved amino acids within the spacer play roles in DNA binding, in subunit interactions, and in the interplay between the pol and exo activities (N. Luo and L.S. Kaguni, submitted for publication). Crystal structures of archaeal DNA polymerases in the family B group (10, 11) suggest that the exonuclease domain and the fingers subdomain of the polymerase may move as a unit during the catalytic cycle, providing a structural explanation for the functional coupling between pol and exo that has been documented in bacteriophage T4 DNA polymerase (53). Although this does not occur in *E. coli* pol I Klenow (12), it will be of substantial interest to dissect structure-function relationships in these regions of the γ subfamily of the family A polymerases.

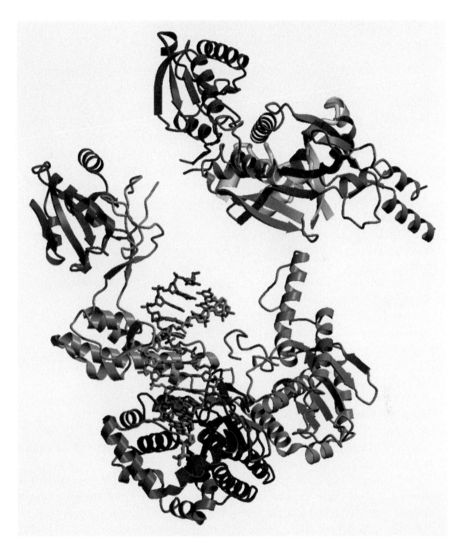

Figure 2 Structural representations of the accessory subunit of mouse pol γ and bacteriophage T7 DNA polymerase. The structure depicted at upper right represents that of mouse pol γ−β (PDB 1G5H), with its C-terminal domain colored in pink and the remainder in light blue. The gray regions represent the sequences deleted in *Drosophila* pol γ−β. The structure depicted at lower left represents the T7 pol in a ternary complex with template primer and ddGTP (PDB 1T7P). T7 pol is colored in fuschia (palm), gold (fingers), and chartreuse (thumb) with the pol active site motifs in orange. The exo domain is gray, and the exo active site motifs are aqua. Thioredoxin is pink, the template strand is medium gray, the primer strand is light green, and the ddGTP is dark green. The figure was rendered by Charles Carter using the programs MOLSCRIPT (158) and Raster3D (159).

Structure and Functional Contributions of the Accessory Subunit

Molecular cloning of the small subunit of *Drosophila* pol γ confirmed our biochemical conclusion that it is a distinct polypeptide and also allowed us to identify mammalian homologs (54), validating our earlier proposal of the two-subunit structure of animal pol γ. Sequence alignments showed that only the C-terminal region of about 120 aa was well conserved between the *Drosophila* polypeptide and its mammalian homologs. Based on a surprising finding of both sequence and structural similarity with class IIa aminoacyl-tRNA synthetases (aaRSs), we used the anticodon-binding domain of *Thermus thermophilus* ProRS as the structural template to build a homology model of the 3D structure of the C-terminal domain of the accessory subunit of *Drosophila* pol γ (55). The model revealed a rare α/β fold comprising a five-stranded mixed beta-sheet surrounded by four α helices that is found only in anticodon binding domains of class IIa aaRSs.

This homology between the C-terminal domain and an RNA-binding domain prompted us to propose a role for the accessory subunit in RNA primer binding in mitochondrial DNA replication (55). Indeed, we subsequently showed by UV cross-linking that the accessory subunit binds RNA in a complex of *Drosophila* pol γ with a synthetic template primer (49). Furthermore, a striking molecular mimicry of the C-terminal domain with the accessory subunit of T7 DNA polymerase, thioredoxin, suggested another possible role in enzyme processivity (55) (see Figure 2). In a concurrent study, the *Xenopus laevis* accessory-subunit homolog was cloned and found to resemble class IIa aaRSs, and it was demonstrated further to increase the processivity of the human catalytic core, which had been produced in a recombinant form in the absence of an accessory subunit (56).

The determination of the crystal structure of the mouse accessory subunit marked a significant advance (57). In parallel biochemical studies of deletion mutants, we concluded that the accessory subunit of *Drosophila* pol γ comprises three domains (49), and the mouse crystal structure demonstrated this. Likewise, our molecular model of the C-terminal domain proved highly accurate. Surprisingly, however, the mouse homolog was found to be a dimer (57), corroborating earlier sedimentation analyses that documented an *S* value of 5.5 for the human accessory-subunit homolog (40). This led to a proposal that mammalian pol γ is a heterotrimer with the structure αβ₂ (57).

The human accessory subunit was shown to bind dsDNA (40), and a subsequent study, which involved specific deletion mutations, demonstrated a requirement for sequences on each of the protomers to do so (58). Notably, this dsDNA-binding property of the accessory subunit is not required for the stimulation of DNA synthesis by the catalytic subunit (58), raising the possibility of alternate roles of the accessory subunit in mtDNA metabolism. Remarkably, the *Drosophila* homolog lacks over 100 of the first 250 residues present in the mouse accessory subunit. These missing residues map to domains 1 and 2, which contain critical amino acid residues for dimerization of the mammalian accessory subunit (see Figure 2). This might suggest that dimerization is a feature of the mammalian form that renders it soluble in the absence of the catalytic core (and perhaps available to serve an alternate function), whereas the accessory subunits from several insect species are all largely insoluble (C.L. Farr and L.S. Kaguni, unpublished observations). At present, the subunit stoichiometry of the native form of mammalian pol γ remains an open issue, though both the available physical and mechanistic data from studies of the reconstituted human holoenzyme would argue it has a heterodimeric structure (40, 41).

In earlier dissociation studies, we showed a strong interaction between the catalytic and accessory subunits of *Drosophila* pol γ and a requirement for the accessory subunit to maintain the structural integrity and/or the catalytic efficiency of the holoenzyme (23). Although the *Drosophila* catalytic subunit alone, produced either in *E. coli* (35) or in the baculovirus system (38), exhibits very low catalytic activity (~ 2% of that of the holoenzyme), reconstitution of the heterodimeric holoenzyme in baculovirus-infected cells produces a form with physical and biochemical properties indistinguishable from the native enzyme from *Drosophila* embryonic mitochondria. In contrast, cloning and biochemical analysis of the human catalytic core showed that under conditions of low ionic strength it retains substantial activity, whereas at more physiological salt concentrations, its very low activity is stimulated markedly by the accessory subunit (36, 37, 40, 41).

DNA Polymerase Activity, and Processivity and Fidelity of Nucleotide Polymerization

TEMPLATE-PRIMER SPECIFICITY   The DNA polymerase activity of pol γ has been assayed historically either on partially degraded, DNase I-activated calf thymus (or salmon sperm) DNA as template primer, or on poly(rA) : oligo(dT). Such assays have provided critical parameters to distinguish the rare mitochondrial enzyme from the multiple, and far more abundant, nuclear DNA polymerases. In both cases, pol γ is generally stimulated by moderate to high salt, sensitive to inhibition by N-ethylmaleiimide and dideoxynucleoside triphosphates, and insensitive to aphidicolin. Pol γ exhibits a broad pH optimum (7.5–9.5), absolutely requires a divalent cation (Mg^{2+} or Mn^{2+}, depending on the template primer used), and has a high affinity for dNTPs.

Among all of the animal cell DNA polymerases, pol γ displays a remarkable ability to utilize diverse template-primer substrates. This is perhaps the case because pol γ is thought to be solely responsible for the DNA synthetic reactions in replication, repair, and recombination in mitochondria, in contrast to the nuclear pols, which have distinct roles in these processes. Furthermore, pol γ is the only cellular DNA polymerase that can utilize effectively a ribohomopolymeric template. Although pol γ was thought not to utilize natural sequence RNA as a template, a recent study shows that it does so with a slightly higher catalytic rate than HIV-1 reverse transcriptase, albeit at a much lower efficiency than it replicates natural sequence DNA (59). The physiological significance that underlies this specialized pol activity remains to be elucidated.

The relative activity of pol γ isolated from multifarious sources varies as much as 10-fold on primed ribo- and deoxyribo-homopolymers [e.g., poly(rA) : oligo(dT) and poly(dA) : oligo(dT)] versus natural sequence DNA, but in general activity is greatest on substrates with a high primer density. Differences in reaction conditions and particularly in the types and concentrations of mono- and divalent salts used are responsible for much of the variability in activity levels reported. Steady-state kinetic constants were determined for native *Drosophila* pol γ on DNase I-activated calf thymus DNA [$K_{m\ (DNA)}$ of 12.5 μM (as nucleotides)], where a k_{cat} of 3.1 nucleotides (nt) s^{-1} was found (60). Notably, pol γ also catalyzes efficient DNA synthesis on template-primer DNAs with low primer density. *Drosophila* pol γ was found to have a high affinity for single-stranded natural DNA as the template [$K_{m_{(DNA)}}$ of 1.1 μM (as nucleotides) on singly primed ϕX174 DNA] on which it exhibits a twofold-lower k_{cat} but a substantially higher substrate specificity.

These catalytic features are generally consistent with the in vivo requirements for mtDNA replication as would be dictated by the model describing a highly asymmetric mode of DNA strand synthesis in which the leading strand is synthesized on a double-stranded DNA template that is unwound just ahead of the fork, whereas at the time of lagging DNA strand synthesis, the template is largely single stranded (61). As discussed below, challenges to this long-standing model prompt investigation of alternative mechanisms of DNA strand synthesis.

CATALYTIC MECHANISM   The processivity of nucleotide polymerization was first analyzed, by both kinetic and gel analyses, for *Drosophila* pol γ on singly primed ϕX174 DNA (60). The kinetic analysis yielded an average processive unit of $\sim$ 30 nt, which was corroborated by the gel analysis that identified a minimum value of 25–40 nt. However, the appearance of a complex mixture of products >200 nt in length at very low extents of DNA synthesis led to the conclusion that a significant number of enzyme-substrate interactions result in processive synthesis catalyzed by pol γ, even under these conditions of moderate ionic strength. In a subsequent study, we showed that, in the absence of other factors, the two-subunit holoenzyme is capable of highly processive DNA synthesis (>1000 nt) only at low ionic strength and that reaction conditions affect greatly both the activity and processivity of nucleotide polymerization by *Drosophila* pol γ (62).

New insights into the mechanism of DNA strand synthesis by pol γ have come from studies of recombinant forms and, in particular, the isolated human catalytic core and the reconstituted holoenzyme. Concurrent studies of the human catalytic core explored the effects of reaction conditions and inhibitors and of the kinetics of nucleotide polymerization (36, 37). The isolated human catalytic core was found to be salt sensitive and resistant to aphidicolin and butyl-phenyl dGTP and to exhibit a processivity of $\sim$ 50 nt on M13 DNA. In single nucleotide incorporation assays using an oligonucleotide template primer at moderate ionic strength, maximum polymerization rates of 3.5–8.7 s^{-1} were reported, with a relatively low affinity for DNA and dNTPs (36). An initial burst of nucleotide incorporation followed by a steady-state rate was interpreted as a fast polymerization step followed by the dissociation and reassociation of the enzyme with another template-primer molecule.

The effects of the accessory subunit on animal pol γ from several sources have been described in recent reports (38, 40, 41, 56). The accessory subunit increases substantially both the catalytic activity and the processivity of the core. These increases are due in part to enhanced DNA binding but, perhaps surprisingly, also to increased nucleotide binding. The latter suggests that the accessory subunit contributes to the structural integrity of, or imparts functionality to, the catalytic subunit, which allows it to bind its substrates more tightly and to catalyze faster polymerization. This interpretation is consistent with those from our earlier dissociation studies of *Drosophila* pol γ (23).

Notably, the enhanced processivity conferred on the pol γ holoenzyme by the accessory subunit does not result from a decreased rate of dissociation from the template primer (41), as was observed for T7 pol (16), but rather from enhanced template primer and dNTP binding and polymerization. These enhancements may derive from multiple contacts between the accessory subunit and the catalytic core of pol γ (49), as compared to the single, flexible tether of thioredoxin to the tip of the thumb subdomain in T7 pol (14). In that regard, it is tempting to consider the superposition of the thioredoxin-like C-terminal domain of the pol γ accessory subunit onto thioredoxin in the T7 pol ternary complex and to imagine the likely numerous interactions between its middle and N-terminal regions and the template-primer binding cleft (see Figure 2). At the same time, one might consider a distinct contrast between the vertebrate pol γs and the *Drosophila* enzyme, based on the observation that processivity in the presence of the accessory subunit is high even at moderate ionic strength for the vertebrate enzymes (40, 41). With the *Drosophila* enzyme, high processivity is achieved only under low ionic strength conditions or in the presence of mitochondrial single-stranded DNA-binding protein (39, 62, 63). It will be of substantial interest to determine how this feature may relate to differences in the overall structure of the vertebrate and insect holoenzymes and/or in their interactions with template primer in ternary complexes with dNTPs.

FIDELITY AND THE ROLE OF ACTIVE SITE RESIDUES   Mitochondrial DNA (mtDNA) evolves at a 5- to 10-fold-higher rate than single-copy nuclear DNA (64). Furthermore, it has been established clearly that mutations in mitochondrial DNA result in human disease (65, 66). These include base-substitution, deletion, and duplication mutations that result in a broad spectrum of degenerative diseases involving the heart, muscle, kidney, liver, and the central nervous and endocrine systems. Each of the numerous mtDNA diseases can result from several different mutations, and each exhibits variable clinical pathologies. These deleterious mutations likely result from replication copy errors, either unprovoked or promoted by DNA damage. Nonetheless, native pol γ purified from chick (67) and *Drosophila* embryos (60) and from pig liver (28) is highly accurate in nucleotide polymerization, with an in vitro error rate of only $\sim$ 1 misincorporated nucleotide per $\sim$ a half-million bases polymerized. It has been shown, however, that replicative bypass in vitro of abasic sites and sites of oxidative damage by *Xenopus* pol γ is error prone (68). The high base-substitution fidelity of pol γ is also compromised by nucleotide pool imbalances (27, 60), a situation that may be physiologically relevant to normal fluctuations in metabolites known to occur in the mitochondrial matrix (69, 70) and in view of the link between intramito-chondrial regulation of dNTP pools and mitochondrial genetic disease (71–73). Nucleotide pools in the mitochondrion are also affected by cellular nucleotide metabolism (74, 75). To date, defects in the nuclear genes for mitochondrial adenine nucleotide translocator 1 (76), deoxyguanosine kinase (72), and thymidine kinase 2 (73) and for cytoplasmic thymidine phosphorylase (74) have been shown to induce pathogenic mutations of mtDNA [reviewed in (77)].

The fidelity of recombinant human pol γ has been a focus in recent studies. In a forward mutational assay that detects a broad spectrum of DNA synthetic errors, human pol γ was found to exhibit a high base-substitution fidelity that derives from high nucleotide selectivity and exonucleolytic proofreading (78–80). Kinetic parameters of nucleotide binding and incorporation with correct and incorrect nucleotides were determined using synthetic template primers and enzyme single-turnover conditions, yielding an average fidelity for the human pol γ holoenzyme of one error in $\sim$ 3 X 10^5 base pairs (79). This value is similar to those reported for the native pol γs from various sources that were studied earlier.

The role of active site residues in governing fidelity has been explored both in vivo and in vitro. These amino acids map to each of the three $3' \rightarrow 5'$ exonuclease and three polymerase motifs (Figure 3). The contribution of $3' \rightarrow 5'$ exonuclease to fidelity will be discussed in a later section. Random mutagenesis of yeast pol γ identified 10 mutator alleles that conferred mitochondrial erythromycin resistance in vivo (81). Three of these mapped within the pol domain (T716I, E724K, P851L) and exhibited mutation frequencies 15- to 50-fold above the spontaneous level in the presence of $3' \rightarrow 5'$ exonuclease function. Interestingly, all three mutations map outside of the Pol A, B, and C active site motifs

Figure 3   Sequence alignment of the $3' \rightarrow 5'$ exonuclease and DNA polymerase active site motifs in family A DNA pols. The sequences of the active site motifs of γ pols were aligned with those of bacteriophage T7 pol and *E. coli* pol I using the Clustal W server (http://clustalw.genome.ad.jp). Boxes frame the sequences that are highly conserved in the family A group, and amino acid residues indicated in bold represent those for which genetic and/or biochemical data are discussed in the text. The abbreviations are *Dm, D. melanogaster; Hs, Homo sapiens; Mm, Mus musculus; Xl, Xenopus laevis; Sc, Saccharomyces cerevisiae*; T7, bacteriophage T7; and Pol I, *E. coli* pol I.

Exo I

```
Dm  177  PlEKGL  VFDVE  VCVsEGqaPvLA
Hs  190  PEERAL  VFDVE  VCLaEGtCPTLA
Mm  173  PEERAL  VFDVE  VCLaEGtCPTLA
Xl  164  PDEKAM  VFDVE  VCVtEGcCPTLA
Sc  163  PDEelV  VFDVE  tlynvsdyPTLA
T7    1     mi   VsDIE  anal
PolI 349  apvf   aFDtE  tdsl
```

Exo II

```
Dm  254  LV  VGHNVSYLR  ARLKEQYLI
Hs  265  LV  VGHNVSFlR  AHIREQYLI
Mm  248  LV  VGHNVSFDR  AHIREQYLI
Xl  239  LV  VGHNVSFDR  AHIREQYLI
Sc  221  VI  IGHNVAYDf  ARVlEEYnf
T7   56  IV  fhnghkYDv  paLtkla
PolI 415 Lk  VGqNLkYDR  GiLanyg
```

Exo III

```
Dm  351  qVRQsFQsLtN  YCAsDVeAT  HrIlrvlYP
Hs  384  DIRENFQDLMQ  YCAqDVwAT  HEVFQqQlP
Mm  366  DIRENFQDLMQ  YCArDVwAT  fEVFQqQlP
Xl  358  DIRteFQELMr  YCAlDVqAT  HEVFQeQFP
Sc  332  tiENFQklVN   YCAtDVtAT  sQVFdeiFP
T7  161      wnfnEEMMd  YnvqDVvvT  KaLlEll
PolI 488     qialEEagr  YaAeDadvT  lQLhlkm
```

Pol A

```
Dm  755  YGAIcPQVvacGTLTRRAMEPTWMTASNSRpDRLGSELRSMVQAP?GYrL  VGADVDSQEL  WIASVLGDAyacGeHgaTplGWMTLsG
Hs  837  YGAILPQVVTAGTITRRAVEPTWLTASNARpDRVGSELKAMVQAPPGYtL  VGADVDSQEL  WIAAVLGDAHFAGMHGCTAFGWMTLQG
Mm  816  YGAILPQVVTAGTITRRAVEPTWLTASNARpDRVGSELKAMVQAPFGYvL  VGADVDSQEL  WIAAVLGDAHFAGMHGCTAFGWMTLQG
Xl  807  YGAILaQVVSAGTITRRAVEPTWLTASNARaDRVGSELKAMVQvPPSYhL  IGADVDSQEL  WIAAILGEAHFAGIHGCTAFGWMTLQG
Sc  640  lAilIPkIVpmGTITRRAVEnaWLTASNAkaNRIGSELKtqVKAPPSYcf  WIASLVGDSiFn-VHGgTAiGWMcLEG
T7  465                              tgkpWvq  aGiDasglEL  rclAhf
PolI 695                             APecYvI  VsADysqiEL  rImAhL
```

Pol B

```
Dm  857  VGIS  RDHAKViNYARIYGAG  QlFAEtLlrQFN
Hs  939  VGIS  REHAKIFNYGRIYGAG  QPFAERLLMQFN
Mm  917  VGIS  REHAKIFNYGRIYGAG  QsFAERLLMQFN
Xl  909  VGIS  REHAKVFNYGRIYGAG  QPFAERLLMQFN
Sc  715  LGcS  RneAKIFNYGRIYGAG  akFAsqLLkrFN
T7  514  elpT  RDnAKtFiYGfLYGAG  dekig
PolI 750 tseq  RrsAKainFGliIYGms afgla
```

Pol C

```
Dm  1032  RFCLSf  HDELRYLV  KEElsp
Hs  1128  RFCISI  HDEVRYLV  REEDRY
Mm  1106  RFCISI  HDEVRYLV  REEDRY
Xl  1097  RFCISI  HDEVRYLV  HskDRY
Sc   859  RlCISI  HDEIRFLV  sEkDKY
T7   647  aYmawV  HDEIqvgc  RtEEia
PolI 875  RmiMqV  HDELvFeV  HrDDvd
```

that are conserved among family A polymerases and likely fall within the palm and finger subdomains. In fact, whereas there is little conservation outside of the Exo I, II, and III active site motifs, the γ pols share extensive amino acid sequence similarity in the pol domain, both surrounding the active site motifs and between them. In that regard, the conserved γ-specific elements γ5 and γ6 show substantially greater sequence similarity than the γ1–4 elements located in the spacer region (see Figure 1).

The importance in the γ pols of conserved amino acids that surround the consensus Pol A motif shared within the larger family A group is highlighted by the identification of two deleterious mutations (G848S and G923D); these result in the human autosomal mitochondrial disorder PEO (51, 52). PEO is characterized by the accumulation of base-substitution mutations and deletions within mtDNA. Notably, sequence alignment shows that two of the yeast mutators (T716I and E724 K) map immediately adjacent to the human G923D mutation (Figure 3). Furthermore, the yeast mutator P851L and a third human mutation associated with PEO (S1176L) map to positions within the γ5 and γ6 conserved sequence elements, respectively, that flank the conserved Pol C active site motif, reflecting the importance of these sequences in the γ subclass.

Three human mutations of the catalytic subunit, associated with PEO, map within the conserved Pol B active site motif (R943H, Y955C, A957S). Tyr^{766} in *E. coli* pol I, equivalent to Tyr^{955} in human pol γ, was shown to be involved in nucleotide selection and binding (82, 83). Its substitution with alanine or serine generates an error-prone DNA polymerase that promotes deletions between direct repeat sequences (84, 85).

A *Drosophila* mutation, E813V, represents the sole genetic mutant of pol γ identified to date that maps to the Pol A motif (50). As with the spacer-region mutant described earlier, this mutation results in developmental lethality that is characterized by defects in mitochondrial morphology and localization (86) and in loss of mtDNA (N. Luo and L.S. Kaguni, unpublished results). There are no known genetic mutants of pol γ that map to the Pol C motif. The Pol A and C motifs in *E. coli* pol I were shown to contain aspartate residues, Asp^{705} and Asp^{882}, that are critical for catalysis (87). The aspartate residues coordinate the two metal ions that stabilize the resulting pentacoordinated transition state (5) and are invariant among family A pols. Likewise, there are two invariant glutamate residues in motifs A and C represented by Glu^{710} and Glu^{883} in pol I respectively (Figure 3). Glu^{813} in *Drosophila* is the equivalent of Glu^{710} in pol I

Guided by studies of conserved active site residues in *E. coli* pol I Klenow and T7 pol, site-directed mutagenesis of residues involved in nucleotide selection and binding in the catalytic subunit of human pol γ has revealed that Tyr^{951} and Tyr^{955} in Pol motif B and Glu^{895} in motif A play critical roles (88). Steady-state kinetic analysis of pol γ derivatives lacking $3' \rightarrow 5'$ exonuclease activity demonstrated that Tyr^{951} is largely responsible for the ability of pol γ to incorporate dideoxynucleotides and other analogs with alterations of the ribose moieties that are used in antiviral nucleotide analog therapy. Changing Tyr^{951} to

Phe, which occurs naturally in *E. coli* pol I and which is resistant to ddNTPs (89), confers resistance in pol γ without affecting pol activity. In contrast, alteration of Tyr^{951} and Glu^{895} to Ala produces dramatic increases in K_m for normal dNTPs and large decreases in k_{cat}. Changing Tyr^{955} to Ala results in a large increase in $K_{m(dNTP)}$ without affecting overall catalytic efficiency. Furthermore, a Y955C derivative, which corresponds to the mutation associated with PEO that was found in the large-scale studies of several independent European families, was shown to exhibit a base-substitution fidelity twofold lower than wild-type pol γ in the presence of $3' \rightarrow 5'$ exonuclease function, with error rates elevated 10- to 100-fold upon inactivation of $3' \rightarrow 5'$ exonuclease (90). This is consistent with the data described earlier for derivatives of the equivalent Tyr^{766} in *E. coli* pol I and with the accumulation of base-substitution mutations in the mtDNA of patients with PEO.

$3' \rightarrow 5'$ Exonuclease Activity and Coordination of Pol and Exo Function

REACTION REQUIREMENTS AND MISPAIR SPECIFICITY $3' \rightarrow 5'$ exonuclease was documented in association with native pol γs from various animal sources (27–30). As with DNA polymerase activity, $3' \rightarrow 5'$ exonuclease activity is generally stimulated by moderate to high salt, exhibits a broad pH optimum, and is absolutely dependent on a divalent metal cation. Though $3' \rightarrow 5'$ exonuclease is frequently assayed on single-stranded DNAs, the $3' \rightarrow 5'$ exonuclease activity of pol γ shows a marked preference for double-stranded DNA containing 3'-terminal mispairs. Mispair specificity is high, with values reported for the native enzymes from various sources ranging from 5- to 34-fold depending on the substrate and the mispair (25, 29, 30, 91). Steady-state kinetic analysis of the porcine pol γ showed little preference among the 16 possible mispairs (25), and no loss of mispair specificity was observed for the *Drosophila* enzyme over a 14-fold range of activity upon titration of KCl concentration (91). *Drosophila* pol γ was shown to exhibit no detectable dNTP turnover ($<0.03\%$) yet, under conditions of DNA synthesis, hydrolyzes quantitatively a mispaired nucleotide prior to polymerization (29). These data would suggest that the cost of proofreading during mitochondrial DNA synthesis is low, a feature that distinguishes the mitochondrial replicase in general from bacterial and nuclear replicative polymerases (92).

COORDINATION OF DNA POLYMERASE AND $3' \rightarrow 5'$ EXONUCLEASE AND CONTRIBUTION TO FIDELITY   The coordination of DNA polymerase and exonuclease function has been evaluated by various experimental approaches. The contribution of $3' \rightarrow 5'$ exonuclease to the fidelity of DNA synthesis by native pol γ was shown in assays in which the $3' \rightarrow 5'$ exonuclease was inhibited by NMPs and/or in reactions with nucleotide pool biases that promote mispair extension, and it was estimated to be $\sim$ 100-fold (27, 28). In the random mutagenesis study of yeast pol γ that was described earlier, 4 of the 10 mutator alleles identified were found to map within the 3 exonuclease active site motifs. Mutations in the Exo I (E173 K), Exo II (G2240, H225D), and Exo III (T351I) motifs increased the mutation frequency by 200- to 500-fold in yeast cells (81). Site-directed mutagenesis of the conserved aspartate residues in each of the three exonuclease motifs (D171G, Exo I; D230A, Exo II; D347A, Exo III) in yeast pol γ, which were shown to be involved in metal ion binding and catalysis in *E. coli* pol I (93), demonstrated their effects both in vivo and in vitro. Mutation frequency was increased several 100-fold for the individual mutants and as much as 1500-fold for the D171G/D230A double mutant in vivo, and exonuclease activity was decreased by 10^4-fold in vitro (33, 34). Three deleterious mutations in the gene encoding the human catalytic subunit map in the region of the exonuclease active site motifs [T251I, L304R, R309L (51, 52)]. Interestingly, all of these are associated with PEO and map between rather than within the active site motifs. All are recessive mutations, in contrast to the polymerase region mutants discussed earlier, that are all dominant mutants and of which three of five map within the pol active site motifs. In consideration of the apparent mutagenic potential of exonuclease versus polymerase active site mutants in yeast, it could be argued that exonuclease active site mutants would be too deleterious and might result in developmental lethality in animals.

Kinetic analysis of 3' → 5' exonuclease in recombinant forms of human pol γ has enhanced our understanding of the interplay between pol and exo function and of the factors that govern fidelity (78–80). Consistent with the mispair specificity documented for native forms of pol γ, the reconstituted human enzyme hydrolyzes mispaired 3'-terminal nucleotides at a rate 20-fold greater than that of correctly paired bases (80). Furthermore, the use of single-turnover kinetics establishes the contribution of 3' → 5' exonuclease to fidelity up to ~ 200-fold. The key mechanistic feature governing pol γ fidelity is apparently the rate-limiting step of mispair extension, which allows the intramolecular transfer of the mispaired 3' terminus from the pol to the exo active site for hydrolysis (80). Alternatively, dissociation from the pol active site is increased 10-fold in the presence of a mispaired versus base paired 3' terminus, providing another opportunity for exonucleolytic hydrolysis upon rebinding of pol γ (79). Overall, pol γ catalysis is accurate, processive, and efficient, and the cost of exonucleolytic proofreading is minimal.

Nucleoside Inhibitors and Mitochondrial Toxicity

A direct correlation has been established between the administration of nucleoside analogs in life-extending antiviral therapy and mitochondrial toxicity [reviewed in (94)]. The medical community and pharmaceutical industry became acutely aware of the hazards of therapeutic approaches using nucleoside analogs when five deaths occurred upon administration of fialuridine to combat chronic hepatitis B (95).

Despite the multiple steps that might mitigate toxicity, e.g., cellular uptake, efficacy of cytoplasmic phosphorylation and mitochondrial transport of phosphorylated nucleosides, and/or mitochondrial phosphorylation, the clear cellular target of these analogs is the mitochondrial replicase. The first report of mitochondrial myopathy followed five years after FDA approval in 1985 of azidothymidine (96), and the cellular target was identified as pol γ in 1994 (97). Although FDA approval for use of the first dideoxynucleoside antiviral inhibitor came in 1991, it was reported as early as 1986 that dideoxynucleotides are potent inhibitors of *Drosophila* pol γ, exhibiting K_i values ranging from 1.0 μM for ddGTP to 3.5 μM for ddTTP, as compared to K_ms of 0.4–1.0 μM for the natural dNTPs (22, 98). Since then it has been demonstrated clearly that pol γ is unique among the cellular DNA polymerases with regard to its strong inhibition by anti-HIV-1 reverse transcriptase drugs, with the general hierarchy of HIV-RT $\gg$ pol γ > pol β> pol α > pol ϵ (99). The general hierarchy for inhibition of pol γ among the FDA approved nucleoside inhibitors (ddC, ddI, D4T > 3TC > AZT > carbovir) is consistent with recent kinetic analyses of recombinant human pol γ (100–102). The kinetic studies show that dideoxynucleotides and D4TTP are incorporated by pol γ as efficiently as natural dNTPs, whereas AZTTP, 3TCTP and CBVTP are moderately inhibitory. Furthermore, dideoxynucleotides are very poorly proofread, as was initially determined for the native porcine pol γ with ddCMP termini (25). AZTMP at physiologically relevant concentrations inhibits exonuclease activity (100).

Mitochondrial toxicity is manifest by cardiac dysfunction, hepatic failure, skeletal myopathy, and lactic acidosis correlating with defective mtDNA replication, mtDNA depletion, and altered mitochondrial ultrastructure [(94) and references therein]. Notably, similar manifestations characterize the development of mitochondrial genetic diseases, and animal models of mitochondrial mutagenesis (see the next section) will likely provide new insight into the development of mitochondrial toxicity promoted by either mechanism. With regard to pol γ mechanism, overall toxicity of nucleoside analogs must be assessed in terms of both its polymerase and exonuclease activities, that is, as a function of frequency of nucleotide incorporation versus excision relative to the rate of mtDNA replication and/or the energetic demands of relevant tissues.

Animal Models of Mitochondrial Mutagenesis and Dysfunction

A current focus in the field is the generation of animal models to evaluate the role of pol γ in mitochondrial mutagenesis and dysfunction, which might serve to mimic the development of human mitochondrial disorders. Mitochondrial content and mtDNA copy number vary greatly among the different tissues of an organism but are maintained relatively constant within specific cell types, implying that mtDNA replication is strictly controlled at the cellular level. The catalytic subunit of human pol γ is expressed constitutively in cultured cells, even in the absence of mtDNA (103). In *Drosophila*, the catalytic and accessory subunits of pol γ are encoded in a compact gene cluster, which suggests the possibility of coordinate regulation (104). However, whereas the catalytic-subunit gene is expressed constitutively at a low level, the accessory-subunit gene is regulated under the control of the transcription factor DREF (105), which also activates the genes for mitochondrial single-stranded DNA-binding protein (mtSSB) (106) and a number of nuclear proteins involved in DNA replication and cell cycle control, such as the catalytic and 73-kDa subunits of pol α, proliferating cell nuclear antigen, cyclin A, and E2F (107–110). We sought to alter the expression of the catalytic subunit of *Drosophila* pol γ to evaluate the cellular and developmental consequences (111). Overexpression of pol γ-α in cultured *Drosophila* cells alters neither mtDNA content nor growth rate. In contrast, overexpression of the catalytic subunit at the same level in transgenic flies produces a significant mtDNA depletion that causes a broad variety of phenotypic effects. These alterations range from pupal lethality to moderate morphological abnormalities in adults, depending on the level and temporal pattern of overexpression. Differences between the rapid and dynamic process of cell proliferation during animal development as compared to the relatively relaxed conditions for cell proliferation in culture might explain the lack of a deleterious effect in the latter situation. We proposed that mitochondrial levels of the catalytic subunit in large excess of its critical accessory subunit likely lead to aberrant mtDNA replication. Such replication would occur either by virtue of the activity of the catalytic subunit alone, which is neither processive nor efficient, or by sequestration of other interacting replication proteins whose physiological levels are more narrowly regulated, thereby preventing their participation at replication forks. This *Drosophila* model of mtDNA depletion may be exploited to gain understanding of the role of mitochondria in cell and tissue differentiation during development and also to understand the progression of human mtDNA depletion disorders (77).

Several models of mitochondrial mutagenesis involving exonuclease-deficient, "mutator" forms of pol γ are under study. In a human cell culture model, the catalytic aspartate Asp[198] in the Exo I active site motif of pol γ-α was converted to alanine (112). Two cell lines stably expressing the exonuclease-deficient catalytic subunit construct during three months of continuous culture were shown to accumulate base-substitution mutations in mtDNA at a frequency of 1:1700 as compared with a frequency of <1:28,000 determined for a cell line expressing the wild-type pol γ-α. The mutator cell lines exhibited only mild impairment of mitochondrial metabolism, consistent with the notion that most mutations are functionally recessive due to the presence of multiple copies of mtDNA in animal cells, such that the mutants are complemented by the wild-type mtDNAs still remaining. Continued growth of the mutator cells for longer periods results either in arrest of cell growth or in a reduced mutational load. The reduced mutation frequency is manifest together with a reduced level of steady-state expression of the transgene at the protein level, which correlates with the accumulation of an R823S substitution in the transgene. Interestingly, R823 lies in the extended Pol A region (Figure 3), very close to the position of the G848S mutation that was found in association with PEO (51, 52).

In the same study, transient expression of constructs carrying substitutions of the conserved aspartates in the pol A (D890N) and pol C (D1135A) active site motifs produced a dominant-negative phenotype (112), demonstrating the importance of these amino acid residues in pol γ function and consistent with our sequestration hypothesis for catalytic-subunit overexpression in the *Drosophila* model (111). Moreover, there was no effect of transient expression of the wild-type pol γ-α in human cells, a result similar to that found in the concurrent study involving stable expression of wild-type pol γ-α in *Drosophila* cell culture (111).

A mouse mutator model in which 3' $\rightarrow$ 5' exonuclease was inactivated at the catalytic aspartate Asp^{181} in the Exo I motif (equivalent to the residue that was altered in the human study) was developed to examine the role of mitochondrial DNA mutations as a causative agent in cardiac dysfunction (113–115). To this end, the mutator pol γ-α was placed under the control of the cardiac-specific promoter for the α-myosin heavy chain. Base-substitution and deletion mutations were detected shortly after birth, and several weeks later the mice developed clinical signs of congestive heart failure. Excised hearts were much larger than those of control mice and showed marked ventricular dilation and atrial enlargement. At one month, when mtDNA mutation frequency was determined to be 0.014%, no effects were observed on mitochondrial DNA or protein content, mitochondrial gene expression, or respiratory function. However, the dilated cardiomyopathy was accompanied by apparent activation of the mitochondrial apoptotic pathway, as evidenced by sporadic myocytic death and release of cytochrome *c*.

Our laboratory developed a *Drosophila* model of mitochondrial mutagenesis to evaluate the effect of mtDNA mutations and mitochondrial dysfunction during development and aging. Using the model, we found that constitutive, low-level expression of pol γ-α carrying an exonuclease I and II motif double substitution (D185A/D263A) results in a generational phenotype. The phenotype progresses from normal development, reproductive capacity, and life span in the first generation to substantial developmental delays, developmental arrest at the stage of pupal eclosion, severe loss of reproductive capacity, and reduction in adult life span from 3 weeks to $\sim$ 10 days in the sixth generation (L.S. Kaguni and R. Garesse, unpublished results). At present, the emergent theme from each of the animal models is that mitochondrial biogenesis drives development, and it is intimately associated with the processes of aging, disease, and apoptosis. Determining the underlying mechanisms relevant to the various processes remains a fascinating challenge.

PROTEINS AT THE MITOCHONDRIAL DNA REPLICATION FORK

Overview of mtDNA Replication and its Regulation

Mitochondrial biogenesis is a key process in animal cell proliferation, and mtDNA replication is an essential component of that process. The hallmark of mtDNA replication might lie in its simplicity. Animal mtDNAs are double-stranded circular molecules that range in size from $\sim$ 16 to 20 kb and contain a limited number of genes, which include those encoding its 13 polypeptides that are essential for energy production via oxidative phosphorylation. Whereas a large number of proteins participate in multicomponent replication machines in bacterial and nuclear systems, a modest group of essential proteins are likely involved in the mtDNA replication process. A long-standing model of mammalian mtDNA replication involves unidirectional and asymmetric synthesis initiating from a single origin, O_H, that proceeds by a displacement-loop mechanism (61). Lagging DNA strand synthesis is initiated at a specific site, O_L, on the displaced parental strand, at a point at which the leading DNA strand is $\sim$ 2/3 complete. This model predicts continuous DNA strand synthesis on both parental strand templates for which a minimal set of proteins would be required. In recent years, a second model has been proposed involving coupled leading and lagging DNA strand synthesis in a symmetric, semidiscontinuous DNA synthetic mode. This model is based on the study of replication intermediates by 2D-gel electrophoresis and on the ribosubstitution pattern in mtDNA (116–118). Though the mode of mtDNA replication remains a controversy (119–121), each of the two proposed models predicts the involvement of a limited group of proteins and enzymatic activities, which represent a focus of current biochemical studies.

A number of proteins have been established firmly by genetic studies to be essential for mitochondrial DNA maintenance. Three of these have been studied in detail with regard to their roles at the mitochondrial DNA replication fork: the catalytic and accessory subunits of pol γ and the mitochondrial single-stranded DNA-binding protein, mtSSB. Thermosensitive mutants of the MIP1 gene encoding yeast pol γ result in loss of mtDNA (122), as do Drosophila mutants in the genes for both the catalytic and accessory subunits of pol γ in which the endpoint is developmental lethality (50, 86). Likewise, a null mutation in the yeast RIM1 gene encoding mtSSB causes loss of mtDNA (123), as does a Drosophila mutant in the corresponding lopo gene in which developmental lethality is concomitant with loss of respiratory function and a dramatic decrease in cell proliferation (124). Recombinant mtSSB proteins carrying mutations in specific amino acids within the conserved DNA-binding domain that is shared with E. coli SSB fail to stimulate the activity of pol γ in vitro (124a). Moreover, their overexpression fails to rescue a defective cell growth phenotype in Drosophila cultured cells that exhibit a mtDNA depletion syndrome, which results from inhibition of endogenous production by double-stranded RNA (dsRNA) interference.

Structure and Function of Mitochondrial Single-Stranded DNA-Binding Protein

Single-stranded DNA-binding proteins (SSBs) serve critical roles in DNA replication, repair, and recombination: They enhance helix destabilization by DNA helicase and stimulate the activity of DNA polymerase, enhancing both the processivity and fidelity of nucleotide polymerization (125). Mitochondrial SSBs have been purified from yeast and from various animal sources (63, 123, 126–128), and the *S. cerevisiae* gene (123) and cDNAs from *X. laevis* oocytes (129), rat and human tissues (130), and *Drosophila melanogaster* ovaries (131) have been cloned and sequenced. Mitochondrial SSB comprises a 13–15 kDa polypeptide with an estimated native molecular mass of 56 kDa and has been shown to be a tetramer in solution (123, 126, 130, 132, 133). It binds cooperatively to DNA with a binding site size of 8–17 nt per protomer and has a high affinity for DNA, similar to that displayed by *E. coli* SSB (63, 127, 134), to which it is ~ 25% identical and 50% similar by deduced amino acid sequence. A crystal structure of human mtSSB that highlights the similarities between the bacterial and mitochondrial forms was determined, and it proposes a model for wrapping of ssDNA around the tetramer through electropositive channels guided by flexible loops (135). Consistent with a central role in mtDNA replication, mtSSB has been shown to coat the displaced ssDNA that is the template for lagging DNA strand synthesis in mtDNA replication (136, 137), and its apparent role in preventing the renaturation of displacement loops likely enhances DNA helicase activity (138). Rat, *X. laevis*, and human mtSSBs were shown to stimulate partially purified forms of pol γ in vitro on a variety of template-primer substrates (134, 139, 140). *X. laevis* mtSSB was also shown to stimulate highly purified pol γ (141) and to enable pol γ to replicate through dT-rich regions that are capable of assuming a triplex structure (142). In an assay that mimics lagging DNA strand synthesis in mitochondrial replication, we demonstrated functional interactions between the near-homogenous *Drosophila* pol γ and both native and recombinant forms of mtSSB, identifying roles for mtSSB in initiation and elongation of DNA strands (39, 63, 143). DNase I footprinting of pol γ : DNA complexes and initial rate measurements show that mtSSB enhances primer recognition and binding, and it stimulates 30-fold the rate of initiation of DNA strands by *Drosophila* pol γ. At the same time, mtSSB stimulates exonuclease activity to a similar extent over a broad range of KCl concentrations. That this occurs without any apparent loss of mispair specificity argues that the enhancement of pol γ catalytic efficiency is likely not accompanied by increased nucleotide turnover and suggests coordination of its DNA polymerase and proofreading activities at the replication fork. Moreover, mtSSB increases severalfold the processivity of *Drosophila* pol γ in DNA strand elongation. Recent experiments provide a physical link with the functional interactions; we find that mtSSB interacts specifically with the catalytic core of *Drosophila* pol γ in the absence of DNA (P.R. Kiefer and L.S. Kaguni, unpublished observations).

Other Proteins

The unidirectional, asymmetric model of mtDNA replication in which leading and lagging DNA strands are synthesized continuously (61) would require a minimal group of proteins at the replication fork. These would include pol γ, mtSSB, a mtDNA helicase, and a mtDNA topoisomerase. The strand-coupled model (116) would require additional enzymatic activities in the initiation and processing of Okazaki fragments on the lagging DNA strand, which include a priming enzyme, an RNase H or FEN1-like enzyme to catalyze primer removal, and a mtDNA ligase.

Two potential players at the mtDNA replication fork, a mtDNA helicase and mtRNase H, represent exciting new discoveries. The gene for a putative mtDNA helicase, *Twinkle*, was identified in individuals with PEO associated with multiple mtDNA deletions from 12 pedigrees of various ethnic origins (144). As with the catalytic cores of pol γ and mtRNA polymerase, *Twinkle* encodes a protein with a deduced amino acid sequence similar to a bacteriophage T7 protein, the T7 gene 4 helicase-primase (144). The human *Twinkle* gene contains the five helicase sequence motifs of the T7 gene 4 protein but lacks its primase-associated sequences. The Twinkle protein colocalizes with mtDNA in mitochondrial nucleoids (144, 145), and a recent study demonstrates that a recombinant form is an NTP-dependent DNA helicase that is stimulated by human mtSSB (146).

In 2003, it was demonstrated that generation of RNase H1 null mice leads to mtDNA depletion and results in apoptotic cell death (147). Although much of the RNase H1 protein localizes to the nucleus, a fraction localizes to the mitochondrion. Thus it is apparent that a mitochondrial form of mouse RNase H1 is required for mtDNA maintenance. The requirement for RNase H1 might be for primer removal at the origin, O_H, and/or at the site of lagging DNA strand initiation, O_L. Additionally, RNase H1 might catalyze RNA primer removal in Okazaki fragment synthesis.

Two other enzymes that likely function at the mtDNA replication fork and their corresponding genes have been identified in the past few years, mtDNA ligase III (148, 149) and mtDNA topoisomerase I (150). *Xenopus* mtDNA ligase III has been shown to function with pol γ in reconstitution of base excision repair in vitro (42). Antisense-mediated decreases in DNA ligase III expression in cultured human cells result in both reduced mtDNA content and numerous single-strand nicks in mtDNA not found in control cells, providing a physiological demonstration of a requirement for DNA ligase III in mtDNA repair (151). Human mtDNA topoisomerase I is a 72 kDa protein that relaxes negative supercoils (150). Because it is a type IB enzyme, it could also catalyze the removal of positive supercoils at the replication fork that result from DNA helicase action (152).

The identification of a priming enzyme in mtDNA replication remains elusive. Although a mtDNA primase was identified in human mitochondria as early as 1985 and partially purified and characterized (153, 154), no further reports of a highly purified form have appeared, nor has a corresponding gene been identified. On the other hand, a processed transcript synthesized by mtRNA pol is thought to provide the primer in the initiation of mtDNA replication [reviewed in (155, 156)]. Additionally, the catalytic core of mtRNA pol has been shown to synthesize short RNAs with little sequence specificity (157). Thus, it is possible that mtRNA pol also serves a role in lagging DNA strand initiation. Alternatively, biochemical features shared between pol γ and HIV-1 reverse transcriptase in binding and copying RNA templates (59) and the tRNA-like binding fold in the accessory subunit of pol γ (55, 57) tempt the speculation that pol γ might use mitochondrial tRNAs as primers. In this regard, 22 tRNAs are produced in both sense and antisense form as a result of the synthesis of genome-length primary transcripts of both mtDNA strands in mammals. Furthermore, the tRNA genes punctuate the 13 protein-coding and 2 rRNA genes, and the full-length transcripts are processed to liberate tRNAs, mRNAs, and rRNAs. At present, determining the requirements for, and the mechanism of, priming at the mtDNA replication fork poses an important challenge for future biochemical and genetic studies.

PERSPECTIVES

Nearly 30 years of research on DNA polymerase γ has elucidated its physical properties, subunit composition, and catalytic activities and documented its key role in mtDNA replication and repair. The three-dimensional structure of pol γ, its subunit-subunit interactions, and specific contacts with its substrates remain to be determined. Although the presence of an additional DNA polymerase in animal mitochondria that functions in replication, e.g., in lagging DNA strand synthesis, remains a possibility, current studies show that pol γ is well suited for the multifarious DNA synthetic reactions that are known to occur in mitochondrial DNA metabolism. Recombination in animal mitochondria is an emergent area, and a role for pol γ is entirely consistent with its catalytic repertoire. The reconstitution of mtDNA replication in vitro will require the identification and validation of new protein players and the crucial physiological evidence to support their roles. The development of animal models of mitochondrial mutagenesis and toxicity may pave the way for understanding the interplay of proteins at the mtDNA replication fork and for the discovery of new antiviral and antimicrobial reagents.

ACKNOWLEDGMENTS

Research from the author's laboratory was supported by grants from the National Institutes of Health (GM34042, GM45295, HL59656), the American Cancer Society (JFRA-144, NP-686), and the National Science Foundation (9600681). I thank the members of my laboratory, past and present, for their research contributions. I acknowledge in particular Carol Farr for her numerous and sustained efforts in the past 15 years. I am grateful to Drs. Robert Bambara, Charles Carter, and Charles McHenry for critical review of the manuscript and to Dr. Charles Carter and Ningguang Luo for help with the figures. I wish to express my gratitude to my long-standing mentors, Dr. Robert Lehman and Dr. Shelagh Ferguson-Miller, for their insights and encouragement, collegiality, and friendship.

The *Annual Review of Biochemistry* is online at http://biochem.annualreviews.org

LITERATURE CITED

1. Weissbach A. 1979. *Arch. Biochem. Biophys.* 198:386–96
2. Wang TS. 1991. *Annu. Rev. Biochem.* 60:513–52
3. Hübscher U, Maga G, Spadari S. 2002. *Annu. Rev. Biochem.* 71:133–63
4. Goodman MF. 2002. *Annu. Rev. Biochem.* 71:17–50
5. Steitz TA. 1999. *J. Biol. Chem.* 274:17395–98
6. Delarue M, Poch O, Tordo N, Moras D, Argos P. 1990. *Protein Eng.* 3:461–67
7. Ito J, Braithwaite DK. 1991. *Nucleic Acids Res.* 19:4045–57
8. Steitz TA, Smerdon SJ, Jager J, Joyce CM. 1994. *Science* 266:2022–25
9. Wang J, Sattar AK, Wang CC, Karam JD, Konigsberg WH, Steitz TA. 1997. *Cell* 89:1087–99
10. Hopfner KP, Eichinger A, Engh RA, Laue F, Ankenbauer W, et al. 1999. *Proc. Natl. Acad. Sci. USA* 96:3600–5
11. Rodriguez AC, Park HW, Mao C, Beese LS. 2000. *J. Mol. Biol.* 299:447–62
12. Ollis DL, Brick P, Hamlin R, Xuong NG, Steitz TA. 1985. *Nature* 313:762–66
13. Kim Y, Eom SH, Wang J, Lee DS, Suh SW, Steitz TA. 1995. *Nature* 376:612–16
14. Doublie S, Tabor S, Long AM, Richardson CC, Ellenberger T. 1998. *Nature* 391:251–58
15. Kiefer JR, Mao C, Braman JC, Beese LS. 1998. *Nature* 391:304–7
16. Huber HE, Tabor S, Richardson CC. 1987. *J. Biol. Chem.* 262:16224–32
17. Ye F, Carrodeguas JA, Bogenhagen DF. 1996. *Nucleic Acids Res.* 24:1481–88
18. Lecrenier N, Van Der Bruggen P, Foury F. 1997. *Gene* 185:147–52
19. Fridlender B, Fry M, Bolden A, Weissbach A. 1972. *Proc. Natl. Acad. Sci. USA* 69:452–55
20. Bolden A, Noy GP, Weissbach A. 1977. *J. Biol. Chem.* 252:3351–56
21. Yamaguchi M, Matsukage A, Takahashi T. 1980. *J. Biol. Chem.* 255:7002–9
22. Wernette CM, Kaguni LS. 1986. *J. Biol. Chem.* 261:14764–70
23. Olson MW, Wang Y, Elder RH, Kaguni LS. 1995. *J. Biol. Chem.* 270:28932–37
24. Insdorf NF, Bogenhagen DF. 1989. *J. Biol. Chem.* 264:21491–97
25. Longley MJ, Mosbaugh DW. 1991. *J. Biol. Chem.* 266:24702–11
26. Gray H, Wong TW. 1992. *J. Biol. Chem.* 267:5835–41
27. Kunkel TA, Soni A. 1988. *J. Biol. Chem.* 263:4450–59
28. Kunkel TA, Mosbaugh DW. 1989. *Biochemistry* 28:988–95
29. Kaguni LS, Olson MW. 1989. *Proc. Natl. Acad. Sci. USA* 86:6469–73
30. Insdorf NF, Bogenhagen DF. 1989. *J. Biol. Chem.* 264:21498–503
31. Foury F. 1989. *J. Biol. Chem.* 264:20552–60
32. Ito J, Braithwaite DK. 1990. *Nucleic Acids Res.* 18:6716
33. Foury F, Vanderstraeten S. 1992. *EMBO J.* 11:2717–26
34. Vanderstraeten S, Van den Brule S, Hu J, Foury F. 1998. *J. Biol. Chem.* 273:23690–97
35. Lewis DL, Farr CL, Wang Y, Lagina AT III, Kaguni LS. 1996. *J. Biol. Chem.* 271:23389–94
36. Graves SW, Johnson AA, Johnson KA. 1998. *Biochemistry* 37:6050–58
37. Longley MJ, Ropp PA, Lim SE, Copeland WC. 1998. *Biochemistry* 37:10529–39
38. Wang Y, Kaguni LS. 1999. *J. Biol. Chem.* 274:28972–77

39. Farr CL, Wang Y, Kaguni LS. 1999. *J. Biol. Chem.* 274:14779–85

40. Lim SE, Longley MJ, Copeland WC. 1999. *J. Biol. Chem.* 274:38197–203

41. Johnson AA, Tsai Y, Graves SW, Johnson KA. 2000. *Biochemistry* 39:1702–8

42. Pinz KG, Bogenhagen DF. 1998. *Mol. Cell. Biol.* 18:1257–65

43. Pinz KG, Bogenhagen DF. 2000. *J. Biol. Chem.* 275:12509–14

44. Longley MJ, Prasad R, Srivastava DK, Wilson SH, Copeland WC. 1998. *Proc. Natl. Acad. Sci. USA* 95:12244–48

45. Bogenhagen DF. 1999. *Am. J. Hum. Genet.* 64:1276–81

46. Copeland WC, Longley MJ. 2003. *ScientificWorldJournal* 3:34–44

47. Ropp PA, Copeland WC. 1995. *Gene* 165:103–7

48. Ropp PA, Copeland WC. 1996. *Genomics* 36:449–58

49. Fan L, Kaguni LS. 2001. *Biochemistry* 40:4780–91

50. Iyengar B, Roote J, Campos AR. 1999. *Genetics* 153:1809–24

51. Van Goethem G, Dermaut B, Lofgren A, Martin JJ, Van Broeckhoven C. 2001. *Nat. Genet.* 28:211–12

52. Lamantea E, Tiranti V, Bordoni A, Toscano A, Bono F, et al. 2002. *Ann. Neurol.* 52:211–19

53. Reha-Krantz LJ, Nonay RL. 1993. *J. Biol. Chem.* 268:27100–8

54. Wang Y, Farr CL, Kaguni LS. 1997. *J. Biol. Chem.* 272:13640–46

55. Fan L, Sanschagrin PC, Kaguni LS, Kuhn LA. 1999. *Proc. Natl. Acad. Sci. USA* 96:9527–32

56. Carrodeguas JA, Kobayashi R, Lim SE, Copeland WC, Bogenhagen DF. 1999. *Mol. Cell. Biol.* 19:4039–46

57. Carrodeguas JA, Theis K, Bogenhagen DF, Kisker C. 2001. *Mol. Cell* 7:43–54

58. Carrodeguas JA, Pinz KG, Bogenhagen DF. 2002. *J. Biol. Chem.* 277:50008–14

59. Murakami E, Feng JY, Lee H, Hanes J, Johnson KA, Anderson KS. 2003. *J. Biol. Chem.* 278:36403–9

60. Wernette CM, Conway MC, Kaguni LS. 1988. *Biochemistry* 27:6046–54

61. Clayton DA. 1982. *Cell* 28:693–705

62. Williams AJ, Wernette CM, Kaguni LS. 1993. *J. Biol. Chem.* 268:24855–62

63. Thommes P, Farr CL, Marton RF, Kaguni LS, Cotterill S. 1995. *J. Biol. Chem.* 270:21137–43

64. Brown WM, George MJ, Wilson AC. 1979. *Proc. Natl. Acad. Sci. USA* 76:1967–71

65. Wallace DC. 1992. *Annu. Rev. Biochem.* 61:1175–212

66. Wallace DC, Shoffner JM, Trounce I, Brown MD, Ballinger SW, et al. 1995. *Biochim. Biophys. Acta* 1271:141–51

67. Kunkel TA. 1985. *J. Biol. Chem.* 260:12866–74

68. Pinz KG, Shibutani S, Bogenhagen DF. 1995. *J. Biol. Chem.* 270:9202–6

69. Hackenbrock CR. 1968. *Proc. Natl. Acad. Sci. USA* 61:598–605

70. Bestwick RK, Mathews CK. 1982. *J. Biol. Chem.* 257:9305–8

71. Rampazzo C, Gallinaro L, Milanesi E, Frigimelica E, Reichard P, Bianchi V. 2000. *Proc. Natl. Acad. Sci. USA* 97:8239–44

72. Salviati L, Sacconi S, Mancuso M, Otaegui D, Camano P, et al. 2002. *Ann. Neurol.* 52:311–17

73. Mancuso M, Salviati L, Sacconi S, Otaegui D, Camano P, et al. 2002. *Neurology* 59:1197–202

74. Nishino I, Spinazzola A, Hirano M. 1999. *Science* 283:689–92

75. Lecrenier N, Foury F. 1995. *Mol. Gen. Genet.* 249:1–7

76. Kaukonen J, Juselius JK, Tiranti V, Kyttala A, Zeviani M, et al. 2000. *Science* 289:782–85

77. Zeviani M, Spinazzola A, Carelli V. 2003. *Curr. Opin. Genet. Dev.* 13:262–70

78. Longley MJ, Nguyen D, Kunkel TA,

Copeland WC. 2001. *J. Biol. Chem.* 276:38555–62

79. Johnson AA, Johnson KA. 2001. *J. Biol. Chem.* 276:38090–96

80. Johnson AA, Johnson KA. 2001. *J. Biol. Chem.* 276:38097–107

81. Hu J, Vanderstraeten S, Foury F. 1995. *Gene* 160:105–10

82. Polesky AH, Steitz TA, Grindley ND, Joyce CM. 1990. *J. Biol. Chem.* 265:14579–91

83. Astatke M, Grindley ND, Joyce CM. 1995. *J. Biol. Chem.* 270:1945–54

84. Carroll SS, Cowart M, Benkovic SJ. 1991. *Biochemistry* 30:804–13

85. Bell JB, Eckert KA, Joyce CM, Kunkel TA. 1997. *J. Biol. Chem.* 272:7345–51

86. Iyengar B, Luo N, Farr CL, Kaguni LS, Campos AR. 2002. *Proc. Natl. Acad. Sci. USA* 99:4483–88

87. Polesky AH, Dahlberg ME, Benkovic SJ, Grindley ND, Joyce CM. 1992. *J. Biol. Chem.* 267:8417–28

88. Lim SE, Ponamarev MV, Longley MJ, Copeland WC. 2003. *J. Mol. Biol.* 329:45–57

89. Astatke M, Grindley ND, Joyce CM. 1998. *J. Mol. Biol.* 278:147–65

90. Ponamarev MV, Longley MJ, Nguyen D, Kunkel TA, Copeland WC. 2002. *J. Biol. Chem.* 277:15225–28

91. Olson MW, Kaguni LS. 1992. *J. Biol. Chem.* 267:23136–42

92. Echols H, Goodman MF. 1991. *Annu. Rev. Biochem.* 60:477–511

93. Derbyshire V, Grindley ND, Joyce CM. 1991. *EMBO J.* 10:17–24

94. Lewis W, Day BJ, Copeland WC. 2003. *Nat. Rev. Drug Discov.* 2:812–22

95. McKenzie R, Fried MW, Sallie R, Conjeevaram H, Di Bisceglie AM, et al. 1995. *N. Engl. J. Med.* 333:1099–105

96. Dalakas MC, Illa I, Pezeshkpour GH, Laukaitis JP, Cohen B, Griffin JL. 1990. *N. Engl. J. Med.* 322:1098–105

97. Lewis W, Simpson JF, Meyer RR. 1994. *Circ. Res.* 74:344–48

98. Kaguni LS, Wernette CM, Conway MC, Yang-Cashman P. 1988. In *Cancer Cells: Eukaryotic DNA Replication*, ed. TJ Kelly, B Stillman, pp. 425–32. Cold Spring Harbor, NY: Cold Spring Harbor Lab.

99. Kakuda TN. 2000. *Clin. Ther.* 22:685–708

100. Lim SE, Copeland WC. 2001. *J. Biol. Chem.* 276:23616–23

101. Feng JY, Johnson AA, Johnson KA, Anderson KS. 2001. *J. Biol. Chem.* 276:23832–37

102. Johnson AA, Ray AS, Hanes J, Suo Z, Colacino JM, et al. 2001. *J. Biol. Chem.* 276:40847–57

103. Davis AF, Ropp PA, Clayton DA, Copeland WC. 1996. *Nucleic Acids Res.* 24:2753–59

104. Lefai E, Fernandez-Moreno MA, Kaguni LS, Garesse R. 2000. *Insect Mol. Biol.* 9:315–22

105. Lefai E, Fernandez-Moreno MA, Alahari A, Kaguni LS, Garesse R. 2000. *J. Biol. Chem.* 275:33123–33

106. Ruiz de Mena I, Lefai E, Garesse R, Kaguni LS. 2000. *J. Biol. Chem.* 275:13628–36

107. Hirose F, Yamaguchi M, Handa H, Inomata Y, Matsukage A. 1993. *J. Biol. Chem.* 268:2092–99

108. Ohno K, Hirose F, Sakaguchi K, Nishida Y, Matsukage A. 1996. *Nucleic Acids Res.* 24:3942–46

109. Takahashi Y, Yamaguchi M, Hirose F, Cotterill S, Kobayashi J, et al. 1996. *J. Biol. Chem.* 271:14541–47

110. Sawado T, Hirose F, Takahashi Y, Sasaki T, Shinomiya T, et al. 1998. *J. Biol. Chem.* 273:26042–51

111. Lefai E, Calleja M, Ruiz de Mena I, Lagina AT III, Kaguni LS, Garesse R. 2000. *Mol. Gen. Genet.* 264:37–46

112. Spelbrink JN, Toivonen JM, Hakkaart GA, Kurkela JM, Cooper HM, et al. 2000. *J. Biol. Chem.* 275:24818–28

113. Mott JL, Zhang D, Farrar PL, Chang SW, Zassenhaus HP. 1999. *Ann. NY Acad. Sci.* 893:353–57

114. Zhang D, Mott JL, Chang SW, Denniger G, Feng Z, Zassenhaus HP. 2000. *Genomics* 69:151–61

115. Zhang D, Mott JL, Farrar P, Ryerse JS, Chang SW, et al. 2003. *Cardiovasc. Res.* 57:147–57

116. Holt IJ, Lorimer HE, Jacobs HT. 2000. *Cell* 100:515–24

117. Yang MY, Bowmaker M, Reyes A, Vergani L, Angeli P, et al. 2002. *Cell* 111:495–505

118. Bowmaker M, Yang MY, Yasukawa T, Reyes A, Jacobs HT, et al. 2003. *J. Biol. Chem.* 278:50961–69

119. Bogenhagen DF, Clayton DA. 2003. *Trends Biochem. Sci.* 28:357–60

120. Holt IJ, Jacobs HT. 2003. *Trends Biochem. Sci.* 28:355–56

121. Bogenhagen DF, Clayton DA. 2003. *Trends Biochem. Sci.* 28:404–5

122. Genga A, Bianchi L, Foury F. 1986. *J. Biol. Chem.* 261:9328–32

123. Van Dyck E, Foury F, Stillman B, Brill SJ. 1992. *EMBO J.* 11:3421–30

124. Maier D, Farr CL, Poeck B, Alahari A, Vogel M, et al. 2001. *Mol. Biol. Cell* 12:821–30

124a. Farr CL, Matsushima Y, Lagina AT III, Luo N, Kaguni LS. 2004. *J. Biol. Chem.* In press

125. Kornberg A, Baker TA. 1992. *DNA Replication.* New York: Freeman

126. Pavco PA, Van Tuyle GC. 1985. *J. Cell Biol.* 100:258–64

127. Mignotte B, Barat M, Mounolou JC. 1985. *Nucleic Acids Res.* 13:1703–16

128. Curth U, Urbanke C, Greipel J, Gerberding H, Tiranti V, Zeviani M. 1994. *Eur. J. Biochem.* 221:435–43

129. Tiranti V, Barat-Gueride M, Bijl J, DiDonato S, Zeviani M. 1991. *Nucleic Acids Res.* 19:4291

130. Tiranti V, Rocchi M, DiDonato S, Zeviani M. 1993. *Gene* 126:219–25

131. Stroumbakis ND, Li Z, Tolias PP. 1994. *Gene* 143:171–77

132. Ghrir R, Lecaer JP, Dufresne C,

Gueride M. 1991. *Arch. Biochem. Biophys.* 291:395–400

133. Li K, Williams RS. 1997. *J. Biol. Chem.* 272:8686–94

134. Hoke GD, Pavco PA, Ledwith BJ, Van Tuyle GC. 1990. *Arch. Biochem. Biophys.* 282:116–24

135. Yang C, Curth U, Urbanke C, Kang C. 1997. *Nat. Struct. Biol.* 4:153–57

136. Van Tuyle GC, Pavco PA. 1985. *J. Cell Biol.* 100:251–57

137. Takamatsu C, Umeda S, Ohsato T, Ohno T, Abe Y, et al. 2002. *EMBO Rep.* 3:451–56

138. Van Tuyle GC, Pavco PA. 1981. *J. Biol. Chem.* 256:12772–79

139. Mignotte B, Marsault J, Barat-Gueride M. 1988. *Eur. J. Biochem.* 174:479–84

140. Genuario R, Wong TW. 1993. *Cell. Mol. Biol. Res.* 39:625–34

141. Mikhailov VS, Bogenhagen DF. 1996. *J. Biol. Chem.* 271:18939–46

142. Mikhailov VS, Bogenhagen DF. 1996. *J. Biol. Chem.* 271:30774–80

143. Williams AJ, Kaguni LS. 1995. *J. Biol. Chem.* 270:860–65

144. Spelbrink JN, Li FY, Tiranti V, Nikali K, Yuan QP, et al. 2001. *Nat. Genet.* 28:223–31

145. Garrido N, Griparic L, Jokitalo E, Wartiovaara J, van der Bliek AM, Spelbrink JN. 2003. *Mol. Biol. Cell* 14:1583–96

146. Korhonen JA, Gaspari M, Falkenberg M. 2003. *J. Biol. Chem.* 278:48627–32

147. Cerritelli SM, Frolova EG, Feng C, Grinberg A, Love PE, Crouch RJ. 2003. *Mol. Cell* 11:807–15

148. Lakshmipathy U, Campbell C. 1999. *Mol. Cell. Biol.* 19:3869–76

149. Perez-Jannotti RM, Klein SM, Bogenhagen DF. 2001. *J. Biol. Chem.* 276:48978–87

150. Zhang H, Barcelo JM, Lee B, Kohlhagen G, Zimonjic DB, et al. 2001. *Proc. Natl. Acad. Sci. USA* 98:10608–13

151. Lakshmipathy U, Campbell C. 2001. *Nucleic Acids Res.* 29:668–76

152. Wang JC. 2002. *Nat. Rev. Mol. Cell Biol.* 3:430–40
153. Wong TW, Clayton DA. 1985. *J. Biol. Chem.* 260:11530–35
154. Wong TW, Clayton DA. 1985. *Cell* 42:951–58
155. Shadel GS, Clayton DA. 1997. *Annu. Rev. Biochem.* 66:409–35
156. Shadel GS. 1999. *Am. J. Hum. Genet.* 65:1230–37
157. Kelly JL, Lehman IR. 1986. *J. Biol. Chem.* 261:10340–47
158. Kraulis PJ. 1991. *J. Appl. Crystallogr.* 24:946–50
159. Merritt EA, Bacon DJ. 1997. *Methods Enzymol.* 277:505–24

Annu. Rev. Biochem. 2004. 73:321–54
doi: 10.1146/annurev.biochem.73.011303.073731
Copyright © 2004 by Annual Reviews. All rights reserved
First published online as a Review in Advance on March 5, 2004

LYSOPHOSPHOLIPID RECEPTORS:
Signaling and Biology

Isao Ishii,[1] Nobuyuki Fukushima,[2] Xiaoqin Ye,[3] and Jerold Chun[3]

[1]*Department of Molecular Genetics, National Institute of Neuroscience, NCNP, Kodaira, Tokyo 187-8502, Japan; email: isao@ncnp.go.jp*
[2]*Department of Molecular Biochemistry, Hokkaido University Graduate School of Medicine, Kita-ku, Sapporo 060-8638, Japan; email: nfuku@med.hokudai.ac.jp*
[3]*Department of Molecular Biology, The Scripps Research Institute, La Jolla, California 92037; email: xiaoqin@scripps.edu, jchun@scripps.edu*

Key Words lysophosphatidic acid, lysophosphatidylcholine, sphingosine 1-phosphate, sphingosylphosphorylcholine, G protein-coupled receptor, LPA, S1P

■ **Abstract** Lysophospholipids (LPs), such as lysophosphatidic acid and sphingosine 1-phosphate, are membrane-derived bioactive lipid mediators. LPs can affect fundamental cellular functions, which include proliferation, differentiation, survival, migration, adhesion, invasion, and morphogenesis. These functions influence many biological processes that include neurogenesis, angiogenesis, wound healing, immunity, and carcinogenesis. In recent years, identification of multiple cognate G protein-coupled receptors has provided a mechanistic framework for understanding how LPs play such diverse roles. Generation of LP receptor-null animals has allowed rigorous examination of receptor-mediated physiological functions in vivo and has identified new functions for LP receptor signaling. Efforts to develop LP receptor subtype-specific agonists/antagonists are in progress and raise expectations for a growing collection of chemical tools and potential therapeutic compounds. The rapidly expanding literature on the LP receptors is herein reviewed.

CONTENTS

INTRODUCTION

Lysophospholipids (LPs) are quantitatively minor lipid species, compared to their major phospholipid counterparts (e.g., phosphatidylcholine, phosphatidylethanolamine, and sphingomyelin) that structurally compose mammalian cell membranes. Prominent among the LPs are lysophosphatidic acid (LPA), sphingosine 1-phosphate (S1P), lysophosphatidylcholine (LPC), and sphingosylphosphorylcholine (SPC); all of which share rather simple chemical structures of a 3-carbon glycerol or sphingoid backbone on which is attached a single acyl chain of varied length and saturation (Figure 1). LPs were initially identified as precursors and metabolites in the de novo biosynthesis of phospholipids. They were subsequently observed to have properties resembling extracellular growth factors or signaling molecules, although the mechanisms of action for LPs remained unclear for decades after description of their bioactivities.

In recent years, LPs have been shown to act through sets of specific G protein-coupled receptors (GPCRs) in an autocrine or paracrine fashion. Since the identification of the first LP receptor, LPA$_1$, in 1996 (1), a growing family of GPCRs has been identified as high-affinity LP receptors in mammals (Table 1), with sequence and functional homologies that extend evolutionarily at least through Amphibia (2). The best characterized receptors have been renamed for their high-affinity ligands and constitute the LPA and S1P receptors (3), whereas other orphan receptor names have been maintained for those receptors with provisional ligand identities. For a given LP ligand, LP receptors generally share high amino acid similarities, although exceptions are also evident (Figure 2). One or more LP receptor genes and gene products are expressed in most mammalian tissues with spatially and temporally regulated expression patterns (Table 2). Many cell types express more than one LP receptor, and each receptor can couple with multiple types of G proteins to activate a range of downstream effectors that mediate a variety of cellular responses upon LP stimulation (Table 2) [reviewed in (4)]. The identification of LP receptors has provided a major focus for understanding not only the signaling pathways activated by LPs but also their biology in vivo. This review focuses on our current knowledge about the LP receptors, with emphasis on their signaling and biological roles. We have deferred detailed discussion of the basic biochemistry and biosynthesis/metabolism of the LPs, along with in-depth specialty areas, to many excellent reviews and the papers cited therein (5–18).

1. LYSOPHOSPHOLIPIDS

LPA (1-acyl-2-hydroxy-*sn*-glycero-3-phosphate) and S1P are in many ways prototypic examples of LPs relevant to this review. Early hints of a role for LPA as a biological effector molecule were recognized during the 1960s (19, 20), and an increasingly diverse range of physiological actions were identified in the ensuing decades, including effects on blood pressure, platelet activation, and smooth muscle contraction (21–23). A wealth of cell biological studies in the mid-1980s defined a variety of cellular effects, which include cell growth, cell rounding, neurite retraction, and actin stress fiber formation [reviewed in (24)]. These findings suggested the existence of specific receptors that could mediate the effects of this small lipid. However, biophysical properties of LPA or the possibility of second messenger activities were also proposed as competing mechanisms for LPA actions, and this mechanistic ambiguity persisted in the absence of identified receptors [reviewed in (4)]. The identification of cloned LPA receptors, combined with molecular genetic strategies to establish their functions, allowed determination of both signaling and biological effects that are dependent on receptor mechanisms. Activation of these LPA receptors demonstrated that a range of downstream signaling cascades mediate LPA signaling. These include mitogen-activated protein kinase (MAPK) activation, adenylyl cyclase (AC) inhibition/activation, phospholipase C (PLC) activation/Ca^{2+} mobilization, arachidonic acid release, Akt/PKB activation, and the activation of small GTPases, Rho, Rac, and Ras (Table 2). It is critical to note that the actual pathway and realized end point are dependent on a range of variables that include receptor usage, cell type, expression level of a receptor or signaling protein, and LPA concentration. Many discrepancies in the literature can be explained by the different experimental conditions employed.

LPA is produced from activated platelets, activated adipocytes, neuronal cells, and other cell types [reviewed in (4, 8)]. Although mechanisms of LPA synthesis in individual cell types remain to be elucidated, serum LPA is produced by multiple enzymatic pathways that involve monoacylglycerol kinase, phospholipase A_1, secretory phospholipase A_2, and lysophospholipase D (lysoPLD), including autotaxin [reviewed in (9, 26)]. Several enzymes are involved in LPA

SPC

S1P

LPC

LPA

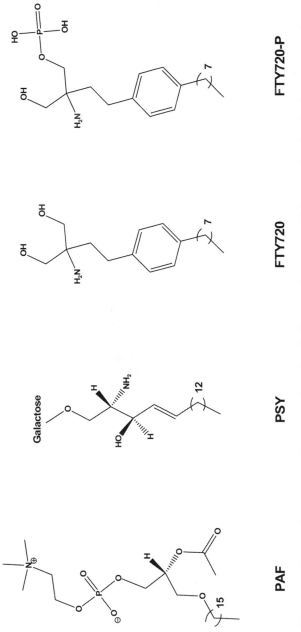

Figure 1 Chemical structures of signaling lysophospholipids and two related bioactive lipids, platelet-activating factor (PAF) and psychosine (PSY). LPA, (1-oleoyl) lysophosphatidic acid; LPC, (1-oleoyl) lysophosphatidylcholine; S1P, sphingosine 1-phosphate; SPC, sphingosylphosphorylcholine; FTY720-P, the phosphorylated form of FTY720.

TABLE 1   Mammalian lysophospholipid receptors

Receptors (Synonyms)	Species	Accession number	Amino acids	Predicted MW (kDa)	Chromosomal location
LPA$_1$	Human	(NM_057159)	364	41.1	9q32
(EDG-2/VZG-1)	Mouse	(NM_010336)	364	41.2	4 B3
	Rat	(NM_053936)	364	41.1	5q22
LPA$_2$	Human	(NM_004720)	351	39.1	19p12
(EDG-4)	Mouse	(XM_193070)	348	39.0	8 B3.3
LPA$_3$	Human	(NM_012152)	353	40.1	1p22.3-p31.1
(EDG-7)	Mouse	(NM_022983)	354	40.3	3 H2
	Rat	(AB051164)	354	40.3	1q55
LPA$_4$	Human	(NM_005296)	370	41.9	Xq13-q21.1
(p2y$_9$/GPR23)	Mouse	(NM_175271)	370	41.9	X D
S1P$_1$	Human	(NM_001400)	382	42.8	1p21
(EDG-1/LP$_{B1}$)	Mouse	(NM_007901)	382	42.6	3 G1
	Rat	(NM_017301)	383	42.7	2q41
S1P$_2$	Human	(NM_004230)	353	38.8	19p13.2
(EDG-5/AGR16/	Mouse	(XM_134731)	352	38.8	9 A3
H218/LP$_{B2}$)	Rat	(NM_017192)	352	38.7	5q36
S1P$_3$	Human	(NM_005226)	378	42.3	9q22.1-q22.2
(EDG-3/LP$_{B3}$)	Mouse	(NM_010101)	378	42.3	13 B1
S1P$_4$	Human	(NM_003775)	384	41.6	19p13.3
(EDG-6/LP$_{C1}$)	Mouse	(NM_010102)	386	42.3	10 C1
S1P$_5$	Human	(NM_030760)	398	41.8	19p13.2
(EDG-8/NRG-1	Mouse	(NM_053190)	400	42.3	9 A3
LP$_{B4}$)	Rat	(NM_021775)	400	42.4	5q36
G2A	Human	(NM_013345)	380	42.5	14q32.3
	Mouse	(NM_019925)	382	42.7	12 F2
OGR1	Human	(NM_003485)	365	41.1	14q31
(GPR68)	Mouse	(NM_175493)	365	41.2	12 E
	Rat	(XM_234483)	365	41.2	6q32
GPR4	Human	(NM_005282)	362	41.0	19q13.3
	Mouse	(NM_175668)	365	41.1	7 A1
	Rat	(NM_218415)	365	41.3	1q21
GPR12	Human	(NM_005288)	334	36.7	13q12
(GPCR01)	Mouse	(NM_008151)	334	36.6	5 G3
	Rat	(NM_030831)	334	36.7	12p11
TDAG8[a]	Human	(NM_003608)	337	39.3	14q31-q32.1
(GPCR25/GPR65)	Mouse	(NM_008152)	337	39.4	12 E
PAFR[a]	Human	(NM_000952)	342	39.2	1p35-p34.3
	Mouse	(D50872)	341	39.1	4 D2.2

[a]Not a lysophospholipid receptor

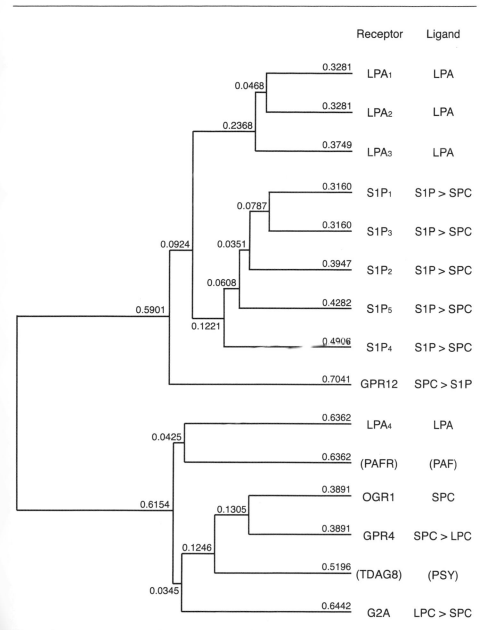

	Receptor	Ligand
0.3281	LPA₁	LPA
0.3281	LPA₂	LPA
0.3749	LPA₃	LPA
0.3160	S1P₁	S1P > SPC
0.3160	S1P₃	S1P > SPC
0.3947	S1P₂	S1P > SPC
0.4282	S1P₅	S1P > SPC
0.4906	S1P₄	S1P > SPC
0.7041	GPR12	SPC > S1P
0.6362	LPA₄	LPA
0.6362	(PAFR)	(PAF)
0.3891	OGR1	SPC
0.3891	GPR4	SPC > LPC
0.5196	(TDAG8)	(PSY)
0.6442	G2A	LPC > SPC

Figure 2   Phylogenic tree of the human lysophospholipid receptor family. For comparison, receptors for two related but distinct lipids are also shown in parentheses, the platelet activating factor (PAF) receptor and the putative psychosine receptor (TDAG8). The tree was derived by the neighbor joining method run on a Genetyx-Mac program (Genetyx Corp., Tokyo, Japan). The amino acid sequence divergence between any pair of sequences is equal to the sum of the lengths of the horizontal branches connecting two sequences. Preferential ligands for each receptor are aligned with the potency orders.

TABLE 2   Lysophospholipid receptor signaling and distribution in mice

Receptors	G-protein coupling	Cellular signaling	Tissue distribution in mice
LPA$_1$	G$_{i/o}$, G$_q$, G$_{12/13}$	[a]DNA ↑, SRE ↑, MAPK ↑, AC ↓, PLC/Ca ↑, Rho ↑, PI3K/Akt ↑	Ubiquitous
LPA$_2$	G$_{i/o}$, G$_q$, G$_{12/13}$	DNA ↑, SRE ↑, MAPK ↑, AC ↓, PLC/Ca ↑, Rho ↑, PI3K/Akt ↑	Ubiquitous
LPA$_3$	G$_{i/o}$, G$_q$, G$_s$	MAPK ↑, AC ↑ ↓, PLC/Ca ↑	Ubiquitous
LPA$_4$	Unknown	Ca ↑, AC ↑	(Ov, Pa, Th in human)
S1P$_1$	G$_{i/o}$	MAPK ↑, AC ↓, PLC/Ca ↑, (Rho ↑), Rac ↑, PI3K/Akt ↑	Ubiquitous
S1P$_2$	G$_{i/o}$, G$_q$, G$_{12/13}$, G$_s$	SRE ↑, MAPK ↑, AC ↑, PLC/Ca ↑, Rho ↑, Rac ↓	Ubiquitous
S1P$_3$	G$_{i/o}$, G$_q$, G$_{12/13}$, G$_s$	SRE ↑, MAPK ↑, AC ↑ ↓, PLC/Ca ↑, Rho ↑, Rac ↑, PI3K/Akt ↑	Ubiquitous
S1P$_4$	G$_{i/o}$, G$_{12/13}$, G$_s$	MAPK ↑, AC ↑, PLC/Ca ↑, Rho ↑	Ln, Sp, Lg, Th
S1P$_5$	G$_{i/o}$, G$_{12/13}$	DNA ↓, MAPK ↓, AC ↓, PLC/Ca ↑	Br, Sk, Sp
G2A	G$_{i/o}$, G$_q$, G$_{12/13}$, G$_s$	SRE ↑, MAPK ↑, AC ↑, PLC/Ca ↑, Rho ↑, Rac ↑, Ras ↑	Th, Sp, Bm
OGR1	G$_{i/o}$, G$_q$	DNA ↓, MAPK ↑, PLC/Ca ↑	(Ubiquitous in human)
GPR4	G$_{i/o}$	DNA ↑, SRE ↑, MAPK ↑, PLC/Ca ↑	(Ubiquitous in human)
GPR12	G$_{i/o}$, G$_s$	AC ↑, PLC/Ca ↑	Br, Ts, Li
TDAG8	G$_q$, G$_z$	AC ↓, PLC/Ca ↑	Th, Sp, Bm

[a]Abbreviations used are: DNA, DNA synthesis (proliferation); SRE, serum-responsive element; MAPK, mitogen-activated protein kinase; AC, adenylyl cyclase; PLC, phospholipase C; PI3K, phosphoinositide 3-kinase; Ov, ovary; Pa, pancreas; Th, thymus; Ln, lymph node; Sp, spleen; Lg, lung; Br, brain; Sk, skin; Ts, testis; Li, liver; and Bm, bone marrow.

degradation: lysophospholipase, lipid phosphate phosphatase, and LPA acyl transferase such as endophilin [reviewed in (9)]. LPA concentrations in human serum are estimated to be 1–5 μM (27). Serum LPA is bound to albumin, low-density lipoproteins, or other proteins, which possibly protect LPA from rapid degradation [reviewed in (9)]. LPA molecular species with different acyl chain lengths and saturation are naturally occurring, including 1-palmitoyl (16:0), 1-palmitoleoyl (16:1), 1-stearoyl (18:0), 1-oleoyl (18:1), 1-linoleoyl (18:2), and 1-arachidonyl (20:4) LPA (27) [reviewed in (8)]. Quantitatively, minor alkyl LPA has biological activities similar to acyl LPA (28), and different

LPA species activate LPA receptor subtypes with varied efficacies (29). Thus far, three LPA receptors (LPA_1-LPA_3) that share high amino acid sequence similarity have been identified and are complemented by a fourth (LPA_4) that is dissimilar.

S1P represents the second major prototype of bioactive LPs whose biological effects were first described in the early 1990s. S1P was found to induce Ca^{2+} mobilization from internal Ca^{2+} stores (30), and it was subsequently observed to stimulate fibroblast proliferation and morphological changes (31). These phenomena were attributed to S1P acting as a second messenger in fibroblast proliferation induced by platelet-derived growth factor (PDGF) and serum (32). Like inositol triphosphate, S1P was proposed to act as a Ca^{2+}-mobilizing second messenger that is produced intracellularly from membranes upon cell activation [reviewed in (6)], and a receptor-independent mechanism remains a viable explanation for some S1P-mediated effects (33). However, following the functional identification of LPA receptors, multiple S1P receptors were also identified based on their shared sequence similarities [reviewed in (4, 6, 34)]. S1P has been shown to exert most of its previously documented effects through cell surface receptors with signaling pathways and cellular/organismal physiologies that are comparable to LPA (Table 2) [reviewed in (4, 6, 8, 35)]. Thus far, five high-affinity receptors ($S1P_1$-$S1P_5$) and a possible low-affinity receptor (GPR12) have been identified in mammals, and additional receptors are likely to exist (Figure 2).

S1P, unlike LPA, is stored in platelets at relatively high concentrations and released from platelets upon activation (36, 37) [reviewed in (10)]. S1P in serum is bound to albumin or lipoproteins, and its concentration is estimated to be 0.5–0.8 μM in human serum and 0.2–0.4 μM in human plasma [reviewed in (11)]. S1P is synthesized exclusively from sphingosine by sphingosine kinases and is degraded either by S1P lyases or by S1P phosphatases [reviewed in (10, 12)]. As a calcium-mobilizing second messenger, intracellular S1P homeostasis maintains low concentrations by appropriately balanced synthesis and degradation. Platelets are the primary source of S1P in serum because of the presence of sphingosine kinase and the absence of S1P lyase [reviewed in (10)]. Recent genetic-null studies in lower organisms indicate that deficiency of S1P lyase leads to abnormal development of *Dictyostelium discoideum*, *Caenorhabditis elegans*, or *Drosophila melanogaster* (38–40). In *Drosophila*, most of the examined 31 genes, whose mammalian homologues were shown to be involved in LPA or S1P metabolism, are expressed in spatio-temporal patterns that suggest physiological roles for LPs during *Drosophila* development (41).

In addition to LPA and S1P, two phosphorylcholine-containing LPs, SPC and LPC, have been shown to induce a variety of biological responses [reviewed in (13, 42)]. SPC induces cell proliferation, migration, and cytoskeletal rearrangement; all are actions shared by the structurally related ligand S1P. SPC may activate high-affinity S1P receptors as a lower affinity ligand (Figure 2), although it can also be converted to S1P by enzymes like autotaxin, underscoring a need for cautious interpretation of observed effects. Moreover, SPC mobilizes Ca^{2+}

from internal stores as does S1P, suggesting common sites of action and mechanisms for SPC and S1P [reviewed in (13, 14)]. It is notable, however, that SPC has also been shown to act differentially from S1P, as observed in human platelets in which S1P initiates, but SPC inhibits, activation (43). SPC can be produced from sphingomyelin by sphingomyelin deacylase, but little is known about the synthesis/degradation of SPC in a physiological context [reviewed in (13)]. Elevated SPC levels were observed in some tissues of Niemann–Pick disease patients, who lack sphingomyelinase that degrades sphingomyelin, but the relevance of SPC to the observed pathological conditions remains unclear [reviewed in (13)].

LPC is present as a component of oxidized low-density lipoprotein (LDL); it has been proposed to be involved in atherosclerosis and inflammatory diseases [reviewed in (14, 42)]. It can be produced from phosphatidylcholine by phospholipase A_2. Compared to other signaling LPs, the physiological concentrations of LPC in body fluids including blood and ascitic fluid can be very high (5–180 μM), and LPC is capable of cell lysis at >30 μM, which suggests the operation of nonreceptor actions (that include toxicity) in addition to receptor-based activities in considering LPC actions [reviewed in (14, 42)]. Possible high-affinity receptors for both lipids have been reported; these include three high-affinity SPC receptors (OGR1, GPR4, and GPR12) and one high-affinity LPC receptor (G2A) (44–47). The identity of OGR1 and GPR4 as bona fide SPC receptors is currently unclear in view of a recent report finding them unresponsive to SPC (48), and this receptor group in general requires additional analysis. Further characterization of these receptors will clarify ambiguities, and they are reviewed while noting these caveats.

2. LYSOPHOSPHOLIPID RECEPTORS

A growing number of receptors are reported to interact with lysophospholipids, albeit with variable apparent affinities. The best characterized are those for LPA (LPA$_1$-LPA$_4$) and S1P (S1P$_1$-S1P$_5$ and possible low-affinity interactions with GPR12). In addition, provisional identifications were reported for SPC receptors (OGR1, GPR4, GPR12, S1P$_1$-S1P$_5$, and G2A) and LPC receptors (G2A and GPR4). The receptors and relative ligand interactions are listed in Figure 2. Historical details on identification of the first LP receptors were previously covered (4, 25). Many of the LP receptor genes have been referred to by different names, such as *EDGs* (*Endothelial Differentiation Genes*). However, this review adopts the recommended IUPHAR nomenclature that is based on the optimal biological ligand for a given receptor (Table 1) (3). The LP receptors generally share high amino acid similarity within a ligand group and also share some similarity with other GPCRs for structurally related bioactive lipids, such as platelet-activating factor (PAF; 1-*O*-alkyl-2-acetyl-*sn*-glycero-3-phosphorylcholine), psychosine (Figure 1), and endogenous cannabinoids [reviewed in (4)].

Here we review the basic characteristics of each LP receptor, which include gene structures, signaling properties, tissue distribution, cellular functions, and in vivo roles, particularly as revealed by genetic-null studies.

2.1 Lysophosphatidic Acid Receptor 1

LPA_1 (previously called VZG-1/EDG-2/mrec1.3) was the first identified, high-affinity receptor for LPA (1) [reviewed in (4)]. The mammalian (human, mouse, and rat) lpa_1 genes encode 41-kDa proteins consisting of 364 amino acids with 7 putative transmembrane domains (Table 1). The mouse receptor is encoded by two of five exons with a characteristic conserved intron in transmembrane domain 6 that is shared with lpa_2 and lpa_3, and amino terminus isoforms generated by alternative exon usage exist (49). Functional analyses with mammalian heterologous receptor expression systems reveal the multi-functionality of this receptor [reviewed in (4, 25)]. LPA_1 couples with three types of G proteins, $G_{i/o}$, G_q, and $G_{12/13}$ (50, 51). Through activation of these G proteins, LPA induces a range of cellular responses through LPA_1: cell proliferation, serum-response element (SRE) activation, MAPK activation, AC inhibition, PLC activation, Ca^{2+} mobilization, Akt activation, and Rho activation (Table 2) [reviewed in (4, 25)].

Wide expression of lpa_1 is observed in adult mice, with clear presence in testis, brain, heart, lung, small intestine, stomach, spleen, thymus, and skeletal muscle (25). Similarly, human tissues also express lpa_1; it is present in brain, heart, placenta, colon, small intestine, prostate, testis, ovary, pancreas, spleen, kidney, skeletal muscle, and thymus (52). In situ hybridization studies reveal varied patterns of expression within a single tissue (1, 53, 54).

The nervous system is a major locus for lpa_1 expression; there it is spatially and temporally regulated throughout brain development [reviewed in (4, 25)]. Its embryonic central nervous system (CNS) expression is restricted to the neocortical neurogenic region called ventricular zone, which disappears at the end of cortical neurogenesis, just before birth (53). During postnatal life, lpa_1 expression is apparent in and around developing white matter tracts, and its expression coincides with the process of myelination (53). In situ hybridization and immunohistochemistry show that oligodendrocytes, the myelinating cells in the CNS, express lpa_1 in mammals (53, 55, 56). In addition, Schwann cells, the myelinating cells of the peripheral nervous system, also express lpa_1 (57), which is involved in regulating Schwann cell survival and morphology (57, 58). These observations identify important functions for receptor-mediated LPA signaling in neurogenesis, cell survival, and myelination.

The targeted disruption of lpa_1 in mice revealed unanticipated in vivo functions of this receptor (59). The $lpa_1^{(-/-)}$ mice show $\sim 50\%$ lethality in the perinatal period in a mixed genetic background. Survivors have reduced body size, craniofacial dysmorphism with flattened facies, and increased apoptosis in sciatic nerve Schwann cells (58, 59). Defective suckling, attributable to olfactory defects, likely accounts for neonatal lethality. Small fractions of $lpa_1^{(-/-)}$ embryos have exencephaly ($\sim 5\%$) or frontal cephalic hemorrhage ($\sim 2.5\%$). Loss of LPA responsivity in embryonic neuroblasts and fibroblasts demonstrates nonredundant functions and roles for lpa_1 in vivo (59, 60).

2.2 Lysophosphatidic Acid Receptor 2

LPA_2 (EDG-4 nonmutant form) was identified from GenBank homology searches of orphan GPCR genes [reviewed in (4)]. A related carboxyl-terminus mutant termed EDG-4 that was isolated from a neoplasm (52) is not present in wild-type genomes, and it should not be confused with wild-type LPA_2. The mouse LPA_2 gene contains three exons with the coding region in exons two and three. As with LPA_1, LPA_2 couples with three types of G proteins, $G_{i/o}$, G_q, and $G_{12/13}$, to mediate LPA-induced cellular signaling (Table 2) [reviewed in (4, 25)]. Expression of lpa_2 is observed in the testis, kidney, lung, thymus, spleen, and stomach of adult mice (25) and in the human testis, pancreas, prostate, thymus, spleen, and peripheral blood leukocytes (52). Expression of lpa_2 is upregulated in various cancer cell lines, and several human lpa_2 transcriptional variants with mutations in the 3′-untranslated region have been observed (see Section 4.5).

Targeted deletion of lpa_2 in mice (60) does not result in any obvious phenotypic abnormalities. However, significant loss of normal LPA signaling (e.g., PLC activation, Ca^{2+} mobilization, and stress fiber formation) is observed in primary cultures of mouse embryonic fibroblasts (MEFs) (60). Creation of $lpa_1^{(-/-)}lpa_2^{(-/-)}$ double-null mice (60) does not reveal obvious additional phenotypic abnormalities beyond those attributable to $lpa_1^{(-/-)}$ except for a higher incidence of frontal cephalic hemorrhage (26% versus 2.5% in $lpa_1^{(-/-)}$ mice); however, more subtle phenotypes may be present (59, 60), along with effects observed under gain-of-function conditions (see Section 4.1). Importantly, many LPA-induced responses, which include cell proliferation, AC inhibition, PLC activation, Ca^{2+} mobilization, JNK and Akt activation, and stress fiber formation, are absent or severely reduced in double-null MEFs. All these responses, except for AC inhibition (AC inhibition is nearly abolished in $lpa_1^{(-/-)}$ MEFs), are only partially affected in either $lpa_1^{(-/-)}$ or $lpa_2^{(-/-)}$ MEFs (60). These results indicate that LPA_2 contributes to normal LPA-mediated signaling responses in at least some cell types.

2.3 Lysophosphatidic Acid Receptor 3

LPA_3 (EDG-7) was isolated as an orphan GPCR gene by degenerate polymerase chain reaction (PCR)-based cloning and homology searches: As with the three exon structure of the LPA_2 gene, LPA_3 is also encoded by exons two and three (61, 62) [reviewed in (4)]. The LPA_3 receptor is distinct from LPA_1 and LPA_2 in its ability to couple with $G_{i/o}$ and G_q but not $G_{12/13}$ (51) and is much less responsive to LPA species with saturated acyl chains (61, 62). Nonetheless, LPA_3 can mediate pleiotropic LPA-induced signaling that includes PLC activation, Ca^{2+} mobilization, AC inhibition/activation, and MAPK activation (51, 61, 62). LPA_3 has variable effects on AC that likely depend on cell type and expression levels.

This could explain the elevated intracellular cAMP levels seen in Sf9 insect cells compared to slightly decreased levels in neuronal cells or no change in cAMP levels in hepatoma cells (51, 61, 62). Additionally, LPA_3 does not couple to actomyosin machinery that produces cell rounding in neuronal cells in which $G_{12/13}$ and Rho are involved (51). Overexpression of LPA_3 in neuroblastoma cells leads, surprisingly, to neurite elongation, whereas that of LPA_1 or LPA_2 results in neurite retraction and cell rounding when stimulated with LPA (51). Null receptor mutations for LPA_3 have not yet been reported. Expression of lpa_3 is observed in adult mouse testis, kidney, lung, small intestine, heart, thymus, and brain (25). In humans, it is found in the heart, pancreas, prostate, testis, lung, ovary, and brain (frontal cortex, hippocampus, and amygdala) (61, 62).

2.4 Lysophosphatidic Acid Receptor 4

LPA_4 (p2y$_9$/GPR23) was identified from orphan GPCR gene libraries within another evolutionary branch of the LP receptor superfamily (Figure 2). Unlike the other three LPA receptors, LPA_4 is encoded by a single exon. The orphan receptor, p2y$_9$/GPR23, is a functional high-affinity LPA receptor (K_d = 45 nM) that is now classified as LPA_4 (63). It is of divergent sequence compared to LPA_1-LPA_3 with closer similarity to the PAF receptor. LPA_4 mediates LPA-induced Ca^{2+} mobilization and cAMP accumulation, and functional coupling to G_s for AC activation is probable, although coupling preferences to other G proteins are currently unknown. Among 16 human tissues tested with quantitative real-time PCR, the lpa_4 gene is expressed at very high levels in the ovary and, to a much lesser extent, in the pancreas, thymus, and human kidney and skeletal muscle (63). Its physiological roles are currently unknown.

2.5 Sphingosine 1-Phosphate Receptor 1

$S1P_1$ (EDG-1/LP_{B1}) was the first identified S1P receptor that was initially isolated as an orphan GPCR in human endothelial cells, and it was later shown to encode a high-affinity (K_d = 8 nM) S1P receptor. The $S1P_1$ gene contains two exons with the coding region entirely on exon two [reviewed in (4)]. The $S1P_1$ receptor primarily couples with PTX-sensitive $G_{i/o}$ proteins and mediates S1P-induced MAPK activation, AC inhibition, PLC activation, Ca^{2+} mobilization, cell aggregation, Rac (and often Rho) activation, and cell migration [reviewed in (4, 15–17)]. Akt-mediated phosphorylation of $S1P_1$ is required for Rac activation, cortical actin assembly, and cell migration, but not for $G_{i/o}$-dependent signaling in endothelial cells (64). $S1P_1$ mutants lacking an intracellular Akt phosphorylation site (Thr^{236} in human, mouse, and rat) act as dominant negatives and inhibit S1P-induced chemotaxis and angiogenesis (64). Computational modeling suggests that basic amino acids, Arg^{120} and Arg^{292}, may form an ion-pair with the phosphate group of S1P, whereas the acidic Glu^{121} residue forms an ion pair with the ammonium moiety of S1P (65).

Several lines of evidence suggest that $S1P_1$ signaling is involved in PDGF-induced cellular responses. First, PDGF-induced Src activation, focal adhesion kinase activation, and cell migration are defective in $slp_1^{(-/-)}$ MEFs, yet neither PDGF receptor autophosphorylation nor DNA synthesis are altered in this setting (66, 67). Second, PDGF activates sphingosine kinase, inducing its membrane translocation (66), which may increase S1P levels in local areas where the $S1P_1$ receptor is present (68, 69). Third, immunoprecipitation experiments suggest possible protein interactions between the PDGF receptor and $S1P_1$ (68, 69), although this finding may not be universal and requires further examination (70).

Expression of slp_1 is pervasive, including spleen, brain, heart, lung, adipose tissues, liver, thymus, kidney, and skeletal muscle (71, 72). The deletion of slp_1 in mice results in embryonic lethality (73). The $slp_1^{(-/-)}$ embryos appear to be normal by E11.5, but they are identifiable at E12.5 by their edematous yolk sac with less blood. All $slp_1^{(-/-)}$ embryos show hemorrhage at E12.5 to E14.5 and fail to survive beyond E14.5. Vasculogenesis and angiogenesis are normal, contrasting with vascular maturation that is incomplete because of defects in surrounding vascular smooth muscle cells (VSMCs)/pericytes in $slp_1^{(-/-)}$ embryos (73). S1P-induced Rac activation and cell migration are both defective in $slp_1^{(-/-)}$ MEFs, suggesting that defects in VSMC/pericyte migration result in vascular immaturity that leads to embryonic death (73). Recent studies utilizing cell type-specific slp_1 deletion in mice via a *Cre/loxP* system show that $S1P_1$ receptors in vascular endothelial cells (VECs), rather than those in VSMCs, are responsible for the constitutive deletion phenotype (74) (see Section 4.2).

2.6 Sphingosine 1-Phosphate Receptor 2

$S1P_2$ (EDG-5/AGR16/H218/LP_{B2}) was first isolated as an orphan GPCR gene from rat cardiovascular and nervous systems. $S1P_2$ was later identified by many groups as a high-affinity ($K_d = 20–27$ nM) S1P receptor and low-affinity SPC receptor, and it is also encoded on a single exon [reviewed in (4)]. It couples with $G_{i/o}$, G_q, $G_{12/13}$, and possibly G_s, and it can mediate S1P-induced cell proliferation, cell survival, cell rounding, SRE activation, MAPK activation, AC activation, PLC activation, Ca^{2+} mobilization, and Rho activation [reviewed in (4, 15–17)]. $S1P_2$ inhibits Rac activity and prevents cell migration (75), contrasting with $S1P_1$ [reviewed in (17)], although it appears that $S1P_2$ can also produce counteracting signals (G_i versus $G_{12/13}$-Rho pathways) to influence cell migration (76).

Expression of slp_2 is widespread; it is present in heart, lung, thymus, brain, liver, kidney, spleen, adipose tissues, and all other tissues tested in adult mouse (71, 72) and in lung, heart, stomach, intestine, and adrenal glands in rat (77). During early stages of rat CNS development, slp_2 may be expressed in young, differentiating neuronal cell bodies and axons (78), although no significant expression is observed during embryonic mouse development (134).

Vertebrate slp_2 genetic models include a zebra fish (*Danio rerio*) mutant (see Section 4.2) as well as slp_2-null mice generated by two independent groups (79,

80). One report indicates that $slp_2^{(-/-)}$ mice do not show anatomical/histological defects but have spontaneous and sporadic seizures, which were occasionally lethal (79). Electroencephalographic abnormalities are also observed both during and between seizures, and whole-cell patch-clamp recording revealed a significant increase in the excitability of neocortical pyramidal neurons (79). Independently generated $slp_2^{(-/-)}$ mice also do not show gross phenotypic abnormalities nor evidence of seizure activity. However, a slight but statistically significant decrease in litter size is observed (80). The $slp_2^{(-/-)}$ mice are born at the expected Mendelian ratios without sexual bias, and they are fertile and healthy. However, a significant loss of S1P-induced intracellular signaling is observed in $slp_2^{(-/-)}$ MEFs. Wild-type MEFs express slp_1, slp_2, and slp_3 but neither slp_4 nor slp_5. The observation that S1P-induced Rho activation is significantly impaired while PLC activation/Ca^{2+} mobilization remains intact in $slp_2^{(-/-)}$ MEFs indicates that S1P$_2$ is critical for S1P-induced Rho activation but not for PLC activation/Ca^{2+} mobilization (80). The discrepancy in phenotypes, observed between two groups of $slp_2^{(-/-)}$ mice, may derive from differences in mouse genetic backgrounds used for analyses. All these results indicate that S1P$_2$ has identifiable signaling properties in MEFs, and it plays a minor yet discernible role in normal mouse development.

2.7 Sphingosine 1-Phosphate Receptor 3

S1P$_3$ (EDG-3/LP$_{B3}$) was isolated as an orphan GPCR gene by degenerate PCR-based cloning from a human genomic DNA library (81). Like S1P$_2$, S1P$_3$ is a high-affinity (K_d = 23–26 nM) S1P receptor and low-affinity SPC receptor and is encoded on a single exon [reviewed in (4)]. S1P$_3$ is evolutionarily more related to S1P$_1$ than to S1P$_2$ (Figure 2), but the intracellular signaling mediated by S1P$_3$ appears to resemble that of S1P$_2$, except for regulation of Rac (75) [reviewed in (17)]. The S1P$_3$ receptor couples with $G_{i/o}$, G_q, $G_{12/13}$, and possibly G_s, and it induces cell proliferation, cell survival, cell rounding, MAPK activation, SRE activation, AC activation/inhibition, PLC activation, Ca^{2+} mobilization, Rho activation, Rac activation, and cell migration [reviewed in (4, 15–17)].

Expression of slp_3 is widespread; it is present in the spleen, heart, lung, thymus, kidney, testis, brain, and skeletal muscle in adult mice (71, 72) and, in humans, in the heart, placenta, kidney, liver, pancreas, skeletal muscle, lung, and brain (81). Targeted disruption of slp_3 in mice results in no obvious abnormality (72). The $slp_3^{(-/-)}$ mice are born at the expected Mendelian ratios without sexual bias, are fertile, and appear healthy. The litter size from $slp_3^{(-/-)}$ crosses is modestly smaller (5.6 pups per litter) than that from $slp_3^{(+/-)}$ × wild-type crosses (7.5 pups per litter), but the reason for this is unknown (72). However, significant loss of S1P signaling is observed in $slp_3^{(-/-)}$ MEFs that normally express only three of the five slp genes ($slp_{1,2,3}$) (80). S1P-induced PLC activation and Ca^{2+} mobilization are abolished, but Rho activation and AC inhibition remain intact in $slp_3^{(-/-)}$ MEFs (72). In vitro functional analyses reveal that S1P$_3$ and S1P$_2$ can mediate many of the same S1P responses except for Rac regulation [reviewed in

(17)]. However, the results from analyses of MEFs demonstrate that $S1P_3$ is indispensable for PLC activation/Ca^{2+} mobilization. Nevertheless, the $s1p_3$ and $s1p_2$ genes are coexpressed in many mouse tissues (72), and functional redundancy might exist in vivo in other cell types. Production of $s1p_2^{(-/-)}s1p_3^{(-/-)}$ double-null mice (80) results in a clear phenotype of reduced litter sizes compared to single-null crosses, and most of the $s1p_2^{(-/-)}s1p_3^{(-/-)}$ pups do not survive beyond the first three postnatal weeks (only 1.2 pups per litter survived beyond three weeks) (80). Surprisingly, the double-null survivors display no obvious phenotype and are fertile. Although $s1p_3$ deletion alone does not significantly affect S1P-induced Rho activation, $s1p_2$ deletion partially impairs it, and deletion of both receptors eliminates it in MEFs (80). These data demonstrate that $S1P_2$ and $S1P_3$ can function redundantly in vivo, whereas elimination of both receptors in mice results in marked perinatal lethality and S1P signaling defects. The reasons for the lethality remain to be elucidated.

2.8 Sphingosine 1-Phosphate Receptor 4

$S1P_4$ (EDG-6/LP_{C1}) was isolated from in vitro differentiated human and murine dendritic cells (82) as an orphan GPCR gene. It was found to encode a high-affinity (K_d = 13–63 nM) receptor for S1P and a low-affinity receptor for SPC (83, 84) and is also encoded on a single exon [reviewed in (4)]. Its comparatively low amino acid sequence similarity compared to the other high-affinity S1P receptors (Figure 2) suggested that this receptor could have a distinct, preferred ligand (4, 85), and indeed, phytosphingosine 1-phosphate (4D-hydroxysphinganine 1-phosphate) has a 50-fold higher affinity for $S1P_4$ (at 1.6 nM) compared to S1P itself in at least one assay system (86). Virtually all other analyses on $S1P_4$ were conducted using S1P as the activating ligand, which could have relevance to reported signaling properties of $S1P_4$ in some biological settings. $S1P_4$ couples with $G_{i/o}$, $G_{12/13,}$ and possibly G_s, and it mediates S1P-induced MAPK activation, PLC activation, Ca^{2+} mobilization, AC activation, Rho activation, cytoskeletal rearrangement (stress fiber formation and cell rounding), and cell motility (72, 83, 84, 87).

Unlike $s1p_1$–$s1p_3$ receptors, $s1p_4$ expression is restricted in human and mouse to lymph node, spleen, lung, and thymus (72, 82). This expression pattern suggests potential roles of $S1P_4$ in the immune system. In vivo roles and functions of $S1P_4$ are still unknown.

2.9 Sphingosine 1-Phosphate Receptor 5

$S1P_5$ (EDG-8/LP_{B4}) was isolated as an orphan GPCR gene from rat pheochromocytoma 12 (PC12) cells (88), and it was later found to encode a high-affinity S1P receptor (K_d = 2–10 nM) and low-affinity SPC receptor (89–91) that also has a single exon coding region [reviewed in (4)]. Expression of $s1p_5$ is restricted to specific tissues: brain, spleen, and peripheral blood leukocytes in human and brain, skin, and spleen in rat and mouse (72, 88, 89). In rat brain, $s1p_5$ is predominantly expressed in white matter tracts and cells of oligodendrocyte lineage (89, 92), suggesting its potential roles in maturation and myelination of oligodendrocytes. $S1P_5$ can couple with $G_{i/o}$ and $G_{12/13}$, and it mediates S1P-induced AC inhibition and Ca^{2+} mobilization like the other S1P receptors. However, unlike the other S1P receptors, it mediates inhibition of MAPK activation/cell proliferation (72, 89, 90, 93). Physiological roles for $S1P_5$ have not been reported in the published literature.

2.10 G2A

The *G2A* gene was isolated as an orphan GPCR gene from mouse bone marrow cells during a search for genes that were induced by BCL-ABL tyrosine kinase oncogene (94). It was named because ectopic *G2A* expression resulted in accumulation of NIH3T3 cells at G2/M cell cycle boundary, which blocked further progression to mitosis (94). G2A has properties of a high-affinity LPC receptor (K_d = 65 nM) and low-affinity SPC receptor (K_d = 230 nM) (46).

G2A appears to couple with $G_{12/13}$ and mediates the activation of small GTPases (Rho, Rac, and Ras), stress fiber formation that requires G_{13} and Rho, and SRE activation (95, 96). It also mediates LPC-induced Ca^{2+} mobilization and MAPK activation in a pertussis toxin (PTX)-sensitive manner, suggesting coupling to $G_{i/o}$. G2A can also mediate LPC induced cell migration and apoptosis (46, 97). A recent report indicates that G2A expression also produces ligand-independent, PTX-insensitive activation of PLC and AC (probably via G_q and G_s, respectively) and that G2A couples with G_q and G_{13} for NF-κB activation (97). This observation has theoretical implications for identifying with certainty the specificity of LPC as the biological, high-affinity ligand for G2A. The current availability of a genetic-null mouse for G2A should help to clarify this issue (98).

G2A expression is restricted to lymphoid tissues, such as thymus, spleen, and bone marrow in mice (95). Overexpression of G2A may antagonize fibroblast transformation by the BCL-ABL (94) or induce it (95). Disruption of *G2A* in mice leads to immunological disorders (98). The $G2A^{(-/-)}$ mice develop enlarged spleens and lymph nodes with abnormal expansion of both T- and B-lymphocytes (98). Older $G2A^{(-/-)}$ mice (>1 year) develop a late-onset autoimmune syndrome, indicating that G2A also plays a role in the control of lymphocyte homeostasis (98).

2.11 Ovarian Cancer G Protein-Coupled Receptor 1

OGR1 was originally isolated as an orphan GPCR gene in HEY human ovarian cancer cells (99) and was later reported to encode a high-affinity (K_d = 33 nM) receptor for SPC (44). The mammalian *OGR1* gene encodes a 41-kDa protein consisting of 365 amino acids with relatively low similarity to any of the *lpa* or *s1p*-type receptors (Figure 2). It is notable that $S1P_1$-$S1P_5$ receptors can be activated by SPC with much lower affinities than by S1P. In contrast, OGR1 is only activated by SPC, not S1P (44). A recent report raises questions about the identity of OGR1 based on the reported inability of SPC to activate it (48).

OGR1 appears to couple with both $G_{i/o}$ and G_q, mediating SPC-induced Ca^{2+} mobilization via $G_{i/o}$ and MAPK activation via G_q. It also mediates SPC inhibition of cell proliferation (44). The *OGR1* gene is expressed in various human tissues: lung, placenta, brain, heart, spleen, testis, small intestine, and peripheral blood lymphocytes (99). Current information on functions and physiological roles of OGR1 is still limited [reviewed in (13, 14)]. Comparatively little independent confirmation of OGR1 as a bona fide SPC receptor exists, and based on the current literature, its identity as an SPC receptor should be considered provisional.

2.12 G Protein-Coupled Receptor 4

The GPR4 gene was originally isolated as an orphan human GPCR (100, 101) and was later reported to encode a high-affinity receptor for SPC (K_d = 36 nM) and low-affinity receptor for LPC (K_d = 159 nM) (45). As with OGR1, the identity of GPR4 is currently unclear in view of its unresponsiveness to SPC in recently reported assays (48).

The GPR4 protein shows the highest homology with OGR1 (~50% amino acid identity) (Figure 2) and appears to couple (at least) with $G_{i/o}$ to mediate cell proliferation, SRE activation, MAPK activation, Ca^{2+} mobilization, and cell motility (45) [reviewed in (13, 14)]. It is widely expressed in human tissues: ovary, liver, lung, kidney, heart, and lymph node (45, 101), and its physiological function remains to be determined. As with OGR1, its identity as a high-affinity SPC receptor should be considered provisional.

2.13 G Protein-Coupled Receptor 12

The GPR12 (GPCR01) gene was isolated from mouse cDNA by degenerate PCR during homology searches for orphan GPCR genes, and it was originally called *GPCR01* (102). GPR12 has been reported to be a moderate-affinity S1P receptor, unresponsive to SPC (103), and as a high-affinity SPC receptor (47). This discrepancy requires clarification.

There is little information regarding GPR12 signaling properties. SPC induces a G protein-gated, inwardly rectifying K^+ channel in *Xenopus* oocytes expressing GPR12 via PTX-sensitive pathways (47). S1P was reported to activate AC via PTX-insensitive pathways and Ca^{2+} mobilization via PTX-sensitive pathways, suggesting coupling with both G_s and $G_{i/o}$ (103). The *GPR12* gene is expressed in mouse brain, testis, and liver (102). In situ hybridization reveals *GPR12* expression in all areas of the developing mouse CNS from E14.5; this occurs especially in regions of neuroblast differentiation but not in areas of neuroblast proliferation. SPC stimulates cell proliferation and clustering in hippocampal HT22 cell lines, and it increases amounts of synaptophysin (a neuronal differentiation marker) in primary rat cortical cells that could be mediated by *GPR12*

(47), suggesting roles for SPC–GPR12 signaling in neuronal differentiation as observed in LPA-LPA receptor signaling (54, 104). However, the identity of this receptor must still be clarified beyond provisional classification as a receptor for S1P and/or SPC.

2.14 Candidates for Additional LP Receptors

A human orphan receptor GPR63 (accession no. NM_030784) was reported to act as a low-affinity receptor for S1P, dihydro-S1P, and dioleoylphosphatidic acid (105). The *GPR63* gene is highly expressed in human brain (especially in thalamus and caudatus), thymus, stomach, and small intestine, and a lower transcript abundance is present in kidney, spleen, pancreas, and heart (105). GPR63 overexpressed in CHO cells mediates S1P-induced Ca^{2+} mobilization and cell proliferation via PTX-insensitive pathways (105). In addition, GPR3 and GPR6, which are similar to GPR12, may mediate S1P-induced Ca^{2+} mobilization and AC activation (103), and further characterization is again necessary for their identification as functional S1P receptors. The orphan GPCRs continue to be an ample source for the discovery of new LP receptor members.

2.15 LP Receptor Agonists and Antagonists

As with many other GPCRs, LP receptors should be amenable to the development of highly specific and potent agonists or antagonists that have favorable pharmacokinetic, bioavailability, and metabolic characteristics. The anthelmintic drug suramin (106), despite its use in many earlier studies as a possible receptor inhibitor, should be considered to have poor specificity for LP receptors. Contrary to earlier speculations, receptor-mediated actions of LPs are stereo-selective (107). Currently available compounds represent a promising start to the development of useful chemical tools, although none can be considered definitive in determining receptor selectivity or biological functions, especially for studies in vivo. With these caveats in mind, a partial list of compounds with reported LP receptor selectivity includes an LPA$_1$ antagonist, 3-(4-[4-([1-(2-chlorophenyl)-ethoxy]carbonylamino)-3-methyl-5-isoxazolyl] benzylsulfanyl) propanoicacid (Ki16425) (108); an LPA$_1$ antagonist that is an ethanolamide derivative (109, 110); LPA$_2$ agonists that are decyl and dodecyl fatty alcohol phosphates (FAP-10 and FAP-12) (111); an LPA$_3$ agonist that is a phosphothionate analog of LPA (112); an LPA$_3$ agonist that is a monofluorinated analog of LPA (107); an LPA$_3$ antagonist DGPP 8:0 that is a diacylglycerol pyrophosphate (110, 113); and an S1P$_2$ antagonist, JTE-013, pyrazolopyridine (114). Perhaps most encouraging are studies on a nonselective S1P receptor agonist prodrug that is an analog of myriocin, called FTY720, and becomes active following phosphorylation by sphingosine kinase (Figure 1), because these studies mark the entry of major pharmaceutical companies into the LP receptor field (108, 115, 116). Appropriately validated compounds are essential for in vivo studies, particularly in view of potential off-target effects. Combining chemical compounds with genetic mutants for receptors and/or related signaling components offers an attractive strategy for validating compounds and revealing new biological functions.

3. NON-GPCR TARGETS FOR LYSOPHOSPHOLIPIDS

Recently, the nuclear hormone receptor peroxisome proliferator-activated receptor γ (PPARγ) was proposed as an intracellular receptor for LPA based on the ability of LPA to displace a synthetic PPARγ agonist (rosiglitazone) (117). The physiological significance of this observation for mechanistically explaining the effects of LPA signaling is currently unclear on the basis of several criteria. Specificity of LPA binding to PPARγ is absent, and numerous biological ligands including eicosanoids, anionic fatty acids, and components of oxidized LDL can also bind PPARγ (118). Similarly, LPA itself clearly acts at other loci beyond PPARγ. Independent molecular genetic studies, in which PPARγ deletion was coupled with lacZ expression (119) or analyzed as chimeras (120), demonstrated restricted expression and effects of PPARγ deletion, primarily in adipose tissues. By contrast, LPA effects are well known in many tissue types that are unaffected by PPARγ elimination or that do not express PPARγ. Morever, the loss of LPA signaling associated with LPA$_1$ and LPA$_2$ receptor deletion, along with the observed null phenotypes, are clearly not rescued by the concomitant expression of PPARγ (59, 60). These data do not detract from the observation that LPA can interact with PPARγ, and future studies should clarify its physiological significance.

It remains possible that other intracellular LPA receptors exist. The LPA$_1$ receptor can apparently be expressed in nuclear membranes to mediate LPA-induced signaling, which leads to proinflammatory gene expression (121). This finding suggests that the GPCR-type LPA receptors could theoretically serve as intracellular receptors. Further, the lipid bilayer of membranes can be the direct target of LPA action. Endophilin I mediates synaptic vesicle formation by its LPA acyl transferase activity that produces phosphatidic acid (PA) from LPA (122). The local balance between PA and LPA concentrations could affect membrane curvature, leading to membrane invagination and synaptic vesicle uncoating (123). Because endophilins are essential for clathrin-mediated endocytosis and form complexes with various signaling molecules, which include cell surface receptors, metalloprotease disintegrins, and germinal center kinase-like kinase, local accumulation of LPA could result in altered cellular signaling [reviewed in (124)].

In comparison to LPA, the second messenger-like actions of S1P were first documented in the early 1990s. Molecular identification of S1P intracellular targets remains to be elucidated. S1P is rapidly produced intracellularly from sphingosine by sphingosine kinase upon cell activation with mitogens, such as PDGF and serum, and mobilizes Ca^{2+} from internal stores via inositol triphosphate (IP$_3$)-independent pathways (32, 125). SPC has been shown to mobilize

Ca^{2+} from internal stores by activating the IP_3 receptor and/or ryanodine receptor. In addition, an SPC-gated Ca^{2+} channel called sphingolipid Ca^{2+} release-mediating protein of endoplasmic reticulum (SCaMPER) (30, 126–128) has also been implicated, although some disagreement exists over the functional distribution of SCaMPER in endoplasmic reticulum (129) [reviewed in (130)]. In cardiomyocytes, however, SCaMPER is localized to the sarcotubular junction on the plasma membrane where connections between the transverse tubules and sarcoplasmic reticulum function to regulate cell calcium levels and mediate SPC-induced Ca^{2+} release (131). S1P and SPC seem to act differentially as possible second messengers. They mobilize Ca^{2+} from different internal pools that could be distinguished by sensitivity to thapsigargin pretreatment (132). S1P has also been proposed as a "calcium influx factor," which links internal calcium store depletion to downstream store-operated calcium entry (133). Superimposed on these studies is the action of both known and perhaps unknown GPCR-type S1P receptors, some of which might function intracellularly; future studies should provide mechanistic clarification.

4. BIOLOGY OF LYSOPHOSPHOLIPIDS

4.1 Nervous System

The nervous system is one of the major loci for LP receptor expression (25, 53, 72, 134). The expression profiles are correlated with neuronal development processes, such as neurogenesis, neuronal migration, neuritogenesis, and myelination [reviewed in (4)]. Exogenous application of LPA or S1P to neural cells induces responses that are relevant to both the development and function of the nervous system. Furthermore, both lipids exist in the brain at relatively high concentrations (28, 135). LPs affect most neural cell types, such as neural cell lines, neural progenitors, primary neurons, oligodendrocytes, Schwann cells, astrocytes, and microglia [reviewed in (136)]. Receptor-mediated LPA signaling induces a variety of cellular responses in all these cell types, and additional functions are being identified, such as roles in some forms of pain (137). By comparison, S1P-induced responses have thus far been demonstrated in only neural cell lines and glia, although a role for S1P signaling in other cells of neural origin is likely because several S1P receptors ($s1p_1$– $s1p_3$, $s1p_5$) are expressed in the developing and mature nervous system (72, 80, 134). Here we briefly review cellular effects of LPA on neural cell lines, neuroblasts, and neurons and discuss their biological relevance. Actions of LPA on other neural cell types, particularly Schwann cells, have been described elsewhere [reviewed in (4, 136)].

Historically, studies using peripheral nervous system cell lines hinted at roles for LPA in the nervous system (138, 139). Exposure of those cells to LPA produced a rapid retraction of their processes resulting in cell rounding, which was, in part, mediated by polymerization of the actin cytoskeleton. Combined with the fact that LPA was present in serum at high concentrations, it has been proposed that LPA may cause neuronal degeneration under pathological conditions when the blood-brain barrier is damaged and serum components leak into the brain (140). LPA receptor gene expression is not clearly detectable in adult neurons (53), and therefore, direct influences of LPA on adult CNS neurons remain unclear, although up-regulation of LPA receptor expression following insults remains a possibility. Immortalized CNS neuroblast cell lines from the cerebral cortex also display retraction responses to LPA exposure through Rho activation and actomyosin interactions (1, 51, 141, 142). Neurite retraction occurs in minutes following LPA exposure; however, an even earlier response has been identified in LPA-responsive cells that involves loss of membrane ruffling associated with actin depolymerization (142). This phenomenon is independent of Rho activation and requires interactions between Ca^{2+} and α-actinin, an actin-cross-linking protein (142). These cellular phenomena may reflect collapse and retraction of basal neuroblast processes that are perhaps involved in the proper reorganization of the actin cytoskeleton following LPA exposure.

Growth cone collapse is a well-known cellular response triggered by extracellular stimuli, occurring during axonal pathfinding or migration of differentiating neurons. LPA signaling might influence neurite formation and migration of differentiating neurons. For example, low concentrations of LPA induce repulsive turning of extending growth cones for *Xenopus* spinal cord neurons (143). Ubiquitin-dependent proteolysis via proteosomes is required for LPA-induced growth cone collapse in retinal neurons in which the apoptotic pathway involving p38 kinase and caspase-3 plays an important role (144, 145). In addition to a classical view of cytoskeletal rearrangement in growth cone motility, these observations raise the intriguing possibility that regulation of protein degradation is involved in not only collapse but also in turning of growth cones that involve LPA signaling.

Neuroprogenitor cells that express lpa_1 appear to include precursor cells of neurons and glia. As observed in neural cell lines, LPA induces cellular and nuclear rounding and migration, accompanied by the formation of fine retraction fibers (54). These morphological changes resemble the well known rounding-up phase of "to-and-fro" nuclear movement present in the cerebral cortical ventricular zone called "interkinetic nuclear migration" (146). LPA receptor-mediated cell proliferation is only modestly increased by LPA and is distinct from the more prominent proliferative responses produced by basic fibroblast growth factor (bFGF) (59). LPA also stimulates depolarizing ionic conductances in cortical neuroblasts (147), consisting of increases in both chloride and nonselective cation conductances.

LPA can be produced by postmitotic cortical neurons (54), and the interaction between neuroblasts and neurons by means of released LPA is likely involved in cortical neurogenic processes. Expression of lpa_2 in cortical plate or differentiating neurons (104, 134) combined with the effects of LPA on their cytoskeleton suggest a role of LPA in neuronal migration and/or neurite outgrowth. In young differentiating cortical neurons, LPA induces retraction of neurites or lamellar structures, which could be related to growing axons, dendrites, or leading processes (104). The overall effects of LPA signaling in the embryonic cerebral cortex have been recently addressed in a culture settting that maintains the normal organization and growth characteristics observed in utero (148). Exogenous LPA exposure produces increases in cell number, width, and cortical folds resembling sulci and gyri. Each of these phenomena is absent in embryonic cerebral cortices from mice that are null for *lpa*$_1$ and *lpa*$_2$, demonstrating the LPA receptor dependence of this marked, growth phenomenon. Collectively, these data demonstrate multiple functions for LPA signaling during embryonic brain development, and it is probable that an equivalent biology exists for S1P (134) and perhaps related ligands.

4.2 Angiogenesis and Cardiovascular Development

Angiogenesis is the formation of new capillary networks from preexisting vasculature by sprouting and/or splitting of capillaries; it involves coordinated proliferation, migration, adhesion, differentiation, and assembly of both VECs and their surrounding VSMCs. This process is also implicated in physiological processes, which include wound healing and myocardial angiogenesis after ischemic injury, and is precisely controlled by (both angiogenic and antiangiogenic) protein growth factors, such as vascular endothelial growth factor (VEGF), bFGF, and PDGF. It is also influenced by the lysophospholipids, LPA and S1P. Dysregulation of angiogenesis can lead to pathological conditions such as atherosclerosis, hypertension, solid tumor growth, rheumatoid arthritis, and diabetic retinopathy [reviewed in (5, 149)].

Several lines of evidence suggest that S1P receptor-mediated signaling plays a major regulatory role in angiogenesis. First, S1P induces both proliferation and migration of VECs (36, 64, 114, 150–152) while inducing proliferation but inhibiting migration of VSMCs (114, 151, 153, 154). VEC migration is inhibited by antisense oligonucleotides against *s1p*$_1$ or *s1p*$_3$ (150) or by *s1p*$_2$ overexpression (151). Using an S1P$_2$-specific antagonist (JTE-013), S1P-induced migration of VECs is enhanced, and inhibition of VSMCs migration is reversed (114). Overexpression of *s1p*$_1$ in VSMCs enhances both mitogenic and migration responses to S1P (151, 153). Furthermore, S1P protects VECs from serum-deprived apoptosis by nitric oxide production through both S1P$_1$ and S1P$_3$ receptors (155). These results implicate S1P receptor signaling in VEC/VSMC proliferation and migration. Second, S1P stimulates the formation and maintenance of VECs assembly/integrity by activating both S1P$_1$ and S1P$_3$. S1P-induced VEC adherens junction assembly and cell barrier integrity are blocked by antisense oligonucleotides against *s1p*$_1$ or *s1p*$_3$ (156, 157). Expression of a dominant negative S1P$_1$ mutant inhibits S1P-induced VEC assembly and migration (64). Although *s1p*$_2$ is not expressed or is expressed at only low levels in VECs (114, 150, 151, 156), overexpression of *s1p*$_2$ augments cell barrier integrity of VECs (157). Third, slp_1– slp_3 transcripts are found in embryonic brain blood vessels (134).

Different S1P responses can be explained by altered S1P receptor expression patterns in VECs and VSMCs and by differential regulation of the two small GTPases, Rho and Rac (Table 2). In general, VEC expresses both slp_1 and slp_3 but not slp_2 (36, 150–152, 154, 156), whereas VSMCs express all three receptors with high slp_2 expression levels (114, 151, 153, 154). Functional analyses revealed that these receptors differentially regulate Rho and Rac; $S1P_1$ mediates the activation of Rac [and often Rho (150)], $S1P_2$ mediates Rho activation and Rac inhibition, and $S1P_3$ mediates the activation of both Rho and Rac (Table 2). Rac activation (via $S1P_1$) is required for migration of both VEC and VSMC (64), whereas both Rac-mediated cortical actin assembly (via $S1P_1$) and Rho-mediated stress-fiber formation (via $S1P_3$) are essential for VEC adherens junction assembly (64, 156, 157). $S1P_1$ plays a primary role in angiogenesis by its potent activation of Rac (64, 153), potentially through the intimate interplay with PDGF (as mentioned in Section 2.5). Negative regulation of Rac by $S1P_2$ in VSMCs underlies the S1P inhibitory response in migration (114, 151).

Although LPA regulates VSMC functions (158, 159), the roles of LPA signaling in angiogenesis appear to be in pathological conditions, such as wound healing (see next section) and atherosclerosis [reviewed in (160, 161)], rather than in a normal or basal conditions. However, frontal cephalic hemorrhages are observed in a significant percentage of $lpa_1^{(-/-)}$ or $lpa_1^{(-/-)}lpa_2^{(-/-)}$ embryos (59, 60) (see Sections 2.1 and 2.2), suggesting potential roles for LPA receptor signaling in some aspect of normal angiogenesis/vascular maturation.

The cardiovascular system is another major locus for LP receptor expression; at least five LP receptor genes (lpa_1, lpa_3, slp_1– slp_3, along with $OGR1$ and $GPR4$) are expressed in mammalian heart (25, 45, 72, 99). Vasoregulatory actions of LPA were described as early as 1978 in which intravenous LPA application produced hypertension in rats/guinea pigs but hypotension in cats/rabbits (21). Later, S1P was also shown to regulate the cardiovascular system; intravenous administration of S1P decreased heart rates, ventricular contraction, and blood pressure in rats (162). The effects of LPA and S1P are predominantly receptor mediated. Direct evidence for S1P receptor signaling in angiogenesis and cardiovascular development comes from the phenotype of two genetic-null studies in mice and zebra fish (73, 163) [reviewed in (5, 6, 35)]. The slp_1-null embryos die in utero because of defective vascular maturation in which VSMCs/pericytes do not migrate to surround the vessels (see Section 2.5). In zebra fish, the homolog of mammalian slp_2 (the *mil* gene) was mutated (163), resulting in a cardiac phenotype. Normally cardiac muscle progenitor cells migrate from bilateral positions toward the dorsal midline and fuse to form a single heart tube. However, Mil mutant progenitors do not migrate to the midline, leading to lethality for lack of proper blood circulation. Transplanted mutant progenitors migrate normally in wild-type embryos, whereas transplanted wild-type progenitors do not migrate in the Mil mutant, suggesting defects in the guidance of progenitor cell migration by surrounding paraxial cells (163). By contrast, deletion of slp_2 in mice does not produce discernible cardiovascular defects, and $slp_2^{(-/-)}$ mice are alive and grossly normal (see Section 2.6) (79, 80). Deletion of both slp_2 and slp_3 in mice leads to marked perinatal lethality, despite absence of gross anatomical and histological defects in the rare surviving double-null mice (see Section 2.7) (80). Deletion of lpa_1, lpa_2, or both in mice does not reveal obvious cardiac defects (59, 60).

4.3 Wound Healing

When wounded, damaged blood vessels activate platelets. The activated platelets play pivotal roles in subsequent repair processes by releasing bioactive mediators to induce cell proliferation, cell migration, blood coagulation, and angiogenesis. LPA and S1P are likely to be such mediators because both are released from activated platelets (36, 37) [reviewed in (10)]; this induces platelet aggregation along with mitogenic/migration effects on the surrounding cells, such as endothelial cells, smooth muscle cells, fibroblasts, and keratinocytes (164). Indeed, topical LPA application to cutaneous wounds in mice promotes repair processes (wound closure and increased neoepithelial thickness) by increasing cell proliferation/migration without affecting secondary inflammation (165, 166). However, normal wound closure is observed in $lpa_1^{(-/-)}lpa_2^{(-/-)}$ mice (60), suggesting the potential involvement of other LPA receptors and/or nonreceptor-mediated mechanisms in this process. S1P has not been reported in wound healing in vivo. S1P may have paradoxical effects on cutaneous wound healing, because S1P induces fibroblast proliferation and keratinocyte migration while inhibiting keratinocyte proliferation, a critical step for reepithelialization of the wound (167). All five high-affinity S1P receptor genes, slp_1– slp_5, are expressed in keratinocytes, and S1P inhibition of keratinocyte proliferation is partially inhibited by PTX pretreatment. In addition, microinjection of S1P inhibits keratinocyte proliferation. These results suggest that both S1P receptor signaling and perhaps intracellular actions may mediate this effect (167). S1P could also modulate actions of other mediators released from platelets (168). In contrast, SPC inhibits platelet activation (not via specific LP receptors) (43), and its effect on wound healing is unknown.

4.4 Immunity

Consistent with their roles as pleiotropic lipid mediators, LPA and S1P have been shown to regulate immunological responses by modulating activities/functions of immune cells such as T-/B-lymphocytes and macrophages [reviewed in (169–172)]. These immune cells and/or other cells involved with their normal function express several LP receptors, and their activities are regulated differentially by the expressed LP receptor subtypes. Furthermore, expression patterns of LP receptors can be altered by cell activation [reviewed in (169–171)]. Through LP receptors, T-cell migration and immune responses can be influenced with high receptor sensitivity (173). LPA and S1P might also protect T cells from apoptosis through LPA_1 in combination with LPA_2, and $S1P_2$ in combination with $S1P_3$, respectively (174). LPA induces migration of and inhibits interleukin-2 (IL-2) production from unstimulated T cells that predominantly express lpa_2. Mitogen activation of T cells leads to down-regulation of lpa_2 as well as up-regulation of lpa_1 expression. Therefore, in activated T cells, LPA inhibits cell migration but activates IL-2 production/cell proliferation through LPA_1 (175, 176). S1P has been reported to stimulate migration of T cells that express $s1p_1$ and $s1p_4$ under some conditions. T cell receptor-mediated activation of T cells suppresses expression of both $s1p_1$ and $s1p_4$, and it has been reported to eliminate their migration responses to S1P (177).

Immunomodulatory actions of S1P on lymphocytes represent a particularly active area of investigation [reviewed in (172, 178)]. The phosphorylated metabolite of FTY720 (Figure 1), a novel immunomodulator that causes lymphopenia, has been shown to act through S1P receptors (115, 116). FTY720 is being evaluated in human transplant studies in which it induces lymphocyte sequestration. Assayed models of autoimmunity and transplantation indicate that sequestered lymphocytes may be prevented from reacting/migrating to inflammatory chemokines at graft sites (172, 179–181). Unlike current immunosuppressive drugs, such as cyclosporine, FTY720 neither inhibits T-cell activation/proliferation nor impairs general immunological responses (181, 182).

FTY720 is phosphorylated by sphingosine kinase in vivo and in vitro, and the phosphorylated form of FTY720 (FTY720-P in Figure 1) acts as a S1P receptor agonist (115, 116). FTY720-P can bind to each of four S1P receptors ($S1P_1$, $S1P_3$–$S1P_5$) and activates them with varied potency and efficacy compared to S1P. Because S1P concentrations are exquisitely regulated both by synthesis and degradation, S1P homeostasis may contribute to the normal status of lymphocyte homing. S1P receptors as well as enzymes involved in S1P synthesis thus represent attractive immunoregulatory targets.

4.5 Ovarian Cancer and Preservation of Female Reproduction

Several lines of evidence suggest that abnormal LPA metabolism/signaling may contribute to the initiation and progression of ovarian cancers. First, LPA is present at significant concentrations (2–80 μM) in the ascitic fluid of ovarian cancer patients (183, 184). Ovarian cancer cells (OCCs) constitutively produce increased amounts of LPA as compared to normal ovarian surface epithelial cells (OSEs), the precursor of ovarian epithelial cancer (185). Elevated LPA levels are also detected in plasma from patients with early-stage ovarian cancers compared with controls, and therefore, the plasma LPA level might represent a potential biomarker for ovarian cancer (186). Second, LPA stimulates proliferation/survival of OCCs (187–189). In OCCs but not OSEs, LPA induces cell proliferation (189) and activates secretion of urokinase plasminogen activator, a critical component of the metastasis cascade (190). Overexpression of LPA-hydrolyzing lipid phosphate phosphohydrolase-3 in OCC lines decreases colony forming activity and tumor growth in vitro and in vivo (191). Third, the expression of lpa_2 is markedly increased in OCCs as compared with OSEs, whereas lpa_1 expression is not consistently different between OCCs and OSEs (189, 190). Overexpression of lpa_1 in OCCs results in apoptosis and anoikis (192). Markedly increased lpa_3 expression is observed in OCCs as compared with OSEs (185, 190, 193), and LPA_3 is required for LPA-induced OCC migration (194). Notably, the newly identified lpa_4 has the highest expression level in ovary among human tissues examined (63). These results indicate that LPA is involved in ovarian carcinogenesis in which LPA_2 and LPA_3 (and possibly LPA_4) could mediate LPA-induced OCC proliferation and possibly metastasis (193–195). It should be noted that the first reported human lpa_2 clone was derived from an ovarian tumor library (52) and contained a frame-shift mutation that produced 31 extra amino acids at its intracellular carboxyl terminal end, which could produce a gain of function mutant (171). Several 3′-untranslated region (or coding region) variants of the lpa_2 transcripts have been found in multiple tumors, suggesting oncogenic potential by altered LPA_2 stability/signaling (196).

The source of LPA in ascites is unclear but may include macrophages, lymphocytes, mesothelial cells, or OCCs themselves (184, 185). Recent studies indicate that human plasma lysoPLD, one of the LPA-producing enzymes, is identical to autotaxin, a cell motility-stimulating ectophosphodiesterase implicated in tumor progression (197, 198) [reviewed in (26)]. Several cancer cell lines express autotaxin and release significant amounts of LPC, a substrate of autotaxin, thus producing LPA in culture media (197). LysoPLD activity in human serum is increased in normal pregnant women at the third trimester of pregnancy and to a higher extent in patients at risk for preterm delivery (198, 199). LPA can also be found in the follicular fluid of healthy individuals (200) and induces MAP kinase activation in ovarian theca cells that express lpa_1 (201). These results suggest that, in addition to its potential roles in ovarian cancer progression, LPA may have physiological functions in normal ovarian as well as in normal reproductive processes such as pregnancy and parturition.

Stimulatory roles of LPA in cancer progression were also described in other cancers. LPA is produced from and induces proliferation of prostate cancer cell lines (202, 203). LPA induces human colon carcinoma DLD1 cell proliferation, migration, adhesion, and secretion of angiogenic factors, possibly through LPA_1 (204). In other human colon carcinoma cells lines (HT29 and WiDR), LPA enhances cell proliferation and secretion of angiogenic factors, possibly through LPA_2 but not cell migration and adhesion (204). The genetic or pharmacological manipulation of LPA metabolism, specific blockade of receptor signaling, and inhibition of downstream signal transduction, represent possible approaches for cancer therapies [reviewed in (18, 205)].

In contrast with tumor-promoting effects of LPA, S1P could protect female germ cells from cancer therapy. Chemotherapy of cancers in young female patients induces oocyte apoptosis that leads to early ovarian failure and premature onset of menopause. Total body irradiation for leukemia or lymphoma before bone marrow transplantation may cause complete oocyte depletion [reviewed in (206–208)]. S1P inhibits chemotherapy-induced oocyte apoptosis (209) and suppresses radiation-induced oocyte loss in vivo without propagating genomic damage in offspring (210), raising the possibility for lipid-based therapy in clinical oocyte preservation.

CONCLUDING REMARKS

The identification of LP receptors over the last decade has provided a mechanistic foundation from which the pleiotropic effects attributed to LPs can be understood. Continued "deorphaning" of GPCRs—and potentially, non-GPCRs—will surely result in new additions to the receptors discussed in this review. How receptor mechanisms interface with LP metabolism, particularly synthesis and degradation, will become increasingly clear with the continued identification of relevant enzymes involved in these processes. Similarly, new aspects of receptor-activated intracellular signaling will also be determined. A major current challenge is understanding the physiological and pathophysiological roles played by single LP receptor subtypes, as well as combinations of receptors, which include those from different ligand classes. This challenge will be aided by the continued generation of receptor mutants, combined with LP receptor-specific agonists and antagonists that have favorable properties allowing their use in vivo. This new information will be important for understanding the fundamental biology of LP receptors and for the development of human therapies based on targeting LP receptors and components of their signaling pathways.

ACKNOWLEDGMENTS

We apologize to our colleagues for being unable to cite many excellent references because of editorial length restrictions. Isao Ishii and Nobuyuki Fukushima contributed equally to this review. Our work was supported by the NIMH and The Helen L. Dorris Institute for the Study of Neurological and Psychiatric Disorders of Children and Adolescents (J.C.), by a Grant-in-Aid for Young Scientists (No. 15790066) from the MEXT of Japan, and by grants from Uehara Memorial Foundation, Yamanouchi Foundation for Research on Metabolic Disorders, and the ONO Medical Research Foundation (I.I.). We thank Dr. Brigitte Anliker and Christine Higgins for reading the manuscript and Dr. Hugh Rosen for helpful comments.

The *Annual Review of Biochemistry* is online at http://biochem.annualreviews.org

LITERATURE CITED

1. Hecht JH, Weiner JA, Post SR, Chun J. 1996. *J. Cell Biol.* 135:1071–83
2. Kimura Y, Schmitt A, Fukushima N, Ishii I, Kimura H, et al. 2001. *J. Biol. Chem.* 276:15208–15
3. Chun J, Goetzl EJ, Hla T, Igarashi Y, Lynch KR, et al. 2002. *Pharmacol. Rev.* 54:265–69
4. Fukushima N, Ishii I, Contos JJA, Weiner JA, Chun J. 2001. *Annu. Rev. Pharmacol. Toxicol.* 41:507–34
5. Osborne N, Stainier DY. 2003. *Annu. Rev. Physiol.* 65:23–43
6. Spiegel S, Milstien S. 2003. *Nat. Rev. Mol. Cell Biol.* 4:397–407
7. Xu Y, Xiao YJ, Zhu K, Baudhuin LM, Lu J, et al. 2003. *Curr. Drug Targets Immune Endocr. Metab. Disord.* 3:23–32
8. Xie Y, Gibbs TC, Meier KE. 2002. *Biochim. Biophys. Acta* 1582:270–81
9. Pages C, Simon MF, Valet P, Saulnier-Blache JS. 2001. *Prostaglandins Other Lipid Mediat.* 64:1–10
10. Yatomi Y, Ozaki Y, Ohmori T, Igarashi Y. 2001. *Prostaglandins Other Lipid Mediat.* 64:107–22
11. Okajima F. 2002. *Biochim. Biophys. Acta* 1582:132–37
12. Spiegel S, Kolesnick R. 2002. *Leukemia* 16:1596–602
13. Heringdorf DMZ, Himmel HM, Jakobs KH. 2002. *Biochim. Biophys. Acta* 1582:178–89
14. Xu Y. 2002. *Biochim. Biophys. Acta* 1582:81–88
15. Kluk MJ, Hla T. 2002. *Biochim. Biophys. Acta* 1582:72–80
16. Siehler S, Manning DR. 2002. *Biochim. Biophys. Acta* 1582:94–99
17. Takuwa Y. 2002. *Biochim. Biophys. Acta* 1582:112–20
18. Mills GB, Moolenaar WH. 2003. *Nat. Rev. Cancer* 3:582–91
19. Vogt W. 1963. *Biochem. Pharmacol.* 12:415–20
20. Pieringer RA, Bonner H Jr, Kunnes RS. 1967. *J. Biol. Chem.* 242:2719–24
21. Tokumura A, Fukuzawa K, Tsukatani H. 1978. *Lipids* 13:572–74
22. Gerrard JM, Kindom SE, Peterson DA, Peller J, Krantz KE, White JG. 1979. *Am. J. Pathol.* 96:423–38
23. Tokumura A, Fukuzawa K, Yamada S, Tsukatani H. 1980. *Arch. Int. Pharmacodyn. Ther.* 245:74–83
24. Moolenaar WH. 2000. *Ann. NY Acad. Sci.* 905:1–10
25. Contos JJA, Ishii I, Chun J. 2000. *Mol. Pharmacol.* 58:1188–96
26. Moolenaar WH. 2002. *J. Cell Biol.* 158:197–99
27. Baker DL, Desiderio DM, Miller DD, Tolley B, Tigyi GJ. 2001. *Anal. Biochem.* 292:287–95
28. Sugiura T, Nakane S, Kishimoto S, Waku K, Yoshioka Y, et al. 1999. *Biochim. Biophys. Acta* 1440:194–204
29. Bandoh K, Aoki J, Taira A, Tsujimoto M, Arai H, Inoue K. 2000. *FEBS Lett.* 478:159–65
30. Ghosh TK, Bian J, Gill DL. 1990. *Science* 248:1653–56
31. Zhang H, Desai NN, Olivera A, Seki T, Brooker G, Spiegel S. 1991. *J. Cell Biol.* 114:155–67
32. Olivera A, Spiegel S. 1993. *Nature* 365:557–60
33. Olivera A, Rosenfeldt HM, Bektas M, Wang F, Ishii I, et al. 2003. *J. Biol. Chem.* 278:46452–60
34. Hla T. 2003. *Pharmacol. Res.* 47:401–7
35. Yang AH, Ishii I, Chun J. 2002. *Biochim. Biophys. Acta* 1582:197–203
36. Yatomi Y, Ohmori T, Rile G, Kazama F, Okamoto H, et al. 2000. *Blood* 96:3431–38
37. Sano T, Baker D, Virag T, Wada A, Yatomi Y, et al. 2002. *J. Biol. Chem.* 277:21197–206
38. Herr DR, Fyrst H, Phan V, Heinecke K,

Georges R, et al. 2003. *Development* 130:2443–53

39. Mendel J, Heinecke K, Fyrst H, Saba JD. 2003. *J. Biol. Chem.* 278:22341–49
40. Li GC, Foote C, Alexander S, Alexander H. 2001. *Development* 128:3473–83
41. Renault AD, Starz-Gaiano M, Lehmann R. 2002. *Gene Expr. Patterns* 2:337–45
42. Kabarowski JH, Xu Y, Witte ON. 2002. *Biochem. Pharmacol.* 64:161–67
43. Altmann C, Heringdorf DMZ, Boyukbas D, Haude M, Jakobs KH, Michel MC. 2003. *Br. J. Pharmacol.* 138:435–44
44. Xu Y, Zhu K, Hong GY, Wu WH, Baudhuin LM, et al. 2000. *Nat. Cell Biol.* 2:261–67
45. Zhu K, Baudhuin LM, Hong G, Williams FS, Cristina KL, et al. 2001. *J. Biol. Chem.* 276:41325–35
46. Kabarowski JH, Zhu K, Le LQ, Witte ON, Xu Y. 2001. *Science* 293:702–5
47. Ignatov A, Lintzel J, Hermans-Borgmeyer I, Kreienkamp HJ, Joost P, et al. 2003. *J. Neurosci.* 23:907–14
48. Ludwig M-G, Vanek M, Guerini D, Gasser JA, Jones CE, et al. 2003. *Nature* 425:94–98
49. Contos JJA, Chun J. 1998. *Genomics* 51:364–78
50. Fukushima N, Kimura Y, Chun J. 1998. *Proc. Natl. Acad. Sci. USA* 95:6151–56
51. Ishii I, Contos JJA, Fukushima N, Chun J. 2000. *Mol. Pharmacol.* 58:895–902
52. An SZ, Bleu T, Hallmark OG, Goetzl EJ. 1998. *J. Biol. Chem.* 273:7906–10
53. Weiner JA, Hecht JH, Chun J. 1998. *J. Comp. Neurol.* 398:587–98
54. Fukushima N, Weiner JA, Chun J. 2000. *Dev. Biol.* 228:6–18
55. Allard J, Barron S, Trottier S, Cervera P, Daumas-Duport C, et al. 1999. *Glia* 26:176–85
56. Handford EJ, Smith D, Hewson L, McAllister G, Beer MS. 2001. *NeuroReport* 12:757–60
57. Weiner JA, Chun J. 1999. *Proc. Natl. Acad. Sci. USA* 96:5233–38
58. Weiner JA, Fukushima N, Contos JJA,

Scherer SS, Chun J. 2001. *J. Neurosci.* 21:7069–78

59. Contos JJA, Fukushima N, Weiner JA, Kaushal D, Chun J. 2000. *Proc. Natl. Acad. Sci. USA* 97:13384–89
60. Contos JJA, Ishii I, Fukushima N, Kingsbury MA, Ye XQ, et al. 2002. *Mol. Cell. Biol.* 22:6921–29
61. Bandoh K, Aoki J, Hosono H, Kobayashi S, Kobayashi T, et al. 1999. *J. Biol. Chem.* 274:27776–85
62. Im DS, Heise CE, Harding MA, George SR, O'Dowd BF, et al. 2000. *Mol. Pharmacol.* 57:753–59
63. Noguchi K, Ishii S, Shimizu T. 2003. *J. Biol. Chem.* 278:25600–6
64. Lee MJ, Thangada S, Paik JH, Sapkota GP, Ancellin N, et al. 2001. *Mol. Cell* 8:693–704
65. Parrill AL, Wang D, Bautista DL, Van Brocklyn JR, Lorincz Z, et al. 2000. *J. Biol. Chem.* 275:39379–84
66. Rosenfeldt HM, Hobson JP, Maceyka M, Olivera A, Nava VE, et al. 2001. *FASEB J.* 15:2649–59
67. Hobson JP, Rosenfeldt HM, Barak LS, Olivera A, Poulton S, et al. 2001. *Science* 291:1800–3
68. Waters C, Sambi B, Kong KC, Thompson D, Pitson SM, et al. 2003. *J. Biol. Chem.* 278:6282–90
69. Alderton F, Rakhit S, Kong KC, Palmer T, Sambi B, et al. 2001. *J. Biol. Chem.* 276:28578–85
70. Kluk MJ, Colmont C, Wu MT, Hla T. 2003. *FEBS Lett.* 533:25–28
71. Zhang GF, Contos JJA, Weiner JA, Fukushima N, Chun J. 1999. *Gene* 227:89–99
72. Ishii I, Friedman B, Ye X, Kawamura S, McGiffert C, et al. 2001. *J. Biol. Chem.* 276:33697–704
73. Liu Y, Wada R, Yamashita T, Mi Y, Deng CX, et al. 2000. *J. Clin. Investig.* 106:951–61
74. Allende ML, Yamashita T, Proia RL. 2003. *Blood* 102:3665–67
75. Okamoto H, Takuwa N, Yokomizo T,

Sugimoto N, Sakurada S, et al. 2000. *Mol. Cell. Biol.* 20:9247–61

76. Sugimoto N, Takuwa N, Okamoto H, Sakurada S, Takuwa Y. 2003. *Mol. Cell. Biol.* 23:1534–45

77. Okazaki H, Ishizaka N, Sakurai T, Kurokawa K, Goto K, et al. 1993. *Biochem. Biophys. Res. Commun.* 190:1104–9

78. MacLennan AJ, Marks L, Gaskin AA, Lee N. 1997. *Neuroscience* 79:217–24

79. MacLennan AJ, Carney PR, Zhu WJ, Chaves AH, Garcia J, et al. 2001. *Eur. J. Neurosci.* 14:203–9

80. Ishii I, Ye XQ, Friedman B, Kawamura S, Contos JJA, et al. 2002. *J. Biol. Chem.* 277:25152–59

81. Yamaguchi F, Tokuda M, Hatase O, Brenner S. 1996. *Biochem. Biophys. Res. Commun.* 227:608–14

82. Graler MH, Bernhardt G, Lipp M. 1998. *Genomics* 53:164–69

83. Yamazaki Y, Kon J, Sato K, Tomura H, Sato M, et al. 2000. *Biochem. Biophys. Res. Commun.* 268:583–89

84. Van Brocklyn JR, Graler MH, Bernhardt G, Hobson JP, Lipp M, Spiegel S. 2000. *Blood* 95:2624–29

85. Contos JJA, Ye XQ, Sah VP, Chun J. 2002. *FEBS Lett.* 531:99–102

86. Candelore MR, Wright MJ, Tota LM, Milligan J, Shei GJ, et al. 2002. *Biochem. Biophys. Res. Commun.* 297: 600–6

87. Graler MH, Grosse R, Kusch A, Kremmer E, Gudermann T, Lipp M. 2003. *J. Cell. Biochem.* 89:507–19

88. Glickman M, Malek RL, Kwitek-Black AE, Jacob HJ, Lee NH. 1999. *Mol. Cell. Neurosci.* 14:141–52

89. Im DS, Heise CE, Ancellin N, O'Dowd BF, Shei GJ, et al. 2000. *J. Biol. Chem.* 275:14281–86

90. Malek RL, Toman RE, Edsall LC, Wong S, Chiu J, et al. 2001. *J. Biol. Chem.* 276: 5692–99

91. Im DS, Clemens J, Macdonald TL, Lynch KR. 2001. *Biochemistry* 40: 14053–60

92. Terai K, Soga T, Takahashi M, Kamohara M, Ohno K, et al. 2003. *Neuroscience* 116:1053–62

93. Niedernberg A, Scherer CR, Busch AE, Kostenis E. 2002. *Biochem. Pharmacol.* 64:1243–50

94. Weng Z, Fluckiger AC, Nisitani S, Wahl MI, Le LQ, et al. 1998. *Proc. Natl. Acad. Sci. USA* 95:12334–39

95. Zohn IE, Klinger M, Karp X, Kirk H, Symons M, et al. 2000. *Oncogene* 19:3866–77

96. Kabarowski JH, Feramisco JD, Le LQ, Gu JL, Luoh SW, et al. 2000. *Proc. Natl. Acad. Sci. USA* 97:12109–14

97. Lin P, Ye RD. 2003. *J. Biol. Chem.* 278: 14379–86

98. Le LQ, Kabarowski JH, Weng Z, Satterthwaite AB, Harvill ET, et al. 2001. *Immunity* 14:561–71

99. Xu Y, Casey G. 1996. *Genomics* 35: 397–402

100. Heiber M, Docherty JM, Shah G, Nguyen T, Cheng R, et al. 1995. *DNA Cell Biol.* 14:25–35

101. Mahadevan MS, Baird S, Bailly JE, Shutler GG, Sabourin LA, et al. 1995. *Genomics* 30:84–88

102. Saeki Y, Ueno S, Mizuno R, Nishimura T, Fujimura H, et al. 1993. *FEBS Lett.* 336:317–22

103. Uhlenbrock K, Gassenhuber H, Kostenis E. 2002. *Cell. Signal.* 14:941–53

104. Fukushima N, Weiner JA, Kaushal D, Contos JJA, Rehen SK, et al. 2002. *Mol. Cell. Neurosci.* 20:271–82

105. Niedernberg A, Tunaru S, Blaukat A, Ardati A, Kostenis E. 2003. *Cell. Signal.* 15:435–46

106. Edwards G, Breckenridge AM. 1988. *Clin. Pharmacokinet.* 15:67–93

107. Xu Y, Qian L, Prestwich GD. 2003. *J. Org. Chem.* 68:5320–30

108. Ohta H, Sato K, Murata N, Damirin A, Malchinkhuu E, et al. 2003. *Mol. Pharmacol.* 64:994–1005

109. Lynch KR, Macdonald TL. 2002. *Biochim. Biophys. Acta* 1582:289–94

110. Sardar VM, Bautista DL, Fischer DJ, Yokoyama K, Nusser N, et al. 2002. *Biochim. Biophys. Acta* 1582:309–17

111. Virag T, Elrod DB, Liliom K, Sardar VM, Parrill AL, et al. 2003. *Mol. Pharmacol.* 63:1032–42

112. Hasegawa Y, Erickson JR, Goddard GJ, Yu S, Liu S, et al. 2003. *J. Biol. Chem.* 278:11962–69

113. Dillon DA, Chen X, Zeimetz GM, Wu WI, Waggoner DW, et al. 1997. *J. Biol. Chem.* 272:10361–66

114. Osada M, Yatomi Y, Ohmori T, Ikeda H, Ozaki Y. 2002. *Biochem. Biophys. Res. Commun.* 299:483–87

115. Mandala S, Hajdu R, Bergstrom J, Quackenbush E, Xie J, et al. 2002. *Science* 296:346–49

116. Brinkmann V, Davis MD, Heise CE, Albert R, Cottens S, et al. 2002. *J. Biol. Chem.* 277:21453–57

117. McIntyre TM, Pontsler AV, Silva AR, St Hilaire A, Xu Y, et al. 2003. *Proc. Natl. Acad. Sci. USA* 100:131–36

118. Rosen ED, Spiegelman BM. 2001. *J. Biol. Chem.* 276:37731–34

119. Barak Y, Nelson MC, Ong ES, Jones YZ, Ruiz-Lozano P, et al. 1999. *Mol. Cell* 4:585–95

120. Rosen ED, Sarraf P, Troy AE, Bradwin G, Moore K, et al. 1999. *Mol. Cell* 4:611–17

121. Gobeil F Jr, Bernier SG, Vazquez-Tello A, Brault S, Beauchamp MH, et al. 2003. *J. Biol. Chem.* 278:38875–83

122. Schmidt A, Wolde M, Thiele C, Fest W, Kratzin H, et al. 1999. *Nature* 401:133–41

123. Kooijman EE, Chupin V, de Kruijff B, Burger KN. 2003. *Traffic* 4:162–74

124. Reutens AT, Begley CG. 2002. *Int. J. Biochem. Cell Biol.* 34:1173–77

125. Mattie M, Brooker G, Spiegel S. 1994. *J. Biol. Chem.* 269:3181–88

126. Ghosh TK, Bian J, Gill DL. 1994. *J. Biol. Chem.* 269:22628–35

127. Betto R, Teresi A, Turcato F, Salviati G, Sabbadini RA, et al. 1997. *Biochem. J.* 322:327–33

128. Mao C, Kim SH, Almenoff JS, Rudner XL, Kearney DM, Kindman LA. 1996. *Proc. Natl. Acad. Sci. USA* 93:1993–96

129. Schnurbus R, Tonelli DD, Grohovaz F, Zacchetti D. 2002. *Biochem. J.* 362:183–89

130. Young KW, Nahorski SR. 2002. *Cell Calcium* 32:335–41

131. Cavalli AL, O'Brien NW, Barlow SB, Betto R, Glembotski CC, et al. 2003. *Am. J. Physiol. Cell Physiol.* 284:C780–90

132. Brailoiu E, Patel S, Dun NJ. 2003. *Biochem. J.* 373:313–18

133. Itagaki K, Hauser CJ. 2003. *J. Biol. Chem.* 278:27540–47

134. McGiffert C, Contos JJA, Friedman B, Chun J. 2002. *FEBS Lett.* 531:103–8

135. Yatomi Y, Igarashi Y, Yang L, Hisano N, Qi R, et al. 1997. *J. Biochem.* 121:969–73

136. Ye X, Fukushima N, Kingsbury MA, Chun J. 2002. *NeuroReport* 13:2169–75

137. Renback K, Inoue M, Yoshida A, Nyberg F, Ueda H. 2000. *Mol. Brain Res.* 75:350–54

138. Jalink K, Eichholtz T, Postma FR, van Corven EJ, Moolenaar WH. 1993. *Cell Growth Differ.* 4:247–55

139. Tigyi G, Miledi R. 1992. *J. Biol. Chem.* 267:21360–67

140. Moolenaar WH. 1995. *Curr. Opin. Cell Biol.* 7:203–10

141. Chun J, Jaenisch R. 1996. *Mol. Cell. Neurosci.* 7:304–21

142. Fukushima N, Ishii I, Habara Y, Allen CB, Chun J. 2002. *Mol. Biol. Cell* 13:2692–705

143. Yuan XB, Jin M, Xu X, Song YQ, Wu CP, et al. 2003. *Nat. Cell Biol.* 5:38–45

144. Campbell DS, Holt CE. 2001. *Neuron* 32:1013–26

145. Campbell DS, Holt CE. 2003. *Neuron* 37:939–52

146. Sauer FC. 1935. *J. Comp. Neurol.* 62:377–405

147. Dubin AE, Bahnson T, Weiner JA, Fukushima N, Chun J. 1999. *J. Neurosci.* 19:1371–81

148. Kingsbury MA, Rehen SK, Contos JJ, Higgins CM, Chun J. 2003. *Nat. Neurosci.* 6:1292–99

149. Levade T, Auge N, Veldman RJ, Cuvillier O, Negre-Salvayre A, Salvayre R. 2001. *Circ. Res.* 89:957–68

150. Paik JH, Chae S, Lee MJ, Thangada S, Hla T. 2001. *J. Biol. Chem.* 276: 11830–37

151. Ryu Y, Takuwa N, Sugimoto N, Sakurada S, Usui S, et al. 2002. *Circ. Res.* 90:325–32

152. Kimura T, Watanabe T, Sato K, Kon J, Tomura H, et al. 2000. *Biochem. J.* 348: 71–76

153. Kluk MJ, Hla T. 2001. *Circ. Res.* 89:496–502

154. Tamama K, Kon J, Sato K, Tomura H, Kuwabara A, et al. 2001. *Biochem. J.* 353:139–46

155. Kwon YG, Min JK, Kim KM, Lee DJ, Billiar TR, Kim YM. 2001. *J. Biol. Chem.* 276:10627–33

156. Lee MJ, Thangada S, Claffey KP, Ancellin N, Liu CH, et al. 1999. *Cell* 99:301–12

157. Garcia JG, Liu F, Verin AD, Birukova A, Dechert MA, et al. 2001. *J. Clin. Investig.* 108:689–701

158. Hayashi K, Takahashi M, Nishida W, Yoshida K, Ohkawa Y, et al. 2001. *Circ. Res.* 89:251–58

159. Cui MZ, Zhao G, Winokur AL, Laag E, Bydash JR, et al. 2003. *Arterioscler. Thromb. Vasc. Biol.* 23:224–30

160. Karliner JS. 2002. *Biochim. Biophys. Acta* 1582:216–21

161. Siess W. 2002. *Biochim. Biophys. Acta* 1582:204–15

162. Sugiyama A, Aye NN, Yatomi Y, Ozaki Y, Hashimoto K. 2000. *Jpn. J. Pharmacol.* 82:338–42

163. Kupperman E, An SZ, Osborne N, Waldron S, Stainier DY. 2000. *Nature* 406: 192–95

164. Lee H, Goetzl EJ, An SZ. 2000. *Am. J. Physiol. Cell Physiol.* 278:C612–18

165. Balazs L, Okolicany J, Ferrebee M, Tolley B, Tigyi G. 2001. *Am. J. Physiol. Regul. Integr. Comp. Physiol.* 280: R466–72

166. Demoyer JS, Skalak TC, Durieux ME. 2000. *Wound Repair Regen.* 8:530–37

167. Vogler R, Sauer B, Kim DS, Schafer-Korting M, Kleuser B. 2003. *J. Investig. Dermatol.* 120:693–700

168. Takeya H, Gabazza EC, Aoki S, Ueno H, Suzuki K. 2003. *Blood* 102:1693–700

169. Lee H, Liao JJ, Graeler M, Huang MC, Goetzl EJ. 2002. *Biochim. Biophys. Acta* 1582:175–77

170. Graler MH, Goetzl EJ. 2002. *Biochim. Biophys. Acta* 1582:168–74

171. Huang MC, Graeler M, Shankar G, Spencer J, Goetzl EJ. 2002. *Biochim. Biophys. Acta* 1582:161–67

172. Rosen H, Sanna G, Alfonso C. 2003. *Immunol. Rev.* 195:160–77

173. Rosen H, Alfonso C, Surh CD, McHeyzer-Williams MG. 2003. *Proc. Natl. Acad. Sci. USA* 100:10907–12

174. Goetzl EJ, Kong Y, Mei B. 1999. *J. Immunol.* 162:2049–56

175. Zheng Y, Voice JK, Kong Y, Goetzl EJ. 2000. *FASEB J.* 14:2387–89

176. Zheng Y, Kong Y, Goetzl EJ. 2001. *J. Immunol.* 166:2317–22

177. Graeler M, Goetzl EJ. 2002. *FASEB J.* 16:1874–78

178. Brinkmann V, Lynch KR. 2002. *Curr. Opin. Immunol.* 14:569–75

179. Chen S, Bacon KB, Garcia G, Liao R, Pan ZK, et al. 2001. *Transplant. Proc.* 33:3057–63

180. Henning G, Ohl L, Junt T, Reiterer P, Brinkmann V, et al. 2001. *J. Exp. Med.* 194:1875–81

181. Brinkmann V, Pinschewer DD, Feng L, Chen S. 2001. *Transplantation* 72: 764–69

182. Xie JH, Nomura N, Koprak SL, Quackenbush EJ, Forrest MJ, Rosen H. 2003. *J. Immunol.* 170:3662–70

183. Xu Y, Gaudette DC, Boynton JD, Frankel A, Fang XJ, et al. 1995. *Clin. Cancer Res.* 1:1223–32

184. Westermann AM, Havik E, Postma FR, Beijnen JH, Dalesio O, et al. 1998. *Ann. Oncol.* 9:437–42

185. Eder AM, Sasagawa T, Mao M, Aoki J, Mills GB. 2000. *Clin. Cancer Res.* 6:2482–91

186. Xu Y, Shen Z, Wiper DW, Wu M, Morton RE, et al. 1998. *JAMA* 280:719–23

187. Xu Y, Fang XJ, Casey G, Mills GB. 1995. *Biochem. J.* 309:933–40

188. Frankel A, Mills GB. 1996. *Clin. Cancer Res.* 2:1307–13

189. Goetzl EJ, Dolezalova H, Kong Y, Hu YL, Jaffe RB, et al. 1999. *Cancer Res.* 59:5370–75

190. Pustilnik TB, Estrella V, Wiener JR, Mao M, Eder A, et al. 1999. *Clin. Cancer Res.* 5:3704–10

191. Tanyi JL, Morris AJ, Wolf JK, Fang X, Hasegawa Y, et al. 2003. *Cancer Res.* 63:1073–82

192. Furui T, LaPushin R, Mao M, Khan H, Watt SR, et al. 1999. *Clin. Cancer Res.* 5:4308–18

193. Fang X, Schummer M, Mao M, Yu S, Tabassam FH, et al. 2002. *Biochim. Biophys. Acta* 1582:257–64

194. Sengupta S, Xiao YJ, Xu Y. 2003. *FASEB J.* 17:1570–72

195. Fujita T, Miyamoto S, Onoyama I, Sonoda K, Mekada E, Nakano H. 2003. *Cancer Lett.* 192:161–69

196. Contos JJA, Chun J. 2000. *Genomics* 64:155–69

197. Umezu-Goto M, Kishi Y, Taira A, Hama K, Dohmae N, et al. 2002. *J. Cell Biol.* 158:227–33

198. Tokumura A, Majima E, Kariya Y, Tominaga K, Kogure K, et al. 2002. *J. Biol. Chem.* 277:39436–42

199. Tokumura A, Kanaya Y, Miyake M, Yamano S, Irahara M, Fukuzawa K. 2002. *Biol. Reprod.* 67:1386–92

200. Budnik LT, Mukhopadhyay AK. 2002. *Biol. Reprod.* 66:859–65

201. Budnik LT, Brunswig-Spickenheier B, Mukhopadhyay AK. 2003. *Mol. Endocrinol.* 17:1593–606

202. Xie Y, Gibbs TC, Mukhin YV, Meier KE. 2002. *J. Biol. Chem.* 277:32516–26

203. Daaka Y. 2002. *Biochim. Biophys. Acta* 1582:265–69

204. Shida D, Kitayama J, Yamaguchi H, Okaji Y, Tsuno NH, et al. 2003. *Cancer Res.* 63:1706–11

205. Mills GB, Fang X, Lu Y, Hasegawa Y, Eder A, et al. 2003. *Gynecol. Oncol.* 88: S3–6, S88–92

206. Casper RF, Jurisicova A. 2000. *Nat. Med.* 6:1100–1

207. Revel A, Laufer N. 2002. *Mol. Cell. Endocrinol.* 187:83–91

208. Tilly JL, Kolesnick RN. 2002. *Biochim. Biophys. Acta* 1585:135–38

209. Morita Y, Perez GI, Paris F, Miranda SR, Ehleiter D, et al. 2000. *Nat. Med.* 6:1109–14

210. Paris F, Perez GI, Fuks Z, Haimovitz-Friedman A, Nguyen H, et al. 2002. *Nat. Med.* 8:901–2

Annu. Rev. Biochem. 2004. 73:355–82
doi: 10.1146/annurev.biochem.73.011303.074118
Copyright © 2004 by Annual Reviews. All rights reserved
First published online as a Review in Advance on March 5, 2004

Protein Modification by SUMO

Erica S. Johnson

*Department of Biochemistry and Molecular Pharmacology, Thomas Jefferson
University, Philadelphia, Pennsylvania 19107; email: erica.johnson@jefferson.edu*

Key Words   post-translational modification, ubiquitin-like protein, PIAS, Ubc9,
Ulp

■ **Abstract**   Small ubiquitin-related modifier (SUMO) family proteins function by
becoming covalently attached to other proteins as post-translational modifications.
SUMO modifies many proteins that participate in diverse cellular processes, including transcriptional regulation, nuclear transport, maintenance of genome integrity,
and signal transduction. Reversible attachment of SUMO is controlled by an enzyme
pathway that is analogous to the ubiquitin pathway. The functional consequences of
SUMO attachment vary greatly from substrate to substrate, and in many cases are not
understood at the molecular level. Frequently SUMO alters interactions of substrates
with other proteins or with DNA, but SUMO can also act by blocking ubiquitin
attachment sites. An unusual feature of SUMO modification is that, for most
substrates, only a small fraction of the substrate is sumoylated at any given time. This
review discusses our current understanding of how SUMO conjugation is controlled,
as well as the roles of SUMO in a number of biological processes.

CONTENTS

INTRODUCTION

Covalent modifications of proteins are rapid, energetically inexpensive mechanisms for reversibly altering protein function, and modifications such as phosphorylation, acetylation, and ubiquitylation participate in most cellular activities. Ubiquitylation, which involves attachment of the 76-residue protein ubiquitin (Ub) to other proteins, often targets the substrate protein for degradation by the proteasome, but it can also have several other functions (1, 2). Recently, several small ubiquitin-like proteins (Ubls) that also act as post-translational modifications on other proteins have been discovered. These Ubls vary widely in their degree of sequence similarity to Ub but share a common chemistry for becoming attached to internal lysine residues in substrate proteins (3). Ubls have a variety of different functions, but they do not target their substrates directly for proteasome-dependent proteolysis. The Ubls with the widest range of functions and the most known substrates are the members of the SUMO (small ubiquitin-related modifier) family. Several previous reviews on SUMO cover earlier work and specific topics in depth (4–8).

SUMOs constitute a highly conserved protein family found in all eukaryotes and are required for viability of most eukaryotic cells, including budding yeast, nematodes, fruit flies, and vertebrate cells in culture (9–13). In multicellular organisms, SUMO conjugation takes place in all tissues at all developmental stages (14–21). Since its discovery in 1996, SUMO has been found covalently attached to more than 50 proteins, which include the androgen receptor, IκBα, c-jun, histone deacetylases (HDACs), p53, and other proteins that participate in transcription, DNA repair, nuclear transport, signal transduction, and the cell cycle. Most SUMO-modified proteins that have been characterized in mammalian systems are involved in transcription, which is often repressed by SUMO conjugation. However, genetic studies in model organisms have pointed to a role for SUMO in chromosome dynamics and higher order chromatin structures, illustrating the diversity of SUMO function.

At this time, only one fairly uninformative generalization about the downstream consequences of SUMO attachment is possible: SUMO alters substrate interactions with other macromolecules. SUMO often has a positive effect on protein-protein interactions, and it promotes assembly of several multi-protein complexes. However, the effects of SUMO on interactions vary for different substrates. For example, sumoylation allows RanGAP1 to bind tightly to the nuclear pore complex protein RanBP2/Nup358 (22, 23), but no other sumoylated proteins participate in a stable complex with RanBP2. SUMO can also act by a completely different mechanism: preventing ubiquitylation of a protein by blocking the lysine where Ub would normally be attached (24–27).

There are several reasons why proteins that have been intensely studied for many years, such as c-jun and the androgen receptor, have only recently been shown to be modified by SUMO. One is that SUMO-cleaving enzymes rapidly desumoylate all conjugates instantly upon cell lysis, unless cells are lysed under denaturing conditions or cleaving enzymes are inhibited. Another is that usually only a small fraction of the substrate, often less than 1%, is sumoylated at any given time. A third reason for the late discovery of SUMO is that, for some sumoylated proteins, eliminating the SUMO attachment site has fairly subtle effects on protein function, so that functional domains containing the attachment sites were not immediately apparent.

However, recent experiments have uncovered a variety of effects that can clearly be attributed to sumoylation of specific proteins at specific sites, and new substrates and functions for SUMO continue to be discovered at a rapid pace.

THE SUMO CONJUGATION PATHWAY

The linkage between SUMO and its substrates is an isopeptide bond between the C-terminal carboxyl group of SUMO and the ϵ-amino group of a lysine residue in the substrate. A three-step enzyme pathway attaches SUMO to specific substrates, and other enzymes cleave SUMO off its targets (Figure 1). The enzymes of the SUMO pathway, although analogous to those of the Ub pathway, are specific for SUMO and have no role in conjugating Ub or any of the other Ubls.

The SUMO pathway begins with a SUMO-activating enzyme (also called an E1), which carries out an ATP-dependent activation of the SUMO C terminus and then transfers activated SUMO to a SUMO-conjugating enzyme (E2) called Ubc9. SUMO is then transferred from Ubc9 to the substrate with the assistance of one of several SUMO-protein ligases (E3s). Ubc9 and the E3s both contribute to substrate specificity. Many of the Lys residues where SUMO becomes attached are in the short consensus sequence ΨKXE, where Ψ is a large hydrophobic amino acid, generally isoleucine, leucine, or valine; K is the lysine residue that is modified; X is any residue; and E is a glutamic acid. This motif is bound directly by Ubc9. E3s probably enhance specificity by interacting with other features of the substrate. Sumoylation is a reversible modification, and removal of SUMO is carried out by enzymes of the Ulp family that specifically cleave at the C terminus of SUMO. Ulps are also required for generating mature SUMO from the SUMO precursor, which contains a short peptide blocking its C terminus.

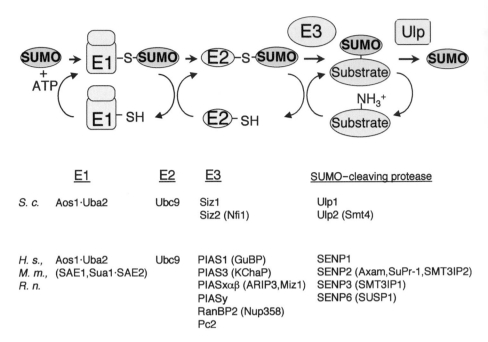

	E1	E2	E3	SUMO–cleaving protease
S. c.	Aos1·Uba2	Ubc9	Siz1	Ulp1
			Siz2 (Nfi1)	Ulp2 (Smt4)
H. s.,	Aos1·Uba2	Ubc9	PIAS1 (GuBP)	SENP1
M. m.,	(SAE1,Sua1·SAE2)		PIAS3 (KChaP)	SENP2 (Axam,SuPr-1,SMT3IP2)
R. n.			PIASxαβ (ARIP3,Miz1)	SENP3 (SMT3IP1)
			PIASy	SENP6 (SUSP1)
			RanBP2 (Nup358)	
			Pc2	

Figure 1   The SUMO conjugation pathway. (*top*) Enzymes and reactions of the SUMO pathway are described in the text. (*bottom*) Enzymes present in *S. cerevisiae* (*S.c.*) and in human (*H.s.*), mouse (*M.m.*), and rat (*R.n.*) are listed. Alternative names and names of splice variants are in parentheses.

SUMO

SUMOs share only ~18% sequence identity with Ub, but the folded structure of the SUMO C-terminal Ub-like domain is virtually superimposable on that of Ub (28) (Figure 2). However, the surface charge topology of SUMO is very different from that of Ub, with distinct positive and negative regions (28). SUMOs are ~11 kDa proteins, but they appear larger on SDS-PAGE and add ~20 kDa to the apparent molecular weight of most substrates. SUMOs are ~20 amino acids longer than Ub, and the extra residues are found in an N-terminal extension, which is flexible in solution. The N-terminal extension of yeast SUMO can be entirely deleted with only modest effects on SUMO function, indicating that the Ub-like domain is sufficient for conjugation to many substrates and for any downstream interactions required for yeast viability (29). All SUMO genes actually encode a precursor bearing a short C-terminal peptide, which is cleaved off by Ulps to produce the mature Gly-Gly C terminus found in most Ubls.

The yeast and invertebrates studied to date contain a single SUMO gene, whereas vertebrates contain three: SUMO-1 (also known as sentrin, PIC1, GMP1, Ubl1, and Smt3c), SUMO-2 (sentrin-3, Smt3a), and SUMO-3 (sentrin-2, Smt3b) (13, 15, 16, 22, 23, 30–33). Plants contain even more SUMO genes, with

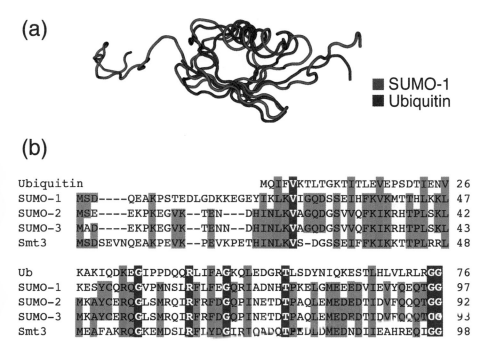

(a)

■ SUMO-1
■ Ubiquitin

(b)

```
Ubiquitin                      MQIFVKTLTGKTITLEVEPSDTIENV  26
SUMO-1     MSD----QEAKPSTEDLGDKKEGEYIKLKVIGQDSSEIHFKVKMTTHLKKL  47
SUMO-2     MSE----EKPKEGVK--TEN---DHINLKVAGQDGSVVQFKIKRHTPLSKL  42
SUMO-3     MAD----EKPKEGVK--TENN--DHINLKVAGQDGSVVQFKIKRHTPLSKL  43
Smt3       MSDSEVNQEAKPEVK--PEVKPETHINLKVS-DGSSEIFFKIKKTTPLRRL  48


Ub         KAKIQDKEGIPPDQQRLIFAGKQLEDGRTLSDYNIQKESTLHLVLRLRGG  76
SUMO-1     KESYCQRQGVPMNSLRFLFEGQRIADNHTPKELGMEEEDVIEVYQEQTGG  97
SUMO-2     MKAYCERQGLSMRQIRFRFDGQPINETDTPAQLEMEDEDTIDVFQQQTGG  92
SUMO-3     MKAYCERQGLSMRQIRFRFDGQPINETDTPAQLEMEDEDTIDVFQQQTGG  93
Smt3       MEAFAKRQGKEMDSLRFIYDGIRTQADQTPEDLDMEDNDIIEAHREQIGG  98
```

Figure 2 Comparison of SUMO and ubiquitin. (*a*) Structural alignment of the backbones of SUMO-1 (*pink*) and ubiquitin (*blue*) is from the VAST database (NCBI) with structures from References 28 and 28a. The N termini are on the left and and the C termini on the right. The SUMO structure is of the precursor and includes the C-terminal tetrapeptide that is cleaved off. (*b*) Sequence alignment of *H. sapiens* Ub, SUMO-1, SUMO-2, and SUMO-3 and the *S. cerevisiae* SUMO protein Smt3 was made using ClustalW. Positions that are identical in all sequences are shaded dark blue, and conserved positions are light blue. Positions that are identical in at least three of the SUMO proteins, but not in Ub, are shaded pink.

eight in *Arabidopsis* (20, 21). The single SUMO genes in the nematode *Caenorhabditis elegans* and the budding yeast *Saccharomyces cerevisiae* are essential for viability, while fission yeast *Schizosaccharomyces pombe* lacking the SUMO gene *pmt3* are barely viable and have severe defects in genome maintenance (13, 32, 33).

Mammalian SUMO-2 and -3 share ~95% sequence identity with each other and are ~50% identical to SUMO-1. Although the same E1 and E2 enzymes activate and conjugate all SUMO isoforms, SUMO-1 appears to have a partially distinct function from SUMO-2 and -3, which are assumed, at present, to be functionally identical. Cells contain a large pool of free, unconjugated SUMO-2/3, but there is virtually no pool of free SUMO-1; at any given time, the vast majority of SUMO-1 is conjugated to other proteins (23, 34). Furthermore, conjugation of SUMO-2/3 is strongly induced in response to various stresses, but

SUMO-1 conjugation is not (34). Plants have a similar pattern of SUMO isoform utilization, with some isoforms conjugated primarily under stress conditions (20, 21). Thus, one function of SUMO-2/3 may be to provide a reservoir of free SUMO for stress responses. There is also evidence that different SUMOs are used preferentially for different substrates. RanGAP1 is the major substrate of SUMO-1, but it is not strongly modified by SUMO-2/3 (34). Other proteins can be modified equally well by SUMO-1 and SUMO-2/3 (35, 36). It is likely that E3s mediate the differential conjugation of the SUMO isoforms (see below).

Another difference between SUMO-1 and SUMO-2/3 is that SUMO-2 and -3 contain ΨKXE sequences in their N-terminal extensions, which can serve as SUMO attachment sites, thereby allowing formation of poly-SUMO chains (37). Yeast SUMO also contains a ΨKXE sequence and can form chains (29, 38, 39). Chain formation by SUMO was a surprise because in vivo most SUMO attachment-site Lys residues bear only a single copy of SUMO, although proteins are often multiply sumoylated by attachment of mono-SUMO at different sites (40, 41). The only protein on which a SUMO-2 chain has been observed in cells is the histone deacetylase HDAC4; it forms a di-sumoylated conjugate that disappears when the SUMO attachment site in SUMO-2 is mutated (37). However, there are intriguing data suggesting that cleavage of the amyloid precursor protein to generate the amyloid β peptide involves SUMO-2/3 chain formation (42). The function of SUMO chains is unclear in yeast, where chain formation can be eliminated without notable effects on either SUMO function or the pattern of conjugates (29, 43).

SUMO-Activating Enzyme (E1)

Like the E1 for Ub, the SUMO-activating enzyme (E1) catalyzes a three-part reaction. First, the C-terminal carboxyl group of SUMO attacks ATP, forming a SUMO C-terminal adenylate and releasing pyrophosphate. Next, the thiol group of the active site cysteine in the E1 attacks the SUMO adenylate, releasing AMP and forming a high-energy thiolester bond between the E1 and the C terminus of SUMO. Finally, the activated SUMO is transferred to a cysteine in the E2. The crystal structure of the related E1 for the Ubl Nedd-8 suggests that three distinct domains catalyze each of the steps (44). Most organisms contain a single SUMO-activating enzyme, which is required for conjugation of all SUMO variants to all substrates. Interestingly, the SUMO E1 is a heterodimer, whereas the Ub E1 is a monomer, but both components of the SUMO enzyme are related to the Ub enzyme. Aos1 (also called SAE1, Sua1) resembles the N terminus of the Ub E1, while Uba2 (SAE2) corresponds to the C terminus and contains the active site cysteine (33, 45, 46). Although the two-subunit structure of the SUMO E1 suggests that Aos1 and Uba2 might function or be regulated separately, all cellular Uba2 and Aos1 is found in the heterodimer (47). However, *Arabidopsis* actually has two *SAE1* (*AOS1*) genes, whose products presumably each partner with the product of the single *SAE2* (*UBA2*) gene (21).

SUMO-Conjugating Enzyme (E2)

In the second step of the pathway, SUMO is transferred from the E1 to the active site cysteine of the SUMO-conjugating enzyme (E2), forming a SUMO-E2 thiolester intermediate. This serves as the SUMO donor in the final reaction in which SUMO is transferred to the amino group of a Lys in the substrate. Ubc9 is the only SUMO-conjugating enzyme in yeast and invertebrates and most likely in vertebrates as well (10, 13, 48, 49). The presence of only one SUMO E2 contrasts with the Ub pathway where multiple E2s participate in ubiquitylating distinct sets of substrates. Ubc9 shares considerable sequence similarity with ubiquitylation E2s and also assumes essentially the same folded structure, although Ubc9 has a strong overall positive charge (50). A patch surrounding the active site cysteine of Ubc9 binds directly to the ΨKXE consensus sequence in the substrate (51, 52). A second region on Ubc9, separate from the active site, binds directly to SUMO and is involved in transfer of SUMO from the E1 (39, 53). Like the genes for SUMO, Aos1, and Uba2, the gene encoding Ubc9 is essential in all organisms tested except *S. pombe*, in which the mutant lacking the Ubc9 gene *hus5* has the same phenotypes as mutants lacking SUMO, Aos1 or Uba2 (11–13, 32, 54–56).

SUMO Ligases (E3s)

Three distinct types of SUMO ligases (E3s) have been discovered recently. One includes members of the PIAS (protein inhibitor of activated STAT) family (57, 58), originally discovered as inhibitors of STAT transcription factors (59); another consists of a domain in the large vertebrate nuclear pore protein RanBP2/Nup358 (60); and the third is the polycomb group protein Pc2 (61). These proteins meet the definition of an E3 in that they (*a*) bind the E2, (*b*) bind the substrate, and (*c*) promote transfer of SUMO from the E2 to the substrate in vitro (1). These SUMO E3s, like the RING domain-containing E3s involved in ubiquitylation, do not form covalent intermediates with SUMO, but instead they appear to act by bringing together Ubc9 and the substrate. They may also activate Ubc9. There was initially some doubt as to whether there would be E3s in the SUMO pathway because SUMO conjugation can take place in vitro in the absence of an E3, and this reaction is specific for the Lys residues that are actually modified in vivo (45, 46). However, the vast majority of sumoylation in yeast is E3-dependent (38, 62), and E3s enhance SUMO attachment in vitro to all substrates that have been tested (38, 60, 63–68). Together these results indicate that E3s participate in at least most of the sumoylation that occurs in cells.

PIAS FAMILY E3s   PIAS proteins share a conserved ~400 residue N-terminal domain that includes several shorter regions of greater similarity, notably a SAP domain (SAR, Acinus, PIAS), which has been implicated in binding AT-rich DNA sequences (64, 69–71), and an SP-RING, which resembles the RING domains found in many ubiquitylation E3s (57, 58). Like RING domains, which bind ubiquitylation E2s, the SP-RING binds directly to Ubc9 and is required for the E3 activity of PIAS proteins, suggesting that it is the critical element for promoting the sumoylation reaction (63–65). PIAS proteins also contain a short motif of hydrophobic amino acids followed by acidic amino acids, called an SXS domain or SIM (SUMO interaction motif), which has been implicated in binding directly to SUMO (72). Deletion of the SIM has little effect on the ability of PIAS proteins to promote SUMO conjugation, but it can affect their localization and transcriptional effects (64, 66). The main differences between PIAS proteins lie in their 100–450 residue C-terminal tails, which share no sequence similarity with each other or with other known proteins. Some PIAS proteins also have splice variants that produce alternative C-terminal tails. It is likely that these C-terminal domains interact with specific substrates.

S. cerevisiae contains two PIAS family proteins, Siz1 and Siz2/Nfi1. Siz1 is required for sumoylation of septin family cytoskeletal proteins and of the replication processivity factor PCNA; whereas Siz2 does not promote septin or PCNA sumoylation but sumoylates other, as yet unidentified, proteins (24, 38, 62). Together, SIZ1 and SIZ2 are required for most sumoylation in yeast, but the siz1Δ siz2Δ double mutant still carries out low levels of SUMO conjugation. This double mutant is also viable, indicating that Siz-independent sumoylation can fulfill the essential functions of SUMO. However, the siz1Δ siz2Δ mutant does have significant growth defects not seen in either single mutant, suggesting that Siz1 and Siz2 have some overlapping functions. Drosophila melanogaster has a single PIAS gene, known as dpias, Su(var)2–10, or zimp, which produces at least two isoforms derived from alternative splicing. dpias is an essential gene that functions in chromosome organization and segregation as well as in blood cell and eye development (73–75).

Four mammalian genes encoding PIAS proteins have been described, PIAS1 (also called GuBP), PIAS3, PIASx, and PIASy (59, 76, 77). PIAS3 has a splice variant called KChAP, and PIASx also produces two isoforms derived from alternative splicing, designated PIASxα (ARIP3) and PIASxβ (Miz1) (78–80). PIAS1 and PIAS3 are found in all cell types, whereas PIASx and PIASy appear to be expressed primarily in testis (76, 81). PIASxα, PIASxβ, PIASy, PIAS1, and PIAS3 all localize to intranuclear dots, which are, at least in part, PML nuclear bodies (see below) (64, 66, 82, 83).

By analogy with the Ub system, the purpose of the different PIAS proteins may be to sumoylate different substrates, but currently the only clear example of this is the specificity of Siz1 for septins and PCNA. Sumoylation of many vertebrate-derived substrates can be stimulated by several different PIAS proteins, upon overexpression both in cells and in vitro. For example, PIAS1, PIAS3, and PIASy can all promote sumoylation of p53 (63, 68). Such a result may suggest either that PIAS proteins have overlapping substrate specificities or that in vitro assays do not faithfully reproduce physiological substrate selection mechanisms. In support of this second possibility, Siz2/Nfi1 can stimulate SUMO attachment to septins in vitro, even though it is incapable of promoting septin sumoylation in vivo (43). However, PIAS proteins do show different substrate specificities with some substrates: PIAS1 and PIASxβ, but not PIASxα, stimulate sumoylation of Mdm2 (82).

Another function of the different PIAS proteins may be to promote attachment of the different SUMO isoforms. PIASy preferentially conjugates SUMO-2, rather than SUMO-1, to the transcription factors LEF1 and GATA-2, and it strongly enhances overall SUMO-2 conjugation (64, 84). It is also not clear that all PIAS effects are mediated by SUMO conjugation. In particular, PIAS proteins inhibit binding of STAT transcription factors to DNA in vitro, and there is no evidence that this effect involves SUMO (76, 77, 85, 86).

RanBP2/Nup358   A second type of SUMO E3 consists of an ~300 residue region in the large vertebrate-specific nuclear pore protein RanBP2 (also called Nup358), which localizes to the cytoplasmic fibrils of the nuclear pore and contains several types of functional domains (60, 87, 88). The E3 domain, called the internal repeat (IR) domain, contains two repeats of an ~50 residue sequence that shares no sequence similarity with any of the known ubiquitylation E3s or any other protein. In addition to having the capacity to act as an E3 in the sumoylation of several proteins, including RanGAP1, the IR domain forms a stable trimeric complex with SUMO-RanGAP1 and Ubc9, and thus it is responsible for the localization of SUMO-RanGAP1 to the nuclear pore (89, 90). RanBP2 itself can also be sumoylated (60, 91). Presumably, sumoylation of nuclear proteins by RanBP2 would have to occur during nuclear import.

Although it has not been demonstrated conclusively that RanBP2 is required in vivo for sumoylation of proteins other than RanGAP1, in vitro results indicate that RanBP2 and PIAS proteins have mostly distinct sets of substrates, suggesting they may have fundamentally different specificities. The IR domain promotes SUMO attachment in vitro to several proteins, including HDAC4, Sp100, and RanGAP1, whose sumoylation is not stimulated by PIAS proteins. Conversely, PIAS proteins, but not RanBP2, stimulate sumoylation of p53 and Sp3 (60, 67, 92). However, other proteins can be sumoylated by either RanBP2 or PIAS proteins (82, 93).

Pc2   A third reported E3 for SUMO is the polycomb group (PcG) protein Pc2 (61). PcG proteins form large multimeric complexes that have histone methylation activity and that participate in transcriptional repression through establishment of epigenetically inherited domains of silent chromatin. The transcriptional corepressor CtBP associates with PcG bodies via Pc2, and Pc2 stimulates sumoylation of CtBP both in vivo and in vitro. Moreover, overexpression of Pc2 in cells causes SUMO and Ubc9 to colocalize at PcG bodies, suggesting that PcG bodies may be major sites of sumoylation. However, the enhancement of CtBP sumoylation by Pc2 in vitro is very modest (61), and PIAS1, PIASxβ, and RanBP2 can also promote CtBP sumoylation (93), suggesting that there may be multiple factors involved in CtBP sumoylation.

SUMO-Cleaving Enzymes

The pattern of SUMO conjugates is dynamic and changes during the cell cycle and in response to various stimuli (94). SUMO-cleaving enzymes (also called isopeptidases) have at least two functions in this process: They remove SUMO from proteins, making the modification reversible, and they also provide a source of free SUMO to be used for conjugation to other proteins. Free SUMO is generated both from newly synthesized SUMO, which must be cleaved to remove a short C-terminal peptide, and from desumoylation of existing conjugates. Both of these sources of free SUMO are likely to be critical for maintaining normal levels of SUMO conjugation because cellular pools of unconjugated SUMO-1 and yeast SUMO are very low (23, 33).

All known SUMO-cleaving enzymes contain an ~200 amino acid C-terminal domain (the Ulp domain), which has the SUMO cleaving activity (95). The Ulp domain does not share sequence similarity with the enzymes that cleave Ub. Instead, it is distantly related to a number of viral proteases (94, 96). The different SUMO-cleaving enzymes have varying N-terminal domains, which are apparently regulatory and target the enzymes to different parts of the cell (97–100). Overexpression of the SUMO cleaving domain of the yeast enzyme Ulp1 is lethal in yeast, consistent with the likelihood that uncontrolled desumoylation is toxic (95).

Two desumoylating enzymes with distinct functions have been described in *S. cerevisiae*. Ulp1 localizes to the nuclear pore complex (NPC) and is required for cleaving both the SUMO precursor and SUMO conjugates to other proteins; whereas Ulp2/Smt4 localizes to the nucleus, does not cleave the precursor, and appears to desumoylate a distinct set of conjugates (94, 98, 101–103). Ulp1 and Ulp2 cannot compensate for each other functionally, as *ulp1Δ* cells are inviable, and *ulp2Δ* cells are stress sensitive and have defects in genome maintenance. The substrate specificity of Ulp1 is controlled by its N-terminal regulatory domain, which targets it to the NPC. Mutants lacking this domain both nonspecifically desumoylate Ulp2 targets and fail to desumoylate the normal targets of Ulp1 (97).

Seven genes in mammalian genomes encode proteins with Ulp domains, but at least one of these cleaves the Ubl Nedd-8 instead of SUMO (104–106). All have divergent N-terminal domains, and those that have been characterized localize to different parts of the cell, suggesting that they may desumoylate different proteins. These enzymes include SENP3 (SMT3IP1), which localizes to the nucleolus (107); SENP6 (SUSP1), found primarily in the cytoplasm (108); SENP1, which localizes to foci in the nucleus and the nuclear rim (109); and SENP2 (Axam, SMT3IP2/Axam2, SuPr-1), which produces at least three different isoforms derived from alternatively spliced mRNAs (110–112). Of these, the SENP2/Axam isoform has an N-terminal extension that allows it to bind the nucleoplasmic side of the nuclear pore complex (99, 100); Axam2/SMT3IP2 has a different N terminus and localizes to the cytoplasm

(112); and SuPr1 lacks these N-terminal domains and localizes to PML nuclear bodies (110).

Substrate Specificity in Sumoylation

SUMO is attached to most substrates at the lysine in a ΨKXE sequence, but there are clearly other determinants involved in substrate selection as well. Of the positions in the consensus sequence, the glutamic acid is the most highly conserved position other than the lysine. In some cases, even a conservative Glu to Asp mutation significantly reduces sumoylation (92, 113), although a few ΨKXD sequences are sumoylated (40). The ΨKXE motif is bound directly by the E2 Ubc9 (114), and this direct interaction explains why so many sumoylation substrates have been identified via their interaction with Ubc9 in the yeast two-hybrid screen and also why the E1 and Ubc9 alone are sufficient to sumoylate many substrates at the correct sites in vitro in the absence of an E3. Remarkably, a ΨKXE sequence and a nuclear localization sequence (NLS) are sufficient to target an artificial substrate for sumoylation, indicating that the requirements for SUMO conjugation can be very simple (113). Most SUMO substrates localize to the nucleus, and many, including Sp100, HDAC4, Mdm2, and Smad4, require their NLSs for sumoylation (26, 67, 82, 115).

The ΨKXE motif is very short and is found in many proteins, most of which are probably not modified by SUMO. For example, out of 5884 open reading frames (ORFs) in *S. cerevisiae*, there are 2799 sequences of the form (IVL)KXE distributed in 1913 different ORFs. Thus, interactions other than those between Ubc9 and the ΨKXE motif are likely to be critical in determining which proteins are sumoylated. Most of these probably involve interactions between an E3 and the substrate or a substrate-associated protein. However, the crystal structure of the RanGAP1-Ubc9 complex shows an additional contact besides the ΨKXE interaction (51), suggesting that other interactions between the substrate and Ubc9 may also participate in substrate selection.

Several proteins are also modified at sites other than ΨKXE. The replication processivity factor PCNA has two sumoylation sites, one conforming to the consensus sequence and the other at a T*K*ET sequence (24). TEL, PML, Smad4, and the Epstein Barr virus BZLF1 protein have reported sumoylation sites at T*K*ED, A*K*CP, V*K*YC, and V*K*FT, respectively, and both lysines in a G*K*VE*K*VD sequence in Axin are sumoylated (116–120). Moreover, some sumoylated proteins, such as Mdm2, Daxx, CREB, and CTBP-2, do not contain a ΨKXE sequence; others are still sumoylated when all consensus sites are mutated (61, 82, 121–124). It is not known how these nonconsensus sites are recognized.

Regulation of SUMO Conjugation

The set of proteins that is modified by SUMO changes during the cell cycle and in response to various conditions, but how SUMO conjugation is regulated is not well understood. In theory, sumoylation could be regulated at the level of either attachment or removal of SUMO; a change in either rate would alter the steady-state amount of protein modified. Some examples of proteins showing regulated SUMO modification are the yeast bud neck-associated septin proteins, which are modified only during mitosis and only on the mother-cell side of the bud neck (40). Septin sumoylation requires the E3 Siz1, which itself localizes to the mother-cell side of the bud neck exclusively during mitosis (38, 62). Thus, it is likely that septin sumoylation is regulated by controlling the localization of Siz1, possibly via phosphorylation of Siz1.

Phosphorylation of several substrates affects their sumoylation, mostly negatively. Phosphorylation of c-jun, PML, and IκBα correlates with reduced SUMO attachment (25, 125, 126). Furthermore, the antagonistic relationship between phosphorylation and sumoylation is involved in activation of the transcription factor Elk-1 by MAP kinases (127). In unstimulated cells, sumoylated Elk-1 represses Elk-1-dependent gene expression (see below). Upon MAPK-dependent phosphorylation, Elk-1 is desumoylated and transcription is activated. However, phosphorylation has the opposite effect on sumoylation of the heat shock transcription factor HSF1, which must be phosphorylated in order to be sumoylated (128, 129).

Because lysines serve as the attachment sites for several modifications, which include Ub, other Ubls, acetylation, and methylation, it is possible that these modifications might regulate each other by competing for the same lysines. In fact, several proteins contain a lysine that can be modified by either Ub or SUMO (see below), and the transcription factor Sp3 contains a lysine that can be either acetylated or sumoylated (92, 130, 131).

Sumoylation of some proteins is regulated by binding interactions with other macromolecules. Sumoylation of Mdm2 and p53 in vivo is enhanced by association with the tumor suppressor ARF (132). In another example, sumoylation of the base excision repair protein thymine DNA glycosylase (TDG) in vitro is stimulated both by DNA and by the downstream enzyme in the repair pathway (36). In contrast, the transcription factor Sp3 is resistant to sumoylation when bound to DNA (92).

BIOLOGICAL FUNCTIONS OF SUMO

Although identification of the enzymes of the SUMO pathway has proceeded rapidly, investigations of how SUMO affects biological processes are only at the early stages. Several features of the SUMO system, including the low levels of modification, the presence of Ulp activity in native lysates, and a number of complex interactions among different enzymes and substrates, combine to make functional analysis challenging. In fact, for some proteins that have been reported to be sumoylated, it is not clear that there is a function, or even that the protein is really sumoylated under endogenous expression levels of SUMO pathway enzymes.

The most important experiment in studying the function of SUMO conjugation to a particular protein is mutational elimination of the SUMO attachment site(s). This is usually done by mutating the attachment-site lysine, but because lysines can also serve as attachment sites for other modifications, the assignment of any effects to SUMO is more convincing if mutations at other positions in the ΨKXE motif show similar effects. Overexpression, dominant negative, or knockdown experiments involving SUMO pathway enzymes can complement these results, but it is imperative that such experiments be done with both wild-type substrate and the substrate that cannot be sumoylated, to confirm that any effects are direct. Often the same effect is seen whether or not the SUMO attachment site in the protein being studied is present, suggesting that the effect involves sumoylation of another protein in the same pathway.

Transcription

Many of the known substrates of SUMO in mammalian systems are involved in gene expression and include transcriptional activators, repressors, coactivators, corepressors, and components of large subnuclear structures called PML nuclear bodies (PML NBs), PODs, or ND10. Two recent reviews address the role of SUMO in transcription (4, 5). For simplicity, the transcriptional effects of SUMO can be divided into two groups: those that are likely to involve sumoylated transcription factors bound to a particular promoter and those that involve PML NBs (4). The activities of many transcription factors are regulated by association with PML NBs, and assembly of PML NBs requires sumoylation of the PML protein. Thus, changes in the level of PML sumoylation have broad effects on transcription by several pathways.

Although direct evidence of sumoylated transcription factors bound to promoters has not yet been obtained, it is becoming clear that the presence of SUMO at a promoter represses trancription. Mutations that prevent SUMO attachment to the transcription factors Elk-1, Sp-3, SREBPs, STAT-1, SRF, c-myb, C/EBPs, to the androgen receptor, or to the coactivator p300 all increase transcription from responsive promoters, consistent with a negative role for SUMO in gene expression (86, 92, 127, 131, 133–138). Some other interesting examples of this effect are several transcription factors with "synergy control motifs," which were originally identified in the glucocorticoid receptor (GR) as peptide motifs that reduce GR-dependent transcription from promoters containing multiple GR binding elements (139). Mutating these motifs does not affect transcription from promoters with a single element. The critical feature of synergy control motifs is a ΨKXE sequence, and these sites are sumoylated, suggesting that SUMO attachment reduces the positive synergistic effect of having multiple receptors bound to the same promoter (139–141). Sumoylation of the progesterone receptor (PR) is also involved in complex negative regulatory interactions in which one isoform of PR, PR-A, can "transrepress" the transcriptional activity of the other isoform PR-B or of the estrogen receptor (142).

A possible clue to the mechanism of SUMO in transcriptional repression is that targeting SUMO itself to a promoter, by fusing it to a DNA binding domain, is sufficient to reduce promoter activity (127, 131). Because it is unlikely that SUMO per se has this activity, this result suggests that SUMO recruits other factors that repress transcription. Candidates for such factors include HDACs, the repressor protein Daxx, the NuRD complex component CHD3/ZFH, and PIAS proteins. HDAC6 binds to the repressor domain of p300 only when it is sumoylated. Furthermore, si-RNA-mediated knockdown of HDAC6 relieves SUMO-dependent transcriptional repression by p300, consistent with a model where SUMO attenuates transcription through recruitment of HDAC6 (138). Daxx and CHD3/ZFH both bind directly to SUMO, and both also associate with HDACs and are involved in transcriptional repression (72, 143–145). Intriguingly, some PIAS proteins interact with HDACs, and PIAS proteins also bind directly to SUMO and sumoylated proteins (72, 92, 146, 147). In fact, PIASy actually binds more tightly to SUMO-Sp3 than it does to unmodified Sp3 (92), which it targets for sumoylation, suggesting that PIASy may also function downstream of Sp3 sumoylation. When different PIAS proteins are tethered to promoters by fusion to DNA binding domains, some have negative effects on transcription, while others have positive effects (148). A distinct mechanism for SUMO in transcriptional repression involves sumoylation of HDACs themselves. HDAC1 and HDAC4 are both sumoylated, and sumoylation enhances their transcriptional repression activities (67, 149, 150).

Although SUMO attachment to most transcription factors results in repression, SUMO apparently has positive effects on transcriptional activation by the heat shock factors HSF1 and HSF2 and the β-catenin activated factor Tcf-4. HSF1 is sumoylated in response to heat shock, coinciding with HSF1 activation (129, 151), and, remarkably, sumoylation promotes binding of both HSF1 and HSF2 to DNA in vitro (151, 152). However, it is not yet clear whether this mechanism operates in vivo (129). Tcf-4-dependent transcription is activated by coexpression of β-catenin and PIASy, and this activation is reduced when Tcf-4 lacks SUMO attachment sites, suggesting that sumoylation activates Tcf-4 (153).

PML Nuclear Bodies

Other effects on transcription are mediated by PML NBs, whose central component is the PML protein. PML was discovered because the t(15;17) chromosomal translocation that causes acute promyelocytic leukemia (APL) generates a fusion between PML and the retinoic acid receptor (RARα) [reviewed in (154, 155)]. Normal interphase cells have 5–10 PML NBs per nucleus, but NBs are disrupted by many viruses and by expression of the PML-RARα fusion. PML -/- mice are viable but vulnerable to infection and to developing tumors, while PML-/- cells in culture are radiation resistant and defective in p53-induced apoptosis. A number of other proteins also localize to PML NBs; these include

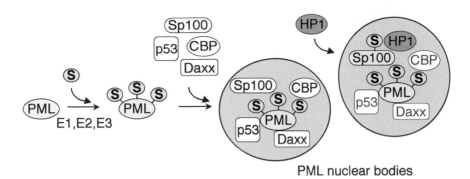

PML nuclear bodies

Figure 3   Assembly of PML nuclear bodies (NBs). Attachment of SUMO (S) to PML promotes formation of PML NBs and recruitment of associated proteins. Sumoylation of associated proteins may allow additional proteins (e.g., HP1) to bind.

the tumor suppressor p53, the Bloom Syndrome gene product BLM, the coactivator CBP, and Daxx, a transcriptional repressor that has been implicated in apoptosis. Two hypotheses regarding the function of NBs are that they are storage depots for nuclear factors or that they are the site of specific activities, such as modification or assembly of transcription factors. For example, there is evidence that acetylation of p53 by CBP takes place in PML NBs (156).

PML is covalently modified by SUMO at three sites (117, 157, 158), and sumoylation of PML is essential for formation of morphologically normal NBs and for recruitment of interacting proteins. When PML lacking SUMO attachment sites is introduced into PML -/- cells, the mutant PML protein forms aggregates, and many of the interacting proteins, including Sp100, CBP, ISG20, Daxx, and SUMO-1, fail to colocalize with either the PML or with each other (117, 157–161) (Figure 3). In several situations, higher levels of PML-SUMO conjugates correlate with enhanced PML NB formation. Arsenic trioxide, which can be used to treat APL, promotes both sumoylation of PML and reorganization of NBs (157). The converse effect is seen early in infection by many viruses, where disruption of PML NBs takes place simultaneously with desumoylation of PML [reviewed in (162, 163)].

Many of the other proteins that localize to PML NBs also become sumoylated. Curiously, most of these proteins still localize to PML NBs even if their sumoylation sites are mutated, suggesting that sumoylation of these proteins has some purpose other than to promote association with NBs. Proteins for which this is true include p53, LEF1, Sp100, Daxx, SRF1, and the cytomegalovirus proteins IE1 and IE2 (35, 64, 115, 121, 137, 164–167). One possible explanation for these results is that sumoylation of different proteins may produce a hierarchy of interactions: Sumoylation of PML could allow binding of one set of proteins, and sumoylation of these proteins could promote binding of another layer of proteins. For example, sumoylation of Sp100 enhances binding to the heterochromatin protein HP1 in vitro, suggesting that Sp100 sumoylation may recruit HP1 to NBs

(168) (Figure 3). A second possibility is that sumoylation of different proteins creates a web of cooperative interactions, and loss of some of them is not sufficient to destabilize the whole structure (4). It is also conceivable that proteins in NBs may be sumoylated somewhat nonspecifically because high levels of sumoylation occur in NBs or because NB proteins are protected against de-sumoylating activities. However, arguing against this, mutant versions of Sp100 and CMV IE1 that do not localize to NBs are still sumoylated (115, 169), suggesting that they are specifically targeted for sumoylation.

Changes in the levels of various components of the SUMO pathway can have dramatic effects on the structure of PML NBs, with correspondingly dramatic effects on transcription, probably through sequestration and release of various NB-associated factors. For example, sequestration of the repressor protein Daxx by conditions that promote PML NB formation leads to activation of promoters that are otherwise repressed by Daxx (170–172). Another example involves c-jun-dependent transcription, which is strongly induced by overexpression of the SuPr-1 isoform of the SUMO isopeptidase SENP-2 (110). This induction does not depend on sumoylation of c-jun, but of PML, and does not take place in cells expressing only unsumoylatable PML. Paradoxically, SuPr-1 reduces PML sumoylation and disrupts PML NBs. This result suggests that c-jun-dependent transcription may be induced by a factor that is activated and sequestered in SUMO-PML-containing NBs but that is then released in greater quantities when PML NBs are disrupted by SuPr-1. In addition, overexpression of PIAS proteins has many transcriptional effects, and although it has not been tested in most cases, it seems likely that some of these effects are mediated by changes in PML NBs.

Chromosome Organization and Function

Genetic studies of SUMO pathway function in model organisms indicate a role for SUMO conjugation in higher-order chromatin structure and in chromosome segregation, but the molecular basis of these effects is largely unknown. *S. pombe* strains lacking SUMO conjugation, although viable, grow very poorly, are sensitive to DNA damaging agents, have a high frequency of chromosome loss and aberrant mitosis, and develop elongated telomeres (32, 54, 56). Furthermore, a mutant in the *D. melanogaster dpias* gene was isolated as a suppressor of position effect variegation, an effect in which heterochromatin induces transcriptional silencing of adjacent loci (74). *dpias* mutants also have chromosome condensation defects, aberrant chromosome segregation, high frequency of chromosome loss, and defects in telomere clustering and telomere-nuclear lamina associations (74). The *S. cerevisiae ulp2Δ* strain also has a number of phenotypes indicating genomic instability and is defective in targeting the condensin complex, which is required for chromosome condensation, to rDNA repeats (96, 102, 103).

Several lines of evidence implicate SUMO in kinetochore function. SUMO was first identified in yeast as a high-copy suppressor of mutations in the *MIF2* gene, which encodes a centromere-binding protein related to vertebrate CENP-C (173). CENP-C mutants are also suppressed by overexpression of SUMO (174). In addition, SUMO localizes at or adjacent to the kinetochore in mammalian cells, and a number of proteins associate with both centromeres and PML NBs, raising the possibility of a common, SUMO-related mechanism (175–178). The best characterized centromere function involves *S. cerevisiae* strains lacking the SUMO isopeptidase Ulp2, which exhibit premature separation of a section of the chromosome near the centromere prior to mitosis (179). *ulp2Δ* strains contain elevated levels of sumoylated topoisomerase II (Top2), and mutating the SUMO attachment sites in Top2 suppresses not only this precocious chromosome separation phenotype but also the temperature sensitivity of *ulp2Δ* mutants, suggesting that these phenotypes result in part from excess SUMO conjugation to Top2.

DNA Repair

Specific roles for SUMO in two DNA repair pathways have been described, and there are indications that SUMO also acts in other repair pathways. An elegant study focuses on the sumoylation of thymine DNA glycosylase (TDG), a base excision repair enzyme that removes thymine or uracil from T-G or U·G mismatched base pairs (36). The product of the TDG reaction is an abasic site, which is then repaired by downstream enzymes. In vitro, unmodified TDG carries out only a single round of base removal because the enzyme binds tightly to the abasic site that is produced by the reaction. In vivo this interaction probably facilitates transfer of the abasic site to the downstream machinery for completion of repair. SUMO-TDG, in contrast, catalyzes multiple rounds of base removal in vitro, indicating that it is not as strongly inhibited by its product as is unmodified TDG. Furthermore, SUMO conjugation to TDG is stimulated by DNA and by APE1, a downstream enzyme that processes abasic sites. These data suggest a model in which unmodified TDG cleaves the mismatched T or U and then, coordinated with recruitment of the downstream enzymes to the site, is sumoylated, released, and then desumoylated, regenerating the high-affinity form to carry out the next cycle of catalysis (36) (Figure 4a).

SUMO may also participate in maintaining the activity of DNA topoisomerase I (TOP1) and topoisomerase II, which are both sumoylated in mammalian cells in response to topoisomerase inhibitors (180, 181). A TOP1 mutant lacking the active site is also constitutively sumoylated in the absence of inhibitors (182), suggesting that some feature of the inactive protein promotes its sumoylation. Upon treatment of cells with the TOP1 inhibitor camptothecin, wild-type TOP1 clears from the nucleoli and disperses throughout the nucleus, whereas TOP1 lacking the SUMO attachment sites remains in the nucleoli (183), indicating that sumoylation may regulate TOP1 localization or may increase its activity.

SUMO also participates in the yeast postreplication repair system, which repairs DNA lesions during the G2 phase of the cell cycle (24). A critical element of this system is the attachment of Ub, either as mono-Ub or as a Ub chain, to

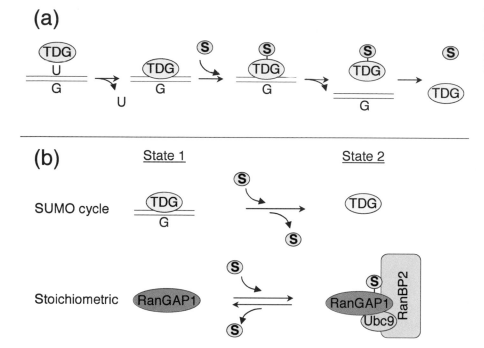

Figure 4 Stoichiometric versus cycling mechanisms for SUMO. (*a*) A model for the role of SUMO (S) in the thymine DNA glycosylase (TDG) reaction (36). (*b*) SUMO may be able to work through a sumoylation-desumoylation cycle in which SUMO promotes a change in the substrate that persists after desumoylation, or stoichiometrically, such that desumoylation restores the original state. Examples are described in the text.

the proliferating cell nuclear antigen (PCNA) at Lys^{164}. SUMO competes for attachment to this lysine and can also be attached at a second site. As would be expected if SUMO is blocking ubiquitylation, genetic evidence indicates that SUMO conjugation inhibits damage-induced DNA repair and mutagenesis (24, 184). PCNA is sumoylated most heavily during the S phase of the cell cycle; this may suggest that sumoylation prevents inappropriate recruitment of postreplication repair enzymes during the wrong phase of the cell cycle. Interestingly, either sumoylation or mono-ubiquitylation of PCNA can participate in spontaneous mutagenesis by this pathway (184), suggesting that SUMO can also affect PCNA function independently of Ub.

Nuclear Transport

Investigators studying nuclear transport were the first to discover that SUMO modifies other proteins when they isolated sumoylated RanGAP1, which is the most abundant SUMO-1 conjugate in vertebrate cells (22, 23). RanGAP1 is the

GTPase activating protein for the small GTPase Ran, which plays a central role in nucleocytoplasmic transport and also participates in several events during mitosis (185). It is not clear what role sumoylation of RanGAP1 plays in nuclear transport. SUMO-RanGAP1 binds tightly to the nuclear pore complex (NPC) by participating in a stable trimeric complex with Ubc9 and the IR domain of RanBP2/Nup358 (22, 89, 90). This tightly bound RanGAP1 is crucial in nuclear import assays in vitro, and soluble RanGAP1 cannot substitute for it (22). However, plant and yeast RanGAPs are not sumoylated, and yeast RanGAP localizes to the cytoplasm, indicating that in yeast, nuclear transport does not depend on RanGAP localization at the NPC. Another possible function of RanGAP1 sumoylation is that it could participate in the mitotic functions of Ran. During mitosis, SUMO-RanGAP1 localizes to the mitotic spindle and associates most strongly with the kinetochores. RanGAP1 that cannot be sumoylated does not associate with spindles (176). These results may indicate a centromere-associated function for SUMO-RanGAP1.

SUMO conjugation to proteins other than RanGAP1 also affects nuclear versus cytoplasmic localization. Even though yeast RanGAP is not sumoylated, nuclear import of certain yeast proteins is impaired in SUMO pathway mutants (186). This effect could involve sumoylation of other nuclear transport factors or of the cargo proteins. In mammalian cells, the presence of the E3 RanBP2 on the cytoplasmic site of the NPC and the SUMO isopeptidase SENP2 on the nucleoplasmic side suggests a model in which proteins might be rapidly sumoylated and desumoylated as they are imported into the nucleus (22, 23, 99, 100). There is no direct evidence for this idea, but the SUMO attachment sites in several proteins are required for their nuclear localization (93, 122, 124, 187). However, it is not clear whether SUMO affects nuclear transport or nuclear retention. The opposite effect on nuclear localization has also been seen: Sumoylation of TEL and *Dictyostelium* MEK1 is associated with their export to the cytoplasm (188, 189).

Sumoylation of Nonnuclear Proteins

Most SUMO conjugates are nuclear proteins, and it is likely that most of the major functions of SUMO take place in the nucleus. However, there are several cytoplasmic SUMO conjugates. The most prominent example is of course SUMO-RanGAP1, which is found on the cytoplasmic fibrils of the nuclear pore complex. Others include the yeast septins, which form a filamentous structure at the yeast bud neck, the glucose transporters GLUT1 and GLUT4 (40, 190, 191), and the signaling proteins IκBα, Axin, and *Dictyostelium* MEK1.

Signal Transduction Pathways

Stimulation of the inflammatory response pathway leads to activation of the transcription factor NFκB by promoting Ub-dependent degradation of the NFκB inhibitor IκB. SUMO conjugation to IκBα can inhibit this step because SUMO is linked to the same Lys where Ub would be attached, thereby preventing IκBα degradation (25). Consistent with a role for SUMO in stabilizing IκBα, SUMO overexpression inhibits NFκB-dependent transcription in mammalian cells (25). Curiously, SUMO has the opposite effect on the orthologous pathway in *Drosophila*, where sumoylation of the NFκB ortholog Dorsal apparently promotes its import into the nucleus and transcriptional activity (192).

Another example of a role for SUMO in signal transduction is found in *Dictyostelium*, where a MAP kinase pathway controls chemotaxis and aggregation in response to extracellular cAMP (189). Within 15 s after cAMP addition, the MAPK kinase MEK1 becomes sumoylated, and the initially nuclear MEK1 and SUMO localize to the plasma membrane. It is not clear whether SUMO enhances nuclear export or plasma membrane association of MEK1. Simultaneously, the downstream MAPK ERK1 relocalizes from the cytoplasm to the plasma membrane, suggesting that ERK1 activation takes place at the plasma membrane. Strikingly, by 3 min after pathway activation, MEK1 has been desumoylated, and MEK1 and SUMO both disappear from the plasma membrane.

The SUMO pathway also affects signaling dependent on Axin, a protein that serves as a scaffold for enzymes in the Wnt pathway and participates in activation of the JNK MAP kinase. Axin is sumoylated at two sites at its extreme C terminus (119), and deletion of these sites eliminates MEKK1-dependent JNK activation but has no effect on Wnt signaling. Axin also interacts with two isoforms of the isopeptidase SENP-2, Axam and Axam2, and expressing either of these inhibits Wnt signaling, although the mechanism is not clear (111, 112).

MECHANISMS OF SUMO ACTION

SUMO's Interactions With the Ub-Proteasome Pathway

One way SUMO affects the function of its substrates is by preventing ubiquitylation at specific lysine residues. PCNA, Smad4, and IκBα are all examples of substrates where a single lysine residue can be either sumoylated or ubiquitylated. However, it is still not clear in these cases whether SUMO exclusively regulates ubiquitylation or whether it also has a distinct function. The model that sumoylation acts solely by blocking ubiquitylation is perplexing, because often very little of the protein is sumoylated. For example, only a small fraction of IκBα is sumoylated in unstimulated cells, so that upon activation of the inflammatory response pathway, most of the NFκB in the cell could still be activated via degradation of the remaining unsumoylated IκBα. One possible answer to this dilemma is that SUMO may act primarily by shutting off the inflammatory response, rather than by modulating its activation (124). Hypoxia induces proinflammatory genes through Ub-dependent degradation of CREB (cAMP response element binding protein). CREB can also be sumoylated and is stabilized by SUMO overexpression. Strikingly, hypoxia induces ubiquitylation of CREB within one hour, but it induces sumoylation of both CREB and IκBα slowly, with maximal sumoylation after 24–48 h (124). This late induction of sumoylation is consistent with a role for SUMO in resolution of the response.

In other cases, SUMO also appears to have a separate function in addition to preventing ubiquitylation. The transcription factor Smad4 is protected from Ub-dependent proteolysis by attachment of SUMO at its ubiquitylation site, but there is also evidence that sumoylation separately promotes nuclear retention of Smad4 (26). In another example, sumoylation of PCNA inhibits Ub-dependent postreplication DNA repair, consistent with a function for SUMO in blocking ubiquitylation. However, sumoylated PCNA can itself promote spontaneous mutagenesis through the postreplication repair pathway (184), indicating an independent role for SUMO.

SUMO also interacts with the Ub-proteasome pathway by other uncharacterized mechanisms. Sumoylation inhibits degradation of c-myb but not by competing for the ubiquitylation site (134). In contrast, SUMO conjugation coincides with degradation of both PML and the PML-RARα fusion protein. Agents such as arsenic trioxide induce both sumoylation and proteasome-dependent degradation of PML and PML-RARα, and the SUMO attachment sites are required for this degradation (161, 193). Of course, if these sites were also used as ubiquitylation sites, the same result would be obtained. Arsenic trioxide also enhances recruitment of the 11S proteasome regulator to PML NBs (161).

Sumoylation Modulates Interactions of Substrate

The most common mode of SUMO action is to alter substrate binding interactions with other macromolecules. Three nonmutually exclusive models for this are (a) the linked SUMO itself could interact with other proteins; (b) both SUMO and the substrate could contribute determinants of the interaction surface; or (c) SUMO could alter the conformation of the substrate, exposing or hiding binding sites within the modified protein. Several of the proteins isolated in the yeast two-hybrid screen with SUMO do bind SUMO noncovalently in pull-down assays. A number of these, including HIPK2 (homeodomain-interacting protein kinase 2), the cytomegalovirus protein IE2, and PIAS proteins, contain a SUMO interacting motif (SIM), which is likely to mediate this interaction. The function of SIMs has not been fully investigated, but the SIM in PIASy is involved in its localization and transcriptional effects (64), and the SIM in HIPK-2 is required for HIPK-2-dependent disruption of PML NBs (194), demonstrating that these motifs have relevant physiological functions. Many of the proteins that interact with SUMO noncovalently, such as TDG, Daxx, CMV IE2, and Dnmt3b, are also covalently modified by SUMO (36, 144, 164, 195). This ability of proteins both to be sumoylated and to interact noncovalently with SUMO may enhance complex formation between various sumoylated proteins, as in PML NBs. However, there are very little data on the prevalence and function of direct noncovalent interactions with conjugated SUMO, and it seems likely that other interactions involving the substrate would also be required for the effects of SUMO to be substrate-specific.

For most of the substrates that have been characterized, changes in binding capabilities are a collaboration between SUMO and the substrate. Sumoylation alters the DNA-binding characteristics of TDG, reducing its affinity for the abasic sites that are the products of its reaction (36). Two ideas for the way this might take place are that SUMO attachment could induce a conformational change in TDG or that SUMO could act more directly, possibly blocking access to the DNA by steric hindrance. In another example, association of the SUMO-RanGAP1 conjugate with RanBP2 requires both SUMO and sequences in RanGAP1. Unsumoylated RanGAP1 does not bind RanBP2, but there is also one RanGAP1 deletion mutant that is sumoylated properly but still does not associate with RanBP2 (90). This result shows that the presence of SUMO is not sufficient for binding to RanBP2; the binding determinant must include sequences in RanGAP1. Supporting this interpretation, free SUMO does not compete with SUMO-RanGAP1 for binding to RanBP2 (22). These results could be explained either by a model in which SUMO induces a conformational change in RanGAP1 to expose a RanBP2 binding-site that is entirely in RanGAP1 or by a model in which RanBP2 interacts with elements in both RanGAP1 and SUMO (90).

Stoichiometric Versus Cycling Roles for SUMO Conjugation

A notable feature of the SUMO system is that SUMO is often attached to only a few percent or less of a given protein. The only clear exception is RanGAP1, which is $\sim$50% modified in most cells (22, 23). Therefore an important unresolved question is how SUMO can affect protein function when only a very small fraction is modified. One possibility is that SUMO could act on a subpopulation of a protein that is different structurally or functionally from the rest of the pool of that protein. For example, it is possible that some transcription factors are preferentially sumoylated when they are bound to certain promoters. Another possibility is that SUMO conjugation could be acting through a cycle of sumoylation and desumoylation, rather than by persistent attachment of SUMO to the substrate. In this model, SUMO attachment would promote a single event, whose consequences would persist after desumoylation. The role that has been proposed for SUMO in TDG function is an example of a such a cycle (36) (Figure 4a). Unmodified TDG removes the thymine or uracil at a mismatched site and then remains bound until it is sumoylated. The sumoylated TDG releases from the abasic site and is then desumoylated to prepare it for the next round of high-affinity binding. This cycle converts the DNA-bound form of TDG, state 1, to the unbound form, state 2, where neither of the two states is modified by SUMO (Figure 4b). In this way, the whole population of a protein could be affected by sumoylation, but very little of it would be modified at a given time. It easy to imagine how a sumoylation-desumoylation cycle could act in other situations as well, possibly by promoting protein interactions, inducing conformational changes, or even stimulating other protein modifications that would then be maintained after removal of SUMO. Many of the functions of Ub, including the proteasome pathway and the sorting of membrane proteins in the endosomal system, are carried out by a cycle of ubiquitylation and deubiquitylation. This SUMO cycle model contrasts with a model in which SUMO acts stoichiometrically (Figure 4b). Here, attachment of SUMO alters the state of the substrate, and desumoylation returns it to its original state. This is likely to be the case with RanGAP1; the unmodified form localizes to the cytoplasm, state 1, while the sumoylated form associates tightly with the nuclear pore, state 2.

CONCLUDING REMARKS

Work over the last several years has shown SUMO to be a remarkably versatile regulator of protein function, both in the number of different biological pathways that it affects and in the different sorts of mechanisms by which it controls the activities of other proteins. Many fundamental questions remain to be answered about both the biological function of SUMO and its mechanism of action. Why is SUMO essential for viability of most eukaryotic cells? What role does it play in maintaining chromosome structure? What are the substrates whose sumoylation participates in these processes? How are sumoylation and desumoylation regulated? How does SUMO alter binding properties of proteins?

There are also fields in which we are catching only our first glimpses of a role for SUMO, as in the pathogenesis of several neurodegenerative diseases. The difficulties associated with detecting SUMO-modified proteins have delayed recognition of the widespread participation of SUMO in cellular processes, and it is likely that as these difficulties are overcome, even more roles for SUMO will be discovered.

ACKNOWLEDGMENTS

I thank G. Bylebyl and A. Reindle for comments on the manuscript. Work in the author's lab is supported by the NIH (GM62268).

The *Annual Review of Biochemistry* is online at http://biochem.annualreviews.org

LITERATURE CITED

1. Hershko A, Ciechanover A. 1998. *Annu. Rev. Biochem.* 67:425–79
2. Pickart CM. 2001. *Annu. Rev. Biochem.* 70:503–33
3. Schwartz DC, Hochstrasser M. 2003. *Trends Biochem. Sci.* 28:321–28
4. Seeler JS, Dejean A. 2003. *Nat. Rev. Mol. Cell Biol.* 4:690–99
5. Verger A, Perdomo J, Crossley M. 2003. *EMBO Rep.* 4:137–42
6. Kim KI, Baek SH, Chung CH. 2002. *J. Cell Physiol.* 191:257–68

7. Müller S, Hoege C, Pyrowolakis G, Jentsch S. 2001. *Nat. Rev. Mol. Cell Biol.* 2:202–10

8. Melchior F. 2000. *Annu. Rev. Cell Dev. Biol.* 16:591–626

9. Fraser AG, Kamath RS, Zipperlen P, Martinez-Campos M, Sohrmann M, Ahringer J. 2000. *Nature* 408:325–30

10. Hayashi T, Seki M, Maeda D, Wang W, Kawabe Y, et al. 2002. *Exp. Cell Res.* 280:212–21

11. Epps JL, Tanda S. 1998. *Curr. Biol.* 8:1277–80

12. Apionishev S, Malhotra D, Raghavachari S, Tanda S, Rasooly RS. 2001. *Genes Cells* 6:215–24

13. Jones D, Crowe E, Stevens TA, Candido EP. 2002. *Genome Biol.* 3:RESEARCH0002

14. Chen A, Mannen H, Li SS. 1998. *Biochem. Mol. Biol. Int.* 46:1161–74

15. Kamitani T, Kito K, Nguyen HP, Fukuda-Kamitani T, Yeh ET. 1998. *J. Biol. Chem.* 273:11349–53

16. Shen Z, Pardington-Purtymun PE, Comeaux JC, Moyzis RK, Chen DJ. 1996. *Genomics* 36:271–79

17. Mannen H, Tseng HM, Cho CL, Li SS. 1996. *Biochem. Biophys. Res. Commun.* 222:178–80

18. Joanisse DR, Inaguma Y, Tanguay RM. 1998. *Biochem. Biophys. Res. Commun.* 244:102–9

19. Howe K, Williamson J, Boddy N, Sheer D, Freemont P, Solomon E. 1998. *Genomics* 47:92–100

20. Lois LM, Lima CD, Chua NH. 2003. *Plant Cell* 15:1347–59

21. Kurepa J, Walker JM, Smalle J, Gosink MM, Davis SJ, et al. 2003. *J. Biol. Chem.* 278:6862–72

22. Mahajan R, Delphin C, Guan T, Gerace L, Melchior F. 1997. *Cell* 88:97–107

23. Matunis MJ, Coutavas E, Blobel G. 1996. *J. Cell Biol.* 135:1457–70

24. Hoege C, Pfander B, Moldovan GL, Pyrowolakis G, Jentsch S. 2002. *Nature* 419:135–41

25. Desterro JM, Rodriguez MS, Hay RT. 1998. *Mol. Cell* 2:233–39

26. Lin X, Liang M, Liang YY, Brunicardi FC, Feng XH. 2003. *J. Biol. Chem.* 278: 31043–48

27. Lee PS, Chang C, Liu D, Derynck R. 2003. *J. Biol. Chem.* 278:27853–63

28. Bayer P, Arndt A, Metzger S, Mahajan R, Melchior F, et al. 1998. *J. Mol. Biol.* 280:275–86

28a. Vijay-Kumar S, Bugg CE, Cook WJ. 1987. *J. Mol. Biol.* 194:531–44

29. Bylebyl GR, Belichenko I, Johnson ES. 2003. *J. Biol. Chem.* 278:44113–20

30. Boddy MN, Howe K, Etkin LD, Solomon E, Freemont PS. 1996. *Oncogene* 13:971–82

31. Kamitani T, Nguyen HP, Yeh ET. 1997. *J. Biol. Chem.* 272:14001–4

32. Tanaka K, Nishide J, Okazaki K, Kato H, Niwa O, et al. 1999. *Mol. Cell. Biol.* 19:8660–72

33. Johnson ES, Schwienhorst I, Dohmen RJ, Blobel G. 1997. *EMBO J.* 16:5509–19

34. Saitoh H, Hinchey J. 2000. *J. Biol. Chem.* 275:6252–58

35. Hofmann H, Floss S, Stamminger T. 2000. *J. Virol.* 74:2510–24

36. Hardeland U, Steinacher R, Jiricny J, Schär P. 2002. *EMBO J.* 21:1456–64

37. Tatham MH, Jaffray E, Vaughan OA, Desterro JM, Botting CH, et al. 2001. *J. Biol. Chem.* 276:35368–74

38. Johnson ES, Gupta AA. 2001. *Cell* 106: 735–44

39. Bencsath KP, Podgorski MS, Pagala VR, Slaughter CA, Schulman BA. 2002. *J. Biol. Chem.* 277:47938–45

40. Johnson ES, Blobel G. 1999. *J. Cell Biol.* 147:981–94

41. Mahajan R, Gerace L, Melchior F. 1998. *J. Cell Biol.* 140:259–70

42. Li Y, Wang H, Wang S, Quon D, Liu YW, Cordell B. 2003. *Proc. Natl. Acad. Sci. USA* 100:259–64

43. Takahashi Y, Toh-e A, Kikuchi Y. 2003. *J. Biochem.* 133:415–22

44. Walden H, Podgorski MS, Schulman BA. 2003. *Nature* 422:330–34
45. Okuma T, Honda R, Ichikawa G, Tsumagari N, Yasuda H. 1999. *Biochem. Biophys. Res. Commun.* 254:693–98
46. Desterro JM, Rodriguez MS, Kemp GD, Hay RT. 1999. *J. Biol. Chem.* 274:10618–24
47. Azuma Y, Tan SH, Cavenagh MM, Ainsztein AM, Saitoh H, Dasso M. 2001. *FASEB J.* 15:1825–27
48. Desterro JM, Thomson J, Hay RT. 1997. *FEBS Lett.* 417:297–300
49. Johnson ES, Blobel G. 1997. *J. Biol. Chem.* 272:26799–802
50. Tong H, Hateboer G, Perrakis A, Bernards R, Sixma TK. 1997. *J. Biol. Chem.* 272:21381–87
51. Bernier-Villamor V, Sampson DA, Matunis MJ, Lima CD. 2002. *Cell* 108:345–56
52. Tatham MH, Chen Y, Hay RT. 2003. *Biochemistry* 42:3168–79
53. Tatham MH, Kim S, Yu B, Jaffray E, Song J, et al. 2003. *Biochemistry* 42:9959–69
54. al-Khodairy F, Enoch T, Hagan IM, Carr AM. 1995. *J. Cell Sci.* 108(Pt. 2):475–86
55. Seufert W, Futcher B, Jentsch S. 1995. *Nature* 373:78–81
56. Shayeghi M, Doe CL, Tavassoli M, Watts FZ. 1997. *Nucleic Acids Res.* 25:1162–69
57. Jackson PK. 2001. *Genes Dev.* 15:3053–58
58. Hochstrasser M. 2001. *Cell* 107:5–8
59. Shuai K. 2000. *Oncogene* 19:2638–44
60. Pichler A, Gast A, Seeler JS, Dejean A, Melchior F. 2002. *Cell* 108:109–20
61. Kagey MH, Melhuish TA, Wotton D. 2003. *Cell* 113:127–37
62. Takahashi Y, Toh-e A, Kikuchi Y. 2001. *Gene* 275:223–31
63. Kahyo T, Nishida T, Yasuda H. 2001. *Mol. Cell* 8:713–18
64. Sachdev S, Bruhn L, Sieber H, Pichler A, Melchior F, Grosschedl R. 2001. *Genes Dev.* 15:3088–103
65. Takahashi Y, Kahyo T, Toh-e A, Yasuda H, Kikuchi Y. 2001. *J. Biol. Chem.* 276:48973–77
66. Kotaja N, Karvonen U, Jänne OA, Palvimo JJ. 2002. *Mol. Cell. Biol.* 22:5222–34
67. Kirsh O, Seeler JS, Pichler A, Gast A, Müller S, et al. 2002. *EMBO J.* 21:2682–91
68. Schmidt D, Müller S. 2002. *Proc. Natl. Acad. Sci. USA* 99:2872–77
69. Aravind L, Koonin EV. 2000. *Trends Biochem. Sci.* 25:112–14
70. Kipp M, Gohring F, Ostendorp T, van Drunen CM, van Driel R, et al. 2000. *Mol. Cell. Biol.* 20:7480–89
71. Tan JA, Hall SH, Hamil KG, Grossman G, Petrusz P, French FS. 2002. *J. Biol. Chem.* 277:16993–7001
72. Minty A, Dumont X, Kaghad M, Caput D. 2000. *J. Biol. Chem.* 275:36316–23
73. Mohr SE, Boswell RE. 1999. *Gene* 229:109–16
74. Hari KL, Cook KR, Karpen GH. 2001. *Genes Dev.* 15:1334–48
75. Betz A, Lampen N, Martinek S, Young MW, Darnell JE Jr. 2001. *Proc. Natl. Acad. Sci. USA* 98:9563–68
76. Chung CD, Liao J, Liu B, Rao X, Jay P, et al. 1997. *Science* 278:1803–5
77. Liu B, Liao JY, Rao XP, Kushner SA, Chung CD, et al. 1998. *Proc. Natl. Acad. Sci. USA* 95:10626–31
78. Wible BA, Wang L, Kuryshev YA, Basu A, Haldar S, Brown AM. 2002. *J. Biol. Chem.* 277:17852–62
79. Wu L, Wu H, Ma L, Sangiorgi F, Wu N, et al. 1997. *Mech. Dev.* 65:3–17
80. Moilanen AM, Karvonen U, Poukka H, Yan W, Toppari J, et al. 1999. *J. Biol. Chem.* 274:3700–4
81. Gross M, Liu B, Tan J, French FS, Carey M, Shuai K. 2001. *Oncogene* 20:3880–87
82. Miyauchi Y, Yogosawa S, Honda R, Nishida T, Yasuda H. 2002. *J. Biol. Chem.* 277:50131–36
83. Liu B, Shuai K. 2001. *J. Biol. Chem.* 276:36624–31
84. Chun TH, Itoh H, Subramanian L,

Iñiguez-Lluhi JA, Nakao K. 2003. *Circ. Res.* 92:1201–8

85. Rogers RS, Horvath CM, Matunis MJ. 2003. *J. Biol. Chem.* 278:30091–97
86. Ungureanu D, Vanhatupa S, Kotaja N, Yang J, Aittomaki S, et al. 2003. *Blood* 102:3311–13
87. Yokoyama N, Hayashi N, Seki T, Pante N, Ohba T, et al. 1995. *Nature* 376:184–88
88. Wu J, Matunis MJ, Kraemer D, Blobel G, Coutavas E. 1995. *J. Biol. Chem.* 270: 14209–13
89. Saitoh H, Pu R, Cavenagh M, Dasso M. 1997. *Proc. Natl. Acad. Sci. USA* 94: 3736–41
90. Matunis MJ, Wu J, Blobel G. 1998. *J. Cell Biol.* 140:499–509
91. Saitoh H, Sparrow DB, Shiomi T, Pu RT, Nishimoto T, et al. 1998. *Curr. Biol.* 8:121–24
92. Sapetschnig A, Rischitor G, Braun H, Doll A, Schergaut M, et al. 2002. *EMBO J.* 21:5206–15
93. Lin X, Sun B, Liang M, Liang YY, Gast A, et al. 2003. *Mol. Cell* 11:1389–96
94. Li SJ, Hochstrasser M. 1999. *Nature* 398:246–51
95. Mossessova E, Lima CD. 2000. *Mol. Cell* 5:865–76
96. Strunnikov AV, Aravind L, Koonin EV. 2001. *Genetics* 158:95–107
97. Li SJ, Hochstrasser M. 2003. *J. Cell Biol.* 160:1069–81
98. Panse VG, Küster B, Gerstberger T, Hurt E. 2003. *Nat. Cell Biol.* 5:21–27
99. Hang J, Dasso M. 2002. *J. Biol. Chem.* 277:19961–66
100. Zhang H, Saitoh H, Matunis MJ. 2002. *Mol. Cell. Biol.* 22:6498–508
101. Takahashi Y, Mizoi J, Toh-e A, Kikuchi Y. 2000. *J. Biochem.* 128:723–25
102. Schwienhorst I, Johnson ES, Dohmen RJ. 2000. *Mol. Gen. Genet.* 263:771–86
103. Li SJ, Hochstrasser M. 2000. *Mol. Cell. Biol.* 20:2367–77
104. Gan-Erdene T, Kolli N, Yin L, Wu K, Pan ZQ, Wilkinson KD. 2003. *J. Biol. Chem.* 278:28892–900
105. Mendoza HM, Shen LN, Botting C, Lewis A, Chen JW, et al. 2003. *J. Biol. Chem.* 278:25637–43
106. Yeh ET, Gong L, Kamitani T. 2000. *Gene* 248:1–14
107. Nishida T, Tanaka H, Yasuda H. 2000. *Eur. J. Biochem.* 267:6423–27
108. Kim KI, Baek SH, Jeon YJ, Nishimori S, Suzuki T, et al. 2000. *J. Biol. Chem.* 275: 14102–6
109. Bailey D, O'Hare P. 2002. *J. Gen. Virol.* 83:2951–64
110. Best JL, Ganiatsas S, Agarwal S, Changou A, Salomoni P, et al. 2002. *Mol. Cell* 10:843–55
111. Kadoya T, Yamamoto H, Suzuki T, Yukita A, Fukui A, et al. 2002. *Mol. Cell. Biol.* 22:3803–19
112. Nishida T, Kaneko F, Kitagawa M, Yasuda H. 2001. *J. Biol. Chem.* 276: 39060–66
113. Rodriguez MS, Dargemont C, Hay RT. 2001. *J. Biol. Chem.* 276:12654–59
114. Sampson DA, Wang M, Matunis MJ. 2001. *J. Biol. Chem.* 276:21664–69
115. Sternsdorf T, Jensen K, Reich B, Will H. 1999. *J. Biol. Chem.* 274:12555–66
116. Adamson AL, Kenney S. 2001. *J. Virol.* 75:2388–99
117. Kamitani T, Kito K, Nguyen HP, Wada H, Fukuda-Kamitani T, Yeh ET. 1998. *J. Biol. Chem.* 273:26675–82
118. Lin X, Liang M, Liang YY, Brunicardi FC, Melchior F, Feng XH. 2003. *J. Biol. Chem.* 278:18714–19
119. Rui HL, Fan E, Zhou HM, Xu Z, Zhang Y, Lin SC. 2002. *J. Biol. Chem.* 277: 42981–86
120. Chakrabarti SR, Sood R, Nandi S, Nucifora G. 2000. *Proc. Natl. Acad. Sci. USA* 97:13281–85
121. Jang MS, Ryu SW, Kim E. 2002. *Biochem. Biophys. Res. Commun.* 295:495–500
122. Rangasamy D, Woytek K, Khan SA, Wilson VG. 2000. *J. Biol. Chem.* 275: 37999–8004
123. Xirodimas DP, Chisholm J, Desterro JM,

Lane DP, Hay RT. 2002. *FEBS Lett.* 528:207–11

124. Comerford KM, Leonard MO, Karhausen J, Carey R, Colgan SP, Taylor CT. 2003. *Proc. Natl. Acad. Sci. USA* 100:986–91

125. Everett RD, Lomonte P, Sternsdorf T, van Driel R, Orr A. 1999. *J. Cell Sci.* 112(Pt. 24):4581–88

126. Müller S, Berger M, Lehembre F, Seeler JS, Haupt Y, Dejean A. 2000. *J. Biol. Chem.* 275:13321–29

127. Yang SH, Jaffray E, Hay RT, Sharrocks AD. 2003. *Mol. Cell* 12:63–74

128. Hilgarth RS, Hong YL, Park-Sarge OK, Sarge KD. 2003. *Biochem. Biophys. Res. Commun.* 303:196–200

129. Hietakangas V, Ahlskog JK, Jakobsson AM, Hellesuo M, Sahlberg NM, et al. 2003. *Mol. Cell. Biol.* 23:2953–68

130. Braun H, Koop R, Ertmer A, Nacht S, Suske G. 2001. *Nucleic Acids Res.* 29:4994–5000

131. Ross S, Best JL, Zon LI, Gill G. 2002. *Mol. Cell* 10:831–42

132. Chen LH, Chen JD. 2003. *Oncogene* 22:5348–57

133. Poukka H, Karvonen U, Jänne OA, Palvimo JJ. 2000. *Proc. Natl. Acad. Sci. USA* 97:14145–50

134. Bies J, Markus J, Wolff L. 2002. *J. Biol. Chem.* 277:8999–9009

135. Kim J, Cantwell CA, Johnson PF, Pfarr CM, Williams SC. 2002. *J. Biol. Chem.* 277:38037–44

136. Hirano Y, Murata S, Tanaka K, Shimizu M, Sato R. 2003. *J. Biol. Chem.* 278:16809–19

137. Matsuzaki K, Minami T, Tojo M, Honda Y, Uchimura Y, et al. 2003. *Biochem. Biophys. Res. Commun.* 306:32–38

138. Girdwood D, Bumpass D, Vaughan OA, Thain A, Anderson LA, et al. 2003. *Mol. Cell* 11:1043–54

139. Iñiguez-Lluhi JA, Pearce D. 2000. *Mol. Cell. Biol.* 20:6040–50

140. Tian S, Poukka H, Palvimo JJ, Jänne OA. 2002. *Biochem. J.* 367:907–11

141. Subramanian L, Benson MD, Iñiguez-Lluhi JA. 2003. *J. Biol. Chem.* 278:9134–41

142. Abdel-Hafiz H, Takimoto GS, Tung L, Horwitz KB. 2002. *J. Biol. Chem.* 277:33950–56

143. Hollenbach AD, McPherson CJ, Mientjes EJ, Iyengar R, Grosveld G. 2002. *J. Cell Sci.* 115:3319–30

144. Ryu SW, Chae SK, Kim E. 2000. *Biochem. Biophys. Res. Commun.* 279:6–10

145. Tong JK, Hassig CA, Schnitzler GR, Kingston RE, Schreiber SL. 1998. *Nature* 395:917–21

146. Long JY, Matsuura I, He DM, Wang GN, Shuai K, Liu F. 2003. *Proc. Natl. Acad. Sci. USA* 100:9791–96

147. Tussie-Luna MI, Bayarsaihan D, Seto E, Ruddle FH, Roy AL. 2002. *Proc. Natl. Acad. Sci. USA* 99:12807–12

148. Kotaja N, Aittomaki S, Silvennoinen O, Palvimo JJ, Jänne OA. 2000. *Mol. Endocrinol.* 14:1986–2000

149. David G, Neptune MA, DePinho RA. 2002. *J. Biol. Chem.* 277:23658–63

150. Petrie K, Guidez F, Howell L, Healy L, Waxman S, et al. 2003. *J. Biol. Chem.* 278:16059–72

151. Hong YL, Rogers R, Matunis MJ, Mayhew CN, Goodson ML, et al. 2001. *J. Biol. Chem.* 276:40263–67

152. Goodson ML, Hong Y, Rogers R, Matunis MJ, Park-Sarge OK, Sarge KD. 2001. *J. Biol. Chem.* 276:18513–18

153. Yamamoto H, Ihara M, Matsuura Y, Kikuchi A. 2003. *EMBO J.* 22:2047–59

154. Borden KL. 2002. *Mol. Cell. Biol.* 22:5259–69

155. Salomoni P, Pandolfi PP. 2002. *Cell* 108:165–70

156. Pearson M, Carbone R, Sebastiani C, Cioce M, Fagioli M, et al. 2000. *Nature* 406:207–10

157. Müller S, Matunis MJ, Dejean A. 1998. *EMBO J.* 17:61–70

158. Duprez E, Saurin AJ, Desterro JM, Lallemand-Breitenbach V, Howe K, et al. 1999. *J. Cell Sci.* 112(Pt. 3):381–93

159. Ishov AM, Sotnikov AG, Negorev D, Vladimirova OV, Neff N, et al. 1999. *J. Cell Biol.* 147:221–34

160. Zhong S, Müller S, Ronchetti S, Freemont PS, Dejean A, Pandolfi PP. 2000. *Blood* 95:2748–52

161. Lallemand-Breitenbach V, Zhu J, Puvion F, Koken M, Honore N, et al. 2001. *J. Exp. Med.* 193:1361–71

162. Regad T, Chelbi-Alix MK. 2001. *Oncogene* 20:7274–86

163. Everett RD. 2001. *Oncogene* 20:7266–73

164. Ahn JH, Xu Y, Jang WJ, Matunis MJ, Hayward GS. 2001. *J. Virol.* 75:3859–72

165. Fogal V, Gostissa M, Sandy P, Zacchi P, Sternsdorf T, et al. 2000. *EMBO J.* 19:6185–95

166. Kwek SS, Derry J, Tyner AL, Shen Z, Gudkov AV. 2001. *Oncogene* 20:2587–99

167. Spengler ML, Kurapatwinski K, Black AR, Azizkhan-Clifford J. 2002. *J. Virol.* 76:2990–96

168. Seeler JS, Marchio A, Losson R, Desterro JM, Hay RT, et al. 2001. *Mol. Cell. Biol.* 21:3314–24

169. Müller S, Dejean A. 1999. *J. Virol.* 73:5137–43

170. Lin DY, Lai MZ, Ann DK, Shih HM. 2003. *J. Biol. Chem.* 278:15958–65

171. Li H, Leo C, Zhu J, Wu X, O'Neil J, et al. 2000. *Mol. Cell. Biol.* 20:1784–96

172. Lehembre F, Müller S, Pandolfi PP, Dejean A. 2001. *Oncogene* 20:1–9

173. Meluh PB, Koshland D. 1995. *Mol. Biol. Cell* 6:793–807

174. Fukagawa T, Regnier V, Ikemura T. 2001. *Nucleic Acids Res.* 29:3796–803

175. Pluta AF, Earnshaw WC, Goldberg IG. 1998. *J. Cell Sci.* 111(Pt. 14):2029–41

176. Joseph J, Tan SH, Karpova TS, McNally JG, Dasso M. 2002. *J. Cell Biol.* 156:595–602

177. Everett RD, Earnshaw WC, Pluta AF, Sternsdorf T, Ainsztein AM, et al. 1999. *J. Cell Sci.* 112(Pt. 20):3443–54

178. Everett RD, Earnshaw WC, Findlay J, Lomonte P. 1999. *EMBO J.* 18:1526–38

179. Bachant J, Alcasabas A, Blat Y, Kleckner N, Elledge SJ. 2002. *Mol. Cell* 9:1169–82

180. Mao Y, Desai SD, Liu LF. 2000. *J. Biol. Chem.* 275:26066–73

181. Mao Y, Sun M, Desai SD, Liu LF. 2000. *Proc. Natl. Acad. Sci. USA* 97:4046–51

182. Horie K, Tomida A, Sugimoto Y, Yasugi T, Yoshikawa H, et al. 2002. *Oncogene* 21:7913–22

183. Rallabhandi P, Hashimoto K, Mo YY, Beck WT, Moitra PK, D'Arpa P. 2002. *J. Biol. Chem.* 277:40020–26

184. Stelter P, Ulrich HD. 2003. *Nature* 425:188–91

185. Weis K. 2003. *Cell* 112:441–51

186. Stade K, Vogel F, Schwienhorst I, Meusser B, Volkwein C, et al. 2002. *J. Biol. Chem.* 277:49554–61

187. Endter C, Kzhyshkowska J, Stauber R, Dobner T. 2001. *Proc. Natl. Acad. Sci. USA* 98:11312–17

188. Wood LD, Irvin BJ, Nucifora G, Luce KS, Hiebert SW. 2003. *Proc. Natl. Acad. Sci. USA* 100:3257–62

189. Sobko A, Ma H, Firtel RA. 2002. *Dev. Cell* 2:745–56

190. Giorgino F, de Robertis O, Laviola L, Montrone C, Perrini S, et al. 2000. *Proc. Natl. Acad. Sci. USA* 97:1125–30

191. Lalioti VS, Vergarajauregui S, Pulido D, Sandoval IV. 2002. *J. Biol. Chem.* 277:19783–91

192. Bhaskar V, Valentine SA, Courey AJ. 2000. *J. Biol. Chem.* 275:4033–40

193. Zhu J, Lallemand-Breitenbach V, de The H. 2001. *Oncogene* 20:7257–65

194. Engelhardt OG, Boutell C, Orr A, Ullrich E, Haller O, Everett RD. 2003. *Exp. Cell Res.* 283:36–50

195. Kang ES, Park CW, Chung JH. 2001. *Biochem. Biophys. Res. Commun.* 289:862–68

Annu. Rev. Biochem. 2004. 73:383–415
doi: 10.1146/annurev.biochem.73.011303.074021
Copyright © 2004 by Annual Reviews. All rights reserved
First published online as a Review in Advance on April 2, 2004

PYRIDOXAL PHOSPHATE ENZYMES:
Mechanistic, Structural, and Evolutionary Considerations

Andrew C. Eliot[1] and Jack F. Kirsch[2]

[1]Department of Chemistry and [2]Departments of Chemistry and Molecular and Cell Biology, University of California, Berkeley, California 94720-3206; email: aeliot@life.uiuc.edu, jfkirsch@uclink.berkeley.edu

Key Words   substrate specificity, reaction type specificity, enzyme inhibition, enzyme mechanism

■ **Abstract**   Pyridoxal phosphate (PLP)-dependent enzymes are unrivaled in the diversity of reactions that they catalyze. New structural data have paved the way for targeted mutagenesis and mechanistic studies and have provided a framework for interpretation of those results. Together, these complementary approaches yield new insight into function, particularly in understanding the origins of substrate and reaction type specificity. The combination of new sequences and structures enables better reconstruction of their evolutionary heritage and illuminates unrecognized similarities within this diverse group of enzymes. The important metabolic roles of many PLP-dependent enzymes drive efforts to design specific inhibitors, which are now guided by the availability of comprehensive structural and functional databases. Better understanding of the function of this important group of enzymes is crucial not only for inhibitor design, but also for the design of improved protein-based catalysts.

CONTENTS

INTRODUCTION

Following its identification in 1951 as one of the active vitamers of vitamin B_6 (1), pyridoxal 5'-phosphate (PLP) has been the subject of extensive research directed toward understanding its unequaled catalytic versatility.[1] As a result, the basic mechanisms of PLP-assisted reactions, both in solution and enzyme-associated, have been well characterized and are now a staple of biochemistry textbooks (for example, see 2, 3). A number of reviews have addressed the subject in greater detail, and readers are directed particularly to the recent articles by Hayashi (4) and John (5). *Transaminases,* edited by Christen & Metzler (6), remains an excellent and thorough source of information about these enzymes. More detailed reviews concentrate on the function of a number of specific enzymes, including tryptophan synthase (7), O-acetylserine sulfhydrylase (8), δ-aminolevulinate synthase (9), serine hydroxymethyltransferase (SHMT) (10), and branched chain amino acid aminotransferase (BCAT) (11).

In addition to their versatility as catalysts, PLP-dependent enzymes have attracted attention because of their widespread involvement in cellular processes. These enzymes are principally involved in the biosynthesis of amino acids and amino acid-derived metabolites, but they are also found in the biosynthetic pathways of amino sugars (12) and other amine-containing compounds. Their importance is further underscored by the number identified as drug targets. For example, inhibitors of γ-aminobutyric acid aminotransferase (GABA ATase) are used in the treatment of epilepsy (13), SHMT has been identified as a target for cancer therapy (14), and inhibitors of ornithine decarboxylase (ODC) are employed in the treatment of African sleeping sickness (15). Functional defects in PLP enzymes have furthermore been implicated in a number of disease pathologies, including homocystinuria, which is most frequently caused by mutations in cystathionine β-synthase (16, 17).

---

[1]Abbreviations: AATase, aspartate aminotransferase; ACC, 1-aminocyclopropane-1-carboxylate; AlaP, 1-aminoethylphosphonate; ALR, alanine racemase; AONS, 8-amino-7-oxononanoate synthase; ATase, aminotransferase; BCAT, branched chain amino acid aminotransferase; DAAT, D-amino acid aminotransferase; DGD, dialkylglycine decarboxylase; GABA, γ-aminobutyric acid; OAT, ornithine aminotransferase; ODC, ornithine decarboxylase; PLP, pyridoxal 5'-phosphate; PMP, pyridoxamine 5'-phosphate; SAM, S-adenosyl-L-methionine; SHMT, serine hydroxymethyl transfease; and TATase, aromatic amino acid aminotransferase.

Despite the long history of research in the field, we are only now beginning to answer some of the most exciting questions. In particular, technological advances that have allowed ever more rapid determination of enzyme structures and the accumulation of large sequence databases have also enhanced our understanding of enzymatic catalysis. Analyses of sequences and structures yield significant insight regarding the evolution of this diverse class of enzymes. Additionally, crystal structures elegantly demonstrate how PLP enzymes harness the potential of the cofactor to accelerate the rates only of specific reactions, and they provide the basis for targeted mutagenesis.

This review describes many of the new insights that have come from recent structural and mechanistic studies, with a primary focus on determinants of substrate specificity and reaction type, as well as the design of inhibitors for this class of enzyme. The mechanisms and structures are briefly discussed to provide background.

MECHANISTIC VERSATILITY OF PLP

PLP-catalyzed reaction types can be divided according to the position at which the net reaction occurs. Reactions at the α position include transamination, decarboxylation, racemization, and elimination and replacement of an electrophilic R group. Those at the β or γ position include elimination or replacement. Examples of each of these reactions are shown in Scheme 1, and the basic mechanisms are shown in Schemes 2 (α) and 3 (β and γ). Exceptions to these common types include the formation of a cyclopropane ring from S-adenosyl-L-methionine (SAM), catalyzed by 1-aminocyclopropane-1-carboxylate (ACC) synthase (18), and the cleavage of ACC to α-ketobutyrate and ammonia, catalyzed by ACC deaminase (19); these are not discussed here. Because many of the reaction pathways share common intermediates, a number of enzymes also catalyze reactions that are combinations of the basic types, such as the decarboxylation-dependent transamination (aminoisobutyrate + pyruvate $\rightarrow$ acetone + CO_2 + L-Ala) catalyzed by dialkylglycine decarboxylase (DGD; 20) or the γ-elimination and β-replacement (O-phospho-L-homoserine + H_2O $\rightarrow$ L-Thr + P_i) catalyzed by threonine synthase (21).

Although the scope of PLP-catalyzed reactions initially appears to be bewilderingly diverse, there is a simple unifying principle. The cofactor in all cases functions to stabilize negative charge development at C_α in the transition state that is formed after condensation of the amino acid substrate with PLP to form a Schiff base (referred to as the external aldimine).[2] The fully formed carbanion

---

[2]There are two exceptions to this basic principle. The glycogen phosphorylase family of enzymes utilizes the phosphate group of PLP for catalysis [reviewed in (106)], and the aminomutase family catalyzes a radical-initiated reaction on PLP-bound amino acid substrates [reviewed in (107)]. Neither of these families will be discussed in this review.

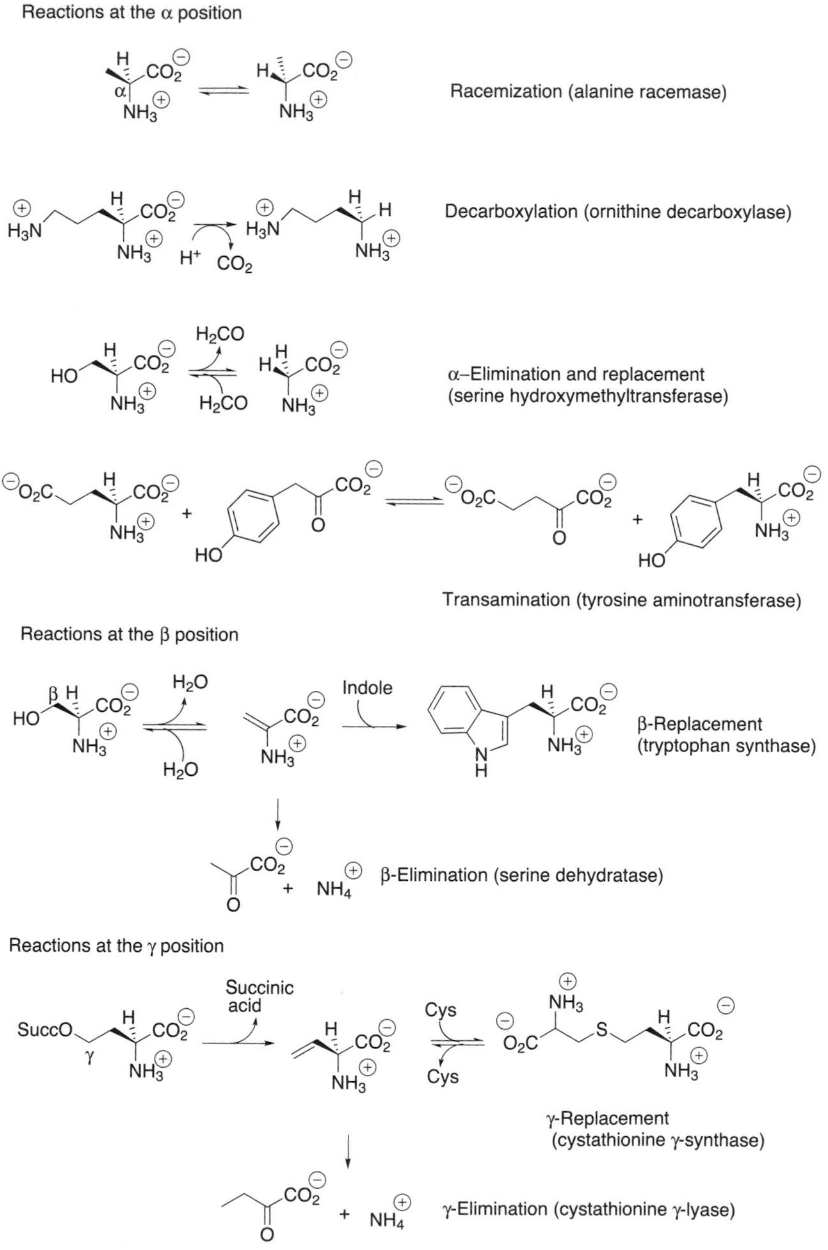
Reactions at the α position
Racemization (alanine racemase)
Decarboxylation (ornithine decarboxylase)
α–Elimination and replacement
(serine hydroxymethyltransferase)
Transamination (tyrosine aminotransferase)
Reactions at the β position
Indole
β-Replacement
(tryptophan synthase)
β-Elimination (serine dehydratase)
Reactions at the γ position
Succinic
acid
SuccO
Cys
γ-Replacement
(cystathionine γ-synthase)
γ-Elimination (cystathionine γ-lyase)

Scheme 1 Examples of each of the common reaction types catalyzed by PLP-dependent enzymes. Net changes at the α carbon (*top*) include racemization, decarboxylation, α-elimination and replacement, and transamination. Net changes at the β carbon (*middle*) include β-elimination or replacement, and those at the γ-carbon (*bottom*) are γ-replacement or elimination. The basic mechanisms for these reactions are shown in Schemes 2 and 3.

is referred to as the quinonoid intermediate (Scheme 4). The pK_a for loss of the C_α-proton from an amino acid in the absence of PLP is ~ 30 (22); therefore, that anion, formation of which is required in order to enable all of the described chemistry, is ordinarily inaccessible under physiological conditions.[3] The stabilization of the C_α-anion is facilitated by delocalization of the negative charge through the pi system of the cofactor, and for this reason PLP is often described as an electron sink. This factor allows PLP in the absence of enzyme to catalyze many of the possible reactions slowly (reviewed in 6). The function of the protein apoenzyme, therefore, is to enhance this innate catalytic potential and to enforce selectivity of substrate binding and reaction type. In most cases, this selectivity is exquisite—potential side reactions are limited to barely detectable levels.

STRUCTURAL DIVERSITY

Although PLP enzymes were for a time underrepresented in protein structure databases (5), the situation has been rectified in recent years, as an abundance of new structures have been solved. It was initially postulated that the structures of PLP enzymes would correlate with the reaction type (24), but it has since been found that each of the major structural classes contains representatives of multiple reaction types, the evolutionary implications of which are discussed below. All PLP enzymes whose structures have been solved to date belong to one of five fold types, which have been described in detail in two recent reviews (25, 26), and therefore are only briefly summarized here. Figure 1 shows single representatives of Fold Types I-IV, whose mechanisms are discussed in this review.

The majority of known structures are of Fold Type I (aspartate aminotransferase family) enzymes, a group that includes many of the best-characterized PLP enzymes. They invariably function as homodimers or higher-order oligomers, with two active sites per dimer. The active sites lie on the dimer interface, and each monomer contributes essential residues to both active sites. In general, the two active sites are independent, but asymmetry has been observed in a few cases. Negative cooperativity, for example, has been reported in GABA ATase, where a dimer with two functional active sites exhibits the same specific activity as a dimer with only one functional active site (27). The two active sites in each dimer of glutamate-1-semialdehyde aminomutase also exhibit different reactivities, as evidenced by the biphasic kinetics of reduction of the enzyme-PLP aldimine by sodium borohydride (28). Each monomer of Fold Type I enzymes has a large and a small domain. In a number of cases [e.g., aspartate aminotransferase (AATase)], these domains move significantly upon association with

---

[3]Certain non-PLP-dependent amino acid racemases, such as glutamate racemase (23), remain puzzling exceptions. If these reactions proceed through a transition state that has substantial anion character, it is not apparent how that state is stabilized.

Racemization (alanine racemase)

L-Ala aldimine   Quinonoid   D-Ala aldimine

Decarboxylation (prokaryotic ornithine decarboxylase)

Ornithine aldimine   Quinonoid   Putrescine aldimine

α–Elimination and replacement (serine hydroxymethyltransferase)

Serine aldimine   Quinonoid   Glycine aldimine

Transamination (tyrosine aminotransferase first half-reaction)

Glutamate aldimine   Quinonoid   Ketimine   PMP substrate, creating a closed conformation that may contribute to the specificity for both the substrate and the reaction type (see below).

The structures of Fold Type II (tryptophan synthase family) enzymes are similar to those of Fold Type I, but the proteins are evolutionarily distinct (29). One significant difference is that the active sites of Fold Type II enzymes are composed entirely of residues from one monomer [first observed in tryptophan synthase (30)]. Nevertheless, the functional form remains a homodimer or higher-order oligomer. These enzymes also differ from those of Fold Type I in that they often contain additional regulatory domains. Examples include threonine synthase (31) and cystathionine β-synthase (32), which are allosterically regulated by SAM, and threonine deaminase, which is regulated by isoleucine and valine (reviewed in 33).

The Fold Type IV (D-amino acid aminotransferase family) enzymes are superficially similar to Fold Types I and II, in that they are also functional homodimers, and the catalytic portion of each monomer is composed of a small and a large domain. The cofactor is bound in a site that is a near mirror image of the Fold Types I and II binding sites, so that the *re* rather than *si* face is solvent exposed (34).

Fold Types III (alanine racemase family) and V (glycogen phosphorylase family) are strikingly different from the other PLP enzymes. The Fold Type V enzymes are mechanistically distinct in utilizing the phosphate group of the cofactor for catalysis and are not considered further. The Fold Type III enzymes consist of a classical α/β barrel and a second β-strand domain. Interestingly, the mode of binding of PLP is similar to that of other fold types, with the phosphate

---

Scheme 2   Examples of mechanisms of reactions involving net change at the α position. All reactions are shown starting with the substrate aldimine, which is formed by transaldimination of the lysine-bound PLP. Racemization: the bacterial alanine racemase utilizes a tyrosine residue (51, 104) to deprotonate L-alanine, forming the quinonoid intermediate, which is reprotonated by a lysine residue on the opposite face of the cofactor to produce D-alanine. Decarboxylation: the reaction begins with loss of CO_2 from the substrate aldimine, producing the quinonoid intermediate. Protonation by an unidentified active site residue in ornithine decarboxylase generates the product aldimine. α-Replacement: the well-studied serine hydroxymethyltransferase (10) initiates the retro-aldol cleavage of serine by deprotonation of the hydroxyl group. Formaldehyde is released to generate the quinonoid intermediate. Protonation of the quinonoid at C_α by the lysine produces the product aldimine of glycine. Transamination: the first half-reaction catalyzed by tyrosine aminotransferase involves initial proton abstraction from the glutamate aldimine at C_α by the active site lysine, yielding the quinonoid intermediate. Reprotonation at C_4' of the cofactor by that lysine generates the ketimine intermediate, which is subsequently hydrolyzed to release α-ketoglutarate, leaving the enzyme in the PMP form. A complete catalytic cycle involves subsequent reaction with hydroxyphenylpyruvate to give tyrosine and to regenerate the PLP form of the enzyme.

β-Replacement (tryptophan synthase)

Serine aldimine   Quinonoid   Aminoacrylate aldimine

Tryptophan aldimine   Quinonoid   Quinonoid

γ-Replacement (cystathionine γ-synthase)

O-Succinylhomoserine   Quinonoid   Ketimine   Ketimine
aldimine

Cystathionine aldimine   Quinonoid   Ketimine   Enamine

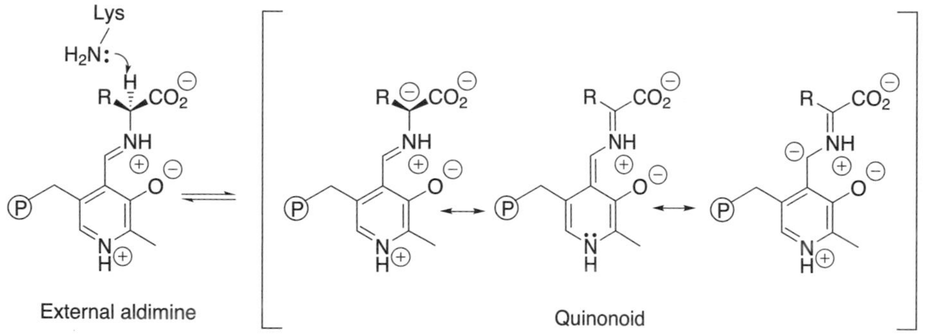

Scheme 4   The primary function of PLP is to stabilize anions generated at C_α. The negative charge is delocalized by resonance in the pi system of the cofactor in the quinonoid intermediate after loss of a proton from the external aldimine.

group anchored at the N terminus of an α−helix, H-bond interactions made to the 3′-OH, and the presence of the ubiquitous lysine Schiff base. Furthermore, these enzymes are also obligate dimers, as each monomer contributes residues to both active sites (35).

Multiple scaffolds have clearly evolved to bind PLP and to assist in catalysis by this cofactor. In no case does the fold type dictate the reaction type, as each fold type contains multiple reaction types, and all common reaction types are found in at least two fold types.[4]

---

[4]Although only one racemase structure has been solved (alanine racemase Fold Type III), serine racemase is predicted to be in Fold Type II based on sequence (36), and the fungal alanine racemase in Fold Type I (37).

←

Scheme 3   Examples of mechanisms of reactions at the β and γ positions. β-Replacement: tryptophan synthase (7) catalyzes a net β-replacement by first deprotonating the serine aldimine at C_α, producing the quinonoid intermediate. Protonation of the hydroxyl group by the active site lysine promotes its elimination, generating the aminoacrylate aldimine. Indole adds to C_β to form a second quinonoid that is subsequently protonated at C_α to generate the product aldimine. γ-Replacement: in the cystathionine γ-synthase-catalyzed γ-replacement reaction (105), O-succinyl homoserine is deprotonated at C_α to produce the quinonoid intermediate that is subsequently protonated at C_4' of the cofactor to give the ketimine intermediate. Proton abstraction at C_β by an unknown active site base results in elimination of the succinyl group, which may occur in either a step-wise or concerted (shown) manner. Michael addition of cysteine to the β,γ-unsaturated ketimine and subsequent proton transfers yield a second quinonoid intermediate that is protonated at C_α to form the product aldimine.

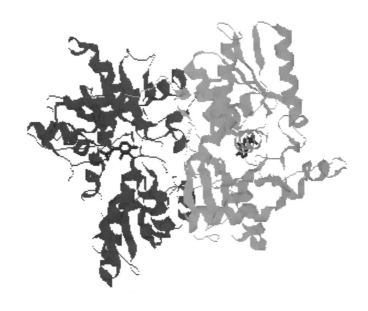

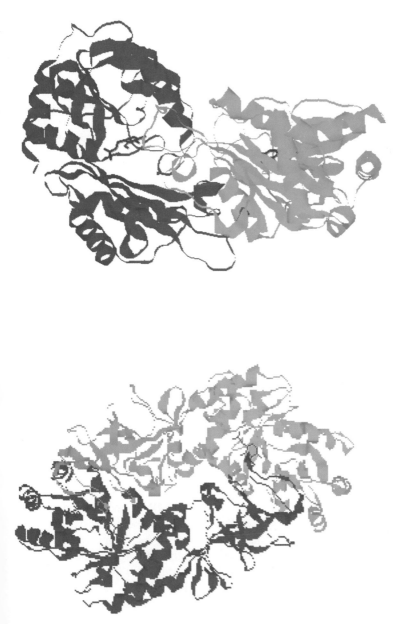

Figure 1   Ribbon diagrams of representative enzymes of Fold Types I - IV. Each structure depicts a homodimer with the individual monomers distinguished by color. The PLP cofactor is shown in red (*top left*). Fold Type I (*E. coli* aspartate aminotransferase; pdb file 1asn) (*top right*). Fold Type II (*Salmonella typhimurium* O-acetylserine sulfhydrylase; 1oas) (*bottom left*). Fold Type III (*Bacillus stearothermophilus* alanine racemase; 1sft) (*bottom right*). Fold Type IV (thermophilic *Bacillus* sp. D-amino acid aminotransferase; 1daa). The figure was prepared with RasMol.

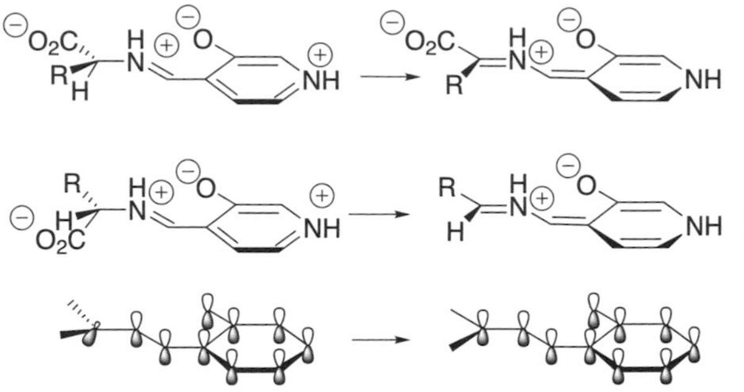

Scheme 5   The Dunathan stereoelectronic hypothesis. Substrates are bound to PLP such that the bond to C_α that is to be broken is aligned with the pi orbitals of the cofactor. Control of the substrate orientation thus enables the enzyme to distinguish between, for example, decarboxylation and deprotonation.

DETERMINANTS OF REACTION TYPE

Dunathan Stereoelectronic Hypothesis

Well before the lack of correlation between fold type and reaction type was recognized, much research was directed toward understanding the determinants of reaction type specificity. Dunathan postulated in 1966 (38) that the topology of the amino acid aldimine determined the bond to C_α that would be broken. He suggested that that bond must be situated so that it will align perpendicularly with the pyridine ring of the cofactor in the transition state of the reaction (Scheme 5). The ensuing carbanion is stabilized by conjugation with the extended pi system. This hypothesis was later confirmed when the structure of the aspartate aminotransferase/phosphopyridoxyl aspartate complex was solved (39). All subsequently determined structures are consistent with this idea. Of particular interest in this respect are enzymes that catalyze C_α-deprotonation and decarboxylation at different points during their catalytic cycle.

8-AMINO-7-OXONONANOATE SYNTHASE (AONS)   AONS, the second of four enzymes in the biotin biosynthetic pathway, catalyzes the decarboxylation and addition of a pimeloyl group to alanine (Scheme 6). Interestingly, the reaction mechanism involves initial deprotonation rather than decarboxylation (40). The quinonoid thus formed attacks pimeloyl-CoA, forming a β-keto acid intermediate, which is decarboxylated and reprotonated to form the product aldimine (Scheme 7). Decarboxylation of a β-keto acid is a facile reaction that may not require the participation of the cofactor, but the observation of the expected quinonoid intermediate indicates that it is likely involved (41). Although no structure is available of the intermediate aldimine formed prior to decarboxylation, the structure of the product aldimine shows that pimeloyl addition occurs

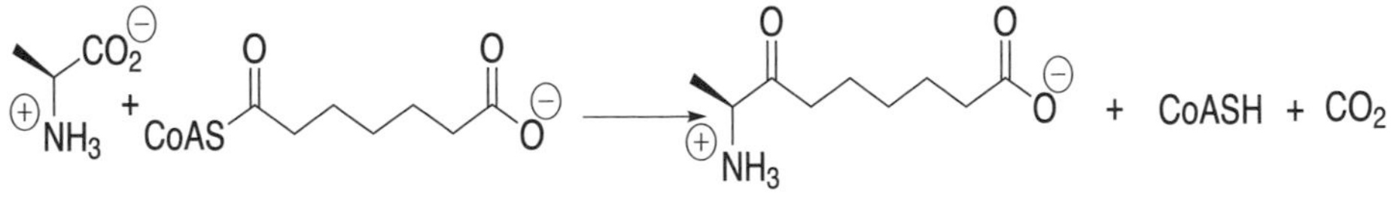

Scheme 6   The reaction catalyzed by 8-amino-7-oxononanoate synthase. The carboxylate group of alanine is replaced by a pimeloyl moiety. The mechanism of this reaction is shown in Scheme 7.

on the face opposite from that of deprotonation, resulting in the carboxylate group occupying nearly the same position as that previously held by the proton (Scheme 7).

DIALKYLGLYCINE DECARBOXYLASE (DGD)   DGD is both an aminotransferase and a decarboxylase. A complete catalytic cycle involves decarboxylation and transamination of dialkylglycine to generate a ketone product and the pyridox-amine phosphate (PMP) form of the enzyme followed by reaction with pyruvate in a typical transamination to generate alanine and to restore the PLP form of the enzyme (Scheme 8). Structural and mechanistic studies demonstrated that the decarboxylation of the dialkyl amino acids is forced by large side chains, which are not accommodated in the same site as the methyl side chain of alanine. They instead occupy the same position as the alanine carboxylate. This reorientation results in the scissile bond being that between C_α and the carboxylate rather than that between C_α and the proton (42) (Scheme 9). Thus decarboxylation is prefered over deprotonation.

These examples demonstrate how PLP enzymes can strictly control the initial bond-breakage step and limit other potential reactions by restricting the bound substrate to a specific orientation. Control of the steps subsequent to the initial bond breakage is less well understood. C_α-deprotonation, for example, can lead to β- or γ-elimination/replacement, racemization, or transamination (Schemes 2, 3). To promote the desired reaction, an enzyme must rigorously govern the electron flow and proton transfers. The structures of PLP enzymes have shown how active site residues are positioned to promote particular reaction types and to restrict possible side reactions. These principles are readily appreciated through a comparison of the structures of enzymes catalyzing transamination to that of a racemase.

Transamination

The first structures of a PLP-dependent enzyme to be determined were of the mitochondrial and cytosolic aspartate aminotransferases (AATase) (39, 43, 44). They explained much of the available mechanistic data and inspired fruitful

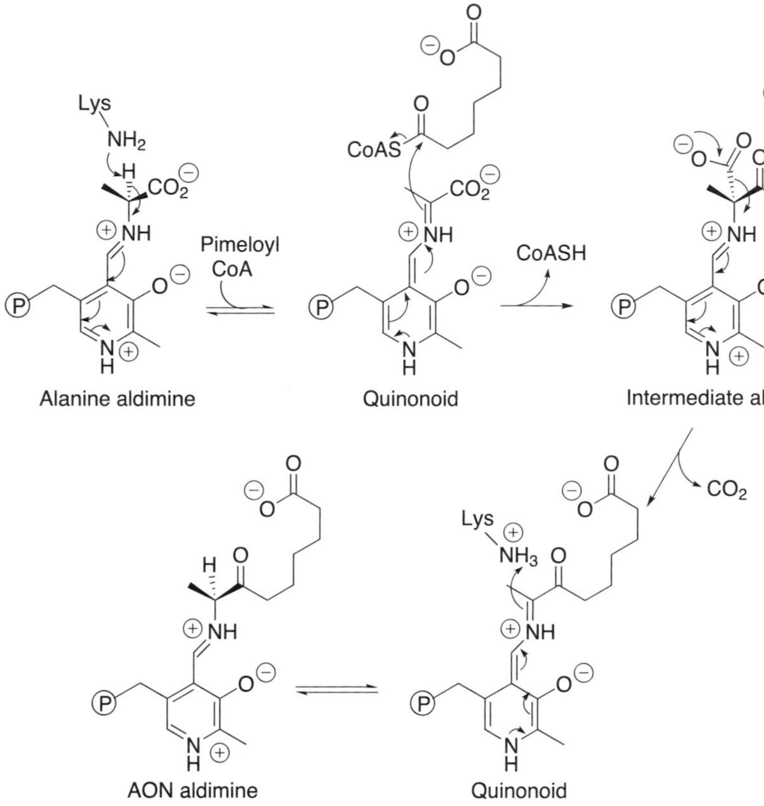

Scheme 7 The mechanism of the 8-amino-7-oxononanoate synthase-catalyzed reaction. The alanine aldimine is initially deprotonated, and the resulting quinonoid attacks the pimeloyl CoA thioester. Addition of the pimeloyl group to the face opposite from the deprotonation event causes the carboxylate to reorient perpendicular to the pyridine ring, resulting in subsequent decarboxylation. Protonation of this second quinonoid produces the product aldimine. Adapted from Reference 41.

targeted mutagenesis investigations (45–49). The electron sink nature of the cofactor is enhanced by a close interaction of the pyridinium nitrogen with an aspartate residue (D222), which acts to maintain the cofactor in the protonated form (Scheme 10). The displaced Schiff base lysine residue (K258) is positioned to transfer a proton to and from C_α and C_4'. Since the substrates of AATase, aspartate and glutamate, cannot undergo β- or γ- elimination, the primary side reaction that must be minimized is racemization. This objective is achieved by closure of the enzyme around the substrate, so that solvent water molecules do not access the quinonoid intermediate to protonate the *re* face (50). Furthermore, there are no active site acids in position to donate a proton to the *re* face of that intermediate. Subsequently solved structures of other aminotransferases con-

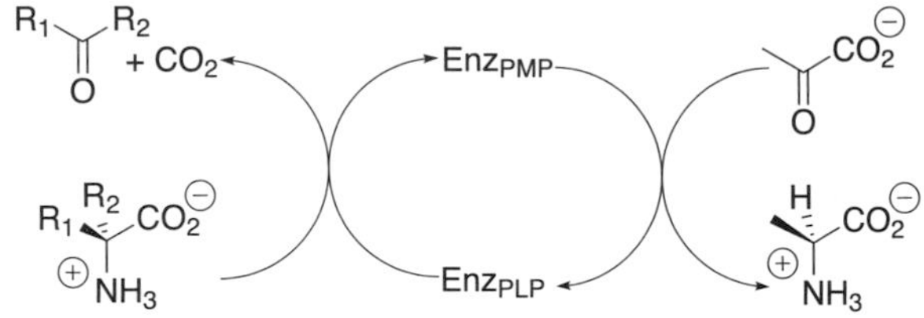

Scheme 8 The reaction catalyzed by dialkylglycine decarboxylase. In the first half-reaction, a dialkyl substrate is decarboxylated and transaminated, producing CO_2, a ketone, and the PMP form of the enzyme. The latter form subsequently reacts with pyruvate in a transamination half-reaction to give alanine and to restore the PLP form of the enzyme.

firmed that these are general properties of the entire class of enzymes. The D-amino acid aminotransferase (Fold Type IV) is particularly interesting because its active site is in part a mirror image of the L-amino acid aminotransferases (34), in that the active site lysine is positioned on the *re* face and the *si* face is solvent exposed, neatly accounting for its opposite stereospecificity.

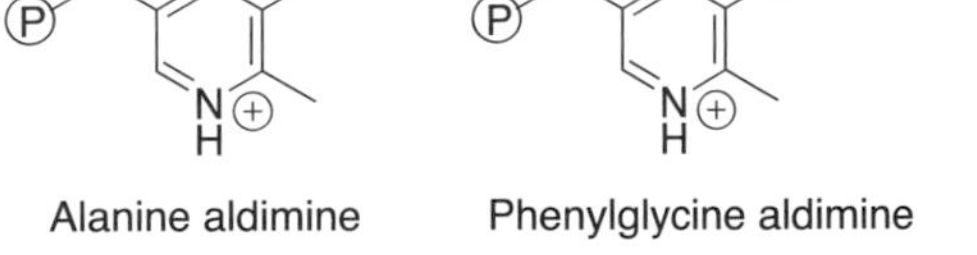

Alanine aldimine   Phenylglycine aldimine

Scheme 9 Steric control of reaction type by dialkylglycine decarboxylase. The three substituents on C_α occupy distinct binding sites, labeled *A*, *B*, and *C*. Whereas the *A* and *B* sites are tolerant of carboxylates and hydrogen or large alkyl groups, respectively, the *C* site only accommodates small alkyl groups. Amino acid substrates with small side chains, such as alanine, bind preferably with their carboxylate moieties in the *B* site, placing the C_α-proton in the reactive *A* site. Substrates with a large side chain (such as the phenylglycine shown), however, bind with that group in the *B* site, forcing the carboxylate into the reactive *A* site. For this reason, the distinction between decarboxylation or deprotonation is a consequence of the substrate structure. Adapted from Reference 42.

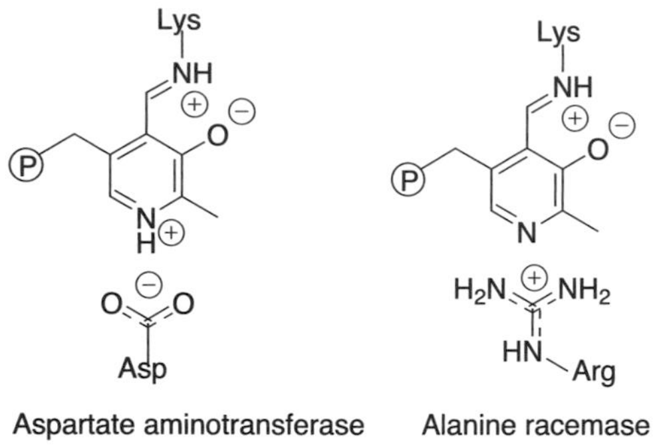

Aspartate aminotransferase   Alanine racemase

Scheme 10 Control of the electron sink properties of PLP by the amino acid side chain positioned nearest to the pyridinium nitrogen atom. An aspartate residue occupies this locus in aspartate aminotransferase and all other known Fold Type I enzymes, thereby maintaining ring protonation. The corresponding residue is an arginine in the bacterial alanine racemase, which is expected to maintain the cofactor in the unprotonated form. Lack of a proton at this position would greatly diminish the electron withdrawing properties of PLP.

Racemization

Although the structure of alanine racemase (ALR) from *Bacillus stearother-mophilus* (51) is the only one presently available, the catalytic mechanism is apparent from the unique architecture of its active site. An active site Brønsted acid is expected to be on the opposite *re* face to complement the lysine on the *si* face, and a tyrosine provides this function here (Scheme 2). Like AATase, ALR acts on substrates incapable of undergoing elimination; therefore, the primary side reaction that must be limited is transamination. A major factor in promoting that restriction is likely the arginine that replaces the aspartate near the pyridinium nitrogen (Scheme 10). The positively charged arginine prevents protonation of the cofactor (51), thereby negating its electron sink properties. Although it might be expected that this interaction would also impair the normal function of the enzyme, the racemization reaction may require less delocalization of electron density into the cofactor. In this regard, it is notable that some known amino acid racemases do not require a cofactor (see Footnote 1). Although racemization is generally presumed to proceed through a fully formed carbanion at C_α (52), the instability of this species elicits consideration of an S_E2 mechanism where the incoming proton develops some bonding with C_α in a concerted transition state. Jencks pointed out (53) that stepwise mechanisms are generally preferred because of entropy considerations, but that mechanisms become concerted when the intermediate that would develop in the stepwise process is too unstable to exist.

Interestingly, some of the insights gained from this structure cannot be generally applied to all PLP-dependent racemases. Paiardini et al. (54) reported an alanine racemase that is predicted to be a member of Fold Type I and to have an aspartate residue that interacts with the pyridinium nitrogen, as do all other Fold Type I enzymes.

Role of the Closed Conformation

Although it is not clear if the observed conformational changes of PLP enzymes contribute to substrate specificity (see below), the existence of the closed conformation undoubtedly contributes to reaction type specificity. Wolfenden pointed out the advantages of a closed conformation that makes more contacts with the bound substrate than are possible in a conformation from which the substrate must be able to dissociate (55). The closed conformation can favor specific reactions by providing greater control over proton transfers and solvent accessibility. A well-studied example is serine hydroxymethyltransferase, which catalyzes a variety of side reactions (transamination, decarboxylation, racemization, etc.) when presented with substrates other than serine. The open form of the enzyme is unable to discriminate between the various reaction types in the way that the closed form can (37, 56). Therefore, these reactions occur, at least in part, because the substrates in question do not induce the closed conformation.

DETERMINANTS OF SUBSTRATE SPECIFICITY

Role of the Closed Conformation

The potential contribution of induced fit to enzymatic substrate specificity has been much debated (for a recent overview, see 57), but it is now accepted that a conformational change induced only by certain substrates does not necessarily result in increased specificity for those substrates. Thus, it has been argued that the conformational change observed in AATase and many other PLP enzymes does not contribute to their substrate specificities (4). Exceptions to the general rule, however, include cases where substrate association or product dissociation is rate-determining for good substrates, whereas chemistry is rate-determining for poor substrates (58). Since it has been shown that release of the product oxalacetate is partially rate determining for the reaction of AATase with aspartate and α-ketoglutarate (αKG) (59), AATase is an example of an induced fit enzyme in which the conformational change may contribute to substrate specificity. It is not yet clear whether this conclusion extends to other PLP enzymes.

Dual Specificity of Aminotransferases

The basis for the dual specificity of aminotransferases has been elucidated by recent structural information. Because the aminotransferase reaction requires two different substrates to bind in succession to the same cofactor in the active site, these enzymes must be able to accommodate both structures while discriminating

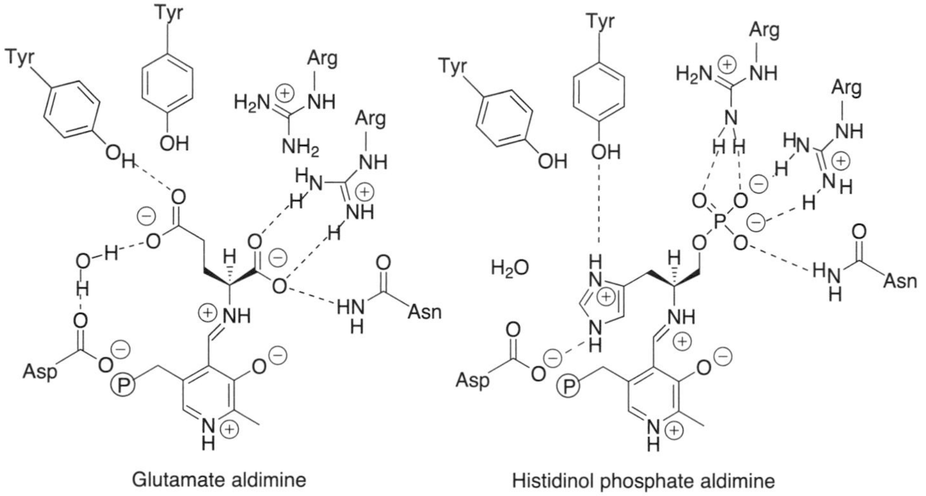

Glutamate aldimine   Histidinol phosphate aldimine

Scheme 11   Dual specificity of histidinol phosphate aminotransferase. The phosphate and carboxylate moieties of glutamate and histidinol phosphate, respectively, bind in the same site, although an additional arginine does interact with the phosphate group. In contrast, the oppositely charged imidazole and carboxylate side chains occupy spatially distinct sites and interact with different active site residues. The enzyme thus recognizes each substrate specifically. Adapted from Reference 60.

against all others. One possible solution would be for the PLP itself to move between two different substrate binding sites, but such movement has never been observed. An alternative is to take advantage of flexible side chains to position the functional groups into exclusive binding sites. This is the case for histidinol phosphate aminotransferase, which reacts with both histidinol phosphate and glutamate. The recently solved structures of substrate complexes of this enzyme (60) show that, although the phosphate and carboxylate groups interact primarily with the same arginine residue, the side chains of the two substrates have separate binding sites (Scheme 11).

Most known aminotransferases, however, adopt strategies for binding both substrates in the same site. Because many use the common substrate glutamate (thereby linking other amino acids to the cellular nitrogen pool), the problem of dual specificity is generally that of accommodating the negatively charged γ-carboxylate of glutamate in a site that must also accept a neutral or positively charged side chain. Two common solutions to this problem have been found: the use of an arginine switch and an extended hydrogen bond network.

ARGININE SWITCHES   The first example of an arginine switch was observed in an engineered enzyme that was constructed by introducing six mutations into AATase. These changes resulted in a substantial increase in activity toward aromatic substrates (61). The subsequent determination of the structure of the

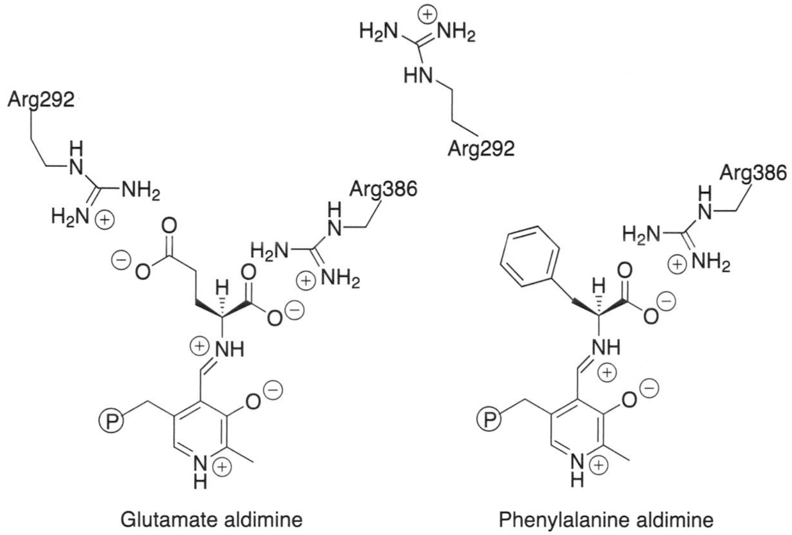

Scheme 12   Schematic of the arginine switch in aminotransferases that react with both dicarboxylic and aromatic amino acids. The γ carboxylate of glutamate (*left*) interacts closely with Arg292. This residue reorients to point out of the active site when aromatic substrates bind (*right*). This movement allows the enzyme to accept both types of substrates. Adapted from Reference 62.

mutant enzyme (62) showed that the large aromatic substrates are accommodated by movement of Arg292 out of the active site. Dicarboxylate substrates bind their β- or γ-carboxylate via direct interaction with this arginine through a bidentate hydrogen bond/ion pair in the canonical AATase conformation (Scheme 12). The position of Arg292 is locked in AATases; therefore, substrates lacking a carboxylate side chain are effectively excluded.

More recently, directed evolution techniques have been utilized to broaden the substrate specificity of AATase to include branched chain (63) or aromatic amino acids (64). Crystal structures of mutants with increased activity toward branched chain amino acids (65, 66) indicated that they also have acquired the ability to switch Arg292 out of the active site when uncharged substrates are bound, indicating that this trait is easily induced in AATase and is crucial for dual specificity.

Arginine switches are not unique to engineered enzymes. Tyrosine (aromatic) aminotransferase (TATase) is a well-characterized Fold Type I enzyme that has natural specificity for the aromatic amino acids tyrosine, phenylalanine, and tryptophan, as well as for the dicarboxylic amino acids aspartate and glutamate. A number of structures are now available of the *Paracoccus denitrificans* TATase that clearly demonstrate the arginine switch (67, 68).

Crystallographic and modeling studies illuminated a similar strategy employed by the GABA (69, 70) and ornithine (71) aminotransferases, which react with both

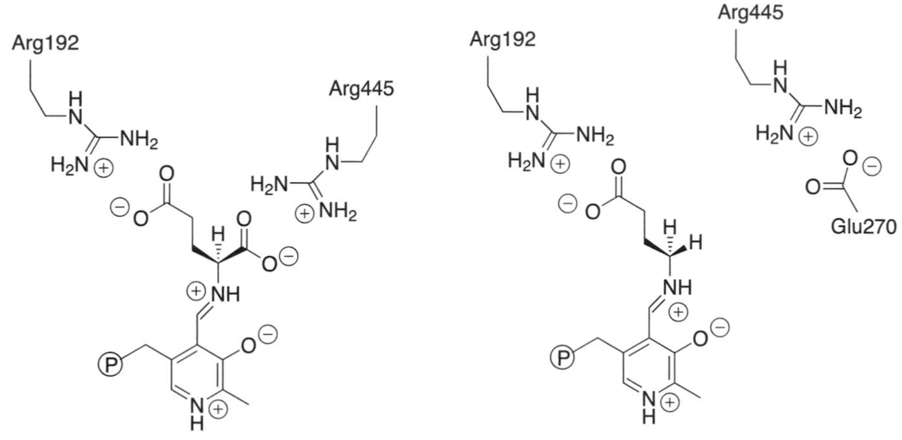

Glutamate aldimine   GABA aldimine

Scheme 13   Schematic of the arginine switch in GABA aminotransferase. As in the case of AATase, GABA ATase binds the dicarboxylic acid substrate glutamate via two conserved arginines. In order to accommodate GABA, Arg445 moves away from the cofactor to engage in a salt bridge with a nearby glutamate residue. Adapted from References 69 and 70.

ω-amino acid substrates and the common substrate glutamate. GABA ATase, like AATase and TATase, binds the dicarboxylic acid substrate via two conserved arginines. The carboxylate of the ω-amino acid, GABA, occupies the same position as the γ-carboxylate of glutamate, thereby taking advantage of the similar distance between the amino and carboxylate groups of GABA and the amino and γ-carboxylate groups of glutamate. The second arginine (equivalent to Arg386 of AATase) moves to interact with a conserved glutamate near the active site (Scheme 13).

HYDROGEN BOND NETWORKS   Arginine switches, however, are not ubiquitous even among TATases. Early sequence alignments indicated that a subgroup of aminotransferases (designated Iγ) lack an Arg292 equivalent (72). The only structure available for a TATase of this group is that of the unliganded *Pyrococcus horikoshii* enzyme (73). Modeling of the substrates in the active site suggests that this enzyme binds glutamate via an extended hydrogen-bonding network, as has been observed in the AATase from this same organism (74) (Scheme 14). The absence of a positively charged residue in this TATase makes it much easier to accommodate the uncharged substrates by simple rearrangement of the hydrogen bond network. Moreover, absence of the flexible arginine side chain allows this enzyme to distinguish between glutamate and aspartate. The specificity ratio $(k_{cat}/K_m^{Glu})/(k_{cat}/K_m^{Asp})$ for the *P. horikoshii* TATase is 3400 (73), compared to 0.27 for the *Escherichia coli* TATase (75), a typical member of the Iα family.

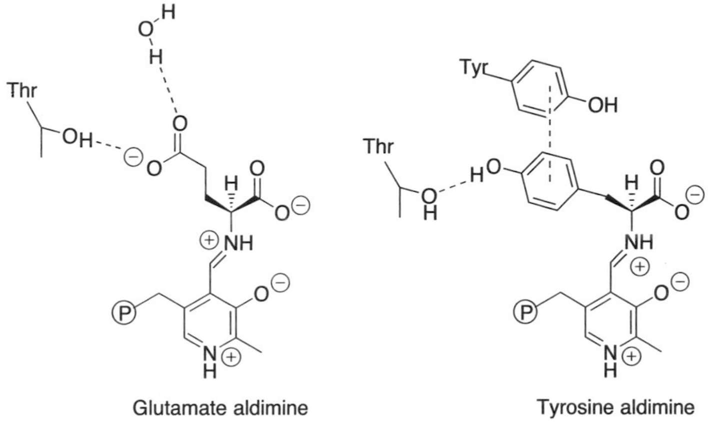

Glutamate aldimine   Tyrosine aldimine

Scheme 14 Dual substrate specificity can also be achieved by hydrogen bond rearrangement. *Pyrococcus horikoshii* TATase binds the γ-carboxylate of glutamate via a hydrogen bond network rather than an arginine residue. Direct interactions are made with a threonine residue and a tightly bound water molecule. Modeling of bound tyrosine suggests that the hydrogen bond network rearranges, so that the aromatic ring stacks against a nearby tyrosine residue, as well as makes a hydrogen bond to the same threonine residue that is involved in glutamate association. Adapted from References 73 and 74.

The recently solved structures of the natural *E. coli* branched chain amino acid aminotransferase (BCAT) (76) show that this enzyme also takes advantage of different hydrogen bond interactions. BCAT reacts with the branched chain aliphatic amino acids isoleucine, leucine, and valine, as well as with glutamate. The hydrophobic substrates and glutamate all bind in the same site, which is a hydrophobic pocket consisting largely of aromatic residues. In contrast to the TATases, there is no large-scale rearrangement of the H-bond network to accommodate the charged substrate. Instead, the glutamate side chain, which is slightly longer than that of the other substrates, extends into a pocket where four residues (Arg97, Tyr31, Tyr129, and the backbone amide of Val109) hydrogen bond to the carboxylate. The arginine in this case forms only a monodentate interaction rather than the bidentate H-bond network seen in other enzymes, and it is extensively H-bonded to neighboring residues, allowing it to remain in position when the small hydrophobic substrates are bound (Scheme 15).

EVOLUTION

Evolutionary relationships among PLP-dependent enzymes have been extensively examined (25, 29, 72). The following discussion is therefore limited to recent insights emanating from the increased number of available structures.

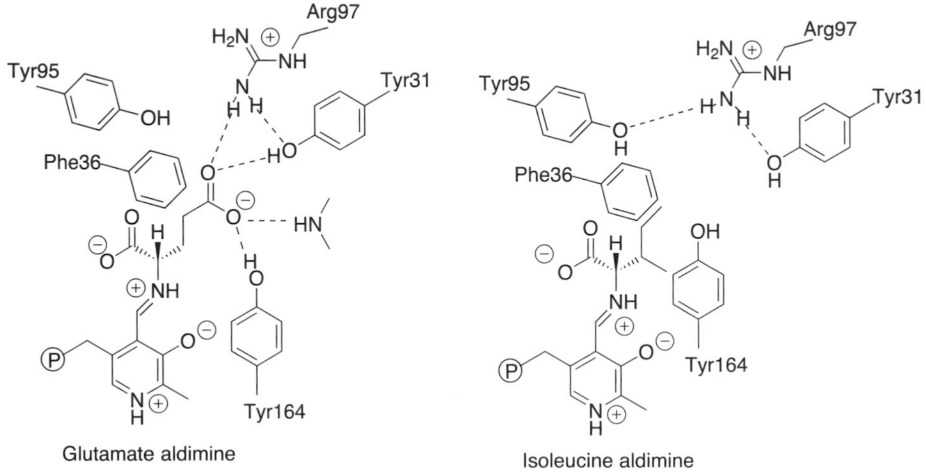

Scheme 15 Dual substrate recognition by branched chain amino acid aminotransferase. The substrate binding pocket is composed primarily of aromatic residues, and the hydrophobic substrate isoleucine is surrounded by five of them, only three of which are shown (Phe36, Tyr164, and Tyr31). The longer glutamate substrate extends far enough to form hydrogen bonds with the hydroxyl groups of two tyrosines and the guanidino group of an arginine residue. Note that the orientation of the substrate C_α-H bond is opposite that found in other fold types. Adapted from Reference 76.

The analyses of sequences and structures that led to the categorization of PLP enzymes into the five recognized fold types also indicated that the fold types are evolutionarily distinct (29). The similarities of the cofactor binding sites thus provide an excellent example of convergent evolution. It is believed that reaction type generally evolved first within each fold type, followed by narrowing substrate specificity (29). A number of enzymes, however, group most closely with those that catalyze reactions of a different type, suggesting that their reaction type-specificity arose later in evolution. In these cases, the change in reaction type can often be explained as a consequence of altered substrate specificity. For example, enzymes catalyzing β-elimination are found in many evolutionary subgroups, often among enzymes that catalyze transamination or γ-elimination. Since β-elimination of a good leaving group is a very facile reaction, it is easy to imagine that the acquisition of improved binding of a substrate with a β leaving group could readily lead to a change in the reaction specificity to favor elimination. Another example of an enzyme where a substrate specificity change effects a change in reaction type is dialkylglycine decarboxylase (see above). DGD is fundamentally an aminotransferase like most of its closest evolutionary relatives, but it also catalyzes decarboxylation of dialkyl substrates that bind in a unique orientation.

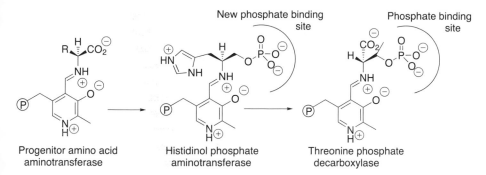

New phosphate binding site

Phosphate binding site

Progenitor amino acid aminotransferase

Histidinol phosphate aminotransferase

Threonine phosphate decarboxylase

Scheme 16 Possible evolutionary route to threonine phosphate decarboxylase. Histidinol phosphate aminotransferase is quite similar to Fold Type I amino acid aminotransferases and must share a common ancestral aminotransferase. However, it acquired additional affinity for a phosphate group. Relatively minor changes are required for that enzyme to accommodate threonine phosphate. With the large phosphate group as an anchor, the carboxylate is forced into the position occupied by the C_α proton in the related aminotransferases, thereby effecting a change in the reaction type from transamination to decarboxylation.

L-Threonine-O-3-phosphate decarboxylase, an enzyme that is also most closely related to aminotransferases, is unusual because the change in reaction type from transamination to decarboxylation requires a different substrate orientation with respect to the cofactor, and therefore mandates substantial rearrangement of the substrate binding mode. How this change was achieved evolutionarily is evident in the fact that the closest evolutionary relative of the enzyme is histidinol-phosphate aminotransferase, which has evolved to bind a phosphate group in the position usually occupied by the substrate carboxylate. One can imagine that from this starting point (or something similar), it is easy to improve binding of threonine-phosphate in the same site. With the phosphate as an anchor point in the usual carboxylate site, the carboxylate is positioned for bond breakage (Scheme 16). It is not clear, however, how decarboxylation-dependent transamination is averted, but presumably the mechanism is similar to those of other decarboxylases and is related to the lack of a proton on the active site lysine after decarboxylation. Solvent exclusion can also prevent hydrolysis of the ketimine intermediate formed by protonation at C_4' (78).

Possibly the most interesting group of PLP enzymes from an evolutionary standpoint is that of Fold Type IV (D-amino acid aminotransferase (DAAT) family). This family is unique in containing both a D-amino acid and an L-amino acid aminotransferase (the previously described BCAT). Since the active site lysine can only catalyze proton transfer on one face (79), DAAT and BCAT must bind their substrates in opposite orientations (Scheme 17), as evidenced in the crystal structures of these enzymes (76, 80). Either the binding mode of the

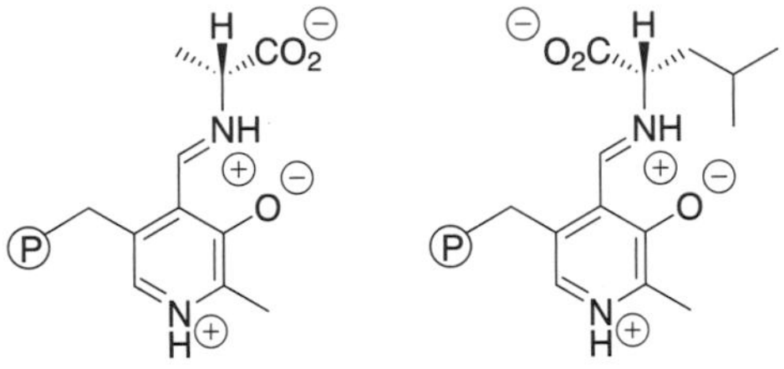

DAAT   BCAT

Scheme 17   Reverse orientation of substrate binding by D-amino acid aminotrans-ferase (DAAT) and branched chain amino acid aminotransferase (BCAT). These closely related enzymes are both in Fold Type IV, where the active site lysine is positioned on the *re* face of the cofactor, opposite from its position in Fold Type I and II enzymes. As a result, D-amino acids (shown bound to DAAT) are bound with the carboxylate pointing away from, and L-amino acids (shown bound to BCAT) are bound with the carboxylate proximal to, the cofactor phosphate group.

substrate was reversed at some point during evolution, or the enzymes share a common ancestor with broad specificity.

MECHANISMS OF INHIBITION

The prominent role of PLP enzymes in metabolism has generated a great deal of interest in the mechanisms of their inhibition. Although they, like nearly all enzymes, are susceptible to simple competitive inhibition, the catalytic versatility of the cofactor enhances their potential susceptibility to natural or designed mechanism-based inhibitors. Because such inhibition is often irreversible, it is of much greater practical utility than competitive inhibition (81). The inhibited complexes are also particularly useful for crystallographic studies, as they often mimic substrate or reaction intermediate complexes. A large number of inhibitors of PLP enzymes have now been identified (for more detailed reviews, see 6, 82), and they have been generally grouped into three categories according to their mode of inactivation.

Noncovalent Inactivation

The simplest mechanism-based inhibitors of PLP enzymes are those that form exceptionally stable complexes that often resemble normal reaction intermedi-ates. Although the inhibitor is covalently bound to the cofactor, the affinity for the protein is through noncovalent interactions. The combined affinity may be very high. A recently reported example is the stable ketimine intermediate formed in the reaction of ACC synthase with L-aminoethoxyvinylglycine (83) (Scheme 18). The ketimine is an intermediate in the reaction catalyzed by

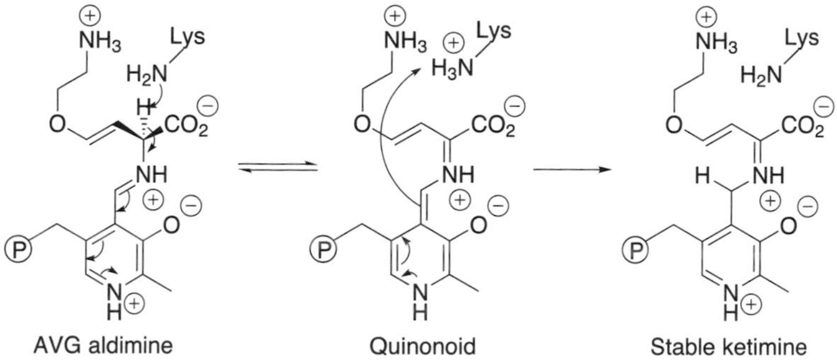

AVG aldimine   Quinonoid   Stable ketimine

Scheme 18 Inhibition of 1-aminocyclopropane-1-carboxylate synthase by amino-ethoxyvinylglycine. The inhibitor reacts to form a stable ketimine that is not hydrolyzed and remains tightly bound to the enzyme. Dissociation by dialysis removes the cofactor together with the covalently bound inhibitor (83).

aminotransferases but not in the ACC synthase-catalyzed elimination reaction; thus the enzyme is unable to catalyze its hydrolysis. Of particular interest is the subset of these inhibitors whose stability is the result of enzyme-catalyzed aromatization. An example is the inhibition of GABA ATase by (S)-4-amino-4,5-dihydro-2-thiophenecarboxylic acid (84) (Scheme 19). Since the products of these reactions are not covalently attached to the enzyme, the inhibited enzymes can often be reactivated by dialysis in the presence of PLP, so that the PLP-inhibitor species is replaced with fresh PLP (for example, see 83), but this mode of reactivation does not generally occur on a physiologically relevant timescale.

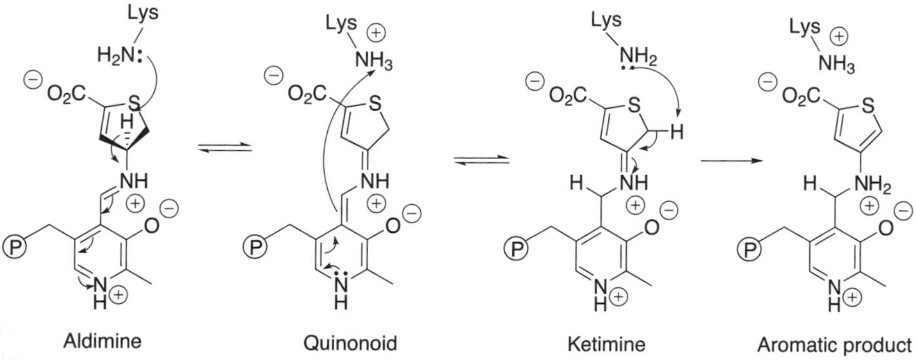

Aldimine   Quinonoid   Ketimine   Aromatic product

Scheme 19 Inhibition of γ-aminobutyrate aminotransferase by (S)-4-amino-4,5-dihydro-2-thiophenecarboxylic acid. The reaction proceeds identically to the aminotransferase reaction up to the formation of the ketimine intermediate. At this point, deprotonation of a β carbon yields a very stable aromatic product that does not react further and remains in the active site (84).

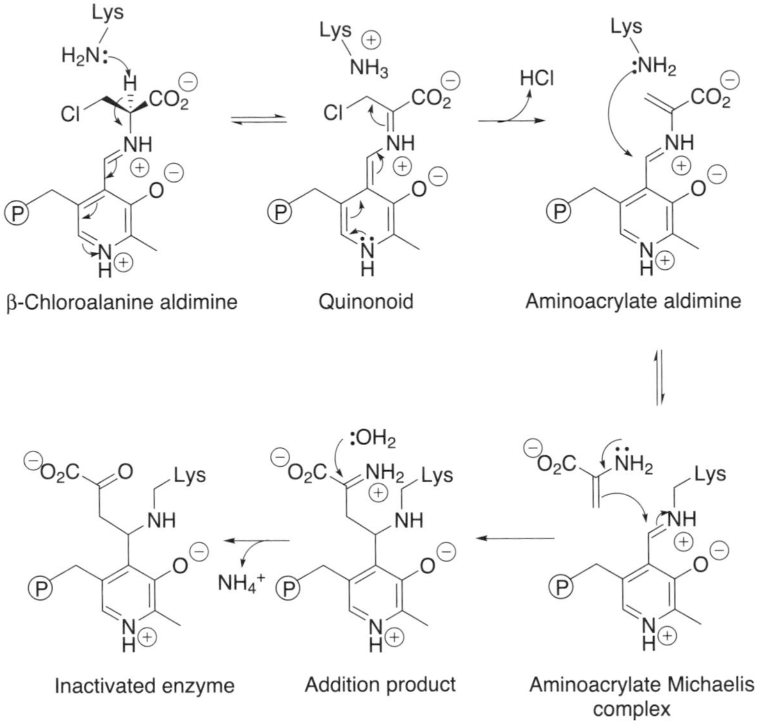

Scheme 20   Mechanism of inactivation of a PLP-dependent enzyme by β-chloroalanine. The reaction follows the normal pathway to β-elimination up to the formation of the aminoacrylate aldimine. From there, the aminoacrylate may be released by transaldimination and may subsequently attack the cofactor nucleophilically. Subsequent hydrolysis of the imine yields the final inactivated enzyme. A number of other mechanisms are possible, as noted in the text.

Activated Nucleophiles

A number of amino acids with good β-leaving groups, such as β-chloroalanine, inhibit several types of PLP enzymes (6). The general mechanism for this reaction is by initial formation and release of an enamine intermediate, which is a potent nucleophile and can attack C_4' of the cofactor (85) (Scheme 20). Inactivation is irreversible, because the cofactor remains covalently bound to the enzyme. An alternative proposed inactivation pathway (86) is by direct Michael addition of an active site nucleophile to the aminoacrylate aldimine. Free aminoacrylate can also diffuse out of the enzyme (as it does in natural β-elimination reactions), where it spontaneously decomposes to pyruvate and ammonia, leaving the enzyme in an active form. Because of this possibility of turnover as well as inhibition, the effectiveness of these inhibitors is often quantitated by the inactivation/turnover ratio.

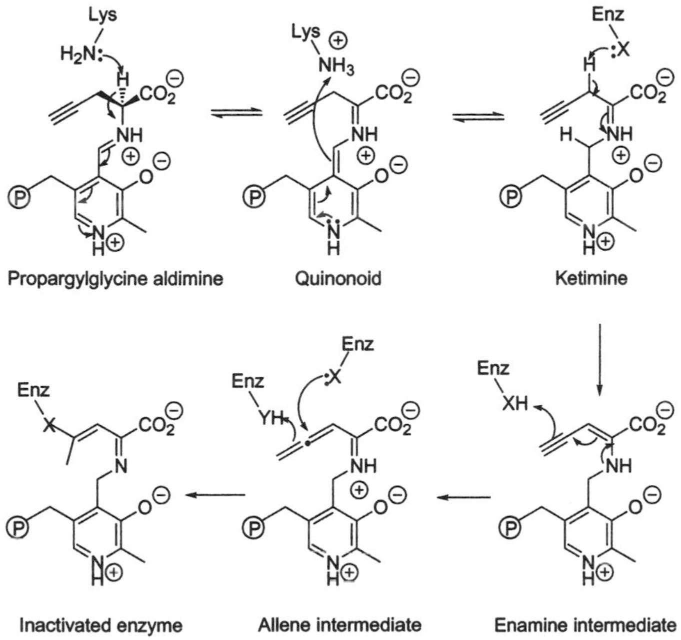

Scheme 21 Mechanism of inactivation by propargylglycine. The reaction parallels a γ-elimination mechanism through the formation of the enamine intermediate. At this point, the acetylene moiety is rearranged to form a highly reactive allene. Nucleophilic attack by an active site residue results in covalent attachment to the enzyme. The mechanism shown is slightly altered from that originally proposed by Abeles & Walsh (87) to include a ketimine intermediate in accord with the most recent proposal for the mechanism of γ-elimination (105).

Activated Electrophiles

Inhibition can also result from rearrangement of the inhibitor to generate an electrophile that subsequently reacts irreversibly with an active site nucleophile (often the active site lysine). Two common types of inhibitors of this group are acetylenic compounds such as propargylglycine, which reacts to form a highly reactive allene intermediate (87) (Scheme 21), and vinylic compounds such as vinylglycine, which reacts to form an β,γ-unsaturated imine Michael acceptor (88) (Scheme 22). The reactivity of the potential intermediates formed from these inhibitors allows for alternative reaction fates in addition to those shown. In the case of vinylglycine, for instance, an aminocrotonate aldimine, a potential Michael acceptor, can be formed from the quinonoid (89). The aminocrotonate may also be released by transaldimination, at which point it can nucleophilically attack the cofactor in the manner described above for aminoacrylate (90). A third possible fate is diffusion off the enzyme, where it decomposes spontaneously to α-ketobutyrate and ammonia (91), leaving the enzyme in the active PLP form.

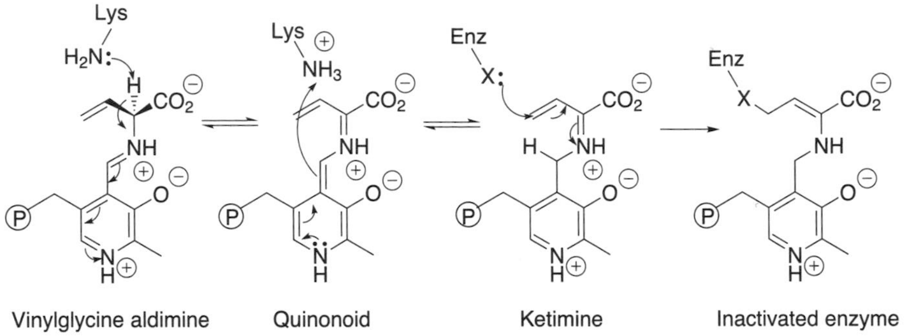

Vinylglycine aldimine   Quinonoid   Ketimine   Inactivated enzyme

Scheme 22   Mechanism of inactivation by vinylglycine. The reaction parallels an aminotransferase mechanism through the formation of the ketimine intermediate. Michael addition by an active site nucleophile to the vinylglycine ketimine results in a covalent adduct.

The vinylglycine ketimine can also be hydrolyzed to release the potentially toxic Michael acceptor 2-ketobut-3-enoic acid, leaving the enzyme in the PMP form (92). Effective inhibitor design requires limiting these possible alternative pathways.

Reversible Competitive Inhibition

Competitive inhibitors have proven to be exceptionally useful in studies of enzyme function and as unreactive substrate mimics in crystallographic studies. There are two general classes of competitive inhibitors—those that bind noncovalently and those that react reversibly with the enzyme to form an aldimine that does not react further. A classical example of the former is maleate, which inhibits AATase by binding in the aspartate site (6). This association induces closure of the enzyme into the active form; thus this inhibitor has proven useful for crystallographic studies of this form of the enzyme.

The most common of the inhibitors that form unreactive aldimines are α-methyl substrate analogs and amino-oxy or hydrazine analogs. The close similarity of the α-methyl compounds to the substrates makes them particularly useful for crystallography, and structures of complexes of AATase and phosphoserine aminotransferase with α-methylaspartate and α-methylglutamate, respectively, have been reported (93–95). Recent examples of the use of amino-oxy compounds for structure determination are those of ACC synthase with the amino-oxy analogue of SAM (96) (Scheme 23) and of ornithine aminotransferase with L-canaline, an analogue of ornithine (97) (Scheme 23). The amino-oxy adducts with PLP are sufficiently stable so that these compounds often need not be substrate analogs. Hydroxylamine itself binds to AATase to form a PLP-oxime whose K_i = 700 nM, a figure less than that for any of the dicarboxylic inhibitor complexes (6). A final example of an inhibitor that forms an unreactive aldimine is that of 1-aminoethyl phosphonate (AlaP), which binds tightly to alanine

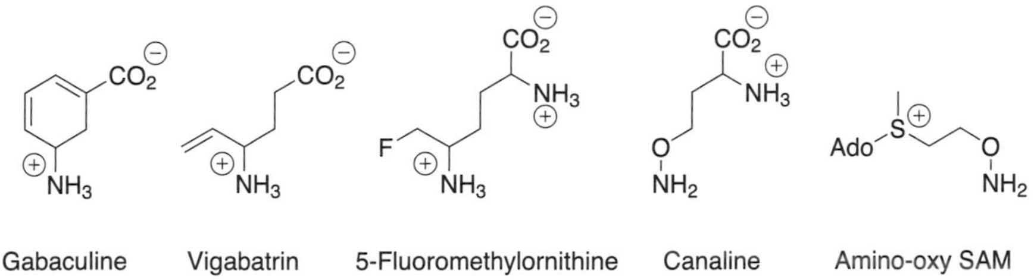

Gabaculine   Vigabatrin   5-Fluoromethylornithine   Canaline   Amino-oxy SAM

Scheme 23   The structures of the inhibitors gabaculine, vigabatrin, 5-fluoromethylornithine, canaline, and amino-oxy SAM.

racemase (98). Although this molecule has a C_α proton, it does not undergo deprotonation. The stability of this complex enabled determination of the enzyme structure with the inhibitor in the active site, providing a mimic of the complex with the natural substrate (99).

The Challenge of Inhibitor Specificity

Although the variety of possible mechanisms of inhibition makes it easy to design effective inhibitors, specificity remains as a major challenge for use in vivo. Gabaculine (Scheme 23), for example, is a potent inhibitor of GABA ATase and also inhibits the closely related ornithine aminotransferase (100), making it unsuitable for pharmaceutical applications. One promising approach is to incorporate the reactive functional groups described above in structures that are very close substrate analogs. An example is γ-vinylGABA (vigabatrin; Scheme 23), a specific inhibitor of GABA ATase, which is used in the treatment of epilepsy (reviewed in 101). In this case, the vinylic group is appended to the natural substrate GABA to form a potential electrophile analogous to vinylglycine. Another example is the ornithine aminotransferase (OAT) inhibitor 5-fluoromethylornithine (102) (Scheme 23). The fluoride ion is susceptible to β-elimination, which generates an enamine capable of nucleophilic attack on the cofactor in the manner described above (71, 103), while the ornithine scaffold provides OAT specificity.

Understanding mechanisms of inhibition is crucial not only to enable the design of better inhibitors, but also to understand the control of reaction mechanisms by this important class of enzymes. It is also of interest to ask how enzymes whose natural reaction pathways include reactive intermediates, such as the aminoacrylate aldimine and vinylglycine ketimine, manage to avoid inactivation.

CONCLUSIONS

The recent effusion of X-ray structures for PLP-dependent enzymes—more than twice as many were deposited in the protein data bank between 1997 and 2001 as in the preceding five years—has done much to provide a visual framework in which the mechanistic concepts secured over the past half-century can be interpreted. The findings served to focus more penetrating targeted mutagenesis experiments and, taken together with those from newer studies capitalizing upon directed evolution methods, helped to elucidate some of the fundamental principles of protein design.

We now have a good understanding of the mechanisms of dual substrate recognition by aminotransferases and of how the fates of the common C_α anion are directed. Many questions remain unanswered, particularly with regard to the control of reaction pathways of both natural substrates and mechanism-based inhibitors. Of the four fold types, only Fold Type I is well represented in the structural database, and many of the reaction types are only represented by one or two structures. We need additional structures of underrepresented fold types in order to direct further functional studies. The understanding of the substrate and reaction type-specificity of PLP enzymes is crucial for the design not only of specific and medicinally useful inhibitors, but also of improved protein-based catalysts.

ACKNOWLEDGMENT

The authors thank Susan Aitken and Kathryn McElroy for critically reviewing the manuscript.

The *Annual Review of Biochemistry* is online at http://biochem.annualreviews.org

LITERATURE CITED

1. Heyl D, Luz E, Harris SA, Folkers K. 1951. *J. Am. Chem. Soc.* 73:3430–33
2. Stryer L. 1995. *Biochemistry*. New York: Freeman
3. Voet D, Voet J. 1995. *Biochemistry*. New York: Wiley
4. Hayashi H. 1995. *J. Biochem.* 118:463–73
5. John RA. 1995. *Biochim. Biophys. Acta* 1248:81–96
6. Christen P, Metzler DE, eds. 1985. *Transaminases*. New York: Wiley
7. Miles EW. 2001. *Chem. Rec.* 1:140–51
8. Tai CH, Cook PF. 2001. *Acc. Chem. Res.* 34:49–59
9. Ferreira GC, Gong J. 1995. *J. Bioenerg. Biomembr.* 27:151–59
10. Rao NA, Talwar R, Savithri HS. 2000. *Int. J. Biochem. Cell Biol.* 32:405–16
11. Hutson S. 2001. *Prog. Nucleic Acid Res. Mol. Biol.* 70:175–206
12. He XM, Liu HW. 2002. *Annu. Rev. Biochem.* 71:701–54
13. Kleppner SR, Tobin AJ. 2001. *Emerging Ther. Targets* 5:219–39
14. Snell K, Riches D. 1989. *Cancer Lett.* 44:217–20
15. Wang CC. 1995. *Annu. Rev. Pharmacol. Toxicol.* 35:93–127
16. Mudd SH, Laster L, Finkelstein JD, Irreverre F. 1964. *Science* 143:1443–45
17. Kraus JP, Janosik M, Kozich V, Mandell R, Shih V, et al. 1999. *Hum. Mutat.* 13:362–75
18. Adams DO, Yang SF. 1979. *Proc. Natl. Acad. Sci. USA* 76:170–74
19. Honma M, Shimomura T. 1978. *Agric. Biol. Chem.* 42:1825–31

20. Keller JW, Baurick KB, Rutt GC, O'Malley MV, Sonafrank NL, et al. 1990. *J. Biol. Chem.* 265:5531–39

21. Watanabe Y, Shimura K. 1956. *J. Biochem.* 43:283–94

22. Rios A, Amyes TL, Richard JP. 2000. *J. Am. Chem. Soc.* 122:9373–85

23. Gallo KA, Knowles JR. 1993. *Biochemistry* 32:3981–90

24. Alexander FW, Sandmeier E, Mehta PK, Christen P. 1994. *Eur. J. Biochem.* 219:953–60

25. Jansonius JN. 1998. *Curr. Opin. Struct. Biol.* 8:759–69

26. Schneider G, Kack H, Lindqvist Y. 2000. *Struct. Fold. Des.* 8:R1–6

27. Churchich JE, Moses U. 1981. *J. Biol. Chem.* 256:1101–4

28. Hennig M, Grimm B, Contestabile R, John RA, Jansonius JN. 1997. *Proc. Natl. Acad. Sci. USA* 94:4866–71

29. Mehta PK, Christen P. 2000. *Adv. Enzymol. Relat. Areas Mol. Biol.* 74:129–84

30. Hyde CC, Ahmed SA, Padlan EA, Miles EW. 1988. *J. Biol. Chem.* 263:17857–71

31. Madison JT, Thompson JF. 1976. *Biophys. Biochem. Res. Commun.* 71:684–91

32. Finkelstein JD, Kyle WE, Martin JJ, Pick AM. 1975. *Biophys. Biochem. Res. Commun.* 66:81–87

33. Gallagher DT, Gilliland GL, Xiao G, Zondlo J, Fisher KE, et al. 1998. *Structure* 6:465–75

34. Sugio S, Petsko GA, Manning JM, Soda K, Ringe D. 1995. *Biochemistry* 34:9661–69

35. Kern AD, Oliveira MA, Coffino P, Hackert ML. 1999. *Struct. Fold. Des.* 7:567–81

36. Wolosker H, Blackshaw S, Snyder SH. 1999. *Proc. Natl. Acad. Sci. USA* 96:13409–14

37. Contestabile R, Paiardini A, Pascarella S, di Salvo ML, D'Aguanno S, Bossa F. 2001. *Eur. J. Biochem.* 268:6508–25

38. Dunathan HC. 1966. *Proc. Natl. Acad. Sci. USA* 55:712–16

39. Kirsch JF, Eichele G, Ford GC, Vincent MG, Jansonius JN, et al. 1984. *J. Mol. Biol.* 174:497–525

40. Ploux O, Marquet A. 1996. *Eur. J. Biochem.* 236:301–8

41. Webster SP, Alexeev D, Campopiano DJ, Watt RM, Alexeeva M, et al. 2000. *Biochemistry* 39:516–28

42. Sun S, Zabinski RF, Toney MD. 1998. *Biochemistry* 37:3865–75

43. Ford GC, Eichele G, Jansonius JN. 1980. *Proc. Natl. Acad. Sci. USA* 77:2559–63

44. Borisov VV, Borisova SN, Sosfenov NI, Vainshtein BK. 1980. *Nature* 284:189–90

45. Kuramitsu S, Inoue Y, Tanase S, Morino Y, Kagamiyama H. 1987. *Biochem. Biophys. Res. Commun.* 146:416–21

46. Kochhar S, Finlayson WL, Kirsch JF, Christen P. 1987. *J. Biol. Chem.* 262:11446–48

47. Toney MD, Kirsch JF. 1987. *J. Biol. Chem.* 262:12403–5

48. Cronin CN, Kirsch JF. 1988. *Biochemistry* 27:4572–79

49. Hayashi H, Kuramitsu S, Inoue Y, Morino Y, Kagamiyama H. 1989. *Biochem. Biophys. Res. Commun.* 159:337–42

50. Kochhar S, Christen P. 1992. *Eur. J. Biochem.* 203:563–69

51. Shaw JP, Petsko GA, Ringe D. 1997. *Biochemistry* 36:1329–42

52. Albery WJ, Knowles JR. 1986. *Biochemistry* 25:2572–77

53. Jencks WP. 1985. *Chem. Rev.* 85:511–27

54. Paiardini A, Contestabile R, D'Aguanno S, Pascarella S, Bossa F. 2003. *Biochim. Biophys. Acta* 1647:214–19

55. Wolfenden R. 1974. *Mol. Cell. Biochem.* 3:207–11

56. Schirch V, Shostak K, Zamora M, Guatam-Basak M. 1991. *J. Biol. Chem.* 266:759–64

57. Pasternak A, White A, Jeffery CJ, Medina N, Cahoon M, et al. 2001. *Protein Sci.* 10:1331–42

58. Herschlag D. 1988. *Bioorg. Chem.* 16:62–96

59. Goldberg JM, Kirsch JF. 1996. *Biochemistry* 35:5280–91

60. Haruyama K, Nakai T, Miyahara I, Hirotsu K, Mizuguchi H, et al. 2001. *Biochemistry* 40:4633–44

61. Onuffer JJ, Kirsch JF. 1995. *Protein Sci.* 4:1750–57

62. Malashkevich VN, Onuffer JJ, Kirsch JF, Jansonius JN. 1995. *Nat. Struct. Biol.* 2:548–53

63. Yano T, Oue S, Kagamiyama H. 1998. *Proc. Natl. Acad. Sci. USA* 95:5511–15

64. Rothman SC, Kirsch JF. 2003. *J. Mol. Biol.* 327:593–608

65. Oue S, Okamoto A, Yano T, Kagamiyama H. 1999. *J. Biol. Chem.* 274:2344–49

66. Oue S, Okamoto A, Yano T, Kagamiyama H. 2000. *J. Biochem.* 127:337–43

67. Okamoto A, Nakai Y, Hayashi H, Hirotsu K, Kagamiyama H. 1998. *J. Mol. Biol.* 280:443–61

68. Okamoto A, Ishii S, Hirotsu K, Kagamiyama H. 1999. *Biochemistry* 38:1176–84

69. Toney MD, Pascarella S, De Biase D. 1995. *Protein Sci.* 4:2366–74

70. Storici P, Capitani G, De Biase D, Moser M, John RA, et al. 1999. *Biochemistry* 38:8628–34

71. Storici P, Capitani G, Muller R, Schirmer T, Jansonius JN. 1999. *J. Mol. Biol.* 285:297–309

72. Jensen RA, Gu W. 1996. *J. Bacteriol.* 178:2161–71

73. Matsui I, Matsui E, Sakai Y, Kikuchi H, Kawarabayasi Y, et al. 2000. *J. Biol. Chem.* 275:4871–79

74. Ura H, Harata K, Matsui I, Kuramitsu S. 2001. *J. Biochem.* 129:173–78

75. Hayashi H, Inoue K, Nagata T, Kuramitsu S, Kagamiyama H. 1993. *Biochemistry* 32:12229–39

76. Goto M, Miyahara I, Hayashi H, Kagamiyama H, Hirotsu K. 2003. *Biochemistry* 42:3725–33

77. Deleted in proof

78. Eliot AC, Kirsch JF. 2003. *Acc. Chem. Res.* 36:757–65

79. Soda K, Yoshimura T, Esaki N. 2001. *Chem. Rec.* 1:373–84

80. Peisach D, Chipman DM, Van Ophem PW, Manning JM, Ringe D. 1998. *Biochemistry* 37:4958–67

81. Silverman RB. 1988. *J. Enzyme Inhib.* 2:73–90

82. Nanavati SM, Silverman RB. 1989. *J. Med. Chem.* 32:2413–21

83. Capitani G, McCarthy DL, Gut H, Grutter MG, Kirsch JF. 2002. *J. Biol. Chem.* 277:49735–42

84. Fu M, Nikolic D, Van Breemen RB, Silverman RB. 1999. *J. Am. Chem. Soc.* 121:7751–59

85. Likos JJ, Ueno H, Feldhaus RW, Metzler DE. 1982. *Biochemistry* 21:4377–86

86. Kishore GM. 1984. *J. Biol. Chem.* 259:10669–74

87. Abeles RH, Walsh CT. 1973. *J. Am. Chem. Soc.* 95:6124–25

88. Rando RR. 1974. *Biochemistry* 13:3859–63

89. Soper TS, Manning JM, Marcotte PA, Walsh CT. 1977. *J. Biol. Chem.* 252:1571–75

90. Nanavati SM, Silverman RB. 1991. *J. Am. Chem. Soc.* 113:9341–49

91. Miles EW. 1975. *Biochem. Biophys. Res. Commun.* 66:94–102

92. Choi S, Storici P, Schirmer M, Silverman RB. 2002. *J. Am. Chem. Soc.* 124:1620–24

93. McPhalen CA, Vincent MG, Picot D, Jansonius JN, Lesk AM, Chothia C. 1992. *J. Mol. Biol.* 227:197–213

94. Okamoto A, Higuchi T, Hirotsu K, Kuramitsu S, Kagamiyama H. 1994. *J. Biochem.* 116:95–107

95. Hester G, Stark W, Moser M, Kallen J, Markovic-Housley Z, Jansonius JN. 1999. *J. Mol. Biol.* 286:829–50

96. Capitani G, Eliot AC, Gut H, Khomutov

RM, Kirsch JF, Grutter MG. 2003. *Biochim. Biophys. Acta* 1647:55–60

97. Shah SA, Shen BW, Brunger AT. 1997. *Structure* 5:1067–75

98. Badet B, Walsh C. 1985. *Biochemistry* 24:1333–41

99. Stamper GF, Morollo AA, Ringe D, Stamper CG. 1998. *Biochemistry* 37:10438–45

100. Jung MJ, Seiler N. 1978. *J. Biol. Chem.* 253:7431–39

101. Mumford JP, Cannon DJ. 1994. *Epilepsia* 35:S25–28

102. Daune G, Gerhart F, Seiler N. 1988. *Biochem. J.* 253:481–88

103. Bolkenius FN, Knodgen B, Seiler N. 1990. *Biochem. J.* 268:409–14

104. Sun S, Toney MD. 1999. *Biochemistry* 38:4058–65

105. Brzovic P, Holbrook EL, Greene RC, Dunn MF. 1990. *Biochemistry* 29:442–51

106. Livanova NB, Chebotareva NA, Eronina TB, Kurganov BI. 2002. *Biochemistry* 67:1089–98

107. Frey PA. 2001. *Annu. Rev. Biochem.* 70:121–48

Annu. Rev. Biochem. 2004. 73:417–35
doi: 10.1146/annurev.biochem.73.011303.073651
Copyright © 2004 by Annual Reviews. All rights reserved
First published online as a Review in Advance on March 18, 2004

THE SIR2 FAMILY OF PROTEIN DEACETYLASES

Gil Blander and Leonard Guarente

*Department of Biology, Massachusetts Institute of Technology, Cambridge,
Massachusetts 02139; email: gblander@mit.edu, leng@mit.edu*

Key Words   Sir2, deacetylase, NAD$^+$, caloric restriction, aging

■ **Abstract**   The yeast SIR protein complex has been implicated in transcription silencing and suppression of recombination. The Sir complex represses transcription at telomeres, mating-type loci, and ribosomal DNA. Unlike *SIR3* and *SIR4*, the *SIR2* gene is highly conserved in organisms ranging from archaea to humans. Interestingly, Sir2 is active as an NAD$^+$-dependent deacetylase, which is broadly conserved from bacteria to higher eukaryotes. In this review, we discuss the role of NAD$^+$, the unusual products of the deacetylation reaction, the Sir2 structure, and the Sir2 chemical inhibitors and activators that were recently identified. We summarize the current knowledge of the Sir2 homologs from different organisms, and finally we discuss the role of Sir2 in caloric restriction and aging.

CONTENTS

THE DISCOVERY AND IDENTIFICATION OF SIR2

Studies in yeast have led to the identification of cellular factors that are required for transcriptional silencing. Among these are the proteins encoded by the yeast SIR genes, which are responsible for silencing at repeated DNA sequences in yeast: mating-type loci, telomeres, and rDNA. *SIR2*, *SIR3*, and *SIR4* genes are required for silencing at mating-type loci (1) and telomeres (2), whereas the *SIR2*, but not the *SIR3* and the *SIR4*, is required for silencing in the rDNA (3, 4). Silencing causes a closed, inaccessible regional chromatin structure, as assayed by various probes of DNA accessibility (5, 6). Silencing requires particular lysine residues in the amino-terminal tail of histones H3 and H4 (7–9). These and other lysine residues of the tail are acetylated in active chromatin but deacetylated in silenced chromatin (9, 10). The deacetylated histones can fold into a more compact, closed nucleosomal structure (11). Due to global deacetylation of yeast histones observed when the SIR2 was overexpressed (10), Sir2 was suggested to be a histone deacetylase. However, early attempts to demonstrate histone deacetylase activity for Sir2 were not successful.

Unlike the *SIR3* or *SIR4*, the *SIR2* gene is highly conserved in the organisms ranging from archaea to humans (12). Studies on a bacterial homolog, *cobB*, led to the conclusion that this gene could substitute for another bacterial gene, *cobT*, in the pathway of cobalamin synthesis (13). The *cobT* encodes an enzyme that transfers a ribose-phosphate moiety from nicotinic acid mononucleotide (NMN) to dimethyl benzimidazole. Thus, it seemed possible that Sir2 proteins might catalyze a similar but different reaction because the cobalamin pathways are not present in yeast or mammals. Frye (14) showed that Sir2 proteins from bacteria, yeast, and mammals were able to transfer an ADP-ribose group from NAD^+ to a protein carrier. Subsequent work by others proved that the Sir2 protein was indeed an ADP-ribosyltransferase in vitro, and its activity was essential for silencing in vivo (15). Imai et al. (16) discovered that the amino-terminal tails of histone H3 or H4 peptides could accept ADP-ribose from NAD^+, but only if the peptides were acetylated. Through mass spectrometry, they showed that the relative molecular weight of the product was actually smaller than that of the substrate by 42 Da, indicating that the major modification catalyzed by Sir2 was deacetylation and not ADP-ribosylation. When NAD^+ was omitted from the reaction, Sir2 exhibited no deacetylase activity. NADH, NADP, and NADPH could not substitute for NAD^+ in this reaction. Landry et al. (17) characterized the role of SIR2 and HST2 in a nicotinamide-NAD^+ exchange reaction. HST2 had NAD^+-dependent deacetylase activity (17). Smith et al. (18) also characterized SIR2 and HST genes as an NAD^+-dependent deacetylase. They also first showed that mutation, which lowered NAD^+ synthesis in vivo, compromised the silencing activity of SIR2p.

CATALYTIC MECHANISM

The Sir2-dependent deacetylation reaction is different from other HDACs. Here we review the role of NAD^+ and the unusual reaction products. Finally, we discuss the possible biological role of those products.

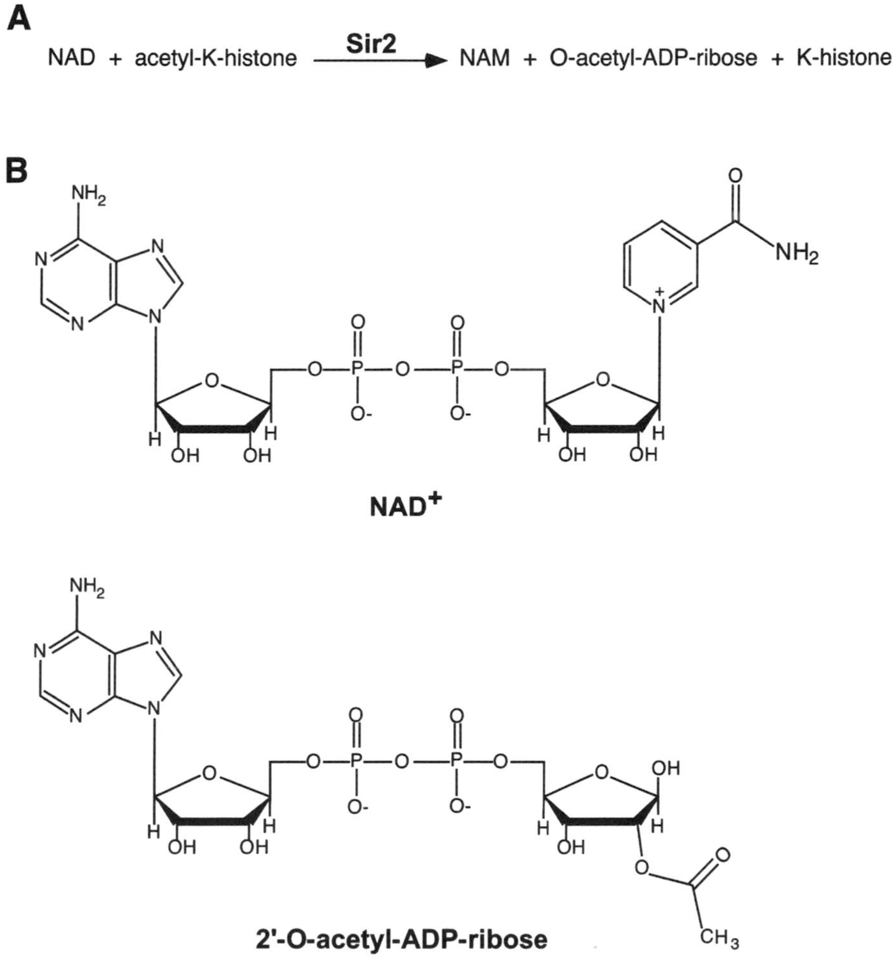

A

NAD + acetyl-K-histone $\xrightarrow{\text{Sir2}}$ NAM + O-acetyl-ADP-ribose + K-histone

B

NAD$^+$

2'-O-acetyl-ADP-ribose

Figure 1 Coupling of deacetylation to NAD cleavage and acetyl transfer from substrate to ADP-ribose by Sir2 and Sir2-like proteins. (A) The overall reaction scheme. (B) Structures of NAD$^+$ and 2'-O-acetyl-ADP-ribose, the novel compound produced by Sir2 and Sir2-like proteins, are shown. NAM = nicotinamide.

An important clue for the molecular mechanism of Sir2 activity came from the stoichiometry between deacetylation and NAD$^+$ breakdown. For each acetyl lysine that was deacetylated by Sir2, one NAD$^+$ molecule was cleaved (17, 19, 20) to produce unexpected products, nicotinamide and O-acetyl-ADP-ribose, instead of the predicted nicotinamide and ADP-ribose (Figure 1) (19, 20). The O-acetyl-ADP-ribose was generated from the transfer of an acetyl group from substrate to the ADP-ribose moiety of NAD$^+$ (19, 20). It was clear that Sir2 has two coupled enzymatic activities, deacetylation and NAD$^+$ breakdown, that produce a new compound, O-acetyl-ADP-ribose. These two enzymatic activities and the novel enzymatic reaction product raise important questions: Why is an energetically favorable reaction (deacetylation) coupled to NAD^+ hydrolysis, and what is the biological role of O-acetyl-ADP-ribose?

The NAD^+ dependence of deacetylation by Sir2 may possibly provide regulation of the biological functions of the Sir2, such as gene silencing, metabolism, and aging. It was proposed that O-acetyl-ADP-ribose has a unique cellular function that may be linked to the Sir2 gene-silencing effect, raising the possibility that this product had an important signaling role as a cofactor for its catalytic activity (20). In support of this notion, a quantitative microinjection of exogenous O-acetyl-ADP-ribose into starfish oocytes delayed oocyte maturation (21). Interestingly, a group of enzymes, the Nudix hydrolase family, can hydrolyze O-acetyl-ADP-ribose (22).

STRUCTURE OF THE SIR2 FAMILY PROTEINS

The structures of Sir2 homologs have been determined (archaeal Sir2 homolog sir2-Af1, human SIRT2, and archaeal sir2-Af2). Here, we discuss the Sir2 structure and the implications of the structure in our understanding of the Sir2 family.

The structure of an archaeal Sir2 homolog sir2-Af1 revealed that the protein consists of a large domain of a classical open α/β, Rossmann-fold structure and a small domain of α three-stranded antiparallel β-sheet, two α-helices, and a long loop (23). The structure of the human SIRT2 was later solved and compared to that of the sir2-Af1. The two enzymes shared a similar domain architecture. The large domains had an identical topology but only part of the Rossmann folds could be aligned (24). One NAD^+ molecule was bound in a pocket between two domains, and it adopted an extended conformation, which was found in Rossmann-fold NAD^+-bound proteins. Moreover, the NAD^+ molecule was oriented in an inverted orientation in contrast to other NAD^+-bound proteins. In the open conformation, the large domain formed the floor, and the small domain formed the ceiling of the NAD^+-binding pocket. The majority of the SIR2 family members contain a motif of Cys-X-X-Cys-(X)15–20-Cys-X-X-Cys in the conserved domain, which binds to Zn^+ ion (23).

The structure of archaeal sir2-Af2 complexed with an acetylated p53 peptide was solved. The p53 peptide lies in the large groove between the Rossmann fold and the small domain. The N terminus of the peptide was close to the zinc-binding domain, and the C terminus was close to the flexible loop region (Figure 2) (25). The key Sir2-Af2 peptide-binding residues were conserved in other Sir2 proteins. The side chain of the acetyl-lysine residue (K-Ac382) fits into a tunnel that led to the NAD^+ binding site. Two amino acids at the N terminus from the Ac-Lysine and five amino acids C-terminal

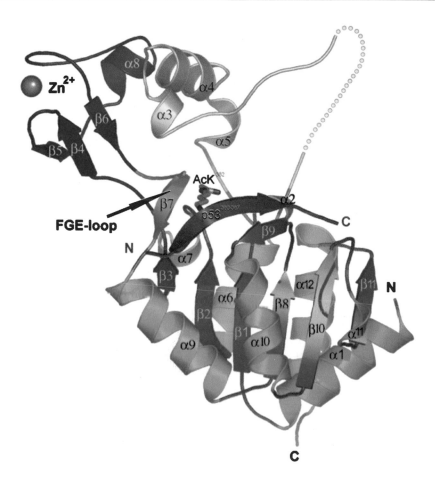

Figure 2   Cartoon representation of the overall structure of the Af2 and p53 acetylated peptide complex. The reverse Rossmann fold is shown in green and orange. The helical module is shown in yellow, and the disordered region represented as small circles. The zinc-binding module is shown in dark blue, the FGE loop in cyan, and the p53-KAc382 peptide in red.

to the Ac-Lysine of p53 peptide were included in the electron density map. The dominant role of peptide backbone hydrogen bound in substrate binding and the limited extent of substrate side chain burial result in only weak selectivity for certain side chains (25).

In conclusion, the Sir2 solved structure adds to our understanding of the Sir2 family. One issue that has not been explained by the structure is the substrate specificity of the Sir2 enzyme. To understand this issue, other techniques, such as identification of the in vivo substrate by two-dimensional acetylated gels or peptide library screening, may be utilized.

INHIBITORS AND ACTIVATORS

Several different chemical compounds have been shown to inhibit the deacetylase activity of the Sir2 family. One inhibitor, splitomicin, was identified in a 6000 chemical compound screen. Treatment with splitomicin disrupted silencing at the HML, HMR, and telomeric loci of the budding yeast. Interestingly, transcriptional profiles of splitomicin-treated cells mimic those of a *sir2* mutant. Splitomicin inhibits the NAD^+-dependent deacetylase activity of Sir2 in vitro (26).

Another inhibitor was identified in a high-throughput phenotypic screen. This inhibitor, sirtinol is derived from 2-hydroxy-1-napthaldehyde. Sirtinol interferes with body axis formation in *Arabidopsis* (27).

Nicotinamide, a product of the Sir2 deacetylation reaction, is an inhibitor of Sir2 activity both in vivo and in vitro. In yeast cells, exogenous nicotinamide derepresses all three Sir2 target loci, increases recombination at the *rDNA* loci, and shortens life span, comparable to that of the *sir2* mutant (28). Nicotinamide has been shown to inhibit a Sir2 homolog, SIRT1, a negative p53 regulator, promoting p53-dependent apoptosis in mammalian cells (29, 30). To prove that Sir2 was regulated by the changes in nicotinamide levels, Anderson et al. (31) showed that the *PNC1*, which encodes an enzyme that deaminates nicotinamide, was both necessary and sufficient for life span extension by depleting nicotinamide (Figure 3, yeast). Characterization of the base-exchange reaction revealed that nicotinamide regulates Sir2 activity by switching between deacetylation and base exchange (32).

Small molecules that stimulate the SIRT1 deacetylase activity were recently identified. Two of these activators, quercetin and piceatannol, are structurally similar to each other. These compounds stimulate SIRT1 activity by more than severalfold. A secondary screen within the quercetin and piceatannol family identified 15 additional SIRT1 activators. The most potent activator was resveratrol, a polyphenol found in red wine, which is implicated in a number of health benefits. In vitro, resveratrol lowers the K_m values for the acetylated peptide and NAD^+ by 35 and 5 times, respectively. At a low concentration, resveratrol increased the yeast life span by 70%, whereas a high concentration had only partial effect. Resveratrol did not further extend the life span of caloric restricted yeast, indicating that they probably act through the same pathway. In human cells, treatment with a low concentration of resveratrol increased cell survival following DNA damage. Moreover, a low resveratrol concentration decreased the acetylation of p53 at the lysine residue 382, a known SIRT1 substrate; however, a high concentration caused the opposite (33).

In conclusion, a large number of chemical activators and inhibitors were identified in the recent years. Those compounds may help us to further understand Sir2 biology.

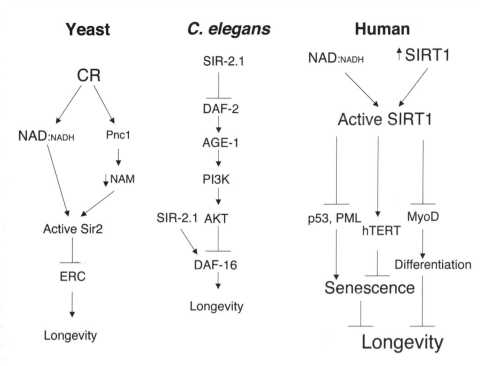

Figure 3 Model showing how Sir2 may extend the life span of yeast, *Caenorhabditis elegans*, and humans. The role of Sir2 in these organisms is discussed in the text.

CELLULAR LOCALIZATION, PROTEIN INTERACTIONS AND PROTEIN SUBSTRATES

The amount of data on the cellular localization, protein interactions, and substrates of the Sir2 family members are rapidly increasing. In Table 1, we summarize the current knowledge of the most important Sir2 family members. In Figure 4, we aligned the sequences of Sir2 substrates, in order to find a consensus sequence.

Yeast Sir2

Sir2 AS A DEACETYLASE   Since Sir2 was discovered to be an NAD^+-dependent deacetylase, significant progress in the Sir2 substrates and Sir2 chromatin binding sites has been made. In this section, we summarize the recent discoveries. To learn more about the yeast Sir2, you may read other recent reviews (34–37).

The Sir2 enzymatic activity is not required for the initial binding, but it is important for the association with the regions distal to the nucleation sites. At the *rDNA* sites, histone H4 was hypoacetylated in a Sir2-dependent manner (38).

With a genome-wide acetylation microarray, Gruastein and colleagues (39) showed that a *SIR2* deletion leads to hyperacetylation of subtelomeric regions,

TABLE 1 Sir2 family members

Gene	Subcellular localization	Interactors	Targets	Activity
Sir2 (yeast)	Nucleolus (86)	RAP1 (87)	Histone H3 (K9) (K14)	Deacetylase (16)
	Telomeres (77)	SIR4 (88)	H4 (K16) (16)	
		Net1 (89)		
CobB			Acetyl-CoA synthetase (44)	Deacetylase (44)
Archaeal Sir2		Alba (46)	Alba (46)	Deacetylase ADP ribosyl transferase (46)
dSIR2 (*Drosophila*)	Nucleus (47)	Hairy (48)	Histone H4 (K5) (K8) (K12) (K16)	Deacetylase (47, 48)
SIRT1	Nucleus (29, 30) PML bodies (57)	p53 (29, 30) PML (57) PCAF/GCN5 (53) CTIP2 (63) HES1 (62)	Histone H3 (K9 & 14) H4 (K16) (16) p53 (k382) (29, 30) TAFI68 (61) PCAF/MyoD (53)	Deacetylase (16)
SIRT2	Cytoplasm (68, 69)	HDAC6 (68)	Tubulin (68)	Deacetylase (68, 69)
SIRT3	Mitochondria (72, 90)			Deacetylase (72, 90)
SIRT4				Not active (68)
SIRT5				Deacetylase (68)
SIRT6				Not active (68)
SIRT7				Not active (68)

less than 4 kb from the telomeric ends, the mating-type loci (*HML* and *HMR*), and the *rDNA* loci.

Deletion of *SIR2* caused an increase in histone H3 and H4 acetylation within the rDNA region. The largest increase in acetylation occurred on histone H3 (40).

Two Sir2-containing protein complexes were identified via biochemical purification. One complex includes Sir4 and is active as an NAD$^+$-dependent histone deacetylase, whereas another complex contains Net1, which is a part of nucleolar silencing and the telophase exit (RENT) complex and recruits Sir2 to the *rDNA* loci, and possesses a deacetylase activity that is only partially dependent on NAD$^+$. Both complexes efficiently bind to nucleosomes (41). All of these data

Sir2 deacetylation consensus sequences

H3(9)	A<u>RTKQTAR</u>**K**STGG**K**APRKQLATKAA
H3(14)	ARTKQTA<u>RKSTGG</u>**K**APR<u>K</u>QLATKAA
H4(16)	AGG<u>KGGKG</u>MGAK<u>VGA</u>**K**RHS
P53(382)	HL<u>KSKKG</u>QSTS<u>R</u>HK**K**LMF<u>K</u>TEGPDSD
TUBULIN	ME<u>RLSVDYG</u>**K**<u>K</u>SKLEFSIYP
Alba	MSSGTPTPSN<u>VV</u>LIG**KK**
Consensus	K(V/S)(V/T)G<u>G</u>**K**KXX<u>K</u>

Figure 4 Sir2 family deacetylation target consensus sequences. The amino acid sequence of the Sir2 targets are shown: H3(9), histone H3 lysine 9; H3(14), histone H3 lysine 14; H4(16), histone H4 lysine 16; p53(382), p53 lysine 382; TUBULIN, tubulin; and Alba, alba. The Sir2 target consensus sequence is indicated.

indicate that, in budding yeast, the histone deacetylation is the major biological function of Sir2.

SCHIZOSACCHAROMYCES POMBE Sir2   The *Schizosaccharomyces pombe* Sir2 is the closest Sir2 homolog. Similar to *Saccharomyces cerevisiae* Sir2, it is an NAD^+-dependent deacetylase with histone H3 lysine 9 and H4 lysine 16 as substrates. In vivo the *S. pombe* Sir2 regulates silencing at the donor mating-type loci, telomeres, and the inner centromeric repeats but not at the *rDNA* loci. These results suggest that the molecular function of the Sir2-dependent silencing involves the deacetylation of histone H3 lysine 9 in chromatin (42).

Bacterial Sir2

It is remarkable that SIR2 genes are conserved not only in all eukaryotic organisms examined, but also in prokaryotes and archaea. Below we describe features of the noneukaryotic SIR2 proteins.

SALMONELLA Sir2 (*CobB*)   *CobB* complements the lack of phosphoribosyltransferase activity in a *cobT* mutant. *CobB* catalyzes the synthesis of N-(5-phospho-α-D-ribosyl)5,6-dimethylbenzimidazole, a cobalamin biosynthetic intermediate, from nicotinate mononucleotide and 5,6-dimethylbenzimidazole (13). Recently, *CobB* was shown to be required for the acetate and propionate activation via the high-affinity acyl-CoA synthetase pathway. The acyl-CoA synthetase is acetylated on lysine 609 (43), which is an essential residue for catalysis (44). *CobB* deacetylates the acyl-CoA synthetase in an NAD^+-dependent manner to activate the enzyme.

ARCHAEAL Sir2   The archaeal *Sulfolobus solfataricus* P2 encodes a single Sir2 homolog, ssSIR2 (45). The ssSIR2 is an active NAD^+-dependent deacetylase as well as a mono-ADP-ribosyl transferase (46). The archaeal protein Alba, which is a major *Sulfolobus* chromatin protein, interacts with the ssSir2. Alba was doubly acetylated at the α-amino group of the N terminus and at either lysine 16 or 17 of the protein. Lysine to alanine mutations at these residues resulted in a significant decrease in the DNA binding affinity of Alba and less repression of the transcription in vitro. The acetylated Alba also has a lower DNA affinity in vivo. ssSIR2 deacetylates Alba, which increases DNA affinity and thereby represses transcription (46).

Drosophila Sir2

Five *Drosophila* genes belong to the Sir2 family. Of these, dSir2 is the closest homolog of the yeast Sir2. dSir2 is an active deacetylase in vitro, capable of deacetylating histone H4 (47–49). In vivo dSir2 was shown to be a requirement for hetrochromatic silencing. Moreover, dSir2 genetically and physically interacted with Hairy, a bHLH euchromatic repressor and key regulator of *Drosophila* development (48). The dSir2 mRNA was detected during the first 2 h of embryogenesis, indicating that the transcripts are maternally derived (49). In the adult fly, the protein is primarily nuclear (49). In the embryo, prior to nuclear cycle 12, it is detected both in the nucleus and in the cytoplasm. By the syncytial blastoderm stage (cycle 13), dSir2 was only in the cytoplasm. As cellularization begins (cycle 14), it appeared again both in the nucleus and the cytoplasm (48). Contrary to what had been reported previously (48), Astrom et al. (50) found that a dSir2 knockout fly has no effect on viability, developmental rate, and sex ratio. In agreement with the previous report, they observed a modest effect on position effect variegation (50). Notably, they showed that dSir2 knockout results in the shortening of the fly life span (50).

Mammalian Sir2

The mammalian Sir2 gene family is comprised of seven members. Each is defined by a conserved core domain, and some contain additional N- or C-terminal sequences. Whereas much is known about SIRT1, less is known about the other six mammalian sir2 homologues.

SIRT1   The human Sir2 ortholog SIRT1 is an NAD^+-dependent deacetylase (16). The SIRT1 protein is localized in the nucleus (29, 30). SIRT1 interacts with and deacetylates a growing number of proteins. A knockout mouse showed that this protein is important for the embryonic development (51, 52), and recently it was shown that it plays a role in muscle differentiation (53). Moreover, when overexpressed, SIRT1 appears to increase the *hTERT* expression (54).

p53 INTERACTION   Following DNA damage, p53 is acetylated and activated by p300 acetyltransferase. (55). Moreover, the HDAC1 is capable of deacetylating and repressing p53 (56). SIRT1 was hypothesized as playing a role in the p53 pathway as well. To test this possibility the physical interaction between SIRT1 and p53 was investigated. Vaziri et al. (29) and Luo et al. (56) showed that p53 and SIRT1 coimmunoprecipitate. DNA-damaging agents augment in vivo interaction (30). In vitro a p53 peptide, acetylated at lysine 382, served as a substrate for SIRT1. Importantly, NAD$^+$ was required for the deacetylation reaction (30, 57). Upon exposure of immortalized human fibroblast to ionizing radiation, a marked increase in the p53 acetylation level was detected. The increase in the acetylation levels was abrogated in the cells that overexpress the *SIRT1* protein (29). In vivo deacetylation of p53 was inhibited by nicotinamide (30). The biological consequences of the deacetylation are the repression of the p53-dependent transcription and apoptosis (29, 30, 57). Also, a SIRT1 point mutation in the conserved deacetylase motif inhibited p53 deactylation as a dominant negative and activated the p53-dependent apoptosis (29). Recently, a SIRT1 knockout mouse was shown to have highly acetylated p53, and it induced apoptosis in thymocytes (52).

PML INTERACTION   The nuclear bodies (NB), often termed promyelocytic leukemia protein (PML) NB, are distinct nuclear substructures that accumulate PML proteins (58). A typical cell contains 10–30 PML NB per nucleus. The PML interacts with CBP and HDAC1, which led the Kouzarides group (57) to explore possible interactions between SIRT1 and PML4. Indeed, endogenous SIRT1 interacts with PML4. When SIRT1 was overexpressed with PML4, it localizes to the PML NB (57). Moreover, SIRT1 and PML4 colocalizes with p53 in the PML NB. Overexpression of PML4 in primary cells leads to an immediate growth arrest. Interestingly, SIRT1 overexpression rescued the cells from the growth arrest (57). Together, these results indicate that SIRT1 may be a positive effector of cell growth that negatively regulates p53 and PML.

BCL6 INTERACTION   BCL6 is a nuclear protein, which belongs to the BTB/POZ family of zinc finger transcription factors. It primarily functions as a transcriptional repressor (59). BCL6 plays a role in the control of lymphocyte activation, differentiation, and apoptosis. p300 binds to and acetylates BCL6, inhibiting its activity as a transcriptional repressor. Inhibition of classes I and II HDACs by trichostatin A and inhibition of SIR2 family (class III HDAC) by nicotinamide leads to an increase in the BCL6 acetylation levels (60). These results suggest that one of the SIR2 family members deacetylates BCL6, represses the transcription of BCL6 target genes, and consequently inhibits B cell differentiation.

TAF$_I$68 INTERACTION   TAF$_I$68 is the second largest subunit of the RNA polymerase I complex that binds to the TATA box and is acetylated by PCAF. In vitro TAF$_I$68 acetylation stimulates RNA polymerase transcription (61). The mouse

SIRT1 deacetylates TAF$_I$68 in vitro, and it may thereby repress the polymerase I transcription apparatus (61).

HES1 INTERACTION   The Hairy related bHLH protein functions as a transcriptional repressor and plays an important role in diverse aspects of development. HES1 and the HEY2 are human Hairy homologs. SIRT1 associates with both HES1 and HEY2 in vivo. In vitro SIRT1 interacts with the bHLH domain of HES1. However, SIRT1 augments only slightly the repression mediated by HES1 and HEY2 (62). The biological implication of this interaction needs to be studied.

CTIP2 INTERACTION   CTIP2 is a sequence-specific DNA binding protein that represses transcription via direct DNA binding. In vivo and in vitro SIRT1 binds to CTIP2 and is recruited to CTIP2 target promoters in a CTIP2-dependent manner. SIRT1 stimulates the repression by CTIP2 and enhances the histone deacetylation of CTIP2 target promoters (63). These data suggest that SIRT1 can be recruited to promoters by specific transcription factors and function to repress the transcription of specific genes.

KNOCKOUT MICE   Two different groups created two independent lines of the *SIRT1* knockout mice. Both groups showed that the SIRT1 knockout mice are viable. However, the birth ratio of the homozygous knockout mice to other animals was lower than expected. The lower survival of the null animals at birth reflects the immediate postnatal loss of abnormal fetuses. The mice were smaller than the wild-type littermates, and most of them died during the first few months after birth (51, 52). One of the most obvious developmental defects was the delay in eyelid opening. In the *SIRT1* null mice, eyelids stayed closed for at least several months after birth. Eyes of the *SIRT1* knockout mice were smaller and irregularly shaped (52). This observation indicates that the *SIRT1* may have an important role in eye development. McBurney et al. (51) found that the lung and pancreas were affected in the mutant mice. The pancreas showed patchy atrophy of the exocrine epithelia (51). The *SIRT1* knockout mice made by the Alt group (52) showed a cardiac defect but did not show lung and pancreas defects. The *SIRT1* transcript is widely expressed in many tissues and particularly evident in testis and ovary. Interestingly, both sexes of the null animals are sterile. The female had smaller ovaries in which corpora lutea were conspicuously absent and the wall of uterus was thin. The authors (51) showed that the sterility was due to a hormonal inadequacy. The male null mice had a dramatically reduced number of mature sperms. Importantly, none of the *SIRT1* null sperms were motile nor had a normal morphology (51). The level of p53 acetylation was much higher in the knockout mice. In contrast to the previous reports, the hyperacetylation was not specific to lysine 379 but also occurred at lysine residues 317 and 370. Following DNA damage, the steady-state level of p53 was not induced, and p21 was not induced in knockout mouse embryonic fibroblasts (MEFs). However, apoptosis in thymocytes was elevated in the mutant mice (52).

REGULATION OF SKELETAL MUSCLE DIFFERENTIATION   The expression of muscle-cell-specific genes is regulated by acetylation and deacetylation of transcription factors (64). The muscle transcriptional regulator MyoD is an acetylated protein. In a recent study, the Sartorelli group (64) showed that the mouse SIRT1 negatively regulates skeletal muscle differentiation. By using a battery of Sir2 family inhibitors, they demonstrated that these inhibitors activated the transcription of muscle-specific reporters. Furthermore, SIRT1 overexpression negatively regulated the transcription of those genes and inhibits differentiation into muscle cells. SIRT1 directly interacted with the PCAF/GCN5 acetyltransferases in vitro and in muscle cells. PCAF mediated the interaction between SIRT1 and MyoD. In vitro SIRT1 deacetylates MyoD and PCAF in an NAD^+-dependent manner. With microarray experiments, the authors (64) showed that myogenin and MEF2C expression were negatively regulated by SIRT1. They further showed that many genes that were activated by MyoD and involved in myogenesis were repressed by SIRT1. In addition, by chromatin immunoprecipitation, they found that the SIRT1 was recruited to the MyoD targets and deacetylased histones in the target promoters. Notably, they explored whether the NAD^+/NADH ratio, which decreases during muscle differentiation, regulates SIRT1 deacetylase activity. Indeed, they found that the change in the ratio regulates SIRT1 enzymatic activity (53).

SIRT2   SIRT2 is another member of the Sir2 family. In the phylogenic tree, it is localized to the same branch as Sir2 and SIRT1 (65). A Northern blot analysis showed that *SIRT2* is highly expressed in heart, brain, testis, and skeletal muscle (66, 67). The protein is cytoplasmic (66–68), and its levels are regulated during the cell cycle. The SIRT2 protein level increases dramatically during mitosis, and it becomes phosphorylated at the G2/M transition (69). SIRT2 overexpression dramatically prolongs the M phase. Moreover, SIRT2 is targeted for degradation by the proteasome (69).

Reversible acetylation of tubulin had been implicated in the regulation of microtubule stability and function (70). Interestingly, the HDAC6 was shown to deacetylate tubulin (71). The Verdin group (68) showed that SIRT2 protein colocalizes with microtubules. They showed that SIRT2 deacetylates lysine 40 of α-tubulin in vivo and in vitro and that SIRT2 *RNAi* results in hyperacetylation of tubulin. They further demonstrated that SIRT2 and HDAC6 coimmunoprecipitate in vivo (68). In contrast to HDAC6, which regulates microtubule-dependent cell motility, the biological implication of tubulin deacetylation by SIRT2 is not known.

SIRT3   SIRT3 is another member of the Sir2 family. In the phylogeny tree, SIRT3 is localized to the same branch as Sir2 and SIRT1, and it is the closest paralog of SIRT2 (65). A Northern blot analysis demonstrated that the gene is highly expressed in brain, heart, liver, kidney, testis, and muscle (66, 72). The SIRT3 protein is localized to the mitochondrial matrix (90). The N terminus of the protein is proteolytically processed in the mitochondrial matrix to form a mature product. The unprocessed SIRT3 is enzymatically inactive, but following signal peptide cleavage, it becomes active as a histone deacetylase (72, 90). Biological targets of this protein are not yet known.

SIRT4   SIRT4 is another member of the Sir2 family. Other than SIRT5, SIRT4 is the most distant from Sir2 and SIRT1 in phylogeny tree (65). RT-PCR analysis of SIRT4 expression from adult and fetal tissues showed that the gene is broadly expressed in all tissues, except leukocytes in the adult and thymus in the fetus (14). SIRT4 may be enzymatically inactive as a histone deacetylase in vitro (68), and a biological role is yet to be discovered.

SIRT5   SIRT5 is is the closest homolog of the bacterial *CobB* and is the most distant from Sir2 and SIRT1 (65). RT-PCR analysis of SIRT5 expression profiles from adult and fetal tissues demonstrated that the gene is broadly expressed (14). SIRT5 is enzymatically active as a histone deacetylase (68), and its biological role is yet to be discovered.

SIRT6   SIRT6 is another member of the Sir2 family, and it is the closest SIRT7 homolog (65). SIRT6 is enzymatically inactive as a histone deacetylase (68), and its biological role is yet to be discovered.

SIRT7   SIRT7 is highly expressed in the spleen, ovary, and thyroid. SIRT7 is also highly expressed in thyroid carcinomas when compared to normal thyroid tissues. In contrast, its expression is almost undetectable in adenomas and normal thyroid tissues (73, 74). SIRT7 is enzymatically inactive as a histone deacetylase (68).

SIR2 AND AGING

The yeast *Saccharomyces cerevisiae* divides asymmetrically to give rise to a larger mother cell and a smaller daughter cell. In this organism, life span can be defined by the number of cell divisions undergone by a mother cell before it stops dividing (75). In a screen to isolate long-living mutants, Kennedy et al. (76) found eight strains that exhibited extended life span. One of them, *SIR4–42*, was a *SIR4* mutant that lacks a C-terminal domain (76). The life span extension by the Sir4–42 allele was dominant, and mutations in the SIR2 or the SIR3 genes, which are part of the SIR complex, abolished the life span extension by the SIR4–42. In a further characterization, the *SIR4–42* gene product changed the localization of the SIR complex from the telomeres to the nucleolus (77). Interestingly, in old yeast mother cells, the Sir3 translocated from the telomeres to the nucleolus (77), suggesting that something in the nucleolus might regulate the yeast life span. Deletion of either the *SIR3* or *SIR4* resulted in a 20% decrease in mean life span, and it was due to simultaneous expression of the two mating-type genes. A *sir2* mutant resulted in much shorter life span (approximately 50% of the control) (78). The life span shortening by *sir2* mutants was due to the increase in homologous recombination at the *rDNA* loci, which results in the formation of an extrachromosomal *rDNA* circle (ERC) in the nucleolus (Figure 3, yeast). Interestingly, introduction of a second copy of the *SIR2* into the yeast genome extended the replicative life span by 30% (78). Therefore, the Sir2 protein is a limiting factor of yeast life span. Mutations that abolish the Sir2 deacetylase activity shorten life span (16). Interestingly, SIR2 is the only yeast Sir complex member that is conserved through evolution.

Tissenbaum & Guarente (79) decided to look at the effect of overexpression of a Sir2 homolog on *Caenorhabditis elegans* life span. The cross of a sir-2.1 overexpression strain with a *daf-16* mutant strain, a downstream target of the insulin pathway that shortens life span when mutated, abolished the life span extension by the sir-2.1. Also, a cross with a daf-2 mutant, an insulin receptor, resulted in no further extension of life span (79) (Figure 3, *C. elegans*). Thus, sir-2.1 overexpression significantly extended the life span for *C. elegans* (79). The life span extension by the *sir-2.1* overexpression is via the insulin pathway, which was already established in regulating its life span [for review, see (80)]. It is remarkable that both replicative aging in yeast and postmitotic aging in the worm are regulated by Sir2. It will be interesting to determine whether Sir2 regulates the aging process in higher eukaryotes as well.

CALORIC RESTRICTION

Caloric restriction (CR) refers to a dietary regime, low in calories without undernutrition. CR extends the life span in many organisms, which include rotifers, spiders, worms, fish, mice, and rats (81). Recent data suggest that it may be true for primates as well (82). Although it has been suggested that CR might work by reducing the levels of reactive oxygen species during respiration, the mechanism of life span extension was uncertain. Lin et al. (83) showed that yeast cells exhibited a longer life span on 0.5% glucose-containing media than on 2% glucose-containing media, suggesting that CR in yeast extends life span. Limiting the availability of glucose by mutating a glucose transporter or blocking its downstream signaling pathway also extends the life span. Thus, reduction in glucose concentration extends life span and provides a model for CR in yeast (83). By using a mutant strain that mimics reduction in glucose concentration, Lin et al. demonstrated that the yeast life span extension by CR requires *SIR2* (83, 84) (Figure 3, yeast). As mentioned before, Sir2 is an NAD^+-dependent enzyme. NAD^+ may be a sensor that activates Sir2 during CR. In fact, *NPT1,* involved in de novo synthesis of NAD^+, was required for CR-dependent life span extension. The life span extension by CR in yeast is likely caused by the reduction in rDNA circles by Sir2 (83).

Next, the mechanism by which CR increases Sir2 activity and extends life span was explored. A clue came from the glucose metabolism. Glucose is metabolized to pyruvate, where the pathway flows into either respiration or fermentation, depending on O_2 availability. Respiration generates 36 ATP molecules per glucose, whereas fermentation generates only two ATP molecules. Indeed, respiration was activated during CR (84). The shift toward respiration was necessary for life span extension (84). Following this shift, the NAD^+/NADH ratio widely changes to favor the activation of Sir2, resulting in the inhibition of recombination at the rDNA locus and the extension of life span (Figure 3, yeast).

CONCLUSION

How do these findings relate to human aging? In mammals, the Sir2 enzymatic activity may be regulated by changes in the steady-state protein levels, the NAD^+:NADH ratio, or the nicotinamide (NAM) levels (Figure 3, human). The Sartorelli group (53) showed that the NAD^+/NADH ratio decreases during muscle differentiation and thus regulates SIRT1 deacetylase activity (Figure 3, human). Activation of SIRT1 could then inhibit cell senescence by repressing p53 (29, 30, 57), repressing PML (57), and activating hTERT (54) (Figure 3, human). In addition, active SIRT1 would stall differentiation of muscle by repressing MyoD (53) (Figure 3, human) and perhaps exert similar effects on other tissues. It will be fascinating to observe additional functions of the mammalian SIR2 homologs when they are discovered.

ACKNOWLEDGMENTS

We thank D. Moazed and C. Wolberger for allowing us to use Figures 1 and 2, respectively. We thank N. Chang, M. Haigis, A. Berdichevsky, and R. Machado de Oliveira for critical reading of the manuscript. Our laboratory is supported by grants from the National Institute of Health. G. Blander is an EMBO Long-Term Postdoctoral Fellow.

The *Annual Review of Biochemistry* is online at http://biochem.annualreviews.org

LITERATURE CITED

1. Rine J, Herskowitz I. 1987. *Genetics* 116: 9–22
2. Gottschling DE, Aparicio OM, Billington BL, Zakian VA. 1990. *Cell* 63:751–62
3. Bryk M, Banerjee M, Murphy M, Knudsen KE, Garfinkel DJ, Curcio MJ. 1997. *Genes Dev.* 11:255–69
4. Smith JS, Boeke JD. 1997. *Genes Dev.* 11:241–54
5. Loo S, Rine J. 1994. *Science* 264: 1768–71
6. Bi X, Broach JR. 1997. *Mol. Cell. Biol.* 17: 7077–87
7. Thompson JS, Ling X, Grunstein M. 1994. *Nature* 369:245–47
8. Hecht A, Laroche T, Strahl-Bolsinger S, Gasser SM, Grunstein M. 1995. *Cell* 80: 583–92

9. Braunstein M, Sobel RE, Allis CD, Turner BM, Broach JR. 1996. *Mol. Cell. Biol.* 16: 4349–56

10. Braunstein M, Rose AB, Holmes SG, Allis CD, Broach JR. 1993. *Genes Dev.* 7:592–604

11. Luger K, Mader AW, Richmond RK, Sargent DF, Richmond TJ. 1997. *Nature* 389: 251–60

12. Brachmann CB, Sherman JM, Devine SE, Cameron EE, Pillus L, Boeke JD. 1995. *Genes Dev.* 9:2888–902

13. Tsang AW, Escalante-Semerena JC. 1998. *J. Biol. Chem.* 273:31788–94

14. Frye RA. 1999. *Biochem. Biophys. Res. Commun.* 260:273–79

15. Tanny JC, Dowd GJ, Huang J, Hilz H, Moazed D. 1999. *Cell* 99:735–45

16. Imai S, Armstrong CM, Kaeberlein M, Guarente L. 2000. *Nature* 403:795–800

17. Landry J, Sutton A, Tafrov ST, Heller RC, Stebbins J, et al. 2000. *Proc. Natl. Acad. Sci. USA* 97:5807–11

18. Smith JS, Brachmann CB, Celic I, Kenna MA, Muhammad S, et al. 2000. *Proc. Natl. Acad. Sci. USA* 97:6658–63

19. Tanny JC, Moazed D. 2001. *Proc. Natl. Acad. Sci. USA* 98:415–20

20. Tanner KG, Landry J, Sternglanz R, Denu JM. 2000. *Proc. Natl. Acad. Sci. USA* 97: 14178–82

21. Borra MT, O'Neill FJ, Jackson MD, Marshall B, Verdin E, et al. 2002. *J. Biol. Chem.* 277:12632–41

22. Rafty LA, Schmidt MT, Perraud AL, Scharenberg AM, Denu JM. 2002. *J. Biol. Chem.* 277:47114–22

23. Min JR, Landry J, Sternglanz R, Xu RM. 2001. *Cell* 105:269–79

24. Finnin MS, Donigian JR, Pavletich NP. 2001. *Nat. Struct. Biol.* 8:621–25

25. Avalos JL, Celic I, Muhammad S, Cosgrove MS, Boeke JD, Wolberger C. 2002. *Mol. Cell* 10:523–35

26. Bedalov A, Gatbonton T, Irvine WP, Gottschling DE, Simon JA. 2001. *Proc. Natl. Acad. Sci. USA* 98:15113–18

27. Grozinger CM, Chao ED, Blackwell HE, Moazed D, Schreiber SL. 2001. *J. Biol. Chem.* 276:38837–43

28. Bitterman KJ, Anderson RM, Cohen HY, Latorre-Esteves M, Sinclair DA. 2002. *J. Biol. Chem.* 277:45099–107

29. Vaziri H, Dessain SK, Eagon EN, Imai SI, Frye RA, et al. 2001. *Cell* 107:149–59

30. Luo JY, Nikolaev AY, Imai S, Chen DL, Su F, et al. 2001. *Cell* 107:137–48

31. Anderson RM, Bitterman KJ, Wood JG, Medvedik O, Sinclair DA. 2003. *Nature* 423:181–85

32. Sauve AA, Celic I, Avalos J, Deng H, Boeke JD, Schramm VL. 2001. *Biochemistry* 40:15456–63

33. Howitz KT, Bitterman KJ, Cohen HY, Lamming DW, Lavu S, et al. 2003. *Nature* 425.191–96

34. Hekimi S, Guarente L. 2003. *Science* 299: 1351–54

35. Sinclair DA. 2002. *Mech. Ageing Dev.* 123: 857–67

36. Gasser SM, Cockell MM. 2001. *Gene* 279: 1–16

37. Guarente L. 2000. *Genes Dev.* 14: 1021–26

38. Hoppe GJ, Tanny JC, Rudner AD, Gerber SA, Danaie S, et al. 2002. *Mol. Cell. Biol.* 22:4167–80

39. Robyr D, Suka Y, Xenarios I, Kurdistani SK, Wang A, et al. 2002. *Cell* 109: 437–46

40. Buck SW, Sandmeier JJ, Smith JS. 2002. *Cell* 111:1003–14

41. Ghidelli S, Donze D, Dhillon N, Kamakaka RT. 2001. *EMBO J.* 20: 4522–35

42. Shankaranarayana GD, Motamedi MR, Moazed D, Grewal SI. 2003. *Curr. Biol.* 13:1240–46

43. Starai VJ, Takahashi H, Boeke JD, Escalante-Semerena JC. 2003. *Genetics* 163:545–55

44. Starai VJ, Celic I, Cole RN, Boeke JD, Escalante-Semerena JC. 2002. *Science* 298: 2390–92

45. She Q, Singh RK, Confalonieri F, Zivanovic Y, Allard G, et al. 2001. *Proc. Natl. Acad. Sci. USA* 98:7835–40

46. Bell SD, Botting CH, Wardleworth BN, Jackson SP, White MF. 2002. *Science* 296: 148–51

47. Barlow AL, van Drunen CM, Johnson CA, Tweedie S, Bird A, Turner BM. 2001. *Exp. Cell Res.* 265:90–103

48. Rosenberg MI, Parkhurst SM. 2002. *Cell* 109:447–58

49. Newman BL, Lundblad JR, Chen Y, Smolik SM. 2002. *Genetics* 162:1675–85

50. Astrom SU, Cline TW, Rine J. 2003. *Genetics* 163:931–37

51. McBurney MW, Yang XF, Jardine K, Hixon M, Boekelheide K, et al. 2003. *Mol. Cell. Biol.* 23:38–54

52. Cheng HL, Mostoslavsky R, Saito S, Manis JP, Gu YS, et al. 2003. *Proc. Natl. Acad. Sci. USA* 100:10794–99

53. Fulco M, Schiltz RL, Iezzi S, King MT, Zhao P, et al. 2003. *Mol. Cell* 12:51–62

54. Lin SY, Elledge SJ. 2003. *Cell* 113: 881–89

55. Abraham J, Kelly J, Thibault P, Benchimol S. 2000. *J. Mol. Biol.* 295:853–64

56. Luo JY, Su F, Chen DL, Shiloh A, Gu W. 2000. *Nature* 408:377–81

57. Langley E, Pearson M, Faretta M, Bauer UM, Frye RA, et al. 2002. *EMBO J.* 21: 2383–96

58. Seeler JS, Dejean A. 1999. *Curr. Opin. Genet. Dev.* 9:362–67

59. Chang CC, Ye BH, Chaganti RS, Dalla-Favera R. 1996. *Proc. Natl. Acad. Sci. USA* 93:6947–52

60. Bereshchenko OR, Gu W, Dalla-Favera R. 2002. *Nat. Genet.* 32:606–13

61. Muth V, Nadaud S, Grummt I, Voit R. 2001. *EMBO J.* 20:1353–62

62. Takata T, Ishikawa F. 2003. *Biochem. Biophys. Res. Commun.* 301:250–57

63. Senawong T, Peterson VJ, Avram D, Shepherd DM, Frye RA, et al. 2003. *J. Biol. Chem.* 278:43041–50

64. Sartorelli V, Puri PL, Chang CC, Ye BH, Chaganti RS, Dalla-Favera R. 2001. *Front. Biosci.* 6:D1024–47

65. Frye RA. 2000. *Biochem. Biophys. Res. Commun.* 273:793–98

66. Yang YH, Chen YH, Zhang CY, Nimmakayalu MA, Ward DC, Weissman S. 2000. *Genomics* 69:355–69

67. Afshar G, Murnane JP. 1999. *Gene* 234: 161–68

68. North BJ, Marshall BL, Borra MT, Denu JM, Verdin E. 2003. *Mol. Cell* 11:437–44

69. Dryden SC, Nahhas FA, Nowak JE, Goustin AS, Tainsky MA. 2003. *Mol. Cell. Biol.* 23:3173–85

70. Piperno G, LeDizet M, Chang XJ. 1987. *J. Cell Biol.* 104:289–302

71. Hubbert C, Guardiola A, Shao R, Kawaguchi Y, Ito A, et al. 2002. *Nature* 417:455–58

72. Onyango P, Celic I, McCaffery JM, Boeke JD, Feinberg AP. 2002. *Proc. Natl. Acad. Sci. USA* 99:13653–58

73. de Nigris F, Cerutti J, Morelli C, Califano D, Chiariotti L, et al. 2002. *Br. J. Cancer* 86:917–23

74. Frye R. 2002. *Br. J. Cancer* 87:1479

75. Muller I, Zimmermann M, Becker D, Flomer M. 1980. *Mech. Ageing Dev.* 12 47–52

76. Kennedy BK, Austriaco NR Jr, Zhang J, Guarente L. 1995. *Cell* 80:485–96

77. Kennedy BK, Gotta M, Sinclair DA, Mills K, McNabb DS, et al. 1997. *Cell* 89 381–91

78. Kaeberlein M, McVey M, Guarente L. 1999. *Genes Dev.* 13:2570–80

79. Tissenbaum HA, Guarente L. 2001. *Nature* 410:227–30

80. Tissenbaum HA, Guarente L. 2002. *Dev. Cell* 2:9–19

81. Weindruch R. 1996. *Toxicol. Pathol.* 24 742–45

82. Lane MA, Black A, Handy A, Tilmor EM, Ingram DK, Roth GS. 2001. *Ann. N Acad. Sci.* 928:287–95

83. Lin SJ, Defossez PA, Guarente L. 2000. *Science* 289:2126–28

84. Lin SJ, Kaeberlein M, Andalis AA, Sturtz LA, Defossez PA, et al. 2002. *Nature* 418: 344–48

85. Deleted in proof

86. Gotta M, Strahl-Bolsinger S, Renauld H, Laroche T, Kennedy BK, et al. 1997. *EMBO J.* 16:3243–55

87. Moretti P, Freeman K, Coodly L, Shore D. 1994. *Genes Dev.* 8:2257–69

88. Strahl-Bolsinger S, Hecht A, Luo K, Grunstein M. 1997. *Genes Dev.* 11: 83–93

89. Straight AF, Shou W, Dowd GJ, Turck CW, Deshaies RJ, et al. 1999. *Cell* 97: 245–56

90. Schwer B, North BJ, Frye RA, Ott M, Verdin E. 2002. *J. Cell Biol.* 158: 647–57

Annu. Rev. Biochem. 2004. 73:437–65
doi: 10.1146/annurev.biochem.73.071403.161303
Copyright © 2004 by Annual Reviews. All rights reserved
First published online as a Review in Advance on March 26, 2004

INOSITOL 1,4,5-TRISPHOSPHATE RECEPTORS AS SIGNAL INTEGRATORS

Randen L. Patterson,[1] Darren Boehning,[1] and Solomon H. Snyder[1,2,3]

Department of Neuroscience,[1] Pharmacology and Molecular Science,[2] and Psychiatry and Behavioral Sciences,[3] Johns Hopkins University, Johns Hopkins Medical School, Baltimore, Maryland 21205; email: rpatter1@jhmi.edu, dboehnin@jhmi.edu, ssnyder@jhmi.edu

Key Words   calcium, kinases, scaffold, entry, release, structure, regulation

■ **Abstract**   The inositol 1,4,5 trisphosphate (IP$_3$) receptor (IP$_3$R) is a Ca^{2+} release channel that responds to the second messenger IP$_3$. Exquisite modulation of intracellular Ca^{2+} release via IP$_3$Rs is achieved by the ability of IP$_3$R to integrate signals from numerous small molecules and proteins including nucleotides, kinases, and phosphatases, as well as nonenzyme proteins. Because the ion conduction pore composes only ~5% of the IP$_3$R, the great bulk of this large protein contains recognition sites for these substances. Through these regulatory mechanisms, IP$_3$R modulates diverse cellular functions, which include, but are not limited to, contraction/excitation, secretion, gene expression, and cellular growth. We review the unique properties of the IP$_3$R that facilitate cell-type and stimulus-dependent control of function, with special emphasis on protein-binding partners.

CONTENTS

INTRODUCTION

Inositol 1,4,5-trisphosphate (IP$_3$) is a second messenger produced primarily by phospholipase C (PLC) metabolism of phosphoinositol-4,5-bisphosphate (PIP$_2$) in response to the stimulation of G-protein-coupled receptors (GPCRs) or receptor tyrosine kinases (RTKs) (1). IP$_3$ rapidly releases Ca^{2+} from intracellular Ca^{2+} pools within the endoplasmic reticulum (ER) and other cellular membranes by binding to the IP$_3$ receptor (IP$_3$R), which amplifies/transduces cellular signals, many of which are generated at the plasma membrane (PM) (1, 2).

Researchers initially pursued study of the IP$_3$R following the discovery that IP$_3$ mediates Ca^{2+} signaling in both excitable and nonexcitable cells (2, 3). [^{3}H] IP$_3$ binding was first reported in membranes from neutrophils (4), hepatocytes, and adrenal cells (5, 6), but low levels of specific binding precluded molecular characterization of the binding entity. Autoradiographic studies revealed extraordinarily high densities of IP$_3$ binding in the cerebellum (7) with levels 100–300 times higher than in peripheral tissues, permitting detailed characterization. IP$_3$R was purified to homogeneity from this tissue by exploiting the ability of the receptor to bind heparin and conconavalin A (8), allowing functional characterization of the purified channel (9, 10). Reconstitution of the purified IP$_3$-binding protein in lipid vesicles established that the protein is an IP$_3$-gated Ca^{2+} channel (11) with the unique ability to release Ca^{2+} in quantal units (12). Cloning of the gene corresponding to the IP$_3$R (13) facilitated structure/function analysis (14, 15). The *IP$_3$R* gene encodes a very large protein with an open reading frame of more than 2700 amino acids (13).

Three distinct *IP$_3$R* genes have been identified with differential expression throughout the body (16). One suggested difference among them involves regulation by Ca^{2+} (17, 18); however, several studies have failed to detect any functional differences between isoforms (19, 20). Although still not definitive, it appears that Ca^{2+} modulation of Ca^{2+} release may be the same for the three isoforms. Evidence includes the functional redundancy uncovered by genetic ablation of individual *IP$_3$R* isoforms in DT40 cells (21) and the lack of a strong phenotype in type 2 and type 3 IP$_3$R knockout mice (K. Mikoshiba, personal communication). However, IP$_3$R1 knockout mice that survive past birth display severe neurological and physiological defects (22). Because the cerebellum is unique in expressing a single isoform, IP$_3$R-1 (23), this phenotype may reflect a major dependence of cerebellar function on IP$_3$R. Additional forms of IP$_3$R are derived by alternative splicing. There are three well-characterized splice variants of the IP$_3$R-1 (SI, SII, SIII), but isolates found in expressed sequence tag (EST libraries predict many others in the human genome, most of which are truncations not containing the pore-forming region of the channel (23, 24). Best characterized is the splicing of IP$_3$R1 into a longer form that predominates in the brain (SII+) and a shorter form (SII-) that is found mostly in peripheral nonneuronal tissues (25). No functional differences have been observed in this splice variant with respect to Ca^{2+} or IP$_3$ regulation (26, 27). The overall structure of IP$_3$R and its splicing sites are depicted in Figure 1.

Purification, cloning, and characterization of IP$_3$R have provided insight into a multiplicity of cellular functions for IP$_3$ (1, 28). Given that even short peptides are capable of transporting calcium ions across lipid bilayers (29), the large size of IP$_3$R ($\sim$1000 kDa) suggests that it does not merely function as a Ca^{2+} channel. In fact, a large number of proteins interact directly with IP$_3$R, which indicates that IP$_3$R acts in many signaling pathways. The role of the IP$_3$R as a signal integration for multiple proteins and small molecules is a major focus of this review.

STRUCTURE AND LOCALIZATION

The IP$_3$R has a tetrameric structure similar to other Ca^{2+} channels. Recent work has helped to define the three dimensional structure of the IP$_3$R, information important to deciphering the exact mechanisms of IP$_3$R Ca^{2+} release. The IP$_3$R has also been localized to numerous organelles within the cell, including the PM. This wide distribution likely reflects the multiplicity of function contained within the IP$_3$R.

Primary Structure

The functional IP$_3$R exists as a tetramer (30) with the Ca^{2+} channel region of the receptor sharing substantial sequence homology with the corresponding region of the ryanodine receptor (RyR). The channel region of IP$_3$R is located at the extreme C-terminal end (14); it is characterized by six membrane-spanning helices with the C terminus projecting into the cytoplasm (30). The pore domain of the channel shares structural similarity to both voltage-gated potassium and calcium channels. The selectivity filter is comprised of the amino acid sequence GVGD [amino acids (aa) 2547–2550, rat sequence] with Ca^{2+} selectivity mediated by Asp2550 as determined by site-directed mutagenesis (31). The IP$_3$ binding site resides in the N-terminal portion of the protein; the core element contains residues 224–578 (32). Mutagenesis studies reveal that multiple sites in the N-terminal portion are required for IP$_3$ binding, demonstrating that the tertiary structure is critical in establishing the IP$_3$-binding pocket (32, 33), a result confirmed by the crystal structure of the ligand-binding region of the IP$_3$R (34). Because the IP$_3$-binding domain is separated from the channel pore by $\sim$2000 aa, IP$_3$ binding presumably elicits large conformational changes that regulate chan-

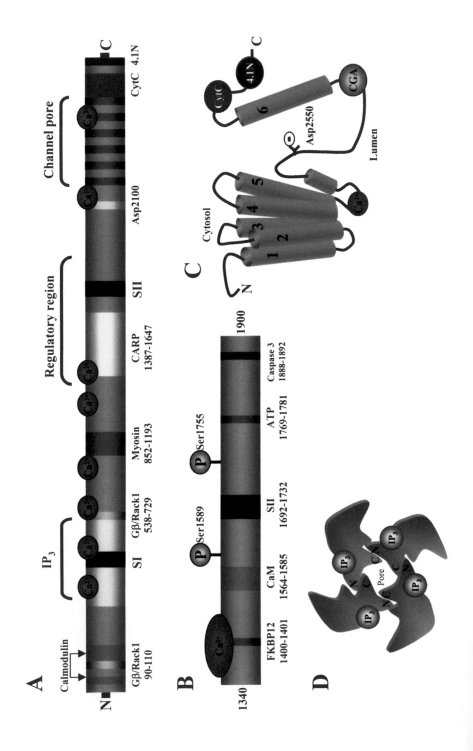

nel gating (28). However, IP_3 does not substantially change the pattern of peptides produced by trypsin digestion (35), and trypsinized IP_3Rs retain the ability to release Ca^{2+} (32), indicating that subtle changes in conformation lead to channel gating. Interestingly, the N terminus of the channel directly binds to and gates the C terminus of an adjacent subunit in a tetramer (36), perhaps explaining why trypsinized channels retain functional activity and providing insight into the three-dimensional organization of the channel (Figure 1C). Close proximity between the IP_3-binding domain and the channel pore is also implied by evidence for interactions of IP_3Rs with synthetic poly(ethylene glycol)-linked dimers of IP_3 (37) and by a recent cryo-electron mycroscopy of the IP_3R structure, which mapped the extreme N terminus to the channel pore (38). Intervening between the ligand-binding domain and the channel pore is the bulk of the receptor, which is loosely referred to as the modulatory domain (Figure 1). This domain, which shares the least amount of homology among the three IP_3R isoforms, contains binding sites for Ca^{2+}, nucleotides, calmodulin, other proteins, and modulatory factors (discussed in more detail below).

Four groups have recently investigated the three-dimensional structure of the IP_3R (30, 38–40). On the basis of negative staining of a partial structure, Hamada et al. (39) demonstrated two distinct conformational states of IP_3R, a windmill and a square, wherein interconversion is regulated by Ca^{2+}. IP_3-binding regions were identified at the tips of the windmill by heparin-gold labeling. Serysheva et al. (38) magnified a cryo-electron microscopic structure to 30 Å, which also demonstrated a windmill shape. This study assigned putative domain organizations on the basis of the crystal structure of the IP_3-binding domain, placing the IP_3-biding domain within a spoke of the windmill. A similar assignment of domains was achieved with a negatively stained structure that was much more compact (40). These structures contrast with the 24 Å cryo-electron structure of Jiang et al. (30), which is primarily globular with only small projections extending from the top and bottom. The substantially divergent structures obtained by the four groups may reflect variations in sample preparation or number of particles analyzed. The structure of Serysheva et al. (38) may be the most accurate one as the result of sample preparation at physiologic pH,

---

←

Figure 1   Structure of the type 1 IP_3 receptor. (*A*) Depiction of the full-length IP_3R. Ca^{2+}-binding sites, splicing regions (SI, SII), and binding sites for Gβ, RACK1, CaM, myosin, CARP, 4.1N protein, and cytochrome *c* (CytC) are demarcated. The core IP_3 binding site, regulatory region, and channel pore segment are also outlined. (*B*) An expanded view of the regulatory region of the IP_3R, showing the binding sites for FKBP12, CaM, and ATP. The PKA/PKG phosphorylation sites are also labeled along with the caspase-3 cleavage site. (*C*) An expanded view of the channel region of the IP_3R. Asp2550 is critical for channel function. (*D*) A pictorial representation of the tetramerization of the IP_3R, representing a combined model of the cryo-electron microscopy data and biochemical data on the IP_3R structure. N and C termini of each subunit are indicated.

mapping the crystal structure of the IP_3-binding domain onto the projection, and the large number of particles analyzed.

Intracellular Localization

IP_3R has been localized to a number of cellular membranes. First identified in the ER, IP_3R also resides in the Golgi apparatus where it colocalizes with calcium-binding nucleobindin and releases Ca^{2+} in response to IP_3 (41). IP_3R occurs in secretory granules (discussed below), in the PM (42), and in a newly discovered organelle designated the nucleoplasmic reticulum (NR) (43, 44). Immunofluorescent studies of the HepG2 liver cell line, which expresses only IP_3R-2 and IP_3R-3, reveals IP_3R-2 in the ER and the nucleus, whereas IP_3R-3 is only present in the ER (43). IP_3 releases nuclear Ca^{2+} with an EC_{50} (64 nM) mirroring its affinity for IP_3R-2 (Kd~58 nM) (17) but not for IP_3R-3 (3.4 μM) (45). Echevarria et al. (44) imaged an NR using two-photon confocal microscopy, differentiating nuclear and cytoplasmic Ca^{2+} signals as well as localizing protein kinase C to the NR in response to Ca^{2+} signals that did not extend outside of the NR.

With the exception of the PM, all IP_3R-containing membranes surround organelles with releasable Ca^{2+} stores. Though IP_3R was localized to the PM in 1992 (46), its PM function remains unclear. Cell-surface biotinylation reveals that 5% to 14% of total IP_3R protein resides in the PM in several cell types (47). Mayrleitner et al. (48) purified IP_3R from the PM and observed Ca^{2+} channel behavior in planar lipid bilayers. However, an IP_3-activated Ca^{2+} conductance in PM-matching IP_3R conductances measured in other organelles has not yet been demonstrated. Perhaps IP_3R in the PM functions as a scaffold for its numerous binding partner proteins. If IP_3R in the PM is a functional Ca^{2+} channel, its conductance/open probability should be decreased by lipids or other IP_3R-interacting proteins. Consistent with this possibility, Lupu et al. (49) detected phosphatidylinositol bisphosphate binding to and inhibition of IP_3R channel activity in planar lipid bilayers.

REGULATION BY SMALL MOLECULES

The complexity and sophistication of IP_3R regulation is exemplified by its regulation by small molecules. Phosphorylation, nucleotides, and the ubiquitous second messenger (Ca^{2+}), which the channel transports, all work in concert to provide precise control over IP_3R function.

Calcium Regulation

IP_3Rs are regulated biphasically by Ca^{2+} (50–52). Ca^{2+} stimulates IP_3R Ca^{2+} conductance in tissue membranes with maximal effects at about 100–300 nM Ca^{2+}(53). Studies of purified IP_3R-1 at the single-channel level in bilayers and of native nuclear membrane patches show a sharply positive cooperative increase in open probability at physiologic concentrations of Ca^{2+} and inhibition at low micromolar concentrations (27, 54). Other studies with purified IP$_3$R did not reveal inhibition by Ca^{2+} (55), suggesting that a protein mediating the Ca^{2+} inhibition of the IP$_3$R was lost during purification. Mikoshiba and associates (56) showed that calmodulin mediates Ca^{2+}-dependent inhibition of IP$_3$R function (discussed in more detail below) and so may be the protein lost during purification. Studies with reconstituted IP$_3$R have revealed multiple direct binding sites for ^{45}Ca^{2+} with affinity of about 800 nM (57) (Figure 1), although no direct functional role for these binding sites has been determined. Miyakawa et al. (51) reported that a single residue, Asp-2100, is required for the Ca^{2+} responsiveness of IP$_3$R because its replacement with Glu causes a 10-fold decrease in Ca^{2+} stimulation. The homologous residue in the ryanodine receptor also mediates Ca^{2+} sensitivity of the channel (58). Type 2 and type 3 IP$_3$Rs are not inhibited by Ca^{2+} when activity is measured in planar lipid bilayers (17, 18), but regulation by Ca^{2+} has indeed been demonstrated with the nuclear patch clamp technique (27), microsomal ^{45}Ca^{2+} flux (19), and in intact cells (59). The exact mechanism of Ca^{2+} regulation is still controversial; however, it is generally accepted that Ca^{2+} initially released by IP$_3$R feeds back to augment further Ca^{2+} release in a positively cooperative fashion. Higher Ca^{2+} concentrations inhibit channel function, creating the classical oscillatory pattern of Ca^{2+} release observed in response to most agonist-mediated pathways.

Phosphorylation

IP$_3$R is phosphorylated by multiple kinases. Cyclic-AMP-dependent protein kinase (PKA) phosphorylation of IP$_3$R regulates channel activity and has been implicated in influences of muscarinic-cholinergic receptor pathways (60), ERK-kinase activation (61), *Drosophila* larval development (62), neuroprotection (63), neostriatal signaling (64), and PLC activation (65). PKA stoichiometrically phosphorylates purified IP$_3$R-1 at two known sites, serine 1589 and 1755 (S1589 and S1755) (66) (Figure 1). IP$_3$R-1(SII-) in peripheral tissues is preferentially phosphorylated by PKA at S1589, whereas neuronal IP$_3$R-1(SII+) is more heavily phosphorylated at S1755 (66). In intact and permeabilized cells, the PKA activators forskolin or dibutyryl cAMP stimulate Ca^{2+} release and PKA-mediated IP$_3$R phosphorylation (60, 67). In pancreatic acinar cells, cholecystokinin-elicited Ca^{2+} release is associated with IP$_3$R-3 phosphorylation (67).

PKA phosphorylation of IP$_3$R varies throughout the brain (63). Antibodies that distinguish between IP$_3$R S1755 phosphorylated and unphosphorylated forms of IP$_3$R1 reveal phosphorylated IP$_3$R enrichment in dendrites, whereas nonphosphorylated forms predominate in cell bodies. Evidently, within areas where rapid signaling and Ca^{2+} release are required, IP$_3$R is phosphorylated to increase its responsiveness. The phosphorylation state of IP$_3$R can be rapidly altered because PKA and protein phosphatases 1 and 2A form a complex with IP$_3$R (Figure 2A), analogous to the complex formed with RyR (64, 68). This

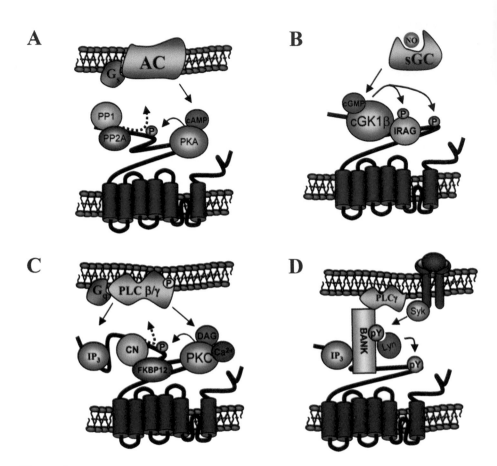

Figure 2   Multi-protein IP$_3$R signaling complexes. (*A*) Through G-protein-coupled signaling (G$_s$) to adenylate cyclase (AC), PKA is activated by cyclic-AMP to phosphorylate the channel and increases activity. PP1 and PP2A are complexed at the N terminus of the IP$_3$R, allowing rapid regulation by dephosphorylation. (*B*) Activation of soluble guanylyl cyclase (sGC) by nitric oxide (NO) leads to the production of cyclic-GMP and the activation of cyclic-GMP kinase 1β (cGK1β). cGK1β phosphorylates the IP$_3$R and IRAG in a trimeric complex, inhibiting channel function. (*C*) Activation of G-protein-coupled signaling pathways leads to the activation of PLCβ/γ for the production of IP$_3$ and DAG. DAG stimulates PKC phosphorylation activity, increasing IP$_3$R channel activity. FKBP12 likely anchors calcineurin to the IP$_3$R near the phosphorylation site to dephosphorylate the channel, allowing for rapid regulation of the phosphorylation state of the IP$_3$R. (*D*) Activation of the B-cell receptor complex leads to the phosphorylation of BANK by the tyrosine kinase Syk. The phosphorylation of BANK then allows the tyrosine kinase Lyn to bind to the IP$_3$R, allowing Lyn to phosphorylate the IP$_3$R and modulating channel activity.

protein complex facilitates rapid changes in IP_3R phosphorylation state and likely represents the functional unit by which cAMP regulates IP_3R function. Isoform-specific modulation of IP_3R function may depend on PKA actions. Thus, serine 1589 and 1755 PKA phosphorylation sequences in IP_3R-1 are absent in IP_3R-2 and IP_3R-3, although these channels are also phosphorylated by PKA (69).

Cyclic GMP-dependent protein kinase (PKG) phosphorylates IP_3R at the same sites as does PKA (70), generating similar Ca^{2+} oscillations as with phosphorylation by PKA (71). In brain slices, PKG and PKA preferentially phosphorylate S1589 and S1755, respectively (72). No function for the direct phosphorylation of IP_3R by PKG has been ascribed, though PKG phosphorylation of proteins that bind IP_3R influences IP_3R functions, discussed below.

Protein kinase C (PKC) and calcium/calmodulin-dependent protein kinase II (CAMK-II) phosphorylate IP_3R at distinct sites (73). PKC phosphorylates IP_3R in vitro (73) and in vivo (74), increasing sensitivity to IP_3-mediated Ca^{2+} release from IP_3R. Calcium directly activates many forms of PKC, providing a positive feedback loop that may influence downstream events, such as long-term depression, neuronal exocytosis, and apoptosis (65, 75, 76). CAMK-II regulates Ca^{2+} oscillations via IP_3R in certain systems (77, 78). He et al. (79) demonstrated that continuous activation of CAMK-II is required for IP_3R-linked Ca^{2+} oscillations that influence neurotransmitter release. Moreover, the IP_3R modulation by CAMK-II creates a positive feedback loop that influences neurotransmitter release at the neuromuscular junction (79).

A consistent pattern in serine/threonine phosphorylation of the IP_3R is that these modifications alter the spatio-temporal Ca^{2+} release properties of the channel. This likely facilitates the regulation of "microdomain" Ca^{2+} in spatially restricted areas, providing signal specificity.

IP_3R can be phosphorylated at tyrosine residues. In activated T cells, the nonreceptor tyrosine kinase, Fyn, binds and phosphorylates IP_3R (80). This phosphorylation is diminished in Fyn -/- mice, which manifest reduced intracellular Ca^{2+} release and defective T-cell signaling. Lyn also tyrosine phosphorylates IP_3R in response to B-cell receptor stimulation, a process facilitated by the B-cell scaffold protein with ankyrin repeats (BANK) (Figure 2D) (81). Because this phosphorylation occurs in response to B-cell receptor stimulation, tyrosine phosphorylation of IP_3R in B cells may be selective for immune responses but not mitogenic signals.

IP_3R is autophosphorylated at the same sites used by PKA and PKC (82). Persistence of the phosphorylation through extensive purification, which includes denaturation and renaturation, apparently rules out actions of contaminating kinases. Autophosphorylation is well known for tyrosine kinase receptors but not for ion channel receptors. The plasma-membrane ion channel TRPM7 is also autophosphorylated in an ATP-dependent manner (83). IP_3R autophosphorylation might provide a sensitization/desensitization mechanism that occurs with other autophosphorylating proteins.

Nucleotides

ATP influences IP_3R function independently of phosphorylation or energy-dependent processes. In reconstituted vesicles containing purified IP_3R, ATP maximally enhances IP_3-induced Ca^{2+} flux at ~ 100 μM ATP; these effects are fairly selective for adenine nucleotides, occur with hydrolysis-resistant analogues, and reflect specific ATP-binding sites on IP_3R (11) (Figure 1). At high concentrations, ATP inhibits IP_3-induced release of Ca^{2+}, probably by competing directly for the ligand-binding pocket. Effects of low ATP concentrations are physiologically relevant because the open probability of IP_3-regulated Ca^{2+} channels in muscle membranes is also augmented by ATP (57). Resting levels of ATP in the cell are about 1 mM. Therefore, physiologic activation of IP_3R may occur when ATP levels are depleted in the microenvironment of IP_3R. Thus, Ca^{2+} released through IP_3R is followed immediately by activation of SERCA to refill stores, and this leads to depletion of ATP in the vicinity of IP_3R, further activating Ca^{2+} release via a feed-forward mechanism. Because the ER containing IP_3R is closely juxtaposed to mitochondria (84), alterations in mitochondrial ATP may regulate IP_3R.

The adenine nucleotide NADH also regulates IP_3R. Physiologic concentrations of NADH augment IP_3R-mediated Ca^{2+} flux in lipid vesicles containing purified IP_3R (85). This action is highly selective for NADH and appears to be physiologically relevant because hypoxia of PC12 cells or cerebellar Purkinje cells raises NADH levels and elicits rapid increases in cytosolic calcium derived from IP_3-sensitive stores (85). Glyceraldehyde-3-phosphate dehydrogenase (GAPDH) binds IP_3R, so that NADH can be physiologically generated in close proximity to the IP_3R (R.L. Patterson and S.H. Snyder, unpublished data). This system may represent a mechanism for rapid adaptation to hypoxia. Ca^{2+} released from IP_3R in close proximity to the mitochondria would stimulate mitochondrial production of ATP in an effort to enhance energy production during hypoxic inhibition of mitochondrial respiration (86).

PROTEIN-PROTEIN INTERACTIONS

IP_3R is regulated by its interactions with a multiplicity of proteins and small molecules. Presumably, the complexity of its control provides extreme precision of signaling that enables the IP_3R to respond accurately to extracellular messages that impact upon IP_3-evoked Ca^{2+} release. To do this, IP_3R must regulate and be regulated by a large number of proteins. To date, more than 25 proteins have been reported to interact with IP_3R, influencing function in a variety of ways.

Scaffolding Proteins

A number of scaffolding proteins associate with the IP_3R, promoting stable localization of the IP_3R within the cell. Mutations of these proteins can result in mislocalization of the IP$_3$R within the cell, leading to pathophysiological cellular misfunction.

ANKYRIN   Ankyrins comprise a family of adapter proteins that link the spectrin-based cytoskeleton to proteins in the PM, ER, and Golgi complex. Ankyrin binds IP$_3$R, inhibiting IP$_3$ binding and thereby blocking IP$_3$- mediated Ca^{2+} release (87, 88). A synthetic peptide, which corresponds to an 11 residue sequence in IP$_3$R (residues 2548–2558 in rat IP$_3$R), competes for full-length ankyrin binding to IP$_3$R and blocks ankyrin-induced inhibition on IP$_3$-induced Ca^{2+} release (89). This sequence constitutes the ion conduction pathway of the channel, so it is an unlikely place for ankyrin binding (28). Because ankyrin binds both resident ER and PM proteins, ankyrin might create microdomains between the two membranes, perhaps using IP$_3$R as an anchor to the ER. Another example of a scaffolding protein linking ER-localized IP$_3$Rs to the PM is B-cell scaffold protein with ankyrin repeats (BANK) (81). BANK indeed contains ankyrin repeats that complex the B-cell receptor to IP$_3$R and the tyrosine kinase Lyn (Figure 2*D*), although it is not clear whether the ankyrin repeats are responsible for this binding. Presumably, this trimeric complex creates a restricted space in the B cell selectively augmenting Ca^{2+} levels at the PM.

The recent discovery that mutations in ankyrin-B cause type 4 long-QT syndrome cardiac arrhythmia and sudden cardiac death suggests a role for IP$_3$R ankyrin interactions in cardiac physiology (90). Most of these patients display a single-point mutation at amino acid 1425 (glutamic acid to glycine), a residue close to the regulatory domain of ankyrin-B. Heterozygous mice transgenic for this mutation display type 4 long-QT syndrome and sudden death after exercise. This mutation causes mislocalization of IP$_3$R, the Na$^+$/Ca^{2+} exchanger, and the SERCA pump (all proteins that associate with ankyrin-B) and alters Ca^{2+} signaling.

Ankyrin-dependent regulation of IP$_3$R localization and function may also be amenable to pharmacologic manipulation based on findings that the sigma-1 receptor forms a heterotrimeric complex with ankyrin and IP$_3$R (91). Sigma-1 receptors were initially identified as targets for the psychomimetic actions of opiates, phencyclidine, and other drugs. The cloned sigma-1 receptor resembles a sterol isomerase but lacks this enzyme activity. Sigma-1 agonists, such as cocaine, dissociate ankyrin from IP$_3$R and potentiate IP$_3$R-mediated Ca^{2+} release.

HOMER   Homer 1a, an immediate early gene rapidly induced by neural stimulation, binds to and regulates metabotropic glutamate receptors (mGluR). Longer forms of Homer, which incorporate a coiled-coil domain at their C termini, constitutively bind to metabotropic glutamate receptors. Worley and associates (92) demonstrated direct binding of Homer to IP$_3$R, which contains the same Homer-binding domain as does mGluR. Homer proteins that contain a coiled-coil domain occur as multimers and can link mGluRs with IP$_3$R, providing regulation of mGluR/IP$_3$R signaling at excitatory synapses (93). Because Homer 1a does

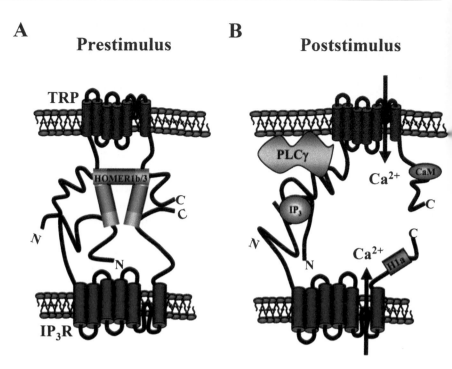

Figure 3   Model of the role for the IP$_3$R in TRPC activation. (*A*) Before agonist stimulation, TRPCs are likely precoupled to the IP$_3$R, possibly through their N terminus, C terminus, or both. For TRPC1 and possibly other TRPC, Homer 1b/3 cross-links N and C termini with the IP$_3$R, keeping the channel in an off state. (*B*) After agonist stimulus, IP$_3$ binds to the IP$_3$R, causing a conformational change in the channel. This leads to dissociation of Homer from TRPC, allowing for activation of the channel in coordination with PLC-gamma binding to the N terminus and CaM binding to the C terminus of TRPC. Long-term modulation of Ca^{2+} entry can be achieved by regulated expression of the dominant-negative Homer isoform 1a (H1a).

not multimerize and blocks the association of a coiled-coil domain containing Homer proteins, it functions as a dominant-negative to disrupt mGluR/IP$_3$R interactions. The rapid induction of Homer 1a following neural activity down-regulates PLC-coupled receptors. It is likely that Homer isoforms, similar to ankyrin, organize microdomains of closely juxtaposed ER and PM, influencing vesicle secretion and Ca^{2+} entry (Figure 3 *A,B*). In support of this hypothesis, Yuan et al. (94) showed that Homer binds directly to the transient receptor potential channel (TRPC1) Ca^{2+} channels in the PM, presumably via its two Homer-binding sites at the C and N termini. When bound to Homer, TRPC1 is unable to open, so that TRPC1 channels deficient for Homer binding are constitutively active. Neurally mediated changes in Homer 1a levels likely represent a regulatory mechanism for Ca^{2+} entry.

4.1N PROTEIN   Using yeast 2-hybrid techniques, two groups independently observed 4.1N interactions with the extreme C terminus of IP₃R, (95, 96). 4.1N is the neuronal-enriched form of 4.1R protein originally isolated from red blood cells (97). 4.1R is also selectively enriched in certain neuronal populations, and its genetic deletion in mice leads to neurobehavioral abnormalities (98). 4.1N has 2 FERM (4.1-ezrin-moesin-radixin) domains as well as a spectrin/actin-binding domain. Binding of 4.1N to IP₃R is required for proper localization of IP₃R to the basolateral membrane in Madin-Darby canine kidney cells (95). In neurons, Maximov et al. (96) describe a ternary complex existing between IP₃R, 4.1N CASK and syndecan 2. This protein complex is likely to help create the ultrastructure at the postsynaptic density.

MYOSIN   Myosins are ubiquitous motor/scaffolding proteins. Both muscle and nonmuscle myosins are involved in the movement and localization of protein complexes. In rodents with mutations in myosin-Va, ER innervation of dendritic spines in Purkinje cells is lost, leading to disturbances of long-term depression (99). The *Caenorhabditis elegans* isoform of IP₃R binds to the heavy chain of several myosin II isoforms (Figure 1) (100), an interaction that may explain abnormalities in myosin-deficient mice. Overexpression of peptides that compete for myosin/IP₃R binding abolishes IP₃R-dependent pharyngeal pumping in *C. elegans*. Although it has been previously demonstrated that pharmacological manipulation of the cytoskeleton can affect IP₃R-mediated Ca^{2+} signaling (101, 102), this is the first example of a physiological role whereby the actin/myosin cytoskeleton can regulate IP₃R function. It seems likely that this type of regulation occurs in mammals, particularly in smooth muscle where the IP₃R is the predominant Ca^{2+} release channel.

IP3 RECEPTOR   IP₃R itself functions as a scaffold as is evident in a recent model (96) with IP₃R as the central core of the mGluR1α/5-Homer-CASK-syndecan-2 signalsome at the postsynaptic density. A similar complex has been proposed for IP₃R coupling G-protein-coupled receptors to Ca^{2+} entry through TRPC channels (103). Immunoprecipitation and immunofluorescence experiments reveal IP₃R complexes with a number of focal contact/scaffolding proteins, such as talin, vinculin, and α-actin (104), suggesting that IP₃R anchors the ER to focal adhesions consistent with the requirement for Ca^{2+} in focal adhesion regulation (105, 106). Just as phospholipase Cγ has lipase-independent activities (107, 108), IP₃R probably exerts nonchannel activities, most of which have yet to be elucidated.

Calcium-Binding Proteins

Because the IP₃R is a Ca^{2+}-carrying ion channel, it is not surprising that a number of Ca^{2+}-binding proteins directly associate with the channel. However, the actual function of these proteins with relationship to the

IP$_3$R is of great debate, in particular calmodulin, caldendrin, chromogranins, and FKBP12.

CALMODULIN   Yamada et al. (109) first demonstrated direct calcium-dependent high-affinity binding of calmodulin to the regulatory domain of IP$_3$R-1. A distinct, calcium-independent calmodulin-binding site with similar affinity occurs on the N terminus. The two binding sites have been mapped; the calcium-dependent binding sequence is at residues 1564–1589 in the type 1 SII-isoform (109), and the N-terminal site is between residues 6–159 (110). The N-terminal site was further mapped by Sienaert et al. (111) to two adjacent regions, residues 49–81 and 106–128 (Figure 1). Calmodulin binding appears to be required for Ca^{2+}-dependent inhibition of IP$_3$R (112, 113). However, in other studies a single-point mutation (W1557A, IP$_3$R SII-isoform) of the calcium-dependent calmodulin-binding site eliminated binding of IP$_3$R to calmodulin in vitro without influencing Ca^{2+}-dependent inhibition of the IP$_3$R in microsomal flux (114) or lipid bilayer assays (115).

Because multiple studies have demonstrated influences of calmodulin on IP$_3$ binding to IP$_3$R in vitro and in vivo [reviewed in Patel et al. (28)], calmodulin is likely a physiologic regulation of IP$_3$R. The N-terminal, Ca^{2+}-independent binding site on IP$_3$R appears to be most physiologically relevant (111, 116). One plausible mechanism is that Ca^{2+}-dependent binding of calmodulin to one lobe of IP$_3$R exposes the other lobe to Ca^{2+}-independent binding of calmodulin, which inhibits the channel. The N terminus of IP$_3$R can be crystallized (117), so cocrystallization of IP$_3$R with calmodulin as well as additional mutational analysis may provide clarification.

Calmodulin probably does not influence Ca^{2+}-dependent inhibition of the IP$_3$R at the high-affinity Ca^{2+}-dependent site because two independent groups (114, 115) have not confirmed the reported inhibitory effects. Calmodulin can sequester plasma-membrane proteins into the cytosol (118), so perhaps it brings proteins from other regions of the cell into close proximity with IP$_3$R (103).

CaBP/CALDENDRIN   CaBPs are neuronal EF-hand containing calcium-binding proteins, which include calsenilin, GCAPs, recoverin, hippocalcin, visinin, and VILIPs (119). They resemble calmodulin in possessing four EF-hand Ca^{2+}-binding motifs. Unlike calmodulin, these proteins often contain one or more nonfunctional EF-hands. The CaBP subfamily contains five members and additional forms derived by alternative splicing, including caldendrin (a spliced variant of CaBP1). Yeast 2-hybrid analysis with the N-terminal portion of IP$_3$R-3 identified interactions of CaBPs with IP$_3$R (120). CaBP1 binds to a portion of IP$_3$R that contains the IP$_3$-binding domain and was reported to gate the IP$_3$R channel in the absence of IP$_3$. The high affinity of CaBP for IP$_3$R (Ka, 25 nM) is enhanced by Ca^{2+} (Ka, 1 nM). Thus, CaBPs may provide a mechanism for opening IP$_3$R channels in response to an increase in Ca^{2+} concentration without augmentation of IP$_3$ levels (121). However, Haynes et al. (122; H.L

Roderick et al., personal communication) reported essentially opposite results with CaBP inhibiting IP$_3$ binding and reducing channel activity. The divergent findings may reflect the different methodologies used: patch-clamping of *Xenopus laevis* nuclei and calcium imaging in intact cells.

CHROMOGRANINS   Chromogranins provide a novel example of direct coupling between an ion channel and the storage protein for the same ion. Chromogranins are Ca^{2+}-binding proteins that are highly enriched (1–2 mM) in secretory granules of neurons, neuroendocrine, and endocrine cells, where they sequester Ca^{2+} to ~40 mM Ca^{2+} content. Yoo and colleagues (123–125) discovered that IP$_3$R binds chromogranins, which enhance IP$_3$-induced Ca^{2+} release from IP$_3$R reconstituted in liposomes and planar lipid bilayers by shifting IP$_3$ concentration-response relationships. Chromogranins may bind IP$_3$R on its intraluminal loops (Figure 1), rendering IP$_3$R in secretory granules more sensitive to IP$_3$ than other cellular organelles, such as the ER and nucleus. Direct evidence for this notion emerges from studies of pancreatic acinar cells in which acetylcholine and cholecystokinin cause Ca^{2+} release initially from the secretory granule area even when the stimulating hormone is selectively applied to the opposite side of the cell (126). Srivastava et al. (127) demonstrated a physiological link between IP$_3$R and chromogranins because mice heterozygous for the *Anx 7* gene, a Ca(2+)/GTPase associated with Ca^{2+}-dependent exocytosis, have defective IP$_3$R expression and chromaffin cell hyperplasia. These cells also have constitutively high levels of chromogranin A expression, presumably as a compensation mechanism. Chromogranins and IP$_3$R are closely opposed at the base of cilia in oviduct, where cilia movement is dependent on Ca^{2+} signaling (119). Thus, secretory granules comprise a previously unrecognized pool of Ca^{2+} stores, which may segregate Ca^{2+} signaling pathways by creating microdomains of IP$_3$-sensitive secretory granules.

FKBP-12

The immunophilin proteins were first identified as receptors for the immunosuppressant drugs cyclosporin A and FK506. These drugs cause immunosuppression by inhibiting calcineurin, which suppresses NFAT translocation to the nucleus and thereby inhibits interleukin-2 production. Cyclosporin A binds to the cyclophilin family of proteins, most of which are small and soluble. FK506 binds to a family of FK506-binding proteins (FKBPs) of which the most prominent is a 12 kDa protein, FKBP12.

FKBP12 binds to the ryanodine receptor (RyR) stabilizing its full conductance state (128). FK506 application dissociates FKBP12 from RyR causing it to be leaky to Ca^{2+}. Interactions of FKBP12 with IP$_3$R are less clear. Cameron et al. (129) demonstrated that FKBP12 copurifies with IP$_3$R and binds to IP$_3$R in yeast 2-hybrid models (Figure 2C) (130). FK506 was shown to augment Ca^{2+} release from purified IP_3R and IP_3R in brain membranes, presumably due to increased conductance of the IP_3R channel (131). Calcineurin was demonstrated to be a component of the IP_3R/FKBP12 complex because the three proteins copurified (129). The yeast 2-hybrid method demonstrated that FKBP12 binds IP_3R through a leucine-proline sequence at residues 1400–1401 and anchors calcineurin to this domain. The leucine-proline dipeptide structurally resembles the FK506 drug, which may explain the unique affinity of FKBP12 for this site. Inhibition of calcineurin in the complex by treatment with cyclosporin A augmented phosphorylation of IP_3R by PKC, but not by PKA or CAMK-II, leading to augmented Ca^{2+} release. These data agree with other evidence that PKC phosphorylation enhances IP_3-induced release of Ca^{2+} (73, 74). The failure of cyclosporin A to influence phosphorylation by other kinases suggests that the sites phosphorylated by these enzymes were not dephosphorylated by calcineurin. PKC is predicted to be a part of the quaternary complex that includes FKBP12, IP_3R, and calcineurin (Figure 2C). In such a complex, cellular activation that stimulates PKC activity would augment IP_3-induced release of calcium, whereas calcineurin in the complex would rapidly dephosphorylate the receptor to the off state, providing a mechanism for oscillatory Ca^{2+} signaling.

Using brain, cardiac, and skeletal muscle microsomes Carmody et al. (132) reported that skeletal microsomes required the application of FK506 to remove FKBP12 from the membranes; however, in brain and cardiac microsomes, FKBP12 could be eliminated by washing. Furthermore, when passed over GST-FKBP12 Sepharose columns, microsomes containing skeletal muscle RyR1 were retained, whereas brain or cardiac microsomes were not retained. Concurrently, using proteolytic fragments of RyR or IP_3R passed over GST-FKBP12 columns, Bultynck et al. (133) reported that RyR but not IP_3R associated with the column; also FKBP12 did not affect release through IP_3R, although caffeine-evoked RyR release was inhibited. Bultynck et al. (134) noted differences in structures of IP_3R and RyR that might affect FKBP12 interactions. Proteolytic mapping and site-directed mutagenesis of RyR revealed that Val 2322 was critical for the binding of FKBP12 to RyR, and when replaced with Leu as in IP_3R, binding affinity was drastically decreased. A chimeric RyR substitution of residues 2318–2328 of IP_3R into RyR retained FKBP12 binding, although binding of native IP_3R to FKBP12 could not be demonstrated. Dargan et al. (135) showed that FKBP12 increases the mean open time of purified IP_3R in lipid bilayers and coordinates the gating of neighboring channels in a manner similar to RyRs.

In summary, while yeast 2-hybrid and copurification studies indicate IP_3R-FKBP12 binding, other approaches do not detect such binding, suggesting that the interactions are weak. Several groups report functional regulation of IP_3R by FKBP12. Moreover, numerous workers have shown that calcineurin interacts with and modifies IP_3R phosphorylation. Newer techniques, such as transiently deleting FKBP12 from cells using siRNA and measuring IP_3R function, protein-protein interacting assays, such as surface plasmon resonance, or fluorescent resonance energy transfer (FRET) may provide clarification.

Apoptotic Proteins

Abundant evidence implicates elevations of intracellular Ca^{2+} in apoptotic cell death. Perturbations that cause significant increases in intracellular Ca^{2+} activate numerous proteins linked to apoptosis. IP$_3$R impacts the apoptotic pathway in several ways. Buffering cytosolic Ca^{2+} or suppressing IP$_3$R levels in T cells or B cells inhibits apoptosis with antisense to IP$_3$R-3 more effectively than does antisense to IP$_3$R-1 (136). In developing chick embryos, IP$_3$R-3 levels increase during developmental apoptosis in early postnatal cerebellar granule cells, dorsal root ganglia, embryonic hair follicles, and intestinal villi destined for programmed cell death, and neurotoxicity elicited by glutamate agonists is associated with increased IP$_3$R-3 expression (137). Moreover, apoptosis of chick dorsal root ganglia elicited by deprivation of nerve growth factor is accompanied by augmented IP$_3$R levels. Apoptosis is selectively prevented by antisense to IP$_3$R-3, once again suggestive of a type-specific function for IP$_3$R (137). Additionally, IP$_3$R levels decline in cells overexpressing the antiapoptotic protein BCL-X$_L$ (138). Genetic deletion of all IP$_3$R subtypes from DT 40 avian B cells inhibits anti-IgM-induced apoptosis by 75% (139).

CASPASE 3   Cleavage of IP$_3$R by caspase 3, first characterized by Hirota et al. (140) involves a caspase consensus site within IP$_3$R (^{1888}DEVD^{1892} in mouse IP$_3$R-1) (Figure 1). IP$_3$R degradation during apoptosis elicited in T cells by staurosporine or Fas ligand is blocked by the caspase inhibitor z-VAD-CH(2)DCB. IP$_3$R is not cleaved during apoptosis in caspase-3-deficient cells, but cleavage is restored by transfection with caspase 3; these findings have been replicated in many cells (141–143). Thus, caspase-3 cleavage of the IP$_3$R appears to provide a universal mechanism for IP$_3$R degradation during apoptosis (Figure 4B). However, early in the apoptotic program, Ca^{2+} release through IP$_3$R promotes cell death, an action that is not blocked by the global caspase inhibitor Z-VAD-fmk (148). Therefore, it is not clear whether caspase cleavage of IP$_3$Rs is causal, participates in apoptosis, or is a secondary consequence of caspase activation.

CALPAIN   Calpain cleaves IP$_3$R, creating 130kDa and 95 kDa fragments, though the exact cleavage sites are unknown (144) (Figure 4b). Calpain degradation of IP$_3$R occurs during apoptosis induced by chronic receptor stimulation (145), cadmium poisoning (142), and TNF-α signaling (141). However, with chronic agonist stimulation, the calpain-like proteolytic activity might reflect proteosomal degradation (145). Although activities of both calpain and caspase 3 increase during apoptosis, IP$_3$R transcription and protein levels are also up-regulated. Why would the cell make more IP$_3$R only to degrade it? Perhaps the N terminus of IP$_3$R acquires novel functions once cleaved from the channel domain of the protein. ESTs contain mRNA transcripts of IP$_3$R lacking the channel domain (24,

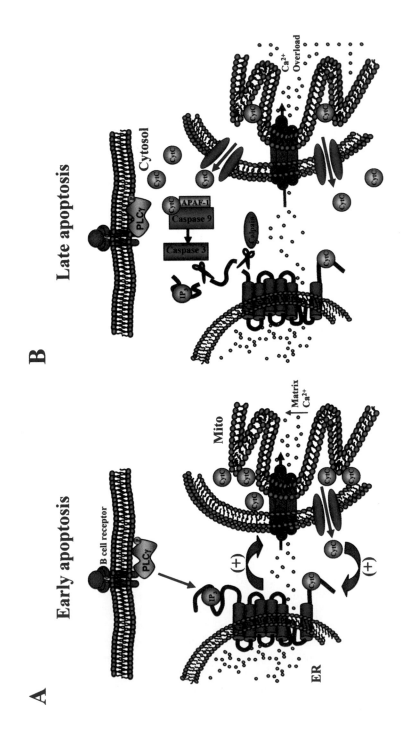

25), some of which may be expressed. One potential physiologic role for IP$_3$R fragments cleaved during apoptosis is the activation of Ca^{2+} entry through TRPC channels (Figure 3). Kiselyov et al. (146, 147) demonstrated that the N terminus of IP$_3$R can gate TRPC3 channels and activate endogenous store-operated Ca^{2+} entry. Because Ca^{2+}entry is inhibited by Ca^{2+}, an active IP$_3$R at the PM associated with Ca^{2+}entry would be predicted to be inhibitory, not stimulatory as determined experimentally. IP$_3$Rs with a mutation eliminating Ca^{2+} permeation activate TRPC3 and endogenous Ca^{2+} entry, an action lacking IP$_3$-binding mutants of IP$_3$R. These data suggest that the N terminus of the IP$_3$R can activate Ca^{2+} entry in the absence of the IP$_3$R channel domain. Calcium entry activated by IP$_3$R N-terminal peptides, generated in response to apoptotic stimuli, would further increase intracellular Ca^{2+} levels and augment apoptosis.

CYTOCHROME c Cytochrome c binds to the C terminus of IP$_3$R in yeast 2-hybrid analysis in vitro and in apoptosis-dependent coimmunoprecipitation models (148) (Figure 1). Cytochrome c blocks calcium-induced inhibition of IP$_3$R with nanomolar affinity. Very early after stimulation with apoptotic agents, cytochrome c translocates from the mitochondria to the ER. FRET analysis using YFP-tagged cytochrome c and CFP-tagged IP$_3$R reveals direct interaction of IP$_3$R with cytochrome c at time points as early as 10 min. In DT-40 cells devoid of IP$_3$R, stimulation of apoptosis increases cytochrome c in the cytosol rather than in the ER, indicating that IP$_3$R is the predominant cytochrome c binding partner within ER membranes. Protracted Ca^{2+} oscillations in cells undergoing apoptotic stimulus are inhibited by overexpression of IP$_3$R's cytochrome c binding domain. Thus, apoptotic stimuli appear to release small amounts of cytochrome c, which sensitize IP$_3$Rs to unregulated Ca^{2+} release, causing more cytochrome c release in a positive feedback loop and eventually leading to massive cell-wide cytochrome c release and cell death (Figure 4 *A,B*).

←───

Figure 4   IP$_3$ receptor functions in apoptosis. (*A*) Early after stimulation with apoptotic agents, such as anti-IgM, cytochrome c translocates from the mitochondria exclusively to the ER, where is binds to IP$_3$R. This sensitizes IP$_3$Rs to unregulated Ca^{2+} release, causing mitochondrial matrix Ca^{2+} levels to rise via uptake by the Ca^{2+} uniporter. This leads to more cytochrome c release via the permeability transition pore in a positive feedback loop. (*B*) Cell-wide CytC release is coordinated by Ca^{2+} release by IP$_3$R. This results in APAF-1/caspase 9/CytC apoptosome formation and caspase-3 activation. Caspase 3 can cleave IP$_3$R, resulting in nonfunctional channels. Because ATP production, a CytC-dependent process, is required for the completion of the apoptosis, it is conceivable that after the large release of CytC IP$_3$Rs are turned off by caspase cleavage, allowing resealing of the permeability transition pore and the resumption of mitochondrial respiration. The Ca^{2+}-activated protease calpain may also help in the degradation of the IP$_3$R because of the increased intracellular Ca^{2+} levels maintained during apoptosis.

Mitochondria actively accumulate Ca^{2+} released by IP_3R during physiologic and pathophysiologic stimulation, leading to activation of Ca^{2+}-sensitive metabolic enzymes (84, 149). Rizzuto et al. (149) demonstrated a close physical association or "privileged communication" between IP_3R and mitochondria, findings which have been confirmed by others (150–152). IP_3R-mediated Ca^{2+} release is associated with activation of mitochondrial enzymes leading to ATP production (153), which is required for apoptosis (154). When ATP is depleted, incipient apoptosis is transformed to necrosis. Apoptotic stimuli can elicit IP_3-mediated Ca^{2+} spikes to trigger mitochondrial permeability transition and cytochrome c release (150, 155). This cytochrome c could then bind IP_3R, increasing Ca^{2+} flow into the mitochondria. The permeability transition pore reseals after cytochrome c release, allowing mitochondrial metabolism to recover and to synthesize the ATP required for full induction of the apoptotic cascade (150, 155).

Plasma-Membrane Calcium Channels

Stimulation of GPCRs and RTKs in the PM and subsequent release of intracellular Ca^{2+} triggers agonist-induced Ca^{2+} entry into cells through the PM (1) (Figure 4). The molecular identity of channels mediating nonvoltage-gated Ca^{2+} entry has not been established (1, 156). The canonical transient receptor potential channels (TRPC) constitute part of agonist-induced Ca^{2+} entry (107, 157). TRPCs are homologous to the TRP channels of *Drosophila*, which are required for fly vision. Seven mammalian isoforms of these channels are expressed ubiquitously throughout the body.

TRPC3   The best-studied TRPC with respect to interaction with the IP_3R is TRPC3. Kiselyov et al. (147) observed that TRPC3 coimmunoprecipitates with IP_3R. In addition, cell-excised patches containing TRPC3 could be activated by microsomes containing IP_3R in the presence of IP_3. Kiselyov and colleagues further demonstrated that the N terminus of the IP_3R alone was capable of gating the activity of TRPC3 (146). Binding involves residues 777–797 of human TRPC3 and residues 638–1183 of rat IP_3R-3 in vitro (158). TRPC3 channel activity in response to agonists is decreased by overexpression of IP_3R fragments corresponding to the binding region of IP_3R for TRPC3 (158). This site also binds calmodulin, suggesting that IP_3R and calmodulin compete for the binding site, providing a regulatory mechanism (159). TRPCs may bind IP_3R at other sites because TRPC4 binds to the C terminus of the IP_3R (160), and TRPC2 binds the extreme N terminus of the IP_3R in yeast 2-hybrid experiments (R.L. Patterson and S.H. Snyder, unpublished data).

TRPC1   TRPC1 binds to IP_3R in vitro and in vivo through its C terminus, enhancing channel activity (159, 161). In salivary glands, overexpression of TRPC1 augments agonist-induced Ca^{2+} entry and increases salivary gland fluid secretion (161, 162). In endothelial cells stimulated by thrombin, RhoA interacts with IP$_3$R and overexpressed TRPC1 increasing endothelial permeability (163). This process results in movement of TRPC1 and IP$_3$R to the PM, increasing agonist-induced Ca^{2+} entry, similar to interactions of Homer with TRPC1 and IP$_3$R (94).

Although TRPC family members bind to varying regions of IP$_3$R in vitro and in vivo through their C terminus, functional relevance is not altogether clear. In overexpression studies, activation of TRPC can occur in the absence of IP$_3$R (165) and IP$_3$ (166). However, these studies employed overexpressed TRPC channels that often act as nonselective cation channels, unlike endogenous TRPC channels, which are Ca^{2+}-selective in response to agonist stimulation (1, 157, 167). Recent work performed by van Rossum et al. (168) in DT40 IP$_3$R triple knockout cells demonstrates how endogenous Ca^{2+} entry channels can be activated by IP$_3$R. Deletion of IP$_3$R leads to loss of Ca^{2+} entry; this can be restored with IP$_3$R mutants that do not permeate Ca^{2+}, but not with IP$_3$ binding mutants of IP$_3$R. In similar experiments by (H.L. Roderick, personal communication), transient deletion of IP$_3$R by siRNA or depletion of IP$_3$ with the IP$_3$ sponge inhibits agonist-induced Ca^{2+} entry.

In summary, there exists a physiologically relevant interaction between IP$_3$R and TRPC. Moreover, TRPCs likely constitute a large portion of agonist-induced Ca^{2+} entry within excitable and nonexcitable cells.

Heterotrimeric G-Protein β and Receptor for Activated C Kinase-1

Zeng et al. (169) reported coimmunoprecipitation of IP$_3$R with G$\beta\gamma$. In the same study, G$\beta\gamma$ activated Ca^{2+} oscillations in pancreatic acinar cells, which were not affected by the PLC inhibitor U73122 or antibody absorption of PIP$_2$. G$\beta\gamma$ appeared to gate IP$_3$R channels directly in the absence of IP$_3$ when tested on IP$_3$R expressed in *Xenopus* nuclei, as in studies with CaBP (120). IP$_3$-dependent Ca^{2+} release by VIP is abolished by 20 nM concentrations of a Gβ scavenger, which does not affect Gβ independent muscarinic cholinergic signaling in the same cells (170).

In yeast 2-hybrid studies, we showed that Gβ and the Gβ homologue receptor for activated C kinase-1 (RACK1) bind to two distinct regions of IP$_3$R (171). Both RACK1 and Gβ bind at residues 90–110 and 578–678 in the N terminus of IP$_3$R, straddling the ligand-binding domain of the channel (Figure 1). Deletion of these sites decreases the affinity of the channel for IP$_3$ by an order of magnitude and eliminates the ability of Gβ to increase the IP$_3$ affinity of the channel. Overexpression of RACK1 in PC12, A7r5, or DT40 cells augments IP$_3$-mediated calcium release through muscarinic, serotonin, and purinergic receptors but has no effect on bradykinin receptor signaling. Transient depletion of RACK1 using siRNA in PC12 cells and A7r5 cells inhibits signaling through muscarinic, serotonin, and purinergic receptors but not bradykinin receptors.

In our studies monitoring microsomal ^{45}Ca^{2+} flux, neither Gβ nor RACK1 activate IP$_3$R in the absence of IP$_3$; this differs from findings of Zeng et al. (120).

However, Gβ and RACK1 could strongly potentiate IP$_3$-mediated Ca^{2+} release. In our experiments, binding of Gβ to IP$_3$R persists after extensive purification of the channel, indicating high-affinity binding in vivo, whereas Zeng et al. (120) observed low-affinity binding of IP$_3$R to Gβ in vitro. Thus, Gβ binds to IP$_3$R in vivo, an interaction that affects IP$_3$ binding to its receptor. Whether Gβ association alone can gate IP$_3$R and other regulatory effects of Gβ on IP$_3$ binding remains to be established.

IP$_3$R-ASSOCIATED cGMP KINASE SUBSTRATE, CARBONIC ANHYDRASE-RELATED PROTEIN, AND IP$_3$R-BINDING PROTEIN RELEASED WITH IP$_3$

As previously mentioned, the large size of the IP$_3$R provides surfaces that allow for a commensurately large number of proteins to make contact with the channel. In our yeast 2-hybrid studies with the IP$_3$R, we have identified well over 300 proteins that appear to interact with the IP$_3$R (R.L. Patterson, S.H. Snyder, unpublished results); many of these proteins are uncharacterized. We review here three novel proteins, which have been determined to bind to the IP$_3$R, but whose functions are unknown or only poorly understood.

IP3R-ASSOCIATED cGMP KINASE SUBSTRATE (IRAG)   IRAG was initially discovered through phosphorylation assays in search of substrates for cyclic-GMP kinase I (cGK1) (172). cGK1 exists in a ternary complex with IP$_3$R and IRAG, and it can phosphorylate both proteins (Figure 2B). Overexpression of cGK1 and IRAG in COS-7 cells does not effect agonist-mediated Ca^{2+} release through IP$_3$R unless cGMP is added, which abolishes IP$_3$R activity. Inhibition of IP$_3$R function occurs when cGK1 phosphorylates IRAG at serine 696 (173). IRAG is widely distributed with notable enrichment in smooth muscle where cGMP levels are regulated by NO signaling. IP$_3$-dependent Ca^{2+} oscillations in hepatocytes are abolished by removal of endothelial cells from the coculture or the use of NO scavengers (174), suggesting NO-Ca^{2+} links.

CARBONIC ANHYDRASE-RELATED PROTEIN (CARP)   In multiple paradigms, CARP binds IP$_3$R at residues 1387–1647, part of the modulatory domain (Figure 1) (175). CARP is highly concentrated in cerebellar Purkinje cells. Because CARP inhibits IP$_3$ binding to IP$_3$R, it may account for the extremely low sensitivity of Purkinje cells to IP$_3$-modulated Ca^{2+} release.

IP3R-BINDING PROTEIN RELEASED WITH IP3 (IRBIT)   Using the yeast 2-hybrid method, Mikoshiba and associates (176) discovered a novel protein, IRBIT, which interacts with the ligand-binding region of IP$_3$R (Figure 1). IRBIT, a 530 amino acid protein, is widely distributed with highest expression in the brain. The IRBIT sequence, which contains homology to S-adenosylhomocysteine hydrolase, includes numerous putative phosphorylation sites. Treatment of cell extracts with alkaline phosphatases inhibits the coimmunoprecipitation of endogenous IP$_3$R with IRBIT, suggesting a role of phosphorylation in their interaction. Association of IRBIT with IP$_3$R is also inhibited by the physiological levels of IP$_3$ in vitro and in vivo.

Signal Integration by the IP$_3$ Receptor

The likelihood that IP$_3$R integrates signals by binding diverse proteins and small molecules is implicit in the very large size of the IP$_3$R modulatory domain. Regulation by the binding partners described in this review impacts an extensive range of cellular processes. Here we mention a few that illustrate key links of IP$_3$R and cellular physiology.

Crucial components of male and female reproductive physiology that depend on IP$_3$R include the acrosome reaction and oocyte fertilization. IP$_3$R localizes to the acrosome cap and provides the Ca^{2+} stores that enable sperm to fuse with the zona pellucida during fertilization (177). These stores are maintained as membrane-attached vesicles at the equatorial segment of the sperm within the cytoplasmic droplet together with the Ca^{2+}-binding protein calreticulin (178). The acrosome reaction requires Ca^{2+} entry (179), which is likely also regulated by IP$_3$R interactions with TRPC (180). The oocyte requires IP$_3$R-derived oscillatory Ca^{2+} release for fertilization, maturation (181), and the formation of cleavage furrows during the first divisions of the oocyte (182).

IP$_3$R is required for neuronal signaling, such as Ca^{2+} release in dendritic spines during long-term depression (99, 183), as well as for neuronal growth cone formation and neurite growth (184). Discrete neuronal microdomains enable bradykinin but not muscarinic activity to activate IP$_3$R, demonstrating the elegant ability of the cell to segregate receptor stimuli through the IP$_3$R signaling pathway (185).

The proliferation of protein-protein interaction research has emphasized the importance of discrete protein targeting in cellular physiology. IP$_3$R is one of the most striking exemplifications of the process, which provides ion channel activity and scaffolding ability, and, most likely, contains enzymatic activities that are, as yet, uncharacterized.

CONCLUSIONS

It is evident that the IP$_3$R binds a multitude of proteins, thereby placing the IP$_3$R at a nodal point in cellular signaling. The proliferation of literature concerning proteins that have multiple binding partners indicates many proteins require discrete targeting for their physiological function. Because a large number of proteins have already been identified to interact with the IP$_3$R, these proteins likely associate with the IP$_3$R, depending upon the microenvironment in which the IP$_3$R resides. This implies the IP$_3$R is a major provider of second messenger responses and signaling cascade liaison activity. To fully understand the intricacies of IP$_3$R signaling, the identity of the "holoenzyme," proteins constitutively bound to the IP$_3$R, needs to be considered and identified. The IP$_3$R may be among a few select proteins that impact the signaling of such a great diversity of intracellular domains. The targeting and function of the IP$_3$R is exquisitely modulated by myriad interactions with multiple proteins and small molecules.

ACKNOWLEDGMENTS

We thank Drs. R.E. Rothe, H.L. Roderick, K. Mikoshiba, and H. Stern for useful discussions. Research was supported by USPHS grants MH-18501 and DA-000266, Research Scientist Award DA-00074 (S.H.S.), National Research Service Award NH65090 (R.L.P.), and NS043850 (D.B.).

The *Annual Review of Biochemistry* is online at http://biochem.annualreviews.org

LITERATURE CITED

1. Berridge MJ, Lipp P, Bootman MD. 2000. *Nat. Rev. Mol. Cell Biol.* 1:11–21
2. Streb H, Irvine RF, Berridge MJ, Schulz I. 1983. *Nature* 306:67–69
3. Streb H, Bayerdorffer E, Haase W, Irvine RF, Schulz I. 1984. *J. Membr. Biol.* 81:241–53
4. Prentki M, Wollheim CB, Lew PD. 1984. *J. Biol. Chem.* 259:13777–82
5. Williamson JR, Cooper RH, Joseph SK, Thomas AP. 1985. *Am. J. Physiol. Cell Physiol.* 248:C203–16
6. Baukal AJ, Guillemette G, Rubin R, Spat A, Catt KJ. 1985. *Biochem. Biophys. Res. Commun.* 133:532–38
7. Worley PF, Baraban JM, Colvin JS, Snyder SH. 1987. *Nature* 325:159–61
8. Supattapone S, Worley PF, Baraban JM, Snyder SH. 1988. *J. Biol. Chem.* 263: 1530–34
9. Ross CA, Meldolesi J, Milner TA, Satoh T, Supattapone S, Snyder SH. 1989. *Nature* 339:468–70
10. Supattapone S, Danoff SK, Theibert A, Joseph SK, Steiner J, Snyder SH. 1988. *Proc. Natl. Acad. Sci. USA* 85:8747–50
11. Ferris CD, Huganir RL, Snyder SH. 1990. *Proc. Natl. Acad. Sci. USA* 87:2147–51
12. Ferris CD, Cameron AM, Huganir RL, Snyder SH. 1992. *Nature* 356:350–52
13. Furuichi T, Yoshikawa S, Miyawaki A, Wada K, Maeda N, Mikoshiba K. 1989. *Nature* 342:32–38
14. Nakade S, Maeda N, Mikoshiba K. 1991. *Biochem. J.* 277(Part 1):125–31
15. Maeda N, Kawasaki T, Nakade S, Yokota N, Taguchi T, et al. 1991. *J. Biol. Chem.* 266:1109–16
16. Nakagawa T, Okano H, Furuichi T, Aruga J, Mikoshiba K. 1991. *Proc. Natl. Acad. Sci. USA* 88:6244–48
17. Ramos-Franco J, Fill M, Mignery GA. 1998. *Biophys. J.* 75:834–39
18. Hagar RE, Burgstahler AD, Nathanson MH, Ehrlich BE. 1998. *Nature* 396: 81–84
19. Boehning D, Joseph SK. 2000. *J. Biol. Chem.* 275:21492–99
20. Mak DOD, McBride S, Foskett AJ. 2001. *J. Gen. Physiol.* 117:435–46
21. Miyakawa T, Maeda A, Yamazawa T,

Hirose K, Kurosaki T, Iino M. 1999. *EMBO J.* 18:1303–8

22. Matsumoto M, Nagata E. 1999. *J. Mol. Med.* 77:406–11

23. Ross CA, Danoff SK, Schell MJ, Snyder SH, Ullrich A. 1992. *Proc. Natl. Acad. Sci. USA* 89:4265–69

24. De Smedt H, Missiaen L, Parys JB, Bootman MD, Mertens L, et al. 1994. *J. Biol. Chem.* 269:21691–98

25. Danoff SK, Ferris CD, Donath C, Fischer GA, Munemitsu S, et al. 1991. *Proc. Natl. Acad. Sci. USA* 88:2951–55

26. Tu HP, Miyakawa T, Wang ZN, Glouchankova L, Iino M, Bezprozvanny I. 2002. *Biophys. J.* 82:1995–2004

27. Boehning D, Joseph SK, Mak DOD, Foskett JK. 2001. *Biophys. J.* 81:117–24

28. Patel S, Joseph SK, Thomas AP. 1999. *Cell Calcium* 25:247–64

29. Hetz C, Bono MR, Barros LF, Lagos R. 2002. *Proc. Natl. Acad. Sci. USA* 99:2696–701

30. Jiang QX, Thrower EC, Chester DW, Ehrlich BE, Sigworth FJ. 2002. *EMBO J.* 21:3575–81

31. Boehning D, Mak DOD, Foskett JK, Joseph SK. 2001. *J. Biol. Chem.* 276: 13509–12

32. Yoshikawa F, Iwasaki H, Michikawa T, Furuichi T, Mikoshiba K. 1999. *J, Biol. Chem.* 274:328–34

33. Yoshikawa F, Morita M, Monkawa T, Michikawa T, Furuichi T, Mikoshiba K. 1996. *J. Biol. Chem.* 271:18277–84

34. Bosanac I, Alattia JR, Mal TK, Chan J, Talarico S, et al. 2002. *Nature* 420: 696–700

35. Yoshikawa F, Iwasaki H, Michikawa T, Furuichi T, Mikoshiba K. 1999. *J. Biol. Chem.* 274:316–27

36. Boehning D, Joseph SK. 2000. *EMBO J.* 19:5450–59

37. Riley AM, Morris SA, Nerou EP, Correa V, Potter BV, Taylor CW. 2002. *J. Biol. Chem.* 277:40290–95

38. Serysheva II, Bare DJ, Ludtke SJ, Ketlun CS, Chiu W, Mignery GA. 2003. *J. Biol. Chem.* 278:21319–22

39. Hamada K, Miyata T, Mayanagi K, Hirota J, Mikoshiba K. 2002. *J. Biol. Chem.* 277:21115–18

40. da Fonseca PC, Morris SA, Nerou EP, Taylor CW, Morris EP. 2003. *Proc. Natl. Acad. Sci. USA* 100:3936–41

41. Lin P, Yao Y, Hofmeister R, Tsien RY, Farquhar MG. 1999. *J. Cell Biol.* 145: 279–89

42. Khan AA, Steiner JP, Snyder SH. 1992. *Proc. Natl. Acad. Sci. USA* 89:2849–53

43. Leite MF, Thrower EC, Echevarria W, Koulen P, Hirata K, et al. 2003. *Proc. Natl. Acad. Sci. USA* 100:2975–80

44. Echevarria W, Leite MF, Guerra MT, Zipfel WR, Nathanson MH. 2003. *Nat. Cell Biol.* 5:440–46

45. Hagar RE, Ehrlich BE. 2000. *Biophys. J.* 79:271–78

46. Khan AA, Steiner JP, Klein MG, Schneider MF, Snyder SH. 1992. *Science* 257: 815–18

47. Tanimura A, Tojyo Y, Turner RJ. 2000. *J. Biol. Chem.* 275:27488–93

48. Mayrleitner M, Schafer R, Fleischer S. 1995. *Cell Calcium* 17:141–53

49. Lupu VD, Kaznacheyeva E, Krishna UM, Falck JR, Bezprozvanny I. 1998. *J. Biol. Chem.* 273:14067–70

50. Bezprozvanny I, Watras J, Ehrlich BE. 1991. *Nature* 351:751–54

51. Miyakawa T, Mizushima A, Hirose K, Yamazawa T, Bezprozvanny I, et al. 2001. *EMBO J.* 20:1674–80

52. Finch EA, Augustine GJ. 1998. *Nature* 396:753–56

53. Worley PF, Baraban JM, Supattapone S, Wilson VS, Snyder SH. 1987. *J. Biol. Chem.* 262:12132–36

54. Kaznacheyeva E, Lupu VD, Bezprozvanny I. 1998. *J. Gen. Physiol.* 111: 847–56

55. Danoff SK, Supattapone S, Snyder SH. 1988. *Biochem. J.* 254:701–5

56. Hirota J, Michikawa T, Natsume T,

Furuichi T, Mikoshiba K. 1999. *FEBS Lett.* 456:322–26

57. Mak DOD, McBride S, Foskett JK. 1999. *J. Biol. Chem.* 274:22231–37

58. Li P, Chen SR. 2001. *J. Gen. Physiol.* 118:33–44

59. Missiaen L, Parys JB, Sienaert I, Maes K, Kunzelmann K, et al. 1998. *J. Biol. Chem.* 273:8983–86

60. Bruce JI, Shuttleworth TJ, Giovannucci DR, Yule DI. 2002. *J. Biol. Chem.* 277: 1340–48

61. Kovalovsky D, Refojo D, Liberman AC, Hochbaum D, Pereda MP, et al. 2002. *Mol. Endocrinol.* 16:1638–51

62. Venkatesh K, Siddhartha G, Joshi R, Patel S, Hasan G. 2001. *Genetics* 158: 309–18

63. Pieper AA, Brat DJ, O'Hearn E, Krug DK, Kaplin AI, et al. 2001. *Neuroscience* 102:433–44

64. Tang TS, Tu HP, Wang ZN, Bezprozvanny I. 2003. *J. Neurosci.* 23:403–15

65. Otani S, Daniel H, Takita M, Crepel F. 2002. *J. Neurosci.* 22:3434–44

66. Ferris CD, Cameron AM, Bredt DS, Huganir RL, Snyder SH. 1991. *Biochem. Biophys. Res. Commun.* 175:192–98

67. Straub SV, Giovannucci DR, Bruce JI, Yule DI. 2002. *J. Biol. Chem.* 277: 31949–56

68. DeSouza N, Reiken S, Ondrias K, Yang YM, Matkovich S, Marks AR. 2002. *J. Biol. Chem.* 277:39397–400

69. Wojcikiewicz RJH, Luo SG. 1998. *J. Biol. Chem.* 273:5670–77

70. Komalavilas P, Lincoln TM. 1994. *J. Biol. Chem.* 269:8701–7

71. Rooney TA, Joseph SK, Queen C, Thomas AP. 1996. *J. Biol. Chem.* 271: 19817–25

72. Haug LS, Jensen V, Hvalby O, Walaas SI, Ostvold AC. 1999. *J. Biol. Chem.* 274:7467–73

73. Ferris CD, Huganir RL, Bredt DS, Cameron AM, Snyder SH. 1991. *Proc. Natl. Acad. Sci. USA* 88:2232–35

74. Matter N, Ritz MF, Freyermuth S, Rogue P, Malviya AN. 1993. *J. Biol. Chem.* 268:732–36

75. Zhu DM, Fang WH, Narla RK, Uckun FM. 1999. *Clin. Cancer Res.* 5:355–60

76. Belmeguenai A, Leprince J, Tonon MC, Vaudry H, Louiset E. 2002. *Eur. J. Neurosci.* 16:1907–16

77. Zhu DM, Tekle E, Chock PB, Huang CY. 1996. *Biochemistry* 35:7214–23

78. Bagni C, Mannucci L, Dotti CG, Amaldi F. 2000. *J. Neurosci.* 20:RC76

79. He XP, Yang F, Xie ZP, Lu B. 2000. *J. Cell Biol.* 149:783–92

80. Jayaraman T, Ondrias K, Ondriasova E, Marks AR. 1996. *Science* 272:1492–94

81. Yokoyama K, Su I-h, Tezuka T, Yasuda T, Mikoshiba K, et al. 2002. *EMBO J.* 21:83–92

82. Ferris CD, Cameron AM, Bredt DS, Huganir RL, Snyder SH. 1992. *J. Biol. Chem.* 267:7036–41

83. Runnels LW, Yue L, Clapham DE. 2002. *Nat. Cell Biol.* 4:329–36

84. Rizzuto R, Brini M, Murgia M, Pozzan T. 1993. *Science* 262:744–47

85. Kaplin AI, Snyder SH, Linden DJ. 1996. *J. Neurosci.* 16:2002–11

86. Szalai G, Krishnamurthy R, Hajnoczky G. 1999. *EMBO J.* 18:6349–61

87. Joseph SK, Samanta S. 1993. *J. Biol. Chem.* 268:6477–86

88. Bourguignon LY, Jin H, Iida N, Brandt NR, Zhang SH. 1993. *J. Biol. Chem.* 268:7290–97

89. Bourguignon LY, Jin H. 1995. *J. Biol. Chem.* 270:7257–60

90. Mohler PJ, Schott JJ, Gramolini AO, Dilly KW, Guatimosim S, et al. 2003. *Nature* 421:634–39

91. Hayashi T, Su TP. 2001. *Proc. Natl. Acad. Sci. USA* 98:491–96

92. Tu JC, Xiao B, Yuan JP, Lanahan AA, Leoffert K, et al. 1998. *Neuron* 21:717–26

93. Xiao B, Tu JC, Petralia RS, Yuan JP, Doan A, et al. 1998. *Neuron* 21:707–16

94. Yuan JP, Kiselyov KI, Shin DM, Chen J,

Shcheynikov N, et al. 2003. *Cell* 114: 777–89

95. Zhang S, Mizutani A, Hisatsune C, Higo T, Bannai H, et al. 2003. *J. Biol. Chem.* 278:4048–56

96. Maximov A, Tang TS, Bezprozvanny I. 2003. *Mol. Cell. Neurosci.* 22:271–83

97. Wang K, Richards FM. 1975. *J. Biol. Chem.* 250:6622–26

98. Walensky LD, Shi ZT, Blackshaw S, DeVries AC, Demas GE, et al. 1998. *Curr. Biol.* 8:1269–72

99. Miyata M, Finch EA, Khiroug L, Hashimoto K, Hayasaka S, et al. 2000. *Neuron* 28:233–44

100. Walker DS, Ly S, Lockwood KC, Baylis HA. 2002. *Curr. Biol.* 12:951–56

101. Patterson RL, van Rossum DB, Gill DL. 1999. *Cell* 98:487–99

102. Rosado JA, Jenner S, Sage SO. 2000. *J. Biol. Chem.* 275:7527–33

103. Zhang ZM, Tang JS, Tikunova S, Johnson JD, Chen ZG, et al. 2001. *Proc. Natl. Acad. Sci. USA* 98:3168–73

104. Sugiyama T, Matsuda Y, Mikoshiba K. 2000. *FEBS Lett.* 466:29–34

105. Masiero L, Lapidos KA, Ambudkar I, Kohn EC. 1999. *J. Cell. Sci.* 112(Part 19):3205–13

106. Perrin BJ, Huttenlocher A. 2002. *Int. J. Biochem. Cell Biol.* 34:722–25

107. Patterson RL, van Rossum DB, Ford DL, Hurt KJ, Bae SS, et al. 2002. *Cell* 111: 529–41

108. Ye K, Aghdasi B, Luo HR, Moriarity JL, Wu FY, et al. 2002. *Nature* 415:541–44

109. Yamada M, Miyawaki A, Saito K, Nakajima T, Yamamoto-Hino M, et al. 1995. *Biochem. J.* 308(Part 1):83–88

110. Adkins CE, Morris SA, De Smedt H, Sienaert I, Torok K, Taylor CW. 2000. *Biochem. J.* 345(Part 2):357–63

111. Sienaert I, Nadif KN, Vanlingen S, Parys JB, Callewaert G, et al. 2002. *Biochem. J.* 365:269–77

112. Michikawa T, Hirota J, Kawano S, Hiraoka M, Yamada M, et al. 1999. *Neuron* 23:799–808

113. Missiaen L, Parys JB, Weidema AF, Sipma H, Vanlingen S, et al. 1999. *J. Biol. Chem.* 274:13748–51

114. Zhang X, Joseph SK. 2001. *Biochem. J.* 360:395–400

115. Nosyreva E, Miyakawa T, Wang Z, Glouchankova L, Mizushima A, et al. 2002. *Biochem. J.* 365:659–67

116. Vanlingen S, Sipma H, De Smet P, Callewaert G, Missiaen L, et al. 2000. *Biochem. J.* 346(Part 2):275–80

117. Bobanovic LK, Laine M, Petersen CC, Bennett DL, Berridge MJ, et al. 1999. *Biochem. J.* 340:593–99

118. Liu M, Yu B, Nakanishi O, Wieland T, Simon M. 1997. *J. Biol. Chem.* 272: 18801–7

119. Haeseleer F, Imanishi Y, Sokal I, Filipek S, Palczewski K. 2002. *Biochem. Biophys. Res. Commun.* 290:615–23

120. Zeng W, Mak DO, Li Q, Shin DM, Foskett JK, Muallem S. 2003. *Curr. Biol.* 13:872–76

121. Bootman MD, Berridge MJ, Roderick HL. 2002. *Proc. Natl. Acad. Sci. USA* 99:7320–22

122. Haynes LP, Tepikin AV, Burgoyne RD. 2004. *J. Biol. Chem.* 279:547–55

123. Yoo SH, Lewis MS. 2000. *J. Biol. Chem.* 275:30293–300

124. Yoo SH, Jeon CJ. 2000. *J. Biol. Chem.* 275:15067–73

125. Thrower EC, Park HY, So SH, Yoo SH, Ehrlich BE. 2002. *J. Biol. Chem.* 277: 15801–6

126. Petersen OH, Burdakov D, Tepikin AV. 1999. *Eur. J. Cell Biol.* 78:221–23

127. Srivastava M, Kumar P, Leighton X, Glasman M, Goping G, et al. 2002. *Ann. NY Acad. Sci.* 971:53–60

128. Brillantes AB, Ondrias K, Scott A, Kobrinsky E, Ondriasova E, et al. 1994. *Cell* 77:513–23

129. Cameron AM, Steiner JP, Roskams AJ, Ali SM, Ronnett GV, Snyder SH. 1995. *Cell* 83:463–72

130. Cameron AM, Nucifora FC Jr, Fung ET,

Livingston DJ, Aldape RA, et al. 1997. *J. Biol. Chem.* 272:27582–88

131. Cameron AM, Steiner JP, Sabatini DM, Kaplin AI, Walensky LD, Snyder SH. 1995. *Proc. Natl. Acad. Sci. USA* 92:1784–88

132. Carmody M, Mackrill JJ, Sorrentino V, O'Neill C. 2001. *FEBS Lett.* 505:97–102

133. Bultynck G, De Smet P, Rossi D, Callewaert G, Missiaen L, et al. 2001. *Biochem. J.* 354:413–22

134. Bultynck G, Rossi D, Callewaert G, Missiaen L, Sorrentino V, et al. 2001. *J. Biol. Chem.* 276:47715–24

135. Dargan SL, Lea EJ, Dawson AP. 2002. *Biochem. J.* 361:401–7

136. Khan AA, Soloski MJ, Sharp AH, Schilling G, Sabatini DM, et al. 1996. *Science* 273:503–7

137. Blackshaw S, Sawa A, Sharp AH, Ross CA, Snyder SH, Khan AA. 2000. *FASEB J.* 14:1375–79

138. Li C, Fox CJ, Master SR, Bindokas VP, Chodosh LA, Thompson CB. 2002. *Proc. Natl. Acad. Sci. USA* 99:9830–35

139. Sugawara H, Kurosaki M, Takata M, Kurosaki T. 1997. *EMBO J.* 16:3078–88

140. Hirota J, Furuichi T, Mikoshiba K. 1999. *J. Biol. Chem.* 274:34433–37

141. Diaz F, Bourguignon LY. 2000. *Cell Calcium* 27:315–28

142. Li M, Kondo T, Zhao QL, Li FJ, Tanabe K, et al. 2000. *J. Biol. Chem.* 275:39702–9

143. Haug LS, Walaas SI, Ostvold AC. 2000. *J. Neurochem.* 75:1852–61

144. Magnusson A, Haug LS, Walaas SI, Ostvold AC. 1993. *FEBS Lett.* 323:229–32

145. Wojcikiewicz RJ, Oberdorf JA. 1996. *J. Biol. Chem.* 271:16652–55

146. Kiselyov KI, Mignery GA, Zhu MX, Muallem S. 1999. *Mol. Cell* 4:423–29

147. Kiselyov KI, Xu X, Mohayeva G, Kuo T, Pessah IN, et al. 1998. *Nature* 396:478–82

148. Boehning D, Patterson RL, Sedaghat L, Glebova NO, Kurosaki T, Snyder SH. 2003. *Nature Cell Biol.* 5:1051–61

149. Rizzuto R, Pinton P, Carrington W, Fay FS, Fogarty KE, et al. 1998. *Science* 280:1763–66

150. Csordas G, Thomas AP, Hajnoczky G. 1999. *EMBO J.* 18:96–108

151. Montero M, Alonso MT, Albillos A, Cuchillo-Ibanez I, Olivares R, et al. 2002. *Biochem. J.* 365:451–59

152. Jaconi M, Bony C, Richards SM, Terzic A, Arnaudeau S, et al. 2000. *Mol. Biol. Cell* 11:1845–58

153. Poitras M, Ribeiro-Do-Valle RM, Poirier SN, Guillemette G. 1995. *Biochemistry* 34:9755–61

154. Liu X, Kim CN, Yang J, Jemmerson R, Wang X. 1996. *Cell* 86:147–57

155. Csordas G, Hajnoczky G. 2001. *Cell Calcium* 29:249–62

156. Venkatachalam K, van Rossum DB, Patterson RL, Ma HT, Gill DL. 2002. *Nat. Cell Biol.* 4:E263–72

157. Zhu X, Jiang M, Peyton M, Boulay G, Hurst R, et al. 1996. *Cell* 85:661–71

158. Boulay G, Brown DM, Qin N, Jiang M, Dietrich A, et al. 1999. *Proc. Natl. Acad. Sci. USA* 96:14955–60

159. Tang J, Lin YK, Zhang ZM, Tikunova S, Birnbaumer L, Zhu MX. 2001. *J. Biol. Chem.* 276:21303–10

160. Mery L, Magnino F, Schmidt K, Krause KH, Dufour JF. 2001. *FEBS Lett.* 487:377–83

161. Singh BB, Liu XB, Ambudkar IS. 2000. *J. Biol. Chem.* 275:36483–86

162. Singh BB, Zheng CY, Liu XB, Lockwich T, Liao D, et al. 2001. *FASEB J.* 15:1652–54

163. Mehta D, Rahman A, Malik AB. 2001. *J. Biol. Chem.* 276:22614–20

164. Deleted in proof

165. Venkatachalam K, Ma HT, Ford DL, Gill DL. 2001. *J. Biol. Chem.* 276:33980–85

166. Trebak M, Bird GSJ, McKay RRM, Birnbaumer L, Putney JW Jr. 2003. *J. Biol. Chem.* 278:16244–52

167. Wu XY, Babnigg G, Zagranichnaya T, Villereal ML. 2002. *J. Biol. Chem.* 277: 13597–608

168. van Rossum DB, Patterson RL, Kiselyov KI, Boehning D, Barrow RK, Gill DL, Snyder SH. 2004. *Proc. Natl. Acad. Sci. USA.* In press

169. Zeng WZ, Mak DOD, Li Q, Shin DM, Foskett JK, Muallem S. 2003. *Curr. Biol.* 13:872–76

170. Zeng W, Xu X, Muallem S. 1996. *J. Biol. Chem.* 271:18520–26

171. Patterson RL, van Rossum DB, Barrow RK, Snyder SH. 2004. *Proc. Natl. Acad. Sci. USA.* In press

172. Schlossmann J, Ammendola A, Ashman K, Zong XG, Huber A, et al. 2000. *Nature* 404:197–201

173. Ammendola A, Geiselhoringer A, Hofmann F, Schlossmann J. 2001. *J. Biol. Chem.* 276:24153–59

174. Patel S, Robb-Gaspers LD, Stellato KA, Shon M, Thomas AP. 1999. *Nat. Cell Biol.* 1:467–71

175. Hirota J, Ando H, Hamada K, Mikoshiba K. 2003. *Biochem. J.* 372:435–41

176. Ando H, Mizutani A, Matsu-ura T, Mikoshiba K. 2003. *J. Biol. Chem.* 278: 10602–12

177. Walensky LD, Snyder SH. 1995. *J. Cell Biol.* 130:857–69

178. Naaby-Hansen S, Wolkowicz MJ, Klotz K, Bush LA, Westbrook VA, et al. 2001. *Mol. Hum. Reprod.* 7:923–33

179. Patrat C, Serres C, Jouannet P. 2000. *Biol. Cell* 92:255–66

180. Jungnickel MK, Marrero H, Birnbaumer L, Lemos JR, Florman HM. 2001. *Nat. Cell Biol.* 3:499–502

181. Iwasaki H, Chiba K, Uchiyama T, Yoshikawa F, Suzuki F, et al. 2002. *J. Biol. Chem.* 277:2763–72

182. Mitsuyama F, Sawai T. 2001. *Int. J. Dev. Biol.* 45:861–68

183. Inoue T, Kato K, Kohda K, Mikoshiba K. 1998. *J. Neurosci.* 18:5366–73

184. Takei K, Shin RM, Inoue T, Kato K, Mikoshiba K. 1998. *Science* 282:1705–8

185. Delmas P, Wanaverbecq N, Abogadie FC, Mistry M, Brown DA. 2002. *Neuron* 34:209–20

Annu. Rev. Biochem. 2004. 73:467–89
doi: 10.1146/annurev.biochem.73.011303.074104

STRUCTURE AND FUNCTION OF TOLC:
The Bacterial Exit Duct for Proteins and Drugs

Vassilis Koronakis, Jeyanthy Eswaran, and Colin Hughes

Department of Pathology, Cambridge University, Cambridge CB2 1QP, United Kingdom; email: vk103@mole.bio.cam.ac.uk, je236@hermes.cam.ac.uk, ch@mole.bio.cam.ac.uk

Key Words   multidrug resistance, toxin secretion, efflux pumps, membrane protein, membrane transport

■ **Abstract**   The bacterial TolC protein plays a common role in the expulsion of diverse molecules, which include protein toxins and antibacterial drugs, from the cell. TolC is a trimeric 12-stranded α/β barrel, comprising an α-helical *trans*-periplasmic tunnel embedded in the outer membrane by a contiguous β-barrel channel. This structure establishes a 140 Å long single pore fundamentally different to other membrane proteins and presents an exit duct to substrates, large and small, engaged at specific inner membrane translocases. TolC is open to the outside medium but is closed at its periplasmic entrance. When TolC is recruited by a substrate-laden translocase, the entrance is opened to allow substrate passage through a contiguous machinery spanning the entire cell envelope, from the cytosol to the external environment. Transition to the transient open state is achieved by an iris-like mechanism in which entrance α-helices undergo an untwisting realignment, thought to be stabilized by interaction with periplasmic helices of the translocase. TolC family proteins are ubiquitous among gram-negative bacteria, and the conserved entrance aperture presents a possible cheomotherapeutic target in multidrug-resistant pathogens.

CONTENTS

0066-4154/04/0707-0467$14.00

INTRODUCTION

It has been known for some time that mutants of *Escherichia coli* lacking the protein TolC exhibit pleiotropic phenotypes, which include tolerance to colicins and bacteriophage (1–3) and heightened sensitivity to environmental stresses [such as detergents, bile salts, and organic solvents (4–7)]. Although tolerance is due to the exploitation of TolC as a cell surface receptor, it is now known that the other phenotypes reflect the role of TolC in the expulsion of a wide range of molecules from gram-negative bacteria. TolC family proteins are central to the export of many large proteins, which include several that aid bacterial survival in mammalian hosts (8–11), and also the efflux of a plethora of small noxious molecules that are inhibitory to the bacteria (10, 12, 13). These efflux substrates have recently been shown to include antibacterial drugs. TolC is therefore a key player in the increasing problem of multidrug resistance and is important to the survival of pathogens during infections (7, 14–19). This review will discuss the structure of TolC and its common function in protein export and multidrug efflux.

TolC-DEPENDENT PROTEIN EXPORT AND DRUG EFFLUX

Proteins destined for the cell surface or the surrounding medium of gram-negative bacteria, such as *E. coli*, must cross both the inner (cytoplasmic) and outer membranes and the intervening periplasmic space (20), which is believed to measure at least 130Å across. Most are secreted by large multiprotein assemblies that either span the periplasm or establish two-step mechanisms employing periplasmic intermediates (21–23). TolC-dependent type I protein export contrasts with these pathways because it bypasses the periplasm but requires only the outer membrane TolC, acting with an inner membrane trans-locase containing a traffic ATPase and an accessory or adaptor protein (11, 17, 24–27). Our work has focused on type I export of the 110 kDa hemolysin toxin (HlyA) (26) common among uropathogenic and enterohemorrhagic *E. coli*, but the same mechanism exports many other proteins, which include toxins, proteases, and lipases from pathogens of humans, animals, and plants (17, 28, 29) (Table 1). Genes encoding the protein substrates are usually located in operons coding for the corresponding export proteins. Export substrates are not subject to N-terminal processing by the signal peptidase; their secretion signal is typically uncleaved, about 50–60 residues long, and located at the extreme C terminus (30–32). There is little primary sequence identity among the signals, but interchangeability of type I export genes suggests that they may contain common higher order structures, such as an amphipathic α-helix (33, 34). C-terminal polypeptides are exported, and they direct export of heterologous proteins through the type I system, albeit weakly (32, 35).

TABLE 1 Substrates of TolC-dependent type I protein export

Substrates[a]	Protein export systems (inner membrane/ outer membrane)[b]	Bacterium
HlyA hemolysin	HlyBD/TolC	*Escherichia coli*
LktA leukotoxin	LktBD/—	*Pasteurella haemolytica*
AaltA leukotoxin	AaltBD/—	*Actinobacillus actinomycetemcomitas*
ApxI/II/III hemolysin	ApxBD/—	*Actinobacillus pleuropneumoniae*
CyaA adenylate cyclase	CyaBD/CyaE	*Bordetella pertussis*
PrtC protease	PrtDE/PrtF	*Erwinia chrysanthemi*
HasA metalloprotease	HasDE/HasF	*Serratia marcescens*
LipA lipase	LipBC/LipD	*Serratia marcescens*
Colicin V	CvaBA/TolC	*Escherichia coli*
Apr alkaline protease	AprDE/AprF	*Pseudomonas aeruginosa*
TliA thermostable lipase	TliDE/TliF	*Pseudomonas fluorescens*
PlyA exopolysaccharide glycanase	PlsDC/—	*Rhizobium leguminosarum*
SapA S-layer protein	SapDE/SapF	*Campylobacter fetus*
RasA S-layer protein	RsaDE/—	*Caulobacter crescentus*

[a]All substrates have a C-terminal uncleaved export signal, except the hemoprotein HasA, which has a cleaved C-terminal signal, and Colicin V, which has an N-terminal signal with similarity to the leader peptides of lantibiotics of gram-positive bacteria.

[b]No data is shown by —.

As in type I protein export, the TolC-dependent efflux of antibacterial drugs and other small inhibitory molecules involves TolC interacting with a translocase/pump of two inner membrane proteins (6, 15, 16). These also comprise a protein of the adaptor family and an energy-providing protein, which is sometimes an ATP-binding cassette (ABC) protein, but more often it is a proton antiporter of either the resistance nodulation division (RND) or major facilitator superfamily (MFS) class (10, 36, 37). Cells typically have several TolC homologues that act in a number of parallel efflux pumps, which typically have broad and sometimes overlapping substrate specificities (Table 2). For example, the *E. coli* genome encodes TolC, three homologues, and about 30 inner membrane translocases of the ABC, MFS (e.g., *E. coli* EmrAB), and RND (e.g., *E. coli* AcrAB) families (38), whereas the opportunistic pathogen *Pseudomonas aeruginosa*, in which multidrug resistance (MDR) poses a particular and growing clinical problem, has four major efflux (Mex) systems containing an RND proton antiporter and one of the three TolC homologues, OprM, OprJ, and OprN (39, 40).

TABLE 2   TolC-dependent drug efflux systems of *E. coli* and *P. aeruginosa*

Substrates[a]	Efflux systems (inner membrane/ outer membrane)
E. coli	
AC, AZ, BL, BS, CH, CM, CV, CP, DOC, EB, ER, FA, FQ, FU, NAL, NV, RF, TC, SDS, TX	AcrAB (RND)/TolC AcrEF (RND)/TolC
LC, CCCP, NAL	EmrAB (MFS)/TolC
ML	MacAB (ABC)/TolC
P. aeruginosa	
AC, AH, BL, CL, CM, CV, EB, FQ, ER, NV, SM, SDS, TL, TP	MexAB (RND)/OprM
AC, AH, CL, CM, CV, EB, ER, FQ, NV, SDS, TC, TP, TS	MexCD (RND)/OprJ
AH, CM, ER, FQ, TP, TS	MexEF (RND)/OprN
AG, ER, FQ, TC	MexXY (RND)/OprN

[a]Substrate abbreviations: AC, acriflavin; AG, aminoglycoside; AH, aromatic hydrocarbons; AZ, azithromycin; BL, β-lactams; BS, bile salts; CCCP, carbonyl cyanide m-chlorophenylhydrazone; CH, cholate; CL, cerulenin; CM, chloramphenicol; CP, ciprofloxacin; CV, crystal violet; DOC, deoxycholate; EB, ethidium bromide; ER, ethyromycin; FA, fatty acids; FQ, fluoroquinolones; FU, fusidic acid; LC, lipophilic cations; ML, macrolides; NAL, nalidixic acid; NV, novobiocin; RF, rifampicin; SDS, sodium dodecyl sulfate; SM, sulphonamides; TC, tetracycline; TL, thiolactomycin; TS, triclosan; TP, trimethoprim; TX, Triton X-100.

In addition to its role in these three-component protein export and drug efflux machineries, *E. coli* TolC is also utilized for the exit of low-molecular-weight peptides, such as the heat-stable enterotoxin, cationic antimicrobial peptides, and microcins, which are transported to the periplasm either by the housekeeping Sec system or other designated inner membrane transport systems (41–43). Little is known of how TolC is accessed by these periplasmic peptides.

THE TolC STRUCTURE: THE KEY TO A COMMON MECHANISM OF EXPORT AND EFFLUX

Biochemical studies have shown that TolC-dependent protein export requires direct contact between outer membrane TolC and substrate-laden translocases in the inner membrane (44). However, it was not evident how substrate engagement at the inner membrane could be coupled, without periplasmic intermediates, to substrate exit through what was imagined to be a simple outer membrane porin-like channel. Reconstitution of purified TolC into phospholipid bilayers

Extracellular

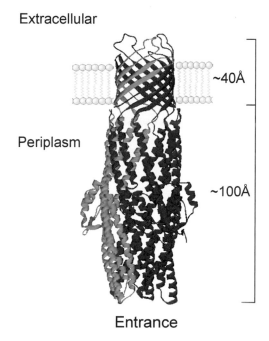

Periplasm

~40Å

~100Å

Entrance

Figure 1 The structure of TolC. Cα trace of the trimer (47). The individual protomers are colored blue, red, and green. The lipid bilayer represents the bacterial outer membrane. The molecular threefold rotation axis is aligned vertically and is assumed to be normal to the plane of the outer membrane. The outer membrane embedded β barrel is open to the extracellular medium, but the coiled coils taper to close the periplasmic entrance of the α-helical barrel.

allowed electron microscopy of two-dimensional crystals of 13Å resolution (11), establishing that the 471 amino acid TolC is trimeric and indicating that it had a novel single pore and possibly an additional domain that could contribute to a periplasmic bypass. Nevertheless, to establish how TolC could mediate export and efflux, a high-resolution structure was needed. Crystallization of TolC was a complex process. Initially, crystals of the intact membrane protein had high mosaicity, and crystallization was further complicated by the nature of TolC, as it turned out to be an atypical membrane protein with a lipid-embedded domain fused to a large extramembranous domain stable in an aqueous environment. Satisfactory crystal packing was only achieved when the flexible C-terminal 43 residues were removed, a truncation that did not attenuate TolC function in *E. coli*. Collection of multiple wavelength anomalous dispersion (MAD) data using selenomethionine (SeMet) derivatives was finally possible when heterogeneity in the oxidation state of selenium was overcome by oxidation of the SeMet residues (45). The resulting TolC crystals were loosely packed with a relatively high solvent content of 70% (46). It is noteworthy that the charged residues (in the

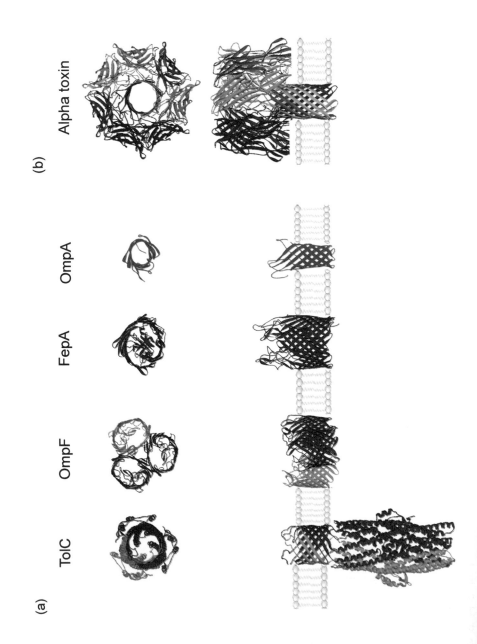

equatorial domain and extracellular loops, see below) that dictated intermolecular contact in the TolC crystal lattice are variable throughout TolC homologues, and therefore they might not be crystallized by the same protocol.

TolC: A *TRANS*-PERIPLASMIC CHANNEL TUNNEL

X-ray crystallography at 2.1Å resolution (47) revealed that TolC is fundamentally different to known outer membrane proteins. The TolC homotrimer is a tapered hollow cylinder 140Å in length; it comprises a 40Å long outer membrane β barrel (the channel domain) that anchors a contiguous 100Å long α-helical barrel projecting across the periplasmic space (the tunnel domain) (Figure 1). A third domain, with a mixed α/β structure, forms a belt around the equator of the tunnel. TolC thus provides a water-filled exit duct with a volume of 43,000Å^3. The average accessible interior diameter of the single TolC channel-tunnel pore is 19.8Å (30Å backbone to backbone) throughout the outer membrane channel and most of the tunnel. TolC is a 12-stranded barrel. Each of the three monomers contributes four antiparallel β-strands and four antiparallel α-helical strands (two continuous long helices and two pairs of shorter helices) to form the channel and the tunnel domains, respectively.

Although β barrels are typical of outer membrane proteins (48–50), the TolC channel domain is distinct because the trimer forms a single β barrel (Figure 2a, Table 3). In TolC, the protomers each contribute four β-strands to the 12-strand β barrel, an architecture distinct from other membrane proteins (51, 52), which form one barrel per monomer (Figure 2a). It is to some degree comparable to that of the *Staphylococcus aureus* alpha-toxin in which seven subunits assemble a single barrel in mammalian target membranes (53) (Figure 2b). The TolC outer membrane β barrel is constitutively open to the external medium; it lacks the inward folded loop that constricts the β barrels of channel-forming proteins like OmpA (54, 55) and does not have a plug domain, such as the one that closes the β barrels of FhuA and FepA (56, 57) (Figure 2a). The small extracellular loops of TolC are the sites of colicin and bacteriophage attachment (3). Notwithstanding these singularities of the TolC β barrel, the most distinctive feature of TolC is the 100Å-long periplasmic tunnel (Figure 1). The TolC α barrel comprises 12

←———————————————————————————————————————

Figure 2   TolC and other pore-forming proteins. Viewed from above the lipid bilayer (*upper*) and through the plane of the membrane (*lower*), (*a*) *E. coli* outer membrane proteins TolC, OmpF, FepA, and OmpA. The porin OmpF is trimeric, but each monomer forms a barrel of 16 β-strands (51). Monomeric OmpA forms the smallest known barrel of 8 β-strands (52), while the large β barrels of the iron siderophore transporters FhuA and FepA comprise 22 β-strands (56, 57). (*b*) The pore-forming *S. aureus* alpha-toxin inserted in a eukaryotic membrane. Seven subunits contribute 14 β-strands to a single barrel (53).

TABLE 3   Structural properties of TolC and other membrane proteins

Properties	*E. coli* TolC	*E. coli* OmpF	*E. coli* FepA	*E. coli* OmpA	*S. aureus* alpha-toxin
Length (Å)	140	35	70	57	100
Radius (Å)	17.5	15.5	19.9	13	8
Constriction diameter (Å)	3.9	11	n/a	n/a	14
Number of pores	1	3	1	1	1
Number of monomers	3	3	1	1	7
β-strands per monomer	4	16	22	8	2
Conductance[a] (pS)	80	840	n/a	n/a	1000

[a]Conductances are measured at 1M KCl or NaCl.

α-helices (4 from each monomer) packing in an antiparallel arrangement. The structural principles underlying how helices are constrained in the α barrel have been described (46, 47, 58). The barrel is assembled by each of the 12 helices packed laterally with two neighboring helices. This is stabilized by intermeshing of side chains, known as "knobs-into-holes" packing. Throughout the α barrel the helices follow a left-handed superhelical twist, but they are underwound in the upper (β-barrel-proximal) half compared to helices in a conventional two-stranded coiled coil (Figure 3). This enables them to lie on a cylindrical surface, possibly further facilitated by bulkier side chains that tend to partition to the outside of the barrel. Assembly of the tunnel is additionally supported by hydrogen bonds within and between the helices, and salt bridges formed at the interface of the monomers may play a role in the trimerization. (It is not known how TolC trimers are assembled in vivo nor whether this is aided by cellular proteins, such as chaperones or enzymes that degrade the peptidoglycan.) In the lower (β-barrel-distal) half of the α barrel, neighboring helices form six pairs of regular two-stranded coiled coils, and at the periplasmic end, one coil from each monomer folds inward. This constricts the periplasmic entrance to a resting closed state, with an effective diameter of ~3.9Å (47), which is consistent with the small (c.80pS) conductance of TolC in planar lipid bilayers (59, 60). The periplasmic entrance is the only constriction in the TolC pore. The structure indicated vividly the means by which TolC allows the exit of a wide range of substrates from the cell, by presenting a common exit duct for substrates engaged by inner membrane translocases.

TOLC IS CONSERVED THROUGHOUT GRAM-NEGATIVE BACTERIA

TolC homologues are seemingly ubiquitous among gram-negative bacteria, as nearly 100 family members have been identified in over 30 bacterial species (10, 61). Primary sequence similarity correlates with the function of the homologues

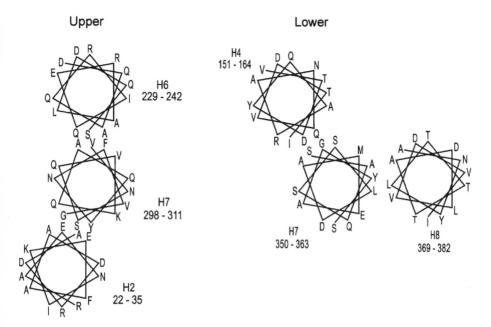

Figure 3   Helix interactions in the ToLC periplasmic α barrel. Helical wheel representations summarizing the inter-helical contacts in the upper part (*left*) and the lower part (*right*) of the α-helical barrel.

in protein export, cation efflux, or multidrug efflux (61, 62) (Figure 4). The sequences and structures of the N- and C-terminal halves of the ToLC monomer are similar to each other (47), and this internal duplication is evident throughout the family. The strongest intramolecular identity is seen in *Bordetella pertussis* CyaE (61) (Table 1), and CyaE is also nearest to the root of the tree, suggesting that it is closest to the family progenitor.

Although homologues vary in sequence length, this is due primarily to extensions at the periplasmic N and C termini (61). Significant sequence gaps or insertions occur only in the equatorial domain outside the α/β-barrel structure and in the extracellular loops. Sequences encoding the α/β barrels do not vary substantially in length, and experimental deletions or insertions in the barrel domains are poorly tolerated (65, 66). Few amino acids are conserved among the ToLC homologues, but those that are seem structurally significant. In particular, transition from the left-twisted α barrel of the periplasmic domain to the right-twisted β barrel of the outer membrane channel domain is accomodated by conserved linkers containing proline and glycine. Glycines facilitate a tight turn between the helices forming the periplasmic tunnel entrance, and small residues, such as alanine and serine, at the interface of tunnel-forming helices allow the dense packing that determines tapering and entrance closure. At the entrance aperture, an electronegative inner surface is a conserved feature, most commonly

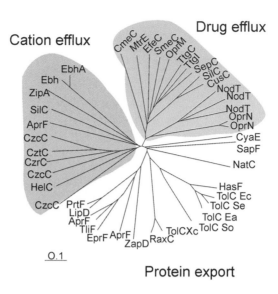

Figure 4   The TolC family. The phylogenetic tree of homologues is derived from primary amino acid sequence alignment (63) and sorted by Treeview (64). TolC homologue function, either known or strongly implicated by the coding gene context, is indicated as cation efflux, drug efflux, or protein export (homologues of unknown function are not included). Scale 0.1 indicates 0.1 nucleotide substitutions.

due to a ring of aspartic acid residues (47) (in *E. coli* TolC Asp^{371} and Asp^{374}). Conserved aromatic residues face outward to form a ring around the channel domain at the base of the β-strands, delimiting the inner edge of the lipid bilayer. This seems to be a universal feature in OM protein structures and possibly performs an anchor function. Conservation of the principle structural elements suggests that the functions of TolC are common to the homologues (61). This is compatible with genetic data indicating that at least some TolC homologues are interchangeable (67–69).

ASSEMBLY OF TolC-DEPENDENT MACHINERIES: RECRUITMENT OF TolC

In vivo cross-linking has defined the sequence of protein-protein interactions underlying export (44). The inner membrane translocase is formed constitutively, i.e., even in the absence of their respective substrates (44, 70, 71). TolC recruitment by the inner membrane translocase is clearly a central step in the mechanism, and it has been shown to occur in response to substrate engagement by the translocase (Figure 5, upper). All three export components undergo conformational changes during the substrate-induced assembly of the machinery

(44). The use of chemical uncouplers has indicated that this pretranslocation assembly of the substrate-bound export complex requires the membrane electrochemical potential but not ATP hydrolysis by the traffic ATPase (72–74). Recruitment is mediated by the adaptor protein, which is a common element in all translocases of TolC-dependent drug efflux and protein export (44, 75). Efflux adaptors are predicted to be anchored to the inner membrane by a transmembrane helix, e.g., MFS EmrA, or by an N-terminal lipid modification, e.g., RND AcrA and MexA (Figure 6) (18), and biophysical studies of AcrA have predicted an extended structure in solution, c.210Å long with an 8:1 axial ratio (76, 77). The adaptor of the *E. coli* type I hemolysin export machinery is HlyD, which comprises a large periplasmic domain (residues 81–478) connected by a single transmembrane helix to a small N-terminal cytosolic domain (residues 1–59) (Figure 6). HlyD seems to assemble to at least a trimer (44, 70), while oligomerization is also indicated by low resolution (20Å) electron microscopy of lipid-reconstituted RND efflux adaptor AcrA (77), although no structure is discernable.

The principle feature of adaptor proteins is a large periplasmic domain, which is predicted to contain long coiled coils that may form an α-helical hairpin (62) (Figure 6). During assembly, the predicted coiled-coil structures of the adaptor could contact the coiled coils of the periplasmic tunnel a barrel, possibly reaching to the equatorial domain to recruit TolC. When the protein substrate is exported, the inner and outer membrane components revert to their resting state (44). It is therefore envisaged that the adaptor has a dynamic function, which effects a transient coupling of the TolC channel tunnel to the energy-providing protein of the cognate translocase.

Notwithstanding the evidence supporting a common role for the adaptors in TolC recruitment, the 3.5Å resolution crystal structure (78) of the *E. coli* RND AcrB [which forms a constitutive inner membrane complex with AcrA (71)] has prompted speculation that this class of antiporter might be able to contact the TolC periplasmic tunnel entrance directly (Figure 5, lower). AcrB is a trimer with a periplasmic domain of length 70Å and an 80Å diameter comprising two large hydrophilic periplasmic loops from each monomer (Figure 7). The contiguous transmembrane domain is composed of 36 α-helices, 12 from each monomer, and is 50Å in length and 100Å in diameter, with an opening at the cytosolic side of the inner membrane (78). As with TolC, there is structural similarity between the N- and C-terminal halves of the 1049 residue AcrB, suggesting that evolution of this protein has also involved a gene duplication event. It is proposed that six hairpins at the top of the AcrB trimer could dock with the six α-helix turn α-helix structures at the base of TolC. Such a docking might not be stable and may be precluded in the protein export pathway and the MFS and ABC efflux machineries because their energy transducing components, traffic ATPases and antiporters, are not predicted to have substantial periplasmic domains (18, 38) (Figure 8). In the various systems, the adaptor may therefore function to stabilize RND antiporter docking, while in all cases effecting the TolC recruitment required to

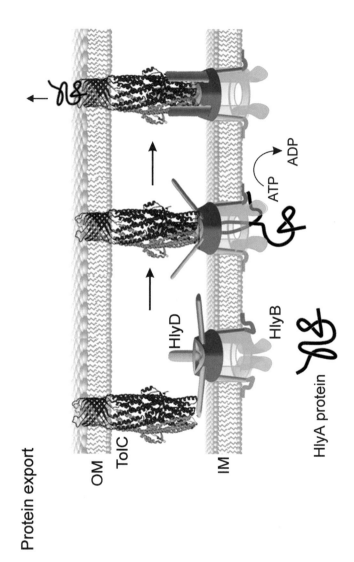

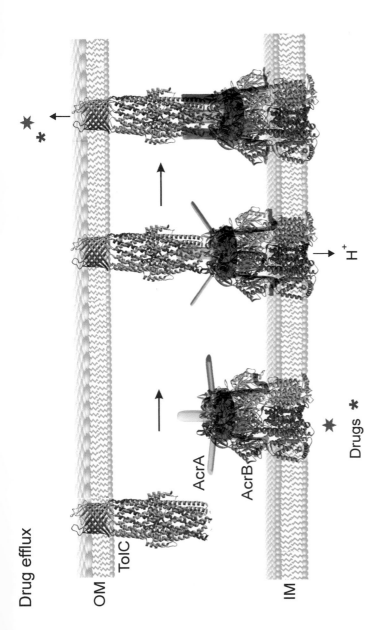

Figure 5   Assembly of TolC-dependent export and efflux machineries. A model indicating reversible interaction of outer membrane TolC or a homologue with substrate-specific inner membrane complexes (translocases) containing an adaptor protein and an energy-providing protein, either a traffic ATPase in protein export (*upper*, indicating the type I hemolysin export) or typically an antiporter in drug efflux (*lower*, indicating the RND pump). In both cases, transport occurs across the inner membrane (IM), outer membrane (OM), and the intervening periplasmic space.

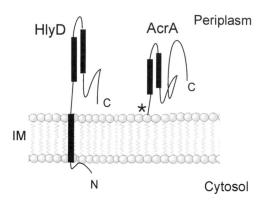

Figure 6 Representation of inner membrane (IM) adaptor proteins. A putative topological representation of the HlyD (type I protein export) and AcrA (RND drug efflux) monomers based on secondary structure predictions. The asterisk indicates lipophilic modification; black boxes are predicted α-helices, which could potentially form coiled coils.

open the periplasmic entrance. The assembled RND drug efflux machinery has not been isolated (37), but recent results show that, though the purified AcrA adaptor has micromolar affinity for the AcrB antiporter and for TolC, no binding was detectable between AcrB and TolC in vitro (V. Koronakis, T. Touzé, J. Eswaran, E. Bokma, E. Koronokis, and C. Hughes, unpublished), possibly reflecting a need to be stabilized by AcrA in vivo. It may be significant that the export adaptor HlyD has additional sequence in the periplasm (62, 75), compared to AcrA; perhaps this compensates for the lack of a periplasmic domain in its translocase partner.

A view of substrate-responsive TolC recruitment by translocases envisages transduction of the substrate-binding signal across the inner membrane from the cytosolic face of the translocase. Protein substrates interact independently with both the traffic ATPase and adaptor in vivo (44, 70), although genetic and biophysical studies suggest that the initial interaction involves the substrate export signal and the (c.250 residue) cytosolic ATPase domain of the traffic ATPase (79, 80). This is predicted to be fused to an N-terminal domain encompassing six transmembrane helices, as exemplified by the 707 residue hemolysin B (HlyB) (81, 82) (Figure 8). However, substrate interaction involves not only this initial signal recognition as the substrate is engaged (70) but also a subsequent step in which substrate binding to the adaptor triggers TolC recruitment. Removal of the small N-terminal cytosolic domain of the export adaptor HlyD abolishes protein export and substrate-adaptor interaction, and small deletions within this domain disable TolC recruitment (and therefore protein export), even though substrate is still engaged. Such mutants thus appear to be defective specifically in triggering recruitment in response to substrate engage-

AcrB

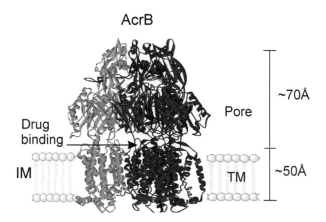

Figure 7   Structure of the inner membrane (IM) drug efflux antiporter AcrB. Ribbon depiction of the crystal structure of trimeric AcrB (78); colors indicate each protomer. AcrB comprises transmembrane (TM) and periplasmic regions, the latter encompassing pore and TolC-docking domains. The arrow indicates a region suggested to bind drug efflux substrates (86).

ment. It is assumed that this engagement signal is transduced by an intramolecular allosteric mechanism to the coiled coils of the adaptor periplasmic domain.

SUBSTRATE TRANSLOCATION THROUGH TolC-DEPENDENT MACHINERIES

Little is known about substrate translocation, although experiments with chemical uncouplers (72) suggest that it does not require the electrochemical potential but is driven solely by traffic ATPase ATP hydrolysis (the HlyB cytosolic domain has an in vitro Vmax of 1 μmolATP/min/mg and a Km of 0.2 mM ATP) (73, 74). Mutations that abolish hydrolysis of ATP bound at the cytosolic domain disable passage of protein through the TolC-dependent system, causing accumu-

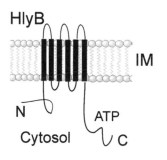

Figure 8   Representation of the inner membrane (IM) traffic ATPase. A putative topology of the HlyB monomer based on secondary structure predictions.

lation of a stalled intermediate complex containing the substrate, translocase and TolC (44). Although this evidence supports a specific role for ATP hydrolysis in substrate translocation through the assembled machinery, little is known yet of how ATP hydrolysis, substrate (un)folding, and substrate movement are coupled (82). It is possible that large protein substrates pass through the translocase in a partially unfolded state, resulting in "racheted" translocation driven by ATP hydrolysis, analogous to that suggested for mitochodrial protein import (83). This, however, is speculation, especially as no contacts have been defined between internal regions of the substrate and any of the export proteins. Although no high-resolution structure of a traffic ATPase has been solved, i.e., complete with its transmembrane domain, the structures of several ATP-binding proteins and isolated domains (84, 85) are compatible with biochemical evidence suggesting they function as homodimers (73, 84, 85). But it is not yet possible to discern the nature of interaction between the ATPase monomers or their transmembrane and cytosolic domains (84).

Drug efflux systems, such as AcrAB-TolC, expel a wide variety of small structurally unrelated compounds (13, 16) (Table 2). It seems likely that these substrates could enter the efflux channel through the opening at the cytosolic face of AcrB (78), but it has also been proposed that substrates might enter AcrB laterally from within the membrane, on both the cytosolic and periplasmic sides (78). The crystal structure of AcrB liganded with diverse substrates revealed binding at nonidentical positions in the transmembrane domain (86). Although this could partly explain how diverse drug efflux substrates are accommodated, hybrid transporter studies indicate that the antiporter periplasmic domain plays the major role in determining substrate specificity (68, 69), and the drug binding sites in the transmembrane domain may be only a snapshot of a transient point in the efflux process. Once past the TolC entrance, the substrates, even large proteins, can readily pass through what is effectively the external environment. Nevertheless, substrates have a wide variation in charge and other physicochemical properties, and it has been proposed that the strikingly electronegative surface of the TolC tunnel could affect movement of substrates out of the bacterium (47).

TolC OPERATION: TRANSITION TO THE ENTRANCE OPEN STATE

In both protein export and drug efflux, substrates are channeled through the translocase to the periplasmic entrance of TolC. It is not known how small molecules, such as drugs, trigger opening of the entrance, but as in protein export, this transition to the open state must occur because the resting state entrance aperture is too small for their passage (47) (Figure 9, left). Opening of the entrance is therefore key to the function of TolC and to the export and efflux

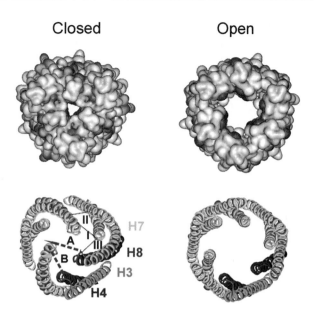

Closed   Open

Figure 9 The closed and modeled open states of TolC. Space-filled (*upper*) and ribbon (*lower*) depictions of the closed (*left*) and modeled open (*right*) states (47) of the tunnel entrance, viewed from the periplasm. The coiled coils of one protomer are colored (H3/4 and H7/8; numbering taken from Reference 47) and show the constraining intramonomer (I and II) and intermonomer (III) links. The crystal structure (*closed*) shows the coiled coils closing the end of the tunnel. The open-state model (*open*) illustrates how the channel may be opened.

machineries. The dense packing of the helices at the periplasmic entrance suggests a very stable structure, and our in vitro analyses of TolC in planar lipid bilayers show that opening can not be induced by high voltage, low pH, or even urea (60). TolC must therefore undergo a conformational change to allow passage of substrate. An allosteric mechanism has been proposed for TolC opening (47). This is based on the observation that the three inner coiled coils (comprising helices H7 and H8) differ from the outer coiled coils (H3/H4) only by small changes in superhelical twist, and it envisages that transition to the open state is achieved by the inner coil of each monomer realigning relative to the outer coil, thereby enlarging the aperture diameter (Figure 9, right). Comparison of the resting closed state of the entrance observed in the crystal structure with the modeled open state (47) identifies inter- and intramolecular bonds that constrain the three inner coils in the closed conformation (Figure 9). Links I and II connect each inner coiled coil to the outer coil of the same monomer by hydrogen bonds between Asp[153]-Tyr[362] and Gln[136]-Glu[359], respectively. Link III connects Arg[367] of each inner coiled coil to the outer coil of the adjacent monomer by a salt bridge to Asp[153] and a hydrogen bond to Thr[152]. In this model of entrance opening, these links must be disrupted for the inner coiled coils to move outward and enlarge the entrance diameter.

This model is supported by data from both in vivo and in vitro experiments. Formation of the salt bridge and hydrogen bonds was prevented by substituting critical residues, and because the periplasmic entrance is the sole constriction of the 140Å long pore (47, 60), change in the diameter of the entrance aperture was monitored as the conductance of purified TolC proteins in black lipid bilayers. Elimination of individual connections I and II caused only small changes in conductivity, whereas significant increases resulted from disruption of the R^{367}-D^{153} salt-bridge of intermonomer connection III (Figure 10). When both components of link III were disrupted simultaneously with the intramonomer link I, there was a synergistic effect, dictating a 6- to 10-fold increase over wild-type conductance (87). This would be compatible with an aperture of 16Å, which corresponds to the modeled open state. These results support a view of transition to the open state by an iris-like realignment of the entrance helices, generating an aperture large enough to allow passage of diverse substrates. Complementary in vivo evidence was obtained by introducing disulphide bonds to constrain the entrance coiled coils in the closed state (88). Type I hemolysin export from E. coli was abolished by introducing intermonomer disulphide bridges cross-linked at the narrowest point of the entrance constriction, either between Asp^{374} of adjacent monomers (link A) or between Asn^{156} and Ala^{375} to connect the inner coil of each monomer to the outer coiled coil of its adjacent monomer (link B). When the TolC entrance was locked and there was no export, the hemolysin protein substrate was still bound at the inner membrane translocase and triggered recruitment of the locked TolC (88). These results confirm that untwisting the entrance helices is essential for TolC function and show that this acts specifically to open the entrance and allow passage of substrate engaged at the inner membrane complex.

In the bacterium, transition to the TolC open state is linked to recruitment of TolC by the inner membrane translocase adaptor, although the recruitment of locked TolC shows that the opening step can be uncoupled. The target for the adaptor interaction could be the TolC entrance itself, but interaction with the periplasmic equatorial domain could also mediate opening. Its strands and helices pack against the inner set of coiled coils, and any change in this relationship, induced by interactions with the adaptor protein, could activate an allosteric transition in the coiled coils. The adaptor coiled-coil domain could repack against the untwisted coiled coils at the base of TolC, stabilizing the open state.

PUMPS AS DRUG TARGETS: IS IT POSSIBLE TO BLOCK THE TolC ENTRANCE?

Knowledge of the structure and function of the machineries will not only further the understanding of the mechanism underlying protein export and drug efflux, but it may permit rational design of potential antibacterial agents for the treatment of

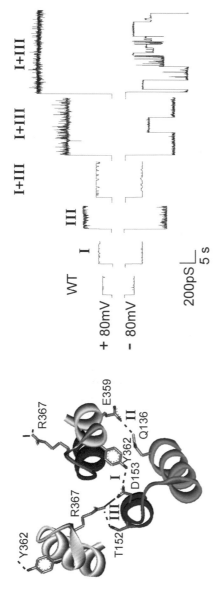

Figure 10 The entrance opening mechanism. Experimental evaluation of the "twist-to-open" mechanism for transition from the closed to open state of TolC. The four helices of one protomer are colored as in Figure 9. Experiments shown are in vitro disruption of entrance constraining links I and III using different amino acid substitutions. The corresponding changes in conductance of the purified TolC derivatives in planar lipid bilayers are compared with the TolC wildtype (87).

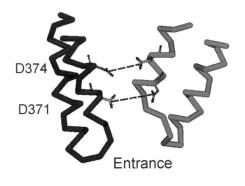

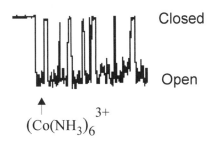

Figure 11 Blocking the electronegative entrance. The electronegative aspartate rings are formed by residues D^{371} and D^{374} of each monomer (two adjacent monomers are shown) (*top*). Conductance in lipid bilayers of wild-type TolC shows single reversible blocking events in the presence of 3 nM hexaamminecobalt trichloride (*bottom*).

multidrug-resistant bacterial infections. The importance of channel tunnels to bacterial survival, especially during infection, suggests they may present a possible target. The periplasmic entrance of TolC is the sole constriction in the exit duct (Figure 1), and the negatively charged residues at the entrance might be liganded to effect irreversible closure of the tunnel, thus reducing virulence and drug resistance. Initial investigation of this possibility shows that TolC pore function in artificial lipid bilayers is severely inhibited by divalent and trivalent cations introduced into the channel from the extracellular side (89). Trivalent cations are most potent, with hexaamminocobalt binding at nanomolar affinity. The TolC entrance constriction is lined by a ring of six aspartate residues, D^{37} and D^{374} from each of the three monomers (Figure 11) (46, 47). When either or both of the entrance aspartates are substituted by alanines, high-affinity binding is abolished, and blocking of the membrane pore is alleviated (89). This is compatible with the inhibitor binding to the entrance aspartate ring, which is also indicated by X-ray crystallography of the liganded TolC (M.H. Higgins and V. Koronakis, unpublished). These results may suggest a strategy to develop bioactive molecules, especially as the electronegative entrance is widely conserved throughout the TolC family of gram-negative bacteria.

ACKNOWLEDGMENTS

We thank all our colleagues who worked with us on this research, and the MRC for its sustained support.

The *Annual Review of Biochemistry* **is online at http://biochem.annualreviews.org**

LITERATURE CITED

1. de Zwaig N, Luria S. 1967. *J. Bacteriol.* 94:1112–23
2. Davies K, Reeves R. 1975. *J. Bacteriol.* 123:102–17
3. German J, Misra R. 2001. *J. Mol. Biol.* 11:579–85
4. Fralick A, Burns-Keliher L. 1994. *J. Bacteriol.* 176:6404–6
5. Nikaido H. 1994. *Science* 264:382–88
6. Aono R, Tsukagoshi N, Yamamoto M. 1998. *J. Bacteriol.* 180:938–44
7. Bina JE, Mekalanos JJ. 2001. *Infect. Immun.* 69:4681–85
8. Glaser P, Sakamoto H, Bellalou J, Ullmann A, Danchin A. 1988. *EMBO J.* 7:3997–1004
9. Wandersman C, Delepelaire P. 1990. *Proc. Natl. Acad. Sci. USA* 87:4776–80
10. Paulsen T, Park H, Choi S, Saier H. 1997. *FEMS Microbiol. Lett.* 156:1–8
11. Koronakis V, Li J, Koronakis E, Stauffer K. 1997. *Mol. Microbiol.* 23:617–26
12. Nikaido H. 1996. *J. Bacteriol.* 178:5853–59
13. Poole K. 2002. *Curr. Pharm. Biotechnol.* 3:77–98
14. Stone BJ, Miller VL. 1995. *Mol. Microbiol* 17:701–12
15. Fralick A. 1996. *J. Bacteriol.* 178:5803–5
16. Sulavik MC, Houseweart C, Cramer C, Jiwani N, Murgolo N, et al. 2001. *Antimicrob. Agents Chemother.* 45:1126–36
17. Delepelaire P, Wandersman C. 1990. *J. Biol. Chem.* 265:17118–25
18. Lewis K. 2000. *Curr. Biol.* 10:678–81
19. Bina JE, Mekalanos JJ. 2001. *Infect. Immun.* 69:4681–85
20. Graham L, Harris R, Villiger W, Beveridge J. 1991. *J. Bacteriol.* 173:1623–33
21. Hultgren J, Abraham S, Caparon M, Falk P, St. Geme W, Normark S. 1993. *Cell* 73:887–901
22. Russel M. 1998. *J. Mol. Biol.* 279:485–99
23. Hueck J. 1998. *Microbiol. Mol. Biol. Rev.* 62:379–433
24. Thanassi DG, Hultgren SJ. 2000. *Curr. Opin. Cell Biol.* 12:420–30
25. Buchanan SK. 2001. *Trends Biochem. Sci.* 26:3–6
26. Koronakis V, Hughes C. 1993. *Semin. Cell Biol.* 4:7–15
27. Wagner W, Vogel M, Gobel W. 1983. *J. Bacteriol.* 154:200–10
28. Scheu K, Economou A, Hong F, Ghelani S, Johnston W, Downie A. 1992. *Mol. Microbiol.* 6:231–38
29. Kawai E, Akatsuka H, Idei A, Shibatani T, Omori K. 1998. *Mol. Microbiol.* 27:941–52
30. Felmlee T, Pellett S, Lee EY, Welch RA. 1985. *J. Bacteriol.* 163:88–93
31. Stanley P, Koronakis V, Hughes C. 1991. *Mol. Microbiol.* 10:2391–403
32. Blight MA, Holland IB. 1994. *Trends Biotechnol.* 12:450–55
33. Zhang F, Yin Y, Arrowsmith CH, Ling V. 1995. *Biochemistry* 4:4193–201
34. Duong F, Lazdunski A, Murgier M. 1996. *Mol. Microbiol.* 21:459–70

35. Gentschev I, Dietrich G, Goebel W. 2002. *Trends Microbiol.* 10:39–45

36. Putman M, van Veen HW, Konings WN. 2000. *Microbiol. Mol. Biol. Rev.* 64: 672–93

37. Zgurskaya H, Nikaido H. 2000. *J. Bacteriol.* 182:4264–67

38. Saier MH, Paulsen IT, Sliwinski MK, Pao SS, Skurray RA, Nikaido H. 1998. *FASEB J.* 12:265–74

39. Nikaido H. 1998. *Curr. Opin. Microbiol.* 1:516–23

40. Hancock RE, Brinkman FS. 2002. *Annu. Rev. Microbiol.* 56:17–38

41. Foreman DT, Martinez Y, Coombs G, Torres A, Kupersztoch YM. 1995. *Mol. Microbiol.* 18:237–45

42. Delgado MA, Solbiati JO, Chiuchiolo MJ, Farias RN, Salomon RA. 1999. *J. Bacteriol.* 181:1968–70

43. Lagos R, Baeza M, Corsini G, Hetz C, Strahsburger E, et al. 2001. *Mol. Microbiol.* 42:229–43

44. Thanabalu T, Koronakis E, Hughes C, Koronakis V. 1998. *EMBO J.* 17:6487–96

45. Sharff AJ, Koronakis E, Luisi B, Koronakis V. 2000. *Acta Crystallogr. D* 56:785–88

46. Koronakis V, Andersen C, Hughes C. 2001. *Curr. Opin. Struct. Biol.* 11:403–7

47. Koronakis V, Sharff A, Koronakis E, Luisi B, Hughes C. 2000. *Nature* 405: 914–19

48. Buchanan SK. 1999. *Curr. Opin. Struct. Biol.* 9:455–61

49. Koebnik R, Locher KP, Van Gelder P. 2000. *Mol. Microbiol.* 37:239–53

50. Postle K, Vakharia H. 2000. *Nat. Struct. Biol.* 7:527–30

51. Cowan SW, Garavito RM, Jansonius JN, Jenkins JA, Karlsson R, et al. 1995. *Structure* 3:1041–50

52. Pautsch A, Schulz GE. 1998. *Nat. Struct. Biol.* 5:1013–17

53. Song LZ, Hobaugh MR, Shustak C, Cheley S, Bayley H, Gouaux JE. 1996. *Science* 274:1859–66

54. Jordy M, Andersen C, Schulein K, Ferenci T, Benz R. 1996. *J. Mol. Biol.* 259: 666–78

55. Saint N, Lou KL, Widmer C, Luckey M, Schirmer T, Rosenbusch JP. 1996. *J. Biol. Chem.* 271:20676–80

56. Locher P, Rees B, Koebnik R, Mitschler A, Moulinier L, et al. 1998. *Cell* 95: 771–78

57. Buchanan SK, Smith BS, Venkatramani L, Xia D, Esser L, et al. 1999. *Nat. Struct. Biol.* 6:56–63

58. Calladine CR, Sharff A, Luisi B. 2001. *J. Mol. Biol.* 305:603–18

59. Benz R, Maier E, Gentschev I. 1993. *Zentralbl. Bakteriol.* 278:187–96

60. Andersen C, Hughes C, Koronakis V. 2002. *J. Membr. Biol.* 185:83–92

61. Andersen C, Hughes C, Koronakis V. 2000. *EMBO Rep.* 1:313–18

62. Johnson JM, Church GM. 1999. *J. Mol. Biol.* 287:695–715

63. Corpet F. 1988. *Nucleic Acids Res.* 16: 10881–90

64. Page RDM. 1996. *Comput. Appl. Biosci.* 12:357–58

65. Li XZ, Poole K. 2001. *J. Bacteriol.* 183: 12–27

66. Wong KK, Brinkman FS, Benz RS, Hancock RE. 2001. *J. Bacteriol.* 183:367–74

67. Letoffe S, Ghigo JM, Wandersman C. 1993. *J. Bacteriol.* 175:7321–28

68. Elkins A, Nikaido H. 2002. *J. Bacteriol.* 184:6490–98

69. Tikhonova EB, Wang QJ, Zgurskaya HI. 2002. *J. Bacteriol.* 184:6499–507

70. Balakrishnan L, Hughes C, Koronakis V. 2001. *J. Mol. Biol.* 313:501–10

71. Zgurskaya H, Nikaido H. 2000. *J. Bacteriol.* 182:4264–67

72. Koronakis V, Hughes C, Koronakis E. 1991. *EMBO J.* 10:3263–72

73. Koronakis V, Hughes C, Koronakis E. 1993. *Mol. Microbiol.* 8:1163–75

74. Koronakis E, Hughes C, Milisav I, Koronakis V. 1995. *Mol. Microbiol.* 16: 87–96

75. Dinh T, Paulsen T, Saier H. 1994. *J. Bacteriol.* 176:3825–31
76. Zgurskaya HI, Nikaido H. 1999. *J. Mol. Biol.* 285:409–20
77. Avila-Sakar J, Misaghi S, Wilson-Kubalek M, Downing H, Zgurskaya H, et al. 2001. *J. Struct. Biol.* 136:81–88
78. Murakami S, Nakashima R, Yamashita E, Yamaguchi A. 2002. *Nature* 419:87–93
79. Sheps JA, Cheung I, Ling V. 1995. *J. Biol. Chem.* 270:14829–34
80. Benabdelhak H, Kiontke S, Horn C, Ernst R, Blight MA, et al. 2003. *J. Mol. Biol.* 327:1169–79
81. Gentschev I, Goebel W. 1992. *Mol. Gen. Genet.* 232:40–48
82. Holland IB, Blight MA. 1999. *J. Mol. Biol.* 293:381–99
83. Schatz G, Dobberstein B. 1996. *Science* 271:1519–26
84. Kerr D. 2002. *Biochim. Biophys. Acta* 1561: 47–64
85. Schmitt L, Benabdelhak H, Blight MA, Holland IB, Stubbs MT. 2003. *J. Mol. Biol.* 330:333–42
86. Yu EW, McDermott G, Zgurskaya H, Nikaido H, Koshland DE. 2003. *Science* 300:976–80
87. Andersen C, Koronakis E, Bokma E, Eswaran J, Humphreys D, et al. 2002. *Proc. Natl. Acad. Sci. USA* 99:11103–8
88. Eswaran J, Hughes C, Koronakis V. 2003. *J. Mol. Biol.* 327:309–15
89. Andersen C, Koronakis E, Hughes C, Koronakis V. 2002. *Mol. Microbiol.* 44: 1131–39

Annu. Rev. Biochem. 2004. 73:491–537
doi: 10.1146/annurev.biochem.73.011303.074043
First published online as a Review in Advance on March 18, 2004

ROLE OF GLYCOSYLATION IN DEVELOPMENT

Robert S. Haltiwanger[1] and John B. Lowe[2]

[1]Department of Biochemistry and Cell Biology, Institute for Cell and Developmental Biology, State University of New York, Stony Brook, New York 11794-5215; email: robert.haltiwanger@stonybrook.edu

[2]Department of Pathology, Life Sciences Institute, and Howard Hughes Medical Institute, University of Michigan, Ann Arbor, Michigan 48109-2216; email: johnlowe@umich.edu

Key Words   N-glycans, O-glycans, O-fucose, O-mannose, glycosaminoglycans, glycosphingolipids

■ **Abstract**   Researchers have long predicted that complex carbohydrates on cell surfaces would play important roles in developmental processes because of the observation that specific carbohydrate structures appear in specific spatial and temporal patterns throughout development. The astounding number and complexity of carbohydrate structures on cell surfaces added support to the concept that glycoconjugates would function in cellular communication during development. Although the structural complexity inherent in glycoconjugates has slowed advances in our understanding of their functions, the complete sequencing of the genomes of organisms classically used in developmental studies (e.g., mice, *Drosophila melanogaster*, and *Caenorhabditis elegans*) has led to demonstration of essential functions for a number of glycoconjugates in developmental processes. Here we present a review of recent studies analyzing function of a variety of glycoconjugates (O-fucose, O-mannose, N-glycans, mucin-type O-glycans, proteoglycans, glycosphingolipids), focusing on lessons learned from human disease and genetic studies in mice, *D. melanogaster*, and *C. elegans*.

CONTENTS

INTRODUCTION

Polymeric glycans cover the surface of virtually all cells that comprise multi-cellular animals with the ability to differentiate and form tissues. At the ultrastructural level, these glycans contribute to the supramembrane surface layer known as the glycocalyx, an osmophilic layer disclosed by staining with ruthenium red (a dye that binds to sialic acid and other acidic molecules) (1). As the most prominent surface by which the cell faces its neighbors and molecules in the extracellular milieu, the components of the glycocalyx are optimally positioned to help the cell communicate with its environment. The amorphous appearance of this layer in the electron microscope belies the extraordinary degree of carefully orchestrated molecular complexity inherent in the glycan components of the glycocalyx. In principle, the combinatorial possibilities for molecular diversity inherent in glycan structure can exceed peptide-based structural diversity by orders of magnitude [reviewed in (2)]. The remarkably elaborate and precise synthetic processes that regulate such diversity parallel this astonishing extent of glycan-based molecular diversity [reviewed in (3)].

Early hints of glycan-based biomolecular complexity are evident from studies in which plant lectins were applied to red blood cells (4). The sugar binding specificities of these proteins, together with biochemical approaches, helped to define the monosaccharide-dependent variation among blood group types [reviewed in (5)]. This work certainly facilitated the disclosure that the glycan-based ABO blood group antigens are expressed in a dynamic and lineage-specific regulated manner during human embryogenesis [see (6–8) for examples]. Lectin-based immunohistochemistry expanded these observations in humans, rodents, and other mammals to disclose that the cells in these organisms elaborate a wide array of glycan structures and control the amount and structure of such glycans with temporal and lineage specificity during development and differentiation [reviewed in (9)]. This carefully orchestrated modulation of glycan diversity during development implies corresponding and important functions for these molecules during this process. In the mouse, such implications were reinforced by early efforts to use monoclonal antibodies to identify developmentally important cell surface molecules whose expression was restricted to specific stages early in embryogenesis. This work generated a series of monoclonal antibodies that did indeed recognize stage specific embryonic antigens (SSEAs). SSEA-1 was the first example (10); reviewed in (11). Subsequent analyses disclosed that the anti-SSEA-1 antibody and many other such monoclonal antibodies bind to glycan epitopes whose expression is stage- and/or lineage-specific [SSEA-1 corresponds to the Lewis x/CD15 epitope (12, 13); SSEA-3 and SSEA-4 bind to globoseries gangliosides (14, 15)].

Descriptive studies in lower metazoans have also indicated that glycan structural diversity is characteristic of developmental events in these animals. In the sea urchin (*Arbacia punctulata*), for example, the observation that cell surface glycoprotein determinants are modulated during gastrulation (16) helped to prompt an examination of requirements for glycosylation during sea urchin embryogenesis (17, 18). Using chemical inhibitors of N-glycosylation, these studies demonstrated that N-glycosylation is apparently dispensable for developmental processes prior to major morphogenic events, but it is absolutely required for development from mesenchyme blastula to the gastrula stage and for postgastrulation development (17, 18). These studies presage recent gene deletion studies in the mouse, discussed below, which disclose a requirement for N-glycosylation in the early postimplantation stages of development.

Descriptive studies in *Drosophila melanogaster* have also implied that glycans may modulate developmental processes. These include lectin-based surveys of variation in glycosylation as a function of organ, cell type, and developmental stage in *D. melanogaster* (19, 20). Inhibition of N-glycosylation in the fly disrupts rhodopsin function (21, 22), suggesting that glycosylation might also contribute to the functions of proteins with important roles in development, though glycosylation inhibition experiments akin to those done in the sea urchin have not been performed. Similar lectin-based surveys in *Caenorhabditis elegans* have not been published, though developmental defects observed in some mutant worms with altered plant lectin binding phenotypes implied roles for glycans in cell migration and morphogenesis (23).

Implicit in many of these studies is the possibility that glycan counterreceptors, corresponding to endogenous lectins, function as partners in decoding the information inherent in developmentally regulated glycan structures. Indeed, this possibility is consistent with the multiplicity of endogenous lectins in mammals (24, 25), as well as in *D. melanogaster* and *C. elegans* (26, 27). Nonetheless, the mere existence of diverse families of endogenous metazoan lectins and the descriptive observations corresponding to glycan structure modulation during development have not afforded solid evidence of function in development. As this review summarizes, however, such evidence is now forthcoming from purposeful and precise perturbation of glycan structure in development. Observations derived from induced mutation of loci that control glycosylation in the mouse, RNAi inhibition of glycosylation in lower metazoans and correlations between developmental phenotypes and genetic deficiencies in glycan expression/structure in humans, mice, *D. melanogaster*, and *C. elegans* now confirm that glycans contribute essentially to development and differentiation events required for the assembly of a complex multicellular body plan.

ROLE OF O-FUCOSE IN DEVELOPMENT

O-fucose modifications were first described as amino acid O-fucosides isolated from human urine nearly 30 years ago (28). The first protein identified bearing an O-fucose modification was urinary-type tissue plasminogen activator (29). Soon afterwards, several serum glycoproteins involved in either blood clot formation or dissolution were demonstrated to be modified; these include factors VII, IX, and XII, and tissue-type plasminogen activator (30). Comparison of the sites of glycosylation on these proteins led to the realization that O-fucose modification exists on serine or threonine residues within an epidermal growth factor-like (EGF) repeat. EGF repeats are small protein domains of ~40 amino acids in length found in numerous cell-surface and secreted proteins (31). They are defined by the presence of six conserved cysteines that form three conserved disulfide bonds. The O-fucose modification site lies adjacent to the third conserved cysteine at the putative consensus sequence C^2-X_{4-5}-(S/T)-C^3 (32, 33). Database searches using this sequence identify dozens of proteins predicted to bear O-fucose (34), several of which have been confirmed, including the Notch family of receptors (35, 36), Notch ligands (Delta, Serrate/Jagged) (33), and Cripto (37, 38). Recently O-fucose modifications were described in a different protein motif, thrombospondin type 1 repeat (TSR), although little is know about the function of O-fucose modifications in this context (39, 40).

O-fucose exists as a monosaccharide that can be elongated to a tetrasaccharide on certain EGF repeats (Figure 1) (32, 41). O-fucose is transferred to EGF repeats from GDP-fucose by protein O-fucosyltransferase I (O-FucT-1, Figure 1). This enzyme was purified and cloned (gene designator *Pofut1*) from Chinese hamster ovary (CHO) cells and subsequently shown to be widely expressed in all metazoans examined to date (42–44). Elongation of O-fucose to the tetrasaccharide is initiated by the action of glycosyltransferases from the Fringe family. Fringe was originally identified in *D. melanogaster* (45), and subsequently three mammalian homologues were isolated: Lunatic fringe, Manic fringe, and Radical fringe (46, 47). Fringe homologues have not been identified in *C. elegans*. The Fringes are O-fucose-specific β1,3 N-acetylglucosaminyltransferases, which transfer GlcNAc from UDP-GlcNAc to O-fucose on specific EGF repeats (Figure

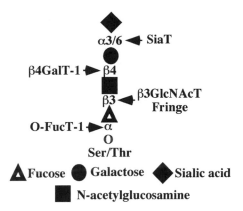

Figure 1 Structures of O-fucose saccharides found on EGF repeats. O-fucose modifications can exist as simple monosaccharides, the tetrasaccharide shown, or intermediates. The symbols used to represent each monosaccharide are identified at the bottom. The anomeric nature of each glycosidic linkage is identified by an α or β symbol, which is adjacent to a number that identifies the carbon atom of the acceptor monosaccharide involved in the glycosidic linkage. The enzymes responsible for formation of glycosidic linkages are indicated with arrows. The abbreviations are SiaT, sialyltransferase; β4GalT-1, β1,4 galactosyltransferase 1; and β3GlcNAcT, β1,3 N-acetylglucosaminyltransferase.

1) (36, 48). *D. melanogaster* Fringe, mouse Lunatic fringe, and mouse Manic fringe have all been demonstrated to catalyze this reaction in vitro. Each Fringe also appears to have somewhat different specificity; this suggests that each will modify an overlapping but slightly different subset of O-fucose bearing EGF repeats (32, 41). Following addition of β1,3-GlcNAc by Fringe, the mature tetrasaccharide is completed by the sequential action of β4GalT-1 and either an α-2,3 or α-2,6 sialyltransferase (Figure 1) (35, 49). Maturation of O-fucose beyond the disaccharide GlcNAc-β1,3-Fucose has not been confirmed to occur on EGF repeats in *D. melanogaster*.

Lessons in O-Fucose Function From Genetic Manipulation of *D. melanogaster*

Because O-FucT-1 appears to be the only enzyme in either *D. melanogaster* or mammals capable of O-fucosylating EGF repeats, studies on elimination of O-FucT-1 have been very revealing regarding the function of O-fucose modifications (44, 50, 51). The role of O-fucose modifications on EGF repeats in *D. melanogaster* has been examined both by reduction of O-FucT-1 levels using RNAi (51) and more recently by isolating a mutant in O-FucT-1 termed *neurotic* (52). In both cases, strong phenotypes resembling those of *Notch* mutants were seen. *Notch* is a genetic locus identified in *D. melanogaster* nearly 100 years ago with an X-linked lethal phenotype. The females have a small notch in their wings, from which the mutant gets its name. The gene encoded by the Notch locus was identified in 1985 and shown to be a large ($\sim$300 kDa) cell surface receptor (53). A large portion of the extracellular domain of Notch is composed of 36 tandem EGF repeats, many of which are modified with O-fucose (Figure 2A) (32, 36, 48). Homologues have been identified in all metazoans, and the Notch signaling pathway is now recognized as a key player in numerous stages of development [for a review, see (54)]. Defects in Notch signaling result in a number of human diseases, including T-cell leukemia, other forms of cancer, and several developmental disorders (e.g., CADASIL, Alagille Syndrome, and spondylocostal dystoses) (55–58). Recent studies have also shown a link between defects in Notch signaling and the pathogenesis of multiple sclerosis (59). Notch activation is described in Figure 2B [see (60) for a review].

The fact that either reduction or elimination of O-FucT-1 causes Notch-like phenotypes strongly suggests that O-fucose modifications are essential for this pathway. Interestingly, both Notch ligands in *D. melanogaster* (Delta and Serrate) are also modified with O-fucose (33). Nonetheless, genetic studies in *D. melanogaster* wing suggest that O-FucT-1 acts cell autonomously with respect to Notch, not the ligands, indicating that O-fucose must be added to Notch for proper function (51, 52). Although many of the O-fucose modification sites on Notch are evolutionarily conserved, functions at individual sites are only just beginning to be unraveled. For instance, recent work suggests that the O-fucose site at EGF repeat 12, a portion of the ligand binding site (Figure 2A), plays an important role in regulating Notch function by Fringe (see below) (60a).

Not only is O-fucose essential for Notch function, but elongation of O-fucose on Notch by Fringe serves as a means to modulate Notch activation (Figure 2B). The *Fringe* gene was originally identified in a *D. melanogaster* mutant screen for genes involved in formation of the dorsal-ventral boundary during wing development (45). *Fringe* mutants show similar defects in boundary formation in several other tissues including eye, leg, and ovary [for reviews, see (61–63)]. Subsequent work demonstrated that Fringe acts through modulation of Notch activation, stimulating activation from Delta but inhibiting activation from Serrate (64, 65). Although Notch ligands are also modified by Fringe (33), Fringe acts cell autonomously with respect to Notch (like O-FucT-1), suggesting it affects Notch directly (64, 65). The demonstration that Fringe catalyzes the transfer of GlcNAc to O-fucose residues on Notch strongly suggested that Fringe alters Notch function by altering glycan structures on Notch (36, 48). Mutation of residues predicted to play a role in catalysis abrogated the ability of Fringe to modulate Notch activity, indicating that the glycosyltransferase activity is essential for the biological function of Fringe (36, 48, 66). The subsequent demonstration that *Fringe connection* encodes a transporter for UDP-GlcNAc (and other nucleotide sugars) into the Golgi and that mutants in *Fringe connection* show Fringe-like phenotypes added further weight to this model (67, 68). Thus, it appears that Fringe modulates Notch activity by altering the structure of

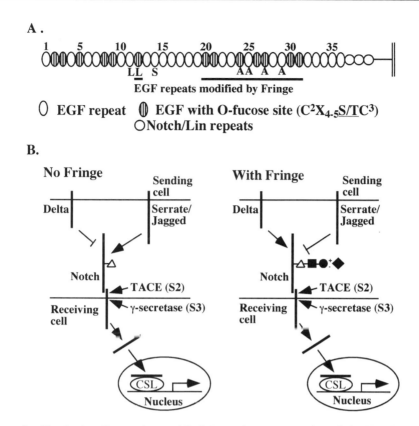

Figure 2   Notch signaling pathway. (*A*) Schematic representation of the Notch extracellular domain. Evolutionarily conserved O-fucose sites were identified by comparing *D. melanogaster* Notch, mouse Notch1 and -2, and human Notch1 and -2. EGF repeats containing putative ligand binding sites (L), the *split* mutation (S), and *Abruptex* mutations (A) are indicated. EGF repeats that are modified by Fringe (Lunatic or Manic) are indicated by a line (based on analysis of mouse Notch1) (32). Adapted from (32). (*B*) Schematic representation of Notch activation in the absence or presence of Fringe. Ligands (Delta and Serrate/Jagged) expressed on the sending cells bind to Notch. Ligand binding is believed to induce a conformational change in the Notch extracellular domain, allowing cleavage by TNFα-converting enzyme (TACE), followed immediately by cleavage by γ-secretase. This releases the cytoplasmic domain from the membrane, allowing it to move to the nucleus where it binds to members of the CSL family of transcriptional regulators and activates transcription. For a detailed review on Notch activation, see (60). Activation in cells with or without Fringe is shown. With Fringe, O-fucose saccharides can be elongated to a tetrasaccharide, as shown in Figure 1. The two forms of Notch (with monosaccharide or elongated forms of O-fucose) respond differently to activation by the ligands as indicated. Inhibition of Jagged1-mediated Notch1 activation by Fringe (Lunatic or Manic) does not require the addition of the sialic acid, as indicated by +/- (49). Elongation of O-fucose beyond GlcNAc has not yet been demonstrated in *D. melanogaster*. Adapted from (36).

O-fucose glycans on Notch. In contrast to the requirement of O-FucT-1 for Notch activity in essentially all contexts (50–52), Fringe only affects a subset of Notch functions, such as those in wing, eye, and leg development (62); this is consistent with the milder phenotypes of Fringe mutants compared to O-FucT-1 mutants.

Lessons in O-Fucose Function From Genetic Manipulation of Mice

O-fucosylation appears to play a similar role in mice, although the system is more complicated. Four mammalian Notch homologues exist (Notch1–4), five mammalian Notch-ligand homologues (Delta homologues Delta-like 1, 3, and 4 and Serrate homologues Jagged1 and -2), and three mammalian Fringe homologues (Lunatic fringe, Manic fringe, and Radical fringe) (54, 60, 69). In contrast, as in *D. melanogaster*, mice produce only a single O-fucosyltransferase for EGF repeats: O-FucT-1 (44). Mouse embryos lacking O-FucT-1 are embryonic lethal; they die at day 9.5 with severe defects in somitogenesis, cardiogenesis, vasculogenesis, and neurogenesis, similar to *Notch* mutants (50). Interestingly, detailed analysis of the *Pofut1$^{-/-}$* embryos shows the phenotype is slightly more severe than any individual *Notch* mutant, suggesting O-fucosylation is essential for all Notch receptors. The phenotype is most similar to mice lacking other core components of the Notch pathway, such as the presenilins (believed to encode a portion of the γ-secretase complex) or CSL proteins. Thus, as in *D. melanogaster*, O-fucose modification appears to be essential for proper Notch function.

Similarly, *Fringe* mutants in mice show much less severe phenotypes than O-FucT-1 knockouts. Targeted mutation of *Lunatic fringe* in mice results in a perinatal lethality with severe defects in somitogenesis (70, 71). Subsequent work has shown that Lunatic fringe is a central player in the "segmentation clock," the Notch-dependent process of budding somites from presomitic mesoderm during development [for a review, see (72)]. Mutation of *Radical fringe* in mice shows no obvious phenotype, and mice lacking both *Radical fringe* and *Lunatic fringe* show phenotypes similar to *Lunatic fringe* alone (73, 74). No report of a *Manic fringe* mutation yet exists. It is not clear whether the phenotype of the *Radical fringe* mutant is too subtle to detect or if it possibly shares some redundant function with *Manic fringe*.

Complexity in the control of development by O-fucosylation in mice is also illustrated by studies in which the synthesis of GDP-fucose has been rendered conditional (75). In mammals, the de novo and primary synthetic pathway for GDP-fucose derives from a two-step enzymatic transformation of GDP-mannose [reviewed in (76)]. The two enzymes, GDP-mannose 4,6-dehydratase (GMD) and GDP 4-keto 6-deoxymannose epimerase-reductase (also known as FX) (77) localize to the cytosol. Their concerted activities normally account for greater than 90% of the de novo formation of GDP-fucose in mammalian cells (78). A salvage pathway for GDP-fucose also exists that can utilize fucose directly. FX (+/-) mice are indistinguishable from wild-type mice. However, the frequency of homozygous FX null progeny derived from intercrosses of such heterozygote can be rare, or essentially Mendelian, depending on the genetic background of the parents (79). On genetic backgrounds where such progeny are rare, FX(-/-) embryos perish throughout the time between postimplantation period and weaning beginning via mechanisms that remain to be determined. FX(-/-) mice are smaller than wild-type mice at birth but are otherwise indistinguishable from wild-type mice when provided with dietary fucose. However, when these mice are deprived of fucose, their cells and tissues are virtually devoid of fucosylation. In this circumstance, they exhibit a number of rather severe phenotypes, which include loss of selectin-dependent leukocyte trafficking that is accounted for by deficiency of the α1-3fucosylation events required for selectin ligand activity (80). FX null mice also exhibit aberrant myelopoiesis, infertility, and inflammatory bowel disease (75). It seems likely that some of these latter phenotypes will be accounted for by faulty Notch-dependent signal transduction events and/or defective Cripto-initiated, Nodal-dependent signaling, given the evidence that O-fucosylation is required for these processes (discussed below).

Because O-FucT-1 uses GDP-fucose as its substrate, the observation that FX(-/-) embryos can survive to term in the absence of the de novo GDP-fucose synthetic pathway is apparently at odds with the embryonic lethal phenotype observed in the *Pofut1*[-/-] embryos. However, there is evidence that at least some fucosylation is maintained during gestation in FX(-/-) embryos (75). This observation suggests the possibility that the GDP-fucose salvage pathway in FX(-/-) embryos is nourished by maternal and/or fraternal glycans or free fucose in utero and is sufficiently active to support O-fucosylation at a level sufficient to sustain viability. This concept is consistent with recent studies that indicate under conditions of limiting GDP-fucose availability Notch O-fucosylation is preserved in the face of apparent loss of terminal fucosylation detected with fucose-selective lectins (81).

In Vitro Studies on the Function of O-Fucose Modifications

Although the genetic studies in *D. melanogaster* and mice suggest that the major target of O-fucosylation in early development is Notch, the fact that many other proteins also bear the modification suggests it may function in other contexts as well. For instance, removal of O-fucose [either chemically or by synthesizing the EGF repeat from urinary-type plasminogen activator (uPA) in bacteria where O-fucosylation does not occur] abolishes its ability to activate the uPA receptor, suggesting that O-fucose plays a role in modulating activation of the receptor (82). In addition, mutation of an O-fucose modification site on Cripto, a member of the EGF-Cripto-FRL-Cryptic (EGF-CFC) family of proteins (which include mammalian Cripto and Cryptic, zebrafish one-eyed pinhead, and *Xenopus* FRL-1), inhibits Nodal-dependent signal transduction (37, 38). Nodal is a member of the TGFβ superfamily and plays a key role in left-right axis formation and mesoderm induction during vertebrate development (83, 84). EGF-CFC proteins are required cofactors for Nodal signaling, and mutations of Cripto in mice show

Nodal-like phenotypes (85). All of the EGF-CFC family members have a predicted O-fucose modification site within their EGF repeat, and elimination of the site in Cripto by mutagenesis prevents O-fucosylation and results in a protein unable to support Nodal-dependent activation in various cell-based assays (37, 38). Although these results are intriguing, the fact that a Cripto/Nodal-like phenotype was not observed in *Pofut1⁻/⁻* mice was surprising because the Cripto/Nodal defects should occur earlier in embryogenesis than the Notch defects (50). Thus, the O-fucose modification on Cripto may only support Nodal signaling in specific contexts. Alternatively, the mutations introduced into Cripto for the in vitro studies may have caused the protein to misfold, abrogating its function.

A number of in vitro studies on the role of O-fucose on Notch have been reported since the demonstration that Fringe can modulate Notch function by altering O-fucose structures. Genetic studies in *D. melanogaster* indicate that both O-FucT-1 and Fringe function prior to the proteolytic cleavages, presumably altering the interaction between Notch and its ligands (51, 64). Accumulating evidence suggests that changes in O-fucose structure do indeed affect these interactions. The most compelling data comes from recent studies utilizing cell-based binding assays taking advantage of the relative simplicity of the *D. melanogaster* system (48, 52, 86). Reduction of O-FucT-1 levels using RNAi in Notch-producing *D. melanogaster* S2 cells severely compromises the ability of either Delta or Serrate to interact with Notch, indicating O-fucose modifications on Notch are essential for binding (52, 86). Expression of Fringe in the Notch-producing cells causes a decrease in Serrate binding and an increase in Delta binding to Notch, consistent with the effects of Fringe on Notch activation in vivo (48, 52, 86). Interestingly, overexpression of O-FucT-1 has opposite effects to Fringe, decreasing Delta binding and increasing Serrate binding, raising the possibility that Notch activity could be controlled not only by Fringe but also by modulating levels of O-FucT-1 (51, 86). Similar cell-based binding studies have been performed utilizing mammalian components, but the results have not been as clear. In one study, expression of Lunatic fringe enhanced the binding of soluble forms of Delta1 and Jagged1 to Notch1 expressed in HEK293 cells (87), though in another study expression of either Lunatic fringe or Manic fringe inhibited binding of soluble Jagged1 to Notch2 expressed in CHO cells (41). These studies have confirmed that expression of enzymes predicted to alter O-fucose structures on Notch results in changes in ligand binding in mammalian systems, but obviously, further work needs to be done in the mammalian system to better understand the complexities of multiple receptors, ligands, and Fringes.

Although these studies suggest that changes in O-fucose modifications alter Notch-ligand binding interactions, the molecular mechanism for this change is not yet clear. Several possible models exist; these include a direct affect of glycans on binding, a conformational change in the Notch extracellular domain caused by glycans, or a change in the interactions with a third partner (e.g., scabrous or cell-autonomous inhibition by a *cis* ligand) (88–90) caused by glycans [for a more detailed discussion of potential models, see (63, 69)]. As an initial attempt at sorting out these models, several sites of O-fucose modification, and those affected by Fringes, have been mapped to specific EGF repeats in the extracellular domain of mouse Notch1 (Figure 2A) (32). Interestingly, an O-fucose site capable of being modified by Fringe is found on EGF repeat 12 of Notch1. EGF repeat 12 is a portion of the proposed ligand-binding site (EGF repeats 11–12, Figure 2A) (91). This O-fucose site is conserved in every Notch homologue found in databases (32, 50), suggesting that O-fucose modifications at this site could easily have a direct effect on Notch-ligand interactions. Recent studies show that mutation of the O-fucose site at EGF repeat 12 of *D. melanogaster* (Figure 2A) has little or no effect on the ability of Delta to bind Notch but enhances the ability of Serrate to bind to Notch in in vitro assays (60a). In addition, the EGF repeat 12 mutant prevents Fringe from inhibiting Serrate binding to Notch but has no effect on Delta. These results suggest that the O-fucose modification at this site does directly affect binding of Serrate to Notch, but they do not explain the effects of O-fucose and Fringe on Delta (60a). Further work is needed to examine whether similar effects will be seen using mammalian components.

In addition to the ligand binding site, sites modified by O-fucose and Fringe were mapped to the *Abruptex* region of Notch, which includes EGF repeats 24–29 (Figure 2A) (92). The *Abruptex* mutations are gain-of-function mutations in *D. melanogaster* Notch that result in highly active Notch. The *Abruptex* region is believed to be inhibitory, possibly acting through association with ligands expressed in the same cell (so-called *cis-* or cell-autonomous inhibition by ligand) (90). The mapping of several O-fucose modifications to the *Abruptex* region suggests that changes in O-fucose structure could affect this inhibitory function. Analysis of mutations in O-fucose sites within this region of *D. melanogaster* Notch has not yet revealed any specific effect (60a).

Finally, recent studies show that the *Split* mutation introduces a novel O-fucose site in EGF repeat 14 of *D. melanogaster* Notch (Figure 2A) (89). The *Split* mutant phenotype is suppressed in the absence of Scabrous, indicating that Scabrous may interact with O-fucose on EGF repeat 14. Scabrous normally interacts with EGF repeats 19–26 of *D. melanogaster* Notch, a region with numerous O-fucose modification sites. Thus, Scabrous may normally mediate its effects on Notch activation through interacting with sugars in this region (88, 89).

In addition to binding assays and mapping of sites, cell-based Notch signaling assays have been used to study the role of O-fucose modifications. These assays typically consist of a sending cell expressing a ligand cocultured with a receiving cell expressing Notch and a luciferase reporter construct (87, 93). A number of key observations have been made by using glycosylation mutants in CHO cells as the receiving cells in these assays. For instance, loss of complex- and hybrid-type N-glycosylation in Lec1 CHO cells (lacking GlcNAcT-I) has no effect on the ability of Lunatic or Manic fringe to inhibit Jagged1-dependent Notch activation. In contrast, loss of fucosylation in Lec13 CHO cells (which have low GDP-fucose levels due to a lack of GDP-4,6-dehydratase) has a significant effect on both Notch activation and the ability of Lunatic or Manic fringe to modulate it (36, 49). In combination, these results suggest that Fringe mediates its effects through O-fucose modifications and not through N-glycans. Further, the ability of Lunatic or Manic fringe to inhibit Jagged1-dependent Notch activation was not impaired in Lec2 CHO cells (which have reduced sialylation due to a defect in CMP-sialic acid transport) but was prevented in both Lec 8 CHO cells (which have reduced galactosylation due to a defect in UDP-galactose transport) and Lec 20 CHO cells (with a defect in β4GalT-1) (49). Complementation of the defect in Lec 20 CHO cells by transfection with wild-type β4GalT-1 restored the ability of Manic and Lunatic fringe to modulate Notch activity. These results demonstrate that the addition of a β1,4-linked galactose to form an O-fucose trisaccharide (Gal-β1,4-GlcNAc-β1,3-Fuc) is required for the inhibition of Jagged1-dependent Notch activation in CHO cells (Figure 2B).

ROLE OF O-MANNOSE IN DEVELOPMENT

O-mannose saccharides were originally thought to exist only in yeast, such as *Saccharomyces cerevisiae*, where they occur as linear chains of one to six mannose units on a wide variety of proteins (94, 95). The sites of modification, which are similar to mucin domains bearing O-GalNAc modifications (see below), occur in regions rich in serine, threonine, and proline residues. The initial reports of O-mannose-linked oligosaccharides in mammals were made by Margolis and coworkers (96, 97) over 20 years ago. A number of structures have been reported, most of which contain a core structure of Gal-β1,4-GlcNAc-β1,2-Man-α-O-Ser/Thr (Figure 3). This core can be modified with sialic acid, fucose, and GalNAc, and branched structures with di-substituted mannose (GlcNAc linked β1,2 and β1,6) have also been reported (98). Although very few proteins from higher eukaryotes have been identified with O-mannose modifications (99, 100), one report states that O-mannose is an abundant modification, suggesting that up to 30% of protein O-glycosylation in the brain is accounted for by O-mannose saccharides (100). One of the proteins identified with an O-mannose modification is α-dystroglycan (101, 102). The middle third of α-dystroglycan contains a mucin domain with over 50 serine or threonine residues available for modification (Figure 4). Several reports have described the structures of O-mannose saccharides on α-dystroglycan isolated from various species [for reviews, see (98, 99, 103)]. The main form appears to be a tetrasaccharide with the structure Sia-α2,3-Gal-β1,4-GlcNAc-β1,2-Man-α-O-Ser/Thr (Figure 3), although truncated forms (e.g., asialo) and fucosylated forms (α1,3-Fuc to GlcNAc) have also been seen. Interestingly, the mucin domain in α-dystroglycan is modified with a mixture of O-mannose saccharides and conventional O-GalNAc-based mucin-type structures (Figure 4).

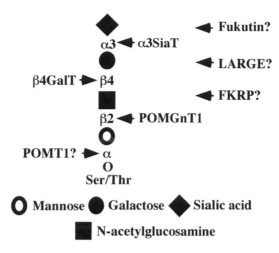

Figure 3   Structure of O-mannose saccharides. A representative structure found on α-dystroglycan is shown. Additional modifications include fucosylation, addition of GalNAc, and branching. The abbreviations are POMT, protein O-mannosyltransferase; POMGnT1, protein O-mannosyl β1,2-N-acetylglucosaminyltransferase 1; β4GalT, β1,4 galactosyltransferase; α3SiaT, α2,3-sialyltransferase; and FKRP, Fukutin-related protein. The symbols representing the carbohydrate structures are as described in the Figure 1 legend.

O-mannose is transferred from dolicholphosphate-mannose to the hydroxyls of serine or threonine residues by protein O-mannosyltransferases (POMTs) in the endoplasmic reticulum. Seven POMTs have been described in *Saccharomyces cerevisiae* [for a review, see (94)]. The large number of isoforms is reminiscent of the polypeptide GalNAc transferases responsible for initiation of mucin-type glycosylation in higher eukaryotes (104). Like the polypeptide GalNAc transferases, each POMT appears to show somewhat different specificity. The POMTs are predicted to have multiple transmembrane helices, an unusual structure for a glycosyltransferase (94). Although homologues to the yeast POMTs have been identified in *D. melanogaster* (*rotated abdomen, rt*) (105) and mammals (*POMT1, POMT2, SDF2L1,* and *SDF2*) (106–109), none of them have been demonstrated either in vivo or in vitro to catalyze O-mannosylation reactions. Recent results consistent with O-mannosyltransferase activity for human POMT1 are discussed below.

In contrast to the POMTs, an enzyme capable of elongating O-mannose has been identified. Protein O-mannose N-acetylglucosaminyltransferase 1 (POMGnT1) catalyzes the transfer of GlcNAc from UDP-GlcNAc to mannose to form GlcNAc-β1,2-Man (Figure 3) (110, 111). It shows a marked preference for mannose O-linked to serine or threonine residues in glycopeptides (110, 111). POMGnT1 was identified because of sequence similarity to GlcNAcT-I, which catalyzes the formation of the same linkage (GlcNAc-β1,2-Man) in the synthesis

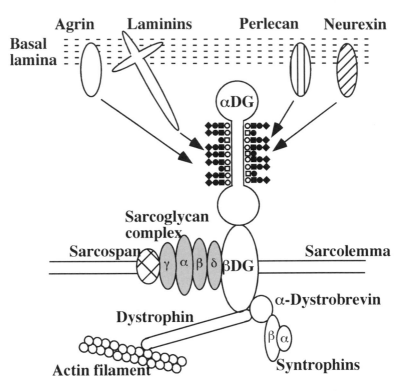

Figure 4 Model of the dystrophin-glycoprotein complex (DGC). A schematic representation of the link between the actin cytoskeleton and components of the extracellular matrix through α-dystroglycan is shown. Representative O-linked structures (both O-mannose and O-GalNAc based) on the mucin-like region of α-dystroglycan are shown. Note that all of the cell surface proteins are also modified with N-glycans, but they are not shown for simplicity. Potential interactions between extracellular matrix components and α-dystroglycan are indicated with arrows. Adapted from (98, 99, 103, 114).

of complex and/or hybrid-type N-glycans (110). The sequence predicts it will have a type II membrane glycoprotein orientation, typical for a Golgi-localized glycosyltransferase. POMGnT1 appears to be widely expressed in mammalian tissues. It is not yet clear whether POMGnT1 is the only enzyme capable of catalyzing this reaction, or whether it will be a member of a family of related enzymes.

None of the enzymes responsible for further elongation of O-mannose saccharides have been identified, although several known glycosyltransferases are capable of forming such structures (β4GalTs, SiaTs). In addition, several other genes (*LARGE, fukutin,* and *fukutin-related protein*), each with sequence similarity to glycosyltransferases, have been identified that appear to affect the function of α-dystroglycan, presumably through alterations in O-mannose oligosaccharides. These will be discussed further below in the context of their role in α-dystroglycan function.

Lessons in O-Mannose Function From Genetic Studies in *D. melanogaster*

The *rotated abdomen* (*rt*) locus was defined in *D. melanogaster* by Bridges & Morgan (112) over 80 years ago as a weakly viable recessive mutation resulting in a 60° to 90° twist in the abdomen relative to the body axis. Detailed analysis of *rt* mutants reveals defects in embryonic muscle development and helical staggering of cuticular patterns in abdominal segments of the adult (105). The gene responsible for the *rt* defects has been cloned and demonstrated to share significant sequence similarity and topology (based on hydropathy analysis) to PMT1 and PMT2 from *S. cerevisiae* (105). The authors proposed that the rotation of body axis is due to defects in muscle architecture, suggesting that O-mannosylation may affect muscle structure and function. Nonetheless, it is not known whether the *rt* gene product has protein O-mannosyltransferase activity, nor what proteins it modifies in *D. melanogaster*.

Lessons in O-Mannose Function From Genetic Studies in Mice and Human Disease

The potential link between muscle structure/function and protein O-mannosylation was greatly strengthened by the identification of O-mannose saccharides on α-dystroglycan (101, 102). α-Dystroglycan is an essential component of the dystrophin-glycoprotein complex (DGC), which functions in linking the actin cytoskeleton to extracellular matrix in muscle and nervous tissue (Figure 4) [for more detailed reviews, see (98, 99, 103, 113, 114)]. Dystroglycan is synthesized as a single polypeptide chain that becomes proteolytically processed during maturation into α and β chains, although the chains remain noncovalently associated at the cell surface. β-Dystroglycan spans the membrane, associates with dystrophin (or utrophin in certain contexts) in the cytoplasm, and provides the link to the actin cytoskeleton. α-Dystroglycan associates with several different extracellular matrix proteins (depending on context), which include laminins, neurexin, agrin, and perlecan. Several other accessory proteins are required to make a functional DGC. Defects in any of these central components can result in congenital muscular dystrophies. For instance, defects in dystrophin cause Duchene muscular dystrophy. Both α- and β-dystroglycan are heavily glycosylated, and studies from the early 1990s suggested that O-linked carbohydrates play a role in the interaction with laminin (115). The subsequent demonstration that α-dystroglycan is heavily modified with O-linked sugars, which include both O-mannose- and O-GalNAc-based structures (101, 102), provided further evidence that O-glycans may be involved in proper DGC function.

A direct link between O-mannose glycans and DGC function came with the discovery by Yoshida et al. (116) that several patients with muscle-eye-brain (MEB) disease, a severe autosomal recessive congenital muscular dystrophy (CMD), harbor mutations in the *POMGnT1* gene. Patients with MEB show CMD, brain malformation, and ocular abnormalities (103, 113). The mutations result in abrogation of POMGnT1 enzymatic activity, suggesting that the primary defect in these patients is the ability to elongate O-mannose residues with a β1,2-linked GlcNAc (Figure 3). A recent study by Michele et al. (117) showed that α-dystroglycan in muscle biopsies from MEB patients migrates at a significantly lower molecular weight than wild-type α-dystroglycan on SDS-PAGE, indicating loss of glycosylation. They also observed that α-dystroglycan from the patients no longer reacted with monoclonal antibodies (VIA4–1, IIH6) believed to be directed against carbohydrate-dependent epitopes on α-dystroglycan. Although the precise nature of the carbohydrate epitope for these antibodies is not known, these data clearly show that mutations in *POMGnT1* have a direct effect on the α-dystroglycan in the affected tissue. Significantly, they also demonstrated that α-dystroglycan from these patients shows significantly reduced binding to laminin, neurexin, or agrin compared to wild-type α-dystroglycan in a variety of assays. These data provide strong support for the hypothesis that O-mannose glycans on α-dystroglycan play a crucial role in the function of the DGC complex.

Further support for this concept has come from the demonstration that mutations in a putative protein O-mannosyltransferase, *POMT1*, are related to another form of CMD, Walker-Warburg syndrome (WWS) (118). WWS is one of the most severe CMDs; patients usually survive less than a year. Many of the defects are similar to those in MEB, with muscular dystrophy, type II lissencephaly, and ocular abnormalities, although some differences exist (103, 113). Mutations in the *POMT1* gene were found in 6 of 30 unrelated patients, indicating that some, but not all, WWS patients have a defect in this gene. This is not surprising; at least three different loci appear to be involved in WWS (118). Because no enzyme assay exists for the protein O-mannosyltransferases, it is not known whether the mutations result in inactive enzyme. Nonetheless, a significant decrease in reactivity of the carbohydrate-dependent monoclonal antibody VIA4–1 with tissues from these patients suggests some change in glycosylation (118). This is accompanied by a less dramatic decrease in levels of α-dystroglycan levels.

In addition to *POMGnT1* and *POMT1*, defects in several other genes with sequence similarity to glycosyltransferases have been shown to result in CMDs. *Fukutin* encodes a protein of unknown function with weak sequence similarity to bacterial and yeast proteins involved in glycosylation reactions (119). Like most Golgi glycosyltransferases, the Fukutin protein is predicted to be a type II membrane glycoprotein, and it contains a DxD motif in the predicted luminal domain (120). Mutations in *Fukutin* result in Fukuyama-type CMD (FCMD), one of the most common forms of muscular dystrophy in Japan (103, 113). Fukutin-related protein (FKRP) also shares sequence features found in glycosyltransferases and has been reported to be Golgi localized (121). Mutations in FKRP are related to congenital muscular dystrophy MDC1C and a milder allelic variant known as Limb-Girdle muscular dystrophy 2I (122, 123). *LARGE* is a mouse gene also predicted to be a glycosyltransferase based on its sequence (124). LARGE appears to be a type II membrane glycoprotein with two potential DxD-containing glycosyltransferase domains separated by a coiled-coil domain in the predicted luminal region. The first glycosyltransferase domain shows sequence similarity to bacterial α-glycosyltransferases, whereas the second glycosyltransferase domain is similar to a human β1,3-N-acetylglucosaminyltransferase (iGlcNAcT) involved in polylactosamine biosynthesis. An internal deletion of the *LARGE* gene is the cause of the defect found in the myodystrophy (*myd*) mouse (124). Homozygous *myd* mice show a severe progressive muscular dystrophy, as well as defects in the central and peripheral nervous systems (114). Very recently, mutations in a human homologue of *LARGE* have been identified in patients with a novel congenital muscular dystrophy, MDC1D (125). Analysis of a muscle biopsies from *myd* mice and patients with FCMD, MDC1C, and MDC1D have shown similar effects to those seen in the MEB biopsies: reduced glycosylation of α-dystroglycan as detected by reduction in molecular weight and lack of reactivity with carbohydrate-dependent antibodies (117, 122, 123). The α-dystroglycan from these biopsies also displays reduced binding to extracellular matrix components, such as laminin, agrin, and neurexin. Thus, although it is not at all clear precisely what the molecular function of the *Fukutin, FKRP*, and *LARGE* gene products are, the data suggest that mutations in these genes have a direct affect on the glycosylation state of α-dystroglycan and that these changes appear to affect the ability of α-dystroglycan to interact with the extracellular matrix.

ROLE OF PROTEOGLYCANS IN DEVELOPMENT

Proteoglycans are an incredibly diverse set of molecules consisting of a glycosaminoglycan chain linked to serine residues of a core protein. Glycosaminoglycans are linear chains of negatively charged (from sulfate and carboxylate side groups) polysaccharides that range in size from ~40 to 400 sugars in heparan sulfate and chondroitin sulfate to as high as 10^5 sugars in hyaluronic acid. They are composed of repeating disaccharide units, and different classes of glycosaminoglycans are defined by the composition of the disaccharides. For instance, heparan sulfate has a repeating disaccharide of GlcNAc-α1,4-GlcA-β1,4, whereas chondroitin sulfate has GalNAc-β1,4-GlcA-β1,3 (Figure 5). Monosaccharides within the polymers can be modified in several ways; these include uronic acid epimerization, GlcNAc N-deacetylation, and N-sulfation and O-sulfation at different positions, resulting in molecules with a high degree of structural diversity. Proteoglycans exist as large macromolecular complexes in

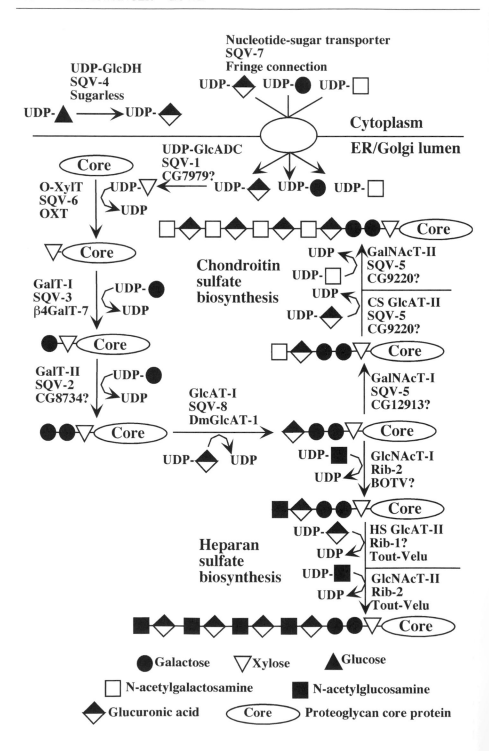

the extracellular matrix or as cell surface molecules. Cell surface proteoglycans fall into two classes: glypicans, which are linked to the plasma membrane through a glycosylphosphatidylinositol anchor and are modified with heparan sulfate glycosaminoglycans, and syndecans, which are transmembrane proteins bearing both heparan sulfate and chondroitin sulfate chains. Most of the enzymes responsible for synthesis of heparan sulfate and chondroitin sulfate chains have been identified (Figure 5). Secreted proteoglycans play a key role in defining the structural properties of extracellular matrix, cartilage, and bone (126). Cell surface heparan sulfate proteoglycans bind growth factors and signaling molecules (e.g., members of the FGF-family, Wnt-family, TGFβ-family, and hedgehog), influencing the ability of these molecules to induce signaling pathways (127, 128). Our understanding of the role of heparan sulfate proteoglycans in signaling and development has exploded over the past several years, largely owing to the identification of mutations in *D. melanogaster* genes encoding glypicans (e.g., *dally* and *dally-like*) and enzymes involved in heparan sulfate biosynthesis (e.g., *sugarless, fringe connection, sulfateless, tout-velu,* and *β4GalT-7*) (Figure 5). Further insight has come from the study of mouse knockouts of genes involved in heparan sulfate proteoglycan biosynthesis and the

---

Figure 5   Synthesis of chondroitin sulfate and heparan sulfate. A schematic representation of the initial steps in chondroitin sulfate and heparan sulfate biosynthesis is shown; these begin in the cytoplasm and move into the lumen of the ER/Golgi. After synthesis of the polymer, further modifications (e.g., de-N-acetylation, N- and O-sulfation, and epimerization, which are not shown) can occur [see (130) for details]. The enzyme/transporter responsible for each step is indicated, followed by the enzymes identified in *C. elegans* (e.g., *sqv* gene products) and *D. melanogaster.* Celera genomics (CG) numbers or names with question marks denote homologues not yet demonstrated to catalyze the appropriate activity. References for the *C. elegans* proteins are given in the text. References for the *D. melanogaster* proteins sugarless, fringe connection, OXT, BOTV (brother of tout-velu), and tout-velu can be found in the following reviews (127, 128, 130). Both *D. melanogaster* β4GalT7 (with associated phenotypes) (265, 266) and DmGlcAT-I (267) have been recently described. For details concerning the corresponding mammalian enzymes, see (130). Abbreviations are as follows: UDP-GlcDH, UDP-glucose dehydrogenase; UDP-GlcADC, UDP-glucuronic acid decarboxylase; O-XylT, polypeptide O-xylosyltransferase; GalT-I, xylose-β1,4-galactosyltransferase (β4GalT-7); GalT-II, galactose-β1,3-galactosyltransferase (β3GalT-6); GlcAT-I, galactose-β1,3-glucuronosyltransferase; GalNAcT-I, glucuronic acid-β1,4-N-acetylgalactosaminyltransferase; CS GlcAT-II, chondroitin sulfate GalNAc-β1,3-glucuronosyltransferase (part of chondroitin sulfate synthase); GalNAcT-II, glucuronic acid-β1,4-N-acetylgalactosaminyltransferase (part of chondroitin sulfate synthase); GlcNAcT-I, glucuronic acid-α1,4-N-acetylglucosaminyltransferase; HS GlcAT-II, heparan sulfate GlcNAc-β1,4-glucuronosyltransferase (part of heparan sulfate copolymerase); and GlcNAcT-II, glucuronic acid α1,4-N-acetylglucosaminyltransferase (part of heparan sulfate copolymerase). Adapted from (142, 201).

recognition that several human diseases arise from defects in their synthesis (129). Because several excellent recent reviews detailing heparan sulfate biosynthesis and function have appeared in this series (130, 131) and in other places [e.g., (127, 128, 132)], we will not cover them further. Instead, we focus on recent studies showing the importance of chondroitin sulfate modifications in the development of *C. elegans*.

Lessons in Chondroitin Sulfate Function From Genetic Studies in *C. elegans*

A genetic screen by Herman et al. (133) for genes involved in vulval morphogenesis in *C. elegans* yielded mutants in eight complementation groups, called *sqv* (for squashed vulva) *1–8*. The vulva is a small invagination, or tube, linking the gonads to the external cuticle. It arises postembryonically, between the third and fourth larval stages [for a review, see (134)]. In homozygous *sqv* mutants, all of the cells necessary for proper vulval formation are present, but the invagination does not fully form. Strong alleles of *sqv* also show a maternal effect lethal phenotype, which arrests development at the one-cell stage (133). Fertilization, pronuclear migration, fusion, and postmitotic nuclear division all appear normal, but the mutants are unable to couple nuclear division with cellular division (135). The defect in cytokinesis is associated with the loss of formation of a fluid-filled space between the egg shell and plasma membrane (135). Loss of this space may be related to the inhibition of cytokinesis. Interestingly, proper vulval formation also requires formation of a fluid-filled space, suggesting that the defects in these two processes may have a similar mechanism (135).

Sequence similarity of *sqv-3* and *sqv-8* to glycosyltransferases, and of *sqv-7* to a nucleotide sugar transporter, led Herman & Horvitz (136) to propose that the *sqv* phenotypes may be caused by a defects in glycosylation reactions, potentially those involved in glycosaminoglycan biosynthesis. Since then, all eight of the *sqv* genes have been demonstrated, either by expression and in vitro assay or by complementation of mutants in mammalian cells, to encode enzymes essential for chondroitin sulfate biosynthesis (Figure 5). SQV-4 has UDP-glucose dehydrogenase activity, capable of converting UDP-Glc to UDP-GlcA (137). It is found in the cytoplasm of numerous cells, which include vulval cells and oocytes. Levels of SQV-4 in vulval cells dramatically increase between the third and fourth larval stages, suggesting formation of UDP-GlcA may play a regulatory role in glycosaminoglycan biosynthesis. SQV-7 is a nucleotide sugar transporter, capable of transporting UDP-GlcA, UDP-GalNAc, and UDP-Gal (138, 139), and it localizes to what is believed to be the Golgi apparatus in many cells (135). SQV-1 appears to be a UDP-GlcA decarboxylase, which converts UDP-GlcA to UDP-Xyl (135). It is predicted to have a type II membrane topology and colocalizes with SQV-7 in cells, suggesting it functions in the lumen of the ER or Golgi apparatus. The remaining proteins (SQV-6, SQV-3, SQV-2, SQV-8, and SQV-5) are also predicted to be type II transmembrane glycoproteins, similar to most Golgi glycosyltransferases (140). SQV-6 is a protein O-xylosyltransferase capable of initiating glycosaminoglycan biosynthesis by transfer of xylose from UDP-xylose to serines on core proteins (141). SQV-3 appears to be GalT-I (β4GalT7), the enzyme responsible for addition of the first β1,4-Gal in the core region (139), and SQV-2 appears to be GalT-II (β3GalT6), responsible for addition of the second galactose (β1,3-linked) in the core (141). SQV-8 seems to be GlcAT-I, responsible for addition of GlcA to complete the core region (139).

Interestingly, each of the SQV proteins described above (SQV 1–4 and 6–8) are involved in synthesis of both heparan sulfate and chondroitin sulfate, making it impossible to determine whether the *sqv* phenotypes are caused by defects in synthesis of heparan sulfate, chondroitin sulfate, or both (Figure 5). The recent demonstration that *sqv-5* encodes a homolog of chondroitin sulfate synthase (ChSy) (142, 143) strongly suggests that the *sqv* phenotypes are the result of defects in chondroitin sulfate synthesis. Chondroitin sulfate synthase is responsible for the alternating, stepwise addition of β1,3-GlcA (GlcAT-II) and β1,4-GalNAc (GalNAcT-II) to form the growing chondroitin sulfate polymer. Homozygous *sqv-5* mutants show loss of chondroitin sulfate synthase activity (both GlcAT-II and GalNAcT-II), but no decrease in GlcNAcT-II activity (specific for heparan sulfate biosynthesis) was observed (142). Interestingly, Mizuguchi et al. (143) were able to detect GalNAcT-II activity but not GlcAT-II activity when expressing a related form (splice variant) of the *sqv-5* gene. Homozygous *sqv-5* mutants also show a decrease in GalNAcT-I activity, suggesting that unlike vertebrate chondroitin sulfate synthases SQV-5 adds the first GalNAc to the core region as well as the GalNAc in the repeating disaccharide. Reduction in SQV-5 levels using RNAi causes a decrease in chondroitin sulfate levels but no decrease in heparan sulfate, adding further support to the idea that the *sqv* phenotypes are caused by a loss of chondroitin sulfate and not heparan sulfate (142, 143).

The mechanism by which a reduction in chondroitin sulfate levels results in the vulval and embryonic cytokinesis phenotypes is not yet clear. It is possible that cell-surface proteoglycans bearing chondroitin sulfate chains may be essential for some as yet undescribed signaling pathway necessary for vulval formation and cytokinesis, although no evidence for such a pathway exists. Interestingly, both vulval formation and cytokinesis in embryos involves formation of a fluid-filled space. Homozygous *sqv* mutants form neither the space resulting in vulval invagination nor the space between the plasma membrane and eggshell, which is believed to be necessary for cytokinesis (135). On the basis of this observation, Hwang et al. (135) have proposed a model suggesting that secretion of chondroitin sulfate proteoglycans is necessary in both of these cases to create an osmotic gradient, which results in the fluid-filled spaces. It remains to be seen whether the lack of chondroitin sulfate biosynthesis abrogates such an osmotic gradient and whether such a gradient is necessary for formation of the fluid-filled spaces in vulval development and early embryonic cytokinesis.

ROLE OF N-GLYCANS IN DEVELOPMENT

A large fraction of membrane-associated and secreted proteins in every eukaryote cell are modified by glycans attached at the reducing terminus through N-acetylglucosamine (GlcNAc) to the amide group of certain asparagine residues (144–146). These so-called N-glycans are typically comprised of seven or more component monosaccharides, and they are complex with respect to the diversity of component monosaccharides, the nature of the linkages among these components, and the processes by which they are elaborated. The details of N-glycan biosynthesis in mammals have been reviewed elsewhere (145, 146) (see Figure 6 for an overview). The N-glycan synthetic scheme has been less well studied in *D. melanogaster* species and in *C. elegans*. Although some differences have been identified between the mammalian N-glycan synthetic pathway and the pathways in the fly and worm, and these will be noted where appropriate, the pathways appear to be quite similar overall.

Lessons in N-glycan Function From Genetic Studies in Mice and Humans

Evidence of major roles for N-glycans in mammalian development derives from the creation and analysis of mice with targeted mutations in glycosyltransferase loci that contribute to N-glycan synthesis (129) and from the study of some types of human congenital disorders of glycosylation (CDG) (147, 148), wherein defective N-glycan synthesis has been assigned a causal role in postnatal pathobiology. There is limited information about early prenatal developmental defects associated with N-glycan synthesis in CDG, in contrast to the situation in mice—for which it is possible to prospectively study developmental processes at the very earliest stages of embryogenesis. In virtually all reports published to date, targeted mutation studies in mice have used null alleles in loci that control N-glycan formation. By contrast, the CDGs are typically caused by missense alleles in loci that control glycosylation. Such mutations leave the corresponding protein with partial function and allow viability at least to term (147, 148).

Developmental Defects in Mice With Targeted Mutations in N-Glycan Biosynthetic Loci

An essential role for N-glycans in cellular viability has been inferred from work with tunicamycin, an analogue of GlcNAc that competitively inhibits the enzyme UDP-GlcNAc: dolichol phosphate N-acetylglucosamine-1-phosphate transferase (GPT) (Figure 6*a*) (149, 150). In cultured cells, tunicamycin depletes the cell of GlcNAc-PP-Dol and thus disables synthesis of the lipid-linked N-glycan precursor $Glc_3Man_9GlcNAc_2$-PP-Dol (151). N-glycan synthesis ceases, in turn, and loss of cellular viability ensues (151, 152). N-glycan-dependent cellular viability depends in large part on contributions made by partially processed N-glycans in the quality control machinery that monitors protein folding in the ER. Glucosidase- and mannosidase-dependent processing of N-glycans in the ER create N-glycan substrates, which support enzymatic reglucosylation when the associated glycoprotein is not yet properly folded (153, 154). The partially folded protein is retained in the folding environment of the ER by calreticulin and calnexin, which recognize glucose-terminated N-glycans. Once properly folded, the N-glycosylated glycoprotein may then proceed to the next steps along the secretory pathway. In the absence of N-glycans that contribute to this process, many N-glycosylated glycoproteins will not fold properly and will not be allowed to traffic to their final membrane-associated or extracellular locale; their loss will contribute to cellular dysfunction or loss of viability. Furthermore, the cellular unfolded protein response (UPR) is strongly triggered in the absence of N-glycan synthesis (155), and prolonged UPR activation leads to programmed cell death (156).

Observations made with tunicamycin have been extended through analysis of mice with a targeted deletion of the *GPT* locus (157, 158). GPT (+/-) mice are viable, fertile, and grossly normal, but homozygous null, full-term progeny are not obtained from intercrosses of heterozygotes. Homozygous GPT null embryonic tissue can be identified prior to E5.5 of gestation in the uteri of GPT heterozygous females mated with GPT heterozygous males, but the tissue is characterized by degeneration of embryonic and extraembryonic cells. Development of homozygous GPT null embryos in vitro up to the blastocyst stage and subsequent implantation is thought to be sustained by maternally derived GPT.

Physiological roles for N-glycans with increasing biosynthetic maturity have been explored using mice with null alleles in loci that encode Golgi-localized N-glycan machinery (159–162). The most biosynthetically proximal of these corresponds to deletion of the *Mgat1*-encoded *GlcNAc-TI* locus, which controls a committed step in the conversion of high-mannose N-glycans to hybrid and complex N-glycans (Figure 6*b*) (163, 164). GlcNAc-TI deficiency per se is not associated with loss of cellular viability (165, 166). *Mgat1* heterozygous null mice are without obvious developmental defects (159–162), but *Mgat1*(+/-) intercrosses do not yield *Mgat1*(-/-) mice at term. Embryonic lethality of *Mgat1* null embryos becomes evident at E10.5 and later in gestation, and it is associated with an amorphic and relative small cephalic area, defective vascularization, an increased frequency of situs inversus of the heart, defective somite formation, and neural tube defects.

At E9.5, *Mgat1* null embryos do not bind to the lectin L-PHA, a lectin with selectivity for complex N-glycans, confirming that GlcNAc-TI is required for complex N-glycan formation. By contrast, *Mgat1* deficient embryos do express complex N-glycans at E3.5, yet expression declines and then is extinguished by E7.5. Maternally derived Mgat1 transcripts in preimplantation embryos may account for maternally determined GlcNAc-TI activity and complex N-glycans that support viability of *Mgat1*-null embryos through E7.5. Developmental contributions by GlcNAc-TI-dependent complex N-glycans have been explored

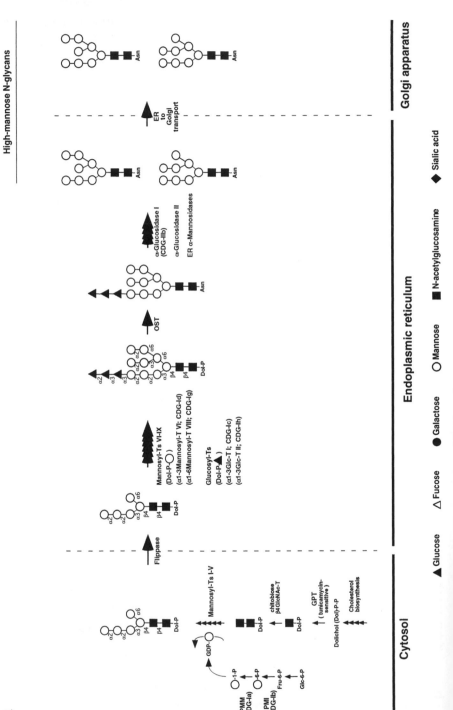

Figure 6 N-glycan synthesis in mammals. Most biosynthetic steps with specific relevance to the text are illustrated. (*a*) Biosynthetic events localized to the cytosol and endoplasmic reticulum (≡R) lumen are shown. (*b*) Biosynthetic steps that localize to the lumen of the Golgi apparatus are shown. General categories of N-glycans (high-mannose, hybrid, and complex) are denoted at the top of each panel. Glycosidic linkages and component monosaccharides are as denoted in Figure 1; symbols for component monosaccharides are also shown at the bottom of each panel. All glycosidic linkages are shown fully in dolichol-P-linked structures in panel *a*, prior to the action the oligosaccharyltransferase complex (OST). Subsequent structures show primarily the linkage(s) created in the preceding step(s). With four exceptions, biosynthetic steps of relevance to N-glycan assembly, where deficiency leads to congenital disorders of glycosylation, are labeled with the corresponding type of CDG (*in parentheses*). The hypothetical complex N-glycan, shown as the final product in panel *b*, illustrates representative monosaccharide linkages that have been described in mammalian N-glycans but are not likely to be found in nature.

TABLE 1 Congenital disorders of glycoylation[a]

Name	Enzyme (*Locus*)	Clinical phenotypes commonly associated with enzyme deficiency
CDG-Ia	Phosphomannomutase 2 (*PMM2*)	Psychomotor retardation, axial hypotonia, and altered levels of serum glycoproteins
CDG-Ib	Phosphomannose isomerase (*PMI*)	Diarrhea, protein-losing enteropathy, thrombosis, and altered serum protein levels Normal mental and motor development
CDG-Ic	Dolichyl-P-Glc:Man$_9$GlcNAc$_2$-PP-dolichyl α-1-3-glucosyltransferase (*ALG6*), (glucosyltransferase I)	Psychomotor retardation, milder than CDG-Ia
CDG-Id	Dolichyl-P-Man:Man$_5$GlcNAc$_2$-PP-dolichyl α-1-3-mannosyltransferase (*ALG3*), (mannosyltransferase VI)	Craniofacial dysmorphology and severe psychomotor retardation
CDG-Ie	Dolichol-P-Man synthase I (*DPM1*)	Craniofacial dysmorphology and severe psychomotor retardation
CDG-If	Dolichol-P-Man utilization defect 1 (*MPDU1*) (*lec35*)	Severe psychomotor retardation and icthyosis
CDG-Ig	Dolichyl-P-Man:Man$_7$GlcNAc$_2$-PP-dolichyl α1-6-mannosyltransferase (*ALG12*), (mannosyltransferase VIII)	Delayed psychomotor development and serum protein abnormalities
CDG-Ih	Dolichyl-P-Glc:Glc$_1$Man$_9$GlcNAc$_2$-PP-dolichyl α-1-3-glucosyltransferase (*ALG6*) (glucosyltransferase II)	Liver disease and protein-losing enteropathy
CDG-IIa	N-acetylglucosaminyltransferase II (GnT II) (*MGAT2*)	Psychomotor retardation, craniofacial dysmorphology, skeletal abnormalities, decreased levels of some serum glycoproteins, gastrointestinal disease, and growth retardation; it has extensive overlap with phenotype of *Mgat2*-null mouse (172)
CDG-IIb	α-1-2-glucosidase I (*GCS1*)	Craniofacial dysmorphology, psychomotor retardation, liver disease

TABLE 1   (*Continued*)

Name	Enzyme (*Locus*)	Clinical phenotypes commonly associated with enzyme deficiency
CDG-IIc	GDP-fucose transporter (*FUCT1*)	Dysmorphology, psychomotor retardation, growth retardation, and leukocyte adhesion deficiency associated with selectin ligand deficiency
CDG-IId	β1-4 GalT-I galactosyltransferase (*β4GalTI*)	CNS pathology, psychomotor retardation; it has some phenotypic overlap with GalT-I null mouse (178–182)

[a]Reviewed in (129, 147, 148, 197)

using chimeric embryos made with *Mgat1*-null embryonic stem cells (166). At E10.5–E16.5, *Mgat1* null cells contribute to several tissues in these chimeras but are absent from the bronchial epithelium. These results imply a requirement for complex N-glycans in the development of this epithelial cell layer. Mechanisms to account for these observations, as well as those in *Mgat1* null embryos, remain to be discovered.

Roles for hybrid and complex N-glycans have been sought in mice with a deficiency in α-mannosidase II (αM-II), historically assigned to a unique role in trimming a single mannose residue from the hybrid N-glycan product of GlcNAcT-1 (Figure 6*b*) (167). αM-II null mice are produced in normal numbers from heterozygous intercrosses and are found to be largely normal with the exception of splenomegaly with a dyserythropoietic anemia (168). Complex N-glycan expression is intact in all tissues except the erythroid lineage, whereas αM-II activity is not detectable. These observations identify a novel alternative pathway for the mannose-trimmed GlcNAcT-II substrate that is dependent upon an α-mannosidase activity termed αM-III (Figure 6*b*). Although the αM-II null mice do not exhibit gross developmental anomalies, many acquire an immune complex glomerulonephritis with autoantibody formation similar to human systemic lupus erythrematosus (169). This observation implies that αM-III is not fully compensating for loss of αM-II activity in these mice, and it suggests that subtle developmental or physiological defects may eventually be identified in these animals.

The alternate pathway for complex N-glycan formation identified in the αM-II null mice is probably catalyzed α mannosidase-IIx (MX) (170). MX deficient mice survive to term without gross developmental defects and have essentially normal complex N-glycans; female MX null mice are grossly normal (171). By contrast, MX null male mice are not fully fertile and exhibit testicular hypoplasia, hypospermatogenesis, and defects in the adherence of spermatogenic cells to Sertoli cells. These adhesion defects are associated with an increase in hybrid-type N-glycans in testicular tissue at the expense of GlcNAc-terminated complex N-glycan. It is not yet known how these glycan alterations may account for the adhesion defects. Studies of combined deficiency of αM-II and MX will be necessary to confirm the hypothesis that MX corresponds to the αM-III activity in αM-II deficient mice.

It seems likely that mice with combined deficiency of αM-II and MX will be developmentally abnormal, as indicated by mice with a deficiency in GlcNAc-TII (Mgat2) (172), the enzyme that catalyzes the post-αM-II/post-MX N-glycan synthetic step (Figure 6b) (173). Mgat2 deletion in C57BL/6 genetic background is associated with partially penetrant embryonic lethal phenotype, which has a loss of viability between E9 and E15. Surviving *Mgat2* null embryos are small but apparently developmentally normal. Most *Mgat2* nulls die during the first postnatal week, all perish within four weeks of birth, and they are deficient in complex N-glycans in most tissues. The phenotype of *Mgat2* null mice overlaps with CDG-IIa (Table 1) caused by mutation of the human locus (*MGAT2*) [reviewed in (147, 148)]. In an outbred strain of mice, (ICR) mice, *Mgat2* deficiency is associated with a less severe phenotype, suggesting the existence of modifier loci that alter the penetrance of the Mgat2-deficienct phenotypes. The phenotypic pleiotropy in GlcNAcT-II deficiency implies that numerous proteins in many tissues are dysfunctional or absent, but a molecular basis for any specific abnormality has yet to be assigned.

Mice with targeted mutation of *Mgat3*-encoded GlcNAcT-III, *Mgat4a*-encoded GlcNAcT-IVa, and *Mgat5*-encoded GlcNAcT-V have been constructed and characterized. In each instance, null progeny are apparently generated from heterozygous crosses with Mendelian frequencies, and developmental defects have not been described in *Mgat3* nulls (174, 175), nor in *Mgat4a* nulls [cited in (129)]. *Mgat5* null mice (176) exhibit an autoimmune syndrome associated with enhanced clustering of the T-cell receptor (177), but with the possible exception of a partially characterized defect in maternal nurturing behavior, developmental abnormalities are not part of the *Mgat5* null phenotype.

The mannose-linked GlcNAc moieties within N-glycans are typically modified by galactose in β1,3 or β1,4 linkage. The galactose residues and the underlying GlcNAc moieties may, in turn, serve as substrates for modification by sialylation, fucosylation, sulfation, and other monosaccharides (Figure 6b). Mice with targeted deletions have been described for several such glycosyltransferase loci. These include one β1-4galactosyltransferase (GalT-1) (178–182), two α1-2fucosyltransferases (Fut1 and Fut2) (183), four sialyltransferases [ST6Gal (184), ST3Gal-III (185), ST3Gal-IV (185, 186), ST8Sia-IV, (187)], and one α1-3galactosyltransferase (α1,3 GalT) (188, 189). Developmental defects identified in GalT-1 and ST8Sia-IV null mice are summarized below. Nondevelopmental phenotypes are characteristic of the other strains and are reviewed elsewhere (129).

Galactosylation of the GlcNAc moieties is catalyzed by members of a family of galactosyltransferases (190). One of these genes (β1-4 galactosyltransferase-1

GalT-1) (191) and its cognate glycan structure have been targeted for deletion in the mouse (178–182). This enzyme is responsible for the rather ubiquitous β1-4-linked galactosylation of GlcNAc moieties on both N-linked and O-linked glycans. A nonenzymatic role (as a cell adhesion molecule) has also been assigned to one isoform of this protein [reviewed in (178, 179)]. Heterozygous GalT-1 null mice are indistinguishable from wild-type mice, whereas homozygous GalT-1 nulls exhibit growth retardation and a markedly shortened life span. Premature death is thought to be due to abnormal epithelial growth and differentiation and to endocrine insufficiency, though underlying mechanisms have not yet been identified. These phenotypes are associated with altered glycan structures on plasma proteins, including a loss of type II chains (Galβ1-4GlcNAc-) together with loss of α2-6-linked sialylation characteristic of type II glycans. In these mice, the glycans on plasma proteins are instead decorated with type I chains (Galβ1-3GlcNAc-) and α2-3-linked sialylation characteristic of type I glycans. It remains to be determined if or how such changes contribute to the phenotype in these mice.

Polysialylation of N-glycans, a characteristic modification of neural cell adhesion molecule (NCAM) is elaborated by several different sialyltransferases, which include the polysialyltransferase ST8Sia-IV (192–194). Polysialic acid (PSA) modification of NCAM modulates the homotypic adhesion properties of this glycoprotein and contributes to the development of the central nervous system (CNS) in utero (195). ST8SiaIV null mice (187) are born at a Mendelian frequency from heterozygous intercrosses. At birth, these mice maintain essentially normal gross and microscopic anatomy and physiology, and they express normal amounts of polysialylated NCAM. However, postnatal development is characterized by loss of polysialylation throughout the brain. PSA deficiency is associated with impaired long-term potentiation and long-term depression in some areas of the hippocampus, though the anatomy remains normal. Mechanisms to account for the defective synaptic plasticity observed in these studies are not yet apparent. Maintenance of normal CNS developmental anatomy in these mice can be explained by redundancy afforded by other polysialyltransferases, especially ST8SiaII. ST8SiaII is likely to be responsible for the wild-type levels of PSA in the ST8SiaIV null neonates. This hypothesis predicts that mice lacking both ST8SiaIV and ST8SiaII will manifest defects in axonal fasciculation and pathfinding that are observed in NCAM-deficient mice (196).

Targeted mutation experiments in mice imply that deficiency in N-glycan structure prior to the formation of hybrid-type structures is incompatible with prenatal development. By contrast, elaboration of high-mannose and hybrid structures in the context of block to formation of complex N-glycans permit some aspects of prenatal development to occur, albeit with various degrees of developmental deficiency. Finally, diversified hybrid and complex-type N-glycans elaborated by enzymes responsible for terminal decoration appear to maintain highly specialized functions postnatally but are apparently not generally necessary for proper development in utero. These conclusions must be tempered by keeping in mind that the set of mutant mice from which such conclusions have been derived is incomplete; numerous glycosyltransferase and glycohydrolase loci remain to be altered by mutagenesis. Moreover, analysis of the available mutant mice, especially with respect to developmental anomalies, remains preliminary in many instances, so additional developmental phenotypes informative for glycan function may yet be identified.

Congenital Disorders of Glycosylation

The general conclusions derived from mouse mutation studies are in large measure consistent with the results from analysis of patients with various forms of congenital disorders of glycosylation (CDG). The literature in this field is extensive and rapidly expanding, in keeping with the estimate that as much as 1% of the human genome may be devoted to glycosylation processes (197). Space considerations permit only a brief overview of the developmental defects and molecular pathology relevant to CDG. Refer to excellent reviews for an in-depth discussion of CDG and for references to the primary literature (147, 148, 197).

CDG is characterized by 12 known defects in N-glycan synthesis and 4 in O-glycan synthesis (147, 148, 197). Diseases caused by defective O-glycan synthesis include two O-mannosylation defects (Walker-Warburg Syndrome and muscle-eye-brain disease, discussed above) and two defects in O-xylosylation [hereditary multiple exostosis and progeroid Ehlers-Danlos syndrome, reviewed elsewhere (130–132)]. N-glycan defects in CDG are assigned to categories characterized by assembly defects (CDG-Ia through CDG-Ih) and by processing defects (CDG-IIa through CDG-IId). Table 1 summarizes the name of each syndrome, its molecular defect and general phenotype, and any relationship to mouse mutational studies. Typically, these are serious multisystem diseases that can include growth retardation, dysmorphology, CNS abnormalities, and dysfunction of specific organs. The biochemical basis for each is reasonably well understood (outlined in Figure 6), but it is not yet possible to assign a molecular explanation to the phenotypes in CDG, except the selectin ligand deficiency observed in CDG-IIc (147).

Lessons in N-Glycan Function From Genetic Studies in C. elegans and D. melanogaster

Although the majority of work on the structure and function of N-glycosylation has been performed in mammalian systems, the power of the genetic manipulations that can be performed in developmental systems such as C. elegans and D. melanogaster has led a number of researchers to begin using them to explore roles for N-glycans. A weakness has been the lack of biochemical knowledge of N-glycan structures in these systems, but several recent reports have added significantly to what is known (198–200). Whereas the mammalian N-glycan biosynthetic pathway appears to be fairly well conserved in both C. elegans and D. melanogaster (201), the resulting structures are somewhat simpler than those seen in mammals. The major N-glycan structures observed in both systems are of the oligomannose variety. Trimming by mannosidases I and II appears to occur, followed by addition of GlcNAc by several GlcNAc transferases (Figure 6). Homologues of GlcNAcT-I and GlcNAcT-II exist in both *C. elegans* and *D. melanogaster* (202–204). Interestingly, *C. elegans* has three GlcNAcT-I homologues, GLY-12, GLY-13, and GLY-14 (203, 205). All three show the appropriate in vitro enzymatic activity, but mutations in GLY-12 and/or GLY-14 show no phenotype. In contrast, GLY-13 mutants are partially lethal. Survivors have a variety of morphological and behavioral abnormalities, indicating that formation of complex and/or hybrid-type N-glycan structures by GlcNAcT-I is essential for proper development in *C. elegans* as it is in mice (159, 160). In addition, a functional GlcNAcT-V homolog has been identified in *C. elegans* (GLY-2), though no overt phenotypes were observed in homozygous mutant *gly-2* animals (206). The conservation of the core N-glycan biosynthetic pathway up to the formation of complex and/or hybrid-type structures indicates that *C. elegans* and *D. melanogaster* are good model systems for examining the function of core N-glycans in development.

Direct evidence for further modification of complex or hybrid-type N-glycan structures with galactose and sialic acids in either *C. elegans* or *D. melanogaster* is only just emerging (201, 206a). Homologues of β4-galactosyltransferases exist in both genomes (199), but it is not yet clear whether any of these modify N-glycans. A *D. melanogaster* α2,6-sialyltransferase has recently been expressed and demonstrated to preferentially modify GalNAc-β1,4-GlcNAc structure (so-called LacDiNAc structure) on N-glycans (206a). This enzyme can modify GalNAc residues on its own N-glycans, providing direct evidence that *D. melanogaster* proteins can be modified with sialic acids. The recent expression and characterization of a β1,4-N-acetylglucosaminyltransferase capable of forming LacDiNAc in *C. elegans* suggests similar structures may be found in worms, but no N-glycans bearing this structure have yet been identified in *C. elegans* (207). In fact, a very recent report indicates that this enzyme (also called BRE-4) is also involved in glycosphingolipid biosynthesis (see below) (208). In addition, a number of fucosylated N-glycans have been reported. Core α1,6-fucosylation is common, and di-fucosylation (both α1,6 and α1,3-fucose on the core GlcNAc) commonly found in insects has been reported in both *C. elegans* and *D. melanogaster* (201, 209). Both the core α1,6- and α1,3-fucosyltransferases have been identified in *D. melanogaster*, and a homologue of the core α1,6-fucosyltransferase exists in *C. elegans* (199, 209). Interestingly, core α1,3-fucosylation appears to be responsible for the reactivity of *D. melanogaster* neuronal proteins with antihorse radish peroxidase antibodies, which have been used as a marker for neurons in *D. melanogaster* (209). Horseradish peroxidase has several unusual carbohydrate epitopes, for example, core α1,3-linked fucose. Multiple fucosylated N-glycan structures have been identified in *C. elegans*; these include some with novel terminal fucose modifications linked to mannose residues (198). A number of genes with homology to fucosyltransferases have been identified in

C. elegans, several of which have been expressed and demonstrated to encode $\alpha 1,2$- and $\alpha 1,3$-fucosyltransferase activities (210–212). Although it is not yet clear whether these modify N-glycans or other types of structures, these results indicate that *C. elegans* will be a rich source for mining functions of terminal fucosylation. In addition to fucose, N-glycans from *C. elegans* can be methylated and modified with phosphocholine (198, 201).

ROLE OF MUCIN-TYPE O-GLYCANS IN DEVELOPMENT

Mucin-type O-glycans correspond to linear and branched glycan chains attached to serine and threonine residues via N-acetylgalactosamine (GalNAc). Mucin-type O-glycans typically decorate segments of membrane-associated or secreted polypeptides that are rich in serine, threonine, and proline residues and are thus believed to assume densely clustered arrays. Synthesis of mucin-type O-glycans differs from N-glycan synthesis in that the process is largely localized to the Golgi and does not require glycosidase-dependent trimming events; the initial synthetic step is controlled by the addition of a monosaccharide directly to the protein. Mucin-type O-glycans are most notably elaborated by epithelia, leukocytes, and vascular endothelia. The widespread distribution and abundance of glycoproteins bearing mucin-type O-glycans imply important functions for these complex carbohydrates, yet much remains to be learned about this subject.

Developmental and Physiological Functions Assigned to Mammalian Mucin-Type O-Glycans

Addition of the initial GalNAc moiety is catalyzed by members of the UDP-N-acetylgalactosamine:polypeptide N-acetylgalactosaminyltransferase (ppGaN-Tase) family [reviewed in (104)] (Figure 7). In mammals, there are at least 12 confirmed members of this family. This is probably an underestimate because database searches predict that the human genome may encode as many as 24 such enzymes (104). It is well established that there is site-specific modification of threonines and serines in proteins that bear mucin-type O-glycans. Nonetheless, a linear consensus sequence for such specificity has not yet been defined, nor has it yet been possible to understand how the three-dimensional configuration of ppGaNTase substrates may dictate specificity. Some of these enzymes are able to add GalNAc directly to threonines or serines in nonglycosylated peptide substrates, whereas others use or even require GalNAc-modified substrates that are the product of prior pp-GaNTase activity. These substrate specificity issues, when considered together with the intricate tissue-specific expression patterns of these genes, imply that control GalNAc-T-dependent O-glycan synthesis is complex and incompletely understood. Redundancy in the number, substrate

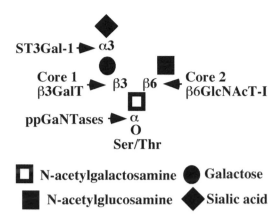

Figure 7 Mucin-type O-glycan structures. Mucin-type O-glycans can exist as simple O-GalNAc monosaccharides or elongated with core 1 and/or core 2 enzymes as shown. More extensive modification (with GalNAc, GlcNAc, fucose, and sialic acid) can occur. Abbreviations are ppGaNTases, polypeptide: O-N-acetylglucos-aminyltransferases; core 1 β3GalT, core 1 β1,3-galactosyltransferase; core 2 β6Glc-NAcT-I, core 2 β1,3-N-acetylglucosaminyltransferase-I; and ST3Gal-I, β-galacto-side: α2,3-sialyltransferase-I. The symbols representing the carbohydrate structures are used as described in the legend for Figure 1.

specificity, and expression patterns of these enzymes has confounded efforts to define mucin-type O-glycan function through targeted deletion approaches in the mouse. For example, mice that are homozygous for a null mutation at the ppGaNTase-T13 locus (revised nomenclature; formerly termed ppGalNAc-T null mice) (213) are developmentally and behaviorally indistinguishable from wild-type mice (214). These mice retain wild-type mucin-type O-glycan expression in most tissues (214) except in the Purkinje cells and the internal granule cell layer in the cerebellum (213). Altered O-glycan expression is also observed in lymphocytes (215). Corresponding functional deficits have not yet been reported. Similarly, unpublished data cited in (104) implies that ppGaNTase-T4 and -T5 null mice are grossly normal, though detailed examination of these mice is undoubtedly underway. The early gene deletion studies, and the properties and extent of redundancy of these enzymes, suggest that abrogation of O-glycan synthesis at the point of GalNAc addition, as a means to uncover developmental contributions for O-glycans, is likely to require mutational inactivation of two or more ppGaNTase loci.

In O-glycan synthesis, GalNAc addition is typically followed by the addition of a β1,3-linked galactose moiety (core 1 branch) and then by the addition of a β1,6-linked N-acetylglucosamine residue (core 2 branch) (Figure 7). Core 3 β1,3-linked N-acetylglucosamine addition followed by core 4 β1,6-linked N-acetylglucosamine addition also occurs. Each branch may be elongated by

GlcNAc and/or Gal addition. Each branch is also subject to sialylation and/or fucosylation [see (216) and references therein]. To date, mice with targeted deletions of genes that contribute to these elongation events include strains deleted for a $\alpha2,3$-sialyltransferase that operates on the core 1 branch (ST3Gal-I) (217, 218), for one of three known $\beta1,6$-GlcNAc transferases (Core 2 GlcNAcT-I) (219–221), and for a pair of $\alpha1,3$-fucosyltransferases (FucT-IV and FucT-VII) (80). In each instance, homozygous null mice are born at the expected Mendelian frequency and do not exhibit major developmental defects at birth. Nonetheless, each such strain exhibits interesting and informative immune defects in the adult animal, which disclose key roles for glycans in thymocyte differentiation (217, 218) and in leukocyte trafficking (80, 219). These mice and their phenotypes are reviewed elsewhere (129, 216).

Lessons in Mucin-Type O-Glycan Function From Genetic Studies in *C. elegans* and *D. melanogaster*

As with N-glycans, analysis of mucin-type O-glycosylation function in *C. elegans* and *D. melanogaster* has only just begun. Very little data on mucin-type O-glycan structures from these organisms exist. Both make mucin-type core 1 structures (Gal-$\beta1,3$-GalNAc-O-Ser/Thr, Figure 7) (222, 223), and both contain a family of ppGaNTase, up to 9 in *C. elegans* and up to 15 in *D. melanogaster*, as found in mammals (104, 224–226). Several of the ppGaNTases from both *C. elegans* and *D. melanogaster* have been expressed and demonstrated to be active in vitro, displaying broad but overlapping peptide substrate specificities, similar to their mammalian counterparts. The importance of mucin-type O-glycosylation for development has recently been illustrated by the fact that mutations in the coding region of one of the ppGaNTases in *D. melanogaster* [*pgant35A* or *l(2)35A*] result in a recessive lethal phenotype, arresting during early pupal development (226, 227). Less is known about the importance of elongating O-GalNAc residues. A preliminary report of a $\beta1,3$-galactosyltransferase from *C. elegans* has appeared (228). In addition, substitution of either the Gal and/or GalNAc of the mucin-type core 1 structures with $\beta1,6$-linked Glc and/or Gal has been reported in *C. elegans* (223). A glucosyltransferse that may be responsible for formation of this linkage has recently been identified (GLY-1) (229). Several apparently nonessential genes with significant similarity to core2/I type N-acetyl-glucosaminyltransferases have also been identified (GLY 15–19), although enzymatic activity has not yet been demonstrated (230). No further information yet exists on the importance of elongating O-GalNAc for development in either *C. elegans* or *D. melanogaster*. Even so, although the mechanism of action is unknown, the fact that a mutation in a specific ppGaNTase results in embryonic lethality in *D. melanogaster* indicates that the mucin-type O-glycosylation is essential for proper development (226, 227).

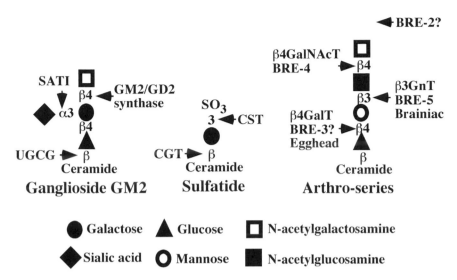

Figure 8 Glycosphingolipid structures. Structures for a representative ganglioside (GM2), sulfatide, and the core structure for the arthroseries of glycosphingolipids found in *C. elegans* and *D. melanogaster* are shown. Mammalian gangliosides exist as a highly diverse set of structures [see (231, 232) for details]. In *C. elegans* and *D. melanogaster*, the arthroseries core can be further modified by the addition of neutral sugars (e.g., GalNAc, GlcNAc, and Gal) and charged groups (phosphorylcholine in *C. elegans*, GlcA, or phosphoethanolamine in *D. melanogaster*). See (248–251) for details. Abbreviations used are UGCG, glucosylceramide synthase; SATI, GM3 synthase; CGT, UDP-galactose:ceramide galactosyltransferase; CST, cerebroside sulfotransferase. For the enzymes involved in synthesis of the arthro-series core, the name of the enzyme is followed by the *C. elegans* (e.g., *bre* gene products) and *D. melanogaster* (Egghead and Brainiac) homologues. Names followed by a question mark have not been demonstrated to catalyze the predicted reaction. The enzymes discussed in the text are identified. The symbols representing the carbohydrate structures are used as described in the legend for Figure 1.

ROLE OF GLYCOSPHINGOLIPIDS IN DEVELOPMENT

Glycolipids are a relatively abundant class of cell surface glycoconjugates (231). These molecules, more properly termed glycosphingolipids, consist of an extracellular glycan component covalently linked to a membrane-embedded ceramide. Glycolipid synthesis is more similar to O-glycan synthesis than to N-glycan synthesis in that glycan chain initiation begins with Golgi-localized addition of a monosaccharide (glucose or galactose) to the ceramide moiety (Figure 8). Subsequent glycosyltransferase-dependent modifications lead to linear and branched complex carbohydrate chains [reviewed in (231, 232)]. In mammals, glycolipids can be neutral, or can be sialylated (gangliosides). Glycolipids have been assigned functions as signal transduction molecules in membrane assembly and maintenance and in cell adhesion processes (231–233).

Glycolipids in Mammalian Development and Physiology

Most mammalian glycolipids are characterized by a glucosylceramide core. This structure is synthesized in the cytosol by glucosylceramide synthase (234). Glucosylceramide then moves to the Golgi, where it is subjected to modification by Golgi-localized glycosyltransferases. Mice that are homozygous for a targeted disruption of the glucosylceramide synthase locus (*Ugcg* locus, Figure 8) die in utero between E6.5 and E7.5. Intrauterine death is characterized by apoptosis in the embryonic ectoderm (235). Embryonic stem cells that are homozygous for the *Ugcg* locus are not distinguishable from wild-type ES cells with respect to growth, morphology, and potential to differentiate in vitro, but they do not properly contribute to the formation of differentiated tissues as do teratomas in mice. These observations imply essential roles for glucosylceramide-dependent glycolipids in organization of different cell types in the context of multicellular differentiation via mechanisms that remain to be defined.

The other, less abundant class of glycolipids is derived from galactosylceramide. These molecules are normally abundant components of myelin in the central and peripheral nervous systems. Galactosylceramide is formed by galactosylation of ceramide by galactosylceramide synthase [UDP-galactose:ceramide galactosyltransferase (CGT), Figure 8] (231, 232). Galactosylceramide can then be sulfated to form sulfatide, its 3-O-sulfated isomer, or can be α2,3-sialylated to form ganglioside GM4. Galactosylceramide, sulfatide, and GM4 are all eliminated in mice with a targeted deletion of the *Cgt* locus (236–238). In the absence of galactosylceramide, the myelin in these mice accumulates a variant of glucosylceramide in which the ceramide contains hydroxyfatty acids. Deficiency of galactosylceramide and sulfatide, perhaps together with the presence of the glucosylceramide isomer, presumably accounts for the localized myelin defects, tremor, and conduction defects that become evident in these mice beginning at postnatal day 12 and that lead to dysfunction of the hind limbs and death between 18 and 90 days. Spermatogenesis is also arrested in these mice, and it is associated with apoptosis of spermatogenic cells and loss of seminolipid (3-sulfogalactosyl-1-alkyl-2-acyl-sn-glycerol) (239). A similar phenotype is observed in cerebroside sulfotransferase (*Cst*) null mice that are defective for sulfatide formation (240), suggesting that the loss of sulfatide in the *Cgt* null mice contributes significantly to these phenotypes. The molecular basis for defective myelin formation or spermatogenetic apoptosis in these two strains of mice is not yet clear, nor is the function of seminolipid understood.

Gene deletion approaches have generated strains of mice in which ganglioside synthesis has been disrupted at three synthetic points distal to glucosylceramide synthesis (241–246). One such strain (generated independently by two groups) is deleted for GM2/GD2 synthase (Figure 8), a β1,4GalNAc transferase that modifies the second intermediate in ganglioside synthesis (lactosylceramide) and its mono-, di-, and trisialylated isomers, all of which can be subsequently elongated by a galactosyltransferase GalT II and at least two sialyltransferases (SAT IV and SAT V) [reviewed in (231, 232)]. These mice are thus deficient in all complex gangliosides (241, 242). Initial evaluations at 10 weeks after birth disclose subtle nerve conduction defects but no histological abnormalities. In older mice, however, progressive axonal degeneration is observed in the central and peripheral nervous system; this degeneration is associated with decreased myelination and with motor and behavioral aberrancies (243). These morphological and functional phenotypes are believed to arise from disruption of interactions between myelin-associated glycoprotein and complex gangliosides that are thought to maintain myelin integrity. Male GM2/GD2 null mice also exhibit testicular pathology that is ascribed to defective testosterone transport (244). Defective IL-2-dependent T-cell proliferation has also been described in these mice (245). A molecular basis for these latter two phenotypes has not been reported.

A second ganglioside mutant strain is deleted for sialyltransferase II (SATII, GD3 synthase), and thus it does not elaborate di- and trisialylated gangliosides. These mice are not distinguishable from wild-type control animals (246). However, mice deficient for GD3 synthase and GM2/GD2 synthase express primarily GM3 (246), and they suffer from lethal audiogenic seizures of an as yet undetermined cause.

A third ganglioside mutant mouse is deficient in the α2,3-sialyltransferase known as sialyltransferase I (SATI, GM3 synthase, encoded by the *Siat9* locus) (Figure 8) (247). This mutant strain is deficient in GM3, the most ubiquitous ganglioside and a precursor for most complex gangliosides. The mice are grossly indistinguishable from wild-type mice, but they maintain enhanced phosphorylation of the insulin receptor in their skeletal muscle with an increased sensitivity to insulin. Mechanisms to account for this phenotype are not yet known.

Considered together, the mouse gene deletion studies summarized above indicate that gangliosides provide essential contributions to developmental processes during embryogenesis. These studies also demonstrate that specific classes of these glycolipids are required for maintaining normal architecture and function of the peripheral and central nervous systems and for certain aspects of male reproductive physiolology. In addition, these studies reinforce previously reported experimental inferences concerning roles for membrane gangliosides in modulating the function of membrane-associated, protein-dependent signal transduction processes.

Lessons in Glycosphingolipid Function From Genetic Studies in *C. elegans* and *D. melanogaster*

As with N-glycans and mucin-type O-glycans, the study of the role of glycosphingolipids in the development of *C. elegans* and *D. melanogaster* is only just beginning. Structural work has revealed that the cores of the major glycosphingolipids in *C. elegans* and *D. melanogaster* are different from those of mammals (248–251). In contrast to the Gal-β1,4-Glc-β1-ceramide core found in most mammalian glycosphingolipids, those in *C. elegans* and *D. melanogaster* are of the arthro-series, which contain a core of Man-β1,4-Glc-β1-ceramide (Figure 8). This core can be extended by the addition of neutral sugars (e.g., GalNAc, GlcNAc, and Gal) and charged groups (phosphorylcholine in *C. elegans* and GlcA or phosphoethanolamine in *D. melanogaster*). The *D. melanogaster egghead* gene has recently been demonstrated to encode a GDP-mannose: βGlc-β1,4-mannosyltransferase, capable of forming the Man-β1,4-Glc linkage in vitro (252). In addition, the *D. melanogaster braoniac* gene encodes a UDP-GlcNAc: βMan-β1,3-N-acetylglucosaminyltransferase, capable of adding a β1,3-GlcNAc to Man-β1,4-Glc-β1-ceramide (253, 254). These studies suggest that the *egghead* and *brainiac* gene products are responsible for the formation of glycosphingolipids in *D. melanogaster*, although no data demonstrating a loss of these structures in either *egghead* or *brainiac* mutants yet exists. Interestingly, both *egghead* and *brainiac* were originally identified because of their effects on epithelial morphogenesis during oogenesis and neurogenesis in *D. melanogaster* (255, 256). They have very similar, nonadditive phenotypes, suggesting that they function in the same pathway. Defects in *egghead* or *brainiac* appear to affect several signaling cascades, which include those involving TGFα/EGF receptor and Delta/Notch (255). Interestingly, *egghead* and *brainiac* exert their effect in the germ line during oogenesis but not in the follicular epithelial cells. Delta is also expressed in the germ line and activates Notch in the epithelial cells (257). These authors proposed that brainiac and egghead function by modulating the ability of Delta to activate Notch. The mechanism by which egghead and brainiac function is not known, although numerous studies in mammalian systems indicate that changes in glycosphingolipids can result in alterations in signaling either through direct interactions with receptors or through modulation of lipid raft function (258, 259). Arthro-series glycosphingolipids have been identified as components of lipid rafts in *D. melanogaster* (260), suggesting that modifications to their structures could modulate signaling events mediated by lipid rafts. Nonetheless, the demonstration that *egghead* and *brainiac* encode glycosyltransferases is additional evidence of the essential role glycoconjugates play in signaling events during development.

Homologues to *egghead* and *brainiac* were identified in *C. elegans* during a screen for mutations that allow survival in the presence of crystal (Cry) proteins from *Bacillus thuringiensis* (208, 261). The Cry proteins are pore-forming toxins that bind to specific receptors in the intestinal epithelium of insects and nematodes and are used widely as insecticides. Griffitts and coworkers (208, 261) have isolated a series of *C. elegans* mutants that are resistant to Cry toxins, termed *bre* (for Bt-toxin resistant). Interestingly, *bre-5* is the *brainiac* homologue and has been demonstrated to complement *brainiac* mutations in *D. melanogaster*. *Bre-3* shares a high degree of sequence similarity to *egghead*, although enzymatic activity has not been demonstrated. BRE-4 is identical to a UDP-GalNAc: βGlcNAc β1,4-N-acetylgalactosaminyltransferase previously identified by Cummings and coworkers in *C. elegans* (207), suggesting it may be responsible for addition of a β1,4-GalNAc to a glycosphingolipid. Finally, BRE-2 shows sequence similarity to the β1,3-glycosyltransferase family, although no specific enzymatic activity has yet been demonstrated. Generation of multiple mutations demonstrates that Bre-2, Bre-3, Bre-4 and Bre-5 all function in the same pathway, suggesting that they function to synthesize a glycosphingolipid functioning as the receptor for the Cry toxins (208). All of the *bre* mutants appear wild type in the absence of Cry toxins, suggesting that these glycosphingolipids do not function in development as they appear to in *D. melanogaster*. Nonetheless, these studies have identified several components of a glycosphingolipid biosynthetic pathway in *C. elegans* (Figure 8).

PERSPECTIVES

The task of assigning functions to glycans in any metazoan biological process is made daunting by several considerations. The structures of these molecules can be complex in the extreme, and they typically vary considerably between species. Complexity in glycan structures is compounded by the commensurately intricate temporal and spatial regulation patterns of these molecules during development and differentiation. Additional levels of difficulty are added by the qualitative and quantitative heterogeneity that characterizes these molecules, even at the level of a specific glycan attachment site on single specific proteins in a clonal cell line [reviewed in (262)], and by the numerous cellular proteins and/or lipids that may be modified by the same class of glycan. Our grasp of the repertoire of enzymes that can determine glycan structures in humans, mice, *D. melanogaster*, *C. elegans*, and other species is perhaps nearing completion with the convergence of genome studies and work that identified these enzymes and their cognate genes via traditional biochemical and molecular studies. Nonetheless, much remains to be learned about how these enzymes work in concert in specific cell types to bring about proper protein- and lipid-linked glycosylation. Finally, even though spectacular progress has been made in developing tools to define the structures of these biopolymers [reviewed in (263)], early experience with assembly of the glycome in *C. elegans* [reviewed in (264)] implies that we have a largely incomplete understanding of the spectrum of glycan structures on most cell types in virtually all metazoans. This may be especially true when considering *D. melanogaster*, *C. elegans*, and other small model organisms because technical approaches for defining glycan structure with small amounts of material have become available relatively recently (263). Nonetheless, there are examples in the mouse where application of more sensitive techniques for assessing glycan structure, in conjunction with gene ablation approaches, identified important yet unappreciated glycan structures, associated biosynthetic pathways, and cognate functions (168, 169, 172).

In the future, it seems likely that much of what we will learn about glycan function in the context of development and other biological processes in metazoans will continue to come from studies in which glycan structure is perturbed by natural mutation or by intentional genetic manipulation. Correlations between glycan structural alterations, corresponding genetic alteration, and phenotype, of the sort summarized earlier in this review, will provide clues to the function of the corresponding glycans. Going forward, technical advances in defining glycan structure with small amounts of material will certainly represent a major contribution to this approach. Nonetheless, the major challenge for the future, as has been the case to date, will be to use the correlations between glycan structure, genetic perturbation, and phenotype to develop a deeper, more molecular understanding of how specific glycans contribute to the function of specific proteins or lipids and, in turn, how these functions contribute to cellular and organismal physiology.

ACKNOWLEDGMENTS

The authors apologize for the inability to cite all of the relevant literature because of space considerations. We thank Jeff Esko and members of the Haltiwanger laboratory for helpful discussions on the manuscript. RSH is supported by grants from the NIH (GM48666) and the Mizutani Foundation for Glycoscience. JBL is supported by grants from the NIH (GM62116 and 1PO1-CA71932) and is an Investigator of the Howard Hughes Medical Institute.

The *Annual Review of Biochemistry* is online at http://biochem.annualreviews.org

LITERATURE CITED

1. Luft JH. 1966. *Fed. Proc.* 25:1773–83
2. Gabius HJ. 2000. *Naturwissenschaften* 87:108–21
3. Varki A, Cummings R, Esko J, Freeze H, Hart G, Marth J, eds. 1999. *Essentials of Glycobiology*. Cold Spring Harbor, NY: Cold Spring Harbor Lab.
4. Morgan WT, Watkins WM. 1953. *Br. J. Exp. Pathol.* 34:94–103
5. Morgan WT, Watkins WM. 2000. *Glycoconj. J.* 17:501–30
6. Szulman AE, Marcus DM. 1973. *Lab. Investig.* 28:565–74
7. Szulman AE. 1964. *J. Exp. Med.* 119:503–16
8. Szulman AE. 1971. *Hum. Pathol.* 2:575–85
9. Oppenheimer SB. 1977. *Curr. Top. Dev. Biol.* 11:1–16
10. Solter D, Knowles BB. 1978. *Proc. Natl. Acad. Sci. USA* 75:5565–69
11. Solter D, Knowles BB. 1979. *Curr. Top. Dev. Biol.* 13(Pt. 1):139–65
12. Gooi HC, Feizi T, Kapadia A, Knowles BB, Solter D, Evans MJ. 1981. *Nature* 292:156–58
13. Knowles BB, Rappaport J, Solter D. 1982. *Dev. Biol.* 93:54–58
14. Kannagi R, Cochran NA, Ishigami F, Hakomori S, Andrews PW, et al. 1983. *EMBO J.* 2:2355–61
15. Tippett P, Andrews PW, Knowles BB, Solter D, Goodfellow PN. 1986. *Vox Sang.* 51:53–56
16. McClay DR, Chambers AF, Warren RH. 1977. *Dev. Biol.* 56:343–55
17. Schneider EG, Nguyen HT, Lennarz WJ. 1978. *J. Biol. Chem.* 253:2348–55

18. Carson DD, Lennarz WJ. 1979. *Proc. Natl. Acad. Sci. USA* 76:5709–13

19. D'Amico P, Jacobs JR. 1995. *Tissue Cell* 27:23–30

20. Fredieu JR, Mahowald AP. 1994. *Acta Anat.* 149:89–99

21. O'Tousa JE. 1992. *Vis. Neurosci.* 8: 385–90

22. Kaushal S, Ridge KD, Khorana HG. 1994. *Proc. Natl. Acad. Sci. USA* 91: 4024–28

23. Link CD, Silverman MA, Breen M, Watt KE, Dames SA. 1992. *Genetics* 131: 867–81

24. Weis WI, Taylor ME, Drickamer K. 1998. *Immunol. Rev.* 163:19–34

25. Cooper DN. 2002. *Biochim. Biophys. Acta* 1572:209–31

26. Dodd RB, Drickamer K. 2001. *Glycobiology* 11:R71–79

27. Haslam SM, Gems D, Morris HR, Dell A. 2002. *Biochem. Soc. Symp.*, pp. 117–34

28. Hallgren P, Lundblad A, Svensson S. 1975. *J. Biol. Chem.* 250:5312–14

29. Kentzer EJ, Buko AM, Menon G, Sarin VK. 1990. *Biochem. Biophys. Res. Commun.* 171:401–6

30. Harris RJ, Spellman MW. 1993. *Glycobiology* 3:219–24

31. Campbell ID, Bork P. 1993. *Curr. Opin. Struct. Biol.* 3:385–92

32. Shao L, Moloney DJ, Haltiwanger RS. 2003. *J. Biol. Chem.* 278:7775–82

33. Panin VM, Shao L, Lei L, Moloney DJ, Irvine KD, Haltiwanger RS. 2002. *J. Biol. Chem.* 277:29945–52

34. Haltiwanger RS. 2002. *Curr. Opin. Struct. Biol.* 12:593–98

35. Moloney DJ, Shair L, Lu FM, Xia J, Locke R, et al. 2000. *J. Biol. Chem.* 275: 9604–11

36. Moloney DJ, Panin VM, Johnston SH, Chen J, Shao L, et al. 2000. *Nature* 406: 369–75

37. Yan YT, Liu JJ, Luo Y, Chaosu E, Haltiwanger RS, et al. 2002. *Mol. Cell. Biol.* 22:4439–49

38. Schiffer SG, Foley S, Kaffashan A, Hronowski X, Zichittella AE, et al. 2001. *J. Biol. Chem.* 276:37769–78

39. de Peredo AG, Klein D, Macek B, Hess D, Peter-Katalinic J, Hofsteenge J. 2002. *Mol. Cell. Proteomics* 1:11–18

40. Hofsteenge J, Huwiler KG, Macek B, Hess D, Lawler J, et al. 2001. *J. Biol. Chem.* 276:6485–98

41. Shimizu K, Chiba S, Saito T, Kumano K, Takahashi T, Hirai H. 2001. *J. Biol. Chem.* 276:25753–58

42. Wang Y, Lee GF, Kelley RF, Spellman MW. 1996. *Glycobiology* 6:837–42

43. Wang Y, Spellman MW. 1998. *J. Biol. Chem.* 273:8112–18

44. Wang Y, Shao L, Shi S, Harris RJ, Spellman MW, et al. 2001. *J. Biol. Chem.* 276:40338–45

45. Irvine KD, Wieschaus E. 1994. *Cell* 79: 595–606

46. Cohen B, Bashirullah A, Dagnino L, Campbell C, Fisher WW, et al. 1997. *Nat. Genet.* 16:283–88

47. Johnston SH, Rauskolb C, Wilson R, Prabhakaran B, Irvine KD, Vogt TF. 1997. *Development* 124:2245–54

48. Bruckner K, Perez L, Clausen H, Cohen S. 2000. *Nature* 406:411–15

49. Chen J, Moloney DJ, Stanley P. 2001. *Proc. Natl. Acad. Sci. USA* 98: 13716–21

50. Shi S, Stanley P. 2003. *Proc. Natl. Acad. Sci. USA* 100:5234–39

51. Okajima T, Irvine KD. 2002. *Cell* 111: 893–904

52. Sasamura T, Sasaki N, Miyashita F, Nakao S, Ishikawa HO, et al. 2003. *Development* 130:4785–95

53. Wharton KA, Johansen KM, Xu T, Artavanis-Tsakonas S. 1985. *Cell* 43: 567–81

54. Artavanis-Tsakonas S, Rand MD, Lake RJ. 1999. *Science* 284:770–76

55. Joutel A, Tournier-Lasserve E. 1998. *Semin. Cell Dev. Biol.* 9:619–25

56. Bulman MP, Kusumi K, Frayling TM,

McKeown C, Garrett C, et al. 2000. *Nat. Genet.* 24:438–41

57. Ellisen LW, Bird J, West DC, Soreng AL, Reynolds TC, et al. 1991. *Cell* 66: 649–61

58. Artavanis-Tsakonas S. 1997. *Nat. Genet.* 16:212–13

59. John GR, Shankar SL, Shafit-Zagardo B, Massimi A, Lee SC, et al. 2002. *Nat. Med.* 8:1115–21

60. Mumm JS, Kopan R. 2000. *Dev. Biol.* 228:151–65

60a. Lei L, Xu A, Panin VM, Irvine KD. 2003. *Development* 130:6411–21

61. Irvine KD, Rauskolb C. 2001. *Annu. Rev. Cell Dev. Biol.* 17:189–214

62. Irvine KD. 1999. *Curr. Opin. Genet. Dev.* 9:434–41

63. Haines N, Irvine KD. 2003. *Nat. Rev. Mol. Cell Biol.* 4:786–97

64. Panin VM, Papayannopoulos V, Wilson R, Irvine KD. 1997. *Nature* 387: 908–12

65. Fleming RJ, Gu Y, Hukriede NA. 1997. *Development* 124:2973–81

66. Munro S, Freeman M. 2000. *Curr. Biol.* 10:813–20

67. Selva EM, Hong K, Baeg GH, Beverley SM, Turco SJ, et al. 2001. *Nat. Cell Biol.* 3:809–15

68. Goto S, Taniguchi M, Muraoka M, Toyoda H, Sado Y, et al. 2001. *Nat. Cell Biol.* 3:816–22

69. Haltiwanger RS, Stanley P. 2002. *Biochim. Biophys. Acta* 1573:328–35

70. Zhang N, Gridley T. 1998. *Nature* 394: 374–77

71. Evrard YA, Lun Y, Aulehla A, Gan L, Johnson RL. 1998. *Nature* 394:377–81

72. Pourquie O. 2003. *Science* 301:328–30

73. Moran JL, Levorse JM, Vogt TF. 1999. *Nature* 399:742–43

74. Zhang N, Norton CR, Gridley T. 2002. *Genesis* 33:21–28

75. Smith PL, Myers JT, Rogers CE, Zhou L, Petryniak B, et al. 2002. *J. Cell Biol.* 158:801–15

76. Becker DJ, Lowe JB. 2003. *Glycobiology* 13:R41–53

77. Tonetti M, Sturla L, Bisso A, Benatti U, De Flora A. 1996. *J. Biol. Chem.* 271: 27274–79

78. Yurchenco PD, Atkinson PH. 1977. *Biochemistry* 16:944–53

79. Becker DJ, Myers JT, Ruff MM, Smith PL, Gillespie BW, et al. 2003. *Mamm. Genome* 14:130–39

80. Homeister JW, Thall AD, Petryniak B, Maly P, Rogers CE, et al. 2001. *Immunity* 15:115–26

81. Sturla L, Rampal R, Haltiwanger RS, Fruscione F, Etzioni A, Tonetti M. 2003. *J. Biol. Chem.* 278:26727–33

82. Rabbani SA, Mazar AP, Bernier SM, Haq M, Bolivar I, et al. 1992. *J. Biol. Chem.* 267:14151–56

83. Schier AF, Shen MM. 2000. *Nature* 403: 385–89

84. Whitman M. 2001. *Dev. Cell* 1:605–17

85. Shen MM, Schier AF. 2000. *Trends Genet.* 16:303–9

86. Okajima T, Xu A, Irvine KD. 2003. *J. Biol. Chem.* 278:42340–45

87. Hicks C, Johnston SH, DiSibio G, Collazo A, Vogt TF, Weinmaster G. 2000. *Nat. Cell Biol.* 2:515–20

88. Li Y, Fetchko M, Lai ZC, Baker NE. 2003. *Development* 130:2819–27

89. Li Y, Lei L, Irvine KD, Baker NE, Li L. 2003. *Development* 130:2829–40

90. De Celis JF, Bray SJ. 2000. *Development* 127:1291–302

91. Rebay I, Fleming RJ, Fehon RG, Cherbas L, Cherbas P, Artavanis-Tsakonas S. 1991. *Cell* 67:687–99

92. Kelley MR, Kidd S, Deutsch WA, Young MW. 1987. *Cell* 51:539–48

93. Lindsell CE, Shawber CJ, Boulter J, Weinmaster G. 1995. *Cell* 80:909–17

94. Strahl-Bolsinger S, Gentzsch M, Tanner W. 1999. *Biochim. Biophys. Acta* 1426:297–307

95. Gemmill TR, Trimble RB. 1999. *Biochim. Biophys. Acta* 1426:227–37

96. Finne J, Krusius T, Margolis RK, Mar-

golis RU. 1979. *J. Biol. Chem.* 254: 10295–300

97. Krusius T, Finne J, Margolis RK, Margolis RU. 1986. *J. Biol. Chem.* 261: 8237–42

98. Martin PT. 2003. *Glycobiology* 13: R55–66

99. Endo T. 1999. *Biochim. Biophys. Acta* 1473:237–46

100. Chai W, Yuen C-T, Kogelberg H, Carruthers RA, Margolis RU, et al. 1999. *Eur. J. Biochem.* 263:879–88

101. Chiba A, Matsumura K, Yamada H, Inazu T, Shimizu T, et al. 1997. *J. Biol. Chem.* 272:2156–62

102. Smalheiser NR, Haslam SM, Sutton-Smith M, Morris HR, Dell A. 1998. *J. Biol. Chem.* 273:23698–703

103. Michele DE, Campbell KP. 2003. *J. Biol. Chem.* 278:15457–60

104. Ten Hagen KG, Fritz TA, Tabak LA. 2003. *Glycobiology* 13:1–16

105. Martin-Blanco E, Garcia-Bellido A. 1996. *Proc. Natl. Acad. Sci. USA* 93: 6048–52

106. Jurado LA, Coloma A, Cruces J. 1999. *Genomics* 58:171–80

107. Willer T, Amselgruber W, Deutzmann R, Strahl S. 2002. *Glycobiology* 12: 771–83

108. Fukuda S, Sumii M, Masuda Y, Takahashi M, Koike N, et al. 2001. *Biochem. Biophys. Res. Commun.* 280:407–14

109. Hamada T, Tashiro K, Tada H, Inazawa J, Shirozu M, et al. 1996. *Gene* 176: 211–14

110. Zhang W, Betel D, Schachter H. 2002. *Biochem. J.* 361:153–62

111. Takahashi S, Sasaki T, Manya H, Chiba Y, Yoshida A, et al. 2001. *Glycobiology* 11:37–45

112. Bridges CB, Morgan TH. 1923. *The Third-Chromosome Group of Mutant Characters of Drosophila melanogaster.* Washington, DC: Carnegie Inst. 251 pp.

113. Martin PT, Freeze HH. 2003. *Glycobiology* 13:R67–75

114. Hewitt JE, Grewal PK. 2003. *Cell. Mol. Life Sci.* 60:251–58

115. Ervasti JM, Campbell KP. 1993. *J. Cell Biol.* 122:809–23

116. Yoshida A, Kobayashi K, Manya H, Taniguchi K, Kano H, et al. 2001. *Dev. Cell* 1:717–24

117. Michele DE, Barresi R, Kanagawa M, Saito F, Cohn RD, et al. 2002. *Nature* 418:417–22

118. de Bernabe DBV, Currier S, Steinbrecher A, Celli J, van Beusekom E, et al. 2002. *Am. J. Hum. Genet.* 71:1033–43

119. Kobayashi K, Nakahori Y, Miyake M, Matsumura K, Kondo-Iida E, et al. 1998. *Nature* 394:388–92

120. Aravind L, Koonin EV. 1999. *Curr. Biol.* 9:R836–37

121. Esapa CT, Benson MA, Schroder JE, Martin-Rendon E, Brockington M, et al. 2002. *Hum. Mol. Genet.* 11:3319–31

122. Brockington M, Yuva Y, Prandini P, Brown SC, Torelli S, et al. 2001. *Hum. Mol. Genet.* 10:2851–59

123. Brockington M, Blake DJ, Prandini P, Brown SC, Torelli S, et al. 2001. *Am. J. Hum. Genet.* 69:1198–209

124. Grewal PK, Holzfeind PJ, Bittner RE, Hewitt JE. 2001. *Nat. Genet.* 28: 151–54

125. Longman C, Brockington M, Torelli S, Jimenez-Mallebrera C, Kennedy C, et al. 2003. *Hum. Mol. Genet.* 12:2853–61

126. Iozzo RV. 1998. *Annu. Rev. Biochem.* 67:609–52

127. Selleck SB. 2000. *Trends Genet.* 16: 206–12

128. Nybakken K, Perrimon N. 2002. *Biochim. Biophys. Acta* 1573:280–91

129. Lowe JB, Marth JD. 2003. *Annu. Rev. Biochem.* 72:643–91

130. Esko JD, Selleck SB. 2002. *Annu. Rev. Biochem.* 71:435–71

131. Bernfield M, Gotte M, Park PW, Reizes O, Fitzgerald ML, et al. 1999. *Annu. Rev. Biochem.* 68:729–77

132. Perrimon N, Berfield M. 2000. *Nature* 404:725–28

133. Herman T, Hartwieg E, Horvitz HR. 1999. *Proc. Natl. Acad. Sci. USA* 96: 968–73

134. Bulik DA, Robbins PW. 2002. *Biochim. Biophys. Acta* 1573:247–57

135. Hwang HY, Horvitz HR. 2002. *Proc. Natl. Acad. Sci. USA* 99:14218–23

136. Herman T, Horvitz HR. 1999. *Proc. Natl. Acad. Sci. USA* 96:974–79

137. Hwang HY, Horvitz HR. 2002. *Proc. Natl. Acad. Sci. USA* 99:14224–29

138. Berninsone P, Hwang HY, Zemtseva I, Horvitz HR, Hirschberg CB. 2001. *Proc. Natl. Acad. Sci. USA* 98:3738–43

139. Bulik DA, Wei G, Toyoda H, Kinoshita-Toyoda A, Waldrip WR, et al. 2000. *Proc. Natl. Acad. Sci. USA* 97: 10838–43

140. Paulson JC, Colley KJ. 1989. *J. Biol. Chem.* 264:17615–18

141. Hwang HY, Olson SK, Brown JR, Esko JD, Horvitz HR. 2003. *J. Biol. Chem.* 278:11735–38

142. Hwang HY, Olson SK, Esko JD, Horvitz HR. 2003. *Nature* 423:439–43

143. Mizuguchi S, Uyama T, Kitagawa H, Nomura KH, Dejima K, et al. 2003. *Nature* 423:443–48

144. Hart GW, Brew K, Grant GA, Bradshaw RA, Lennarz WJ. 1979. *J. Biol. Chem.* 254:9747–53

145. Herscovics A. 1999. *Biochim. Biophys. Acta* 1473:96–107

146. Kornfeld R, Kornfeld S. 1985. *Annu. Rev. Biochem.* 54:631–64

147. Marquardt T, Denecke J. 2003. *Eur. J. Pediatr.* 162:359–79

148. Jaeken J, Matthijs G. 2001. *Annu. Rev. Genomics Hum. Genet.* 2:129–51

149. Lehrman MA. 1994. *Glycobiology* 4: 768–71

150. Takasuki A, Tamura G. 1971. *J. Antibiot.* 24:785–94

151. Elbein AD. 1984. *CRC Crit. Rev. Biochem.* 16:21–49

152. Surani MA. 1979. *Cell* 18:217–27

153. Ellgaard L, Molinari M, Helenius A. 1999. *Science* 286:1882–88

154. Trombetta ES, Parodi AJ. 2001. *Adv. Protein Chem.* 59:303–44

155. Kaufman RJ, Scheuner D, Schroder M, Shen XH, Lee K, et al. 2002. *Nat. Rev. Mol. Cell Biol.* 3:411–21

156. Kaufman RJ. 2002. *J. Clin. Investig.* 110: 1389–98

157. Rajput B, Muniappa N, Vijay IK. 1994. *J. Biol. Chem.* 269:16054–61

158. Marek KW, Vijay IK, Marth JD. 1999. *Glycobiology* 9:1263–71

159. Ioffe E, Stanley P. 1994. *Proc. Natl. Acad. Sci. USA* 91:728–32

160. Metzler M, Gertz A, Sarkar M, Schachter H, Schrader JW, Marth JD. 1994. *EMBO J.* 13:2056–65

161. Campbell RM, Metzler M, Granovsky M, Dennis JW, Marth JD. 1995. *Glycobiology* 5:535–43

162. Ioffe E, Liu Y, Stanley P. 1997. *Glycobiology* 7:913–19

163. Kumar R, Yang J, Larsen RD, Stanley P. 1990. *Proc. Natl. Acad. Sci. USA* 87: 9948–52

164. Sarkar M, Hull E, Nishikawa Y, Simpson RJ, Moritz RL, et al. 1991. *Proc. Natl. Acad. Sci. USA* 88:234–38

165. Kumar R, Stanley P. 1989. *Mol. Cell. Biol.* 9:5713–17

166. Ioffe E, Liu Y, Stanley P. 1996. *Proc. Natl. Acad. Sci. USA* 93:11041–46

167. Moremen KW, Robbins PW. 1991. *J. Cell Biol.* 115:1521–34

168. Chui D, Oh-Eda M, Liao YF, Panneerselvam K, Lal A, et al. 1997. *Cell* 90: 157–67

169. Chui D, Sellakumar G, Green R, Sutton-Smith M, McQuistan T, et al. 2001. *Proc. Natl. Acad. Sci. USA* 98:1142–47

170. Misago M, Liao YF, Kudo S, Eto S, Mattei MG, et al. 1995. *Proc. Natl. Acad. Sci. USA* 92:11766–70

171. Akama TO, Nakagawa H, Sugihara K, Narisawa S, Ohyama C, et al. 2002. *Science* 295:124–27

172. Wang Y, Tan J, Sutton-Smith M, Ditto D, Panico M, et al. 2001. *Glycobiology* 11:1051–70

173. Bendiak B, Schachter H. 1987. *J. Biol. Chem.* 262:5775–83
174. Priatel JJ, Sarkar M, Schachter H, Marth JD. 1997. *Glycobiology* 7:45–56
175. Bhattacharyya R, Bhaumik M, Raju TS, Stanley P. 2002. *J. Biol. Chem.* 277:26300–9
176. Granovsky M, Fata J, Pawling J, Muller WJ, Khokha R, Dennis JW. 2000. *Nat. Med.* 6:306–12
177. Demetriou M, Granovsky M, Quaggin S, Dennis JW. 2001. *Nature* 409:733–39
178. Lu QX, Hasty P, Shur BD. 1997. *Dev. Biol.* 181:257–67
179. Lu QX, Shur BD. 1997. *Development* 124:4121–31
180. Asano M, Furukawa K, Kido M, Matsumoto S, Umesaki Y, et al. 1997. *EMBO J.* 16:1850–57
181. Kotani N, Asano M, Iwakura Y, Takasaki S. 2001. *Biochem. J.* 357:827–34
182. Kotani N, Asano M, Iwakura Y, Takasaki S. 1999. *Biochem. Biophys. Res. Commun.* 260:94–98
183. Domino SE, Zhang L, Gillespie PJ, Saunders TL, Lowe JB. 2001. *Mol. Cell. Biol.* 21:8336–45
184. Hennet T, Chui D, Paulson JC, Marth JD. 1998. *Proc. Natl. Acad. Sci. USA* 95:4504–9
185. Ellies LG, Sperandio M, Underhill GH, Yousif J, Smith M, et al. 2002. *Blood* 100:3618–25
186. Ellies LG, Ditto D, Levy GG, Wahrenbrock M, Ginsburg D, et al. 2002. *Proc. Natl. Acad. Sci. USA* 99:10042–47
187. Eckhardt M, Bukalo O, Chazal G, Wang L, Goridis C, et al. 2000. *J. Neurosci.* 20:5234–44
188. Thall AD, Maly P, Lowe JB. 1995. *J. Biol. Chem.* 270:21437–40
189. Tearle RG, Tange MJ, Zannettino ZL, Katerelos M, Shinkel TA, et al. 1996. *Transplantation* 61:13–19
190. Hennet T. 2002. *Cell. Mol. Life Sci.* 59:1081–95

191. Shaper NL, Hollis GF, Douglas JG, Kirsch IR, Shaper JH. 1988. *J. Biol. Chem.* 263:10420–28
192. Eckhardt M, Mühlenhoff M, Bethe A, Koopman J, Frosch M, Gerardy-Schahn R. 1995. *Nature* 373:715–18
193. Yoshida Y, Kojima N, Tsuji S. 1995. *J. Biochem.* 118:658–64
194. Nakayama J, Fukuda MN, Fredette B, Ranscht B, Fukuda M. 1995. *Proc. Natl. Acad. Sci. USA* 92:7031–35
195. Muhlenhoff M, Eckhardt M, Gerardy-Schahn R. 1998. *Curr. Opin. Struct. Biol.* 8:558–64
196. Cremer H, Chazal G, Goridis C, Represa A. 1997. *Mol. Cell. Neurosci.* 8:323–35
197. Freeze HH. 1998. *J. Pediatr.* 133:593–600
198. Cipollo JF, Costello CE, Hirschberg CB. 2002. *J. Biol. Chem.* 277:49143–57
199. Altmann F, Fabini G, Ahorn H, Wilson IB. 2001. *Biochimie* 83:703–12
200. Natsuka S, Adachi J, Kawaguchi M, Nakakita S, Hase S, et al. 2002. *J. Biochem.* 131:807–13
201. Wilson IB. 2002. *Curr. Opin. Struct. Biol.* 12:569–77
202. Sarkar M, Schachter H. 2001. *Biol. Chem.* 382:209–17
203. Chen S, Spence AM, Schachter H. 2003. *Biochimie* 85:391–401
204. Chen S, Tan J, Reinhold VN, Spence AM, Schachter H. 2002. *Biochim. Biophys. Acta* 1573:271–79
205. Chen S, Zhou S, Sarkar M, Spence AM, Schachter H. 1999. *J. Biol. Chem.* 274:288–97
206. Warren CE, Krizus A, Roy PJ, Culotti JG, Dennis JW. 2002. *J. Biol. Chem.* 277:22829–38
206a. Koles K, Irvine KD, Panin VM. 2004. *J. Biol. Chem.* 279:4346–57
207. Kawar ZS, Van Die I, Cummings RD. 2002. *J. Biol. Chem.* 277:34924–32
208. Griffitts JS, Huffman DL, Whitacre JL, Barrows BD, Marroquin LD, et al. 2003. *J. Biol. Chem.* 278:45594–602

209. Fabini G, Freilinger A, Altmann F, Wilson IB. 2001. *J. Biol. Chem.* 276:28058–67

210. Oriol R, Mollicone R, Cailleau A, Balanzino L, Breton C. 1999. *Glycobiology* 9:323–34

211. Zheng Q, Van Die I, Cummings RD. 2002. *J. Biol. Chem.* 277:39823–32

212. DeBose-Boyd RA, Nyame AK, Cummings RD. 1998. *Glycobiology* 8:905–17

213. Zhang Y, Iwasaki H, Wang H, Kudo T, Kalka TB, et al. 2003. *J. Biol. Chem.* 278:573–84

214. Hennet T, Hagen FK, Tabak LA, Marth JD. 1995. *Proc. Natl. Acad. Sci. USA* 92:12070–74

215. Westerman EL, Ellies LG, Hagen FK, Marek KW, Sutton-Smith M, et al. 1999. *Glycobiology* 9:1121 (Abstr.)

216. Lowe JB. 2002. *Immunol. Rev.* 186:19–36

217. Priatel JJ, Chui D, Hiraoka N, Simmons CJ, Richardson KB, et al. 2000. *Immunity* 12:273–83

218. Moody AM, Chui D, Reche PA, Priatel JJ, Marth JD, Reinherz EL. 2001. *Cell* 107:501–12

219. Ellies LG, Tsuboi S, Petryniak B, Lowe JB, Fukuda M, Marth JD. 1998. *Immunity* 9:881–90

220. Snapp KR, Heitzig CE, Ellies LG, Marth JD, Kansas GS. 2001. *Blood* 97:3806–11

221. Sperandio M, Thatte A, Foy D, Ellies LG, Marth JD, Ley K. 2001. *Blood* 97:3812–19

222. Kramerov AA, Arbatsky NP, Rozovsky YM, Mikhaleva EA, Polesskaya OO, et al. 1996. *FEBS Lett.* 378:213–18

223. Guerardel Y, Balanzino L, Maes E, Leroy Y, Coddeville B, et al. 2001. *Biochem. J.* 357:167–82

224. Hagen FK, Nehrke K. 1998. *J. Biol. Chem.* 273:8268–77

225. Ten Hagen KG, Tran DT, Gerken TA, Stein DS, Zhang Z. 2003. *J. Biol. Chem.* 278:35039–48

226. Schwientek T, Bennett EP, Flores C, Thacker J, Hollmann M, et al. 2002. *J. Biol. Chem.* 277:22623–38

227. Ten Hagen KG, Tran DT. 2002. *J. Biol. Chem.* 277:22616–22

228. Ju T, Canfield WM, Cummings RD. 2001. *Glycobiology* 11:886 (Abstr.)

229. Warren CE, Krizus A, Partridge EA, Dennis JW. 2001. *Glycobiology* 12:G8–9

230. Warren CE, Krizus A, Dennis JW. 2001. *Glycobiology* 11:979–88

231. Sandhoff K, Kolter T. 2003. *Philos. Trans. R. Soc. London Ser. B* 358:847–61

232. Kolter T, Proia RL, Sandhoff K. 2002. *J. Biol. Chem.* 277:25859–62

233. van Meer G, Wolthoorn J, Degroote S. 2003. *Philos. Trans. R. Soc. London Ser. B* 358:869–73

234. Ichikawa S, Ozawa K, Hirabayashi Y. 1998. *Biochem. Biophys. Res. Commun.* 253:707–11

235. Yamashita T, Wada R, Sasaki T, Deng C, Bierfreund U, et al. 1999. *Proc. Natl. Acad. Sci. USA* 96:9142–47

236. Coetzee T, Fujita N, Dupree J, Shi R, Blight A, et al. 1996. *Cell* 86:209–19

237. Bosio A, Binczek E, Stoffel W. 1996. *Proc. Natl. Acad. Sci. USA* 93:13280–85

238. Dupree JL, Coetzee T, Blight A, Suzuki K, Popko B. 1998. *J. Neurosci.* 18:1642–49

239. Fujimoto H, Tadano-Aritomi K, Tokumasu A, Ito K, Hikita T, et al. 2000. *J. Biol. Chem.* 275:22623–26

240. Honke K, Hirahara Y, Dupree J, Suzuki K, Popko B, et al. 2002. *Proc. Natl. Acad. Sci. USA* 99:4227–32

241. Takamiya K, Yamamoto A, Furukawa K, Yamashiro S, Shin M, et al. 1996. *Proc. Natl. Acad. Sci. USA* 93:10662–67

242. Sheikh KA, Sun J, Liu Y, Kawai H, Crawford TO, et al. 1999. *Proc. Natl. Acad. Sci. USA* 96:7532–37

243. Chiavegatto S, Sun J, Nelson RJ,

Schnaar RL. 2000. *Exp. Neurol.* 166: 227–34

244. Takamiya K, Yamamoto A, Furukawa K, Zhao J, Fukumoto S, et al. 1998. *Proc. Natl. Acad. Sci. USA* 95: 12147–52

245. Zhao J, Furukawa K, Fukumoto S, Okada M, Furugen R, et al. 1999. *J. Biol. Chem.* 274:13744–47

246. Kawai H, Allende ML, Wada R, Kono M, Sango K, et al. 2001. *J. Biol. Chem.* 276:6885–88

247. Yamashita T, Hashiramoto A, Haluzik M, Mizukami H, Beck S, et al. 2003. *Proc. Natl. Acad. Sci. USA* 100: 3445–49

248. Gerdt S, Dennis RD, Borgonie G, Schnabel R, Geyer R. 1999. *Eur. J. Biochem.* 266:952–63

249. Gerdt S, Lochnit G, Dennis RD, Geyer R. 1997. *Glycobiology* 7:265–75

250. Seppo A, Tiemeyer M. 2000. *Glycobiology* 10:751–60

251. Seppo A, Moreland M, Schweingruber H, Tiemeyer M. 2000. *Eur. J. Biochem.* 267:3549–58

252. Wandall HH, Pedersen JW, Park C, Levery SB, Pizette S, et al. 2003. *J. Biol. Chem.* 278:1411–14

253. Schwientek T, Keck B, Levery SB, Jensen MA, Pedersen JW, et al. 2002. *J. Biol. Chem.* 277:32421–29

254. Muller R, Altmann F, Zhou D, Hennet T. 2002. *J. Biol. Chem.* 277:32417–20

255. Goode S, Melnick M, Chou TB, Perrimon N. 1996. *Development* 122: 3863–79

256. Goode S, Morgan M, Liang YP, Mahowald AP. 1996. *Dev. Biol.* 178: 35–50

257. Lopez-Schier H, St Johnston D. 2001. *Genes Dev.* 15:1393–405

258. Hakomori S. 2000. *Glycoconj. J.* 17: 627–47

259. Hakomori SI. 2002. *Proc. Natl. Acad. Sci. USA* 99:225–32

260. Rietveld A, Neutz S, Simons K, Eaton S. 1999. *J. Biol. Chem.* 274:12049–54

261. Griffitts JS, Whitacre JL, Stevens DE, Aroian RV. 2001. *Science* 293:860–64

262. Rudd PM, Dwek RA. 1997. *Crit. Rev. Biochem. Mol. Biol.* 32:1–100

263. Dell A, Morris HR. 2001. *Science* 291: 2351–56

264. Haslam SM, Dell A. 2003. *Biochimie* 85:25–32

265. Nakamura Y, Haines N, Chen J, Okajima T, Furukawa K, et al. 2002. *J. Biol. Chem.* 277:46280–88

266. Vadaie N, Hulinsky RS, Jarvis DL. 2002. *Glycobiology* 12:589–97

267. Kim BT, Tsuchida K, Lincecum J, Kitagawa H, Bernfield M, Sugahara K. 2003. *J. Biol. Chem.* 278:9116–24

Annu. Rev. Biochem. 2004. 73:539–57
doi: 10.1146/annurev.biochem.73.011303.074048
Copyright © 2004 by Annual Reviews. All rights reserved
First published online as a Review in Advance on March 9, 2004

Structural Insights Into the Signal Recognition Particle

Jennifer A. Doudna[1] and Robert T. Batey[2]

[1]Department of Molecular and Cell Biology and Department of Chemistry, Howard Hughes Medical Institute, University of California at Berkeley, Berkeley, California 94705; email: doudna@uclink.berkeley.edu
[2]Department of Chemistry and Biochemistry, University of Colorado at Boulder, Boulder, Colorado 80309; email: robert.batey@colorado.edu

Key Words SRP, signal sequence, SRP54, SRP19, SRP receptor, SRP RNA

■ **Abstract** The signal recognition particle (SRP) directs integral membrane and secretory proteins to the cellular protein translocation machinery during translation. The SRP is an evolutionarily conserved RNA protein complex whose activities are regulated by GTP hydrolysis. Recent structural investigations of SRP functional domains and interactions provide new insights into the mechanisms of SRP activity in all cells, leading toward a comprehensive understanding of protein trafficking by this elegant pathway.

CONTENTS

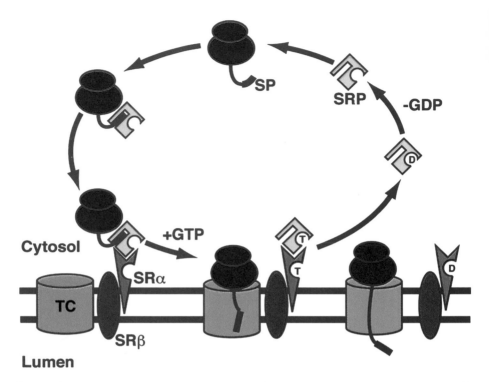

Figure 1   The eukaryotic SRP-dependent protein targeting cycle. A nascent polypeptide chain that is being actively translated by the ribosome (*red*) contains an amino-terminal signal sequence (SP) that is bound by the SRP (*yellow*), which arrests translation. The SRP-ribosome complex is targeted to the translocational complex (TC) embedded in the endoplasmic reticulum membrane via an interaction with the membrane-bound receptor complex SRα/SRβ. Following docking of the ribosome with the translocon, the signal sequence is released from the SRP, and the SRP is released from the SRP receptor in a GTP-dependent fashion.

INTRODUCTION

In all cells, the signal recognition particle (SRP) targets proteins destined for secretion or membrane insertion by binding to hydrophobic signal sequences at the N terminus of polypeptides as they emerge from ribosomes. In a process central to the ability of cells to communicate with other cells and the environment, the SRP recognizes ribosome-nascent chain complexes, docks with a specific membrane-bound SRP receptor, releases the ribosome-associated polypeptide into a translocon channel in the membrane, and dissociates from the SRP receptor primed for another cycle of protein targeting (Figure 1). GTP binding and hydrolysis by both the SRP and its receptor coordinate this process, suggesting that induced conformational changes enable ordered binding and release of the signal peptide, the ribosome, the SRP receptor, and the translocon. Interestingly, all cytoplasmic SRPs are ribonucleoproteins that consist of one RNA molecule and up to six proteins, a subset of which share sequence, structural, and functional homology. Structural studies of SRP components and complexes, together with genetic and biochemical experiments, have provided significant insights into the mechanism of SRP-mediated protein trafficking as well as the pathway of SRP assembly (reviewed in 1–5). Here we discuss recent advances in understanding how both the bacterial and mammalian SRP particles form functional complexes and how they bind specifically to SRP receptor proteins and the ribosome. These data make possible future experiments to understand the specificity of signal peptide recognition and the coupling of nucleotide hydrolysis with structural dynamics that drive the protein translocation cycle.

MECHANISM OF COTRANSLATIONAL PROTEIN TRAFFICKING

All cells localize secretory and integral membrane proteins, whose biosynthesis begins in the cytoplasm, to specialized pores that enable cotranslational protein export. The SRP is an RNA-protein complex that provides an evolutionarily conserved mechanism for protein trafficking by recognizing the hydrophobic signal sequence found on the N terminus of targeted proteins.

Phylogenetic Conservation

In mammalian cells, the SRP includes one RNA molecule (7S RNA) and six proteins named according to their molecular weight: SRP72, SRP68, SRP54, SRP19, SRP14, and SRP9. SRP54 and one stem-loop, helix 8 of the 7S RNA, are conserved in archaea and eubacteria, and together with a second RNA helical region (helix 6) constitute the signal peptide binding domain (S domain) of the SRP. SRP54 also includes a GTPase domain and an N-terminal helical domain that play roles in communicating the peptide-bound state of the SRP to the SRP receptor, the ribosome, and the translocon pore. SRP19 induces a structural change in the S domain of 7S RNA required for SRP54 binding. SRP72 and SRP68 form a heterodimeric subcomplex that binds the middle segment of the 7S RNA, whereas SRP14 and SRP9 bind cooperatively to the end opposite helix 8 in the SRP RNA to form the Alu domain responsible for transient translational arrest during protein targeting.

In eubacteria, SRP comprises the smaller 4.5S RNA and an SRP54 homolog called Ffh (Fifty-four homolog). All SRP RNAs include the highly conserved binding site for the SRP54 protein, called domain IV in bacterial SRP RNA; however, the less well conserved Alu domain is missing in gram-negative bacteria. Both Ffh and 4.5S RNA are essential genes in *Escherichia coli*, and

SRP54 and Ffh have been shown to be functional homologs. Furthermore, a truncated form of 4.5S RNA, which includes just the Ffh binding site, supports growth in a 4.5S RNA-depleted strain, showing that the critical function of the RNA is contained within the peptide-recognition domain of the ribonucleoprotein complex.

The SRP receptor consists of a conserved protein, SRα in mammals and FtsY in bacteria, with GTPase activity, sequence, and structural homology to the GTPase domain of SRP54/Ffh. In mammalian cells, SRα binds a second subunit, SRβ, containing a single transmembrane region. SRβ is also a GTPase but has only distant homology to the GTPases of SRP54 and SRα. Interestingly, mutation of the SRβ GTPase domain, but not deletion of the transmembrane domain, disrupts signal recognition (SR) function in vivo (6). FtsY weakly associates with the bacterial inner membrane, perhaps through direct interaction with membrane phospholipids (7–11).

Structural Features of the Conserved SRP Components

Crystal structures of the GTPase domains of *Thermus aquaticus* Ffh and *E. coli* FtsY provided the first structural insights into GTPase function in SRP (12, 13). Ffh and FtsY each contain three domains, two of which, the N and G domains, are related at both the sequence and structural level and comprise the GTPase of each protein (Figure 2*a*). The G domain adopts a classical GTPase fold in which four conserved sequence motifs (I-IV) are organized around the nucleotide binding site. Motif II is part of a sequence unique to the SRP GTPases called the insertion-box domain (IBD) that extends by two strands the central β-sheet of the domain. The amino-terminal N domain, a four-helix bundle, packs against the G domain to form a contiguous unit referred to as the NG domain (Figure 2*b*). Side chains from the C-terminal end of the G domain contribute to the hydrophobic core of the N domain, an interface conserved in Ffh and FtsY that creates an axis about which the relative orientations of the N and G domains vary. Nucleotide-dependent changes in the relative N and G domain positions are proposed to enable N domain detection of the GTP-bound state of the G domain (14–16).

Crystal structures of full-length *T. aquaticus* Ffh as well as the human SRP54 M domain revealed an all-helical domain featuring a prominent hydrophobic cleft comprising helices αM1, αM2, and αM4 and an extended flexible loop, the "finger loop," connecting αM1 and αM2 (Figure 3) (17, 18). Adjacent to this cleft, helices αM3 and αM4 form a classical helix-turn-helix (HTH) motif that contains a conserved sequence of serine, arginine, and glycine residues essential for high-affinity binding to SRP RNA. The crystal structure of the *E. coli* Ffh M domain bound to the phylogenetically conserved region of SRP RNA revealed that the HTH motif binds the distorted minor groove of the RNA (Figure 4*a*) (19). In the protein-RNA complex, nucleotides in the asymmetric loop of the RNA wrap around the outside of the helix and make specific contacts to the M domain. Comparison of the structure to that of the unbound RNA shows that a

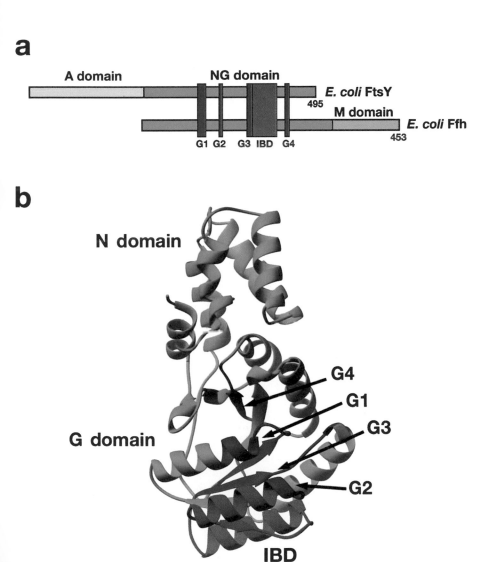

Figure 2 (*a*) Cartoon of the alignment of the *E. coli* FtsY and Ffh proteins. The figure emphasizes the conservation of the NG domain of each and includes the four conserved sequence motifs (G1-G4) and an insertion element (IBD that is unique to the SRP-associated ras-type GTPases. (*b*) Structure of the *E. coli* NG domain of the SRP receptor protein FtsY (PDB ID: 1FTS). The N domain represents the four helix bundle on the top and the ras-type GTPase domain (G domain) on the bottom. The four conserved sequence motifs in the Ffh/FtsY family of proteins (G1–G4) are highlighted in blue and the insertion-box domain (IBD) are highlighted in magenta.

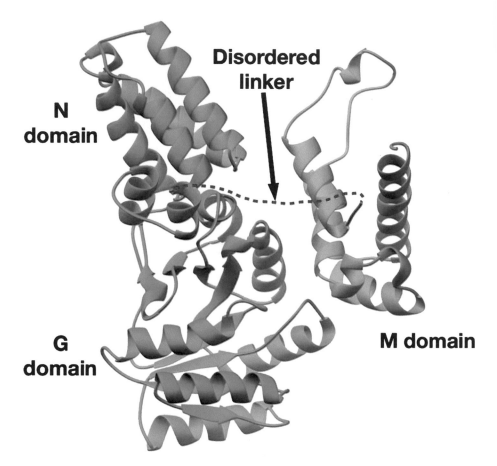

Figure 3   Crystal structure of the *Thermus aquaticus* Ffh protein (PDB ID: 2FFH). The two principal domains, the NG domain (*green*) and the M domain (*blue*), are shown. Ffh crystallized as a trimer of proteins in which a flexible linker between the two domains (residues 308–318) (shown as a *red dashed line*) was disordered, and thus the orientation of the two domains is ambiguous. Only one of the three possible pairs is shown.

significant conformational change is induced upon M domain binding, coupled to the ordered binding of metal ions and waters in the complex (19–23).

Role of GTP Hydrolysis

The protein targeting cycle is regulated by the coordinated action of GTPases, SRP54 and SRα/SRβ in eukaryotes and Ffh/FtsY in bacteria, that control signal peptide binding and release. In the GTP-bound state, SRP binds to a nascent signal peptide and the SRP receptor, leading to localization of the associated ribosome on the translocon. The GTPases in both SRP54/Ffh and SRα/FtsY are

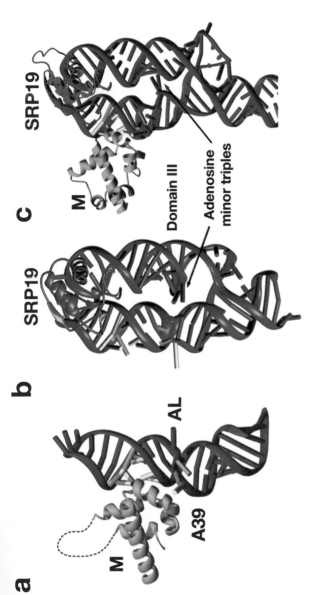

Figure 4 Structures of the signal recognition domain of the SRP from the three kingdoms of life. (*a*) Structure of the conserved domain of the *E. coli* 4.5S RNA (corresponding to domain IV of the eukaryotic SRP RNA) bound to Ffh M domain (PDB ID: 1DUL). A 33-amino acid segment between helices one and two of the M domain (*light blue*) was disordered in the structure (*dashed red line*). Nucleotides in the 4.5S RNA that are highly (*green*) and universally (*yellow*) conserved are highlighted along with A39 of the asymmetric internal loop (AL). (*b*) Structure of the *Methanococcus jannaschii* S domain RNA in complex with SRP19 (*magenta*) (PDB ID: 1LNG). Two adenosine residues in domain III, highlighted in red, contact the asymmetric loop of domain IV by forming A-minor base triples that potentially stabilize a conformation productive for M domain binding. (*c*) Structure of the ternary complex between the human S domain RNA/SRP19/SRP54 M domain (PDB ID: 1MFQ). The adenosines from the asymmetric loop that form A-minor triples with domain III are highlighted in red.

mutually stimulated upon complex formation and have been proposed to act as GAPs (GTPase activating proteins) for each other (24, 25). Following release of the signal peptide from SRP54/Ffh into the translocon, hydrolysis of SRP- and SR-bound GTP molecules causes dissociation of the SRP-SR complex and resumption of ribosome-catalyzed polypeptide synthesis. In eukaryotes, GTP hydrolysis by SRβ leads to dissociation of its complex with SRα. Key questions about the mechanism of SRP center on the role of GTP hydrolysis in coordinating and controlling the timing of these events in the cell.

The role of SRβ and why it is required in eukaryotic cells remains unclear, but some clues were provided by the crystal structure of SRβ bound to GTP and the N-terminal interaction domain of SRα (26). SRβ has nanomolar affinity for GTP (27), and in its GTP-bound state, it is catalytically inert when bound to SRα, requiring a GAP and a GEF (guanine-nucleotide exchange factor) to function as a GTPase switch for release of the SRα subunit (28). In the structure, an extensive intermolecular interface includes both polar and hydrophobic contacts with interdigitated side chains that produce a high-affinity complex (26). Recent data from a fluorescence nucleotide exchange assay show that the β subunit of the protein-conducting channel in the endoplasmic reticulum functions as the GEF for SRβ in yeast, and by analogy in other organisms as well (29). The nature of the SRβ GAP activity and how it correlates with ribosome and translocon binding and release await further biochemical and structural studies.

SIGNAL RECOGNITION PARTICLE ASSEMBLY

SRP comprises discrete domains corresponding to the translational arrest and peptide recognition functions of the particle. Results of structural and biochemical experiments highlight the topology of SRP as well as the hierarchy of interactions that result in a functional complex.

Structural and Functional Domains

Electron microscopy (EM) revealed that the eukaryotic SRP has an elongated rod-like structure ($\sim$240 x 60 Å) comprising 3 distinct regions (30, 31). The Alu domain and the S domain, responsible for translation arrest and signal sequence binding, respectively, lie at opposite ends of the complex. Connecting these functional centers is a low-mass region thought to be a flexible RNA linker. This global organization is consistent with simultaneous binding of the elongation arrest domain at the ribosomal subunit interface and binding of the signal recognition domain near the peptidyl exit site on the large ribosomal subunit. In crystal structures of the murine Alu domain, which include the Alu region of the SRP RNA bound to the SRP9/14 heterodimer, the RNA forms a U-turn that connects two helical stacks (32, 33). In contrast to earlier models, however, this structure does not resemble that of tRNA, implying that molecular mimicry is no the mechanism of ribosome binding or elongation arrest. Part of the structural organization of the S domain of the SRP has recently been revealed in several crystal structures, which culminate with the human S domain RNA/SRP19/SRP54 M domain complex (34–36). The cryo-EM reconstruction of the mammalian SRP bound to the 80S ribosome is eagerly awaited to reveal the mechanism of simultaneous translational arrest and signal peptide binding by the SRP (89).

Assembly of the Signal Sequence Binding Domain

The universally conserved ribonucleoprotein core of the cytosolic SRP comprises the SRP54/Ffh protein and domain IV of the SRP RNA. In bacteria, these two components interact through the methionine-rich M domain of SRP54/Ffh to form a functional enzyme. The crystal structure of an *E. coli* SRP RNA/Ffh M domain complex (19) revealed contacts between two internal loops in the RNA and a series of strictly conserved amino acids in the M domain (Figure 4a). Strikingly, the crystal structure of a ternary complex between the human M-domain/SRP-19/S-domain RNA showed nearly identical architecture in the protein and RNA at this site of contact as well as an identical set of contacts between the two (36). Clearly, this interface remained constant over evolutionary time.

The first internal loop, symmetric in all SRP RNAs, contains six universally conserved nucleotides that form three noncanonical base pairs. Functional group mutagenesis showed that disruption of either of two of these pairs, a sheared G-G pair or a reverse Hoogsteen A-C pair, abolishes protein binding (22, 37); similarly, a single point mutation in this loop is lethal in *E. coli*. In contrast to the symmetric loop, the sequence of the asymmetric internal loop of the SRP RNA is more variable (38, 39). In both the *E. coli* and human SRP complexes, the asymmetric loop presents a 5′-side adenosine base to the M domain for extensive recognition by three universally conserved amino acids (19, 36). Because all phylogenetic variants of the SRP RNA have at least one adenosine on the 5′ side of the asymmetric loop, this set of contacts is probably universally conserved.

To overcome unfavorable electrostatic consequences of extruding the adenosine, a series of metal ions interact specifically with the major groove of the asymmetric internal loop (23). An extensive hydrogen bonding network between water, three cations, and the RNA, observed crystallographically (19), is essential for stability; removal of all metal cations from the binding reaction reduced the binding affinity of the complex by at least 10^6-fold (23). Although metal ions were observed throughout the major groove of the RNA, the metal ions in the asymmetric loop appear to stabilize the protein-RNA complex. For example, Mn^{2+} preferentially stabilizes the complex compared to Mg^{2+} by enabling formation of an additional A-A pair that likely further stabilizes the bound conformation (23). Additionally, Cs^+ enhances stability relative to other monovalent cations, in contrast to its behavior in other RNAs (40) and protein-RNA complexes (41), because Cs^+ binds at the site where the backbone comes into close contact with itself and presumably alleviates unfavorable electrostatic interactions (23). Thus, the two metals that most stabilize the protein-RNA complex both affect the structure of the asymmetric internal loop.

Archaea and eukarya require the presence of a second protein, SRP19, for the efficient binding of SRP54 to the SRP RNA. This protein binds to another domain of the SRP RNA, domain III, and to the conserved GNRA tetraloop of domain IV (35) (Figure 4b). A solution structure of SRP19 alone (42), as well as crystal structures the S-domain RNA bound to SRP 19 (34, 35, 43), and SRP19/SRP54-M domain (36), show that cooperative assembly occurs through the stabilization of the bound form of the asymmetric internal loop motif via A-minor base triples (37, 44) with a second RNA helix (helix 6) (Figure 4b,c) (45–47). As in the *E. coli* structure, magnesium ions bind the asymmetric loop, presumably further stabilizing this region of the RNA. Thus, cooperative assembly of this RNP appears to be driven by the formation of new RNA-RNA contacts rather than protein-protein contacts, similar to the assembly of the central domain of the 30S ribosomal subunit (48, 49).

Although their interactions with SRP RNA are nearly identical, Ffh and SRP54 in the bacterial and human SRP complexes, respectively, use different strategies to stabilize the critical asymmetric loop. This begs the question as to why the SRP19/domain III-mediated assembly arose. Currently no other role for SRP19 in SRP function has been determined, suggesting that the primary role for this protein in the SRP is facilitating its assembly and stabilizing the intact particle. Thus, the domain III/SRP19 extension may be an evolutionary adaptation of the SRP to enhance or control the kinetics of assembly that cannot be achieved by metals alone.

THE SIGNAL RECOGNITION PARTICLE CYCLE

Nascent protein targeting involves a choreographed cycle of signal peptide binding and release coupled to GTP hydrolysis and interactions with the ribosome and translocon pore. Exciting recent advances in understanding GTPase activity and SRP-ribosome interactions have come from x-ray crystallography, cryo-electron microscopy and mutagenesis studies.

Signal Sequence Recognition

Though a number of structures of various domains of the SRP have revealed a wealth of information about protein-RNA, protein-protein, and protein-nucleotide interactions, few insights have emerged as to how the most critical ligand of all—the signal sequence of the protein to be targeted—is specifically recognized. A typical signal sequence comprises 9–12 large hydrophobic residues in a row (50) that adopt an α-helical conformation (51–53). In eukarya, these signals are typically found at the N terminus of the protein; however, in *E. coli* the SRP-dependent signal is often a transmembrane helix within inner membrane proteins (54–57). The SRP appears to recognize any sequence that bears a critical level of hydrophobicity, though flanking basic residues are also important (52, 58). Currently unanswered is the fascinating question of how the SRP recognizes and productively binds almost any such hydrophobic α-helix.

Early studies of target recognition by the SRP suggested that nascent polypeptide chains bearing signal sequences cross-linked to SRP54/Ffh through the M domain and that the M domain was sufficient to mediate this interaction (59, 60). From these data, the "methionine bristle" hypothesis was proposed in which a flexible, methionine-rich pocket is used by the SRP to recognize almost any given signal sequence (61). Consistent with this hypothesis, structures of the M domain in the free and RNA-bound states show that the methionine-rich region of the protein, the presumed signal binding site, is conformationally flexible (Figure 5). In the structure of the *T. aquaticus* M domain (17), this region is involved in extensive crystal packing contacts with the signal binding site of an adjacent protein in the crystal lattice (Figure 5*b*). Consequently, it takes on a β-hairpin like structure, called the finger loop, that is partially inserted into a neighboring hydrophobic groove. Because authentic signal sequences are probably α-helical, the authors contend that the structure illustrates the inherent flexibility rather than the signal sequence binding mechanism of the finger loop. Underscoring its flexibility, this region of the M domain was entirely disordered in the *E. coli* M domain-4.5S RNA complex structure (Figure 5*a*) (19, 22). This may represent the true state of the signal recognition site in the SRP in solution, consistent with the conformational flexibility required for productive binding of heterogeneous targets.

Crystal structures of the human M domain present a different picture of how the signal sequence potentially interacts with the SRP. In the absence (18) and presence of SRP-RNA (36), the first helix of the M domain undergoes a "domain swap" with an adjacent protein (Figure 5*c*). This has been observed in the crystal structures of other proteins (62, 63) and suggests weak interactions between this helix and the rest of the M domain. The result of the domain swap is that helix one (h1′, Figure 5*c*) packs into a shallow, moderately hydrophobic groove of an adjacent molecule. Thus, in an alternative model of how the M domain may recognize signal sequences, helix 1′ (h1′) occupies nearly the same position as helix 1 in the *E. coli* and *T. aquaticus* structures, although helix 1 represents the signal peptide (Figure 5*d*).

These models for peptide recognition leave unanswered the question concerning the involvement of the NG domain and the RNA in target recognition by the SRP. Even though the M domain can bind the signal on its own or in complex with SRP RNA, the affinity of the interaction was improved by the NG domain, and the NG domain can be cross-linked to signal peptides in solution (64, 65). Whereas biochemical and crystallographic studies indicate that the NG and M domains of Ffh are loosely associated in the absence of a bound signal sequence and/or GTP (17, 66), a recent crystal structure of the intact SRP54 protein

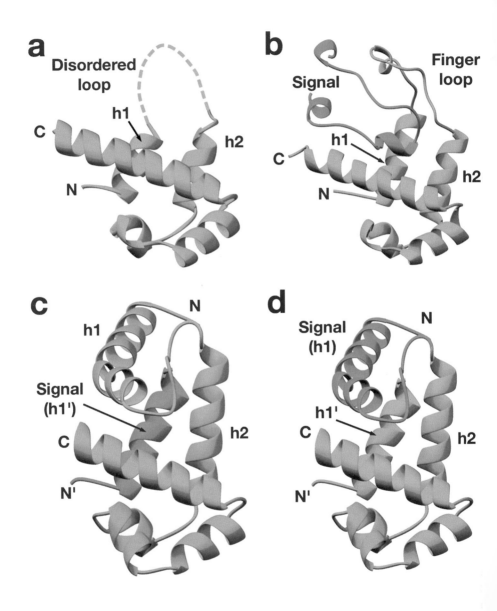

provides clues to interdomain communication that may occur upon signal binding (see below). It has also been proposed that the SRP RNA has a role in signal recognition via electrostatic interactions between the backbone of the RNA and positively charged residues adjacent to the hydrophobic sequence (19), a hypothesis supported by mutagenesis and in vivo experiments (58, 67). Although these data provide tantalizing clues to the mechanism of protein targeting by the SRP, a structure of the SRP-signal complex will be an important step toward an atomic-level understanding of SRP function.

Structure of the SRP54-RNA Complex

SRP54 is the only protein subunit conserved in all SRPs and controls communication with the SRP receptor, the ribosome, and the translocon. In the crystal structure of *T. aquaticus* SRP54, the linker region between the G and M domains was disordered and hence provided no information about the three-dimensional domain arrangement of SRP54 or its organization in complex with SRP RNA. Recent determination of structures of SRP54 from the archaeon *Sulfolobus solfataricus* alone and complexed with helix 8 of SRP RNA reveal the architecture of the complex and a hydrophobic contact between the M and N domains, suggesting a possible mechanism for interdomain communication (67a). The structures, solved by molecular replacement at a resolution of ~4 Å, reveal an L-shaped protein in which the NG domain represents the long arm and the M domain represents the short arm. Helix 8 of SRP RNA lies parallel to the long axis of the NG domain, giving the complex an overall U shape (Figure 6). Only one region of interaction, which involves a short stretch of hydrophobic contacts between the loop connecting helices αN3 and αN4 at the distal end of the N domain, the N-terminal region of αML, and the C-terminal region of the short α helix αM1b adjacent to the finger loop, is observed between the N and M domains (Figure 6). The high degree of evolutionary conservation of these

---

Figure 5 Potential models of signal sequence recognition by the SRP54/Ffh M domain. (*a*) The hydrophobic, methionine-rich pocket in the *E. coli* M domain (PDB ID: 1DUL) is significantly disordered (*dashed line*), suggesting a conformationally dynamic binding site that can recognize a variety of sequences. (*b*) The conformation of the finger loop between helices one and two of the *T. aquaticus* M domain structure (PDB ID: 2FFH) is stabilized through its interaction with a significantly hydrophobic groove of an adjacent M domain in the crystal (note that the signal corresponds to the finger loop). (*c*) A model of signal recognition proposed from the structure of the human SRP M domain (PDB ID: 1QB2). Helix one (h1′) of an adjacent M domain in the crystal structure packs into a shallow, moderately hydrophobic groove formed between helices one and two (h1, h2). (*d*) An alternative model for signal sequence binding by the human M domain in which h1′ is considered part of the domain structure by virtue of its close superposition with helix 1 of the *E. coli* and *T. aquaticus* structures and helix 1 represents the signal sequence.

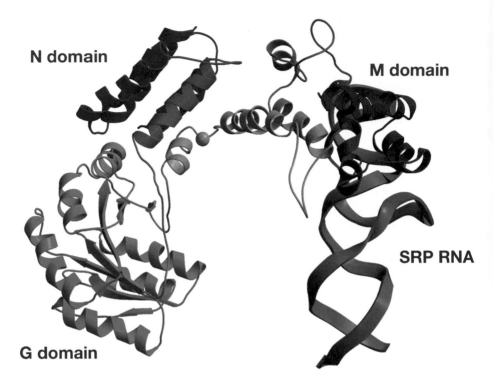

Figure 6   Crystal structure of a complex between *Sulfolobus solfataricus* SRP domain IV RNA and SRP54 (67a). The N domain of SRP54 (*blue*) contacts the N-terminal helix of the M domain (*red*), while the M domain forms a complex with the SRP RNA (*gray*) in a fashion similar to that of the bacterial and eukaryal variants. Figure courtesy of I. Sinning.

residues suggests a functional role of the contact, perhaps as a greasy hinge enabling interdomain flexibility in SRP54. Although the relative orientations of the NG and M domains are not affected by RNA binding, shape and charge complementarity between the phosphate backbone of the RNA minor groove and residues 121–126 of the G domain suggest a possible contact that might be induced at some stage in the SRP cycle; this would enable communication between the peptide binding and GTPase functionalities of the particle.

GTPase Stimulation in the SRP-SR Complex

Two central questions about the mechanism of SRP-mediated protein targeting are how GTPase activities in the SRP54/Ffh and SRα/FtsY proteins are enhanced upon heterodimerization of the SRP with its receptor, and how GTP hydrolysis is coupled to peptide binding and release. Crystal structures of the homologous GTPase domains of these two proteins, determined in their apo- and nucleotide-bound states, revealed a common two-domain NG fold comprising an α-helical

N domain packed against a G domain with a fold similar to those of other members of the GTPase superfamily. Structural similarity between the G domains of Ffh and FtsY and dimeric ATP-utilizing proteins led to a model for the Ffh-FtsY complex in which the G domains dimerize in an antiparallel orientation (68). Similar to other GTPases, four conserved motifs, I-IV, comprise the SRP GTPases and include residues directly involved in GTP binding and hydrolysis. In addition, the G domains of both Ffh and FtsY contain a subdomain, termed the insertion box domain (IBD), that extends the core α/β fold along the face distal to the N domain and was thought to provide the site of interaction in the targeting complex (12, 14, 69). Despite their similarities, the SRP GTPases exhibit several distinct properties relative to other GTPases, which include low affinity and rapid exchange of both GDP and GTP (70). GTP hydrolysis in an Ffh/FtsY complex is stimulated about 10-fold above that observed in either protein alone.

A significant advance in understanding GTPase activation has come from recent structure determinations of a complex between the Ffh and FtsY NG domains (70a, 70b). The two proteins associate longitudinally along the N and G domains, bringing the two active sites into direct contact to form a contiguous catalytic chamber that contains two bound GMPPCP molecules, two hydrated magnesium ions, and several waters. The contact region between the domains comprises most of the conserved sequence motifs of the SRP GTPases and buries ~ 1800 Å^2 on each protein surface. Mutation of residues across the interface surface in FtsY disrupts GTPase activity in the complex, confirming the importance of the crystallographically observed interaction (70a). Formation of the Ffh-FtsY interface requires several conformational changes relative to the structures of the domains alone, including a rigid body motion of the N and G domains that translates the distal loops of the N domain by ~ 11–12 Å. This movement is accompanied by a substantial shift in the position of the C-terminal helix, leading to speculation that the N domain functions as a sensor of targeting complex formation by triggering a change in the relative orientations of the M-domain and A-domain that are found on the C-terminal ends of the full-length Ffh and FtsY proteins, respectively. Conserved arginine sidechains in each protein that might serve as arginine fingers, analogous to the Ras/GAP structure, are oriented asymmetrically such that only the Arg in Ffh is positioned like the Arg supplied by the RasGAP. Possibly, the arrangement of arginines alternates within the chamber, perhaps contributing sequentially to hydrolysis of the two bound nucleotides.

Ribosome Interactions

In the mammalian system, SRP54 cross-links to two ribosomal proteins, L23a and L35, located near the polypeptide exit site on the ribosome (71). A similar interaction is observed in eubacteria, where Ffh/4.5S RNA complexes can be cross-linked to L23 (72, 73). Interestingly, protein L23 also cross-links to a chaperone protein, trigger factor (TF), though L23-SRP and L23-TF interactions appear to be mutually exclusive. These data suggest that L23 may play a role in directing nascent polypeptides into the translocation machinery, though this hypothesis remains to be tested in vivo. The ribosome somehow induces a structural change in SRP54 that leads to increased GTP binding affinity, as well as increased affinity for the ribosome itself (74, 75). One possibility, suggested by the SRP54/RNA structure, is that this structural rearrangement corresponds to rotation of the NG domain with respect to the M domain that may occur upon signal peptide binding, with consequent opening of the finger loop similar to the conformation observed in the *T. aquaticus* Ffh structure (17). Signal peptide binding to the M domain might therefore result in a similar structural rearrangement in the NG domain interface as occurs upon GTP binding to the G domain, effectively linking signal sequence binding to the M domain with GTP binding to the G domain.

Nucleotide Exchange

The molecular mechanism of reciprocal GTPase activity in SRP and its receptor remains poorly understood. Recent evidence indicates that structural changes induced in the bacterial SRP receptor, FtsY, upon formation of the SRP-FtsY complex enhance nucleotide binding specificity in FtsY (76). Mutagenesis studies support a similar weak-binding affinity in the eukaryotic homolog SRα (77, 78). Why might this occur? One idea is that loosely bound GTP in free FtsY would prevent futile cycles of GTP hydrolysis in the substantial fraction of free FtsY in the cytosol (79). In this way, nucleotide hydrolysis would be coupled to binding of SRP and presumably to nascent signal peptides.

FUNCTIONS OF SIGNAL RECOGNITION PARTICLE RNA

Why does cellular SRP include an essential RNA, and what does the RNA contribute to SRP function as well as to other physiological activities in the cell? Most of the evidence to date addressing these questions comes from studies in the *E. coli* system. In vitro, the 4.5S RNA appears to stabilize the structure of the Ffh M domain, as indicated by circular dichroism and proteolysis experiments (66, 80), and studies on the kinetics of Ffh-FtsY complex formation show that 4.5S RNA enhances both association and dissociation of the complex (81, 82). Intriguingly, several lines of evidence support an additional role for 4.5S RNA in translation on the ribosome, because of the observation that the deleterious effects of 4.5S RNA depletion can be suppressed by mutations in translation factor EF-G or in the 16S or 23S ribosomal RNAs (83–86). Recent evidence implies that although the SRP RNA interaction with EF-G homologs is conserved in archaea, the essential activity of SRP RNA is in fact as part of the signal-sequence binding particle rather than on the ribosome (90). It is intriguing to note that in chloroplasts the SRP RNA has apparently been replaced by a protein, indicating that study of this SRP may provide clues to the role of the RNA (87, 88).

FUTURE DIRECTIONS

With many structures of individual components of the SRP pathway now known, attention is focused on understanding how these molecules interact to enable efficient protein targeting. How signal peptide binding is achieved, how peptide binding and release is controlled, and how the SRP coordinates interactions with its receptor, the ribosome, and the translocon remain fascinating unanswered questions. A combination of genetic, biochemical, and structural approaches will be required to address these issues and fully illuminate the function of one of the most ancient of the cellular ribonucleoproteins.

ACKNOWLEDGMENTS

We thank Irmgard Sinning for sharing unpublished data and for preparation of Figure 6. This work was supported in part by the NIH (GM 22778 to J.A.D.).

The *Annual Review of Biochemistry* is online at http://biochem.annualreviews.org

LITERATURE CITED

1. Keenan RJ, Freymann DM, Stroud RM, Walter P. 2001. *Annu. Rev. Biochem.* 70: 755–75
2. Driessen AJ, Manting EH, van der Does C. 2001. *Nat. Struct. Biol.* 8: 492–98
3. Eichler J, Moll R. 2001. *Trends Microbiol.* 9:130–36
4. Sauer-Eriksson AE, Hainzl T. 2003. *Curr. Opin. Struct. Biol.* 13:64–70
5. Nagai K, Oubridge C, Kuglstatter A, Menichelli E, Isel C, Jovine L. 2003. *EMBO J.* 22:3479–85
6. Fulga TA, Sinning I, Dobberstein B, Pool MR. 2001. *EMBO J.* 20:2338–47
7. de Leeuw E, Poland D, Mol O, Sinning I, ten Hagen-Jongman CM, et al. 1997. *FEBS Lett.* 416:225–29
8. de Leeuw E, te Kaat K, Moser C, Menestrina G, Demel R, et al. 2000. *EMBO J.* 19:531–41
9. Herskovits AA, Bibi E. 2000. *Proc. Natl. Acad. Sci. USA* 97:4621–26
10. Herskovits AA, Seluanov A, Rajsbaum R, ten Hagen-Jongman CM, Henrichs T, et al. 2001. *EMBO Rep.* 2:1040–46
11. Moll RG. 2003. *Biochem. J.* 374:247–54
12. Freymann DM, Keenan RJ, Stroud RM, Walter P. 1997. *Nature* 385:361–64
13. Montoya G, Svensson C, Luirink J, Sinning I. 1997. *Nature* 385:365–68
14. Freymann DM, Keenan RJ, Stroud RM, Walter P. 1999. *Nat. Struct. Biol.* 6: 793–801
15. Ramirez UD, Minasov G, Focia PJ, Stroud RM, Walter P, et al. 2002. *J. Mol. Biol.* 320:783–99
16. Shepotinovskaya IV, Freymann DM. 2002. *Biochim. Biophys. Acta* 1597: 107–14
17. Keenan RJ, Freymann DM, Walter P, Stroud RM. 1998. *Cell* 94:181–91

18. Clemons WM Jr, Gowda K, Black SD, Zwieb C, Ramakrishnan V. 1999. *J. Mol. Biol.* 292:697–705

19. Batey RT, Rambo RP, Lucast L, Rha B, Doudna JA. 2000. *Science* 287:1232–39

20. Schmitz U, James TL, Lukavsky P, Walter P. 1999. *Nat. Struct. Biol.* 6:634–38

21. Jovine L, Hainzl T, Oubridge C, Scott WG, Li J, et al. 2000. *Struct. Fold. Des.* 8:527–40

22. Batey RT, Sagar MB, Doudna JA. 2001. *J. Mol. Biol.* 307:229–46

23. Batey RT, Doudna JA. 2002. *Biochemistry* 41:11703–10

24. Miller JD, Wilhelm H, Gierasch L, Gilmore R, Walter P. 1993. *Nature* 366:351–54

25. Powers T, Walter P. 1995. *Science* 269:1422–24

26. Schwartz T, Blobel G. 2003. *Cell* 112:793–803

27. Bacher G, Pool M, Dobberstein B. 1999. *J. Cell Biol.* 146:723–30

28. Legate KR, Falcone D, Andrews DW. 2000. *J. Biol. Chem.* 275:27439–46

29. Helmers J, Schmidt D, Glavy JS, Blobel G, Schwartz T. 2003. *J. Biol. Chem.* 278:23686–90

30. Andrews DW, Walter P, Ottensmeyer FP. 1985. *Proc. Natl. Acad. Sci. USA* 82:785–89

31. Andrews DW, Walter P, Ottensmeyer FP. 1987. *EMBO J.* 6:3471–77

32. Weichenrieder O, Wild K, Strub K, Cusack S. 2000. *Nature* 408:167–73

33. Weichenrieder O, Stehlin C, Kapp U, Birse DE, Timmins PA, et al. 2001. *RNA* 7:731–40

34. Wild K, Sinning I, Cusack S. 2001. *Science* 294:598–601

35. Hainzl T, Huang S, Sauer-Eriksson AE. 2002. *Nature* 417:767–71

36. Kuglstatter A, Oubridge C, Nagai K. 2002. *Nat. Struct. Biol.* 9:740–44

37. Doherty EA, Batey RT, Masquida B, Doudna JA. 2001. *Nat. Struct. Biol.* 8:339–43

38. Regalia M, Rosenblad MA, Samuelsson T. 2002. *Nucleic Acids Res.* 30:3368–77

39. Rosenblad MA, Gorodkin J, Knudsen B, Zwieb C, Samuelsson T. 2003. *Nucleic Acids Res.* 31:363–64

40. Basu S, Rambo RP, Strauss-Soukup J, Cate JH, Ferre-D'Amare AR, et al. 1998. *Nat. Struct. Biol.* 5:986–92

41. Conn GL, Gittis AG, Lattman EE, Misra VK, Draper DE. 2002. *J. Mol. Biol.* 318:963–73

42. Pakhomova ON, Deep S, Huang Q, Zwieb C, Hinck AP. 2002. *J. Mol. Biol.* 317:145–58

43. Oubridge C, Kuglstatter A, Jovine L, Nagai K. 2002. *Mol. Cell* 9:1251–61

44. Nissen P, Ippolito JA, Ban N, Moore PB, Steitz TA. 2001. *Proc. Natl. Acad. Sci. USA* 98:4899–903

45. Diener JL, Wilson C. 2000. *Biochemistry* 39:12862–74

46. Rose MA, Weeks KM. 2001. *Nat. Struct. Biol.* 8:515–20

47. Yin J, Yang CH, Zwieb C. 2001. *RNA* 7:1389–96

48. Agalarov SC, Williamson JR. 2000. *RNA* 6:402–8

49. Agalarov SC, Prasad GS, Funke PM, Stout CD, Williamson JR. 2000. *Science* 288:107–13

50. Valent QA, Kendall DA, High S, Kusters R, Oudega B, Luirink J. 1995. *EMBO J.* 14:5494–505

51. McKnight CJ, Rafalski M, Gierasch LM. 1991. *Biochemistry* 30:6241–46

52. Lee HC, Bernstein HD. 2001. *Proc. Natl. Acad. Sci. USA* 98:3471–76

53. Adams H, Scotti PA, De Cock H, Luirink J, Tommassen J. 2002. *Eur. J. Biochem.* 269:5564–71

54. Ulbrandt ND, Newitt JA, Bernstein HD. 1997. *Cell* 88:187–96

55. Beck K, Wu LF, Brunner J, Muller M. 2000. *EMBO J.* 19:134–43

56. Park SK, Jiang F, Dalbey RE, Phillips GJ. 2002. *J. Bacteriol.* 184:2642–53

57. Sijbrandi R, Urbanus ML, ten Hagen-

Jongman CM, Bernstein HD, Oudega B, et al. 2003. *J. Biol. Chem.* 278:4654–59

58. Peterson JH, Woolhead CA, Bernstein HD. 2003. *J. Biol. Chem.* 278:46155–62

59. Romisch K, Webb J, Lingelbach K, Gausepohl H, Dobberstein B. 1990. *J. Cell Biol.* 111:1793–802

60. Zopf D, Bernstein HD, Johnson AE, Walter P. 1990. *EMBO J.* 9:4511–17

61. Bernstein HD, Poritz MA, Strub K, Hoben PJ, Brenner S, Walter P. 1989. *Nature* 340:482–86

62. Clemons WM Jr, Davies C, White SW, Ramakrishnan V. 1998. *Structure* 6:429–38

63. Liu Y, Eisenberg D. 2002. *Protein Sci.* 11:1285–99

64. Newitt JA, Bernstein HD. 1997. *Eur. J. Biochem.* 245:720–29

65. Cleverley RM, Gierasch LM. 2002. *J. Biol. Chem.* 277:46763–68

66. Zheng N, Gierasch LM. 1997. *Mol. Cell* 1:79–87

67. Huang Q, Abdulrahman S, Yin J, Zwieb C. 2002. *Biochemistry* 41:11362–71

67a. Rosendal, KR, Wild, K, Montoya, G, Sinning, I. 2003. *Proc. Natl. Acad. Sci. USA* 100:14701–6

68. Montoya G, Kaat K, Moll R, Schafer G, Sinning I. 2000. *Struct. Fold. Des.* 8:515–25

69. Padmanabhan S, Freymann DM. 2001. *Structure* 9:859–67

70. Lu Y, Qi HY, Hyndman JB, Ulbrandt ND, Teplyakov A, et al. 2001. *EMBO J.* 20:6724–34

70a. Egea, PF, Shan, S, Napetschnig, J, Savage, DF, Walter, P, Stroud, RM. 2004. *Nature* 427:215–21

70b. Focia, PJ, Shepotinovskaya, IV, Seidler, JA, Freymann, DM. 2004. *Science* 303:373–7

71. Pool MR, Stumm J, Fulga TA, Sinning I, Dobberstein B. 2002. *Science* 297:1345–48

72. Gu SQ, Peske F, Wieden HJ, Rodnina MV, Wintermeyer W. 2003. *RNA* 9:566–73

73. Ullers RS, Houben EN, Raine A, ten Hagen-Jongman CM, Ehrenberg M, et al. 2003. *J. Cell Biol.* 161:679–84

74. Bacher G, Lutcke H, Jungnickel B, Rapoport TA, Dobberstein B. 1996. *Nature* 381:248–51

75. Flanagan JJ, Chen JC, Miao Y, Shao Y, Lin J, et al. 2003. *J. Biol. Chem.* 278:18628–37

76. Shan SO, Walter P. 2003. *Proc. Natl. Acad. Sci. USA* 100:4480–85

77. Kusters R, Lentzen G, Eppens E, van Geel A, van der Weijden CC, et al. 1995. *FEBS Lett.* 372:253–58

78. Rapiejko PJ, Gilmore R. 1997. *Cell* 89:703–13

79. Luirink J, ten Hagen-Jongman CM, van der Weijden CC, Oudega B, High S, et al. 1994. *EMBO J.* 13:2289–96

80. Cleverley RM, Zheng N, Gierasch LM. 2001. *J. Biol. Chem.* 276:19327–31

81. Peluso P, Herschlag D, Nock S, Freymann DM, Johnson AE, Walter P. 2000. *Science* 288:1640–43

82. Peluso P, Shan SO, Nock S, Herschlag D, Walter P. 2001. *Biochemistry* 40:15224–33

83. Brown S. 1987. *Cell* 49:825–33

84. Brown S, Thon G, Tolentino E. 1989. *J. Bacteriol.* 171:6517–20

85. Brunelli CA, O'Connor M, Dahlberg AE. 2002. *FEBS Lett.* 514:44–48

86. Rinke-Appel J, Osswald M, von Knoblauch K, Mueller F, Brimacombe R, et al. 2002. *RNA* 8:612–25

87. Eichacker LA, Henry R. 2001. *Biochim. Biophys. Acta* 1541:120–34

88. Groves MR, Mant A, Kuhn A, Koch J, Dubel S, et al. 2001. *J. Biol. Chem.* 276:27778–86

89. Halic M, Becker T, Pool MR, Spahn CM, Grassucci RA, Frank J, Beckmann R. 2004. *Nature* 427:808–14

90. Sagar MB, Lucast L, Doudna JA. 2004. *RNA*. In press

Annu. Rev. Biochem. 2004. 73:559–87
doi: 10.1146/annurev.biochem.73.011303.073954
First published online as a Review in Advance on March 9, 2004

PALMITOYLATION OF INTRACELLULAR SIGNALING PROTEINS: Regulation and Function

Jessica E. Smotrys and Maurine E. Linder

Department of Cell Biology and Physiology, Washington University School of Medicine, St. Louis, Missouri 63110; email: jessica.smotrys@cellbiology.wustl.edu, mlinder@cellbiology.wustl.edu

Key Words thioacylation, protein acyltransferase, lipid raft, lipid modification, acyl-protein thioesterase

■ **Abstract** Protein S-palmitoylation is the thioester linkage of long-chain fatty acids to cysteine residues in proteins. Addition of palmitate to proteins facilitates their membrane interactions and trafficking, and it modulates protein-protein interactions and enzyme activity. The reversibility of palmitoylation makes it an attractive mechanism for regulating protein activity, and this feature has generated intensive investigation of this modification. The regulation of palmitoylation occurs through the actions of protein acyltransferases and protein acylthioesterases. Identification of the protein acyltransferases Erf2/Erf4 and Akr1 in yeast has provided new insight into the palmitoylation reaction. These molecules work in concert with thioesterases, such as acyl-protein thioesterase 1, to regulate the palmitoylation status of numerous signaling molecules, ultimately influencing their function. This review discusses the function and regulation of protein palmitoylation, focusing on intracellular proteins that participate in cell signaling or protein trafficking.

CONTENTS

BACKGROUND AND SCOPE

Protein modification with fatty acids is a universal feature of eukaryotic cells. Historically, fatty acylation of eukaryotic proteins was divided into two classes: cotranslational addition of myristate to N-terminal glycine through amide linkage (myristoylation) and posttranslational addition of palmitate through thioester linkage to cysteine (palmitoylation) (1). However, fatty acylation of eukaryotic proteins is more diverse, and this categorization has been expanded to include modification of other amino acid residues with amide- and ester-linked fatty acids.

Protein N-myristoylation is the covalent attachment of myristate, a 14-carbon saturated fatty acid, to N-terminal glycine residues. The reaction is catalyzed by N-myristoyltransferase. Typically this occurs cotranslationally after the initiating methionine is cleaved by an aminopeptidase, but it can occur posttranslationally when an internal glycine residue is exposed by proteolytic cleavage (2). The enzymology and function of N-myristoylation have been well characterized, and a number of reviews are available (3–5).

S-palmitoylation refers to the addition of palmitate (C16:0) to cysteine residues through thioester linkage (6). Fatty acids thioesterified to proteins include fatty acids other than palmitate. These can be saturated, monounsaturated, and polyunsaturated species of varying chain lengths (C14 or longer). Thus, S-acylation and thioacylation are more general terms used to describe this process. In this review, the traditional term palmitoylation will be used for S-acylation. The enzymology of protein S-acylation is an emerging field and is covered in this review.

N-palmitoylation was first described for Sonic hedgehog, a secreted signaling protein (7). Palmitate modifies Hedgehog at the N-terminal cysteine residue through an amide linkage. An acyltransferase that palmitoylates Hedgehog has recently been identified in *Drosophila* (8–10). Lipid modifications of Hedgehog and other secreted signaling proteins are reviewed elsewhere in this volume (11).

Fatty acylation also occurs through oxyester attachment of palmitate or other fatty acids to serine or threonine. Oxyester-linked octanoate (C8:0) is found attached to serine in ghrelin, a growth-hormone-releasing peptide from stomach (12). O-palmitoyl threonine is found at the carboxyl terminus of an insect toxin derived from the spider *Plectreurys tristes* (13).

Finally, several secreted proteins are modified at the ϵ-amino group of lysine with myristate or palmitate through an amide linkage. Fatty acylation at lysine occurs both in prokaryotic and eukaryotic cells. Lysine palmitoylation is found in bacterial toxins (14), whereas lysine myristoylation has been reported for two cytokine precursor proteins, tumor necrosis factor (15) and interleukin 1α (16). Although a bacterial acyltransferase for lysine palmitoylation has been identified (14), there is little information about the enzymes that acylate serine, threonine, and lysine residues in eukaryotic cells.

Palmitoylation is a widespread modification found almost exclusively on membrane proteins. The family of proteins modified with thioester-linked palmitate is large and diverse. It includes transmembrane-spanning proteins and proteins that are synthesized on soluble ribosomes. The focus of this review is on the latter with the role of palmitoylation in the function and regulation of intracellular signaling molecules emphasized. The diverse contexts and functions of palmitoylation demonstrate the versatility of this modification. In general, palmitoylation increases affinity of proteins for membranes and thereby affects protein localization and function. Several surprising functions that go beyond membrane affinity have also been described. Selected examples will be explored in detail to illustrate both common and unique roles of palmitoylation. Several other recent reviews on protein palmitoylation are available (6, 17–19).

STRUCTURAL REQUIREMENTS FOR PROTEIN PALMITOYLATION

An examination of the palmitoylated sequence motifs in Table 1 reveals that there is no well-defined consensus sequence for this modification other than a requirement for cysteine. Very few S-palmitoylation sites in proteins have been mapped using mass spectroscopy or other physical techniques. In most cases, palmitoylation sites are deduced by mutagenesis of candidate cysteine residues. As palmitoylated proteins have been characterized over the years, several sequence contexts have emerged; representative proteins harboring these motifs are shown in Table 1. The list is by no means exhaustive. Palmitoylated proteins that are synthesized on soluble ribosomes either undergo sequential modification with different lipids or are exclusively S-palmitoylated.

Sequential lipid modification is a property of several families of signaling proteins (5, 20). A number of members of the Ras superfamily are prenylated and palmitoylated (21). The substrate recognition motif for a major class of prenylated proteins is a C-terminal –CaaX sequence in which C represents cysteine, a represents an aliphatic amino acid, and X represents the amino acid specifying farnesylation or geranylgeranylation (22). Prenylation at the cysteine residue is followed by proteolytic cleavage of the C-terminal three amino acids and by carboxyl-methylation of the prenylated cysteine. Cysteine residues immediately upstream of a prenylated cysteine are likely to be modified with palmitate. This modification has been well documented for H- and N-Ras proteins, which are farnesylated and palmitoylated (Table 1). There is evidence to suggest that palmitoylation is also found near geranylgeranylated cysteines. On the basis of sequence, R-Ras, RhoB, TC21, TCL, and Rap2b are predicted to be palmitoylated and geranylgeranylated. Radiolabeled palmitate incorporation into R-Ras and RhoB has been documented (23, 24). The prenyl modification on RhoB has been characterized and is unexpectedly heterogeneous with both farnesylated and

TABLE 1 Palmitoylation motifs

Protein class Examples	Sequence[a]	Reference
I. Dually lipidated proteins		
A. -Prenylation and S-palmitoylation		
Ras GTPases		
H-Ras	–GCMSCKCfarn	(165)
N-Ras	–GCMGLPCfarn	(165)
R-Ras	–KGGGCPCgg	(24)
Ras2 (*S. cerevisiae*)	–KSGSGGCCfarn	(166)
Rho GTPases		
RhoB	–QNGCINCCfarn/gg	(25)
TC10	–GSRCINCCfarn	(23, 113)
B. N-myristoylation and S-palmitoylation		
p59*fyn*	myr-N-GCVQCKDKE–	(167, 168)
p56*lck*	myr-N-GCGCSSHPE–	(83, 168)
G$_{i\alpha 1}$	myr-N-GCTLSAEDK–	(169)
eNOS	myr-N-GNLKSVAQEPGPPCGLGLGLGLGLCG–	(27)
Gpal (*S. cerevisiae*)	myr-N-GCTVST–	(121)
Vac8 (*S. cerevisiae*)	myr-N-GSCCSCLKD–	(141)
C. N-palmitoylation and S-palmitoylation		
G$_{s\alpha}$	palm-N-**GCL**GNSKTE–	(30, 149)
II. Cytoplasmic proteins—exclusively palmitoylated		
A. N-terminal motifs		
GAP-43	[1]MLCCM–	(170)
PSD-95	[1]MDCLCITT–	(171)
SCG10	[1]M–[19]SLICSCFYP[27]–	(114)
GAD65	[1]M–[27]RAWCQVAQKFTGGIGNKLCALL[48]–	(118)
G$_{q\alpha}$	[1]MTLESIMACC–	(75)
RGS4	[1]MCKGLAGLPASCLR–	(126)
RGS16	[1]MCRTLAAFPTTCLE–	(127)
B. C-terminal motif		
Yck2 (*S. cerevisiae*)	–FFSKLGCC-COOH	(37)
AtRac8 (*Arabidopsis thaliana*)	–GCLSNILCGKN-COOH	(38)
C. Cysteine string motifs		
SNAP-25b	–[83]KFCGLCVCPCNKL[95]–	(36)
Cysteine string protein 1	–[116]LTCCYCCCCLCCCFNCCCGKCKPK[139]–	(173)
GAIP (RGS19)	–[36]RNPCCLCWCCCCSCSW[51]–	(174)

TABLE 1 (*Continued*)

Protein class Examples	Sequence[a]	Reference
III. Pleckstrin homology domains		
Phospholipase D1	$-^{236}$PGLN**CC**GQGR$^{245}-$	(39)
Phospholipase D2 (*Rattus norvegicus*)	$-^{219}$PGFT**CC**GRDQ$^{228}-$	(40)
VI. RGS core domains		
RGS4	$-^{91}$FWIS**C**EEYKKI$^{103}-$	(128)
RGS10	$-^{56}$FWLA**C**EDFKKM$^{66}-$	(128)
RGS16	$-^{94}$FWLA**C**EEFKKI$^{104}-$	(129)

[a]S-palmitoylated cysteines are underlined and in boldface type (**C**). The superscript numbers refer to the amino acid position in the protein sequence. Amino acid sequences are human unless indicated otherwise. Abbreviations used are farn, farnesyl isoprenoid; gg, geranylgeranyl isoprenoid. Modified with permission from Biochemistry (6). Copyright (2003) American Chemical Society.

geranylgeranylated forms (25). It seems likely that palmitoylation is found in the context of geranylgeranyl groups, but this remains to be demonstrated directly.

N-myristoylated proteins that are palmitoylated have one or more cysteine residues adjacent to or nearby the N-myristoylated glycine (5, 20). All members of the Src family of protein tyrosine kinases and the $G_{i\alpha}$ subfamily are N-myristoylated, and the majority of these proteins are also palmitoylated (see Table 1 for selected examples). A palmitoylated cysteine residue is often adjacent to the N-myristoylated glycine, but palmitoylation occurs at cysteine residues up to 20 amino acids distant from the N terminus (26). Endothelial nitric oxide synthase (eNOS), which is N-myristoylated and palmitoylated at Cys15 and 26, has an unusual sequence requirement for palmitoylation (27). There are five GL repeats between the two palmitoylated cysteine residues that are required for palmitoylation (28). In most cases, however, the sequence context surrounding the palmitoylated cysteines has not been shown to be important for palmitoylation in a N-myristoylated protein (26, 29).

$G_{s\alpha}$ is an unusual dually lipidated protein in that it is N-palmitoylated at the amino-terminal glycine (30) and S-palmitoylated at the adjacent cysteine (31) (Table 1). Although $G_{s\alpha}$ has a glycine at position 2, similar to $G_{i\alpha}$ proteins, it is not a substrate for N-myristoylation (32). S-palmitoylation at Cys3 is conserved in $G_{s\alpha}$ and may contribute to modification at Gly2. This hypothesis is suggested by studies of the N-palmitoylated Hedgehog protein (7). In Hedgehog, N-palmitoylation requires a cysteine side chain; a serine substitution blocks modification (7). This observation suggests a mechanism by which palmitate is first added to the free thiol on the cysteine side chain; then it is transferred to the α-amino group of the same residue (8). A similar mechanism may operate in $G_{s\alpha}$, with palmitate transfer from the cysteine side chain to the free amino group on the glycine residue. $G_{i\alpha}$ proteins would be resistant to this mechanism because the α-amino group of the glycine is occupied with myristate.

Proteins that are exclusively palmitoylated include a group that is modified at cysteine residue(s) within the N-terminal 25 amino acids of the protein. A number of examples are listed in Table 1. Inspection of the sequences does not reveal any obvious motif that will direct palmitoylation other than the presence of the cysteines. Of these proteins, the sequence requirements for palmitoylation of PSD-95 have been studied most extensively using site-directed mutagenesis and deletion mapping (33). PSD-95 (postsynaptic density protein of 95 kDa) is a PDZ-containing protein that clusters proteins into complexes at the postsynaptic density. The first 13 amino acids of PSD-95 are sufficient for palmitoylation with modification occurring at Cys3 and Cys5. Palmitoylation depends on five consecutive hydrophobic amino acids (^{3}CLCIV^7) that reside near the N terminus of the protein.

Palmitoylation of cysteine string motif proteins has been documented, but the requirements for palmitoylation have been studied only for SNAP-25. SNAP-25 is a neuronal SNARE protein found at the plasma membrane that mediates synaptic vesicle exocytosis. The cluster of four palmitoylated cysteines is found in the interhelical domain that connects the two SNARE complex-forming domains (34) (Table 1). Residues 85–120 of the interhelical domain are sufficient for palmitoylation and plasma membrane targeting of a heterologous soluble protein (35). A five amino acid motif (QPARV) that is 25 amino acids downstream of the cysteine motif is required for efficient palmitoylation and plasma membrane targeting (35). This sequence may mediate a protein interaction that is a prerequisite for palmitoylation. Palmitoylation of SNAP-25 requires at least three cysteine residues; mutation of two of the four cysteines results in a complete loss of radiolabeling (36). The requirement for three cysteine residues is interesting and may suggest a distinct mechanism (or protein acyltransferase) for proteins with cysteine clusters versus proteins that are modified at single or double sites.

The C-terminal palmitoylation motif of Yck2 (yeast casein kinase 2) ends with a cysteine doublet (Table 1). This motif is typically recognized by geranylgeranyltransferase type II with both cysteines modified by a geranylgeranyl isoprenoid (22). However, Yck2 is modified with palmitate at these cysteine residues (37). Other canonical prenylation sites may be modified with palmitate instead of prenyl groups (38). A Rho GTPase in plants (AtRac8) has a *−CaaX* motif but does not appear to be prenylated. Instead, AtRac8 is palmitoylated near the C terminus, and its plasma membrane localization is dependent upon this modification (38). These studies underscore the limitations of sequence motifs as predictors of the lipidation state of a protein.

The diversity of palmitoylation motifs is evident from the discovery of sites within a well-characterized protein domain. Phospholipases D1 and D2 (PLD) are palmitoylated at cysteine residues within a pleckstrin homology (PH) domain (39, 40) (Table 1). Modeling of the PLD1 PH domain places the palmitoylated cysteines in the loop connecting the first two beta strands of the domain (39). There is a complex interplay between palmitoylation and the phosphoinositide binding PH and PX domains in PLD that regulates its dynamic localization in the plasma membrane and Golgi (41, 42).

MECHANISMS OF PROTEIN PALMITOYLATION

Since the discovery of S-palmitoylation, much effort has gone into understanding how palmitate is transferred onto proteins. Two mechanisms have been proposed. The first is through the action of an enzyme generically referred to as protein acyltransferase (PAT). The second mechanism is nonenyzmatic: spontaneous autoacylation of a protein in the presence of long-chain acyl-coenzyme As (CoAs). Genetic strategies in yeast have recently yielded the identity of two PATs (37, 43). Thus it is clear that some palmitoylation reactions are mediated by enzymes. Autoacylation of two mitochondrial enzymes is important for their regulation (44, 45), suggesting that both mechanisms are operational in cells.

Autoacylation

Acylation of peptides and proteins can occur in vitro in the absence of an enzyme. Quesnel & Silvius (46) utilized short lipid-modified peptides mimicking a myristoylated G_α N terminus or the prenylated and carboxymethylated C terminus of H-Ras. The cysteine residues in these peptides are acylated in the presence of lipid vesicles and acyl-CoAs (46). The presence of a membrane surface and affinity for that surface (as provided by prior lipid modification) were both found to be important for this reaction (46).

The amino acid context of cysteine residues is important for the rate of autoacylation in vitro. Surrounding amino acids influence the proximity and orientation of the cysteine to a membrane as well as the ability of that cysteine's thiol group to ionize. Using a peptide derived from the β_2-adrenergic receptor (β_2AR), Belanger et al. (47) demonstrated that basic and hydrophobic residues promote autoacylation, whereas acidic residues are inhibitory. In a separate study, basic and aromatic residues were found to decrease the pK_a of peptide cysteine residues, which correlated with increased autoacylation (48). The pH affects the rate of palmitoylation of a myristoylated peptide in vitro (46), again indicating that autoacylation is more efficient under conditions that favor formation of thiolate ion in the cysteine residue. The requirement for specific sequences to yield palmitoylation of a protein in cells is often used as an argument against autoacylation as a physiological mechanism. However, these studies demonstrate that sequence context influences autoacylation.

Proteins as well as peptides undergo autoacylation. For example, myristoylated $G_{i\alpha}$ reacts with palmitoyl CoA (palm-CoA) in vitro, resulting in nearly stoichiometric palmitoylation (49). Autoacylation of free $G_{i\alpha}$ requires myristoylation, and the rate of palmitoylation is accelerated by $G_{\beta\gamma}$ binding (49). Both of these properties parallel what is observed in vivo. Not all palmitoylated proteins examined were found to autoacylate: Under the same conditions that $G_{i\alpha}$ and $G_{o\alpha}$ subunits autoacylate, $G_{s\alpha}$, $G_{q\alpha}$, GAP43, p59^{fyn}, and SNAP-25 (but see 50) do not (49).

Palmitoylation of peptides and proteins clearly can occur in the absence of enzyme in vitro, but is autoacylation likely to be relevant in vivo? An important consideration is the available concentration of acyl-CoAs. In cells, acyl-CoAs are bound to acyl-CoA binding protein (ACBP), effectively reducing their concentration at the membrane surface. Inclusion of ACBP in peptide or protein palmitoylation reactions decreases autoacylation in a concentration-dependent manner (51, 52). At physiological concentrations of lipids, acyl-CoA, and ACBP, Leventis et al. (51) calculated that autoacylation in cells would occur with halftimes on the order of days. This is in sharp contrast to the short palmitoylation halftimes observed in vivo for some proteins ($<$ 1 h for $G_{s\alpha}$) (53) and suggests that autoacylation is not likely to be responsible for palmitoylation of proteins where palmitate is rapidly cycling (51). It should be noted that estimates of the concentration of acyl-CoAs in cells vary widely (54). Better information about the concentration and spatial distribution of acyl-CoAs in cells would facilitate our understanding of enzymatic and nonenzymatic mechanisms of fatty acylation in vivo.

One area where autoacylation is likely to play an important physiological role is in the regulation of mitochondrial enzymes. Methylmalonate semialdehyde dehydrogenase is a mitochondrial enzyme that is acylated on its active site cysteine, presumably preventing this residue from interacting with substrate (44). Another mitochondrial enzyme, carbamoyl-phosphate synthetase 1 (CPS 1), is also palmitoylated within its active site, resulting in inhibition of enzymatic activity (45). Palmitoylation of purified CPS 1 occurs at palm-CoA concentrations within the physiological range found in mitochondria, indicating that autoacylation may also occur in vivo (45). This phenomenon may be a general mechanism for regulation of mitochondrial metabolic enzymes by fatty acids.

DHHC Cysteine-Rich Domain Proteins Are Protein Acyltransferases

PAT activities have been characterized in tissues and cell lines, but purification of the activities has not been sufficient to unveil their molecular identity (6). To date, two PATs for intracellular proteins have been identified, both in yeast: Erf2/Erf4 and Akr1. Erf2/Erf4 complex stimulates palmitoylation of Ras2 substrates, whereas Akr1 palmitoylates Yck2 (37, 43). Erf2 and Akr1 each have a DHHC-CRD (Asp-His-His-Cys-cysteine-rich domain), a protein module that is a distinct variation of cysteine-rich zinc-finger domains (55, 56). The DHHC-CRD is hypothesized to be a palmitoyltransferase domain (37, 43).

Genetics in yeast led to both PAT discoveries. Yeast Ras2, similar to mammalian H-Ras and N-Ras, is palmitoylated on a cysteine adjacent to a -*CaaX* motif in which the cysteine is prenylated and carboxymethylated (Table 1). Erf2 (effect on ras function) was identified, along with its binding partner Erf4, in a genetic screen for mutations that were synthetically lethal with a palmitoylation-dependent mutant allele of Ras2 (57). Erf4 had previously been identified in a screen for genes affecting Ras function and named Shr5 (suppressor of hyperactive ras) (58). In cells lacking Erf2 or Erf4, palmitate labeling of Ras2 is reduced, and Ras2 is partially mislocalized from the plasma membrane (57, 58). When the Erf2/Erf4 complex is partially purified and incubated with Ras2 and palm-CoA, palmitoylation of Ras2 is stimulated 160-fold over the spontaneous rate, indicating that the complex is a PAT for Ras2 (43).

The other PAT enzyme-substrate pair that has been identified is Akr1 and Yck2. Akr1 is an ankyrin-repeat-containing protein with a DHHC-CRD that is involved in mating, cell morphology, and regulation of G-protein-coupled-receptor (GPCR) trafficking (59). The connection between Akr1 and its substrate Yck2 was identified by phenotypic analysis. Loss of either Akr1 or Yck2 function results in defective ligand-mediated endocytosis of the GPCR Ste2. Further study revealed that Akr1 is required for localization of Yck2 to the plasma membrane: In the absence of Akr1, Yck2 is mislocalized to the cytosol (60). Yck2 terminates with a dicysteine motif that is palmitoylated, and loss of Akr1 in cells abolishes in vivo palmitate labeling of Yck2 (37). Roth et al. (37) purified Akr1 to apparent homogeneity and demonstrated that Akr1 palmitoylates Yck2 in vitro.

Erf2 and Akr1 both have several predicted transmembrane (TM) domains and behave as integral membrane proteins (37, 57). In contrast to Erf2 and Akr1, Erf4 does not contain any recognizable domains. Erf4 is tightly membrane associated, but the mode of association is not readily apparent because it is not predicted to contain TM domains or lipid modifications. A weak hydrophobic cluster present in the C-terminal half of the protein is essential for Erf4 interaction with Erf2 (61). However, Erf4, which lacks the hydrophobic cluster, is still tightly membrane associated, indicating that Erf2 is not mediating Erf4 membrane association (61).

The PAT activity generated by the Erf2/Erf4 complex has specificity toward Ras2 substrates (43). Erf2/Erf4 palmitoylated yeast Ras2 much better than mammalian H-Ras or myristoylated $G_{i\alpha}$. Within the Ras2 protein, the hypervariable domain is important for recognition by Erf2/Erf4. GST fused to a sequence, which consists of just the palmitoylated cysteine and the -CaaX motif, is weakly palmitoylated by Erf2/Erf4. Addition of 28 residues of the hypervariable domain upstream of the -CaaX motif results in a 10-fold increase in palmitoylation (43). Further analysis revealed that basic residues within the hypervariable domain, especially Arg297, are important for palmitoylation of Ras2 by Erf2/Erf4, but the role of these residues is unknown (62). The specific sequence requirements for substrate recognition displayed by Erf2/Erf4 is consistent with the existence of a large family of PATs, each with a unique set of substrates.

The diversity of fatty acids found on palmitoylated proteins in vivo suggests that a PAT should not be highly selective for acyl-CoAs. This specificity appears to be true for Erf2/Erf4. Long-chain acyl-CoAs (C16, C16:1, C18, and C18:1) are effective competitors for palm-CoA, whereas shorter acyl-CoAs are less so (C<14) (43). One activator of DHHC-CRD PAT activity has been reported: Addition of ATP to Akr1 PAT assays has a stimulatory effect on both Akr1 autoacylation and Yck2 palmitoylation (37). The mechanism of this stimulation is unknown.

The common element shared by Akr1 and Erf2 is the DHHC-CRD. The critical role of this domain in PAT activity has been documented both genetically and biochemically. In the synthetic lethal screen in which *ERF2* mutations were isolated, several point mutations within the DHHC-CRD were found (57). Directed mutagenesis demonstrated that two residues of the DHHC motif are essential (His201 and Cys203) for PAT activity of the purified Erf2/Erf4 complex (43). Similar mutations in Akr1 also abolish transferase activity (37).

Further insight into the importance of this domain is provided by the observation that both Erf2 and Akr1 are themselves palmitoylated during in vitro reactions (37, 43), and Akr1 is palmitoylated in vivo (37). Palmitoylation may occur on the cysteine residue within the DHHC motif because mutation of that residue blocks autoacylation of either protein (37, 43). However, this cysteine residue could be required for palmitoylation at another site. There are at least two possible roles for DHHC-CRD palmitoylation. First, DHHC-CRD palmitoylation may represent an acyl intermediate in the reaction with subsequent transfer of the fatty acid to the substrate (37, 43). Peptide studies reveal that acylated peptides can efficiently donate their acyl group to an acceptor cysteine (46). The histidine in the DHHC motif is also required for PAT activity in vitro, suggesting a possible role for this residue in the transfer of palmitate from the DHHC-CRD to the substrate (37, 43). An alternative possibility is that DHHC-CRD palmitoylation is required for the protein to adopt a conformation that permits palmitate transfer through another mechanism. These possibilities have yet to be explored.

DHHC-CRDs are found throughout the eukaryotic kingdom. There are seven DHHC-CRD-containing proteins in *Saccharomyces cerevisiae*. Of these, five are similar in overall organization to Erf2: They are approximately the same size (39–44 kDa), and contain four predicted TM domains. The DHHC-CRD is between the second and third TM domains extending into the third predicted TM. This group includes Erf2, Swf1 (spore wall formation), Yol003c, Ydr459c, and Ynl326c. The sequence similarity is generally restricted to within the DHHC-CRD. Other than Erf2, very little is known about these proteins. None of the DHHC-CRD genes is essential (63). Diploid cells lacking Swf1 sporulate very inefficiently due to a defect in formation of the spore wall (64) and have a strong vacuolar protein-sorting defect (65). The role of Swf1 in either of these processes is unknown.

Akr1 and Akr2 make up the remainder of the family in yeast. These proteins contain six predicted TMs and several N-terminal ankyrin repeats in addition to the DHHC-CRD. Ankyrin repeats are 33 amino acid conserved domains found in numerous proteins. Ankyrin repeats are often involved in protein-protein interactions, but how or if the ankyrin repeats have a role in palmitoylation of Yck2 by Akr1 is unknown. One possibility is that the ankyrin repeats bind Yck2 to bring it into proximity to the DHHC-CRD for palmitoylation. Whereas Akr1 is involved in mating, cell morphology, and endocytosis (59), no phenotypes have been described for cells lacking Akr2. Neither gene is essential (63).

In three of the DHHC-CRD proteins (Ydr459c, Akr1, and Akr2) the second histidine of the DHHC motif is replaced with tyrosine (DHYC). The CRDs present in these proteins are also more distantly related to the rest of the family in terms of cysteine conservation. The canonical domain contains seven cysteines with conserved spacing. Ydr459c has four of these cysteines, whereas Akr1 and Akr2 contain only two. Despite the divergence in the DHHC-CRD domains, Akr1 and Erf2 both harbor PAT activity and share essential features of the domain required for palmitate transfer.

Genomic databases indicate that human and mouse genomes each contain over 20 DHHC-CRD proteins (6). In contrast, Erf4 homologs are not readily identifiable in organisms other than fungi. To date, a few mammalian DHHC-CRDs have been cloned and initially characterized. A human DHHC-CRD-containing protein, huntingtin interacting protein 14 (HIP14), interacts with huntingtin (htt), the protein in which polyglutamine expansion leads to Huntington's disease (66). Some characteristics of HIP14 indicate that it may be involved in the pathogenesis of Huntington's disease: HIP14 interaction with htt is decreased by polyglutamine expansion in htt, and HIP14 expression is limited to the neural cells affected in Huntington's disease (66). Similar to Akr1, HIP14 contains ankyrin repeats. The human HIP14 cDNA complements the temperature-sensitive growth and endocytic defects of *akr1* null yeast cells (66). Auto-acylation or PAT activity of HIP14 remains to be explored. However, the ability of HIP14 to rescue the endocytic defect of *akr1* null cells suggests that it may be able to palmitoylate Yck2 in yeast. Other cloned mammalian proteins that contain a DHHC-CRD include Golgi-specific DHHC zinc-finger protein (67), a proapoptotic c-Abl-interacting protein named Aph2 (68), and Sertoli cell gene with a zinc-finger domain (69). Experiments to determine if these proteins are involved in protein palmitoylation are eagerly awaited.

FUNCTIONS OF PALMITOYLATION

Similar to other lipid modifications, palmitoylation promotes membrane association of otherwise soluble proteins. The function of palmitoylation, however, ranges beyond that of a simple membrane anchor. Trafficking of lipidated proteins from the early secretory pathway to the plasma membrane is dependent upon palmitoylation in many cases. In addition, modification with fatty acids impacts the lateral distribution of proteins on the plasma membrane by targeting them to lipid rafts. Palmitoylation works in concert with other lipid modifications and protein motifs to facilitate targeting to the appropriate cellular destination through mechanisms that are just beginning to be defined.

Membrane Association

Palmitoylation occurs at membranes. Thus, soluble proteins must interact with membranes at least transiently to be palmitoylated. Once added, palmitate promotes stable membrane association. The two-signal hypothesis was developed for mammalian Ras proteins when it was discovered that plasma membrane targeting required a component in addition to farnesylation (70). This second signal could either be another lipid modification (palmitate) or an electrostatic interaction (basic amino acids that interact with the negatively charged phospholipid head groups at the plasma membrane). The model has been extended to other proteins and is supported by biophysical studies described below. For many proteins, however, the mechanisms that underlie initial membrane targeting to permit palmitoylation remain elusive.

S-palmitoylation of dually modified proteins is dependent upon prior modification with either the prenyl or N-myristoyl group (5, 20). Current evidence favors a model in which the requirement for the first lipid modification is membrane targeting rather than providing a specific recognition determinant for a PAT (71). The model emerged from measurements of the relative affinities of lipidated peptides for model membranes. These studies demonstrated that a myristate or farnesyl group provides only a weak affinity for membranes, whereas tandem lipid modifications are sufficient for high-affinity interactions (71, 72). The rates of interbilayer transfer of singly and dually lipid-modified peptides showed that dual lipid modifications resulted in essentially permanent association with the membrane (71). This led to the kinetic membrane trapping model, which states that proteins modified solely with myristate or farnesyl will cycle on and off membranes until they encounter a membrane with an appropriate membrane-targeting receptor. This interaction leads to the addition of the second lipid modification and results in essentially permanent membrane association. The protein interaction that permits palmitoylation need only be transient. Thus, a PAT is an excellent candidate for the membrane-targeting receptor. In the absence of the N-myristoylation or prenylation, other mechanisms that provide affinity for membranes can substitute for the addition of the first lipid.

G-protein α subunits of the G_q and G_{12} families are modified only with thioester-linked palmitate (31, 73). Their ability to be brought to a membrane compartment for palmitoylation involves interactions with $G_{\beta\gamma}$ subunits (74). However, other mechanisms must exist because constitutively active forms of $G_{q\alpha}$ and $G_{13\alpha}$ have low affinity for $G_{\beta\gamma}$, but are palmitoylated when expressed at high levels in mammalian cells (75, 76). Kosloff et al. (77) recently proposed on the basis of homology modeling that all G-protein α subunits have a positively charged structural motif near the N terminus that could facilitate membrane interactions and palmitoylation. The model may be applicable for other proteins, which include a subset of the regulators of G-protein signaling (RGS) proteins. RGS4, -5, and -16 have amphipathic α-helical domains near the palmitoylated residues in the N terminus (Table 1) that mediate membrane binding, in part through electrostatic interactions (78, 79).

Palmitoylation and Lipid Rafts

Palmitoylation targets proteins to lipid rafts, subdomains of the plasma membrane that are enriched in sphingolipids and cholesterol (80, 81). This characteristic lipid composition confers resistance to detergent extractions and allows the isolation of rafts as low-buoyant-density membranes. Raft lipids are organized in the liquid-ordered (l_o) phase in which the acyl chains of lipids are tightly packed, highly ordered, and extended (80, 82). Proteins modified with saturated acyl chains are predicted to have high affinity for an ordered lipid environment (82). Indeed, the content of lipid rafts is rich in proteins modified with fatty acyl chains. These are proteins attached to the cell surface through glycosylphosphatidylinositol (GPI) anchors (which contain predominately saturated fatty acids) and proteins anchored to the cytoplasmic inner leaflet through tandem myristoylation and palmitoylation or tandem palmitoylation.

Several lines of evidence support the importance of palmitoylation in targeting proteins to lipid rafts. Blocking palmitoylation through site-directed mutagenesis (83–86) or the use of the palmitoylation inhibitor 2-bromopalmitate (87) delocalizes proteins from rafts. Although palmitoylation increases affinity for rafts, not all palmitoylated proteins, e.g., many palmitoylated TM proteins, are residents of rafts (88). Targeting proteins to rafts via palmitoylation relies on the ability of the saturated fatty acid to fit well in an ordered lipid environment, and is not simply a function of hydrophobicity. Proteins modified with bulkier lipids like unsaturated fatty acids or prenyl groups are predicted to have low affinity for rafts (82). This has been demonstrated in model membranes (89, 90) and in cells (91, 92). Cholesterol is essential for organizing lipid rafts. Depletion of cell-surface cholesterol with β-methyl cyclodextran or other cholesterol-sequestering reagents disrupts lipid rafts and changes the lateral distribution of proteins within the plasma membrane (81, 93, 94).

The in vivo correlate of lipid rafts has been difficult to pin down, and estimates of the size and number of rafts in cells vary widely, depending upon the technique used (95). It is clear that proteins dynamically associate with lipid rafts. This is exemplified by GPI-anchored proteins, which are uniformly distributed on the cell surface until clustered with cross-linking antibodies (80, 96). Reversible palmitoylation may represent a second mechanism for dynamic raft association, although to date this has not been demonstrated for a protein in response to a physiological signal.

Morphological studies support the organization of palmitoylated signaling proteins in membrane subdomains. Myristoylated, palmitoylated $G_{i\alpha}$ subunits have a punctate appearance in plasma membrane fragments when detected by immunofluorescence and have a clustered appearance when decorated with gold particles in electron micrographs of plasma membrane (97, 98). Green fluorescent protein (GFP) tagged with raft targeting sequences (tandem myristoylation and palmitoylation or tandem palmitoylation) exhibits plasma membrane clustering in cells as assessed by fluorescence resonance energy transfer (92). Prior et al. (93) used statistical methods to analyze the distribution of H- and K-Ras proteins and derivatives thereof in electron micrographs of the plasma membrane. The study revealed the existence of multiple, nonoverlapping microdomains within the plasma membrane that would allow for a different spatial distribution of Ras isoforms. The distinctive lipid modifications of the Ras isoforms are an important (but not exclusive) determinant of targeting to these subdomains (21, 99).

Localization of signaling proteins in rafts appears to be important for their function. This has been studied most extensively in immune cells (96, 100). Palmitoylated proteins associated with early events in T-cell signaling include nonreceptor tyrosine kinases (NRTKs) and the TM protein LAT (linker for activation of T cells). Mutation of the palmitoylation sites in NRTKs (101) or LAT (102, 103) results in attenuated T-cell responses, demonstrating the necessity for the palmitoylated cysteines in signaling. Treatment of T cells with polyunsaturated fatty acids or 2-bromopalmitate inhibits T-cell signaling and palmitoylation of LAT and NRTK, and it disrupts raft structure (87, 91, 104), strengthening the correlation between palmitoylation, raft targeting, and productive signaling. The precise role of lipid rafts in immune cell signaling is controversial (100). Nonetheless, the data strongly imply a role for palmitoylation in organizing protein interactions needed for signal propagation.

Most of the work on lipid rafts has focused on how they organize proteins on the plasma membrane. An open question is whether proteins are sorted into lipid rafts on intracellular membranes as a mechanism for transporting them to the cell surface. It has been proposed that cholesterol-sphingolipid rafts first assemble in the Golgi (80, 81). Acquisition of detergent resistance for a newly synthesized GPI-anchored protein occurs in the medial Golgi, correlating with the presence of significant amounts of sphingolipid and cholesterol in the Golgi membranes. Sphingolipids and GPI-anchored proteins are components of the outer leaflet of the plasma membrane. Much less is understood about how lipid rafts are organized on the cytoplasmic leaflet of membranes and how or if the raft domains in both leaflets are coupled (81). This is an important question with respect to the trafficking of fatty acylated proteins that associate with the cytoplasmic leaflet. As described below, palmitoylation represents an important signal for transport to the plasma membrane and to specialized domains in neurons. Sorting of palmitoylated proteins into cholesterol-rich subdomains that form on intracellular membranes represents one model of how this could occur.

Trafficking of Lipid-Modified Signal Transducers

Our understanding of the trafficking of lipidated proteins has been enhanced by information about the location of the modifying enzymes and improved visualization of proteins using GFP. Ras proteins are among the best studied (105, 106). All Ras proteins are synthesized on soluble ribosomes and farnesylated in the cytoplasm by farnesyltransferase. The enzyme may transport Ras to the endoplasmic reticulum (ER) where it encounters the next two processing enzymes, the -CaaX protease Ras-converting enzyme 1 (107) and the isoprenyl-cysteine carboxyl methyltransferase (108). At this point, Ras isoforms diverge in their trafficking pathways. K-Ras4B, a nonpalmitoylated form of Ras, trafficks to the plasma membrane through a pathway that is independent of the classical secretory pathway. H- and N-Ras are further modified by palmitoylation in the early secretory pathway. The mammalian Ras PAT has not been identified, but the *S. cerevisiae* Ras PAT Erf2/Erf4 is localized in the ER (57, 61). When palmitoylation of H- or N-Ras is blocked by mutation or treatment of cells with 2-bromopalmitate, Ras accumulates in the early secretory pathway. This suggests that palmitoylation is a signal for exit from the early secretory pathway. Trafficking of Ras through intracellular membranes is not simply a biosynthetic journey. Contrary to the long-standing dogma that Ras signals exclusively from the plasma membrane, recent studies demonstrate that Ras is also actively signaling on intracellular membranes (109). If as the data suggest, palmitoylation controls exit from the early secretory pathway, it could have an important regulatory role in controlling where Ras is available for signaling.

Newly synthesized heterotrimeric G proteins appear to use a pathway similar to that of palmitoylated Ras proteins to get to the plasma membrane (110, 111). G proteins are functional heterodimers of the GTP-binding α subunit and a tightly associated $\beta\gamma$ complex. G-protein α subunits are modified with amide-linked myristate, thioester-linked palmitate, or both. The γ subunit is modified by prenylation and carboxylmethylation at the C terminus. G-protein β and γ subunits assemble in the cytoplasm after synthesis (112). Following prenylation in the cytoplasm, the complex moves to the ER where subsequent C-terminal processing of the G_γ prenylated cysteine occurs. The heterotrimer appears to assemble on Golgi membranes and then transit to the plasma membrane. The model is based on studies of exogenously expressed proteins in which $G_{\beta\gamma}$ subunits accumulate in the ER and Golgi in the absence of G_α expression (110, 111). Engineering a palmitoylation site into G_γ (111) or coexpression of palmitoylated G_α subunits (110) rescues plasma membrane localization of $G_{\beta\gamma}$. Both G_α and $G_{\beta\gamma}$ accumulate on intracellular membranes when palmitoylation–defective G_α is expressed (110, 111). Thus, palmitoylation again appears to be important for exit from the Golgi and transport to the plasma membrane.

The studies just described for H-Ras and G-protein heterotrimers are consistent with a model in which they move to the plasma membrane through the classical secretory pathway tethered to the cytosolic surface of transport vesicles.

A close examination of this question was recently performed in yeast with respect to the trafficking of Ras2 (62). In yeast, the classical secretory pathway can be inactivated genetically with temperature-sensitive mutations in *SEC* genes or with the inhibitor Brefeldin A (BFA). Perturbation of the classical secretory pathway does not affect the ability of Ras2 to localize to the plasma membrane, indicating that the secretory pathway is not required for trafficking (62). With the exception of Erf2/Erf4, the components of the alternative trafficking pathway in yeast are a mystery (61, 62). Interestingly, the alternative pathway does not appear to require vesicular transport because Ras2 can proceed to the plasma membrane even when Sec18, the yeast homolog of N-ethylmaleimide-sensitive factor (NSF), is inactivated (62).

Does an equivalent alternative trafficking pathway for palmitoylated proteins exist in mammalian cells? TC10 is a mammalian GTPase that is farnesylated and palmitoylated similar to H-Ras (Table 1). Disruption of the secretory pathway in adipocytes by BFA or by a temperature block does not affect plasma membrane localization of TC10 (or H-Ras); this is consistent with the existence of an alternative trafficking pathway (113). The results obtained with H-Ras in adipocytes (113) are inconsistent with those obtained in BHK cells (106), suggesting that the pathways taken by proteins may vary by cell type.

Targeting of Palmitoylated Proteins in Neurons

In neurons, newly synthesized proteins in the cell body must be transported to their appropriate destination, sometimes at great distances. Proteins are sorted at the trans-Golgi network into vesicles for transport to the axon, dendrites, or growth cones. PSD-95, localized prominently in dendrites, requires palmitoylation to associate with perinuclear vesiculotubular structures that bud transport intermediates destined for postsynaptic sites (33). Similarly, SCG-10, a microtubule-destabilizing protein, requires palmitoylation for targeting to Golgi membranes for transport to neuronal growth cones (114, 115). In other cases, proteins synthesized on soluble ribosomes use a hierarchy of signals to be transported to the appropriate location. One such example is the enzyme glutamic acid decarboxylase 65 kDa (GAD65) that synthesizes the neurotransmitter γ-aminobutyric acid. GAD65 is localized in synaptic vesicles at the nerve terminal and synaptic-like vesicles in pancreatic β cells (116). The N-terminal 72 amino acids of GAD65 are sufficient to target GFP to presynaptic clusters in nerve terminals and for its relative exclusion from dendrites (94). GAD65 is palmitoylated at residues 30 and 45 (117) (Table 1). The enzyme is synthesized as a soluble protein and associates with membranes through a palmitoylation-independent mechanism that requires residues 24–31 (118). Membrane targeting specifically to the Golgi requires residues 1–24, again in a palmitoylation-independent fashion (119). Palmitoylation is essential for post-Golgi trafficking to presynaptic clusters in a cholesterol-dependent manner (94). Thus, for GAD65, the primary function of palmitoylation is to facilitate sorting of GAD65 into cholesterol-rich membrane microdomains, rather than initial targeting of the protein to Golgi membranes.

Regulation of Protein Activity

The requirement for protein palmitoylation in a signaling pathway or other cellular process is often associated with localization of the modified protein. However, palmitoylation also functions in the regulation of protein activity. Although the structural basis of this regulation is not well understood, in vitro assays are revealing the importance of palmitoylation in promoting or inhibiting protein interactions.

G–PROTEIN SIGNALING   Palmitoylation regulates protein activity at multiple levels of G-protein-mediated signaling, starting with localization of heterotrimeric G proteins at the plasma membrane. Interfering with palmitoylation of G_α subunits has variable consequences on signal transduction, depending upon the G_α in question. Palmitoylation-defective G_s and G_q poorly transduce signals from receptor to effector in cell-based assays of second-messenger production (75). However, $G_{i\alpha}$ family members that are N-myristoylated, but are not palmitoylated, retain some ability to signal (120, 121). Effects on signaling can be correlated with mislocalization. However, the possibility that loss of palmitoylation also has an effect on receptor or effector interactions cannot be excluded. Furthermore, because many assays to assess the function of palmitoylation use replacement of cysteine residues, caution must be used when attributing the loss of function to a deficit in palmitoylation rather than a change in the amino acids. This possibility was elegantly demonstrated by Hepler et al. (122) who showed that depalmitoylated $G_{q\alpha}$ functioned normally in a purified reconstituted system of M_1-muscarinic receptor G_q and phospholipase $C\beta$. A palmitoylation-defective mutant $G_{q\alpha}$ (C9S, C10S) was nonfunctional in the same assay, demonstrating the importance of the cysteine residues themselves, rather than the modification for receptor/effector coupling (122).

Palmitoylation also impacts the interactions of G-protein α subunits and RGS proteins. RGS proteins constitute a large family with a shared RGS domain that harbors GTPase activating protein (GAP) activity toward G_α subunits (123). In vitro palmitoylated $G_{i\alpha}$ proteins resist the GAP activity of RGS proteins (124). Palmitoylation of $G_{z\alpha}$, a $G_{i\alpha}$ family member, decreased the affinity of its cognate RGS protein RGSZ1 for the GTP-bound form of $G_{z\alpha}$ by at least 90% and decreased the maximum rate of GTP hydrolysis. Other G_α-RGS pairs exhibited similar regulation, suggesting that palmitoylation of $G_{i\alpha}$ family members and perhaps other G proteins may be an important determinant in terminating signaling through GPCR-coupled pathways (124).

A number of RGS proteins are themselves substrates for palmitoylation (125). The functional consequences of RGS palmitoylation have been studied best for RGS4 and RGS16. These proteins share an N-terminal membrane interaction domain that includes palmitoylated cysteines at positions 2 and 12 (126, 127) (Table 1). Modification of the N-terminal residues precedes palmitoylation at a cysteine residue buried in the RGS core domain (Cys98 in RGS16 and Cys95 in

RGS4) (128, 129). Palmitoylation of the N-terminal domain is not necessary for plasma membrane localization (126, 127) but is important for localization in lipid rafts (129). In cells, raft localization appears to facilitate palmitoylation of RGS16 on the core domain cysteine, but it is not required for RGS16 activity in signaling (129).

In vitro multisite palmitoylation of RGS proteins has profound and complex effects on GAP activity (128). Palmitoylation of RGS4 at the N terminus potentiates GAP activity when assayed at a membrane surface but has no effect in solution-based turnover assays. Palmitoylation at the internal cysteine residue is inhibitory for GAP activity when assayed in solution and when reconstituted into lipid vesicles with receptor and G protein. These data suggest that palmitoylation of the core domain has a negative regulatory role in RGS GAP activity. The core domain cysteine is highly conserved among RGS proteins, and this may represent a common regulatory mechanism. However, in the same study, analysis of a second RGS protein yielded different results. RGS10 is palmitoylated at the internal cysteine in the core domain, but lacks the N-terminal cysteines found in RGS4 and RGS16 (Table 1). Palmitoylation of the RGS10 core domain is inhibitory in the solution-based assay, as was the case for RGS4, but stimulates GAP activity greater than 20-fold in receptor–G-protein proteoliposomes (128).

Studies of RGS16 and RGS10 core domain palmitoylation in cells lend support for a positive regulatory role for this modification (129–131). RGS10 and RGS16 inhibit signaling through G_q- and G_i-mediated pathways in cells. When the core domain cysteine is mutated, RGS10 and RGS16 are less effective regulators of the signaling pathway, suggesting that the cysteine residue plays a positive role in regulating RGS activity (130, 131). Osterhout et al. (131) ruled out mislocalization of the RGS16 mutant as an explanation for the diminished activity of the core domain palmitoylation mutant in cells and demonstrated that GAP activity of RGS4 and RGS16 was increased when the core domain cysteine was palmitoylated. In their study, GAP activity was measured using mammalian cell membranes expressing a functional fusion of the 5-hydroxytryptamine-1A receptor with $G_{o\alpha 1}$ and stoichiometrically palmitoylated RGS4. These results are in opposition to what was measured for RGS4 in receptor–G-protein proteoliposomes (128). However, the discrepancy may be explained by the different lipid composition or additional protein factors that are present in cell membranes (131).

It is clear from these and other studies (132, 133) that RGS protein activity is exquisitely sensitive to its membrane environment. Not unexpectedly, palmitoylation of RGS4 and RGS16 in the N-terminal domain contributes to interactions at membrane surfaces in vitro and lipid raft localization in vivo. In addition to membrane interactions, RGS palmitoylation might also exert effects through conformational changes. In RGS4, the core domain cysteine is found in the middle of helix 4 in the four-helix bundle that makes contact with $G_{i\alpha}$. The presence of palmitate at this site could shift the packing of the helices, thereby altering the domain that binds G_α-GTP and accordingly affects GAP activity (128).

VACUOLAR MEMBRANE FUSION   Recent studies implicate protein palmitoylation as an essential step in the process of vacuolar membrane fusion (134, 135). The yeast vacuole is equivalent to the mammalian lysosome and has been exploited as a model to understand organelle biogenesis and membrane trafficking (136). The discovery of a large number of genes and their role in vacuolar inheritance, homeostasis, and membrane trafficking has been complemented by biochemical reconstitution of a vacuolar function, homotypic vacuole fusion (137). Similar to fusion of Golgi transport vesicles (138, 139), vacuolar fusion is strongly stimulated by palm-CoA (140). The effector of palm-CoA in these assays has not been identified but is presumed to be a protein. A candidate for the palm-CoA effector is the protein product of *VAC8*, which was uncovered in a screen for genes involved in vacuolar inheritance (141).

Vac8 is an armadillo repeat protein with an N-terminal sequence that directs both N-myristoylation and palmitoylation (141). Both modifications are required for complete localization of Vac8 at the vacuolar membrane (141). By contrast to most proteins that are dually modified with myristate and palmitate, N-myristoylation is dispensable for function (vacuolar inheritance), whereas S-palmitoylation is not. Thus, myristoylation and palmitoylation are uncoupled in Vac8, suggesting that palmitoylation of Vac8 might serve functions in addition to membrane binding. Vac8 is necessary for vacuolar fusion in vivo and in vitro. Notably, fusion reactions with vacuoles lacking Vac8 or harboring a palmitoylation-defective Vac8 are not stimulated by palm-CoA (134, 135). Palmitoylation of Vac8 occurs early during the vacuolar fusion reaction with palm-CoA serving as the palmitate donor (135, 142). Vac8 acts at a late step in the fusion reaction, following formation of *trans*-SNARE pairs (134, 135). Weisman and coworkers (134) proposed that Vac8p may bind the fusion machinery through its armadillo repeats and that palmitoylation brings this machinery to specialized lipid raft domains that facilitate bilayer mixing. The enrichment of Vac8 at sites of vacuole-vacuole contact in vivo is consistent with this model (143, 144). Although the precise role of Vac8 palmitoylation remains to be elucidated, it is clear that it plays an essential role in the process of vacuole fusion. Mammalian homologues of Vac8 have not been identified, but the requirement for palm-CoA in mammalian Golgi transport assays suggests that the mechanisms discovered here might represent a more general role in membrane fusion (134).

PALMITATE TURNOVER AND ITS REGULATION

Removal of palmitate from proteins occurs both constitutively and in response to signals. Depalmitoylation is carried out by thioesterases. Palmitoyl-protein thioesterase I (PPT1) and PPT2 are lysosomal hydrolases involved in the cleavage of acyl-cysteine linkages during the process of protein degradation (145, 146). These enzymes have been recently reviewed elsewhere (6). Acyl-protein thioesterase 1 (APT1) is a cytosolic thioesterase that depalmitoylates several signaling molecules and may have a role in regulation of their activity. APT1 has a wide range of substrates in vitro, but the cellular roles of APT1 are just beginning to emerge.

Palmitate Turnover

Palmitate is a dynamic modification that is continually turning over on cellular proteins. An example of this phenomenon is the mannose-6-phosphate receptor. The $t_{1/2}$ of the receptor protein is 40 h, but the $t_{1/2}$ of the attached palmitate is only 2 h (147). This suggests that a single protein undergoes several cycles of palmitoylation and depalmitoylation during its lifetime.

The mannose-6-phosphate receptor undergoes constitutive palmitate turnover. Depalmitoylation can also be a regulated process within a cell. For example, in response to activation of β-ARs with the agonist isoproterenol, palmitate turn-over on $G_{s\alpha}$ increases, as revealed by increased palmitate incorporation (53, 148, 149). The increased turnover is due to an increase in the rate of depalmitoylation as revealed by pulse-chase analysis (53, 149). Jones et al. (150) determined that ~70% of endogenous $G_{s\alpha}$ is palmitoylated in COS or S49 lymphoma cells. Following isoproterenol treatment of these cells, the percentage of palmitoylated $G_{s\alpha}$ does not change significantly (150). This suggests that repalmitoylation occurs quickly because unpalmitoylated $G_{s\alpha}$ does not accumulate. Mutationally activated $G_{s\alpha}$ incorporates much less palmitate in cells than does wild-type $G_{s\alpha}$, again linking activation to depalmitoylation (53). This may be a general G_α phenomenon, as palmitate turnover on $G_{i\alpha}$ is also affected by activation of the 5-hydroxytryptamine-1A receptor (151).

In addition to G_α subunits, β_2AR and m_2 muscarinic acetylcholine receptors show enhanced palmitate turnover after stimulation with agonists (152, 153). The enhancement of palmitate turnover on both $G_{s\alpha}$ and β_2AR depends directly on their activation rather than on feedback from downstream components of the pathway. This was demonstrated utilizing a β_2AR-$G_{s\alpha}$ fusion protein in which both receptor and $G_{s\alpha}$ are depalmitoylated in response to treatment with agonist (154). Treatment of cells with forskolin, which mimics downstream components of the pathway and bypasses the receptor and G protein, did not increase depalmitoylation of this fusion protein (154).

The cycle of palmitate turnover on a GPCR and heterotrimeric G protein in response to pathway activation is illustrated in Figure 1. In the basal state (A), both the GPCR and G_α are palmitoylated. G_α also contains a stable lipid modification, either amide-linked palmitate ($G_{s\alpha}$) (30) or myristate ($G_{i\alpha}$) (32), and $G_{\beta\gamma}$ is prenylated (20). In response to ligand (B), G_α exchanges GDP for GTP and dissociates from $G_{\beta\gamma}$. Free G_α is a substrate for the thioesterase APT1 (see below) and is depalmitoylated (155–157). The GPCR is also depalmitoy-lated, but the identity of the thioesterase is unknown. Following activation and depalmitoylation (C), G_α distribution may be altered. One group reports that $G_{s\alpha}$ moves into the cytoplasm following activation (158). However, Huang et al. (98) demonstrated that overexpressed $G_{s\alpha}$ and endogenous $G_{i\alpha}$ remain membrane

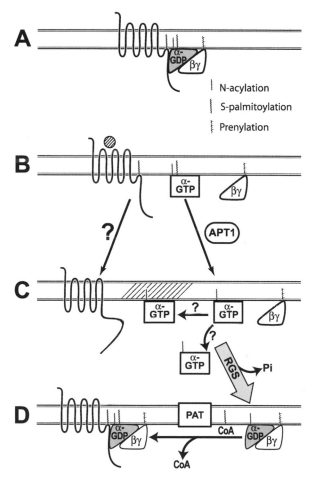

Figure 1   Palmitate turnover on G_α and GPCR is regulated during the G-protein activation cycle.

associated following in vitro depalmitoylation by APT1. Instead, depalmitoylation may result in movement of G_α into a different membrane subdomain. Visualization of endogenous $G_{i\alpha}$ revealed a subtle redistribution within the membrane in the presence of activating GTPγS (98). In the depalmitoylated state, G_α is susceptible to the GAP activity of RGS proteins (124). RGS proteins promote GTP hydrolysis, and G_α reassociates with $G_{\beta\gamma}$. Once the heterotrimer is reformed (*D*), G_α is a substrate for an unidentified PAT in the plasma membrane (159). The GPCR is also repalmitoylated.

The palmitoylation status of other proteins has also been linked to their activation. Depalmitoylation of eNOS increases in response to treatment of cells with bradykinin (160). Receptor-mediated increases in turnover have not been reported for Ras proteins; however, the rate of palmitate turnover on H-Ras does correlate with its activation state. Activated H-Ras has faster palmitate turnover than nonactivated H-Ras in vivo, again because of an increased rate of depalmitoylation (161). Palmitate turnover has also been described for the postsynaptic density protein PSD-95. Glutamate receptor activity triggers depalmitoylation of PSD-95 in neurons (162). How glutamate receptor activity regulates palmitoylation of PSD-95 is unknown, but it depends on the influx of extracellular calcium (162).

As mentioned above, depalmitoylation has a subtle effect on membrane distribution of $G_{i\alpha}$ (98). Because palmitoylation confers affinity for lipid rafts, depalmitoylation may also affect partitioning within these membrane subdomains. In the case of H-Ras, lateral movements within the membrane appear to coincide with activation: GTP-bound activated H-Ras is excluded from rafts, whereas GDP-bound inactive H-Ras partitions into rafts (99). Knowing that GTP-bound Ras has faster palmitate turnover, it seems possible that there is a connection between activation, depalmitoylation, and lateral movement within the membrane.

Palmitate cycling on PSD-95 may have a role in regulating synaptic strength. Blocking repalmitoylation by treatment of neurons with 2-bromopalmitate results in dispersal of PSD-95 from synaptic clusters (162). The effect of 2-bromopalmitate indicates that palmitate is turning over on PSD-95 and that repalmitoylation is necessary to keep it localized at the synapse. 2-Bromopalmitate also declusters receptors for α-amino-3-hydroxy-5-methylisoxazole-4-propionic acid (AMPA) and reduces the amplitude and frequency of AMPA-mediated miniature excitatory postsynaptic currents (162). The role of palmitate turnover becomes clear when examining the internalization of AMPA receptors following exposure to glutamate. Internalization was enhanced with overexpression of PSD-95 but abolished when cells expressed PSD-95 prenyl, which is membrane targeted but does not undergo lipid turnover (162). Thus, palmitate turnover on PSD-95 is critical for downregulation of AMPA receptors in response to glutamate, which in turn may regulate synaptic strength.

Acyl-Protein Thioesterase 1

One enzyme has been identified that removes palmitate from proteins on the cytosolic surface of membranes. APT1 was purified from rat liver cytosol because of its ability to remove palmitate from $G_{i\alpha}$ (156). APT1 is a 29 kDa protein that was previously characterized as having lysophospholipase activity toward palmitoylated glycerol-3-phosphocholine (163). Duncan & Gilman (156, 157) demonstrated that palmitoylated proteins are in fact the preferred substrate of APT1 with both a lower K_m and higher catalytic efficiency toward protein substrates. APT1 is widely conserved from yeast to humans (156).

Several palmitoylated proteins are substrates in vitro for mammalian APT1 in addition to $G_{i\alpha}$ (156). These are $G_{s\alpha}$, RGS4, H-Ras, and eNOS (156, 164). The catalytic efficiency of rat APT1 toward H-Ras is about one third of that observed for $G_{i\alpha}$ (156). In contrast to the rat enzyme, Apt1 from *S. cerevisiae* does not depalmitoylate RGS4 or H-Ras, but mammalian $G_{i\alpha}$ is a substrate (157). There is also evidence indicating that G_α subunits and eNOS are substrates for APT1 in cells. Heterologous expression of rat APT1 in mammalian cells leads to increased loss of palmitate from $G_{s\alpha}$ (156) and from eNOS (164). In yeast, deletion of the *APT1* gene results in decreased turnover of palmitate on Gpa1 (Table 1), a heterotrimeric G-protein α subunit, in the unstimulated state (157).

As described in the previous section, activation of some signaling proteins correlates with increased palmitate turnover. The observation that APT1 prefers activated proteins as substrates for its depalmitoylating activity is consistent with those studies. G_α is a better substrate for APT1 in the absence of $G_{\beta\gamma}$ (in the GDP- or GTP-bound state) or when activated with AlF_4^- to release $G_{\beta\gamma}$ from the heterotrimeric complex (156, 157). Similarly, eNOS is a better substrate for APT1 in vitro in the presence of Ca^{2+}-calmodulin, which converts eNOS into an active conformation (164). The same appears to hold true with Ras proteins. H-Ras proteins that are predominantly inactive show a slower palmitate turnover in vivo than do activated oncogenic forms (161). This finding again suggests that the activated protein is a better substrate for depalmitoylation, although APT1 has not been specifically implicated in the latter reaction.

The variety of activated substrates indicates that APT1 does not recognize a specific protein sequence determinant. A conformational change (or dissociation of a binding partner) in the vicinity of the palmitoyl group may make it more accessible for cleavage, regardless of protein sequence context. Another possibility concerns the proximity of APT1 to substrate. As mentioned, APT1 was purified from cytosol, but its substrates are membrane associated. It is unknown if APT1 is recruited to membranes following signal activation or if there is simply enough present near the membrane to accomplish depalmitoylation.

The effect of APT1 deficiency in mammalian cells has yet to be explored, but yeast *apt1* null cells have been generated. Deletion of *APT1* results in loss of thioesterase activity of yeast cell extracts toward lipid-modified mammalian $G_{i\alpha}$ but not H-Ras (157). If these results are confirmed with the yeast orthologs, Gpa1 and Ras2, then they suggest that another acyl-protein thioesterase activity may be present in yeast (157). Yeast cells lacking Apt1 are viable and do not have defects in growth rate or lipid metabolism (157). Despite the fact that Apt1 affects palmitate turnover on Gpa1, this does not translate into a gross signaling defect in the pheromone response pathway (157). Further studies will be necessary to uncover the signaling function of Apt1 in vivo.

CONCLUSIONS AND PERSPECTIVES

It is clear that protein palmitoylation plays important and diverse roles in eukaryotic cells. It is also evident that much remains to be understood about this lipid modification. Although palmitoylation contributes to initial membrane binding, it is probably more important for trafficking between membrane compartments or subdomains within the same membrane than serving as a stable membrane anchor. A key area for future investigation is how protein-bound palmitate cooperates with other protein determinants and membrane lipids to direct proteins to their appropriate membrane compartment.

The discovery of two *S. cerevisiae* PATs will invigorate the study of protein palmitoylation. Both of these PATs, Erf2/Erf4 and Akr1, contain a DHHC-CRD that may function generally as a PAT domain. Identification of mammalian counterparts of these proteins is an important goal. Much remains to be learned about the mechanism of the PAT reaction itself, and this may lead to the development of specific inhibitors. The molecular identification of PAT activities provides the opportunity to manipulate protein expression using gene knockouts or mRNA knockdowns. Similar strategies can be applied to APT1. The field is now positioned to make significant advances in understanding the mechanism and function of protein palmitoylation.

ACKNOWLEDGMENTS

Work in the authors' laboratory is supported by grants from the National Institutes of Health and the American Cancer Society. M.E.L. is an Established Investigator of the American Heart Association. J.E.S. is a predoctoral fellow of the Howard Hughes Medical Institute and the beneficiary of an Olin Fellowship for Women. We thank Dr. Deborah Brown and members of our laboratory, Dr. John Swarthout, Vivek Mittal, and Andrew Grillo-Hill for comments on the manuscript.

The *Annual Review of Biochemistry* is online at http://biochem.annualreviews.org

LITERATURE CITED

1. Magee AI, Courtneidge SA. 1985. *EMBO J.* 4:1137–44
2. Zha J, Weiler S, Oh KJ, Wei MC, Korsmeyer SJ. 2000. *Science* 290:1761–65
3. Farazi TA, Waksman G, Gordon JI. 2001. *J. Biol. Chem.* 276:39501–4
4. Johnson D, Bhatnagar R, Knoll L, Gordon J. 1994. *Annu. Rev. Biochem.* 63:869–914
5. Resh MD. 1999. *Biochim. Biophys. Acta* 1451:1–16
6. Linder ME, Deschenes RJ. 2003. *Biochemistry* 42:4311–20
7. Pepinsky RB, Zeng C, Wen D, Rayhorn P, Baker DP, et al. 1998. *J. Biol. Chem.* 273:14037–45
8. Chamoun Z, Mann RK, Nellen D, von Kessler DP, Bellotto M, et al. 2001. *Science* 293:2080–84
9. Lee JD, Treisman JE. 2001. *Curr. Biol.* 11:1147–52
9a. Amanai K, Jiang J. 2001. Development 128: 5119–27
10. Micchelli CA, The I, Selva E, Mogila V, Perrimon N. 2002. *Development* 129: 843–51
11. Mann RK, Beachy PA. 2004. *Annu. Rev. Biochem.* 73:891–923
12. Kojima M, Hosoda H, Date Y, Nakazato M, Matsuo H, Kangawa K. 1999. *Nature* 402:656–60

13. Branton WD, Rudnick MS, Zhou Y, Eccleston ED, Fields GB, Bowers LD. 1993. *Nature* 365:496–97

14. Stanley P, Koronakis V, Hughes C. 1998. *Microbiol. Mol. Biol. Rev.* 62: 309–33

15. Stevenson FT, Bursten SL, Locksley RM, Lovett DH. 1992. *J. Exp. Med.* 176: 1053–62

16. Stevenson FT, Bursten SL, Fanton C, Locksley RM, Lovett DH. 1993. *Proc. Natl. Acad. Sci. USA* 90:7245–49

17. Bijlmakers MJ, Marsh M. 2003. *Trends Cell Biol.* 13:32–42

18. El-Husseini AE-D, Bredt DS. 2002. *Nat. Rev. Neurosci.* 3:791–802

19. Qanbar R, Bouvier M. 2003. *Pharmacol. Ther.* 97:1–33

20. Dunphy JT, Linder ME. 1998. *Biochim. Biophys. Acta* 1436:245–61

21. Hancock JF. 2003. *Nat. Rev. Mol. Cell Biol.* 4:373–84

22. Zhang FL, Casey PJ. 1996. *Annu. Rev. Biochem.* 65:241–69

23. Michaelson D, Silletti J, Murphy G, D'Eustachio P, Rush M, Philips MR. 2001. *J. Cell Biol.* 152:111–26

24. Furuhjelm J, Peranen J. 2003. *J. Cell Sci.* 116:3729–38

25. Adamson P, Marshall CJ, Hall A, Tilbrook PA. 1992. *J. Biol. Chem.* 267: 20033–38

26. Navarro-Lerida I, Alvarez-Barrientos A, Gavilanes F, Rodriguez-Crespo I. 2002. *J. Cell Sci.* 115:3119–30

27. Robinson LJ, Michel T. 1995. *Proc. Natl. Acad. Sci. USA* 92:11776–80

28. Liu J, Hughes TE, Sessa WC. 1997. *J. Cell Biol.* 137:1525–35

29. Schroeder H, Leventis R, Shahinian S, Walton PA, Silvius JR. 1996. *J. Cell Biol.* 134:647–60

30. Kleuss C, Krause E. 2003. *EMBO J.* 22: 826–32

31. Linder ME, Middleton P, Hepler JR, Taussig R, Gilman AG, Mumby SM. 1993. *Proc. Natl. Acad. Sci. USA* 90: 3675–79

32. Mumby SM, Heukeroth RO, Gordon JI, Gilman AG. 1990. *Proc. Natl. Acad. Sci. USA* 87:728–32

33. El-Husseini AE, Craven SE, Chetkovich DM, Firestein BL, Schnell E, et al. 2000. *J. Cell Biol.* 148:159–72

34. Sutton RB, Fasshauer D, Jahn R, Brunger AT. 1998. *Nature* 395:347–53

35. Gonzalo S, Greentree WK, Linder ME. 1999. *J. Biol. Chem.* 274:21313–18

36. Lane SR, Liu Y. 1997. *J. Neurochem.* 69:1864–69

37. Roth AF, Feng Y, Chen L, Davis NG. 2002. *J. Cell Biol.* 159:23–28

38. Lavy M, Bracha-Drori K, Sternberg H, Yalovsky S. 2002. *Plant Cell* 14: 2431–50

39. Sugars JM, Cellek S, Manifava M, Coadwell J, Ktistakis NT. 1999. *J. Biol. Chem.* 274:30023–27

40. Xie Z, Ho WT, Exton JH. 2002. *Biochim. Biophys. Acta* 1580:9–21

41. Sugars JM, Cellek S, Manifava M, Coadwell J, Ktistakis NT. 2002. *J. Biol. Chem.* 277:29152–61

42. Du G, Altshuller YM, Vitale N, Huang P, Chasserot-Golaz S, et al. 2003. *J. Cell Biol.* 162:305–15

43. Lobo S, Greentree WK, Linder ME, Deschenes RJ. 2002. *J. Biol. Chem.* 277: 41268–73

44. Berthiaume L, Deichaite I, Peseckis S, Resh MD. 1994. *J. Biol. Chem.* 269: 6498–505

45. Corvi MM, Soltys CL, Berthiaume LG. 2001. *J. Biol. Chem.* 276:45704–12

46. Quesnel S, Silvius JR. 1994. *Biochemistry* 33:13340–48

47. Belanger C, Ansanay H, Qanbar R, Bouvier M. 2001. *FEBS Lett.* 499:59–64

48. Bizzozero OA, Bixler HA, Pastuszyn A. 2001. *Biochim. Biophys. Acta* 1545: 278–88

49. Duncan JA, Gilman AG. 1996. *J. Biol. Chem.* 271:23594–600

50. Veit M. 2000. *Biochem. J.* 345: 145–51

51. Leventis R, Juel G, Knudsen JK, Silvius JR. 1997. *Biochemistry* 36:5546–53

52. Dunphy JT, Schroeder H, Leventis R, Greentree WK, Knudsen JK, et al. 2000. *Biochim. Biophys. Acta* 1485:185–98

53. Wedegaertner PB, Bourne HR. 1994. *Cell* 77:1063–70

54. Gossett RE, Frolov AA, Roths JB, Behnke WD, Kier AB, Schroeder F. 1996. *Lipids* 31:895–918

55. Bohm S, Frishman D, Mewes HW. 1997. *Nucleic Acids Res.* 25:2464–69

56. Putilina T, Wong P, Gentleman S. 1999. *Mol. Cell. Biochem.* 195:219–26

57. Bartels DJ, Mitchell DA, Dong XW, Deschenes RJ. 1999. *Mol. Cell. Biol.* 19: 6775–87

58. Jung V, Chen L, Hofmann SL, Wigler M, Powers S. 1995. *Mol. Cell. Biol.* 15: 1333–42

59. Dohlman HG, Thorner JW. 2001. *Annu. Rev. Biochem.* 70:703–54

60. Feng Y, Davis NG. 2000. *Mol. Cell. Biol.* 20:5350–59

61. Zhao L, Lobo S, Dong X, Ault AD, Deschenes RJ. 2002. *J. Biol. Chem.* 277: 49352–59

62. Dong X, Mitchell DA, Lobo S, Zhao L, Bartels DJ, Deschenes RJ. 2003. *Mol. Cell. Biol.* 23:6574–84

63. Winzeler EA, Shoemaker DD, Astromoff A, Liang H, Anderson K, et al. 1999. *Science* 285:901–6

64. Enyenihi AH, Saunders WS. 2003. *Genetics* 163:47–54

65. Bonangelino CJ, Chavez EM, Bonifacino JS. 2002. *Mol. Biol. Cell* 13: 2486–501

66. Singaraja RR, Hadano S, Metzler M, Givan S, Wellington CL, et al. 2002. *Hum. Mol. Genet.* 11:2815–28

67. Uemura T, Mori H, Mishina M. 2002. *Biochem. Biophys. Res. Commun.* 296: 492–96

68. Li B, Cong F, Tan CP, Wang SX, Goff SP. 2002. *J. Biol. Chem.* 277:28870–76

69. Chaudhary J, Skinner MK. 2002. *Endocrinology* 143:426–35

70. Hancock JF, Paterson H, Marshall CJ. 1990. *Cell* 63:133–39

71. Shahinian S, Silvius JR. 1995. *Biochemistry* 34:3813–22

72. Peitzsch RM, McLaughlin S. 1993. *Biochemistry* 32:10436–43

73. Veit M, Nurnberg B, Spicher K, Harteneck C, Ponimaskin E, et al. 1994. *FEBS Lett.* 339:160–64

74. Evanko DS, Thiyagarajan MM, Wedegaertner PB. 2000. *J. Biol. Chem.* 275: 1327–36

75. Wedegaertner PB, Chu DH, Wilson PT, Levis MJ, Bourne HR. 1993. *J. Biol. Chem.* 268:25001–8

76. Bhattacharyya R, Wedegaertner PB. 2000. *J. Biol. Chem.* 275:14992–99

77. Kosloff M, Elia N, Selinger Z. 2002. *Biochemistry* 41:14518–23

78. Chen C, Seow KT, Guo K, Yaw LP, Lin SC. 1999. *J. Biol. Chem.* 274: 19799–806

79. Bernstein LS, Grillo AG, Loranger SS, Linder ME. 2000. *J. Biol. Chem.* 275: 18520–26

80. Brown DA, London E. 1998. *Annu. Rev. Cell Dev. Biol.* 14:111–36

81. Edidin M. 2003. *Annu. Rev. Biophys. Biomol. Struct.* 32:257–83

82. Schroeder R, London E, Brown D. 1994. *Proc. Natl. Acad. Sci. USA* 91:12130–34

83. Shenoy-Scaria AM, Dietzen DJ, Kwong J, Link DC, Lublin DM. 1994. *J. Cell Biol.* 126:353–63

84. Robbins SM, Quintrell NA, Bishop JM. 1995. *Mol. Cell. Biol.* 15:3507–15

85. Guzzi F, Zanchetta D, Chini B, Parenti M. 2001. *Biochem. J.* 355:323–31

86. Arni S, Keilbaugh SA, Ostermeyer AG, Brown DA. 1998. *J. Biol. Chem.* 273: 28478–85

87. Webb Y, Hermida-Matsumoto L, Resh MD. 2000. *J. Biol. Chem.* 275:261–70

88. Melkonian KA, Ostermeyer AG, Chen JZ, Roth MG, Brown DA. 1999. *J. Biol. Chem.* 274:3910–17

89. Moffett S, Brown DA, Linder ME. 2000. *J. Biol. Chem.* 275:2191–98

90. Wang TY, Leventis R, Silvius JR. 2001. *Biochemistry* 40:13031–40

91. Liang X, Nazarian A, Erdjument-Bromage H, Bornmann W, Tempst P, Resh MD. 2001. *J. Biol. Chem.* 276:30987–94

92. Zacharias DA, Violin JD, Newton AC, Tsien RY. 2002. *Science* 296:913–16

93. Prior IA, Muncke C, Parton RG, Hancock JF. 2003. *J. Cell Biol.* 160:165–70

94. Kanaani J, El-Husseini AE-D, Aguilera-Moreno A, Diacovo JM, Bredt DS, Baekkeskov S. 2002. *J. Cell Biol.* 158:1229–38

95. Anderson RG, Jacobson K. 2002. *Science* 296:1821–25

96. Dykstra M, Cherukuri A, Sohn HW, Tzeng SJ, Pierce SK. 2003. *Annu. Rev. Immunol.* 21:457–81

97. Huang C, Hepler JR, Chen LT, Gilman AG, Anderson RG, Mumby SM. 1997. *Mol. Biol. Cell* 8:2365–78

98. Huang C, Duncan JA, Gilman AG, Mumby SM. 1999. *Proc. Natl. Acad. Sci. USA* 96:412–17

99. Prior IA, Harding A, Yan J, Sluimer J, Parton RG, Hancock JF. 2001. *Nat. Cell Biol.* 3:368–75

100. Pizzo P, Viola A. 2003. *Curr. Opin. Immunol.* 15:255–60

101. Kabouridis PS, Magee AI, Ley SC. 1997. *EMBO J.* 16:4983–98

102. Zhang W, Trible RP, Samelson LE. 1998. *Immunity* 9:239–46

103. Lin J, Weiss A, Finco TS. 1999. *J. Biol. Chem.* 274:28861–64

104. Zeyda M, Staffler G, Horejsi V, Waldhausl W, Stulnig TM. 2002. *J. Biol. Chem.* 277:28418–23

105. Choy E, Chiu VK, Silletti J, Feoktistov M, Morimoto T, et al. 1999. *Cell* 98:69–80

106. Apolloni A, Prior IA, Lindsay M, Parton RG, Hancock JF. 2000. *Mol. Cell. Biol.* 20:2475–87

107. Schmidt WK, Tam A, Fujimura-Kamada K, Michaelis S. 1998. *Proc. Natl. Acad. Sci. USA* 95:11175–80

108. Dai Q, Choy E, Chiu V, Romano J, Slivka SR, et al. 1998. *J. Biol. Chem.* 273:15030–34

109. Chiu VK, Bivona T, Hach A, Sajous JB, Silletti J, et al. 2002. *Nat. Cell Biol.* 4:343–50

110. Michaelson D, Ahearn I, Bergo M, Young S, Philips M. 2002. *Mol. Biol. Cell* 13:3294–302

111. Takida S, Wedegaertner PB. 2003. *J. Biol. Chem.* 278:17284–90

112. Rehm A, Ploegh HL. 1997. *J. Cell Biol.* 137:305–17

113. Watson RT, Furukawa M, Chiang SH, Boeglin D, Kanzaki M, et al. 2003. *Mol. Cell. Biol.* 23:961–74

114. Di Paolo G, Lutjens R, Pellier V, Stimpson SA, Beuchat MH, et al. 1997. *J. Biol. Chem.* 272:5175–82

115. Lutjens R, Igarashi M, Pellier V, Blasey H, Di Paolo G, et al. 2000. *Eur. J. Neurosci.* 12:2224–34

116. Christgau S, Schierbeck H, Aanstoot HJ, Aagaard L, Begley K, et al. 1991. *J. Biol. Chem.* 266:21257–64

117. Christgau S, Aanstoot HJ, Schierbeck H, Begley K, Tullin S, et al. 1992. *J. Cell Biol.* 118:309–20

118. Shi Y, Veit B, Baekkeskov S. 1994. *J. Cell Biol.* 124:927–34

119. Solimena M, Dirkx R Jr, Radzynski M, Mundigl O, De Camilli P. 1994. *J. Cell Biol.* 126:331–41

120. Morales J, Fishburn CS, Wilson PT, Bourne HR. 1998. *Mol. Biol. Cell* 9:1–14

121. Song J, Dohlman HG. 1996. *Biochemistry* 35:14806–17

122. Hepler JR, Biddlecome GH, Kleuss C, Camp LA, Hofmann SL, et al. 1996. *J. Biol. Chem.* 271:496–504

123. Ross EM, Wilkie TM. 2000. *Annu. Rev. Biochem.* 69:795–827

124. Tu YP, Wang J, Ross EM. 1997. *Science* 278:1132–35

125. Hollinger S, Hepler JR. 2002. *Pharmacol. Rev.* 54:527–59

126. Srinivasa SP, Bernstein LS, Blumer KJ, Linder ME. 1998. *Proc. Natl. Acad. Sci. USA* 95:5584–89

127. Druey KM, Ugur O, Caron JM, Chen CK, Backlund PS, Jones TL. 1999. *J. Biol. Chem.* 274:18836–42

128. Tu YP, Popov S, Slaughter C, Ross EM. 1999. *J. Biol. Chem.* 274:38260–67

129. Hiol A, Davey PC, Osterhout JL, Waheed AA, Fischer ER, et al. 2003. *J. Biol. Chem.* 278:19301–8

130. Castro-Fernandez C, Janovick JA, Brothers SP, Fisher RA, Ji TH, Conn PM. 2002. *Endocrinology* 143:1310–17

131. Osterhout JL, Waheed AA, Hiol A, Ward RJ, Davey PC, et al. 2003. *J. Biol. Chem.* 278:19309–16

132. Tu YP, Woodson J, Ross EM. 2001. *J. Biol. Chem.* 276:20160–66

133. Popov SG, Krishna UM, Falck JR, Wilkie TM. 2000. *J. Biol. Chem.* 275:18962–68

134. Wang YX, Kauffman EJ, Duex JE, Weisman LS. 2001. *J. Biol. Chem.* 276:35133–40

135. Veit M, Laage R, Dietrich L, Wang L, Ungermann C. 2001. *EMBO J.* 20:3145–55

136. Bryant NJ, Stevens TH. 1998. *Microbiol. Mol. Biol. Rev.* 62:230–47

137. Wickner W. 2002. *EMBO J.* 21:1241–47

138. Glick BS, Rothman JE. 1987. *Nature* 326:309–12

139. Pfanner N, Glick BS, Arden SR, Rothman JE. 1990. *J. Cell Biol.* 110:955–61

140. Haas A, Wickner W. 1996. *EMBO J.* 15:3296–305

141. Wang YX, Catlett NL, Weisman LS. 1998. *J. Cell Biol.* 140:1063–74

142. Veit M, Dietrich LE, Ungermann C. 2003. *FEBS Lett.* 540:101–5

143. Pan X, Goldfarb DS. 1998. *J. Cell Sci.* 111:2137–47

144. Fleckenstein D, Rohde M, Klionsky DJ, Rudiger M. 1998. *J. Cell Sci.* 111:3109–18

145. Soyombo AA, Hofmann SL. 1997. *J. Biol. Chem.* 272:27456–63

146. Verkruyse LA, Hofmann SL. 1996. *J. Biol. Chem.* 271:15831–36

147. Schweizer A, Kornfeld S, Rohrer J. 1996. *J. Cell Biol.* 132:577–84

148. Degtyarev MY, Spiegel AM, Jones TL. 1993. *J. Biol. Chem.* 268:23769–72

149. Mumby SM, Kleuss C, Gilman AG. 1994. *Proc. Natl. Acad. Sci. USA* 91:2800–4

150. Jones TL, Degtyarev MY, Backlund PS Jr. 1997. *Biochemistry* 36:7185–91

151. Chen CA, Manning DR. 2000. *J. Biol. Chem.* 275:23516–22

152. Hayashi MK, Haga T. 1997. *Arch. Biochem. Biophys.* 340:376–82

153. Loisel TP, Adam L, Hebert TE, Bouvier M. 1996. *Biochemistry* 35:15923–32

154. Loisel TP, Ansanay H, Adam L, Marullo S, Seifert R, et al. 1999. *J. Biol. Chem.* 274:31014–19

155. Iiri T, Backlund PS Jr, Jones TL, Wedegaertner PB, Bourne HR. 1996. *Proc. Natl. Acad. Sci. USA* 93:14592–97

156. Duncan JA, Gilman AG. 1998. *J. Biol. Chem.* 273:15830–37

157. Duncan JA, Gilman AG. 2002. *J. Biol. Chem.* 277:31740–52

158. Wedegaertner PB, Bourne HR, von Zastrow M. 1996. *Mol. Biol. Cell* 7:1225–33

159. Dunphy JT, Greentree WK, Manahan CL, Linder ME. 1996. *J. Biol. Chem.* 271:7154–59

160. Robinson LJ, Busconi L, Michel T. 1995. *J. Biol. Chem.* 270:995–98

161. Baker TL, Zheng H, Walker J, Coloff JL, Buss JE. 2003. *J. Biol. Chem.* 278:19292–300

162. El-Husseini AE-D, Schnell E, Dakoji S, Sweeney N, Zhou Q, et al. 2002. *Cell* 108:849–63

163. Sugimoto H, Hayashi H, Yamashita S. 1996. *J. Biol. Chem.* 271:7705–11

164. Yeh DC, Duncan JA, Yamashita S, Michel T. 1999. *J. Biol. Chem.* 274:33148–54

165. Hancock JF, Magee AI, Childs JE, Marshall CJ. 1989. *Cell* 57:1167–77

166. Deschenes RJ, Broach JR. 1987. *Mol. Cell. Biol.* 7:2344–51

167. Alland L, Peseckis SM, Atherton RE,

Berthiaume L, Resh MD. 1994. *J. Biol. Chem.* 269:16701–5

168. Koegl M, Zlatkine P, Ley SC, Courtneidge SA, Magee AI. 1994. *Biochem. J.* 303:749–53
169. Degtyarev MY, Spiegel AM, Jones TL. 1994. *J. Biol. Chem.* 269:30898–903
170. Liu Y, Fisher DA, Storm DR. 1993. *Biochemistry* 32:10714–19
171. Topinka JR, Bredt DS. 1998. *Neuron* 20: 125–34
172. Deleted in proof
173. Gundersen CB, Mastrogiacomo A, Faull K, Umbach JA. 1994. *J. Biol. Chem.* 269: 19197–99
174. De Vries L, Elenko E, Hubler L, Jones TL, Farquhar MG. 1996. *Proc. Natl. Acad. Sci. USA* 93:15203–8

Annu. Rev. Biochem. 2004. 73:589–615
doi: 10.1146/annurev.biochem.73.012803.092453
Copyright © 2004 by Annual Reviews. All rights reserved
First published online as a Review in Advance on March 18, 2004

FLAP ENDONUCLEASE 1: A Central Component of DNA Metabolism

Yuan Liu, Hui-I Kao, and Robert A. Bambara

Department of Biochemistry and Biophysics, University of Rochester School of Medicine and Dentistry, Rochester, New York 14642; email: liu14@niehs.nih.gov, huii_kao@urmc.rochester.edu, robert_bambara@urmc.rochester.edu

Key Words   FEN1, structure-specific endonuclease, DNA replication, DNA long-patch base excision repair, genomic stability

■ **Abstract**   One strand of cellular DNA is generated as RNA-initiated discontinuous segments called Okazaki fragments that later are joined. The RNA terminated region is displaced into a 5′ single-stranded flap, which is removed by the structure-specific flap endonuclease 1 (FEN1), leaving a nick for ligation. Similarly, in long-patch base excision repair, a damaged nucleotide is displaced into a flap and removed by FEN1. FEN1 is a genome stabilization factor that prevents flaps from equilibrating into structures that lead to duplications and deletions. As an endonuclease, FEN1 enters the flap from the 5′ end and then tracks to cleave the flap base. Cleavage is oriented by the formation of a double flap. Analyses of FEN1 crystal structures suggest mechanisms for tracking and cleavage. Some flaps can form self-annealed and template bubble structures that interfere with FEN1. FEN1 interacts with other nucleases and helicases that allow it to act efficiently on structured flaps. Genetic and biochemical analyses continue to reveal many roles of FEN1.

CONTENTS

INTRODUCTION

Replication of double-stranded DNA involves opening of the helix to create a replication fork. Because of the 5′ to 3′ directionality of DNA synthesis and the antiparallel structure of the double helix, one strand of the nascent DNA, the leading strand, is synthesized continuously in the direction of the opening fork. The other strand is extended discontinuously in the opposite direction (1, 2). This lagging strand is initially made as a series of short segments, named Okazaki fragments, which are later joined. In eukaryotes, the size of Okazaki fragments is 100–150 nucleotides (nt) (3–7). The synthesis of an Okazaki fragment is initiated by the DNA polymerase α/primase complex (pol α). The primase domain of the complex produces an 8–12 oligoribonucleotide primer, and then the polymerase extends the primer by adding another 20 deoxyribonucleotides (dNTPs) (8). Subsequently, an accessory protein, replication factor C (RFC), initiates a process called polymerase switching (9, 10) in which RFC displaces pol α and loads a toroidal protein, the proliferating cell nuclear antigen (PCNA), onto the duplex DNA encircling the double helix at the 3′ terminus of the primer. PCNA then acts as a sliding clamp to tether DNA polymerase δ (pol δ) onto the 3′ end of the primer. This event initiates highly processive DNA polymerization (2).

When synthesis of one fragment encounters the next, the elongating DNA strand displaces the 5′-end region of the downstream segment. A 5′-unannealed flap is created that contains the initiator RNA primer. The primer must be removed prior to ligation of the fragments to allow completion of lagging-strand synthesis. All the proposed models for primer removal, discussed below, contain a common key nuclease at the step preceeding the ligation, the flap endonuclease 1 (FEN1).

FEN1 is a structure-specific 5′ endo/exonuclease. As an endonuclease, FEN1 specifically recognizes a double-stranded DNA with a 5′-unannealed flap and makes an endonucleolytic cleavage at the base of the flap (11, 12). As a 5′ exonuclease, it degrades nucleotides from a nick or a gap progressively (11, 12).

THE ROLES OF FEN1 IN OKAZAKI FRAGMENT MATURATION PATHWAYS

Participation of FEN1 in mammalian Okazaki fragment processing was initially examined using the simian virus 40 (SV40) (13) and circular plasmid (14) systems reconstituted from purified proteins. In both systems, FEN1 is required to degrade RNA primers in Okazaki fragments (13–15). Analyses using Okazaki fragment primer-template model substrates have shown that RNase H could remove most of the initiator RNA but leaves a residual ribonucleotide upstream of the RNA-DNA junction, which is subsequently processed by FEN1 exonucleolytically (16, 17). The cooperation between RNase H and FEN1 generates a nick for DNA ligase I to form an intact DNA strand (16). FEN1 could aid this process by cleaving endonucleolytically within and beyond the RNA region (18) after displacement into a flap (Figure 1A). A genetic study has shown that RNase H is not essential for Okazaki fragment processing, suggesting that FEN1 alone could remove the initiator RNA (19). In this scenario, after strand-displacement synthesis by DNA polymerase pol α, δ, or ϵ, FEN1 alone removes the flap containing the RNA primer (2, 12, 17, 20–23) (Figure 1B).

Recently, another endonuclease, named Dna2p, has also been implicated in Okazaki fragment processing. Dna2p physically interacts with the yeast FEN1 homologue, Rad27(ScFEN1;RTH1)p (24). It can cleave a 5′ flap containing an RNA primer (7), although only within the DNA region (7). Cleavage is stimulated by the RNA segment at the 5′ end and by the single-stranded DNA-binding protein, replication protein A (RPA) (7, 25). In one model, RPA plays a critical role in governing endonuclease switching (25). Displacement synthesis produces a flap long enough to be bound by RPA. RPA coating stimulates cleavage by Dna2p but inhibits FEN1. Dna2p makes a cut within the DNA beyond the initiator RNA, leaving a shorter DNA flap that is no longer bound by RPA. Subsequently, FEN1 loads onto the trimmed flap and cuts it at the base (25) (Figure 1C). This pathway may be particularly important when FEN1 activity is compromised because pol δ and FEN1 are very efficient at creating and removing flaps (26). Different proposed pathways for Okazaki fragment processing are presented schematically in Figure 1. FEN1 activity is critical in all three models.

In vivo studies in mammals, yeast, and bacteria also attest to the importance of FEN1 in DNA replication and repair. Deletion of both copies of FEN1 genes leads to mouse embryonic lethality (27). A haploinsufficiency of FEN1 function may play a role in tumor progression, and loss of FEN1 causes genomic instability (27). FEN1 also appears to be critical for normal cell cycle progression in mouse embryo. Mouse FEN1 null blastocysts (Fen1$^{-/-}$) cannot enter S phase to carry out normal DNA synthesis and are arrested in the endocycle (28). Fen1$^{-/-}$ cells also undergo extensive apoptosis when they are exposed to reactive oxygen species and ionizing radiation (28), indicating that these cells fail to maintain normal DNA replication and repair.

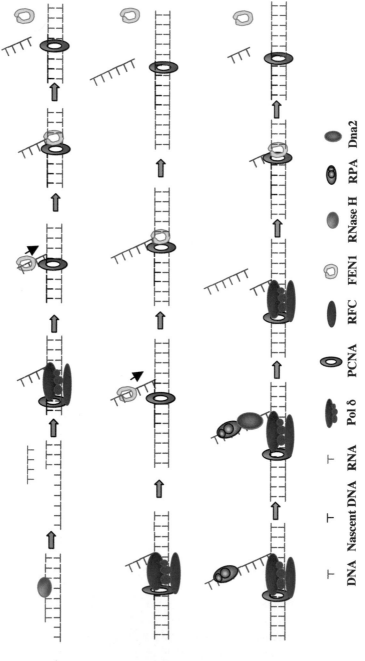

Figure 1 Models for Okazaki fragment maturation. *(A)* RNase H/FEN1 pathway. *(B)* FEN1-only pathway. *(C)* Dna2p/RPA/FEN1 pathway.

Although FEN1 is not essential for survival in *Saccharomyces cerevisiae* and *Schizosaccharomyces pombe*, it is critical for normal cell growth and proliferation. The growth rate of the *rad27(scfen1)*Δ mutant was significantly reduced even at the permissive temperatures (16°C, 23°C, and 30°C) (29, 30). At 37°C, *rad27(scfen1)*Δ strains exhibited cell cycle arrest in the late S phase or G2 phase, and DNA replication was halted (29–31). The appearance is characterized by the formation of a large cell with a nonseparated daughter cell, and a nucleus wedged inside the mother cell or in the junction between mother and daughter cells (29, 30). These phenotypes resemble those of other mutants, such as *cdc2* and *cdc9*, known to be defective in DNA replication. In fact, the accumulation of replication forks, bubbles, and flaps in a *rad27(scfen1)* mutant strain was implied by the characteristics of chromosome banding in a pulse-field gel (29).

Mutations of the 5'-3' exonuclease (FEN1) domain in *Escherichia coli* polymerase I (pol I) lead to temperature sensitivity (32). At a nonpermissive temperature, mutants cannot complete Okazaki fragment maturation, which prevents the sealing of short fragments into large DNA (32). It appears that inability of the mutant protein to remove initiator RNAs in Okazaki fragments is the cause of these phenotypes.

DISCOVERY OF FEN1—A STRUCTURE-SPECIFIC ENDONUCLEASE

The first FEN1 homologue was discovered in 1968 as a 5' exonuclease activity associated with highly purified *E. coli* pol I (33a). Later, Kornberg and colleagues (33) found that the 5'-3' nuclease of *E. coli* pol I possesses an endonucleolytic cleavage activity that can excise mismatched sequences and hydrolyze distorted regions such as irradiated poly-thymine segments to release thymine dimers from duplex DNA. This implies an important role for this nuclease in both DNA replication and repair. Around the same time, Lindahl et al. (34) reported that an exonuclease, isolated from rabbit tissues, could cleave the 5' ends of a double-stranded DNA. This DNase IV was shown to have cleavage properties similar to that of the pol I nuclease (34). One year later, Setlow et al. (35) defined the unique functions of the large and small subtilisin fragments of pol I. They found that the small fragment, a 5'-3' exonuclease, could make oligonucleotide products. Moreover, a mixture of both small and large fragments also generated oligomers (35). This result implied that DNA synthesis catalyzed by the polymerase displaced the 5' end of the downstream strand to create a "frayed" single-stranded DNA (35), which was then removed endonucleolytically.

In the early 1970s, Kelly et al. (36a) reported a unique function of *E. coli* pol I, that they called "nick translation." On nicked DNA, the extension of the 3' hydroxyl terminus and the 5' cleavage of the downstream strand occur simultaneously. This results in the movement of the nick along the DNA duplex in the 5'-3' direction but with no net synthesis. Soon after, Masamune & Richardson showed that *E. coli* pol I could also produce net synthesis on a nicked substrate. They postulated a mechanism involving branched intermediates (36b). In the 1980s, Lundquist & Olivera determined that the preferred substrate for the 5'-3' exonuclease of *E. coli* pol I is a displaced single-stranded DNA. The polymerase and nuclease functions act on the same substrate, creating transient overhangs, which may become intermediates in general recombination (36).

During the late 1980s and early 1990s, several mammalian 5'-3' exonuclease homologues were isolated and characterized (13–15, 21, 37, 38). Guggenheimer et al. (37) purified a 39 ± 2 kDa protein from human HeLa cell nuclear extracts. The protein, designated factor pL, was essential for initiation of adenovirus DNA replication (37). Further biochemical characterization of factor pL indicates that it was actually a 5'-3' exonuclease that can degrade a 5'-end displaced single-stranded DNA of the adenovirus replication origin (38). Later, Ishimi et al. (13) also obtained a 44 kDa 5'-3' exonuclease from the same cell extracts and demonstrated that it was essential for removing RNA primers for the joining of adjacent Okazaki fragments. Goulian et al. (14) also isolated a 50 kDa 5'-3' exonuclease, which was required to remove RNA primers from DNA segments resulting from discontinuous synthesis prior to their joining, from mouse cell extracts. They named this protein circle closing activity (cca)/exonuclease because it is essential for converting a circular single-stranded DNA template to a covalently closed duplex circle. Later, Siegal et al. (20) identified a 56 kDa protein, named calf 5' to 3' exonuclease, that functionally interacted with DNA polymerase ϵ (pol ϵ). It cleaved a downstream DNA fragment in a length corresponding to the number of nucleotides incorporated at the upstream primer (20). Considered retrospectively, this finding indicated that polymerization created the flap structure in the downstream primer and that mammalian nick translation proceeded by a strand displacement-flap cleavage mechanism.

In 1994, Lieber and colleagues (11) cloned a mouse nuclease and named the DNA structure, containing both a duplex DNA and 5'-displaced single-stranded DNA, a DNA flap. The nuclease that can specifically cleave the flap was named FEN1 (11). They identified a high similarity in DNA sequence, protein structure, and enzyme activities between the mouse FEN1 and the *S. cerevisiae* YKL510 open reading frame (11, 39), indicating that YKL510 encoded a form of FEN1 (40). Later, YKL510 was renamed *RAD27* (29), consistent with nomenclature for a radiation-sensitive mutant in yeast, because the *rad27Δ* strain is sensitive to DNA damaging reagents and UV radiation (29). Sommers et al. (30) named the same gene RTH1 (RAD two homolog) because of the high homology between the YKL510 encoding region and the RAD2 nuclease family, indicating a structural connection between the two classes of nucleases. Mass spectrometry analysis on the tryptic fragments of the DNase IV enzyme revealed that it is actually a homologue of Rad2 protein in *S. pombe* (42). The protein sequence also exhibits significant homology with that of the 5'-3' exonuclease of *E. coli* pol I. In the early 1990s, FEN1 homologues in bacteriophage and archea were also identified. These findings were followed by identification of FEN1 in plants (43,

44) and *Xenopus* (45). Based on the similar molecular weight, substrate specificity, and enzymatic reaction optimi among the 5'-3' exonucleases isolated from various mammalian tissues, it was evident that all are forms of FEN1 (11, 13–15, 20, 34, 38, 46).

Eukaryotic FEN1 proteins consist of three domains, defined as N-terminal (N), intermediate (I), and C-terminal (C) (40, 47, 48), whereas the prokaryotic versions only contain the N and I domains (40, 47, 48). Both the N and I domains share high homology among various FEN1 species in bacteriophage, bacteria, eubacteria, archeabacteria, yeast, and mammals. Some proteins that share high homology in the N and I domains with FEN1 homologues are also classified as members of the FEN1 family. Among them, Exo I (49), XPG (50), and ERCC5 (40, 41, 47, 48, 50) are examples. These proteins are involved in mismatch repair (51) and nucleotide excision repair (50). Thus far, the FEN1 family consists of more than 10 members. The key amino acid residues that are involved in enzyme catalysis and substrate binding are highly conserved (48), indicating that these FEN1 homologues employ similar mechanisms. The motifs for substrate binding and catalysis are present within the N and I domains of FEN1. The C domain is responsible for FEN1 interacting with other proteins (48). The consensus sequence for human FEN1 to interact with other proteins is XXFXFF. However, one study suggests that the region around this motif also participates in substrate binding (52).

BIOCHEMICAL PROPERTIES, SUBSTRATE SPECIFICITY, ENZYME-SUBSTRATE INTERACTION, AND ENZYMATIC MECHANISM OF FEN1

The cleavage reaction of FEN1 requires divalent cations (11, 14, 38). Efficient cleavage activity is maintained in the presence of 1 mM to 10 mM Mg^{2+} (11). Mn^{2+} actually accelerates the cleavage rate. However, Ca^{2+} and Zn^{2+} cannot substitute for Mg^{2+}. Cleavage activity is maximal in the absence of salt and greatly reduced at 50 mM NaCl (11) or KCl (11, 14). The pH for efficient enzyme cleavage is broad, ranging from 7.5 to 9.5 (14). The optimal pH for mammalian FEN1 is 8 (11).

FEN1 endonucleolytically cleaves a 5'-unannealed flap (11, 12). It is capable of cleaving both RNA and DNA flaps (11, 12, 18). FEN1 can cleave a flap structure lacking the adjacent upstream primer (pseudo Y structure) but with low efficiency (11, 12). The nuclease is inert on single-stranded DNA, double-stranded DNA, a heterologous loop, a D loop, a Holliday junction, and both 3' and 5' overhangs (11). Besides its endonucleolytic activity, FEN1 also has a low efficiency exonuclease that cleaves a nick, a gap, or a recessed 5' end of double-stranded DNA (11, 12, 53). Among various 5' flaps, FEN1 removes a flap with its base directly adjacent to an upstream primer (nick flap). FEN1 endonucleolytic cleavage on nick-flap substrates generally results in two kinds of products, a nick that can be ligated and a 1-nt gap (11, 12).

Recent studies have shown that a double-flap structure with 1-nt 3' tail is the optimal substrate for FEN1 homologues in archeabacteria (54), yeast (55), bacteria (57) and human (56, 58). Cleavage of the optimal double-flap makes only a single product, a nick that can be ligated, and this structure is proposed to be created in vivo by a transient flap equilibration after strand-displacement synthesis (55). FEN1 cuts this substrate one nucleotide into a duplex DNA downstream of the bifurcated junction, allowing the 3' tail to anneal to form one side of the nick (54, 55, 57). The cleavage efficiency and specificity with the optimal double-flap substrate correlate with higher binding affinity of the nuclease compared to the nick flap (59). Thus the substrate binding (59), and specificity of FEN1 on the double-flap with 1-nt 3' tail, is significantly improved suggesting that this structure is the biological substrate for FEN1 in vivo (55). In all of these substrates cleavage is structure but not sequence specific (11, 55).

FEN1 interacts with both the flap and the duplex region of its substrate (60). The substrate binding regions of human FEN1 have been mapped using micrococcal nuclease (60) and *E. coli* exonuclease III (J. Qiu, R. Liu, M. Sherman, and B. Shen, unpublished data). The bound nuclease occupies about 25 nts of a flap adjacent to the bifurcated junction (60). The footprint of FEN1 on the template was shown to be 13 nt downstream and 15 nt upstream, respectively, from this junction (61; J. Qiu, R. Liu, M. Sherman, and B. Shen, unpublished data).

Catalysis of cleavage by FEN1 resembles the two metal ion mechanism proposed for Klenow 3'-5' exonuclease (62). In this model, one of the metal ions facilitates the formation of a hydroxyl group that can perform nucleophilic attack on a scissile phosphodiester bond, likely through a metal-bound hydroxide (63). The other ion stabilizes both the oxyanion of the leaving group and the pentavalent species during the transition state, promoting the leaving of this group. This mechanism requires the distance between the metal ions to be less than 4Å. In all resolved FEN1 homologue crystals, however, this distance is longer than 4Å. Thus, substrate-induced conformational changes must occur in FEN1 before the cleavage to bring the two metal ion sites close enough for cleavage (63). The occurrence of this conformational change in human FEN1 was detected by small-angle X-ray scattering (64) and Fourier transform infrared spectroscopy (65).

FEN1 TRACKING MECHANISM

FEN1 Employs a Unique Tracking Mechanism to Remove a Flap

FEN1 has been proposed to load from the 5' end of a flap and traverse along its entire length to the junction between the unannealed 5' flap and the downstream duplex DNA (23, 66). Results, supporting this conclusion, show that a region of double-stranded DNA anywhere on the flap is inhibitory to cleavage (23, 66). A free 5' end is absolutely required for human FEN1 to load onto a flap because binding of streptavidin at the 5' end of even very long flaps prevents FEN1 binding and cleavage at the base of the flap (23, 67). Also, FEN1 cannot cleave bubble substrates (68, 69). In such substrates, the 5' end region of the flap is annealed to the template, creating a region of single strands flanked by double strands.

Implications of FEN1 Structure for Tracking

The crystal structures of FEN1 homologues such as T5 5'-3' exonuclease (70) and *Methanococcus jannaschii* FEN1 (71) exhibit either a preformed loop (71) or a helical arch (70) above their catalytic sites. The 5' flap was proposed to thread through the hole in the protein created by these structures (70, 71). However, use of these fully encircling structures is inconsistent with the ability of the nuclease to accommodate large modifications of the flaps, which include branch structures and chemical adducts (72). These results prompted a proposal that the flexible loop region undergoes a conformational change allowing it to fold into a C clamp with an open end (72). The latter model is supported by a resolved *Pyrococcus furiosus* FEN1 crystal structure that contains a helical clamp (73). Upon binding to a substrate, the enzyme undergoes a conformational change that closes the clamp, effectively enclosing the single-stranded flap (73). The substrate-induced conformational change identified in human FEN1 (64, 65) is consistent with this mechanism. Thus, a tracking, instead of a threading, mechanism has been implicated for FEN1 cleavage.

The FEN1 homologues Taq polymerase (74) and T4 RNase H (75) do not contain any preformed loop, arch, or clamp in their crystal structures. However, tracking is still required for Taq polymerase to cleave a flap (66). This nuclease may employ a different structural approach to clamp onto the flap for tracking.

ELEMENTS OF FEN1 STRUCTURE FOR SUBSTRATE RECOGNITION AND CLEAVAGE

The crystal structures of FEN1 homologues from bacteriophage, bacteria, eubacteria, and archeabacteria exhibit several common structures for enzyme catalysis and substrate binding. Beyond the helical loop implicated in tracking, there is a catalytic groove that contains two metal ion binding sites below the helical loop. The core of this groove is composed of a bundle of β sheets surrounded by α helices on both sides (70, 71, 73–75). A motif, named the helix-3 turn-helix (H3TH), which resembles the helix-hairpin-helix motif in some transcription factors, is present in most FEN1 structures (70, 71, 73–75). Because the helix-hairpin helix is a double-stranded DNA binding motif, the H3TH motif is proposed to be responsible for interacting with the downstream double-stranded DNA of a flap substrate (76).

In the catalytic groove of various FEN1 homologues, two clusters of acidic amino acids, mainly aspartic and glutamic acids, comprise the metal ion binding sites (70, 71, 73–75, 77, 78). Mutational analysis identified the essential amino acids for catalysis and substrate binding as being the residues that coordinate with the metal ions in human FEN1 (58, 77–80). D34, D86, and D181 are responsible for cleavage, whereas E156, D158, D179, G231, and D233 are involved in substrate binding (58, 77, 78, 80). Because these amino acids are predicted to coordinate with distinct metal ions, it is proposed that one metal ion may directly participate in enzyme catalysis. The other is responsible for substrate binding (78). R47, R70, R192, R200, and R201, located near the catalytic center of human FEN1, are also involved in substrate binding and may interact with the upstream primer of flap substrates (81; J. Qiu, R. Liu, M. Sherman, and B. Shen, unpublished data). Amino acids located within the flexible loop, such as K93, L97, and L111, are critical for human FEN1 cleavage but not binding (58) (J. Qiu, R. Liu, M. Sherman, and B. Shen, unpublished data). R104, R129, and K132 on the same loop are also important for interaction with the flap (J. Qiu, R. Liu, M. Sherman, and B. Shen, unpublished data). Possibly these residues participate in enzyme tracking and orienting the flap to the catalytic site of FEN1 (58).

ROLES OF FEN1 IN DNA BASE EXCISION REPAIR

Cells can suffer extensive DNA damage from exposure to either endogenous reactive metabolites or exogenous DNA damaging agents that oxidize or alkylate DNA bases (82, 83). It has been estimated that one of the oxidative damage products, 8-oxo G, is generated at a frequency of 10^4 to 10^5 per cell per day (84). The major system for correcting DNA base damage is the DNA base excision repair (BER) pathway. This process is initiated by a DNA glycosylase that recognizes a damaged base and cleaves a glycosidic bond linking the base to a sugar phosphate backbone to create an apurinic/apyrimidinic or abasic (AP) site. Subsequently, AP endonuclease 1 (APE1) cleaves the DNA backbone at the 5' end of the AP site. Depending on the type of DNA damage, the abasic site is repaired by either the short (SP-BER) or long (LP-BER) patch base excision repair pathway. SP-BER only involves the repair of one base in a nucleotide that has not sustained damage on the sugar moiety. In mammals, DNA polymerase β (β-pol) is responsible for removing the abasic sugar residue from the nicked AP site by a β-elimination reaction and then for filling in the resultant 1-nt gap (83, 85). However, an oxidized or a reduced base cannot be removed by SP-BER. The lesion is subjected to LP-BER, a process involving a displacement of the damaged base into a flap intermediate 2 to 12 nts in length by either polymerase

β, δ, or ϵ (86, 87). Subsequently, FEN1 removes the damage by cleaving off the flap endonucleolytically.

The relative use of each BER pathway is different among species. In mammals, SP-BER has been proposed as the major repair pathway. However, LP-BER still accounts for some DNA damage repair. In bacteria, LP-BER is a minor back-up pathway (88). In *S. cerevisiae* and malaria parasite, SP-BER is absent, and LP-BER is the major repair pathway (89–92).

The importance of FEN1 in LP-BER has been demonstrated both in vitro and in vivo. In vitro reconstitution of BER has shown that the 5'-3' exonuclease of *E. coli* pol I is required in LP-BER but not in SP-BER (88). In *S. cerevisiae* the *rad27(scfen1)Δ* mutant is sensitive to damage resulting from the DNA alkylating agent, methylmethane sulfonate (MMS) (29, 30). Deletion of both the *APE1* and *RAD27(ScFEN1)* genes allowed cells to be more resistant to MMS than those with disruption of FEN1 alone (93), suggesting that the products from APE1 endonucleolytic cleavage are subjected to a repair pathway that requires FEN1. The *rad27(scfen1)* mutants bearing a partial functional defect also exhibited MMS sensitivity (94). In addition, knockout of both copies of chicken FEN1 leads to hypersensitivity of cells to methylating agents and hydrogen peroxide (95). Human carcinoma cell lines expressing a dominant-negative FEN1 mutant also exhibit high sensitivity to MMS (96). FEN1 cannot remove an abasic sugar residue directly. It initiates an endonucleolytic cleavage only if the 5'-AP-site-terminated flap is at least 1 nt long (97). With increased flap length, the enzyme efficiently removes DNA damage by cleaving within the AP site-terminated flap (97). Thus, the nuclease relies on strand-displacement synthesis to create a damage-containing flap for cleavage.

In both *S. cerevisiae* and *S. pombe*, the FEN1 null mutant is also sensitive to UV irradiation (29, 30, 41). The UV sensitivity can be complemented by human FEN1 (41), indicating that FEN1 is also responsible for repairing UV-induced DNA damage. Deletion of both *S. pombe* UV damage endonuclease (UVDE) and *RAD2(SpFEN1)* makes cells more resistant to UV than the strain with deletion of *RAD2(SpFEN1)* alone (98). This result indicates that the UV-damaged DNA, 5'-incised by UVDE, is subsequently processed by FEN1. Indeed, FEN1 homologues from human, *S. cerevisiae*, and *S. pombe* can endonucleolytically remove photoproducts resulting from UV irradiation. These photoproducts include cyclobutane pyrimidine dimers (CPDs), (6–4) photoproducts, and thymine dimers nicked by a UVDE 5' to the damaged nucleotide (99). These results highlight the importance of FEN1 in the DNA LP-BER pathway of bacteria, yeast, vertebrates, and mammals. Interestingly, FEN1 mutants are fairly resistant to ionizing irradiation (29, 95, 96). In addition, other FEN1 homologues, such as Exo I, Din 7, and Yen, cannot substitute for FEN1 in the repair of alkylation damage (100). This finding indicates that FEN1 is specifically involved in the LP-BER pathway but not other repair pathways.

In *Xenopus* and mammals, LP-BER was found to be PCNA-dependent (101, 102). Depletion of PCNA completely abolishes LP-BER (103). A defective interaction between FEN1 and PCNA reduces FEN1 cleavage activity in BER reactions thereby compromising the efficiency of damage repair (104, 105). PCNA may support high efficiency of LP-BER by stimulating both FEN1 cleavage activity (67, 106–110) and processivity of pol δ and ε, which perform strand-displacement synthesis (111).

Recent studies also indicate a PCNA-independent LP-BER pathway that is mediated by FEN1 and β-pol. β-pol plays a significant role in LP-BER by mediating strand-displacement synthesis (107, 110, 112, 113). This synthesis was initially thought to be PCNA-independent (107, 110). Interestingly, FEN1 can stimulate a β-pol-dependent strand-displacement synthesis to produce a 5′-deoxyribose phosphate trinucleotide (dRP-N$_3$) flap for its own cleavage (114). In addition, poly(ADP-ribose) polymerase (PARP), a nick surveillance protein, has been shown to stimulate both FEN1 and β-pol. Its stimulation of β-pol requires the presence of FEN1. The biological relevance of interaction among FEN1, β-pol, and PARP is unclear. The advantage might be that these proteins form a complex, or repairosome, allowing the proteins to achieve an optimal activity in LP-BER (115). Alternatively to strand displacement, β-pol may add the first nucleotide and dissociate, allowing pol δ and ε to continue further strand-displacement synthesis (116).

FEN1 interacts with APE1 physically and functionally to achieve an efficient cleavage. Both FEN1 endo- and exonucleolytic cleavage activities are stimulated by APE1 (117, 118). A physical interaction was also identified between the two enzymes by immunoprecipitation (117). In LP-BER, APE1 is proposed to bind a nicked AP site and protects this site until a DNA polymerase displaces it during creation of the flap for FEN1 cleavage. Interestingly, APE1 can stimulate LP-BER and partially compensate for p21 inhibition of the pathway (119). Possibly when PCNA-dependent LP-BER is inhibited by p21, APE1 serves as a substitute for PCNA to stimulate FEN1 activity (117–119).

ROLE OF FEN1 IN MAINTAINING GENOME STABILITY

FEN1 Functional Defects Destabilize the Genome

The *rad27(scfen1)*Δ and other mutations lead to a high frequency of chromosome loss as measured by either a loss of a centromeric plasmid (29), a minichromosome (41), or *MAT* gene (31). In addition, *rad27(scfen1)* mutants demonstrate an increased rate of recombination that results in loss of heterozygosity at *MAT* (31) or mitotic crossover (120). Functionally defective mutations of *RAD27(ScFEN1)* enhance inter- and intra-chromosomal recombination. When combined with a nuclease deficiency mutation in the 3′-5′ exonuclease domain of pol δ, recombination is further increased (104, 121). This result suggests that uncleaved flaps are potentially lethal and that cells have developed mechanisms to back up this removal process.

Yeast genetic studies have suggested several pathways to resolve the accumulated flaps when FEN1 function is defective. *S. pombe rad2(spfen1)*Δ or *S. cerevisiae rad27(scfen1)*Δ causes lethality in combination with a deletion of either the *RAD51* (*RHP51* in *S. pombe*) or *RAD52* genes (41, 120, 122). These genes are important in double-stranded DNA break repair (122), suggesting that this pathway is a backup for flap removal. Furthermore, a *rad27(scfen1*Δ*)/pol3-exo* double mutant is synthetically lethal, which indicates that either the 3'-5' exonuclease decreases displacement synthesis and prevents further flap formation, or 5' flaps can equilibrate into 3' flaps that are the substrates for the 3'-5' exonuclease (121).

Defects in FEN1 increase the rate of spontaneous mutations. The *rad27(scfen1)*Δ and FEN1 nuclease defective mutations enhanced the rates of reversion mutations (120, 121), frameshifts (104), and forward mutations of the *CAN1* gene by 10- to 100-fold (94, 104, 120, 121, 123). Detailed analysis of the mutation spectra revealed that FEN1 mutant strains allowed sequence insertions or duplications (94, 100, 120). Interestingly, in the FEN1 null strain, the duplicated fragments, resulting from Can^r forward mutations or Lys$^+$ reversion mutations, exhibit a unique pattern. They have a duplication of a unique DNA sequence ranging from 5 to 108 bps in size and a flanking region of direct repeats ranging from 3 to 12 bps in length (120). This implies that an uncleaved flap reanneals to the template in a misaligned configuration to form a bubble structure. Evidently annealing of 3 to 12 bps is sufficient to allow the flap to be ligated with the upstream DNA strand. In fact, ligation of similar structures has been observed in vitro (69). As a result, this event causes integration of the duplicated segment into the genome. Proper FEN1 function, therefore, prevents sequence duplication and repeat sequence expansion (124).

FEN1 AND REPEAT SEQUENCE EXPANSIONS

Proposed Models for Repeat Sequence Expansion

Simple repetitive DNA sequences occur commonly in the eukaryotic genome (125, 126). In the human population, they are highly polymorphic, allowing them to be used as markers for genetic mapping and medicine (127). However, they also underlie a characteristic type of mutation that causes human diseases (127). Two major repeat sequence expansions that are associated with human diseases have been identified to date. Dinucleotide repeat (DNR) sequence expansion has been implicated in some human cancers (128, 129), and trinucleotide repeat (TNR) expansion is responsible for human neurodegenerative diseases (125, 130–133), such as Huntington's disease and myotonic dystrophy (125, 130). Thus far, there are 15 human neurodegenerative diseases exclusively caused by CTG, CAG, and CCG repeat expansion (131, 133). TNR expansion and disease correlate with the propensity of self-base pairing of these specific TNRs (132,

134, 135). The TNRs, CTG/CAG and CGG/CCG, that can form secondary structures, such as a hairpin, G-quartets, or a triplex on a flap, readily undergo triplet repeat expansion (134). However, TNRs, such as GAT/CTA, which are not capable of base pairing, fail to expand (135). Self-annealing repeats inhibit the activity of DNA polymerases and other replication proteins (135). Thus, two replication-related models for TNR expansion have been proposed.

In the DNA slippage model, the blockage of strand-displacement synthesis across the TNR region by DNA polymerases leads to a transient dissociation of newly synthesized DNA from its template. This dissociation subsequently causes misalignment between these DNA strands, termed DNA slippage (134). If the nascent DNA strand forms a hairpin or loop structure, resumption of DNA synthesis can bypass these structures and lead to TNR expansion. This model is supported by the studies in vitro that indicate DNA synthesis is blocked by various structures formed by triplet repeats (135–137). Another expansion model that involves FEN1 will be discussed in detail below. Alternative models involving mechanisms, such as homologous recombination (138, 139), gene conversion (140), and base excision repair (141), are also proposed to explain why triplet repeats expand. In addition, recent studies have indicated that another mechanism, gap repair, may cause CAG repeat expansion in the mammalian germ cell (142, 143). It is proposed that the expansion arises from a gap-filling synthesis when a CAG hairpin generated from a normal postmeiotic DNA strand break is integrated into a DNA strand (142, 143).

Sequence Expansion Model Involving FEN1

In a DNA segment containing repeat units, creation of flaps during discontinuous replication offers an opportunity for DNA slippage-induced repeat expansion. The DNA replication enzymes involved in lagging-strand synthesis, termed trans-elements, may have a role in suppressing sequence expansion. A useful model has been proposed to describe how FEN1 relates to repeat sequence expansion (144). In this model, FEN1 removes a nonrepeat-containing flap resulting from strand-displacement synthesis. However, a structured flap, such as one with a self-complementary repeat sequence, will inhibit cleavage by blocking the loading of FEN1 from the 5′ end. The long-lived flap could then equilibrate into a bubble intermediate, caused by misaligned annealing to the template, that can be ligated into an expanded strand (144). This model points out the importance of both *cis* and *trans* elements of DNA sequence and FEN1, respectively, in determining the occurrence of sequence expansion.

The model is supported by evidence both in vitro and in vivo. TNR flaps readily form stable secondary structures that significantly reduce FEN1 binding and cleavage (68, 145). With an increase in the number of CTG repeats, removal of a CTG flap was found to be significantly reduced (68, 146). In addition, bubble structures that simulate the critical step of the expansion completely inhibited FEN1 endonucleolytic cleavage (68, 69), thereby allowing TNR expansion (69).

A yeast genetic study has also demonstrated that preformed large CTG bubbles or loops are inefficiently repaired in vivo by wild-type DNA replication and repair enzymes, including FEN1 (147). The model also suggests that once a flap folds into a stable secondary structure and before FEN1 can remove it the probability of sequence expansion at that site is enhanced.

FEN1 Prevents Repeat Sequence Expansion

Under normal circumstances, wild-type FEN1 exhibits a proficient capacity to prevent di- or trinucleotide repeat expansion. In *E. coli*, a mutant strain lacking 5'-3' exonuclease showed a frequency of dinucleotide repeat sequence expansion 13- to 30-fold higher than the wild type (148). This mutation mainly caused base insertion and sequence duplications that contained clusters of direct repeats or imperfect inverted repeats (149). The mutation pattern is similar to the one in *rad27(scfen1)Δ* (120), indicating that the sequence duplication results from integration of a newly synthesized flap that cannot be processed by the deficient 5'-3' exonuclease.

Due to the association of sequence expansion and human diseases, there has been interest in determining the roles of Rad27(ScFEN1)p in the stability of both di- and trinucleotide repeats in *S. cerevisiae* (128, 150). In a *rad27(scfen1)Δ* strain, the rate of dinucleotide repeat expansion is 100- 160-fold of that found in the wild-type strain (123, 151). Partial defects of the FEN1 also led to instability of mono- or dinucleotide repeats (94, 152). In all mutant strains, the changes in dinucleotide repeat sequences are predominantly expansions (94, 123, 151), suggesting that flap misalignment is the major pathway for sequence expansion in the absence of FEN1.

Deletion of *RAD27(ScFEN1)* also greatly enhances triplet repeat instability, predominantly causing expansion (145, 153, 154). TNR expansion appears to specifically correlate with the loss of functional RAD27(ScFEN1)p but does not relate to the proteins involved in the mismatch repair pathway (153, 154). Furthermore, triplet repeats, such as CGG and CTG, exhibit orientation-dependent instability (154, 155) in a *rad27(scfen1)Δ* strain with a newly synthesized lagging strand constituted by CGG or CTG repeats readily experiencing sequence expansion (154, 155). Thus, it appears that the same mechanism as with DNR expansion is involved in TNR expansion when FEN1 function is defective. Additional evidence for a misalignment mechanism is that CTG and CGG repeat expansion exclusively results from increases of small numbers of repeat units ranging from 5 to 40 repeats (154, 155). These expansions are mainly duplications within the repeat tracts (154, 155) and are consistent with joining an unresolved flap to an upstream Okazaki fragment. This observation argues against significant involvement of other mechanisms, such as recombination, which should allow expansion of a large segment of repeat tracts. Long TNR expansions were occasionally observed in *rad27(scfen1)Δ*. However, this could be due to multiple rounds of joining events of TNR repeat flaps.

Interestingly, FEN1 cannot alter a minimum length threshold of CTG repeats that allows expansion (156), suggesting that TNRs must grow long enough to form a stable secondary structure. This structure subsequently leads to expansion. Evidently the sequence expansion threshold is mainly determined by the nature of triplet repeat structural intermediates and not by FEN1.

A recent study has demonstrated that FEN1 haploinsufficiency results in expansion of CAG repeats in Huntington's disease model mice (157). FEN1 is implicated in suppressing expansion in developing germ cells (157). In human cells, expression of a nuclease-deficient FEN1 (D181A) protein that retains a normal substrate binding ability also causes instability of CAG repeats (157). It appears that FEN1 is also essential for the stability of triplet repeats in humans (157). These results support an important role for FEN1 in maintaining TNR stability that is conserved from yeast to humans.

A linkage associating Huntington's disease and a FEN1 functional defect in human population has not been established (158). More evidence is needed to draw confident conclusions about the exact relationship between FEN1 and triplet repeat expansion in humans.

Mechanisms by Which FEN1 Inhibits Repeat Sequence Expansion

It is possible that in vivo FEN1 effectively resolves a repeat flap before it can fold into a secondary structure. Alternatively, in vivo FEN1 evolution has developed a strategy to deal with various structure-containing flaps, allowing appropriate cleavage. In vitro analysis has indicated that FEN1 competes with DNA ligase I at the last step of Okazaki fragment processing to inhibit sequence expansion (159). Experiments characterizing the yeast rad27(scfen1)-G67Sp and rad27(scfen1)-G240Dp (94) and human hfen1-G66Sp and hfen1-G242Dp (69), all exhibiting a partial endonucleolytic and severe exonucleolytic defects and sequence expansion phenotypes, have been utilized to address how the endonuclease activity of FEN1 competes with DNA ligase I to prevent expansion. It was found that the severe exonuclease defect of these mutants correlates well with their sequence expansion mutator phenotypes. However, reconstituting the last step of sequence expansion demonstrated that even wild-type FEN1 exonuclease cannot efficiently compete with DNA ligase I to suppress sequence expansion (69). Instead, FEN1 employs its endonuclease activity to remove a repeat flap before it can equilibrate into a bubble intermediate that allows sequence expansion (69). These observations indicate that in vivo the nuclease may remove triplet repeat flaps early in their formation before they form expandable structures.

Other Connections Between FEN1 and Genomic Integrity

Yeast genetic studies demonstrate that the human minisatellite DNAs, such as MS1, MS32, MS305, and CEB1, which contain various lengths of repeat units ranging from 9 to 54 bps in size, are destabilized by the *rad27(scfen1)*Δ mutation when they are integrated into the yeast genome (160, 161). In the absence of *RAD27(ScFEN1)*, minisatellites, such as CEB1, undergo sequence duplication, deletion, and also double-stranded DNA breaks that subsequently initiate homologous recombination (160). In addition, disruption of *RAD27(ScFEN1)* significantly increased recombination resulting from Alu inverted repeats (162).

FEN1 may also be involved in maintaining stability of telomeres. In a *rad27(scfen1)*Δ strain, the telomeric repeats became very heterogeneous (163, 164), particularly when mutant cells were grown at 37°C. In addition, the mutant strain accumulated a single-strand G–tail, which is indicative of a defect in synthesis of the lagging-strand C tail (163, 164). Because incomplete C strands and single-stranded DNA breaks can subsequently cause telomere shortening, strains having a deletion of *RAD27(ScFEN1)* undergo senescence earlier than the wild type (164). These observations imply an important role of FEN1 in telomere stability and in prevention of aging.

FEN1 has also been implicated in nonhomologous end joining by yeast genetic studies (165, 166). Although *RAD27(ScFEN1)* has been proposed to facilitate nonhomologous end joining by creating a blunt ended double-stranded DNA (165), the relative importance of this reaction compared to the functions of other protein players in this pathway is not clear.

The endonuclease activity of eukaryotic FEN1 is directly involved in restricting homologous recombination between short sequences, a process that causes genome rearrangement (166). FEN1 efficiently removes a short 5' flap resulting from unwinding of a heteroduplex formed between the ends of a short DNA fragment and its target genomic sequence (166). In this manner, FEN1 endonuclease activity disrupts the insertion of the short DNA fragment into its genome target (166).

Recent work demonstrated that *RAD27(ScFEN1)* maintains genome stability by inhibiting Ty1 mobility that could lead to chromosomal rearrangement and affect genome integrity (167). *RAD27(ScFEN1)* can reduce the stability of Ty1 cDNA, a rate-limiting component required to produce high-level transposition of Ty1 elements (167). Thus far, it is unknown what mechanism FEN1 employs in this process.

A very recent study indicates that FEN1 may participate in resolving a Holliday junction by cooperating with Werner protein (WRN), a member of the RecQ family (168). WRN is the gene product altered in patients with Werner's syndrome, which produces symptoms of premature aging (169). WRN exhibits both 5'-3' helicase activity and 3'-5' exonuclease activity (170, 171). Its helicase activity, rather than exonuclease activity, is required to stimulate FEN1 cleavage activity on a Holliday junction (168). FEN1 alone will not cleave this junction structure. However, WRN helicase unwinds a Holliday junction to create a pseudo Y structure that serves as the substrate for FEN1 (168). In addition, WRN recruits FEN1 onto the junction to facilitate resolution of this structure (168).

FEN1 is also involved in creation of double-stranded DNA breaks when mammalian cells are subjected to X-ray irradiation (172). Double-stranded DNA breakage can occur when FEN1 cleaves transient flaps formed by two single-stranded DNA breaks located on opposite DNA strands, resulting from ionizing radiation. In this scenario, double-stranded DNA breakage is a by-product of normal FEN1 flap cleavage (172). Thus, overexpression of FEN1 may enhance sensitivity of cells to ionizing radiation. This principle might be useful for developing a gene therapy strategy to improve treatment for cancers that are resistant to radiation therapy.

INTERACTIONS BETWEEN FEN1 AND OTHER PROTEINS

FEN1 interacts with other replication and repair proteins to achieve an optimal activity in DNA replication and repair. Thus far, the proteins identified to functionally and/or physically interact with FEN1 include PCNA (67, 173–175), Dna2 (24), WRN (176), RPA (177), AP endonuclease 1 (117), β-pol (114), p300 (178), chromatin proteins (179), and others involved in cell cycle control (180) as well as apoptosis (181).

FEN1-PCNA interaction has been extensively studied because the proteins are the common players in both DNA replication and repair. Both FEN1 endo- and exonucleolytic activities are stimulated by PCNA from 5- to 10-fold (173, 174). PCNA stimulates FEN1 cleavage on a 5′ flap, a nick, and a gap but not a bubble, loop structure, or a flap with its 5′ end blocked with a primer or a biotin-streptavidin complex (67, 174). The FEN1 stimulation is flap length independent (67). Therefore, PCNA can neither alter substrate specificity of FEN1 nor bypass its requirement for loading from the 5′ end of a flap and for tracking to the base of the flap for cleavage (67). Kinetic and substrate binding analyses indicate that PCNA stabilizes FEN1 binding onto its substrates rather than changing its catalytic properties (67) by reducing the Km value up to 11- to 12-fold (67).

The physical interaction between PCNA and FEN1 has been identified by immunoprecipitation using cell extracts (96, 173, 175), purified proteins (110, 174), and the yeast two-hybrid assay (174, 175). Three molecules of FEN1 bind to one PCNA trimer with each FEN1 molecule binding to one PCNA monomer (175). FEN1 interacts with the domain connecting loop (182) of PCNA via a consensus fragment, QGRLDG/D/SFF, designated the PCNA binding motif (182–184). PCNA is loaded onto the double-stranded DNA upstream of a flap with its C-terminal side facing in the direction of DNA synthesis (182). Thus, PCNA is located below the flap to stabilize FEN1 binding to its cleavage site (182). Interestingly, in budding yeast, FEN1 employs two different binding modes to PCNA (184). In the presence of substrates, FEN1 interacts with the PCNA C terminus and is stimulated. In the absence of substrates, FEN1 interacts with the PCNA interdomain connector loop in solution. Deletion of this interdomain connector loop produces an altered PCNA that still supports stimulation of FEN1 on substrates, suggesting that the solution binding mode is not involved in the stimulation (184, 185). Evidence for solution binding, however, suggests that FEN1 and PCNA may form a complex prior to substrate binding. Prebinding of FEN1 to PCNA implies a mega protein complex, which consists of all the replication enzymes at the fork. Perhaps this interaction will be switched to the other binding mode induced by conformational change upon substrate recognition to allow processing of the flap and stimulation by PCNA.

An interaction of FEN1 with WRN stimulates FEN1 cleavage activity (176, 186). WRN alters the Vmax of FEN1 cleavage but not substrate binding (176), and the stimulation is independent of WRN ATPase/helicase function (176). Therefore, the WRN stimulation mechanism is different from that employed by PCNA. However, WRN cannot help FEN1 to resolve a flap that folds back into a hairpin (176), suggesting that the stimulation of FEN1 by WRN is also mediated through the FEN1 tracking mechanism. The C terminus of WRN, containing amino acid residues 949–1092, is required for its interaction with FEN1 (176). It appears that interaction between FEN1 and WRN enhances the activity of FEN1 in processing Okazaki fragments. Recently a physical interaction between FEN1 and Bloom protein (BLM), another RecQ family helicase, has been indicated by coimmunoprecipitation (187). Mutation of BLM has been identified in patients with Bloom's syndrome, a disorder of cancer predisposition due to high frequency of sister chromatid exchange (188). Both WRN and BLM have been proposed to be important genomic stability factors because the defects of either protein cause hyperrecombination phenotypes (189). Because FEN1 has a role in preventing sequence expansions, the FEN1-WRN and FEN1-BLM functional interactions may synergize the effects of each protein.

FEN1 also interacts with chromatin to accomplish its roles in DNA replication. FEN1 is able to make an efficient cleavage on flap structures and DNA damage sites assembled into nucleosomes independent of chromatin remodeling (179). FEN1 may be recruited to a nucleosome through binding to a core histone tail for an access of a flap substrate (179). FEN1-nucleosome interaction may help FEN1 to accommodate flaps oriented either away from or toward the surface of the histone octamers (179). This interaction provides the flexibility for FEN1 to capture all the cleavable flaps.

Direct interactions between FEN1 and the Cdk1-cyclin A and Cdk2-cyclin A complexes (180) have been identified in vitro and in vivo. These interactions allow FEN1 to be a substrate for phosphorylation by the Cdk1or Cdk2-cyclin A complex (180). Besides phosphorylation, the C terminus of FEN1 also interacts with a transcription coactivator p300 (178). The interaction is substrate independent (178). FEN1-p300 interaction allows acetylation of FEN1 by p300 in vitro and in vivo (178). Because these modifications inhibit FEN1 substrate binding and cleavage, the data suggest that both the phosphorylation and acetylation of FEN1 are important for downregulation of its activity in mammalian cells (180).

C. elegans FEN1, CRN-1, physically interacts with a nuclease named CPS-6, which cleaves single-stranded DNA endonucleolytically to initiate programmed cell death or apoptosis (181). CRN-1(CeFEN1) acts as a cofactor to stimulate CPS-6 nuclease activity, whereas CPS-6 enhances 5'-3' exonuclease of CRN-1(CeFEN1) and its gap-dependent 5'-3' endonuclease activity (181). This latter activity presumably reflects the ability of the *C. elegans* FEN1 to access single-stranded DNA without the need to load from a free 5' end. It is proposed that the two proteins cooperate to initiate DNA fragmentation with the creation of a nick on a double-stranded DNA by CPS-6, followed by conversion of the nick into a gap by CRN-1(CeFEN1) 5'-3' exonucleolytic cleavage (181). Subsequently, double-stranded DNA breaks are produced by CRN-1(CeFEN1) gap-dependent 5'-3'endonuclease to enter the apoptosis.

A recent study demonstrates that FEN1 can physically interact with nuclear ribonucleoprotein A1 (hnRNP A1), a protein that participates in RNA metabolism (Q. Chai, L. Zheng, M. Zhou, J. Turchi, and B. Shen, unpublished data). The C-terminal glycine-rich domain of hnRNP A1 is responsible for the interaction (Q. Chai, L. Zheng, M. Zhou, J. Turchi, and B. Shen, unpublished data). FEN1 endonucleolytic cleavage activity is stimulated significantly by hnRNP A1 (Q. Chai, L. Zheng, M. Zhou, J. Turchi, and B. Shen, unpublished data). Kinetic analysis demonstrates that hnRNP A1 enhances substrate binding and catalysis by FEN1 (Q. Chai, L. Zheng, M. Zhou, J. Turchi, and B. Shen, unpublished data), suggesting that hnRNP A1 helps FEN1 loading onto a substrate and tracking to its cleavage site. The biological significance of this interaction needs to be further explored.

Even though coating of RPA on a flap substrate inhibits FEN1 cleavage activity, a moderate level of yeast RPA has been demonstrated to slightly stimulate FEN1 endonucleolytic cleavage activity (177). FEN1 may coordinate with RPA in a multiple protein complex located on a replication fork.

Both genetic and physical interactions between FEN1 and Dna2 have also been identified (24, 190). The interactions suggest cooperation between these enzymes in DNA replication and repair as shown in the Dna2 replication model (Figure 1C).

In summary, through a multiplicity of protein-protein interactions, FEN1 acts as a nexus that links the pathways of DNA replication, base excision repair, cell cycle control, and apoptosis.

EXPRESSION OF FEN1 AND REGULATION OF ITS ACTIVITY

FEN1 activity is regulated at both the transcriptional level and posttranslational level. The latter is mediated by posttranslational modifications, such as phosphorylation and acetylation. The nuclease is an abundant enzyme relative to other DNA replication-associated proteins (41). It is constitutively expressed in normal cells (29, 31, 41, 158, 191) and is localized within the nucleus (41, 192, 193). FEN1 has been identified throughout all human tissues with highest expression levels in testes, thymus, and bone marrow (158). The abundance, tissue distribution, and subcellular localization of FEN1 are consistent with its critical role in DNA replication during normal cell growth and proliferation.

In normal proliferating cells and malignant cell lines, FEN1 expression is upregulated (158, 192). However, the expression of FEN1 is downregulated during cell differentiation (193). The enzyme level is also cell cycle regulated with an increase in early G1 phase and a peak in late G1 phase (29, 31), suggesting that an accumulation of FEN1 will be needed for DNA synthesis during S phase. In addition, FEN1 expression can be induced, independent of position in the cell cycle, by DNA damaging agents such as MMS (31). This expression pattern and regulation of FEN1 correlates with its important roles in DNA replication and repair.

Four lysine residues, K354, K375, K377, and K380 located near the C terminus of FEN1 can be acetylated by the histone acetyl transferase domain of p300 (194). The acetylated FEN1 exhibits reduced endo- and exonuclease activity resulting from reduced substrate binding ability (194). This implies that FEN1-mediated DNA metabolism needs to be downregulated when cells undergo transcription. The acetylation does not affect FEN1-PCNA interaction (194), although PCNA binding sites are adjacent to the acetylated sites (194). The potential phosphorylation sites of FEN1 have been identified as three serine residues, S16, S157, and S187 (180). These sites can be phosphorylated by Cdk1-Cyclin A, Cdk2-Cyclin A, and Cdk2-Cyclin E complexes (180). The phosphorylation of FEN1 increases at the end of S phase. Phosphorylated FEN1 exhibits reduced endo- and exonucleolytic cleavage activity but retains normal substrate binding ability (180). Moreover, FEN1 phosphorylation abolishes interaction with PCNA and thereby prevents PCNA stimulation of FEN1 (180). Disruption of one of the phosphorylation sites, S187, led to a delay in S phase, indicating that FEN1 phosphorylation may be a critical step for normal cells to exit S phase (180). Thus, it appears that FEN1 plays a role in cell cycle regulation. It has been proposed that FEN1 phosphorylation occurs after it removes a flap, which may inhibit the interaction between PCNA and FEN1. This would allow more free PCNA to interact with DNA ligase and stimulate sealing of the Okazaki fragments (67). Additional studies are needed to further explore the biological significance of FEN1 acetylation and phosphorylation.

CONCLUDING REMARKS

FEN1 functions as a critical enzyme in lagging-strand DNA synthesis, LP-BER, and maintaining genome stability. Moreover, recent studies have demonstrated that its role may not be limited to these functions. FEN1 also appears to work with apoptosis and cell cycle control proteins, thereby having a wide-ranging influence on cell development and growth. These characteristics ensure an expansion of future FEN1 research.

ACKNOWLEDGMENTS

We thank Dr. B. Shen for communicating results prior to publication. We are sorry that space limitations prevented us from citing all references in the large FEN1 field. The work in the laboratory of R.A.B. is supported by grant GM24441 from the National Institutes of Health.

The *Annual Review of Biochemistry* is online at http://biochem.annualreviews.org

LITERATURE CITED

1. Kornberg A, Baker TA. 1992. *DNA Replication*. New York: Freeman
2. Bambara RA, Murante RS, Henricksen LA. 1997. *J. Biol. Chem.* 272:4647–50
3. Anderson S, DePamphilis ML. 1979. *J. Biol. Chem.* 254:11495–504
4. Murakami Y, Eki T, Hurwitz J. 1992. *Proc. Natl. Acad. Sci. USA* 89:952–56
5. Nethanel T, Reisfeld S, Dinter-Gottlieb G, Kaufmann G. 1988. *J. Virol.* 62:2867–73
6. Nethanel T, Zlotkin T, Kaufmann G. 1992. *J. Virol.* 66:6634–40
7. Bae SH, Seo YS. 2000. *J. Biol. Chem.* 275:38022–31
8. Arezi B, Kuchta RD. 2000. *Trends Biochem. Sci.* 25:572–76
9. Tsurimoto T, Stillman B. 1991. *J. Biol. Chem.* 266:1961–68
10. Waga S, Stillman B. 1994. *Nature* 369:207–12
11. Harrington JJ, Lieber MR. 1994. *EMBO J.* 13:1235–46
12. Murante RS, Huang L, Turchi JJ, Bambara RA. 1994. *J. Biol. Chem.* 269:1191–96
13. Ishimi Y, Claude A, Bullock P, Hurwitz J. 1988. *J. Biol. Chem.* 263:19723–33
14. Goulian M, Richards SH, Heard CJ, Bigsby BM. 1990. *J. Biol. Chem.* 265:18461–71
15. Waga S, Bauer G, Stillman B. 1994. *J. Biol. Chem.* 269:10923–34
16. Turchi JJ, Huang L, Murante RS, Kim Y, Bambara RA. 1994. *Proc. Natl. Acad. Sci. USA* 91:9803–7
17. Huang L, Rumbaugh JA, Murante RS, Lin RJR, Rust L, Bambara RA. 1996. *Biochemistry* 35:9266–77
18. Murante RS, Rumbaugh JA, Barnes CJ, Norton JR, Bambara RA. 1996. *J. Biol. Chem.* 271:25888–97
19. Qiu J, Qian Y, Frank P, Wintersberger U, Shen BH. 1999. *Mol. Cell. Biol.* 19:8361–67
20. Siegal G, Turchi JJ, Jessee CB, Mallaber LM, Bambara RA, Myers TW. 1992. *J. Biol. Chem.* 267:3991–99
21. Siegal G, Turchi JJ, Myers TW, Bambara RA. 1992. *Proc. Natl. Acad. Sci. USA* 89:9377–81
22. Turchi JJ, Bambara RA. 1993. *J. Biol. Chem.* 268:15136–41
23. Murante RS, Rust L, Bambara RA. 1995. *J. Biol. Chem.* 270:30377–83
24. Budd ME, Campbell JL. 1997. *Mol. Cell. Biol.* 17:2136–42
25. Bae SH, Bae KH, Kim JA, Seo YS. 2001. *Nature* 412:456–61
26. Ayyagari R, Gomes XV, Gordenin DA, Burgers PM. 2003. *J. Biol. Chem.* 278:1618–25
27. Kucherlapati M, Yang K, Kuraguchi M, Zhao J, Lia M, et al. 2002. *Proc. Natl. Acad. Sci. USA* 99:9924–29
28. Larsen E, Gran C, Saether B, Seeberg E,

Klungland A. 2003. *Mol. Cell. Biol.* 23: 5346–53

29. Reagan MS, Pittenger C, Siede W, Friedberg EC. 1995. *J. Bacteriol.* 177:364–71

30. Sommers CH, Miller EJ, Dujon B, Prakash S, Prakash L. 1995. *J. Biol. Chem.* 270:4193–96

31. Vallen EA, Cross FR. 1995. *Mol. Cell. Biol.* 15:4291–302

32. Konrad EB, Lehman IR. 1974. *Proc. Natl. Acad. Sci. USA* 71:2048–51

33a. Klett RP, Cerami A, Reich E. 1968. *Proc. Natl. Acad. Sci. USA* 60:943–50

33. Kelly RB, Atkinson MR, Huberman JA, Kornberg A. 1969. *Nature* 224:495–501

34. Lindahl T, Gally JA, Edelman GM. 1969. *Proc. Natl. Acad. Sci. USA* 62: 597–603

35. Setlow P, Brutlag D, Kornberg A. 1972. *J. Biol. Chem.* 247:224–31

36a. Kelly RB, Cozzarelli NR, Deutscher MP, Lehman IR, Kornberg A. 1970. *J. Biol. Chem.* 245:39–45

36b. Masamune Y, Richardson CC. 1971. *J. Biol. Chem.* 246:2692–701

36. Lundquist RC, Olivera BM. 1982. *Cell* 31:53–60

37. Guggenheimer RA, Nagata K, Kenny M, Hurwitz J. 1984. *J. Biol. Chem.* 259: 7815–25

38. Kenny MK, Balogh LA, Hurwitz J. 1988. *J. Biol. Chem.* 263:9801–8

39. Jacquier A, Legrain P, Dujon B. 1992. *Yeast* 8:121–32

40. Harrington JJ, Lieber MR. 1994. *Genes Dev.* 8:1344–55

41. Murray JM, Tavassoli M, al-Harithy R, Sheldrick KS, Lehmann AR, et al. 1994. *Mol. Cell. Biol.* 14:4878–88

42. Robins P, Pappin DJ, Wood RD, Lindahl T. 1994. *J. Biol. Chem.* 269:28535–38

43. Kimura S, Kai M, Kobayashi H, Suzuki A, Morioka H, et al. 1997. *Nucleic Acids Res.* 25:4970–76

44. Kimura S, Ueda T, Hatanaka M, Takenouchi M, Hashimoto J, Sakaguchi K. 2000. *Plant Mol. Biol.* 42:415–27

45. Bibikova M, Wu B, Chi E, Kim KH,

Trautman JK, Carroll D. 1998. *J. Biol. Chem.* 273:34222–29

46. Lindahl T. 1970. *Methods Enzymol.* 21: 148–53

47. Lieber MR. 1997. *BioEssays* 19:233–40

48. Shen B, Qiu J, Hosfield D, Tainer JA. 1998. *Trends Biochem. Sci.* 23:171–73

49. Zankasi P, Smith GR. 1992. *J. Biol. Chem.* 267:3014–23

50. Scherly D, Nouspikel T, Corlet J, Ucla C, Bairoch A, Clarkson SG. 1993. *Nature* 363:182–85

51. Genschel J, Bazemore L, Modrich P. 2002. *J. Biol. Chem.* 277:13302–11

52. Stucki M, Jonsson ZO, Hübscher U. 2001. *J. Biol. Chem.* 276:7843–49

53. Lindahl T. 1971. *Eur. J. Biochem.* 18: 407–14

54. Kaiser MW, Lyamicheva N, Ma W, Miller C, Neri B, et al. 1999. *J. Biol. Chem.* 274:21387–94

55. Kao HI, Henricksen LA, Liu Y, Bambara RA. 2002. *J. Biol. Chem.* 277:14379–89

56. Friedrich-Heineken E, Henneke G, Ferrari E, Hübscher U. 2003. *J. Mol. Biol.* 328:73–84

57. Xu Y, Grindley ND, Joyce CM. 2000. *J. Biol. Chem.* 275:20949–55

58. Storici F, Henneke G, Ferrari E, Gordenin D, Hübscher U, Resnick M. 2002. *EMBO J.* 21:5930–42

59. Harrington C, Perrino FW. 1995. *J. Biol. Chem.* 270:26664–69

60. Barnes CJ, Wahl AF, Shen B, Park MS, Bambara RA. 1996. *J. Biol. Chem.* 271: 29624–31

61. Allawi H, Kaiser M, Onufriev A, Ma W, Brogaard A, et al. 2003. *J. Mol. Biol.* 328: 537–54

62. Beese L, Steitz T. 1991. *EMBO J.* 10: 25–33

63. Tock MR, Frary E, Sayers JR, Grasby JA. 2003. *EMBO J.* 22:995–1004

64. Kim CY, Shen BH, Park MS, Olah GA. 1999. *J. Biol. Chem.* 274:1233–39

65. Kim CY, Park MS, Dyer RB. 2001. *Biochemistry* 40:3208–14

66. Lyamichev V, Brow MA, Dahlberg JE. 1993. *Science* 260:778–83

67. Tom S, Henricksen LA, Bambara RA. 2000. *J. Biol. Chem.* 275:10498–505

68. Henricksen LA, Tom S, Liu Y, Bambara RA. 2000. *J. Biol. Chem.* 275:16420–27

69. Liu Y, Bambara RA. 2003. *J. Biol. Chem.* 278:13728–39

70. Ceska TA, Sayers JR, Stier G, Suck D. 1996. *Nature* 382:90–93

71. Hwang KY, Baek K, Kim HY, Cho Y. 1998. *Nat. Struct. Biol.* 5:707–13

72. Bornarth CJ, Ranalli TA, Henricksen LA, Wahl AF, Bambara RA. 1999. *Biochemistry* 38:13347–54

73. Hosfield DJ, Mol CD, Shen BH, Tainer JA. 1998. *Cell* 95:135–46

74. Kim Y, Eom SH, Wang J, Lee DS, Suh SW, Steitz TA. 1995. *Nature* 376: 612–16

75. Mueser TC, Nossal NG, Hyde CC. 1996. *Cell* 85:1101–12

76. Dervan J, Feng M, Patel D, Grasby J, Artymiuk P, et al. 2002. *Proc. Natl. Acad. Sci. USA* 99:8542–47

77. Shen BH, Nolan JP, Sklar LA, Park MS. 1996. *J. Biol. Chem.* 271:9173–76

78. Shen B, Nolan JP, Sklar LA, Park MS. 1997. *Nucleic Acids Res.* 25:3332–38

79. Frank G, Qiu J, Somsouk M, Weng Y, Somsouk L, et al. 1998. *J. Biol. Chem.* 273:33064–72

80. Busen W, Frank P. 1998. *Ribonucleases H*. Paris: INSERM. 146 pp.

81. Qiu J, Bimston D, Partikian A, Shen BH. 2002. *J. Biol. Chem.* 277:24659–66

82. Lindahl T. 1993. *Nature* 362:709–15

83. Matsumoto Y, Kim K. 1995. *Science* 269: 699–702

84. Ames B, Shigenaga M, Hagen T. 1993. *Proc. Natl. Acad. Sci. USA* 90:7915–22

85. Singhal RK, Prasad R, Wilson SH. 1995. *J. Biol. Chem.* 270:949–57

86. Memisoglu A, Samson L. 2000. *Mutat. Res.* 451:39–51

87. Sattler U, Frit P, Salles B, Calsou P. 2003. *EMBO Rep.* 4:363–67

88. Dianov G, Lindahl T. 1994. *Curr. Biol.* 4:1069–76

89. Budd ME, Campbell JL. 1997. *Mutat. Res.* 384:157–67

90. Budd ME, Campbell JL. 1995. *Methods Enzymol.* 262:108–30

91. Haltiwanger B, Matsumoto Y, Nicolas E, Dianov G, Bohr V, Taraschi T. 2000. *Biochemistry* 39:763–72

92. Prakash S, Sung P, Prakash L. 1993. *Annu. Rev. Genet.* 27:33–70

93. Wu XH, Wang ZG. 1999. *Nucleic Acids Res.* 27:956–62

94. Xie YL, Liu Y, Argueso JL, Henricksen LA, Kao HI, et al. 2001. *Mol. Cell. Biol.* 21:4889–99

95. Matsuzaki Y, Adachi N, Koyama H. 2002. *Nucleic Acids Res.* 30:3273–77

96. Shibata Y, Nakamura T. 2002. *J. Biol. Chem.* 277:746–54

97. DeMott MS, Shen BH, Park MS, Bambara RA, Zigman S. 1996. *J. Biol. Chem.* 271:30068–76

98. Yonemasu R, McCready SJ, Murray JM, Osman F, Takao M, et al. 1997. *Nucleic Acids Res.* 25:1553–58

99. Yoon JH, Swiderski PM, Kaplan BE, Takao M, Yasui A, et al. 1999. *Biochemistry* 38:4809–17

100. Johnson RE, Kovvali GK, Prakash L, Prakash S. 1998. *Curr. Genet.* 34:21–29

101. Fortini P, Pascucci B, Parlanti E, Sobol RW, Wilson SH, Dogliotti E. 1998. *Biochemistry* 37:3575–80

102. Matsumoto Y, Kim K, Bogenhagen DF. 1994. *Mol. Cell. Biol.* 14:6187–97

103. Biade S, Sobol RW, Wilson SH, Matsumoto Y. 1998. *J. Biol. Chem.* 273: 898–902

104. Gary R, Park MS, Nolan JP, Cornelius HL, Kozyreva OG, et al. 1999. *Mol. Cell. Biol.* 19:5373–82

105. Matsumoto Y, Kim K, Hurwitz J, Gary R, Levin DS, et al. 1999. *J. Biol. Chem.* 274:33703–8

106. Frosina G, Fortini P, Rossi O, Carrozzino F, Raspaglio G, et al. 1996. *J. Biol. Chem.* 271:9573–78

107. Kim K, Biade S, Matsumoto Y. 1998. *J. Biol. Chem.* 273:8842–48

108. Pascucci B, Stucki M, Jonsson ZO, Dogliotti E, Hübscher U. 1999. *J. Biol. Chem.* 274:33696–702

109. Wu X, Lieber MR. 1996. *Mol. Cell. Biol.* 16:5186–93

110. Gary R, Kim K, Cornelius HL, Park MS, Matsumoto Y. 1999. *J. Biol. Chem.* 274:4354–63

111. Stucki M, Pascucci B, Parlanti E, Fortini P, Wilson SH, et al. 1998. *Oncogene* 17:835–43

112. Klungland A, Lindahl T. 1997. *EMBO J.* 16:3341–48

113. Dianov GL, Prasad R, Wilson SH, Bohr VA. 1999. *J. Biol. Chem.* 274:13741–3

114. Prasad R, Dianov GL, Bohr VA, Wilson SH. 2000. *J. Biol. Chem.* 275:4460–66

115. Lavrik OI, Prasad R, Sobol RW, Horton JK, Ackerman EJ, Wilson SH. 2001. *J. Biol. Chem.* 276:25541–48

116. Podlutsky AJ, Dianova II, Podust VN, Bohr VA, Dianov GL. 2001. *EMBO J.* 20:1477–82

117. Dianova I, Bohr VA, Dianov GL. 2001. *Biochemistry* 40:12639–44

118. Ranalli TA, DeMott MS, Bambara RA. 2002. *J. Biol. Chem.* 277:1719–27

119. Tom S, Ranalli TA, Podust VN, Bambara RA. 2001. *J. Biol. Chem.* 276:48781–89

120. Tishkoff DX, Filosi N, Gaida GM, Kolodner RD. 1997. *Cell* 88:253–63

121. Jin YH, Obert R, Burgers PM, Kunkel TA, Resnick MA, Gordenin DA. 2001. *Proc. Natl. Acad. Sci. USA* 98:5122–27

122. Symington LS. 1998. *Nucleic Acids Res.* 26:5589–95

123. Johnson RE, Kovvali GK, Prakash L, Prakash S. 1995. *Science* 269:238–40

124. Kunkel T, Resnick M, Gordenin D. 1997. *Cell* 88:155–58

125. Paulson HL, Fischbeck KH. 1996. *Annu. Rev. Neurosci.* 19:79–107

126. Debrauwere H, Gendrel CG, Lechat S, Dutreix M. 1997. *Biochimie* 79:577–86

127. Sutherland GR, Richards RI. 1995. *Proc. Natl. Acad. Sci. USA* 92:3636–41

128. Ionov Y, Peinado MA, Malkhosyan S, Shibata D, Perucho M. 1993. *Nature* 363:558–61

129. Merlo A, Mabry M, Gabrielson E, Vollmer R, Baylin SB, Sidransky D. 1994. *Cancer Res.* 54:2098–101

130. Bates G, Lehrach H. 1994. *BioEssays* 16:277–84

131. Miret JJ, Pessoa-Brandao L, Lahue RS. 1998. *Proc. Natl. Acad. Sci. USA* 95:12438–43

132. Usdin K, Grabczyk E. 2000. *Cell. Mol. Life Sci.* 57:914–31

133. Mirkin S, Smirnova E. 2002. *Nat. Genet.* 31:5–6

134. Wells RD. 1996. *J. Biol. Chem.* 271:2875–78

135. McMurray CT. 1999. *Proc. Natl. Acad. Sci. USA* 96:1823–25

136. Usdin K, Woodford KJ. 1995. *Nucleic Acids Res.* 23:4202–9

137. Kang S, Jaworski A, Ohshima K, Wells RD. 1995. *Nat. Genet.* 10:213–18

138. Cemal CK, Huxley C, Chamberlain S. 1999. *Gene* 236:53–61

139. Jakupciak JP, Wells RD. 1999. *J. Biol. Chem.* 274:23468–79

140. Richard GF, Goellner GM, McMurray CT, Haber JE. 2000. *EMBO J.* 19:2381–90

141. Lyons-Darden T, Topal MD. 1999. *J. Biol. Chem.* 274:25975–78

142. Kovtun IV, McMurray CT. 2001. *Nat. Genet.* 27:407–11

143. McMurray CT, Kortun IV. 2003. *Chromosoma* 111:505–8

144. Gordenin DA, Kunkel TA, Resnick MA. 1997. *Nat. Genet.* 16:116–18

145. Spiro C, Pelletier R, Rolfsmeier ML, Dixon MJ, Lahue RS, et al. 1999. *Mol. Cell* 4:1079–85

146. Lee S, Park MS. 2002. *Exp. Mol. Med.* 34:313–17

147. Moore H, Greenwell PW, Liu CP, Arnheim N, Petes TD. 1999. *Proc. Natl. Acad. Sci. USA* 96:1504–9

148. Morel P, Reverdy C, Michel B, Ehrlich SD, Cassuto E. 1998. *Proc. Natl. Acad. Sci. USA* 95:10003–8

149. Nagata Y, Mashimo K, Kawata M, Yamamoto K. 2002. *Genetics* 160:13–23

150. Thibodeau SN, Bren G, Schaid D. 1993. *Science* 260:816–19

151. Kokoska RJ, Stefanovic L, Tran HT, Resnick MA, Gordenin DA, Petes TD. 1998. *Mol. Cell. Biol.* 18:2779–88

152. Greene A, Snipe JR, Gordenin DA, Resnick MA. 1999. *Hum. Mol. Genet.* 8: 2263–73

153. Freudenreich CH, Kantrow SM, Zakian VA. 1998. *Science* 279:853–56

154. White PJ, Borts RH, Hirst MC. 1999. *Mol. Cell. Biol.* 19:5675–84

155. Schweitzer JK, Livingston DM. 1998. *Hum. Mol. Genet.* 7:69–74

156. Rolfsmeier ML, Dixon MJ, Pessoa-Brandao L, Pelletier R, Miret JJ, Lahue RS. 2001. *Genetics* 157:1569–79

157. Spiro C, McMurray CT. 2003. *Mol. Cell. Biol.* 23:6063–74

158. Otto CJ, Almqvist E, Hayden MR, Andrew SE. 2001. *Clin. Genet.* 59:122–27

159. Henricksen L, Veeraraghavan J, Chafin DR, Bambara RA. 2002. *J. Biol. Chem.* 277:22361–69

160. Lopes J, Debrauwere H, Buard J, Nicolas A. 2002. *EMBO J.* 21:3201–11

161. Maleki S, Cederberg H, Rannug U. 2002. *Curr. Genet.* 41:333–41

162. Lobachev K, Stenger J, Kozyreva O, Jurka J, Gordenin D, Resnick M. 2000. *EMBO J.* 19:3822–30

163. Parenteau J, Wellinger RJ. 1999. *Mol. Cell. Biol.* 19:4143–52

164. Parenteau J, Wellinger RJ. 2002. *Genetics* 162:1583–94

165. Wu X, Wilson TE, Lieber MR. 1999. *Proc. Natl. Acad. Sci. USA* 96:1303–8

166. Negritto MC, Qiu J, Ratay DO, Shen B, Bailis AM. 2001. *Mol. Cell. Biol.* 21: 2349–58

167. Sundararajan A, Lee B, Garfinkel D. 2003. *Genetics* 163:55–67

168. Sharma S, Otterlei M, Sommers JA, Driscoll HC, Dianov GL, et al. 2004. *Mol. Biol. Cell* 15:734–50

169. Yu CE, Oshima J, Fu YH, Wijsman EM, Hisama F, et al. 1996. *Science* 272: 258–62

170. Huang S, Li B, Gray MD, Oshima J, Mian IS, Campisi J. 1998. *Nat. Genet.* 20:114–16

171. Kamath-Loeb AS, Shen JC, Loeb LA, Fry M. 1998. *J. Biol. Chem.* 273: 34145–50

172. Vispe S, Ho ELY, Yung TMC, Satoh MS. 2003. *J. Biol. Chem.* 278:35279–85

173. Li X, Li J, Harrington J, Lieber MR, Burgers PM. 1995. *J. Biol. Chem.* 270: 22109–12

174. Wu X, Li J, Li X, Hsieh CL, Burgers PM, Lieber MR. 1996. *Nucleic Acids Res.* 24:2036–43

175. Chen JJ, Chen S, Saha P, Dutta A. 1996. *Proc. Natl. Acad. Sci. USA* 93: 11597–602

176. Brosh RJ, Driscoll H, Dianov G, Sommers J. 2002. *Biochemistry* 41:12204–16

177. Biswas EE, Zhu FX, Biswas SB. 1997. *Biochemistry* 36:5955–62

178. Hasan S, Stucki M, Hassa PO, Imhof R, Gehrig P, et al. 2001. *Mol. Cell* 7:1221–31

179. Huggins CF, Chafin DR, Aoyagi S, Henricksen LA, Bambara RA, Hayes JJ. 2002. *Mol. Cell* 10:1201–11

180. Henneke G, Koundrioukoff S, Hübscher U. 2003. *Oncogene* 22:4301–13

181. Parrish JZ, Yang CL, Shen BH, Xue D. 2003. *EMBO J.* 22:3451–60

182. Jonsson ZO, Hindges R, Hübscher U. 1998. *EMBO J.* 17:2412–25

183. Warbrick E, Lane D, Glover D, Cox L. 1997. *Oncogene* 14:2313–21

184. Gomes XV, Burgers PM. 2000. *EMBO J.* 19:3811–21

185. Frank G, Qiu JZ, Zheng L, Shen BH. 2001. *J. Biol. Chem.* 276:36295–302

186. Brosh RJ, von Kobbe C, Sommers J, Karmakar P, Opresko P, et al. 2001. *EMBO J.* 20:5791–801

187. Imamura O, Campbell JL. 2003. *Proc. Natl. Acad. Sci. USA* 100:8193–98

188. Hickson ID. 2003. *Nat. Rev. Cancer* 3: 169–78

189. Nakayama H. 2002. *Oncogene* 21: 9008–21

190. Budd ME, Campbell JL. 1995. *Proc. Natl. Acad. Sci. USA* 92:7642–46

191. Kim NK, Lee SH, Sohn TJ, Roy R, Mitra S, et al. 2000. *Anticancer Res.* 20:3037–43

192. Warbrick E, Coates PJ, Hall PA. 1998. *J. Pathol.* 186:319–24

193. Kim I, Lee M, Lee I, Shin S, Lee S. 2000. *Biochim. Biophys. Acta* 1496: 333–40

194. Hasan S, Stucki M, Hassa PO, Imhof R, Gehrig P, et al. 2001. *Mol. Cell* 7: 1221–31

Annu. Rev. Biochem. 2004. 73:617–56
doi: 10.1146/annurev.biochem.72.121801.161837
Copyright © 2004 by Annual Reviews. All rights reserved
First published online as a Review in Advance on March 26, 2004

EMERGING PRINCIPLES OF CONFORMATION-BASED PRION INHERITANCE

Peter Chien, Jonathan S. Weissman, and Angela H. DePace

Graduate Group in Biophysics, Howard Hughes Medical Institute, Department of Cellular and Molecular Pharmacology, University of California, San Francisco, California 94107-2240; email: pchien@fas.harvard.edu, jsw1@itsa.ucsf.edu, adepace@uclink4.berkeley.edu

Key Words   PSI, amyloid, prion strains, species barrier, misfolding

■ **Abstract**   The prion hypothesis proposes that proteins can act as infectious agents. Originally formulated to explain transmissible spongiform encephalopathies (TSEs), the prion hypothesis has been extended with the finding that several non-Mendelian traits in fungi are due to heritable changes in protein conformation, which may in some cases be beneficial. Although much remains to be learned about the specific role of cellular cofactors, mechanistic parallels between the mammalian and yeast prion phenomena point to universal features of conformation-based infection and inheritance involving propagation of ordered β-sheet-rich protein aggregates commonly referred to as amyloid. Here we focus on two such features and discuss recent efforts to explain them in terms of the physical properties of amyloid-like aggregates. The first is prion strains, wherein chemically identical infectious particles cause distinct phenotypes. The second is barriers that often prohibit prion transmission between different species. There is increasing evidence suggesting that both of these can be manifestations of the same phenomenon: the ability of a protein to misfold into multiple self-propagating conformations. Even single mutations can change the spectrum of favored misfolded conformations. In turn, changes in amyloid conformation can shift the specificity of propagation and alter strain phenotypes. This model helps explain many common and otherwise puzzling features of prion inheritance as well as aspects of noninfectious diseases involving toxic misfolded proteins.

CONTENTS

INTRODUCTION

The idea that a protein conformation can replicate itself and therefore serve as a genetic element was first formalized by the prion hypothesis, which seeks to explain an unusual set of neurodegenerative diseases known as the transmissible spongiform encephalopathies (TSEs). These devastating diseases result in progressive cognitive and motor impairment and are characterized by the accumulation of proteinaceous brain lesions or plaques (1). Sheep scrapie was the first of these diseases to be recognized, but subsequently a set of human diseases, such as kuru and Creutzfeldt-Jacob Disease (CJD), was shown to have similar clinical and pathological features. TSEs have now been identified in a wide range of mammals, including cats, cows, mink, deer, and elk (2).

Though the TSEs can arise spontaneously or be inherited, they are also infectious (3). The earliest illustrations of infectivity were accidental; sheep scrapie was transmitted to an entire flock during routine vaccination, and kuru was transmitted through ritual cannibalism practiced by a tribe in New Guinea.

Subsequent experiments showed that human disease could be transmitted to primates and surprisingly indicated that the infectious agent was resistant to classic methods for inactivating nucleic acid. The purification of the infectious agent responsible for scrapie led to the remarkable discovery that it was composed primarily, if not entirely, of protein. On the basis of this observation, Stanley Prusiner proposed that a novel proteinaceous infectious agent, termed a "prion," was responsible for these diseases [reviewed in (4)]. It was later found that the infectious protein is a ubiquitous endogenous cellular protein, termed "PrP" for prion protein.

How might an endogenous protein be infectious? In a prescient argument prompted by the need to reconcile the failure to detect nucleic acids in the infectious agent responsible for scrapie with the newly emerging central dogma of molecular biology, Griffith (5) described three general mechanisms for replication of a protein so that "the occurrence of a protein agent would not necessarily be embarrassing." In the first mechanism, a transcriptional activator could be infectious if it were to turn on a normally quiescent gene that participated in a positive feedback loop driving its own production. The second mechanism postulated a change in either protein conformation or multimeric state that cannot occur without a catalyst, such as a preformed multimeric nucleus. The final mechanism invoked an immune response feedback loop. Another mechanism has recently been described by Wickner (6): A zymogen, or self-activating enzyme, can be infectious if an active form is introduced into a pool of otherwise stably inactive proteins.

One of above mechanisms, propagation of conformational change, appears to underlie the mammalian TSEs. During purification of the infectious scrapie agent, a β-sheet-rich insoluble protease-resistant fragment of PrP was associated with highly infectious preparations. Surprisingly this form is covalently identical to the normal cellular form of PrP, but in uninfected animals PrP is alpha-helical, soluble, and protease sensitive (Figure 1). A variety of observations now support a model where the scrapie-associated form of PrP, termed "PrP^{SC}," is transmitted by conformational conversion of the normal cellular form, called PrP^{C}. Though two Nobel prizes have been awarded for research on TSEs, the biology of mammalian prion diseases is still hotly contested (7). Specifically, the size and nature of the infectious particle remain unresolved due to experimental limitations. First, the specific activity of purified material is extremely low (8). Second, infectious preparations without protease-resistant PrP^{SC} have been found (9). Third, and most challenging, recombinant infectious material has not yet been produced in vitro to provide the formal proof of the protein-only hypothesis. Together, these technical limitations have left lingering questions whether other components, such as chaperones, small molecules, or even RNAs (10) could play a role in prion infection.

Despite these questions about the mechanisms of mammalian TSEs, it has become clear that proteins can serve as genetic elements and that prions are more widespread in biology than previously thought. In 1994, Wickner (11) proposed

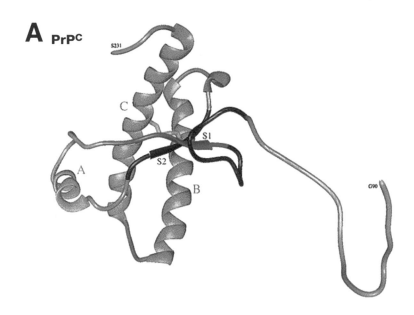

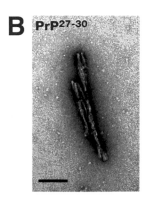

C Conformational Differences
Between PrP^C and PrP^SC

PrP^C	PrP^SC
monomeric	multimeric
soluble	insoluble
protease sensitive	protease resistant
predominantly α-helical	predominantly β-sheet

Figure 1 PrP^C and PrP^{SC} are conformationally distinct. (*A*) Solution NMR structure of Syrian hamster PrP^C, residues 90–231. The structure is predominantly alpha helical with an unstructured amino terminus (199). (*B*) Negative stain EM of Syrian hamster PrP$^{27–30}$ (Sc237 strain), stained with uranyl acetate. The material is in insoluble protease-resistant high molecular weight aggregates that are predominantly β-sheet. Scale bar is 100 nm. Image courtesy of Dr. Holger Wille (unpublished material). (*C*) Summary of differences between PrP^C and PrP^{SC}.

that the behavior of two non-Mendelian cytoplasmically inherited traits in *Saccharomyces cerevisiae*, [*PSI*$^+$] and [*URE3*], could be explained by a prion-like mechanism where an alternate protein form does not cause disease but does "infect" daughter cells as they bud from the mother. This model was based on three remarkable features shared by [URE3] and [PSI⁺]. One, propagation of [URE3] and [PSI⁺] is dependent on the continuous expression of an associated gene, URE2 and SUP35, respectively, yet their phenotypes mimic loss-of-function mutations in these genes. Two, [URE3] and [PSI⁺] can be cured by growth on guanidine hydrochloride and can return to the prion state without any changes in the genome. Three, overexpression of Ure2p and Sup35p increases the frequency of de novo [URE3] and [PSI⁺] appearance. Wickner's model elegantly explained these observations by postulating that overexpression results in a novel prion form of the protein. The prion form is self-propagating, which allows inheritance, and inactivating, which results in the apparent loss-of-function phenotype. This model has been confirmed and expanded by the work of multiple labs, and it is now established that both [PSI⁺] and [URE3] are due to the self-propagating aggregation of Sup35p and Ure2p, respectively (12, 13). More fungal prion domains were subsequently discovered, including [RNQ⁺], also known as [PIN⁺], [NU⁺], and [Het-s]. These have been comprehensively reviewed elsewhere (14, 15).

The yeast prions have provided genetically and biochemically tractable systems for studying prion behavior, greatly facilitating studies on the mechanism of conformation-based inheritance and infection (12, 13). [URE3] and [PSI⁺] are the best characterized, and both offer accessible in vivo and in vitro experimental systems. In vivo, genetic screens can exploit the nitrogen uptake phenotype of [URE3] yeast or the nonsense suppression phenotype of [PSI⁺] yeast (Figure 2). In vitro, propagation of [URE3] and [PSI⁺] are modeled by the formation of amyloid fibers. Both Ure2p and Sup35p are modular proteins with their prion activity localized to an amino-terminal glutamine/asparagine-rich domain separable from domains responsible for their normal cellular function (Figure 2). These purified prion domains spontaneously form amyloid fibers only after a characteristic lag phase that can be eliminated by the addition of preformed seeds, mimicking propagation in vivo. Importantly, formal proof of the prion hypothesis has come from studies with [PSI⁺] and [Het-s]. When introduced into cells, amyloid seeds generated in vitro from purified recombinant Sup35p, or Het-s* are able to cause de novo formation of the [PSI⁺] and [Het-s] states, respectively (16, 17).

Although there are critical differences in the cellular location and phenotypic consequences between mammalian and yeast prions, they share a remarkable number of common mechanistic features. In this review, we compare and contrast these systems in an effort to build a general model for conformation-based infection and inheritance. We first consider that both mammalian and yeast prions appear to be due to the propagation of β-sheet-rich aggregates that resemble amyloid fibers. In contrast to disordered amorphous aggregates, amyloids are highly ordered fibrillar structures, formed by a wide variety of polypeptides with no homology in either their native structures or in their amino acid sequence (Figure 3). In many cases these amyloids are self-propagating; in

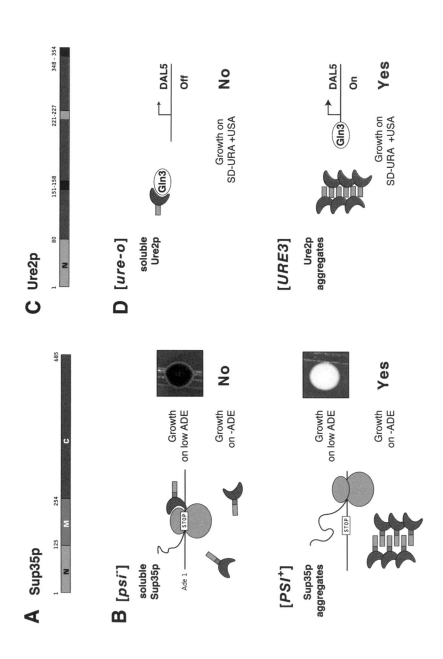

general, however, amyloids are not infectious. Thus, an unresolved question is—what distinguishes prions from this larger class of misfolded proteins?

We next consider that both mammalian and yeast prions display multiple strains in which infectious particles composed of the same protein give rise to distinct phenotypes. This strain phenomenon has been difficult to reconcile with the protein-only hypothesis, but evidence is accumulating from both the prion and amyloid fields that a single polypeptide can form multiple distinct conformations, which may provide the structural basis for strain diversity. We then examine the sequence specificity of prion propagation, which manifests in both the mammalian and yeast systems as a "species barrier," inhibiting transmission between even highly related species. Finally, we review the evidence that strains and species barriers can result from the same underlying process, namely that a single polypeptide can form multiple self-propagating states. These different conformations can lead to distinct strain phenotypes and can determine the sequence specificity of prion propagation.

Figure 2   The yeast prions [*PSI*⁺] and [*URE3*] are the result of self-propagating protein conformations. (*A*) Sup35p is a modular protein involved in translation termination; self-propagating aggregation is responsible for the [*PSI*⁺] phenotype. The amino-terminal prion-forming domain, N (*green*), is glutamine- and asparagine-rich. The middle domain, M (*blue*), is rich in charged residues. The carboxy-terminal domain, C (*orange*), contains the essential translation-termination function of the protein. (*B*) Sup35p is soluble in [*psi*⁻] yeast and able to facilitate translation termination while in [*PSI*⁺] yeast; Sup35p is aggregated, resulting in suppression of nonsense codons. Translation termination can be monitored using an *ADE1* reporter harboring a premature stop codon. [*PSI*⁺] cells are white and capable of growth on media lacking adenine, whereas [*psi*⁻] yeast accumulate a red pigment caused by lack of Ade1p and are incapable of growth on adenine-less media. (*C*) Ure2p is a modular protein involved in regulation of nitrogen catabolism; self-propagating aggregation of Ure2p is responsible for the [*URE3*] phenotype. In addition to the glutamine/asparagine-rich amino terminus (*green*), Ure2p also contains another region that facilitates prion behavor (*green*) and portions that antagonize prion formation (*black*). The remainder of the protein (*orange*) resembles glutathione S-transferase and is necessary for Ure2p signaling of the presence of high-quality nitrogen sources through Gln3p. (*D*) Normally Ure2p binds the transcription factor Gln3p, preventing the upregulation of genes, such as *DAL5*, required for uptake of poor nitrogen sources. Serendipitously, Dal5p imports not only the poor nitrogen source allantoate, but also USA (*n*-carbamyl aspartate), an intermediate in uracil biosynthesis. Thus [*ure-o*] yeast cannot grow on ureidosuccinate (USA) medium lacking uracil. In [*URE3*] yeast, Ure2p is aggregated and inactive, leading to constitutive activation of Dal5p and enabling growth on USA media lacking uracil.

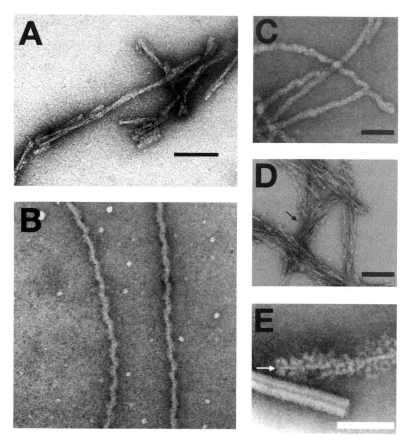

Figure 3   Amyloid-like fibers are formed by a variety of prion proteins. (*A*) EM of Syrian hamster PrP$^{27-30}$ (Sc237 strain), stained with uranyl acetate. Bar = 100 nm. Image courtesy of Dr. Holger Wille (unpublished material). (*B*) Amyloid fibers formed by Sup35NM, stained with uranyl acetate. Sup35NM fibers are on average 5–10 nm in diameter. (*C, D*) EM of full-length Ure2p fibers stained with uranyl acetate before (*C*) and after (*D*) digestion with proteinase K. Arrow in *D* indicates position of a single fiber. Bar = 100 nm (89). (*E*) Amyloid fibers formed by full-length Ure2p, stained with vanadate and visualized by dark-field scanning transmission electron microscopy (STEM). Arrow indicates the core of the fiber. Bar = 50 nm (89).

AMYLOID-LIKE SELF-PROPAGATING PROTEIN AGGREGATES UNDERLIE PRION INHERITANCE

The fundamental requirement of the prion hypothesis is that a protein be capable of adopting a state that can initiate and sustain its own replication. Although the cellular machinery for transcription and translation are used to generate new polypeptides, the infectious protein must contain enough information to direct production of the prion rather than the normal cellular form. Abundant evidence has accumulated that both mammalian and yeast prions accomplish this by directing conformational change of a normal cellular host protein into an alternate prion conformation. These alternate conformations are β-sheet-rich multimers and resemble a broader class of ordered protein aggregates termed amyloids. Amyloids are associated with a variety of noninfectious neurodegenerative diseases, such as Alzheimer's and Parkinson's diseases, as well as a range of systemic amyloidoses (18). Here we review the evidence that prions operate by directing conformational change of a host protein and what is known about the formation and structure of these alternate conformations. Finally, we explore the steps of prion replication to explain why prions are an infectious subset of the larger class of proteins that misfold into amyloid. Specifically, we will focus on the infectivity requirements beyond simple self-propagating protein structures, a feature shared by many amyloids.

Evidence for Conformational Changes in Mammalian Prions

The first evidence indicating that conformational change was involved in prion diseases arose during the purification of the infectious scrapie agent. A protease resistant protein fragment was found to copurify along with infectivity (8, 19). Subsequent cloning of this fragment revealed that it was part of a larger 33–35 kDa host glycoprotein, encoded by the *PRNP* gene (20–22). The normal cellular version of this protein, PrP^C, is distributed throughout many visceral tissues and is both soluble and highly sensitive to proteinase K digestion. However, in infected animals, an insoluble form of PrP is also present, PrP^{SC}. PrP^{SC} accumulates in aggregates and plaques in the brain. Digestion with proteinase K cleaves only its first 66 amino-terminal residues, leaving a fragment referred to as PrP$^{SC27-30}$ with an SDS-PAGE mobility of 27–30 kDa.

The difference between PrP^C and PrP^{SC} appears to reside completely in their conformations. Though mutations in the nucleic acid genome can increase rates of spontaneous disease (2), the infectious disease occurs in the absence of such mutations. Moreover, systematic analysis of posttranslational modifications have failed to find any evidence that covalent modifications underlie formation of the infectious form (23, 24). By contrast, extensive evidence argues that PrP^C and PrP^{SC} adopt distinct conformations. For example, in addition to the protease resistance and solubility mentioned above, the two conformers vary in the exposure of a number of different epitopes (25) and have dramatically different thermodynamic stabilities (26) and secondary structure content. The structures of human, hamster, bovine, and mouse PrP^C have been solved by NMR (27–30), and all are highly similar, predominantly alpha-helical folds. PrP^{SC} on the other hand is predominantly β-sheet, as revealed by Fourier transform infrared spectroscopy studies (31).

Extensive evidence reviewed elsewhere implicates the conversion of PrP^C to PrP^{SC} in disease progression (32). In vivo, PrP$^{0/0}$ mice are not susceptible to prion infection, arguing that conversion of the endogenous protein is required to develop disease (33, 34). Furthermore, the lag time before developing disease is dependent on the concentration of PrP in the host (34–37). The infectious process can be recapitulated in cell culture using a neuroblastoma N2a cell line (38). In vitro extracts enriched in PrP^{SC} can convert recombinant PrP^C to a protease resistant form called PrP^{RES}, and this material exhibits similar specificity to that seen in vivo (39, 40). It has also been reported that shearing aggregates during the polymerization reaction increases the yield of protease-resistant material (41, 42). Nonetheless, to date de novo infectious material has failed to be created in vitro. A second caveat is that infectious prion diseases have been observed in the absence of detectable protease-resistant PrP^{SC} aggregates (9, 43, 44). However, the question remains whether this absence is due to a titer of aggregates below detection limits, to a formation of an infectious conformation that is genuinely protease-sensitive, or to some more radical departure from the idea that conformational changes are necessary for generating infectivity.

Evidence for Conformational Changes in Fungal Prions

Yeast prions, like mammalian prions, are characterized by the presence of an alternate conformation of a normal cellular protein. All fungal prion proteins identified have been shown using either differential sedimentation or size-exclusion chromatography to form high-molecular-weight complexes specifically in prion-containing cells (45–53). However, the degree of aggregation in vivo can vary with genetic background (54, 55) or with the expression level of the cognate prion protein (56). Aggregated protein can be visualized in intact cells by generating prion-GFP fusion proteins that are soluble and distributed evenly throughout the cytoplasm in wild-type cells, but they are organized into punctate foci stainable by the amyloid-specific dye, thioflavin-S, in prion-containing cells (46, 49, 50, 57–59). Ure2p has also been visualized by thin-section EM followed by immunogold staining and shown to form short cytoplasmic fibrils specifically in [URE3] yeast (60). In most cases, these aggregates have been shown to be highly stable and to have altered resistance to protease digestion (45–47, 51). Finally, de novo formation of these aggregates is slow, but once formed, they are stably inherited by daughter cells during mitosis.

The fungal prions have proven to be far more amenable to reconstitution in vitro than the mammalian prion system. Extracts from [PSI$^+$] yeast can catalyze conversion of soluble Sup35p, whereas extracts from [psi$^-$] yeast do not have this activity (61). Moreover, for Ure2p, Sup35p and HET-s, inheritance can be modeled using purified protein. Following a characteristic lag phase, these proteins spontaneously form amyloid-like aggregates. Importantly, the lag phase can be eliminated by the addition of small amounts of preformed fiber seed (62–64). A number of lines of experiments argue that this seeding effect underlies prion inheritance in vivo. For example, mutations in Sup35, which affect aggregation in vivo, have parallel effects on the in vitro reaction (46, 65, 66). More directly for Sup35p and HET-s, it has been possible to create aggregates in vitro from recombinant protein and use these to convert wild-type cells to the prion state (16, 17, 66a). These experiments have provided the most complete evidence to date for the protein-only hypothesis.

Beyond supporting the prion hypothesis, this facile in vitro system allows more detailed mechanistic studies of prion conversion. Three questions stand out. One, what is the aggregation state of the infectious material? Specifically, are fibers necessary for infection or are they merely an assembly by-product of conformational conversion? What is the minimum size of an infectious particle? Two, when does conformational conversion occur? Monomers or oligomers could undergo spontaneous conformational conversions in solution that are subsequently stabilized by assembly into polymers, or conformational conversion could be driven by the assembly process itself. Three, what is the rate-limiting step in prion formation? Nucleated polymerization models argue that the formation of a multimeric nucleus is the slow step, whereas templated assembly models argue that conformational conversion is rate-limiting, though these two are not necessarily mutually exclusive. Detailed coverage of the literature addressing the conversion reaction is beyond the scope of this review, but we encourage interested readers to consult recent reviews (67–69) and research papers addressing the subject (70–74).

Prion Aggregates Resemble Amyloid Fibers

Recently it has become clear that a wide range of unrelated proteins form structurally similar β-sheet-rich aggregates, often referred to as amyloids. Amyloids have received an enormous amount of attention caused by their association with a wide variety of protein misfolding disorders, including neurodegenerative diseases such as Alzheimer's, Parkinson's, and Huntington's (75). They are also found in a number of systemic amyloidoses, characterized by peripheral deposition of a number of aggregated proteins, such as lysozyme, transthyretin (TTR), immunoglobulin light chain, β2-microglobulin, and islet amyloid protein or amylin (76). Furthermore, a number of nondisease-associated proteins, for example acylphosphatase and the SH3 domain from PIP_3 kinase, have been shown to form amyloid under mildly denaturing conditions (77, 78). The ability of such diverse polypeptides to form amyloid argues that this fold is generally accessible to polymers of amino acids, perhaps because it is stabilized by main chain rather than side chain interactions (18).

Despite the variety in amyloid-prone proteins and their aggregated states, amyloids share similarities that make it useful to discuss them as a family of related structures (Figure 3). Amyloid fibers are characterized by a set of fiber diffraction reflections indicating that the β-sheets are organized in a cross-β fold where the strands of the sheets run perpendicular to the fiber axis while the sheets run parallel to it (79–81). The repeating β-sheet structure allows the binding of the hydrophobic dyes thioflavin-T and Congo Red, both of which are commonly

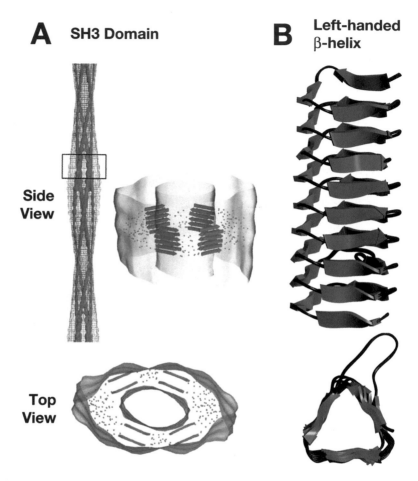

Figure 4 Two models for amyloid structure. Both fulfill the requirements of the cross-β fold in which individual β-strands are oriented perpendicular to the fiber axis, whereas β-sheets are oriented parallel to it. (*A*) Model from cryo-EM studies of amyloid formed by the SH3 domain from PIP_3 kinase (83). (*B*) An example of a left-handed β-helix (from UDP-N-acetylglucosamine pyrophosphorylase of *Streptococcus pneumoniae*, PDB ID 1G97), which has been proposed to resemble PrP^{SC} (86). Image is courtesy of Dr. Cedric Govaerts (unpublished material).

used to monitor amyloid formation in vitro. Amyloids are often composed of multiple thin protofilaments that can associate in a variety of ways to create mature fibers with a range of diameters and helical twists [reviewed in (82)]. Currently, cryo-electron microscopy has yielded the most detailed structural model of an amyloid and indicates that protofilaments can be arranged around a hollow core (83) (Figure 4). Multiple folds can satisfy the constraints of a cross-β-sheet structure (81), such as the β-helix shown in Figure 4. Most generally, amyloids are uncapped β-sheets that can incorporate new protein on their edges, leading to a fiber of defined diameter but unlimited length. In fact, well-behaved β-sheet-rich proteins appear to avoid aggregation by protecting their β-sheet edges with a variety of strategies (84).

Though the heterogeneity and insolubility of prion aggregates have made high-resolution structural studies difficult, they are known to share many features with amyloid (62–64, 85). Recently, reconstruction from electron micrographs of two-dimensional crystals of PrP^{SC} present in infectious preparations has provided enough constraints to propose structural models (86). Additionally, a crystal structure of a PrP dimer has been solved, and these findings suggest how subunits might assemble into a fiber (87). Sup35p fibers give rise to the stereotypical amyloid cross-β diffraction pattern when subjected to fiber diffraction (71), and Ure2p fibers give rise to this pattern after being subjected to heat, though Ure2p may assemble into native-like filaments under physiological conditions (88). Both Sup35p and Ure2p fibers appear to consist of a central core made up of the prion domain with globular domains corresponding to the remainder of the protein decorating the periphery (62, 89) (Figure 3). Taken together with the self-propagating behavior of amyloid, the structural similarities between amyloid and prion aggregates suggest that propagation of β-sheet-rich amyloid-like core could provide the molecular mechanism responsible for prion growth.

Self-Propagating Aggregates Are Not Sufficient for Infection/Inheritance

Amyloid fibers formed by many proteins are self-seeding (90–94), but few are infectious. For example, Aβ, the peptide whose aggregation is intimately correlated with Alzheimer's disease (95), exhibits stereotypical self-propagating behavior in vitro, forming an amyloid after a lag phase that can be eliminated by the addition of preformed fibers (67). Yet, Alzheimer's disease is not transmissible to primates or rodents (96, 97). What then is unique among the prion-associated amyloids that allows them to be infectious? We consider the steps of aggregation and transmission in Figure 5, comparing PrP and [*PSI*$^+$].

Initially, a self-propagating aggregate must form spontaneously. This is a step common to all of the amyloid diseases; in fact, most cases of Alzheimer's and Creutzfeldt-Jacob occur spontaneously in patients without any genetic predisposition to the disease (3). Yeast prions rarely occur spontaneously but are stable once formed (14). In mammals, mutations can accelerate the rate of spontaneous aggregation, as can overexpression of the aggregation-prone protein (34–37). In the case of yeast, truncations and expansions can accelerate the rate of spontaneous occurrence (66), and overexpression greatly increases the rate of prion formation (98). Finally, exposure to environmental factors, such as metals and pesticides, may also facilitate protein aggregation (99, 100).

Next, the newly formed prion must replicate itself. This involves two separable steps: growth of the infectious particle by addition to the aggregate and amplification of the number of infectious particles. Growth of the infectious

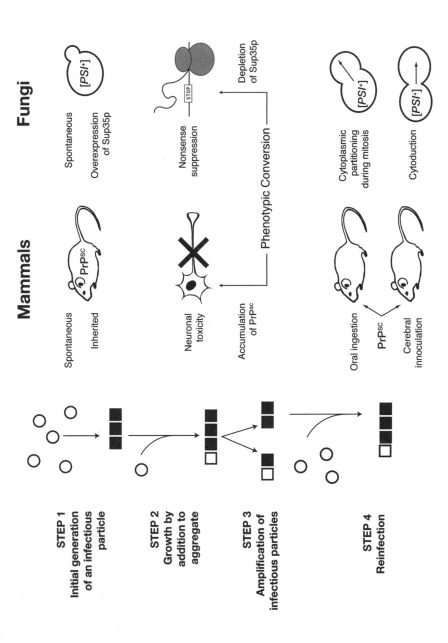

Figure 5 Steps in prion transmission. (*left*) A general replication cycle for self-propagating conformationally based prion protein is shown. (*right*) Corresponding steps during prion infection in mammals and prion inheritance in fungi are shown.

particle comes about through recruitment and assembly of new protein onto the prion. However, this process alone would only lead to an increase in mass of protein in the prion form without net increase in the number of catalytic surfaces. Therefore, new infectious particles must somehow be released from the aggregate, either by spontaneous shedding or division by a cellular factor. Though how division is accomplished by mammalian prions is unclear, division of [PSI^+] aggregates appears to require the chaperone Hsp104p. A key piece of evidence for this model is the peculiar relationship between [PSI^+] and Hsp104p in which both deletion and overexpression of *HSP104* interferes with [PSI^+] propagation (101). Hsp104p is normally involved in the rescue of aggregated protein with the help of Hsp70p and Hsp40p (102), and indeed it may play a general role in prion propagation because deletion of *HSP104* cures all known yeast prions (14, 15). Division of aggregates could be another step differentiating transmissible and nontransmissible aggregates; if aggregates are too stable to either release small units or to be degraded by chaperones, they would never exponentially amplify during infection or inheritance.

Finally, these aggregates must be transmitted into a naive host and reach a pool of substrate protein. Mammalian prions can be ingested orally, as evidenced by the Mad Cow epidemic and by the transmission of kuru through ritualistic cannibalism. The prion infection then reaches the central nervous system (CNS) apparently through the lymphoid tissue. Once in the CNS, prions are able to spread from one cell to another presumably due to the presence of PrP^C, which is exposed on the surface of the cell (103). Yeast prions are transmitted naturally during cell division or experimentally using cytoplasmic mixing. This is clearly a critical step in differentiating between infectious and nontransmissible amyloids (104). Protein aggregates could vary in their ability to circumvent the body's defenses by being differentially susceptible to degradation and/or transport. There is recent evidence that systemic amyloids can also be transmissible when administered either orally or intraveneously after an inflammatory stimulus, arguing that under the right conditions more aggregates may prove to be infectious (105).

In addition to this growth and replication cycle, prion aggegates also cause a phenotypic change in their hosts. The need to distinguish phenotypic output from prion replication is emphasized by recent work that shows high titers of prions can exist without development of clinical symptoms (106). In the case of amyloid-related neurodegenerative diseases, the mechanism of toxicity and their tissue specificity must also be determined (107). In systemic amyloidoses, disease may be caused by mechanical disruption due to enormous amyloid burden because simple removal of amyloid deposits alleviates symptoms (108). In some cases, it may be that amyloid fibers are not toxic but instead are an inert repository for improperly folded proteins. In this case, the intermediates along the pathway to amyloid formation would be the neurotoxic species (107). Indeed, partially unstructured oligomers of both Aβ and α-synuclein are toxic to cells (109, 110), as are partially unstructured oligomers of an SH3 domain, which is not associated with any known disease (111). For [*URE3*] and [*PSI⁺*], the relationship to phenotype is more straightforward; sequestration of the prion protein in aggregates leads to a state similar to a loss-of-function phenotype (11, 112). In fact, the prion domain from Sup35p can be fused to other proteins to create novel prion elements with phenotypes caused by inactivation of the fusion protein (49, 66). However, a simple inactivation model is not sufficient to explain all fungal prions because [Het-s] and [*PIN⁺*] lead to a gain-of-function phenotypes (51, 113, 114). More recently it has been shown that a neuronal member of the CPEB family shows prion-like properties in yeast and it is the prion-like form that has the greatest capacity to enhance translation of CPEB-regulated mRNA (114a). These results suggest a remarkable model in which conversion of CPEB to a prion-like state in stimulated synapses helps to maintain long-term synaptic changes associated with memory storage.

PRION STRAIN VARIATION

One of the most fascinating and perplexing features of prion biology is the existence of multiple prion strains, wherein infectious particles composed of the same protein give rise to a range of prion states that vary in incubation time, pathology, and other phenotypic aspects. Observation of strain variability preceded the prion hypothesis, and in fact, it was originally used as evidence for the existence of a nucleic acid genome in the infectious particle. Strain variation was postulated to be caused by mutations in this genome. In the context of the proposal that transmissible encephalopathies result from propagating conformational changes in a prion protein, one must postulate that a single polypeptide can misfold into multiple infectious conformations, at least one for each phenotype. As disconcerting as this idea may be, there is increasing evidence from studies of both fungal and mammalian prions that it is indeed true. Nonetheless, there remain many unresolved questions regarding the origin of prion strains and their relationship to phenotype. For instance, what roles do cellular factors play (10, 115)? Do prion conformational differences lead to strain variation or simply reflect some other mechanism that actually encodes strain diversity? Formal proof of the conformational basis for prion strains has very recently been provided for the yeast [*PSI⁺*] prion. Here it has been possible to fold the Sup35p prion protein into distinct infectious conformations (66a, 115a). Remarkably, infection of yeast with these different Sup35p conformations leads to distinct and heritable differences in [*PSI⁺*] prion strains.

Strain Variability in Mammalian Prions

Strain variability has always been closely associated with transmissible spongiform encephalopathies. Classic experiments in transmission of sheep scrapie to goats led researchers to group isolates according to clinical syndromes, such as "drowsy" and "scratchy" strains (116). Material derived from these animals could infect mice in which these strains would propagate with distinct clinical and pathological parameters, such as patterns of brain lesions (117) and lag in incubation times. Use of isogenic mouse models made it unlikely that this variation arose from host genome polymorphisms (118).

With the identification of the PrP protein as the core component of the infectious particle, classification of strains could focus on molecular analysis of differences in the prion protein. Differences in secondary structure content (119), thermal stability (26, 120), and epitope exposure (121) of PrP^{sc} isolates can be used to distinguish prion strains. Posttranslational modifications, such as glycosylation and attachment of GPI anchors, also show differences among known prion strains (2, 122, 123). Whether these covalent modifications modulate prion strains or reflect an inherent diversity among strains is still unknown.

Conformational Differences Distinguish Mammalian Prion Strains

In light of the hypothesis that prions result from propagation of an infectious conformation, much of the effort in analyzing strains has focused on identifying strain-specific conformational differences in the prion protein. Initial evidence for such differences came from strains of transmissible mink encepholopathy (TME) PrP^{SC} accumulated within the brains of infected minks and showed distinct proteolysis patterns and glycosylation profiles that correlated with different strain types. Upon injection of this material into naive hosts, not only did the newly infected animals exhibit strain-specific brain lesions and incubation times, but the converted PrP^{SC} retained the proteolytic digestion pattern of the inoculum (124). Similarly, transmission of human-derived infectious material into transgenic mice expressing a human-mouse chimera produces PrP^{SC} with hallmarks of the original strain, including protease sensitivity and glycosylation patterns (125).

A series of cell-free experiments have provided evidence that these protein conformations are sufficient to mediate their own propagation. Caughey and coworkers (39) developed an in vitro system in which brain-derived PrP^{SC} mixed with PrP^{C} converts PrP^{C} to a protease-resistant PrP^{SC}-like state, called PrP^{RES} (42). Paralleling the in vivo experiments, TME prion strains convert PrP^{C} to a PrP^{SC} similar to the initial strain, as defined by proteolysis and extent of glycosylation (126). Although other cellular factors, such as chaperones or a potential Protein X (127, 128), may be required for robust propagation of strain differences in vivo, the above observations suggest that the particular prion conformation can mediate strain-specific conversion of PrP^{C}.

Strain Variability in Yeast Prions

The existence of strains appears to be a ubiquitous feature of prions, independent of the specific prion protein, the types of posttranslational modifications, or the cellular site of conversion. Strain variability in fungal prions affects a range of different properties, including the strength of the associated phenotype, mitotic stability, and the dependence on molecular chaperones. Fungal prion strains were discovered during analysis of de novo induction of [*PSI*⁺] by overexpression of Sup35p. Remarkably, inductants showed clear and heritable differences in color phenotype, caused by differences in the strength of nonsense suppression (98, 113). Genomic mutations cannot account for these differences; once a particular [*PSI*⁺] variant was cured, the full spectrum of strains was reproduced upon reinduction. This variation among prion states has also been documented in Sup35p derived from other species. For instance, [*PSI*⁺] elements arising in *S. cerevisiae* expressing the *Pichia methanolica SUP35* showed phenotypic variation, and they can be distinguished by their differential sensitivity to a host of chaperones, such as Hsp70p and Hsp40p family members (53, 129, 130). Chaperone discrimination is also seen with a chimeric prion domain derived from *Candida albicans* and *S. cerevisiae* Sup35p; overexpression of Hsp104p results in differential curing of these prion variants (P. Chien and J. Weissman, unpublished information).

Although [*PSI*⁺] prion variants are the best characterized, similar variants have been seen in all yeast prions examined so far. De novo induction of the [*URE3*] prion results in variants distinguished by the strength of the associated phenotype and their susceptibility to curing by expression of an inhibitory fragment of Ure2p (48). The [*PIN*⁺] element, mediated by self-propagating aggregates of the Rnq1p protein, also shows phenotypic variation. Unlike [*PSI*⁺] and [*URE3*], the [*PIN*⁺] phenotype is caused by a gain of function of the protein aggregate. [*PIN*⁺] is required for efficient induction of [*PSI*⁺] by overexpression of Sup35p (15, 49, 131, 132), and deletion of *RNQ1* does not mimic this phenotype. Strong [*PIN*⁺] elements can generate high numbers of [*PSI*⁺] cells upon overexpression of Sup35p, whereas weak variants are not as efficient at conversion (114).

Yeast Prion Strains Modulate Solubility of the Prion Protein

An important link between yeast prion strain phenotypes and the conformation of the prion protein came from studies of variant-specific differences in the solubility of the endogenous prion protein (113, 133, 134). This was first shown using [*ETA*], a non-Mendelian genetic element isolated through synthetic lethality with particular alleles of translation release factors (135). Elegant experiments by Zhou et al. (133) showed that [*ETA*] was a weak variant of [*PSI*⁺] and was distinguished primarily by a reduced level of Sup35p aggregation relative to strong [*PSI*⁺] strains. Other [*PSI*⁺] variants have now also been characterized and been shown to have similar differences in the degree of aggregation of Sup35p. Importantly, these variants propagate faithfully and are largely independent of the yeast genetic background (134). The variants of [*PIN*⁺], which show differential ability to promote [*PSI*⁺], also show differences in the amount of aggregated Rnq1p, but there is no clear correlation between that phenotype and the degree of aggregation (114). Therefore changes in the relative fraction of aggregated protein can result in prion variants, but it is not the only possible mechanism for phenotypic diversity.

In Vitro Analysis of Yeast Prion Strains

Further evidence that [*PSI*⁺] variants are encoded by different prion conformations came from two lines of experiments. The first took advantage of an extract-based system in which [*PSI*⁺] extracts containing aggregated Sup35p were mixed with [*psi*⁻] extracts containing only soluble Sup35p that was converted to an insoluble form after incubation (61). Conformational differences between variants could be propagated in a cell-free system. When extracts from [*PSI*⁺] variants showing differential sedimentation profiles of Sup35p were used to seed [*psi*⁻] extracts, the newly aggregated material showed the same sedimentation as the original variants (134). Further experiments validate this notion because Sup35p aggregate-containing extracts generated from either strong or weak [*PSI*⁺] variants showed different seeding efficiencies in in vitro polymerization reactions. However, this difference was lost when the newly polymerized material was used as seeds for secondary rounds of in vitro reactions, raising the possibility that faithful propagation of different Sup35p conformations in vivo depends on host factors (136).

Yeast Prion Proteins Adopt Multiple Self-Propagating Forms

Work with pure protein has established that both Ure2p and Sup35p are able to adopt multiple self-propagating conformations. Spontaneous polymerization of either Sup35NM (the amino-terminal domain of Sup35p) (62) or Ure2p protein produces a range of amyloid fiber types (137). Even though the specific conformational differences between fiber types have not yet been determined, characteristics correlated with strain phenotypes, such as kinetics and seeding specificity, have been measured for Sup35NM fibers. When an atomic force microscopy (AFM)-based assay was used to measure growth from individual Sup35NM fibers, it was found that the purified polypeptide spontaneously forms multiple kinetically distinguishable fiber types. These could be sorted into a discrete number of classes on the basis of their growth polarity and elongation rate (Figure 6). Both the number of distinguishable fiber types and their relationship to protein aggregation rates suggest that these differences are well suited to account for [*PSI*⁺] strain variation in vivo (138). Other in vitro work with a chimeric Sup35p system demonstrates that a single protein can form multiple biochemically distinguishable conformations with properties that reflect their in vivo strain phenotypes (Figure 7) (139, 140).

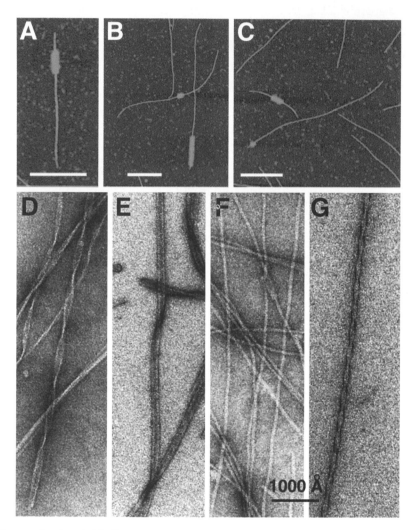

Figure 6 Amyloid fibers adopt multiple distinguishable structures. (*A, B, C*) Amyloid fibers formed spontaneously by Sup35NM vary in their growth patterns, including overall rate and polarity of growth (138). Four kinetic fiber types visualized by an AFM single fiber growth assay are shown. The original seed is labeled with antibody and is therefore wider than the new growth extending from its ends. Note the presence of long and short symmetric and asymmetric fibers. Scale bar is 500 nm. (*D, E, F, G*) Negative stain EM of amyloid fibers formed spontaneously by the SH3 domain from PIP_3 kinase illustrates that they vary in the number of protofilaments and helical pitch (83). Scale bar is 100 nm.

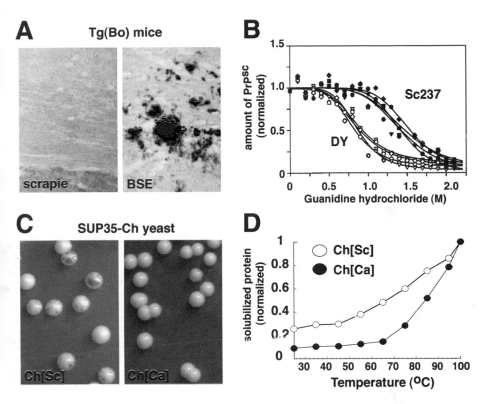

Figure 7   Strain phenotypes in vivo correlate with in vitro differences in prion protein. (*A*) Subcallosal plaques caused in transgenic mice expressing bovine *PRNP* characteristic of infection by the indicated prion strain (200). (*B*) Denaturation profile of indicated strains of PrP^{SC} showing stability differences (26). (*C*) Yeast harboring *SUP35* with Ch prion domain, induced to prion state by either Sc (Ch[Sc]) or Ca overexpression (Ch[Ca]) (139) with associated differences in phenotype. (*D*) Thermal denaturation of Ch fibers seeded by either Sc or Ca fibers (140).

Structural Polymorphism Is a Common Feature of Amyloid Aggregates

Although the notion that prion strains are due to multiple infectious conformations was radical when first proposed, it is now clear that many proteins misfold into a variety of aggregates. This is especially true of amyloids; in this case, even during the same polymerization reaction a single polypeptide can adopt multiple fiber types, distinguishable by their ultrastructural properties, such as number of protofilaments and helical pitch as shown in Figure 6 (141–145). Just as crystal growth is intimately dependent on the nature of the solution, changing reaction conditions can shift the relative populations of morphologically distinct fibers. For example, amyloid forms of an SH3 domain were shown to be highly sensitive to pH (83, 146). Aggregation of an Alzheimer's Aβ-derived peptide also showed strong dependence on pH, forming thicker ribbon-like fibers at higher pH (147). Changes in temperature can also alter the range of fiber morphologies as illustrated by work with polyglutamine peptide aggregates (74) and yeast prion amyloid fibers (140).

In addition to reaction conditions, covalent changes in the polypeptide, such as mutations or chemical modifications, can also modulate the spectrum of misfolded protein conformations. Mutations in Aβ affect the ultrastructural packing of amyloid protofilaments and overall length of fibers when compared to wildtype (141, 148–150). Mutations in light chain domains also alter the morphology of amyloid fibers (151). Formation of nonfibrillar intermediates along the pathway to amyloid formation can also be influenced by changes in primary structure. For example, mutations in α-synuclein correlating with early onset of Parkinson's disease have been linked to accelerated formation of toxic protofilament structures that are normally not present in wild-type polymerization reactions (152–154). Finally, the propagating form of yeast prion proteins can also be influenced by mutations (140), resulting in a shift in aggregate stability and in species-specific seeding processes (see below for discussion).

How do even relatively small changes in environment or in primary structure cause shifts in final aggregate morphology? Models for amyloid structure may shed light on the origin of this effect. The core of the amyloid structure is thought to consist of multiple layers of closely packed β-sheets. Morphological variants could arise from differences in side chain packing, register or topology of β-sheets, or quarternary structures. Though amyloid fibers are notoriously difficult to study using classical methods such as X-ray crystallography or NMR, high-resolution structural studies are clearly needed.

SEQUENCE-DEPENDENT PRION TRANSMISSION: THE SPECIES BARRIER

Passage of transmissible spongiform encephalopathies between species has long been known to be limited by species barriers (155) and analogous barriers to propagation exist in the yeast prion systems (52, 53, 129, 156, 157). The primary structure of the prion protein is a critical determinant of the specificity of propagation because the inhibition of cross-species infectivity is intimately dependent on the degree of similarity between the sequences of the two prion proteins (2, 158, 159). Indeed, even point mutations or allelic variants can have dramatic effects on the specificity of prion propagation (160–164). Although host factors may play an important role, increasing evidence from in vitro studies argues that the growth of amyloid-like aggregates can account for much of the observed specificity. In general, prion infectivity is also highly dependent on the prion strain in question; we address this feature in the following section but focus now on primary structure differences.

Mammalian Species Barriers In Vivo

Species barriers are common among the TSEs. Sheep scrapie isolates are delayed in transmission to goats (155), and human TSEs do not easily infect laboratory mice (165). Systematic exploration of this phenomenon has been greatly facilitated by the establishment of scrapie in transgenic mice in which a species barrier greatly slows prion transmission between Syrian hamsters and mice. From these studies, the sequence of the PrP protein has emerged as a critical determinant of cross-species transmission (158, 159). When PrP^{SC} isolated from Syrian hamsters was intracerebrally injected into hamster hosts, the animals rapidly came down with disease, whereas mouse hosts showed no clinical symptoms after inoculation with the same material. A transgenic mouse expressing a copy of the Syrian hamster *PRNP* gene in addition to the endogenous copy was now highly susceptible to both hamster and mouse innoculum (158). Strikingly, inoculation with mouse-derived prions resulted in formation of exclusively mouse prions, and innoculation with hamster prions resulted in exclusive formation of hamster prions (159). However, interpretation of these results is complicated somewhat by the recent finding that high prion titers can exist in the absence of clinical disease features (106).

Since these classic studies, several transgenic experiments have confirmed the intimate relationship between the sequence of the prion protein and specificity of transmission (2, 166, 167). Nonetheless, other studies establish that in some contexts It is not the sole determinant. For example, transgenic mice expressing a human copy of *PRNP* in addition to their endogenous mouse copy [Tg(Hu) mice] are immune to human prions (127). Ablation of the mouse *PRNP* gene in Tg(Hu) mice makes them susceptible to human prions, whereas mice expressing a mouse-human chimera PrP [Tg(MH2 M) mice] are susceptible to human prions independent of the presence of the endogenous mouse copy (128). These data led to the suggestion that a species-specific factor (known as protein X) is necessary for prion susceptibility. In Tg(Hu) mice, this factor would bind selectively to the wild-type mouse PrP^C protein, preventing proper conversion of the human PrP^C. On the other hand, Tg(MH2 M) mice were postulated to have both the human-derived sequence necessary for conversion and the recognition epitopes required for binding the prion-promoting factor (128).

Role of Polymorphisms in Prion Transmission

Even within a single species, allelic variants of PrP affect mammalian prion transmission. The effects of genetic background on scrapie susceptibility were observed as early as 1959 by Gordon (see 128a) who found that some breeds of sheep were particularly sensitive to scrapie. In humans, familial forms of prion diseases are often associated with particular alleles of *PRNP* (2). Although mutations can result in general acceleration of prion onset and transmission, there is also a potential role for *PRNP* alleles to modulate prion transmission specificity. These observations have now been more extensively studied using transgenic mice (161, 168), cell culture (169), and cell extract systems (164), which showed that single substitutions in primary structure can determine susceptibility and specificity to prion infection.

Sequence-Specific Mammalian Prion Replication In Vitro

Conversion experiments in cell extract systems have helped define the molecular nature of species specificity in prion transmission (40, 42, 170). Incubation of mouse or hamster PrP^{SC} extracts with recombinant PrP^{C} protein from the same species resulted in conversion of the PrP^{C} to a protease resistant form remiscent of PrP^{SC}. However, hamster PrP^{SC} could not convert mouse PrP^{C}, suggesting that the specificity of prion propagation resulted from the ability of the infectious particle to bind to and convert soluble PrP^{C}(40). A similar in vitro result was also described for species-specific transmission of chronic wasting disease from cervids to other mammals (170). Although other factors necessary for prion replication (127, 128) may be present in these extracts, it seems likely that sequence-specific and direct interactions between PrP^{SC} and PrP^{C} underlie much of the prion species barrier.

Transmission of [*PSI*$^+$] Is Highly Sequence Specific

Barriers inhibiting yeast prion transmission have been extensively studied using the yeast prion [*PSI*$^+$]. Cloning of *SUP35* genes from a broad range of budding yeast revealed that although the exact sequence of the amino-terminal domain varies, the features thought to be important for prion propagation, such as high glutamine/asparagine content, are preserved (52, 53, 129, 156, 157). Moreover, these domains can support prion states when expressed in a heterologous *S. cerevisiae* system (52, 53, 129) and, in one case examined, in the original yeast species (*Kluyveromyces lactis*) from which it was derived (156). The conservation of the prion-forming abilities of Sup35p together with the observation that presence of the prion can provide a selective advantage in certain conditions (15, 171, 172) suggests that rather than being a pathogen, [*PSI*$^+$] may represent a beneficial and conserved epigenetic mechanism for regulating protein function.

Analogous to the mammalian species barrier that limits induction and transmission, [*PSI*$^+$] prions are typically species specific (52, 53, 129, 156, 157). A particularly robust barrier exists between [*PSI*$^+$] prions formed from *S. cerevisiae*– and *C. albicans*–derived *SUP35* prion domains. Although these organisms would not naturally interact, species specificity can be studied using genetically manipulated yeast. Overexpression of *S. cerevisiae* Sup35p induces [*PSI*$^+$] in wild-type *S. cerevisiae* but not in yeast where the *SUP35* gene encodes for the *C. albicans* prion domain and vice versa (52). Even a single point mutation within the *S. cerevisiae* *SUP35* sequence is sufficient to confer specificity (65). However, in other cases, cross transmission between different *SUP35* sequences is possible albeit with reduced efficiency (53, 156). Such cross transmission could arise directly from some propensity of those Sup35p to be recruited into heterologous prions or indirectly through interactions with cellular machinery, such as chaperones.

Specificity of Transmission in Other Yeast Prions

Barriers to transmission between different yeast prions have also been observed. For example overexpression of New1p induces $[NU^+]$ but not $[PSI^+]$, whereas overexpression of Sup35p induces $[PSI^+]$ but not $[NU^+]$ (49) or $[URE3]$. Finally, transient expression of heterologous species of Ure2p rarely induced $[URE3]$ formation in *S. cerevisiae* even though similar expression of the *S. cerevisiae* Ure2p generated $[URE3]$-containing cells (173, 174). However, an important caveat is that it has not yet been shown that these alternate species of Ure2p can even form self-propagating prion states. If they cannot, then the lack of induction can be easily explained by the inability to form any type of infectious particles, rather than reflecting a specific transmission barrier between prions.

Antagonism and Cooperation Between Yeast Prions

Even when a barrier prevents transmission of prion states between two different prion proteins, the presence of one prion can strongly influence both induction and propagation of a second. This influence can be positive, such as in the well-characterized $[PSI^+]$-inducibility ($[PIN^+]$) effect , where de novo induction of $[PSI^+]$ by Sup35p overexpression only occurs in yeast harboring a second prion (49, 131, 132). Alternatively, prions can interfere with each other's propagation. For example, the $[URE3]$ state is not inherited stably in $[PSI^+]$ cells and vice versa (132, 175). The molecular bases of the above phenomena are poorly understood. In particular, a major open question is the extent to which this represents mixed polymers or an indirect effect, such as modulation of aggregation by chaperones.

On a related note, this effect of protein aggregates affecting de novo appearance of other aggregates seems to be a general effect, at least in yeast. Recent experiments demonstrated that aggregation of polyglutamine proteins is sensitive to the presence of other yeast prions (49), even though the polyglutamine proteins themselves cannot support prion inheritance in yeast. A mutant allele of the Machado-Joseph Disease (MJD) protein, containing an expanded polyglutamine tract fused to GFP, was used as a fluorescent reporter of aggregation. Aggregates of the Rnq1p or New1p prion domain were sufficient to promote aggregation of the mutant MJD protein, though in the absence of these aggregates, the reporter construct remained soluble (49).

In Vitro Evidence for a Molecular Basis of Yeast Prion Specificity

Complementing in vivo observations of yeast prion specificity, it has been possible to recapitulate the sequence-specific propagation of the $[PSI^+]$ prion in vitro. Extracts of $[PSI^+]$ cells expressing *S. cerevisae* Sup35p can induce aggregation of Sup35p present in $[psi^-]$ extracts from cells expressing *S. cerevisiae* but not *Pichia methanolica* Sup35p (129). An obligatory role for other cellular factors can be eliminated using an in vitro polymerization reaction with only purified recombinant prion domains (62, 176). Both *S. cerevisiae–* and *C. albicans–*derived prion domains form amyloid fibers after characteristic lag times; addition of preformed fibers of *S. cerevisiae* Sup35p prion domains efficiently seeds polymerization of *S. cerevisiae* Sup35p prion domains but not domains derived from *C. albicans* and vice versa. Remarkably, even when present together in a mixture, these two species of prion domains show exquisite sequence specificity and form homopolymeric fibers (52).

Sequence-Specific Amyloid Propagation

The sequence specificity seen in the purified Sup35p amyloid system is a common property of amyloid fibers, even of those not involved in prion phenomenon. Recent work with polyglutamine-containing proteins showed that formation of detergent-resistant amyloid aggregates is highly protein specific with coaggregation limited to proteins that share sequence homology outside the polyglutamine tract region (177). Peptides derived from the PrP protein also show preferential formation of homogeneous amyloids (178, 179), and polymers of Aβ have stereochemical specificity for aggregate formation (180). Quantitative analyses of seeding efficacy also revealed that Aβ was poorly seeded by fibers composed of completely unrelated peptides as well by fibers made up of a short amyloidogenic peptide, islet amyloid polypeptide (IAPP), despite the high degree of sequence similarity between Aβ and IAPP (180a). Moreover, as was seen with the Sup35 prion protein, Aβ appeared to be able to misfold into more than one amyloidogenic conformation as detected by seeding efficiencies. Finally, in vivo specificity is also seen during inclusion body formation and aggresome assembly (181, 182).

RELATIONSHIP BETWEEN PRION STRAINS AND SPECIES BARRIERS

The phenomenological connection between strains and species barriers has long been appreciated. Even before the identification of an infectious agent responsible for TSEs, it was known that scrapie strains played a strong role in determining specificity of transmission (155, 160, 183). Indeed, these observations led to the concept of a transmission barrier, rather than a species barrier, to reflect the role of features other than simple sequence homology in determining prion infectivity (2). Understanding this relationship between prion strains and interspecies transmission has become especially relevant with the finding that the recent appearance of new variant CJD (nvCJD) seems to have resulted from the transmission of the prion strain responsible for Mad Cow Disease or bovine spongiform encephalopathy (BSE). BSE appears to be an especially promiscuous prion type, capable of crossing the species barrier that normally prevents transmission of animal prions, such as scrapie, to humans (2, 32, 184–186). Remarkably, the link between strains and species barriers seems to be general because strain variants of yeast prions can differ widely in specificity of transmission (139, 140, 187, 188). Below, we review the evidence linking strains, species barriers, and protein conformational changes in various prion systems. A synthesis of these observations suggests a model in which prion strains and transmission barriers are in large part manifestations of the same phenomenon: the ability of proteins to misfold into multiple amyloid-like conformations. This model helps explain several characteristic features of prion strains and species barriers.

Prion Strains Affect Interspecies Mammalian Prion Transmission

During early studies of TSE infectivity, isolates of sheep scrapie were found to vary in their ability to infect goats, mice, and other laboratory animals (189). In one case, two sheep prion strains were investigated, a clinical isolate of naturally occurring sheep scrapie and a strain generated through experimental passage of BSE through sheep. The results were striking: Successful transmission of sheep scrapie to laboratory mice took ~800 days, whereas the mice inoculated with sheep-passaged BSE strains showed clinical signs in half that time (160). Later these experiments were refined through the use of isogenic transgenic mice. For instance, a transgenic mouse expressing a chimeric human-mouse PrP showed different susceptibility to two hamster-derived prion strains, even though they were composed of the same prion protein (190). Altogether these data argue that the nature of the prion strain is a key component of determining transmission across a species barrier.

Passage Through a Species Barrier Modulates Prion Strains

The relationship between strains and species barrier is reciprocal: Just as strains show differing ability to cross between species, crossing a species barrier can result in a shift in strain characteristics. For example, clinical features and pathological hallmarks of scrapie were altered upon inoculation of goats with sheep scrapie. However, the infection of other sheep did not show this shift in scrapie disease profile (191). In studies of transmissible mink encephalopathies, researchers found that transmission of mink-derived drowsy prion strains into hamsters resulted in formation of both drowsy and hyper prion strains in a titer-dependent fashion (191a). Polymorphisms in the PrP gene present in a single species can also modulate the transmission of prion strains. Passage of BSE through transgenic mice expressing human PrP homozygous for valine at codon 129 does not affect strain type (165, 185). In contrast, BSE transmission to mice expressing human PrP homozygous for methionine at codon 129 resulted in mixture of both the parental BSE/nvCJD strain and new types similar to spontaneous CJD (192). An important caveat is that the existence of a species barrier does not necessitate a change in strain type. For example, in one study, large species barrier effects were observed upon mouse-to-hamster and upon mouse-to-rat transmission when using a particular mouse-derived prion strain. When the material was then inoculated into mice, the mouse-to-hamster passaged strain showed significantly different properties as compared to the parent, but the mouse-to-rat isolate appeared unchanged (191).

Emergence of Prion Strains Is Accompanied by a Change in Conformation

Experiments by Peretz and colleagues (190) helped define a molecular mechanism for the link between species barriers and strains by showing that emergence of new prion strains following interspecies transmission is accompanied by changes in prion conformations. These studies used two hamster prion strains, drowsy (DY) and Sc237, which could be distinguished by relative stability as measured by chemical denaturation (Figure 7) (26). The prion strains were administered to a line of transgenic mice in which a chimeric hamster/mouse PrP gene replaced the wild-type mouse allele. Inoculation with the DY strain resulted in rapid onset of disease and a characteristic DY-specific clinical phenotype. Consistent with this observation, the newly converted host PrP^{SC} retained the conformational stability associated with the parent strain. In contrast, the Sc237 hamster prion strain exhibited a delayed transmission characteristic of a species barrier. The passage resulted in the emergence of a strain with significantly different clinical features and conformational stability than the original Sc237 strain. The new strain propagated faithfully in transgenic mice with a fixed period of latency distinct from the Sc237 strain (190). Thus both the clinical features and conformational hallmarks of a prion strain can change upon transmission across a species barrier.

Yeast Prion Strains and Sequence-Dependent Transmission

A number of in vivo and in vitro experiments have pointed to an intimate link among strains, sequence, and conformational differences in the [PSI$^+$] prion systems. An early example of this came from studies of a Sup35p mutant [glycine at residue 58 to aspartic acid—known as PNM2 (193, 194)], which in some contexts is defective in yeast prion propagation (188). As mentioned previously, weak [PSI$^+$] variants exhibit mitotic instability and have lower levels of termination suppression when compared to strong [PSI$^+$] variants. Paradoxically, expression of PNM2 interfered with the suppression phenotype of strong [PSI$^+$] strains, whereas expression of PNM2 in a weak [PSI$^+$] strain actually enhanced the suppression phenotype (188). Because both weak and strong [PSI$^+$] variants were in genetically identical backgrounds, the clearest interpretation was that the variants consisted of distinct propagating forms of Sup35p that could be differentially influenced by expression of the mutant protein.

Elegant experiments by King (187) further investigated this link between yeast prion strains and sequence specificity. King explored the ability of three different [PSI$^+$] variants to recruit a panel of Sup35p mutants, as monitored by both suppression phenotype and by recruitment of GFP fusions. He found that certain mutants could be preferentially recruited by some of the variants, but other mutants could not be recruited by any of the variants. Furthermore, coexpression of mutant prion domains cured the [*PSI*⁺] variants to different degrees. These data led to the conclusion that the [*PSI*⁺] strain variants were caused by structurally different Sup35p aggregates, each exposing different regions of the polypeptide. The ability to interact with a mutant would then be determined by the surface presented by a particular aggregate.

Work from our own lab, using a combination of in vivo and in vitro studies, has directly established that a single polypeptide can form more than one self-propagating amyloid conformation and that these conformations can determine the specificity of prion propagation. Moreover, we demonstrated that point mutations in a prion protein, by changing the spectrum of favored conformations, generate a de novo species barrier (139, 140). These experiments used a chimeric prion domain (known as Ch) composed of the first 40 amino acids of the *S. cerevisiae* Sup35p fused to the remainder of the prion domain from *C. albicans*. Whereas a barrier normally inhibits transmission between *S. cerevisiae* and *C. albicans* SUP35, the Ch prion domain is able to bridge this barrier. In vivo, Ch formed distinct prion strains with markedly different strengths and specificities upon induction by different Sup35p species (Figure 7). Similarly, when seeded with different species of Sup35p fibers in vitro, the purified Ch protein forms two distinct self-propagating amyloid forms. These conformations dictate seeding specificity: Ch seeded by *S. cerevisiae* Sup35p fibers efficiently catalyzes conversion of *S. cerevisiae* Sup35p (Sc) but not *C. albicans* Sup35p (Ca), and vice versa (139). These observations indicated that Ch bridges the species barrier by adopting two conformations (Figure 8), one that is specific for Sc (Ch[Sc]) and the other specific for Ca (Ch[Ca]).

This work was extended by looking at the effect of mutations in the Ch protein that were chosen to specifically disfavor Ch[Sc] or Ch[Ca]. Mutations that inhibited formation of the Ch[Ca] state, both in vivo and in vitro, prevented transmission between Ch and Ca without disrupting transmission to Sc. Conversely, mutants disfavoring Ch[Sc] were incapable of transmitting to Sc but remained susceptible to Ca. Interestingly, modulation of temperature also strongly influenced the preference for forming Ch[Sc] and Ch[Ca] (140). These observations indicate how changes in the sequence of a prion or changes in the environment can affect the specificity of a prion by modulating conformations.

MODEL INTEGRATING PRION STRAINS, SPECIES BARRIERS, AND PRINCIPLES OF AMYLOID FORMATION

Tenets of the Model

A synthesis of the experimental observations above suggests the following tenets linking prion strains, species barriers, and the physical principles that govern

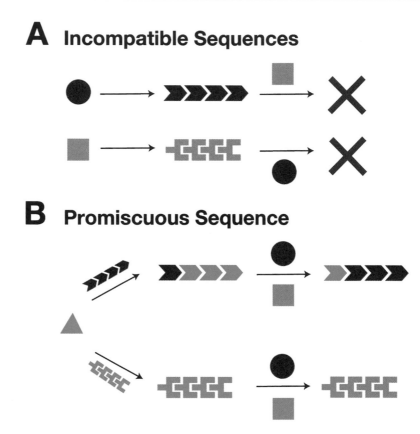

Figure 8 Models depicting the relationship between transmission specificity and conformation. (*A*) Robust species barrier between two variants of prion protein that do not form compatible conformations and thus do not cross seed. (*B*) A single polypeptide, which can adopt two distinct conformations that allow assembly onto two otherwise incompatible prions.

protein misfolding. For the most part, these tenets have substantial experimental support and can also serve to guide the direction of future experiments.

1. Self-propagation of amyloid-like protein aggregates underlies prion growth.

2. A single protein can often misfold into multiple different amyloid conformations.

3. The phenotypic consequences resulting from an aggregated protein are highly dependent on the specific amyloid conformation.

4. The particular amyloid conformation that a protein adopts determines the specificity of growth.

5. Changes in protein sequence can modulate the spectrum of favored amyloid conformations.

Relationship Between Conformation, Strains and Species Barriers

Based on these tenets, a model emerges in which prion strains and transmission barriers are in large part two different manifestations of the same phenomenon, the ability of a protein to misfold into multiple amyloid conformations. These conformations in turn determine both the specificity of growth and the phenotypic consequences of harboring a prion. Changes in sequence alter the range of preferred amyloid conformations thereby modulating transmission barriers and strain phenotypes.

Though this model, based on the propagation of amyloid-like structures, can account for many observed prion phenomena, it is clear that host cellular factors, such as chaperones or degradative machinery, can play a significant role in both the phenotype and propagation of prions. These factors are also likely to contribute to strains and species barriers by mechanisms other than changes in conformation. Furthermore, the simple ability to form a self-propagating aggregated state does not guarantee that a protein will be infectious. Although amyloid-like aggregation forms a physical basis for the propagation of prions, true understanding of what makes a prion more than an aggregated protein remains a central challenge. Nonetheless, our model suggests explanations for several features of prion inheritance.

STRAINS ARE A COMMON FEATURE OF PRION INHERITANCE   Extensive evidence from both mammalian and yeast prion systems show that different propagating amyloid-like conformations are strongly correlated with distinct prion strains. This may be a specific case of the more general ability of amyloid fibers to form a range of self-propagating conformations. The ability of each conformation to robustly propagate differences, such as distinct fiber morphologies and assembly kinetics, could lead directly to heritable variation in phenotypes. For example, if the phenotype is due to the amount of soluble protein, variations in the aggregation rate will directly influence phenotype. Alternately, cellular factors may interact differently with the various conformations, leading to a distinct physiological outcome for different prion aggregates. A major goal is to elucidate the mechanism by which alternate prion conformations can cause different strain phenotypes.

TRANSMISSION BARRIERS ARE COMMON AND APPARENTLY EASY TO GENERATE   Not only are species-specific transmission barriers a ubiquitous feature of prion propagation, they also arise rapidly, as evidenced by the small number of amino acid changes required to inhibit transmission between prions. This phenomenon can be explained by the ability of changes in polypeptide primary structure to alter the range of preferred amyloid fiber conformations. Even single point

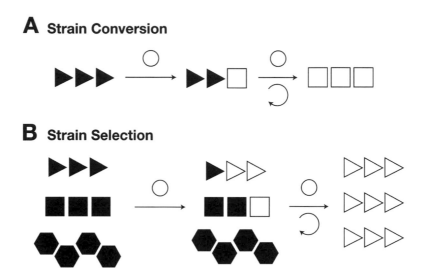

Figure 9   Two models for strain switching upon passage through a species barrier. (*A*) In strain conversion, heterologous protein adopts a new conformation upon incorporation into prion seeds. (*B*) In strain selection, host protein selects a compatible conformation from a heterogeneous innoculum. Over multiple rounds of prion replication, the distribution of conformations changes.

mutations can in some cases shift the fiber conformation, resulting in a novel self-specific aggregate that is incompatible with the original parent sequence.

STRAINS DETERMINE TRANSMISSION SPECIFICITY   Because amyloid conformations vary in their ability to recruit heterologous proteins, prion particles composed of the same protein but differing in their strain conformation will differ in their compatibility with the corresponding prion protein from another species. Variation in conformational compatibility between prion proteins thereby contributes to strain-specific transmission across a species barrier.

STRAINS SWITCH UPON TRANSMISSION ACROSS A SPECIES BARRIER   Crossing a species barrier, though inefficient, is possible if a compatible prion conformation can be found and amplified. Observed switches in strains therefore may result from such an amplification of a conformation compatible with cross-species transmission. Although there are a variety of models that may explain this effect, two major possibilities stand out (2, 190) (Figure 9). In the first model, upon interspecies passage, assembly of the new polypeptide onto the infectious seed results in a new conformation. In the second model, new protein selectively grows on the subset of compatible seeds. In this interpretation, the transmission barrier acts as a sieve by selectively amplifying one component of a pool of conformations. This model demands that the initial strain actually consists of a number of subtypes and that the biological strain phenotype reflects this collection.

PERSPECTIVE

The ability of proteins to adopt multiple amyloid forms indicates a fundamental difference between the rules of protein folding and misfolding. Globular folds are stabilized by multiple cooperative interactions between specific side chains, resulting in unique well-defined structures. By contrast, recent studies indicate that amyloid formation is driven predominantly by main chain interactions, which can be locally favored or disfavored by specific side chains (195). As a consequence, a polypeptide can adopt multiple amyloid forms differing in their quaternary or possibly tertiary structures with specific side chains disfavoring a particular subset of structures without preventing amyloid formation altogether. Small differences in the rates of forming the various conformations will be reinforced by the self-propagating nature of amyloid formation; once a stable nucleus of a given conformation is formed, it rapidly dominates the reaction as it grows exponentially. Because of this, the reaction will be under kinetic control with the final conformation choice determined by the specific conditions of polymerization rather than the global thermodynamic minimum.

The fact that one polypeptide can misfold into multiple self-propagating forms helps explain a range of observations regarding prion inheritance. For example, the existence of transmission barriers between highly related species can be explained by the fact that the infectious conformation is sensitive to small changes in primary structure. Mutations affect the initial choice of conformation during de novo prion formation, and in turn, the conformation of the prion will determine which sequences can be recruited. A robust transmission barrier will therefore arise when the range of conformations adopted by two sequences are incompatible (2, 52, 53, 129, 156). Similarly, if the conformation affects phenotype as well as specificity, then changes in the primary structure of a prion protein could also alter the strain phenotype by shifting the infectious conformations (161, 168). It has been reported that crossing a transmission barrier can result in a change in the prion strain. Because conformations vary in their specificity, the transmission barrier could act as a sieve, selectively amplifying infectious forms compatible with the recipient prion sequence (2, 190). Finally, the failure to create transmissible forms of the mammalian prion protein (PrP) in vitro, despite a number of reports demonstrating production of self-propagating or protease-resistant PrP states (196–198), could be due to the preferential formation of noninfectious conformations outside the normal cellular context.

The degeneracy of amyloid formation may be important for understanding a range of protein misfolding disorders. There is increasing evidence that the toxicity of the different misfolded forms varies greatly, with some species being highly pathogenic, though others might even be protective (107, 109, 111). Given the strong propensity of nonnative proteins to aggregate, therapeutic strategies designed to promote formation of nontoxic conformations rather than preventing amyloid formation altogether may be more tractable. In addition to selective pressure for function, this variability suggests that polypeptide sequences that form less toxic conformations when they do misfold will be preferred. More generally, any analysis of formation and consequences of amyloid-like aggregates needs to be tempered with the knowledge that there exists a range of conformationally distinct subtypes that will influence their biological effects.

ACKNOWLEDGMENTS

We acknowledge Drs. Reed Wickner, Helen Saibil, Holger Willie, and Cedric Govaerts for providing images and Dr. Holly Field, Kim Tipton, Sean Collins, Anna Weissman, and members of the Weissman lab for helpful comments. P.C., now at Harvard University, was supported by a National Science Foundation Graduate Fellowship and was a Scholar of the Achievement Rewards for College Scientists (ARCS) Foundation. A.H.D., now at the University of California, Berkeley, was supported by a Howard Hughes Medical Institute Predoctoral Fellowship and funding from the National Institutes of Aging. J.S.W. was funded through the Howard Hughes Medical Institute, the Packard Foundation, the Searle Scholars Program, and the NIH.

The *Annual Review of Biochemistry* is online at http://biochem.annualreviews.org

LITERATURE CITED

1. Prusiner SB, Scott MR, DeArmond SJ, Cohen FE. 1998. *Cell* 93:337–48
2. Collinge J. 2001. *Annu. Rev. Neurosci.* 24:519–50
3. Cohen FE, Prusiner SB. 1998. *Annu. Rev. Biochem.* 67:793–819
4. Horwich AL, Weissman JS. 1997. *Cell* 89:499–510
5. Griffith JS. 1967. *Nature* 215:1043–44
6. Roberts BT, Wickner RB. 2003. *Genes Dev.* 17:2083–87
7. Priola SA, Chesebro B, Caughey B. 2003. *Science* 300:917–19
8. Bolton DC, McKinley MP, Prusiner SB. 1982. *Science* 218:1309–11
9. Lasmezas CI, Deslys JP, Robain O, Jaegly A, Beringue V, et al. 1997. *Science* 275:402–5
10. Deleault NR, Lucassen RW, Supattapone S. 2003. *Nature* 425:717–20
11. Wickner RB. 1994. *Science* 264:566–69
12. Wickner RB, Taylor KL, Edskes HK, Maddelein ML, Moriyama H, Roberts BT. 2001. *Adv. Protein Chem.* 57:313–34
13. Serio TR, Lindquist SL. 2001. *Adv. Protein Chem.* 57:335–66
14. Uptain SM, Lindquist S. 2002. *Annu. Rev. Microbiol.* 56:703–41
15. Osherovich LZ, Weissman JS. 2002. *Dev. Cell* 2:143–51
16. Sparrer HE, Santoso A, Szoka FC, Weissman JS. 2000. *Science* 289:595–99
17. Maddelein ML, Dos Reis S, Duvezin-Caubet S, Coulary-Salin B, Saupe SJ. 2002. *Proc. Natl. Acad. Sci. USA* 99:7402–7

18. Dobson CM. 1999. *Trends Biochem. Sci.* 24:329–32
19. Prusiner SB, Bolton DC, Groth DF, Bowman KA, Cochran SP, McKinley MP. 1982. *Biochemistry* 21:6942–50
20. Prusiner SB, Groth DF, Bolton DC, Kent SB, Hood LE. 1984. *Cell* 38: 127–34
21. Oesch B, Westaway D, Walchli M, McKinley MP, Kent SB, et al. 1985. *Cell* 40:735–46
22. Chesebro B, Race R, Wehrly K, Nishio J, Bloom M, et al. 1985. *Nature* 315: 331–33
23. Stahl N, Baldwin MA, Teplow DB, Hood L, Gibson BW, et al. 1993. *Biochemistry* 32:1991–2002
24. Baldwin MA. 2001. *Adv. Protein Chem.* 57:29–54
25. Peretz D, Williamson RA, Matsunaga Y, Serban H, Pinilla C, et al. 1997. *J. Mol. Biol.* 273:614–22
26. Peretz D, Scott MR, Groth D, Williamson RA, Burton DR, et al. 2001. *Protein Sci.* 10:854–63
27. James TL, Liu H, Ulyanov NB, Farr-Jones S, Zhang H, et al. 1997. *Proc. Natl. Acad. Sci. USA* 94:10086–91
28. Hosszu LL, Baxter NJ, Jackson GS, Power A, Clarke AR, et al. 1999. *Nat. Struct. Biol.* 6:740–43
29. Riek R, Hornemann S, Wider G, Billeter M, Glockshuber R, Wuthrich K. 1996. *Nature* 382:180–82
30. Lopez Garcia F, Zahn R, Riek R, Wuthrich K. 2000. *Proc. Natl. Acad. Sci. USA* 97:8334–39
31. Pan KM, Baldwin M, Nguyen J, Gasset M, Serban A, et al. 1993. *Proc. Natl. Acad. Sci. USA* 90:10962–66
32. Prusiner SB. 1998. *Proc. Natl. Acad. Sci. USA* 95:13363–83
33. Bueler H, Aguzzi A, Sailer A, Greiner RA, Autenried P, et al. 1993. *Cell* 73: 1339–47
34. Prusiner SB, Groth D, Serban A, Koehler R, Foster D, et al. 1993. *Proc. Natl. Acad. Sci. USA* 90:10608–12
35. Bueler H, Raeber A, Sailer A, Fischer M, Aguzzi A, Weissmann C. 1994. *Mol. Med.* 1:19–30
36. Carlson GA, Ebeling C, Yang SL, Telling G, Torchia M, et al. 1994. *Proc. Natl. Acad. Sci. USA* 91:5690–94
37. Manson JC, Clarke AR, McBride PA, McConnell I, Hope J. 1994. *Neurodegeneration* 3:331–40
38. Enari M, Flechsig E, Weissmann C. 2001. *Proc. Natl. Acad. Sci. USA* 98: 9295–99
39. Kocisko DA, Come JH, Priola SA, Chesebro B, Raymond GJ, et al. 1994. *Nature* 370:471–74
40. Kocisko DA, Priola SA, Raymond GJ, Chesebro B, Lansbury PT Jr, Caughey B. 1995. *Proc. Natl. Acad. Sci. USA* 92: 3923–27
41. Saborio GP, Permanne B, Soto C. 2001. *Nature* 411:810–13
42. Lucassen R, Nishina K, Supattapone S. 2003. *Biochemistry* 42:4127–35
43. Wille H, Zhang GF, Baldwin MA, Cohen FE, Prusiner SB. 1996. *J. Mol. Biol.* 259:608–21
44. Tzaban S, Friedlander G, Schonberger O, Horonchik L, Yedidia Y, et al. 2002. *Biochemistry* 41:12868–75
45. Masison DC, Wickner RB. 1995. *Science* 270:93–95
46. Patino MM, Liu JJ, Glover JR, Lindquist S. 1996. *Science* 273:622–26
47. Paushkin SV, Kushnirov VV, Smirnov VN, Ter-Avanesyan MD. 1996. *EMBO J.* 15:3127–34
48. Schlumpberger M, Prusiner SB, Herskowitz I. 2001. *Mol. Cell. Biol.* 21: 7035–46
49. Osherovich LZ, Weissman JS. 2001. *Cell* 106:183–94
50. Sondheimer N, Lindquist S. 2000. *Mol. Cell* 5:163–72
51. Coustou V, Deleu C, Saupe S, Begueret J. 1997. *Proc. Natl. Acad. Sci. USA* 94: 9773–78
52. Santoso A, Chien P, Osherovich LZ, Weissman JS. 2000. *Cell* 100:277–88

53. Chernoff YO, Galkin AP, Lewitin E, Chernova TA, Newnam GP, Belenkiy SM. 2000. *Mol. Microbiol.* 35:865–76

54. Fernandez-Bellot E, Guillemet E, Cullin C. 2000. *EMBO J.* 19:3215–22

55. Fernandez-Bellot E, Guillemet E, Ness F, Baudin-Baillieu A, Ripaud L, et al. 2002. *EMBO Rep.* 3:76–81

56. Coustou-Linares V, Maddelein ML, Begueret J, Saupe SJ. 2001. *Mol. Microbiol.* 42:1325–35

57. Zhou P, Derkatch IL, Liebman SW. 2001. *Mol. Microbiol.* 39:37–46

58. Ripaud L, Maillet L, Cullin C. 2003. *EMBO J.* 22:5251–59

59. Kimura Y, Koitabashi S, Fujita T. 2003. *Cell Struct. Funct.* 28:187–93

60. Speransky VV, Taylor KL, Edskes HK, Wickner RB, Steven AC. 2001. *J. Cell Biol.* 153:1327–36

61. Paushkin SV, Kushnirov VV, Smirnov VN, Ter-Avanesyan MD. 1997. *Science* 277:381–83

62. Glover JR, Kowal AS, Schirmer EC, Patino MM, Liu JJ, Lindquist S. 1997. *Cell* 89:811–19

63. Taylor KL, Cheng N, Williams RW, Steven AC, Wickner RB. 1999. *Science* 283:1339–43

64. Dos Reis S, Coulary-Salin B, Forge V, Lascu I, Begueret J, Saupe SJ. 2002. *J. Biol. Chem.* 277:5703–6

65. DePace AH, Santoso A, Hillner P, Weissman JS. 1998. *Cell* 93:1241–52

66. Li L, Lindquist S. 2000. *Science* 287:661–64

66a. Tanaka M, Chien P, Nariman N, Cooke R, Weissman JS. 2004. *Nature* 428:323–28

67. Harper JD, Lansbury PT Jr. 1997. *Annu. Rev. Biochem.* 66:385–407

68. Rochet JC, Lansbury PT Jr. 2000. *Curr. Opin. Struct. Biol.* 10:60–68

69. Thirumalai D, Klimov DK, Dima RI. 2003. *Curr. Opin. Struct. Biol.* 13:146–59

70. Masel J, Jansen VA. 1999. *Proc. R. Soc. London Ser. B* 266:1927–31

71. Serio TR, Cashikar AG, Kowal AS, Sawicki GJ, Moslehi JJ, et al. 2000. *Science* 289:1317–21

72. Scheibel T, Lindquist SL. 2001. *Nat. Struct. Biol.* 8:958–62

73. Padrick SB, Miranker AD. 2002. *Biochemistry* 41:4694–703

74. Chen S, Berthelier V, Hamilton JB, O'Nuallain B, Wetzel R. 2002. *Biochemistry* 41:7391–99

75. Taylor JP, Hardy J, Fischbeck KH. 2002. *Science* 296:1991–95

76. Kelly JW. 1996. *Curr. Opin. Struct. Biol.* 6:11–17

77. Chiti F, Webster P, Taddei N, Clark A, Stefani M, et al. 1999. *Proc. Natl. Acad. Sci. USA* 96:3590–94

78. Guijarro JI, Sunde M, Jones JA, Campbell ID, Dobson CM. 1998. *Proc. Natl. Acad. Sci. USA* 95:4224–28

79. Sunde M, Serpell LC, Bartlam M, Fraser PE, Pepys MB, Blake CC. 1997. *J. Mol. Biol.* 273:729–39

80. Perutz MF, Finch JT, Berriman J, Lesk A. 2002. *Proc. Natl. Acad. Sci. USA* 99:5591–95

81. Wetzel R. 2002. *Structure* 10:1031–36

82. Serpell LC, Sunde M, Benson MD, Tennent GA, Pepys MB, Fraser PE. 2000. *J. Mol. Biol.* 300:1033–39

83. Jimenez JL, Guijarro JI, Orlova E, Zurdo J, Dobson CM, et al. 1999. *EMBO J.* 18:815–21

84. Richardson JS, Richardson DC. 2002. *Proc. Natl. Acad. Sci. USA* 99:2754–59

85. Prusiner SB, McKinley MP, Bowman KA, Bolton DC, Bendheim PE, et al. 1983. *Cell* 35:349–58

86. Wille H, Michelitsch MD, Guenebaut V, Supattapone S, Serban A, et al. 2002. *Proc. Natl. Acad. Sci. USA* 99:3563–68

87. Knaus KJ, Morillas M, Swietnicki W, Malone M, Surewicz WK, Yee VC. 2001. *Nat. Struct. Biol.* 8:770–74

88. Bousset L, Briki F, Doucet J, Melki R. 2003. *J. Struct. Biol.* 141:132–42

89. Baxa U, Taylor KL, Wall JS, Simon

MN, Cheng N, et al. 2003. *J. Biol. Chem.* 278:43717–27

90. Kayed R, Bernhagen J, Greenfield N, Sweimeh K, Brunner H, et al. 1999. *J. Mol. Biol.* 287:781–96

91. Wood SJ, Wypych J, Steavenson S, Louis JC, Citron M, Biere AL. 1999. *J. Biol. Chem.* 274:19509–12

92. Friedhoff P, von Bergen M, Mandelkow EM, Davies P, Mandelkow E. 1998. *Proc. Natl. Acad. Sci. USA* 95:15712–17

93. Scherzinger E, Sittler A, Schweiger K, Heiser V, Lurz R, et al. 1999. *Proc. Natl. Acad. Sci. USA* 96:4604–9

94. Morozova-Roche LA, Zurdo J, Spencer A, Noppe W, Receveur V, et al. 2000. *J. Struct. Biol.* 130:339–51

95. Selkoe DJ. 1999. *Nature* 399:A23–31

96. Goudsmit J, Morrow CH, Asher DM, Yanagihara RT, Masters CL, et al. 1980. *Neurology* 30:945–50

97. Godec MS, Asher DM, Kozachuk WE, Masters CL, Rubi JU, et al. 1994. *Neurology* 44:1111–15

98. Chernoff YO, Derkach IL, Inge-Vechtomov SG. 1993. *Curr. Genet.* 24:268–70

99. Yamin G, Glaser CB, Uversky VN, Fink AL. 2003. *J. Biol. Chem.* 278:27630–35

100. Uversky VN, Li J, Bower K, Fink AL. 2002. *Neurotoxicology* 23:527–36

101. Chernoff YO, Lindquist SL, Ono B, Inge-Vechtomov SG, Liebman SW. 1995. *Science* 268:880–84

102. Glover JR, Lindquist S. 1998. *Cell* 94:73–82

103. Aguzzi A, Heppner FL, Heikenwalder M, Prinz M, Mertz K, et al. 2003. *Br. Med. Bull.* 66:141–59

104. Borchsenius AS, Wegrzyn RD, Newnam GP, Inge-Vechtomov SG, Chernoff YO. 2001. *EMBO J.* 20:6683–91

105. Lundmark K, Westermark GT, Nystrom S, Murphy CL, Solomon A, Wes-

termark P. 2002. *Proc. Natl. Acad. Sci. USA* 99:6979–84

106. Hill AF, Joiner S, Linehan J, Desbruslais M, Lantos PL, Collinge J. 2000. *Proc. Natl. Acad. Sci. USA* 97:10248–53

107. Caughey B, Lansbury PT Jr. 2003. *Annu. Rev. Neurosci.* 26:267–98

108. Pepys MB, Herbert J, Hutchinson WL, Tennent GA, Lachmann HJ, et al. 2002. *Nature* 417:254–59

109. Walsh DM, Klyubin I, Fadeeva JV, Cullen WK, Anwyl R, et al. 2002. *Nature* 416:535–39

110. Conway KA, Rochet JC, Bieganski RM, Lansbury PT Jr. 2001. *Science* 294:1346–49

111. Bucciantini M, Giannoni E, Chiti F, Baroni F, Formigli L, et al. 2002. *Nature* 416:507–11

112. Ter-Avanesyan MD, Dagkcsamanskaya AR, Kushnirov VV, Smirnov VN. 1994. *Genetics* 137:671–76

113. Derkatch IL, Chernoff YO, Kushnirov VV, Inge-Vechtomov SG, Liebman SW. 1996. *Genetics* 144:1375–86

114. Bradley ME, Edskes HK, Hong JY, Wickner RB, Liebman SW. 2002. *Proc. Natl. Acad. Sci. USA* 99(Suppl. 4):16392–99

114a. Si K, Lindquist SL, Kandel ER. 2003. *Cell* 115:879–91

115. Weissmann C. 1991. *Nature* 352:679–83

115a. King CY, Diaz-Avalos R. 2004. *Nature* 428:319–23

116. Pattison I, Millson GC. 1961. *J. Comp. Pathol.* 71:101–8

117. Fraser H, Dickinson AG. 1968. *J. Comp. Pathol.* 78:301–11

118. Bruce ME, McConnell I, Fraser H, Dickinson AG. 1991. *J. Gen. Virol.* 72(Pt. 3):595–603

119. Caughey B, Raymond GJ, Bessen RA. 1998. *J. Biol. Chem.* 273:32230–35

120. Somerville RA, Oberthur RC, Havekost U, MacDonald F, Taylor DM, Dickin-

son AG. 2002. *J. Biol. Chem.* 277: 11084–89

121. Safar J, Wille H, Itri V, Groth D, Serban H, et al. 1998. *Nat. Med.* 4: 1157–65

122. Priola SA, Lawson VA. 2001. *EMBO J.* 20:6692–99

123. Vorberg I, Priola SA. 2002. *J. Biol. Chem.* 277:36775–81

124. Bessen RA, Marsh RF. 1992. *J. Virol.* 66:2096–101

125. Telling GC, Parchi P, DeArmond SJ, Cortelli P, Montagna P, et al. 1996. *Science* 274:2079–82

126. Bessen RA, Kocisko DA, Raymond GJ, Nandan S, Lansbury PT, Caughey B. 1995. *Nature* 375:698–700

127. Telling GC, Scott M, Hsiao KK, Foster D, Yang SL, et al. 1994. *Proc. Natl. Acad. Sci. USA* 91:9936–40

128. Telling GC, Scott M, Mastrianni J, Gabizon R, Torchia M, et al. 1995. *Cell* 83:79–90

128a. Tranulis MA. 2002. *APMIS* 110:33–43

129. Kushnirov VV, Kochneva-Pervukhova NV, Chechenova MB, Frolova NS, Ter-Avanesyan MD. 2000. *EMBO J.* 19: 324–31

130. Kushnirov VV, Kryndushkin DS, Boguta M, Smirnov VN, Ter-Avanesyan MD. 2000. *Curr. Biol.* 10: 1443–46

131. Derkatch IL, Bradley ME, Masse SV, Zadorsky SP, Polozkov GV, et al. 2000. *EMBO J.* 19:1942–52

132. Derkatch IL, Bradley ME, Hong JY, Liebman SW. 2001. *Cell* 106:171–82

133. Zhou P, Derkatch IL, Uptain SM, Patino MM, Lindquist S, Liebman SW. 1999. *EMBO J.* 18:1182–91

134. Kochneva-Pervukhova NV, Chechenova MB, Valouev IA, Kushnirov VV, Smirnov VN, Ter-Avanesyan MD. 2001. *Yeast* 18:489–97

135. All-Robyn JA, Kelley-Geraghty D, Griffin E, Brown N, Liebman SW. 1990. *Genetics* 124:505–14

136. Uptain SM, Sawicki GJ, Caughey B,

Lindquist S. 2001. *EMBO J.* 20: 6236–45

137. Baxa U, Speransky V, Steven AC, Wickner RB. 2002. *Proc. Natl. Acad. Sci. USA* 99:5253–60

138. DePace AH, Weissman JS. 2002. *Nat. Struct. Biol.* 9:389–96

139. Chien P, Weissman JS. 2001. *Nature* 410:223–27

140. Chien P, DePace AH, Collins SR, Weissman JS. 2003. *Nature* 424: 948–51

141. Kirschner DA, Inouye H, Duffy LK, Sinclair A, Lind M, Selkoe DJ. 1987. *Proc. Natl. Acad. Sci. USA* 84:6953–57

142. Halverson K, Fraser PE, Kirschner DA, Lansbury PT Jr. 1990. *Biochemistry* 29: 2639–44

143. Goldsbury CS, Cooper GJ, Goldie KN, Muller SA, Saafi EL, et al. 1997. *J. Struct. Biol.* 119:17–27

144. Goldsbury C, Goldie K, Pellaud J, Seelig J, Frey P, et al. 2000. *J. Struct. Biol.* 130:352–62

145. Kad NM, Thomson NH, Smith DP, Smith DA, Radford SE. 2001. *J. Mol. Biol.* 313:559–71

146. Zurdo J, Guijarro JI, Jimenez JL, Saibil HR, Dobson CM. 2001. *J. Mol. Biol.* 311:325–40

147. Abe H, Kawasaki K, Nakanishi H. 2002. *J. Biochem.* 132:863–74

148. Miravalle L, Tokuda T, Chiarle R, Giaccone G, Bugiani O, et al. 2000. *J. Biol. Chem.* 275:27110–16

149. Fraser PE, Duffy LK, O'Malley MB, Nguyen J, Inouye H, Kirschner DA. 1991. *J. Neurosci. Res.* 28:474–85

150. Fraser PE, McLachlan DR, Surewicz WK, Mizzen CA, Snow AD, et al. 1994. *J. Mol. Biol.* 244:64–73

151. Helms LR, Wetzel R. 1996. *J. Mol. Biol.* 257:77–86

152. Li J, Uversky VN, Fink AL. 2001. *Biochemistry* 40:11604–13

153. Lashuel HA, Hartley D, Petre BM, Walz T, Lansbury PT Jr. 2002. *Nature* 418:291

154. Li J, Uversky VN, Fink AL. 2002. *Neurotoxicology* 23:553–67
155. Pattison IH. 1965. See Ref. 201, pp. 249–57
156. Nakayashiki T, Ebihara K, Bannai H, Nakamura Y. 2001. *Mol. Cell* 7: 1121–30
157. Resende C, Parham SN, Tinsley C, Ferreira P, Duarte JA, Tuite MF. 2002. *Microbiology* 148:1049–60
158. Scott M, Foster D, Mirenda C, Serban D, Coufal F, et al. 1989. *Cell* 59: 847–57
159. Prusiner SB, Scott M, Foster D, Pan KM, Groth D, et al. 1990. *Cell* 63: 673–86
160. Bruce M, Chree A, McConnell I, Foster J, Pearson G, Fraser H. 1994. *Philos. Trans. R. Soc. London Ser. B* 343: 405–11
161. Mastrianni JA, Capellari S, Telling GC, Han D, Bosque P, et al. 2001. *Neurology* 57:2198–205
162. Manson JC, Jamieson E, Baybutt H, Tuzi NL, Barron R, et al. 1999. *EMBO J.* 18:6855–64
163. Manson JC, Barron R, Jamieson E, Baybutt H, Tuzi N, et al. 2000. *Arch. Virol.* Suppl. 16:95–102
164. Bossers A, Belt P, Raymond GJ, Caughey B, de Vries R, Smits MA. 1997. *Proc. Natl. Acad. Sci. USA* 94: 4931–36
165. Collinge J, Palmer MS, Sidle KC, Hill AF, Gowland I, et al. 1995. *Nature* 378: 779–83
166. Supattapone S, Bosque P, Muramoto T, Wille H, Aagaard C, et al. 1999. *Cell* 96:869–78
167. Scott MR, Safar J, Telling G, Nguyen O, Groth D, et al. 1997. *Proc. Natl. Acad. Sci. USA* 94:14279–84
168. Barron RM, Thomson V, Jamieson E, Melton DW, Ironside J, et al. 2001. *EMBO J.* 20:5070–78
169. Priola SA, Chesebro B. 1995. *J. Virol.* 69:7754–58
170. Raymond GJ, Bossers A, Raymond LD,

O'Rourke KI, McHolland LE, et al. 2000. *EMBO J.* 19:4425–30
171. True HL, Lindquist SL. 2000. *Nature* 407:477–83
172. Eaglestone SS, Cox BS, Tuite MF. 1999. *EMBO J.* 18:1974–81
173. Baudin-Baillieu A, Fernandez-Bellot E, Reine F, Coissac E, Cullin C. 2003. *Mol. Biol. Cell* 14:3449–58
174. Edskes HK, Wickner RB. 2002. *Proc. Natl. Acad. Sci. USA* 99(Suppl. 4): 16384–91
175. Schwimmer C, Masison DC. 2002. *Mol. Cell. Biol.* 22:3590–98
176. King CY, Tittmann P, Gross H, Gebert R, Aebi M, Wuthrich K. 1997. *Proc. Natl. Acad. Sci. USA* 94:6618–22
177. Busch A, Engemann S, Lurz R, Okazawa H, Lehrach H, Wanker EE. 2003. *J. Biol. Chem.* 278:41452–61
178. Come JH, Fraser PE, Lansbury PT Jr. 1993. *Proc. Natl. Acad. Sci. USA* 90: 5959–63
179. Kundu B, Maiti NR, Jones EM, Surewicz KA, Vanik DL, Surewicz WK. 2003. *Proc. Natl. Acad. Sci. USA* 100:12069–74
180. Esler WP, Stimson ER, Fishman JB, Ghilardi JR, Vinters HV, et al. 1999. *Biopolymers* 49:505–14
180a. O'Nuallain B, Williams AD, Westermark P, Wetzel R. 2004. *J. Biol Chem.* PMID:14752113
181. Speed MA, Wang DI, King J. 1996. *Nat. Biotechnol.* 14:1283–87
182. Rajan RS, Illing ME, Bence NF, Kopito RR. 2001. *Proc. Natl. Acad. Sci. USA* 98:13060–65
183. Dickinson AG, Fraser H. 1969. *Nature* 222:892–93
184. Collinge J, Sidle KC, Meads J, Ironside J, Hill AF. 1996. *Nature* 383:685–90
185. Hill AF, Desbruslais M, Joiner S, Sidle KC, Gowland I, et al. 1997. *Nature* 389: 448–50, 526
186. Will RG, Ironside JW, Zeidler M, Cousens SN, Estibeiro K, et al. 1996. *Lancet* 347:921–25

187. King CY. 2001. *J. Mol. Biol.* 307: 1247–60

188. Derkatch IL, Bradley ME, Zhou P, Liebman SW. 1999. *Curr. Genet.* 35: 59–67

189. Zlotnick I. 1965. See Ref. 201, pp. 237–48

190. Peretz D, Williamson RA, Legname G, Matsunaga Y, Vergara J, et al. 2002. *Neuron* 34:921–32

191. Kimberlin RH, Cole S, Walker CA. 1987. *J. Gen. Virol.* 68(Pt. 7):1875–81

191a. Bartz JC, Bessen RA, McKenzie D, Marsh RF, Aiken JM. 2000 *J. Virol.* 74(12):5542–47

192. Asante EA, Linehan JM, Desbruslais M, Joiner S, Gowland I, et al. 2002. *EMBO J.* 21:6358–66

193. Doel SM, McCready SJ, Nierras CR, Cox BS. 1994. *Genetics* 137:659–70

194. Kochneva-Pervukhova NV, Paushkin SV, Kushnirov VV, Cox BS, Tuite MF, Ter-Avanesyan MD. 1998. *EMBO J.* 17: 5805–10

195. Fandrich M, Dobson CM. 2002. *EMBO J.* 21:5682–90

196. Hill AF, Antoniou M, Collinge J. 1999. *J. Gen. Virol.* 80:11–14

197. Post K, Brown DR, Groschup M, Kretzschmar HA, Riesner D. 2000. *Arch. Virol.* Suppl. 16:265–73

198. Baskakov IV, Legname G, Baldwin MA, Prusiner SB, Cohen FE. 2002. *J. Biol. Chem.* 277:21140–48

199. Liu H, Farr-Jones S, Ulyanov NB, Llinas M, Marqusee S, et al. 1999. *Biochemistry* 38:5362–77

200. Scott MR, Will R, Ironside J, Nguyen HO, Tremblay P, et al. 1999. *Proc. Natl. Acad. Sci. USA* 96:15137–42

201. Gadjusek DC, Gibbs CJ Jr, Alpers MP, eds. 1965. *Slow, Latent, and Temperature Virus Infections,* NINDB Monogr. 2. Washington, DC: US GPO

Annu. Rev. Biochem. 2004. 73:657–704
doi: 10.1146/annurev.biochem.73.030403.080419
Copyright © 2004 by Annual Reviews. All rights reserved
First published online as a Review in Advance on March 25, 2004

THE MOLECULAR MECHANICS OF EUKARYOTIC TRANSLATION

Lee D. Kapp and Jon R. Lorsch

*Department of Biophysics and Biophysical Chemistry, Johns Hopkins University
School of Medicine, 725 North Wolfe Street, Baltimore, Maryland 21205-2185; email:
lkapp@jhmi.edu, jlorsch@jhmi.edu*

Key Words   protein synthesis, initiation, elongation, termination, recycling

■ **Abstract**   Great advances have been made in the past three decades in understanding the molecular mechanics underlying protein synthesis in bacteria, but our understanding of the corresponding events in eukaryotic organisms is only beginning to catch up. In this review we describe the current state of our knowledge and ignorance of the molecular mechanics underlying eukaryotic translation. We discuss the mechanisms conserved across the three kingdoms of life as well as the important divergences that have taken place in the pathway.

CONTENTS

INTRODUCTION

This chapter reviews what we think we know about the molecular mechanisms underlying protein synthesis in eukaryotic organisms and outlines what we do not know. Most molecular biology textbooks contain beautiful cartoons of the current model of the pathway of translation in eukaryotes. These models, albeit useful for teaching purposes and to guide the thinking of researchers, have also lulled many in the biological community into a false sense of the level of our understanding. It is not uncommon to hear, either explicitly or implicitly, the statement that "translation is a solved problem." Therefore, we examine the current models of the steps of eukaryotic translation to determine where they are soft and where they are firm. The enormity of this task compels us to limit our discussion to the fundamental mechanisms underlying eukaryotic translation. Translation of specific mRNAs, ER-associated translation, translational control, and special cases of translation such as internal ribosome entry, frame-shifting, ribosome shunting, etc., are not discussed, except as an occasional example in which these events have shed light on the underlying mechanics of eukaryotic protein synthesis.

Overview of the Steps of Translation

Translation can be broken into four stages: initiation, elongation, termination, and recycling. In initiation, the ribosome is assembled at the initiation codon in the mRNA with a methionyl initiator tRNA bound (presumably) in its peptidyl (P) site. In elongation, aminoacyl tRNAs enter the acceptor (A) site where decoding takes place. If they are the correct (cognate) tRNA, the ribosome catalyzes the formation of a peptide bond. After the tRNAs and mRNA are translocated such that the next codon is moved into the A site, the process is repeated. Termination takes place when a stop codon is encountered and the finished peptide is released from the ribosome. In the final stage, recycling, the ribosomal subunits are dissociated, releasing the mRNA and deacylated tRNA and setting the stage for another round of initiation.

The above outline of the stages of translation describes the fundamental events in the process that occur throughout all kingdoms of life. Although these fundamental events are the same, how they are achieved in each kingdom sometimes differs. The differences in the processes required to make a protein in each of the three kingdoms may well contain important information about the underlying molecular mechanics of the processes in each kingdom. Therefore, our discussion of each phase of eukaryotic translation begins with a comparison of the steps and components used to accomplish it in each kingdom.

INITIATION

Summary of the Current Model of the Steps of Eukaryotic Translation Initiation

The current model of the steps of eukaryotic translation initiation is shown in Figure 1 (1). The first step in the initiation pathway is the assembly of the eIF2·GTP·Met-tRNA$_i$ ternary complex. Because eIF2 has ~100-fold higher affinity for GDP than for GTP (2–5) and the rate constant for GDP release is slow (2, 6, 7), the eIF2·GDP complex that results from each completed round of translation initiation must be recycled to eIF2·GTP before a new round can begin. This exchange reaction is facilitated by eIF2B. After its formation, the ternary complex binds to the small (40S) ribosomal subunit. This binding is facilitated by (at least) eIFs 1, 1A, and 3. The resulting complex is called the 43S complex. The eIF4F complex assembles on the 5′-cap of the mRNA and unwinds structures found in the 5′-untranslated region (UTR). This is accomplished through the ATP-dependent action of eIF4A assisted by the RNA-binding proteins eIF4B and, in mammals, eIF4H. eIF4F, in conjunction with eIF3 and the poly(A) binding protein (PAB) bound to the 3′-poly(A) tail, loads the mRNA onto the 43S complex. The 43S complex then begins scanning down the message in the 5′ to 3′ direction, looking for the initiation codon. This scanning process is thought to require ATP hydrolysis, although the ATPase involved has not been identified. eIF1 and eIF1A may play a role in this scanning process as well. When the 43S complex encounters an AUG codon that is embedded in a favorable sequence context (e.g., the Kozak sequence), usually the first AUG, codon-anticodon base pairing takes place between the initiation codon and the initiator tRNA in the ternary complex. This then triggers GTP hydrolysis by eIF2, a reaction facilitated by the GTPase-activating protein (GAP) eIF5. eIF1 also plays a role in the detection of the correct initiation codon. After GTP hydrolysis by eIF2, eIF2·GDP releases the Met-tRNA$_i$ into the P site of the 40S subunit and then dissociates from the complex. In the current model, eIFs 1, 1A, 3, and 5 also dissociate at this stage. At some point either before or after GTP hydrolysis by eIF2, eIF5B·GTP binds to the complex. Once the other factors are gone, it facilitates the joining of the large (60S) ribosomal subunit to the 40S·Met-tRNA$_i$·mRNA complex. This event triggers GTP hydrolysis by eIF5B, and, because the eIF5B·GDP complex has a low affinity for the ribosome, it dissociates from the complex. This is thought to be the end of translation initiation, although other steps may be required before the complex is fully active to make a peptide bond (8, 9).

Comparative Translation Initiation

BACTERIA   Translation initiation in bacteria involves three initiation factors: IF1, IF2 and IF3. IF1 binds over the A site of the small ribosomal subunit (10, 11) and is thought to prevent the initiator tRNA from binding to the A site,

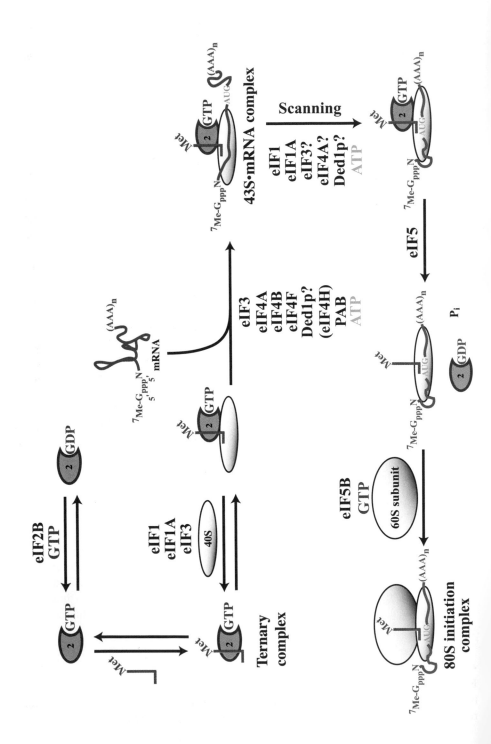

instead promoting its binding to the P site (12). IF2 is a GTPase that enhances binding of the initiator tRNA to the small ribosomal subunit and likely also facilitates the subunit joining step (12–14). IF3 appears to have at least four functions (12): to ensure the fidelity of initiation site selection; along with IF2 to help select the formylated methionyl initiator tRNA (fMet-tRNA$_i$) for use in initiation rather than the elongator methionyl tRNA used to insert methionine residues into polypeptides during elongation; to dissociate 70S ribosomes into 30S and 50S subunits and likely prevent premature association of the subunits during initiation; and to act during ribosome recycling to remove the deacylated tRNA from the P site of the 30S subunit (15).

The recruitment of the mRNA to the small ribosomal subunit is accomplished in bacteria via a base-pairing interaction between the 3′-end of the 16S rRNA (the anti-Shine-Dalgarno) and the purine-rich Shine-Dalgarno sequence located upstream of the initiation codon in the mRNA. Because the Shine-Dalgarno sequence is usually ∼10 bases upstream of the initiation codon, it makes the placement of the initiation codon in or near the P site of the small ribosomal subunit a relatively straightforward process.

EUKARYOTES   Although the fundamental goals of translation initiation are the same in bacteria and eukaryotes—to get the initiator tRNA and the mRNA onto the small ribosomal subunit, to find the AUG codon, and to join the ribosomal subunits to form an initiation complex—the steps and machinery required to accomplish these goals are much more complicated in eukaryotes than in bacteria. The free amino group on the methionine moiety attached to the initiator tRNA, which is formylated in bacteria, is free in eukaryotes, although the functional significance of this difference is not yet clear. Perhaps most remarkably, the three translation initiation factors of bacteria are replaced by at least 12 in eukaryotes, which are comprised of at least 23 different polypeptides. Two of the bacterial initiation factors have clear orthologs in eukaryotes (16). The naming of eukaryotic initiation factors is somewhat confusing. The ortholog of bacterial IF1 is not eIF1 (the *e* stands for eukaryotic) but is instead eIF1A, and the ortholog of IF2 is eIF5B, not eIF2 (17–19). At the sequence level, there does not appear to be a clear counterpart to IF3 in eukaryotes. However, based on functional and structural similarities, eIF1 is suggested to be the eukaryotic ortholog of IF3 (20, 21). [Oddly, a few bacteria have a gene encoding a clear eIF1 ortholog, although its function in these organisms is not known (16).]

Functionally, the story is slightly more complex. Mammalian eIF1A has been reported to prevent spurious association of ribosomal subunits (22) just as its ortholog IF1 does. The large, multisubunit factor eIF3 also has antisubunit-

Figure 1   Cartoon of our interpretation of the current model of the steps in eukaryotic translation initiation and the roles of the factors. This is just a model and should not be regarded as factual. For clarity, the 5′- and 3′-ends of the mRNA are not shown interacting.

association activity (23–25) and thus eukaryotes may have multiple means for dissociating prematurely coupled ribosomal subunits and/or preventing their premature association. Although eIF5B is the clear ortholog of bacterial IF2, some of the proposed roles of IF2 appear to be performed in eukaryotes by eIF2 rather than by eIF5B. In its GTP-bound form, eIF2 forms a complex with Met-tRNA$_i$ and carries the initiator tRNA onto the small ribosomal subunit, similar to the initiator tRNA loading function of IF2 in bacteria [although it is still controversial whether IF2 forms a complex with initiator tRNA in bacteria prior to binding to the 30S subunit (26, 27)]. In contrast, eIF5B does not bind initiator tRNA detectably in solution (28; L.K. & J.R.L., unpublished), although it stabilizes initiator tRNA binding to the ribosome (17, 29). eIF5B does, however, play a critical role in facilitating the joining of the two ribosomal subunits following location of the initiation codon (30), just as IF2 is proposed to do (14). Finally, as mentioned above, it is not known whether there is an actual ortholog for IF3 in eukaryotes, but, like IF3, eIF1 is clearly involved in ensuring the fidelity of initiation site selection (21, 31, 32), suggesting they may at least be functional orthologs (21).

The other translation initiation factors in eukaryotes do not appear to have any counterparts in bacteria. eIF2, for which the closest relative in bacteria is the elongation factor EF-Tu (33), functions as previously described to bring the initiator tRNA onto the ribosome. It may also be directly involved in locating the initiation codon (34). eIF3 is a giant heteromultimeric complex comprised of five core subunits in yeast (35) and at least ten subunits in mammals (36–39) and plants (40). With a mol wt of 360 kDa in yeast, it is roughly one quarter of the size of the small ribosomal subunit. The evidence to date suggests a role for eIF3 in facilitating the binding of both the ternary complex and the mRNA to the 40S subunit (38, 41–44). eIF3 also binds to most other initiation factors plus the small ribosomal subunit, suggesting that it may function as a central hub in the assembly of the translation initiation complex. In general, the function of this factor is still not entirely clear, as discussed in more detail below.

Location of the initiation codon in eukaryotes is a very different process from that in bacteria. Although the 18S rRNA of the eukaryotic small ribosomal subunit is very similar to bacterial 16S rRNA over most of its length, the anti-Shine-Dalgarno sequence found in 16S rRNA is missing in 18S rRNA (45). Correspondingly, eukaryotic mRNAs do not have a Shine-Dalgarno sequence that allows easy identification of the initiation codon, but rather have a 5′-7-methylguanosine cap structure that identifies the 5′-end of the message. The 43S complex is proposed to be loaded onto the 5′-end of the mRNA via the cap and to then locate the initiation codon via the scanning process (46). As long as the 43S complex can be faithfully loaded onto the message at its 5′-end and then scan along linearly without hopping about, this method should allow the identification of the correct (usually the first) initiation codon.

The initial loading of the mRNA onto the 43S complex itself presents several problems for the eukaryotic translational apparatus. In bacteria, transcription and translation are coupled processes and thus as soon as the Shine-Dalgarno sequence emerges from the transcriptional apparatus it can be bound by the small ribosomal subunit, whereas in eukaryotes, the mRNA substrate for translation initiation is thought to be a complete, fully processed messenger RNA, potentially rife with secondary and tertiary structures and coated with RNA-binding proteins such as hnRNPs. [However, the first round of translation initiation on an mRNA may well occur concomitantly with export of the mRNA from the nucleus. Thus, a free, full-length mRNA may not be the substrate for initial initiation (47).] The eIF4 group of initiation factors, which are not present in bacteria, appears to be required because of the uncoupling of transcription and translation and the lack of the Shine-Dalgarno initiation codon identification system in eukaryotes. eIF4A is a DEAD-box RNA-dependent ATPase (48) that can unwind RNA duplexes in vitro and thus has been proposed to be an RNA helicase that functions to unwind regions of secondary and tertiary structure in the 5'-ends of mRNAs. eIF4B is an RNA-binding protein that stimulates the RNA-unwinding activity of eIF4A in vitro (49–52) and in vivo suppresses both mutations in eIF4A (53) and the inhibitory effect of secondary structure in the 5'-untranslated region of a reporter gene (54). It is therefore thought that eIF4B assists eIF4A in unwinding structures in mRNAs. A second RNA-binding protein with similar functions, called eIF4H, has also been identified in mammals (55–57). eIF4E binds to the 7-methylguanosine cap on mRNAs, thus locating their 5'-ends. Both eIF4E and eIF4A bind to eIF4G, which, like eIF3, is thought to act as a multifaceted adapter protein—a sort of central hub for interactions among factors in the initiation apparatus (58). Together, eIFs 4A, 4E, and 4G are called eIF4F (59). The proposed role of the eIF4F complex is to bind to the 5'-end of the mRNA, unwind any structures found there, and then facilitate the loading of the 43S complex onto the now unstructured 5'-untranslated region (5'-UTR). Since the RNA-binding proteins that coat an mRNA probably must also be removed prior to 43S complex loading, it may be that eIF4F and/or eIF4A perform this task as well.

In addition to the 7-methylguanosine cap structure on their 5'-ends, eukaryotic mRNAs have 3'-poly(A) tails that are bound by the poly(A)-binding protein, or PAB. PAB interacts with eIF4G, and this interaction is thought to lead to the circularization of eukaryotic mRNAs, which in turn stimulates translation (60–64). The prevailing notion is that this interaction facilitates binding of the 43S complex to the mRNA, although this is most likely not the whole story (see discussion below). This system provides a quality control mechanism: If the mRNA has been partially degraded and has lost its 3'-end, it will not be translated efficiently. Thus, the system helps guard against the synthesis of truncated proteins that could be toxic to the cell. Translation of degraded mRNAs is somewhat less of a problem in bacteria because of the coupling between transcription and translation—as the 3'-end of the message is being made, ribosomes are rapidly moving down from the 5'-end to translate it. Of course, the presence of tmRNAs and the *trans*-translation system in bacteria that can rescue ribosomes stuck at the end of partially degraded massages lacking stop codons (65) indicates that 3′-degradation of mRNAs is also important in bacteria. One interesting difference between the two kingdoms in this regard is that it is apparently sufficient for bacteria to deal with the problem after the fact (i.e., release the stalled ribosome and tag the incomplete protein for degradation), whereas eukaryotes have had to develop a system to prevent the translation of truncated mRNAs from happening in the first place.

In addition to the core group of initiation factors, a number of proteins have been suggested to be involved in the initiation process, although their roles, if any, are not yet clear. For example, eIF5A was initially isolated as a protein from rabbit reticulocyte lysates that increased the efficiency of assays measuring the formation of the first peptide bond from 80S initiation complexes (66, 67). Although it is an essential protein in yeast, shutting off expression of an unstable form of eIF5A has only a small ($\leq$ twofold) effect on overall protein synthesis in vivo, suggesting it may not in fact be a general translation initiation factor (68). Other data suggest a role for eIF5A in mRNA turnover (69), in ribosome biogenesis, and in the maintenance of cell wall integrity (70). Other potential factors such as eIF6, eIF2A (71), eIF2C (72), and an ATP-binding protein called ABC50 (73) have all been implicated in translation initiation, although eIF6 and eIF2C have also been implicated in other processes, 60S subunit biogenesis (74, 75), and RNA interference (e.g., 76), respectively.

The differences between the required components for eukaryotic and prokaryotic translation might be construed as excess factors required solely for regulatory purposes. However, most of the authentic initiation factors absent in bacteria but present in eukaryotes appear to be involved in central aspects of the mechanism of translation rather than serving peripheral, albeit important, roles. Furthermore, eIFs 1, 1A, 2, 2B, 3, 4A, 4E, 4G, and 5 as well as ded1p are all essential in yeast, further arguing against their being merely ornamentation on top of the core mechanism of bacterial translation. One of the few factors that is not absolutely essential in yeast is eIF5B, the ortholog of bacterial IF2, although its deletion produces a severe slow-growth phenotype (17).

ARCHAEA   Translation initiation in archaea appears to constitute an evolutionary rest-stop between the bacterial and eukaryotic systems. Archaeal mRNAs are not capped and most contain Shine-Dalgarno-like sequences upstream of their initiation codons (33). Accordingly, in many cases a prokaryotic-like mode of initiation codon recognition appears to be used in archaea (33), i.e., via base pairing between the 16S rRNA and the Shine-Dalgarno sequence upstream of the initiation codon. This is consistent with the lack of a nucleus in archaea and the presumed coupling between transcription and translation. However, a significant number of archaeal mRNAs lack appropriately spaced Shine-Dalgarno sequences upstream of their initiation codons, and thus other mechanisms of initiation site selection, possibly akin to eukaryotic scanning, may also take place (33, 77).

Like both bacteria and eukaryotes, archaea have IF1/eIF1A and IF2/eIF5B orthologs. [To avoid confusion, we adopt the nomenclature of Dever (20) and call the archaeal factors by the names of their eukaryotic counterparts, for instance, a-eIF5B.] Unlike translation initiation in bacteria, the methionine on the initiator tRNA does not appear to be formylated in archaea (78). The initiator tRNA itself is a mix of eukaryotic and bacterial features; several of the conserved initiator tRNA identity elements are the same as those found in the bacterial initiator tRNA, whereas others, including the important A1:U72 base pair at the end of the acceptor stem, are the same as those found in the eukaryotic tRNA (78). Although all three kingdoms contain IF1/eIF1A counterparts (16, 79), the eukaryotic and archaeal versions are more similar to each other than are the archaeal and bacterial versions (16). In addition, archaea contain an eIF2 ortholog (80) and an eIF1 ortholog (16). They also have proteins related to two of the five subunits of the GDP:GTP-exchange factor for eIF2, eIF2B (81), although it is unlikely that they actually perform this function because the subunits to which they are homologous are not the catalytic subunits of the complex (82). Thus it appears that archaea possess some of the same translation initiation machinery as eukaryotes, albeit in a bare-bones version.

There is probably some significance to which components of the translation initiation machinery are common to eukaryotes and archaea and which are not. First, the core components, IF1/eIF1A, IF2/eIF5B, and IF3/eIF1 (the last being a tenuous homology at this point), must ensure that the initiator tRNA gets into the right place on the small ribosomal subunit, then that a start codon is also in the right place and that subunit joining happens at the right time. Because eIF1 and eIF1A may both interact with the initiator tRNA during initiation, the reason why archaea have more eukaryotic-like versions of these factors may be because the eukaryotic and archaeal initiator tRNAs are more similar than are the archaeal and bacterial tRNAs (e.g., the A1:U72 base pair and the lack of formylation of the methionine). Also, eIF1A and eIF1 may have evolved to work in regions of the small ribosomal subunits of eukaryotes and archaea that are structurally distinct from the same regions in the bacterial 30S subunit; at least in terms of their complements of ribosomal proteins, the archaeal ribosome is more similar to the eukaryotic ribosome than to the bacterial ribosome (83).

Differences between the archaeal and eukaryotic translation initiation apparatus are also potentially informative. For instance, as previously pointed out by Dever (20), it is striking that archaea possess an eIF2 but not an eIF5 ortholog, given that eIF5 is thought to act as a GTPase activating protein (GAP) for eIF2. Thus, a-eIF2 must either be capable of hydrolyzing GTP without assistance from a GAP, or a GAP that is not homologous to eIF5 must exist in archaea. If there is no eIF5 equivalent, this could be because of the different modes of initiation site selection in archaea and eukaryotes, which would in turn suggest that eIF5 plays a direct role in location of the initiation codon during scanning, rather than acting simply as a GAP. The lack of eIF3 in archaea is also striking because eIF3 is thought to be a centrally important factor in eukaryotes, acting as an interaction hub for most of the components of the machinery and facilitating binding of the eIF2·GTP·Met-tRNA$_i$ ternary complex and mRNA to the small ribosomal subunit. Why would eukaryotes need a very large factor to help binding of the ternary complex to the small ribosomal subunit but archaea would not? Perhaps facilitating ternary complex binding is not actually one of eIF3's essential functions in eukaryotes but rather the consequence of mutual interaction among eIF3, the 40S subunit, and eIF2. If one of eIF3's main roles is to aid in mRNA loading onto the 43S complex and initiation site location, it might then make sense that archaea lack the factor given their apparently different mode of initiation site selection.

By the same logic, it makes some sense that archaea lack the eIF4 complement of factors. It has been suggested, however, that there might in fact be one or more orthologs of eIF4A in archaea (77, 84), although this assignment is based entirely on sequence similarity, and DEAD box proteins in general share significant sequence similarity with eIF4A [the canonical DEAD box protein (85)]. The sequence identity between yeast eIF4A and the putative eIF4A orthologs in archaea is ~36%. For comparison, the sequence identity between yeast eIF4A and Fal1p, a nucleolar DEAD box protein involved in ribosome biogenesis and not translation initiation (86), is 57%, and the sequence identity between yeast eIF4A and the 50S-subunit associated protein DbpA from *Escherichia coli* is 34%. Thus an identity of 36% does not imply a common function for DEAD box proteins. It has also been suggested that the *E. coli* cold-shock protein DeaD/CsdA is a bacterial version of eIF4A (87). While this suggestion suffers from the same sequence similarity constraints just discussed, it was strengthened somewhat by data showing that the protein could promote the translation of a structured mRNA in vitro (87). This effect, however, did not depend on ATP hydrolysis, as expected for the action of a true eIF4A ortholog. Furthermore, it was later shown that DeaD/CsdA can function in ribosome biogenesis in vivo (88), casting some doubt on its role in translation initiation. In general, there is no convincing evidence to date of eIF4A counterparts acting in translation initiation either in archaea or bacteria.

The Order of Events

Much effort has gone into defining the order of the steps in the pathway of eukaryotic translation initiation. An understanding of this order is a necessary foundation for attempts to understand the molecular mechanics of each step in the pathway. Assignment of the order, however, has often been complicated by confusion over the difference between kinetics and thermodynamics. Thermodynamics can tell us nothing about the order of events; this is solely the domain of kinetics. For example, the determination that A binds more tightly to B than C does provides no evidence that A binds to B before C.

There is considerable evidence behind the assignment of the order of several steps. For example, ternary complex clearly must be formed at the beginning of the pathway and bind to the 40S subunit prior to initiation codon location. In contrast, the evidence supporting the assigned order of other steps is less compelling. For example, in the current model, ternary complex binds to the 40S subunit prior to binding of the mRNA. The assignment of this order of binding does not appear to be strong, however. Although ternary complexes can bind to 40S subunits in the absence of mRNA, small model mRNAs can also bind to naked 40S ribosomal subunits with K_ds in the high nM to low μM range (89; D. Maag & J.R.L., unpublished), suggesting that the ribosome has a functional mRNA-binding site in the absence of ternary complex and other factors. A priori there does not seem to be any reason why unwinding of the mRNA's 5'-UTR and loading of the message onto the 40S subunit by the factors could not take place, at least some of the time, prior to binding of the ternary complex. In fact, on the GCN4 mRNA in yeast, after a short open reading frame upstream of the GCN4 ORF has been translated, the 40S subunit lacking ternary complex appears to scan along the mRNA on its own before rebinding another ternary complex and recommencing initiation (90), consistent with the idea that mRNA can be bound to the 40S subunit in the absence of a ternary complex. A similar model was proposed to explain why lowering the concentration of eIF2 in an in vitro system increased the efficiency with which the second of two AUG codons was used to initiate translation: The 40S subunit without the ternary complex binds mRNA and scans past the first AUG before binding ternary complex and becoming competent to locate the (second) initiation codon (91). Resolution of the order of ternary complex and mRNA binding awaits detailed kinetic analysis of these steps.

The prevailing model regarding the timing of the release of the initiation factors is that eIFs 1, 1A, 3, and 5 are released from the 40S subunit along with eIF2·GDP following initiation codon location. Much of the data for this part of the model comes from experiments in which preinitiation complexes were isolated by centrifugation on density gradients and then the fractions subjected to Western blot analysis to determine whether the factors were still associated with ribosomal complexes. This technique, although somewhat informative about the stability of the factor-ribosome complexes, lacks kinetic and thermodynamic resolution. Even complexes that are very stable on the time scale and/or under the concentrations of translation initiation can dissociate from the preinitiation complexes during the lengthy process of gradient fractionation. Therefore, until a thorough kinetic dissection of the dissociation of the factors from actively initiating ribosomal complexes can be performed, determination of when most of the factors leave the ribosome will remain unresolved.

Complex Complexes: Interactions Among the Factors

Considerable attention has been directed in the last five years toward determining all of the interactions among the components of the initiation machinery. Initially, the degree of separation between any two components in the system— how many intermediate components bridged the interaction between any two factors—appeared to be amazingly small, perhaps only one or two. As more

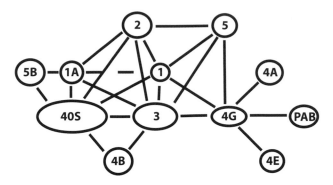

Figure 2   Map of known interactions among eukaryotic translation initiation factors and the 40S ribosomal subunit. The dashed line between eIFs 1 and 1A indicates that this interaction might be indirect (mediated by conformational changes in the 40S subunit).

interactions are discovered and mapped, the real answer may in fact turn out to be closer to zero; that is, everything interacts directly with everything else. As can be seen from Figure 2, not many more interactions need to be found before this becomes true.

THE MULTIFACTOR COMPLEX(ES)   Clearly one hub of interactions is eIF3. There are data indicating that one or more of the subunits of eIF3 interact with eIFs 1 (35, 92), 1A (93), 2 (42), 4B (94, 95), 4G (96, 97), and 5 (35, 44, 97), as well as with the 40S subunit (39, 98, 99). In addition, various subunits of eIF3 have been shown to interact with a wide variety of other proteins in the cell, most notably with cytoskeletal proteins (100–104) and components of the proteasome (reviewed in 105). These interactions suggest that eIF3 may function not only as a central organizer of the translation initiation machinery but also as a key link between the initiation machinery and the rest of the cell.

A variety of experiments have shown that a "multifactor complex" (MFC) consisting of at least eIFs 1, 2, 3, 5, and Met-tRNA$_i$ (presumably bound to eIF2) exists stably in vivo (106) and that the integrity of this complex is important for translation initiation (42, 92, 97, 106, 107). Thus, at least some components of the translational machinery may function in vivo as a unit that binds *en masse* to the 40S ribosomal subunit and the mRNA, coordinately facilitates the loading of the mRNA and location of the initiation codon, and then dissociates from the ribosome as a single complex. This idea is reminiscent of the recent proposal in the splicing field, based on isolation of a similar stable multifactor spliceosomal complex, that most of the components of the spliceosome may function in splicing as one unit, rather than coming on and off the pre-mRNA at different times (108). This stable multifactor arrangement could even be a general feature of how complex biological processes operate within the cell. It would have the advantage that it reduces the number of assembly and disassembly steps required for each round of the process in question and it ensures that all the required components are always in the same place at the same time. This could be particularly important inside a cell where macromolecular diffusion can be slow (109). Another advantage might be that once the complex was assembled correctly, its constituent parts would be resistant to competing inhibitory interactions, either spurious or regulatory, with other components of the cell [this assumes a slow dissociation rate for the correct components within the complex—which seems to be the case for the MFC (106)].

At least some of the interactions among the factors may be dynamic and only occur at certain points in the initiation pathway. Several observations support this notion. First, some components of the system appear to stay associated with the 40S subunit after other components have dissociated. For example, as described above, eIF2, at least, has most likely left the 40S subunit before eIF5B facilitates subunit joining. Furthermore, in the cell the concentrations of some of the components of the multifactor complex are higher than others, suggesting pools of certain factors that are associated with the complex and other pools that are not. In yeast, careful measurements of the concentrations of most of the factors has revealed that there is approximately threefold more eIF1 than eIF3 and ~50-fold more eIF4A than eIF4G (110), suggesting the existence of pools of eIF1 not associated with the multifactor complex and large pools of eIF4A not associated with eIF4F. It seems unlikely that these pools of factors not associated with multifactor complexes play no roles in initiation. Finally, in several cases, interactions between two factors and a third factor are mutually exclusive, suggesting that if they occur as part of the initiation pathway, they do so at different times. For example, the interactions of eIFs 1 and 5 with eIF4G are mutually exclusive (111), as are the interactions of eIFs 2 and 4G with eIF5 (97) and the interactions of eIFs 5 and 2B with eIF2 (112).

MOVING MOUNTAINS AND LOWERING ENERGY BARRIERS: MODULATING ENERGETICS IN COMPLEX SYSTEMS   In addition to colocalizing the factors and coordinating their binding and release, interactions among factors could induce conformational changes that modulate their behavior at different points in the pathway. For example, binding of factor A to factor B could enhance or reduce factor B's affinity for another component of the system. Fundamentally, such changes in thermodynamic and kinetic parameters as the processes proceed from one step to the next allow complex biological systems to operate. Although our knowledge of the physical basis for each of the steps in eukaryotic translation initiation is limited, some studies are suggesting the interconnectedness of the energetics in the system. For example, studies have indicated that eIF4E's affinity for the 7-methylguanosine cap structure is enhanced by interaction between eIF4E and eIF4G, and that the interaction between eIF4G and PAB increases eIF4E's cap affinity still further (113–115). Data also suggest that eIF4A and eIF3 bind to eIF4G cooperatively (116) and since there is no known interaction between eIF4A and eIF3, this effect could be the result of a conformational change in one or more factors. The same could be true for eIFs 1 and 1A, which bind cooperatively to the 40S subunit (117). These studies are most likely just beginning to elucidate the modulation in the energetics of different interactions that takes place during the course of translation initiation.

COMPLEX STRUCTURES   Structures of complexes are needed to make molecular sense of changes in thermodynamic and kinetic parameters that occur from step to step in the initiation process. Progress has been made in the past few years toward understanding the molecular basis for several interfactor interactions.

One complex of central importance is the cap-binding complex eIF4F. Although the entire structure of eIF4F has not been determined, the structures of parts of the complex have been elucidated. The crystal structure of a segment of human eIF4G that binds to eIF4A has been determined, revealing a surface that interacts with eIF4A (118). The crystal structure of the ternary complex of eIF4E, an analog (7-methyl-GDP) of the 5′-cap and a peptide that corresponds to a conserved eIF4E-binding site in eIF4G, has also been determined (119). The ternary complex of eIF4E, 7-methyl-GDP, and a homologous peptide from eIF4E-binding protein (4E-BP), a regulatory protein that binds to and inhibits eIF4E's interaction with the cap, was determined in the same study. In both cases, the peptides undergo a transition from a random coil in solution to an α-helix when bound to eIF4E. While it is not yet clear whether the eIF4G peptide is structured or unstructured in the context of full-length eIF4G [although a 98-amino acid fragment of the eIF4E binding site of yeast eIF4G was shown to undergo a similar unfolded-to-folded transition upon eIF4E binding (120)], circular dichroism and NMR studies both indicate that full-length 4E-BP is not a structured protein on its own, suggesting that the unstructured-to-structured transition upon binding occurs in this context as well (119, 121, 122).

Such binding-induced unstructured-to-structured transitions may be common features of interactions between eukaryotic translation initiation factors. The interaction between eIF1A and eIF5B also involves an initially unstructured region of eIF1A (the C terminus) binding to the structured C-terminal domain of eIF5B, and becoming structured in the process (93, 123, 124). eIF1A also has a long, unstructured N-terminal tail (125) that is involved in interacting with eIFs 2 and 3 (93), and eIF1's N terminus is also long and unstructured (126), suggesting that they may undergo similar binding-induced ordering. Although it is not yet clear what these structural transitions are used for, if anything, the long, unstructured tails may be convenient to span the long distances between factors imposed by the size of the ribosome, allowing factors to communicate with one another using a minimal amount of excess protein (essentially polypeptide telephone lines). As more structures of complexes in these systems become available, it should be more obvious whether this really is a general feature of factor-factor interactions and what its purpose is.

The Molecular Mechanics of the Steps

In the following sections we discuss the current state of knowledge (and lack thereof) of the molecular mechanics underlying several key steps in eukaryotic translation initiation. Space and stamina constraints led us to choose topics of particular interest to us rather than try to survey the entire field.

TERNARY COMPLEX FORMATION AND DISSOLUTION   eIF2 is also a heteromultimer, made up of three different subunits with a total mol wt of ~125 kDa. A variety of biochemical and genetic experiments have indicated that the γ subunit is the site of both GTP and Met-tRNA$_i$ binding (6, 81, 127) and consistent with this, the γ subunit is homologous to the elongation factor eEF1A/EF-Tu (80), which is also a GTP-dependent carrier of aminoacylated tRNAs. The crystal structure of the γ subunit of archaeal eIF2 has confirmed that the protein's structure is similar to EF-Tu/eEF1A (8). Unlike eEF1A/EF-Tu, however, which must bind all elongator aminoacyl tRNAs with roughly equal affinity, eIF2 must bind only the methionylated initiator tRNA. The elements in the initiator tRNA that make this discrimination possible include three consecutive G:C base pairs in the anticodon stem, some conserved sequences in and around the T loop, a phosphoribosyl modification at position 64, and an A1:U72 base pair at the end of the acceptor stem (128). This last feature is particularly important for identifying the tRNA as an initiator tRNA (2, 129, 130). Several of the other identity elements, most notably the phosphoribosyl modification at position 64, prevent binding of the initiator tRNA to eEF1A (128, 131–133), although whether they also play nonessential roles in initiation is open to question. It was recently shown that eIF2 in its GTP-bound form has a positive contact with the methionine on the initiator tRNA and that this contact is disrupted when GTP is exchanged for GDP (e.g., upon GTP hydrolysis), whereas contacts to the body of the tRNA are not altered (2). This GTP-dependent recognition of the methionine moiety may in part prevent unacylated tRNA$_i$ from entering the initiation pathway and is likely to be an important part of the mechanism of release of the initiator tRNA from eIF2 upon initiation codon recognition.

LOADING THE TERNARY COMPLEX ONTO THE 40S SUBUNIT   A number of components of the initiation machinery seem to be involved in facilitating 43S complex formation. Early on, eIFs 1A and 3 were implicated in enhancing ternary complex binding to the 40S subunit in a reconstituted mammalian initiation system (41, 134). More recently, using the same system, it has been found that eIFs 1, 1A, and 3 are minimally required to observe stable binding of the ternary complex to the 40S subunit (43, 135, 136), whereas in a yeast-based reconstituted initiation system, only eIFs 1 and 1A are critical (137), although eIF3 does stabilize ternary complex binding in this system approximately twofold (M. Algire & J.R.L., unpublished). In most cases, it is difficult to disentangle effects on ternary complex and mRNA binding in vivo because the two are strongly linked (38), but experiments monitoring the translationally controlled production of the transcription factor GCN4 have allowed the binding of ternary complex to be uncoupled from mRNA binding (reviewed in 90, 138). The efficiency with which the GCN4 ORF is translated is inversely proportional to the efficiency with which ternary complex binds to a 40S subunit that is already on the mRNA (because it has already translated an ORF upstream of GCN4's). Studies using this system have shown that in vivo both eIF1A (93) and eIF3 (42) affect ternary complex binding. In addition, in vitro translation experiments using yeast cell extracts indicate that eIF5 affects ternary complex recruitment and/or mRNA binding to the 40S subunit, although given the coupling between the two events, it is uncertain which step is affected. However, because eIF5 is a central component of the multifactor complex, interacting with the ternary complex in addition to eIFs 1 (111) and 3, both of which bind the 40S subunit, eIF5 is likely at least to stabilize ternary complex binding to the 40S subunit via the chelate effect.

How then do these factors facilitate 43S complex formation? Some, as suggested for eIF5, might stabilize ternary complex binding via simultaneous interactions with the ternary complex and the 40S subunit. Others could alter the conformation of the ribosome either locally or globally in such a way as to promote ternary complex binding. Small factors such as eIFs 1 and 1A might alter the local conformation of the eIF2 binding site, which is thought to be over and around the P site (139–141), whereas a large factor such as eIF3 might distort the conformation of the entire 40S subunit to allow easier access of eIF2 with its attached Met-tRNA$_i$. This latter possibility might explain the puzzle of why eIF3 is so big.

MOLECULAR SNAKE HANDLING: GETTING THE mRNA ON THE RIBOSOME   The presumed first step in loading of the mRNA onto the small ribosomal subunit is the recognition of the mRNA's 5'-cap. The structure of eIF4E in complex with the cap analog 7-methyl-GDP has been determined by both X-ray crystallography (142, 143) and NMR (144), revealing the molecular basis for the factor's specificity for 7-methylguanosine: base stacking of the positively charged 7meG between two conserved electron-rich tryptophans and a network of other interactions. The importance of the eIF4E-cap interaction has been demonstrated both in vivo and in vitro. For example, in vitro translation of capped mRNAs is enhanced 3- to 30-fold relative to the uncapped message in both yeast (61, 63, 145) and mammalian (62, 145–147) extracts. Depletion of active eIF4E in the extracts abrogates the effect of the mRNA cap on translational efficiency (61), as expected if eIF4E mediates the interaction of the cap with the translational machinery. At high concentrations, uncapped mRNA is translated as efficiently as capped mRNA in extracts (147), consistent with the idea that the role of the cap is to enhance binding of the mRNA to the preinitiation complex. Electroporation of capped and uncapped mRNAs into plant, yeast, and mammalian cells has also shown that the cap enhances translation in vivo 3- to 20-fold (62, 63, 148, 149), consistent with the in vitro results.

Careful measurements of the affinity of eIF4E for a variety of cap analogs showed a strong correlation between the analogs' affinities and their abilities to inhibit translation in vitro (150), suggesting that modulation of the strength of the cap-eIF4E interaction could be important for controlling rates of translation. For example, mRNAs with 5'-caps partially occluded by structures in their 5'-UTRs would be expected to have an intrinsically weaker cap-eIF4E interaction than messages with unobscured cap structures, i.e., some mRNAs are "strong" and can compete well for binding to eIF4E, whereas others are "weak" and compete poorly. However, this notion is as yet unproven.

In general, a stable secondary structure near the 5'-cap seems to be more inhibitory than is a structure farther from the 5'-end, consistent with the strong versus weak mRNA model (151–155), although several other models could also account for this effect. It is not clear that a simple hairpin structure, as used in most of the studies cited, would seriously occlude the cap. UV cross-linking studies have, in fact, suggested that the cap remains fully accessible to eIF4E even when structures are present near to it (156, 157). Based on these experiments, a model was proposed in which eIF4E can associate with the cap even when the 5'-end of the mRNA has a stable structure in it, but functional engagement requires association of one or more of the other subunits of eIF4F (or factors associated with eIF4F) with the 5'-end of the mRNA, and it is this interaction that is inhibited by secondary structures. This model was supported by a study showing that the binding of iron regulatory protein-1 (IRP-1) to its hairpin binding site in the 5'-UTR prevented stable association of eIF3 with the mRNA, but did not affect binding of the eIF4F complex (158). According to this model, the IRP-1-stabilized structure prevents eIF4G from recruiting eIF3, which in turn prevents recruitment of the 40S subunit to the mRNA. Some structures in the 5'-UTR may also prevent binding of the 40S subunit but have no effect on initiation factor binding.

The role of the eIF4F-cap interaction is proposed to be twofold. First, it brings eIF4A to the 5'-end of the mRNA where it can unwind secondary structures found there and make a suitable binding site for the 43S complex. Second, it brings factors such as eIF3 to the 5'-end of the mRNA that help the 43S complex bind to the newly ironed out mRNA (159). Data supporting the first role of eIF4F have been obtained using in vitro initiation systems. Initiation on small, unstructured model mRNAs does not require any of the components of the eIF4F complex, ATP or a 5'-cap, nor is it stimulated by them (9, 41, 137). Furthermore, a natural mRNA with an unstructured 5'-UTR can be recruited to a 43S complex without the need of the eIF4F complex, ATP or a 5'-cap, whereas these components are required if the 5'-UTR has significant secondary structure (21). In addition, a dominant negative mutant of eIF4A lacking in vitro ATPase and RNA helicase activities was less inhibitory to translation initiation on an mRNA with a low amount of secondary structure in its 5'-UTR than on an mRNA with a very structured 5′-UTR (160). As the mutant eIF4A was incorporated into the eIF4F complex even more efficiently than the wild-type factor, these data suggest that eIF4F functions in a step slowed by increasing secondary structure in the 5′-end of the mRNA, consistent with the idea that eIF4A within the eIF4F complex unwinds these structures.

What is known about eIF4A's RNA helicase activity? eIF4A is an RNA-dependent ATPase (161) that can unwind RNA duplexes in vitro (50–52, 162). Its ATPase activity is required for duplex unwinding in vitro (162) and translation initiation in vivo (163), consistent with the notion that it uses energy from ATP hydrolysis to disrupt RNA structure. The enzyme undergoes a cycle of changes in its conformation and RNA affinity as it binds ATP, hydrolyzes it, and releases ADP (164, 165). These changes in RNA affinity and conformation could be used to transduce the energy from ATP hydrolysis into the breaking of RNA base pairs. The enzyme's structure is similar to that of a dumbbell, with a compact N-terminal domain connected to a compact C-terminal domain by an unstructured 11-amino acid linker (166–168). The highly conserved Walker A and B motifs involved in ATP binding and hydrolysis are found in the N-terminal domain, but other conserved residues known to be involved in ATP hydrolysis and in transducing RNA binding into ATPase activation are in the C-terminal domain (162, 167, 169), which suggests that some of the conformational changes observed involve engagement and disengagement between the two compact domains of the protein.

Unlike DNA helicases, eIF4A has no preference for unwinding duplexes with either a 5′ or a 3′ single-stranded overhang (50, 170, 171), and it has recently been shown that the protein can in fact unwind blunt-ended RNA duplexes with nearly the same efficiency as it unwinds duplexes with single-stranded overhangs (170). One possible interpretation of these results is that eIF4A, unlike most DNA helicases, does not translocate along the single-stranded region of the RNA as part of its unwinding mechanism; directional binding and processive translocation along single-stranded nucleic acids is thought to give DNA helicases their specificity for duplexes with 3′ or 5′ single-stranded overhangs (172). In support of this view, recent kinetic analyses of the RNA-dependent ATPase and RNA duplex unwinding activities of eIF4A strongly suggests that eIF4A does not function in a processive manner in vitro (57, 164, 171). In vitro at least, eIF4A is ineffective in unwinding RNA duplexes longer than ~10 base pairs and its activity drops off dramatically even for these short duplexes if the stability is increased by increasing the G:C content (170, 171). This low processivity of eIF4A might actually make physiological sense because it is unlikely that in the cell the factor would often encounter long stretches of duplex within an mRNA. Thus, unwinding short regions of secondary structure might be exactly what eIF4A has evolved to do.

Incorporation of eIF4A into the eIF4F complex stimulates its RNA-unwinding activity several fold (50, 57). eIF4B also stimulates the in vitro RNA-unwinding activity of eIF4A (57, 171). There is evidence that eIF4B affects the affinity of eIF4A for its RNA (173) and ATP (174) substrates, and thus interaction of the two proteins (never directly detected) might alter the conformation of eIF4A in mechanistically important ways. A second RNA-binding protein recently discovered in mammalian systems, called eIF4H, has a similar effect on eIF4A's RNA-unwinding activity (55–57). In fact, the functions of the two factors seem to be the same: Once enough eIF4B is added to the reaction to achieve maximal stimulation of eIF4A's unwinding activity, addition of eIF4H has no additional effect, and vice versa. Similar results were found for the effects of eIF4B and eIF4H on in vitro translation (55). These data suggest that eIFs 4B and 4H may play functionally redundant roles in translation initiation.

Though it is unclear how eIF4A functions to unwind RNA duplexes, the general model currently (57, 174) is that eIF4A can interact with the RNA duplex in its ATP-bound form [although this interaction has not been directly detected (164)] and then upon ATP hydrolysis disrupts some of the base pairs within the helix. After a round of helix disruption, it moves along, either actively (i.e., using the energy from ATP hydrolysis) or passively (diffusion) to the next part of the helix, which it proceeds to unwind. eIF4A might simply translocate along single-stranded RNA and thus disrupt the duplexes encountered along the way, but if so, its processivity is very low. In the models, eIF4B and 4H function to capture regions of single-stranded RNA produced by eIF4A. The eIF4G subunit of eIF4F might perform a similar function via its three RNA-binding sites, which are required for efficient translation initiation (175).

Another model is related to one originally proposed by Sonenberg in 1988 (176), but that has since been largely ignored. It seems consistent with most of the current data that eIF4A might function by transiently polymerizing along single-stranded RNA and in the process disrupt RNA duplexes. The factor's ATPase activity could affect the polymer lifetime and influence the stability of the RNA duplex by modulating protein-protein and protein-RNA contacts [it is known to do the latter; (164)]. Consistent with this model, the concentration of eIF4A in yeast and mammalian cells has been estimated to be ~50 μM (110, 177), comparable to that of actin (178). In in vitro unwinding reactions, eIF4A must always be present at high concentrations (μM) and superstoichiometric ratios relative to the RNA substrate (typically >500:1) in order to observe activity. Both observations are consistent with the need to drive an unstable binding equilibrium involving the factor. Also consistent with the polymerization hypothesis is the fact that when the rate of RNA unwinding is measured as a function of the concentration of eIF4A, a lag is often observed at low concentrations of the factor (51, 174; J.R.L. & D. Herschlag, unpublished), suggesting a higher-order assembly process might be taking place (i.e., the curve could be sigmoidal). This model would also explain the apparent bidirectionality of unwinding: Polymerization could take place from either end once the factor binds to the single-stranded RNA, so it could disrupt a duplex on either side. The eIF4F complex might serve as a nucleation center for the polymerization of eIF4A, which could be especially important since nucleation is often the rate-limiting step in polymerization processes. It might also explain why a number of ATPase-deficient mutants that bind eIF4G quite well produce dominant negative phenotypes (160, 179)—polymer poisoning by mutant subunits is a well-established phenomenon. eIF4B might serve to capture the unwound RNA, as in the current model, and then hand it off to the ribosome (180).

Another possibility is that eIF4A might not be an RNA helicase in vivo. For example, it could function to remove proteins associated with the mRNA prior to translation. In fact, it was recently shown that inhibition of 43S·mRNA complex formation in vitro by the mRNA-binding protein p50/YB-i can be partially reversed by addition of eIF4F (181). Also possible is that eIF4A acts to unwind or rearrange RNA structures, but rearranges rRNA rather than mRNA, perhaps to cause a conformational change in the 40S subunit that facilitates mRNA loading. Another RNA-dependent ATPase of the DEAD box family, Ded1p, unwinds RNA helices in vitro (182) and is involved in an early step of translation initiation (183, 184). Ded1p may well unwind structures in the 5'-UTR of mRNAs or perform any of the other possible functions mentioned above. Overexpression of Ded1p suppresses the effect of a mutation in eIF4E, suggesting that the factor may be involved in the mRNA recruitment step (184). Consistent with this proposal, a mutation in Ded1p is synthetically lethal with mutations in eIFs 4A, 4E and 4G. However, eIF4A and Ded1p appear not to be functionally redundant (183, 184).

The giant multisubunit factor eIF3 also may play a role in mRNA loading (38, 41, 44). eIF3 binds both the 40S ribosomal subunit and the ternary complex and several of its subunits bind RNA (185, 186). It might serve to coordinate the complicated arrangement of components required to put both the ternary complex and mRNA on the ribosome, while preventing the large ribosomal subunit from prematurely joining to the small subunit. As suggested in the previous section, eIF3 might serve as a scaffold to alter the conformation of the 40S subunit, allowing easier access for mRNA and/or ternary complex.

In addition to the events taking place at the 5'-end of the mRNA, strong evidence indicates that the 3'-end of the message also takes part in facilitating loading of the mRNA onto the 40S subunit. The key players in these 3'-end-mediated events are the 3'-poly(A) tail and the poly(A) binding protein, PAB. PAB contains four RNA recognition motifs (RRMs) and has a minimal RNA site size of ~10-nucleotides (187). Multiple molecules of PAB will bind to a poly(A) tail with approximately one PAB/27 nucleotides (188, 189). Given that the average length of poly(A) tails in yeast is ~70 nucleotides and is several fold higher in mammals, between two and ten PAB molecules will generally be bound to the end of an mRNA. The crystal structure of the two N-terminal RRMs of human PAB in complex with an 11-nucleotide long poly(A) molecule was recently determined (187). The protein resembles a trough, with the two RRMs forming a single, contiguous RNA-binding site. Specific recognition of the adenine bases is mediated by conserved residues in the protein via a complex assortment of hydrogen bonds, van der Waals interactions, and aromatic stacking. RRM2 appears to contain the binding site for eIF4G (190) and the structure reveals a strip of highly conserved residues suggestive of a binding site on the side of this domain opposite the RNA-binding cleft. Thus the picture that arises of the poly(A) tail of an mRNA is that it is coated with PAB protomers that form a multivalent attachment site for eIF4G.

The 3'-poly(A) tail stimulates translation initiation both in vivo and in vitro (61–63, 146), just as the 5'-cap does, and this effect is mediated by PAB (61, 191). In vivo, the effects of the cap and poly(A) tail are generally synergistic rather than additive (62). For example, addition of a cap stimulates translation of luciferase mRNA fourfold in yeast when no poly(A) tail is present on the message, but 24-fold when the poly(A) tail is present (62). The effect of the poly(A) tail is likewise increased by the presence of the cap structure. A number of controls indicated that the stability of the mRNA was not affected by addition of the cap or poly(A) tail and thus their effects were ascribed to differences in translation rather than mRNA stability.

Experiments using yeast extracts have provided evidence that the poly(A) tail facilitates the recruitment of the mRNA to the 43S complex (61). The discovery that eIF4G and PAB bind to each other (60) led to the proposal that this interaction circularizes eukaryotic mRNAs. In support of this notion, it was shown that the eIF4E and 4G subunits of eIF4F together with PAB could circularize capped and polyadenylated nucleic acids in vitro (192). Based on these observations, a working model has been developed in which recognition of the poly(A) tail by PAB acts as a signal that the mRNA has not been degraded and thus is fit for translation. The interaction between PAB and eIF4F synergistically promotes binding of the 43S complex to the mRNA. This synergism could be due to something as simple as the chelate effect—one interaction pays the entropic price of reduction in degrees of freedom upon binding, allowing more benefit to be gained from the second interaction. Alternatively, or, more likely, in addition, the effect could be due to conformational changes in PAB and eIF4F induced by binding to one another that have positive effects on the steps facilitated by these factors. There is evidence to suggest that binding of PAB to eIF4F affects eIF4F's cap-binding and ATPase activities (115, 193–195). PAB and eIF4F could also affect the kinetics of independent steps in the pathway, leading to synergism that does not depend on cooperative binding or mutually induced conformational changes (196).

Further support for the proposal that the poly(A) tail is involved in recruiting the 43S complex to the mRNA was derived from the observation that even though initiation on polyadenylated mRNAs lacking a 5'-cap is more efficient than on their nonpolyadenylated counterparts, these initiation events frequently take place at internal AUG codons, which is not true when the mRNAs are capped (63). The interpretation of these data was that the poly(A) tail can facilitate binding of the 43S complex to the mRNA, but it cannot efficiently direct it to the 5'-end in the absence of the eIF4F-5'-cap complex. This result provides compelling evidence that the 5'-cap helps enforce the use of only the very 5'-end of the mRNA as the binding site for the 40S subunit. To observe similar synergistic stimulation of initiation between the cap and poly(A) tail in cell extracts as was observed in vivo, it was necessary either to increase the concentration of competitor RNA in the extracts (63) or to decrease the concentrations of ribosomes and associated factors (146). In addition, poly(A)⁻ mRNAs can be efficiently translated in vivo as long as the ribosome:mRNA concentration ratio is high (197). These results could be interpreted as suggesting a connection between the poly(A)-tail and binding of mRNA to the ribosome because either addition of competitor RNA or a decrease in ribosome concentration might be expected to adversely affect the binding of the mRNA and the 40S subunit, in the first case by competition for binding and in the second by mass action. The data are not, however, inconsistent with other possible roles for the poly(A) tail.

Whereas much of the data has suggested that the main role of the poly(A) tail and PAB is to cooperate with eIF4F and the 5'-cap to facilitate loading of the 43S complex onto the 5'-end of the mRNA, other data suggest that the poly(A) tail also affects different steps in the pathway. First, experiments in mammalian cell extracts suggested that the poly(A) tail might stimulate the subunit joining step of initiation (198, 199) rather than mRNA binding to the 43S complex. Mutations that lower the concentration of 60S subunits in vivo specifically decrease the efficiency with which poly(A)⁻ mRNAs are translated, a result hard to reconcile with the proposal that the main role of the poly(A) tail is to promote mRNA recruitment to the 43S complex (197). Decreasing the concentration of 40S subunits also adversely affected translation of poly(A)⁻ mRNA relative to poly(A)⁺ mRNA, however, consistent with a role for the poly(A) tail in 43S binding as well. Additionally, when yeast bearing a temperature-sensitive allele of PAB are shifted to the nonpermissive temperature, they accumulate free 60S subunits (200), as one might expect if PAB is involved in the subunit joining step. It has also been suggested that PAB may interact with the 60S subunit directly (200, 201). Thus there is now considerable evidence that the poly(A) tail and PAB may influence subunit joining in addition to its effects on mRNA binding to the 43S subunit.

The subunit joining story is not without its complications either. For example, while lowering the concentration of 60S subunits in vivo suppressed translation of poly(A)⁻ mRNA relative to poly(A)⁺ mRNA (197), it also obviated the need for PAB itself (200), a seemingly contradictory result if PAB and the poly(A) tail conspire to facilitate subunit joining. Furthermore, deletion of two nonessential putative RNA helicases, Ski2p and Slh1p, dramatically enhances translation of poly(A)⁻ mRNA, bringing it to levels comparable to translation of poly(A)⁺ mRNA (202, 203). Thus it has been suggested that these proteins are enforcers of discrimination against the translation of mRNAs lacking poly(A) tails. This may be a mechanism for preventing translation of nonpolyadenylated viral mRNAs (203), but it may also play a role in mediating poly(A) effects on native mRNAs. Finally, the poly(A) tail and PAB (as well as the Ski genes) have roles in mRNA degradation and stability and, while in most of the experiments discussed above care was taken to control for effects on mRNA degradation rates, some of the effects observed might still be due to alteration of mRNA stability rather than the rates of steps in the initiation pathway.

Even the appealing closed loop model of eukaryotic mRNAs has recently been questioned. Experiments in mammalian cell extracts have demonstrated that addition of exogenous poly(A) [but not poly(dA) or poly(U)] can stimulate translation of both poly(A)$^+$ and poly(A)$^-$ capped mRNA (194). The stimulation was inhibited by adding a viral protein that is able to disrupt the eIF4G-PAB interaction, suggesting that the (poly(A)·PAB)-eIF4G interaction itself and not the circularization of the mRNA is responsible for the enhancement of translation initiation provided by the poly(A) tail and the cap. Adding PAB to yeast extracts stimulates translation of capped mRNAs lacking poly(A) tails (204), either via the *trans*-stimulatory effects just described or because PAB can perform functions without being bound to poly(A). Disruption of the PAB-eIF4G interaction did not prevent this poly(A) tail-independent effect of PAB, however, suggesting that it may be distinct from stimulation by exogenous poly(A). Finally, a homologue of the central region of eIF4G, called PABP-interacting protein-1 (PAIP1), that binds eIF4A and PAB but not eIF4E, has been discovered in mammals (205). Overexpression of PAIP1 stimulates translation, suggesting that this protein can enhance initiation of polyadenylated mRNAs without joining the 5'- and 3'-ends.

IN SEARCH OF THE INITIATION CODON: SCANNING AND AUG RECOGNITION   The scanning model proposes that once a 43S complex is loaded onto the 5'-end of an mRNA it moves along the message toward the 3'-end looking for the initiation codon (46). Placing stable secondary structures in the middle of the 5'-UTR in the path of the ribosome can lower the rate of 80S complex assembly without diminishing the efficiency with which the 43S complex binds to the 5'-end of the message, consistent with the predictions of the model (206, 207). Toe-printing experiments in a reconstituted mammalian initiation system have shown that the 43S complex can assemble on the 5'-end of the mRNA (albeit in a nonfunctional and unstable form) in the absence of eIFs 1 and 1A, but that it will not reach the initiation codon unless the two factors are included (208). These data indicate that eIFs 1 and 1A are required for the formation of a 43S complex capable of scanning the mRNA and further support the notion that binding of the 43S complex to the mRNA and location of the initiation codon can be decoupled.

It has also been reported that the scanning process requires the hydrolysis of ATP (209), although it is not clear what this energy is used for. One possibility is that a molecular motor (e.g., eIF4A or Ded1p) uses the energy derived from ATP hydrolysis to actively translocate the ribosome in the 5' to 3' direction. The actual movement of the ribosome may occur in a passive manner via diffusion, and the ATP requirement may reflect the need for unwinding of structures in the mRNA, possibly by the action of eIF4A and/or Ded1p, in order to allow diffusive movement. Also possible is a Brownian ratchet mechanism in which an ATP-dependent RNA unwindase located on the front (3′) side of the 40S subunit unwinds structures in the mRNA, allowing the ribosome to slide past via diffusion. When the ribosome randomly slides in the 3′ direction over the unwound structures, they reform behind it, preventing backsliding. In support of the diffusion-based models, it was recently shown in a reconstituted mammalian system that 43S complexes can reach the AUG codon in an mRNA in the absence of ATP and eIF4A if the 5′-UTR is unstructured but not if it contains significant secondary structure (21). These data indicate that the search for the AUG codon does not inherently require the hydrolysis of ATP and suggest that energy input is required to get through structured regions in the message. It is not yet clear, however, if the rate at which the AUG codon in an unstructured mRNA is located is affected by ATP hydrolysis or by the addition of factors not present in these experiments (e.g., Ded1p or as-yet undiscovered factors).

The molecular events that take place when the scanning 43S complex reaches an AUG codon are also unclear. Aside from the initiator tRNA, which seems to be the key point of contact with the mRNA in the decoding process (210), three players appear to be central to initiation codon location: eIF1, eIF2, and eIF5. Each has been found in genetic screens to affect the fidelity of initiation codon selection (31, 32, 34, 211–214), allowing initiation at a UUG codon (but oddly in most cases, not at other near-cognate codons). The mutations in eIF1 that affect the fidelity of initiation site selection all cluster at the top of a strip on the surface of the molecule that is made up of highly conserved residues and is suggestive of the binding site for another component of the system (126). A role for eIF1 in initiation codon selection has also been demonstrated in vitro (21).

The mechanism of action employed by eIF1 during initiation codon recognition is not known. It may directly interrogate the anticodon-mRNA pairings and when three perfect base pairs are formed, interact with the resulting duplex and reduce the energy of the complex enough to stop scanning. It might also then send a signal to eIF2 and/or eIF5, initiating GTP hydrolysis. In this model, eIF1 might play a similar role to the decoding bases A1492, A1493, and G530 in the bacterial 30S ribosomal subunit that swing out to interact with the codon-anticodon duplex in the A site during tRNA selection in elongation, thus (presumably) signaling to EF-Tu that the cognate tRNA has been found (215). For most of the time in translation, all decoding happens in the A site and thus it may require an accessory factor (eIF1) in the P site to allow it to perform this function during initiation. The proposal that eIF1 binds in or near the ribosomal P site comes mainly from the experiments showing that bacterial IF3, a possible homologue of eIF1, binds in the P site, although IF3 does not appear to be close enough to the anticodon end of the tRNA to monitor its base-pairing interactions (216). [Note: The binding site of eIF1 was recently determined by footprinting and found to be similar to that of IF3, as predicted (216a).] A related model is that eIF1 alters the conformation of the 40S subunit's P site in such a way as to allow decoding to take place (21). eIF1 might also detect the formation of the codon-anticodon interaction indirectly, for example by interacting with the body of the initiator tRNA and responding to conformational changes in it when the initiation codon is reached. Such conformational changes in the tRNA have been proposed to be important for sensing the cognate codon-anticodon base pairing during A-site decoding (217, 218).

The function of eIF2 in the decoding process is to respond to the discovery of the initiation codon by hydrolyzing GTP and then, at the appropriate time, release the Met-tRNA$_i$ into the P site. Mutations that reduce initiation site selection fidelity have been isolated in all three subunits of eIF2 (34, 211, 212, 214). The initiation codon fidelity mutation in the γ-subunit is in the GTP-binding site and appears to increase both the eIF5-independent GTPase activity of eIF2 and the rate of dissociation of Met-tRNA$_i$ from ternary complex (34). One of the mutations in the β-subunit was found to cause similar effects. Based on these data, it was proposed that these mutations allow premature release of the initiator tRNA into the P site of the 40S subunit in response to an anticodon-UUG mismatch, although the fact that they still confer a preference for UUG over other near-cognate codons suggests that it is not as simple as spontaneous GTP hydrolysis and tRNA dissociation. Instead, the barrier to these events has more likely been lowered such that UUG can now trigger them, but for whatever reasons, other near-AUG codons do not do so as efficiently. Mutations that weaken the interface between the β- and γ-subunits of eIF2 also lead to a decrease in the fidelity of initiation codon selection, and these mutations are synthetically lethal with the mutation mentioned above in the β-subunit that appears to increase the intrinsic GTPase activity of the factor (214). Thus there appears to be a complex interplay between the tertiary and quaternary structure of the factor, its Met-tRNA$_i$ binding and GTPase activities, and its ability to respond to the location of the correct initiation codon.

eIF5 stimulates the GTPase activity of eIF2 (219). eIF5 can bind directly to the ternary complex via the β-subunit of eIF2, and this interaction is critical for eIF5's function (97, 220). In addition, eIF5 interacts more strongly in solution with eIF2·GDP·AlF$_4^-$, a putative transition-state analogue of the GTP hydrolysis reaction, than with eIF2·GDP (221), consistent with its proposed role as a GAP. However, constitutively activating eIF2 would serve no useful purpose, and thus the activity of eIF5 must be regulated in response to a signal that the AUG codon has been found. This signal could be direct, e.g., eIF5 could monitor the codon-anticodon pairings itself, or it could be indirect, mediated through conformational changes in eIF1 or the 40S subunit. As with two of the eIF2 mutations described above, the mutation in eIF5 that reduces the fidelity of initiation site selection in vivo appears to result in hyperstimulation of eIF2's GTPase activity (34). However, in vivo this mutant allele produces a strong ($\geq$100-fold) preference for initiation at UUG versus other one-base changes from AUG (AUU, GUG, CUG) and thus its effect cannot be as simple as unregulated overstimulation of eIF2's GTPase activity. Instead, the barrier to its being triggered to activate eIF2 for GTP hydrolysis has probably been lowered in such a way as to allow it to be activated by the anticodon-UUG mismatch in addition to the anticodon-AUG match. Again, the fact that it does not respond to the other near-AUG codons is probably significant.

While eIFs 1, 2, and 5 appear to be the core of the AUG-recognition apparatus, at least as far as initiation factors are concerned, evidence is mounting that other factors may have a hand in the process as well. For example, a mutation in eIF4G that weakens its interaction with eIF1 produces a modest initiation site selection fidelity phenotype (111). Remarkably, this mutation, like those in eIFs 1, 2, and 5 described above, increases initiation at a UUG codon but not at AUU or UUA codons. Overexpression of eIF1 could suppress the phenotype, consistent with the interpretation that the effect results from a decrease in the binding of eIF1 to eIF4G. One of the initiation codon-selection fidelity mutants in eIF1 weakened binding of the factor to eIF4G, further suggesting that this interaction is important in initiation codon location (111). One possible interpretation of these data is that eIF4G remains bound to the 43S complex as it scans the mRNA in search of the initiation codon and that it helps to organize the complex to respond appropriately to the AUG's identification. Because eIF3 interacts with eIFs 1, 2, and 5, it has been proposed that it too may play a role in locating the initiation codon (35, 92).

A wide variety of experiments have established that in mammals the sequences surrounding AUG codons play a role in specifying which one is used as the initiation site (222). The first AUG in the mRNA can be bypassed if it is in an unfavorable sequence context and a downstream AUG imbedded in a favorable context will be used instead. The mechanism through which this Kozak sequence exerts its effect remains mysterious, however. Attempts to find sequences in the 18S rRNA that interact with these regions, analogous to the role of the anti-Shine-Dalgarno in 16S rRNA, have thus far proved futile (45). Recent in vitro experiments, however, have suggested that eIF1 might play a role in discriminating between AUGs in favorable and unfavorable contexts by destabilizing preinitiation complexes on "incorrect" AUGs (21). Thus something about the sequence around an AUG codon may allow it to either engage or not engage correctly with the apparatus that is searching for the initiation codon in the P site, although how is unclear. Also possible is that some part of the preinitiation complex (e.g., eIF1) directly monitors the sequences around the AUG codon.

By contrast, yeast and plants have very different sequences around their initiation sites [AAAAAUGUCU and AA(A/C)AAUGGC, respectively; (223–225)] than do mammals (GCCACCAUGG), the only common nucleotide being the A at position −3 (with the A of AUG as 0). A variety of experiments have shown that in yeast the consensus sequence has only a small effect on the efficiency of use of an AUG codon as the initiation site (154, 226–229), in contrast to the larger effects usually seen in mammalian systems. This apparent lack of an important role for the consensus sequence in yeast mRNAs may be message dependent, as there has been at least one report in which it influences the choice of initiation codons by up to tenfold (230). Also unclear is whether the context of the AUG codon plays a significant role on initiation site selection in plants (225). Given the differences among the sequences surrounding the initiation codons in these diverse eukaryotes, the mechanism of action of these nucleotides in mammals may not be general to all eukaryotes. Finally, even the situation in mammals appears to be more complicated than originally thought. An analysis of genes from the human genome has indicated that as many as 40% of mRNAs have AUG codons upstream of the initiation site and that the consensus sequence around the initiation codon may not be as strongly conserved as originally proposed (231). Based on these data, it has been suggested that initiation of translation on many mRNAs occurs via nonstandard pathways, for example internal ribosome entry or leaky scanning past upstream AUGs (231), a controversial conclusion, however, in part because the analysis requires that the initiation site be known for each mRNA, which is not always a straightforward matter (232).

BRINGING THINGS TOGETHER: SUBUNIT JOINING   The final (known) step of eukaryotic translation initiation is the joining of the 40S and 60S subunits following initiation codon recognition and deposition of the initiator tRNA into the P site of the 40S subunit. Considerable progress has been made in the past few years concerning the mechanics of this step. It was long thought that only a single GTP hydrolysis event was required for eukaryotic translation initiation (233). However, the discovery that a yeast homologue of bacterial IF2, eIF5B, was a GTPase and a central initiation factor (17) and a kinetic dissection of the initiation process in an in vitro mammalian system (9) both indicated that two GTP hydrolysis events are required, one catalyzed by eIF2 upon initiation codon recognition and another at the end of the pathway, after 80S complex formation (9). This requirement was subsequently confirmed both in vivo and in vitro (18, 29, 234). eIF5B, originally called eIF5 (41, 235) before being rediscovered as a separate initiation factor, was subsequently shown to facilitate subunit joining (30). eIF5B's GTPase activity is not actually required to bring the subunits together. Instead, GTP hydrolysis promotes the release of the factor from the 80S complex once the subunit joining step has been completed (29, 30, 234).

The crystal structure of an archaeal eIF5B has been determined without nucleotide and bound to GDP and GDPNP (28). These structures reveal an overall shape resembling a chalice, with three domains on top, including the GTPase domain, connected to a fourth domain on the bottom via a long α-helix. The top and bottom of the chalice undergo small movements relative to each other when GDPNP is replaced by GDP, suggesting that GTP hydrolysis is used to induce conformational changes in the factor. These changes could potentially be employed for mechanical work on the 80S complex. For example, eIF5B might move the initiator tRNA into its final position in the P site using the energy from GTP hydrolysis (8), similar to proposals for the role of EF-G/EF2 during translation elongation (236). However, the idea that the essential role of eIF5B's GTP hydrolysis activity was to reorganize the initiation complex was cast into doubt when it was found that the key translational functions of a GTPase-deficient mutant of eIF5B could be restored by a second mutation that reduced the affinity of the factor for the ribosome without restoring its GTPase activity (29). These and other data have indicated that the GTPase activity serves as an on/off switch: In its GTP-bound form, eIF5B has a high affinity for the ribosome and in its GDP-bound form, a low affinity. If the affinity is decreased by a mutation such that it can dissociate from the ribosome at a reasonable rate in its GTP-bound form, the factor can still perform all of its essential functions even without being able to hydrolyze GTP. The main role of the conformational change in eIF5B, then, appears to be to change the affinity of the factor for the ribosome. This could easily be achieved by the relatively small changes observed, either by creating a steric clash between the factor and the ribosome or by breaking important positive interactions.

Experiments testing the effects of the GTPase-deficient mutants of eIF5B on the translation of the GCN4 mRNA (138) indicated that while eIF5B's GTPase activity is not required for the core process of translation or for maintaining reasonably healthy yeast cells, its disruption increases the frequency with which ribosomes fail to begin translation at upstream initiation codons and proceed downstream (29). This suggests that the timing of GTP hydrolysis plays a regulatory function and that premature release of the second-site-suppressed eIF5B mutant from the 80S complex leads either to movement of the complex past the AUG where it was located or to premature dissociation of the Met-tRNA$_i$ followed by downstream scanning of the message and reinitiation. This model predicts that GTP hydrolysis by eIF5B is not triggered until the factor somehow senses that the complex is correctly set up to begin elongation. This signal could be that additional factors have been released or that the initiator tRNA is in the right place within the complex. An alternative model that still cannot be ruled out is that GTP hydrolysis by eIF5B, while not being required in most instances to mechanically reorganize the initiation complex, can jiggle into place incorrectly set-up complexes that occur occasionally during normal initiation. The failure by the second-site-suppressed GTPase-deficient mutant to jiggle these aberrant complexes might lead to the missed-AUG effects observed.

Despite recent advances in understanding the subunit joining step of eukaryotic translation initiation, the molecular mechanisms used by eIF5B to facilitate the process remain to be elucidated. First, it is not yet clear whether eIF5B binds to the 40S subunit before the 60S subunit or vice versa. The exact timing of GTP hydrolysis by eIF2 and eIF5B relative to binding of eIF5B to the ribosomal subunits and subunit joining have also not yet been fully explored. How does eIF5B actually facilitate the joining of the subunits? Binding of eIF5B to the 40S and/or 60S subunit could well alter the subunit's conformation in such a way as to increase the rate constant for the joining reaction. Also possible is that binding of eIF5B to one subunit presents a new molecular surface that facilitates the docking of the other subunit. What actually triggers eIF5B's GTPase activity? eIF5B is suggested to facilitate the release of other factors, such as eIF1A, from either the 40S subunit or final 80S complex (93, 123), although if so and how remain to be explored. Finally, as discussed above, understanding how eIF5B checks or otherwise ensures that the final 80S complex is properly set up to elongate a polypeptide may provide important insights into how the translational machinery maintains quality control in protein synthesis.

ELONGATION

In contrast to the initiation and termination stages of translation, the machinery used during the course of translation elongation has been highly conserved across the three kingdoms of life. Because of this conservation it is assumed that the mechanisms underlying elongation are the same in eukaryotes as they are in bacteria and archaea (reviewed in 237, 238). We therefore outline only briefly the current state of the field and then focus on one intriguing aspect that appears to be unique to one branch of the eukaryotes.

Peptide chain elongation begins with a peptidyl tRNA in the ribosomal P site next to a vacant A site (Figure 3). An aminoacyl tRNA is carried to the A site as part of a ternary complex with GTP and the elongation factor 1A (eEF1A; EF-Tu in bacteria). eEF1A·GTP·aa-tRNA ternary complexes with either the cognate or noncognate aminoacyl tRNAs can bind to the ribosomal A site. However, several steps involving codon-anticodon base pairing between the mRNA and the tRNA, conformational changes in the decoding center of the small ribosomal subunit, and GTP hydrolysis by eEF1A/EF-Tu ensure that only the cognate tRNA is selected for entry into the next stage of elongation (reviewed in 239). Codon-anticodon base pairing induces three bases in the small ribosomal subunit's rRNA to swing out and interact with the resulting mRNA-tRNA duplex (215). This in turn appears to activate eEF1A/EF-Tu's GTPase activity. eEF1A·GDP releases the aminoacyl tRNA into the A site in a form that can continue with peptide bond formation.

The ribosomal peptidyl transferase center then catalyzes the formation of a peptide bond between the incoming amino acid and the peptidyl tRNA (240). The result is a deacylated tRNA in a hybrid state with its acceptor end in the exit (E) site of the large ribosomal subunit and its anticodon end in the P site of the small subunit (241). The peptidyl-tRNA is in a similar hybrid situation with its acceptor end in the P site of the large subunit and its anticodon end in the A site of the small subunit. This complex must be translocated such that the deacylated tRNA is completely in the E site, the peptidyl tRNA completely in the P site, and the mRNA moved by three nucleotides to place the next codon of the mRNA into the A site. This task is accomplished by elongation factor 2 (EF-G in bacteria), which hydrolyzes GTP as it facilitates translocation (242). This cycle is repeated until a stop codon is encountered and the process of termination is initiated.

Following the hydrolysis of GTP and the release of aminoacyl tRNA onto the ribosome, eEF1A·GDP is released and must be recycled to its GTP-bound form

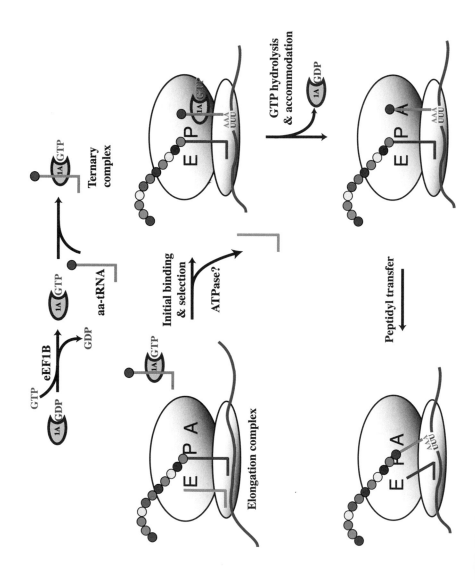

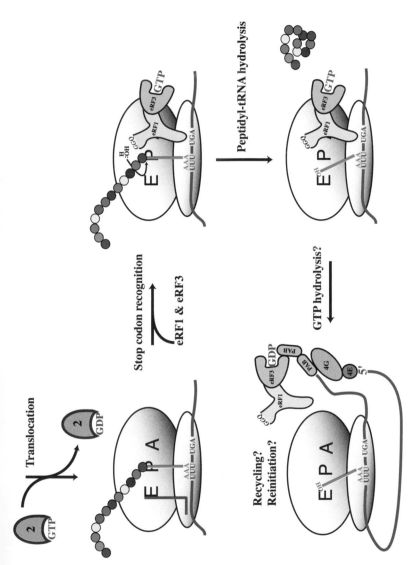

Figure 3   Cartoon depicting the current models for translation elongation and termination in eukaryotes. Not all steps are shown and several steps have been condensed. It has also been assumed that the molecular mechanics of elongation are conserved between bacteria and eukaryotes. The lines between the UUU and UGA codons are not meant to represent mRNA, but are merely for clarity.

so that it may participate in successive rounds of polypeptide elongation. A multifactor complex called eEF1B (eEF1Bα plus eEF1Bβ, formerly called eEF1βγ) catalyzes this exchange. Currently, no known guanine-nucleotide exchange factor (GEF) exists for either EF-G or eEF2.

Human eEF1A and *E. coli* EF-Tu are 33% identical overall, with even greater identity between their GTP-binding domains (243). The archaeal EF1A from *Halobacterium halobium* is 49% identical to yeast eEF1A and 36% identical to *E. coli* EF-Tu (244). Among EF2 homologs, aEF2 from *H. halobium* is 36% identical to eEF2 from *Saccharomyces cerevisiae* and *Drosophila melanogaster* and 30% identical to *E. coli* EF-G (245). Aside from the extensive sequence similarity, X-ray crystallographic studies and cryo-EM reconstructions have demonstrated remarkable structural similarity between EF1A and EF2 homologs on and off the ribosome (246–249). One interesting exception to these similarities is a variety of posttranslational modifications of both factors in eukaryotes (250), including the conversion of a histidine residue in eEF2 to a diphthamide (251), which is the target of ADP-ribosylation by diphtheria toxin.

Perhaps the most significant exception to the rule of evolutionary conservation in translation elongation is the existence of elongation factor 3 (eEF3) exclusively in fungi. eEF3 was shown to be required when yeast ribosomes are used in poly-Phe synthesis assays in vitro employing yeast eEF1A and eEF2 but not when mammalian ribosomes are used in the same system (252). Furthermore, the gene encoding eEF3 has been shown to be essential for viability in yeast (253). eEF3 possesses ribosome-dependent ATPase and GTPase activities (254) and contains a duplicated nucleotide binding motif homologous to those found in the typically membrane-associated ATP Binding Cassette (ABC) proteins (253), such as the cystic fibrosis transmembrane conductance regulator. eEF3 is found primarily associated with translating cytosolic ribosomes, mostly in polysome fractions, and is required for each round of peptide bond formation. It interacts with eEF1A (255) and this interaction is important for protein synthesis in vivo (256). eEF3 is thought to facilitate release of the E site-deacylated tRNA and to enable efficient binding of the eEF1A·GTP·aa-tRNA ternary complex to the A site (257). The data also suggest that binding of tRNA to the A and E sites is anticoupled. ATP hydrolysis by eEF3 appears to be required for its effects.

Why do fungal ribosomes require a distinct elongation factor to facilitate E site clearance and A site loading when other eukaryotic, bacterial, and archaeal ribosomes apparently do not? It has been suggested that mammalian ribosomes actually possess an intrinsic eEF3-like activity. Very stable ATPase and GTPase activities, lost only after exposure to 4 M LiCl, are associated with mammalian ribosomes (258). ATP hydrolysis has also been reported to stimulate release of E-site-bound tRNA from mammalian ribosomes (259), and occupancy of the A site appears to stimulate the intrinsic ATPase activity (258). However, no candidate in mammalian ribosomes has yet been identified as the homologue of fungal eEF3. *E. coli* possesses a soluble ribosome-dependent ATPase, RbbA, that stimulates poly-Phe synthesis in vitro, requires the hydrolysis of ATP for its effects, and has some sequence similarity to eEF3 (260, 261). Thus it remains possible that bacteria may have an eEF3-like activity as well.

TERMINATION

The termination of translation (reviewed in 262–267) occurs in response to the presence of a stop codon in the ribosomal A site. The end result of this process is the release of the completed polypeptide following the hydrolysis of the ester bond linking the polypeptide chain to the P site tRNA. The peptidyl transferase center of the ribosome is believed to catalyze the hydrolysis reaction (268–273), in response to the activity of class 1 release factors, which decode stop codons presented in the A site. Class 2 release factors are GTPases that stimulate the activity of class 1 release factors regardless of which stop codon the class 1 factor has engaged (Figure 3).

Comparative Termination

BACTERIA   Bacteria possess two class 1 release factors, RF1 and RF2, with overlapping codon specificity, such that each responds to UAA, whereas UAG is decoded only by RF1 and UGA is decoded only by RF2 (274). The swapping of RF1- and RF2-specific domains among RF1/RF2 chimeric factors led to the identification of a putative peptide "anticodon" within each factor; the sequences PAT in RF1 and SPF in RF2 were capable of imposing either RF1 or RF2 specificity on the chimeric factor (275). Thus, it was concluded that the prokaryotic factors employ a localized linear amino acid sequence for the recognition of a stop codon. The peptide anticodon and the stop codon are thought to form an interaction that is analogous to the codon:anticodon pairing between mRNA and tRNA. These data have greatly influenced the study of eukaryotic translation termination, as discussed below.

However, swapping the discriminatory domain between native factors, consisting exclusively of RF1 or RF2 domains, as opposed to a particular mixture of RF1 and RF2 domains, resulted in an inactive release factor. This suggests that the codon:anticodon base-pairing analogy is an oversimplification of the actual mechanism of stop codon recognition by RFs 1 and 2. Structures present in other domains most likely help define the codon recognition ability of each factor, perhaps by influencing the structure of the discriminator region.

In bacteria RF3 is the class 2 release factor, which not only stimulates the activities of RF1 and RF2, but is also required to eject them from the ribosome following peptidyl-tRNA hydrolysis (276–278).

EUKARYOTES   In contrast to the prokaryotic factors, only one class 1 release factor is present in eukaryotes. Accordingly, eRF1 has an omnipotent decoding capacity and can promote the hydrolysis of peptidyl-tRNA in response to any of the three stop codons, UAA, UAG, or UGA (279–281). In organisms that utilize variant genetic codes, such as the ciliates, in which either UGA or both UAA and UAG have been reassigned as sense codons, eRF1 decoding potential is appropriately restricted by mechanisms that are currently not well understood.

Eukaryotes also possess a single class 2 release factor, eRF3 (282, 283). In contrast to RF3, a role for eRF3 in triggering the release of eRF1 from the ribosome following peptidyl-tRNA hydrolysis has yet to be experimentally verified. Furthermore, eRF3 is an essential protein in eukaryotes, whereas RF3 is dispensable for viability in bacteria (276, 277, 284). eRF1 and eRF3 bind to each other in the absence of the ribosome, and this interaction is required for optimum efficiency of termination in *S. cerevisiae*. In contrast, no such cytosolic complex between RFs 1 or 2 and RF3 has been observed (285–288), although transient interactions on the ribosome between the class 1 factors and RF3 have been proposed (273).

ARCHAEA   Translation termination, as far as stop codon recognition and peptidyl-tRNA hydrolysis are concerned, appears to be similar between archaea and eukaryotes based on the homology of aRF1 and eRF1 as well as the discovery that *Methanococcus jannaschii* aRF1 is fully functional in an in vitro release factor (RF) assay employing mammalian ribosomes (281). Surprisingly, no gene encoding an archaeal homologue of eRF3 has been identified, although this does not necessarily mean that there is no archaeal counterpart to eRF3. It has been suggested, however, that a process of reductive evolution of the translational apparatus occurred during the divergence of archaea (83), which may have made such a factor unnecessary.

IS THE MECHANISM OF TERMINATION CONSERVED?   The lack of sequence homology between bacterial and eukaryotic factors might indicate independent origins of the process of termination (280, 282). Alternatively, these factors may have diverged so much that their lack of resemblance belies a true evolutionary relationship. In fact, there appears to be one universally conserved sequence motif among class 1 release factors, the GGQ motif, which is required for the activation of peptidyl-tRNA hydrolysis (272, 289). The universal conservation of the core ribosomal proteins and rRNA (83, 237) also suggests that the molecular mechanism of the peptidyl-tRNA hydrolysis reaction, which is catalyzed by the ribosome, is likely conserved. However, as there are no other apparent regions of similarity aside from the GGQ motif in class 1 release factors from bacteria and eukaryotes/archaea, the mechanism of stop codon decoding may not be conserved. Furthermore, the fact that there are two class 1 factors in bacteria but only one in eukaryotes and archaea also suggests important differences in decoding mechanisms. One similarity, however, is that the termination signal in both bacteria and eukaryotes consists not of a three-base codon, but includes the +4 position as well (290–293), suggesting vestiges of a primordial termination codon-recognition mechanism. Finally, there are no homologous sequences between RF3 and eRF3 aside from their GTP-binding motifs (282).

The Structure of eRF1

The three-dimensional structure of eRF1 has been determined (294). The overall shape of the protein resembles the letter "Y," each arm being formed by a distinct domain. Domains 2 and 3 form the branches of the Y and domain 1 forms its base. Together, domains 1 and 2 are thought to constitute the core of eRF1 responsible for ribosome binding, stop codon recognition, and the activation of peptidyl-tRNA hydrolysis (295). Domain 3 is required for the binding of eRF3 (296, 297), while all three domains are required to activate eRF3's GTPase activity, as it is the ribosome-bound eRF1 that functions as the GAP (283).

Interestingly, the crystal structure reported for *E. coli* RF2 differs significantly from that of eRF1, raising doubts, as mentioned above, as to the evolutionary conservation of stop codon decoding (294, 298). In this structure, the GGQ motif and the putative SPF tripeptide anticodon were only 23Å apart, much too short to bridge the gap between the peptidyl transferase center and the mRNA. However, a more recent cryo-EM structure revealed that RF2 undergoes significant reorganization when it binds to the ribosome, such that the helix contiguous with the GGQ loop rotates outward, thereby increasing the distance to nearly 60Å (299, 300). An extended conformation on the ribosome was also indicated by hydroxyl radical probing studies (301). As it now appears that RF2 on the ribosome adopts a structure similar to that of eRF1, these factors may be evolutionarily related after all. Alternatively, this could be an example of convergent evolution dictated by the constraints imposed by the ribosomal A site.

Domain 1 of eRF1: The Mechanism of Stop Codon Decoding

Several mutations in eRF1 were isolated by a genetic screen for a unipotent suppressor phenotype; the suppression of only one of the three stop codons rather than a general stop codon recognition defect (302). These mutations are localized near the interface between two α-helices and an opposed β sheet in domain 1. The van der Waals surface of this interface constitutes a continuous groove hypothesized to be the site of mRNA interaction. Because all mutant eRF1s isolated by the screen also displayed weak omnipotent suppression phenotypes, it was suggested that the decoding of stop codons in eukaryotes is a holistic process; that is, recognition of one stop codon cannot be completely decoupled from recognition of the others. This "articulated coupling" hypothesis advocates that the decoding of stop codons in eukaryotes is more complicated than, and might not be directly analogous to, the decoding of sense codons by tRNA (302).

Additional data concerning the molecular basis for stop codon discrimination have been obtained from experiments involving eRF1 proteins from variant genetic code organisms that have reassigned either UAG or both UAA and UAG to function as sense codons (reviewed in 303). For these changes to be effective, the reassigned codons must no longer be recognized by eRF1 as signals for the termination of translation. It has been proposed that the residues that determine the specificity of eRF1 for particular stop codons can be identified using the sequence divergence of variant and universal code eRF1s (304). Based on these sequence comparisons, several hypotheses to explain decoding have been proposed (303, 305–307). Unfortunately, experiments have not yet provided support for them or revealed any patterns of discriminatory residues in eRF1 as obvious as those apparently utilized by RF1 and RF2. A general finding of many experiments has been that stop codon recognition cannot be attributed to a single linear sequence of residues in eRF1, but likely involves more complicated structural motifs (308–311).

The Mechanism of Peptidyl-tRNA Hydrolysis

It was proposed based on the crystal structure of eRF1 that the terminal glutamine of the universally conserved GGQ motif at the tip of domain 2 might coordinate a water molecule that serves as the nucleophile for the hydrolysis of the peptidyl-tRNA ester bond (294). However, mutation of this glutamine to a glycine does not affect the ability of the mutant eRF1 to stimulate peptidyl tRNA hydrolysis (272, 294). It has also been suggested that, because they lack side chains, the glycine residues allow the entry of a water molecule into the peptidyl transferase center (289). The termination reaction is quite distinct from polypeptide elongation in that it relies on the inclusion of water in the peptidyl transferase center, whereas an elongating ribosome must employ mechanisms to keep water out during amide bond formation, lest premature peptidyl-tRNA hydrolysis occurs. On the other hand, the function of eRF1 may not simply be to provide a channel for water but to facilitate some conformational change in the ribosome that permits the entry of water into its active site as well as the activation of this water and of the ester bond, resulting in hydrolysis of the peptidyl-tRNA.

That the peptidyl transferase center catalyzes the hydrolysis reaction was supported by experiments conducted decades ago, which revealed that in the presence of 20% ethanol a deacylated tRNA or a CCA trinucleotide in the A site can stimulate peptidyl tRNA hydrolysis nearly as efficiently as *E.coli* class 1 RFs (273). The peptidyl transferase center thus appears to be activated for peptidyl tRNA hydrolysis by the GGQ motif in a manner similar to its activation for peptide elongation by tRNA, although the molecular details of this activation and the subsequent hydrolysis reaction remain unclear.

It's Not All About Factors: The Role of the Ribosome

Experiments involving eRF1s of variant genetic code organisms have suggested that the structure of eRF1 is the sole determinant of its decoding function (308, 312). Nevertheless, the evidence that the peptidyl transferase center, rather than the release factor, catalyzes peptidyl-tRNA hydrolysis implies that there must be some communication between eRF1 and the ribosome. Work in both bacteria and eukaryotes indicates that the ribosome is not simply a bystander in the process of stop codon recognition either. Several groups have reported the isolation of mutations in both the 16S and 23S rRNAs from *E. coli* and in yeast 18S rRNA that influence the fidelity of translation termination (269, 270, 313–316; reviewed in 271, 317). For example, some mutations of base 1054 in the decoding center of yeast 18S rRNA facilitate stop codon read-through, whereas different changes to the same base compensate for the defects associated with mutant eRF1 and eRF3 proteins that themselves display ominpotent suppressor phenotypes (314). Such rRNA mutations were termed antisuppressors, because they restored fidelity to the termination reaction in the presence of the mutant RFs, suggesting that rRNA plays an important role in maintaining the efficiency of translation termination, just as it plays a central role in the decoding process during polypeptide elongation (318, 319).

Characterization of eRF3

eRF1 and eRF3 form a complex mediated by their C termini (282, 296, 297). Addition of eRF3 reduces the concentration of stop codon tetraplets required for eRF1-mediated translation termination in vitro. eRF3 binds GTP independently of eRF1, and stimulation of its GTPase activity requires eRF1 and the ribosome, but not the presence of a stop codon (282, 283).

The mechanism of eRF3-mediated stimulation of eRF1 activity is not known, nor is the function of GTP hydrolysis understood. The precise signal that results in the activation of GTP hydrolysis also has yet to be identified, in contrast to what has been learned about its bacterial counterpart. RF3 binds to the ribosome in its GDP-bound form and undergoes nucleotide exchange. The posttermination ribosome, still bound by RF1 or RF2 in the presence of a stop codon, was shown to be the GEF. Nucleotide exchange is activated following release of the peptide from the peptidyl-tRNA by a class 1 release factor. RF3•GTP destabilizes the binding of RFs 1 and 2, causing them to dissociate from the ribosome. GTP hydrolysis by RF3 appears to be required for its release at the conclusion of termination (278, 320). It will be interesting to determine how similar the mechanism of eRF3 is to that of RF3.

RECYCLING

The fourth stage of translation is the recycling of the ribosomal subunits so that they can be used in another round of initiation. Significant information regarding the steps in this process is available only for bacteria. At the end of the termination stage the ribosome is left on the mRNA with a deacylated tRNA, presumably in a P/E hybrid state in which the acceptor end has moved into the E site of the 50S subunit while the anticodon end remains in the P site of the 30S subunit (321). In bacteria, this complex is recognized by ribosome release factor (RRF). RRF was initially assumed to bind in the vacant A site because of its striking structural similarity to a tRNA (322), but recent footprinting experiments have indicated that it binds across the A site, almost orthogonally to the orientation of an A-site tRNA (323). EF-G·GTP and IF3 then assist RRF in disassembling the posttermination complex (15). The exact roles of these factors are still not entirely clear, although one model (15, 323) posits that RRF in conjunction with EF-G alters the ribosome's structure to destabilize the binding of tRNA and mRNA as well as the subunit interface. IF3 then binds and facilitates complete subunit dissociation and release of the tRNA and mRNA. GTP hydrolysis by EF-G could be used actively to facilitate these disruptions or could be used to trigger release of the factors at the appropriate time.

The events and players involved in ribosome recycling in eukaryotes and archaea are largely mysterious. There does not appear to be an RRF ortholog in either kingdom. eIF1A (134), eIF3 (23, 25, 325), and eIF6 (326, 327) all have ribosome antiassociation activity in vitro, and eIF1 might possibly have a similar activity, given that it may be the ortholog of bacterial IF3, which is involved in recycling. However, it is not clear whether these activities are relevant for ribosome recycling or if they are instead used to prevent premature association of the subunits during initiation. Antiassociation activity is defined as shifting the equilibrium from associated to dissociated subunits, which could be important for recycling, but does not a priori have to be so. For ribosome recycling, a factor or factors that increase the rate of dissociation of the subunits, mRNA and deacylated tRNA would more likely be required, as is the case in the bacterial system (15). Furthermore, one might expect that factors that bind to the interface side of one of the ribosomal subunits would prevent subunit association because they sterically block joining. Whether this "activity" is relevant to the mechanism of translation, however, is another matter. For example, eIF6 has antiassociation activity and was believed to be involved in ribosome recycling (326, 327). It was later shown, however, that eIF6 is required for ribosome biogenesis, and cell extracts depleted in eIF6 did not show a defect in in vitro translation assays (74, 75), suggesting its antiassociation activity may not be relevant for ribosome recycling but is instead a consequence of the fact that it interacts tightly with the 60S subunit as part of its role in other pathways. [Note: Recent work has indicated that phosphorylation of eIF6 regulates the level of free 60S subunits in vivo (327a).]

Of the factors proposed to be involved in ribosome recycling in eukaryotes, eIF3 is perhaps the most intriguing. A variety of experiments including electron microscopic studies have suggested that eIF3 binds to the side of the 40S subunit opposite the interface (99, 139, 328). If this is so, it indicates that the antiassociation activity of the factor is due to its induction of a conformational change in the 40S subunit rather than sterically preventing binding of the 60S subunit. Inducing such a conformational change could potentially increase the rate of subunit dissociation as well as lowering the rate of association.

Finally, the closed-loop model of eukaryotic mRNAs has suggested the possibility that termination and recycling may not release the 40S subunit back into the cytoplasm. Instead, the 40S subunit may be shuttled across or over the poly(A) tail back to the 5'-end of the mRNA via the 5'- and 3'-end-associated factors. In this model, the closed loop serves to facilitate reinitiation of translation rather than (or in addition to) the first initiation event. This proposal was recently bolstered by the finding that eRF3 and PAB interact with each other (329), connecting the termination apparatus to the poly(A) tail. Overexpression of PAB was found to suppress effects associated with mutant and aberrantly folded (the prion-like PSI^+ form) eRF3 in vivo (330), suggesting that this interaction is functionally important for termination. In addition, it was recently shown that disruption of the PAB-eRF3 interaction inhibits translation and that addition of eRF3 to in vitro translation assays stimulates the initiation process (331). Recent modeling studies have also suggested that closed-loop reinitiation could substantially increase the rate of initiation (332). At this stage, however, this model remains unproven but intriguing.

ACKNOWLEDGMENTS

We thank members of our lab for comments on the manuscript, Mikkel Algire for making Figures 1 and 2, Katsura Asano for communicating results prior to publication, and Tom Dever, Alan Hinnebusch, and Reed Wickner for discussions. We apologize for omitting discussions of many interesting topics from the original manuscript due to lack of space. Work in our lab is funded by grants from the NIH/NIGMS, the American Cancer Society, and the American Heart Association.

The *Annual Review of Biochemistry* is online at http://biochem.annualreviews.org

LITERATURE CITED

1. Hershey JWB, Merrick WC. 2000. See Ref. 333, pp. 33–88
2. Kapp LD, Lorsch JR. 2004. *J. Mol. Biol.* 335:923–36
3. Kimball SR, Everson WV, Myers LM, Jefferson LS. 1987. *J. Biol. Chem.* 262: 2220–27
4. Panniers R, Rowlands AG, Henshaw EC. 1988. *J. Biol. Chem.* 263:5519–25
5. Walton GM, Gill GN. 1975. *Biochim. Biophys. Acta* 390:231–45
6. Erickson FL, Hannig EM. 1996. *EMBO J.* 15:6311–20
7. Nika J, Yang W, Pavitt GD, Hinne-busch AG, Hannig EM. 2000. *J. Biol. Chem.* 275:26011–17
8. Schmitt E, Blanquet S, Mechulam Y. 2002. *EMBO J.* 21:1821–32
9. Lorsch JR, Herschlag D. 1999. *EMBO J.* 18:6705–17
10. Moazed D, Samaha RR, Gualerzi C, Noller HF. 1995. *J. Mol. Biol.* 248: 207–10
11. Carter AP, Clemons WM Jr, Brodersen DE, Morgan-Warren RJ, Hartsch T, et al. 2001. *Science* 291:498–501
12. Gualerzi CO, Pon CL. 1990. *Biochemistry* 29:5881–89

13. Luchin S, Putzer H, Hershey JW, Cenatiempo Y, Grunberg-Manago M, Laalami S. 1999. *J. Biol. Chem.* 274:6074–79

14. Grunberg-Manago M, Dessen P, Pantaloni D, Godefroy-Colburn T, Wolfe AD, Dondon J. 1975. *J. Mol. Biol.* 94:461–78

15. Karimi R, Pavlov MY, Buckingham RH, Ehrenberg M. 1999. *Mol. Cell* 3:601–9

16. Kyrpides NC, Woese CR. 1998. *Proc. Natl. Acad. Sci. USA* 95:224–28

17. Choi SK, Lee JH, Zoll WL, Merrick WC, Dever TE. 1998. *Science* 280:1757–60

18. Lee JH, Choi SK, Roll-Mecak A, Burley SK, Dever TE. 1999. *Proc. Natl. Acad. Sci. USA* 96:4342–47

19. Sorensen HP, Hedegaard J, Sperling-Petersen HU, Mortensen KK. 2001. *IUBMB Life* 51:321–27

20. Dever TE. 2002. *Cell* 108:545–56

21. Pestova TV, Kolupaeva VG. 2002. *Genes Dev.* 16:2906–22

22. Goumans H, Thomas A, Verhoeven A, Voorma HO, Benne R. 1980. *Biochim. Biophys. Acta* 608:39–46

23. Goss DJ, Rounds DJ. 1988. *Biochemistry* 27:3610–13

24. Nakaya K, Ranu RS, Wool IG. 1973. *Biochem. Biophys. Res. Commun.* 54:246–55

25. Goss DJ, Rounds D, Harrigan T, Woodley CL, Wahba AJ. 1988. *Biochemistry* 27:1489–94

26. Gualerzi C, Wintermeyer W. 1986. *FEBS Lett.* 202:1–6

27. Szkaradkiewicz K, Zuleeg T, Limmer S, Sprinzl M. 2000. *Eur. J. Biochem.* 267:4290–99

28. Roll-Mecak A, Cao C, Dever TE, Burley SK. 2000. *Cell* 103:781–92

29. Shin B-S, Maag D, Roll-Mecak A, Arefin MS, Burley SK, et al. 2002. *Cell* 111:1015–25

30. Pestova TV, Lomakin IB, Lee JH, Choi SK, Dever TE, Hellen CU. 2000. *Nature* 403:332–35

31. Cui Y, Dinman JD, Kinzy TG, Peltz SW. 1998. *Mol. Cell Biol.* 18:1506–16

32. Yoon HJ, Donahue TF. 1992. *Mol. Cell Biol.* 12:248–60

33. Keeling PJ, Doolittle WF. 1995. *Proc. Natl. Acad. Sci. USA* 92:5761–64

34. Huang H, Yoon H, Hannig EM, Donahue TF. 1997. *Genes Dev.* 11:2396–413

35. Phan L, Zhang XL, Asano K, Anderson J, Vornlocher HP, et al. 1998. *Mol. Cell Biol.* 18:4935–46

36. Merrick WC. 1979. *Methods Enzymol.* 60:108–23

37. Merrick WC. 1979. *Methods Enzymol.* 60:101–6

38. Trachsel H, Erni B, Schreier MH, Staehelin T. 1977. *J. Mol. Biol.* 116:755–67

39. Benne R, Hershey JW. 1976. *Proc. Natl. Acad. Sci. USA* 73:3005–9

40. Burks EA, Bezerra PP, Le H, Gallie DR, Browning KS. 2001. *J. Biol. Chem.* 276:2122–31

41. Benne R, Hershey JWB. 1978. *J. Biol. Chem.* 253:3078–87

42. Valasek L, Nielsen KH, Hinnebusch AG. 2002. *EMBO J.* 21:5886–98

43. Chaudhuri J, Chowdhury D, Maitra U. 1999. *J. Biol. Chem.* 273:17975–80

44. Phan L, Schoenfeld LW, Valasek L, Nielsen KH, Hinnebusch AG. 2001. *EMBO J.* 20:2954–65

45. Jackson RJ. 1996. See Ref. 334, pp. 71–112

46. Kozak M. 2002. *Gene* 299:1–34

47. Ishigaki Y, Li XJ, Serin G, Maquat LE. 2001. *Cell* 106:607–17

48. Linder P. 1992. *Antonie van Leeuwenhoek J. Microbiol. Serol.* 62:47–62

49. Grifo JA, Tahara SM, Leis JP, Morgan MA, Shatkin AJ, Merrick WC. 1982. *J. Biol. Chem.* 257:5246–52

50. Rozen F, Edery I, Meerovitch K, Dever TE, Merrick WC, Sonenberg N. 1990. *Mol. Cell Biol.* 10:1134–44

51. Lawson TG, Lee KA, Maimone MM,

Abramson RD, Dever TE, et al. 1989. *Biochemistry* 28:4729–34

52. Ray BK, Lawson TG, Kramer JC, Cladaras MH, Grifo JA, et al. 1985. *J. Biol. Chem.* 260:7651–58

53. Coppolecchia R, Buser P, Stotz A, Linder P. 1993. *EMBO J.* 12:4005–11

54. Altmann M, Muller PP, Wittmer B, Ruchti F, Lanker S, Trachsel H. 1993. *EMBO J.* 12:3997–4003

55. Richter-Cook NJ, Dever TE, Hensold JO, Merrick WC. 1998. *J. Biol. Chem.* 273:7579–87

56. Richter NJ, Rogers GW Jr, Hensold JO, Merrick WC. 1999. *J. Biol. Chem.* 274: 35415–24

57. Rogers GW Jr, Richter NJ, Lima WF, Merrick WC. 2001. *J. Biol. Chem.* 276: 30914–22

58. Hentze MW. 1997. *Science* 275:500–1

59. Grifo JA, Tahara SM, Morgan MA, Shatkin AJ, Merrick WC. 1983. *J. Biol. Chem.* 258:5804–10

60. Tarun SZ Jr, Sachs AB. 1996. *EMBO J.* 15:7168–77

61. Tarun SZ Jr, Sachs AB. 1995. *Genes Dev.* 9:2997–3007

62. Gallie DR. 1991. *Genes Dev.* 5: 2108–16

63. Preiss T, Hentze MW. 1998. *Nature* 392: 516–20

64. Munroe D, Jacobson A. 1990. *Mol. Cell Biol.* 10:3441–55

65. Withey JH, Friedman DI. 2003. *Annu. Rev. Microbiol.* 57:101–23

66. Benne R, Brown-Luedi ML, Hershey JW. 1978. *J. Biol. Chem.* 253:3070–77

67. Kemper WM, Berry KW, Merrick WC. 1976. *J. Biol. Chem.* 251:5551–57

68. Kang HA, Hershey JWB. 1994. *J. Biol. Chem.* 269:3934–40

69. Zuk D, Jacobson A. 1998. *EMBO J.* 17: 2914–25

70. Valentini SR, Casolari JM, Oliveira CC, Silver PA, McBride AE. 2002. *Genetics* 160:393–405

71. Zoll WL, Horton LE, Komar AA, Hensold JO, Merrick WC. 2002. *J. Biol. Chem.* 277:37079–87

72. Chakravarty I, Bagchi MK, Roy R, Banerjee AC, Gupta NK. 1985. *J. Biol. Chem.* 260:6945–49

73. Tyzack JK, Wang X, Belsham GJ, Proud CG. 2000. *J. Biol. Chem.* 275: 34131–39

74. Si K, Maitra U. 1999. *Mol. Cell Biol.* 19:1416–26

75. Basu U, Si K, Warner JR, Maitra U. 2001. *Mol. Cell Biol.* 21:1453–62

76. Hutvagner G, Zamore PD. 2002. *Science* 297:2056–60

77. Dennis PP. 1997. *Cell* 89:1007–10

78. Ramesh V, RajBhandary UL. 2001. *J. Biol. Chem.* 276:3660–65

79. Keeling PJ, Doolittle WF. 1995. *Mol. Microbiol.* 17:399–400

80. Kyrpides NC, Wocse CR. 1998. *Proc. Natl. Acad. Sci. USA* 95:3726–30

81. Kurzchalia TV, Bommer UA, Babkina GT, Karpova GG. 1984. *FEBS Lett.* 175: 313–16

82. Pavitt GD, Ramaiah KV, Kimball SR, Hinnebusch AG. 1998. *Genes Dev.* 12: 514–26

83. Lecompte O, Ripp R, Thierry JC, Moras D, Poch O. 2002. *Nucleic Acids Res.* 30:5382–90

84. Ganoza MC, Kiel MC, Aoki H. 2002. *Microbiol. Mol. Biol. Rev.* 66:460–85

85. Tanner NK, Linder P. 2001. *Mol. Cell* 8:251–62

86. Kressler D, de la Cruz J, Rojo M, Linder P. 1997. *Mol. Cell Biol.* 17: 7283–94

87. Lu J, Aoki H, Ganoza MC. 1999. *Int. J. Biochem. Cell Biol.* 31:215–29

88. Moll I, Grill S, Grundling A, Blasi U. 2002. *Mol. Microbiol.* 44:1387–96

89. Parkhurst KM, Hileman RE, Saha D, Gupta NK, Parkhurst LJ. 1994. *Biochemistry* 33:15168–77

90. Hinnebusch AG. 2000. See Ref. 333, pp. 185–244

91. Dasso MC, Milburn SC, Hershey JW,

Jackson RJ. 1990. *Eur. J. Biochem.* 187: 361–71

92. Naranda T, MacMillan SE, Donahue TF, Hershey JW. 1996. *Mol. Cell Biol.* 16:2307–13

93. Olsen DS, Savner EM, Mathew A, Zhang F, Krishnamoorthy T, et al. 2003. *EMBO J.* 22:193–204

94. Methot N, Song MS, Sonenberg N. 1996. *Mol. Cell Biol.* 16:5328–34

95. Vornlocher HP, Hanachi P, Ribeiro S, Hershey JW. 1999. *J. Biol. Chem.* 274: 16802–12

96. Imataka H, Sonenberg N. 1997. *Mol. Cell Biol.* 17:6940–47

97. Asano K, Shalev A, Phan L, Nielsen K, Clayton J, et al. 2001. *EMBO J.* 20: 2326–37

98. Peterson DT, Merrick WC, Safer B. 1979. *J. Biol. Chem.* 254:2509–16

99. Valasek L, Mathew AA, Shin BS, Nielsen KH, Szamecz B, Hinnebusch AG. 2003. *Genes Dev.* 17:786–99

100. Pincheira R, Chen Q, Huang Z, Zhang JT. 2001. *Eur. J. Cell Biol.* 80:410–18

101. Hasek J, Kovarik P, Valasek L, Malinska K, Schneider J, et al. 2000. *Cell Motil. Cytoskelet.* 45:235–46

102. Hou CL, Tang C, Roffler SR, Tang TK. 2000. *Blood* 96:747–53

103. Palecek J, Hasek J, Ruis H. 2001. *Biochem. Biophys. Res. Commun.* 282: 1244–50

104. Lin L, Holbro T, Alonso G, Gerosa D, Burger MM. 2001. *J. Cell Biochem.* 80: 483–90

105. von Arnim AG, Chamovitz DA. 2003. *Curr. Biol.* 13:R323–25

106. Asano K, Clayton J, Shalev A, Hinnebusch AG. 2000. *Genes Dev.* 14: 2534–46

107. Valasek L, Phan L, Schoenfeld LW, Valaskova V, Hinnebusch AG. 2001. *EMBO J.* 20:891–904

108. Stevens SW, Ryan DE, Ge HY, Moore RE, Young MK, et al. 2002. *Mol. Cell* 9:31–44

109. Ellis RJ. 2001. *Curr. Opin. Struct. Biol.* 11:114–19

110. von der Haar T, McCarthy JEG. 2002. *Mol. Microbiol.* 46:531–44

111. He H, von der Haar T, Singh CR, Ii M, Li B, et al. 2003. *Mol. Cell Biol.* 23: 5431–45

112. Asano K, Krishnamoorthy T, Phan L, Pavitt GD, Hinnebusch AG. 1999. *EMBO J.* 18:1673–88

113. von der Haar T, Ball PD, McCarthy JEG. 2000. *J. Biol. Chem.* 275: 30551–55

114. Wei C, Balasta ML, Ren J, Goss DJ. 1998. *Biochemistry* 37:1910–16

115. Luo Y, Goss DJ. 2001. *J. Biol. Chem.* 276:43083–86

116. Korneeva NL, Lamphear BJ, Hennigan FL, Rhoads RE. 2000. *J. Biol. Chem.* 275:41369–76

117. Maag D, Lorsch JR. 2003. *J. Mol. Biol.* 330:917–24

118. Marcotrigiano J, Lomakin IB, Sonenberg N, Pestova TV, Hellen CU, Burley SK. 2001. *Mol. Cell* 7:193–203

119. Marcotrigiano J, Gingras AC, Sonenberg N, Burley SK. 1999. *Mol. Cell* 3: 707–16

120. Hershey PEC, McWhirter SM, Gross JD, Wagner G, Alber T, Sachs AB. 1999. *J. Biol. Chem.* 274:21297–304

121. Fletcher CM, Wagner G. 1998. *Protein Sci.* 7:1639–42

122. Fletcher CM, McGuire AM, Gingras AC, Li H, Matsuo H, et al. 1998. *Biochemistry* 37:9–15

123. Choi SK, Olsen DS, Roll-Mecak A, Martung A, Remo KL, et al. 2000. *Mol. Cell Biol.* 20:7183–91

124. Marintchev A, Kolupaeva VG, Pestova TV, Wagner G. 2003. *Proc. Natl. Acad. Sci. USA* 100:1535–40

125. Battiste JL, Pestova TV, Hellen CU, Wagner G. 2000. *Mol. Cell* 5:109–19

126. Fletcher CM, Pestova TV, Hellen CU, Wagner G. 1999. *EMBO J.* 18:2631–37

127. Naranda T, Sirangelo I, Fabbri BJ, Hershey JW. 1995. *FEBS Lett.* 372:249–52

128. Astrom SU, von Pawel-Rammingen U, Bystrom AS. 1993. *J. Mol. Biol.* 233: 43–58

129. Farruggio D, Chaudhuri J, Maitra U, Rajbhandary UL. 1996. *Mol. Cell Biol.* 16:4248–56

130. von Pawel-Rammingen U, Astrom S, Bystrom AS. 1992. *Mol. Cell Biol.* 12: 1432–42

131. Forster C, Chakraburtty K, Sprinzl M. 1993. *Nucleic Acids Res.* 21:5679–83

132. Astrom SU, Bystrom AS. 1994. *Cell* 79: 535–46

133. Drabkin HJ, Estrella M, Rajbhandary UL. 1998. *Mol. Cell Biol.* 18:1459–66

134. Thomas A, Goumans H, Voorma HO, Benne R. 1980. *Eur. J. Biochem.* 107: 39–45

135. Chaudhuri J, Si K, Maitra U. 1997. *J. Biol. Chem.* 272:7883–91

136. Majumdar R, Bandyopadhyay A, Maitra U. 2003. *J. Biol. Chem.* 278: 6580-87

137. Algire MA, Maag D, Savio P, Acker MG, Tarun SZ Jr, et al. 2002. *RNA* 8: 382–97

138. Hinnebusch AG. 1996. See Ref. 334, pp. 199–244

139. Bommer UA, Lutsch G, Stahl J, Bielka H. 1991. *Biochimie* 73:1007–19

140. Bommer UA, Lutsch G, Behlke J, Stahl J, Nesytova N, et al. 1988. *Eur. J. Biochem.* 172:653–62

141. Bommer UA, Stahl J, Henske A, Lutsch G, Bielka H. 1988. *FEBS Lett.* 233: 114–18

142. Marcotrigiano J, Gingras AC, Sonenberg N, Burley SK. 1997. *Cell* 89: 951–61

143. Tomoo K, Shen X, Okabe K, Nozoe Y, Fukuhara S, et al. 2002. *Biochem. J.* 362: 539–44

144. Matsuo H, McGuire AM, Fletcher CM, Gingras AC, Sonenberg N, Wagner G. 1997. *Nat. Struct. Biol.* 4:717–24

145. Iizuka N, Najita L, Franzusoff A, Sarnow P. 1994. *Mol. Cell Biol.* 7322–30

146. Michel YM, Poncet D, Piron M, Kean KM, Borman AM. 2000. *J. Biol. Chem.* 275:32268–76

147. Svitkin YV, Ovchinnikov LP, Dreyfuss G, Sonenberg N. 1996. *EMBO J.* 15: 7147–55

148. Lang V, Zanchin NI, Lunsdorf H, Tuite M, McCarthy JE. 1994. *J. Biol. Chem.* 269:6117–23

149. Vasilescu S, Ptushkina M, Linz B, Muller PP, McCarthy JE. 1996. *J. Biol. Chem.* 271:7030–37

150. Niedzwiecka A, Marcotrigiano J, Stepinski J, Jankowska-Anyszka M, Wyslouch-Cieszynska A, et al. 2002. *J. Mol. Biol.* 319:615–35

151. Vega Laso MR, Zhu D, Sagliocco F, Brown AJ, Tuite MF, McCarthy JE. 1993. *J. Biol. Chem.* 268:6453–62

152. Pelletier J, Sonenberg N. 1985. *Cell* 40: 515–26

153. Kozak M. 1986. *Proc. Natl. Acad. Sci. USA* 83:2850–54

154. Baim SB, Sherman F. 1988. *Mol. Cell Biol.* 8:1591–601

155. Kozak M. 1989. *Mol. Cell Biol.* 9: 5134–42

156. Lee KA, Guertin D, Sonenberg N. 1983. *J. Biol. Chem.* 258:707–10

157. Lawson TG, Ray BK, Dodds JT, Grifo JA, Abramson RD, et al. 1986. *J. Biol. Chem.* 261:13979–89

158. Muckenthaler M, Gray NK, Hentze MW. 1998. *Mol. Cell* 2:383–88

159. Lamphear BJ, Kirchweger R, Skern T, Rhoads RE. 1995. *J. Biol. Chem.* 270: 21975–83

160. Svitkin YV, Pause A, Haghighat A, Pyronnet S, Witherell G, et al. 2001. *RNA* 7:382–94

161. Grifo JA, Abramson RD, Satler CA, Merrick WC. 1984. *J. Biol. Chem.* 259: 8648–54

162. Pause A, Sonenberg N. 1992. *EMBO J.* 11:2643–54

163. Blum S, Schmid SR, Pause A, Buser P, Linder P, et al. 1992. *Proc. Natl. Acad. Sci. USA* 89:7664–68

164. Lorsch JR, Herschlag D. 1998. *Biochemistry* 37:2180–93

165. Lorsch JR, Herschlag D. 1998. *Biochemistry* 37:2194–206

166. Johnson ER, McKay DB. 1999. *RNA* 5: 1526–34

167. Caruthers JM, Johnson ER, McKay DB. 2000. *Proc. Natl. Acad. Sci. USA* 97: 13080–85

168. Benz J, Trachsel H, Baumann U. 1999. *Struct. Fold. Des.* 7:671–79

169. Pause A, Methot N, Sonenberg N. 1993. *Mol. Cell Biol.* 13:6789–98

170. Rogers GW Jr, Lima WF, Merrick WC. 2001. *J. Biol. Chem.* 276:12598–608

171. Rogers GW Jr, Richter NJ, Merrick WC. 1999. *J. Biol. Chem.* 274: 12236–44

172. Lohman TM, Bjornson KP. 1996. *Annu. Rev. Biochem.* 65:169–214

173. Abramson RD, Dever TE, Merrick WC. 1988. *J. Biol. Chem.* 263:6016–19

174. Bi X, Ren J, Goss DJ. 2000. *Biochemistry* 39:5758–65

175. Berset C, Zurbriggen A, Djafarzadeh S, Altmann M, Trachsel H. 2003. *RNA* 9: 871–80

176. Sonenberg N. 1988. *Prog. Nucleic Acid Res. Mol. Biol.* 35:173–207

177. Duncan R, Milburn SC, Hershey JW. 1987. *J. Biol. Chem.* 262:380–88

178. Pollard TD, Cooper JA. 1986. *Annu. Rev. Biochem.* 55:987–1035

179. Pause A, Methot N, Svitkin Y, Merrick WC, Sonenberg N. 1994. *EMBO J.* 13: 1205–15

180. Altmann M, Wittmer B, Methot N, Sonenberg N, Trachsel H. 1995. *EMBO J.* 14:3820–27

181. Pisarev AV, Skabkin MA, Thomas AA, Merrick WC, Ovchinnikov LP, Shatsky IN. 2002. *J. Biol. Chem.* 277:15445–51

182. Iost I, Dreyfus M, Linder P. 1999. *J. Biol. Chem.* 274:17677–83

183. Chuang RY, Weaver PL, Liu Z, Chang TH. 1997. *Science* 275:1468–71

184. de la Cruz J, Iost I, Kressler D, Linder P. 1997. *Proc. Natl. Acad. Sci. USA* 94: 5201–6

185. Verlhac MH, Chen RH, Hanachi P, Hershey JW, Derynck R. 1997. *EMBO J.* 16:6812–22

186. Naranda T, MacMillan SE, Hershey JW. 1994. *J. Biol. Chem.* 269: 32286–92

187. Deo RC, Bonanno JB, Sonenberg N, Burley SK. 1999. *Cell* 98:835–45

188. Baer BW, Kornberg RD. 1980. *Proc. Natl. Acad. Sci. USA* 77:1890–92

189. Baer BW, Kornberg RD. 1983. *J. Cell Biol.* 96:717–21

190. Kessler SH, Sachs AB. 1998. *Mol. Cell Biol.* 18:51–57

191. Gray NK, Coller JM, Dickson KS, Wickens M. 2000. *EMBO J.* 19: 4723–33

192. Wells SE, Hillner PE, Vale RD, Sachs AB. 1998. *Mol. Cell* 2:135–40

193. Borman AM, Michel YM, Kean KM. 2000. *Nucleic Acids Res.* 28:4068–75

194. Borman AM, Michel YM, Malnou CE, Kean KM. 2002. *J. Biol. Chem.* 277: 36818–24

195. Bi XP, Goss DJ. 2000. *J. Biol. Chem.* 275:17740–46

196. Herschlag D, Johnson FB. 1993. *Genes Dev.* 7:173–79

197. Proweller A, Butler JS. 1997. *J. Biol. Chem.* 272:6004–10

198. Munroe D, Jacobson A. 1990. In *The Ribosome*, ed. WE Hill, A Dahlberg, RA Garrett, PB Moore, D Schlessinger, JR Warner, pp. 299–305. Washington, DC: ASM Press

199. Jacobson A. 1996. See Ref. 334, pp. 459–80

200. Sachs AB, Davis RW. 1989. *Cell* 58: 857–67

201. Proweller A, Butler JS. 1996. *J. Biol. Chem.* 271:10859–65

202. Searfoss A, Dever TE, Wickner R. 2001. *Mol. Cell Biol.* 21:4900–8

203. Searfoss AW, Wickner RB. 2000. *Proc. Natl. Acad. Sci. USA* 97:9133–37

204. Otero LJ, Ashe MP, Sachs AB. 1999. *EMBO J.* 18:3153–63
205. Craig AW, Haghighat A, Yu AT, Sonenberg N. 1998. *Nature* 392:520–23
206. Paraskeva E, Gray NK, Schlager B, Wehr K, Hentze MW. 1999. *Mol. Cell Biol.* 19:807–16
207. Hanson S, Berthelot K, Fink B, McCarthy JEG, Suess B. 2003. *Mol. Microbiol.* 49:1627–37
208. Pestova TV, Borukhov SI, Hellen CUT. 1998. *Nature* 394:854–59
209. Kozak M. 1980. *Cell* 22:459–67
210. Cigan AM, Feng L, Donahue TF. 1988. *Science* 242:93–97
211. Donahue TF, Cigan AM, Pabich EK, Valavicius BC. 1988. *Cell* 54:621–32
212. Cigan AM, Pabich EK, Feng L, Donahue TF. 1989. *Proc. Natl. Acad. Sci. USA* 86:2784–88
213. Castilho-Valavicius B, Yoon H, Donahue TF. 1990. *Genetics* 124:483–95
214. Hashimoto NN, Carnevalli LS, Castilho BA. 2002. *Biochem. J.* 367:359–68
215. Ogle JM, Brodersen DE, Clemons WM Jr, Tarry MJ, Carter AP, Ramakrishnan V. 2001. *Science* 292:902
216. Dallas A, Noller HF. 2001. *Mol. Cell* 8:855–64
216a. Lomakin IB, Kolupaeva VG, Marintchev A, Wagner G, Pestova TV. 2003. *Genes Dev.* 17:2786–97
217. Piepenburg O, Pape T, Pleiss JA, Wintermeyer W, Uhlenbeck OC, Rodnina MV. 2000. *Biochemistry* 39:1734–38
218. Yarus M, Smith D. 1995. In *tRNA: Structure, Biosynthesis and Function*, ed. D Soll, UC Rajbhandary, pp. 443–70. Washington, DC: ASM Press
219. Das S, Maitra U. 2001. *Prog. Nucleic Acid Res. Mol. Biol.* 70:207–31
220. Das S, Maitra U. 2000. *Mol. Cell Biol.* 20:3942–50
221. Paulin FE, Campbell LE, O'Brien K, Loughlin J, Proud CG. 2001. *Curr. Biol.* 11:55–59
222. Kozak M. 1994. *Biochimie* 76:815–21
223. Cigan AM, Donahue TF. 1987. *Gene* 59:1–18
224. Hamilton R, Watanabe CK, de Boer HA. 1987. *Nucleic Acids Res.* 15:3581–93
225. Joshi CP, Zhou H, Huang X, Chiang VL. 1997. *Plant Mol. Biol.* 35:993–1001
226. Yun DF, Laz TM, Clements JM, Sherman F. 1996. *Mol. Microbiol.* 19:1225–39
227. Donahue TF, Cigan AM. 1988. *Mol. Cell Biol.* 8:2955–63
228. Looman AC, Kuivenhoven JA. 1993. *Nucleic Acids Res.* 21:4268–71
229. Cigan AM, Pabich EK, Donahue TF. 1988. *Mol. Cell Biol.* 8:2964–75
230. Slusher LB, Gillman EC, Martin NC, Hopper AK. 1991. *Proc. Natl. Acad. Sci. USA* 88:9789–93
231. Peri S, Pandey A. 2001. *Trends Genet.* 17:685–87
232. Kozak M. 2000. *Genomics* 70:396–406
233. Merrick WC. 1979. *J. Biol. Chem.* 254:3708–11
234. Lee JH, Pestova TV, Shin BS, Cao C, Choi SK, Dever TE. 2002. *Proc. Natl. Acad. Sci. USA* 99:16689–94
235. Merrick WC, Kemper WM, Anderson WF. 1975. *J. Biol. Chem.* 250:5556–62
236. Rodnina MV, Savelsbergh A, Katunin VI, Wintermeyer W. 1997. *Nature* 385:37–41
237. Ramakrishnan V. 2002. *Cell* 108:557–72
238. Spahn CM, Beckmann R, Eswar N, Penczek PA, Sali A, et al. 2001. *Cell* 107:373–86
239. Rodnina MV, Wintermeyer W. 2001. *Annu. Rev. Biochem.* 70:415–35
240. Moore PB, Steitz TA. 2003. *RNA* 9:155–59
241. Green R, Noller HF. 1997. *Annu. Rev. Biochem.* 66:679–716
242. Wintermeyer W, Savelsbergh A, Semenkov YP, Katunin VI, Rodnina

MV. 2001. *Cold Spring Harbor Symp. Quant. Biol.* 66:449–58

243. Cavallius J, Zoll W, Chakraburtty K, Merrick WC. 1993. *Biochim. Biophys. Acta* 1163:75–80

244. Fujita T, Itoh T. 1995. *Biochem. Mol. Biol. Int.* 37:107–15

245. De Vendittis E, Amatruda MR, Masullo M, Bocchini V. 1993. *Gene* 136:41–48

246. Andersen GR, Valente L, Pedersen L, Kinzy TG, Nyborg J. 2001. *Nat. Struct. Biol.* 8:531–34

247. Valle M, Sengupta J, Swami NK, Grassucci RA, Burkhardt N, et al. 2002. *EMBO J.* 21:3557–67

248. Stark H, Rodnina MV, Wieden HJ, Zemlin F, Wintermeyer W, van Heel M. 2002. *Nat. Struct. Biol.* 9:849–54

249. Jorgensen R, Ortiz PA, Carr-Schmid A, Nissen P, Kinzy TG, Andersen GR. 2003. *Nat. Struct. Biol.* 10:379–85

250. Dever TE, Costello CE, Owens CL, Rosenberry TL, Merrick WC. 1989. *J. Biol. Chem.* 264:20518–25

251. Van Ness BG, Howard JB, Bodley JW. 1978. *J. Biol. Chem.* 253:8687–90

252. Skogerson L, Engelhardt D. 1977. *J. Biol. Chem.* 252:1471–75

253. Qin SL, Xie AG, Bonato MC, McLaughlin CS. 1990. *J. Biol. Chem.* 265:1903–12

254. Dasmahapatra B, Chakraburtty K. 1981. *J. Biol. Chem.* 256:9999–10004

255. Kovalchuke O, Kambampati R, Pladies E, Chakraburtty K. 1998. *Eur. J. Biochem.* 258:986–93

256. Anand M, Chakraburtty K, Marton MJ, Hinnebusch AG, Kinzy TG. 2003. *J. Biol. Chem.* 278:6985–91

257. Triana-Alonso FJ, Chakraburtty K, Nierhaus KH. 1995. *J. Biol. Chem.* 270:20473–78

258. Rodnina MV, Serebryanik AI, Ovcharenko GV, El'Skaya AV. 1994. *Eur. J. Biochem.* 225:305–10

259. El'Skaya AV, Ovcharenko GV, Palchevskii SS, Petrushenko ZM, Triana-

Alonso FJ, Nierhaus KH. 1997. *Biochemistry* 36:10492–97

260. Kiel MC, Ganoza MC. 2001. *Eur. J. Biochem.* 268:278–86

261. Kiel MC, Aoki H, Ganoza MC. 1999. *Biochimie* 81:1097–108

262. Inge-Vechtomov S, Zhouravleva G, Philippe M. 2003. *Biol. Cell* 95:195–209

263. Kisselev L, Ehrenberg M, Frolova L. 2003. *EMBO J.* 22:175–82

264. Nakamura Y, Ito K. 2003. *Trends Biochem. Sci.* 28:99–105

265. Bertram G, Innes S, Minella O, Richardson J, Stansfield I. 2001. *Microbiology* 147:255–69

266. Kisselev LL, Buckingham RH. 2000. *Trends Biochem. Sci.* 25:561–66

267. Poole E, Tate W. 2000. *Biochim. Biophys. Acta* 1493:1–11

268. Caskey CT, Beaudet AL, Scolnick EM, Rosman M. 1971. *Proc. Natl. Acad. Sci. USA* 68:3163–67

269. Arkov AL, Freistroffer DV, Ehrenberg M, Murgola EJ. 1998. *EMBO J.* 17:1507–14

270. Arkov AL, Hedenstierna KO, Murgola EJ. 2002. *J. Bacteriol.* 184:5052–57

271. Arkov AL, Murgola EJ. 1999. *Biochemistry* 64:1354–59

272. Seit-Nebi A, Frolova L, Justesen J, Kisselev L. 2001. *Nucleic Acids Res.* 29:3982–87

273. Zavialov AV, Mora L, Buckingham RH, Ehrenberg M. 2002. *Mol. Cell* 10:789–98

274. Scolnick E, Tompkins R, Caskey T, Nirenberg M. 1968. *Proc. Natl. Acad. Sci. USA* 61:768–74

275. Ito K, Uno M, Nakamura Y. 2000. *Nature* 403:680–84

276. Grentzmann G, Brechemier-Baey D, Heurgue V, Mora L, Buckingham RH. 1994. *Proc. Natl. Acad. Sci. USA* 91:5848–52

277. Mikuni O, Ito K, Moffat J, Matsumura K, McCaughan K, et al. 1994. *Proc. Natl. Acad. Sci. USA* 91:5798–802

278. Freistroffer DV, Pavlov MY, MacDougall J, Buckingham RH, Ehrenberg M. 1997. *EMBO J.* 16:4126–33

279. Konecki DS, Aune KC, Tate W, Caskey CT. 1977. *J. Biol. Chem.* 252:4514–20

280. Frolova L, Le Goff X, Rasmussen HH, Cheperegin S, Drugeon G, et al. 1994. *Nature* 372:701–3

281. Dontsova M, Frolova L, Vassilieva J, Piendl W, Kisselev L, Garber M. 2000. *FEBS Lett.* 472:213–16

282. Zhouravleva G, Frolova L, Le Goff X, Le Guellec R, Inge-Vechtomov S, et al. 1995. *EMBO J.* 14:4065–72

283. Frolova L, Le Goff X, Zhouravleva G, Davydova E, Philippe M, Kisselev L. 1996. *RNA* 2:334–41

284. Wilson PG, Culbertson MR. 1988. *J. Mol. Biol.* 199:559–73

285. Stansfield I, Jones KM, Kushnirov VV, Dagkesamanskaya AR, Poznyakovski AI, et al. 1995. *EMBO J.* 14:4365–73

286. Le Goff X, Philippe M, Jean-Jean O. 1997. *Mol. Cell Biol.* 17:3164–72

287. Ito K, Ebihara K, Uno M, Nakamura Y. 1996. *Proc. Natl. Acad. Sci. USA* 93: 5443–48

288. Pel HJ, Moffat JG, Ito K, Nakamura Y, Tate WP. 1998. *RNA* 4:47–54

289. Frolova LY, Tsivkovskii RY, Sivolobova GF, Oparina NY, Serpinsky OI, et al. 1999. *RNA* 5:1014–20

290. Brown CM, Stockwell PA, Trotman CN, Tate WP. 1990. *Nucleic Acids Res.* 18:6339–45

291. Brown CM, Stockwell PA, Trotman CN, Tate WP. 1990. *Nucleic Acids Res.* 18:2079–86

292. Poole ES, Major LL, Mannering SA, Tate WP. 1998. *Nucleic Acids Res.* 26: 954–60

293. Ozawa Y, Hanaoka S, Saito R, Washio T, Nakano S, et al. 2002. *Gene* 300: 79–87

294. Song H, Mugnier P, Das AK, Webb HM, Evans DR, et al. 2000. *Cell* 100: 311–21

295. Frolova LY, Merkulova TI, Kisselev LL. 2000. *RNA* 6:381–90

296. Merkulova TI, Frolova LY, Lazar M, Camonis J, Kisselev LL. 1999. *FEBS Lett.* 443:41–47

297. Eurwilaichitr L, Graves FM, Stansfield I, Tuite MF. 1999. *Mol. Microbiol.* 32: 485–96

298. Vestergaard B, Van LB, Andersen GR, Nyborg J, Buckingham RH, Kjeldgaard M. 2001. *Mol. Cell* 8:1375–82

299. Klaholz BP, Pape T, Zavialov AV, Myasnikov AG, Orlova EV, et al. 2003. *Nature* 421:90–94

300. Rawat UB, Zavialov AV, Sengupta J, Valle M, Grassucci RA, et al. 2003. *Nature* 421:87–90

301. Scarlett DJ, McCaughan KK, Wilson DN, Tate WP. 2003. *J. Biol. Chem.* 278: 15095–104

302. Bertram G, Bell HA, Ritchie DW, Fullerton G, Stansfield I. 2000. *RNA* 6: 1236–47

303. Lozupone CA, Knight RD, Landweber LF. 2001. *Curr. Biol.* 11:65–74

304. Karamyshev AL, Ito K, Nakamura Y. 1999. *FEBS Lett.* 457:483–88

305. Inagaki Y, Doolittle WF. 2001. *Nucleic Acids Res.* 29:921–27

306. Inagaki Y, Blouin C, Doolittle WF, Roger AJ. 2002. *Nucleic Acids Res.* 30: 532–44

307. Muramatsu T, Heckmann K, Kitanaka C, Kuchino Y. 2001. *FEBS Lett.* 488: 105–9

308. Ito K, Frolova L, Seit-Nebi A, Karamyshev A, Kisselev L, Nakamura Y. 2002. *Proc. Natl. Acad. Sci. USA* 99:8494–99

309. Frolova L, Seit-Nebi A, Kisselev L. 2002. *RNA* 8:129–36

310. Seit-Nebi A, Frolova L, Kisselev L. 2002. *EMBO Rep.* 3:881–86

311. Chavatte L, Kervestin S, Favre A, Jean-Jean O. 2003. *EMBO J.* 22:1644–53

312. Kervestin S, Frolova L, Kisselev L, Jean-Jean O. 2001. *EMBO Rep.* 2: 680–84

313. Velichutina IV, Dresios J, Hong JY, Li

C, Mankin A, et al. 2000. *RNA* 6: 1174–84

314. Chernoff YO, Newnam GP, Liebman SW. 1996. *Proc. Natl. Acad. Sci. USA* 93:2517–22

315. Chernoff YO, Vincent A, Liebman SW. 1994. *EMBO J.* 13:906–13

316. Velichutina IV, Hong JY, Mesecar AD, Chernoff YO, Liebman SW. 2001. *J. Mol. Biol.* 305:715–27

317. Liebman SW, Chernoff YO, Liu R. 1995. *Biochem. Cell Biol.* 73:1141–49

318. Ogle JM, Brodersen DE, Clemons WM Jr, Tarry MJ, Carter AP, Ramakrishnan V. 2001. *Science* 292:897–902

319. Ogle JM, Murphy FV, Tarry MJ, Ramakrishnan V. 2002. *Cell* 111: 721–32

320. Zavialov AV, Buckingham RH, Ehrenberg M. 2001. *Cell* 107:115–24

321. Moazed D, Noller HF. 1989. *Nature* 342: 142–48

322. Selmer M, Al-Karadaghi S, Hirokawa G, Kaji A, Liljas A. 1999. *Science* 286: 2349–52

323. Lancaster L, Kiel MC, Kaji A, Noller HF. 2002. *Cell* 111:129–40

324. Deleted in proof

325. Thompson HA, Sadnik I, Scheinbuks J, Moldave K. 1977. *Biochemistry* 16: 2221–30

326. Russell DW, Spremulli LL. 1979. *J. Biol. Chem.* 254:8796–800

327. Valenzuela DM, Chaudhuri A, Maitra U. 1982. *J. Biol. Chem.* 257:7712–19

327a. Ceci M, Gaviraghi C, Corrini C, Sala LA, Offenhauser N, et al. 2003. *Nature* 426:579–84

328. Srivastava S, Verschoor A, Frank J. 1992. *J. Mol. Biol.* 226:301–4

329. Hoshino S, Imai M, Kobayashi T, Uchida N, Katada T. 1999. *J. Biol. Chem.* 274:16677–80

330. Cosson B, Couturier A, Chabelskaya S, Kiktev D, Inge-Vechtomov S, et al. 2002. *Mol. Cell Biol.* 22:3301–15

331. Uchida N, Hoshino S, Imataka H, Sonenberg N, Katada T. 2002. *J. Biol. Chem.* 277:50286–92

332. Chou T. 2003. *Biophys. J.* 85:755–73

333. Sonenberg N, Hershey JWB, Mathews MB, eds. 2000. *Translational Control of Gene Expression.* Cold Spring Harbor, NY: Cold Spring Harbor Lab. Press

334. Hershey JWB, Mathews MB, Sonenberg N, eds. 1996. *Translational Control.* Cold Spring Harbor, NY: Cold Spring Harbor Lab. Press

Annu. Rev. Biochem. 2004. 73:705–48
doi: 10.1146/annurev.biochem.72.121801.161542

MECHANICAL PROCESSES IN BIOCHEMISTRY

Carlos Bustamante,[1-3] Yann R. Chemla,[3] Nancy R. Forde,[1] and David Izhaky[1]

[1]Howard Hughes Medical Institute and the Departments of [2]Molecular and Cell Biology, and [3]Physics, University of California, Berkeley, California 94720-3206; email: carlos@alice.berkeley.edu, ychemla@socrates.berkeley.edu, nforde@alice.berkeley.edu, izhaky@alice.berkeley.edu

Key Words mechanical forces, single-molecule manipulation, molecular motors, mechanical unfolding, enzyme catalysis

■ **Abstract** Mechanical processes are involved in nearly every facet of the cell cycle. Mechanical forces are generated in the cell during processes as diverse as chromosomal segregation, replication, transcription, translation, translocation of proteins across membranes, cell locomotion, and catalyzed protein and nucleic acid folding and unfolding, among others. Because force is a product of all these reactions, biochemists are beginning to directly apply external forces to these processes to alter the extent or even the fate of these reactions hoping to reveal their underlying molecular mechanisms. This review provides the conceptual framework to understand the role of mechanical force in biochemistry.

CONTENTS

PROLOGUE

Fifty years ago, biochemists described cells as small vessels that contain a complex mixture of chemical species undergoing reactions through diffusion and random collision. This description was satisfactory inasmuch as the intricate pathways of metabolism and, later, the basic mechanisms of gene regulation and signal transduction were still being unraveled. Gradually, and in part as a result of the parallel growth in our structural understanding of the molecular components of the cell, the limitations of this "chemical reactor" view of the cell became plain. Armed with a more precise knowledge of the structural bases of molecular interactions, the focus shifted more and more to the mechanisms by which these molecular components recognize and react with each other. Moreover, it also became clear that cells are polar structures and that the cell interior is neither isotropic nor homogeneous; that many of the essential processes of the cell, such as chromosomal segregation, translocation of organelles from one part of the cell to another, protein import into organelles, or the maintenance of a voltage across the membrane, all involve directional movement and transport of chemical species, in some cases against electrochemical gradients. Processes such as replication, transcription, and translation require directional readout of the information encoded in the sequence of linear polymers. Slowly, the old paradigm was replaced by one of a small "factory" of complex molecular structures that behave in machine-like fashion to carry out highly specialized and coordinated processes. These molecular machines are often complex assemblies of many proteins and contain parts with specialized functions, for example, as energy transducers or molecular motors, converting chemical energy (either in the form of binding energy, chemical bond hydrolysis, or electrochemical gradients) into mechanical work through conformational changes and displacements.

To understand the behavior of this molecular machinery requires a fundamental change in our conceptual and practical approaches to biochemical research. The cell, it appears, resembles more a small clockwork device than a reaction vessel of soluble components. Many of the functions of this device (which besides replication, transcription, translation, and organelle transport, include cell crawling, cell adhesion, protein folding, protein and nucleic acid unfolding, protein degradation, and protein and nucleic acid splicing) are indeed mechanical processes, and basic physical concepts such as force, torque, work, energy conversion efficiency, mechanical advantage, etc., are needed to describe them. The recent development of experimental methods that permit the direct mechanical manipulation of single molecules now allows many of these mechanical processes to be investigated directly and in real-time fashion. Analyses of the data so obtained also require the reformulation of many of the traditional concepts of thermodynamics and kinetics to incorporate terms corresponding to forces and torques.

This article attempts to critically review the most recent conceptual and experimental developments in the mechanical characterization of biochemical processes. In the following section we reformulate some of the main results of thermodynamics and kinetics in terms of the effect of mechanical force to provide a conceptual framework for the interpretation of the results presented later in this review. In the third section, we review the use of mechanical force to unfold proteins and nucleic acids. Here, as in the following sections, we describe and illustrate the important new information that can be derived from the mechanical characterization of molecules and molecular processes rather than providing an exhaustive guide to the literature. In the fourth section we describe the mechanical properties of molecular motors, illustrate how mechanical force is used to investigate their mechanisms of mechanochemical transduction, and discuss the many cellular functions now known to be mechanical processes. The final section presents our current understanding of the importance of force and strain in enzyme catalysis: how an otherwise silent form of chemical energy (that associated with binding interactions) can be and is used by enzymes to accelerate the rate of chemical reactions in the cell.

The mechanical characterization of the cellular "factory" is just beginning. Many more mechanical cellular functions are likely to be discovered in the future. This exciting new aspect of the inner workings of the cell challenges us to learn to think in terms of concepts heretofore alien to the trained biochemist. We hope that this review will be a helping step in that direction.

THE EFFECT OF FORCE ON THE THERMODYNAMICS AND KINETICS OF CHEMICAL REACTIONS

Introduction

Many biochemical reactions proceed via large conformational changes within or between interacting molecules. Such conformational changes, which may involve a combination of linear and rotational degrees of freedom, provide convenient, well-defined mechanical reaction coordinates that can be used to follow the progress of the reaction. Examples of these reaction coordinates are the end-to-end distance of a molecule as it is being stretched, the position of a molecular motor as it moves along a track, the angle of a rotary motor's shaft, or the deformation of an enzyme's binding pocket when it binds the substrate. The effect of an applied force can yield valuable information about the free energy surface of the reaction. In this section, we describe, using basic thermodynamic and kinetic relationships, how an applied mechanical force affects the free energy, equilibrium, and rate of a reaction occurring along a mechanical reaction coordinate (for a more detailed treatment, see 1). For brevity, and because of its direct relevance to the majority of the examples in this review, we limit our discussion to linear mechanical reaction coordinates. By replacing linear distance by angle and force by torque, it is possible to derive analogous expressions for the effect of torsion.

From thermodynamics, the energy change of a system (e.g., a molecule being stretched, a motor moving along a track, etc.) can be separated into components related to the heat exchanged and the work done on or performed by the system. When the energy changes slowly enough that the system remains in quasi-static equilibrium, these quantities are the reversible exchanged heat and the reversible work (pressure-volume work and mechanical work):

$$
\begin{aligned}
dE &= dq_{rev} + dw_{rev} \\
&= (TdS) + (-PdV + \int F \cdot dx).
\end{aligned}
\tag{1.}
$$

In practice, the temperature and pressure are usually the independent variables in an experiment, and the Gibbs free energy ($G = E - TS + PV$) provides a more relevant expression:

$$
dG = -SdT + VdP + Fdx.
\tag{2.}
$$

At constant temperature and pressure, the work required to extend the system by an amount Δx is

$$
W = \int_{x_0}^{x_0 + \Delta x} Fdx.
\tag{3.}
$$

If the extension is carried out slowly enough that the system remains in quasi-static equilibrium, the work in Equation 3 is reversible and is equal to the free energy change of the system, $\Delta G_{stretch}(\Delta x)$. The work done to stretch the molecule is positive, as both F and Δx are positive. Positive work means that the surroundings have done work on the system; the free energy of the molecule is increased.

Work is required to stretch a DNA molecule by its ends, and the amount of force required to extend its ends a given distance is described by the worm-like chain model of polymer elasticity (2, 3). When the molecule is relaxed, the tension in the DNA molecule decreases and the molecule does work on its surroundings. The molecule follows the same force-extension curve upon stretching as relaxation, implying that these processes are occurring at equilibrium. Thus, DNA acts as a reversible spring, storing elastic energy as it is stretched and restoring that same energy to the surrounding bath as it relaxes. In other words mechanical stretching and relaxation of DNA are 100% efficient, when all of input (or stored) work is converted to a free energy change of the molecule and none is dissipated as heat. By integrating the area under the force-extension curve of DNA, the free energy change in the molecule due to stretching [$\Delta G_{stretch}(x)$] can thus be determined (Equation 3). Furthermore, by measuring the free energy change as a function of temperature, it is possible to determine the change in entropy, $\Delta S_{stretch}$, and the change in enthalpy, $\Delta H_{stretch}$, upon stretching the molecule. Recalling that the enthalpy $H = G\text{-}TS$ and using Equation 2, it can be shown that:

$$\Delta S_{stretch}(x) = -\left(\frac{\partial \Delta G_{stretch}(x)}{\partial T}\right)_{P,x}, \quad \Delta H_{stretch}(x) = \left(\frac{\partial (\Delta G_{stretch}(x)/T)}{\partial (1/T)}\right)_{P,x}. \qquad 4.$$

Effect of Force on the Free Energy of a Reaction

The effect of force on a reaction in which A is converted into B is illustrated by the two-state system depicted in Figure 1. Here, A and B are distinct observable states of the system; each occupies a local free energy minimum, at position x_A and x_B, respectively, separated by a distance Δx along the mechanical reaction coordinate shown in Figure 1. A and B could represent the states of a motor in sequential locations along its track, or folded and unfolded states of a protein. At zero force, the free energy difference between A and B is simply

$$\Delta G(F = 0) = \Delta G^0 + k_B T \ln \frac{[B]}{[A]}, \qquad 5.$$

where ΔG^0 is the standard state free energy, and [A] and [B] are the concentrations (more appropriately, the activities) of the molecule in states A and B. Since concentrations are only well-defined in bulk measurements, [A] and [B] represent probabilities of populating states A and B in single-molecule experiments. To a first approximation, an applied force "tilts" the free energy surface along the mechanical reaction coordinate by an amount linearly dependent on distance (Equation 2), such that

$$\Delta G(F) = \Delta G^0 - F(x_B - x_A) + k_B T \ln \frac{[B]}{[A]}. \qquad 6.$$

At equilibrium, $\Delta G = 0$ and

$$\Delta G^0 - F\Delta x = -k_B T \ln \frac{[B]_{eq}(F)}{[A]_{eq}(F)} \qquad 7.$$
$$= -k_B T \ln K_{eq}(F).$$

Thus, the equilibrium constant $K_{eq}(F)$ depends exponentially on the applied force. By applying a force assisting ($F > 0$) or opposing ($F < 0$) the transition, we can alter the equilibrium of the reaction, increasing or reducing the population

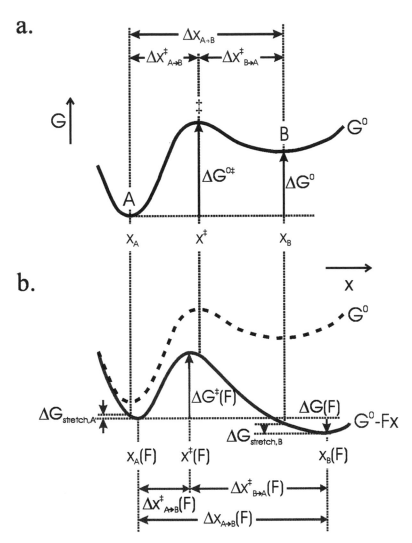

Figure 1 The effect of force on the free energy of a two-state system, where x represents the mechanical reaction coordinate. (*a*) No applied force. (*b*) Solid curve: positive applied force. Dashed curve: no applied force. The application of force lowers the energy of both the transition state ‡ and state B relative to state A ($\Delta G^{0\ddagger}$ and ΔG^0), which increases the rate of the forward reaction and the population of state B, respectively. The positions of the free energy minima (x_A and x_B) and maximum ($x^\ddagger$) shift to longer and shorter x, respectively, with a positive applied force. Their relative shifts in position depend on the local curvature of the free energy surface. The free energy change of states A and B upon stretching is $\Delta G_{stretch}$; see text.

of state B relative to state A, respectively. Furthermore, from Equation 7, one can determine the separation, Δx, between states A and B from the slope of the plot of $\ln K_{eq}(F)$ against the force.

For simplicity, we have assumed that the positions of states A and B, x_A and x_B, are unaffected by an applied force. In general, this assumption is not valid. As an example of the reaction A→B, consider the mechanical unfolding of a protein by its ends. As shown in Figure 1, force not only shifts the equilibrium toward state B (the unfolded state), but also increases the average end-to-end distance x_B of the unfolded molecule from that of a zero-force, random-coil configuration $x_B(F = 0)$ to that of an extended polypeptide chain $x_B(F)$. It is clear from Figure 1 that the shifts in x_A and x_B with force depend inversely on the local curvature of the potential: the steeper the potential well, the more "localized" the state, and the lesser the effect of force on its position. (For a harmonic potential well, the minimum shifts by F/G'', where G'' is the second derivative of the free energy surface at the minimum.) In Figure 1, the end-to-end distance of the folded protein (x_A) is less shifted by force than that of the unfolded molecule (x_B).

Because the free energy of the reaction A→B under an applied force F must be measured between the new free energy minima at $x_B(F)$ and $x_A(F)$, Equation 7 must be corrected to account for the small free energy change due to this shift in minima:

$$\Delta G^0 - F\Delta x + \Delta G^{A \to B}_{stretch}(F) = -k_B T \ln K_{eq}(F) \qquad 8.$$

where $\Delta x = x_B(F = 0) - x_A(F = 0)$ and $\Delta G^{A \to B}_{stretch}(F)$ is given by

$$\Delta G^{A \to B}_{stretch}(F) = \Delta G_{stretch,B}(F) - \Delta G_{stretch,A}(F)$$

$$= \int_{x_B(F = 0)}^{x_B(F)} F_B dx_B - \int_{x_A(F = 0)}^{x_A(F)} F_A dx_A. \qquad 9.$$

The two terms in Equation 9 are the free energy differences due to the shift in the minimum at state B [i.e., the free energy of stretching the molecule from $x_B(0)$ to $x_B(F)$] and at state A [stretching from $x_A(0)$ to $x_A(F)$], respectively (see Figure 1). If states A and B have the same curvature, their minima are shifted by the same amounts. In this situation, the two terms in Equation 9 cancel out exactly, and

$$\Delta G^{A \to B}_{stretch}(F) = 0.$$

This would be the case for a molecular motor moving along a homogeneous track, where states A and B represent sequential positions on the track.

We have seen that force can shift the equilibrium of a reaction. In principle, it is possible to apply a force, $F_{1/2}$, such that the equilibrium constant $K_{eq} = 1$. At $F_{1/2}$, the molecule has an equal probability of being in state B or A, and will

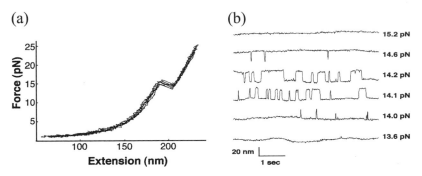

Figure 2 Mechanical unfolding of the 22-bp P5ab RNA hairpin, a domain of the group I intron of *T. thermophila* (4). (*a*) The force-extension curve shows a discontinuity at ~15 pN, due to the hairpin unfolding. The transition is occurring reversibly, as evidenced by the perfect overlap between the stretching and relaxing curves. (*b*) If the molecule is maintained at a constant force near the transition, it "hops" between the folded and unfolded states. Increasing the force through the transition, the hairpin ranges from being predominantly folded (13.6 pN, *bottom curve*), to being predominantly unfolded (15.2 pN, *top*). At the midway point, at a force of ~14.5 pN ($F_{1/2}$), the molecule spends half its time in the folded state and half in the unfolded state.

thus spend half its time in each state. From Equation 8, it is evident that $F_{1/2}$ provides a direct measure of the standard state free energy:

$$F_{1/2} \cdot \Delta x = \Delta G^0 + \Delta G_{stretch}^{A \to B}(F), \qquad 10.$$

where here we include the contribution of the free energy of stretching, $\Delta G_{stretch}$. Furthermore, by taking derivatives of Equation 10 with respect to temperature and inverse temperature (as in Equation 4), one can determine the entropy and enthalpy, respectively, of the reaction:

$$\frac{\partial(F_{1/2} \cdot \Delta x)}{\partial T} = -\Delta S^0 - \Delta S_{stretch}^{A \to B}(F),$$

$$\frac{\partial(F_{1/2} \cdot \Delta x/T)}{\partial(1/T)} = \Delta H^0 + \Delta H_{stretch}^{A \to B}(F). \qquad 11.$$

The effect of force on the free energy of a reaction is illustrated experimentally by the mechanical unfolding of a simple RNA hairpin (4, 5). The force-extension curve of the molecule (Figure 2*a*) shows a discontinuity at ~15 pN, where the length suddenly increases, due to the hairpin unfolding. Upon relaxation, the molecule follows the same force-extension curve, indicating that the unfolding process is occurring reversibly. From the area under the transition, Equation 3 can be used to calculate the free energy of unfolding the molecule mechanically, $\Delta G^0 + \Delta G_{stretch}$. By correcting this value for $\Delta G_{stretch}$, the standard free energy of unfolding was determined, giving $\Delta G^0 = 113 \pm 30$ kJ/mol, a value that agrees well with the predicted free energy value of 147 kJ/mol for unfolding in solution (4).

Effect of Force on the Kinetics of a Reaction

The effect of a mechanical force on the kinetics of a reaction was first described by Bell in 1978, in the context of cellular adhesion (6). Here we apply the concept of a tilted free energy surface to describe the force dependence of kinetic rates. For reactions occurring in solution where inertial forces are negligible, the theory of Kramers (7) gives the rates of transitions between states A and B as

$$k_{A \rightarrow B} = \frac{\omega_A \omega^{\ddagger}}{2\pi\gamma/m} e^{-\Delta G^{0\ddagger}/k_B T}, \quad k_{B \rightarrow A} = \frac{\omega_B \omega^{\ddagger}}{2\pi\gamma/m} e^{-(\Delta G^0 + \Delta G^{0\ddagger})/k_B T}. \qquad 12.$$

The rates depend exponentially on the activation free energies (differences between free energies of the transition state $\ddagger$ and states A and B, for the forward and reverse reactions, respectively). The pre-exponential factors in Equation 12 are related to the rates at which the molecule diffuses[1] to the transition state from states A and B, respectively; they depend on the characteristic frequency ω_A (ω_B) of the harmonic potential well at state A (B), which sets the rate at which the molecule "attempts" to overcome the barrier, $\omega^{\ddagger}$, which sets the rate of passage over the transition state once it has been reached, and the ratio of the friction coefficient experienced by the molecule to its mass, γ/m, which is the damping rate. The characteristic frequencies depend on the curvature of the free energy surface at each state ($\omega_A^2 = G''(x_A)/m$, etc.). Because force "tilts" the free energy along the mechanical reaction coordinate, if an external force assists the forward transition A→B, then the transition state free energy relative to that of state A is lowered by an amount $F\Delta x^{\ddagger}_{A \rightarrow B}$, where $\Delta x^{\ddagger}_{A \rightarrow B} = x^{\ddagger} - x_A$ (see Figure 1). Conversely, the free energy difference between state B and the transition state $\ddagger$ is increased by $F\Delta x^{\ddagger}_{B \rightarrow A} = F(\Delta x_{A \rightarrow B} - \Delta x^{\ddagger}_{A \rightarrow B}) = F(x_B - x^{\ddagger})$. As a result, the forward and reverse rates are modified exponentially by the external force:

$$k_{A \rightarrow B}(F) \sim e^{-\left(\Delta G^{0\ddagger} - F\Delta x^{\ddagger}\right)/k_B T}, \quad k_{B \rightarrow A}(F) \sim e^{-\left(\Delta G^{0\ddagger} - \Delta G^0 + F\left(\Delta x - \Delta x^{\ddagger}\right)\right)/k_B T} \qquad 13.$$

Here, we have again assumed that the locations of A, B, and the transition state are force independent, which as discussed above is not true in general. Positive force shifts the positions of minima to longer extensions, while the positions of

---

[1]Global protein displacements, such as the movement of a molecular motor along its track (discussed below), are expected to occur in the overdamped limit and are well described by diffusion. In cases where conformational changes are underdamped, the prefactors in Equation 12 are modified and do not depend on the damping rate (7). In the Eyring model (8), applicable when covalent bonds are made or broken, the prefactor corresponds to a single quantum mechanical vibrational frequency of the molecule. In all of these models, however, the exponential dependence of the transition rate on the barrier free energy is maintained.

maxima are shifted to shorter extensions. Thus, the distance changes, $\Delta x^{\ddagger}_{A \to B}$ and $\Delta x^{\ddagger}_{B \to A}$, that affect the rates are altered with applied force, no matter what the local curvature of the free energy surface. When states A, B, and $\ddagger$ have steep curvature, these shifts are negligible. For molecular unfolding, this is not the case, as is discussed in the following section.

Returning to our example of the RNA hairpin, we illustrate what can be learned by studying the force-dependent kinetics of the folding-unfolding reaction. Remarkably, by holding the force constant at a value near 15 pN, the hairpin is seen to "hop" between its folded and unfolded states (Figure 2b). Thus, this experiment makes it possible to follow the reversible unfolding of a single molecule in real time. No intermediates are observed and the reaction can be treated as a cooperative, two-state process. The distributions of dwell times in the folded state and in the unfolded state give the unfolding and refolding rate coefficients, respectively, whose force dependence can be determined. As shown in Equation 13, the force dependence of the rate coefficient for unfolding gives the distance from the folded to the transition state $\Delta x^{\ddagger}_{f \to u}$ along the mechanical reaction coordinate:

$$k_{f \to u}(F) = k^0_{f \to u} \; exp \frac{F \Delta x^{\ddagger}_{f \to u}}{k_B T} . \qquad 14.$$

where $k_{f \to u}{}^0$ is the unfolding rate constant along this pathway at zero force. For this molecule, the transition state is found to be midway between the folded and unfolded states ($\Delta x^{\ddagger}_{f \to u} = 11.9$ nm and $\Delta x^{\ddagger}_{u \to f} = 11.5$ nm).

The equilibrium constant for the $f \rightleftharpoons u$ reaction and its force dependence can also be determined from the ratio of dwell times of the molecule in the unfolded and folded states at any given force:

$$K_{eq}(F) = \tau_u(F) / \tau_f(F) . \qquad 15.$$

As shown in Figure 2b, the hopping of the hairpin from the folded to the unfolded state depends on the force. By determining the probability of populating the folded and unfolded states as a function of force, the midpoint of the transition is found to occur at $F_{1/2} = 14.5$ pN. At this force, the molecule spends half its time folded and half unfolded, and $K_{eq} = 1$. From Equation 10, a value of 149 ± 16 kJ/mol is determined for the free energy of unfolding, ΔG^0. $K_{eq}(F)$ is found to depend exponentially on applied force, as predicted by Equation 7 (4). From the force dependence of the equilibrium constant, the distance $\Delta x_{f \to u}$ between the folded and unfolded states is calculated to be 23 ± 4 nm, which agrees well with the expected length increase upon opening a 22-bp hairpin. By extrapolating $K_{eq}(F)$ to zero force, ΔG^0 can also be calculated. The value obtained by this method, 156 ± 8 kJ/mol, agrees well with that found from the area under the force-extension curve and from $F_{1/2}$. In general, because the kinetics of a reaction is pathway-dependent, extrapolation of kinetically determined parameters to zero force can give misleading results. We return to this point in the following section.

MECHANICAL UNFOLDING

Introduction

Many biological molecules have a defined mechanical function. For these molecules, their resistance to unfolding in response to an applied mechanical force—their mechanical stability—is of critical physiological importance. For example, titin is the protein responsible for passive elasticity in the skeletal and cardiac muscle sarcomere, where it functions as a molecular spring and ensures the return of the sarcomere to its initial dimensions after muscle relaxation (9–11). Fibronectin and tenascin are components of the extracellular matrix, where they extend and contract to facilitate, for example, cell migration and adhesion (12, 13). Regulation of the latter is thought to be controlled by a force-dependent recognition site in fibronectin (14). These mechanical proteins have in common a tandem arrangement of β-barrel domains, linked by segments of unstructured polypeptides. In contrast, spectrin is an α-helical protein that plays a central role in the mechanical properties of erythrocytes, which must deform and squeeze through narrow blood vessels during flow (37).

Whereas the examples above demonstrate the importance of a molecule's mechanical stability to its own function, many cellular processes involve the unfolding of macromolecules. For instance, secondary and higher-order nucleic acid structures have to be disrupted to permit translocation by RNA and DNA polymerases, the ribosome, and DNA and RNA helicases. An increasing body of evidence indicates that these machine-like molecules exert mechanical force on their substrates to perform their cellular functions. Similarly, examples of protein "unfoldases" include the import machinery of organelles (15), proteasomes (15, 16), and chaperonins (17–18), all of which use chemical energy from ATP to actively unfold (or fold) proteins. Many of these unfolding processes are likely to be mechanical in nature, although their direct characterization is only now becoming experimentally possible.

While direct measurements of forces in vivo during these types of cellular processes await future technical developments, much can be learned by studying well-defined model systems in vitro and characterizing their responses to force. The development of techniques for manipulating and exerting force on individual molecules enables us to define, for the first time, the conditions under which a molecule unfolds in response to an applied mechanical force. These early studies have revealed a broad distribution of mechanical stabilities among macromolecules, and have found that a molecule's *mechanical* stability cannot be predicted from its thermodynamic stability. Thus, mechanical stability is a property not directly accessible through bulk experiments, and must be determined by direct mechanical measurements. Here, we discuss the mechanical unfolding of proteins and nucleic acids, and how parameters such as the magnitude, direction, and time-dependence of the applied force affect the mechanical stability of these macromolecules.

Irreversibility in Mechanical Unfolding Experiments

When the mechanical unfolding of a macromolecule occurs at equilibrium, it is possible to determine directly the free energy, equilibrium constant and kinetics of the reaction and their dependence on force. We discussed in the previous section how this information can be obtained from reversible, coincident folding and unfolding curves for the example of the mechanical unfolding of a simple RNA hairpin. When the extension and relaxation curves do not overlap, folding/unfolding transitions are not occurring reversibly. From the second law of thermodynamics, the average work done to mechanically unfold the molecule is greater than the free energy of unfolding:

$$W_{irrev} > \Delta G. \qquad \qquad 16.$$

Mechanical unfolding under these conditions is less than 100% efficient because not all of the mechanical work put in to unfold the molecule is converted to a change in the free energy of the molecule. However, for a two-state system, it has recently been demonstrated that it is possible to recover the free energy of unfolding even when the reaction is not occurring at equilibrium (19–21). This result takes advantage of the ability of single-molecule experiments to provide the distribution of unfolding forces (and hence, work done), rather than just the mean value.

More often than not in mechanical unfolding experiments, hysteresis is observed between the extension and relaxation curves, indicating that the molecule is being extended or relaxed at a rate faster than its rate of equilibration. For the molecule to equilibrate during stretching or relaxation, the total change in force applied to the molecule during its slowest relaxation time τ must be less than the root-mean-square force fluctuations it would experience at equilibrium, i.e., $r\tau \leq \Delta F_{rms}$, where the loading rate $r \equiv dF/dt$ (pN/sec) (1). This is the requirement for quasistatic equilibrium during stretching or relaxation. Although most single-molecule mechanical unfolding experiments are performed under nonequilibrium conditions, the observed unfolding force distribution can provide useful information about the free energy surface, such as the position of the transition state. The observed unfolding force distribution is peaked, and the most probable unfolding force F_u^* increases with loading rate as (1)

$$F_u^* = \frac{k_B T}{\Delta x_{f \to u}^{\ddagger}} \ln\left(\frac{r \Delta x_{f \to u}^{\ddagger}}{k_{f \to u}^0 k_B T}\right). \qquad \qquad 17.$$

This maximum arises from the convolution of two competing trends: the probability that a domain remains folded decreases with time (and hence with force, since typically $F = In(r)$, whereas the probability of unfolding increases with force (22). The slope of a plot of $F_u{}^*$ versus $1n(r)$ yields $\Delta x^{\ddagger}_{f \to u}$, whereas the intercept gives the unfolding rate along the mechanical reaction coordinate at zero force $k^0_{f \to u}$.

Why the unfolding force depends on loading rate can be understood by considering the rate of energy input that drives molecular unfolding. Recalling that $k^0_{f \to u} = A \exp(-\Delta G^{\ddagger}/k_B T)$ (Equation 12, where here we denote with A the exponential prefactor), we can rewrite Equation 17 as

$$F_u^* = \frac{\Delta G^{\ddagger}}{\Delta x^{\ddagger}_{f \to u}} + \frac{k_B T}{\Delta x^{\ddagger}_{f \to u}} ln\left(\frac{r \Delta x^{\ddagger}_{f \to u}}{A k_B T}\right).$$

18.

The second term vanishes whenever

$$r \Delta x^{\ddagger}_{f \to u} = A k_B T.$$

19.

The term on the left represents the rate of energy delivery into the system from the pulling process; the term on the right represents the rate of energy exchange with the surrounding thermal bath. Under balanced energy exchange conditions, where these are equal, the unfolding force is equal to the ratio of activation energy to the distance to the barrier. If $r \Delta x^{\ddagger}_{f \to u} \ll A k_B T$, by contrast, F_u^* $< \Delta G^{\ddagger}/\Delta x^{\ddagger}_{f \to u}$ and the process is largely thermally activated. In most unfolding experiments, however, we are far from this limit, and $r \gg A k_B T/\Delta x^{\ddagger}_{f \to u}$, energy is put into the system faster than it can be dissipated, and $F_u^* > \Delta G^{\ddagger}/\Delta x^{\ddagger}_{f \to u}$.

Although pulling at a fixed loading rate is experimentally possible (23), most unfolding experiments instead have stretched the molecule at a constant speed. Because the stiffness of the molecule depends on the applied force, the loading rate varies as the molecule is stretched and Equation 17 cannot be used directly (22, 24). Instead, values of $k^0_{f \to u}$ and $\Delta x^{\ddagger}_{f \to u}$ are typically determined with the help of Monte Carlo simulations, which mimic the stochastic nature of thermally driven unfolding for a molecule stretched at a constant rate. Values for $k^0_{f \to u}$ are less well determined than those of $\Delta x^{\ddagger}_{f \to u}$ because the former depend exponentially on $\Delta x^{\ddagger}_{f \to u}$ (25, 26). Representative values for F_u (at a particular pulling speed), $k^0_{f \to u}$ and $\Delta x^{\ddagger}_{f \to u}$ are listed in Table 1 for various proteins.

The Unfolding Pathway

Because the location of the transition state along the mechanical reaction coordinate can be determined in these experiments, it should be possible to test for the presence of intermediates along the reaction pathway. In a two-state system, the sum of the distances to the unfolding and refolding transition state should equal the distance along the unfolding reaction coordinate between native and denatured states:

$$\Delta x^{\ddagger}_{f \to u} + \Delta x^{\ddagger}_{u \to f} = \Delta x_{f \to u}.$$

20.

TABLE 1 Mechanical properties of selected molecules

Molecule	ΔG^0 (kcal/mol)[a]	F_u (pN)[b]	$\Delta X^{\ddagger}_{F \to u}$ (nm)	$k^0_{F \to u}$ (s^{-1})	Structure at "breakpoint"	Reference
dsDNA	1.5–3 (per base pair)	9–20[c]	—	—	DNA base pairs	(110, 111)
P5ab hairpin	37.5 ± 4.8	14.5 ± 0.4[d]	11.9	—	RNA base pairs	(4)
P5abcΔA three-helix junction − Mg^{2+}	34.4 ± 4.8	11.4 ± 0.5[d]	~12	—	RNA base pairs	(4)
P5abc three-helix junction + Mg^{2+}	36.3[e]	19 ± 3 (at 3 pN/s with optical tweezers)	1.6 ± 0.1	2 × 10^{-4}	Hairpin-bulge	(4)
127 domain from titin (unfolding from native state)	7.6	~100	0.59	1 × 10^{-6}	Parallel β-sheet shear	(27)
127 domain from titin (unfolding from intermediate)	—	204 ± 26	0.25	3.3 × 10^{-4}	Parallel β-sheet shear	(112)
128 domain from titin	3.0	257 ± 27	0.25	2.8 × 10^{-5} or 1.9 × 10^{-6}	Parallel β-sheet shear	(23, 46)
^{10}FNIII domain from fibronectin	—	74 ± 20	0.38	2.0 × 10^{-2}	Antiparallel β-sheet shear	(14, 35)
C2A domain	—	60	—	—	β-sheet zipper	(42)
Spectrin	4.8 ± 0.5[f]	25–35 (at 300 nm/s)	1.7 ± 0.5	3.3 × 10^{-5}	α-helix bundle	(37)
Ubiquitin (N-C linked)	6.7[g]	203 ± 35 (at 400 nm/s)	0.25	4 × 10^{-4}	β-sheet shear	(38)
Ubiquitin (K48-C linked)	6.7[g]	85 ± 20 (at 300 nm/s)	0.63	4 × 10^{-4}	β-sheet shear	(38)

[a]From solution measurements, where available. Note that thermodynamic stability ΔG^0 is not a mechanical property, nor does it correlate with mechanical stability, F_u.

[b]All unfolding forces are quoted for pulling speeds in the AFM of 600 nm/s, unless otherwise noted.

[c]Forces obtained vary from 9 pN for AT base pairs to 20 pN for GC base pairs.

[d]F_u quoted for these secondary structures represents $F_{1/2}$, the force at which the molecule was equally likely to populate the folded and unfolded states. These experiments were done in the optical tweezers under equilibrium, constant force conditions.

[e]Theoretical prediction from Reference 4.

[f]For the α16 domain at 25°C (113).

[g]Reference 114.

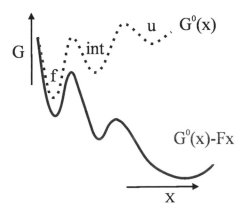

Figure 3 The effect of force on a three-state system, where f, int, and u represent the folded, intermediate and unfolded states of the molecule, respectively. Note how the rate-limiting barrier changes with applied force.

See also Figure 1. The P5ab RNA hairpin (discussed in the previous section), which exhibits no folding intermediates, is described reasonably well by this relation ($\Delta x^{\ddagger}_{f \to u} + \Delta x^{\ddagger}_{u \to f} = 11.5 + 11.9$ nm $= 23.4$ nm; $\Delta x_{f \to u} = 23$ nm). By contrast, a finding of $\Delta x^{\ddagger}_{f \to u} + \Delta x^{\ddagger}_{u \to f} < \Delta x_{f \to u}$ may indicate the presence of intermediates along the reaction pathway. Are there other conditions under which the equality of Equation 20 may not hold? We consider the assumptions that go into determining the distances to the transition state. Equations 13 and 17 assume that the distance to the unfolding transition state from the folded state, $\Delta x^{\ddagger}_{f \to u}$, is independent of force. Because this distance is in general so short and the well and barrier are relatively steep, this is a reasonable assumption. By contrast, the separation between the denatured state and the barrier to refolding, $\Delta x^{\ddagger}_{u \to f}$, is not independent of force, and using the force-dependent refolding kinetics (Equation 13) to determine a fixed value of $\Delta x^{\ddagger}_{u \to f}$ is incorrect (4). This is because the location of the free energy minimum for the unfolded state shifts considerably with force, as described by the worm-like chain equation (2, 3). The distances obtained for P5ab show reasonable agreement with Equation 20 because they were determined over a small range of forces: From 13 to 16 pN, the end-to-end distance of the unfolded RNA chain changes by only 5%, so the distance between the unfolded state minimum and the transition state can be treated as fixed within this force range. Over larger ranges of force, however, this distance cannot be treated as constant.

For reactions that possess intermediates (non-two-state), mechanical unfolding occurs along a complex free energy surface with multiple energetic barriers and may exhibit different rate-limiting transitions in different ranges of force (27, 28). Figure 3 illustrates how force can affect the relative height of barriers along a three-state unfolding reaction coordinate. The energies and locations of the transition state barriers and of the intermediate structure strongly influence the

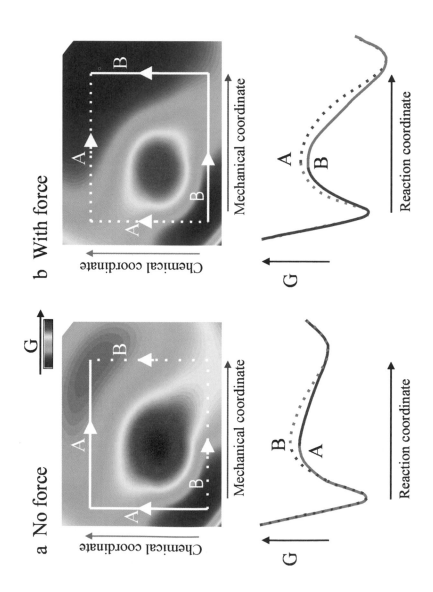

unfolding pathway of a molecule under force, since force reduces the free energy to a greater extent at positions further along the mechanical reaction coordinate than at positions closer to the folded structure. By mechanically unfolding mutant and wild-type I27 domains, and carefully analyzing the unfolding kinetics within the constraints of a three-state system, Clarke and coworkers identified distinct unfolding pathways in three different force ranges (27, 29). At high forces, an intermediate state is rapidly attained and unfolding occurs from this state; at lower forces, unfolding occurs directly from the native state; and at forces below ~43 pN, unfolding follows a distinct "solution" pathway that has a different (lower-energy) transition state than the mechanical transition state. This study demonstrates the difficulties inherent in extrapolating unfolding rates to zero force.

How are the results of mechanical unfolding experiments related to those of bulk unfolding experiments? Thermodynamic properties such as the free energy of unfolding are state functions and depend only on the initial and final states of a process; thus, comparisons between single-molecule mechanical studies and bulk biochemical assays should give identical results (after correcting for the entropic contribution of tethered ends). The kinetics of a reaction is, however, pathway dependent. Because single-molecule unfolding experiments impose a reaction coordinate different from that of bulk experiments, rates of unfolding obtained with these two different approaches will generally differ. Figure 4 illustrates the effect of force on a two-dimensional free energy surface, where one of the axes represents the mechanical reaction coordinate and the other an orthogonal "chemical" coordinate. From this simple depiction, it is clear how the unfolding pathway can change with force: an applied force lowers the high-energy barrier located far along the mechanical reaction coordinate. Above a given force, this barrier becomes lower than the rate-limiting barrier at zero force (which is not located along the mechanical coordinate and is unaffected by force), creating a more energetically favorable trajectory. Thus, under the influence of mechanical force, the unfolding reaction may follow an entirely different trajectory than it does when free in solution (5, 27, 30), as the barrier to unfolding along the mechanical pathway becomes lower in free energy than the chemical barrier.

Figure 4 Force can affect the reaction pathway along a two-dimensional free energy surface. Two paths (*A* and *B*) are shown connecting the folded state minimum (*lower left*) to the unfolded state minimum (*upper right*) of the free energy contour plot; the preferred path is indicated by the solid line. The lower panels illustrate the free energy along these paths, where blue sections occur along the chemical coordinate and those in red occur along the mechanical coordinate. (*a*) No applied force: the reaction following pathway *A* encounters a lower activation barrier than the reaction following pathway *B*. (*b*) An applied force tilts the free energy surface along the mechanical coordinate, lowering the barrier along pathway *B* below the barrier along pathway *A*.

Relating Mechanical Stability to Local Molecular Structure

The mechanical unfolding pathways of RNA and proteins are fundamentally different. Complex RNA structures unfold in a hierarchical manner, with stable secondary domains remaining after tertiary interactions have ruptured. This hierarchical behavior is rooted in the separability between the energy contributions from the secondary interactions of each individual subdomain and the contribution of its tertiary contacts (4, 31, 32). In other words, secondary and tertiary interaction energies are independent and additive, and coupling energy terms can be neglected, to a first approximation:

$$\Delta G = \Delta G_{2^\circ} + \Delta G_{3^\circ} \, (+\Delta G_{coupling}) \qquad\qquad 21.$$

In agreement with this analysis, single-molecule mechanical unfolding experiments have shown that a given secondary RNA domain exhibits the same distribution of unfolding forces and lengths in isolation as when it is involved in tertiary interactions in a larger molecule, meaning that kinetic barriers associated with isolated structural subdomains are maintained within the larger molecule (32). Thus, by pulling on progressively larger pieces of the *Tetrahymena thermophila* L21 ribozyme, it was possible to map completely the unfolding pathway of the molecule (32). Because tertiary structures have shorter distances to their transition states than secondary structures, tertiary interactions tend to be brittle (they break at high forces after small deformation), whereas secondary interactions are compliant (they break at low forces and after large deformations) (4). In addition, tertiary interactions equilibrate over a slower timescale than secondary interactions and their presence is more often than not accompanied by hysteresis in the pulling-relaxation cycle. By pulling the ribozyme many times, it was possible to identify the alternative unfolding pathways of the molecule and their relative probability of occurrence (32). The approach described above suggests that by characterizing the interaction energies of the various tertiary motifs (helix-helix, kissing loops, loop-helix, etc.) it should be possible to develop an *Aufbau* algorithm to solve the RNA folding problem, i.e., to predict the tertiary structures of RNA molecules from their sequence using a semiempirical approach (31): First the most probable secondary structure is predicted from the sequence, and this structure is then used to predict its most probable tertiary fold.

In contrast to RNA, proteins appear to unfold in a highly cooperative manner, and secondary structures are not stable independently of their tertiary context. Mechanical unfolding intermediates have been identified for a few proteins (33–36), in which most intermediate structures involve little disruption of the core domain of the protein, and instead involve peeling off a single external β-strand from a large β-barrel or β-sandwich structure. In the immunoglobin I27 domain, for example, low forces break two hydrogen bonds between outlying antiparallel β-strands, while the core of the protein domain remains folded until higher forces are attained (27, 33). The cooperativity of protein unfolding most likely arises from the high degree of connectivity involved in the interactions that maintain protein structure.

As seen in Table 1, mechanical stability (given by the most probable unfolding force, F_u*) is not correlated with thermodynamic stability (ΔG), nor can it be predicted from the melting temperature ($T_m = \Delta H/\Delta S$) (35, 37–39). Instead, as predicted by Equation 17, the mechanical stability of the molecule depends on the location of the transition state along a specific mechanical reaction coordinate, the height of the barrier (which contributes to k_u^0), and the loading rate. As discussed earlier, force affects most strongly positions furthest from the folded, native state: To mechanically eliminate a $10\text{-}k_BT$ barrier located 0.3 nm from the folded state requires ~ 140 pN of force (when the molecule is stretched under conditions of quasi-static equilibrium), whereas to eliminate a barrier of the same height but located 1.5 nm from the folded state along the mechanical reaction coordinate requires only 28 pN under the same conditions. Many mechanical proteins consist of tandem arrays of domains with different transition state locations and hence differing mechanical stability. The more compliant domains (having longer distances to their transition states) stretch in response to low applied forces, while the more brittle domains (with shorter distances to their transition states) maintain their structure to higher applied forces and only under extreme force conditions would they unfold (35, 36).

The molecular structure provides physical insight into the location of the transition state barrier along the mechanical reaction coordinate. Proteins that contain β-sheets tend to unfold at much higher forces and exhibit less compliance than proteins that are predominantly α-helical (see Table 1) (40). Gaub and coworkers have suggested that this trend can be explained in broad terms by the difference in forces maintaining tertiary structure (37): β-barrel domains are held together by intrasheet hydrogen bonds, which are short-range interactions with short distances to their transition state, whereas tertiary interactions maintaining α-helical bundles are hydrophobic, are more delocalized, and have associated longer distances to their transition state. However, due to the local action of applied mechanical force, the unfolding force is most strongly influenced not by global domain structure but by the local structure near the mechanical "breakpoint."

The strongest protein domain examined thus far, the I27 immunoglobin domain from titin, has a β-sandwich structure. Structural considerations, molecular dynamics simulations, and mutational analysis have been combined to demonstrate that the mechanical breakpoint of this domain at the strongest applied forces consists of a cluster of six hydrogen bonds between parallel β-strands experiencing shear forces (33, 41). Not only is the energy barrier for breaking many bonds in a concerted fashion much greater than the individual energy barriers for breaking individual bonds sequentially, but the stiffness of the connection between the two strands is also greater for shearing six parallel hydrogen bonds than for breaking a single bond (or six bonds in series). The higher transition-state energy barrier requires greater forces to lower it; the increased stiffness implies steeper walls about the native state free energy minimum, which result in a shorter distance to the transition state for the concerted versus sequential cleavage of six hydrogen bonds. Similarly short distances to the transition state were found for other domains experiencing shearing forces between clusters of hydrogen bonds (Table 1). By contrast, sequential breakage of single hydrogen bonds, occurring when β-strands of a protein are "unzipped" or when an RNA hairpin is mechanically unfolded, requires lower energy, and because the potential energy well is shallower, the transition state is located at a longer extension. The result is that sequential cleavage of multiple bonds requires less energy and applied force than concerted cleavage of the same set of bonds. Accordingly, while the structure at the mechanical breakpoint is important for mechanical stability, the direction of applied force can significantly change the force at which the breakpoint ruptures (38–40, 42, 43).

Extending Single-Molecule Mechanical Properties to the Cellular Level

The results of mechanical experiments on individual domains of titin, on single molecules of the entire protein, and on the muscle sarcomere illustrate the convergence, and in some cases complementarity, of information obtained by studying the mechanical properties of proteins at an increasing level of complexity. The similarity of the force-extension curves of single molecules of titin and of the sarcomere (when extrapolated to the single-molecule level) confirm that single-molecule experiments on titin reproduce the physiological elastic response of muscle (36, 44, 45). Muscle elasticity can be described at a still more fundamental level: By determining the mechanical unfolding properties of its individual domains, Fernandez and coworkers have reconstructed the force-extension curve of the complete titin molecule (36). They attribute muscle elasticity at low force (< 4 pN) to the compliance of titin's PEVK and N2B regions, while the inclusion of the more brittle Ig domains in the tandem array provides titin the "mechanical buffer" necessary to react to potentially damaging higher forces or to forces applied for long periods of time (36).

The ability to predict, from studies of individual protein domains, the mechanical behavior of the entire molecule suggests that, to a first approximation, the domains of titin can be treated as independent structural entities. This result appears to be true for most domains of tandem proteins, in which the mechanical unfolding properties of individual domains are independent of their neighbors. Some domains, however, exhibit different unfolding behavior depending on their context within a tandem array (35, 46, 47). Although the characterization of individual domains provides valuable information on the mechanical properties of the parent protein, the mechanical behavior of the entire molecule must also be studied to determine the role of interdomain interactions and other sources of higher-order behavior. By repeatedly stretching both individual titin molecules and the muscle sarcomere, for example, Kellermayer et al. observed mechanical fatigue, where the system became more compliant and extended to greater lengths on consecutive pulls (45). This behavior has not been observed in studies of individual domains of titin, and Kellermayer et al. attribute it to nonspecific intrachain crosslinks, suggesting that the physiological significance of fatigue may be to dynamically modulate the mechanical properties of the muscle in response to its recent mechanical history (45). These studies on the sarcomere, titin, and its constituent domains demonstrate how careful experiments on systems of varying degrees of complexity can provide insight into the physiological mechanical behavior of proteins.

MOLECULAR MOTORS

Introduction

To carry out processes as diverse as cell movement, organelle transport, ion gradient generation, molecular transport across membranes, protein folding and unfolding, and others, cells possess molecular structures that behave as tiny machine-like devices. These molecular machines must use external energy sources to drive directed motion and operate as molecular motors, converting chemical energy into mechanical work.

The myosin, kinesin, and dynein families of molecular motors, whose best-characterized members are most closely associated with muscle contraction, organelle transport, and ciliary beating, respectively, use ATP hydrolysis as a source of energy to step along a track—actin filaments, in the case of myosin, or microtubules, for kinesin and dynein. In replication, transcription, and translation, molecular motors must utilize part of the chemical energy derived from the polymerization reaction to move along the DNA or RNA in a unidirectional manner, generating forces and torques against hydrodynamic drag and/or mechanical roadblocks (48, 49). Many other enzymes that bind to and act on DNA or RNA are molecular motors. Helicases hydrolyze ATP to translocate along DNA, unwinding it into its complementary strands (50, 51). Type II topoisomerases exert forces to pass DNA duplexes through double-strand breaks in DNA and to reseal these breaks (52–54). Many protein translocases are also likely to operate as molecular motors using ATP hydrolysis to mechanically pull polypeptide chains across membranes (15). Finally, during the replication cycle of many dsDNA bacteriophages and viruses, a molecular motor must package the DNA of the virus into newly self-assembled capsids against considerable entropic, electrostatic, and elastic forces (55). Motors not only move along linear tracks (microtubules, polymerized actin, DNA, RNA, or polypeptide chain) exerting forces, but can also operate in a rotary fashion, generating torque. F_1F_0 ATP synthase and the motor at the base of the prokaryotic flagellum utilize the electrochemical energy of a transmembrane proton gradient to generate torque and synthesize ATP, and to propel the bacterium, respectively (56, 57).

Many approaches have been used to study the mechanisms of molecular motors. Structural studies provide detailed information on the various conformations of the motor, but these pictures are static. Biochemical assays of motility and kinetics provide a more dynamic picture, yet they involve averages over large numbers of molecules. Because molecular populations are often heterogeneous, the ensemble-averaged properties measured in these studies may not be representative of individual molecules. The recent development of single-molecule techniques has made it possible to probe the individual molecules that make up the ensemble and, for the first time, allow direct determination of intrinsic molecular motor properties such as efficiency, stall force, and motor step size. In the following section, we discuss in detail the significance of these properties.

Mechanical Properties of Molecular Motors

EFFICIENCY AND COUPLING CONSTANT   All molecular motor reactions are exergonic ($\Delta G < 0$), meaning that they occur energetically "downhill." The free energy ΔG that drives the mechanical motion is derived from chemical sources, for instance, the energy of ATP hydrolysis, the electrochemical energy from a transmembrane ion gradient, or the polymerization energy derived from bond breaking and forming in DNA, RNA, or polypeptides. As in the case for mechanical unfolding, it is possible to define an efficiency—a ratio of output work to input energy—of a molecular motor. The thermodynamic efficiency η_{TD} of a motor is defined as the ratio per step between the work done by the motor against a conservative external force[2] and the free energy associated with the reaction that powers the motor, ΔG: $\eta_{TD} = F\delta/\Delta G$. Here, F is the external mechanical force and δ is the step size of the motor (see below for more on step size). The thermodynamic efficiency must be less than unity, since the motor cannot, on average, do more work than the free energy supply ΔG it is given. Furthermore, since η_{TD} increases with the force, it follows that the motor attains its highest efficiency at the maximum force against which it can work. As discussed below, at this force the motor stalls.

Table 2 lists the thermodynamic efficiencies of a few molecular motors at stall.[2] The values vary between 15% to 100%. The observed variation in molecular motor efficiency may reflect the large uncertainties still associated with the determination of the stall force and the step size using methods of single-molecule manipulation (see below). An observed mechanical efficiency <100% suggests that part of the energy of the reaction is either dissipated as heat or utilized to perform work along a reaction coordinate orthogonal to the direction of the applied mechanical load. RNA polymerases, for example, may

---

[2]A distinction must be made between the case in which the motor works against a conservative force and a dissipative viscous force. The efficiency of the F_1-ATPase (70) was determined from experiments in which the rotary motor operated against the viscous drag of a long actin filament. Here, a different efficiency from that described above must be used: the Stokes efficiency (121).

TABLE 2 Mechanochemical properties of a few molecular motors

Motor	Average speed (nm/s)	Average stall force (pN)	Step size (nm)	Efficiency[a]
F_1-ATPase[b]	4 rps[h]	40 pN·nm	120°	100%
RNA polymerase[c]	6.8	15–25	0.34	9–22%
DNA polymerase[d]	38	34	0.34	23%
Myosin II[e]	8000	3–5	5.3	12–40%
Kinesin[f]	840	7	8	40–60%
Phage ϕ29[g]	34	57	0.68	30%

[a]Calculated from $F_{stall}·\delta/\Delta G$ where δ is the (putative) step size, and ΔG is the free energy change for the reaction.
[b]References 70, 115.
[c]The step size of RNA polymerase has not been measured directly but is expected to be one base pair (64, 74).
[d]The step size of DNA polymerase has not been measured directly but is expected to be one base pair (49).
[e]References 66, 116–118.
[f]The stall force for kinesin can vary between 3–7 pN (63, 119, 120).
[g]The step size of ϕ29 has not been measured directly but is thought to be two base pairs (55, 72).
[h]Revolutions per second (rps).

have the ability to generate torque to overcome the torsional stress built up in the DNA molecule during transcription. The packaging motor in bacteriophage ϕ29 may use part of the energy to twist the DNA and facilitate its arrangement inside the capsid.

A related concept that quantifies the conversion of chemical energy into mechanical motion is the coupling constant ξ, defined as the probability that the motor takes a mechanical step per chemical reaction. If one step is taken per catalytic cycle ($\xi = 1$), the motor is tight coupled; if less than one step is taken ($\xi < 1$), the motor is loose coupled. Still other coupling mechanisms may exist in which many steps are taken per catalytic cycle ($\xi > 1$), so-called one-to-many coupling schemes.

There is no simple correspondence between the coupling ξ and efficiency η. A motor could be tight coupled but have a low efficiency, as appears to be the case for kinesin (58, 59). In principle, a loose-coupled motor could also have a high efficiency. In other words, a motor may step only once per several chemical reactions, but when it does, utilizes all of the chemical energy available to it.

STALL FORCE   The stall force of a molecular motor, F_{stall}, is the force at which the velocity of the motor reduces to zero, and thus it is equal to the maximum force that the motor itself can generate during its mechanical cycle. Stall forces vary greatly from motor to motor (Table 2), depending on the motor's speed of operation and step size, which are ultimately dictated by its biological function. For example, during transcription elongation, RNA polymerases must locally unwind the DNA template, work against torsionally constrained DNA, and possibly disrupt protein roadblocks such as nucleosomes that impede its translocation. To perform these tasks, *Escherichia coli* RNA polymerase generates forces up to 25 pN (48, 60), sufficient to mechanically unzip dsDNA ($\sim$15pN; see Table 1). The connector motor at the base of bacteriophage ϕ29's capsid, on the other hand, must pack the phage DNA in the capsid against the build-up of an internal pressure that reaches a value of nearly 6 megaPascals (MPa) at the end of packaging. To perform this task, the motor is capable of exerting forces as high as $\sim$60 pN (55).

What factors influence the stall force? The stall force is that at which the transition of a motor between states at sequential positions along its track occurs at equilibrium. In other words, the motor oscillates between these states (A and B) so that the net displacement is zero. Because A and B represent sequential positions along a periodic track, the local curvatures of the free energy surface at these positions are identical, in contrast to those of an unfolding molecule (see previous section). As a result, Equation 7 can be used to determine the external opposing force that stalls the motor (see 61):

$$F_{stall} = \frac{\Delta G^0}{\delta} + \frac{k_B T}{\delta} ln\frac{[B]}{[A]}. \qquad 22.$$

[A] and [B] are the populations of states A and B, respectively, ΔG^0 is the standard free energy of the reaction, and δ is the distance translocated along the track: the step size of the motor (see below). Thus, for given values of ΔG^0, [A] and [B], the smaller the step size, the larger the force required to stall the motor. Implicit in Equation 22 is the assumption that the motor is 100% efficient. Indeed, this expression represents the maximum stall force; if the motor utilizes only a fraction of the reaction free energy, Equation 22 should be scaled accordingly. Because the magnitude of the stall force depends on the relative populations of states A and B, albeit in a weak manner, it is important to specify the concentrations of products and reactants under which the stall force was determined. For a motor that hydrolyzes ATP, for instance, [B]/[A] in Equation 22 is replaced by [ADP][P_i]/[ATP]. A dependence of the stall force on product and reactant concentrations was observed for kinesin (62, 63), but not for *E. coli* RNA polymerase (48, 64).

When the population of state B is zero (equivalently, when the product concentration is zero), Equation 22 predicts an infinite stall force. Because the reaction is now irreversible, the motor can never step backward, even under a large force. The motor will wait until it experiences a thermal fluctuation large enough to carry it forward. As a result, the velocity becomes exponentially small at high forces, but will remain positive. In practice, however, there is little experimental distinction between an exponentially small and a zero velocity. Thus, many measurements of the stall force likely underestimate the true stall force (as well as the efficiency), as formally defined in Equation 22. Ultimately, at high enough forces, the motor or its track will deform, rendering it inactive.

If the reaction is reversible such that the stall force is finite, a motor can in principle be run backward, turning products into substrate, by applying sufficiently large forces. A possible example of this is ATP synthase in which the direction of rotation of the F_1 motor—by itself an ATPase—is thought to be reversed by the counter-rotating F_0 motor in order to drive ATP synthesis (65). The mechanical reversibility of the F_1 motor has recently been demonstrated in elegant experiments by Kinosita and coworkers (65a). They mechanically counter-rotated the F_1 motor and found that the hydrolysis reaction was also reversed, synthesizing ATP from ADP and P_i. The work required to reverse a molecular motor must clearly equal or exceed the free energy of the reaction. If a motor is 100% efficient, reversal occurs at forces just beyond the stall force as defined in Equation 22, infinitely slowly, but with 100% efficiency. At larger forces, the process is faster but less efficient.

STEP SIZE   The step size, δ, of a motor is defined as its net displacement during one catalytic cycle (see Table 2). Clearly, the step size is best determined from direct observations. Several groups have employed single-molecule techniques to observe individual steps of myosin II (66, 67) and kinesin (68) that compare well to the known periodicity of their tracks (5.5 nm for actin filaments, 8 nm for microtubules). Recent experiments (69) suggest that the 8-nm displacement of kinesin is comprised of two smaller substeps. However, hydrolysis of one ATP molecule drives the motor the entire 8-nm distance (58, 59).

As discussed above, the distance between two states can vary with force in a manner that depends on the local curvatures of the free energy surface. Because the motor's track is periodic, we expect sequential positions to have identical free energy curvatures, and thus, we expect the step size to be independent of force. This prediction has been confirmed in the case of kinesin (69) and F_1-ATPase (70, 71).

In many cases, the step size has not been observed directly. The difficulty lies in the extraordinary spatial resolution required to make such observations. Optical trap techniques do not currently have sensitivity sufficient to resolve subnanometer displacements. Nevertheless, thermodynamic arguments can be made to place an upper bound on the step size. Since the thermodynamic efficiency of a motor must be less than 100%, from the free energy of NTP binding and hydrolysis and the measured stall force a maximum allowable step size of $\sim$2 basepairs (1 bp = 3.4Å) can be estimated for RNA polymerase (48). For the packaging motor of bacteriophage ϕ29, arguments based on motor efficiency and single-molecule measurements of the maximum stall force ($\sim$70pN) suggest that the step size must be smaller than $\sim$5 base pairs (55). One biochemical bulk study that measured the amount of DNA packaged and ATP hydrolyzed suggests that the movement is $\sim$2 bps per ATP (72).

Mechanochemistry

The distinguishing feature of a molecular motor is the generation of force to produce the mechanical motion that accompanies the reaction. How motors

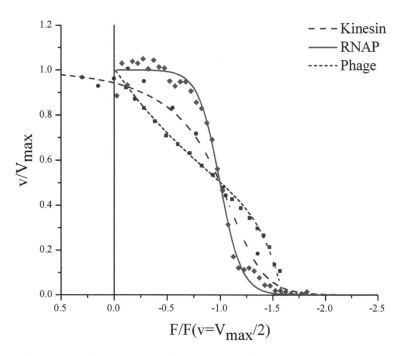

Figure 5   Force-velocity behavior for kinesin (73), RNA polymerase (48), and the bacteriophage ϕ29 packaging motor (55) under saturating conditions. Data were normalized in order to appear on the same graph. Velocities are scaled to their maximum values, and forces are scaled to those at which the velocities are half-maximal, $F(v=V_{max}/2)$. The different shapes of the force-velocity curves imply distinct mechanisms.

convert chemical energy into mechanical movement—the mechanochemistry of the motor—is discussed in this section. In a sense, force can be considered a product of the chemical reaction. Thus, an external force that opposes the motor can function as an inhibitor of the reaction, and one in the aiding direction can promote the reaction and act as an activator. As a result, the velocity of the motor often depends on the external force and does so in a manner dictated by the mechanism of motor operation. Force-velocity behavior can thus provide much insight into a motor's mechanochemical conversion. In general, the shape of the force-velocity relationship will vary depending on the conditions (concentration of reactants and products of the hydrolysis reaction) under which the motor is tested. The rate of a molecular motor will be force dependent if the conditions of the experiment are such that movement itself is the rate-determining step.

Figure 5 shows the normalized force-velocity behavior of three molecular motors—kinesin (73), phage ϕ29 DNA packaging motor (55), and *E. coli* RNA polymerase (48)—at saturating substrate concentrations. The range in behavior among motors is striking, and hints at the underlying differences in their respective mechanisms. The velocity of the $\phi 29$ packaging motor decreases linearly with force over practically the entire range of forces, suggesting that the rate-limiting step of the reaction, in the conditions in which the data were obtained, is DNA translocation, even at low forces. Kinesin exhibits similar behavior, with somewhat less force dependence at low forces than near stall. The velocity was practically independent of force when a constant force was applied assisting the movement of kinesin, suggesting that the rate-limiting step is not translocation for assisting forces (73). In marked contrast to those enzymes, the velocity of RNA polymerase is practically force independent, except near stall, indicating that the movement step is not the rate-limiting step in this case.

The force-velocity relationship of a molecular motor under various concentrations of substrate and its hydrolysis products provide quantitative information about the mechanochemical cycle of the motor. This cycle consists of catalytic steps that connect distinct chemical states and mechanical steps that connect different conformational states. A reaction pathway may typically include a substrate binding step, reaction steps, product release steps, and mechanical steps (which may coincide with chemical steps). Provided the reaction is irreversible (as when the product concentration is zero, for instance), the rate k_t at which the enzyme turns over is given by an expression with the general form of the Michaelis-Menten Equation:

$$k_t = \frac{k_{cat}[S]}{K_M + [S]},$$
23.

where $[S]$ is the substrate concentration, and where k_{cat}, the maximum rate in units of moles·sec^{-1}, and K_M, the Michaelis constant, are determined by the individual transition rates connecting the various states of the motor during its mechanochemical cycle (see below). The velocity of the motor is simply given by $v = \xi \delta \cdot k_t$, where δ is the step size of the motor, and ξ is the coupling constant (which we assume is 1 for the following discussion).

Force-velocity curves alone cannot reveal the location of the movement step in the cycle of the motor. As is illustrated below, it is necessary to determine the force dependence of the parameters of the Michaelis-Menten Equation, $V_{max} = \delta \cdot k_{cat}$ and K_M (61). Consider a general N-step kinetic scheme with two irreversible steps,

$$M_1 + S \underset{k_{\pm 1}}{\longleftrightarrow} M_2 \underset{k_{\pm 2}}{\longleftrightarrow} \ldots M_j \xrightarrow{k_j} M_{j+1} \underset{k_{\pm(j+1)}}{\longleftrightarrow} \ldots M_N \xrightarrow{k_N} M_1$$
24.

where step 1 is the substrate binding step. Because the cycle is irreversible, the velocity obeys the Michaelis-Menten Equation 23. At saturating substrate levels ($[S] \gg K_M$), the velocity $v \sim V_{max} = \delta \cdot k_{cat}$ is independent of substrate. At low concentrations ($[S] < K_M$), the velocity is limited by substrate binding: $v \sim V_{max}[S]/K_M = \delta \cdot k_{cat}/K_M[S]$, where k_{cat}/K_M is an effective second-order binding rate constant. It can be shown that for the scheme in Equation 24, V_{max} is a function of all rate constants $k_{\pm 2}, k_{\pm 3}, \ldots k_N$ except rate constants $k_{\pm 1}$

TABLE 3   Dependence of parameters from the Michaelis-Menten Equation on opposing force

Case	Force-dependent rates	V_{max}	k_{cat}/K_M	K_M
1	$k_{\pm 1}$	–	↓	↑
2	$k_{\pm(j+1)}, k_{\pm(j+2)}, \ldots, k_N$	↓	–	↓
3	$k_{\pm 2}, k_{\pm 3}, \ldots, k_j$	↓	↓	↓ or ↑

associated with the substrate binding step. On the other hand, k_{cat}/K_M depends on all of the rate constants that connect enzyme states reversibly to substrate binding (step 1), and on the forward rate for the first irreversible step that follows binding, k_j (i.e., $k_{\pm 1}$, $k_{\pm 2}$,... k_j).

An interesting consequence of this result is that the location of a force-dependent step in the cycle dictates how the parameters V_{max} and k_{cat}/K_M are affected by force (see Table 3). There are three possible cases depicted in Figure 6 for a simplified reaction cycle, which are discussed below.

THE MOVEMENT STEP COINCIDES WITH BINDING ($k_{\pm 1}$).   This case results in a force-dependent k_{cat}/K_M and a force-independent V_{max}. Here, an opposing force acts like a competitive inhibitor to the substrate, shifting the equilibrium toward the free enzyme state (M_1), reducing the effective binding rate constant k_{cat}/K_M. However, the addition of more substrate outcompetes this effect, shifting the equilibrium back to the substrate-bound state (M_2). As a result, the velocity at infinite substrate concentration, V_{max}, is unaffected by force. K_M increases with opposing force as more substrate is necessary to counteract the effect of force.

Figure 6   Schematic representation of the free energy surface of a molecular motor along a generalized reaction coordinate. Sections marked in blue occur along a chemical coordinate and are independent of force; those in red occur along the mechanical coordinate, and hence depend on force. We assume that the force-dependent step is rate-limiting in each case. For simplicity, we consider the minimal three-state kinetic cycle

$$M + S \xrightleftharpoons[]{k_{\pm 1}} M \cdot S \xrightarrow{k_2} M \cdot P \xrightarrow{k_3} M + P,$$

in which substrate binding is followed by two irreversible steps (depicted by two large free energy drops). The effective second-order binding rate k_{cat}/K_M is the rate at which the motor reaches the state M·P through the first two steps. Once in that state, the motor is committed to proceeding forward, so that binding cannot be affected by any subsequent transition. V_{max}, on the other hand, depends on the second and third transitions, which are independent of substrate concentration. (a) In case 1, the first step is force dependent; k_{cat}/K_M decreases with force, whereas V_{max} is force independent. (b) In case 2, the third transition is force dependent; V_{max} decreases with force, and k_{cat}/K_M is force independent. (c) In case 3, the second step is force dependent; both k_{cat}/K_M and V_{max} decrease with force.

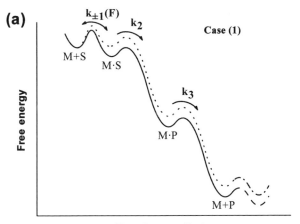

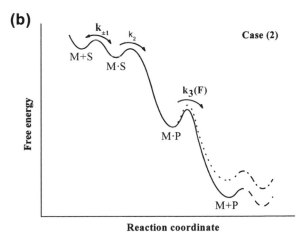

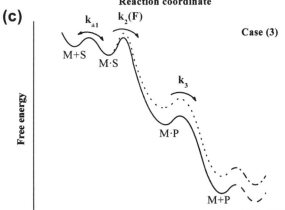

Recent experiments have shown that the force-velocity behavior of T7 RNA polymerase is consistent with a force-dependent rate-limiting step at subsaturating NTP concentrations (75). This motor exhibits a force-independent velocity V_{max} at forces below stall, while K_M increases with opposing force (75). Thus, T7 RNA polymerase appears to follow case 1. The authors have proposed a model for translocation in which the movement step occurs in equilibrium with NTP binding, and hydrolysis and incorporation of NTP serve to rectify translocation in a unidirectional manner.

THE MOVEMENT STEP FOLLOWS THE FIRST IRREVERSIBLE STEP AND PRECEDES THE NEXT BINDING EVENT ($k_{\pm(j+1)}, k_{\pm(j+2)}, \ldots k_N$).   In this case, the force dependence is reversed: V_{max} depends on force, k_{cat}/K_M does not. An opposing force affects the equilibrium between states not connected to substrate binding, like an uncompetitive inhibitor. As a result, when substrate binding is rate-limiting ($[S] < K_M$), the velocity ($v \sim \delta \cdot k_{cat}/K_M[S]$) does not depend on force. In the other extreme ($[S] \gg K_M$), the velocity ($v \sim V_{max}$) is reduced by an opposing force, provided the force-dependent step (translocation) is rate-limiting. Note that K_M must have the same force-dependence as V_{max} in this case to yield a force independent k_{cat}/K_M.

As discussed previously, the force-velocity behavior of the $\phi 29$ packaging motor at saturating ATP indicates that the rate-limiting step is force dependent. At limiting ATP concentrations, however, it was found that the velocity is practically independent of force except close to stall (Y.R. Chemla, A. Karunakaran, J. Michaelis & C. Bustamante, unpublished data). Measurements of V_{max} and K_M from the ATP dependence of the velocity at various forces show that both decrease with opposing force in the same manner, such that k_{cat}/K_M is force independent (Y.R. Chemla, A. Karunakaran, J. Michaelis & C. Bustamante, unpublished data). Thus, $\phi 29$ appears to follow case 2. A putative minimal kinetic scheme for this motor can then be proposed:

$$M_1 + ATP \underset{k_{\pm 1}}{\longleftrightarrow} M_2 \underset{k_2}{\longrightarrow} M_3 \underset{k_3(F)}{\longrightarrow} M_4 \underset{k_{\pm 4}}{\longleftrightarrow} M_5 \underset{k_{\pm 5}}{\longleftrightarrow} M_1.$$

The movement step (step 4) occurs after product release.

THE MOVEMENT STEP OCCURS AFTER BINDING IN ANY STEP UP TO AND INCLUDING THE FIRST IRREVERSIBLE STEP ($k_{\pm 2}, k_{\pm 3}, \ldots k_j$).   In this case, force affects the equilibrium between states indirectly connected to binding, so that both V_{max} and k_{cat}/K_M are force dependent. Here, an opposing force acts essentially as a mixed (noncompetitive) inhibitor. It favors the free enzyme state, reducing the effective binding rate k_{cat}/K_M (as in case 1), but cannot be outcompeted with substrate, so that V_{max} is still reduced (as in case 2). Note that depending on the relative strengths of V_{max} and k_{cat}/K_M, K_M may increase or decrease as a function of force.

Two experiments by Block and coworkers (63, 73) have shown that for kinesin V_{max} and k_{cat}/K_M decrease with opposing force, whereas K_M increases, consistent with case 3. [By contrast, Nishiyama et al. (69), in an experiment based on measuring times between kinesin steps, did not observe any force dependence of K_M.] For assisting forces, other transitions in the cycle are rate-limiting, and little force dependence is observed in V_{max} and k_{cat}/K_M. As a result of these findings, Block et al. (73) have proposed the following minimal reaction pathway for kinesin:

$$M_1 + ATP \underset{k_{\pm 1}}{\longleftrightarrow} M_2 \underset{k_{\pm 2}(F)}{\longleftrightarrow} M_3 \underset{k_3}{\longrightarrow} M_4 \underset{k_4(F_1)}{\longrightarrow} M_5 \underset{k_5(F_1)}{\longrightarrow} M_1.$$

Step 2 is the primary movement step and is rate-limiting for opposing forces. For assisting forces, step 3 is rate-limiting. Steps 4 and 5 are weakly dependent on forces applied perpendicularly to the direction of motion and only affect the maximum velocity V_{max}, not the effective binding rate k_{cat}/K_M. Furthermore, by studying the effect of force on ADP binding to kinesin, Uemura et al. (76) showed that assisting force increases the binding affinity of the motor for ADP, whereas an opposing force decreases it. This observation strongly suggests a reciprocal coupling between the direction of force (whether assisting or opposing) and the enzymatic activity of the motor. The release of ADP increases the binding affinity of the kinesin head to the microtubule (77). Uemura et al. (76) postulate that an internal strain may modulate the binding affinity for ADP and hence may control the motor's processivity.

One of the movement steps of F_1-ATPase occurs during binding of ATP (78). Although this appears consistent with case 1, experiments show that V_{max} decreases with torque, an observation more consistent with cases 2 and 3 (70). The reason for this discrepancy is that our treatment of ATP binding as a single-step, second-order process is an oversimplification of the true binding mechanism. More generally, binding may consist of a second-order "docking" process in which the substrate comes into loose contact with the binding site, followed by one or more first-order "accommodation" steps in which it becomes progressively more tightly bound to the enzyme. Provided these docking steps are reversible, this situation is analogous to case 3. An illustration of this more complex mechanism is precisely the "binding zipper" in F_1-ATPase, in which ATP binding to the catalytic site has been postulated to occur via a zippering of hydrogen bonds, generating a constant torque (79). This example illustrates the need to exercise caution when generating mechanistic models. Force measurements at various ATP, ADP, and P_i concentrations often need to be supplemented with kinetic measurements and structural data to form a complete and accurate picture of motor mechanism.

STRAIN IN ENZYME CATALYSIS

Introduction

In the previous section, we have seen that molecular motors have evolved as energy transducers to convert chemical energy into mechanical work. Binding energy represents a ubiquitous but otherwise "silent" form of energy in the cell that can be considered as one of the main sources of potential energy to perform various molecular tasks. In this section, we generalize the idea of the energy transducer and discuss how the binding energy between an enzyme and its substrate can be utilized to accelerate the rate of chemical reactions.

Chemical catalysis is essential for many critical biological processes to proceed at useful rates under physiological conditions. Thus, a molecular description of the mechanisms by which an enzyme achieves high catalytic efficiency is of crucial importance to understand the control and coordination of the complex biological reactions in the cell. Studies over the past century have greatly improved our understanding of the factors that contribute to the catalytic efficiency of enzymes (80, 81). These factors include, among others, the specificity of an enzyme for its substrate, the correct orientation of its reacting groups at the active site relative to the substrate, and its ability to provide electrostatic shielding of charged intermediates in the reaction. In addition, many mechanisms of rate acceleration have been shown to play an important role in catalysis, such as general acid-base catalysis, and covalent catalysis. These mechanisms are, however, not sufficient to rationalize the extraordinary rate enhancement provided by enzymes. Although other mechanisms have been considered in the past 70 years, most have remained in the realm of hypotheses, due, in part, to the difficulty of designing experiments to quantify their importance. One such mechanism is the hypothesis known as strain-induced catalysis. Viewed from the transition state theory of chemical reactions (82, 83), it is well accepted that to catalyze biochemical reactions, enzymes must lower the activation barrier for the substrate to reach the transition state. Due to the intrinsically different structures of the substrate and the transition state, it is almost certain that the enzyme active center cannot be perfectly complementary to both at the same time. In the 1930s, J.B.S. Haldane suggested that enzymes cannot be efficient catalysts if they are fully complementary to the substrate. Rather, he suggested, enzymes must exert strain on the substrate upon binding. He wrote, ". . . the key does not fit the lock perfectly but exercises a certain strain on it" (84). This idea was elaborated by Pauling in 1946. He pointed out that to accelerate the rate of a reaction, enzymes must display complementarity to the transition state. He wrote, "The only reasonable picture of the catalytic activity of enzymes is that which involves an active region of the surface of the enzyme which is closely complementary in structure not to the substrate molecule itself in its normal configuration, but rather to the substrate molecule in a strained configuration, corresponding to the 'activated complex' for the reaction catalyzed by the enzyme" (85). Combined, these two conjectures are known as the Haldane-Pauling hypothesis.

The Haldane-Pauling hypothesis implies the utilization of binding energy to bring about the acceleration of the rate of the reaction, i.e., strain-induced catalysis. To see how the generation of strain could lead to acceleration of the reaction, let us assume that both the enzyme and the substrate are compliant and can deform upon binding. Let the enzyme's active site be complementary not to the shape of the substrate but to that of the transition state of the reaction. When the substrate first binds, the enzyme is not in a configuration that is catalytically productive. However, because the substrate is not fully complementary to the active site of the enzyme, and the full complement of binding interactions are only realized when the substrate attains the transition state, the enzyme exerts strain on the substrate and this strain favors the attainment of the transition state. It is this gradient of stabilizing binding interactions existing between the substrate and its transition state along the reaction coordinate that amounts to the generation of a force acting on the substrate that ultimately pushes along the reaction. In symbols we can write

$$\frac{\partial E_{bind}}{\partial x_r} = \frac{\partial E_{bind}}{\partial x} \cdot \frac{\partial x}{\partial x_r},$$

25.

where E_{bind} is the binding energy gained through complementarity to the enzyme along the reaction coordinate, x_r, and x is the mechanical coordinate through which the substrate is strained. The second factor on the right-hand side expresses the decrease of substrate strain as it moves along the reaction coordinate toward the transition state, whereas the first represents the effective mechanical force exerted by the enzyme on the substrate during this process:

$$F \equiv \frac{\partial E_{bind}}{\partial x}.$$

26.

These expressions show that whether the enzyme undergoes a conformational change during the process that "drives" the substrate into the transition state or simply provides a gradient of binding interactions along the reaction coordinate that preferentially stabilizes the transition state, the result is the same; in both cases, one can think of the enzyme as exerting a force on the substrate that catalyzes the reaction.

In some cases, the enzyme active site may be more compliant than the substrate and thus, upon binding, the substrate actually induces strain in the enzyme due to its nonideal fit (86). However, such strain will generate an elastic restoring force in the enzyme that will be exerted on the substrate and that, in turn, will tend to distort the substrate to bring about the transition state. In other words, the strain in the enzyme increases the free energy of the enzyme-substrate complex along the reaction coordinate and facilitates the crossing of the transition state beyond which such strain is fully relieved.

The Catalytic Advantage of Transition-State Complementarity

In this model, binding of the substrate to the enzyme induces strain in the complex, and complementarity between the enzyme and substrate is fully realized only at the transition state. We illustrate below the advantage of transition-state complementarity in catalysis.

In Michaelis-Menten kinetics, k_{cat} is a first-order rate constant associated with the rate of decomposition of the enzyme-substrate complex into enzyme and products, whereas k_{cat}/K_M is an apparent second-order rate constant for the generation of products starting from the free enzyme and the free substrate (80), and is a measure of the catalytic efficiency of the enzyme. In symbols,

$$E + S \underset{\Delta G_S}{\overset{K_M}{\xleftrightarrow{\hspace{1cm}}}} \underset{\Delta G^{\ddagger}}{\overset{k_{cat}}{ES \xrightarrow{\hspace{1cm}}}} ES^{\ddagger} ; \quad E + S \underset{\Delta G_T^{\ddagger}}{\overset{k_{cat}/K_M}{\xrightarrow{\hspace{1.5cm}}}} ES^{\ddagger}$$

where ΔG_S is the binding free energy in the association of the enzyme and the substrate, and $\Delta G^{\ddagger}$ and $\Delta G_T^{\ddagger}$ are the activation free energies from the ES complex and from the unbound $E + S$, respectively (Figure 7).

The importance of transition-state complementarity is illustrated in Figure 7 under conditions in which $[S] \gg K_M$ and $[S] < K_M$, and when the active site is complementary to the substrate and to the transition state of the reaction (see 80). Here, we assume that the maximum possible free energy of association between enzyme and substrate is $\Delta G_{S,max}$, which arises only when there is full complementarity between enzyme and substrate. If instead, the enzyme is ideally complementary to the transition state $ES^{\ddagger}$, an adverse free energy ΔG_{strain} arising from the strain between the enzyme and substrate contributes to substrate binding: $\Delta G_S = \Delta G_{S,max} + \Delta G_{strain}$. ($\Delta G_S$, $\Delta G_{S,max} < 0$ when $[S] \gg K_M$; and ΔG_S, $\Delta G_{S,max} > 0$ when $[S] < K_M$.) In addition, the free energy of the transition state is lowered by strain. Note that the expressions for the rate of the reaction depend on whether or not the enzyme is saturated with the substrate. However, in both cases, it is catalytically advantageous for the enzyme to be fully complementary to the transition state (thus inducing strain in the substrate) because the effect is to raise both K_M and k_{cat}. Furthermore, transition-state complementarity maximizes the value of k_{cat}/K_M, the catalytic efficiency of the enzyme. This conclusion is valid regardless of whether the enzyme is saturated with the substrate or not.

Although it is difficult to determine the substrate concentrations in vivo, the cases that have been studied, such as those of the glycolytic pathway (87), indicate that enzymes mostly operate under subsaturation conditions (88), i.e.,

---

Figure 7 Illustration of the advantage of transition state complementarity as discussed in the text. Dashed lines illustrate the free energy when the enzyme binding pocket is complementary to the substrate; solid lines illustrate the free energy for the case of transition state complementarity. Algebraically positive (negative) energies are indicated by arrows pointing upward (downward). (a) $[S] \gg K_M$. K_M, k_{cat}, and k_{cat}/K_M are increased by transition state complementarity, and the rate of catalysis increases with k_{cat}. (b) $[S] < K_M$. K_M, k_{cat}, and k_{cat}/K_M are increased by transition-state complementarity, and the rate of catalysis increases with k_{cat}/K_M.

a. $[S] \gg K_M$

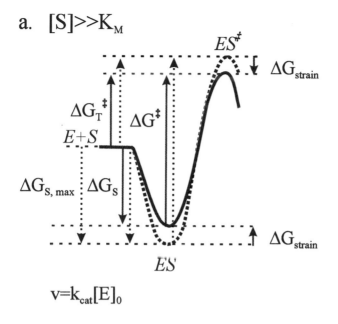

$$v = k_{cat}[E]_0$$

b. $[S] < K_M$

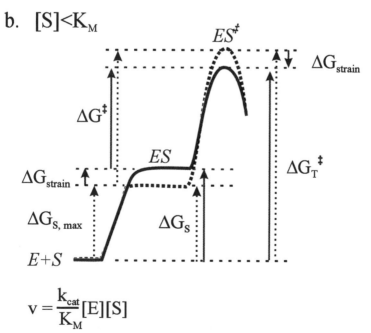

$$v = \frac{k_{cat}}{K_M}[E][S]$$

where substrate concentrations are significantly below their corresponding K_Ms (see Figure 7b). Thus, maximum rates of activity can be obtained if these enzymes have evolved to display full complementarity to the transition state such that k_{cat}/K_M is maximized. Furthermore, a high K_M maximizes the concentration of free enzyme $[E]$. These two conditions ensure the maximum efficiency under subsaturating conditions, i.e., $v = k_{cat}/K_M[E][S]$ (80).

Experimental Evidence for Strain-Induced Catalysis

It has been difficult to demonstrate unequivocally the importance of strain in enzyme catalysis. This is in part due to the lack of methods to directly monitor this elusive mechanical property. Even in cases in which enzymes have been cocrystallized with their substrates or with their substrates near or at the transition state, it is difficult to deduce a priori the energies and forces involved in the enzyme/substrate complex.

Structural studies of enzymes with and without their substrates show clear conformational changes of the enzyme and/or its substrate. A well-known example is hexokinase. Hexokinase catalyzes the transfer of a phosphoryl group from ATP to a substrate, such as glucose or mannose. X-ray crystallographic studies of yeast hexokinase showed that the binding of glucose induces a large conformational change in the enzyme (89, 90). Hexokinase comprises two lobes that form a cleft. Upon binding of the glucose, the cleft between the lobes closes, and the bound glucose becomes surrounded by protein. The substrate-induced closing of the cleft in hexokinase provides an excellent example of the induced-fit mechanism (91, 92).

A complementary technique that probes the dynamics of the enzyme-substrate interaction is fluorescence resonance energy transfer (FRET) (93, 94). A recent FRET study of the enzyme EcoRI, a type II restriction endonuclease, resolved large conformational changes in the N terminus, a region essential for tight binding of the DNA. The conformational changes revealed by the FRET experiments are not visible in any of the crystal structures (95).

More experimental evidence for the development of strain during enzyme catalysis has been indirect through the demonstration of two related issues: enzyme complementarity to the transition state, and the utilization of binding energy in catalysis. The former argument dates back to the 1940s when Pauling suggested that complementarity implied that enzymes should bind more strongly to analogs of the transition state than to the free form of their substrates. Pauling's ideas have been confirmed experimentally and further supported by the generation of catalytic antibodies (81, 96, 97).

Yin et al. (98) recently obtained direct physical evidence for substrate strain in antibody catalysis. These authors raised an antibody to a strained mimic of a substrate of the enzyme ferrochelatase. This mimic has a pyrrole nitrogen pushed out of planarity, a distortion thought to duplicate that involved in the transition state of the reaction (99). Not only does the crystal structure of the antibody complexed with the unstrained substrate show the same distortion, but the antibody also catalyzes its metallation by Zn^{2+} with a catalytic efficiency comparable to that of ferrochelatase. Hokenson et al. (100) studied strain in the carbonyl bond of the substrate of beta-lactamase. Here, a shift in the carbonyl stretch frequency of the beta-lactam substrate upon binding indicated that this bond is stretched in the ES complex; from the frequency shift, the strain in the carbonyl bond can be calculated to be about 3%. Values of the strain of about twice this value have been found in similar systems (101). A final example shows how release of enzyme strain appears to be involved in the catalysis of the reaction by the enzyme-coenzyme complex between aspartate aminotransferase and pyridoxal-5'-phosphate (PLP). Here, crystallographic studies indicated that steric interactions result in the critical hydrogen bond of the protonated Schiff base between the enzyme and PLP being strained in an out-of-plane conformation (102). Because this bond is broken in all ES intermediates and in the transition state, the activation energy to attain the transition state from the unbound E and S is decreased by 16 kJ/mol relative to the unstrained situation, thus decreasing the activation energy of k_{cat}/K_M whose value is increased 10^3-fold. Note that strain in the enzyme, as opposed to the more traditional strain in the substrate, is used in the catalysis by this enzyme.

The Magnitude of the Enzyme-Substrate Forces

The three-dimensional structure of proteins is maintained by a large number of weak interactions and thus, there is a limit to the magnitude of forces and strains that can develop between the enzyme and its substrate in the course of a reaction while maintaining the enzyme's structural integrity. Moreover, the mechanical properties of enzymes, such as their elastic modulus and compressibility, are expected to be anisotropic and inhomogeneous, i.e., to vary with their location within the enzyme and to depend on the direction along which they are measured. Thus, it is difficult to treat this problem in all generality without considering the specific interactions between a particular enzyme and its substrate. Nonetheless, general concepts from the basic physics of deformation can give some insight into the magnitude of the forces and strains that can result from the formation of the ES complex.

It is instructive to consider a simple, elastic, and linear model of the generation of strain in the interaction between enzyme and substrate. Here, we follow the treatment of Gavish (103). Let the complementary surfaces of the enzyme and substrate be depicted as in Figure 8. Here L and l are the length dimensions of the protein and the substrate before binding interactions are turned on, and x and y are the change in these quantities as a result of these interactions. Also, κ_E and κ_S are the spring constants of the enzyme wall and substrate, respectively. Finally,

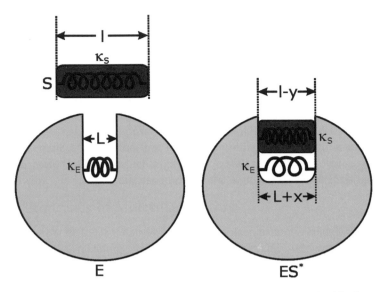

Figure 8   Pictorial representation of the strain induced between the binding pocket of the enzyme and its substrate. (*left*) The enzyme and substrate in their unbound conformations with initial lengths L and l, respectively. (*right*) Distortion is induced upon binding. The enzyme and substrate distort to achieve a net length change (l-L), where the distortion is shared between the enzyme and the substrate according to their relative stiffnesses κ_E and κ_S, as discussed in the text.

F_S and E_{strain} are the force acting on the substrate and the total elastic energy stored in both substrate and enzyme due to the strain. Then it is easy to show that

$$y = \frac{\kappa_E}{\kappa_S + \kappa_E}(l - L), \quad F_S = \frac{\kappa_S \kappa_E}{\kappa_S + \kappa_E}(l - L), \text{ and } E_{strain} = \frac{1}{2}\frac{\kappa_S \kappa_E}{\kappa_S + \kappa_E}(l - L)^2.$$

27.

Note that when $\kappa_E \sim \kappa_S$ the distortion y is half of its maximum possible value ($l - L$). Under these conditions, the strain and stress on the substrate can be written (103)

$$Strain = \frac{n\Delta A(l - L)}{2V_S}; \quad Stress = -\frac{1}{\beta}\frac{n\Delta A(l - L)}{2V_S}$$

28.

where n is the number of interactions between the substrate and the enzyme, ΔA is the area of interaction in the substrate, β the isothermal compressibility of the protein, and V_S is the volume of the substrate. Using $l - L \sim 2$ Å, $\Delta A \sim 10$ Å^2, $V_S \sim 200$ Å^3 and $\beta \sim 10^{-6}$ atm^{-1}, Gavish obtained values of strain $\sim 4\%$, stress $\sim 2 \times 10^4$ atm, and a stored strain energy of ~ 1 kcal/mol per interaction. These values correspond to a force of ~ 50 pN exerted through each interaction on the substrate. For multiple enzyme-substrate interactions, the stored strain energy becomes considerable, and for four interactions would give rise to a catalytic rate enhancement of $\sim 10^3$.

This simple model and calculation show that several interactions working on the substrate can lead to the development of significant stress and strain in the substrate. Indeed, the enzyme and substrate are engaged through multiple interactions, and it is the sum of all these interactions, each of them weak if compared to a covalent bond, that ultimately determines the degree of strain induced in the substrate. Moreover, an enzyme may distort a substrate by pulling in some locations and pushing in others, thus amplifying the mechanical gain (its mechanical advantage) by using a lever design.

Proteins are likely to be more difficult to compress than to stretch. As discussed above, the maximum force that a protein can support before unfolding in either case will depend on the rate at which the force is applied. As a result, it is possible that larger forces than those calculated above can develop in the formation of ES complexes if the force is the result of compression of the substrate and if it is applied only transiently during the catalytic cycle. Such a scenario may occur when the enzyme possesses a cleft that closes upon substrate binding, as is the case for lysozyme (104–106), for example. In fact, many authors have speculated that one reason why enzymes may have evolved to be so large is to increase their mechanical rigidity (107) and thus to be able to generate larger stresses and strains on substrates by compression.

Enzymes as Mechanical Devices

Here, we pose the following questions: Are enzymes mechanical devices capable of actively exerting force on their substrates through a coordinated set of motions akin to the movements of levers and arms of macroscopic machines? Can enzyme movements direct, guide, and mechanically promote the flux and orientation of chemical groups inside the active site as the reaction proceeds (108)? What would be required for these types of machine-like motions to occur in catalysis?

Enzymes undergo concerted conformational changes upon substrate binding, as postulated by the induced-fit hypothesis (91). In this view of enzyme catalysis, the enzyme, fueled by a sequence of successive binding interactions, undergoes a series of motions to bring a substrate into the transition state and to facilitate and push the catalytic process along its reaction coordinate. Indeed, by studying molecular motors, we have learned a great deal about how a series of concerted mechanical motions can be coupled to energy sources such as the binding and hydrolysis of fuel molecules. Like molecular motors, enzymes should be able to function as mechanical devices undergoing a series of concerted movements by utilizing the potential energy stored in the binding of their substrates or cofactors or in the release of the product of the reaction.

Imagine the binding of a ligand on the surface of a globular protein. Locally, the target region on the macromolecule is formed with the participation of groups from regions both close and far removed from each other in the sequence. Maintaining the architecture of this region of the surface, as that of any other part of the molecule, requires the appropriate balance of forces and interactions between the groups that form the target binding area and between groups located farther away at some radial distance from this area. To gain insight into the ability of proteins to behave as mechanical devices capable of "pushing along" the reaction, we need to know how efficiently the strain in one part of the molecule transfers to another because this mechanical communication between adjacent regions of the molecule would be needed for the enzyme to carry out a series of concerted conformational changes. Unfortunately, such information is not yet available. Some evidence exists that steric strain occurs predominantly in regions involved with function, presumably because of the more stringent precision needed for ligand binding and catalysis (109). However, much less is known about changes in strain that result from ligand binding, how far the perturbation brought about by this binding propagates into the macromolecule, or the contribution of side groups to the strain in the structure. Direct experimental measurements of mechanical forces have provided much insight into the mechanical stability of proteins and into the mechanisms by which molecular motors can generate force. We anticipate that similar approaches will be used to gain greater insight into the ability of enzymes to release stored substrate-binding energy (in the form of mechanical strain) to catalyze their reactions.

Single-molecule manipulation methods provide the possibility of directly testing some of the main ideas in strain-induced enzymatic catalysis. The mechanics of the enzyme-substrate interaction can be probed by exerting external mechanical force on the enzyme, the substrate, or both. For example, by applying external mechanical force to the two lobes of hexokinase, it should be possible to directly affect either the formation of the ES complex from the free reactants in solution or, possibly, the attainment of the transition state. If the closure of the cleft is rate-limiting, the external force will significantly affect the rate of the reaction, providing direct evidence for the effect of strain in enzyme catalysis. Moreover, by varying the magnitude of the external force applied to the enzyme-substrate complex, it should be possible to estimate the maximum force generated within the complex.

EPILOGUE

One of the main differences between biological macromolecules and their small organic and inorganic counterparts lies in the number, strength, and nature of the interactions that maintain their three-dimensional structures. The large number of relatively weak interactions that stabilize the tertiary structures of macromolecules also confer on these molecules their unique structural adaptability and flexibility in the cell. Often, this structural pliability is manifested in the form of a conformational change that ensues from the binding interactions of macromolecules with other macromolecules or with small ligands. The concept of conformational change that has played a central role in traditional biochemical studies is also essential to the new description proposed in this review: The energy released through binding interactions or bond hydrolysis leads to molecular displacements and to the generation of forces and mechanical work along a coordinate. The significance of this new description resides in the fact that it makes it possible to directly associate energies in the form of mechanical work along a particular coordinate with the corresponding changes in structure. In the next few years the basic mechanical nature of many more essential biochemical processes will likely be recognized and studied in this fashion. We hope that these processes will be amenable to study and characterization by some of the same concepts and experimental methods described in this review.

The *Annual Review of Biochemistry* is online at http://biochem.annualreviews.org

LITERATURE CITED

1. Tinoco I, Bustamante C. 2002. *Biophys. Chem.* 101:513–33
2. Bustamante C, Marko JF, Siggia ED, Smith S. 1994. *Science* 265:1599–600
3. Marko JF, Siggia ED. 1995. *Macromolecules* 28:8759–70
4. Liphardt J, Onoa B, Smith SB, Tinoco I, Bustamante C. 2001. *Science* 292:733–37
5. Tinoco I. 2004. *Annu. Rev. Biophys. Biomol. Struct.* 33:363–85
6. Bell GI. 1978. *Science* 200:618–27
7. Kramers HA. 1940. *Physica* 7:284–304
8. Eyring H. 1935. *J. Chem. Phys.* 3:107–15
9. Gautel M, Goulding D. 1996. *FEBS Lett.* 385:11–14
10. Linke WA, Ivemeyer M, Olivieri N, Kolmerer B, Ruegg JC, Labeit S. 1996. *J. Mol. Biol.* 261:62–71
11. Linke WA, Rudy DE, Centner T, Gautel M, Witt C, et al. 1999. *J. Cell Biol.* 146:631–44
12. Hynes RO. 1990. *Fibronectins.* New York: Springer
13. Shrestha P, Mori M. 1997. *Tenascin: An Extracellular Matrix Protein in Cell Growth, Adhesion and Cancer.* Austin, Texas: Landes Biosci.
14. Krammer A, Lu H, Isralewitz B, Schulten K, Vogel V. 1999. *Proc. Natl. Acad. Sci. USA* 96:1351–56
15. Matouschek A. 2003. *Curr. Opin. Struct. Biol.* 13:98–109
16. Kenniston JA, Baker TA, Fernandez JM, Sauer RT. 2003. *Cell* 114:511–20
17. Saibil HR, Ranson NA. 2002. *Trends Biochem. Sci.* 27:627–32
17a. Valpuesta JM, Martin-Benito J, Gomez-Puertas P, Carrascosa JL, Willison KR. 2002. *FEBS Lett.* 529:11–16
18. Shtilerman M, Lorimer GH, Englander WS. 1999. *Science* 284:822–25
19. Liphardt J, Dumont S, Smith SB, Tinoco I, Bustamante C. 2002. *Science* 296:1832–35
20. Jarzynski C. 1997. *Phys. Rev. Lett.* 78:2690–93
21. Hummer G, Szabo A. 2001. *Proc. Natl. Acad. Sci. USA* 98:3658–61
22. Evans E, Ritchie K. 1999. *Biophys. J.* 76:2439–47
23. Oberhauser AF, Hansma PK, Carrion-Vazquez M, Fernandez JM. 2001. *Proc. Natl. Acad. Sci. USA* 98:468–72
24. Friedsam C, Wehle AK, Kuhner F, Gaub HE. 2003. *J. Phys. Condens. Matter* 15:S1709–23

25. Best RB, Fowler SB, Toca-Herrera JL, Clarke J. 2002. *Proc. Natl. Acad. Sci. USA* 99:12143–48

26. Best RB, Brockwell DJ, Toca-Herrera JL, Blake AW, Smith DA, et al. 2003. *Anal. Chim. Acta* 479:87–105

27. Williams PM, Fowler SB, Best RB, Toca-Herrera JL, Scott KA, et al. 2003. *Nature* 422:446–49

28. Evans E. 1998. *Faraday Discuss.* 111: 1–16

29. Fowler SB, Best RB, Herrera JLT, Rutherford TJ, Steward A, et al. 2002. *J. Mol. Biol.* 322:841–49

30. Paci E, Karplus M. 2000. *Proc. Natl. Acad. Sci. USA* 97:6521–26

31. Tinoco I, Bustamante C. 1999. *J. Mol. Biol.* 293:271–81

32. Onoa B, Dumont S, Liphardt J, Smith SB, Tinoco I, Bustamante C. 2003. *Science* 299:1892–95

33. Marszalek PE, Lu H, Li HB, Carrion-Vazquez M, Oberhauser AF, et al. 1999. *Nature* 402:100–3

34. Lenne P-F, Raae AJ, Altmann SM, Saraste M, Horber JKH. 2000. *FEBS Lett.* 476:124–28

35. Oberhauser AF, Badilla-Fernandez C, Carrion-Vazquez M, Fernandez JM. 2002. *J. Mol. Biol.* 319:433–47

36. Li H, Linke WA, Oberhauser AF, Carrion-Vazquez M, Kerkvliet JG, et al. 2002. *Nature* 418:998–1002

37. Rief M, Pascual J, Saraste M, Gaub HE. 1999. *J. Mol. Biol.* 286:553–61

38. Carrion-Vazquez M, Li H, Lu H, Marszalek PE, Oberhauser AF, Fernandez JM. 2003. *Nat. Struct. Biol.* 10: 738–43

39. Brockwell DJ, Paci E, Zinober RC, Beddard GS, Olmsted PD, et al. 2003. *Nat. Struct. Biol.* 10:731–37

40. Lu H, Schulten K. 1999. *Proteins: Struct. Funct. Genet.* 35:453–63

41. Li HB, Carrion-Vazquez M, Oberhauser AF, Marszalek PE, Fernandez JM. 2000. *Nat. Struct. Biol.* 7:1117–20

42. Carrion-Vazquez M, Oberhauser AF,

Fisher TE, Marszalek PE, Li HB, Fernandez JM. 2000. *Prog. Biophys. Mol. Biol.* 74:63–91

43. Matouschek A, Bustamante C. 2003. *Nat. Struct. Biol.* 10:674–76

44. Kellermayer MSZ, Smith SB, Granzier HL, Bustamante C. 1997. *Science* 276: 1112–16

45. Kellermayer MSZ, Smith SB, Bustamante C, Granzier HL. 2001. *Biophys. J.* 80:852–63

46. Li HB, Oberhauser AF, Fowler SB, Clarke J, Fernandez JM. 2000. *Proc. Natl. Acad. Sci. USA* 97:6527–31

47. Law R, Carl P, Harper S, Dalhaimer P, Speicher DW, Discher DE. 2003. *Biophys. J.* 84:533–44

48. Wang MD, Schnitzer MJ, Yin H, Landick R, Gelles J, Block SM. 1998. *Science* 282:902–7

49. Wuite GJL, Smith SB, Young M, Keller D, Bustamante C. 2000. *Nature* 404: 103–6

50. Delagoutte E, von Hippel PH. 2003. *Q. Rev. Biophys.* 36:1–69

51. Delagoutte E, von Hippel PH. 2002. *Q. Rev. Biophys.* 35:431–78

52. Strick TR, Croquette V, Bensimon D. 2000. *Nature* 404:901–4

53. Dekker NH, Rybenkov VV, Duguet M, Crisona NJ, Cozzarelli NR, et al. 2002. *Proc. Natl. Acad. Sci. USA* 99:12126–31

54. Stone MD, Bryant Z, Crisona NJ, Smith SB, Vologodskii A, et al. 2003. *Proc. Natl. Acad. Sci. USA* 100:8654–59

55. Smith DE, Tans SJ, Smith SB, Grimes S, Anderson DL, Bustamante C. 2001. *Nature* 413:748–52

56. Boyer PD. 2001. *Biochemistry* 66: 1058–66

57. Berg HC. 2003. *Annu. Rev. Biochem.* 72: 19–54

58. Hua W, Young EC, Fleming ML, Gelles J. 1997. *Nature* 388:390–93

59. Schnitzer MJ, Block SM. 1997. *Nature* 388:386–90

60. Davenport RJ, Wuite GJL, Landick R,

Bustamante C. 2000. *Science* 287: 2497–500

61. Keller D, Bustamante C. 2000. *Biophys. J.* 78:541–56

62. Coppin CM, Pierce DW, Hsu L, Vale RD. 1997. *Proc. Natl. Acad. Sci. USA* 94:8539–44

63. Visscher K, Schnitzer MJ, Block SM. 1999. *Nature* 400:184–89

64. Yin H, Wang MD, Svoboda K, Landick R, Block SM, Gelles J. 1995. *Science* 270: 1653–57

65. Oster G, Wang H. 2003. In *Molecular Motors*, ed. M. Schliwa, pp. 207–27. Weinheim: Wiley-VCH Verlag GmbH

65a. Itoh H, Takahashi A, Adachi K, Noji H, Yasuda R, et al. 2004. *Nature* 427: 465–68

66. Finer JT, Simmons RM, Spudich JA. 1994. *Nature* 368:113–19

67. Molloy JE, Burns JE, Kendrickjones J, Tregear RT, White DCS. 1995. *Nature* 378:209–12

68. Svoboda K, Schmidt CF, Schnapp BJ, Block SM. 1993. *Nature* 365:721–27

69. Nishiyama M, Muto E, Inoue Y, Yanagida T, Higuchi H. 2001. *Nat. Cell Biol.* 3:425–28

70. Yasuda R, Noji H, Kinosita K, Yoshida M. 1998. *Cell* 93:1117–24

71. Adachi K, Noji H, Kinosita K. 2003. *Methods Enzymol.* (Pt. B) 361:211–27

72. Guo P, Peterson C, Anderson D. 1987. *J.Mol. Biol.* 197:229–36

73. Block SM, Asbury CL, Shaevitz JW, Lang MJ. 2003. *Proc. Natl. Acad. Sci. USA* 100:2351–56

74. Wang HY, Elston T, Mogilner A, Oster G. 1998. *Biophys. J.* 74:1186–202

75. Thomen P, Lopez PJ, Heslot F. 2004. Submitted

76. Uemura S, Ishiwata S. 2003. *Nat. Struct. Biol.* 10:308–11

77. Uemura S, Kawaguchi K, Yajima J, Edamatsu M, Toyoshima YY, Ishiwata S. 2002. *Proc. Natl. Acad. Sci. USA* 99: 5977–81

78. Dimroth P, Wang HY, Grabe M, Oster G. 1999. *Proc. Natl. Acad. Sci. USA* 96: 4924–29

79. Oster G, Wang H, Grabe M. 2000. *Philos. Trans. R. Soc. London Ser. B* 355: 523–28

80. Fersht A. 1985. *Enzyme Structure and Mechanism.* New York: Freeman.

81. Jencks WP. 1987. *Catalysis in Chemistry and Enzymology.* New York: Dover.

82. Eyring H. 1935. *Chem. Rev.* 17:65–77

83. Pelzer H, Wigner E. 1932. *Z. Phys. Chem. B* 15:445

84. Haldane JBS. 1965. *Enzymes.* Cambridge: M.I.T. Press.

85. Pauling L. 1946. *Chem. Eng. News* 24: 1375–77

86. Bernstein BE, Michels PA, Hol WG. 1997. *Nature* 385:275–78

87. Lowry OH, Passonneau JV. 1964. *J. Biol. Chem.* 239:31–42

88. Atkinson DE. 1969. In *Current Topics in Cellular Regulation*, ed. ER Stadtman, BL Horecker, pp. 29–43. New York: Academic

89. Bennett WS Jr, Steitz TA. 1978. *Proc. Natl. Acad. Sci. USA* 75:4848–52

90. Bennett WS Jr, Steitz TA. 1980. *J. Mol. Biol.* 140:211–30

91. Koshland DE. 1958. *Proc. Natl. Acad. Sci. USA* 44:98–104

92. Koshland DE. 1963. *Cold Spring Harbor Symp. Quant. Biol.* 28:473–82

93. Selvin PR. 2000. *Nat. Struct. Biol.* 7: 730–34

94. Selvin PR. 1995. *Methods Enzymol.* 246: 300–34

95. Watrob H, Liu W, Chen Y, Bartlett SG, Jen-Jacobson L, Barkley MD. 2001. *Biochemistry* 40:683–92

96. Tramontano A, Janda KD, Lerner RA. 1986. *Science* 234:1566–70

97. Pollack SJ, Jacobs JW, Schultz PG. 1986. *Science* 234:1570–73

98. Yin J, Andryski SE, Beuscher AE, Stevens RC, Schultz PG. 2003. *Proc. Natl. Acad. Sci. USA* 100:856–61

99. McLaughlin GM. 1974. *J. Chem. Soc.Perkin Trans.* 2:136–40

748   BUSTAMANTE ET AL.

100. Hokenson MJ, Cope GA, Lewis ER, Oberg KA, Fink AL. 2000. *Biochemistry* 39:6538–45
101. Chittock RS, Ward S, Wilkinson AS, Caspers P, Mensch B, et al. 1999. *Biochem J.* 338 (Pt. 1):153–59
102. Rhee S, Silva MM, Hyde CC, Rogers PH, Metzler CM, et al. 1997. *J. Biol. Chem.* 272:17293–302
103. Gavish B. 1986. In *The Fluctuating Enzyme*, ed. GR Welch, pp. 263–339. New York: Wiley
104. Kelly JA, Sielecki AR, Sykes BD, James MN, Phillips DC. 1979. *Nature* 282: 875–78
105. Blake CC, Johnson LN, Mair GA, North AC, Phillips DC, Sarma VR. 1967. *Proc. R. Soc. London Ser. B* 167:378–88
106. Blake CC, Koenig DF, Mair GA, North AC, Phillips DC, Sarma VR. 1965. *Nature* 206:757–61
107. Narlikar GJ, Herschlag D. 1997. *Annu. Rev. Biochem.* 66:19–59
108. Williams RJP. 1993. *Trends Biochem. Sci.* 18:115–17
109. Herzberg O, Moult J. 1991. *Proteins* 11: 223–29
110. Essevaz-Roulet B, Bockelmann U, Heslot F. 1997. *Proc. Natl. Acad. Sci. USA* 94:11935–40
111. Rief M, Clausen-Schaumann H, Gaub HE. 1999. *Nat. Struct. Biol.* 6:346–49
112. Carrion-Vazquez M, Oberhauser AF, Fowler SB, Marszalek PE, Broedel SE, et al. 1999. *Proc. Natl. Acad. Sci. USA* 96:3694–99
113. Pantazatos DP, MacDonald RI. 1997. *J. Biol. Chem.* 272:21052–59
114. Khorasanizadeh S, Peters ID, Butt TR, Roder H. 1993. *Biochemistry* 32: 7054–63
115. Noji H, Yasuda R, Yoshida M, Kinosita K. 1997. *Nature* 386:299–302
116. Cooke R, Franks K, Luciani G, Pate E. 1988. *J. Physiol.* 395:77–97
117. Mehta AD, Finer JT, Spudich JA. 1997. *Proc. Natl. Acad. Sci. USA* 94:7927–31
118. Veigel C, Bartoo ML, White DCS, Sparrow JC, Molloy JE. 1998. *Biophys. J.* 75: 1424–38
119. Svoboda K, Block SM. 1994. *Cell* 77: 773–84
120. Kojima H, Muto E, Higuchi H, Yanagida T. 1997. *Biophys. J.* 73:2012–22
121. Wang H, Oster G. 2002. *Europhys. Lett.* 57:134–40

Annu. Rev. Biochem. 2004. 73:749–89
doi: 10.1146/annurev.biochem.73.011303.073823
First published online as a Review in Advance on April 8, 2004

INTERMEDIATE FILAMENTS: Molecular Structure, Assembly Mechanism, and Integration Into Functionally Distinct Intracellular Scaffolds

Harald Herrmann[1] and Ueli Aebi[2]

[1]Department of Cell Biology, German Cancer Research Center, D-69120 Heidelberg, Germany; email: h.herrmann@dkfz.de
[2]Maurice E. Müller Institute for Structural Biology, Biozentrum, CH-4056 Basel, Switzerland; email: ueli.aebi@unibas.ch

Key Words   atomic structure, lamins, vimentin, keratins, coiled coil

■ **Abstract**   The superfamily of intermediate filament (IF) proteins contains at least 65 distinct proteins in man, which all assemble into ~10 nm wide filaments and are principal structural elements both in the nucleus and the cytoplasm with essential scaffolding functions in metazoan cells. At present, we have only circumstantial evidence of how the highly divergent primary sequences of IF proteins lead to the formation of seemingly similar polymers and how this correlates with their function in individual cells and tissues. Point mutations in IF proteins, particularly in lamins, have been demonstrated to lead to severe, inheritable multi-systemic diseases, thus underlining their importance at several functional levels. Recent structural work has now begun to shed some light onto the complex fine tuning of structure and function in these fibrous, coiled coil forming multidomain proteins and their contribution to cellular physiology and gene regulation.

CONTENTS

INTRODUCTION

Background

Without any doubt, intermediate filaments (IFs) are the "stress-buffering" elements of metazoan cells, constituting a principal cytoskeletal moiety as well as a major part of the nucleoskeleton, the nuclear lamina. Compared to microtubules (MTs) and microfilaments (MFs), the other two principal structural elements of the eukaryotic cell cytoskeleton, IFs are rather flexible, and in contrast to MTs and MFs, which break when subjected to shear stress, they become more viscoelastic (1). Tubulin and actin, the molecular building blocks of MTs and MFs, are globular proteins with nucleotidase activity. MTs are polar tubular assemblies built from 13 to 15 laterally associated protofilaments, whereas MFs are polar helical filaments made of two linear strands of actin subunits twisted around each other with an axial stagger of half a subunit. For both types of filaments, atomic models have been built from the crystal structures of their globular subunits (2–4).

In contrast, the molecular building blocks of IFs are fibrous proteins that consist of long, often uninterrupted segments of α-helices. Because single long α-helical chains are unstable in aequous solutions, they usually adopt a rope-like structure by forming multistranded left-handed coiled coils (5). This organizational principle is adopted by many proteins, including keratins, myosin, epidermin, and fibrinogen, i.e., the classical k-m-e-f group (6). By electron microscopy (EM) coiled-coil IF protein dimers are depicted as distinct, 45–50 nm long rod-like molecules (7). To understand the molecular mechanisms by which these rod-like IF protein dimers assemble into 10-nm filaments and associate further into distinct higher-order structures is a formidable challenge and calls for an interdisciplinary experimental approach with biochemists and structural, molecular, cell, and developmental biologists joining forces.

The concept of intermediate filaments being ubiquitous constituents of the structural scaffold within higher eukaryotic cells was put forward by Lazarides (8) who described them "as mechanical integrators of cellular space." Two of their prime biochemical characteristics were their pronounced insolubility and the fact that they could be reassembled in vitro in physiological buffers, after complete denaturation in urea, with no need for any cofactors (9–11). Their widespread occurrence within practically all cell types of the human body and, in particular, their developmentally regulated expression called for their specific function(s) in epithelia (keratins), muscle (desmin), mesenchymal (vimentin), glia cells (GFAP), and neurons (neurofilament triplet proteins) (12).

Despite their similar appearance by EM, it was noted that IF proteins are chemically quite heterogenous, although the accumulation of sequence data allowed some generalization of their domain structure and their distinction into four sequence homology classes (SHCs): acidic keratins (SHC I), basic keratins (SHC II), desmin-/vimentin-type proteins (SHC III), and neurofilament proteins (SHC IV) (13–15). Moreover, the importance of the highly variable non-α-helical end domains for their specific cellular functions was noted (15, 16). A major step forward in understanding IFs was the discovery that the insoluble proteins constituting the nuclear lamina were IF-type proteins too, which were grouped together as SHC V (17, 18). With the cloning of IF genes, a more systematic comparison of the individual domains was possible. Thus, a distinct tripartite molecular organization of all IF proteins emerged: A central α-helical domain (the "rod") with a distinct number of equally sized coiled-coil forming segments is flanked by non-α-helical N-terminal (the "head") and C-terminal (the "tail") end domains of highly varying size and sequence (14, 15). The central rod domain is $\sim$310 amino acids long for all vertebrate cytoplasmic IF proteins, and it is 350 amino acids for the nuclear lamins and the lower invertebrate cytoplasmic IF proteins. In comparison, the tail domain of human and rat nestin is more than 1300 and 1500 amino acids long, respectively (19). The gene structure of various IF members revealed that the assumed functional domains are not contiguous with exons; however, in vertebrates the number and position of introns is largely conserved and splicing does not occur except for one "exotic" member of the cytoplasmic IF family, i.e., synemin, and for lamin A (20, 21). Furthermore, this extensive sequencing work allowed for some speculation on the evolution of the IF gene family. In particular, it explained the distinct positioning of a discontinuity of the heptad polarity (i.e., a "stutter") near the middle of coil 2B and the insertion of three linkers in the rod domain in all IF proteins. Moreover, it was noted that the nuclear lamins most likely represent the primordial form of the cytoplasmic IF proteins (22–25). Despite a large sequence variability between the five SHCs, there exist two IF consensus motifs, one at the N-terminal and the other at the C-terminal end of the central α-helical rod domain; the latter has striking sequence identity even between proteins as distantly related as a nuclear lamin of *Hydra vulgaris* and human hair keratins (26).

A major breakthrough for a concept on the differential function of IFs emerged when (*a*) point mutations in keratin genes were identified that caused severe epidermolytic diseases, and (*b*) it was found that the dynamics of IFs are drastically influenced by phosphorylation [reviewed in (27, 28)]. In addition, by chemical cross-linking of various types of IFs, it was documented that similar interchain interactions may occur within the different cytoplasmic IFs [reviewed in (29)]. Moreover, mass-per-length (MPL) measurements by scanning transmission electron microscopy (STEM) revealed IFs to be rather polymorphic in terms of their number of polypeptides per filament cross section, i.e., ranging between 24 and 40 molecules or 6 to 10 tetramers per cross section [reviewed in (27), see also (30, 31)]. This MPL polymorphism was not just found between different IFs but also along one and the same IF.

With the discovery of IF diseases, such as the various skin fragility disorders (see above), and the laminopathies that are caused by point mutations in the lamin A gene (32–35), IF proteins have moved to center stage of cellular physiology, integrating the structural and functional aspects of IF proteins as constituents of soluble complexes and of larger IF assemblies, such as the nuclear lamina and the IF cytoskeleton (36–39).

The Scope of this Review

In this review, we focus on the molecular properties and structural features of IF proteins that make them unique among the cytoskeletal and nuclear proteins. For example, a rational understanding of the spatial and temporal dynamics of nuclear architecture in interphase versus mitosis has to include the nuclear lamins and most likely the cytoplasmic IF system as well. Similarly, the IF genes are a primary target for point mutations, which eventually lead to various types of morphological and functional tissue damage. Because the IF field has developed so rapidly, this review primarily focuses on the biochemical, structural, and assembly aspects of cytoplasmic and nuclear IF proteins. In addition, the structural and molecular aspects important for the coordinated integration of IF proteins into distinct cellular scaffolds will be discussed. The widespread occurrence of lamins and lamin mutations (37, 40–45), the specific properties of keratins (46) and neurofilaments (47, 48), the structural features related to IF diseases, and the targeted inactivation of IF genes (21, 29, 49–52a) have been the primary topics of many recent reviews. Similarly, the complex family of IF-associated proteins, including motor proteins, as well as the connection of the various IF systems to cellular signaling have been extensively covered (53–58).

THE BIOCHEMISTRY OF INTERMEDIATE FILAMENT PROTEINS

Solubility

A hallmark feature of IF proteins is their insolubility in buffers of physiological ionic strength and pH. Moreover, IFs are completely resistant to extraction with buffers of high ionic strength and high concentrations of nonionic detergents, such as 1.5 M sodium chloride and 1% Triton X-100, and it requires rather drastic conditions to solubilize them, for example, the inclusion of 8 M urea or 3 M guanidinium hydrochloride into the extraction buffer (10, 17). In addition, they also become soluble in buffers of low ionic strength and/or high pH. Therefore, filaments of the mesenchymal IF protein vimentin can readily be dissolved in 5 mM Tris-HCl, pH 7.5, and only when more than 20 mM sodium chloride is added, do they stay intact. These are, however, clearly assembly conditions for keratins, and one has to employ even lower ionic strength and higher pH, i.e., 2 mM Tris-HCl, pH 9, to obtain tetramers (59). Moreover, there are rather big differences between simple epithelial keratins (K8 and K18) and epidermal keratins (K5 and K14). Although K8/18 are completely soluble in 2 mM Tris-HCl, pH 9, K5/14 already exhibit extensive filament formation under these conditions (60). Finally, lamins are soluble both at high pH and high salt conditions, such as 200 to 300 mM sodium chloride, pH 8.0 to 9.0 (17, 61), and in 5 mM Tris-HCl, pH 7.5 (D. Lotsch and H. Herrmann, unpublished observations). These few examples exemplify how strikingly single IF proteins differ with respect to their solubility behavior. One reason is the intricate interplay of the acidic α-helical rods with the basic non-α-helical head domains. Removal of the head domain yields molecules that are soluble under physiological conditions, i.e., they do not associate significantly above tetramers (30, 62). Correspondingly, the introduction of negatively charged groups by phosphorylation of serine residues in the head domain causes the disassembly of filaments (63–65). Another type of posttranslational modification, ADP-ribosylation, leads also to the inhibition of desmin assembly (66). For lamins, it has been demonstrated that three sites immediately adjacent to the rod are substrates for cyclin-dependent kinase 1 (cdk1), and their phosphorylation is essential for the disassembly of the lamin scaffold, both in vivo and in vitro (67, 68).

Analytical Ultracentrifugation

Sedimentation analysis of both authentic and recombinant IF proteins has been, corresponding to their distinct solubility properties, carried out under a variety of conditions (30, 69–71). For desmin isolated from chicken gizzard, a value of $S_{20} = 5.2$ was obtained in 10 mM Tris acetate, pH 8.5 (69), which is in complete agreement with those obtained for recombinant human vimentin in 5 mM Tris-HCl, 1 mM EDTA, pH 8.4 (30). Sedimentation equilibrium analysis for recombinant human vimentin revealed an average M_r value of 158,000 to 370,000, which was interpreted to represent a mixture of coexisting tetramers and octamers, possibly with small amounts of dimers and hexamers (30). In contrast, it was reported that vimentin assembles to a significant part into stable hexameric complexes, which was determined by transient electrical birefringence measurements (72, 73). These measurements were carried out in 0.7 mM sodium phosphate buffers, pH 6.8 to 7.5 (73) and, hence, under conditions that may favor unraveling reactions (74). In stark contrast, analytical ultracentrifugation, employing a global fit software (75, 76) and done in 0.7 mM sodium phosphate, pH 7.5, revealed vimentin to be present in tetrameric form exclusively (77). It was noted in this study that inclusion of 1 mM EDTA into a 5 mM Tris-HCl buffer (pH 8.4) is sufficient to cause oligomerization of the tetramers. Therefore, this system is indeed at the border of assembly.

The neuronal IF protein α-internexin was documented to yield dimeric complexes in the presence of 4 M urea, whereas vimentin forms tetramers under these conditions (78). When dialyzed into 10 mM Tris-HCl, pH 8.0, a sedimentation coefficient $S_{20,w}$ of 10.5 S was obtained for α-internexin. Sedimentation equilibrium centrifugation experiments, carried out in the presence of 2% sucrose to provide a stabilizing density gradient at equilibrium, revealed that under these conditions α-internexin already formed 16-mers. Corresponding EM experiments characterized these complexes as rod-like particles with a length of 68 nm and a width of 11 nm. By adding 10 mM NaCl, the $S_{20,w}$ value increased to 16 S and sedimentation equilibrium measurements exhibited 22- to 62-mers. By EM, these complexes corresponded to full width IFs with a discrete length of 70 nm and 120 nm. Hence, depending on the ionic conditions, various types of α-internexin assemblies can be stably obtained. With keratins, tetrameric complexes prevail even at 7 M urea or in 2 M guanidinium hydrochloride (79).

Sequence Gazing and Evolutionary Considerations

DISTRIBUTION OF CHARGE   A comparative analysis of the amino acid sequences immediately reveals that the α-helical rods of IF proteins are highly charged (Figure 1). For example, the 310–amino acid long α-helical rod domain of human vimentin contains 116 charged amino acids, 70 acidic and 46 basic [see (29)]. Therefore, there is a surplus of 24 acidic residues per subunit or 48 per dimer. In lamin B_1, there are 44 acidic residues more per dimer with an equal surplus number in coil 1B (10 residues) and coil 2B (9 residues). Interestingly, the keratins K5/14 only have a surplus of 32 acidic residues per dimer—just as lamin B_2. Lamin A is even less acidic with 24 extra acidic residues per dimer. This results in an excess of two acidic residues every 2 nm along the rod for vimentin, every 3 nm for lamin B_2, and every 4 nm for lamin A, respectively.

The acidity of the rod domain is contrasted by the high basic character of the head domain. Whereas vimentin contains 12 arginines in its head domain, lamin B_2 harbors only 3. This correlates with the rather different lengths of the two head domains, which are 102 residues (including 20 amino acids preceding coil 1A without any proline and with high probability for α-helix) in vimentin but only 25 in lamin B_2 (Table 1). Even a "short head" version of a cytoplasmic IF protein, such as GFAP, still contains 9 basic residues in its 68–amino acid long head domain. We assume that this indicates a different kind of assembly principle. Notably, both lamin A and lamin B_2 contain 9 basic residues, lamin B_1 has 10 within a stretch of ~40 amino acids of the tail immediately following the IF consensus motif at the end of coil 2 with no acidic residue in it. The four consecutive basic residues at the end of this segment represent the nuclear localization signal [for comparison see (24)]. This structural principle is evolutionarily highly conserved because the lamins of lower vertebrates and invertebrates have short heads and a basic domain of up to 40 amino acids after the IF consensus motif of coil 2. For example, the B-type lamin of *Drosophila melanogaster* has 9 basic and no acidic residues in this region, whereas the single B-type lamin of *Caenorhabditis elegans* differs slightly with only 6 basic and one acidic residue. In the head domain, the *C. elegans* protein differs considerably because it harbors 4 acidic amino acids in addition to its 6 basic ones. Other lamins, such as the human and frog lamin A as well as the B-type lamins of *Hydra*, goldfish, and frog, exhibit one acidic residue in the head. In summary, this "low resolution" sequence comparison suggests that nuclear and cytoplasmic IF proteins use their main structural elements for assembly, i.e., the rod, the head, and the tail, in quite a different way. Note, the tail is not essential for assembly of cytoplasmic IFs but plays a significant role in filament width control (30).

SEQUENCE SPECIFICITY WITHIN THE ROD DOMAIN   The rod domain is regarded as the principal building block of IF proteins; however, primary amino acid sequence identity between the members of the five sequence homology classes (SHC) is quite low (Figures 1 and 2). Even the A- and B-type lamins differ considerably except for the ends of their rod. Rather large stretches in coil 1B and coil 2B have only 3 or less residues in common within consecutive heptads (Figure 2). Vimentin and the keratins have for the most part of their rods no more than two amino acids per heptad that are identical with lamin B_2. Hence, despite the conservation of the general building plan of IF proteins, the coiled-coil forming parts are apparently tailored according to specific functions.

A- and B-type lamins are within the α-helical rod absolutely conserved with respect to the number of amino acids. They differ considerably, however, in individual sequence motifs, particularly in coil 1B, which is, at least for cytoplasmic IF proteins, of high importance for dimer-dimer interactions. Accordingly, the observed segregation of A- and B-type lamins during mitosis indicates that they do not form heterodimers (80). Whether this also applies to higher macromolecular complexes, or even to the entire filament, has to be investigated further, and it is indeed one of the most challenging questions about the assembly mechanism of lamins. The answer may explain too how lamins are involved in DNA replication (81) and how mutations in A-type lamins lead to disease. One attractive hypothesis is that both types of lamins form independent polymers that are integrated within the lamina by the growing number of associated proteins. Future biochemical analyses are needed to elucidate whether A- and B-type lamin molecules are able to form heterodimers. If so, then one would have to assume that they are prevented from forming heterodimers during translation in vivo. If not, we may have to conclude that those parts of the molecules, which are triggers for dimerization, are exactly those that differ between A- and B-type lamins. For example, the sixth heptad of coil 1B is very unique. It reads **FKELKARN** in lamin A and **HDQLLLNY** in lamin B_1 (hydrophobic positions in bold). Even in lamin B_2, which is rather similar to B_1, this motif deviates considerably reading **LDEVNKSA**. In this heptad, no position is identical in the three lamins (Figure 2). Strangely enough, some published

```
           <<— HEAD <<—|              coil 1A                          L1
     gabcdefgabcdefgabcdefgabcdefgabcdefgabcdefgabcdefgabcdefgabcd_____ab
LaB1 -VPPRMGSRAGGPTTPLSPTRLSRLQEKEELRELNDRLAVYIDKVRSLETENSALQLQVTEREEVRGRELTGLRA
LaA  -QRRATRSGAQASSTPLSPTRITRLQEKEDLQELNDRLAVYIDKVRSLETENAGLRLRITESEEVVSREVSGIKA
Vim  -RLLQDSVDFSLADAINTEFKNTRTNEKVELQELNDRFANYIDKVRFLEQQNKILLAELEQLKGQ---GKSRLGD
NFL  -MPSLENLDLSQVAAISNDLKSIRTQEKAQLQDLNDRFASFIEKVHELEQQNKVLEAELLVLRQKHS-EPSRFKA
K8   -VTVNQSLLSPLVLEVDPNIQAVRTQEKEQIKTLNNKFASFIDKVRFLEQQNKMLETKWSLLQQQKT-ARSNMDN
K18  -RGGMGSGGLATGIAGGLAGMGGIQNEKETMQSLNDRLASYLDKVRSLETENRRLESKIREHLEKKG-PQVRDWS
     .    ..    ... .  . x.x**:.:x::**xx.*.xxxx*x.**..*:.*. .:... ... ... ...
                                                                    coil 1B
     cdefgabcdefgabcdefgabcdefgabcdefgabcdefgabcdefgabcdefgabcdefgabcdefgabcde
     LYETELADARRALDDTARERAKLQIELGKCKAEHDQLLLNYAKKESDLNGAQIKLREYEAAALNSRDAALATALGDKKSLE
     AYEAELGDARKTLDSVAKERARLQLELSKVREEFKELKARNTKKEGDLIAAQARLKDLEALLNSKEAALSTALSEKRTLE
     LYEEEMRELRRQVDQLTNDKARVEVERDNLAEDIMRLREK----------------------------LQEEMLQREEAE
     LYEQEIRDLRLAAEDATNEKQALRGEREGLEETLRNLQAR----------------------------YEEEVLSREDAE
     MFESYINNLRRQLETLGQEKLKLEAELGNMQGLVEDFKNK----------------------------YEDEINKRTEME
     HYFKIIEDLRAQIFANTVDNARIVLQIDNARLAADDFRVK----------------------------YETELAMRQSVE
     :xx  :x ::*:.:: ...x::x .x:..........:. :              ...:x  x ..*
                                                                 L12
     fgabcdefgabcdefgabcdefgabcdefgabcdefgabcdefgabcdefgabc_____
     GDLEDLKDQIAQLEASLAAAKKQLADETLLKVDLENRCQSLTEDLEFRKSMYEEEINETRRKHETRLVEVDSGRQIEYEY
     GELHDLRGQVAKLEAALGEAKKQLQDEMLRRVDAENRLQTMKEELDFQKNIYSEELRETKRRHETRLVEIDNGKQREFES
     NTLQSF--------------RQDVDNASLARLDLERKVESLQEEIAFLKKLHEEEIQELQAQIQEQHVQIDVDVSK---P
     GRLME---------------RKGADEAALARAELEKRIDSLMDEISFLKKVHEEEIAELQAQIQYAQISVEMDVTK---P
     NEFVLI--------------KKDVDEAYMNKVELESRLEGLTDEINFLRQLYEEEIRELQSQISDTSVVLSMDNSRS--L
     NDIHGL--------------RKVIDDTNITRLQLETEIEALKEELLFMKKNHEEEVKGLQAQIASSGLTVEVDAPKS--Q
     ..: .:           xx :::. x x::x* ::::x xxx *:x.:.x**x.x::.::  .  x x:.: .:
```

```
     coil 2A           L2                                     coil 2B
     cdefgabcdefgabcdefg_____defgabcdefgabcdefgabcdefgabcdefgabcdefgabcdefg
     KLAQALHEMREQHDAQVRLYKEELEQTYHAKLENARLSSEMNTSTVNSAREELMESRMRIESLSSQLSNLQKESRACLER
     RLADALQELRAQHEDQVEQYKKELEKTYSAKLDNARQSAERNSNLVGAAHEELQQSRIRIDSLSAQLSQLQKQLAAKEAK
     DLTAALRDVRQQYESVAAKNLQEAEEWYKSKFADLSEAANRNNDALRQAKQESTEYRRQVQSLTCEVDALKGTNESLERQ
     DLSAALKDIRAQYEKLAAKNMQNAEEWFKSRFTVLTESAAKNTDAVRAAKDEVSESRRLLKAKTLEIEACRGMNEALEKQ
     DMDSIIAEVKAQYEDIANRSRAEAESMYQIKYEELQSLAGKHGDDLRRTKTEISEMNRNISRLQAEIEGLKGQRASLEAA
     DLAKIMADIRAQYDELARKNREELDKYWSQQIEESTTVVTTQSAEVGAAETTLTELRRTVQSLEIDLDSMRNLKASLENS
     :x.  :x xxx:*:x..:  :.. x.x  :  :.:.... .: .:. x..x:.x:.x.x: x :x. :x..:. ...:x
```

```
     ↓stutter                              intra inter
                                           +...- -....+      |—>> TAIL —>>
     abcdefgdefgabcdefgabcdefgabcdefgabcdefgabcdefgabcdefgabcdefgab
     IQELEDLLAKEKDNSRFMLTDKEREMAEIRDQMQQQLNDYEQLLDVKLALDMEISAYRKLLE-GEEERLKLSPSP- LaB1
     LRDLEDSLARERDTSRRLLAEKEREMAEMRARMQQQLDEYQELLDIKLALDMEIHAYRKLLE-GEEERLRLSPSP- LaA
     MREMEENFAVEAANYQDTIGKLQDEIQNMKEEMARHLREYQDLLNVKMAALDIEIATYRKLLE-GEESRISL-PLP- Vim
     LQELEDKQNADISAMQDTINKLENELRTTKSEMARYLKEYQDLLNVKMALDIEIAAYRKLLE-GEETRLSFTSVG- NFL
     IADAEQRGELAIKDANAKLSELEAALQRAKQDMARQLREYQELMNVKLALDIEIATYRKLLE-GEESRLESGMQN- K8
     LREVEARYALQMEQLNGILLHLESELAQTRAEGQRQAQEYEALLNIKVKLEAEIATYRRLLEDGEDFNLGDALDS- K18
     x.x:*:. :.:.... . .x .:x xx. .x :x.::x.x**:.*x:x*xx*x:**:.**x*** **x.xx ... .
```

sequences of invertebrate lamins, such as *D. melanogaster* lamin C and the single *C. elegans* lamin, respectively, show prolines in this heptad (82, 83). Also, both the lamin of the starfish, *Asterias rubens* and *Astropecten brasiliensis*, have prolines in the very same place, i.e., the *f* position of the sixth heptad. A "helix breaker" in this position, however, calls for new structural concepts regarding this part of coil 1B in these invertebrate lamins. How this correlates to the structure in the corresponding part of vertebrate lamins is entirely elusive. The color code in Figure 2 indicates more areas of interesting differences between A- and B-type lamins as well as between lamins and cytoplasmic IF proteins; these differences may be crucial for recognition of individual chains by either one another or by chaperone-type molecules assisting coiled-coil formation in vivo.

Cytoplasmic chordate IF proteins have lost 42 amino acids in coil 1B. Whether this has occurred by a single deletion step or by two independent ones is not known. However, sequence alignment by conventional computer programs reveals, as one possibility, two deletions of four and two heptads, respectively. This kind of tinkering at the two-heptad level is indeed possible without impeding assembly properties completely as shown by Ip and coworkers (84) for vimentin. Some conspicuous sequence conservation is observed in the first three and the last two heptads of the rod, the "highly conserved end segments." Otherwise, sequence identity is low, in particular in keratin 18. Notably, one heptad at the end of coil 1B in vimentin is very similar to lamin B_2 (5 of 7 amino acids are identical), although it does not exhibit a distinct sequence motif, such as the triple Glu sequence close to the end of coil 1B (Figure 1). Half a heptad further down, lamins have a triple basic amino acid cluster, RRK in lamin B_1, RRR in lamin B_2, and KRR in lamin A. In the cytoplasmic IF proteins, this stretch is principally conserved as QAQ except for keratin 8, which has QSQ.

Figure 1 Amino acid sequence comparison of the α-helical rod domain of human IF proteins from the five sequence homology classes (SHC). The designations above the one-letter-coded sequences indicate the borders of the principal domains. The abbreviations are HEAD, non-α-helical amino-terminal domain; coil 1A, coil 1B, coil 2A, and coil 2B, consecutive coiled-coil forming α-helical segments of the rod; TAIL, non-α-helical carboxy-terminal domain; L1, L12, and L2, connecting segments of unspecified structure or linkers (see text); LaB1, lamin B_1 (SHC5); LaA, lamin A (SHC5); Vim, vimentin (SHC3); Des, desmin (SHC3); NFL, low molecular weight neurofilament triplet protein (SHC4); K8, keratin 8 (SHC2); K18, keratin 18 (SHC1). Lowercase letters designate putative heptad position *abcdefg*. The *a* and *d* positions are highlighted (*yellow*). A discontinuity in the heptad pattern in coil 2B is indicated as "stutter." At the end of coil 2B, the positions of an intrahelical salt bridge (*intra*) and an interhelical salt bridge (*inter*) are indicated. The symbols below the sequences indicate the sequence similarity scores of a particular residue in the six proteins: *, identical in all seven compared proteins, s = 1.0; ×, $0.75 \leq s < 1.0$; ;, $0.5 \leq s < 0.75$; ., $0.25 \leq s < 0.5$ [for details see (97)]. The boxes mark segments with unusually high sequence identity, i.e., the IF consensus motifs. Prolines are marked in gray.

TABLE 1 Amino acid composition of the non-α-helical head domains of nuclear and cytoplasmic IF proteins[a]

Protein		Total	Basic	Acidic	Proline	Hydroxyl	Aliphatic	Aromatic
Lamins								
Human	A	30	5	1	3	10	6	2
	B_1	31	4	—	6	8	7	—
Xenopus	A	26	5	1	3	8	4	—
	B_1	31	5	—	4	11	5	—
Goldfish	B_2	29	4	1	3	9	8	—
Drosophila	Dm_0	53	6	—	12	17	7	—
	C	55	7	—	3	19	9	—
Caenorhabditis		44	6	4	0	15	8	1
Priapulus		36	6	—	5	13	6	—
Hydra		24	4	1	3	5	4	2
Cytoplasmic IF Proteins[b]								
Vimentin[c]		82	11	—	5	31	7	6
GFAP[c]		48	9	1	6	7	12	1
Keratin 18		79	6	—	2	21	19	6
Keratin 8		90	9	1	4	25	21	6

[a]The number of the respective amino acids is presented. Hydroxyl includes serine and threonine; aliphatic includes leucine, isoleucine, valine, and alanine; and aromatic includes phenylalanine and tyrosine.

[b]Values for human proteins are shown. The sequence of the goldfish lamin is from *Carassius aurata*. The other species are *Xenopus laevis, Drosophila melanogaster, Caenorhabditis elegans, Hydra vulgaris,* and the living fossil *Priapulus caudatus*. For sequence information see (139).

[c]SHC3 and SHC4 proteins contain a 20-amino acid segment preceding coil 1A predicted to form an α-helix, which is therefore not considered a head here.

The next amino acid is an isoleucine (a leucine in desmin) for the cytoplasmic IF proteins and, hence, may be an extended "stutter" segment of the coiled coils. These few examples may suffice to demonstrate how little is known about the translation of individual primary sequence motifs into distinct structure and ultimate function.

Structural Principles

It is routine practice for IF biologists to analyze sequences of new, suspected IF proteins for uninterrupted amino acid stretches compatible with the α-helical fold and to distinguish them from those containing a proline or multiple glycines. The second parameter to identify is the heptad repeat pattern of the form $(abcdefg)_n$ where a and d are hydrophobic and $b, c,$ as well as e, f, g positions are often

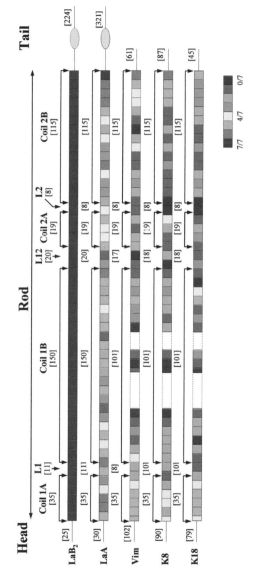

Figure 2 Schematic representation of heptad identity in the coiled-coil forming central α-helical rod of human IF proteins. The number of identical amino acids within a heptad is color coded as indicated in the bottom right. The number in brackets indicates the number of amino acids of the corresponding subdomain (labeled as in Figure 1). The blue ellipsoid in the lamin tail domains represents the Ig-fold (see below).

occupied by charged residues (29, 85). In this way, generalized models, such as the one shown in Figures 1 and 2, have been derived (29). They infer the occurrence of a series of α-helices able to form coiled coils of 35 (coil 1A), 101 (coil 1B), 19 (coil 2A), and 115 (coil 2B) amino acids in length. Lamins are different because they have a "long rod," i.e., exhibit 150 amino acids in their coil 1B. These coiled-coil forming fragments are connected by so-called linkers, i.e., L1 (between coils 1A and 1B), L12 (between coils 1B and 2A), and L2 between coils 2A and 2B) as depicted in Figure 2. L2 is remarkable in that it is made of 8 amino acids in all members of the five sequence homology classes. In particular, this number is also absolutely conserved in the more than 20 known lamins, from *Hydra* to man, and most importantly, it contains no proline. In vertebrate lamins, the first amino acid of this linker is a tyrosine, the fourth is a leucine, and the first amino acid of the presumed start of coil 2B is a tyrosine, assumed to be in a *d* position. Different from previous structural predictions, one may reconsider this linker as a longer structural unit connecting two normal *a* to *g* heptads as an α-helix of the type $(abcd)_3$ where *a* is hydrophobic. This type of "extended stutter" reads by consensus in vertebrates:

$$(abcdefg)\ \textbf{YKEELEQTYQAK}\ (abcdefg),$$

and it may indeed introduce a noncoiled-coil part into the rod with increased flexibility gained by the parallel helices. In cytoplasmic IF proteins, the number of amino acids is again absolutely conserved, but the nature of the amino acids differs significantly from that of vertebrate lamins as shown here for vimentin (see Figure 1):

$$(abcdefg)\ \textbf{NLQEAEEWYKSK}\ (abcdefg).$$

Whereas the first half of this row is very acidic (three glutamic acids) in the fourth, sixth, and seventh positions, this is shielded then by a tryptophan from a very basic charge (two lysines). This sequence is highly conserved in vertebrates, i.e., in both vimentin and desmin from shark to man, even though these proteins differ considerably in assembly properties, including their temperature-dependence of assembly. In particular, the last 9 amino acids EAEEWYKSK are 100% identical from shark to human vimentin as well as shark to human desmin. In lamins, the charges are more balanced within this sequence, and instead of a bulky tryptophan, there is a threonine. However, the heptad before L2 also differs considerably between nuclear and cytoplasmic IF proteins: His and Glu are found in the *a* and *d* position of lamins, which are rather suboptimal amino acids for hydrophobic interaction, whereas the Tyr and Leu (Iso) in the corresponding *a* and *d* positions of all cytoplasmic IF proteins are indeed good ones for coiled-coil formation (86). To more rationally understand these sequence differences, atomic structure is urgently needed for this part of the rod domain. The same applies for linker L1 and L12, where L1 is highly conserved by number in all lamin sequences just as L12, except for the *Hydra* lamin, which has one amino acid more, and the *Tealia* lamin, which has three amino acids more in L12. The strong conservation of the number of amino acids constituting linker L12 calls, however, for independent sequence confirmation of these cDNA clones.

BACTERIAL IF PROTEINS   For both tubulin and actin, bacterial proteins have been recently described with structure and function that resemble the well-characterized components of microtubules and microfilaments (86a). In contrast, it has been a matter of debate for quite some time whether plants, lower eukaryotes, and bacteria express IF-like proteins. To this end, the yeast protein MDM1 has been proposed to be IF related (86b). Although this protein exhibited considerable potential α-helical coiled-coil forming segments and self-assembled into filaments, its primary sequence and its overall structural organization did not really suggest an evolutionary relationship to IF proteins (28). Most recently, a bacterial protein, crescentin, was identified that is involved in the establishment of bacterial cell architecture by mediating the vibrioid and helical cell shapes of *Caulobacter crescentus* (86c). Indeed, its primary sequence organization resembles that of IF proteins, with four α-helical segments, interrupted by non-α-helical linker domains and a stutter in the second half of the fourth α-helical segment, although this stutter is not located in exactly the same relative position as in metazoan IF proteins. Whereas this IF-like rod domain is flanked by a head and a tail domain, the lengths of the individual segments that may correspond to coil 1A, 1B, 2A, and 2B are not directly comparable to those of the metazoan IF proteins. Moreover, the conspicuous IF-specific consensus sequence elements at either end of the rod, which are conserved from *Hydra* lamin to human hair keratins, are completely absent from crescentin. Whether this protein is an independent bacterial invention or the result of a strong sequence drift cannot be decided at present. It should be noted, however, that some invertebrate IF proteins exhibit considerable alterations with respect to the IF domain structure principle, such as truncations within coil 2B or the tail domain (82). Moreover, mammalian IF proteins, such as the beaded filament proteins phakinin and filensin, have acquired strong alterations within their sequence, in particular in the coil 2B IF consensus motif. This modified consensus sequence has recently been found to be conserved in phakinin of the fish *Fugu rubripens*, demonstrating that IF sequences generally do not change at a high rate (86d). In any case, it will be interesting to investigate in more detail the structure and assembly mechanism of crescentin in relation to known IF proteins.

STRUCTURE DETERMINATION OF INTERMEDIATE FILAMENT PROTEINS

X-Ray Crystallography of the Vimentin Rod

IF proteins were among the first proteins to be analyzed by X-ray diffraction methods. In the 1930s, William Astbury used keratins, proteins arranged in a highly ordered manner within hair and nails, to find regularities in the fold of protein chains [reviewed in (87)]. However, it was Francis Crick, who 20 years later, after Linus Pauling's description of the α-helix (88), interpreted the old diffraction patterns correctly in "The Packing of α-Helices: Simple Coiled Coils" (5). In the following years, hundreds of atomic structures of proteins were solved, but progress in the determination of structure in any IF protein was nil. The reason for this lies simply in the fact that larger pieces of coiled-coil structure do not grow into larger, usable crystals. For this very reason, more detailed knowledge on coiled coils stems from investigations of tropomyosin, myosin, and peptides, such as the coiled-coil fragment of the yeast transcription activator GCN4 (89–91).

Crystallization of the oligomerization domain of cortexillin I, an actin-bundling protein from the slime mold *Dictyostelium discoideum* (92), marked a big leap forward. This recombinant 126–amino acid fragment, representing 18 contiguous heptads, grew to thin plates up to a size of $370 \times 370 \times 30\ \mu m$ within 6 months. In addition to a strong "hydrophobic seam" holding the two super-coiled α-helices together via leucines and isoleucines in heptad a and d positions, the crystal structure validated the concept of a coiled-coil "trigger site" (93), which involved a distinct network of intra- and interhelical salt bridges. Corresponding trigger sites have, in a slightly deviated form, also been found in IF proteins (94).

Early attempts to crystallize fragments of vimentin, i.e., the complete rod as well as coil 1 and coil 2 separately, failed (S.V. Strelkov and H. Herrmann, unpublished observations). Therefore, we followed a divide-and-conquer strategy by expressing 25–100–amino acid long fragments of the human vimentin rod in *Escherichia coli* (95). The first crystals obtained were from a synthetic peptide consisting of a 29–amino acid fragment containing the carboxy-terminal IF consensus motif, which was fused in frame to the carboxy terminus of the 31–amino acid leucine zipper domain of GCN4. The rationale for this was that the GCN4 oligomerization domain, demonstrated to adopt a dimeric coiled-coil conformation (96), would force the vimentin rod fragment into a dimeric coiled coil too. The resulting 1.9 Å resolution crystal structure confirmed this assumption and, in addition, revealed three important features (26)

1. The acidic end segment of helix 2B, LEGEESRI, was not in a coiled-coil conformation, but starting with isoleucine 397, the two α-helices gradually separated and past leucine 405 bent away from the coiled-coil axis. Hence, with leucine 404 in an a position, the spacing of the hydrophobic seam was 9.4 Å compared to the typical spacing in a coiled coil of $\sim$6.3 Å.

2. Arginine 401, residing in an e position on one chain, forms an interhelical salt bridge with glutamic acid 396, which is in a g position on the other chain. Evidently, only one such salt bridge is formed because Glu 396 and Arg 401 on the opposite chains are in unfavorable positions to specifically interact.

3. The highly conserved lysine 390 forms an *a*-to-*e* type intrahelical salt bridge with aspartic acid 394 in both α-helices. In some more distantly related IF proteins, this bridge may be formed by a Lys-to-Glu or an Arg-to-Glu pair (26).

This structure was later confirmed by crystallization of a larger coil 2B fragment, i.e., Cys 328 to Ile 411. This 84–amino acid fragment was obtained by chemical cysteine-specific cleavage of recombinant human tailless vimentin. It contained the so-called stutter region where the heptad pattern of coil 2B is disrupted, i.e., either missing three amino acids or harboring four extra amino acids (97). Evidently, this stutter, the first one solved at atomic detail, is tolerated without destroying the coiled-coil geometry. As documented in Figure 3*B*, it represents a segment within the vimentin dimer in which the two supercoiled α-helices run almost parallel to each other without interrupting the coiled-coil geometry. This is an important finding because all IF proteins appear to have retained this heptad discontinuity, which is spaced exactly six heptads away from the evolutionarily conserved IF consensus motif that marks the C-terminal end of coil 2B. At this time, the functional significance of this highly conserved stutter remains elusive. The corresponding dimeric fragment of lamin A exhibits a very similar fold like vimentin (Figure 3*C*). In particular, the structure around the stutter is identical. At the end of the rod, the individual chains of lamin A diverge even more strongly away from the axis of the coiled coil, probably due to an additional glutamic acid at this site (S.V. Strelkov, J. Schumacher, U. Aebi and H. Herrmann, in preparation).

The crystal structure of coil 1A was also solved. Interestingly, although strictly α-helical, the corresponding 39–amino acid fragment was monomeric in solution as well as in the crystal (95). However, rather than being straight, the α-helix was slightly bent with its hydrophobic seam lining the concave face. In fact, the backbone of this 39–amino acid coil 1A structure could be superimposed with good precision onto the GCN4 leucine zipper backbone so that a two-stranded coiled coil of this IF fragment could be built (Figure 3*A*). In addition to four distinct intrahelical salt bridges, this "synthetic" coil 1A dimer reveals a strong *g*-*e*'-type interhelical salt bridge between Lys 120 of one helix and Glu 125 of the other helix. Evidently, the monomeric state of this coiled-coil forming α-helical fragment is caused by its charged N and C termini, because it was derived by thrombin cleavage of a recombinant fusion protein. In contrast, the corresponding chemically synthesized coil 1A peptide with an acetylated N and an amidated C terminus did form a significant amount of coiled-coil dimer in solution (S. Strelkov, personal communication).

By itself the coil 1A dimer appears relatively unstable, but in the complete dimer, it may be strongly stabilized by interacting with the highly basic head domains. Notably, leucines and aromatic residues are prevalent and highly conserved in *a* and *d* positions of coil 1A, which, in turn, may cause this instability (see Figure 3*A*). Moreover, all six human IF proteins aligned in Figure

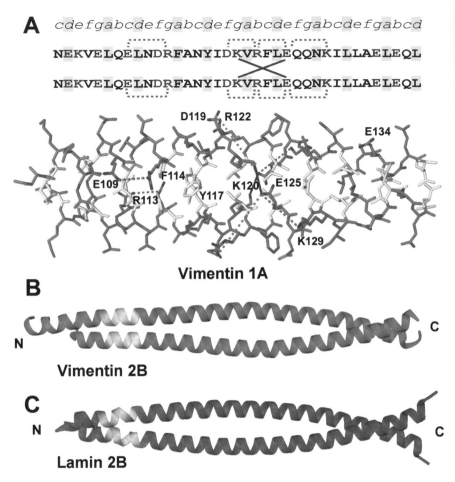

A *cdefgabcdefgabcdefgabcdefgabcdefgabcd*

NEKVELQELNDRFANYIDKVRFLEQQNKILLAELEQL

NEKVELQELNDRFANYIDKVRFLEQQNKILLAELEQL

Vimentin 1A

B

Vimentin 2B

C

Lamin 2B

Figure 3 Atomic model of major parts of the coiled-coil dimer of human vimentin and lamin A. (*A*) Crystal structure of vimentin coil 1A. The position of potential intra- (*green dotted lines*) and interhelical (*red straight lines*) salt bridges are indicated both in the primary sequence (*top*) and the atomic model (*bottom*). (*B*) Crystal structure of the second half of coil 2B of vimentin. (*C*) Crystal structure of the second half of coil 2B from human lamin A. In (*B*) and (*C*) the region encompassing the "stutter" is lighter.

1 harbor an asparagine in the *a* position of the fourth heptad of coil 1A. This is even more conserved than the Asn-Asp-Arg motif (in *efg* positions) in the first complete heptad that, when mutated, causes severe human diseases. Preceding this asparagine, lamins and keratin K18 reveal a Glu-Thr-Glu motif. All the other IF proteins have Glu-Gln-Gln, but in the heptad following it, their primary sequences are very different. This is just one example that documents how little

we presently understand about coiled coils in the context of supramolecular assemblies, such as an IF. Do changes in primary sequence affect the stiffness and/or stability of the filament? What is their role in tetramer formation or filament assembly? Or do coiled-coil interactions mediate "strain hardening" (98) of IFs?

Coiled coils are highly versatile oligomerization motifs not only in IF proteins but also in viral and bacterial proteins, as well as in vertebrate kinesin, myosin, and DNA repair enzymes that can be fine-tuned for a wide range of functional tasks (99–104). In this sense, α-helical coiled coils, e.g., in muscle proteins, exemplify simplicity and economy of protein design: Small variations in sequence lead to remarkable diversity in cellular functions (105). Moreover, the ability of α-helices to form hetero-oligomeric coiled coils of differential inter-active strength renders them ideal constituents of the transcriptional regulation machinery. For example, in the basic region leucine zipper (bZIP) transcription factors considerable partner selectivity was found in the pairing of the 49 human and 10 yeast bZIP domains (106). These domains are typically 30–40 amino acids long, i.e., the size of coil 1A of IF proteins. During development, the specific heterodimer formation of three such transcription factors may determine if a cell will divide and proliferate (Myc-Max) or differentiate and become quiescent (Mad-Max) (107). This brief detour into some of the known functional tasks of α-helical coiled coils may give a flavor as to what remains to be unveiled about the over 300–amino acid long segmented rod of a particular IF protein.

Atomic Structure of the Lamin A Tail Domain

The non-α-helical tail domain can vary drastically between different IF proteins. In the lens protein phakinin or CP49, for example, it is completely missing. Instead, filensin, the obligatory assembly partner of phakinin, harbors a very long tail (108), which may be involved in specific interactions with distinct cellular structures, one being the membrane skeleton (109). Similarly, the two neurofila-ment proteins NF-M and NF-H have long tails with highly charged domains and multiple phosphorylation sites. These domains have been shown to radially project out from the filament backbone when NF-M and NF-H are coassembled with the "short-tail" neurofilament protein NF-L, thereby appearing similar to a millipede (110). In contrast, self-assembly of NF-L yields normal looking 10-nm filaments.

Most significantly, A- and B-type lamins have relatively long tails that carry an evolutionarily conserved 105-amino acid "homology box" (111). Because several mutations were described that give rise to muscular dystrophies and lipodistrophies in this domain of the human lamin A molecule, its atomic structure was solved both by X-ray crystallography (112) and NMR spectroscopy (113). It adopts a novel, all-beta immunoglobulin-like fold (Figure 4), and this compact domain corresponds to the globular appearance in EM (Figure 5A). The structure immediately indicates that the mutations leading to muscle-specific diseases or to lipodistrophy are found in completely different locations. Whereas

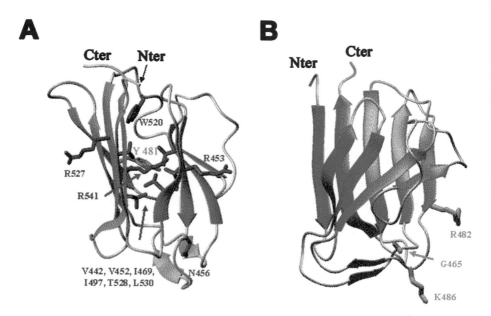

Figure 4   Atomic model of the evolutionarily conserved 105–amino acid box of the non-α-helical carboxy-terminal domain of human lamin A. (*A*) Ribbon diagram highlighting the position of amino acid changes causing cardiac and skeletal muscle diseases: Emery-Dreyfuss muscular dystrophy (*red*) and limb girdle muscular dystrophy (*mustard*). (*B*) Mutations causing Dunnigan-type familial partial lipodistrophy are indicated (*green*). The ribbon diagram is oriented as in (*A*) but rotated by 90° around the vertical axis. Micrographs provided by Sophie Zinn-Justin. From (113), reproduced with permission.

the lipodistrophy mutations are clustered in the corner of the Ig-fold, indicating they mediate interactions with neighboring molecules, the muscle-specific mutations reside within and on the surface of the globular fold and are most likely engaged in the proper stabilization of the entire domain in addition to "cross talk" with other nuclear proteins (Figure 4).

MOLECULAR MECHANISMS OF FILAMENT ASSEMBLY

Three Assembly Groups of Vertebrate IF Proteins

In order to investigate how an IF is formed, one must (*a*) obtain soluble building blocks of defined biophysical nature and (*b*) develop meaningful assays to monitor their assembly. As outlined above, the biochemical properties of the many known IF proteins vary considerably, even within one species, and extensive investigations with representatives of the five sequence homology classes (SHCs) gave rise to the distinction of three principal assembly modes.

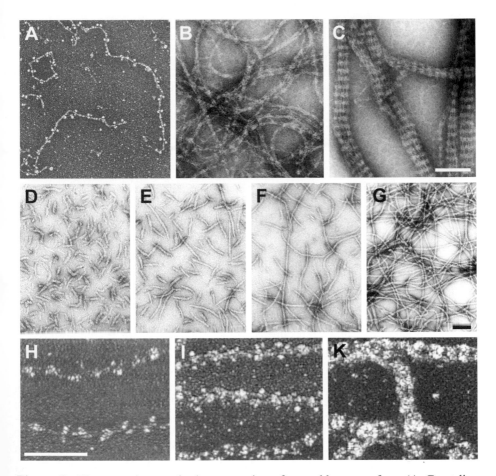

Figure 5 Electron microscopic documentation of assembly stages from (*A–C*) rat liver lamin A/C, (*D–G*) human vimentin, and (*H–K*) mixtures from *Ascaris* high and low IF proteins. (*A–C*) Lamin A/C dialyzed into pH 6.5/150 mM NaCl buffer yielding (*A*) head-to-tail fibers, (*B*) beaded filaments that, particularly in the presence of Ca^{2+}, (*C*) extensively laterally associate to eventually yield paracrystals (17, 118). (*D–G*) Assembly of recombinant human vimentin was initiated by addition of filament buffer and fixed after (*G*) 10 s, (*H*) 1 min, (*I*) 5 min, (*K*) 1 h by addition of glutaraldehyde to 0.1%. (*H–K*) Recombinant *Ascaris* protein assembles, dependent on the ionic conditions, into (*H*) protofilaments, (*I*) octameric protofibrils, or (*K*) full-width filaments. Note the regularly spaced globular domains (140). Bars are 100 nm.

One may therefore categorize IF proteins into the following three assembly groups: assembly group 1, SHC I and II; assembly group 2, SHC III and IV; and assembly group 3, SHC V (36). These three assembly groups can coexist as three separate IF systems within one and the same cell. How this segregation is achieved in vivo is not known, although their tendency to copolymerize in vitro is indeed low (114, 115). The members of two of the five SHCs, i.e., the keratins (SHC I and II), are obligate heteropolymers. Two other classes, the desmin-/vimentin-type (SHC III) and the neurofilament triplet proteins (SHC IV), are able to form homopolymers in vitro, but in the cell, they frequently appear as copolymers. Finally, the nuclear lamins (SHC V) do not assemble in the cytoplasm; instead the majority is recruited into a distinct "nuclear lamina" that is attached to the inner nuclear membrane.

IF assembly already begins in urea-containing buffers. For example, vimentin dimerizes upon lowering the urea concentration to about 6 M (30). In contrast, the epidermal keratin pair K5/14 forms heterodimers even in 9.5 M urea (116). In 4.5 M urea, vimentin exists almost completely in tetrameric complexes, and further dialysis into physiological buffers yields long, uniform IFs, whereas dialysis into low ionic strength buffers (cf. 5 mM Tris-HCl, pH 8.4, or 2 mM sodium phosphate, pH 7.5) freezes vimentin in the tetrameric state.

LAMINS (ASSEMBLY GROUP 3)   In contrast to the cytoplasmic IF proteins, nuclear lamins form dimers in high-pH buffers at medium ionic strength, i.e., in 5 mM Tris-HCl, pH 8–9, and 150–300 mM NaCl (17, 117, 118). By EM of glycerol-sprayed/metal-shadowed specimens, lamin dimers appear as ~50-nm long rods with two globular domains at one end, which adopt an Ig-like fold as outlined above (see Figure 4). By dialysis into more physiological buffers (i.e., pH 6, 300 mM NaCl), lamin dimers associate head-to-tail into protofilaments (Figure 5A), which further laterally associate into extended fibrils exhibiting a pronounced "beading" (Figure 5B) (17, 119). In contrast, dialysis into high-pH buffers (pH 9.0) in the presence of 20–25 mM $CaCl_2$ yields several μm-long paracrystalline fibers with a pronounced 24.5-nm repeat light-dark transverse banding pattern (early forms are shown in Figure 5C) (17, 118, 120). Although recombinant vertebrate lamin polypeptides do not form stable bona fide 10-nm filaments in vitro, authentic rat liver lamin A as well as the single B-type lamin of C. elegans, under certain conditions, yield relatively uniform 10-nm filaments that appear indistinguishable from cytoplasmic IFs by EM (17, 121). There remains, however, the concern that the assembly products observed by negative stain EM, because of selective adsorption to the EM grid, may in fact only represent a minor species (122). Moreover, kinetic intermediates, which have been very informative with cytoplasmic IF assembly (see below), have hardly been explored with lamins. Despite these difficulties in documenting significant formation of stable 10-nm filaments in vitro, in situ lamins clearly do assemble into long, uniform cytoplasmic-like IFs.

An unresolved issue concerns the coassembly of A- an B-type lamins into heteropolymers. In vivo observations clearly favor the segregation of A- and B-type lamins into homopolymers (80). This segregation may be controlled at the translational level, i.e., by selective degradation of heterodimers over homodimers prior to their assembly into lamin polymers. A similar mechanism has been reported to assure the selective supply of α-/β-spectrin heterodimers by rapid proteolytic degradation of possible α-/α- and β-/β-homodimers (123).

In a physiological environment lamins are integrated into a complex supramolecular network containing a wealth of inner nuclear membrane and chromatin binding proteins (see below). Several of these, for example the lamina-associated proteins (LAPs), selectively bind to either A- or B-type lamins before they integrate into the lamina network (38). Consequently, in vitro assembly experiments with pure lamins may only capture their homotypic interactions, which, in the absence of the appropriate binding partners, may produce structures that are never found in the cell except in case of their ectopic or overexpression. One such missense product may be the lamin paracrystals. These assemblies are indeed observed in cultured insect cells after overexpression of *Xenopus* lamin A using the baculovirus system (124). Thus, the in vitro formation of paracrystals represents a lamin-lamin association mode that is attenuated or suppressed by cellular factors in vivo.

DESMIN-LIKE PROTEINS (ASSEMBLY GROUP 2)   Vimentin and desmin dimers rapidly assemble laterally when the ionic strength is raised, e.g., by addition of an equal volume of 100 mM NaCl or KCl (30, 125). This process can be followed by EM because it is arrested instantaneously by the addition of an equal volume of 0.2% glutaraldehyde in filament buffer (30, 31). Therefore, assembly intermediates can be monitored in the subsecond range. After one second of assembly, the structures revealed by negative stain EM (Figure 5D) are short, uniform filaments with a diameter of 16 nm and a length of $\sim$60 nm, i.e., the length of antiparallel, approximately half-staggered vimentin tetramers (see Figure 6B) (30). Because of their similar, albeit somewhat larger diameter than IFs, these assembly intermediates were termed unit-length filaments (ULFs) (30). Following this first phase of lateral association of dimers into ULFs, a second phase of longitudinal annealing of ULFs was observed (see Figure 6B), yielding filaments averaging $\sim$250 nm in length after 60 s (Figure 5E). Concomitantly with the further longitudinal growth of the filaments, the soluble material, very likely tetramers/octamers, was dramatically reduced (Figure 5F). In a third phase, the still $\sim$15-nm diameter, irregular filaments underwent a cooperative radial compaction yielding a smaller, $\sim$11 nm, more uniform diameter (Figure 5G). The same compaction event was also observed with desmin, although desmin IFs are clearly distinguishable from vimentin IFs (see below) (31). Qualitatively, radial compaction is relatively independent of the buffer conditions employed. With five different assembly regimes, Stromer et al. (126) consistently observed a 1-nm reduction of the desmin IF diameter occurring between 5 and 60 min after initiation of assembly. With our regime, the major structural rearrangements, resulting in reduction of the average filament diameter from 16 to 11 nm were nearly complete after 5 min (30, 31). Nevertheless, the early findings by Stromer and coworkers are qualitatively consistent with ours, although the authors did not fully rationalize the significance of this "maturation" step. Meanwhile, this 3-step

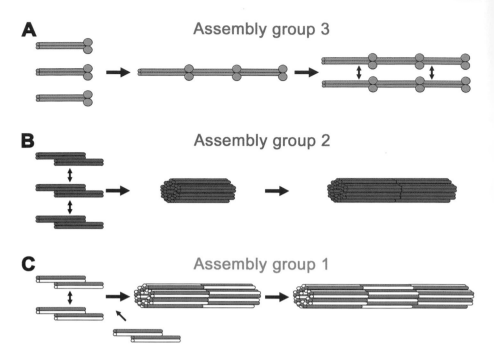

Figure 6 Schematic model of the prime association reactions occurring between dimers and double dimers, respectively, of the three major IF assembly groups. (*A*) Assembly group 3. Lamin dimers associate first into head-to-tail filaments that later laterally associate. The orientation of the two associating filaments is arbitrary. (*B*) Assembly group 2. Vimentin-type assembly starts from antiparallel, half-staggered double dimers (or tetramers) to form full-width, unit-length filaments (ULF). (*C*) Assembly group 1. Keratins assemble from heterodimeric tetramers by lateral and nearly concomitant longitudinal assembly into heterogenous full-width filaments. Only at very low protein concentrations are ULF found (60, 187).

assembly scheme has been documented for all assembly group 2 IF proteins, including vimentin and desmin from shark to man (21, 31, 127, 128), α-internexin, a neuronal protein (78), and the low molecular weight neurofilament triplet protein NF-L (31).

The assembly process of human vimentin has also been followed by atomic force microscopy (AFM). Whereas for EM, the specimen has to be dehydrated and contrasted by heavy metal; for AFM, it can be imaged in buffer solution without a contrasting agent (128a). As documented in Figure 7, corresponding AFM images reveal remarkable surface details of ULFs and IFs. Whereas ULFs measured ~70 nm in length and displayed a ~20-nm axial repeat pattern (panels *A* and *C*), individual protofilaments with a right-handed twist could be resolved within mature IFs (panels *B* and *D*). Hopefully, in the future, it will be possible

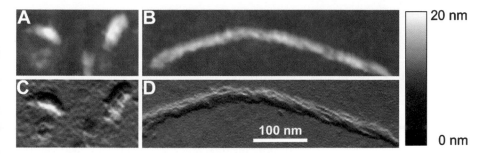

Figure 7   Atomic force microscopy (AFM) of human vimentin assembly intermediates as observed (*A, C*) 5 s and (*B, D*) 5 min after initiation of assembly (the AFM was operated in the tapping mode). Glutaraldehyde-fixed structures were adsorbed to highly oriented pyrolitic graphite and analyzed in solution. (*A, B*) Height images of ULFs and an extended IF, respectively. Note the calibration bar at the right. (*C, D*) Deflection images on the structures visualized in (*A, B*) (micrographs provided by Laurent Kreplak).

to directly follow the assembly of IF dimers into distict oligomers and filamentous polymers by time-lapse AFM.

KERATINS (ASSEMBLY GROUP 1)   A key feature of keratins is their formation of obligate heterodimers from one SHC I and one SHC II group member. The affinity and specificity for their interaction is so strong that even pairs from distantly related species will form heterodimers. The sequence motifs that mediate this recognition have not yet been identified at the molecular level. However, they are so strong and specific that it is possible to determine the SHC-type of a new keratin by interaction screening rather than sequence comparison (129). This type of interaction screening has also been performed on blotted proteins after separation by two-dimensional gel electrophoresis in buffers containing 5 M urea (130, 131). The interaction of soluble keratin tetramers with one another is so strong that no clear ULF phase can be distinguished during filament assembly (see Figure 6*C*) even at low ionic strength, such as 10 mM Tris-HCl, pH 7.5 (31, 116). Only when decreasing the protein concentration far below 0.1 mg/ml are ULFs observed during assembly (31, 60). Evidently, under physiological conditions, keratins very rapidly assemble into polymers, indicating that lateral association and longitudinal annealing are of similar speed, which is distinct from desmin-like proteins in which lateral association of tetramers into ULFs is much faster than their longitudinal annealing (Figure 6*B,C*) (60).

In summary, vertebrate cytoplasmic IF protein dimers spontaneously associate laterally into antiparallel, half-staggered tetramers that, upon raising the ionic strength, associate further into ULFs before longitudinal annealing occurs. In contrast, lamin dimers follow a different assembly path with the head-to-tail association of dimers dominating their lateral association (Figure 6). It was no surprise that trials to produce ULFs from lamins employing vimentin-type assembly regimes have been unsuccessful (D. Lotsch and H. Herrmann, unpublished observations).

Invertebrate Cytoplasmic IF Proteins

Early on, invertebrate organisms, such as squid and snails, were found to contain significant amounts of IF-like structures in epithelia and axons, and their molecular constituents were soon after identified as IF proteins (132–134). Moreover, the nonneuronal IF proteins of snails and nematodes turned out to be closely related to nuclear lamins, i.e., exhibiting a 42–amino acid insertion in coil 1B relative to vertebrate cytoplasmic IF proteins but missing a nuclear localization signal [for an overview see (135)]. In addition, they retained the 105–amino acid homology box in the non-α-helical tail domain that for human lamin A has recently been determined to assume an Ig-like fold (see Figure 4). The same holds true for the invertebrate neuronal IF proteins and all eight described IF proteins of *C. elegans* (136). As an exception, the low- and high-molecular weight neurofilament proteins of the squid lack the globular homology box in their tail domain. In stark contrast, the cephalochordate *Branchiostoma* and the primitive, nonvertebrate chordate *Styela* both harbor the short coil 1B form and miss the homology box in their tail domain (137, 138). As would be expected evolutionarily, the hemichordate *Saccoglossus* exhibits both the coil 1B 42–amino acid insertion and the lamin homology box in its tail domain. In summary, the loss of the 42–amino acid coil 1B extra segment and the loss of the Ig-like homology box are restricted to chordates. Moreover, because no chordate IF protein contains any of these two segments, the cytoplasmic IF protein families must have expanded, by gene duplications, only after the loss of both domains, and therefore, they have most probably been derived from the lamins (139).

Some metazoans, such as insects, may never have acquired a cytoplasmic IF protein or lost it early in evolution. The complexity of organisms and the huge number of specific cell types that express certain IF proteins, in parallel with cellular programs of differentiation, make it difficult to assume one type of function for these many different proteins and also to expect a general filament assembly path. A breakthrough was the dissection by EM of the assembly of the major IF proteins from the snail *Helix pomatia* (i.e., *Helix*-high and *Helix*-low) and the nematode *Ascaris suum* (i.e., *Ascaris*-high and *Ascaris*-low) into filaments (140). Both proteins oligomerize under low ionic strength conditions into dimers, with some tetramers and octamers present. For *Helix*, this condition is 20 mM Tris-HCl, pH 8.5; for *Ascaris*, it is 10 mM Tris-HCl, pH 7.5, 170 mM NaCl. A shift to low ionic strength buffers mediates polymerization; for *Helix*, it is 1 mM Tris-HCl, pH 8.0, and for *Ascaris*, it is 10 mM Tris-HCl, pH 7.0. Intermediate assembly stages were depicted in 10 mM Tris-HCl, pH 8.0, for the *Ascaris* protein, whereas *Helix* required much more nonphysiological conditions, such as 20 mM Tris-HCl, pH 8.5, 250 mM NaCl (140). Taken together, the following filament assembly path was obtained for both *Helix* and *Ascaris*: (*a*) dimer formation with two globular domains at one end; (*b*) formation of antiparallel, half-staggered tetramers with a pair of globules at either end; (*c*) longitudinal annealing of tetramers, first into octamers and eventually into long protofilaments (Figure 5*H*); (*d*) lateral association of the tetrameric protofilaments into octameric protofibrils (Figure 5*I*); and (*e*) octameric protofibrils associate laterally into mature filaments (Figure 5*K*). These filaments are distinctly polymorphic as revealed by STEM mass-per-length (MPL) measurements of unstained specimens. The MPL histogram of the *Ascaris*-high plus -low filaments was fitted by four Gaussian curves peaking at 32, 48, 64, and 80 polypeptides per filament cross section. This indicates that invertebrate cytoplasmic IFs exhibit a relaxed width control and that filaments are able to grow radially by lateral annealing of octameric or even 16-mer-type protofibrils. This assembly behavior contrasts distinctly from that of vertebrate cytoplasmic IFs and, to some extent, resembles that of lamin filament formation.

Toward the Molecular Mechanism of Cytoplasmic IF Assembly

THE ROLE OF THE HEAD DOMAIN   As a result of raising the ionic strength over a certain threshold, tetrameric complexes laterally associate in the subsecond range. Because head-truncated vimentin as well as keratins do not assemble into IFs, their interaction with the highly acidic rods appears to be decisive for this first phase of assembly. Although some kind of tinkering within this domain is tolerated for in vitro assembly, IF formation in vivo is much more restricted (141–143). Moreover, the head domain is essential for the formation of tetramers (30, 62). This was also demonstrated in more detail by stepwise deletion of the amino terminus. Judged by analytical ultracentrifugation and EM of glycerol-sprayed/rotary metal-shadowed specimens, removal of the first 42 amino acids of the 100–amino acid long vimentin head yielded equilibrium mixtures of dimers and tetramers (144). Further deletion, i.e., removal of the first 50 residues, yielded predominantly dimers in low salt buffer. Although vimentin with its first 20 amino acids removed was still able to form IFs, removal of 4 more residues, while assembling into tetramers, yielded neither ULFs nor IFs. Taken together, whereas the region between amino acids 24 and 42 appears to be necessary for directing dimers into tetramers, amino acids 20 to 24, including Arg 23, are critical for the lateral aggregation of tetramers into ULFs. To ultimately localize the active amino acid residues in the head, the arginines in this part of the head domain were replaced one at a time by lysine, and their interaction with the rod monitored by chemical cross-linking, a technique introduced and perfected by Steinert's laboratory (29). Using this approach, it was shown that residues 20 to 45 of the head domain are engaged in both intramolecular interactions, i.e., within dimers, as well as intermolecular, i.e., between two dimers (Figure 8*B*) (145). More specifically, the primary intermolecular target of this part of the head domain is a segment around linker L2, because Arg 23, 28, 36, and 45 all

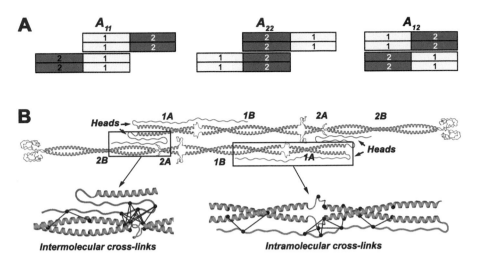

Figure 8 Orientation of the non-α-helical head domains within the tetramer of vimentin. (*A*) Schematic model of dimer-dimer orientation as revealed from chemically cross-linked IFs (29). (*B*) Atomic model of a tetramer in the A_{11} configuration with the head domains depicted according to the identification of cross-linking products (T. Wedig, L. M. Marekov, P. M. Steinert, H. Herrmann, unpublished observations). The enlarged areas depict intermolecular (*red, bottom left*) and intramolecular (*blue, bottom right*) cross-links. The dots indicate authentic lysines (in the coil) and arginines mutationally replaced by lysines (in the head).

cross-linked to residues within this segment when individually mutated to lysines (see Figure 8*B*, *left* inset).

A conclusion concerning a mechanistic model would be that added monovalent ions interfere with salt bridges formed between the arginines of the heads and acidic residues of the rods and thereby liberate the heads from positions to which they had firmly bound during dialysis from urea into low salt buffer. Furthermore, this would imply that the heads should be released from this block by specific peptides competing with these interactions. Indeed, 10-fold molar excess of a 10-mer peptide from the amino terminus of vimentin is able to force the wild-type protein (at 4 μm in 5 mM Tris-HCl, pH 8.4) to assemble into ULF-like polymers [see (142) and references therein]. This peptide motif is evolutionarily highly conserved in SHC III and to some degree also in SHC IV IF proteins (146). The center sequence, SYRRMF, is as potent as the 10-mer peptide in interfering with assembly of wild-type vimentin under standard assembly conditions. The change of one of its two arginines to threonine abolished the effect completely, and IFs formed even in the presence of a 100-fold molar excess of the mutated peptide. A larger peptide consisting of the first 23 amino acids of *Xenopus laevis* vimentin, which was added to the proteins at 20-fold molar excess, was able to push not only wild-type but also headless

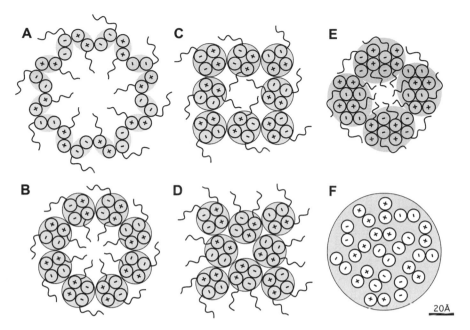

Figure 9   Schematic cross-sectional view of possible dimer/tetramer arrangements in cytoplasmic IF. (*A*) Open, equal dimer; (*B*) microtubule type; (*C*) square; (*D*) compact core type; (*E*) four protofibril; (*F*) random dimer orientation. The plus (+) and minus (-) indicate equally oriented monomers forming a coiled-coil dimer. Only half of the tails are shown (wavy lines).

vimentin and keratins into extensive networks of polymer under low salt/high pH buffers (60, 142). Again, a peptide with the arginine(s) in the conserved motif replaced by threonines was ineffective.

Taken together, if both tetramer formation and association of tetramers to ULFs is mediated by the interaction of arginines in the head domains with acidic residues in the rod domain of the coiled-coil dimers, it is conceivable that in the process of lateral association this network of charge interactions is entirely reorganized also between dimers. As a consequence, the origin of an individual dimer within an ULF from a particular type of tetramer, be it in an A_{11}, A_{22}, or A_{12} configuration (see Figure 8*A*), could no longer be tracked. This, in turn, becomes important when building models that attempt to group the fibrous protein chains with respect to one another in cross section (see Figure 9). In the past, various models for the packing of dimers/tetramers have been presented (28, 29, 147–150). Some of these models did not consider the space requirements of a coiled coil or assumed, for unexplained reasons, an independent positioning of individual dimers (151). Hence, if one translates the events occurring during IF assembly, the first step may involve a symmetrical interaction of tetramers in a circular fashion thereby yielding a relatively open polymer and ULF (Figure

9*A*) that may spread flat upon adsorption to an EM grid and therefore appear relatively wide, i.e., 16 nm or more in diameter (cf. Figure 5*D*) (30, 31). The reorganization that occurs during longitudinal annealing and radial compaction may constitute tetrameric units in various relative positions, thereby giving rise to microtubular (Figure 9*B*), square (Figure 9*C*), core units versus peripheral units (Figure 9*D*), or cross sections in which tetrameric units might be formed by different pairs of neighboring dimers. Alternatively, octameric protofibrils may form an even more compact arrangement (Figure 9*E*). Last but not least, dimers may in fact group independently (Figure 9*F*) such that the acidic dimeric rods are elastically connected to one another by a reorganization event of the β-turn rich head domains via the basic residues and probably also engaging hydrophobic side chains, such as tyrosines, phenylalanines, and aliphatic residues (see Table 1). This view is supported by recent assembly experiments employing electron paramagnetic resonance spectroscopy (152). It was observed in this study that the rod domain structure exhibited little change when tetrameric protofilaments in low ionic strength buffers were compared with mature IFs, indicating that filament formation did not produce strong changes in the tertiary structure of the rod domains with respect to one another.

MUTATIONS INTERFERING WITH THE THREE PHASES OF ASSEMBLY   One of the most prominent interactions in the IF apparently occurs between a highly conserved arginine in an *g* position in coil 1A (ND**R**, see Figure 1) of one dimer with a neighboring dimer. Judged from the crystal structure of the vimentin coil 1A (see Figure 3*A*) (97), mutation of this arginine to a cysteine or histidine should have little or no influence on the coiled-coil geometry or stability. By analytical ultracentrifugation, the corresponding vimentin mutants formed tetrameric complexes in low salt buffers (H. Herrmann and N. Mücke, unpublished data). However, the stability of these tetrameric complexes may be considerably compromised (153). Accordingly, the effect of this point mutation on vimentin or desmin IF assembly was dramatic (Figure 10*A,B*) (154). Instead of 10-nm filaments, distinct roundish aggregates of relatively constant size and shape persisted from 1 s on for at least 1 h. On occasion, individual accumulations appeared somehow to contact each other, but fusion-type reactions were not observed. In contrast, with the heteropolymeric keratins, such severe disturbance was not observed. Whereas the keratin K8/18 R89C pair formed filaments that by negative stain EM slightly differed from the wild-type K8/18 pair (Figure 10*C*), the corresponding mutation R125H in keratin K14 produced IFs that were indistinguishable from those formed by the wild-type K5/14 pair (Figure 10*D*) (60). This morphological phenotype is surprising because both types of mutations have been demonstrated to cause severe disease (28, 155). Very notably, the K14 R125C mutant, which causes the most severe form of Epidermolysis bullosa simplex, coassembles with K5 similar to the wild-type protein, yet the resulting 10-nm filaments exhibit somewhat altered bundling properties. This was shown by rheology, particle tracking, and visualization by differential interference

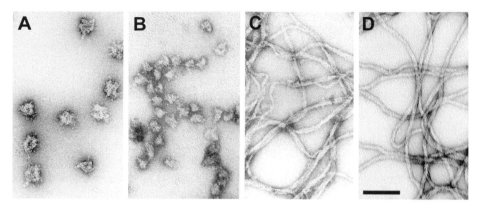

Figure 10   Assembly products of recombinant rat and human IF proteins mutated at the highly conserved arginine of the consensus motif from coil 1A. (*A*) Rat desmin R117C, (*B*) human vimentin R113C, (*C*) human keratin 18 R89C with keratin 8, and (*D*) human keratin 14 R125H with keratin 5. Electron microscopy (EM) of negatively stained specimen. Bar is 100 nm.

contrast microscopy (156). Taken together, morphologically this arginine to cysteine/histidine point mutation is much better tolerated in the keratin hetero-polymers than in the vimentin and desmin homopolymers, but it affects the integration of the mutant keratin polymer into the IF cytoskeleton profoundly, as evidenced by its severe disease phenotype.

Similar to the first phase of assembly, ULF formation, certain other IF protein mutations have been demonstrated to interfere distinctly with either one of the two next phases. Removal of the stutter, by engineering three amino acids into this motif in the middle of coil 2B, yielded a vimentin polypeptide that massively formed ULFs with the same kinetics as wild-type vimentin. However, further elongation occurred only rarely such that after 1 h of assembly most structures seen were ULFs. This lack of assembly above the ULF-state was also clearly demonstrated by viscometric assays. Moreover, the change of lysine 139 to cysteine in linker L1 of human vimentin generated a protein that at room temperature assembled only into ULFs. Upon shifting the temperature to 37°C, however, bona fide IFs were rapidly formed (T. Wedig and H. Herrmann, unpublished observations). Finally, the compaction reaction was severely disturbed in IF proteins missing the tail together with the coil 2 IF consensus motif (26).

Polymorphism

By EM, IFs exhibit diameters ranging between 7 and 11 nm [for an overview see (157)]. There was, however, always some controversy about a potential tubular organization of IFs, i.e., the existence of a "hollow core;" this, in turn, reflects the

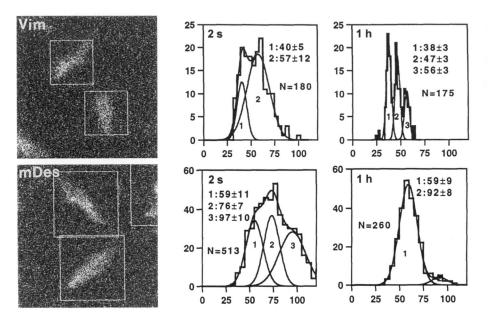

Figure 11   EM and histograms of mass-per-length (MPL) measurements of ULFs formed from human vimentin (*upper panels*) and mouse desmin (*lower panels*). The left panels show STEM dark-field pictures of ULFs fixed 2 s after initiation of assembly. The boxed areas demarcate areas used for quantification of mass. Gaussian curves were fitted into the histograms in order to identify presumptive major peaks, and peak positions and their corresponding standard deviations are indicated. N is the number of segments measured. Ordinate, number of segments; abscissa, MPL value in kDa/nm.

fact that we do not know how dimers are packed within an IF (see Figure 9 and above). Radial mass density profiles computed from STEM images of unstained IFs have documented that the IF core does contain mass, albeit less than anticipated from a compact filament (158–160). Hence an increase in IF diameter from 7 to 11 nm would reflect an increase in cross-sectional area from 38 nm^2 to 95 nm^2 and, by assuming a homogenous mass distribution, a 2.5-fold increase in mass. Moreover, it was demonstrated by STEM that the mass along one and the same IF varies considerably depending on the assembly regime employed (30). Comparison of the MPL of different types of IFs revealed that keratin IFs contain fewer subunits per filament cross section, 16 and 21, than vimentin filaments, 30 and 37 (see Figure 11). Even more striking, desmin, which exhibits high sequence identity with vimentin and which is able to copolymerize with vimentin (161), contains predominantly 47 subunits per filament cross section (see Figure 11) (31). This difference is already evident in ULFs formed after 2 s of assembly (Figure 11). Very interestingly, the MPL profiles of both vimentin and desmin IFs exhibit much sharper MPL peaks 1 h after assembly than after 2 s. At this stage of assembly vimentin IFs yield three distinct peaks at 38, 47, and 56 kDa/nm compared to desmin IFs with just one major peak at 59 kDa/nm. The desmin particle species observed at 2 s with 76 and 97 kDa/nm evidently disappear with time (see Figure 11).

In this context, it is interesting to note that between 5 and 10 min after initiation of assembly unraveled filaments are sometimes seen in significant amounts. Hence, super mass filaments may be unstable and dissolve during the compaction phase (H. Herrmann and U. Aebi, unpublished observations). Most significantly, the same kind of broad MPL distribution was obtained with authentic chicken desmin IFs prepared by dialysis. Therefore, also with a nonrecombinant desmin and an assembly procedure giving molecules more time to explore and establish stable interactions, major MPL peaks were at 52 and 63 kDa/nm and thus considerably larger than those obtained with vimentin. An extreme form of polymorphism was observed during assembly of IF proteins in the presence of phosphate and Mg ions, as used for the transient birefringence experiments described above. Under these conditions, vimentin forms rigid fibrils with 71 and 106 kDa/nm corresponding to 56 and 84 molecules per IF cross section (31). Segments with different types of molecular packing may occur side by side or even within one and the same filament, indicating that these different types of assemblies also elongate by longitudinal annealing.

In summary, MPL polymorphism represents an inherent property of IF assembly whose magnitude strongly depends on the IF protein species and the assembly conditions. It reflects the versatility of the interactions between IF subunit proteins. These interactions occur in very specific, yet multiple ways, and their balance can be fine-tuned by many factors. How this polymorphism is exploited, or controlled, in vivo is completely unknown at present.

INTEGRATION INTO CELLULAR SCAFFOLDS

Nuclear Scaffolds

The establishment of cell type specific nuclear architecture is critically dependent on the nuclear lamina. This proteinaceous network is located at the interface of both the inner nuclear membrane (INM) and the surface of interphase chromosomes (162, 163). Hence, the lamin system is assembled onto and between two opposing, principally different surfaces, one consisting of phopholipids with embedded transmembrane proteins, such as the lamin B receptor (LBR), whereas the other is dominated by basic histones, DNA, and chromatin factors, such as heterochromatin protein 1, also termed HP1 (Figure 12). Because LBR binds also to linker DNA and to histones H3/4, which are in a complex with HP1, lamin B may indeed act as a physical mediator of the chromatin-INM interaction (164, 165). Moreover, the Ig-like fold of a major part of the tail of lamins appears to be a versatile structural module to enable multiple interactions. Although this structure has been solved only for lamin A, the degree of sequence identity and

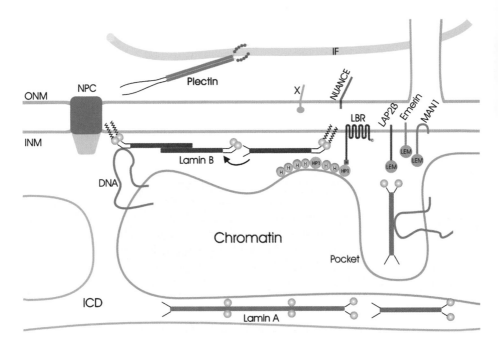

Figure 12 Hypothetical view depicting a minimal complement for the organization of lamins in nuclear envelope reconstitution after mitosis. B-type lamins are anchored to the inner nuclear membrane (INM) via their isoprenyl tails, which form dimers, tetramers, or higher oligomeric complexes. A-type lamins may be associated with chromatin or present as head-to-tail chains in the interchromosomal domain (ICD) compartment (188). INM proteins such as the lamin B receptor (LBR), in complex with heterochromatin protein 1 (HP1) and histones (H), and LEM-domain proteins, such as LAP2 β, emerin, and MAN1, are ready to recruit lamins. Transmembrane proteins present in the outer nuclear membrane (ONM), such as NUANCE (174), may connect the nucleus to the cytoplasmic actin system. Other putative receptors (X) may bind to cytoplasmic IFs or to the cytolinker plectin, thus dynamically integrating the ONM to the cytoskeleton.

the globular appearance of the lamin B dimers upon glycerol spraying/rotary metal shadowing speaks for a similar fold to be formed by B-type lamins. The lamin system is completely solubilized to the dimer/tetramer level during mitosis as determined by centrifugation experiments (17, 166), and therefore its integration into the growing lamina at the end of mitosis may indeed occur from dimeric and/or tetrameric subunits. If no chaperones are engaged, this may represent a diffusion-capture-type mechanism whereby "in place" dephosphorylation may be the switch to generate assembly competent subunits directly within the associated factor ensemble.

A further central issue of this network of associations is encountered during cell proliferation when the nuclear volume is increased and correspondingly the nuclear envelope enlarges (22). Are existing lamin filaments elongated or are new filaments formed, for instance, by nucleation at nuclear pore complexes (NPC)? Indeed, recent experiments underline the importance of NPC factors, such as Nup153, in the dynamics of the nuclear envelope (162, 167). But this may only be one of many factors engaged. Using proteomic approaches, the number of nuclear envelope components has recently been demonstrated to be much higher than previously expected (168, 168a). Finally, during differentiation of myeloid stem cells to granulocytes, the surface of the nuclei increases dramatically in parallel with major alterations of chromatin architecture. The biochemical basis of these processes, including signaling, coordination, and restructuring of the nucleoplasm, is widely unknown (169).

The Cytoskeleton

IF systems of the three general assembly groups are organized into discrete systems within a cell without major interference. Lamins are apparently exported into the nucleus in such a way that heterodimerization with keratins or vimentin, for example, is not a significant problem. Even a chimeric vimentin, ectopically expressed in the nucleus with a lamin tail, is kept outside of the lamin system, and only in the *Xenopus* in vitro nuclear assembly system is it recruited to and integrated into the nuclear rim (170). In addition, vimentin and keratin segregate within the cytoplasm. In epithelial cells, which in culture often coexpress vimentin, the two systems are distributed entirely independently. Hence, vimentin filaments do not contact desmosomes and are excluded from keratin bundles. This is still somehow surprising because vimentin filaments do, in certain tissues, such as in the arachnoid of the brain, bind to desmosomes. Moreover, the IF-binding domain is conserved in plakins such as desmoplakin and plectin. The binding site for IFs has been mapped to the globular carboxy-terminal domains (171), and recent structural data of domains B and C revealed a conserved basic groove that may represent an IF-binding site (172). However, with different cell types, completely different types of crossbridging and anchoring systems are employed, which will probably use different parts of the respective IF for interaction (54). The structural integration of the nucleus into the cytoskeleton is in part mediated by the IF system that has been shown many times to closely approach the outer nuclear envelope. Factors, such as plectin and/or transmembrane proteins indicated by the X in Figure 12, could serve as connectors of IFs and IF bundles [reviewed in (173)]. Moreover, recently identified actin-binding proteins, for example NUANCE, are apparently involved in the interplay of the cytoplasmic actin system with the nucleus (174). Finally, the formation of a superstructure or cytoskeleton may depend heavily on the presence of efficient chaperone systems (175).

Similar to the situation in the nucleus, it is at present not known how the various IF systems are established in the context of the cytoarchitecture in general. During mitosis, some cytoplasmic IF systems are disintegrated into granular, non-IF-type structures [see (176) and references therein]. For its reestablishment, microtubule-dependent motors are used, although the nature of the cargo and the loading and targeting principles are still not well understood (177). However, complex IF systems can be formed even in the complete absence of microtubules, as shown by ectopic expression of vimentin in the nucleus using this cellular compartment as a biochemical test tube (21). But in either situation, the artifical as well as the authentic cellular one, it is entirely elusive how these processes are regulated and coordinated with the general signaling systems of the cell (56).

CONCLUSIONS AND PERSPECTIVES

If cells are the home of an elaborate collection of protein complexes or machines to fulfill its many tasks, we then may ask, how do these complexes have to be placed in space and time to perform their respective functions optimally and to avoid chaotic encounters of the "fast and furious" kind? For a biochemist it may seem best to immobilize them on a solid support that is accessible to an intricate set of transport devices. Although one may argue that internal membranes, with and without lipid rafts, are ideal scaffolds for protein complexes, such as are realized within mitochondria, filament networks appear to be even more versatile because architectural networks could dynamically support the transient establishment of functional molecular networks and thereby gain plasticity for cells in ever changing physiological situations (178). The surface area of the 3 cytoskeletal filament systems exceeds that of all internal membranes 10-fold or more, and furthermore, the "surface code" of individual filaments may be much more versatile than that of membranes (179). Seen in this context, it appears highly desirable to acquire more systematic insights into the binding properties of cytoskeletal filament surfaces for interacting proteins or protein complexes.

Such considerations have gained momentum by the recently discovered mutations in human lamin A that inherit several kinds of rare diseases (35). These single amino acid mutations are spread rather homogenously over the entire molecule, and most of them do not predict obvious structural defects. Moreover, many of the lamin A mutations yield full-blown clinical symptoms only later in life. Hence, they can only be understood in the context of a functional relationship with interacting partners. The importance of the genetic background has recently been documented for a mutated sodium channel modifier: It is the expression level of a putative RNA splicing factor, and not the sodium channel modifier itself, that ultimately modulates disease severity in transgenic mice and decides about life or early death of these animals (180). Therefore, new research activities are needed with IF proteins in order to unravel the relation of their specific structural features to cellular biochemistry.

Obviously, lamins participate in a broad variety of nuclear activities, such as chromatin organization, DNA replication, and transcription. However, how gene regulation depends on lamin levels and/or the presence of specific lamin complexes (i.e., in what way the functional architecture of the nucleus relies on lamins) has not been defined in molecular terms. The same applies to the physiological importance of desmin for muscle or keratins for skin. How little we know about the physiological role of IFs has also been demonstrated, in a certain sense, by corresponding gene targeting experiments [reviewed in (21)].

Yeast two-hybrid experiments revealed new interaction partners of IFs, and subtractive proteomics approaches may become even more successful (168). The transmembrane protein MICAL is a recent example of a completely unexpected binding partner revealed by the two-hybrid approach; MICAL has been shown to bind vimentin and is of utmost importance for kidney function (181). However, one major problem with such "fishing" experiments is that many meaningful but weak interactions may be discarded early on. Moreover, other interactions, not expected to be of physiological relevance at first thought, may be discarded because of conservative reasoning. An interaction, such as that of plectin and lamin B (182), may seem to be odd at first glance because both proteins are harbored by different compartments. However, during mitosis when the nuclear envelope is broken down, the cytoskeletal networks may take over entirely new functions, e.g., binding of nuclear factors not needed for the mitotic process or even disturbing it. To what extent the kinome may reprogram the function of most or all cellular proteins during mitosis is not known at present. However, IF protein complexes may be structurally remodeled by phosphorylation into new types of IF-like or non-IF assemblies to take over new roles (183–185). Recently developed mass spectrometric methods may eventually allow us to follow the type and amount of phosphorylation as well as the absolute amount of protein present in the respective state quantitatively and thus give us more insight into the physiological function of proteins. This was recently achieved for separase (186). With atomic detail of the structure, we may then be able to provide more specific answers regarding IF function(s).

ACKNOWLEDGMENTS

The technical assistance of Monika Brettel and Tatjana Wedig over the last years is gratefully acknowledged. We thank our colleagues Sergei V. Strelkov, Michaela Reichenzeller, Laurent Kreplak, and Jens Schumacher for help with the figures and Sophie Zinn-Justin for supplying Figure 4. Moreover, we thank Eva Gundel for expert secretarial assistance. Support came from the Deutsche Forschungsgemeinschaft to H.H. and from the Swiss National Science Foundation, the Canton Basel-Stadt, and the M.E. Müller Foundation of Switzerland to U.A. Last but not least, we thank Werner W. Franke for continuous interest and support.

The *Annual Review of Biochemistry* is online at http://biochem.annualreviews.org

LITERATURE CITED

1. Janmey PA, Shah JV, Janssen KP, Schliwa M. 1998. *Subcell. Biochem.* 31: 381–97
2. Holmes KC, Popp D, Gebhard W, Kabsch W. 1990. *Nature* 347:44–49
3. Steinmetz MO, Stoffler D, Hoenger A, Bremer A, Aebi U. 1997. *J. Struct. Biol.* 119:295–320
4. Nogales E, Whittaker M, Milligan RA, Downing KH. 1999. *Cell* 96:79–88
5. Crick FHC. 1953. *Acta Crystallogr.* 6: 689–97
6. Squire JM, Vibert PJ. 1987. In *Fibrous Protein Structure*, ed. JM Squire, PJ Vibert, pp. 15–22. London: Academic
7. Aebi U, Häner M, Troncoso J, Eichner R, Engel A. 1988. *Protoplasma* 145: 73–81
8. Lazarides E. 1980. *Nature* 293:249–56
9. Steinert PM, Idler WW, Zimmerman SB. 1976. *J. Mol. Biol.* 108:547–67
10. Starger JM, Brown WE, Goldman AE, Goldman RD. 1978. *J. Cell Biol.* 78: 93–109
11. Renner W, Franke WW, Schmid E, Geisler N, Weber K, Mandelkow E. 1981. *J. Mol. Biol.* 149:285–306
12. Lazarides E. 1982. *Annu. Rev. Biochem.* 51:219–50
13. Geisler N, Kaufmann E, Weber K. 1982. *Cell* 30:277–86
14. Weber K, Geisler N. 1984. In *Cancer Cells 1, The Transformed Phenotype*, ed. AJ Levine, GF Van de Woude, WC Topp, JD Watson, pp. 153–59. Cold Spring Harbor, NY: Cold Spring Harbor Lab.
15. Steinert PM, Parry DAD. 1985. *Annu. Rev. Cell Biol.* 1:41–65
16. Heins S, Aebi U. 1994. *Curr. Opin. Cell Biol.* 6:25–33
17. Aebi U, Cohn J, Buhle L, Gerace L. 1986. *Nature* 323:560–64
18. McKeon FD, Kirschner MW, Caput D. 1986. *Nature* 319:463–68
19. Dahlstrand J, Zimmerman LB, McKay RD, Lendahl U. 1992. *J. Cell Sci.* 103: 589–97
20. Steinert PM, Roop DR. 1988. *Annu. Rev. Biochem.* 57:593–625
21. Herrmann H, Hesse M, Reichenzeller M, Aebi U, Magin TM. 2003. *Int. Rev. Cytol.* 223:83–175
22. Gerace L. 1986. *Trends Biochem. Sci.* 11:443–46
23. Franke WW. 1987. *Cell* 48:3–4
24. Erber A, Riemer D, Hofemeister H, Bovenschulte M, Stick R, et al. 1999. *J. Mol. Evol.* 49:260–71
25. Weber K. 1999. In *Guidebook to the Cytoskeletal and Motor Proteins*, ed. T Kreis, R Vale, pp. 291–93. Cold Spring Harbor, NY: Cold Spring Harbor Lab.
26. Herrmann H, Strelkov SV, Feja B, Rogers KR, Brettel M, et al. 2000. *J. Mol. Biol.* 298:817–32
27. Fuchs E. 1996. *Annu. Rev. Genet.* 30: 197–231
28. Fuchs E, Weber K. 1994. *Annu. Rev. Biochem.* 63:345–82
29. Parry DAD, Steinert PM. 1995. *Intermediate Filament Structure*. Austin: Landes. 183 pp.
30. Herrmann H, Häner M, Brettel M, Müller SA, Goldi KN, et al. 1996. *J. Mol. Biol.* 264:933–53
31. Herrmann H, Häner M, Brettel M, Ku NO, Aebi U. 1999. *J. Mol. Biol.* 286: 1403–20
32. Bonne G, Di Barletta MR, Varnous S, Bécane H-M, Hammouda E-H, et al. 1999. *Nat. Genet.* 21:285–88
33. Fatkin D, MacRae C, Sasaki T, Wolff MR, Porcu M, et al. 1999. *N. Engl. J. Med.* 341:1715–24
34. Worman HJ, Courvalin JC. 2002. *Trends Cell. Biol.* 12:591–98

35. Mounkes L, Kozov S, Burke B, Stewart CL. 2003. *Curr. Opin. Genet. Dev.* 13: 223–30

36. Herrmann H, Aebi U. 2000. *Curr. Opin. Cell Biol.* 12:79–90

37. Goldman RD, Gruenbaum Y, Moir RD, Shumaker DK, Spann TP. 2002. *Genes Dev.* 16:533–47

38. Herrmann H, Foisner R. 2003. *Cell. Mol. Life Sci.* 60:1607–12

39. Owens DW, Lane EB. 2003. *BioEssays* 25:748–58

40. Burke B, Stewart CL. 2001. *Nat. Rev. Mol. Cell Biol.* 3:575–85

41. Hutchison CJ, Alvarez-Reyes M, Vaughan OA. 2001. *J. Cell Sci.* 114: 9–19

42. Moir RD, Spann TP. 2001. *Cell. Mol. Life Sci.* 58:1748–57

43. Wilson KL, Zastrow MS, Lee KK. 2001. *Cell* 104:647–50

44. Gruenbaum Y, Goldman RD, Meyuhas R, Molls E, Margalit A, et al. 2003. *Int. Rev. Cytol.* 226:1–62

45. Shumaker DK, Kuczmarski ER, Goldman RD. 2003. *Curr. Opin. Cell Biol.* 15:358–66

46. Coulombe PA, Omary MB. 2002. *Curr. Opin. Cell Biol.* 14:110–22

47. Lee MK, Cleveland DW. 1996. *Annu. Rev. Neurosci.* 19:187–217

48. Janmey PA, Leterrier J-F, Herrmann H. 2003. *Curr. Opin. Colloid Interface Sci.* 8:40–47

49. Magin TM. 1998. *Subcell. Biochem.* 31: 141–72

50. Magin TM, Hesse M, Schröder R. 2000. *Protoplasma* 211:140–50

51. Kirfel J, Magin TM, Reichelt J. 2003. *Cell. Mol. Life Sci.* 60:56–71

52. Strelkov SV, Herrmann H, Aebi U. 2003. *BioEssays* 25:243–51

52a. Liem RKH, Leung CL. 2003. *Exp. Neurol.* 184:3–8

53. Wiche G. 1998. *J. Cell Sci.* 111: 2477–86

54. Fuchs E, Karakesisoglou I. 2001. *Genes Dev.* 15:1–14

55. Leung CL, Green KJ, Liem RK. 2002. *Trends Cell. Biol.* 12:37–45

56. Paramio JM, Jorcano JL. 2002. *BioEssays* 24:836–44

57. Röper K, Gregory SL, Brown NH. 2002. *J. Cell Sci.* 115:4215–25

58. Helfand BT, Chang L, Goldman RD. 2003. *Annu. Rev. Cell Dev. Biol.* 19: 445–67

59. Paladini RD, Takahashi K, Bravo NS, Coulombe PA. 1996. *J. Cell Biol.* 132: 381–97

60. Herrmann H, Wedig T, Porter RM, Lane EB, Aebi U. 2002. *J. Struct. Biol.* 137:82–96

61. Moir RD, Quinlan RA, Stewart M. 1990. *FEBS Lett.* 268:301–5

62. Hatzfeld M, Burba M. 1994. *J. Cell Sci.* 107:1959–72

63. Inagaki M, Nishi Y, Nishizawa K, Matsuyama M, Sato C. 1987. *Nature* 328: 649–52

64. Geisler N, Weber K. 1988. *EMBO J.* 7:15–20

65. Omary MB, Ku NO, Liao J, Price D. 1998. *Subcell. Biochem.* 31:105–40

66. Huang HY, Graves DJ, Robson RM, Huiatt TW. 1993. *Biochem. Biophys. Res. Commun.* 197:570–77

67. Heald R, McKeon F. 1990. *Cell* 61: 579–89

68. Peter M, Heitlinger E, Häner M, Aebi U, Nigg EA. 1991. *EMBO J.* 10: 1535–44

69. Huiatt TW, Robson RM, Arakawa N, Stromer MH. 1980. *J. Biol. Chem.* 255: 6981–89

70. Quinlan RA, Hatzfeld M, Franke WW, Lustig A, Schulthess T, Engel J. 1986. *J. Mol. Biol.* 192:337–49

71. Hofmann I, Herrmann H. 1992. *J. Cell Sci.* 101:687–700

72. Kooijman M, Bloemendal M, van Amerongen H, Traub P, van Grondelle R. 1994. *J. Mol. Biol.* 236:1241–49

73. Kooijman M, Bloemendal M, Traub P, van Grondelle R, van Amerongen H. 1997. *J. Biol. Chem.* 272:22548–55

74. Aebi U, Fowler WE, Rew P, Sun TT. 1983. *J. Cell Biol.* 97:1131–43

75. Stafford WF 3rd. 1992. *Anal. Biochem.* 203:295–301

76. Philo JS. 2000. *Anal. Biochem.* 279: 151–63

77. Mücke N, Wedig T, Buerer A, Marekov L, Steinert PM. 2001. *Mol. Biol. Cell* 12:55a (Abstr.)

78. Abumuhor IA, Spencer PH, Cohlberg JA. 1998. *J. Struct. Biol.* 123:187–98

79. Quinlan RA, Cohlberg JA, Schiller DL, Hatzfeld M, Franke WW. 1984. *J. Mol. Biol.* 178:365–88

80. Moir RD, Yoon M, Khuon S, Goldman RD. 2000. *J. Cell Biol.* 151:1155–68

81. Moir RD, Spann TP, Herrmann H, Goldman RD. 2000. *J. Cell Biol.* 149: 1179–91

82. Riemer D, Dodemont H, Weber K. 1993. *Eur. J. Cell Biol.* 62:214–23

83. Riemer D, Weber K. 1994. *Eur. J. Cell Biol.* 63:299–306

84. Meng JJ, Khan S, Ip W. 1994. *J. Biol. Chem.* 269:18679–85

85. Letai A, Fuchs E. 1995. *Proc. Natl. Acad. Sci. USA* 92:92–96

86. Lupas A. 1996. *Trends Biochem. Sci.* 21:375–82

86a. van den Ent F, Amos LA, Lowe J. 2001. *Nature* 413:39–44

86b. McConnell SJ, Yaffe MP. 1993. *Science* 260: 687–89

86c. Ausmees N, Kuhn JR, Jacobs-Wagner C. 2003. *Cell* 115:705–13

86d. Zimek A, Stick R, Weber K. 2003. *J. Cell Sci.* 116:2295–302

87. Cohen C. 1998. *J. Struct. Biol.* 122: 3–16

88. Eisenberg D. 2003. *Proc. Natl. Acad. Sci. USA* 100:11207–10

89. Caspar DL, Cohen C, Longley W. 1969. *J. Mol. Biol.* 41:87–107

90. Phillips GN Jr, Fillers JP, Cohen C. 1986. *J. Mol. Biol.* 192:111–31

91. O'Shea EK, Rutkowski R, Kim PS. 1991. *Science* 243:538–42

92. Burkhard P, Kammerer RA, Steinmetz MO, Bourenkov GP, Aebi U. 2000. *Structure* 8:223–30

93. Kammerer RA, Schulthess T, Landwehr R, Lustig A, Engel J, et al. 1998. *Proc. Natl. Acad. Sci. USA* 95: 13419–24

94. Wu KC, Bryan JT, Morasso MI, Jang S-I, Lee J-H, et al. 2000. *Mol. Biol. Cell* 11:3539–58

95. Strelkov SV, Herrmann H, Geisler N, Lustig A, Ivaninskii S, et al. 2001. *J. Mol. Biol.* 306:773–81

96. O'Shea EK, Klemm JD, Kim PS, Alber T. 1991. *Science* 254:539–44

97. Strelkov SV, Herrmann H, Geisler N, Wedig T, Zimbelmann R, et al. 2002. *EMBO J.* 21:1255–66

98. Janmey PA, Euteneuer U, Traub P, Schliwa P. 1991. *J. Cell Biol.* 113: 155–60

99. Burkhard P, Stetefeld J, Strelkov SV. 2001. *Trends Cell. Biol.* 11:82–88

100. Chakrabarty T, Xiao M, Cooke R, Selvin RP. 2002. *Proc. Natl. Acad. Sci. USA* 99:6011–16

101. Delahay RM, Frankel G. 2002. *Mol. Microbiol.* 45:905–16

102. Cohen-Krausz S, Trachtenberg S. 2003. *J. Mol. Biol.* 331:1093–108

103. Kumar PR, Singhal PK, Vinod SS, Mahalingam S. 2003. *J. Mol. Biol.* 331: 1141–56

104. van Noort J, van der Heijden T, de Jager M, Wyman C, Kanaar R, Dekker C. 2003. *Proc. Natl. Acad. Sci. USA* 100: 7581–86

105. Li Y, Brown JH, Reshetnikova L, Blazsek A, Farkas L, et al. 2003. *Nature* 424:341–45

106. Newman JRS, Keating AE. 2003. *Science* 300:2097–101

107. Nair SK, Burley SK. 2003. *Cell* 193–205

108. Goulielmos G, Gounari F, Remington S, Müller S, Häner M, et al. 1996. *J. Cell Biol.* 132:643–55

109. Fischer RS, Quinlan RA, Fowler VM. 2003. *FEBS Lett.* 547:228–32

110. Hisanaga S, Hirokawa N. 1988. *J. Mol. Biol.* 202:297–305

111. Riemer D, Wang J, Zimek A, Swalla BJ, Weber K. 2000. *Gene* 255:317–25

112. Dhe-Paganon S, Werner ED, Chi YI, Shoelson SE. 2002. *J. Biol. Chem.* 277: 17381–84

113. Krimm I, Ostlund C, Gilquin B, Couprie J, Hossenlopp P, et al. 2002. *Structure* 10:811–23

114. Steinert PM, Idler WW, Aynardi-Whitman M, Zackroff RV, Goldman RD. 1982. *Cold Spring Harbor Symp. Quant. Biol.* 46:465–74

115. Steinert PM, Marekov LN, Parry DAD. 1993. *J. Biol. Chem.* 265:24916–25

116. Coulombe PA, Fuchs E. 1990. *J. Cell Biol.* 111:153–69

117. Gieffers C, Krohne G. 1991. *Eur. J. Cell Biol.* 55:191–99

118. Heitlinger E, Peter M, Häner M, Lustig A, Aebi A, Nigg EA. 1991. *J. Cell Biol.* 113:485–95

119. Stuurman N, Heins S, Aebi U. 1998. *J. Struct. Biol.* 122:42–66

120. Sasse B, Lustig A, Aebi U, Stuurman N. 1997. *Eur. J. Biochem.* 250:30–38

121. Karabinos A, Schünemann J, Meyer M, Aebi U, Weber K. 2003. *J. Mol. Biol.* 325:241–47

122. Bremer A, Henn C, Engel A, Baumeister W, Aebi U. 1982. *Ultramicroscopy* 46:85–111

123. Woods CM, Lazarides E. 1986. *Nature* 321:85–89

124. Klapper M, Exner K, Kempf A, Gehrig C, Stuurman N, et al. 1997. *J. Cell Sci.* 110:2519–32

125. Ip W, Hartzer MK, Pang YY, Robson RM. 1985. *J. Mol. Biol.* 183:365–75

126. Stromer MH, Ritter MA, Pang YY, Robson RM. 1987. *Biochem. J.* 246: 75–81

127. Cerdà J, Conrad M, Markl J, Brand M, Herrmann H. 1998. *Eur. J. Cell Biol.* 77:175–87

128. Schaffeld M, Herrmann H, Schultess J, Markl J. 2001. *Eur. J. Cell Biol.* 80: 692–702

128a. Mücke N, Kreplak L, Kirmse R, Wedig T, Herrmann H, et al. 2004. *J. Mol. Biol.* 335:1241–50

129. Karabinos A, Schünemann J, Parry DAD, Weber K. 2002. *J. Mol. Biol.* 316: 127–37

130. Hatzfeld M, Maier G, Franke WW. 1987. *J. Mol. Biol.* 197:237–55

131. Fouquet B, Herrmann H, Franz JK, Franke WW. 1988. *Development* 104: 533–48

132. Bartnik E, Osborn M, Weber K. 1985. *J. Cell Biol.* 101:427–40

133. Szaro BG, Pant HC, Way J, Battey J. 1991. *J. Biol. Chem.* 266:15035–41

134. Way J, Hellmich MR, Jaffe H, Szaro B, Pant HC, et al. 1992. *Proc. Natl. Acad. Sci. USA* 89:6963–67

135. Erber A, Riemer D, Bovenschulte M, Weber K. 1998. *J. Mol. Evol.* 47: 751–62

136. Dodemont H, Riemer D, Ledger N, Weber K. 1994. *EMBO J.* 13: 2625–38

137. Riemer D, Weber K. 1998. *J. Cell Sci.* 111:2967–75

138. Riemer D, Karabinos A, Weber K. 1998. *Gene* 211:361–73

139. Zimek A, Weber K. 2002. *Gene* 288: 187–93

140. Geisler N, Schünemann J, Weber K, Häner M, Aebi U. 1998. *J. Mol. Biol.* 282:601–17

141. Quinlan RA, Moir RD, Stewart M. 1989. *J. Cell Sci.* 93:71–83

142. Herrmann H, Aebi U. 1998. *Subcell. Biochem.* 31:319–62

143. Shoeman RL, Hartig R, Berthel M, Traub P. 2002. *Exp. Cell Res.* 279: 344–53

144. Wedig T, Muecke N, Aebi U, Herrmann H. 2002. *Mol. Biol. Cell* 13:61a (Abstr.)

145. Herrmann H, Wedig T, Aebi U, Parry DA, Marekov LN, Steinert PM. 2002. *Mol. Biol. Cell* 13:61a (Abstr.)

146. Herrmann H, Fouquet B, Franke WW. 1989. *Development* 105:299–307
147. Eichner R, Rew P, Engel A, Aebi U. 1985. *Ann. NY Acad. Sci.* 455: 381–402
148. Ip W, Heuser JE, Pang Y-YS, Hartzer MK, Robson RM. 1985. *Ann. NY Acad. Sci.* 455:185–99
149. Fraser RDB, MacRae TP, Parry DAD. 1990. In *Cellular and Molecular Biology of Intermediate Filaments*, ed. RD Goldman, PM Steinert, pp. 205–31. New York: Plenum
150. Heins S, Wong PC, Müller S, Goldie K, Cleveland DW, Aebi U. 1993. *J. Cell Biol.* 123:1517–33
151. Alberts B, Johnson A, Lewis J, Raff M, Roberts K, Walter P. 2002. *Molecular Biology of the Cell*. New York: Garland Sci. 1463 pp. 4th ed.
152. Hess JF, Voss JC, FitzGerald PG. 2002. *J. Biol. Chem.* 277:35516–22
153. Mehrani T, Wu KC, Morasso MI, Bryan JT, Marekov LN, et al. 2001. *J. Biol. Chem.* 276:2088–97
154. Haubold K, Herrmann H, Langen SJ, Evans RM, Leinwand LA, Klymkowsky MW. 2003. *Cell Motil. Cytoskelet.* 54:105–21
155. Ku NO, Darling JM, Krams SM, Esquivel CO, Keeffe EB, et al. 2003. *Proc. Natl. Acad. Sci. USA* 100: 6063–68
156. Ma L, Yamada S, Wirtz D, Coulombe PA. 2001. *Nat. Cell Biol.* 3:503–6
157. Franke WW, Zerban H, Grund C, Schmid E. 1981. *Biol. Cell* 41: 173–78
158. Engel A, Eichner R, Aebi U. 1985. *J. Ultrastruct. Res.* 90:323–35
159. Steven AC. 1990. See Ref. 149, pp. 233–63
160. Watts NR, Jones LN, Cheng N, Wall JS, Parry DAD, Steven AC. 2002. *J. Struct. Biol.* 137:109–18
161. Quinlan RA, Franke WW. 1982. *Proc. Natl. Acad. Sci. USA* 79:3452–56
162. Hutchison CJ. 2002. *Nat. Rev. Mol. Cell Biol.* 3:848–58
163. Foisner R. 2003. *ScientificWorldJournal* 3:1–20
164. Duband-Coulet I, Courvalin JC. 2000. *Biochemistry* 39:6483–88
165. Polioudaki H, Kourmouli N, Drosou V, Bakou A, Theodoropoulos PA, et al. 2001. *EMBO Rep.* 2:920–25
166. Gerace L, Blobel G. 1980. *Cell* 19: 277–87
167. Liu J, Prunuske AJ, Fager AM, Ullman KS. 2003. *Dev. Cell* 5:487–98
168. Dreger M, Bengtsson L, Schoneberg T, Otto H, Hucho F. 2001. *Proc. Natl. Acad. Sci. USA* 98:11943–48
168a. Schirmer EC, Florens L, Guan T, Yates JR III, Gerace L. 2003. *Science* 310: 1380–82
169. Olins AL, Buendia B, Herrmann H, Lichter P, Olins DE. 1998. *Exp. Cell Res.* 245:91–104
170. Dreger C, König AR, Spring H, Lichter P, Herrmann H. 2002. *J. Struct. Biol.* 140:100–15
171. Nikolic B, Mac Nulty E, Mir B, Wiche G. 1996. *J. Cell Biol.* 134:1455–67
172. Choi HJ, Park-Snyder S, Pascoe LT, Green KJ, Weis WI. 2002. *Nat. Struct. Biol.* 9:612–20
173. Goldman RD, Zackroff RV, Steinert PM. 1990. See Ref. 149, pp. 3–17
174. Zhen YY, Libotte T, Munck M, Noegel AA, Korenbaum E. 2002. *J. Cell Sci.* 115:3207–22
175. Quinlan RA. 2002. In *Progress in Molecular and Subcellular Biology*, Vol. 28, ed. A-P Arrigo, WEG Müller, pp. 219–33. Berlin: Springer-Verlag
176. Chou YH, Khuon S, Herrmann H, Goldman RD. 2003. *Mol. Biol. Cell* 14: 1468–78
177. Chou YH, Helfand BT, Goldman RD. 2001. *Curr. Opin. Cell Biol.* 13: 106 9
178. Bray D. 2003. *Science* 301:1864–65
179. Bray D. 2001. *Cell Movements:From*

Molecules to Motility. New York: Garland. 372 pp. 2nd ed.

180. Buchner DA, Trudeau M, Meisler MH. 2003. *Science* 301:967–69

181. Suzuki T, Nakamoto T, Oawa S, Seo S, Matsumura T, et al. 2002. *J. Biol. Chem.* 277:14933–41

182. Foisner R, Traub P, Wiche G. 1991. *Proc. Natl. Acad. Sci. USA* 88:3812–16

183. Manning G, Whyte DB, Martinez R, Hunter T, Sudarsanam S. 2002. *Science* 298:1912–34

184. Cheng TJ, Tseng YF, Chang WM, Chang MD, Lai YK. 2003. *J. Cell. Biochem.* 89:589–602

185. Meggio F, Pinna LA. 2003. *FASEB J.* 17:349–68

186. Gerber SA, Rush J, Stemman O, Kirschner MW, Gygi SP. 2003. *Proc. Natl. Acad. Sci. USA* 100:6940–45

187. Herrmann H, Aebi U. 1998. *Curr. Opin. Struct. Biol.* 8:177–85

188. Herrmann H, Lichter P. 1999. *Protoplasma* 209:157–65

Annu. Rev. Biochem. 2004. 73:791–836
doi: 10.1146/annurev.biochem.73.011303.073717
First published online as a Review in Advance on March 18, 2004

Directed Evolution of Nucleic Acid Enzymes

Gerald F. Joyce

Departments of Chemistry and Molecular Biology and The Skaggs Institute for Chemical Biology, The Scripps Research Institute, La Jolla, California 92037; e-mail: gjoyce@scripps.edu

Key Words   combinatorial library, DNA enzyme, in vitro evolution, in vitro selection, ribozyme

■ **Abstract**   Just as Darwinian evolution in nature has led to the development of many sophisticated enzymes, Darwinian evolution in vitro has proven to be a powerful approach for obtaining similar results in the laboratory. This review focuses on the development of nucleic acid enzymes starting from a population of random-sequence RNA or DNA molecules. In order to illustrate the principles and practice of in vitro evolution, two especially well-studied categories of catalytic nucleic acid are considered: RNA enzymes that catalyze the template-directed ligation of RNA and DNA enzymes that catalyze the cleavage of RNA. The former reaction, which involves attack of a 2′- or 3′-hydroxyl on the α-phosphate of a 5′-triphosphate, is more difficult. It requires a comparatively larger catalytic motif, containing more nucleotides than can be sampled exhaustively within a starting population of random-sequence RNAs. The latter reaction involves deprotonation of the 2′-hydroxyl adjacent to the cleavage site, resulting in cleaved products that bear a 2′,3′-cyclic phosphate and 5′-hydroxyl. The difficulty of this reaction, and therefore the complexity of the corresponding DNA enzyme, depends on whether a catalytic cofactor, such as a divalent metal cation or small molecule, is present in the reaction mixture.

CONTENTS

INTRODUCTION

All of the enzymes that exist in biology are the product of Darwinian evolution based on natural selection. The processes of Darwinian evolution that occur in nature can be recapitulated in the laboratory, leading to the development of artificial enzymes with catalytic properties that conform to selection constraints imposed by the experimenter. Over the past decade this mode of "evolutionary engineering" has become more sophisticated, yet more routine. Many investigators have employed in vitro evolution methods to probe the structure and function of existing enzymes or to develop enzymes with novel catalytic properties. Others have used these methods to study Darwinian evolution itself, taking advantage of the controlled laboratory setting, which provides access to all genotypic and phenotypic parameters.

This review focuses on a subset of activity in the area of directed evolution that is limited to nucleic acid enzymes obtained from a population of random-sequence RNA or DNA molecules. Catalytic nucleic acids have both genotypic and phenotypic properties, greatly simplifying the procedures needed to carry out Darwinian evolution. The same nucleic acid molecules that can be amplified and mutated on the basis of their genetic information content also can be selected on the basis of their functional properties. By starting with a population of random-sequence nucleic acid molecules, one makes no assumptions about the nucleotide composition that is required to achieve a particular catalytic function, allowing an unbiased exploration of "sequence space." Once molecules with crude catalytic activity have been obtained, they can be used as a starting point for further directed evolution experiments, seeking either to refine that activity or to develop molecules with somewhat different catalytic behavior.

Only a few nucleic acid enzymes are known to occur in nature; all of which are RNA enzymes (ribozymes). These include the self-splicing group I and group II introns (1, 2), the RNA component of RNase P, which cleaves precursor tRNAs to generate mature tRNAs (3), and the various self-cleaving RNAs, including the "hammerhead," "hairpin," HDV (hepatitis delta virus), and VS (*Neurospora* Varkud satellite) motifs (4–7). In addition, the RNA component of the large ribosomal subunit is now recognized to be a ribozyme that catalyzes the peptidyl transferase step of translation (8). Ribosomal RNA is unique among known naturally occurring ribozymes because it contains several modified nucleotides, although it is not clear if these modifications are essential for catalysis. The U2/U6 snRNA complex within the eukaryotic spliceosome also may be a ribozyme (9). These two RNA molecules in isolation catalyze an unusual phosphoryl transfer reaction, but the reaction occurs very slowly and may not reflect the natural catalytic activity of the spliceosome.

There are no known DNA enzymes in biology and little opportunity for such catalysts to arise because most biological DNAs are double stranded. Directed evolution has led to the development of a variety of DNA enzymes (DNAzymes) and an even larger number of ribozymes. It also has provided examples of both RNA and DNA enzymes that contain modified nucleotides. Some of these in vitro–evolved nucleic acid enzymes have a functional counterpart among the naturally occurring RNA enzymes, but most catalyze reactions that previously had only been observed among protein enzymes. The chemical diversity and therefore the catalytic potential of nucleic acids are limited compared to that of proteins. Nucleic acids lack a general acid base with a pK_a that is near neutral (as in histidine), a primary alkyl amine (as in lysine), a carboxylate (as in aspartate), and a sulfhydryl (as in cysteine). However, the missing functionallty can be supplied by employing modified nucleotides or adding a small-molecule cofactor. The directed evolution approach allows one to explore a wide range of chemical possibilities, not necessarily with the aim of imitating natural enzymes, but with the broader goal of understanding what is biochemically feasible.

DIRECTED EVOLUTION METHODS

Selection Based on Catalytic Function

Directed evolution involves first establishing a heterogeneous population of macromolecules and then carrying out repeated rounds of selective amplification based on the physical or biochemical properties of those molecules. In vitro selection is distinguished from in vitro evolution in that the latter also involves the continued introduction of genetic variation through mutation and/or recombination. A critical aspect of directed evolution is the application of an appropriate selection scheme to stringently distinguish molecules that have the desired properties from those that do not. The selection scheme also should be highly sensitive so that even if the desired molecules are very rare in the population, they can be recovered and amplified to give rise to "progeny" molecules. The requirements for stringent selectivity and high sensitivity often are at cross-purposes, challenging the experimenter to devise an appropriate combination of selection pressures to achieve the desired result.

Most of the reactions that have been carried out with in vitro–evolved nucleic acid enzymes involve either bond forming or bond breaking with one reactive group attached to the population of potential catalysts and the other located on a separate substrate or product molecule. Typically the separate molecule also contains a chemical tag, so that a bond-forming reaction results in joining the tag to the catalyst, whereas a bond-breaking reaction results in detaching the tag from the catalyst. A selection procedure is applied that is tag specific, retaining tagged molecules in the case of a bond-forming reaction or rejecting tagged molecules in the case of a bond-breaking reaction.

A simple tag that commonly is employed in the directed evolution of nucleic acid enzymes is an oligonucleotide that can be distinguished on the basis of its size or ability to hybridize to a complementary nucleic acid. As a consequence of a bond-forming reaction, the oligonucleotide becomes attached to the catalyst, allowing active catalysts to be recovered on the basis of their slower gel electrophoretic mobility, hybridization to an oligonucleotide affinity column, or hybridization to a primer that is extended by a polymerase. Conversely, a bond-breaking reaction causes the oligonucleotide tag to be released from the catalyst, allowing active catalysts to be selected on the basis of their faster electrophoretic mobility or detachment from an oligonucleotide affinity column. The stringency of these methods is less than ideal because, for example, anomalous electrophoretic mobility may be indistinguishable from the gain or loss of an oligonucleotide tag. Inventing a false tag or masking an existing tag could have the same consequence as acquiring or detaching a tag, respectively.

Biotin tags also are commonly employed in the directed evolution of nucleic acids. In a bond-forming reaction, the active catalysts become biotinylated, allowing them to be captured using a streptavidin-coated support. In a bond-breaking reaction, the molecules are prebound to streptavidin, and active catalysts become detached from the biotin tag and therefore released from the support. Similar capture and release techniques have been employed for other chemical tags that can be recognized through either covalent or strong noncovalent interactions. There is a risk of false positives with these techniques because the nucleic acid might bind directly to the streptavidin-coated support (or other capture agent) rather than perform the chemistry needed to acquire the tag. The nucleic acid also might bind the tagged molecule rather than react with it, thus becoming captured despite a lack of catalytic activity. Such false positives can be reduced by performing the capture step under denaturing conditions.

Even when the selective tagging scheme is behaving exactly as intended, there still is a risk that the tag will be acquired or lost through a chemical process other than the one that was intended. For example, RNAs selected for the ability to cleave a particular bond that links a biotin tag to their own 5′ end may instead promote cleavage of an RNA phosphodiester that lies just downstream from the 5′ end, thus becoming detached from streptavidin and exhibiting electrophoretic mobility similar to that of molecules that performed the desired reaction. These "accidental catalysts" may be interesting in their own right. More often, however, one must apply appropriate countermeasures to prevent the selection of molecules with undesirable traits.

The selected phenotype entails not just the catalytic activity per se but also the range of biochemical properties associated with that activity, including catalytic rate, substrate binding affinity, substrate specificity, and preferred reaction conditions. All of these traits are selectable and can be refined by imposing the appropriate selection constraints. An adage in directed evolution is "you get what you select for," which refers to the totality of selection pressures, both intended and unintended, that are imposed on the evolving population of molecules. By choosing a particular substrate, substrate concentration, temperature, pH, salt concentration, and reaction time, one favors the development of molecules that are optimized for that situation. As evolution proceeds, one may choose to alter the reaction parameters, driving the population toward altered biochemical properties. Commonly, the reaction time is decreased progressively to favor the development of molecules with faster catalytic rates. In addition, the substrate concentration may be reduced to favor an improvement in K_m, or the reaction conditions may be altered to favor the development of molecules that are adapted to the modified conditions.

Nucleic Acid Amplification

Once a subset of the population has been selected, the selected molecules must be amplified to generate a progeny population that can be subjected to further rounds of selective amplification. Nucleic acid molecules are amplified most conveniently by the polymerase chain reaction (PCR), employing two oligonucleotide primers and a thermostable DNA polymerase. RNA molecules first must be reverse transcribed to a complementary DNA (cDNA), and the PCR products must be transcribed to generate the progeny RNAs. DNA molecules can be amplified directly, but the double-stranded PCR products must be separated, and the functional strand must be isolated by either gel electrophoresis or affinity chromatography.

Nucleic acids also can be amplified in an isothermal reaction that employs both a reverse transcriptase and a DNA-dependent RNA polymerase (10). This method utilizes two oligonucleotide primers, one that binds to the 3′ end of the RNA and initiates cDNA synthesis and another that binds to the 3′ end of the cDNA and introduces a promoter sequence for the DNA-dependent RNA polymerase. Isothermal amplification is especially advantageous for copying RNA because it entails repeated cycles of reverse and forward transcription in a common reaction mixture. Unlike PCR amplification, however, isothermal amplification does not generate double-stranded DNA products that are directly amenable to cloning.

In principle, a third method for nucleic acid amplification would be to clone the selected molecules into bacterial (or eukaryotic) cells, allow the cells to reproduce, and then harvest the amplified DNA (which may be transcribed to RNA). This is a common approach for the directed evolution of protein enzymes, but it is unnecessarily cumbersome for nucleic acids and limits the population size to the number of transformed cells, typically 10^6–10^8 individuals. By contrast, in vitro amplification methods are applied routinely to populations of 10^{12}–10^{14} individuals and require only about one hour to perform.

Any amplification method risks the introduction of uncontrolled selection pressures, for example, the pressure to avoid any sequence that is not copied efficiently by the polymerase or that causes mishybridization of one of the primers. By defining particular primer binding sites at the ends of the nucleic acid molecules, one can change the outcome of a directed evolution experiment. Some molecules will arise that utilize these defined residues as either part of their catalytic apparatus or to help achieve their folded structure. Other potential catalysts will be excluded because they engage in an undesirable interaction involving the defined residues. There are methods for detaching the primer binding sites prior to the reaction and for reattaching them prior to amplification, but these methods are tedious and not routinely practiced. More commonly, if one wishes to avoid catalysts that are dependent on a particular primer binding site, the sequence of that site can be altered during the course of evolution.

It is important to recognize that nucleic acid amplification in the context of directed evolution has a different character than when it is carried out as a standard analytical or preparative procedure. In a directed evolution experiment, one typically carries out 10–20 successive rounds of selective amplification, each involving about 10^6-fold amplification. In one extreme case, an evolving population of RNA enzymes underwent an overall amplification of more than 10^{700}-fold. Iterative amplification of this magnitude normally would give rise to overwhelming artifacts. It is not unusual to carry out a "nested" PCR, involving two successive PCR amplifications, but no one would be so foolhardy as to attempt 10–20 successive PCR amplifications. Directed evolution experiments are different because a selection step is imposed between successive amplification steps, thereby ridding the population of artifacts that do not conform to the selection criteria. In practice, however, directed evolution involves an ongoing battle between the generation of amplification artifacts that subvert the selection process and the development of molecules that exhibit the desired catalytic properties.

Introduction of Genetic Diversity

RANDOM-SEQUENCE LIBRARIES   Genetic diversity is introduced at the start of a directed evolution experiment by creating a combinatorial library of macromolecules. Diversity also can be introduced at any round of evolution by performing amplification under error-prone conditions or by carrying out a separate mutagenesis or recombination procedure. The simplest way to generate a combinatorial library of nucleic acids is to prepare random-sequence DNA molecules using an automated DNA synthesizer. Ideally the four deoxynucleoside phosphoramidites should be premixed at a ratio that compensates for their differential coupling efficiency, ensuring their equal representation in the synthesized library. RNA molecules can be transcribed from random-sequence DNA templates, which results in 10- to 1000-fold amplification due to the ability of RNA polymerase to produce multiple RNA transcripts per DNA template. Even when generating a library of DNA molecules, one typically carries out a few cycles of PCR amplification, thus replacing the synthetic DNA with biochemically prepared DNA that does not contain residual protecting groups or other chemical lesions.

The number of possible sequences for a nucleic acid of length n is 4^n ($\sim 10^{0.6n}$). Thus a combinatorial library of one copy each of all possible 25mers would contain about 10^{15} molecules, or about 1 nmol of material. This is close to the limit of what can be produced using conventional methods, especially if one wishes to have an average of 10–100 copies of each sequence so that, allowing for sampling statistics, almost all sequences are represented in the population. It may not be possible, however, to achieve certain catalytic functions with nucleic acid molecules that contain only 25 residues. Thus combinatorial libraries of larger random-sequence molecules often are prepared with the realization that such libraries contain only a very sparse sampling of all possible sequences. One proceeds on the assumption that there are many possible sequences with the desired activity, so that even a sparse sampling will contain at least a few of them.

Modern automated DNA synthesizers allow the synthesis of up to about 120mers, although the yield of amplifiable DNA molecules deteriorates significantly for lengths above 100. Thus, allowing for primer binding sites of 15–20 nucleotides at each end of the DNA, the maximum number of random nucleotides that can be included within a synthetic DNA molecule is about 80. Two or more such DNA molecules can be assembled by overlap extension or ligation, although this usually requires a region of fixed-sequence nucleotides at the site of overlap or surrounding the ligation junction. There are other methods for generating populations of long, random-sequence nucleic acids that surprisingly have not been applied in directed evolution. For example, one could extend a DNA primer in a template-independent polymerization reaction using terminal transferase and the four deoxynucleoside 5'-triphosphates (dNTPs). Alternatively, one could generate random-sequence RNAs employing polynucleotide phosphorylase and the four ribonucleoside 5'-diphosphates.

MUTAGENESIS   In vitro evolution requires the continued introduction of genetic diversity to compensate for the loss of diversity due to selection. Maintaining diversity allows for a broader exploration of potentially advantageous sequences. A more important but subtle point is that the introduction of diversity will be biased by what has proven selectively advantageous during previous rounds of evolution, and assuming that related sequences tend to encode related behaviors, this will lead to the preferential exploration of nucleic acid sequences that are more likely to result in an improved phenotype. The more advantageous sequences will tend to be present in greater copy number because a larger fraction of the copies of the parental molecules will have performed the target reaction and thus give rise to a larger number of corresponding progeny. If the generation of progeny molecules is accompanied by the introduction of genetic diversity, then the exploration of novel variants will be biased toward the neighborhood of more abundant and therefore more advantageous molecules.

An extreme version of historical bias would be to choose the one sequence that has proven most advantageous during previous rounds of evolution, perhaps by conducting a high-throughput screen, and use it as the sole breeding stock to generate variant progeny. This is colloquially referred to as "declaring wild type." It allows one to prepare a new combinatorial library by automated DNA synthesis or some other means. The problem with this approach is that the correlation between related genotypes and related phenotypes is not absolute. The most advantageous individual at present may reflect a local optimum that does not provide access to other still more advantageous individuals, whereas one of the less advantageous individuals at present may offer the best route to the evolution of the desired phenotype. Of course one should not place all bets on a less advantageous individual because such an individual is even less likely than the most advantageous individual to lead to molecules with improved catalytic behavior. Instead, one should place bets across the entire population with the size of the bet for each sequence proportional to its demonstrated selective advantage. This occurs automatically when the entire population is mutagenized because of the proportional relationship between selective advantage and copy number.

The most common method for introducing genetic diversity during in vitro evolution is error-prone PCR. There are several standardized protocols that utilize either altered reaction conditions or a mutant thermostable DNA polymerase to promote the incorporation and extension of mismatched nucleotides. One procedure for mutagenic PCR achieves an error rate of 0.66% per nucleotide position with roughly equal probability of all possible transitions and transversions (11). Compared to standard PCR, this procedure employs an increased concentration of *Taq* DNA polymerase, added $MnCl_2$, increased concentration of $MgCl_2$, and unbalanced concentrations of the four dNTPs. There also is a "hypermutagenic" PCR procedure that achieves an error rate of about 10% per nucleotide position with only modest sequence bias, but it has a greatly increased risk of introducing amplification artifacts (12). This method also employs added $MnCl_2$ and increased concentration of $MgCl_2$ but relies on more heavily unbalanced concentrations of the four dNTPs and 50 rather than 30 temperature cycles. Yet another procedure utilizes standard PCR conditions but a mutant thermostable DNA polymerase (Mutazyme™) that provides an error rate of up to 0.7% per nucleotide position over the course of the PCR (13). It generates a broad spectrum of mutations, although with a substantial preference for transitions over transversions.

The distribution of mutations following a mutagenesis procedure can be estimated on the basis of the binomial distribution. For a sequence of length n that is mutagenized with an error rate ε, the probability of introducing k mutations is given by:

$$P(k, n, \varepsilon) = (n!/[(n-k)!\, k!])\, \varepsilon^k\, (1-\varepsilon)^{n-k}.$$

For example, a target sequence of 100 nucleotides that is subjected to mutagenic PCR ($\varepsilon = 0.0066$) would give rise to about 52% wild-type sequences, 34% one-error mutants, 11% two-error mutants, 2% three-error mutants, and so on. The number of different sequences in each error class is given by

$$N_k = (n!/[(n-k)!\, k!])\, 3^k.$$

Thus for the target sequence of 100 nucleotides, there are 300 different one-error mutants, 44,550 two-error mutants, 4,365,900 three-error mutants, and so on. A mutagenized population of 10^{15} molecules will contain about 10^{12} copies of each of the possible one-error mutants, 10^9 copies of each of the two-error mutants, 10^7 copies of each of the three-error mutants, and so on. All possible mutants containing five errors or less would be represented in the population, but only about 6% of the six-error mutants would be present as even a single copy, with ever sparser representation of the higher error classes.

RECOMBINATION   A different approach for the introduction of genetic diversity involves in vitro recombination. It has long been possible to shuffle segments of a gene or gene family by taking advantage of restriction sites or primer binding sites that flank those segments. Through either restriction digestion or primer extension at internal sites, the gene can be broken down into component segments, which then can be reassembled in various combinations by either ligation or polymerase-mediated overlap extension. An important advance in recombination technology was the development of "sexual PCR," which involves fragmenting the gene in a random manner, then allowing the fragments to serve as primers for each other to reassemble shuffled versions of the gene (14). This technique does not require knowledge of the gene sequence and thus can be applied to entire populations of nucleic acid molecules.

Sexual PCR has been widely used in the directed evolution of proteins but almost neglected in the directed evolution of nucleic acids. Recombination appears to be more important for proteins than nucleic acids because proteins have an inherent domain structure and are composed of 20 different subunits that prevent complete representation of mutants beyond the three-error class. By this same argument, however, recombination might be useful for the evolution of large nucleic acid enzymes that are composed of multiple subdomains. It also could be used as a technique for genetic backcrossing, recombining an evolved population with its wild-type progenitor to weed out neutral mutations and retain those that confer selective advantage. Some recombination occurs inevitably during nucleic acid amplification due to template switching by the polymerase or the formation of partial extension products that subsequently are completed on a different template molecule. Such spontaneous recombinants have been observed among in vitro–evolved nucleic acid enzymes, suggesting that it may be useful to carry out recombination in a more purposeful manner.

Analysis of Populations of Evolved Enzymes

Darwinian evolution is an open-ended process, and in that sense, the directed evolution of a nucleic acid enzyme is never finished. In practice, a directed evolution experiment is considered completed when the desired catalytic properties have been attained or when the selection process fails to bring about further improvement in phenotype. One usually carries out a tracking assay to monitor the behavior of the evolving population. Such assays can be misleading, however, and until individual molecules have been cloned from the population and studied in isolation, the observed phenotype should be viewed with skepticism.

The isolated individuals are sequenced and tested for their catalytic activity. There usually is a strong correlation between the abundance of a particular sequence and its associated level of catalytic activity, although this relationship may be disrupted by unintended selection pressures that do not pertain to catalytic function. A sequence alignment usually helps to define a consensus sequence and to indicate conserved residues that are important for catalysis. In some cases, sequence covariation may be observed that is indicative of conserved secondary structural elements. Hypotheses regarding the essential composition of the catalytic motif can be tested by site-directed mutagenesis and deletion analysis.

Most investigators carry out a reselection procedure in which a dominant clone or a molecule that corresponds to the consensus sequence is partially randomized and subjected to additional rounds of selective amplification. This is an example of declaring wild type, although the goal is to define more precisely the catalytic motif based on allowable sequence variation, rather than to seek further improvement in catalytic activity. An exception is when additional rounds of selective amplification that are applied to the entire population fail to give further improvement in phenotype because individuals exist within the population that circumvent the selection constraints. In that case, it may be preferable to start fresh with a single, well-behaved individual.

Once the catalytic motif has been defined, rational design approaches often are used to convert the "*cis*-acting" catalyst to a true enzyme that can operate with multiple turnovers on a separate substrate. Recall that during the evolution process one of the reactive groups for either bond forming or bond breaking is attached to the catalyst. This allows reactive molecules to be self-modifying so that they can be distinguished for selective amplification. A true enzyme, however, must be left unchanged following the reaction so that it can enter another catalytic cycle. It usually is straightforward to convert a self-modifying catalytic nucleic acid to one that operates on a separate substrate simply by dividing the catalyst at an appropriate phosphodiester linkage and leaving a few nucleotides attached to the reactive group. It is less trivial to separate the reactive group so that it has no attached nucleotides. This outcome can be made more likely by varying the sequence of the nucleotides adjacent to the reactive group during the course of in vitro evolution.

RNA LIGASE RIBOZYMES EVOLVED FROM RANDOM-SEQUENCE RNAS

The principles of directed evolution described above have been applied to the development of many different nucleic acid enzymes. This section considers one category of in vitro–evolved ribozymes, which catalyze the template-directed ligation of RNA. There are several examples of RNA ligase ribozymes, some of which have been studied in considerable detail. Unfortunately, there is no reported atomic-resolution structure of any of these (or any other) in vitro–evolved nucleic acid enzymes.

The Class I, II, and III Ligase Ribozymes

The first examples of ribozymes evolved from random-sequence RNAs were reported more than 10 years ago (15). They catalyze the ligation of an oligonucleotide substrate to the 5′ end of the ribozyme, directed by an internal template region. The reaction involves attack of the terminal 2′- or 3′-hydroxyl of the substrate on the α-phosphate of the 5′-triphosphate of the ribozyme, forming a phosphodiester linkage and releasing inorganic pyrophosphate. This reaction has special interest because it is analogous to the reaction carried out by an RNA-dependent RNA polymerase.

The evolutionary search for RNA ligase ribozymes began with a starting population of about 10 copies each of 10^{15} different RNA molecules. All of the molecules contained a central region of 220 random-sequence nucleotides flanked by fixed primer binding sites (15). There are 4^{220} ($= 10^{132}$) possible 220mer sequences. Thus only a minute fraction (10^{-117}) of all possible sequences were represented in the starting population, although those molecules contained all possible subsequences up to a length of 28 nucleotides. The starting population was constructed from three synthetic 110mer DNAs that were amplified by PCR, trimmed by restriction digestion, ligated together, subjected to four cycles of PCR amplification, and then transcribed to generate RNA.

The selection scheme was chosen on the basis of acquisition of a tag sequence through ligation of the oligonucleotide substrate to the 5′ end of the ribozyme (15). The tag was recognized first by hybridization to an oligonucleotide affinity column and then, following elution from the column and reverse transcription, by hybridization of a PCR primer to corresponding nucleotides within the cDNA. Random mutations were introduced by mutagenic PCR following the fourth, fifth, and sixth rounds of selective amplification. Selection pressure was maintained by progressively decreasing the reaction time, starting with 16 h in the first round and ending with 10 min in the tenth round. In addition, the concentration of $MgCl_2$ in the reaction mixture was decreased during the last three rounds.

Following the tenth round, individuals were cloned from the population, sequenced, and tested for catalytic activity (16). About two thirds of these fell into a single family of closely related sequences, which was recognized to be part of a larger class of ribozymes (class II ligases) that dominated the population. All of the members of this class share a conserved secondary (and presumed tertiary) structure based on about 50 nucleotides. They catalyze formation of a $2',5'$-phosphodiester linkage with an observed rate of about 0.1 min^{-1}. Another class of ribozymes (class III ligases) has a different conserved structure based on about 80 nucleotides. They too catalyze formation of a $2',5'$-phosphodiester linkage but with an observed rate of only 0.005 min^{-1}. A third class of ribozymes (class I ligases) constitutes less than 10% of the population and forms yet another conserved structure based on about 110 nucleotides. Members of this class catalyze formation of a $3',5'$-phosphodiester linkage with an observed rate of about 0.2 min^{-1}. Because they form the biologically relevant $3',5'$ linkage with a rate enhancement of about 10^6-fold compared to the uncatalyzed reaction, the class I ligases have been the subject of many further studies, which will be discussed below.

In retrospect, there was little selection pressure favoring RNAs that catalyze formation of a $3',5'$ compared to a $2',5'$ linkage because reverse transcriptase can traverse a single $2',5'$ linkage without difficulty. Even during the last rounds of selective amplification, when the reaction time was reduced to 10 min, there would have been only a slight selective advantage for class I compared to class II ligases. Thus domination of the final population by class II ligases can be rationalized on the basis of their relatively simple composition, which makes them more evolutionarily accessible compared to the more complex class I ligases. It is difficult, however, to explain why the weakly reactive class III ligases were not weeded out during the latter rounds of selection. Presumably they enjoyed some other selective advantage not directly attributable to catalytic performance, such as a propensity to be amplified very efficiently.

A clonal isolate of the original class I ligase was truncated at both ends to give a 186-nucleotide RNA that retained a catalytic rate of 0.029 min^{-1} (measured in the presence of 60 mM $MgCl_2$ and 200 mM KCl at pH 7.4 and 22°C) (16). This truncated molecule was used to declare wild type (Figure 1) and to generate a new starting population of 10^{14} variants that contained random mutations at a frequency of about 20% per nucleotide position (17). These were subjected to four additional rounds of selective amplification, using the same selection scheme as before, then again cloned, sequenced, and tested for catalytic activity. On the basis of the consensus sequence of the isolated clones, various constructs were designed and tested, one of which (designated b1-207) exhibited a catalytic rate of 14.4 min^{-1} (measured under the same conditions as above). The catalytic rate of this ribozyme increases log-linearly with increasing pH over the range of 5.7–7.0 (18). Its activity is strongly dependent on the presence of Mg^{2+} with an apparent requirement for at least five Mg^{2+} ions (19). Most of these are bound with a K_d of <2 mM, but there is one Mg^{2+} binding site with an apparent K_d of 70–100 mM.

The b1-207 ligase ribozyme was used to declare wild type once again (Figure 1). Two different research groups then initiated in vitro evolution experiments

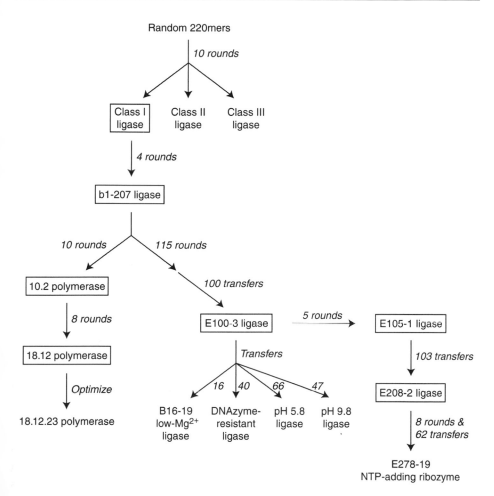

Figure 1 Genealogy of the class I ligase ribozyme, beginning with a population of random-sequence RNAs. Arrows are labeled with either the number of rounds of stepwise evolution or the number of transfers of continuous evolution. Boxes indicate instances of declaring wild type before proceeding with a single selected individual.

beginning with a population of randomized variants of this ribozyme. In one case, the goal was to establish a system for continuous in vitro evolution of catalytic function (20). In contrast to standard stepwise evolution, continuous evolution combines the processes of selection, amplification, and diversification within a common reaction mixture. The other research group sought to use in vitro evolution to convert the RNA ligase to an RNA polymerase that operates on an external RNA template (21), an activity that is central to the notion of RNA-based life (the so-called RNA world). Both efforts were successful and led to several interesting follow-up studies.

Continuous In Vitro Evolution

The continuous evolution system requires the ribozyme to ligate a chimeric DNA-RNA substrate that has the sequence of the promoter element for a DNA-dependent RNA polymerase (20). Following reverse transcription, cDNAs derived from reacted (but not unreacted) ribozymes contain a functional promoter element that, in the presence of the cognate RNA polymerase, gives rise to progeny RNAs. The ligation junction at the RNA level corresponds to the transcription start site at the DNA level, and the inorganic pyrophosphate released at the time of ligation is restored as the β- and γ-phosphates of the NTP that becomes the first residue of the ribozyme. Amplification results from the ability of the RNA polymerase to generate multiple copies of RNA per copy of DNA template. All of these events occur under a common set of conditions and can be repeated indefinitely, so long as there is a supply of the substrate and other components of the evolution system.

The b1-207 ribozyme was unable to ligate the chimeric DNA-RNA substrate, even under optimal reaction conditions. This ribozyme was used to generate a population of 10^{14} variants that contained random mutations at a frequency of about 8% per nucleotide position (20). Fifteen rounds of stepwise evolution were carried out, selecting for activity with the chimeric substrate in the presence of 60 mM $MgCl_2$ and 200 mM KCl at pH 8.5 and 24°C. The first round employed a biotinylated substrate that allowed reacted molecules to be selected on the basis of binding to a streptavidin-coated support. In subsequent rounds, the reacted molecules were selected simply by reverse transcription followed by forward transcription, exploiting the fact that only ligated molecules contained the sequence of the promoter required for transcription. As a precaution against false positives, the full-length cDNA was isolated by polyacrylamide gel electrophoresis during the fifth, tenth, and fifteenth rounds; then it was PCR amplified and transcribed rather than being amplified by transcription alone.

Individual clones isolated following the fifteenth round were able to react with the chimeric substrate at a rate of about 0.1 min^{-1} (measured under the same reaction conditions employed during in vitro evolution) (20). Their activity was substantially reduced, however, under the conditions required for continuous evolution. The population then was subjected to 100 rounds of reaction followed by selective amplification, while closing the gap between the reaction conditions used previously and those required for activity of the reverse transcriptase and RNA polymerase enzymes employed during continuous evolution (25 mM $MgCl_2$ and 50 mM KCl at pH 8.5 and 37°C). Following this procedure, the population as a whole was able to react with the chimeric substrate under the desired reaction conditions with an observed rate of >1 min^{-1}. A typical clone isolated from the population contained 17 mutations relative to the b1-207 ribozyme and had an observed rate of about 10 min^{-1}.

The population of ribozymes that resulted from the procedure described above was used to initiate continuous evolution (20). These molecules were incubated together with the chimeric substrate, the primer used to initiate cDNA synthesis, reverse transcriptase, the four dNTPs, RNA polymerase, and the four NTPs. In order to be amplified, the ribozyme molecules were required to carry out the ligation reaction before they were rendered inactive by reverse transcription. After one hour, a small aliquot of the continuous evolution mixture was transferred to a fresh reaction vessel that contained all of the above components, seeded by whatever ribozyme molecules were present in the aliquot. This is a much more powerful mode of selection compared to what can be achieved with stepwise evolution because a successful catalyst has the opportunity to give rise to several successive generations of progeny prior to each transfer step. The faster the catalyst, the shorter its generation time, providing an exponential advantage in copy number compared to slower catalysts.

The first continuous evolution experiment involved 100 successive transfers carried out over a period of 52 h, using 0.1% of each mixture to seed the next (20). During this time, $\sim$300 successive catalytic events occurred, and there was a substantial improvement in the catalytic efficiency, and therefore the "growth rate," of the evolving population. Following the one hundreth transfer, individual clones were isolated, sequenced, and tested for catalytic activity. A typical clone (designated E100-3) contained 29 mutations relative to the b1-207 ligase ribozyme and exhibited a k_{cat} of 21 min^{-1} and K_m of 1.7 μM (measured in the presence of 25 mM MgCl$_2$ and 50 mM KCl at pH 8.5 and 37°C). This ribozyme has a doubling time of 1.5 min in the continuous evolution mixture, increasing its copy number by one log-unit every 5 min.

Further continuous evolution experiments have been carried out, starting with either randomized variants of the E100-3 ligase or the entire population of ribozymes obtained after the one hundredth transfer (Figure 1). In one study, the E100-3 ligase was randomized at a frequency of 8% per nucleotide position, then subjected to 16 successive transfers of continuous evolution with a 1000-fold dilution between transfers (22). During this time, the concentration of MgCl$_2$ in the evolution mixture was steadily reduced, starting with 25 mM and ending with 15 mM. Individual clones were isolated following the sixteenth transfer, and one sequence was found to dominate the population (designated B16-19). It contains nine mutations relative to the E100-3 ligase and exhibits improved activity, in the presence of either 10 or 25 mM MgCl$_2$, compared to the E100-3 ligase. This improvement was due largely to improved folding characteristics of the B16-19 ribozyme, especially at lower concentrations of MgCl$_2$.

Another study began with variants of the E100-3 ligase that were randomized at a frequency of 10% per nucleotide position, then evolved in a continuous manner for the ability to withstand attack by an RNA-cleaving DNA enzyme (23). The DNA enzyme was directed to cleave the internal template region of the ribozyme, thus preventing the ribozyme from carrying out the ligation reaction. In the presence of 0.1 μM DNA enzyme, most of the ribozyme population was destroyed, but some survivors were carried forward in the 0.1% aliquot that was used to seed the next evolution mixture. Continuous evolution was carried out for

40 successive transfers, progressively increasing the concentration of the attacking DNA enzyme. By the end of this process the ribozymes were able to react and amplify efficiently in the presence of 10 μM DNA enzyme. They could amplify as efficiently in the presence of 10 μM DNA enzyme as the E100-3 ligase could in its absence.

The biochemical mechanism for the development of resistance to the DNA enzyme can be understood by examining the sequence and kinetic properties of the evolved ribozymes (23). They all contain three mutations (plus two compensatory mutations) located on either side of the internal template region, which cause the DNA enzyme to operate with a 10-fold lower k_{cat} and 200-fold higher K_m. Some of the evolved ribozymes also contain a constellation of mutations that caused a threefold improvement in the K_m of the ribozyme for its own RNA substrate. This allows the ribozymes to be more completely saturated with the substrate, protecting them from being bound at the same nucleotide positions by the DNA enzyme.

Another continuous evolution experiment sought to develop RNA ligase ribozymes that could operate under conditions of extreme pH (24). A previous study had shown that randomized variants of the b1-207 ribozyme could be evolved to operate at pH 4.0, albeit with a catalytic rate of only 10^{-4} min^{-1} (25). The E100-3 ribozyme was mutagenized at a frequency of 10% per nucleotide position, and two evolutionary lineages were initiated, progressing gradually to either higher- or lower-pH conditions (24). Over the course of 66 transfers, the low-pH lineage reached pH 5.8, while over the course of 47 transfers the high-pH lineage reached pH 9.8. More extreme conditions could not be explored because they would have been incompatible with the activity of the reverse transcriptase and RNA polymerase used in the continuous evolution system. An evolved low-pH ribozyme exhibited a catalytic rate of 3.3 min^{-1} at pH 5.8, compared with a rate of 0.31 min^{-1} for the E100-3 ribozyme. An evolved high-pH ribozyme retained its structural integrity and activity under such strongly denaturing conditions, whereas the E100-3 ribozyme became denatured and had no detectable activity.

A final example of continuous in vitro evolution involved experiments that sought to address the classic question of recurrence in evolutionary biology (26). If one carries out multiple replicates of the same evolution experiment, will the results be similar, or will it prove impossible to replay the tape? Once again, the starting population consisted of variants of the E100-3 ribozyme that had been randomized at a frequency of 8% per nucleotide position. Thirteen parallel lineages were initiated, four in which the concentration of MgCl$_2$ in the evolution mixture was maintained at 25 mM, five in which it was decreased progressively from 25 to 15 mM, and four in which it was decreased progressively from 25 to 15 mM, and the concentration of substrate was reduced fourfold. Each lineage was carried for 15 successive transfers with a 1000-fold dilution between transfers. Surprisingly, the same constellation of nine mutations that were seen in the B16-19 ligase ribozyme (discussed above) arose in all 13 lineages and eventually came to dominate the population. These mutations occurred as a group, accompanied by several low-frequency mutations that appeared variably among the clones.

The B16-19 ribozyme represents a deep local fitness optimum. Seven of the nine mutations that it contains are present in the DNA-enzyme-resistant ribozymes, and eight of the nine mutations are present in both the low-pH and high-pH ribozymes, although these ribozymes contain other critical mutations that are not present in the B16-19 ribozyme. As noted above, the nine recurrent mutations improve the ability of the ligase ribozyme to fold into an active conformation, which confers a selective advantage under a broad range of reaction conditions.

The above examples illustrate both the strengths and weaknesses of continuous in vitro evolution. It is a powerful method that allows one to carry out tens to hundreds of generations per day. It is methodologically very simple, allowing one to "culture" populations of ribozymes in the same way that one cultures populations of bacteria. In fact, it is provided as a laboratory exercise in the Biological Sciences Curriculum Study undergraduate college curriculum program (27). However, the method only applies to ribozyme-catalyzed reactions that result in attachment of a particular oligonucleotide to the 5' end of the ribozyme via a linkage that can be traversed by reverse transcriptase. There may be chemistries other than RNA ligation that meet this requirement, but so far none have been demonstrated in the context of continuous evolution. The method also is limited to reaction conditions that are compatible with the operation of reverse transcriptase and RNA polymerase. Perhaps the greatest value of continuous in vitro evolution is that it allows one to address fundamental issues in evolutionary biology, such as the development of resistance or the possibility of recurrence. One can control all of the relevant evolutionary parameters, such as population size, sequence diversity, and selection pressure. In addition, one has complete access to the genotype and phenotype of individuals in the population throughout the course of evolution.

From Ligase to Polymerase

As noted previously, there is special interest in the template-directed ligation of RNA because of its relevance to the activity of an RNA-dependent RNA polymerase. The class I ligase and its descendents catalyze what can be regarded as the template-directed extension of an oligonucleotide primer through the addition of an NTP, where the NTP is represented by the first residue of the ribozyme. Indeed, the guanosine 5'-triphosphate (GTP) at the 5' end of the ribozyme can be replaced by free GTP, which is added to the 3' end of the template-bound substrate with a k_{cat} of 0.3 min^{-1} and K_m of 5 mM (28). The three noncomplementary NTPs are added about 1000-fold less efficiently, but if the templating residue of the ribozyme is changed to A, G, or U, then the reaction becomes specific for UTP, CTP, or ATP, respectively. The efficiency of nucleotide addition is lowest for UTP, presumably due to its weaker base pairing and base stacking interactions compared to the other three NTPs. The specificity of addition is lowest for ATP because of the competing addition of GTP that occurs as a result of wobble pairing with a templating U residue.

Surprisingly, the template region of the ribozyme could be expanded by two residues, allowing the template-directed addition of three successive NTPs (28). This was achieved without any further in vitro evolution. Various combinations of NTPs could be added, demonstrating the ability of the ribozyme to operate in a generalized manner within the context of Watson-Crick pairing between the template and incoming nucleotide. However, the second and third NTP additions were much slower than the first, and it was not possible to expand the template region by more than two added residues.

The road to a more robust RNA polymerase ribozyme required 18 additional rounds of in vitro evolution and considerable screening of individual clones (21). The starting point was the b1-207 form of the class I ligase, which was modified in several ways (Figure 2). First, 24 nucleotides, including the internal template region, were deleted from the 5′ end of the ribozyme. The template instead was supplied as a separate oligonucleotide, and another oligonucleotide was added to replace seven deleted residues that normally form part of an essential stem structure. Second, nucleotides that were intended to serve as the primer were tethered to the 5′ end of the ribozyme via an oligonucleotide linker and an unusual 5′,5′-phosphodiester linkage. This made the 3′ end of the primer available for RNA-catalyzed extension. Third, 76 random-sequence nucleotides were attached to the 3′ end of the ribozyme, providing an opportunity to develop an accessory domain that would facilitate RNA polymerization. Fourth, two nonconserved loops within the ligase were replaced by random-sequence nucleotides. Fifth, the remainder of the ribozyme was either left unchanged or mutagenized at a degeneracy of 3%, 10%, or 20%, and a starting population of 10^{15} individuals was constructed by mixing equal portions of these four pools of molecules.

The selection scheme required the ribozyme to bind the external template through interaction with the attached primer, then catalyze the addition of one or more NTPs onto the 3′ end of the primer (21). The NTPs were tagged either as 4-thiouracil or N6-biotinyl-adenine, which could be recognized by their interaction with N-acryloyl-aminophenylmercuric acetate or streptavidin, respectively. Ten rounds of in vitro evolution were carried out, employing two different primer-template combinations and three different downstream template sequences in an effort to maintain sequence generality. The reaction time was reduced progressively from 36 h to 1 h over the first five rounds, during which the ribozyme was required to add a single 4-thiouracil residue. During the sixth through eighth rounds, the reaction time was increased, and the molecules were challenged to add both 4-thiouracil and N6-biotinyl-adenine residues. During the tenth round, the ribozymes were allowed 20 h to add a single 4-thio-uracil residue.

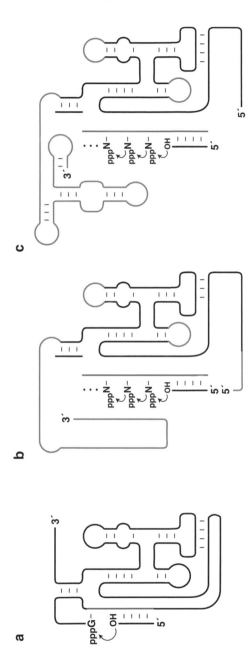

Figure 2 Conversion of the class I ligase ribozyme to a polymerase ribozyme through in vitro evolution. (*a*) Secondary structure of the b1-207 form of the class I ligase. The internal template region is shown in blue and the complementary substrate is shown in red. (*b*) Modified form of the ribozyme that was used to initiate evolution. The template (*blue*) was supplied as a separate oligonucleotide, the primer (*red*) was tethered to the 5' end of the ribozyme via a 5',5'-phosphodiester linkage, and one or more of the NTPs (pppN) were tagged to allow selection of active polymerase ribozymes. Random-sequence nucleotides (*green*) were attached to the 3' end of the ribozyme and in place of two of the internal loop regions. (*c*) Secondary structure of the resulting RNA polymerase ribozyme.

Individuals were cloned from the population following the eighth, ninth, and tenth rounds, sequenced, and tested for catalytic activity. Among 74 clones that were sequenced, 23 different sequence families were identified. Remarkably, only one of these (designated 10.2) had the ability to extend the primer efficiently using both UTP and ATP (21). Most of the cloned individuals operated in a template-independent manner or became tagged by NTP addition at a position other than the 3′ end of the primer. The one desirable individual could operate on a variety of primer-template combinations, even when the primer was detached from the ribozyme. It does not contain any mutations within the original ligase domain, but it had acquired an accessory domain, essential for its catalytic activity, derived from the 76 random-sequence nucleotides. Surprisingly, the 10.2 ribozyme is much less efficient in incorporating N6-biotinyl-ATP compared to unmodified ATP, even though biotinylated ATP was employed during in vitro evolution.

In retrospect, the sixth through eighth rounds of evolution may have been counterproductive to the goal of developing a robust RNA polymerase ribozyme. The greatly increased reaction time and the requirement to acquire a biotin tag by whatever means possible may have allowed rogue molecules to become dominant in the population. Fortunately, a single desirable individual was isolated following the tenth round. At that point there was little choice but to declare wild type before proceeding with additional rounds of evolution.

The 10.2 ribozyme was randomly mutagenized to construct a population of 10^{15} variants (21). The original ligase domain was either left unchanged or mutagenized at a degeneracy of 3%; the accessory domain and two nonconserved loops within the original ligase domain were mutagenized at a degeneracy of 20%, and 17 nucleotides were added to the 3′ terminus to provide a novel binding site for the primer used to initiate cDNA synthesis. Eight more rounds of in vitro evolution were carried out, requiring the molecules to add at least two 4-thio-uracil residues. Three different primer-template combinations and four different downstream template sequences were employed in an effort to maintain sequence generality. During the last four rounds, the concentration of 4-thio-UTP was reduced from 1 to 0.1 mM, and ATP, CTP, and GTP were added at 2 mM each. No attempt was made to tag any NTP other than UTP.

Individuals were isolated from the population following the fourth and eighth rounds. Most were found to have no detectable polymerase activity, but some showed substantially improved activity compared to the 10.2 ribozyme. One individual in particular (designated 18.12) was highly adept at extending a primer on a long template (21). On the basis of sequence comparison with other active individuals, the 18.12 ribozyme was modified in various ways. A variant was constructed (designated 18.12.23) that contains 189 nucleotides, 31 fewer than the 18.12 ribozyme, and has even greater polymerase activity. It can extend a detached primer on an external template molecule by up to 14 nucleotides, exhibiting both sequence generality and high fidelity. The overall fidelity of polymerization is 96.7% in the presence of 4 mM each of the four NTPs. The fidelity is lowest for ATP addition because of competing addition of GTP due to wobble pairing.

The 18.12.23 ribozyme can be regarded as a general, RNA-dependent RNA polymerase; one of the greatest achievements in directed molecular evolution. Its catalytic rate for NTP addition is at least 1.5 min^{-1}. However, the ribozyme has weak affinity for the primer-template complex with a K_d of 700 μM or higher, depending on the choice of primer and template (29). For one well-studied primer-template combination, the on rate is at least 6400 M$^{-1} \cdot$ min^{-1} and the off rate is at least 4.5 min^{-1}. Normally, the ribozyme is made to operate under conditions that are greatly subsaturating with the ribozyme, primer, and template all present at about 1 μM concentration. Under these conditions, the $t_{1/2}$ for binding of the primer-template complex by the ribozyme is about 2 h. Once the primer-template is bound, NTP addition occurs at roughly the same rate as the rate of primer-template release. Interestingly, the ribozyme exhibits a low level of processivity with a 1% to 90% chance (depending on sequence context) of adding a second NTP before releasing the primer-template (29). If the processivity and/or the on rate of the primer-template complex could be increased, the 18.12.23 ribozyme has the potential to evolve to be an RNA replicase ribozyme. Such a molecule could be used to carry out the amplification step of in vitro RNA evolution, perhaps allowing the RNA-directed evolution of the replicase itself.

There is another ribozyme, also a descendant of the class I ligase, that catalyzes the template-directed addition of NTPs onto the end of an RNA primer (30). As discussed previously, the class I ligase has been made to undergo continuous in vitro evolution, leading to the E100-3 ligase. This in turn was evolved to catalyze two successive NTP additions followed by RNA ligation (Figure 1). The E100-3 ligase was randomized at a frequency of 8% per nucleotide position, generating a population of about 10^{14} variants. These were provided with a chimeric DNA-RNA substrate containing the sequence of an RNA polymerase promoter element but lacking the 3′-terminal nucleotide of that sequence. In order to be eligible for amplification, the ribozymes were required to complete the promoter sequence by adding the appropriate NTP and then catalyze ligation of the extended substrate onto their own 5′ end.

The ribozymes first were subjected to five rounds of stepwise evolution. Then individuals were isolated from the population and tested for the ability to undergo continuous evolution employing the shortened substrate (30). One especially well-behaved individual was identified (designated E105-1), randomized at a frequency of 10% per nucleotide position, and then used to initiate continuous evolution. Over a period of 78 h, 103 successive transfers of continuous evolution were carried out, typically with 1000-fold dilution between transfers. Individuals again were isolated from the population and tested for catalytic activity. Most had the ability to catalyze the template-directed addition of an NTP onto the 3′ end of the oligonucleotide substrate. One such individual (designated E208-2) contained 25 mutations relative to the E100-3 ribozyme and was chosen to begin the next phase of in vitro evolution.

The E208-2 ribozyme was randomized at a frequency of 10% per nucleotide position, subjected to two more transfers of continuous evolution as before, and then made to react with an oligonucleotide substrate that lacked two nucleotides at the 3' end of the promoter sequence (30). The ribozymes were required to catalyze two NTP additions followed by ligation of the extended substrate. After 50 more transfers of continuous evolution, followed by a combination of 8 rounds of stepwise evolution and 10 transfers of continuous evolution, molecules with the desired activity were obtained. The combination of stepwise and continuous evolution was necessary because with continuous evolution alone the ribozymes became dependent on the presence of the RNA polymerase protein for their ability to extend the substrate. This dependency was eliminated by occasionally requiring the ribozymes to perform the extension reaction in the absence of the polymerase protein. It proved impossible to use continuous evolution to obtain ribozymes that catalyze three successive NTP additions followed by ligation because ersatz promoter sequences arose instead, allowing transcription to occur at a low level even after only two NTP additions.

Following the last transfer of continuous evolution, individuals were isolated from the population and characterized (30). A typical well-behaved individual (designated E278-19) contained five additional mutations compared to those present in the E208-2 ribozyme. The E278-19 ribozyme catalyzes the template-directed addition of two NTPs followed by ligation with an overall apparent rate of 3×10^{-4} min^{-1} in the presence of 2 mM each of the two NTPs. It also catalyzes direct ligation of the oligonucleotide substrate, without the addition of NTPs, at about a 10-fold faster rate. The products of direct ligation are not amplified but continue to form during each round of in vitro evolution.

Other In Vitro-Evolved Ligase Ribozymes

There are several other reported examples of RNA ligase ribozymes that were evolved starting from random-sequence RNAs. Only three of these (and their descendants) catalyze the template-directed ligation of an oligonucleotide 3'-hydroxyl and an oligonucleotide 5'-triphosphate to form a 3',5'-phosphodiester linkage. They are the L1, hc, and R3C ligases, which will be discussed below. There are other in vitro–evolved ligases that catalyze formation of a 2',5'-phosphodiester, as first demonstrated by the class II and class III ligases.

The simplest of the 2',5' ligases contains only 29 nucleotides; 26 of which form Watson-Crick or wobble pairs with the oligonucleotide substrates (31). This "mini ligase" has a catalytic rate of only 5.4×10^{-4} min^{-1}, which is about 10^4-fold faster than the uncatalyzed rate of reaction in the presence of a complementary template. Another 2',5' ligase was derived from the E100-3 ligase by carrying out stepwise evolution in the presence of sodium bisulfite, which selected strongly against the presence of cytidine residues (32). A total of 24 rounds of evolution were carried out, leading to a molecule that completely lacks cytidine yet has a k_{cat} of 0.006 min^{-1} and K_m of 0.23 μM. The cytidine-free and E100-3 ligases have 45% sequence similarity (despite the difference of 37 cytidines), but they have very different secondary structures and form $2',5'$ versus $3',5'$ linkages, respectively.

Another RNA ligase ribozyme evolved from random-sequence RNA operates by a somewhat different chemical mechanism with adenylate rather than inorganic pyrophosphate as the leaving group (33). The reaction takes place between the $3'$-hydroxyl of a template-bound oligonucleotide and the α-phosphate of an adenosine-$5',5'$-pyrophosphate located at the $5'$ end of the ribozyme. This is the same mechanism employed by various RNA ligase proteins. The starting population contained about 10^{15} different RNAs, each with 210 random-sequence nucleotides surrounding an ATP-binding domain that was mutagenized at a degeneracy of 15%. The ATP-binding domain was obtained previously by in vitro selection and was included to facilitate the emergence of ligase ribozymes that utilize adenylate as the leaving group.

Following ten rounds of in vitro evolution, individuals were isolated from the population and characterized (33). One of these individuals was subjected to $3'$-terminal deletion and mutagenesis based on the sequences of the other isolated individuals, resulting in a variant (designated CM2g) that has a catalytic rate of 0.0063 min^{-1}. This ribozyme forms a $3',5'$-phosphodiester linkage with adenylate as the leaving group. If the reactive adenosine-$5',5'$-pyrophosphate is replaced by guanosine-$5',5'$-pyrophosphate, the rate decreases by only threefold, but if it is replaced by $5'$-triphosphate, the rate decreases by more than 4000-fold. It is unclear whether inclusion of the ATP-binding domain had a beneficial effect. The CM2g ribozyme contains only one critical mutation within this domain, but it is not known whether the domain is involved in binding the adenosine-$5',5'$-pyrophosphate.

There is no reported example of an RNA ligase ribozyme that can also ligate a DNA substrate. However, there is one example of a ribozyme that catalyzes ligation of an oligodeoxynucleotide $3'$-hydroxyl and the $5'$-triphosphate of the ribozyme in the presence of an external template (34). This ligase is unusual in that it was isolated from a population of random-sequence RNAs that contained N6-(6-aminohexyl)-adenosine in place of adenosine. It has a catalytic rate of only 8.5×10^{-7} min^{-1}, which is less than 100-fold faster than the uncatalyzed rate of reaction in the presence of a complementary template. Its rate is reduced by about 20-fold when the N6-(6-aminohexyl)-adenosine residues are replaced by adenosines.

THE L1 LIGASE   The same method that was used to evolve the class I, II, and III ligase ribozymes also was used to obtain an RNA ligase ribozyme that is dependent on the presence of external "effector" molecules for its catalytic activity (35). The latter ribozyme was derived from a starting population of 10^{15} RNA molecules that contained 90 random-sequence nucleotides. After five rounds of in vitro evolution, individual ribozymes were isolated from the population and found to have very little sequence variation. One of the cloned individuals was randomized at an expected frequency of 30% per nucleotide position, and the resulting population was subjected to five additional rounds of in vitro evolution (35, 36). During these rounds, the reaction time was reduced progressively from 16 h to 5 min, and the concentration of $MgCl_2$ in the reaction mixture was reduced from 60 to 10 mM.

The most active ribozyme isolated from the final selected population (designated L1) has a catalytic rate of 0.093 min^{-1} and forms a 3′,5′-phosphodiester linkage (35, 36). Its activity is strongly dependent on the presence of a separate oligodeoxynucleotide, which was included in the reaction mixture during the evolution process and used to prime cDNA synthesis. This oligodeoxynucleotide forms a stem structure with 18 nucleotides at the 3′ end of the ribozyme. In its absence, the catalytic rate of the ribozyme is reduced by about 10^4-fold.

The original isolated form of the L1 ligase contains 130 nucleotides (35). It adopts a simple three-way junction structure that can be trimmed substantially without loss of catalytic activity (36). A 74-nucleotide version of this ribozyme was completely randomized at either 15 or 20 internal nucleotide positions, and the two resulting populations were subjected to either 7 or 8 rounds of in vitro evolution, respectively (37). The population that was randomized at 20 positions showed the greatest improvement in catalytic activity, presumably because it contained more sequence diversity. The most active individual isolated from the final selected population has a catalytic rate of 0.37 min^{-1} and still requires the separate oligodeoxynucleotide for its activity.

The L1 ligase was made to operate in a manner that is dependent on other effector molecules, in addition to the oligodeoxynucleotide effector (35, 36, 38). Ligand-dependent activation of a nucleic acid enzyme was demonstrated previously for the hammerhead ribozyme, which catalyzes the sequence-specific cleavage of RNA. As a prelude to that demonstration, various nucleotides within the catalytic core of the hammerhead were replaced by abasic nucleotide analogs, and the missing base was supplied as a separate molecule (39). For example, replacement of a critical adenylate by an abasic nucleotide analog reduced the catalytic rate of the hammerhead by 40,000-fold, but when 3 mM adenine was added to the reaction mixture, the rate improved by 300-fold. In later experiments, an internal stem-loop was replaced by an ATP-binding domain (40). In the absence of ATP, this domain is unstructured, but upon binding ATP, it adopts a well-defined structure that supports the active conformation of the hammerhead. In still other experiments, the internal stem-loop was replaced by random nucleotides and in vitro selection was carried out to obtain ribozymes that depend on various small molecules for their activity (41).

Similar methods were used to develop ligand-activated forms of the L1 ligase. In one case, the distal portion of an internal stem-loop was replaced by an ATP-binding domain, resulting in 30-fold activation in the presence of 1 mM ATP compared to its absence (35). Optimization of this construct improved ATP-dependent activation to a level of 830-fold. In another case, a FMN-binding domain was introduced and connected to the remainder of the ribozyme via a short stretch of random nucleotides (36). Seven rounds of in vitro selection were carried out, in each round employing a procedure that first selected against activity in the absence of FMN and then selected for activity in the presence of FMN. This resulted in various FMN-activated ribozymes, the best of which has a 260-fold faster catalytic rate in the presence of 0.1 mM FMN compared to its absence. More recently, variants of the L1 ligase were developed that are dependent on the presence of a particular protein, either *Neurospora* mitochondrial tyrosyl tRNA synthetase or chicken egg-white lysozyme, for their catalytic activity (38). The entire central stem-loop of the ribozyme was replaced by 50 random-sequence nucleotides and multiple rounds of in vitro selection were carried out. The selection pressure was increased during successive rounds by increasing the time allowed for the reaction in the absence of the protein (negative selection) and by decreasing the time allowed for the reaction in the presence of the protein (positive selection). After nine rounds, a ribozyme was isolated that had a catalytic rate of 0.035 min^{-1} in the presence of 1 μM tyrosyl tRNA synthetase and a rate of only 4×10^{-7} min^{-1} in its absence, corresponding to 94,000-fold activation. After 11 rounds in the other lineage, a ribozyme was obtained that had a catalytic rate of 0.01 min^{-1} in the presence of 1 μM egg-white lysozyme and a rate of only 3×10^{-6} min^{-1} in its absence, corresponding to 3100-fold activation. Not surprisingly, each ribozyme was highly specific for its cognate protein. Ligand-activated ribozymes such as these have potential applications as biosensors and in proteomics studies (42–45).

THE HC LIGASE   Another example of an RNA ligase ribozyme that catalyzes formation of a 3′,5′-phosphodiester linkage was obtained, not from purely random-sequence RNAs, but from 85 random nucleotides that were attached to a structural scaffold (46). This scaffold was derived from the P4–P6 domain of the *Tetrahymena* group I ribozyme. The *Tetrahymena* ribozyme normally catalyzes two successive phosphoester transfer reactions that result in self-splicing of a precursor rRNA, but the P4–P6 domain alone has no catalytic activity. This domain assumes a well-defined tertiary structure (47, 48) and forms nonsequence-specific tertiary contacts with double-stranded RNA (49, 50). Thus it was hypothesized that by attaching random nucleotides to the P4–P6 scaffold one might take advantage of the structural preorganization of this domain and its propensity to bind double-stranded RNA, leading to a generalized RNA ligase.

A population of 10^{16} different RNAs was constructed. Each molecule contained a 5′-terminal hairpin that provided a template region for binding a complementary oligonucleotide substrate adjacent to the 5′-triphosphate of the ribozyme (46). Ten rounds of in vitro evolution were carried out, alternating two different substrates and changing the sequence of the 5′-terminal hairpin in order to maintain sequence generality. During these rounds, the reaction time was reduced progressively from 24 h to 1 min, and the concentration of MgCl$_2$ in the reaction mixture was reduced from 200 to 4 mM. Mutagenic PCR was performed following the fourth and sixth rounds. After the tenth round, individuals were isolated from the population and found to fall into two different sequence families. A member of the dominant family that exhibited efficient ligase activity (designated hc16) was shown to operate in a generalized manner with regard to the sequence of the template-substrate duplex (hc refers to helical context).

The hc16 ligase ribozyme forms a $3',5'$-phosphodiester with a catalytic rate of 0.26 min^{-1}, measured in the presence of 50 mM MgCl$_2$ and 200 mM KCl at pH 7.5 and 50°C (46). Its rate is about fivefold slower at 30°C. The ribozyme retains the P4–P6 domain but with six point mutations. Based on chemical modification analysis, this domain has the same structure in the hc16 ligase as it does in the *Tetrahymena* group I ribozyme. The regions of random-sequence nucleotides gave rise to two additional stem-loop elements that are essential for the newly developed ligase activity.

A shortcoming of the hc16 ligase is that both the template and $5'$-triphospho-rylated substrate domain are tethered to the ribozyme as part of the $5'$-terminal hairpin. If the hairpin is detached from the ribozyme, the reaction still proceeds but with a k_{cat} of only 0.0035 min^{-1} and K_m of 12 μM (measured in the presence of 100 mM MgCl$_2$ at pH 7.5 and 37°C). Additional rounds of in vitro evolution were carried out in an effort to improve the intermolecular reaction (51). A pseudo-intermolecular reaction format was established by connecting the hairpin to the $5'$ end of the ribozyme via a flexible linker of 28 uridylates. This construct was subjected to mutagenic PCR; then seven rounds of in vitro evolution were carried out, employing two different RNA substrates.

After the seventh round a well-behaved individual was isolated from the population, subjected to mutagenic PCR, and used to begin 11 more rounds of in vitro evolution, reducing the reaction time to only 30 s (51). After the last round, an especially active individual, which contained 15 mutations relative to the hc16 ribozyme, was identified and designated hc18-2. When made to operate on a separate template-substrate hairpin, it had a k_{cat} of 0.074 min^{-1} and K_m of 3.9 μM (measured under the same reaction conditions as above). The sequence of the hairpin can be altered without loss of catalytic activity, as long as Watson-Crick pairing is maintained within the stem region. Activity is substantially reduced, however, if the hairpin loop is removed to provide a tetramolecular reaction format involving the ribozyme, template, and two oligonucleotide substrates. In that case, k_{cat} is only 0.0049 min^{-1} and K_m is 34 μM, measured under multiple-turnover conditions. The ribozyme also is able to catalyze NTP addition on an external RNA template, albeit with very low efficiency.

It is interesting to compare the 18-2 variant of the hc ligase and the 18.12.23 variant of the class I ligase. Both catalyze the formation of a $3',5'$-phosphodiester linkage between two RNA molecules that are bound to a separate RNA template. The class I 18.12.23 ribozyme is able to carry out multiple successive NTP additions with an inherent catalytic rate of at least 1.5 min^{-1} (21, 29). It is general in respect to the sequence of the template-substrate complex but has weak affinity for that complex. In contrast, the catalytic rate of the hc 18-2 ribozyme is much slower, especially for NTP addition, and it is unable to add more than two NTPs in succession (51). However, the hc 18-2 ribozyme has good affinity for the template-substrate complex, even when the hairpin loop is removed. There is no practical method for cross breeding two ribozymes that have no sequence or structural similarity, but it is appealing to contemplate the possibility of a ribozyme that would exhibit the best properties of both the 18.12.23 and 18-2 ribozymes.

THE R3 LIGASE   The final example of an RNA ligase ribozyme that catalyzes formation of a $3',5'$-phosphodiester has an unusual pedigree. It was evolved starting from a population of random-sequence RNAs that contained only adenosine, guanosine, and uridine residues (52). As mentioned previously, a cytidine-free ligase ribozyme was evolved from a starting population of variants of the class I ligase, but that cytidine-free ribozyme catalyzes formation of a $2',5'$-phosphodiester with a catalytic rate of only 0.006 min^{-1} (32). Perhaps its evolutionary development was constrained by the particular starting point that was chosen.

As an alternative approach, a population of 10^{14} cytidine-free RNAs was constructed; each molecule contained two regions of 64 random-sequence nucleotides flanking an internal template (52). Five rounds of in vitro evolution were carried out, progressively decreasing the reaction time from 18 h to 10 s. Following the fifth round, seven individuals were isolated from the population and tested for catalytic activity. Six of these catalyzed formation of a $2',5'$-phosphodiester and one catalyzed formation of a $3',5'$-phosphodiester. The latter individual was randomized at a frequency of 8% per nucleotide position and then was subjected to five more rounds of in vitro evolution. Following the last round, nine individuals were isolated from the population, all of which retained the ability to catalyze formation of a $3',5'$-phosphodiester. The most active of these individuals was chosen for further study.

It was found that all of the nucleotides located on the $3'$ side of the internal template could be deleted without reducing catalytic activity (52). This resulted in a 74-nucleotide ribozyme (designated R3) that has a k_{cat} of 0.013 min^{-1} and K_m of 6.2 μM. The R3 ligase adopts a simple three-way junction structure, reminiscent of the structure of the L1 ligase. These two ribozymes have several critical distinctions, however, most notably in the region surrounding the ligation junction and at the three-way junction.

The R3 ligase, which was evolved in a world free of cytidine, was doped with cytidine at a frequency of 1% per nucleotide position and subjected to hypermutagenic PCR. The resulting population was used to initiate six rounds of in vitro evolution, which led to a population of cytidine-containing molecules that had very little sequence diversity (52). The dominant individual in this population (designated R3C) contained 12 mutations relative to the R3 ligase, including 6 U→C changes and 1 A→C change. It retains $3',5'$-regiospecificity and operates with a k_{cat} of 0.32 min^{-1} and K_m of 0.4 μM. The three-way junction is somewhat remodeled in the R3C compared to the R3 ligase, which required stabilization of the proximal portion of the adjoining stems through replacement of A-U and G-U pairs by G-C pairs. The distal portions of the three stems can have almost any sequence, so long as the stability of the stems is maintained.

More perversely, the R3 ligase was evolved to a form that contains only two different nucleotide subunits, the minimum necessary to specify genetic information (53). This was achieved in two stages. First, all of the adenosine residues were replaced by 2,6-diaminopurine (D) nucleosides and all of the G-U wobble pairs were replaced by D-U pairs. (D-U pairs are intermediate in stability between G-C and A-U pairs.) Second, the remaining G residues were replaced by either D or U, and in vitro evolution was carried out to optimize catalytic activity. After five rounds of evolution, individuals were isolated from the population and sequenced. The two most active individuals, both of which contained only D and U residues, were randomly mutated at a frequency of 12% per nucleotide position; then five more rounds of in vitro evolution were carried out. Again individuals were isolated from the population, and the most active (designated R2) was chosen for further study.

The R2 ligase contains 66 nucleotides (39 D and 27 U residues) plus the 3′-terminal template region of variable composition (53). This ribozyme reacts with a substrate that contains only D and U residues, forming a 3′,5′-phosphodiester linkage with a k_{cat} of 0.0011 min^{-1} and K_m of 1.6 nM. The catalytic rate is about 36,000-fold faster than the uncatalyzed rate of reaction, but 12-fold slower than that of the R3 ligase, which in turn is 25-fold slower than that of the R3C ligase. The R2, R3, and R3C ligases all have the same general architecture, although the R2 ligase is most susceptible to adopting alternative conformations and reacts to a maximum extent of only 6% to 8%. It lies at the edge of the minimum compositional complexity needed to specify catalytic function.

The R3C ligase is the most robust member of the R3 family. It was used to construct a self-replicating ribozyme that operates, not by template-directed polymerization of NTPs, but by RNA-catalyzed ligation of two RNA substrates to form additional copies of itself (54). In the absence of any preformed ribozyme, the two substrates are ligated very slowly with a second-order rate constant of 3.3×10^{-11} M$^{-1} \cdot$ min^{-1}. Upon addition of the ribozyme the reaction becomes autocatalytic with an autocatalytic rate constant of 0.011 min^{-1}. The initial rate of formation of new ribozymes is linearly proportional to the starting concentration of ribozymes with a reaction order of 1.0, corresponding to exponential growth. Dissociation of two ribozyme molecules is not rate limiting. Exponential growth is limited, however, because the two substrates can form a nonproductive complex that prevents them from binding readily to the ribozyme. Nonetheless, it is possible to exceed breakeven, whereby the amount of newly synthesized ribozyme is greater than the starting amount of ribozyme.

The self-replicating ligase ribozyme operates with a very restricted set of component substrates and does not broadly enable the RNA-catalyzed replication of RNA. It is not capable of undergoing Darwinian evolution because it does not allow for the possibility of heritable mutations. Replication in this case involves a single joining reaction, and thus it has low information content despite the complexity of the two components. Perhaps it is possible to develop a ribozyme that ligates several component modules to form additional copies of itself. Competition for a limited supply of modules among a heterogeneous population of self-replicating ribozymes might provide the basis for Darwinian evolution. In the foreseeable future, however, the in vitro evolution of ribozymes will continue to require the use of protein polymerases to carry out the crucial step of amplification.

RNA-CLEAVING DNA ENZYMES EVOLVED FROM RANDOM-SEQUENCE DNAs

It is instructive to contrast the in vitro evolution of RNA ligase ribozymes and RNA-cleaving DNA enzymes. The latter catalyze a bond-breaking reaction, requiring selection based on detachment of a chemical tag. The selected molecules, composed of DNA, can be PCR amplified directly, but the functional strand of the double-stranded PCR products must be isolated prior to the next round of selective amplification. Compared to RNA ligation involving a 3'-hydroxyl and 5'-triphosphate, the cleavage of an RNA phosphodiester is a relatively facile reaction with an uncatalyzed rate of about 10^{-7} min^{-1} in the presence of 2 mM $MgCl_2$ at pH 7.5 and 37°C (55). This makes it relatively easy to discover catalysts, although their enrichment may be more difficult due to the high background resulting from uncatalyzed cleavage events. Fewer nucleotides are required to specify a motif that catalyzes RNA cleavage compared to one that catalyzes RNA ligation. Thus, RNA-cleaving DNA enzymes often exist within a starting population of all possible 25mer DNAs, allowing them to be obtained by in vitro selection without mutagenesis.

The First DNA Enzymes

Shortly after the discovery of ribozymes, there was speculation concerning the possibility of catalytic DNA molecules (56). It was argued that the 2'-hydroxyl group of RNA is important for catalysis and for defining the tertiary structure of RNA, and that even if DNA could be a catalyst, it might be much less efficient than RNA. This view was supported by studies of the hammerhead ribozyme in which component ribonucleotides were replaced systematically by deoxynucleotides, demonstrating that several ribonucleotides are essential for the catalytic activity of this ribozyme (57–61). To this day, no ribozyme has been converted to a DNA enzyme, either by rational design or by in vitro evolution. All known DNA enzymes have been obtained starting from a population of random-sequence DNAs.

The first reported DNA enzyme catalyzes the Pb^{2+}-dependent cleavage of a single RNA phosphodiester that is embedded within an otherwise all-DNA substrate (62). This is a very simple reaction, requiring the proper positioning of a Pb^{2+} ion to assist in deprotonation of the 2'-hydroxyl that lies adjacent to the cleavage site. The uncatalyzed rate of reaction is about 10^{-4} min^{-1}, measured in the presence of 1 mM PbOAc and 10 mM $MgCl_2$ at pH 7.0 and 23°C. A starting population of 10^{14} DNAs was constructed; each molecule contained 50 random-sequence residues flanked by primer binding sites. A substrate domain was placed at the 5' end, which included the target ribonucleotide and a 5'-terminal biotin moiety. The molecules were bound to a streptavidin-coated support, washed, and then triggered for catalysis by addition of 1 mM PbOAc. Any molecule that underwent cleavage following the addition of PbOAc became detached from the support and collected in the eluate. The collected molecules then were PCR amplified, the substrate domain was reattached to the progeny molecules, and the entire procedure was repeated until the desired catalysts were obtained.

Five rounds of selective amplification were carried out, progressively decreasing the reaction time from 1 h to 1 min (62). In principle, any means by which the molecules could become detached from the support following addition of PbOAc would result in their selective amplification. As expected, however, the most expedient means of detachment was DNA-catalyzed cleavage of the target RNA phosphodiester, leaving the 5' product still attached to the support and releasing the 3' product into solution. Uncatalyzed cleavage occurred at a rate of about 10^{-4} min^{-1}, but by the fifth round of selective amplification the population as a whole underwent Pb^{2+}-dependent cleavage at a rate of about 0.1 min^{-1}.

Twenty individuals were isolated from the final selected population and found to conform to a common consensus sequence (62). The catalytic motif was recognized to contain regions of Watson-Crick pairing on either side of the cleavage site, three unpaired substrate nucleotides immediately downstream from the cleavage site, and a catalytic core of 15 deoxynucleotides located opposite the cleavage site. It was straightforward to separate the substrate and catalytic domains to produce a 38mer DNA enzyme that cleaves a 19mer substrate with multiple turnover with a k_{cat} of 1 min^{-1} and K_m of 2 μM. The Pb^{2+}-dependent DNA enzyme is unable to cleave an all-RNA substrate.

The decision to employ a substrate domain that contained a single ribonucleotide embedded within DNA was based on two factors. First, it was feared that if additional ribonucleotides were present in the substrate it would cloud the issue of which residues were responsible for catalysis. It has been shown, for example, that ribonucleotides essential for the activity of the hammerhead ribozyme can be placed within an RNA molecule that also contains the cleavage site, and when supplied with an oligodeoxynucleotide to complete the hammerhead motif, the bimolecular complex undergoes RNA cleavage (63). In this case, of course, most of the catalytic residues lie within the so-called substrate. By restricting the RNA content to the single ribonucleotide that defines the cleavage site, such ambiguity could be avoided. Second, as a practical matter, the uncatalyzed reaction might have been too prominent if multiple ribonucleotides were present, which would have made the selective enrichment of desired catalysts more difficult.

Contemporaneous with the development of the Pb^{2+}-dependent, RNA-cleaving DNA enzyme, in vitro selection was used to obtain a Zn^{2+}/Cu^{2+}-dependent DNA enzyme that catalyzes the template-dependent ligation of an oligodeoxynucleotide 5'-hydroxyl and an oligodeoxynucleotide 3'-phosphorimidazolide (64). This too is a simple reaction with an uncatalyzed rate of 2×10^{-5} min^{-1}. Thus the question arose as to whether DNA could catalyze more difficult chemical transformations that have a much lower uncatalyzed rate of reaction.

The same approach used to obtain the Pb^{2+}-dependent DNA enzyme also was used to obtain an RNA-cleaving DNA enzyme that depends on Mg^{2+} for its catalytic activity (65). The uncatalyzed rate of RNA cleavage is about 10^3-fold slower in the presence of Mg^{2+} compared to Pb^{2+}, which is attributable to the difference in pK_a of the corresponding metal hydrate (11.4 versus 7.7, respectively). A starting population of 10^{13} different DNAs was constructed; each molecule contained 40 random-sequence nucleotides flanked by short regions that could engage in Watson-Crick pairing with the substrate domain. Six rounds of in vitro selection were carried out, triggering the reaction by addition of 1 mM $MgCl_2$. After the sixth round, individuals were isolated from the population and found to have very similar sequences. The most active of these was mutagenized at a degeneracy of 15% and subjected to seven additional rounds of selective amplification, which led to improved catalytic activity and helped to define the catalytic motif.

Thirty individuals, all with similar sequences, were isolated from the final selected population (65). On the basis of analysis of sequence covariation among these individuals, the catalytic motif was recognized to contain regions of Watson-Crick pairing on either side of the cleavage site, extending the paired regions that had been engineered in the starting population of molecules. The catalytic core consists of an internal stem-loop of variable sequence that is connected to the two regions of Watson-Crick pairing by a total of 15 highly conserved nucleotides. The substrate and catalytic domains were separated, resulting in a 36mer DNA enzyme that cleaves a 15mer substrate with a k_{cat} of 0.039 min^{-1} and K_m of 13 μM, measured in the presence of 10 mM $MgCl_2$ and 1 M NaCl at pH 7.0 and 23°C. This corresponds to a rate enhancement of about 10^5-fold compared to the uncatalyzed rate of reaction.

In three other studies, attempts were made to develop DNA enzymes that, without the assistance of a divalent metal, cleave a single RNA phosphodiester embedded within DNA. All three studies employed the same selection scheme as above but triggered the reaction by the addition of different compounds. In one case, the trigger was a mixture of 20 mM L-histidine and 0.5 mM $MgCl_2$ with the aim of developing histidine-dependent catalysts (66). After seven rounds of in vitro selection, catalysts were obtained that depend on Mg^{2+} but have no activity in the presence of histidine alone. After three more rounds, individuals were isolated from the population and characterized. The dominant motif (designated Mg5) has Mg^{2+}-dependent, RNA-cleavage activity, but surprisingly it is about 10-fold more active in the presence of Ca^{2+}. Following separation of the substrate and catalytic domains, the DNA enzyme was found to have a k_{cat} of 0.1 min^{-1} and K_m of 6.4 μM, measured in the presence of 0.5 mM $CaCl_2$ at pH 7.0 and 37°C.

In another study, RNA cleavage was triggered by the addition of 1 M NaCl but without divalent metal (67). The starting population was identical to that used to obtain the Mg^{2+}-dependent DNA enzyme (65). After 12 rounds of in vitro selection, individuals were isolated from the population and found to undergo self-cleavage in the presence of 1 M NaCl (at pH 7.0 and 23°C) with an observed rate of about 10^{-3} min^{-1}. The reaction rate is not diminished by adding divalent-metal-chelating compounds, by replacing Na^+ with a different mono-valent cation, or by changing the buffer used in the reaction. One of the isolated individuals was mutagenized at a degeneracy of 15% and subjected to six additional rounds of selective amplification. This led to an improved version of the Na^+-dependent catalytic motif that has an observed rate of 6.7×10^{-3} min^{-1}.

In the third study, RNA cleavage was triggered by the addition of 50 mM L-histidine and no divalent metal (68). Again the starting population of DNAs contained 40 random-sequence nucleotides flanked by regions that could engage in Watson-Crick pairing with the substrate domain. After 11 rounds of in vitro selection, the population as a whole exhibited RNA-cleavage activity in either the presence or absence of histidine. Individuals were isolated from the population, and a subset of these were found to depend on histidine for their activity. One such individual (designated HD1) was mutagenized at a degeneracy of 21% and subjected to five additional rounds of selective amplification, reducing the triggering concentration of histidine to 5 mM. After the last round, individuals again were isolated from the population, and one of these (designated HD2) was studied in detail.

Both HD1 and HD2 were converted to an intermolecular reaction format by separating the substrate and catalytic domains (68). In the presence of 100 mM histidine at pH 7.5 and 23°C, the HD1 and HD2 DNA enzymes have an observed rate of 0.0047 and 0.2 min^{-1}, respectively. HD1 can be saturated with histidine with an apparent K_d of 25 mM, whereas HD2 is not saturated even at very high concentrations of cofactor. This is somewhat surprising considering that HD2 was derived from HD1 by continuing in vitro selection using a lower concentration of histidine. However, both enzymes are highly specific for L-histidine with about 10^3-fold lower activity, for example, in the presence of D-histidine. The catalytic rate of the HD2 enzyme as a function of pH has a bell-shaped profile with an apparent pK_a of about 6. This is consistent with the imidazole group of histidine acting as a general base to assist in deprotonation of the 2'-hydroxyl adjacent to the cleavage site.

In summary, DNA can be an enzyme with a catalytic rate enhancement similar to that of typical RNA enzymes. DNA enzymes benefit from the assistance of a catalytic cofactor, which may be a divalent metal, histidine, or perhaps some other small molecule. It even is possible for DNA to function without the assistance of a catalytic cofactor, as exemplified by the Na^+-dependent DNA enzyme. It seems likely, however, that cofactors will be required to achieve substantial catalytic rate enhancements, especially for difficult chemical transformations.

Practical Application of RNA-Cleaving DNA Enzymes

With the principle of DNA catalysis firmly established, attention could turn to the development of RNA-cleaving DNA enzymes that might have application in biology or medicine. An important aim was to cleave an all-RNA substrate under physiological conditions with high catalytic efficiency, high substrate sequence specificity, and generality for almost any substrate sequence. DNA enzymes with these properties might be used to cleave, and therefore inactivate, target RNAs in cells or whole organisms. Such applications had already been demonstrated for modified forms of naturally occurring ribozymes, especially the hammerhead and hairpin motifs [for reviews see (69, 70)].

A general-purpose, RNA-cleaving DNA enzyme was obtained by in vitro selection starting from a population of 10^{14} DNAs that contained 50 random-sequence residues (71). The selection scheme was the same as the one used to obtain DNA enzymes that cleave a single ribonucleotide embedded within DNA, except the substrate domain contained 12 ribonucleotides. The reaction was triggered by the addition of 10 mM $MgCl_2$ and 1 M NaCl at pH 7.5 and 37°C. The uncatalyzed rate of cleavage under these conditions, summing over all 12 RNA phosphodiester linkages, is about 10^{-5} min^{-1} (55). This level of background is manageable, but if the substrate domain had included an entire mRNA, uncatalyzed cleavage would have been so prominent that it likely would have been difficult to isolate catalysts.

Ten rounds of in vitro selection were carried out with a reaction time of 60 min in each round (71). A total of 62 individuals were isolated from the population after the eighth and tenth rounds. These were highly variable in sequence, but all were capable of cleaving the attached RNA substrate. Two individuals (designated 8-17 and 10-23) were chosen for further study on the basis of their high level of activity and ability to bind the substrate domain through Watson-Crick pairing. Each individual was randomized at a frequency of 25% per nucleotide position (excluding the substrate-binding regions) and used to initiate six different lineages involving a total of 52 rounds of selective amplification. The reaction times and selection protocols differed among the various lineages, providing a thorough search for variants that might help to refine the two catalytic motifs.

Following separation of the substrate and catalytic domains, the 8-17 and 10-23 DNA enzymes were obtained (71). Both are capable of binding an all-RNA substrate through regions of Watson-Crick pairing that surround the cleavage site. The substrate must contain an unpaired purine nucleotide immediately preceding the cleavage site, located opposite the catalytic core of the enzyme, which contains either 13 or 15 nucleotides for the 8-17 and 10-23 enzymes, respectively. Both enzymes can operate with multiple turnover with a rate that depends on pH, temperature, and the concentration of Mg^{2+} in the reaction mixture. Under simulated physiological conditions (1–3 mM $MgCl_2$, 150–200 mM total ionic strength, pH 7.4–7.6, 37°C), the catalytic rate of both enzymes is about 0.1 min^{-1} (71–73).

The 8-17 motif has been isolated in several different in vitro selection experiments, employing either single or · multiple ribonucleotides within the substrate domain, different number and arrangement of random nucleotides within the starting population, various divalent metals to trigger the reaction, and different reaction conditions. For example, the Mg5 DNA enzyme discussed above, which cleaves a single ribonucleotide in the presence of either Mg^{2+} or Ca^{2+} (66), was later recognized to contain the 8-17 motif (72). When the Mg5 enzyme was trimmed to only the 8-17 portion, its catalytic rate improved by about 20-fold. The 8-17 motif also was obtained when selecting for cleavage of a single ribonucleotide in the presence of Zn^{2+} (74). This motif was found to be about 30-fold more active in the presence of Zn^{2+} compared to Mg^{2+}, and it also has a high level of activity in the presence of Pb^{2+}, Mn^{2+}, or Co^{2+} (73–75). Similar to the B16-19 RNA ligase ribozyme discussed previously, the 8-17 DNA enzyme represents a deep local fitness optimum. Because the 8-17 enzyme has a catalytic core of only 13 nucleotides, it is well represented within a starting population of random-sequence DNA molecules and thus is easily rediscovered.

The 8-17 DNA enzyme has been used as a biosensor for Pb^{2+} ions with practical application for the measurement of the lead content of paint chips (76, 77). When directed to cleave a single ribonucleotide within an otherwise all-DNA substrate, the 8-17 enzyme has a catalytic rate of 5.8 min^{-1} in the presence of 0.1 mM Pb^{2+} at pH 6.0 and 28°C. When the pH is increased to 7.5, the catalytic rate increases to about 220 min^{-1} (73). The apparent K_d for Pb^{2+} (at pH 6.0) is 14 μM, compared to 1–10 mM for other divalent metals. Thus the amount of Pb^{2+} in a sample can be detected over a broad concentration range, even against a substantial background of other metal ions (76). A simple colorimetric assay was developed for Pb^{2+} by linking the substrate to gold particles that form aggregates. The aggregates are disrupted by DNA-catalyzed cleavage, resulting in a blue-to-red color change that can be visualized directly (77). In this way, it is possible to detect the lead content of paint chips down to a level of 0.05% (100 nM in solution), which is 10-fold below the threshold for health risk to children as defined by the U.S. Environmental Protection Agency (78).

The 8-17 DNA enzyme has been applied to a limited extent to the cleavage of biological RNAs (79, 80), although most such applications have involved the 10-23 DNA enzyme. The latter enzyme has relatively higher catalytic efficiency under physiological conditions with a k_{cat} of 0.1 min^{-1} and K_m of 1 nM, corresponding to a k_{cat}/K_m of 10^8 $M^{-1} \cdot min^{-1}$ (71). The catalytic rate of the 10-23 enzyme increases log-linearly with increasing pH and is greater in the presence of either Mn^{2+} or Pb^{2+} compared to Mg^{2+} (81). However, most biochemical studies involving this enzyme have been carried out under simulated physiological conditions. Under those conditions, the rate of formation of the enzyme-substrate complex is determined by the rate of duplex formation, the rate of the chemical step is limiting under saturating conditions, and dissociation of the enzyme-product complex is not rate limiting. The bound substrate dissociates about 100-fold slower than it undergoes cleavage, and the rate of the reverse reaction (ligation of the cleaved products) is 450-fold slower than that of the forward reaction. Thus the enzyme undergoes efficient catalytic turnover until the substrate is almost completely converted to cleaved products.

When optimally configured, the 10-23 DNA enzyme exhibits high substrate sequence specificity, requiring near-perfect complementarity in both of the substrate-binding domains (81, 82). The optimal length of each domain is 7–10 base pairs, corresponding to a predicted $^-\Delta G°_{37}$ for each domain of 8–10 kcal/mol (83). This is a sufficient number of nucleotides to specify a unique target sequence within total cellular mRNA. The DNA enzyme is unable to overcome stable secondary structures within the substrate RNA. This property has been exploited to probe the secondary structure of folded RNAs, immobilizing DNA enzymes on a surface plasmon resonance chip (84). In most cases, however, one must choose a target site that is readily accessible for hybridization to the DNA enzyme (85).

There are many published examples describing the use of the 10-23 DNA enzyme to reduce expression of a target mRNA in vivo [for reviews see (86, 87)]. Discussion of this work is beyond the scope of the present review, but in general, it is possible to reduce expression of a particular cellular mRNA and its corresponding protein by 70% to 90% (80, 85, 88–93). There are a few studies in which the DNA enzyme was administered to whole animals, demonstrating similar efficacy. In one case it was directed to cleave mRNA encoding the transcriptional regulator Egr-1, which mediates a proliferative response of vascular smooth muscle following balloon angioplasty, resulting in restenosis. Administration of the DNA enzyme into either the rat carotid or pig coronary artery at the time of angioplasty resulted in >50% inhibition of thickening of the inner vessel wall compared to untreated controls (94, 95). In another study, the DNA enzyme was directed to cleave TNF-α mRNA, which contributes to the development of congestive heart failure following acute myocardial infarction. The enzyme was administered to rats that had undergone myocardial infarction caused by stricture of the left coronary artery, resulting in 90% improvement of relevant hemodynamic parameters compared to untreated controls (96). In a third study, the DNA enzyme was directed to cleave VEGFR2 mRNA, which mediates angiogenesis associated with tumor growth and metastasis. The enzyme was injected into mice that had been implanted with human breast carcinoma cells, resulting in 75% reduction of tumor size compared to untreated controls (97).

The 10-23 DNA enzyme also has been applied in a diagnostic context as part of the "DzyNA-PCR" method for quantitative PCR (98, 99). This method involves exponential production of copies of the DNA enzyme during amplification of a target nucleic acid. The complement of the DNA enzyme is placed at the 5' end of one of the two PCR primers, so that when the opposing strand is synthesized, it contains a DNA enzyme at its 3' end. The enzyme cleaves a reporter oligonucleotide that is present within the amplification mixture, separating a fluorescent dye and quencher that lie on either side of the cleavage site and giving rise to a fluorescent signal. DzyNA-PCR has been used, for example, to monitor the clinical course of acute promyelocytic leukemia by measuring genetic rearrangements associated with the disease (100).

RNA-cleaving DNA enzymes also have been used to probe for oligonucleotides of a particular sequence. Analogous to molecular beacons (101), a stem-loop structure is placed at the 5' end of the DNA enzyme. Upon hybridization with the target oligonucleotide, the stem-loop is opened, activating the DNA enzyme and allowing it to cleave a reporter oligonucleotide to generate a fluorescent signal (102). Alternatively, one of the substrate-binding domains of the DNA enzyme can be replaced by an oligonucleotide-recognition domain, arranged such that the target oligonucleotide must bind to both the enzyme and substrate in order to activate the enzyme (103). This mode of ligand-dependent activation has been demonstrated for both the 8-17 and 10-23 DNA enzymes (104), as well as for a third Mg^{2+}-dependent, RNA-cleaving DNA enzyme that was obtained more recently (105).

Finally, there have been some clever applications of the 10-23 DNA enzyme in the area of molecular computing. Two or more DNA enzymes can be designed such that the product of one DNA-catalyzed reaction serves as the effector for another (106). The effector may be a positive regulator (logical YES) or a negative regulator (logical NOT). Two effectors may be used as alternative activators (logical OR), mutual activators (logical AND), or mutually exclusive activators (logical XOR). Utilizing the appropriate combination of DNA-enzyme-based logic gates, one can perform more complex computations, such as arithmetic operations (107), and even a game of ticktacktoe against a human opponent in which the DNA enzymes never lose (108). In the special case in which the substrate is an inactive form of the DNA enzyme that becomes activated upon DNA-catalyzed cleavage, one can achieve exponential amplification of the number of active molecules (109). In principle, this operation could be used to add gain to DNA-enzyme-based logic circuits.

DNA Enzymes with Expanded Functionality

On the basis of the examples cited above, it is clear that the efficient, metal-dependent cleavage of RNA is well within the capabilities of DNA. Nonetheless, RNA-cleaving DNA enzymes generally have much slower catalytic rates compared to their protein counterparts. Ribonuclease A, for example, operates with a k_{cat} of 10^4–10^5 min^{-1} (110). This has prompted some investigators to explore chemically modified DNA enzymes that contain additional functional groups, such as those found in protein enzymes. There are a variety of dNTP analogs that

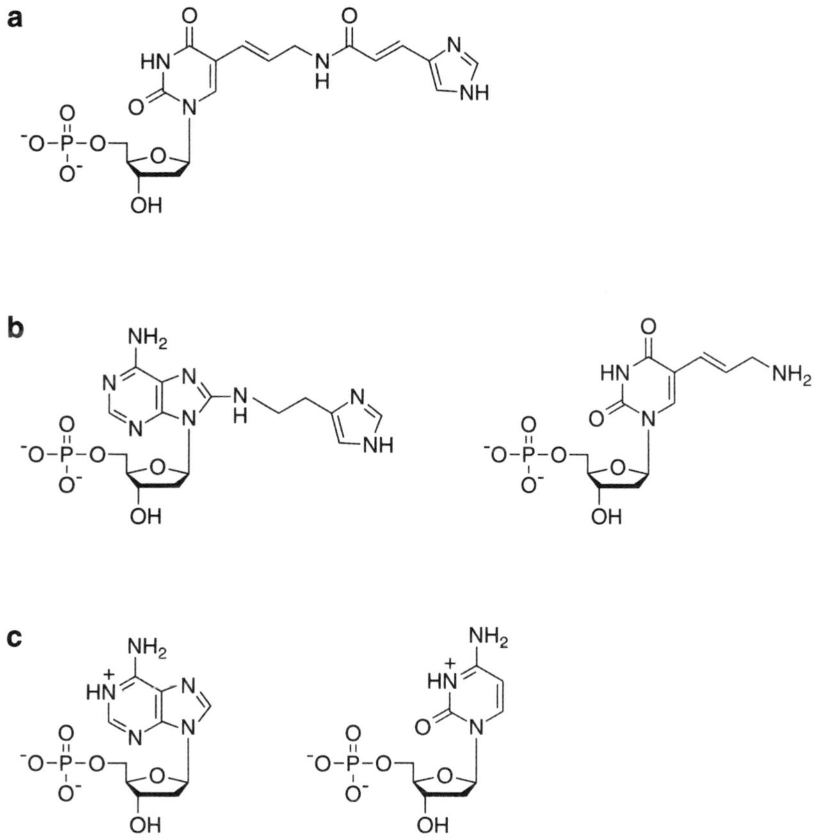

Figure 3 Deoxynucleotide analogs employed in the development of DNA enzymes with expanded chemical functionality. (*a*) C5-imidazole-deoxyuridylate; (*b*) C8-imidazole-deoxyadenylate (*left*) and C5-aminoallyl-deoxyuridylate (*right*); and (*c*) protonated deoxyadenylate (*left*) and protonated deoxycytidylate (*right*), as would occur under low-pH conditions.

can be incorporated efficiently into DNA by a polymerase (111–115). This makes it possible to expand the functionality of DNA by replacing one or more of the standard nucleotides with a suitable nucleotide analog.

In one study, a population of 10^{15} random-sequence DNAs was constructed, each molecule containing C5-imidazole-deoxyuridine (Figure 3*a*) in place of thymidine (116). This analog consists of a 4-imidazoleacrylic acid moiety that is joined by an amide bond to the primary amine of 5-(3-aminopropenyl)-deoxyuridine (112). The imidazole group of this compound has a pK_a of 6.1, which is similar to the pK_a of the imidazole group of histidine. The design of the starting population and the selection scheme were similar to those used to obtain the 8-17 and 10-23 DNA enzymes (81), except that the reaction was triggered by the addition of 10 μM ZnCl$_2$, 2 mM MgCl$_2$ and 150 mM NaCl rather than 10 mM MgCl$_2$ and 1 M NaCl (116).

Sixteen rounds of in vitro evolution were carried out, progressively decreasing the reaction time from 1 h to 10 s (116). Random mutations were introduced at a frequency of 10% per nucleotide position following the tenth and thirteenth rounds. During the last six rounds, the number of ribonucleotides in the substrate domain was increased from 12 to 34, and the reaction was carried out in solution rather than with the molecules bound to a streptavidin-coated support. Individuals were isolated from the population following the sixteenth round, sequenced, and tested for catalytic activity. They were found to represent six different sequence families, all containing a common consensus sequence of about 20 nucleotides. This sequence encodes a catalytic core of 10 nucleotides that form a short hairpin structure flanked by nucleotides that are complementary to those on either side of the cleavage site. The most active individual (designated 16.2-11) was chosen for further study.

It was straightforward to separate the substrate and catalytic domains, resulting in a 32mer DNA enzyme that cleaves a 23mer RNA substrate with multiple turnover (116). This enzyme can be made to cleave almost any RNA substrate, provided that it contains the sequence 5'-AUG-3' located immediately downstream from the cleavage site. There are three C5-imidazole-deoxyuridine residues that are essential for catalysis, one that pairs with the adenosine immediately downstream from the cleavage site, and two others located within the stem portion of the catalytic core. The 16.2-11 enzyme requires Zn^{2+} for its catalytic activity with a maximum catalytic rate of 3.1 min^{-1} in the presence of 30 μM Zn^{2+} at pH 7.5 and 37°C. It has 100- to 1000-fold lower activity in the presence of optimal concentrations of Cd^{2+}, Mn^{2+}, or Mg^{2+}.

The catalytic rate of the 16.2-11 DNA enzyme increases log-linearly with increasing pH over the range of 5.9–7.9 (116). This is not what one would expect if an imidazole group was acting as a general base (assuming that the measured catalytic rate reflects the chemical step of the reaction). Instead, one or more of the imidazole groups may bind a Zn^{2+} ion that assists in deprotonation of the 2'-hydroxyl adjacent to the cleavage site. Compared to most unmodified RNA-cleaving DNA enzymes, the imidazole modification makes it possible to achieve a faster catalytic rate with fewer nucleotides in the catalytic core and a much lower concentration of metal cofactor. Compared to most ribonuclease proteins, however, the imidazole-DNA enzyme has a much slower catalytic rate and a less sophisticated catalytic mechanism.

Another study involving functionally enhanced DNA enzymes sought to mimic more explicitly the catalytic mechanism of ribonuclease A (117). This mechanism involves two histidines, a lysine, and an aspartate at the active site that bring about a catalytic rate enhancement of about 10^{11}-fold (118). A population of 10^{12} random-sequence DNAs was constructed, each molecule containing C8-imidazole-deoxyadenosine in place of deoxyadenosine and C5-aminoallyl-deoxyuridine in place of thymidine (Figure 3b). These analogs were designed to provide the functionality of histidine and lysine, respectively. Each molecule contained only 20 random nucleotides, and the substrate domain contained a single ribonucleotide embedded within DNA. The reaction was triggered by the addition of 200 mM NaCl but without a divalent metal.

Nine rounds of in vitro selection were carried out, progressively decreasing the reaction time from 60 to 5 min (117). Twenty-five individuals were isolated from the final selected population, and one of these (designated 9_{25}-11) was chosen for further study. It has a catalytic rate of 0.044 min^{-1}, measured in the presence of 200 mM NaCl at pH 7.4 and 37°C. The rate does not increase upon addition of Mg^{2+} or other divalent metal ions. The pH optimum of the reaction is 7.4 with decreased activity at lower and higher pH, consistent with one or more imidazole groups acting as a general acid base. Activity is lost when all of the modified nucleotides are replaced by standard nucleotides, but it is not known which modified residues in particular are essential for catalysis.

It was possible to separate the substrate and catalytic domains of the 9_{25}-11 molecule, resulting in a 31mer modified DNA enzyme that cleaves a 15mer RNA substrate with multiple turnover with a catalytic rate of about 0.009 min^{-1} (measured in the presence of 500 mM NaCl at pH 7.9 and 27°C) (119). As was the case with the Na$^+$-dependent, RNA-cleaving DNA enzyme (67), this doubly modified DNA enzyme demonstrates that it is possible to achieve RNA-cleavage activity at near-neutral pH without the benefit of a divalent metal ion. However, the catalytic rate of the modified DNA enzyme remains disappointingly slow, especially in comparison to protein ribonucleases.

The ability of DNA to catalyze RNA cleavage under conditions of varying pH was evaluated by carrying out five parallel lineages of in vitro evolution, selecting for activity at pH 3.0, 4.0, 5.0, 6.0, or 7.0 (120). The starting population contained 70 random-sequence nucleotides, and the substrate domain contained a single ribonucleotide. No modified nucleotides were employed, but under the lower-pH conditions deoxyadenosine and deoxycytidine residues will tend to be protonated at N1 and N3, respectively, in effect behaving as modified nucleotides (Figure 3c). The reaction was triggered by adding a cocktail of metal ions that included 100 mM KCl, 400 mM NaCl, 8.5 mM MgCl$_2$, 5 mM MnCl$_2$, 1.25 mM CdCl$_2$, and 0.25 mM NiCl$_2$.

Seven rounds of in vitro selection were carried out at pH 4.0; then the population was split into five different lineages, one for each of the different pH conditions (120). Nine more rounds of selective amplification were carried out for the pH 3.0 and 4.0 lineages, and 17 more rounds were carried out for the pH 5.0, 6.0, and 7.0 lineages, introducing random mutations at a frequency of 10% per nucleotide position prior to rounds 8 through 12 in each case. It is likely that the diversity of the population was severely restricted by funneling all of the lineages through the initial seven rounds of selection at pH 4.0. This was compensated, however, by the high level of mutagenesis that was performed in the five subsequent rounds. The reaction times were decreased progressively in the latter rounds to maintain stringent selection pressure on the populations.

Individuals were isolated from each of the final populations. Catalytic motifs were recognized that were unique to each lineage, and characterized with regard to their catalytic properties (120). The pH-3.0 catalyst requires only monovalent ions and has a catalytic rate of 0.023 min^{-1} in the presence of 500 mM Na$^+$ at pH 3.0 and 23°C. Its catalytic rate drops off rapidly at higher or lower pH. The pH-4.0, pH-5.0, and pH-6.0 catalysts all require one or more divalent metals for their activity and have a pH optimum within 0.2 units of the condition under which they were selected. In optimum conditions, the catalytic rate of these molecules is 1.1, 0.72, and 0.25 min^{-1}, respectively. Only the pH-7.0 catalyst exhibited more conventional behavior with a log-linear increase in rate with increasing pH and dependence on a Mn^{2+} cofactor that likely assists in deprotonation of the 2'-hydroxyl adjacent to the cleavage site.

The limited exploration of functionally enhanced DNA enzymes that has taken place so far has not led to a dramatic increase in catalytic rates. Perhaps nucleic acids are lacking something in addition to the broader range of functional groups that occur in proteins. It is difficult, for example, for nucleic acids to establish a deep hydrophobic pocket that is shielded from bulk water. Nucleic acids may be hampered by their highly polyanionic nature, which requires extensive charge neutralization in order to draw several subunits into close proximity. They also have a lower density of functional groups compared to proteins; this makes it more difficult to form a compact active site. Yet, combining the methods of in vitro evolution with the ability to synthesize nucleotide analogs might help address these limitations, and the repertoire of nucleic-acid-based catalytic function will continue to expand, certainly going well beyond what has been observed among the naturally occurring nucleic acid enzymes.

LESSONS LEARNED

Evolution Follows the Most Expedient Pathway

As mentioned previously, an adage in directed evolution is "you get what you select for," which refers to both the intended and unintended selection pressures that are imposed on an evolving population of molecules. As a corollary, the solution obtained is the one that is easiest to realize in the short run. This may be advantageous, provided that the selection constraints lead most expeditiously to the desired phenotype. However, difficult phenotypes may be inaccessible because they are eclipsed by a simpler way of meeting the selection constraints. For example, in the study that sought to obtain histidine-dependent, RNA-cleaving DNA enzymes by triggering the reaction by the addition of 20 mM L-histidine and 0.5 mM MgCl$_2$, the resulting catalysts utilized Mg^{2+} rather than histidine as a catalytic cofactor (66). The 40-fold concentration advantage in favor of histidine was insufficient to overcome the greater difficulty of positioning histidine rather than Mg^{2+} to assist in catalysis. Histidine-dependent enzymes could be obtained only when divalent metals were excluded from the reaction mixture (68).

The problem of expediency becomes more challenging when a difficult phenotype must be accessed through several successive in vitro evolution experiments. One aims to discover a series of evolutionary intermediates along the route to the desired phenotype. However, Darwinian evolution has no obligation to follow the planned route and responds only to the task at hand. The best intermediate solutions may in fact preclude the opportunity to reach the final desired behavior. This was nearly the case in the tortuous path that led from the class I ligase to the 18.12.23 polymerase ribozyme (15, 16, 17, 21). Fortunately, a narrow passageway was discovered; at one point, it involved identification of only 1 of 74 cloned individuals that had the desired properties. A good evolutionary engineer seeks to establish a succession of selection constraints that lead inexorably to the desired phenotype, although in practice this may be difficult to achieve.

Beware the Tyranny of the Small Motif

A smaller motif will be present in greater copy number compared to a larger one within a population of random-sequence nucleic acids. Unless the larger motif has a substantial selective advantage, it will not overcome the disadvantage in copy number during successive rounds of selective amplification. This was the case with the 8-17 RNA-cleaving DNA enzyme, which has a catalytic core of only 13 nucleotides and was rediscovered in several independent in vitro selection experiments. Within the catalytic core, there are several permissible sequence variations, adding to the effective copy number of this motif within the starting population. It may be desirable to obtain the smallest motif that can catalyze a particular reaction. However, a small motif is more constrained than a large one with regard to subsequent evolutionary opportunities because it contains fewer nucleotides that can vary in a combinatorial manner. One remedy is to attach a random-sequence domain to a preexisting small motif, thereby providing a more diverse set of permissible sequences.

The tyranny of the small motif often is abetted by the timidity of the experimenter. It is common practice to employ less stringent selection constraints during the early rounds of in vitro evolution to ensure that there will be many survivors. The selection constraints then are made more stringent to weed out the less advantageous individuals. This approach has two important advantages. First, it maintains greater genetic diversity in the population, which provides greater opportunity to discover novel variants that have even more advantageous behavior. Second, it provides a more robust signal of functional molecules that can be distinguished from the inevitable noise that arises during selective amplification. However, the advantage of greater genetic diversity only applies if one carries out mutagenesis and/or recombination to explore variants of the selected individuals. Also, the desire to obtain a robust signal usually is motivated by anxiety regarding the progress of the experiment rather than an awareness of signal versus noise. The main disadvantage of being too lax during the early rounds of in vitro evolution is that it allows small motifs to gain in copy number, making it even more difficult to supplant them later should a more advantageous variant arise.

Declaring Wild Type Is Inadvisable but Often Unavoidable

An ideal in vitro evolution experiment would involve stringent selection pressure, even in the early rounds, compensated by the continued introduction of genetic variation. As discussed above, it is important to maintain genetic diversity in the population to enable a broad exploration of potentially advantageous sequences, biased by what has proven successful in the past. By restricting the population to a single individual, no matter how advantageous, the fine structure of the population is lost. Yet there often are practical reasons for declaring wild type, such as when the desired phenotype is rare and no longer being enriched or when the population is becoming overwhelmed with PCR artifacts. During the evolutionary development of the 18.12.23 polymerase ribozyme, for example, it was necessary to declare wild type on three occasions (15–17, 21).

In some instances, one declares wild type not out of necessity but because an especially interesting individual has been identified as a result of a screening procedure. Surprisingly, no one has taken this to the extreme and carried out high-throughput screens to survey very large numbers of individuals that have been isolated from an evolved population of nucleic acids. High-throughput screening is a mature technology in the areas of combinatorial chemistry and protein engineering, where in vitro selection and evolution are more problematic. Similar procedures could be applied to catalytic nucleic acids, not as a replacement for selection and evolution, but as a way to augment the process of declaring wild type, should it be necessary.

Diversity Management Needs to Be Considered More Rigorously

Most investigators recognize the importance of maintaining diversity during the course of an in vitro evolution experiment to conduct a broad search for novel variants that might have improved catalytic properties. However, the introduction of diversity, usually through error-prone PCR, tends to be carried out in an ad hoc manner. The diversity of the population is seldom measured prior to mutagenesis, and the timing and degree of mutagenesis usually is chosen on the basis of habit or instinct. There is a well-established theoretical framework to guide efforts to balance the loss of diversity due to selection with the introduction of diversity due to mutagenesis (121, 122). Ideally, diversity should be introduced following each round of selective amplification with the degree of diversification determined by the distribution of genotypes and phenotypes in the population.

Methods for diversity management need to be developed, especially those involving high-throughput sampling and profiling of the evolving population. It is straightforward to characterize the aggregate genotype of the population, for example, by digesting with a cocktail of restriction enzymes (15) or by ensemble sequencing. It would be more informative to employ modern DNA sequencing methods to determine the genotype of a representative number of individuals isolated from the population. This would be useful not only for diversity management but also for recognizing sequence covariation and conserved nucleotide positions that help to define a catalytic motif.

Serendipity Counts

There are many examples of evolved catalytic properties that are not what was explicitly selected but that are desirable nonetheless. The 8-17 DNA enzyme, which was selected for RNA-cleavage activity in the presence of millimolar concentrations of Mg^{2+}, turned out to be more active in the presence of micromolar concentrations of Pb^{2+} (73). The latter property was useful for the development of Pb^{2+} sensors (77). The 10-23 DNA enzyme was not selected for high-substrate sequence specificity or rapid product release, but it was found to have these useful properties once the catalytic and substrate domains were separated. These gifts of Darwinian evolution should be welcomed. One should not expect to obtain a property that is not beholding to the selection constraints, but analysis of the evolved individuals may reveal some interesting phenotypes. You get what you select for. . . and sometimes a whole lot more.

ACKNOWLEDGMENTS

This work was supported by the National Aeronautics and Space Administration, the National Institutes of Health, Johnson & Johnson Research, and The Skaggs Institute for Chemical Biology at The Scripps Research Institute.

The *Annual Review of Biochemistry* is online at http://biochem.annualreviews.org

LITERATURE CITED

1. Kruger K, Grabowski PJ, Zaug AJ, Sands J, Gottschling DE, Cech TR. 1982. *Cell* 31:147–57
2. Peebles CL, Perlman PS, Mecklenburg KL, Petrillo ML, Tabor JH, et al. 1986. *Cell* 44:213–33
3. Guerrier-Takada C, Gardiner K, Marsh T, Pace N, Altman S. 1983. *Cell* 35: 849–57
4. Prody GA, Bakos JT, Buzayan JM, Schneider IR, Bruening G. 1986. *Science* 231:1577–80
5. Buzayan JM, Gerlach WL, Bruening G. 1986. *Nature* 323:349–53
6. Sharmeen L, Kuo MYP, Dinter-Gottlieb G, Taylor J. 1988. *J. Virol.* 62:2674–79
7. Saville BJ, Collins RA. 1990. *Cell* 61: 685–96
8. Nissen P, Hansen J, Ban N, Moore PB, Steitz TA. 2000. *Science* 289:920–30

9. Valadkhan S, Manley JL. 2001. *Nature* 413:701–7
10. Guatelli JC, Whitfield KM, Kwoh DY, Barringer KJ, Richman DD, Gingeras TR. 1990. *Proc. Natl. Acad. Sci. USA* 87: 1874–78
11. Cadwell RC, Joyce GF. 1992. *PCR Methods Appl.* 2:28–33
12. Vartanian J-P, Henry M, Wain-Hobson S. 1996. *Nucleic Acids Res.* 24:2627–31
13. Cline J, Hogrefe H. 2000. *Strategies* 13: 157–62
14. Stemmer WPC. 1994. *Nature* 370: 389–91
15. Bartel DP, Szostak JW. 1993. *Science* 261: 1411–18
16. Ekland EH, Szostak JW, Bartel DP. 1995. *Science* 269:364–70
17. Ekland EH, Bartel DP. 1995. *Nucleic Acids Res.* 23:3231–38
18. Bergman NH, Johnston WK, Bartel DP. 2000. *Biochemistry* 39:3115–23
19. Glasner ME, Bergman NH, Bartel DP. 2002. *Biochemistry* 41:8103–12
20. Wright MC, Joyce GF. 1997. *Science* 276: 614–17
21. Johnston WK, Unrau PJ, Lawrence MS, Glasner ME, Bartel DP. 2001. *Science* 292:1319–25
22. Schmitt T, Lehman N. 1999. *Chem. Biol.* 6:857–69
23. Ordoukhanian P, Joyce GF. 1999. *Chem. Biol.* 6:881–89
24. Kühne H, Joyce GF. 2003. *J. Mol. Evol.* 57:292–98
25. Miyamoto Y, Teramoto N, Imanishi Y, Ito Y. 2001. *Biotechnol. Bioeng.* 75: 590–96
26. Lehman N. 2004. *Artif. Life* 10:1–22
27. Atkins JF, Ellington A, Friedman BE, Gesteland RF, Noller HF, et al. 2000. *Bringing RNA into View: RNA and Its Roles in Biology.* Colorado Springs, CO: Biol. Sci. Curric. Study
28. Ekland EH, Bartel DP. 1996. *Nature* 382: 373–76
29. Lawrence MS, Bartel DP. 2003. *Biochemistry* 42:8748–55
30. McGinness KE, Wright MC, Joyce GF. 2002. *Chem. Biol.* 9:585–96
31. Landweber LF, Pokrovskaya ID. 1999. *Proc. Natl. Acad. Sci. USA* 96:173–78
32. Rogers J, Joyce GF. 1999. *Nature* 402: 323–25
33. Hager AJ, Szostak JW. 1997. *Chem. Biol.* 4:607–17
34. Teramoto N, Imanishi Y, Ito Y. 2000. *Bioconjug. Chem.* 11:744–48
35. Robertson MP, Ellington AD. 1999. *Nat. Biotechnol.* 17:62–66
36. Robertson MP, Ellington AD. 2000. *Nucleic Acids Res.* 28:1751–59
37. Robertson MP, Hesselberth JR, Ellington AD. 2001. *RNA* 7:513–23
38. Robertson MP, Ellington AD. 2001. *Nat. Biotechnol.* 19:650–55
39. Peracchi A, Beigelman L, Usman N, Herschlag D. 1996. *Proc. Natl. Acad. Sci. USA* 93:11522–27
40. Tang J, Breaker RR. 1997. *Chem. Biol.* 4:453–59
41. Koizumi M, Soukup GA, Kerr JNQ, Breaker RR. 1999. *Nat. Struct. Biol.* 6: 1062–71
42. Seetharaman S, Zivarts M, Sudarsan N, Breaker RR. 2001. *Nat. Biotechnol.* 19: 336–41
43. Hartig JS, Najafi-Shoushtari SN, Grüne I, Yan A, Ellington AD, Famulok M. 2002. *Nat. Biotechnol.* 20:717–22
44. Vaish NK, Dong F, Andrews L, Schweppe RE, Ahn NG, et al. 2002. *Nat. Biotechnol.* 20:810–15
45. Hesselberth JR, Robertson MP, Knudsen SM, Ellington AD. 2003. *Anal. Biochem.* 312:106–12
46. Jaeger L, Wright MC, Joyce GF. 1999. *Proc. Natl. Acad. Sci. USA* 96:14712–17
47. Murphy FL, Cech TR. 1993. *Biochemistry* 32:5291–300
48. Cate JH, Gooding AR, Podell E, Zhou K, Golden BL, et al. 1996. *Science* 273: 1678–85
49. Strobel SA, Cech TR. 1993. *Biochemistry* 32:13593–604
50. Strobel SA, Ortoleva-Donnelly L, Ryder

SP, Cate JH, Moncoeur E. 1998. *Nat. Struct. Biol.* 5:60–66

51. McGinness KE, Joyce GF. 2002. *Chem. Biol.* 9:297–307
52. Rogers J, Joyce GF. 2001. *RNA* 7: 395–404
53. Reader JS, Joyce GF. 2002. *Nature* 420: 841–44
54. Paul N, Joyce GF. 2002. *Proc. Natl. Acad. Sci. USA* 99:12733–40
55. Li Y, Breaker RR. 1999. *J. Am. Chem. Soc.* 121:5364–72
56. Cech TR. 1987. *Science* 236:1532–39
57. Perreault J-P, Wu T, Cousineau B, Ogilvie KK, Cedergren R. 1990. *Nature* 344: 565–67
58. Perreault J-P, Labuda D, Usman N, Yang J-H, Cedergren R. 1991. *Biochemistry* 30: 4020–25
59. Olsen DB, Benseler F, Aurup H, Pieken WA, Eckstein F. 1991. *Biochemistry* 30: 9735–41
60. Williams DM, Pieken WA, Eckstein F. 1992. *Proc. Natl. Acad. Sci. USA* 89: 918–21
61. Yang J-H, Usman N, Chartrand P, Cedergren R. 1992. *Biochemistry* 31:5005–9
62. Breaker RR, Joyce GF. 1994. *Chem. Biol.* 1:223–29
63. Chartrand P, Harvey SC, Ferbeyre G, Usman N, Cedergren R. 1995. *Nucleic Acids Res.* 23:4092–96
64. Cuenoud B, Szostak JW. 1995. *Nature* 375:611–14
65. Breaker RR, Joyce GF. 1995. *Chem. Biol.* 2:655–60
66. Faulhammer D, Famulok M. 1996. *Angew. Chem.* 35:2837–41
67. Geyer CR, Sen D. 1997. *Chem. Biol.* 4: 579–93
68. Roth A, Breaker RR. 1998. *Proc. Natl. Acad. Sci. USA* 95:6027–31
69. Christoffersen RE, Marr JJ. 1995. *J. Med. Chem.* 38:2023–37
70. Sullenger BA, Gilboa E. 2002. *Nature* 418: 252–58
71. Santoro SW, Joyce GF. 1997. *Proc. Natl. Acad. Sci. USA* 94:4262–66

72. Peracchi A. 2000. *J. Biol. Chem.* 275: 11693–97
73. Brown AK, Li J, Pavot CM-B, Lu Y. 2003. *Biochemistry* 42:7152–61
74. Li J, Zheng W, Kwon AH, Lu Y. 2000. *Nucleic Acids Res.* 28:481–88
75. Ferrari D, Peracchi A. 2002. *Nucleic Acids Res.* 30:e112
76. Li J, Lu Y. 2000. *J. Am. Chem. Soc.* 122: 10466–67
77. Liu J, Lu Y. 2003. *J. Am. Chem. Soc.* 125:6642–43
78. Schmehl RL, Cox DC, Dewalt FG, Haugen MM, Koyak RA, et al. 1999. *Am. Ind. Hyg. Assoc. J.* 60:444–51
79. Kuwabara T, Warashina M, Tanabe T, Tani K, Asano S, Taira K. 1997. *Nucleic Acids Res.* 25:3074–81
80. Wu Y, Yu L, McMahon R, Rossi JJ, Forman SJ, Snyder DS. 1999. *Hum. Gene Ther.* 10:2847–57
81. Santoro SW, Joyce GF. 1998. *Biochemistry* 37:13330–42
82. Cairns MJ, Hopkins TM, Witherington C, Sun L-Q. 2000. *Antisense Nucleic Acid Drug Dev.* 10:323–32
83. Joyce GF. 2001. *Methods Enzymol.* 341: 503–17
84. Okumoto Y, Ohmichi T, Sugimoto N. 2002. *Biochemistry* 41:2769–73
85. Cairns MJ, Hopkins TM, Witherington C, Wang L, Sun L-Q. 1999. *Nat. Biotechnol.* 17:480–86
86. Cairns MJ, Saravolac EG, Sun L-Q. 2002. *Curr. Drug Targets* 3:269–79
87. Khachigian LM. 2000. *J. Clin. Invest.* 106: 1189–95
88. Warashina M, Kuwabara T, Nakamatsu Y, Taira K. 1999. *Chem. Biol.* 6:237–50
89. Yen L, Strittmatter SM, Kalb RG. 1999. *Ann. Neurol.* 46:366–73
90. Oketani M, Asahina Y, Wu CH, Wu GY. 1999. *J. Hepatol.* 31:628–34
91. Sioud M, Leirdal M. 2000. *J. Mol. Biol.* 296:937–47
92. Goila R, Banerjea AC. 2001. *Biochem. J.* 353:701–8
93. Liu C, Cheng R, Sun L-Q, Tien P. 2001.

Biochem. Biophys. Res. Commun. 284: 1077–82

94. Santiago FS, Lowe HC, Kavurma MM, Chesterman CN, Baker A, et al. 1999. *Nat. Med.* 5:1264–69

95. Lowe HC, Fahmy RG, Kavurma MM, Baker A, Chesterman CN, Khachigian LM. 2001. *Circ. Res.* 89:670–77

96. Iversen PO, Nicolaysen G, Sioud M. 2001. *Am. J. Physiol. Heart Circ. Physiol.* 281:H2211–17

97. Zhang L, Gasper WJ, Stass SA, Ioffe OB, Davis MA, Mixson AJ. 2002. *Cancer Res.* 62:5463–69

98. Todd AV, Fuery CJ, Impey HL, Applegate TL, Haughton MA. 2000. *Clin. Chem.* 46:625–30

99. Impey HL, Applegate TL, Haughton MA, Fuery CJ, King JE, Todd AV. 2000. *Anal. Biochem.* 286:300–3

100. Applegate TL, Iland HJ, Mokany E, Todd AV. 2002. *Clin. Chem.* 48: 1338–43

101. Tyagi S, Kramer FR. 1996. *Nat. Biotechnol.* 14:303–8

102. Stojanovic MN, de Prada P, Landry DW. 2001. *ChemBioChem* 2:411–15

103. Wang DY, Sen D. 2001. *J. Mol. Biol.* 310:723–34

104. Wang DY, Lai BHY, Feldman AR, Sen D. 2002. *Nucleic Acids Res.* 30:1735–42

105. Feldman AR, Sen D. 2001. *J. Mol. Biol.* 313:283–94

106. Stojanovic MN, Mitchell TE, Stefanovic D. 2002. *J. Am. Chem. Soc.* 124: 3555–61

107. Stojanovic MN, Stefanovic D. 2003. *J. Am. Chem. Soc.* 125:6673–76

108. Stojanovic MN, Stefanovic D. 2003. *Nat. Biotechnol.* 21:1069–74

109. Levy M, Ellington AD. 2003. *Proc. Natl. Acad. Sci. USA* 100:6416–21

110. delCardayré SB, Raines RT. 1994. *Biochemistry* 33:6031–37

111. Benner SA, Battersby TR, Eschgfäller B, Hutter D, Kodra JT, et al. 1998. *Pure Appl. Chem.* 70:263–66

112. Sakthivel K, Barbas CF. 1998. *Angew. Chem.* 37:2872–75

113. Perrin DM, Garestier T, Helene C. 1999. *Nucleosides Nucleotides* 18:377–91

114. Gourlain T, Sidorov A, Mignet N, Thorpe SJ, Lee SE, et al. 2001. *Nucleic Acids Res.* 29:1898–905

115. Thum O, Jäger S, Famulok M. 2001. *Angew. Chem.* 40:3990–93

116. Santoro SW, Joyce GF, Sakthivel K, Gramatikova S, Barbas CF. 2000. *J. Am. Chem. Soc.* 122:2433–39

117. Perrin DM, Garestier T, Helene C. 2001. *J. Am. Chem. Soc.* 123:1556–63

118. Raines RT. 1998. *Chem. Rev.* 98: 1045–65

119. Lermer L, Roupioz Y, Ting R, Perrin DM. 2002. *J. Am. Chem. Soc.* 124: 9960–61

120. Liu Z, Mei SHJ, Brennan JD, Li Y. 2003. *J. Am. Chem. Soc.* 125:7539–45

121. Eigen M. 1971. *Naturwissenschaften* 58: 465–523

122. Fontana W, Schnabl W, Schuster P. 1989. *Phys. Rev. A* 40:3301–21

Annu. Rev. Biochem. 2004. 73:837–59
doi: 10.1146/annurev.biochem.73.011303.073904

USING PROTEIN FOLDING RATES TO TEST PROTEIN FOLDING THEORIES

Blake Gillespie[1] and Kevin W. Plaxco[1,2]

[1]Department of Chemistry and Biochemistry, [2]Interdepartmental Program in Biomolecular Science and Engineering, University of California, Santa Barbara, Santa Barbara, California 93106; e-mail: blakeg@chem.ucsb.edu, kwp@chem.ucsb.edu

Key Words   simulation, nucleation, topomer search

■ **Abstract**   The fastest simple, kinetically two-state protein folds a million times more rapidly than the slowest. Here we review many recent theories of protein folding kinetics in terms of their ability to qualitatively rationalize, if not quantitatively predict, this fundamental experimental observation.

CONTENTS

Figure 1 The simple, single-domain proteins cytochrome b_{562} and muscle acylphosphatase are of similar size and stability, yet the former folds in microseconds, and the latter folds in seconds (2, 3). Here we critically review the ability of various theories of protein folding to rationalize qualitatively, if not predict quantitatively, this six order of magnitude range of rates.

INTRODUCTION: TWO-STATE FOLDING RATES AS AN EXPERIMENTAL BENCHMARK

Simple, single-domain proteins typically fold via a process that lacks well-populated intermediates [reviewed in (1)]. Despite the potential simplification afforded by the absence of kinetic intermediates, reported two-state folding rates span a remarkable six order of magnitude range (Figure 1) (2, 3). When coupled with the appealing simplicity of two-state behavior, the broad range of two-state rates provides a potentially straightforward and quantitative opportunity to test theories of the folding process.

Theoretical models of protein folding kinetics can, admittedly somewhat artificially, be grouped into two broad classes. Perhaps the more prominent class consists of theories emerging from observations of the simulated folding of simple on- and off-lattice models *in silico*. The second class consists of theoretical models emerging from the experimental observation of protein folding in vitro. In this article, we broadly review many recent simulation- and experiment-based theories of folding kinetics and critically evaluate these theories in terms of their ability to qualitatively rationalize, if not quantitatively predict, the vast range of folding rates observed for two-state proteins.

SIMULATION-DERIVED THEORIES OF FOLDING

Perhaps the majority of contemporary theories of protein folding kinetics are based on observations of the simulated folding of simple, computational models. In an ideal world, the relevant simulations would entail a level of detail commensurate with the complex, atomistic structure of a fully solvated polypeptide. In reality, however, the simulation of protein folding in atomistic detail has proven computationally overwhelming. The difficulty is twofold. First, computational times scale strongly with the number of atoms, and even the smallest proteins are composed of thousands of atoms solvated by thousands of water molecules. Second, even the most rapidly folding proteins fold extremely slowly relative to the femtosecond time step of fully detailed molecular dynamics simulations [for a recent, partial solution to this dilemma, see Pande and coworkers folding@home project (4)]. Faced with this computational obstacle, the large majority of the theoretical literature in protein folding has been based on observations not of fully detailed protein models but on highly simplified representations of the polypeptide chain. Lattice polymers, perhaps the most popular computational model, simplify the description of the polypeptide chain by distilling each amino acid into a single bead and simplify folding dynamics by limiting the moves of each bead to hops between discrete points on a coarse lattice.

Although lattice polymers and many off-lattice computational models are highly simplified, they capture many of the potentially relevant aspects of real proteins. For example, lattice polymers, similar to proteins, are sequence-specific heteropolymers that can encode a unique native fold, and though the coarse lattice significantly reduces the entropy of the unfolded state, its entropy is still sufficiently large that folding would be slow were it a fully random search process. Similarly, some lattice polymer sequences surmount this entropic barrier (the Levinthal Paradox) much more efficiently than others, and thus lattice polymers exhibit a wide range of folding rates. To date a major goal of computational folding studies has been the identification of the equilibrium properties that uniquely identify those rare lattice polymer sequences that fold rapidly, under the assumption that similar behavior will underlie the rapid folding of real proteins. The criteria predicted to distinguish between the rapidly and slowly folding lattice polymer sequences thus provide a clear opportunity for evaluating the correspondence between theory and experiment in protein folding.

In the last decade alone, more than 700 papers on the folding kinetics of simple on- and off-lattice protein models have appeared in the literature. Although it is impossible to accurately distill such a large, diverse, and often conflicting literature into a few brief conclusions, much of this literature can be classified by the criterion that is predicted to separate rapidly folding sequences from slowly folding sequences. Here we briefly describe the three major criteria that have been suggested as potential determinants of the relative folding rates of simplified computational models and review the experimental literature for clues

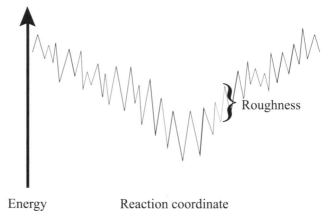

Figure 2   Rapid folding will occur when the energy of each unfolded or partially folded conformation decreases more or less monotonically [along some simple, but often difficult to define, order parameter(s)] as conformations become more native-like (5, 6). Roughness on this energy landscape can produce complex, stretched-exponential kinetics in lattice polymers (9) and may also account for part of the broad range of observed two-state folding rates.

as to whether similar criteria are responsible for the broad range of rates observed for the folding of two-state proteins.

Smooth Energy Landscapes

A large body of theoretical work suggests that the rapid folding of lattice polymers (5, 6) and simplified off-lattice polymers (7, 8) is associated with smooth, funnel-shaped energy landscapes (Figure 2). That is, rapid folding will occur when the energy of each unfolded or partially folded conformation decreases more or less monotonically as conformations become more native-like along some reaction coordinate(s). If this energetic guidance does not occur, or if the landscape is rough (contains local minima deeper than a few k_BT that act as traps), folding becomes glassy, dominated by multiple kinetic traps, and slows dramatically. Differences in the roughness of the energy landscape can lead to orders of magnitude changes in the folding rates of simplified computational models (9).

How can we determine if variations in energy-landscape roughness contribute significantly to the relative folding rates of small proteins? If energetic roughness dominates the energy landscape, slow, nonsingle-exponential kinetics will be observed (5, 9, 10). That is, at temperatures at which k_BT is small relative to the myriad kinetic barriers on a rough landscape (near the so-called glass transition temperature, T_g), the myriad of local kinetic barriers will begin to retard the folding process, switching the kinetics from single exponential ($h = 1$) to a slower, stretched exponential ($h > 1$):

$$S(t) = S_n + A_0 \exp(-(k_f t)^{1/h}),$$ 1.

where k_f is the folding rate, A_0 denotes the amplitude change upon folding, and S_n and $S(t)$ respectively denote the signal of the native state and that observed at time t (5, 8–10). We can determine the extent to which roughness defines relative folding rates in the laboratory by employing this metric to measure the relative energy-landscape roughness of both rapidly and slowly folding proteins.

Do differences in energy-landscape roughness account for a significant fraction of the 1,000,000-fold range of rates observed for two-state protein folding? Although indications of landscape roughness have been reported for the single-domain protein ubiquitin [(11), but see commentary in (12)] at low temperatures, energetic roughness does not generally appear to account for the vast range of folding rates observed under more physiological conditions. Once the complication of proline isomerization is taken into account, the large majority of simple, single domain proteins appear to fold with single exponential kinetics (1, 11, 12). Examples include the folding of the 62-residue protein L, which appears perfectly single exponential down to the experimental limit of $-15°C$ (12), and the pI3k-SH3 domain, which also exhibits a smooth energy landscape despite being the second most slowly folding two-state protein reported to date (13) (Figure 3). It thus appears that although theory is correct in associating smooth energy landscapes with the rapid folding of naturally occurring proteins (5–8), differences in the roughness of the energy landscape do not play a significant role in defining the six order of magnitude range of observed two-state folding rates.

The Energy Gap Hypothesis

Karplus and coworkers have reported that a "necessary and sufficient criterion to ensure [rapid] folding is that the ground state be a pronounced energy minimum (14)" relative to all other maximally compact states (Figure 4). More recently these and other researchers have identified a number of related measures of energetic gap between the ground state and other maximally compact states. These include the Z score, which is a measure of the statistical significance of the size of the gap between the energy of the native state and the mean energy of all other maximally compact states (15). This measure of the energy gap is reported to correlate significantly with the folding rates of a number of simple lattice polymers (16) and may account for the vast dispersion observed in the folding rates of real proteins.

Unfortunately, we cannot measure the size of the energy gap experimentally, and thus, we cannot directly determine whether differences in the energy gap account for the wide range of observed protein folding rates. The difficulty is that, as demonstrated by both solution-phase and crystallographic structural studies, the vast majority of proteins populate only a single maximally compact state, the native state, and therefore, the energy gap between the ground state and all other maximally compact states is too large to measure. An indirect experimental test of the energy gap hypothesis may be provided, however, by the

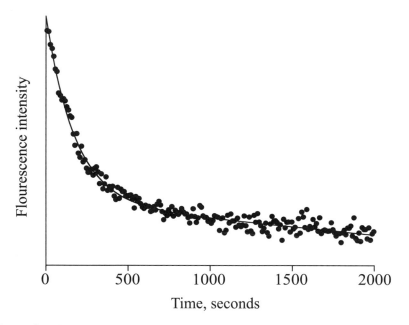

Figure 3 The folding energy landscapes of even the most slowly folding single-domain proteins are exceptionally smooth. Shown here is the refolding of the pI3 k-SH3 domain, one of the most slowly folding of all two-state proteins (13), after folding is initiated by rapid dilution at 5°C. There is no statistically significant evidence of deviation from the fitted single-exponential kinetics (B. Gillespie and K.W. Plaxco, unpublished observations), even at this low temperature, where $k_B T$ is reduced and the effects of landscape roughness enhanced. It appears that energetic roughness plays no significant role in slowing the folding of even the most slowly folding two-state proteins (12).

empirical observation that, for some lattice polymer models, the melting temperature (T_m, often denoted T_f in the theoretical literature) is correlated with the magnitude of the energy gap. This leads to the (indirect, empirically based) prediction that folding rates should also correlate with T_m (16). To the extent that this has been monitored experimentally, however, it appears that folding rates are generally uncorrelated with T_m. For example, although folding rates may be modestly correlated with T_m across a set of mutants of a single protein (A. Fersht, personal communication), a survey of the experimental literature suggests that T_m is effectively completely uncorrelated with folding rates across a set of nonhomologous proteins (Figure 5). Even for sets of closely related proteins, the correlation between T_m and folding rates is often nonexistent. A set of homeodomain sequences provides an extreme example: Despite T_m ranging from 54°C to 116°C, the folding rates of the three characterized homeodomains are effectively within error of one another (12, 17; B. Gillespie and K.W. Plaxco,

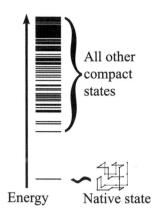

Figure 4   Karplus and coworkers argue that a large gap between the energy of the native state and the second-lowest-energy, maximally compact state is a "necessary and sufficient condition for [rapid] folding" (14) and that the size of this gap is correlated with folding rates (16).

unpublished data). Similarly, there is no correlation between T_m and folding rates across mesophile-, thermophile-, and hyperthermophile-derived cold shock proteins (18). Thus, although theory is correct in associating a large energy gap with the rapid folding of naturally occurring proteins, it appears that there is no evidence in favor of the hypothesis that the size of the energy gap is a significant determinant of relative protein folding rates.

Collapse Cooperativity

Thirumalai and coworkers (19–24) have demonstrated that the relative folding rates of many simple on- and off-lattice polymer models are defined by a dimensionless, equilibrium parameter termed σ. σ is a measure of the ease with which the denatured state undergoes nonspecific, Flory-type coil-to-globule collapse relative to the ease with which the protein folds as temperature or solvent quality (e.g., denaturant concentration) is reduced (Figure 6). Two equivalent measures of σ can be derived. The first, σ_T, is defined in terms of thermal unfolding and is related to the melting (T_m) and collapse (T_θ) temperatures by

$$\sigma_T = 1 - T_m/T_\theta, \qquad\qquad 2.$$

where T_m is the temperature (in Kelvins) at which half of a population of polymers is folded, and T_θ is the temperature at which the mean dimensions of the ensemble are midway between those of the folded and unfolded states. The second, σ_D, is reported to be an equivalent parameter obtained via chemical denaturation experiments (22; D. Thirumalai and D. Klimov, personal communication). It is defined by

$$\sigma_D = 1 - C_m/C_\theta, \qquad\qquad 3.$$

where C_m is the denaturant concentration at which half of the population of polymers is folded, and C_θ is the denaturant concentration at which the mean dimensions of the ensemble are midway between those of the fully folded and

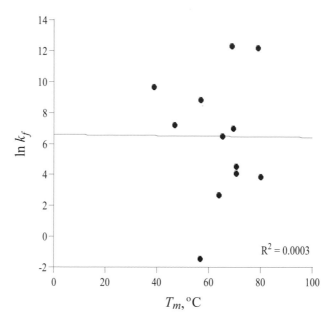

Figure 5 Although it is not possible to test the energy gap hypothesis directly (because the energy gap does not correspond to any experimental observable), an indirect experimental test is provided by the empirical observation that the melting temperatures (T_m) of lattice polymers are highly correlated with both the size of their energy gap and their folding rates (16). No significant correlation between T_m and folding rates is observed, however, when a set of unrelated, single-domain proteins is characterized in the laboratory. Shown are data from 12 nonhomologous two-state proteins randomly selected from the literature (91, 93–105).

unfolded ensembles. Both definitions of σ are thus equilibrium measures of the cooperativity of the collapse of the unfolded state to form the native protein.

Two research groups have explored the relationship between σ and the folding kinetics of simplified computational models. Thirumalai and coworkers (24) have demonstrated that on- and off-lattice polymer sequences with high σ fold orders of magnitude more slowly than sequences that contract only at conditions nearer the midpoint of the equilibrium folding transition. Karplus and coworkers (16) have also investigated the relationship between σ and folding rates. Although these authors report that folding kinetics correlate more strongly with the energy gap criterion than with σ (and, indeed, raise questions about the definition of the parameter), they nevertheless observe a statistically significant correlation between σ and folding rates across several lattice models. Simulations suggest that six order of magnitude change in folding rate should correspond to changing by σ more than 0.5 (20, 24, 25), a difference well within reach of experimental verification.

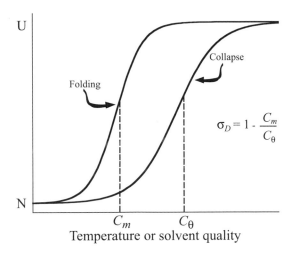

Figure 6 Thirumalai and coworkers (19–24) have reported that the dimensionless, equilibrium collapse parameter σ is highly correlated with the relative folding rates of simple on- and off-lattice polymers. σ is a measure of the point at which the unfolded polymer undergoes nonspecific coil-to-globule collapse relative to the point at which it folds as the solvent quality or temperature is reduced.

And is σ correlated with folding rates in the laboratory? To date σ_D has been experimentally defined for four simple, single-domain proteins spanning a greater than 30,000-fold range of rates (26). All four proteins exhibit σ_D effectively indistinguishable from zero (Figure 7). Similarly, we can employ previously reported data on the thermal unfolding of cytochrome c to determine that, for that protein, σ_T is also ~ 0 (27). Thus in keeping with theory, near zero values of σ are associated with the rapid folding of naturally occurring, two-state proteins. In contrast to the behavior of lattice and simple off-lattice polymers, however, the relative folding rates of simple proteins are not defined by this equilibrium measure of collapse cooperativity.

All Folding Criteria Optimal?

Experimental studies indicate that the criteria that determine folding rates in simple computational models do not account for the range of folding rates observed for simple, single-domain proteins. Indeed, the energy gaps of nearly all two-state proteins are unmeasurably large and both energetic roughness and σ typically adopt the unmeasurably small values that are associated only with the most rapidly folding lattice polymers. Thus, it appears that although biologically relevant folding rates are associated with a smooth landscape/large energy gap/low σ, differences in these parameters do not contribute significantly to the six order of magnitude range of two-state folding rates observed in the laboratory.

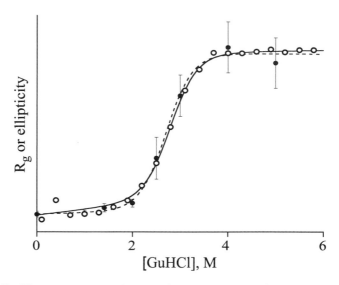

Figure 7 We can measure the putative kinetic determinant σ by monitoring collapse via small angle X-ray scattering and folding via near UV-circular dichroism. Shown is the equilibrium folding and collapse of the single-domain protein acyl phosphatase. As for all two-state proteins characterized to date (26), the σ of this protein is within error of zero, and thus this parameter is not correlated with relative folding rates in the laboratory.

EXPERIMENTALLY MOTIVATED THEORIES OF PROTEIN FOLDING KINETICS

In contrast to simulation-based theories of protein folding, several theories have arisen directly from the study of experimentally observed folding rates. More precisely, these theories have emerged from studies of how folding rates change with changing solvent conditions, under the influence of mutations, and across nonhomologous proteins. Two of these, the nucleation-condensation and topomer search models, specifically address the issue of why some proteins fold more rapidly than others. Here we describe these two models in detail, placing emphasis on their ability to explain or predict the vast range of experimentally observed folding rates.

Nucleation-Condensation Model

It is well established that native-like interactions are formed in the rate-limiting step of folding. For example, the perfectly exponential denaturant dependencies of folding rates (termed linear chevron behavior) demonstrate that the folding transition state contains interactions similar to those that stabilize the native state [reviewed in (28)]. The role that native interactions play in defining folding rates

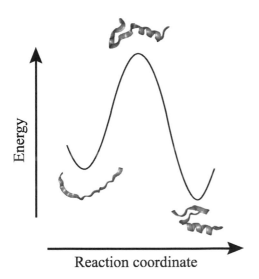

Reaction coordinate

Figure 8 Mutagenesis studies [termed ϕ-value analysis, reviewed in (34)] and other, less direct experimental evidence suggest that the folding transition state contains a nucleus of native-like structure. This has led to the hypothesis that the formation of this nucleus is a necessary and sufficient condition to ensure that the rate-limiting step in folding has been surmounted and that differences in the stability of the folding nucleus account for differences in folding rates (35).

is further supported by reports that native-state stability is an important determinant of the relative folding rates of topologically similar proteins (13, 29) and, often, of point mutants of a single protein (29–32).

Protein engineering studies, termed ϕ-value analysis (33), have further clarified the nature of this native-like transition-state structure and led to the development of the nucleation-condensation model of folding (Figure 8). This theory is based on the observation that a subset of all mutations destabilize the folding transition state approximately as much as they destabilize the native state, suggesting that the mutated residues are in a near-native environment during the rate-limiting step in folding [reviewed in (34)]. Although these structured transition-state residues are frequently distant in the protein sequence, they often cluster when mapped onto the native fold (33). Taken together, these observations support the hypothesis that folding is akin to nucleation in a phase transition. That is, a necessary and sufficient condition to surmount the rate-limiting step in folding is the formation of a small, specific nucleus of native structure upon which the remaining structure condenses, and differences in the stability of this nucleus contribute to relative folding rates (35).

The evidence in support of the nucleation-condensation model of folding is, however, rather qualitative; although native-like structures are clearly present in the folding transition states of many proteins, their quantitative contribution to relative folding rates has not been established. Indeed, much recent evidence suggests that differences in the stability of the folding nucleus contribute comparatively little to the six order of magnitude range of two-state folding rates. This evidence includes the observation that the vast majority of point mutations [reviewed in (28)] and even much larger sequence changes (36–38) produce less than one order of magnitude changes in folding rates. Similarly, circular permu-

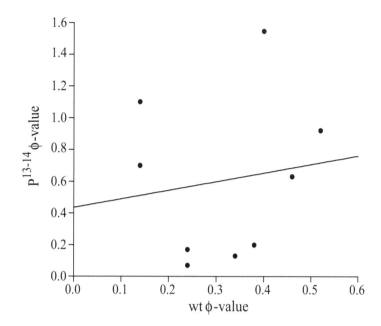

Figure 9 Mutations that dramatically alter the structure of the ϕ-value-defined folding nucleus typically do not significantly alter folding rates. Shown is the correlation between a residue's ϕ-value in the wild-type protein S6 and a circularly permuted variant [data adopted from (40)]. The statistically insignificant correlation demonstrates that the mutation entirely disrupts the wild-type nucleus. Despite completely rearranging the folding nucleus, the mutation accelerates the folding rate by a relatively minor factor of ∼2 (40).

tations (39, 40) and covalent circularizations (41) that largely or entirely disrupt the ϕ-value-defined nucleus produce only rather minor, three- to sevenfold changes in rate (Figure 9). Therefore, though it is extremely well established that a nucleus of native-like structure is formed in the rate-limiting step of folding, it appears that even large-scale alteration of this structure—and thus presumably its thermodynamics—typically fails to change folding rates significantly, suggesting in turn that the precise details of the folding nucleus are not the major determinant of relative folding rates.

Topology as a Determinant of Rates

In contrast to the suggestion that the thermodynamics of the transition state are dominated by a small subset of native interactions, in the late 1990s, it was reported that folding rates are strongly correlated with an empirical metric of global native-state topological complexity termed contact order (42). More recently a number of additional, empirical measures of topology have been reported, including the number of sequence-distant contacts per residue (43), the fraction of contacts that are sequence local (35), the total contact distance (44), and linear models of secondary structure content (45). All of these metrics predict folding rates approximately equally accurately, suggesting they reflect a common underlying physics. Because all four metrics reflect empirical observations rather than theoretical models, they do not, however, directly define the mechanistic origins of this physics.

Because contact order was the first topological metric reported to correlate with folding rates, it has received the most attention vis-à-vis efforts to reconcile the empirically observed topology-rate relationships with a quantitative, mechanistic model of folding. For example, because contact order is related to the sequence separation between contacting residues it has been suggested that it relates to the entropic cost of the loop closures required to surmount the rate-limiting step in folding (42, 46–48). Unfortunately, however, loop-closure entropy is proportional to the logarithm of loop length rather than loop length per se (49), and the average log (separation) between contacting residues is more poorly correlated with rates than is contact order as originally defined (K.W. Plaxco, unpublished observations). Perhaps because of this, several theoretical models of folding based on balancing loop-closure entropy with favorable side chain interactions exhibit poorer correlations with folding rates than the original, empirical observation itself (46, 47). Similarly, relative contact order (the average contact separation in terms of fraction of total peptide length) predicts rates significantly more accurately than absolute measures of the average sequence separation of contacting residues (50, 51). This produces the counter-intuitive result that, of two proteins with the same average contact separation, the longer protein folds faster. Observations such as these lead inevitably to the possibility that contact order predicts rates not because it is directly related to the underlying mechanism of folding but because it is a proxy for some other physically more reasonable parameter.

The first hint as to what this more mechanistically relevant parameter is came from the work of Goddard and coworkers (52), who related folding times to the number of gross topologies a protein must sample in order to enter the native energy well. In this topomer sampling model, the first stage of folding is dynamic as the unfolded chain diffuses between distinct gross topologies (topomers) (Figure 10, *B* to *C* transition). When the native topomer is found, stabilizing interactions can rapidly accrete, trapping the protein in its native conformation (Figure 10, *C* to *E* transition). Two properties of simple, single-domain proteins suggest that this type of search could dominate two-state folding rates. First, the formation of local structures, such as helices, hairpins, and loops, is orders of magnitude more rapid than the rate-limiting step in folding [reviewed in (53)]. Second, the folding free energies of such isolated structural elements, indeed, of almost all of the partially folded and misfolded states of single-domain proteins, are near or above zero (54–56). If proteins fold via the diffusional rearrangement

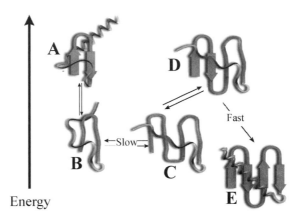

Figure 10   The essence of the topomer search model is that the rate with which an unfolded polymer diffuses between distinct topologies is much slower than the rate with which local structural elements zipper (and, critically, unzip) [reviewed in (59)]. It is well established that the formation of helices, loops, and other sequence-local structural elements (e.g., *B* to *A* transition) is significantly faster than the rate-limiting step in folding. Because the free energy of these partially folded states (e.g., *A*) is almost invariably ≥ 0 for two-state proteins, their disruption (e.g., the *A* to *B* transition) is more rapid still. Given these constraints, the rate-limiting step in folding might be the slow, large-scale diffusion to find the set of conformations (e.g., *C*) close enough to the native topology that they can zipper deeply into the stable, native well (*E*) without requiring slow, large-scale topological rearrangements. Central to this argument is the suggestion that, while the formation of specific, native-like interactions may be necessary in order to surmount the rate-limiting step (*D*), they are neither sufficient to ensure folding (e.g., *A*) nor a major determinant of relative barrier heights.

of these unstable, rapidly interconverting elements, then the rates will depend on the difficulty of finding the native topomer among all other, incorrect topomers. If all topomers are equally well populated, folding rates (for a given protein length) will depend only on the topomer sampling rate (52). If, in contrast, topologically distinct topomers are not equally well populated, rates will be determined by the topomer search rate, which is a function of both the topomer sampling rate and the fraction of unfolded conformations that are in the native topomer (57–59). The latter model predicts that relative folding rates are proportional to the probability of achieving the native topomer and that differences in the probability of achieving various native topomers could account for the vastly differing folding rates observed for topologically distinct proteins.

The topomer search model provides a rationalization for the wide range of two-state folding rates, but does it quantitatively account for that range? We have recently developed a simple method of estimating the probability of an unfolded chain adopting any given topomer that accurately predicts the folding kinetics of single-domain proteins (58, 59). The method emerged from simulations of inert,

Gaussian chains, which suggested that the probability of adopting a given topomer is quantifiable via a straightforward approximation arising from two simplifying effects. First, because the locations of residues that are close in sequence are highly correlated, the probability of achieving a given topomer is dominated by pairs of residues that are distant in the sequence. The second is that the probability of ordering the chain is well described by a mean-field approximation: Once a sufficient number of sequence-distant pairs of residues are brought into proximity, the probability of each of the remaining ordering events becomes independent of all other ordering events and approximately constant. The probability of forming the native topomer, $P(Q_D)$, is approximated by replacing the unique probability of ordering each specific pair with the average probability of ordering a pair (58):

$$P(Q_D) \propto \langle K \rangle^{Q_D},\qquad\qquad 4.$$

where Q_D is the number of sequence-distant pairs whose proximity defines the topomer, and $\langle K \rangle$ is the average equilibrium constant for residue pairs being in proximity (and is less than unity). Folding rates should, in turn, be proportional $P(Q_D)$.

The prediction that folding rates relate to Q_D provides a means of testing the topomer search model. In order to perform this test, however, Q_D must be defined. The definition of Q_D assumes that any pair of residues separated by more than l_c residues along the sequence and within distance r_c in the native state will be in proximity in the native topomer. The model is surprisingly insensitive to the precise values of these parameters; over a wide range of l_c and r_c, Q_D is correlated with $\log(k_f)$ with a correlation coefficient of $r^2 > 0.75$ (Figure 11). Thus this simple model captures in excess of 3/4 of the variance in our kinetic data set using only the fitted parameters $\langle K \rangle$ and the proportionality constant.

The predictive value of the topomer search model can be improved by introducing a length dependence; the extent to which a protein's folding rate differs from that predicted by the topomer search model is strongly correlated with chain length (59). This correlation, however, suggests that, for two proteins with the same Q_D, the longer protein folds more rapidly. This statistically robust observation, which is counter to the predictions of many simulation-based theories (52, 60, 61), is thought to arise due to crowding stemming from excluded volume and persistence length effects (59). With the addition of this crowding effect via the equation

$$k_f \propto QD \cdot J^{Q_{D/N}},\qquad\qquad 5.$$

where J is a constant ($J < 1$) that is analogous to $\langle K \rangle$, the correlation coefficient for the topomer search model increases to $r^2 = 0.85$ (59). The ability of this simple, near first-principles model of folding kinetics to capture 85% of the variance in two-state folding rates further supports the hypothesis that the topomer search process dominates relative barrier heights.

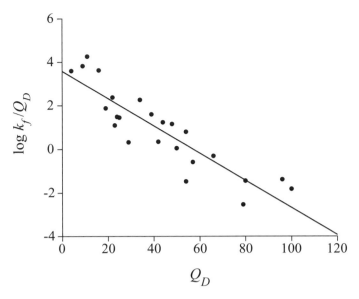

Figure 11   A quantitative version of the topomer search model predicts a simple relationship between folding rates and the number of sequence-distant native contacts (Q_D). As illustrated here, this relationship holds across a dataset of some two-dozen nonhomologous two-state proteins with a correlation coefficient of $r^2 > 0.75$, suggesting that this simple model captures more than 3/4 of the variance in two-state folding rates (59).

A Simple, Predictive Toy Model of Two-State Folding?

The limiting version of the topomer search model assumes that achieving the native topomer is the rate-limiting step in folding and that side-chain interactions do not contribute significantly to the folding barrier (59). In contrast, the original nucleation-condensation model stipulated that the folding barrier is surmounted when a nucleus of native structure is formed (62), presumably irrespective of the topology of the remainder of the chain. Reality clearly lies somewhere on the spectrum between these two extremes, and both a nucleus of native-like interactions and a grossly native topology are formed in the rate-limiting step in folding (58, 63). The observation that the simplest, most extreme version of the topomer search model captures 75% to 85% of the variance in two-state folding rates, however, suggests that this toy model of folding may be a reasonable starting point for further advances in theory. We are optimistic that when specific chemical details (e.g., nucleating interactions) are added to the topomer search model, its predictive value will be further improved, and we note that potentially promising advances in a similar direction have already been reported (64, 65).

RECONCILING THE BEHAVIORS OF LATTICE POLYMERS AND PROTEINS

In thinking of the minimal toy model of folding that captures enough of the relevant physics to recapitulate experimentally observed folding behaviors, we come to the question of how and why the folding of lattice polymers differs from that of real proteins. To address these questions, we must consider that simple computational models typically employed in folding simulations differ from proteins in at least two significant aspects. First, while the energy landscapes of most polymer models are usually quite rugged; those of almost all simple, single domain proteins are extremely smooth. Second, in contrast to lattice polymers, for which rates are not defined by "the number of short- versus long-range contacts in the native state" (14), topological effects dominate the relative folding rates of simple, single-domain proteins. Recent experimental and simulations-based studies are providing insights into the origins of these important discrepancies.

The Origins of Optimal Energy Landscapes

A possible origin of the near-universal observation of smooth landscapes/large energy gaps/low σ among naturally occurring proteins is that these properties are extremely common features of thermodynamically foldable polymers. Lattice polymer simulations suggest, however, that this is not the case. For example, it has been established that lattice polymer sequences with a pronounced energy gap are relatively rare; simulations of randomly selected 27-mer lattice polymers indicate that only 3% to 15% of such sequences encode a sufficiently large energy gap to ensure rapid folding (14, 66, 67), and the fraction of rapidly folding sequences may be yet smaller for 125-mer lattice systems (68). Near zero σ, which equates to rapid, two-state folding, is similarly a relatively rare property of randomly selected lattice and simplified off-lattice polymer sequences (24). Thus simulation studies do not appear to support the hypothesis that smooth energy landscapes/large energy gaps/low σ are a common property of thermodynamically foldable polymers; we must look elsewhere for reasons underlying the ubiquity of rapid folding among naturally occurring proteins.

Alternatively, smooth energy landscapes/large energy gaps/low σ may in fact be rare, but evolutionary pressures aimed specifically at ensuring rapid folding guarantee that naturally occurring proteins are selected from the small subset of sequences that exhibit these critical properties. This hypothesis is supported by studies suggesting that rapidly folding sequences will be produced only rarely in the absence of selective pressures or design constraints aimed at ensuring rapid folding (69–73, 74). Experimental tests of this hypothesis, however, are to be found in observations of the folding kinetics of a number of de novo designed proteins. The algorithms by which these molecules have been designed utterly ignore folding kinetics [(75–77); also see the critical commentary in (78)]. Nevertheless, all of the nearly one dozen de novo designed proteins characterized to date fold approximately as fast as or faster than the analogous naturally occurring proteins (38, 79; F. Gai, personal communication; M. Scalley-Kim & D. Baker, personal communication), suggesting that even in the absence of explicit kinetic selections the large majority of thermodynamically foldable sequences fold rapidly. Rapid folding therefore appears to be relatively common even among sequences that have not been the subject of explicit selection or design aimed at optimizing the energy landscape/energy gap/σ.

The discrepancy in the frequency with which rapid folding arises among protein sequences relative to the paucity of rapidly folding lattice-polymer sequences may be a function of the topologies that natural selection and human design have achieved. Exhaustive simulations demonstrate a close relationship between the native structure of a polymer and the properties of its energy landscape. These simulations suggest that, although most structures are the unique ground state of only a few sequences, a small subset of all structures are encoded by a very large number of sequences (80–83). Naturally occurring protein folds and existing de novo designed proteins are thought to represent such highly designable structures (80, 83). Critically, simulation studies indicate that highly designable structures almost invariably exhibit the smooth landscapes and large energy gaps that theory associates with rapid folding (82, 83). Though smooth landscapes/large gaps/low σ may be rare overall, they may be common among sequences that encode naturally occurring or otherwise designable topologies, thus explaining why, in contrast to randomly selected lattice polymer sequences, energy-landscape issues do not dominate the folding of proteins.

The Origins of Topology-Dependent Rates

The observation that the large majority of naturally occurring or designed proteins exhibit smooth energy landscapes also provides a potential explanation of why the folding kinetics of simple proteins are dominated by topological constraints, rather than the energy-landscape effects that dominate computational models. Namely, it is possible that subtle topological effects will become apparent only in the absence of potentially more dominant energy-landscape issues. Gō polymers, which have very smooth energy landscapes due to the absence of nonnative stabilizing interactions, provide a means of testing this hypothesis (84). Extensive folding simulations of these systems, however, indicate that smooth energy landscapes alone are not sufficient to generate strongly topology-dependent folding rates; on-lattice Gō polymer folding rates are effectively uncorrelated with topology (85, 86). The folding rates of more sophisticated, off-lattice Gō polymers likewise exhibit little (87) if any (88) topology dependence. No matter how smooth their energy landscapes, none of the traditional polymer models exhibit the dramatic topology dependence observed in two-state protein folding.

The topomer search model provides a rationale for the lack of topology dependence among even lattice polymers with smooth energy landscapes. The topomer search dominates folding kinetics because the folding of small proteins is extremely cooperative; breaking noncovalent native-state interactions lowers the energetic barrier to the breaking of additional interactions, producing a nonlinear relationship between free energy and the number of native contacts [for an interesting discussion of cooperativity and its relationship to two-state folding, see (89)]. Because of this cooperativity, an excess of 90% of the native structure is required for the free energy of a typical single-domain protein to drop below zero (54–56). The topomer search dominates rates because this level of cooperativity ensures that only unfolded molecules in the native topomer can zipper sufficient contacts to fold productively, and thus the entropic cost of finding the native topomer is a major contributor to the folding barrier. In comparison to proteins, the folding of traditional Gō polymers is noncooperative (90); the free energy of partially folded Gō lattice polymers is a relatively weak, linear function of the total number of interactions present (85). Gō polymers can, however, be forced to adopt a degree of cooperativity by defining the energy of a given conformation as a nonlinear function of the number of interactions present. Consistent with the predictions of the topomer search model, the introduction of such cooperativity leads to a highly significant relationship between topology and lattice-polymer folding rates (85, 86). If folding is cooperative and the energy landscape is smooth (i.e., lacking nonnative traps), only those unfolded conformations in the native topomer can fold productively, and for this reason, the entropic cost of the topomer search will dominate relative folding rates.

Simulations thus support the suggestion that the topology-dependent folding rates observed for simple, single-domain proteins are an unavoidable consequence of their highly cooperative folding. In addition to producing topology-dependent kinetics, however, the addition of cooperativity also decelerates the folding of Gō polymers (85, 86, 90). Does cooperativity also decelerate the folding rates of real proteins? In net it may not; we have speculated that the net effect of cooperativity is to actually accelerate folding by destabilizing partially structured, misfolded states relative to the native fold (smoothing the landscape/increasing the gap/lowering σ) to a greater extent than it decelerates folding by destabilizing potentially productive intermediates (85). This in turn suggests that the observed dominance of topology in defining folding rates is a consequence of the cooperativity necessary to ensure the formation of an energy landscape that can support rapid folding.

CONCLUSIONS

The criteria that distinguish rapidly folding computational models from those that fold more slowly do not account for the broad range of rates observed for the folding of two-state proteins; the experimentally measurable criteria associated with rapidly folding lattice and off-lattice computational models appear to be perfectly optimized for many naturally occurring proteins. The rapid folding of de novo designed proteins also suggests that this optimization does not require explicitly kinetic selective pressures but is associated with designability. And although native-like interactions are clearly formed during the rate-limiting step in folding, they are apparently not a dominant contributor to relative barrier heights. Finally, a simple mathematical description of the diffusive search for the correct overall topology, the topomer search model, accurately predicts relative folding rates. We believe that the topomer search dominates folding of two-state proteins because their equilibrium folding is so cooperative that only unfolded conformations in the native topomer can zipper sufficient structure for the free energy to drop below zero, and thus the entropic cost of finding the native topomer is a major contributor to the folding barrier. We speculate that this cooperativity provides a means of destabilizing partially structured, misfolded states relative to the native fold, thereby accelerating folding rates more by destabilizing traps (smoothing the landscape/improving the energy gap/decreasing σ) than it decelerates them by destabilizing potentially productive intermediates. This suggests that the observed dominance of topology in defining folding rates is a consequence of the cooperativity necessary to ensure the formation of an energy landscape that can support rapid folding.

ACKNOWLEDGMENTS

In addition to thanking David Baker, Michelle Scalley-Kim, Alan Fersht, and Feng Gai for communicating important unpublished results, the authors wish to thank our many collaborators over the years without whom our contributions to the above described research would not have been possible. Portions of this work were funded by the NIH (R01GM62868-01A1), BioSTAR (s97-79) and an ACS junior postdoctoral research fellowship (ACS CD INC 2-5-00) to BG.

The *Annual Review of Biochemistry* is online at http://biochem.annualreviews.org

LITERATURE CITED

1. Jackson SE. 1998. *Fold. Des.* 3:R81–91
2. van Nuland NAJ, Chiti F, Taddei N, Raugei G, Ramponi G, et al. 1998. *J. Mol. Biol.* 283:883–91
3. Wittung-Stafshede P, Lee JC, Winkler JR, Gray HB. 1999. *Proc. Natl. Acad. Sci. USA* 96:6587–90
4. Zagrovic B, Snow CD, Shirts MR, Pande VS. 2002. *J. Mol. Biol.* 323:927–37
5. Bryngelson JD, Wolynes PG. 1987. *Proc. Natl. Acad. Sci. USA* 84:7524–28
6. Onuchic JN, Wolynes PG, Luthey-Schulten Z, Socci ND. 1995. *Proc. Natl. Acad. Sci. USA* 92:3626 30
7. Thirumalai D, Ashwin V, Bhattacharjee JK. 1996. *Phys. Rev. Lett.* 77:5385–88

8. Nymeyer H, García AE, Onuchic JN. 1998. *Proc. Natl. Acad. Sci. USA* 95:5921–28
9. Onuchic JN, Luthey-Schulten Z, Wolynes PG. 1997. *Annu. Rev. Phys. Chem.* 48:545–600
10. Socci ND, Onuchic JN, Wolynes PG. 1998. *Proteins: Struct. Funct. Genet.* 32:136–58
11. Sabelko J, Ervin J, Gruebele M. 1999. *Proc. Natl. Acad. Sci. USA* 96:6031–36
12. Gillespie B, Plaxco KW. 2000. *Proc. Natl. Acad. Sci. USA* 97:12014–19
13. Guijarro JI, Morton CJ, Plaxco KW, Campbell ID, Dobson CM. 1998. *J. Mol. Biol.* 275:657–67

14. Sali A, Shacknovich E, Karplus M. 1994. *Nature* 369:248–51
15. Gutin AM, Abkevich VI, Shakhnovich EI. 1995. *Proc. Natl. Acad. Sci. USA* 92:1282–86
16. Dinner AR, Abkevich V, Shakhnovich E, Karplus M. 1999. *Proteins Struct. Funct. Genet.* 35:34–40
17. Mayor U, Johnson CM, Daggett V, Fersht AR. 2000. *Proc. Natl. Acad. Sci. USA* 97:13518–22
18. Perl D, Welker C, Schindler T, Schroder K, Marahiel MA, et al. 1998. *Nat. Struct. Biol.* 5:229–35
19. Camacho CJ, Thirumalai D. 1996. *Europhys. Lett.* 35:627–32
20. Klimov DK, Thirumalai D. 1996. *Phys. Rev. Lett.* 76:4070–73
21. Klimov DK, Thirumalai D. 1996. *Proteins Struct. Funct. Genet.* 26:411–41
22. Klimov DK, Thirumalai D. 1998. *Fold. Des.* 3:127–39
23. Klimov DK, Thirumalai D. 1998. *J. Chem. Phys.* 109:4119–25
24. Thirumalai D, Klimov DK. 1999. *Curr. Opin. Struct. Biol.* 9:197–207
25. Veitshans T, Klimov D, Thirumalai D. 1996. *Fold. Des.* 2:1–22
26. Millet IS, Townsley L, Chiti F, Doniach S, Plaxco KW. 2002. *Biochemistry* 41:321–25
27. Hagihara Y, Hoshino M, Hamada D, Kataoka M, Goto Y. 1998. *Fold. Des.* 3:195–201
28. Plaxco KW, Simons KT, Ruczinski I, Baker D. 2000. *Biochemistry* 39:11177–83
29. Clarke J, Cota E, Fowler SB, Hamill SJ. 1999. *Struct. Fold. Des.* 7:1145–53
30. Riddle DS, Grantcharova VP, Santiago JV, Alm E, Ruczinski I, et al. 1999. *Nat. Struct. Biol.* 6:1016–24
31. Villegas V, Martinez JC, Aviles FX, Serrano L. 1998. *J. Mol. Biol.* 6:1027–36
32. Martinez JC, Serrano L. 1999. *Nat. Struct. Biol.* 6:1010–16
33. Jackson SE, elMasry N, Fersht AR. 1993. *Biochemistry* 32:11270–78
34. Fersht AR. 1997. *Curr. Opin. Struct. Biol.* 7:3–9
35. Mirny L, Shakhnovich E. 2001. *Annu. Rev. Biophys. Biomol. Struct.* 30:361–66
36. Riddle DS, Santiago JV, Bray-Hall ST, Doshi N, Grantcharova VP, et al. 1997. *Nat. Struct. Biol.* 4:805–9
37. Kim DE, Gu HD, Baker D. 1998. *Proc. Natl. Acad. Sci. USA* 95:4982–86
38. Gillespie B, Vu D, Shah PS, Marshall S, Dyer RB, et al. 2003. *J. Mol. Biol.* 330:813–19
39. Viguera AR, Serrano L, Wilmanns M. 1996. *Nat. Struct. Biol.* 3:874–80
40. Lindberg M, Tangrot J, Oliveberg M. 2002. *Nat. Struct. Biol.* 9:818–22
41. Grantcharova VP, Baker D. 2001. *J. Mol. Biol.* 306:555–63
42. Plaxco KW, Simons KT, Baker D. 1998. *J. Mol. Biol.* 277:985–94
43. Gromiha MM, Selvaraj S. 2001. *J. Mol. Biol.* 310:27–32
44. Zhou HY, Zhou YQ. 2002. *Biophys. J.* 82:458–63
45. Gong HP, Isom DG, Srinivasan R, Rose GD. 2003. *J. Mol. Biol.* 327:1149–54
46. Alm E, Baker D. 1999. *Proc. Natl. Acad. Sci. USA* 96:11305–10
47. Muñoz V, Eaton WA. 1999. *Proc. Natl. Acad. Sci. USA* 96:11311–16
48. Fersht AR. 2000. *Proc. Natl. Acad. Sci. USA* 97:1525–29
49. Jacobson H, Stockmayer WH. 1950. *J. Chem. Phys.* 18:1600–6
50. Grantcharova VP, Alm EJ, Baker D, Horowitz AL. 2001. *Curr. Opin. Struct. Biol.* 11:70–82
51. Ivankov DN, Garbuzynskiy SO, Alm E, Plaxco KW, Baker D, Finkelstein AV. 2003. *Protein Sci.* 12:2057–62
52. Debe DA, Carlson MJ, Goddard WA. 1999. *Proc. Natl. Acad. Sci. USA* 96:2596–601
53. Eaton WA, Munoz V, Hagen SJ, Jas GS, Lapidus LJ, et al. 2000. *Annu. Rev. Biophys. Biomol. Struct.* 29:327–59
54. Flanagan JM, Kataoka M, Shortle D,

Engelman DM. 1992. *Proc. Natl. Acad. Sci. USA* 89:748–52

55. Ladurner AG, Itzhaki LS, Gay GD, Fersht AR. 1997. *J. Mol. Biol.* 273:317–29

56. Camarero JA, Fushman D, Sato S, Giriat I, Cowburn D, et al. 2001. *J. Mol. Biol.* 308:1045–62

57. Debe DA, Goddard WA. 1999. *J. Mol. Biol.* 294:619–25

58. Makarov DE, Keller CA, Plaxco KW, Metiu H. 2002. *Proc. Natl. Acad. Sci. USA* 99:3535–39

59. Makarov DE, Plaxco KW. 2003. *Protein Sci.* 12:17–26

60. Gutin AM, Abkevich VI, Shakhnovich EI. 1996. *Phys. Rev. Lett.* 77:5433–36

61. Li MS, Klimov DK, Thirumalai D. 2002. *J. Phys. Chem. B* 106:8302–5

62. Fersht AR. 1995. *Proc. Natl. Acad. Sci. USA* 92:10869–73

63. Vendruscolo M, Paci E, Dobson CM, Karplus M. 2001. *Nature* 409:641–45

64. Myers JK, Oas TG. 2001. *Nat. Struct. Biol.* 8:552–58

65. Islam SA, Karplus M, Weaver DL. 2002. *J. Mol. Biol.* 318:199–215

66. Shakhnovich E, Farztdinov G, Gutin AM, Karplus M. 1991. *Phys. Rev. Lett.* 67:1665–67

67. Sali A, Shakhnovich E, Karplus M. 1994. *J. Mol. Biol.* 235:1614–36

68. Dinner AR, So S-S, Karplus M. 1998. *Proteins Struct. Funct. Genet.* 33:177–201

69. Guo Z, Thirumalai D. 1996. *J. Mol. Biol.* 263:323–43

70. Mirny LA, Abkevich VI, Shakhnovich EI. 1998. *Proc. Natl. Acad. Sci. USA* 95:4976–81

71. Gutin AM, Abkevich VI, Shakhnovich EI. 1998. *Fold. Des.* 3:183–94

72. Li L, Mirny LA, Shakhnovich EI. 2000. *Nat. Struct. Biol.* 7:336–42

73. Betancourt M, Thirumalai D. 2002. *J. Phys. Chem. B* 106:599–609

74. Larson S, Ruczinski I, Davidson AR, Baker D, Plaxco KW. 2002. *J. Mol. Biol.* 316:225–33

75. Dahiyat BI, Mayo SL. 1996. *Protein Sci.* 5:895–903

76. Walsh STR, Cheng H, Bryson JW, Roder H, DeGrado WF. 1999. *Proc. Natl. Acad. Sci. USA* 96:5486–91

77. Dantas G, Kuhlman B, Callender D, Wong M, Baker D. 2003. *J. Mol. Biol.* 332:449–60

78. Jin W, Kambara O, Sasakawa H, Tamura A, Takada S. 2003. *Structure* 11:581–90

79. Hill RB, Bracken C, DeGrado WF, Palmer AGI. 2000. *J. Am. Chem. Soc.* 122:11610–19

80. Govindarajan S, Goldstein RA. 1995. *Biopolymers* 36:43–51

81. Li H, Helling R, Tang C, Wingreen N. 1996. *Science* 273:666–69

82. Buchler NEG, Goldstein RA. 2000. *J. Chem. Phys.* 112:2533–47

83. Miller J, Zeng C, Wingreen NS, Tang C. 2002. *Proteins Struct. Funct. Genet.* 47:506–12

84. Abe H, Gō N. 1981. *Biopolymers* 20:10113–31

85. Jewett AI, Pande VS, Plaxco KW. 2003. *J. Mol. Biol.* 326:247–53

86. Kaya H, Chan HS. 2003. *Proteins* 52:524–33

87. Koga N, Takada S. 2001. *J. Mol. Biol.* 313:171–80

88. Cieplak M, Hoang TX. 2002. *Biophys. J.* 84:475–88

89. Kaya H, Chan HS. 2002. *J. Mol. Biol.* 315:899–909

90. Eastwood MP, Wolynes PG. 2001. *J. Chem. Phys.* 114:4702–16

91. Filimonov VV, Azuaga AI, Viguera AR, Serrano L, Mateo PL. 1999. *Biophys. Chem.* 77:195–208

92. Deleted in proof

93. Chiti F, van Nuland NAJ, Taddei N, Magherini F, Stefani M, et al. 1998. *Biochemistry* 37:1447–55

94. Wittung-Stafshede P. 1999. *Biochim. Biophys. Acta* 1432:401–5

95. Huang G, Oas TG. 1995. *Proc. Natl. Acad. Sci. USA* 92:6878–82

96. Huang G, Oas TG. 1996. *Biochemistry* 35:6173–80
97. Jacob M, Geeves M, Holtermann G, Schmid FX. 1999. *Nat. Struct. Biol.* 6: 923–26
98. Jackson SE, Fersht A. 1991. *Biochemistry* 30:10428–35
99. van Nuland NAJ, Meijberg W, Warner J, Forge V, Scheek RM, et al. 1998. *Biochemistry* 37:622–37
100. Horng JC, Moroz V, Raleigh DP. 2003. *J. Mol. Biol.* 326:1261–70
101. Scalley M, Baker D. 1997. *Proc. Natl. Acad. Sci. USA* 94:10637–40
102. Spector S, Raleigh DP. 1999. *J. Mol. Biol.* 293:763–68
103. Knapp S, Karshikoff A, Berndt KD, Christova P, Atanasov B, et al. 1996. *J. Mol. Biol.* 264:1132–44
104. Kubelka J, Eaton WA, Hofrichter J. 2003. *J. Mol. Biol.* 329:625–30
105. Lindberg MO, Tangrot J, Otzen DE, Dolgikh DA, Finkelstein AV, et al. 2001. *J. Mol. Biol.* 314:891–900

Annu. Rev. Biochem. 2004. 73:861–90
doi: 10.1146/annurev.biochem.73.011303.074032
Copyright © 2004 by Annual Reviews. All rights reserved
First published online as a Review in Advance on March 18, 2004

EUKARYOTIC mRNA DECAPPING

Jeff Coller and Roy Parker

*Howard Hughes Medical Institute, Department of Molecular and Cellular Biology,
University of Arizona, Tucson, Arizona 85721; e-mail: jmcoller@u.arizona.edu,
rrparker@u.arizona.edu*

Key Words mRNA turnover, mRNA stability, mRNA decay deadenylation

■ **Abstract**   Eukaryotic mRNAs are primarily degraded by removal of the 3′
poly(A) tail, followed either by cleavage of the 5′ cap structure (decapping) and
5′->3′ exonucleolytic digestion, or by 3′ to 5′ degradation. mRNA decapping
represents a critical step in turnover because this permits the degradation of the
mRNA and is a site of numerous control inputs. Recent analyses suggest decapping
of an mRNA consists of four central and related events. These include removal, or
inactivation, of the poly(A) tail as an inhibitor of decapping, exit from active
translation, assembly of a decapping complex on the mRNA, and sequestration of the
mRNA into discrete cytoplasmic foci where decapping can occur. Each of these steps
is a demonstrated, or potential, site for the regulation of mRNA decay. We discuss the
decapping process in the light of these central properties, which also suggest
fundamental aspects of cytoplasmic mRNA physiology that connect decapping,
translation, and storage of mRNA.

CONTENTS

0066-4154/04/0707-0861$14.00   **861**

THE BIOLOGICAL ROLE OF mRNA TURNOVER

The process of mRNA turnover is important in numerous aspects of eukaryotic mRNA physiology. For example, mRNA turnover plays a key role in the control of gene expression both by setting the basal level of gene expression and as a site of regulatory responses. More recently, it has become clear that mRNA decay mechanisms play an important role in antiviral defenses. This antiviral role can include functions of the basic mRNA decay machinery as well as specialized systems, such as the response to dsRNA (RNA interference) [for reviews, see (113, 118)]. Finally, specialized mRNA turnover systems exist that recognize and degrade aberrant mRNAs, thereby increasing the quality control of mRNA biogenesis (72). Given these functions, it is important to understand how decay rates of different mRNAs are controlled, how aberrant mRNAs are targeted for destruction, and how viral mRNAs are distinguished from normal cellular mRNAs.

PATHWAYS OF mRNA DECAY

In every eukaryote, the regulated destruction of mRNAs can occur via four general mechanisms: deadenylation dependent 3′ to 5′ decay, endonucleolytic digestion, or by specialized quality control pathways. In most species, multiple mechanisms exist. The major pathway for degradation may vary between species, but all organisms seem to have multiple mechanisms for turnover in place. In the following section, we introduce each pathway and briefly discuss key enzymes required for each (Table 1).

Deadenylation Dependent Exonucleolytic Digestion

There are two general pathways by which polyadenylated mRNAs can be degraded in eukaryotic cells (Figure 1). In both cases, the degradation of the transcript begins with the shortening of the poly(A) tail at the 3′ end of the mRNA (77, 99). Three mRNA deadenylases have been identified. In yeast, the predominant deadenylase complex contains two nucleases, Ccr4p and Pop2p,

TABLE 1   Enzymes involved in mRNA decay

Protein	Function	Features
Deadenylation		
Ccr4p	Catalytic subunit of the deadenylase	Homology to Mg2+ dependent endonucleases
Pop2p	Regulator of deadenylation, may also have deadenylase activity	Homology to RNaseD
Pan2p/Pan3p	Minor deadenylases, required for poly(A) length control	Pan2p has homology to RNaseD
PARN	Mammalian deadenylase	Homology to RNaseD. Can bind to 5′ cap, which stimulates deadenylase activity
Decapping		
Dcp1p	Major component of decapping holoenzyme	EVH1/WH1 domain
Dcp2p	Catalytic subunit of decapping holoenzyme	NUDIX motif, conserved Box A and Box B motif
Exonucleases		
Xrn1p	Major cytoplasmic 5′→3′ exonuclease	
Rat1p	Nuclear 5′→3′ exonuclease	
Rrp4, Rrp40p, Rrp41p, Rrp42p, Rrp43p, Rrp44p, Rrp45p, Rrp46p, Mtr3p, Cs14p	A complex of 3′→5′ exonucleases termed the exosome	Domain organization similar to that of bacterial PNPase

and several accessory proteins, Not1-Not5p, Caf4p, Caf16p, Caf40p, and Caf130p (25, 31, 108, 109). Additional deadenylases include the Pan2p/Pan3p complex (13, 18, 109) and the RNase D homolog PARN, which is found in some but not all eukaryotes (61, 62). An unresolved issue is the relative importance of these individual deadenylases in different organisms and for different mRNAs [see (86) for more discussion].

In yeast, shortening of the poly(A) tail primarily leads to removal of the 5′ cap structure (decapping), thereby exposing the transcript to digestion by a 5′ to 3′ exonuclease (34, 51, 80). Several observations suggest that mRNA decapping is a decay mechanism in other eukaryotic cells. First, the decapping enzymes, Dcp1p and Dcp2p, are conserved among eukaryotes (36, 71, 110, 116). Second, analysis of the 5′ and 3′ ends of mammalian mRNAs demonstrated that decapped mRNAs are specifically detected on transcripts with short poly(A) tails (28).

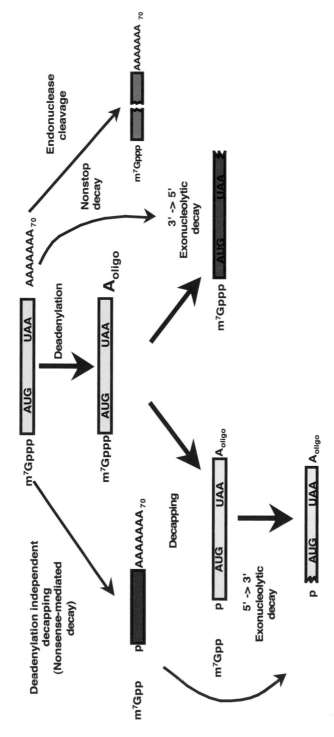

Figure 1 Pathways by which eukaryotic mRNAs are degraded.

Lastly, the aggregation of decapping factors into discrete subcellular structures is also conserved (5, 53, 71, 98, 110).

mRNAs can also be degraded in a 3' to 5' direction following deadenylation (Figure 1) (79). 3' to 5' degradation of mRNAs is catalyzed by the exosome (54, 81, 92), which is a large complex of 3' to 5' exonucleases functioning in several RNA degradative and processing events [for reviews, see (19, 74, 114)]. For the yeast mRNAs that have been studied, the process of 3' to 5' decay is slower than the decapping dependent 5' to 3' decay pathway (23). However, it is likely that for some yeast mRNAs, or in other eukaryotic cells, 3' to 5' degradation will be the primary mechanism of mRNA degradation following shortening of the poly(A) tail (46). Consistent with that view, recent results suggest that degradation of some mammalian mRNAs containing AU-rich destabilizing elements (AREs) may be predominantly in a 3' to 5' direction (25, 81, 117).

Decay of Eukaryotic mRNA via Endonucleolytic Cleavage

Eukaryotic mRNAs can also be degraded via endonucleolytic cleavage prior to deadenylation (Figure 1). Evidence for this mechanism comes from the analysis of transcripts, such as mammalian 9E3, transferrin receptor, *c-myc*, insulin-like growth factor II, serum albumin, vitellogenin mRNA, and Xenopus β-globin mRNA. In these cases, mRNA fragments that correspond to the 5' and/or 3' portions of the transcript are detected in vivo and are consistent with internal cleavage within the mRNA (11, 17, 29, 43, 103, 112). Because there does not appear to be any similarity between the cleavage sites in these mRNAs, a wide variety of endonucleases may exist with different cleavage specificities. For example, the endonuclease suggested to be responsible for β-globin decay appears to cleave at UG and UC dinucleotides (17), whereas decay of the eNOS pre-mRNA appears to occur by cleavage at CA repeats (52).

Endonucleolytic cleavage is also paramount in RNA-mediated gene silencing (RNAi) [for review, see (107)]. The process of RNAi appears to defend the genome against viruses and transposons as well as control gene expression of some endogenous mRNAs. An RNAi response is initiated by the recognition of a dsRNA sequence by the ATP-dependent endonuclease *Dicer. Dicer* cleaves the dsRNA to generate 21–23 nucleotide RNA species that are used to target the RNAi-induced silencing complex (RISC) to the complementary mRNA. In some cases, the RISC complex facilitates a second endonucleolytic event, which triggers the destruction of the mRNA. A simple hypothesis is that this initial cleavage leads to the 5' and 3' portions of the mRNA being degraded by Xrn1p and/or the exosome.

Specialized Decay Pathways for mRNA Quality Control

Eukaryotic cells have evolved quality control mechanisms to monitor mRNA biogenesis (Figure 2). These mechanisms take the form of specialized pathways for the rapid degradation of aberrant mRNAs. For example, aberrant mRNAs

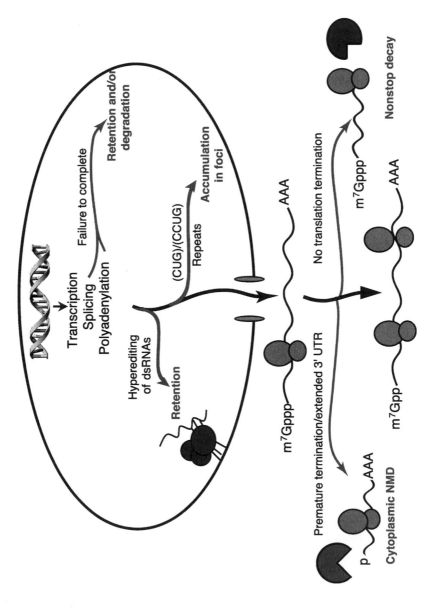

Figure 2   Mechanism of mRNA quality control.

containing a premature translational stop codon are decapped without prior poly(A) shortening (22, 78). This rapid degradation of aberrant transcripts, referred to as nonsense-mediated decay (NMD), is a specialized mRNA turnover process the cell uses to reduce the production of truncated proteins, which can have deleterious consequences to the cell (20). Another aberrancy monitored by specialized decay machinery is the absence of a stop codon, referred to as nonstop decay. In these cases, the mRNA is targeted to the cytoplasmic exosome by a specific adaptor protein, Ski7p, which is proposed to interact with the stalled ribosome (39, 123).

Quality control systems also exist within the nucleus to degrade inappropriately spliced mRNAs and other abnormal transcripts [for review, see (75)]. Nuclear decay appears to predominately occur via the exosome, although there may be some contributions by the nuclear Xrn1p homolog, Rat1p (15, 41). Other aberrancies monitored within the nucleus include hyperediting of dsRNA (122), the presence of CUG expansions (32), and incomplete transcriptional processes, such as polyadenylation and splicing (Figure 2) (30, 35, 47). A subset of these events results in exosome-mediated retention of the mRNA at the site of transcription [for review, see (56)].

Two observations suggest that the frequency with which aberrant mRNAs are produced is quite high. First, analysis of alternatively spliced mRNAs in mammalian cells suggests that one third of all splicing events produce mRNAs with premature translation termination codons, which would then be recognized and degraded by NMD (69). This indicates that splicing produces large amounts of errors, which are then rapidly degraded. Similarly, expressed sequence tags in which premature polyadenylation has occurred within the open reading frame make up ~1% of yeast libraries. Such mRNAs are degraded at least 10× faster than normal mRNAs because of nonstop decay, thus suggesting that errors during 3' end processing may be as high as 10% for all transcripts even in a simple organism like yeast (A. van Hoof and R. Parker, unpublished observations).

The infidelity of mRNA processing may serve as an evolutionary capacitor, allowing rapid responses to changing environmental conditions. However, the accumulation of aberrant and potentially toxic mRNA products is a consequence of this plasticity. Therefore, it appears that one major function of mRNA decay in the cell is quality control, which ensures that only mRNA having an appropriate structure, i.e., cap, start codon, stop codon, and poly(A) tail, survive. Thus the cell has evolved "proofreading" mechanisms to distinguish between normal and abnormal mRNAs. Understanding how these distinctions are made will be important in improving our understanding of how turnover is regulated.

THE mRNA DECAPPING MACHINERY

The bulk of this review focuses on the regulation and execution of the decapping reaction. In this section, we introduce the various factors shown to be required for decapping and provide a description of their characteristics and demonstrated, or

TABLE 2   Regulators of yeast mRNA decapping

Protein	Properties	Function	Significant interactions
Pab1p	Contains four N-terminal RRM domains and a proline-rich C terminus	Major protein associated with poly(A) tail. Blocks mRNA decapping and stimulates translation. Primary coupler of deadenylation and decapping.	eIF-4G, eRF3, Pan2/3p
eIF-4E	Cap-binding protein	Component of the eukaryotic translational initiation complex, eIF-4F. Blocks mRNA decapping by competing with Dcp1/2p for access to the cap.	Lsm7p, eIF-4G, eIF-4A, eIF-4B, Pab1p
Lsm1–7p	Sm-like proteins	Required for the efficiency of decapping in vivo. Forms a heteroheptameric ring complex and interacts with the mRNA after deadenylation. May facilitate the assembly of the decapping complex.	Dcp1p, Dcp2p, Dhh1p, Pat1p, Xrn1p, Upf1p
Pat1p	88kDa protein with no recognizable sequence motifs	Interacts with both polyadenylated and deadenylated transcripts. Required for efficiency of both decapping and formation of P bodies in vivo. May "seed" the decapping complex on the mRNA.	Dcp1p, Dcp2p, Lsm1–7p, Dhh1p, Xrn1p, Crm1p
Dhh1p	Member of the ATP-dependent DExD/H box helicase family	Required for the efficiency of decapping in vivo. Homologs across species are required for translational repression during mRNA storage events.	Dcp1p, Dcp2p, Lsm1–7p, Ccr4p, Pop2p, Caf17p, Pbp1p, Edc3p
Edc1p, Edc2p	Small, basic proteins with weak homology to each other	Required for efficient decapping in vitro. Directly binds to the mRNA substrate.	Dcp1p, Dcp2p
Edc3p	Contains five conserved domains	A general and mRNA-specific regulator of decapping. Regulates the decapping of the RPS28a mRNA.	Dcp1p, Dcp2p, Dhh1p, Crm1p, Rps28ap, Nup157p, Lsm8p
Puf3p	Pumillo-like protein, contains eight PUF repeats	Messages specific activator of mRNA deadenylation and decapping. Homologs facilitate translational repression. Regulates the decapping of the COX17 mRNA.	
Upf1p, Upf2p, Upf3p	Upf1p is an ATP-dependent RNA helicase	Required for non-sense-mediated decapping.	eRF1, eRF3, Dcp2p, Upf2p, Lsm1p
eIF-5a, Vps16p, Mrt4p, Sla2p, Gcr5p, Ths1p		Additional proteins suggested to be involved in mRNA turnover but functions remain unclear.	

hypothesized, biochemical functions (summarized in Table 2). An emerging theme is that several different types of protein complexes can be seen interacting with mRNAs and with the decapping enzyme, thereby providing a possible mechanism by which such proteins can affect decapping. A second theme emerging is that several of these factors appear to have additional roles in the cell and may function to integrate mRNA decapping with translation, transport, and transcription.

The mRNA Decapping Enzyme

Several observations suggest that two proteins, Dcp1p and Dcp2p, function together as a decapping holoenzyme with Dcp2p as the catalytic subunit. First, in both yeast and mammals, these proteins copurify (36, 37, 71). Second, copurification of recombinant yeast Dcp1p and Dcp2p from *Escherichia coli* yields active decapping enzyme under a variety of conditions (100). Third, yeast Dcp2p alone can have decapping activity in the presence of manganese (100) or in high magnesium (110). In contrast, recombinant mammalian Dcp2p is a robust decapping enzyme by itself (110, 116). Fourth, Dcp2p contains a NUDIX motif, which is found in a class of pyrophosphatases (10, 60), and mutations in the Dcp2p NUDIX motif inactivate decapping activity in vivo and in vitro in both the yeast and mammalian enzymes (36, 71, 100, 116). The simplest interpretation of these observations is that Dcp2p is the catalytic subunit of the decapping complex, and Dcp1p primarily functions to enhance Dcp2p activity by a currently unresolved mechanism. It is unclear whether previous results suggesting gel purified or recombinant Dcp1p could have catalytic activity were in error or whether Dcp1p can also have decapping activity (67, 115).

THE CHEMISTRY OF DECAPPING   Biochemical analysis of the Dcp1/Dcp2p decapping activity demonstrated that the m^7GpppX of mRNAs is cleaved to yield the products m^7GDP and a 5′-monophosphate mRNA (Figure 1) (67, 102). Production of a transcript with a 5′-monophosphate is functionally significant in order to ensure rapid degradation by Xrn1p, the 5′ to 3′ exonuclease. Xrn1p is blocked by a cap structure (51) and preferentially degrades substrates containing a 5′-monophosphate end compared to those with a 5′-triphosphate end (101). The decapping enzyme presumably functions in vivo as it does in vitro, yielding an mRNA possessing a monophosphate at its 5′ end, which is then rapidly degraded by Xrn1p.

INTERACTION OF Dcp1/2p AND THE mRNA SUBSTRATE   The *DCP1/DCP2* gene products demonstrate specificity in vitro consistent with an activity that primarily decaps mRNAs. For example, the 7-methyl group of the cap structure contributes to substrate specificity of the enzyme (67, 116). Interestingly, addition of the cap analog m^7GpppG_{OH} fails to inhibit the decapping reaction to an appreciable extent (67, 116). This suggested that the decapping enzyme might also interact with the body of the transcript or that RNA binding may allosterically regulate decapping. This interpretation has been supported by three additional observations. First, Dcp1/2p was effectively inhibited in vitro by uncapped mRNAs (67). Second, Dcp1/Dcp2p or Dcp2p alone prefers substrates that are ≥ 25 nucleotides in length, with a preference for longer mRNA substrates (67, 88, 100, 116). Third, mammalian and yeast Dcp2p binds to RNA (88; C.J. Decker and R. Parker, unpublished observation). These properties suggest that Dcp2p recognizes the mRNA substrate by interactions with both the cap and the RNA moiety. This preference for longer substrates may have biological significance because the 5′ cap structure is often complexed with a set of proteins involved in translation initiation, referred to as the cap-binding complex or eIF-4F. A requirement for a significant length of RNA in addition to the cap structure for substrate recognition by Dcp1/2p may prevent decapping of mRNAs on which translation initiation complexes are assembled (see below).

FEATURES OF Dcp2p   In addition to the NUDIX domain, which appears to be the active site for decapping, Dcp2p also has two additional regions that are highly conserved, termed Box A and Box B. These two regions have no homology to other functional motifs, but it has recently been shown that Box B is required for both RNA binding and decapping activity in vitro (88). The importance of Box A is unknown, but it may serve to facilitate the interaction between Dcp1p and Dcp2p. In yeast, Dcp2p has an extensively long C terminus, which is not seen in other homologs.

FEATURES OF Dcp1p   The Dcp1p protein is also conserved in eukaryotes, and two human homologs, hDcp1a and hDcp1b, have been identified (71). The human homologs appear functional because hDcp1a copurifies with decapping activity (71). Experimental and modeling approaches show that the N-terminal of hDcp1a belongs to a new class of functional EVH1/WH1 domains (Ena/VASP homology 1/Wiskott-Aldrich syndrome protein homology 1). This structural feature was first suggested by modeling of one of the human Dcp1p homologs, referred to as SMIF (21). More recently, a high-resolution X-ray structure for the yeast Dcp1p has confirmed and extended this analysis (97). EVH1/WH1 domains are protein-protein interaction modules that interact with their proline-rich containing ligands, thereby providing essential links for their host proteins to various signal transduction pathways, such as actin filament assembly, synaptic transmission, and Ras signaling (4, 9, 89). Comparison of the proline-rich sequence (PRS) binding sites in this family of proteins with Dcp1p indicates that Dcp1p belongs to a novel class of EVH1 domains. Mapping the sequence conservation on the molecular surface of Dcp1p reveals two prominent sites, one of which is required for function of the Dcp1p/Dcp2p complex, and a second, corresponding to the PRS binding site, that is likely to be a binding site for decapping regulatory proteins. Moreover, a conserved hydrophobic patch is revealed to be critical for decapping.

Dcp1p may have additional functions within the cell. Indeed, hDcp1a may play a role in SMAD-mediated TGFβ signaling, having been shown to interact with Smad4 and to partially translocate to the nucleus in response to transformation growth factor β stimulation (3). The significance of this relocalization is not known but could affect both cytoplasmic degradation and nuclear functions of Dcp1p.

The Scavenger Decapping Enzyme

Eukaryotic cells also contain a second type of decapping enzyme referred to as the scavenger decapping enzyme. These enzymes were first described over thirty years ago and decap short substrates, such as the dinucleotide cap structure or a capped oligonucleotide (64, 83, 84). Recent experiments indicate that the scavenger decapping enzyme acts to decap the capped oligonucleotides produced by exosome mediated 3′ to 5′ degradation of mRNA (117). Consistent with this role, DcpS coimmunoprecipitates with the exosome from mammalian cells (117). DcpS also has a second function in hydrolyzing the m7GDP produced by mRNA decapping to m7GMP and phosphate (111).

DcpS proteins are members of the HIT family of pyrophosphatases and use a histidine triad to perform catalysis (70). Interestingly, DcpS is unable to decap long substrates (70). This suggests that either long mRNAs cannot fit into the active site of DcpS or that RNA is a negative allosteric regulator of this enzyme. The inability to cleave long mRNAs may be biologically important to prevent DcpS from prematurely decapping mRNAs not targeted for degradation.

Regulators of Decapping

Several proteins have been identified that function to accelerate or decelerate the decapping process (see Table 2). One class of proteins includes those that inhibit the decapping process. In particular, the poly(A)-binding protein (Pab1p), is an important member of this group because it is required to couple decapping to prior deadenylation (24, 76). In addition, components of the translation initiation complex also impede decapping (94, 95). Most notably, the cap-binding protein, eIF-4E, inhibits decapping both in vitro and in vivo.

Several proteins are required for the efficient decapping of most normal mRNAs in vivo, but they are not absolutely required for decapping per se. These proteins include the *LSM1–7* complex, which is a presumed RNA-binding complex, Pat1p, which is of unknown biochemical properties, and Dhh1p, which is a member of the DEAD box ATPase family (12, 16, 27, 38, 44, 107). A related class of proteins that affects decapping, referred to as enhancers of decapping, includes Edc1p, Edc2p, and Edc3p (37, 63, 94). Lesions in these genes do not affect general decapping rates unless decapping is already compromised. However, Edc1–3p may be required for the decapping of specific mRNAs. For example, Edc3p specifically affects the decapping rate of the *RPS28a* mRNA (A. Jacquier, personal communication).

Two other classes of proteins are specifically required for the decapping of subsets of mRNAs. For example, specific mRNA-binding proteins, such as the PUF protein family members, can bind individual mRNAs and control their rate of decapping [(85), reviewed in (119)]. Another group of mRNA-specific regulators are Upf1p, Upf2p, and Upf3p, which are primarily required for the recognition and rapid decapping induced by NMD (72).

Important issues for understanding the process of decapping are (*a*) determining how proteins, such as Pab1p, inhibit decapping and (*b*) defining the roles of proteins in activating decapping in specific substeps in the decapping pathway.

OVERVIEW OF DECAPPING

In the following sections, we discuss the process of mRNA decapping and the role of the proteins involved. To introduce this section, we describe a working model for the process of mRNA decapping with three general points. First, a key step in decapping is the removal or functional inactivation of the poly(A). Second, for decapping to be efficient, mRNAs need to exit translation and assemble into specific mRNP structures, referred to as P bodies, where mRNA decapping occurs. Third, specific mRNA-protein complexes form and may function, in part, to recruit the decapping enzyme to the mRNA.

An overall, and currently speculative, model of mRNA decapping based on these three steps is diagrammed in Figure 3 and is discussed in more detail below. First, poly(A) shortening can occur while the mRNA is polysome associated and actively translating, although the process of deadenylation may also be connected to P-body formation (see below). Second, once the mRNA is deadenylated, defective translation initiation events, or stochastic variations in translation initiation, lead to the mRNA becoming ribosome free because of continued elongation in the absence of new initiation events. Third, the combination of being ribosome free and deadenylated leads to a series of mRNP transitions wherein the translation initiation complex disassembles and the decapping complex begins assembly, perhaps nucleated by Pat1p previously bound to the mRNA. Fourth, the assembly of the decapping complex leads to an mRNP structure that aggregates into P bodies wherein decapping complex assembly is completed, decapping occurs, and mRNP proteins are released. Key steps addressed in the following sections are the control of decapping by poly(A) tails, the relationship between translation and decapping, the function and role of P bodies in decapping, and the assembly of mRNA decapping complexes on mRNA.

POLY(A) TAILS AS AN INHIBITOR OF DECAPPING

Several observations indicate that the poly(A) tail can serve as an inhibitor of decapping. First, for those mRNAs examined, the time required before decapping takes place correlates to the time that is required for the mRNAs to deadenylate (34). Second, deadenylation still precedes decapping even when the rate of deadenylation has been either increased or decreased for an individual transcript

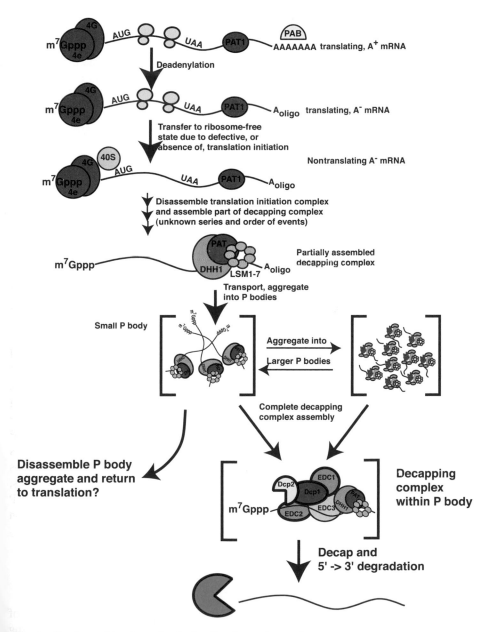

Figure 3   A working hypothesis for the decapping pathway.

by changes in the 3′ UTR sequence (34, 66). Third, decay intermediates produced by the 5′ decapping reaction only appear after at least some of the population of mRNA has undergone deadenylation (34, 79, 80). In addition, these decay intermediates possess only oligo (A)-length tails, suggesting that only deadenylated mRNAs are substrate for decapping (34, 79, 80). Finally, mRNAs with poly(A) tails are resistant to decapping in cell-free extracts in a Pab1p-dependent manner (121). The ability of the poly(A) tail to inhibit decapping adds to the function of poly(A) tails in promoting translation initiation [for review, see (55)] and suggests possible mechanisms by which decapping and translation are connected (see below).

The Poly(A)-Binding Protein Acts as a Block to Decapping

Several observations argue that the ability of the poly(A) tail to inhibit decapping is primarily mediated through Pab1p. First, it has been demonstrated in yeast that decapping occurs when the poly(A) tail has been shortened to an oligo (A) length of ~12 residues (34). This is approximately the minimum length required for Pab1p binding (93). Second, in *pab1* mutant strains, decapping is uncoupled from deadenylation (24, 76). In this case, intermediates in mRNA decay trapped by inhibiting 5′ to 3′ degradation in *cis* with strong secondary structures are produced as decapped mRNA fragments with long poly(A) species (24). This indicates that in the absence of Pab1p the requirement for prior deadenylation before decapping is relieved. It should be noted that poly(A) tails may also inhibit decapping independent of Pab1p. Cell extracts depleted of Pab1p were observed to still show some poly(A) tail inhibition of decapping (121), although the mechanism and in vivo significance of this observation remains to be established.

The block to decapping by Pab1p does not require the presence of the poly(A) tail. If Pab1p is tethered to an mRNA by the RNA-binding domain of the MS2 coat protein, its ability to block decapping becomes independent of its binding to poly(A) (26). This allows Pab1p to bind mRNAs via a MS2-binding sequence in the 3′ UTR, independent of poly(A). These mRNAs accumulate as full-length, capped, and deadenylated messages. This argues that the role of the poly(A) tail is to recruit Pab1p to the transcript and that mRNA-associated Pab1p mediates access to the cap by the decapping machinery. Given this, the process of deadenylation can be considered a process where the binding sites for Pab1p are removed from the transcript, relieving Pab1p-dependent inhibition of decapping.

How Does PAB Block mRNA Decapping?

One unresolved issue is the manner in which Pab1p inhibits decapping and leads to coupling of deadenylation and decapping. Given our current knowledge of mRNA turnover, there are three general, and overlapping, mechanisms by which Pab1p might inhibit decapping (cartooned in Figure 4). Given that no single mechanism is sufficient to explain the effect of Pab1p, it suggests that Pab1p

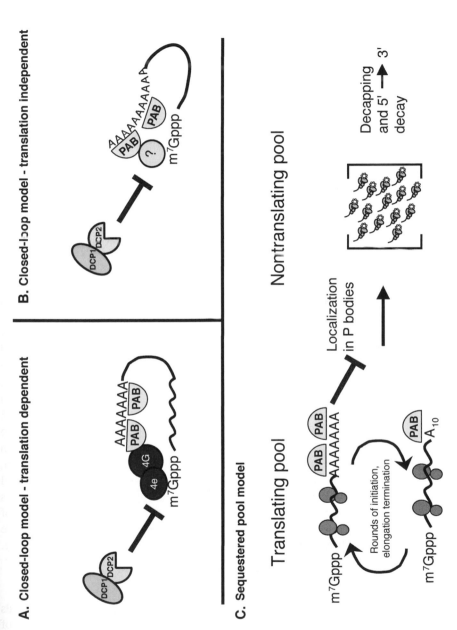

Figure 4   Possible mechanisms for the inhibition of decapping by Pab1p.

inhibits decapping by several mechanisms possibly connecting multiple processes, such as translation, localization, and turnover.

One likely mechanism for Pab1p inhibition of decapping is through interactions with the cap-binding complex (Figure 4A). The closed-loop model is the result of observations that in vitro Pab1p directly interacts with the cap-binding complex via protein-protein interactions with eIF-4G (104). This suggests a simple model in which Pab1p binds the poly(A) tail, then binds eIF-4G, which stabilizes eIF-4E on the mRNA. Such an interaction could inhibit decapping both by keeping the cap protected from Dcp1p/Dcp2p by eIF-4E and by increasing translation of the mRNA, which would have an indirect effect on blocking decapping (see below).

However, several observations suggest that the interaction of Pab1p with eIF-4G, though it may contribute, cannot solely explain the Pab1p inhibition of decapping. First, deletion of the eIF-4G interacting domain in Pab1p has no effect on the ability of tethered Pab1p to block decapping (26). This indicates that the region within the Pab1p that stabilizes mRNAs is distinct from the region that binds eIF-4G. Second, strains carrying mutations in translation initiation factors still exhibit dedeadenylation-dependent mRNA decapping even though translational initiation is severely compromised (95). Third, expression of a PAB1 gene from *Arabidopsis* in a *pab1Δ* yeast strain can complement the translation defect but not the mRNA turnover defects (8). This would suggest that the decay and translation functions of the Pab1p are distinct.

A second manner by which Pab1p could inhibit decapping is through interactions that are independent of translation initiation complex (Figure 4B). This is suggested by the observation that mutations in critical components of the translation initiation complex, including temperature-sensitive eIF-4E, eIF-4A, and deletion of eIF-4G, increase the rate of decapping. However, this decapping process is still dependent on prior deadenylation (95). One possible mechanism in this case would include protein-protein interaction through which Pab1p could directly inhibit the decapping enzyme. Alternatively, on the basis of in vitro studies, Pab1p may directly bind the cap and inhibit decapping through steric exclusion of the decapping enzyme (58).

A third manner by which Pab1p could inhibit decapping is by affecting the spatial location of transcripts within the cell (Figure 4C). For example, the observation that decapping can occur in P bodies suggests translating and degrading pools of mRNAs are spatially distinct (98). Thus, Pab1p might inhibit decapping by preventing mRNAs from entering P bodies or by promoting their rapid exit from P bodies back into the translating pool when Pab1p is bound. This is also consistent with the observation that mRNAs trapped in polysomes (i.e., by treating cells with cyclohexamide) are resistant to decapping and not found in P bodies (6, 98).

The possibility that Pab1p affects the spatial localization of mRNAs and thereby their fate resonates with other observations in the literature. For example, poly(A) has been shown to be preferentially colocalized with microfilaments, suggesting translating mRNAs might be tethered to the cytoskeleton [for review, see (33)]. In addition, a possible role for Pab1p in recruiting mRNAs out of P bodies is analogous to the recruitment of stored mRNAs into a translating pool by cytoplasmic adenylation. Such processes occur in a variety of contexts, which include regulation of the cell cycle, activation of maternal mRNAs, and localized translation in neurons [for reviews, see (90, 91)].

Decapping of Nonsense mRNAs: Does a Proper Translation Termination Event Establish the Pab1p-Dependent Block to Decapping?

One interesting possibility is that the ability of Pab1p to inhibit decapping is dependent on proper translation termination. This idea results from the observation that nonsense-containing mRNAs, which terminate translation prematurely, undergo rapid decapping independently of deadenylation (22, 78). Moreover, such nonsense mRNAs also show enhanced rates of deadenylation by Ccr4p (22), which is inhibited by Pab1p (108). These two consequences of premature translation termination can be explained by inactivation of Pab1p function.

Additional evidence that proper translation termination may affect Pab1p function comes from a variety of observations. For example, the ability of tethered Pab1p to block decapping requires translation of the mRNA (26). Moreover, recent results using tethered Pab1p indicate that a proper termination event requires only that Pab1p be in close proximity to the stop codon. The critical observations are first that tethering Pab1p to the 3'UTR of a nonsense *PGK1* mRNA at a substantial distance from the termination codon has no effect on the deadenylation-independent decay of this transcript (26). However, moving the binding site for tethered Pab1p close to the early stop codon restores the deadenylation dependence for decapping on this message (A. Jacobson, unpublished data). The implication of these results indicates that Pab1p, proximal to a terminating ribosome, is sufficient to initiate the 5' block to decapping; however, it does not address the requirements for maintenance of this inhibition. Interestingly, the yeast and mammalian polypeptide chain release factor (eRF3/GSPT) that functions in translation termination has been shown to physically interact with Pab1p (48, 49). This suggests that possibility that proper translation termination leads to enhancement of Pab1p function in inhibiting decapping, perhaps by affecting the dynamics of Pab1p-mRNA interactions such that Pab1p stays bound for prolonged periods.

DECAPPING REQUIRES A TRANSITION TO A NONTRANSLATION COMPETENT mRNP

Several observations argue that translation and decapping of the mRNA are in competition (Figure 5). For example, mutations in translation initiation factors that decrease translation rates increase the rate of decapping (95). Similarly,

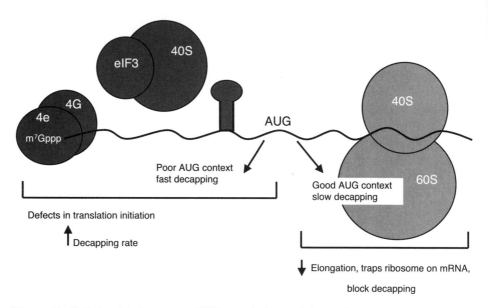

Figure 5   Relationship between mRNA translation and decapping rate.

decreasing translation initiation by placing strong secondary structures within 5′UTRs or with a poor AUG context leads to faster decapping (66, 79). Conversely, inhibition of translation elongation by mutations or by elongation inhibitors (e.g., cyclohexamide) significantly decreased the rates of decapping (6, 87). In addition, increased AUG recognition and an increased length of open reading frame (at least in some cases) led to slower rates of decapping (22, 66). These observations demonstrate a correlation wherein factors that increase an mRNA's association with ribosomes inhibit decapping rates, and factors that decrease an mRNA's association with ribosomes increase decapping rates (Figure 5).

A Critical mRNP Transition From Translation to Decay

An unresolved issue is how active translation of a transcript decreases its rate of decapping. One possibility is that protection from decapping may be due in part to a direct competition between translation and decay machineries for access to the 5′ cap structure. Translation is initiated via recognition of the 5′ m7GpppG cap by eIF-4E, part of the eIF-4F initiation complex. In vitro addition of purified eIF-4E inhibits decapping (96, 121). Consistent with this observation, addition of cap analog enhances decapping rate in mammalian extracts (40). In vivo loss of 4E function stimulates decapping, and lesions in eIF-4E suppress the decapping defect of a *dcp1–1* allele (96). Thus, a translating mRNP is protected from decapping in part by the tight association of the initiation complex with the cap. This conclusion has two clear implications. First, an a priori step in decay must be dissolution of the translating mRNP and deposition of the decapping machinery, although the specific steps of this exchange are unclear. Second, poorly translated mRNAs must have a faster rate of loss of the cap-binding complex because of some difference in their mRNP dynamics.

Additional evidence for a specific transition in mRNP organization leading to decapping has come from using coimmunoprecipitation to examine the interactions between the mRNA decay factors, the mRNA, the cytoplasmic cap-binding complex (eIF-4E and eIF-4G), and Pab1p. After deadenylation, Pat1p, the LSM complex, and the decapping enzyme (Dcp1/2p) can coimmunoprecipitate with the mRNA. In addition, this mRNP complex does not contain eIF-4E, eIF-4G, and Pab1p (106). Although it has not been directly tested, it is likely this decay mRNP also contains other decapping factors, which include Dhh1p, Edc1p, Edc2, Edc3p, and possibly Pop2p (27). This identifies a critical transition in mRNP organization following deadenylation that leads to decapping and degradation of yeast mRNAs.

The mechanics that cause this mRNP transition are unknown. However, at least some components of the decapping machinery may assemble on the mRNP prior to completion of translation, perhaps to initiate this transition. For example, Pat1p coimmunoprecipitates with Pab1p and eIF-4E in an RNA-dependent manner (106). This suggests that Pat1p can interact with the transcript while it is still translationally competent. Consistent with that view, Pat1p can be detected in polysome fractions (14). Given the association of Pat1p with mRNAs early in the decay process, Pat1p may function to facilitate the removal of the cap-binding complex and subsequent assembly of the decapping complex following deadenylation.

Do Elongating Ribosomes Protect mRNAs From Decapping?

Another possible overlapping manner in which translation and decapping could be related is through mRNAs having different properties when associated with elongating ribosomes. This view is suggested by the several observations. First, all treatments that decrease ribosome loading increase decapping (66, 79, 96). Second, all manipulations that increase ribosome loading decrease decapping rates (6, 22, 66, 87). Third, decapping can occur in P bodies (see below), and polysome-bound mRNAs are protected from decapping and excluded from P bodies (6, 98). On the basis of these observations, we propose a hypothesis wherein elongating ribosomes protect mRNAs from decapping, and decreases in translation initiation increase decapping either by direct recognition of a defective translation initiation event or by simply reducing the probability of having an elongating ribosome present on the mRNA. The relationship between ribosome dynamics and mRNP transitions leading to decapping is unclear but may be related to the cytoplasmic localization of the mRNA (see below).

DECAPPING CAN OCCUR IN SPECIALIZED CYTOPLASMIC STRUCTURES TERMED P BODIES

The proteins involved in mRNA decapping and 5′ to 3′ exonucleolytic decay are found in specific cytoplasmic foci, referred to as P bodies. In yeast, GFP-tagged Dcp1p, Dcp2p, Lsm1p, Pat1p, Dhh1p, and Xrn1p have been localized to P bodies (98). In contrast, Ccr4p, Ski7p (a cytoplasmic component of the exosome), Puf3p (an mRNA-binding protein), and translation factors are not concentrated or show small concentrations in P bodies (98; M. Brengues and R. Parker, unpublished observations). In mammalian cells, the Lsm1–7 complex, Xrn1p, Dcp1p, and Dcp2p have been localized to analogous structures (5, 53, 71, 110).

Two experimental observations in yeast cells argue that P bodies are specific sites wherein mRNAs can be decapped and degraded 5′ to 3′ (98). First, the size and number of P bodies varies in a manner correlating with the flux of mRNA molecules through the decapping step. For example, inhibiting mRNA decay at the deadenylation step in a *ccr4Δ* strain leads to a reduction in P-body size and number. Similarly, inhibiting decapping by deleting the *PAT1* gene or by adding cyclohexamide (6) leads to a reduction or loss of P bodies. In contrast, inhibiting the enzymatic steps of decapping or 5′ to 3′ exonuclease digestion leads to an increase in the size and number of P bodies (98). The second key observation is that mRNA decay intermediates, trapped either by the insertion of strong secondary structures or by deletion of the gene for the 5′ to 3′ exonuclease Xrn1p, can be specifically localized to P bodies. The simplest interpretation of these observations is that P bodies are sites of mRNA decapping and 5′ to 3′ exonucleolytic decay. However, because the mRNA decapping factors are also found distributed throughout the cytoplasm, decapping and degradation may also occur outside of P bodies. A reasonable working hypothesis is that mRNAs undergo a transition to an mRNP state (as discussed above) that is both an mRNP precursor to the decapping reaction and has the ability to aggregate into larger structures (Figure 3). Given this, mRNA decapping and decay might take place outside of large P bodies, possibly in smaller-scale aggregates of the same biochemical nature, but too small to be easily observed in the light microscope. It should be noted that the conservation of these structures from yeast to mammals suggests they have functional significance and that decapping is occurring in mammalian cells.

P bodies are dynamic structures and can vary in size and number under different conditions. One striking example of this phenomenon is that P bodies are affected by changes in the translation status of the cell. Specifically, when cells are treated with translation elongation inhibitors, which trap mRNAs in polysomes, P bodies disappear within 5 min (98). In contrast, when mRNAs are driven off polysomes by conditions reducing translation initiation, such as glucose deprivation (1), P bodies rapidly increase in number and size (D. Teixeira and R. Parker, unpublished results). These results are consistent with the polysome and P-body pool of mRNAs being spatially distinct. Moreover, a requirement for entry into P bodies for optimal rates of decapping could explain why elongating ribosomes protect mRNAs from decapping.

P bodies show some remarkable similarities to another form of mRNA containing cytoplasmic particles, referred to as a stress granule. Stress granules form in response to decreased translation initiation and contain poly(A)+ mRNA, translation initiation factors, specific RNA-binding proteins TIA and TIA-R, and 40S subunits [for review, see (57)]. Stress granules and P bodies are similar in their dynamics because both are increased by decreasing translation initiation and both decline when mRNAs are driven into polysomes. At this time, however, stress granules and P bodies appear to have distinct protein compositions and to physically differ. Intriguing issues for future work are (*a*) the relationship between stress granule formation and P bodies, (*b*) finding whether these represent different possible fates of nontranslating mRNAs, and (*c*) finding if there are different steps along a pathway of mRNA sequestration and degradation.

The Relationship Between Deadenylation/Decapping Control and P Bodies

An unresolved issue is the relationship between the processes that control deadenylation and decapping. Several observations argue that there are commonalities in the control of these two mechanisms and that their regulation may be intertwined. First, inhibition of translation initiation by mutations in translation initiation factors, poor AUG context, or insertion of strong secondary structures leads to increases in the rates of both deadenylation and decapping (66, 79, 95). Second, many sequence elements that accelerate mRNA degradation in yeast increase the rates of both deadenylation and decapping. These include the *MFA2* 3′ UTR, the MIE in the *MATα1* mRNA, and the binding of Puf3p to the *COX17* mRNA (24, 77, 85). Similarly, recognition of mRNAs as nonsense-containing by the Upf proteins triggers both accelerated deadenylation and decapping (22). Finally, Dhh1p physically binds to both the decapping enzyme and the deadenylase, possibly mediating the communication between these two events (27). One interpretation of these observations is that transcripts that are poorly translated have increased mRNP dynamics and cycle rapidly through states, possibly nontranslating pools of mRNAs, where they are subject to either mRNA deadenylation or decapping.

Given the similarity in factors affecting both deadenylation and decapping, one question is whether deadenylation can or needs to occur within P bodies. Multiple lines of evidence suggest that, though deadenylation probably occurs in P bodies to some extent, deadenylation can also occur outside of these structures. Evidence that P bodies are not required for deadenylation is twofold. First, treatment of yeast cells with cyclohexamide causes rapid loss of P bodies, but deadenylation of *MFA2* and *PGK1* transcripts proceeds normally (6; D. Muhlrad and R. Parker, unpublished observations). Second, Ccr4p, a component of the major yeast deadenylase, is not primarily found in P bodies when cells are grown in glucose media under standard conditions (98).

Despite these results, several other observations argue that deadenylation may be related to P bodies. First, under some conditions, Ccr4p, and additional components of the deadenylase complex, can be concentrated in P bodies (D. Teixiera and R. Parker, unpublished observations). Similarly, when the flux of mRNAs through P bodies is altered by inhibition of decapping, Ccr4p can be detected in P bodies. Interestingly, cyclohexamide treatment of mammalian cells inhibits the deadenylation of *c-myc* mRNA and raises the possibility that this specific type of deadenylation may be sensitive to P-body structure (68). These results suggest that an understanding of the mRNP dynamics and their spatial location should yield insight into both the control of deadenylation and decapping.

A Conserved Model for mRNA Decapping and Other mRNA Storage Events

The hypothesis that mRNAs enter a nontranslating state following deadenylation and prior to decapping is analogous to the storage of mRNA in numerous biological contexts where deadenylated mRNAs are translationally repressed prior to their later activation (120). Such storage occurs during early development, in neurons, and under conditions of stress in which the regulated repression and derepression of mRNA is paramount for controlling gene expression (50, 57, 59). In essence, the storage that occurs under these situations and the formation of a decapping complex may be manifestations of the same event (Figure 6).

There are three striking similarities between the formation of an untranslated mRNP prior to decapping and the storage of maternal mRNAs. First, in both cases, the mRNA is deadenylated and then enters a translational repressed state. Second, in both cases, similar proteins are involved. For example, the *Drosophila* homolog of Dhh1p, *Me31b*, is required for the masking of *bicoid* and *oskar* mRNA, and the Xenopus homolog, Xp54, is a major component of maternal mRNA storage particles and appears to directly repress mRNA translation (65, 73, 82). Third, both processes occur within distinct cytoplasmic granules (50, 98). For example, maternal mRNA storage is hallmarked by the formation of granular-like structures commonly referred to as polar granules, P granules, or germinal granules (120). In neurons, cytoplasmic structures, termed transport granules, contain stored mRNA that are transported to synaptic junctions (91).

This suggests the hypothesis that cells utilize a conserved mechanism by which mRNAs exit translation and enter a quiescent state with different fates of that quiescent mRNA utilized in different biological contexts. The primordial role of the decapping machinery may be that of mRNA storage following translational repression. In organisms such as yeast in which the decapping rate is high, translational repression following deadenylation would lead to destruction of the mRNA. In oocytes and neurons, deadenylation would lead to the storage of mRNAs as metabolically inert species until activated by cytoplasmic polyadenylation (42, 91). Given this, an understanding of the events that facilitate the assembly of the decapping complex may in fact provide insights into other areas in which mRNAs are posttranscriptionally regulated.

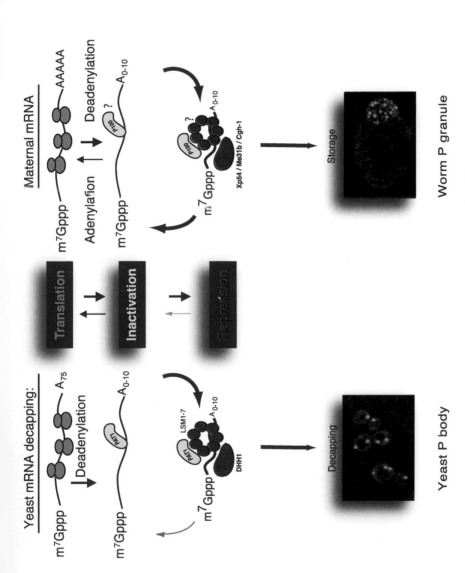

Figure 6   Possible relationship between mRNA decay and maternal mRNA storage.

Advantages of Sequestration of Decapping Factors

The observation that mRNA decapping at least primarily occurs within P bodies raises an obvious question: Why sequester mRNAs for decapping into discrete cytoplasmic foci? There are three potential reasons why sequestration of mRNAs and the decapping machinery would be beneficial to the cell. First, by sequestering the decapping machinery, or dictating it is only active when assembled into P bodies, the cell effectively partitions the degradation machinery away from the functional mRNAs. This partitioning could prevent the premature decapping of mRNAs not yet targeted for decay. In addition, by regulating delivery to the complex, the cell adds an additional layer of control, distinct from the enzymatic steps of degradation, which can be regulated. In this light, it is worth noting that most degradative processes are compartmentalized either within membrane organelles (e.g., the lysosome) or larger structures (e.g., the proteosome). This common compartmentalization must reflect functional advantages.

Another possible rationale for compartmentalization of these mRNAs in both stress granules and P bodies is that it reflects a fundamental buffering system within cells for maintaining a proper ratio of translation capacity to the pool of mRNAs that are translating (Figure 7). An excessive amount of mRNAs within the translating pool may compete for limiting translation factors. Thus, no mRNA would receive its full complement of proteins required for appropriate initiation, and overall translation rates of all mRNAs would decline. By sequestering nontranslating mRNAs in a manner that is not in competition with translation, the cell may be able to successfully translate the mRNAs remaining in the translation pool (Figure 7). In this view, the role of mRNA sequestration into stress granules and/or P bodies may be an ancient system for buffering translation capacity, which has then been co-opted by evolution for various other uses, such as the control of mRNA degradation and storage of mRNAs.

A third hypothesis for these particles is that translational repression and/or targeting a mRNA for decay may require two distinct phases (Figure 6). The first phase would simply be a slowing of translational initiation rate. This could occur quite passively by the loss of the poly(A) tail, or it may be an active consequence of binding a repressor protein. In the latter case, the ability to stay repressed is a function of the dissociation rate of the repressor with its target message. Because mRNP complexes can be highly dynamic, this particular state has the potential to reenter the translational pool, even if at a fairly low level. In order to prevent promiscuous expression of certain messages, a second state may exist in which repression is achieved and maintained. This may be a function of assembling mRNAs into large aggregated complexes. By assembling mRNA into large complexes, the dissociation rate of the binding no longer is the driving force influencing cycles of repression and derepression. Rather, it is the ability to deliver the mRNA to the repressive center (i.e., P bodies) that becomes the rate-limiting step.

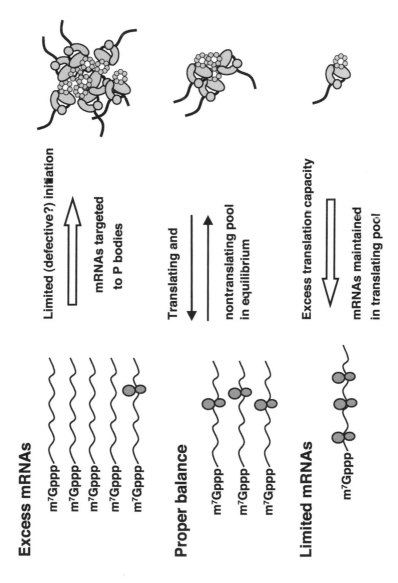

Figure 7   Model showing how compartmentalization can provide buffering to translational capacity.

A. Normal decapping

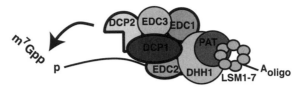

B. Nonsense-mediated decay

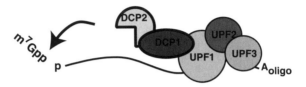

C. Regulation of RPS28a mRNA

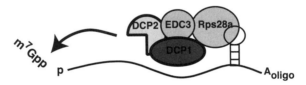

Figure 8   Illustration of three distinct decapping complexes.

ASSEMBLY OF DECAPPING COMPLEXES

Decapping of mRNAs also involves the assembly of different complexes of proteins on the mRNA that may function in vivo to enhance the interaction of the decapping enzyme with mRNA. In yeast, where several proteins affecting decapping have been identified, the majority of mRNAs appear to assemble a decapping complex consisting of the Lsm1 to Lsm7 proteins, Pat1p, Dhh1p, and Dcp1p/Dcp2p (Figure 8A) (7, 12, 16, 27, 38, 44, 105). In contrast, physical interactions between Upf1p and Dcp2p suggest that the process of NMD involves the assembly of a decapping complex wherein the decapping enzyme is recruited to the mRNA through interactions with Upf1p (Figure 8B) (45, 71). Another specialized decapping complex appears to occur on the *RPS28A* mRNA, where in an autoregulatory process, the Rps28a protein binds a stem-loop in the 3′ UTR and then interacts with Edc3p, which interacts with Dcp1p/Dcp2p and enhances their functions (Figure 8C) (A. Jacquier, personal communication). Thus, this set of physical interactions could recruit the decapping enzyme to the Rps28a. One should expect to see more such complexes identified as work in this area progresses. For example, the two small RNA-binding proteins, Edc1p and Edc2p, have been shown to bind RNA and recruit Dcp1/Dcp2p for decapping in vitro. This suggests that Edc1p and Edc2p may also nucleate mRNA-specific decapping complexes on specific mRNAs yet to be identified.

FUTURE DIRECTIONS

In the past few years, decapping has emerged as an important step in the process of mRNA degradation. Moreover, the enzymes and numerous proteins that modulate decapping have been identified. However, little is known about how each factor specifically functions. A deeper understanding into the biochemical properties of the decapping enzyme and its regulators should yield increased insight into the actual process of decapping and its control.

This basic understanding should also set the stage for a mechanistic understanding of how the decay rates for individual mRNAs are initially specified and then regulated in response to environmental cues. Progress in this area should come from understanding mRNA-specific binding proteins and their mechanisms of function. In addition, this area will be aided by a basic understanding of the various mRNP states and the dynamics that exist between a translating mRNA and an mRNA associated with the decapping complex. By understanding how a translating mRNP protects an mRNA from decapping, we shall also gain insight into the process of translational control, the role of Pab1p, and how Pab1p couples deadenylation and decapping.

A new challenge for the field is understanding the role of compartmentalization of mRNAs into discrete cytoplasmic structures, such as P bodies and stress granules. Despite the conservation of these structures, their biological significance is still largely unclear. This area is only in its infancy, and we should expect to see additional biological functions connected to these structures in the future. Insight into the function of these structures will require a biochemical and genetic analysis of their assembly/disassembly and relationships with each other and other cellular structures.

ACKNOWLEDGMENTS

We thank Drs. Allan Jacobson and Alain Jacquier for sharing results prior to publication. In addition, we thank the members of the Parker lab for sharing their time, talent, friendship, and advice over the course of this work. Support was received by grants from both HHMI and NIH.

The *Annual Review of Biochemistry* is online at http://biochem.annualreviews.org

LITERATURE CITED

1. Ashe MP, De Long SK, Sachs AB. 2000. *Mol. Biol. Cell* 11:833–48
2. Deleted in proof
3. Bai RY, Koester C, Ouyang T, Hahn SA, Hammerschmidt M, et al. 2002. *Nat. Cell Biol.* 4(3):181–90
4. Ball LJ, Jarchau T, Oschkinat H, Walter U. 2002. *FEBS Lett.* 513(1):45–52
5. Bashkirov VI, Scherthan H, Solinger JA, Buerstedde JM, Heyer WD. 1997. *J. Cell Biol.* 136(4):761–73
6. Beelman CA, Parker R. 1994. *J. Biol. Chem.* 269(13):9687–92
7. Beelman CA, Stevens A, Caponigro G, Lagrandeur TE, Hatfield L, et al. 1996. *Nature* 382:642–46
8. Belostotsky DA, Meagher RB. 1996. *Plant Cell* 8(8):1261–75
9. Beneken J, Tu JC, Xiao B, Nuriya M, Yuan JP, et al. 2000. *Neuron* 26(1):143–54
10. Bessman M, Frick D, O'Handley S. 1996. *J. Biol. Chem.* 271:25059–62
11. Binder R, Horowitz JA, Basilion JP, Koeller DM, Klausner RD, Harford JB. 1994. *EMBO J.* 13(8):1969–80
12. Boeck R, Lapeyre B, Brown CE, Sachs AB. 1998. *Mol. Cell. Biol.* 18(9):5062–72
13. Boeck R, Tarun S Jr, Rieger M, Deardorff JA, Muller-Auer S, Sachs AB. 1996. *J. Biol. Chem.* 271(1):432–38
14. Bonnerot C, Boeck R, Lapeyre B. 2000. *Mol. Cell. Biol.* (16):5939–46
15. Bousquet-Antonelli C, Presutti C, Tollervey D. 2000. *Cell* 102(6):765–75
16. Bouveret E, Rigaut G, Shevchenko A, Wilm M, Séraphin B. 2000. *EMBO J.* 19:1661–71
17. Bremer KA, Stevens A, Schoenberg DR. 2003. *RNA* 9(9):1157–67
18. Brown CE, Tarun SZ Jr, Boeck R, Sachs AB. 1996. *Mol. Cell. Biol.* 16(10):5744–53
19. Butler JS. 2002. *Trends Cell Biol.* 12(2):90–96
20. Cali BM, Anderson P. 1998. *Mol. Gen. Genet.* 260(2–3):176–84
21. Callebaut I. 2002. *FEBS Lett.* 519(1–3):178–80
22. Cao D, Parker R. 2003. *Cell* 113(4):533–45
23. Cao D, Parker R. 2001. *RNA* 7:1192–212
24. Caponigro G, Parker R. 1995. *Genes Dev.* 9(19):2421–32
25. Chen CY, Gherzi R, Ong SE, Chan EL, Raijmakers R, et al. 2001. *Cell* 107(4):451–64
26. Coller JM, Gray NK, Wickens MP. 1998. *Genes Dev.* 12(20):3226–35
27. Coller JM, Tucker M, Sheth U, Valencia-Sanchez MA, Parker R. 2001. *RNA* 7(12):1717–27
28. Couttet P, Fromont-Racine M, Steel D, Pictet R, Grange T. 1997. *Proc. Natl. Acad. Sci. USA* 94(11):5628–33
29. Cunningham KS, Dodson RE, Nagel MA, Shapiro DJ, Schoenberg DR. 2000. *Proc. Natl. Acad. Sci. USA* 97(23):12498–502
30. Custodio N, Carmo-Fonseca M, Geraghty F, Pereira HS, Grosveld F, Antoniou M. 1999. *EMBO J.* 18(10):2855–66
31. Daugeron MC, Mauxion F, Seraphin B. 2001. *Nucleic Acids Res.* 29(12):2448–55
32. Davis BM, McCurrach ME, Taneja KL, Singer RH, Housman DE. 1997. *Proc. Natl. Acad. Sci. USA* 94(14):7388–93
33. Decker CJ, Parker R. 1995. *Curr. Opin. Cell Biol.* 7(3):386–92
34. Decker CJ, Parker R. 1993. *Genes Dev.* 7:1632–43
35. Dower K, Rosbash M. 2002. *RNA* 8(5):686–97
36. Dunckley T, Parker R. 1999. *EMBO J.* 18:5411–22
37. Dunckley T, Tucker M, Parker R. 2001. *Genetics* 157:27–37
38. Fischer N, Weis K. 2002. *EMBO J.* 21(11):2788–97
39. Frischmeyer PA, van Hoof A, O'Donnel K, Guerrerio AL, Parker R, Dietz HC. 2002. *Science* 295:2258–61

40. Gao M, Wilusz CJ, Peltz SW, Wilusz J. 2001. *EMBO J.* 20(5):1134–43

41. Geerlings TH, Vos JC, Raue HA. 2000. *RNA* 6(12):1698–703

42. Gray NK, Wickens M. 1998. *Annu. Rev. Cell Dev. Biol.* 14:399–458

43. Hanson MN, Schoenberg DR. 2001. *J. Biol. Chem.* 276(15):12331–37

44. Hatfield L, Beelman CA, Stevens A, Parker R. 1996. *Mol. Cell. Biol.* 16:5830–38

45. He F, Jacobson A. 1995. *Genes Dev.* 9(4):437–54

46. Higgs DC, Colbert JT. 1994. *Plant Cell* 6(7):1007–19

47. Hilleren P, McCarthy T, Rosbash M, Parker R, Jensen TH. 2001. *Nature* 413(6855):538–42

48. Hoshino S, Imai M, Kobayashi T, Uchida N, Katada T. 1999. *J. Biol. Chem.* 274(24):16677–80

49. Hosoda N, Kobayashi T, Uchida N, Funakoshi Y, Kikuchi Y, et al. 2003. *J. Biol. Chem.* 278(40):38287–91

50. Houston DW, King ML. 2000. *Curr. Top. Dev. Biol.* 50:155–81

51. Hsu CL, Stevens A. 1993. *Mol. Cell. Biol.* 13:4826–35

52. Hui J, Reither G, Bindereif A. 2003. *RNA* 9(8):931–36

53. Ingelfinger D, Arndt-Jovin DJ, Luhrmann R, Achsel T. 2002. *RNA* 8(12):1489–501

54. Jacobs JS, Anderson AR, Parker RP. 1998. *EMBO J.* 17(5):1497–506

55. Jacobson A, Peltz SW. 1996. *Annu. Rev. Biochem.* 65:693–739

56. Jensen TH, Dower K, Libri D, Rosbash M. 2003. *Mol. Cell* 11(5):1129–38

57. Kedersha N, Anderson P. 2002. *Biochem. Soc. Trans.* 30:963–69

58. Khanna R, Kiledjian M. 2004. *Genes Dev.* In press

59. Kiebler MA, DesGroseillers L. 2000. *Neuron* 1:19–28

60. Koonin EV. 1993. *Nucleic Acids Res.* 21(20):4847

61. Korner CG, Wahle E. 1997. *J. Biol. Chem.* 272(16):10448–56

62. Korner CG, Wormington M, Muckenthaler M, Schneider S, Dehlin E, Wahle E. 1998. *EMBO J.* 17(18):5427–37

63. Kshirsagar M, Parker R. 2004. *Genetics* In press

64. Kumagai H, Kon R, Hoshino T, Aramaki T, Nishikawa M, et al. 1992. *Biochim. Biophys. Acta* 1119(1):45–51

65. Ladomery M, Wade E, Sommerville J. 1997. *Nucleic Acids Res.* 25:965–73

66. LaGrandeur T, Parker R. 1999. *RNA* 5:420–33

67. LaGrandeur TE, Parker R. 1998. *EMBO J.* 17(5):1487–96

68. Laird-Offringa IA, de Wit CL, Elfferich P, van der Eb AJ. 1990. *Mol. Cell. Biol.* 10(12):6132–40

69. Lewis BP, Green RE, Brenner SE. 2003. *Proc. Natl. Acad. Sci. USA* 100(1):189–92

70. Liu HD, Rodgers ND, Jiao X, Kiledjian M. 2002. *EMBO J.* 21(17):4699–708

71. Lykke-Andersen J. 2002. *Mol. Cell. Biol.* 22(23):8114–21

72. Maquat LE, Carmichael GG. 2001. *Cell* 104(2):173–76

73. Minshall N, Thom G, Standart N. 2001. *RNA* (12):1728–42

74. Mitchell P, Tollervey D. 2000. *Curr. Opin. Genet. Dev.* 10(2):193–98

75. Moore MJ. 2002. *Science* 298(5592):370–71

76. Morrissey JP, Deardorff JA, Hebron C, Sachs AB. 1999. *Yeast* 15(8):687–702

77. Muhlrad D, Parker R. 1992. *Genes Dev.* 6:2100–11

78. Muhlrad D, Parker R. 1994. *Nature* 370(6490):578–81

79. Muhlrad D, Decker CJ, Parker R. 1995. *Mol. Cell. Biol.* 15:2145–56

80. Muhlrad D, Decker CJ, Parker R. 1994. *Genes Dev.* 8:855–66

81. Mukherjee D, Gao M, O'Connor JP, Raijmakers R, Pruijn G, et al. 2002. *EMBO J.* 21(1–2):165–74

82. Nakamura A, Amikura R, Hanyu K, Kobayashi S. 2001. *Development* 128(17):3233–42
83. Nuss DL, Furuichi Y. 1977. *J. Biol. Chem.* 252(9):2815–21
84. Nuss DL, Furuichi Y, Koch G, Shatkin AJ. 1975. *Cell* 6(1):21–27
85. Olivas W, Parker R. 2000. *EMBO J.* 19(23):6602–11
86. Parker R, Song H. 2004. *Nat. Struct. Biol.* 11(2):121–27
87. Peltz SW, Donahue JL, Jacobson A. 1992. *Mol. Cell. Biol.* 12(12):5778–84
88. Piccirillo C, Khanna R, Kiledjian M. 2003. *RNA* 9(9):1138–47
89. Renfranz PJ, Beckerle MC. 2002. *Curr. Opin. Cell Biol.* 14(1):88–103
90. Richter JD. 1999. *Microbiol. Mol. Biol. Rev.* 63(2):446–56
91. Richter JD. 2001. *Proc. Natl. Acad. Sci. USA* 98(13):7069–71
92. Rodgers ND, Wang Z, Kiledjian M. 2002. *RNA* 8(12):1526–37
93. Sachs AB, Davis RW, Kornberg RD. 1987. *Mol. Cell. Biol.* 7(9):3268–76
94. Schwartz DC, Decker CJ, Parker R. 2003. *RNA* 9(2):239–51
95. Schwartz DC, Parker R. 1999. *Mol. Cell. Biol.* 19(8):5247–56
96. Schwartz DC, Parker R. 2000. *Mol. Cell. Biol.* 20(21):7933–42
97. She M, Decker C, Liu Y, Sundramurthy K, Parker R, Song H. 2004. *Nat. Struct. Mol. Biol.* In press
98. Sheth U, Parker R. 2003. *Science* 300(5620):805–8
99. Shyu AB, Belasco JG, Greenberg ME. 1991. *Genes Dev.* 5(2):221–31
100. Steiger M, Carr-Schmid A, Schwartz DC, Kiledjian M, Parker R. 2003. *RNA* 9(2):231–38
101. Stevens A, Maupin MK. 1987. *Arch. Biochem. Biophys.* 252(2):339–47
102. Stevens A. 1980. *J. Biol. Chem.* 255(7):3080–85
103. Stoeckle MY, Hanafusa H. 1989. *Mol. Cell. Biol.* 9(11):4738–45
104. Tarun S, Sachs A. 1996. *EMBO J.* 15:7168–77
105. Tharun S, He W, Mayes AE, Lennertz P, Beggs JD, Parker R. 2000. *Nature* 404(6777):515–18
106. Tharun S, Parker R. 2001. *Mol. Cell* 8(5):1075–83
107. Tijsterman M, Ketting RF, Plasterk RH. 2002. *Annu. Rev. Genet.* 36:489–519
108. Tucker M, Staples RR, Valencia-Sanchez MA, Muhlrad D, Parker R. 2002. *EMBO J.* 21(6):1427–36
109. Tucker M, Valencia-Sanchez MA, Staples RR, Chen J, Denis CL, Parker R. 2001. *Cell* 104:377–86
110. van Dijk E, Cougot N, Meyer S, Babajko S, Wahle E, Seraphin B. 2002. *EMBO J.* 21(24):6915–24
111. van Dijk E, Le Hir H, Seraphin B. 2003. *Proc. Natl. Acad. Sci. USA* 100(21):12081–86
112. van Dijk EL, Sussenbach JS, Holthuizen PE. 2001. *Nucleic Acids Res.* 29(17):3477–86
113. van Hoof A, Parker R. 2002. *Curr. Biol.* 12(8):285–87
114. van Hoof A, Parker R. 1999. *Cell.* 99(4):347–50
115. Vilela C, Velasco C, Ptushkina M, McCarthy JE. 2000. *EMBO J.* 19(16):4372–82
116. Wang ZR, Jiao XF, Carr-Schmid A, Kiledjian M. 2002. *Proc. Natl. Acad. Sci. USA* 99(20):12663–68
117. Wang ZR, Kiledjian M. 2001. *Cell* 107(6):751–62
118. Waterhouse PM, Wang MB, Lough T. 2001. *Nature* 411(6839):834–42
119. Wickens M, Bernstein DS, Kimble J, Parker R. 2002. *Trends Genet.* 18(3):150–57
120. Wickens M, Goldstrohm A. 2003. *Science* 300(5620):753–55
121. Wilusz CJ, Gao M, Jones CL, Wilusz J, Peltz SW. 2001. *RNA* 7(10):1416–24
122. Zhang Z, Carmichael GG. 2001. *Cell* 106(4):465–75
123. van Hoof A, Frischmeyer PA, Dietz HC, Parker R. 2002. *Science* 295(5563):2262–64

Annu. Rev. Biochem. 2004. 73:891–923
doi: 10.1146/annurev.biochem.73.011303.073933
Copyright © 2004 by Annual Reviews. All rights reserved
First published online as a Review in Advance on March 26, 2004

NOVEL LIPID MODIFICATIONS OF SECRETED PROTEIN SIGNALS

Randall K. Mann and Philip A. Beachy

*Department of Molecular Biology and Genetics and Howard Hughes Medical Institute,
Johns Hopkins University School of Medicine, Baltimore, Maryland 21205; e-mail:
rkmann@jhmi.edu, pbeachy@jhmi.edu*

Key Words Hedgehog proteins, cholesterol, autoproteolysis, Wnt proteins, protein acylation

■ **Abstract** Secreted signaling proteins function in a diverse array of essential patterning events during metazoan development, ranging from embryonic segmentation in insects to neural tube differentiation in vertebrates. These proteins generally are expressed in a localized manner, and they may elicit distinct concentration-dependent responses in the cells of surrounding tissues and structures, thus functioning as morphogens that specify the pattern of cellular responses by their tissue distribution. Given the importance of signal distribution, it is notable that the Hedgehog (Hh) and Wnt proteins, two of the most important families of such signals, are known to be covalently modified by lipid moieties, the membrane-anchoring properties of which are not consistent with passive models of protein mobilization within tissues. This review focuses on the mechanisms underlying biogenesis of the mature Hh proteins, which are dually modified by cholesteryl and palmitoyl adducts, as well as on the relationship between Hh proteins and the self-splicing proteins (i.e., proteins containing inteins) and the Hh-like proteins of nematodes. We further discuss the cellular mechanisms that have evolved to handle lipidated Hh proteins in the spatial deployment of the signal in developing tissues and the more recent findings that implicate palmitate modification as an important feature of Wnt signaling proteins.

CONTENTS

INTRODUCTION

Pattern formation during embryonic development is directed by several families of secreted protein signals. Whereas some of these signals are limited to local action at emerging tissue interfaces, others exert their influence over a range of distances. The patterning effects of these signaling proteins result from their action as morphogens, that is, by eliciting concentration-dependent responses in cells surrounding a localized site of signal production and release. The patterns of cell proliferation and differentiation resulting from the action of such a morphogen depend critically upon the response of target cells within a tissue and, of course, upon the tissue distribution of the active signal itself. We focus on lipid modification as a factor that critically influences the distribution and activity of Hedgehog (Hh) and Wnt signaling proteins, which act as morphogens in specifying the patterns of proliferation and differentiation in many tissues and structures during embryogenesis.

First discovered in *Drosophila*, the Hh family of secreted signaling proteins has been studied in a wide array of metazoan organisms. The cellular machinery that generates, distributes, transduces, and ultimately implements a cellular response to Hh signals is deployed repeatedly throughout development, and this pathway directly or indirectly influences the development of many tissue and organ systems in vertebrates [for recent reviews, see (1, 2)]. Wnt proteins similarly act in the embryonic patterning of many tissues and structures [see (3–5)]. More recently, the Hh and Wnt signaling pathways are emerging as playing homeostatic roles in the maintenance of postembryonic tissues. Such roles include stem cell maintenance in particular tissues (6–10) and possible participation in tissue repair in response to injury (11). Finally, both the Hh and Wnt pathways have emerged as playing an important role in a growing number of types of cancer (11–13) in which a general feature is unregulated activation of these pathways [see (14–16)].

The normal and abnormal function of these pathways is associated with diverse biological phenomena of considerable importance and interest. Our discussion, however, will focus on the distribution and activity of these signaling proteins, which are critical determinants of their biological effects, and in particular on the mechanisms by which these proteins are lipid modified and how these modifications impact their biological function. Well-known mechanisms that influence the distribution and activity of extracellular protein signals include interactions with the cell matrix (e.g., FGF) (17), release in latent form (e.g., TGFβ) (18), or interaction with secreted inhibitors (e.g., Wnts and BMPs) (19, 20). The lipid modifications undergone by Hh proteins generally are considered to be membrane anchoring, raising the question as to how signaling responses in distant cells are elicited. We discuss the following:

- the biochemical mechanisms of these modifications, which include auto-processing and cholesterol modification of the Hh signal, and further acylation of the Hh protein;

- the more recent finding of Wnt protein acylation, which may occur by a mechanism related to that of Hh protein acylation;

- the relationship between Hh and self-splicing proteins, a well-studied class of autoprocessing proteins, as well as the Hh-like proteins of nematodes; and

- finally, the dual lipid modification of Hh proteins in the context of lipid cell biology and metabolism and, largely in the context of *Drosophila* development, how these hydrophobic modifications target Hh protein to a multicomponent system for distribution of the signaling activity in its appropriate pattern within tissues.

BIOGENESIS OF HEDGEHOG SIGNALING PROTEINS

Newly synthesized Hh proteins undergo a series of posttranslational processing reactions within the secretory pathway that result in the formation and cell surface presentation of the species active in signaling. Although elements of the reaction mechanisms employed are also represented in the metabolism of other proteins, Hedgehog family members are the only examples of signaling proteins known to be covalently modified by cholesterol. In this section, we describe the biogenesis of the Hedgehog signal as compared to that of other autoprocessed proteins. In addition, we review the biochemistry and function of Hedgehog palmitoylation.

Autocatalytic Processing and Cholesterol Modification of Hedgehog Proteins

Whereas many features of Hh autoprocessing [reviewed in (21)] were elucidated from studies of the *Drosophila* protein, the biochemical and functional mechanisms likely apply to Hh proteins from all species. Following cleavage of an amino-terminal signal sequence upon entry into the secretory pathway, the Hh protein undergoes an autocatalytic processing reaction that involves internal cleavage between Gly-Cys residues that form part of an absolutely conserved Gly-Cys-Phe tripeptide (Figure 1A) (22, 23). The amino-terminal product of this

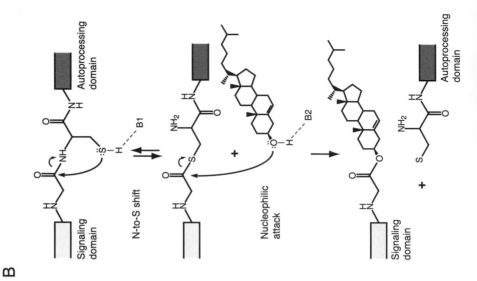

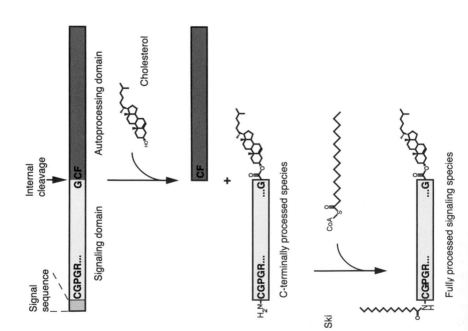

cleavage receives a covalent cholesteryl adduct (24) and is the species active in signaling (23, 25–31) (see Figure 1*A*). Constructs encoding Hh proteins truncated at the normal site of internal cleavage produce proteins that can have signaling activity, but studies in *Drosophila* have demonstrated that such proteins are not appropriately restricted spatially and therefore cause gross mispatterning and lethality in embryos (32). The cholesteryl moiety not only restricts spatial deployment of the mature signal via insertion into the lipid bilayer of the cell membrane, thus influencing the pattern of cellular responses in developing tissues, but it may also function as an essential molecular handle for proper intracellular and extracellular trafficking and localization of the signal (see below). The autoprocessing reaction thus is required not only to release the active signal from precursor but also to specify the properties of this signal within cells and tissues.

The autoprocessing reaction is mediated by the carboxy-terminal domain of the Hh precursor, which has no known additional function. This reaction proceeds by two sequential nucleophilic displacements (Figure 1*B*): The first of which is a rearrangement to replace the main chain peptide linkage between Gly-Cys with a thioester involving the Cys side chain (24, 32). The second step of the Hh autoprocessing reaction involves attack upon the same carbonyl by a second nucleophile, displacing the sulfur and severing the link between Hh-N and Hh-C. The requirement for a second nucleophile in vitro can be met by a high concentration either of a thiol-containing molecule or of another small molecule with nucleophilic properties at neutral pH; these small nucleophiles can be shown to form covalent adducts to the amino-terminal product of the in vitro cleavage reaction (32). This second nucleophile can be provided by the thiol of a cysteine-initiated peptide, leading to a thioester linkage that undergoes further rearrangement via an S-to-N shift to form an amide (32). This reaction represents a variation on the theme of synthetic thioesters used for chemical ligation of peptides (33). Related protein ligation strategies are now used for synthesis of proteins containing specific modified or unnatural amino acids [see (34)].

The importance of autocatalytic processing in biogenesis of active Hh proteins is highlighted by the types of missense mutations that occur in the *Drosophila hh* gene (23). One class of mutations affects amino-terminal coding sequences without affecting the ability of the protein to undergo processing, and these

←

Figure 1 Processing of Hedgehog proteins. (*A*) *Hedgehog* genes encode precursor polypeptides of ~45 kDa that undergo both N-terminal signal sequence trimming and acylation as well as internal proteolysis at a conserved sequence. Endoproteolytic cleavage at the GCF sequence is catalyzed by the processing activity associated with the C-terminal domain and produces an ~19 kDa segment with which all known signaling activities are associated. During cleavage, the signaling domain is modified at its carboxy-terminal glycine by cholesterol; the N-terminal cysteine also becomes palmitoylated (see text for details). (*B*) Mechanism of Hedgehog endoproteolysis (see text for details).

alterations affect either the secretion or the activity of the signaling domain. A second class of mutations comprises alterations of the carboxy-terminal domain without changing amino-terminal sequences, and these alterations can be shown to affect processing, thus demonstrating the requirement for processing in release of the active signal. Missense mutations in the human *Shh* gene associated with holoprosencephaly (see below) also can be classified in this manner, with alterations either in the amino-terminal signaling domain or the carboxy-terminal processing domain (35, 36).

Whereas cholesterol is a prominent constituent of animal cell membranes (as much as 30% of the plasma membrane lipid content in some tissues) and was identified as the modifying lipid in cell-derived Hh-Np (24), other steroidal compounds can substitute for cholesterol in Hh processing reactions performed in vitro (37). Among the compounds tested (see Figure 2), structural variables included the orientation and availability of the 3β hydroxyl, additional specific hydroxylations, the olefin of the cholestene backbone motif, as well as presence and structure of the isooctyl side chain. In this semiquantitative analysis, it is clear that the most important structural feature is the C3 hydroxyl moiety. Not only must the group be free of esterified adducts (no activity with cholesteryl acetate), but there is an absolute requirement for the β orientation of the alcohol (no activity with epicholesterol). Whereas the side chain of these sterols is not essential for activity (see 5-androsten-3β-ol), hydroxylations within the chain can decrease a compound's effectiveness (20-, 22-, 25-hydroxycholesterol). Hydroxylations at other positions within the ring can have no deleterious effect (7β-hydroxycholesterol) or can reduce activity (19-hydroxycholesterol). Finally, changing the cholestene to either a cholestan (coprostan-3-ol) or an ergostatrien (ergosterol) resulted in decreased activity, but this indicates that the olefin at C5 of the backbone is not essential and that conjugation does not prevent the esterification reaction.

The availability of alternative sterols, supplied nutritionally or biosynthetically, and the lack of stringent selectivity of the *Drosophila* Hh processing domain raise the possibility that hedgehog signaling proteins may be modified in vivo, not only by cholesterol, but by other endogenous steroidal nucleophiles. Although this remains to be demonstrated experimentally, it is worth noting that among Hh orthologs there is considerable sequence diversity within the subdomain that in *Drosophila* is known to be required for mediating cholesterol addition (see below).

Evidence for Additional Cholesterol-Modified Proteins

In animal cells, the processed form of overexpressed Sonic hedgehog signaling protein can be detected readily in simple metabolic labeling experiments using radioactive cholesterol (24). In addition to the strong signal due to incorporation within Shh-Np, however, several additional proteins can be detected. Although these proteins and the nature of the chemical linkage remain uncharacterized, this observation suggests that cholesterol modification of polypeptides, perhaps by

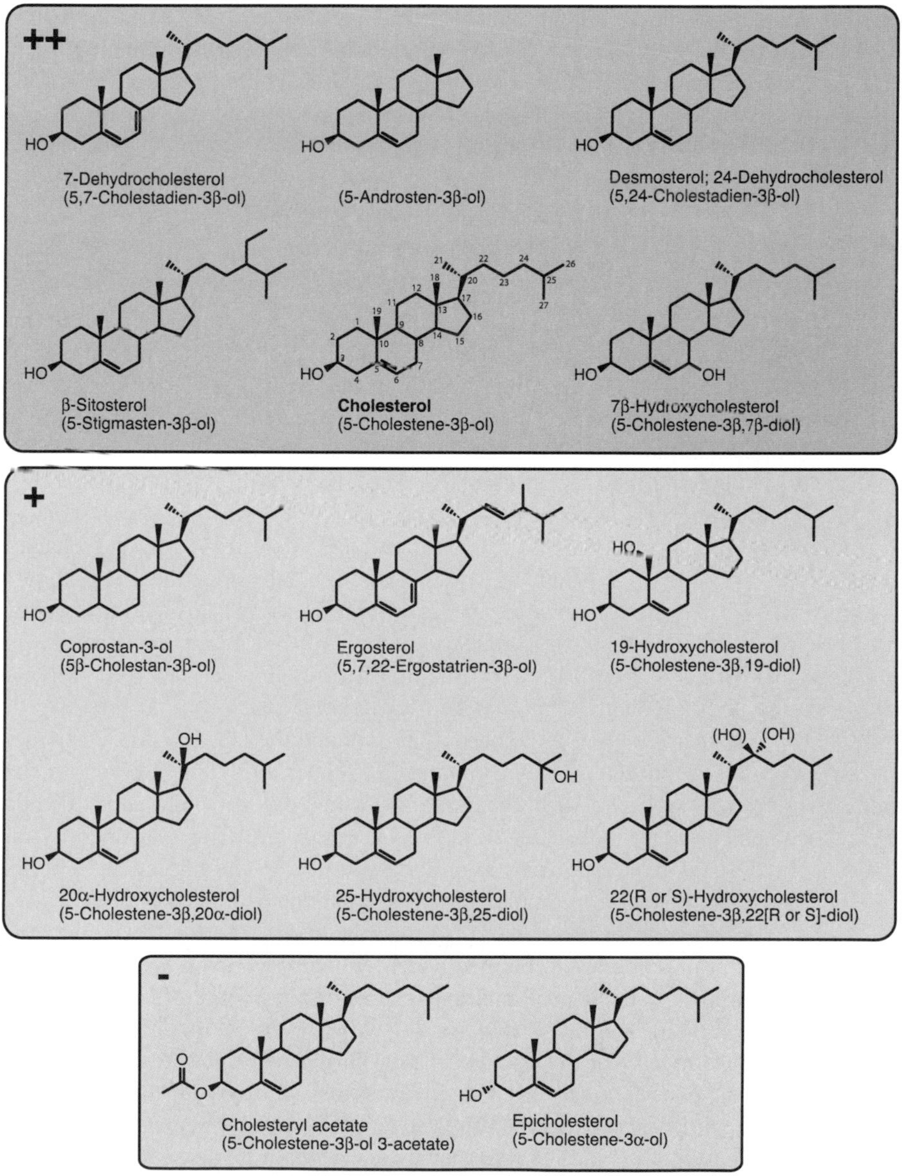

Figure 2   Sterol selectivity in the Hedgehog processing domain. Using a semiquantitative in vitro processing assay (37), the depicted sterols were tested for their ability to substitute for cholesterol in the processing domain-mediated transfer reaction. Full activity (+ +) is observed for compounds that most closely resemble cholesterol, with 5-androstene-3β-ol perhaps representing the minimum structural requirement for a fully active substrate. Various structural changes (see text) allow for partial activity (+), whereas esterification or inversion (to the α orientation) of the 3β hydroxyl completely block participation (-).

esterification, is not unique to the Hh proteins, and it may also be a employed as a means of directing other proteins to membranes or other hydrophobic targets. Alternatively, cholesterol labeling of these proteins could be caused by capture of oxidized sterol intermediates, as suggested for aldehyde dehydrogenase class 1 in bovine lens epithelial cells (38).

Thioesters as Intermediates in Protein Modification

The use of a Cys-derived thioester as an intermediate is a theme common to several other acyl transfers that result in covalent modifications of proteins [reviewed by (39) and (40)]. Following formation of the initial thioester in these systems, the acyl portion of the thioester (the acceptor, corresponding to Hh-N) can receive the final modification directly or alternatively may be transferred to other thiols in one or more subsequent steps before receiving the final modification. The ubiquitin cascade represents such a reaction with multiple intermediates, whose role is to attach ubiquitin to proteins destined for degradation by the proteasome (41). The acyl group for these thioesters is supplied by the carboxy-terminal Gly of ubiquitin, and the thiols come from Cys side chains in three distinct classes of enzymes. The first of these, E1, forms the initial thioester in an ATP-consuming reaction. Then, through transthioesterification reactions, the ubiquitin forms thioesters sequentially with E2 and E3 enzymes before final transfer to the ϵ amine of a Lys side chain. The protein receiving ubiquitin in the resulting amide linkage is thus marked for degradation.

The α2-macroglobulin proteinase inhibitors and the C3, C4, and C5 complement proteins represent members of an ancient superfamily that use an intrachain thioester as a "spring loaded" functionality, which can be triggered for covalent attachment to target molecules (42). The intrachain thioester is formed by thiol attack of a Cys side chain on the amido group of a Gln side chain. The final adducts in the case of the complement proteins are nucleophiles on the surface of cells to be targeted for lysis. In the α2-macroglobulin case, the final adduct is a nucleophile on a protease to be inactivated, which is targeted to α2-macroglobulin through the presence of multiple cleavage sites for proteases of various specificities.

In the examples just discussed, the acyl group contributing to the thioester intermediate derives either from another protein or from an amino acid side chain. In contrast, the acyl group in the Hh thioester intermediate is linked to a main chain carbonyl, and the thioester therefore replaces an amide bond within the peptide backbone. Other proteins likely to utilize main chain ester or thioester intermediates in autoprocessing reactions include prohistidine decarboxylase (43) and members of the Ntn hydrolase family that are processed by an intramolecular mechanism (44, 45). The Ntn (N-terminal nucleophile) family hydrolases, which include proteases active in the proteasome, are autoprocessed with internal cleavage, leaving the active site nucleophile as the amino-terminal residue. The role of these reactions appears to be activation of a precursor protein and takes place without net addition of a modifying adduct. There is no evidence of any evolutionary relationship between Hh autoprocessing domains and either prohistidine decarboxylase or Ntn hydrolase proteins.

Ester Intermediates in Proteins Containing the Hint Domain

Two other groups of proteins that are evolutionarily related to Hh proteins are the self-splicing proteins and a group of novel nematode proteins containing Hh-C-like sequences. The self-splicing proteins undergo a reaction in which an internal portion of the protein, termed an intein, is excised and amino- and carboxy-terminal flanking regions, termed exteins, are ligated to form the mature protein (46, 47). Inteins are found inserted into a wide variety of archaeal, bacterial, chloroplast, and yeast proteins. The intein portion mediates the protein splicing reaction and typically also contains an endonuclease thought to act at the DNA level in mediating movement of intein coding sequences. Similar to Hh auto-processing, the protein splicing reaction is initiated by intramolecular attack of a hydroxyl or thiol upon the preceding carbonyl, and the resulting ester or thioester intermediate replaces the peptide bond at the amino-terminal extein/intein boundary (48) (Figure 3). Unlike Hh proteins, the second nucleophilic attack in the protein self-splicing reaction involves the side chain of another Ser or Cys residue several hundred residues downstream. The resulting branched protein intermediate ultimately resolves into the ligated exteins and the free intein protein (Figure 3).

Nematode proteins with Hh-C-like sequences were identified by searching for homology within the *Caenorhabditis elegans* genomic sequence database. Within this genome, the sequencing of which is now complete (49), 10 putative proteins with homology to the Hh-C autoprocessing domain have been identified (32, 50–52; R. Mann, X. Wang, and P.A. Beachy, unpublished data). As in the Hh family, the Hh-C-like domains, with one exception, are located at the carboxy termini of these proteins and are preceded by an amino-terminal domain bearing a signal sequence. These nematode proteins resemble each other more than they do any other database sequence and can be grouped into three families based on their amino-terminal sequences: Wart, Ground, and a third family identified by the trivial name of its single member, M110. The structures of these proteins suggest the possibility that they are secreted and undergo autoprocessing; a preliminary study of one family member in *Drosophila* cultured cells indeed demonstrates cleavage at the junction between amino- and carboxy-terminal domains (24). Whereas there is no obvious similarity between the amino-terminal domains of the nematode and Hedgehog protein families, a short, shared sequence motif has been noted (52) that, in the context of cysteine conservation at flanking sites nearby, suggests they may share a common ancestor. In the absence of known structural conservation, however, the proposed evolutionary homology must be considered provisional.

The level of amino acid sequence identity between the nematode and Hh processing domains ranges from 24% to 32% in a region approximately corre-

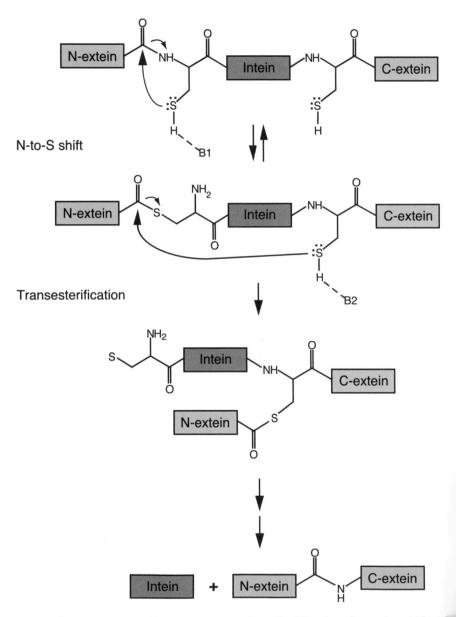

Figure 3 Mechanism of intein autoprocessing. The Hint domain-catalyzed N-to-S shift follows that of Hedgehog. However, the subsequent nucleophilic displacement is intramolecular, resulting in a branched polypeptide intermediate. This structure resolves into an excised intein sequence and a ligated amino- and carboxy-terminal segment [see (40, 46, 47) for recent reviews and reference to the intein database InBase].

sponding to the amino-terminal 2/3 of Hh-C. This same region of Hh-C also can be aligned with inteins, although the alignment is complicated by the presence of sequences corresponding to the endonuclease domain as well as a DNA recognition region domain that is thought to aid in DNA binding (51, 53, 54). The strongest evidence of a common evolutionary origin for these protein families is the presence of a domain with a common fold in the crystal structures of Hh-C and the 454 residue intein protein PI-SceI (Figure 4A) (51, 55). Remarkably, both the endonuclease (Endo) and DNA recognition region (DRR) domains are inserted into peripheral loops of the common domain with little apparent effect upon its three-dimensional fold. Structure-based alignment of Hh-C and PI-SceI intein sequences (Figure 4B; T. Hall and D. Leahy, personal communication), with intein endonuclease and DRR regions removed, reveals a low level of amino acid identity (15/94 residues aligned, ~16%), but the root-mean-square distance for Cα positions is 1.401 Å, and most of the residues known to be essential for Hh-C processing activity are conserved. The crystallized Hh fragment contains the 151 amino-terminal residues of Hh-C of which the first 145 residues are well ordered in the crystal structure; these residues correspond to the region conserved in the nematode proteins (9 identities among various Hh and the Wart/Ground proteins and most other positions featuring similar residues) [see (52)]. This domain alone suffices for thioester formation, as indicated by the ability of a Hh protein truncated after this point to undergo cleavage in the presence of DTT (51; X. Wang, G. Seydoux, and P.A. Beachy, unpublished information), and this domain has been referred to as the Hint module (Hedgehog, intein).

Although the Hint module in Hh-C suffices for the first step of autoprocessing, at least some part of the 63 carboxy-terminal residues missing in the crystallized fragment is required for the second step of cholesterol addition (51). Because of its apparent role in sterol addition, this 63 residue region is referred to as SRR, for sterol recognition region. Whereas residue identities between Hh SRRs and corresponding sequences within the nematode family are limited, the use of a short Hint sequence "anchor," as well as a gap to accommodate sequence insertions, particularly within the Shh and *Drosophila* proteins, reveals a significant degree of sequence similarity, most notably in the spacing of hydrophobic clusters (see Figure 5). Because of their corresponding position with respect to the SRR of Hh, the sequences in these nematode proteins that extend carboxy-terminal to the Hint domain are tentatively designated ARR, for adduct recognition region. Despite the above-mentioned similarities, the overall sequence diversity between the SRR of Hh proteins and the ARR regions of nematode gene family members raises the possibility that molecules other than cholesterol may participate in the processing reaction and form novel protein-modifying adducts.

From the structure of the Hint modules in Hh-C and the PI-SceI intein, and from sequence relationships between these proteins and the nematode proteins, a plausible evolutionary history can be constructed in which an ancestral Hint module evolved and gave rise to all three protein groups (see Figure 6). The evolution of the Hint module is revealed by pseudo twofold symmetry with

A

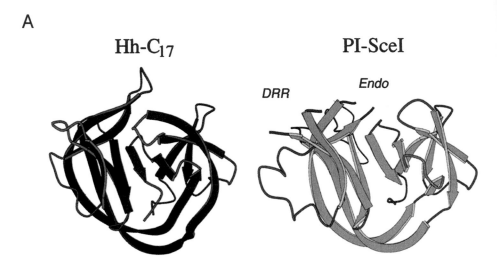

Hh-C$_{17}$ **PI-SceI**

DRR   *Endo*

15/94 matched residues identical
1.40 Å rmsd Cα

Figure 4 (*A*) Hint domain structures of Hh and an intein protein, PI-SceI. Crystal structures of both the Hedgehog processing domain (Hh-C$_{17}$) (51) as well as the PI-SceI autoprocessing protein (55) have been solved, and the Hint domain ribbon diagrams are presented for comparison. Structural similarities are easily detectable after removal of the DRR and endonuclease (Endo) domains of PI-SceI [see (51) for discussion of structural similarities as well as the interesting pseudo twofold axis of symmetry found in both structures]. (*B*) Structure-based alignment of Hint domain sequences from Hh and PI-SceI. The alignment of corresponding structures of the two Hint domains and their associated sequences allows a more accurate assessment of relatedness and reveals a greater degree of conservation. The analysis demonstrates that 94 amino acids are in matched positions, that ~16% of matched residues are identical, and that similar residues are found at most of the other positions. The root-mean-square distance for Cα positions is 1.401 Å.

superimposable subdomains in the crystal structure of the Hh-C protein, which suggests that the ancestral Hint domain arose by gene duplication (see Figure 4*A*) [see (51)]. The duplicated subdomains are interlinked by extended loop-like secondary structure elements that mirror each other in associating primarily with the bulk of the other subdomain, suggesting that following duplication, these secondary structure elements exchanged interactions with their own subdomain for similar interactions with the other subdomain [a "loop swap"; see (56) and Figure 6]. Following establishment of the Hint module, the ancestral intein evolved in one branch by insertion of an endonuclease into a Hint domain and by adjustment (or preservation) of the chemistry to insure that the second nucleophilic attack is made intramolecularly by the side chain of a downstream residue. In a second branch, Hh proteins were formed by association of a Hint domain

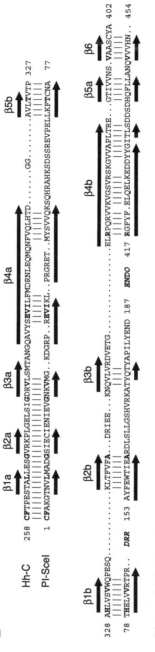

Figure 4 *(Continued)*

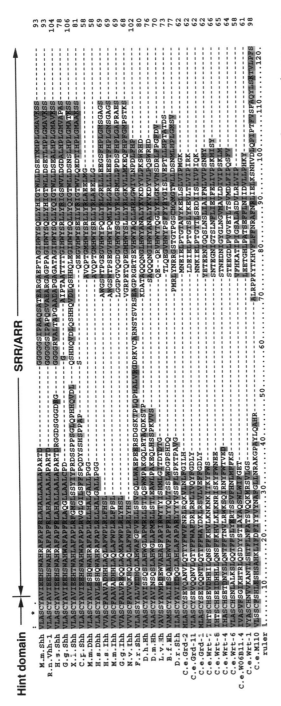

Figure 5 Sequence comparison of Hedgehog SRRs and the ARRs of nematodes. Using several residues of conserved sequence as an alignment anchor, the extreme C-terminal sequences of Hedgehog and nematode processing domains (i.e., the SRR and ARR domains, respectively) were aligned using the ClustalX algorithm (137). Species represented are, alphabetically, B.f., *Branchiostoma floridae* (Amphioxus); C.e., *Caenorhabditis elegans*; C.p., *Cynops pyrrhogaster*; D.h., *Drosophila hydei*; D.m., *Drosophila melanogaster*; D.r., *Danio rerio*; F.r., *Fugu rubripes*; G.g., *Gallus gallus*; H.s., *Homo sapiens*; L.v., *Lytechinus variegatus*; M.m., *Mus musculus*; N.v., *Notophthalmus viridescens*; R.n., *Rattus norvegicus*; and X.l., *Xenopus laevis*. Despite the lack of striking sequence conservation in corresponding segments from nematodes and the Hh proteins, some similarities, most notably the clustering and spacing of hydrophobic residues, are clear. This partial conservation, together with the ability of the *Drosophila* processing domain to utilize alternative steroidal compounds in the in vitro reaction, suggests that other sterols could be used in vivo to modify processed proteins. See Aspock et al. (52) for alignments of entire processing domains of Hedgehog and nematode proteins. Abbreviations are Shh, Sonic hedgehog; Ihh, Indian hedgehog; Dhh, Desert hedgehog; Grd, Ground; and Wrt, Wart.

with the amino-terminal domains of the Hh and nematode proteins. The sequence of events leading to formation of these proteins is not known. One possibility is that the Hint and SRR modules may have been assembled into a cholesterol transfer unit prior to association with the Hh signaling domain; alternatively, the Hint module might have been inserted within a preassembled protein comprising a signaling domain and the SRR precursor. In the second scenario, the SRR precursor in the preassembled protein might have served some function related to sterol recognition, such as membrane association. Similarly, several scenarios are possible in assembly of the nematode proteins. The possibility also exists that additional proteins will be found in which the Hint module initiates novel splicing or transfer reactions.

Amino-Terminal Acylation of Hedgehog Signaling Proteins

A second lipophilic modification of the Hedgehog signaling protein was more recently found to occur on a large proportion of the amino-terminal signaling domain of human Sonic Hedgehog upon expression in either insect or mammalian cell lines (57). This additional modifying adduct is a fatty acid, usually palmitate, and is found in an amide linkage with the amino-terminal cysteine that is exposed by signal sequence cleavage (see Figure 1A). Because this Cys residue is the first of a pentapeptide, CGPGR, that is widely conserved among species, there is a possibility that these residues and others nearby may constitute an important determinant for the palmitoylation reaction. The fatty acylation is proposed to occur via a thioester intermediate involving the side chain of the amino-terminal cysteine, followed by a spontaneous rearrangement to form the amide. The efficiency and specificity of this modification appears to depend in part upon prior cholesterol modification because the level of acylation is reduced and the types of modifying fatty acids are varied when the Shh protein is produced from a truncated construct lacking the processing sequences. When Shh is expressed in cultured mammalian cells using a moderately active (nonviral) promoter, we have observed that the majority of Shh-Np is found in a doubly lipidated form (58; R.K. Mann and P.A. Beachy, unpublished information), providing additional evidence that this is the predominant form of the signaling molecule in vivo.

Whereas N-terminal palmitate is dispensable in some assays of Hh signaling activity, it is now clear, from both animal models and cultured cell-based in vitro assays, that this modification critically contributes to full signal potency. In the responsive chondrogenic cell line C3H10T1/2, recombinant versions of human Shh-N featuring fatty acyl adducts of intermediate chain length were found to be 40- to 160-fold more potent than the corresponding cysteine-initiated protein lacking an N-terminal adduct (59). Replacing the initiating cysteine with the structural cognate serine produces an even weaker signal (10-fold lower than unmodified) and removal of 5–10 N-terminal residues severely impairs signaling potency (>500-fold lower) (60). These results suggest the possibility that some degree of acylation can occur after addition of protein to target cells if the cysteine target for acylation remains in place. Notably, it was also observed that replacing cysteine with hydrophobic residues alone conferred enhanced potency (up to eightfold greater than unmodified) (59). This result, combined with others involving a variety of nonfatty acyl N-terminal adducts indicates that the enhancement of signaling activity can be attributed to a general hydrophobic effect rather than exclusive specificity for palmitate or even long-chain fatty acids. It is also interesting that, at least in this assay system, the fatty acylated versions of recombinant Shh lacking C-terminal cholesterol modification feature potencies comparable to that of the doubly-lipidated Shh-Np (produced from the full-length gene in metazoan cell lines).

The importance of Hedgehog palmitoylation has also been demonstrated in several animal models of development wherein modification site mutants were found to have little or no activity in vivo. Lee et al. (61) found that a transgene encoding a serine substitution at the palmitoylation site in *Drosophila* Hedgehog not only nullified its ability to complement loss-of-function alleles responsible for embryonic and larval defects but also caused a dominant negative effect with wild-type endogenous Hedgehog. A separate study (62) employing ectopic expression in mouse as well as tissue explant assays confirmed the necessity of N-terminal acylation and demonstrated that mutation of the acylation site caused a loss of Shh signaling activity, consistent with cultured cell-based in vitro observations. Furthermore, a *Shh* mutant allele encoding a truncated protein results in reduced signaling and defective limb patterning upon expression within the developing limb (63). The precisely truncated Shh-N protein produced by this allele is not autoprocessed, consequently lacks cholesterol, and has reduced

---

Figure 6 Possible evolutionary history of inteins, Hh proteins, and nematode proteins. Schematic drawings, based on the crystal structure of the Hedgehog C-terminal autoprocessing domain (51), illustrate an ancestral structure that gave rise to the Hint domain proteins through gene duplication, domain swapping, and insertion events, which may have occurred during evolution. The proteins formed by ligation of the N and C extein domains of intein family proteins are of many types with diverse biological activity. The Hh proteins feature a conserved N-terminal signaling domain that is related to a bacterial cell wall enzyme [see (138)], whereas the N-terminal domains of the nematode proteins are novel and sort into three groups (52). Module assembly refers to evolutionary pathways wherein Hint sequences, either alone or carrying an appended domain (e.g., an endonuclease), are inserted into an existing protein. The sequence of depicted events should be considered speculative. Possible scenarios for the assembly include prior association of the SRR with the Hint module to form an independent cholesterol transfer entity. Alternatively, the Hedgehog signaling domain may have evolved with the hydrophobic SRR domain, between which the Hint domain was later inserted. See text and (51) for discussion. An evolutionary dendrogram relating the various Hedgehogs and nematode proteins, which are more closely related, is presented in Aspock et al. (52). Abbreviations are SRR, sterol recognition region; Hint, Hedgehog intein; and DRR, DNA recognition region.

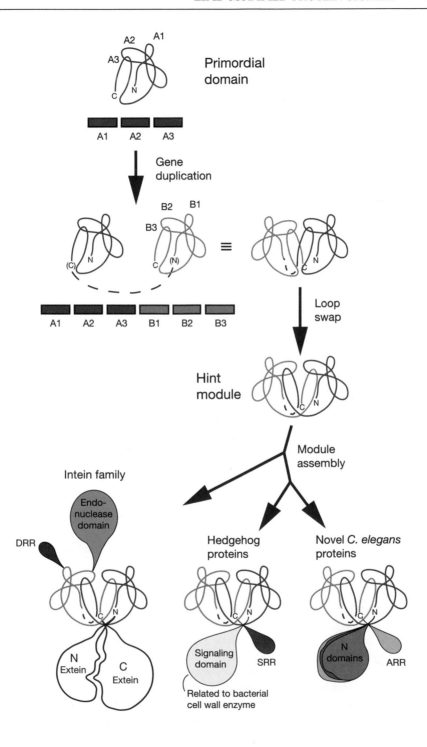

amino-terminal acylation (see below); a reduction in signaling and patterning activity of this mutant is consistent with a requirement for palmitoylation in producing a fully potent Shh signal.

Mechanism and Specificity of Hedgehog Palmitoylation

Most of what is known about the mechanism, specificity, and function of protein S-acylation derives from the numerous and well-documented examples of this dynamic cytoplasmic phenomenon that targets many key intracellular and transmembrane signaling proteins [see (64) and (65) for excellent recent reviews; also see Smotrys and Linder in this volume]. Palmitoylation of Hh proteins, however, is distinctive in that it takes place within the secretory pathway, presumably isolated from the machinery that generates and regulates other S-acylated proteins, and it resolves into a stable amide linkage (see below). Whereas Hedgehog palmitoylation is not the first example of protein S-acylation within the secretory pathway [apolipoprotein B palmitoylation has been documented and shown to be required for assembly of various serum lipoprotein particles; see (66)], it is especially intriguing given the complex biogenesis and signaling activities of Hh proteins.

Although nonenzymatic protein S-acylation has been observed in vitro and may have a specialized role within mitochondria (67), the likelihood of its significance as a general mechanism is minimal given the activity of acyl-CoA binding proteins that reduce the concentration of free acyl-CoA species to levels inadequate for such reactions [(68); also see (64)]. In addition, the recent purification and cloning of two protein acyltransferases (PATs) from *Saccharomyces cerevisiae* that act on cytoplasmic substrates have demonstrated that the acylated components of certain pathways (e.g., the Ras pathway in yeast) have dedicated enzyme-based machinery that is utilized in these reactions (69, 70). Whereas derivatization, including palmitoylation, of N-terminal cysteinyl residues of Hh proteins can also be achieved in vitro [see (59)], autoacylation of Hh in vivo is, likewise, unlikely to happen to any appreciable extent. Recent genetic screens for new patterning mutations in *Drosophila* have supported the idea of enzyme-catalyzed acylation by identifying a single, novel gene that is required for production of an active Hedgehog signal in vivo (71–74). Molecular characterization of this gene revealed that it encodes a protein belonging to the MBOAT family of enzymes (membrane-bound O-acyltransferases), some of which are known to catalyze esterification reactions involving, principally, lipids and other relatively small molecules (75). This insight lead to the obvious proposal that the gene, variously termed *sightless* (*sit*), *skinny hedgehog* (*ski*), *central missing* (*cmn*) and *rasp*, is responsible for the previously identified N-terminal palmitoylation. Genetic knockdown of the activity of *ski* (as we shall refer to this gene) by RNAi causes a reduction in overall Hedgehog protein hydrophobicity, as compared to the fully modified species, and to a level that matches that of the unacylated version (72). Although this reaction, thus far, has not been reconstituted in vitro, these results suggest that this putative acyltransferase is responsible for catalyzing the acylation of Hh proteins in vivo. Vertebrate forms of *ski* have been identified, and their activities, doubtless, will be examined in other models of development and evaluated as to whether they are required in different signaling pathways. In tests of acyltransferase activity with model, non-Hedgehog substrates, the Ski protein has, thus far, proven inactive (R.K. Mann and P.A. Beachy, unpublished information).

The precise mechanism whereby a putative O-acyltransferase accomplishes an amide-linked lipidation is still undefined, but a strict requirement for a cysteine at the site of modification is indicative of, in the early stages at least, a conventional protein S-acylation. In contrast to protein N-myristoylation, enzyme-mediated protein S-acylation, as understood for cytoplasmic proteins, takes place posttranslationally, requires at a minimum a free sulfhydryl acceptor moiety (i.e., an accessible cysteine), and, generally, a prior membrane-anchoring of the target protein through a region of the protein that is typically near the site of the modification. These membrane anchors can take many forms and include transmembrane polypeptide helices, N-terminal myristoyl moieties, C-terminal polyisoprenoid adducts, clusters of hydrophobic residues, as well as spans of basic residues (see above-mentioned reviews). Prior cysteinyl-isoprenylation at the C-terminal CaaX motif of Ras proteins, for example, is required prior to S-acylation at adjacent cysteines [(76), also see (77)]. These membrane anchors are thought to recruit the acylation targets to cellular membranes where most PATs have been shown to localize [see (65)]. Depending on the sequence of processing events, a number of motifs within Hh proteins might satisfy a membrane anchor requirement and include the cholesterol adduct itself (assuming prior internal processing), the polybasic cluster adjacent to the site of palmitoylation (see Figure 7), or the hydrophobic motif (see above) at the extreme C terminus of the Hedgehog processing domain (if N-terminal processing precedes internal). The hydrophobic signal sequence is unlikely to provide the anchoring activity because this is lost early due to cotranslational processing. As mentioned above, both the efficiency and the fatty acyl selectivity of Hedgehog S-acylation depends, to some degree, on the presence of the cholesteryl adduct because expression of Hedgehog signaling proteins lacking the cholesterol transferase domain causes a reduced level of palmitoylation as well as a greater variety of fatty acyl groups to be incorporated. This result suggests that the cholesteryl moiety facilitates membrane association and proximity to the Ski protein and that it does so in an organelle or compartment where palmitoyl-CoA is the predominant fatty acyl donor.

Regardless of the sequence, cysteine-initiated polypeptides, like Hh proteins after signal sequence cleavage, feature a free primary amine that is in an ideal position to attack the thioester and produce the more stable amide (see Figure 7). This S-to-N shift, which proceeds via a cyclic intermediate, has not only been observed in vitro but serves as the basis of expressed protein ligation [see above and (34)].

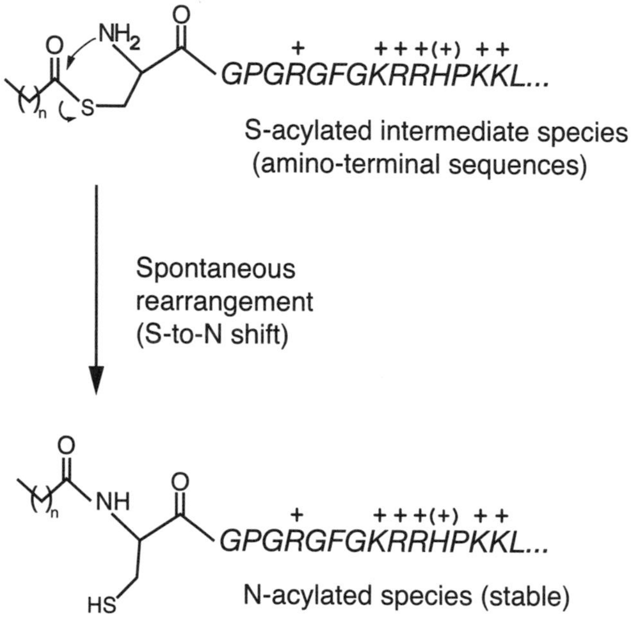

S-acylated intermediate species
(amino-terminal sequences)

Spontaneous
rearrangement
(S-to-N shift)

N-acylated species (stable)

Figure 7 Model of the spontaneous acyl rearrangement at the amino terminus of Hh proteins. Palmitoylation of the amino-terminal cysteine residue likely occurs by conventional side chain S-acylation followed by a spontaneous S-to-N shift that results in the stable amide linkage. A polybasic cluster near the amino terminus of Hh proteins (the mouse Sonic Hedgehog sequence is shown) may serve to facilitate membrane association required during biogenesis and for distribution and activity of the signal. See text for details.

Functional Consequences of Palmitoylation

Although the determinants of protein S-acylation are still poorly defined, some of the consequences of these dynamic modifications are well documented. S-acylation, either on its own (at multiple sites) or in combination with other hydrophobic motifs (posttranslational modifications, hydrophobic amino acid residues, or hydrophobic associated proteins) serves a major role in targeting modified cytosolic proteins to specific membranes [see (64, 65, 77, 78)]. N-myristoylation of many Src family kinases alone, for example, provides only partial and transient association with cellular membranes. The addition of a palmitoyl moiety to the polypeptides (or, in the case of Src itself, the presence of a polybasic cluster), however, confers a strong membrane affinity in general and one that directs them specifically to the plasma membrane. Interestingly, this multiplicity of membrane targeting motifs appears to be the rule rather than the exception among anchored proteins. It is now generally accepted that any single acylation or prenylation is unable to confer stable membrane localization [(77–

79) and references therein], indicating that an additive or cooperative effect between intrinsic anchoring motifs drives the membrane localization. Even cholesterol-modified Hh proteins appear to be subject to the rule of cooperativity. Reducing N-terminal acylation of *Drosophila* Hedgehog in vivo (by RNAi-mediated knockdown of Ski activity; see above) causes a reduction in plasma membrane association as indicated by a significant increase in release of Hedgehog protein into the culture medium (72). This result has implications not only for Hedgehog protein generation and packaging but also in the distribution and reception of the Hedgehog signal (see below).

LIPID MODIFICATION IN THE WNT SIGNALING PATHWAY

The results of computational motif searching (75) used to implicate *ski* as a member of the MBOAT acyltransferase family also revealed the Wnt pathway gene *porcupine (porc)* as a member of this family. As shown for *ski*, *porc* is required in Wnt-producing cells for generating the fully functional protein signal (80, 81). Early reports suggested that Porc is required for Wnt protein secretion, and others have demonstrated that the activity influences the N-linked glycosylation status of the protein (82, 83). Willert et al. (84) recently reported the first purification of biologically active Wnt proteins and found, by metabolic labeling and with mass spectrometric-based mapping analyses, that both vertebrate and *Drosophila* Wnt proteins are modified by palmitate at a conserved cysteine residue. Mutation of this cysteine causes a loss of palmitate incorporation, and this mutation as well as enzymatic removal of the palmitate both reduce the biological activity of the protein. It remains to be demonstrated that Porcupine is directly responsible for Wnt protein palmitoylation, but considering available evidence, the connection appears likely.

It is possible that lipid modification will make another interesting entry into the Wnt pathway, this one at the level of extracellular signaling interactions. Among the many secreted antagonists of Wnt proteins [see (19)], the Dickkopf proteins have been found to possess a structure closely related to that of colipases (85). These proteins act as essential cofactors in the duodenal digestion of nutritional triglycerides by pancreatic lipase and bile salts. Colipases bind lipases, confer enhanced hydrophobicity, and are thought to recruit the enzyme complex to lipid-water interfaces [see (86)]. Although Dickkopf proteins have not been found in *Drosophila* and recruitment of a lipase to a Wnt-Dickkopf complex has yet to be demonstrated in vertebrate systems, hydrolysis of a lipid adduct from the Wnt protein could be the mechanism of Dickkopf inhibition of Wnt signaling.

HEDGEHOG TISSUE DISTRIBUTION

In addition to features intrinsic to the Hh proteins, a number of factors within both signal-generating and receiving tissues have been shown to influence the distribution and activity of the protein signal. Although dual lipidation of Hedgehog promotes membrane affinity, accessory proteins have evolved to deal with this fully processed form of the Hedgehog signal and are essential for its deployment in developing tissues.

The Role of Processing and Cholesterol Modification in Hedgehog Tissue Distribution

Despite its importance in embryonic pattern formation and in facilitating N-terminal acylation, the cholesterol adduct on Hh proteins is, paradoxically, not required for signal transduction through its receptors. Although aberrant in distribution, Hedgehog variants lacking the cholesterol moiety are able to signal to responsive tissues both close to and far from the source (32, 87). Similarly, as mentioned, some derivatives of recombinant Hedgehog that are only N-terminally modified are as active as dually lipidated forms (59). What is the mechanism whereby this adduct influences the proper deployment of Hedgehog signals? Given the overwhelming hydrocarbon content of cholesterol, coupled with the loss of its lone free polar moiety, one would expect that peptide cholesterylation would restrict the complex to the normal residence of a cholesterol molecule, i.e., the various membranes of the cell. This expectation has been borne out in several studies wherein processed, cholesterol-modified forms of Hedgehog have been found to be predominantly membrane-associated, and those lacking the adduct more freely dissociate from cells after secretion (22, 23, 57). The cholesteryl adduct thus functions as a lipid anchor that restricts the spatial mobility of this secreted signal. But processed Hedgehog protein travels beyond the cells in which it is produced and signals over many cell diameters [see (1, 88–91)]. This signaling is direct, rather than through an intermediary signal (92, 93), suggesting that cellular mechanisms may exist for the handling and delivery of cholesterol-modified proteins.

The Contrasting Roles of Dispatched and Patched in Tissue Distribution of the Hh Signal

Specific cellular activities dedicated to the handling of cholesterol-modified Hh proteins indeed have been found and feature distinct activities required in signal-generating and in signal-receiving cells. One of these activities was identified in a genetic screen for new mutations affecting Hedgehog signaling in *Drosophila* embryos and imaginal discs. The function of this gene is required exclusively in Hh-producing cells for release of a fully functional signal (87). In *dispatched* mutants Hh protein production and processing appear to be uncompromised, but the signal generated within mutant cells accumulates and does not travel to distant targets. Interestingly, the gene is not required for release of a truncated form of Hh (i.e., one lacking the cholesterol adduct), suggesting that one of its normal functions is to make the cholesterol-modified version mobile. Dispatched was recently shown to be required in the *Drosophila* embryo for generating apically-localized, Hedgehog-containing punctate structures that are likely involved in many of the embryonic patterning activities of Hedgehog (94). Again, it is the cholesterol-modified forms of Hedgehog that are subject to the activity of Dispatched, confirming its role in mobilizing the anchored form of the signal. The Dispatched protein is predicted to contain 12 transmembrane segments and is related throughout the transmembrane region to a group of proteins that include bacterial transmembrane transporters and Patched, a receptor for the Hh protein (see below). A 5-transmembrane subset of this homology region is conserved in certain proteins that sense and regulate sterol homeostasis and is known as a sterol sensing domain (SSD) (see below).

The murine and human genomes contain two *Dispatched* homologs, and genetic studies in mice support an essential role for the function of one of these genes (*mDispA*) in Hh signaling (95). Homozygous loss-of-function *mDispA* mutations are embryonic lethal, and mutant embryos display an array of phenotypes consistent with a complete loss of Hedgehog pathway patterning activity (95–97). The severity of this phenotype suggests that Hh signaling function can be assigned predominantly to *mDispA*; this is consistent with the ability of *mDispA* but not *mDispB* to rescue *Drosophila disp* mutants.

Some insight into the biochemical function of Dispatched has been derived from cultured cell-based models of Hedgehog protein signal generation (95). Whereas expression of full-length Hh proteins normally results in a strong membrane association, coexpression with *Dispatched,* or *mDispA,* but not with *mDispB* results in a significant increase in Hh-Np and Shh-Np protein levels in the culture medium, indicative of an activity of Disp proteins in release of lipid-modified Hh proteins. In another study, a soluble form of modified Shh-Np was found to be released from mammalian cultured cells and suggested on the basis of gel filtration studies to exist in an aggregate of ∼5–6 molecules (98), but the question of whether Dispatched activity was involved in producing this soluble aggregate was not addressed.

Dispatched as well as Patched proteins display topological and sequence similarity to the RND (for resistance, nodulation, division) family of bacterial transmembrane transporters. These permeases utilize a proton electrochemical gradient to function as antiporters in extruding from bacterial cells a variety of substrates that include heavy metals, hydrophobic drugs, and endogenous compounds [see (99)]. The structure of one member of this transporter family, AcrB, has been solved to atomic resolution (100), and the monomer comprises 12 transmembrane spans that appear to have arisen through a tandem duplication of a 6 transmembrane unit (Figure 8A). The similarity of Disp and Ptc to these transporters extends throughout the transmembrane region and is particularly striking in TM4, which contains a Gly-X-X-X-Asp motif (Figure 8B) that is

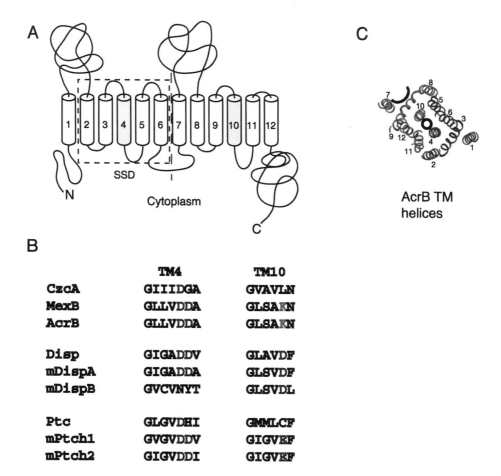

A

SSD

N

Cytoplasm

C

C

AcrB TM
helices

B

	TM4	**TM10**
CzcA	GIIIDGA	GVAVLN
MexB	GLLVDDA	GLSAKN
AcrB	GLLVDDA	GLSAKN
Disp	GIGADDV	GLAVDF
mDispA	GIGADDA	GLSVDF
mDispB	GVCVNYT	GLSVDL
Ptc	GLGVDHI	GMMLCF
mPtch1	GVGVDDV	GIGVEF
mPtch2	GIGVDDI	GIGVEF

Figure 8 Structure and sequence conservation between bacterial RND transporters, Dispatched and Patched proteins. (*A*) Predicted topology of Dispatched proteins. This 12-span polytopic transmembrane model is also representative of other members of a class that includes Patched and the bacterial RND family of transport proteins. The five adjacent transmembrane segments that have been found to confer a sterol sensing activity (e.g., in HMG-CoA reductase and SCAP) are delimited by a dashed box. The depicted topology and domain structure is from sequence-based predictions reported by Burke et al. (87) and Ma et al. (95). Relative loop domain lengths are approximate; loop drawings are stylized and do not reflect known or predicted folds. A red dashed line divides the protein into two homologous spans that are likely the result of gene duplication. (*B*) Conserved sequences within TM4 and TM10 domains of Dispatched and Patched proteins as well as several bacterial transporters, which include AcrB. Note that TM4 of *mDispB*, which fails to rescue Drosophila *disp* mutations, does not contain the conserved Asp (D) residue present in other family members. (*C*) Cross section of transmembrane helices from the crystal structure of AcrB (100); only one subunit of the trimeric structure is shown. It has been proposed that the opening between the central TM4 and TM10 domains (note drawn *circle*) serves as the pore for proton translocation.

critical for function of the transporter in biochemical reconstitution experiments (101–103). TM4 and TM10 are positioned inside a transmembrane helix bundle (see Figure 8C) with the TM4 Asp residue forming a salt bridge with a Lys at the corresponding position in TM10, and these helices are proposed as candidates for the proton-translocating pathway.

Mutations of charged residues at these positions in TM4 and TM10 of Disp (95) and Ptc (see below) disrupt biological function, reinforcing the suggestion from sequence conservation that Disp and Ptc are functionally related to RND transporters and may act by similar mechanisms. Although the proposed export of a lipoprotein would represent a novel activity for an RND transporter family member, there is precedent for this function in members of the structurally distinct ATP binding cassette (ABC) family of exporters [see (104)]. It is also intriguing that AcrB appears to function as a trimer, raising the possibility that multimeric action of Disp protein may produce an aggregated, soluble form of the lipid-modified Hh proteins, perhaps a micelle-like structure with interactions between lipids forming a lipophilic interior. It remains to be seen whether the Dispatched-dependent export of Hh proteins is powered by a proton motive force, like the activities of other RND transporters, or by some other electrochemical gradient.

With possible relevance to Dispatched function, it is interesting to note that Hh-Np from *Drosophila* embryos partitions into detergent-insoluble glycolipid enriched complexes, as has been shown for other raft-associated proteins (105). In addition, murine Shh-Np also partitions with such complexes (58), suggesting that raft association is a general property of the processed Hh signal. It is not known whether raft association is conferred by cholesterylation or palmitoylation alone, or whether both modifications are required. Irrespective of how raft association is specified, it has potential consequences both for signal packaging from signal-producing cells and for signal response in target cells. It is possible that a raft-based process, conceivably involving Dispatched (see above), may operate in signal packaging and secretion of raft-targeted Hh proteins for long-range signaling. With regard to rafts in target cells, *dally-like* (*dlp*) encodes a GPI-linked protein required for response in target cells (106–108) and presumably would be found in rafts, as is the case for other GPI-linked proteins. Perturbations of cholesterol homeostasis, whether genetic or pharmacologic, can also disrupt Hh signaling through components that are unlikely to be involved with the production or distribution of the Hh signal (109). We have thus found that such perturbations disrupt signal response in receiving cells at the level of the seven transmembrane component Smoothened (110).

The Ptc protein itself plays a role in sequestration of the Hh signal within tissues, which is opposite to the role of Disp and is distinct from the better-known role of Ptc in regulating signal transduction as a component of the Hh receptor mechanism. The role of Ptc in Hh signal transduction is to functionally antagonize the activity of Smoothened, a multi-pass transmembrane protein, which in the absence of Ptc function constitutively activates the Hh pathway (111–114).

Ptc acts catalytically in its suppression of Smo activity (115), and disruption of Ptc function by mutation of the Gly-X-X-X-Asp motif in TM4 suggests that this catalysis may occur via a transporter-like activity. Ptc suppression of Smo activity is cell autonomous, and this suppression is alleviated by the presence of the Hh signal, which interacts directly with the Ptc protein (116–118).

A second activity of Ptc, which is more directly relevant to the current discussion, is its role in sequestration of the Hh signal within tissues, a cell non-autonomous activity that restricts the spatial extent of Hh signaling (112). The transduction and tissue distribution activities can be genetically uncoupled, as demonstrated by mutant proteins that retain either function in isolation. A mutant form of Ptc thus has been described that retains the sequestration function but is unable to suppress the Hh pathway (112, 119–121). Conversely, a mutant Ptc protein that fails to bind Hh but retains the ability to suppress Smo can be produced by deletion of a portion of the extracytoplasmic loop between TM7 and TM8 (93, 115). The sequestration and tissue restriction of a ligand by its receptor also has been noted for the Torso pathway in *Drosophila* (122). The sequestering action of Ptc appears to require a processed Hh protein as indicated by the observation that the spatial extent of signaling is greater when a truncated form of the Hh protein is produced at higher levels in its normal location (32). Ptc and Disp thus are proteins with homology throughout their predicted 12 transmembrane spans, and both proteins are required for appropriate tissue distribution of the processed, and therefore cholesterol-modified, form of the Hh protein. However, Disp mobilizes Hh-Np from signal-generating cells, whereas Ptc expression in adjacent tissues causes a striking limitation of that mobility (Figure 9).

The specific requirement of both Ptc and Disp for appropriate tissue distribution of the cholesterol-modified form of the Hh protein is an observation made more striking by the fact that both proteins contain a SSD (Figure 8A), which is a subset of the 12 transmembrane domains conserved between Ptc, Disp, and the RND transporters. The mechanism of sterol sensing remains unclear, but the

---

Figure 9 Short- and long-range Hedgehog signaling. Hedgehog-producing cells are known to signal to both adjacent as well as distant tissues. The described activities that impinge on short- and long-distance signaling, especially those pertaining to generation and distribution of cholesterol-modified Hh proteins, are depicted schematically. The figure integrates both genetic and biochemical relationships. The depicted multimeric form of secreted Hh protein has been proposed by Zeng et al. (98), but a requirement for Dispatched in its biogenesis has not been established. Darker portions of membrane illustrations suggest potential lipid raft/microdomain involvement. Hip1 (not shown) is a membrane-associated protein that is found only in vertebrates and, like Patched, is both induced by Hedgehog and known to restrict the movement of Hh protein by sequestration (139). Other abbreviations are Disp, Dispatched; Dlp, Dally-like protein; Ptc, Patched; Smo, Smoothened; Ttv, Tout velu; and EXT, Exostosin. See text for details.

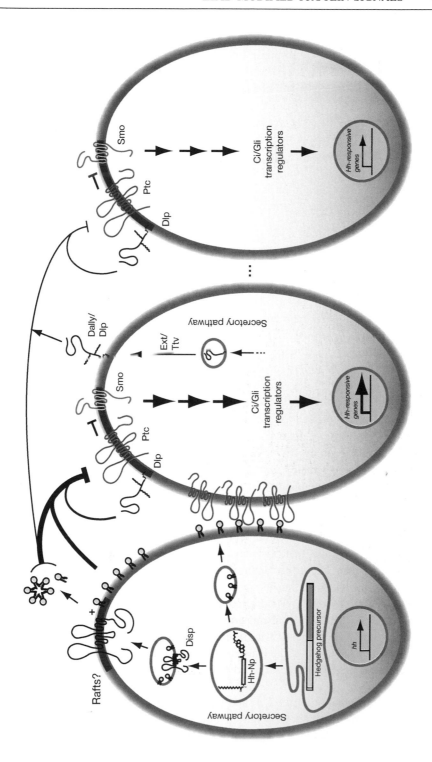

SSDs of HMG-CoA reductase and SCAP allow these proteins to regulate their associated activities in response to cellular sterol levels (123–128). The obvious suggestion that SSDs may aid in binding of the cholesteryl adduct of Hh proteins is belied by the observation that the affinity of Shh-Np binding to Ptc is not significantly different from that of Shh-N, which lacks the cholesteryl adduct (57). Because Ptc and Disp appear to display a more extensive similarity to RND transporters throughout the transmembrane region, it appears likely that the SSD represents a functional subunit, perhaps a conformationally dynamic subdomain, of this larger region. In bacteria, which lack cholesterol, this subunit (the SSD) would have no function on its own, but in HMG-CoA reductase and SCAP, the SSD has perhaps been geared to respond to sterols.

The Role of Tout velu in Hedgehog Tissue Distribution

In contrast to the requirement for Dispatched in signal-generating tissues, a second gene, from yet another genetic screen in *Drosophila*, also causes a *hedgehog*-like phenotype but is required for signal transmission (94, 129, 130). This mutation, termed *tout velu* (or *ttv*, French for hairy), prevents normal propagation of the Hh signal in target tissue and allows signaling only within cells that directly abut the Hh-producing tissue. The normally processed form of Hh is restricted in the *ttv* mutants, whereas overexpression of a truncated, unmodified form of Hh is not, suggesting that the *ttv* gene product is required specifically to enable the transmission of the cholesterol-modified form of Hedgehog to distant targets.

Molecular cloning of *ttv* (130) revealed that it is a homolog of vertebrate *EXT-1*, a gene associated with hereditary multiple exostoses. EXT-1 and EXT-2 (another family member) are type II transmembrane proteins that form a Golgi-localized hetero-oligomeric complex, which has heparan sulfate copoly-merase activity (131–133), and cells lacking EXT-1 function are defective in heparan sulfate proteoglycan biosynthesis. Confirming the structural similarity, it has also been shown that glycosaminoglycan, including heparan sulfate, biosynthesis is disrupted in *ttv* mutant animals (134–136). More recently, a requirement for Ttv activity in normal biosynthesis of the Dally and Dlp proteins has been demonstrated, and Dally and Dlp furthermore have been shown to function in mediating extracellular transmission of the Hh signal to distant targets (108). The role of Ttv in Hh signal distribution thus seems likely to be mediated through its action in biosynthesis of the Dally and Dlp proteins. Current information does not distinguish between direct function of the GAG chains elaborated by Ttv in Hh signal transmission or, instead, a requirement for GAG chain addition for proper surface presentation of the Dlp and Dally proteins. Interestingly, the Dlp protein appears to play roles both in autonomous cellular response to the Hh signal and in transmission of the Hh signal to distant sites (106–108).

PERSPECTIVE ON THE FUNCTION OF CHOLESTEROL MODIFICATION

Hh proteins are deployed in graded concentrations that are dependent not only on proximity to the source but also on influences from producing and receiving tissues. Although the precise mode of transmission through tissues is still unknown, it is likely that Hh proteins retain both of their relatively stable lipid adducts while en route to target tissues. Lipid modification would be expected to affect tissue distribution of Hh protein signals. Analysis of these effects with the use of truncated proteins, which lack cholesterol and therefore might be expected to travel more freely through tissues, is complicated by the accompanying reduction of amino-terminal acylation, which in turn reduces signaling potency (see above). However, when this reduction in signaling potency is compensated for by a higher-level expression of the truncated protein, it is evident that, in comparison to similarly expressed modified protein, Hedgehog protein unanchored by cholesterol is more mobile and acts over a greater range than the modified protein (32). Thus it seems clear that in addition to facilitating amino-terminal acylation and thus indirectly stimulating activity, the primary role of the cholesteryl adduct is to direct the mature signal to a set of cellular components that operate in concert to produce a precisely regulated distribution of Hh signals in responsive tissues. Paramount among a group of unresolved issues are the detailed roles of these lipid modifications in the intra- and extracellular packaging and handling of modified Hh protein, the mechanism by which these modifications modulate signal potency, and the role of such modifications in other extracellular signaling pathways, such as the Wnt pathway.

ACKNOWLEDGMENTS

We thank Daniel Leahy for providing images for Figure 4, Yong Ma for assistance with Figure 8, and Maurine Linder for critical review of the manuscript. Philip Beachy is an investigator of the Howard Hughes Medical Institute.

The *Annual Review of Biochemistry* is online at http://biochem.annualreviews.org

LITERATURE CITED

1. Ingham PW, McMahon AP. 2001. *Genes Dev.* 15:3059–87
2. Muenke M, Beachy PA. 2001. In *The Metabolic and Molecular Bases of Inherited Disease*, ed. CR Scriver, AL Beaudet, WS Sly, D Valle, B Childs, et al., pp. 6203–30. New York: McGraw-Hill
3. Wodarz A, Nusse R. 1998. *Annu. Rev. Cell Dev. Biol.* 14:59–88
4. Miller JR. 2002. *Genome Biol.* 3:1–15
5. Nusse R. 2003. *The Wnt gene homepage.* www.stanford.edu/~rnusse/wntwindow. html
6. Batlle E, Henderson JT, Beghtel H, van den

Born MM, Sancho E, et al. 2002. *Cell* 111: 251–63

7. van de Wetering M, Sancho E, Verweij C, de Lau W, Oving I, et al. 2002. *Cell* 111:241–50

8. Machold R, Hayashi S, Rutlin M, Muzumdar MD, Nery S, et al. 2003. *Neuron* 39:937–50

9. Pinto D, Gregorieff A, Begthel H, Clevers H. 2003. *Genes Dev.* 17:1709–13

10. Reya T, Duncan AW, Ailles L, Domen J, Scherer DC, et al. 2003. *Nature* 423: 409–14

11. Watkins DN, Berman DM, Burkholder SG, Wang BL, Beachy PA, Baylin SB. 2003. *Nature* 422:313–17

12. Berman DM, Karhadkar SS, Maitra A, de Oca RM, Gerstenblith MR, et al. 2003. *Nature* 425:846–51

13. Thayer SP, Di Magliano MP, Heiser PW, Nielsen CM, Roberts DJ, et al. 2003. *Nature* 425:851–56

14. Taipale J, Beachy PA. 2001. *Nature* 411: 349–54

15. Reya T, Morrison SJ, Clarke MF, Weissman IL. 2001. *Nature* 414:105–11

16. Giles RH, van Es JH, Clevers H. 2003. *Biochim. Biophys. Acta* 1653:1–24

17. Ornitz DM, Itoh N. 2001. *Genome Biol.* 2:REVIEWS3005. pp. 1–12; PMID: 11276432

18. Annes JP, Munger JS, Rifkin DB. 2003. *J. Cell Sci.* 116:217–24

19. Kawano Y, Kypta R. 2003. *J. Cell Sci.* 116:2627–34

20. Smith WC. 1999. *Trends Genet.* 15:3–5

21. Mann RK, Beachy PA. 2000. *Biochim. Biophys. Acta* 1529:188–202

22. Lee JJ, Ekker SC, von Kessler DP, Porter JA, Sun BI, Beachy PA. 1994. *Science* 266:1528–37

23. Porter JA, von Kessler DP, Ekker SC, Young KE, Lee JJ, et al. 1995. *Nature* 374:363–66

24. Porter JA, Young KE, Beachy PA. 1996. *Science* 274:255–59

25. Ekker SC, Ungar AR, Greenstein P, von

Kessler DP, Porter JA, et al. 1995. *Curr. Biol.* 5:944–55

26. Fan CM, Porter JA, Chiang C, Chang DT, Beachy PA, Tessier-Lavigne M. 1995. *Cell* 81:457–65

27. Hynes M, Porter JA, Chiang C, Chang DT, Tessier-Lavigne M, et al. 1995. *Neuron* 1:35–44

28. Lai C-J, Ekker SC, Beachy PA, Moon RT. 1995. *Development* 121:2349–60

29. López-Martínez A, Chang DT, Chiang C, Porter JA, Ros MA, et al. 1995. *Curr. Biol.* 5:791–96

30. Marti E, Bumcrot DA, Takada R, McMahon AP. 1995. *Nature* 375:322–25

31. Roelink H, Porter JA, Chiang C, Tanabe Y, Chang DT, et al. 1995. *Cell* 81: 445–55

32. Porter JA, Ekker SC, Park WJ, von Kessler DP, Young KE, et al. 1996. *Cell* 86:21–34

33. Dawson PE, Muir TW, Clark-Lewis I, Kent SB. 1994. *Science* 266:776–79

34. Muir TW. 2003. *Annu. Rev. Biochem.* 72: 249–89

35. Roessler E, Belloni E, Gaudenz K, Jay P, Berta P, et al. 1996. *Nat. Genet.* 14: 357–60

36. Roessler E, Belloni E, Gaudenz K, Vargas F, Scherer SW, et al. 1997. *Hum. Mol. Genet.* 6:1847–53

37. Cooper MK, Porter JA, Young KE, Beachy PA. 1998. *Science* 280:1603–7

38. Cenedella RJ. 2001. *Ophthalmic Res.* 33: 210–16

39. Beachy PA, Cooper MK, Young KE, von Kessler DP, Park WJ, et al. 1997. *Cold Spring Harbor Symp. Quant. Biol.* 62:191–204

40. Perler FB. 1998. *Nat. Struct. Biol.* 5: 249–52

41. Hochstrasser M. 1996. *Annu. Rev. Genet.* 30:405–39

42. Chu CT, Pizzo SV. 1994. *Lab. Investig.* 71:792–812

43. van Poelje PD, Snell EE. 1990. *Annu. Rev. Biochem.* 59:29–59

44. Brannigan JA, Dodson G, Duggleby HJ,

Moody PC, Smith JL, et al. 1995. *Nature* 378:416–19

45. Guan C, Cui T, Rao V, Liao W, Benner J, et al. 1996. *J. Biol. Chem.* 271: 1732–37

46. Perler FB. 2002. *Nucleic Acids Res.* 30: 383–84

47. Gogarten JP, Senejani AG, Zhaxybayeva O, Olendzenski L, Hilario E. 2002. *Annu. Rev. Microbiol.* 56:263–87

48. Xu MQ, Perler FB. 1996. *EMBO J.* 15: 5146–53

49. *C. elegans* Sequencing Consort. 1998. *Science* 282:2012–18

50. Burglin TR. 1996. *Curr. Biol.* 6:1047–50

51. Hall TM, Porter JA, Young KE, Koonin EV, Beachy PA, Leahy DJ. 1997. *Cell* 91:85–97

52. Aspock G, Kagoshima H, Niklaus G, Burglin TR. 1999. *Genome Res.* 9: 909–23

53. Dalgaard JZ, Moser MJ, Hughey R, Mian IS. 1997. *J. Comput. Biol.* 4: 193–214

54. Pietrokovski S, Henikoff S. 1997. *Mol. Gen. Genet.* 254:689–95

55. Duan XQ, Gimble FS, Quiocho FA. 1997. *Cell* 89:555–64

56. Bennett MJ, Schlunegger MP, Eisenberg D. 1995. *Protein Sci.* 4:2455–68

57. Pepinsky RB, Zeng CH, Wen DY, Rayhorn P, Baker DP, et al. 1998. *J. Biol. Chem.* 273:14037–45

58. Taipale J, Chen JK, Cooper MK, Wang B, Mann RK, et al. 2000. *Nature* 406: 1005–9

59. Taylor FR, Wen DY, Garber EA, Carmillo AN, Baker DP, et al. 2001. *Biochemistry* 40:4359–71

60. Williams KP, Rayhorn P, Chi-Rosso G, Garber EA, Strauch KL, et al. 1999. *J. Cell Sci.* 112(Part 23):4405–14

61. Lee JD, Kraus P, Gaiano N, Nery S, Kohtz J, et al. 2001. *Dev. Biol.* 233: 122–36

62. Kohtz JD, Lee HY, Gaiano N, Segal J, Ng E, et al. 2001. *Development* 128: 2351–63

63. Lewis PM, Dunn MP, McMahon JA, Logan M, Martin JF, et al. 2001. *Cell* 105:599–612

64. Bijlmakers MJ, Marsh M. 2003. *Trends Cell Biol.* 13:32–42

65. Linder ME, Deschenes RJ. 2003. *Biochemistry* 42:4311–20

66. Vilas GL, Berthiaume LG. 2004. *Biochem. J.* 377(Part 1):121–30

67. Corvi MM, Soltys CL, Berthiaume LG. 2001. *J. Biol. Chem.* 276:45704–12

68. Dunphy JT, Schroeder H, Leventis R, Greentree WK, Knudsen JK, et al. 2000. *Biochim. Biophys. Acta* 1485:185–98

69. Lobo S, Greentree WK, Linder ME, Deschenes RJ. 2002. *J. Biol. Chem.* 277: 41268–73

70. Roth AF, Feng Y, Chen L, Davis NG. 2002. *J. Cell Biol.* 159:23–28

71. Lee JD, Treisman JE. 2001. *Curr. Biol.* 11:1147–52

72. Chamoun Z, Mann RK, Nellen D, von Kessler DP, Bellotto M, et al. 2001. *Science* 293:2080–84

73. Amanai K, Jiang J. 2001. *Development* 128:5119–27

74. Micchelli CA, The I, Selva E, Mogila V, Perrimon N. 2002. *Development* 129: 843–51

75. Hofmann K. 2000. *Trends Biochem. Sci.* 25:111–12

76. Hancock JF, Magee AI, Childs JE, Marshall CJ. 1989. *Cell* 57:1167–77

77. Resh MD. 1999. *Biochim. Biophys. Acta* 1451:1–16

78. Berthiaume LG. 2002. *Science STKE*152:PE41. PMID:12359913

79. Shahinian S, Silvius JR. 1995. *Biochemistry* 34:3813–22

80. van den Heuvel M, Harryman-Samos C, Klingensmith J, Perrimon N, Nusse R. 1993. *EMBO J.* 12:5293–302

81. Kadowaki T, Wilder E, Klingensmith J, Zachary K, Perrimon N. 1996. *Genes Dev.* 10:3116–28

82. Tanaka K, Okabayashi K, Asashima M, Perrimon N, Kadowaki T. 2000. *Eur. J. Biochem.* 267:4300–11

83. Tanaka K, Kitagawa Y, Kadowaki T. 2002. *J. Biol. Chem.* 277:12816–23

84. Willert K, Brown JD, Danenberg E, Duncan AW, Weissman IL, et al. 2003. *Nature* 423:448–52

85. Aravind L, Koonin EV. 1998. *Curr. Biol.* 8:R477–78

86. van Tilbeurgh H, Bezzine S, Cambillau C, Verger R, Carriere F. 1999. *Biochim. Biophys. Acta* 1441:173–84

87. Burke R, Nellen D, Bellotto M, Hafen E, Senti KA, et al. 1999. *Cell* 99:803–15

88. Johnson RL, Tabin C. 1995. *Cell* 81:313–16

89. Hammerschmidt M, Brook A, McMahon AP. 1997. *Trends Genet.* 13:14–21

90. Ingham PW. 1998. *EMBO J.* 17:3505–11

91. Goetz JA, Suber LM, Zeng X, Robbins DJ. 2002. *BioEssays* 24:157–65

92. Wang B, Fallon JF, Beachy PA. 2000. *Cell* 100:423–34

93. Briscoe J, Chen Y, Jessell TM, Struhl G. 2001. *Mol. Cell* 7:1279–91

94. Gallet A, Rodriguez R, Ruel L, Therond PP. 2003. *Dev. Cell* 4:191–204

95. Ma Y, Erkner A, Gong R, Yao S, Taipale J, et al. 2002. *Cell* 111:63–75

96. Caspary T, Garcia-Garcia MJ, Huangfu D, Eggenschwiler JT, Wyler MR, et al. 2002. *Curr. Biol.* 12:1628–32

97. Kawakami T, Kawcak T, Li YJ, Zhang W, Hu Y, Chuang PT. 2002. *Development* 129:5753–65

98. Zeng X, Goetz JA, Suber LM, Scott WJ Jr, Schreiner CM, Robbins DJ. 2001. *Nature* 411:716–20

99. McKeegan KS, Borges-Walmsley MI, Walmsley AR. 2003. *Trends Microbiol.* 11:21–29

100. Murakami S, Nakashima R, Yamashita E, Yamaguchi A. 2002. *Nature* 419:587–93

101. Goldberg M, Pribyl T, Juhnke S, Nies DH. 1999. *J. Biol. Chem.* 274:26065–70

102. Guan L, Nakae T. 2001. *J. Bacteriol.* 183:1734–39

103. Murakami S, Yamaguchi A. 2003. *Curr. Opin. Struct. Biol.* 13:443–52

104. Yakushi T, Masuda K, Narita S, Matsuyama S, Tokuda H. 2000. *Nat. Cell Biol.* 2:212–18

105. Rietveld A, Neutz S, Simons K, Eaton S. 1999. *J. Biol. Chem.* 274:12049–54

106. Lum L, Yao S, Mozer B, Rovescalli A, Von Kessler D, Nirenberg M, Beachy PA. 2003. *Science* 299:2039–45

107. Desbordes SC, Sanson B. 2003. *Development* 130:6245–55

108. Han C, Belenkaya TY, Wang B, Lin X. 2004. *Development* 131:601–11

109. Cooper MK, Wassif CA, Krakowiak PA, Taipale J, Gong R, et al. 2003. *Nat. Genet.* 33:508–13

110. Chen JK, Taipale J, Cooper MK, Beachy PA. 2002. *Genes Dev.* 16:2743–48

111. van den Heuvel M, Ingham PW. 1996. *Nature* 382:547–51

112. Chen Y, Struhl G. 1996. *Cell* 87:553–63

113. Chen Y, Struhl G. 1998. *Development* 125:4943–48

114. Murone M, Rosenthal A, de Sauvage FJ. 1999. *Curr. Biol.* 9:76–84

115. Taipale J, Cooper MK, Maiti T, Beachy PA. 2002. *Nature* 418:892–97

116. Marigo V, Davey RA, Zuo Y, Cunningham JM, Tabin CJ. 1996. *Nature* 384:176–79

117. Stone DM, Hynes M, Armanini M, Swanson TA, Gu QLJR, et al. 1996. *Nature* 384:129–34

118. Fuse N, Maiti T, Wang B, Porter JA, Hall TM, et al. 1999. *Proc. Natl. Acad. Sci. USA* 96:10992–99

119. Martin V, Carrillo G, Torroja C, Guerrero I. 2001. *Curr. Biol.* 11:601–7

120. Strutt H, Thomas C, Nakano Y, Stark D, Neave B, et al. 2001. *Curr. Biol.* 11:608–13

121. Johnson RL, Zhou L, Bailey EC. 2002. *Dev. Biol.* 242:224–35

122. Casanova J, Struhl G. 1993. *Nature* 362:152–55

123. Gil G, Faust JR, Chin DJ, Goldstein JL, Brown MS. 1985. *Cell* 41:249–58

124. Hua X, Nohturfft A, Goldstein JL, Brown MS. 1996. *Cell* 87:415–26

125. Nohturfft A, Brown MS, Goldstein JL. 1998. *J. Biol. Chem.* 273:17243–50

126. Brown MS, Goldstein JL. 1999. *Proc. Natl. Acad. Sci. USA* 96:11041–48

127. Edwards PA, Tabor D, Kast HR, Venkateswaran A. 2000. *Biochim. Biophys. Acta* 1529:103–13

128. Kuwabara PE, Labouesse M. 2002. *Trends Genet.* 18:193–201

129. Perrimon N. 1996. *Cell* 86:513–16

130. Bellaiche Y, The I, Perrimon N. 1998. *Nature* 394:85–88

131. Lind T, Tufaro F, McCormick C, Lindahl U, Lidholt K. 1998. *J. Biol. Chem.* 273:26265–68

132. McCormick C, Leduc Y, Martindale D, Mattison K, Esford LE, et al. 1998. *Nat. Genet.* 19:158–61

133. McCormick C, Duncan G, Goutsos KT, Tufaro F. 2000. *Proc. Natl. Acad. Sci. USA* 97:668–73

134. Toyoda H, Kinoshita-Toyoda A, Selleck SB. 2000. *J. Biol. Chem.* 275:2269–75

135. Toyoda H, Kinoshita-Toyoda A, Fox B, Selleck SB. 2000. *J. Biol. Chem.* 275:21856–61

136. The I, Bellaiche Y, Perrimon N. 1999. *Mol. Cell* 4:633–39

137. Thompson JD, Gibson TJ, Plewniak F, Jeanmougin F, Higgins DG. 1997. *Nucleic Acids Res.* 25:4876–82

138. Hall TM, Porter JA, Beachy PA, Leahy DJ. 1995. *Nature* 378:212–16

139. Chuang PT, McMahon AP. 1999. *Nature* 397:617–21

Annu. Rev. Biochem. 2004. 73:925–51
doi: 10.1146/annurev.biochem.73.011303.073756
First published online as a Review in Advance on March 18, 2004

RETURN OF THE GDI: The GoLoco Motif in Cell Division

Francis S. Willard, Randall J. Kimple, and
David P. Siderovski

*Department of Pharmacology, Lineberger Comprehensive Cancer Center, and UNC
Neuroscience Center, The University of North Carolina at Chapel Hill, Chapel Hill,
North Carolina, 27599-7365; e-mail: fwillard@med.unc.edu, kimplera@med.unc.edu,
dsiderov@med.unc.edu*

Key Words   asymmetric cell division, guanine nucleotide dissociation inhibitors,
heterotrimeric G proteins, RGS proteins

■ **Abstract**   The GoLoco motif is a 19-amino-acid sequence with guanine nucleotide dissociation inhibitor activity against G-alpha subunits of the adenylyl-cyclase-
inhibitory subclass. The GoLoco motif is present as an independent element within
multidomain signaling regulators, such as Loco, RGS12, RGS14, and Rap1GAP, as
well as in tandem arrays in proteins, such as AGS3, G18, LGN, Pcp-2/L7, and Partner
of Inscuteable (Pins/Rapsynoid). Here we discuss the biochemical mechanisms of
GoLoco motif action on G-alpha subunits in light of the recent crystal structure of
G-alpha-i1 bound to the RGS14 GoLoco motif. Currently, there is sparse evidence for
GoLoco motif regulation of canonical G-protein–coupled receptor signaling. Rather,
studies of asymmetric cell division in *Drosophila* and *Caenorhabditis elegans*, as
well as mammalian mitosis, implicate GoLoco proteins, such as Pins, GPR-1/GPR-2,
LGN, and RGS14, in mitotic spindle organization and force generation. We discuss
potential mechanisms by which GoLoco/Gα complexes might modulate spindle
dynamics.

CONTENTS

INTRODUCTION

Signal transduction via heterotrimeric G-protein–coupled receptors (GPCRs) typically evokes a switch in the status of the G-protein-alpha subunit (Gα), the guanine nucleotide-binding component of the G$\alpha\beta\gamma$ heterotrimer. Although normally tightly bound to its G$\beta\gamma$ partner when in the inactive, guanosine diphosphate (GDP)-bound state, Gα is converted to its active, guanosine triphosphate (GTP)-bound form via the guanine nucleotide exchange factor (GEF) activity of ligand-occupied GPCRs. GTP-bound Gα dissociates from G$\beta\gamma$, and thus both moieties become free to modulate the actions of a multitude of intracellular "effector" enzymes and ion channels (1, 2). Intrinsic guanosine triphosphate phosphohydrolase (GTPase) activity of Gα reverts the subunit back to the GDP-containing, G$\beta\gamma$-complexed form. The return of Gα to its ground state is dramatically hastened by "regulator of G-protein signaling" (RGS) proteins that serve as selective GTPase-accelerating proteins (GAPs) for various Gα subtypes (3, 4).

A common feature of RGS proteins is their possession of additional protein-protein interaction domains beyond their signature "RGS box" that exerts Gα-directed GAP activity (5, 6). In the original cloning of *loco*, the *Drosophila melanogaster* orthologue of *Rgs12*, Granderath, Klämbt, and colleagues identified a second, distinct interaction site for Gα subunits (region D) C-terminal to the RGS box of the encoded protein (7). Bioinformatic analyses of the region-D sequences from Loco and RGS12 led to our realization (8) that several other proteins, each previously identified as binding alpha subunits of the adenylyl-cyclase-inhibitory or Gi subclass (Gα_{i1-i3}, Gα_o, Gα_z), all harbored a highly conserved 19-amino acid polypeptide. Ponting (9) came to the same conclusion independently. We named this conserved polypeptide the GoLoco motif as an acronym for the G$\alpha_{i/o}$-Loco interaction (8). Several groups have since shown that binding of a Gα·GDP subunit to the GoLoco motif slows spontaneous nucleotide release (10–13). Hence, GoLoco motif-containing proteins are considered guanine nucleotide dissociation inhibitors (GDIs) for Gi-subclass alpha subunits.

Contemporaneously with the *in silico* discovery of the GoLoco motif, Cismowski, Lanier, and colleagues (14) identified three rat brain cDNAs in a yeast-based screen for receptor-independent activators of the G$\beta\gamma$-dependent pheromone signaling pathway. The third member of this disparate set of activator of G-protein signaling proteins (AGS3) was found to sequester GDP-bound Gα_{i2}

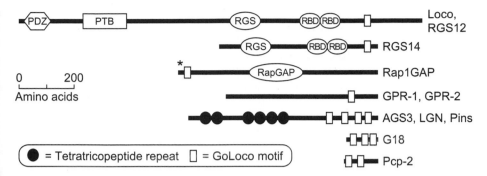

Figure 1 The $G\alpha_{i/o}$-Loco interaction, or GoLoco motif, is found singly, or in tandem arrays, in a number of different proteins. Domain abbreviations are PDZ, PSD-95/Discs large/ZO-1 homology domain; PTB, phosphotyrosine-binding domain; RGS, regulator of G-protein signaling box; RBD, Ras-binding domain; and RapGAP, Rap-specific GTPase-activating protein domain. Asterisk denotes N-terminal variation in GoLoco motif sequence between isoforms I and II of Rap1GAP.

in yeast, resulting in pheromone- and receptor-independent activation of $G\beta\gamma$-dependent responses (15). Lanier and colleagues (15) coined an alternative acronym, GPR for G-protein regulatory, to describe the highly conserved $G\alpha$·GDP-binding site found repeated four times within the C terminus of AGS3. However, the term GPR is already employed in the G-protein signaling field as one naming convention for orphan GPCRs (16). We will use the designation GoLoco for the rest of this review. [In the primary literature, the motif has never been called Loco homology domain contrary to use of that terminology in a recent review in this series (17)].

The GoLoco motif has now been identified in several distinct classes of proteins encoded in metazoan genomes (Figure 1), including modulators of heterotrimeric and Ras family G-protein signaling (RGS12, RGS14, Rap1GAP), several variations on the tetratricopeptide repeat (TPR), multi-GoLoco architecture of AGS3 (LGN, Pins, GPR-1/-2), and two short polypeptides with multiple GoLoco motifs (G18, Pcp-2/L7). The GoLoco motif was first discovered in the context of plasma membrane-delimited GPCR signaling, and its ability to bind $G\alpha$·GDP to the exclusion of $G\beta\gamma$ is proving a useful tool in examining receptor/G-protein/effector coupling. However, a central role has recently emerged for GoLoco motif-containing proteins in otherwise unexpected arenas, the control of mitotic spindle force generation and the act of cell division (18).

STRUCTURE AND FUNCTION OF THE GoLoco MOTIF

An alignment of all currently known GoLoco motifs is presented in Figure 2. The entire motif was predicted to fold as an amphipathic α-helix (10, 15). However, in the structure of the RGS14 GoLoco motif (amino acids 496–531) bound to $G\alpha_{i1}$·GDP

Name	Sequence	Range / Accession
zAGS3.GL4	EDFFSLIQKVQSK.RMDEQR	80- 98/ens|ENSDARP00000026753
(f,h,m,r)AGS3.GL4	EDFFSLIQRVQAK.RMDEQR	579- 597/ens|SINFRUP00000127441
(m,r)LGN.GL4	EDFFSLILRSQAK.RMDEQR	622- 640/ gb|AAL87447
(f1,z)LGN.GL4	EDFFSLIMRSQAK.RMDEQR	626- 644/ens|SINFRUP00000160450
hLGN.GL4	EDFFSLILRSQCK.RMDEQR	622- 640/ gb|AAN01266
fLGN2.GL4	DDFFSLILRSQSN.RMEEQR	597- 615/ens|SINFRUP00000146023
aPins.GL4	EDFFSLIMRLQCG.RMEDQR	653- 671/ gb|EEA14880
dPins.GL3	EDFFSLIMKVQSG.RMEDQR	613- 631/ gb|AAF64499
(m,r)G18.GL2	EQLYSTILSHQCQ.RIEAQR	104- 122/ gb|AAH21942
hG18.GL2	EQLYSTILSHQCQ.RMEAQR	105- 123/ gb|AAF67476
aPins.GL3	DAFLDMLMRCQCS.RIEEQR	599- 617/ gb|EEA14880
dPins.GL2	DDFLDMLMRCQGS.RLEEQR	552- 570/ gb|AAF64499
zAGS3.GL3	DEFFNMLIKYQSS.RINDQR	51- 69/ens|ENSDARP00000026753
(m,r)AGS3.GL3	DEFFNMLIKYQSS.RIDDQR	595- 613/ref|NP_700459
hAGS3.GL3	DDFFNMLIKYQSS.RIDDQR	574- 592/ gb|AAO17260
fAGS3.GL3	DDFFNMLIKCQSS.RIDDQR	545- 563/ens|SINFRUP00000127441
fLGN2.GL3	DVFFDMLVKCQCS.RLDDQR	563- 581/ens|SINFRUP00000146023
fLGN1.GL3	DDFFDMLVKCQGS.RLDDQR	592- 610/ens|SINFRUP00000160450
zLGN.GL3	DQFFDMLVKCQCS.RLEDQR	589- 607/ens|ENSDARP00000026291
(h,m,r)LGN.GL3	EDFFDILVKCQGS.RLDDQR	588- 606/ gb|AAL87447
rRap1GAP2	AEFFEMLEKMQCI.KLEEQR	163- 181/ref|XP_220692
rPcp-2.GL1	EGFFNLLSHVQGD.RMEEQR	24- 42/ref|XP_221787
hPcp-2.GL1	EGFFNLLSHVQGD.RMEGQR	8- 26/ gb|AAN52488
mPcp-2.GL1	EGFFNLLTHVQGD.RMEEQR	8- 26/ gb|AAN52485
fLGN2.GL1	DGFFELLSRFQGN.RLDDQR	450- 468/ens|SINFRUP00000146023
(h,m)G18.GL3	QELLELLLRVQGGRMEEQR	133- 152/ gb|AAF67476
rG18.GL3	QELLELLLRVQGGRMEDQR	131- 150/ref|XP_215346
aLoco	DELLEGLKRAQRS.RLEDQR	639- 657/ gb|EAA08111
dLoco	DELLEGLKRAQLA.RLEDQR	1355-1373/ gb|AAM50799
CbAGS3.GL3	DHLVEWLMRVQSQ.RLDDQR	521- 539/ens|ENSCBRP00000006892
CeAGS3.GL3	EHLVEWLMRVQGE.RLDEQR	522- 540/ gb|AAL27247
(h,m,r)RGS14	EGLVELLNRVQSS.GAHDQR	498- 516/ sp|O08773
dRap1GAP	QDLFELLERVQCS.RLDDQR	45- 63/ gb|AAF52527
(h,m,r)RGS12	EEFFELLSKAQSN.RADDQR	1188-1206/ sp|O08874
fRGS12	EEFFELLSRAQSA.RANDQR	503- 521/ens|SINFRUP00000158915
(f1,h,z)LGN.GL1	EGFFDLLSRFQSN.RMDDQR	476- 494/ens|SINFRUP00000160450
mPins/mLGN.GL1	EGFFDLLRRFQSN.RMDDQR	484- 502/ gb|AAL87447
rLGN.GL1	EGFFDLLRRLQSS.RMEDQR	481- 499/ens|ENSRNOP00000016617
(m,r)AGS3.GL1	ECFFDLLSKFQSS.RMDDQR	471- 489/ gb|AAF08683
fAGS3.GL1	DCFFDLLSKFQSS.RMDDQR	445- 463/ens|SINFRUP00000127441
hAGS3.GL1	ECFFDLLTKFQSS.RMDDQR	473- 491/ gb|AAO17260
(Cb,Ce)AGS3.GL1	EEFFDMLAKLQSK.RMNDQR	427- 445/ gb|AAL27247
aPins.GL1	EDFFDLLTRSQSK.RMDDQR	484- 502/ gb|EEA14880
dPins.GL1	DDFFEMLSRSQSK.RMDDQR	468- 486/ gb|AAF64499
(Cb,Ce)AGS3.GL4	EDVTAIVMRMQAG.RLEDQR	560- 578/ gb|AAL27247
CeRap1GAP	EDFLNMIERMQSN.RLDDQR	80- 98/ gb|AAK71368
CeGPR-1/-2	VDMMDLIFSM.SS.RMDDQR	425- 442/ref|NP_498900;NP_499066
CbGPR	MDFMDLICKM.NS.RMDDQR	421- 438/ens|ENSCBRP00000007225
(h,m,r)G18.GL1	ELLLDLVAEAQSR.RLEEQR	62- 80/ gb|AAH21942
zLGN.GL2	EPFLRLLANAQGR.RLDEQR	531- 549/ens|ENSDARP00000026291
(h,m,r)LGN.GL2	DEFLDLIASSQSR.RLDDQR	537- 555/ gb|AAL87447
fLGN2.GL2	GHFLELLASSQAR.RLDDQR	506- 524/ens|SINFRUP00000146023
(h,m,r,z)AGS3.GL2	EELFDLIASSQSR.RLDDQR	3- 21/ens|ENSDARP00000026753
fAGS3.GL2	EELFDLIASSQSR.RLDDQR	497- 515/ens|SINFRUP00000127441
CbAGS3.GL2	EVLIDLLLNAQER.RMDDQR	474- 492/ens|ENSCBRP00000006892
CeAGS3.GL2	EVLIDLLLNAQGR.RMDDQR	475- 493/ gb|AAL27247
mPcp-2.GL2	DNLMDMLVNTQGR.RMDDQR	27- 45/ sp|P12660
rPcp-2.GL2	DNLMDMLANTQGR.RMDDQR	64- 82/ref|XP_221787
hPcp-2.GL2	DSLMDMLASTQGR.RMDDQR	27- 45/ sp|Q8IVA1
(h,m,r)Rap1GAPII	TDLFEMIEKMQGS.RMDEQR	27- 45/ref|XP_233608
fRap1GAP	TELFEIIEKLQGS.RIDEQR	2- 20/ens|SINFRUP00000130548
aPins.GL2	NVLLEMIAHFQSE.RMDEQR	536- 554/ gb|EEA14880

GoLoco motif consensus: --ΦΦ-ΨΨ + Qπ   RΨ--QR

Alpha-helical   Triad

(19), only the first 13 residues (aa 496–508) adopted an α-helical configuration. This N-terminal α-helix is sandwiched within the Ras-like domain of the Gα subunit between the α3-helix and switch II (Figure 3), the latter being one of three flexible "switch" regions in Gα that adopts nucleotide-state-dependent conformations (20). Binding of the GoLoco peptide results in a significant displacement of switch II away from the α3-helix (19), thus deforming positions within GDP-bound Gα that normally serve as critical contact sites for Gβ in the G$\alpha\beta\gamma$ heterotrimer (21, 22). This observation supports early findings that formation of Gα·GDP-G$\beta\gamma$ and Gα·GDP-GoLoco complexes are mutually exclusive events (13, 15, 23).

The end of the GoLoco α-helix is anchored by burial into Gα of the nearly invariant glutamine found in the middle of the motif (Figure 2). C-terminal to this middle glutamine residue, the GoLoco peptide makes a more relaxed meander across the nucleotide-binding pocket and forms extensive contacts with the Gα all-helical domain (Figure 3C), contacts that are critical to Gα-binding selectivity (19). The spatial relationship between the N-terminal α-helix and the rest of the GoLoco motif is of critical importance to function because insertion of alanines between these two elements destroys all GDI activity (24).

Role of the Acidic-Glutamine-Arginine Triad

Of all the positions within the GoLoco motif, the most highly conserved are the final three residues that comprise an acidic-glutamine-arginine triad (Figure 2). The structure of the RGS14 GoLoco-Gα_{i1}·GDP complex identified the final residue of this triad as an arginine "finger" (19) that is positioned by the

---

←

Figure 2 Comprehensive multiple sequence alignment of all known GoLoco motifs. Individual motifs from multi-GoLoco proteins are numbered from N to C terminus as GL#. The N-terminal α-helix and C-terminal acidic-glutamine-arginine triad, key features of the GoLoco motif fold as defined by the crystal structure of the RGS14 GoLoco/Gα_{i1}·GDP complex (PDB accession 1KJY), are underlined below the GoLoco motif consensus. Species abbreviations are a, *Anopheles gambiae* (mosquito); Cb, *Caenorhabditis briggsae*; Ce, *Caenorhabditis elegans*; d, *Drosophila*; f, *Fugu rubripes*; h, human; m, mouse; r, rat; and z, *Danio rerio* (zebrafish). Sequence ranges and accession numbers from the Ensembl (ens), GenBank (gb), NCBI RefSeq (ref), and Swiss-Prot (sp) databases are denoted to the right of the sequence. Residues are colored according to side chain chemistry using Clustal-X defaults. Consensus symbols for amino acid character are hyphen (-), acidic; Φ, hydrophobic; Ψ, large aliphatic; (+), basic; and π, small side chain. Note that the GoLoco motif sequence denoted rRap1GAP2 is derived from a hitherto unpublished paralogue of Rap1GAP in the rat genome. The species abbreviation "f1" denotes sequence derived from the first of two LGN paralogues present in the *Fugu rubripes* genome (Figure 5). Based on sequence similarity and evolutionary relationships (Figure 5), mouse Pins should be called mouse LGN.

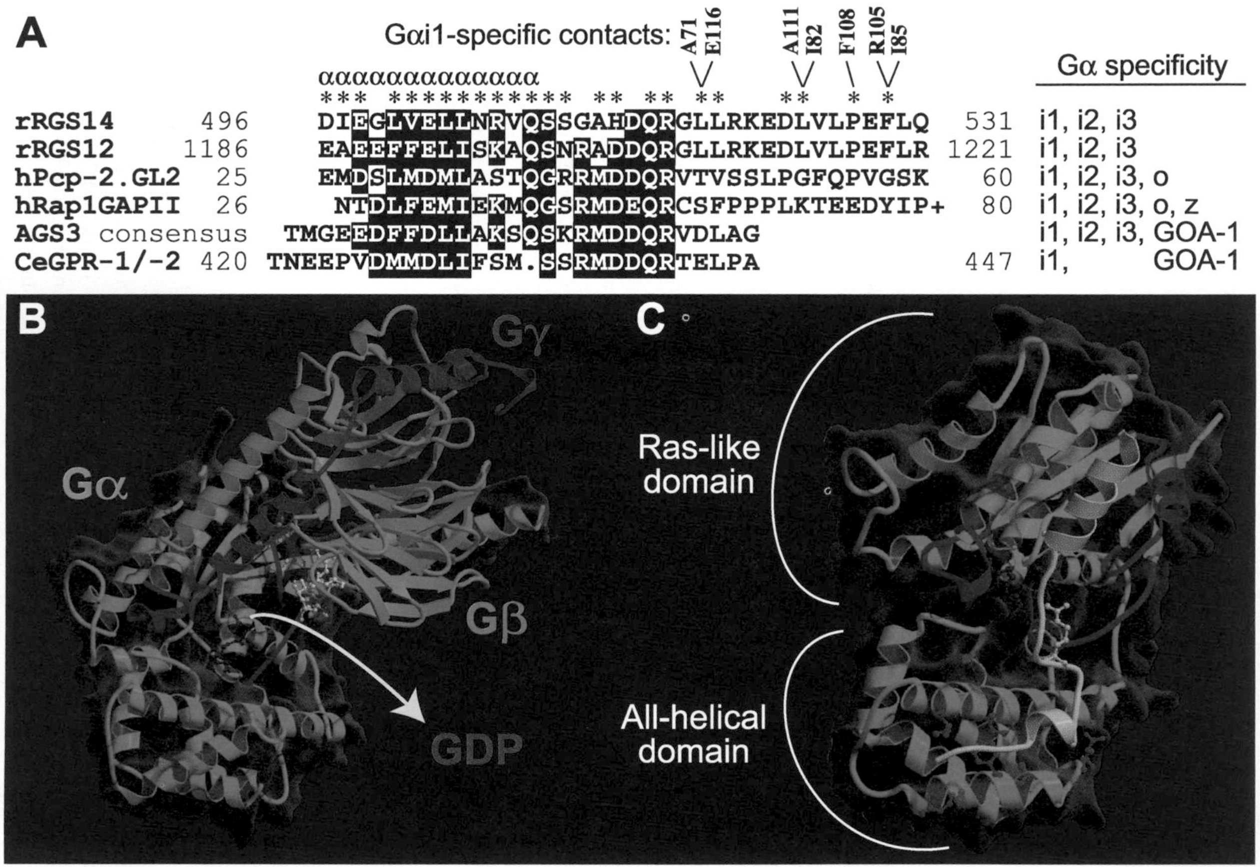

A

Gαi1-specific contacts:

```
                      αααααααααααααα
                      ****    ***********   **    **      **  ** *     *  *     *
rRGS14      496       DIEGLVELINRVQSSGAHDQRGLLRKEDLVLPEFLQ    531    i1, i2, i3
rRGS12      1186      EAEFFFELISKAQSNRADDQRGLLRKEDLVLPEFLR    1221   i1, i2, i3
hPcp-2.GL2  25        EMDSLMDMLASTQGRRMDDQRVTVSSLPGFQPVGSK    60     i1, i2, i3, o
hRap1GAPII  26        NTDLFEMIEKMQGSRMDEQRCSFPPPLKTEEDYIP+   80     i1, i2, i3, o, z
AGS3 consensus        TMGEEDFFDLLAKSQSKRMDDQRVDLAG                  i1, i2, i3, GOA-1
CeGPR-1/-2  420       TNEEPVDMMDLIFSM.SSRMDDQRTELPA          447    i1,        GOA-1
```

Gα specificity

B

Gα   Gγ   Gβ   GDP

C

Ras-like domain   All-helical domain

preceding two residues to reach into the nucleotide-binding pocket of Gα and contact the alpha- and beta-phosphate groups of GDP (Figure 4). This positioning of an arginine side chain in *trans* to contact the guanine nucleotide is reminiscent of the catalytic arginine fingers employed by GAPs for Ras- and Rho-family GTPases (25, 26). Arg-178 within switch I of Gα$_{i1}$ works in *cis* like the arginine finger of a RasGAP or RhoGAP and stabilizes the developing negative charge on the gamma-phosphate leaving group during GTP hydrolysis (27). In the GDP-bound ground state of uncomplexed Gα, the Arg-178 guanidinium group contributes to GDP binding by forming hydrogen bonds with the alpha- and beta-phosphate oxygens (28). However, within GoLoco-bound Gα$_{i1}$, the Arg-178 side chain is displaced and instead contacts the 3′ hydroxyl of the GDP ribose sugar moiety (Figure 4), and forms a salt bridge interaction with the side chain of Glu-43. [This pairing of Arg-178 and Glu-43 is normally seen in Gαβγ heterotrimers but not in the uncomplexed state of Gα·GDP (22).]

A principal aspect of GoLoco-mediated GDI activity therefore appears to be the creation of new contacts to GDP by remodeling Arg-178 and adding a second arginine in *trans* from the GoLoco motif triad. Replacing the triad arginine with a bulky phenylalanine eliminates all GDI activity of the fourth GoLoco motif of AGS3 (23) and of an AGS3-derived GoLoco consensus peptide (11); substitution with phenylalanine also completely eliminates the ability to bind Gi-class α subunits (15). Mutation of the triad arginine to less bulky alanine or leucine

Figure 3 Residues C-terminal to the core GoLoco motif contact the Gα all-helical domain and are important determinants of Gα binding specificity and GDI activity. (*A*) Alignment and sequence ranges of minimal functional GoLoco-motif polypeptides with their known Gα binding specificity are indicated on the right. The core GoLoco motif is boxed in black. Alpha symbols (α) and asterisks (*) denote alpha-helical secondary structure and GoLoco contacts with Gα, respectively, as observed in the 2.7 Å crystal structure of the RGS14 GoLoco/Gα$_{i1}$·GDP complex (1KJY). Contacts to Gα$_{i1}$-specific residues in the all-helical domain are identified by connecting lines. The plus sign (+) after residue 80 in the human Rap1GAP isoform II sequence denotes the fact that the minimal functional Gα$_z$-interacting domain of Rap1GAP isoform I (40) has only been minimized to the first 74 amino acids of that isoform (which starts at methionine 32 relative to the illustrated Rap1GAPII sequence). Other abbreviations follow from Figure 2. (*B*) Model of the Gβγ lever hypothesis, as proposed by Rondard et al. (32), to explain GPCR-mediated guanine nucleotide exchange activity. Gα$_{i1}$ (*green* with translucent space-filling shell; switch regions in *blue*) makes side chain contacts (*white*) with residues in Gβ (*cyan*) of the Gβγ dimer (Gγ in *purple*). Outward movement of switch I and switch II by receptor-induced rotation of Gβγ is proposed to allow the egress of GDP (*brown*) from the nucleotide-binding pocket (32). (*C*) The RGS14 GoLoco-motif peptide (aa 496–531; *yellow*) binds across the Ras-like and all-helical domains of Gα$_{i1}$, trapping GDP within. Note the relative position of the GoLoco triad arginine finger (ball-and-stick representation). Gα$_{i1}$-specific contacts to the GoLoco peptide, as denoted in panel (*A*), are illustrated in pink.

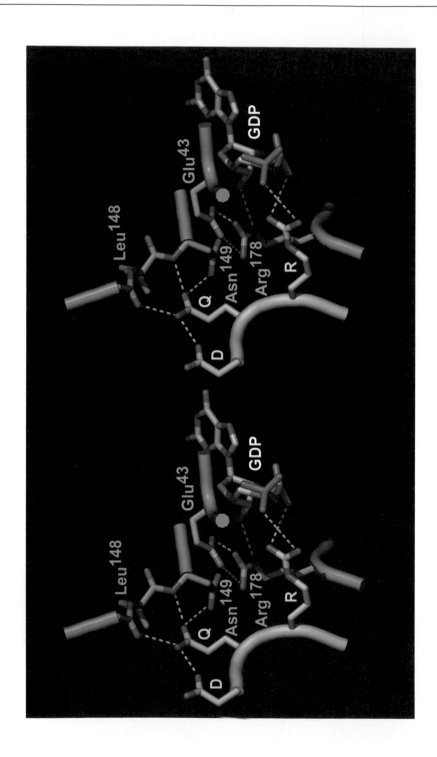

causes, in contrast, a significant decrease in the GDI activity of the RGS14 GoLoco motif without a concomitant decrease in binding affinity for $G\alpha$ (19). Since weak but measureable GDI activity still remains upon substituting the triad arginine with alanine or leucine (19, 24), the GoLoco arginine finger should be considered a principal, but not an absolute, determinant of guanine nucleotide dissociation inhibitor activity.

The invariant glutamine residue that just precedes the arginine finger in the triad (Figure 2) points away from the GDP binding pocket and makes extensive side chain and backbone interactions with Gln-147 and Asn-149 of $G\alpha_{i1}$ (19), thus "kinking" the GoLoco peptide backbone and allowing full extension of the arginine side chain into the nucleotide-binding pocket (Figure 3C and 4). These interactions between triad glutamine and $G\alpha_{i1}$ are critical for GDI activity. Replacing the triad glutamine with alanine eliminates the $G\alpha$-binding and GDI functions of the AGS3 consensus peptide (24), and replacing Asn-149 of $G\alpha_{i1}$ with isoleucine leads to an insensitivity to GDI activity normally exerted by AGS3 and Pcp-2 (29). The latter mutation to $G\alpha_{i1}$ was made in the context of a three-position exchange with $G\alpha_s$ residues (Arg-144 to Asn, Asn-149 to Ile, and Ser-151 to Cys) and was originally interpreted as reflecting direct GoLoco motif interactions with the all-helical domain/switch III interface of $G\alpha_{i1}$ (29). Reappraisal of this conclusion in light of the RGS14-$G\alpha_{i1}$·GDP crystal structure suggests that only the Asn 149 to isoleucine substitution is responsible for the observed GoLoco insensitivity. Arg-144 is far removed from the RGS14 GoLoco peptide-binding site, and the side chain of Ser-151 is involved in hydrogen bonds with the ribose sugar hydroxyl groups of the bound GDP (19).

Preceding the invariant glutamine of the GoLoco motif triad are two acidic residues (Figure 2). Only two GoLoco motifs lack an aspartic acid or glutamic acid at the position immediately adjacent to the invariant glutamine: the second motif of G18 (30) and the first motif of human Pcp-2 (31). Within the asymmetric unit of the RGS14/$G\alpha_{i1}$ crystal (PDB accession number 1KJY), one of the two GoLoco/$G\alpha$ dimers contains this acidic residue in a side chain hydrogen bond with the side chain of the following glutamine (Figure 4). This bond helps anchor the triad glutamine and supports the positioning of the arginine finger. This acidic residue of the GoLoco triad is also important for function; the second GoLoco motif of the triple-motif protein G18, which has an alanine residue at this position

---

Figure 4   Stereo view of the contacts made by the GoLoco motif acidic-glutamine-arginine triad to $G\alpha$ and guanosine diphosphate (GDP), as determined by the crystal structure of the RGS14 GoLoco/$G\alpha_{i1}$·GDP complex (PDB accession 1KJY). RGS14 triad residues aspartate-514 (D), glutamine-515 (Q), and arginine-516 (R) are drawn in yellow; $G\alpha_{i1}$ side chains (Glu-43, Leu-148, Asn-149, and Arg-178) and the intervening backbone are in green; and GDP is rendered in the CPK (Corey-Pauling-Koltun) color scheme: carbon in white, nitrogen in blue, oxygen in red, and phosphorus in magenta. Dotted yellow lines represent hydrogen bonds.

(Ala-121) (Figure 2), neither exhibits GDI activity nor binds $G\alpha$ subunits in vitro, and replacement with aspartate results in a robust gain of $G\alpha_{i1}$-binding and GDI activities (30). Similarly, increased activity is seen for the first GoLoco motif of human Pcp2 upon replacing the glycine residue at this triad position (Gly-24) (Figure 2) with glutamate (our unpublished observations), the amino acid normally present in this position in the rat and mouse orthologues.

Role of Residues C-Terminal to the Conserved Motif

The minimal conserved GoLoco motif ends with the acidic-glutamine-arginine triad and thus is generally 19 residues long. Variants include the single, 18-residue GoLoco motifs present in *C. elegans* and *Caenorhabditis briggsae* GPR proteins, which lack the α-helix-ending central glutamine, and the 20-residue third GoLoco motif within G18, which possesses three glycine residues between the N-terminal α-helix and conserved triad regions (Figure 2). However, and importantly, this minimal conserved sequence is generally not sufficient to mediate interaction with $G\alpha$ subunits. Residues C-terminal to the highly conserved motif (Figure 3A) are poorly conserved yet required for robust GDI activity. [One study reports in vitro GDI activity exerted by an internal, 19-residue span of the AGS3 consensus peptide (24), GFFDLLAKSQSKRMD-DQRV. However, this activity was observed at an inordinately high molar excess of peptide (100 μM) versus $G\alpha$ target (100 nM), and neither IC_{50} nor dissociation constant values for this interaction were reported.]

The structure of the RGS14 GoLoco/$G\alpha_{i1}$·GDP complex suggests a reason for the requirement of residues C terminal to the acidic-Glu-Arg triad in GoLoco motif function—these residues make extensive contacts with the all-helical domain of $G\alpha_{i1}$ (19), affording GoLoco motif proteins the ability to cross over the nucleotide-binding pocket and span both lobes of the $G\alpha$ structure (Figure 3C). One proposed mechanism of GPCR-mediated nucleotide release involves agonist-bound receptor using $G\beta\gamma$ as its lever (32) to induce switch I and switch II of $G\alpha$ to peel back and open the lip of the nucleotide-binding pocket found at the cleft between its Ras-like and all-helical domains (Figure 3B). Hence, beyond direct GDP contact by the arginine finger, other likely aspects of GoLoco-mediated GDI activity include blocking the route of nucleotide egress (29) and restricting any interdomain movement within $G\alpha$ that might be necessary for nucleotide ejection (Figure 3B versus 3C).

Although the GoLoco regions of RGS12 and RGS14 preferentially interact with $G\alpha_{i1}$, $G\alpha_{i2}$, and $G\alpha_{i3}$, and not $G\alpha_o$ (12, 19), other GoLoco-containing proteins can also interact with $G\alpha_o$ (AGS3, LGN, Pcp-2, Rap1GAP isoform II) (13, 19, 23, 33–38), $G\alpha_t$ (AGS3) (23, 39), and $G\alpha_z$ (Rap1GAP isoform I) (40, 41). The RGS14 GoLoco-$G\alpha_{i1}$·GDP structure revealed that the primary determinants of $G\alpha$ selectivity reside within the contacts made between the $G\alpha$ all-helical domain and residues C-terminal to the 19-residue core conserved GoLoco motif (Figure 3A and 3C). Indeed, RGS14 can exert GDI activity on a chimeric $G\alpha_o$ subunit containing the all-helical domain from $G\alpha_{i1}$, yet the same

RGS14 GoLoco peptide is unable to act as a GDI for a chimeric $G\alpha_{i1}$ containing the all-helical domain from $G\alpha_o$ (19). Moreover, exchanging the C-terminal residues (aa 496–531) (Figure 3A) of the RGS14 GoLoco peptide for those found C-terminal to the second GoLoco motif of Pcp-2 (aa 46–60) (Figure 3A) creates a chimeric GoLoco peptide with GDI activity on wild-type $G\alpha_o$ nearly equal in potency to that of wild-type Pcp-2; conversely, the reciprocal chimera, containing the conserved second GoLoco motif of Pcp-2 with the C-terminal residues from RGS14, fails to interact with $G\alpha_o$ (19).

Modulation of GDI Activity by Phosphorylation

Little is currently known about how GoLoco motif activity is regulated in vivo. Two recent reports have proposed that phosphorylation of GoLoco-motif proteins might be one mechanism by which GDI activity can be modulated. Hollinger and colleagues (42) found that RGS14 is phosphorylated in rat B35 neuroblastoma cells by cAMP-dependent protein kinase (PKA). In vitro phosphorylation of recombinant RGS14 protein by PKA occurs at two sites, Ser-258 and Thr-494; the latter site is just N-terminal to the start of the GoLoco motif. Mimicking PKA phosphorylation of Thr-494 by mutation to aspartate or glutamate increases nearly threefold the in vitro GDI activity of RGS14 toward $G\alpha_{i1}$ (42). It is unfortunate that the structure of the RGS14 GoLoco motif peptide (aa 496–531) does not extend N-terminal to the Thr-494 residue (19); thus it remains conjecture whether phosphorylation at this site contributes directly to the interaction with $G\alpha$ or results in structural changes within RGS14 that increase GoLoco motif accessibility. Increased cellular PKA activity is the principal outcome of G_s-coupled receptor stimulation (via adenylyl cyclase activation and the accumulation of cyclic AMP); hence, enhancement of $G\alpha_i$-directed GDI activity mediated by PKA phosphorylation could play a role in cellular cross-modulation of adenylyl cyclase-stimulatory (G_s) and adenylyl cyclase-inhibitory (G_i) GPCR signaling pathways, either by decoupling G_i-linked receptors and/or augmenting effector modulation by $G\beta\gamma$ subunits freed from G_i heterotrimers.

In a yeast two-hybrid screen for AGS3 interactors, Blumer and coworkers (43) identified LKB1/STK11, the mammalian homologue of serine/threonine kinases in *C. elegans* (PAR-4) and *Drosophila* (LKB1) required for establishing early embryonic anterior-posterior axis formation (44, 45). Immunoprecipitated LKB1 was found to phosphorylate a recombinant protein containing the four GoLoco-motif C-terminal region of AGS3 (aa 463–650), a region containing 24 serine and threonine residues, only 9 of which are present within the conserved GoLoco motifs (AGS3.GL1–4) (Figure 2). It is currently unknown which specific serine/threonine residue(s) within AGS3 is phosphorylated by LKB1. Nevertheless, Blumer and colleagues (43) chose to phosphorylate the AGS3 consensus peptide (Figure 3A) at Ser-16 (C-terminal to the nearly invariant middle glutamine) and reported that phosphorylation at this site diminishes GDI activity in vitro. The physiological relevance of this finding is unknown in the absence of evidence that this serine is actually targeted for phosphorylation in vivo.

GoLoco MOTIF PROTEINS IN GPCR PATHWAY MODULATION

GoLoco Peptides as Tools to Uncouple GPCRs

The ability of the GoLoco motif to bind $G\alpha_{i1}$·GDP subunits and prevent concomitant $G\beta\gamma$ association has motivated investigations of whether GoLoco-motif proteins play a role in modulating cellular GPCR signaling pathways. It has been demonstrated, for example, that AGS3 can attenuate rhodopsin-catalyzed activation of transducin and high-affinity agonist binding to 5-HT_{1A} receptors (11, 38, 39). These studies were performed in cell-free systems that rely on reconstitution of GPCR signaling with purified or semipurified components; thus, it cannot be directly inferred that retinal phototransduction and/or serotonin signaling pathways are the direct targets of GoLoco protein modulation.

These studies do, however, highlight the utility of the GoLoco motif as a tool for selective decoupling of G_i-linked GPCRs. For example, intracellular micro-injection of GoLoco peptides into AtT20 mouse pituitary corticotroph cells can selectively antagonize G_i-linked dopamine D2 receptor-mediated enhancement of the G-protein gated inward rectifier potassium (GIRK/K_{ir}3.x) current without affecting somatostatin-induced (G_o-linked) GIRK current activation (46). In this system, the initial GIRK current response to quinpirole application remains unaffected by GoLoco peptide injection, yet subsequent applications of agonist elicit progressively reduced potassium currents. Mutation of the triad arginine within the GoLoco peptide to phenylalanine abrogates its ability to uncouple the D2 receptor. Hence, it appears that GoLoco-mediated decoupling of D2 receptors in this system depends on the GoLoco peptide having access to free $G\alpha$·GDP subunits, as afforded by agonist stimulation and the resultant cycle of receptor GEF activity, heterotrimer separation, and GTP hydrolysis.

These results in AtT20 cells argue against GoLoco motif peptides having any innate ability to displace the $G\beta\gamma$ subunit from a preformed receptor-coupled $G\alpha\beta\gamma$ heterotrimer in vivo. However, a recent, provocative report by Ghosh and coworkers (47) proposes just such an activity for the AGS3 consensus peptide. At a 20,000-fold molar excess of GoLoco peptide (1 μM) over heterotrimer (50 pM), Ghosh et al. observe an accelerated rate of G-protein subunit dissociation in vitro. It remains untested whether scrambled or loss-of-function point-mutant variants of the AGS3 consensus peptide also share this activity (47); moreover, it remains unknown whether such a high relative ratio of GoLoco motif to heterotrimer is ever achieved in a normal cellular context.

Loco and RGS12

Although the existence of the GoLoco motif was first inferred from the detection of a second $G\alpha_i$-binding site within the *Drosophila* protein Loco (7), the physiological role of the GoLoco motif within Loco has not yet been determined. Embryos lacking the *loco* gene display defects in glial cell differentiation and consequently fail to hatch; rare adult escapers that do eclose show severe impairment of spontaneous locomotor activity (7). Loco was also found to be required for dorsal-ventral pattern formation in the *Drosophila* embryo (48). Yet despite this knowledge, no specific G-protein modulatory functions have been ascribed to Loco to help explain its particular role in these processes, nor has a GPCR been identified as having modulatory activity on Loco function.

In contrast, the avian Loco orthologue, RGS12, functions in the context of a GPCR-initiated signal transduction cascade: RGS12 controls the rate of desensitization from γ-aminobutyric acid (GABA)-mediated inhibition of the N-type calcium channel ($Ca_v2.2$) in chick dorsal root ganglion neurons (49). RGS12 is recruited, via its phosphotyrosine-binding (PTB) domain, to the α_{1B} pore-forming subunit of $Ca_v2.2$ in a tyrosine kinase-dependent manner (49). It is currently unknown, however, what specific role the C-terminal GoLoco motif within RGS12 plays in modulating $Ca_v2.2$ inhibition by the G_o-linked $GABA_B$ receptor.

Rap1GAP Isoforms

Currently, the clearest demonstration of a GoLoco/$G\alpha$ interaction being involved in GPCR-mediated modulation of a cellular signaling pathway comes from studies of Rap1GAP, a negative regulator of the Ras-related GTPase, Rap1 (50). Meng and coworkers (40) previously found that Rap1GAP (isoform I) binds $G\alpha_z$ in its activated, GTP-loaded form (a departure from the normal GoLoco motif requirement for a GDP-bound $G\alpha$ subunit). $G\alpha_z$ activation in PC12 cells, via agonist stimulation of α_{2A}-adrenergic receptors, was subsequently shown to recruit Rap1GAP to the plasma membrane (41). $G\alpha_z$-mediated recruitment of Rap1GAP attenuated Rap1-mediated ERK activation and neurite development (41), suggesting that G_z-linked GPCR signaling can antagonize the Rap1/B-Raf/ERK signal transduction cascade in PC12 cells via Rap1GAP translocation to the plasma membrane.

Similarly, Mochizuki and coworkers (37) have found that an N-terminally extended variant of Rap1GAP (isoform II) binds to activated $G\alpha_{i1}$ and $G\alpha_{i2}$ subunits. Activation of the Gi-linked M_2 muscarinic acetylcholine receptor was shown to recruit Rap1GAPII to the plasma membrane and lower cellular levels of GTP-loaded Rap1 (37). However, this reduction in activated Rap1 correlated with an increase in ERK activation. Differences between the findings of Meng et al. and Mochizuki et al. could be the result of differing Rap1GAP isoforms examined and/or differing operative Rap1-effector pathways in the cell lines used: PC12 rat pheochromocytomas (41) versus human embryonic kidney 293T fibroblasts (37).

GoLoco MOTIF PROTEINS IN CELL DIVISION

The seminal observation that first placed a GoLoco motif protein within an unconventional G-protein-signaling paradigm was the finding that a complex comprising $G\alpha$ and the multi-GoLoco protein, Pins, is a crucial component for dictating asymmetric cell division in *Drosophila* neuroblasts (51). Asymmetric cell division (ACD) is used by many organisms during development to generate cellular diversity [reviewed by Knoblich (52)]. Conventional cell division produces two identical daughters, whereas in ACD, RNAs and proteins that determine cell fate are asymmetrically segregated into the two daughter cells. Consequently, daughter cells derived from ACD acquire different developmental potentials. The first step in ACD requires the establishment of an axis of polarity. The second step involves unequal distribution of cell fate determinants along this axis. The third and final step is the orientation of the mitotic spindle along this axis so that cell division segregates these cell fate determinants unequally to produce different daughter cells.

Asymmetric Cell Division in *Drosophila* Neuroblasts

Drosophila neuroblast mitosis is a commonly studied example of ACD (53, 54). Neuroblast cells delaminate from the ventral neuroectoderm and adopt an apical-basal axis of polarity. Subsequent asymmetric division produces a large apical neuroblast and a smaller ganglion mother cell (GMC). Apical neuroblasts can undergo further asymmetric divisions, whereas GMCs are committed to differentiating into neurons or glia. The apical determinants Bazooka (PAR-3), DmPAR-6, and atypical protein kinase C (DaPKC) form a protein complex that recruits the Inscuteable (Insc) protein to the apical cell cortex (55, 56), directs spindle orientation, and helps segregate the basal determinants Numb (a PTB domain-containing protein), Miranda (a coiled-coil protein), and Prospero (a transcription factor).

Insc is expressed during neuroblast delamination and localizes to the apical cell cortex; loss of Insc perturbs spindle orientation and randomizes axes of cell division (57). Deletional analysis has defined a central "asymmetry domain" within Insc as being sufficient to mediate all known functions of the full-length protein (58, 59). In the search for asymmetry domain-interacting proteins as potential Insc effectors, three groups independently discovered the Partner of Inscuteable (Pins/Rapsynoid) protein (51, 60, 61). Pins is the archetypal member of an evolutionarily conserved class of TPR and GoLoco motif-containing proteins involved in cell division processes (Figure 5). Pins is recruited by Insc to the apical cell cortex of delaminated neuroblasts; ablation of maternal and zygotic Pins results in defective spindle orientation, a failure to segregate determinants asymmetrically, and a limited asymmetry in the neuroblast division (51, 60, 61). These phenotypes are equivalent to those observed in *inscuteable*-deficient embryos (57), suggesting that Pins is the predominant effector for Insc function.

Heterotrimeric G-Protein Involvement in *Drosophila* ACD

Insc, Pins, and Gαi form an apical protein complex (Figure 6) essential for ACD in *Drosophila* neuroblasts (51, 62, 63). Pins is selective for GDP-bound Gαi, like

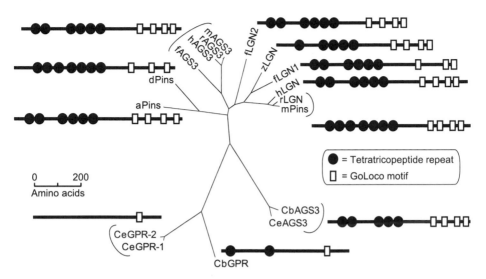

Figure 5   Phylogenic relationship and multidomain architectures of the Pins family of GoLoco proteins. Tetratricopeptide repeat (TPR) regions are illustrated based on detection by Pfam (http://pfam.wustl.edu/) or SMART (http://smart.embl-heidelberg.de/) hidden Markov models. The N termini of *C. elegans* GPR-1 and GPR-2 proteins are predicted to form stable all-alpha-helical folds comprised of TPRs via protein fold recognition algorithms, as implemented by the 3D-PSSM web server (http://www.sbg.bio.ic.ac.uk/~3dpssm). Species abbreviations follow from Figure 2.

other GoLoco proteins (excluding Rap1GAP). Pins is also able to bind Gαo in vitro, but its partner in vivo appears to be Gαi (51, 62). Moreover, elements within the Insc/Pins/Gαi·GDP ternary complex appear mutually codependent for apical localization. Overexpression of wild-type Gαi inhibits polarization of asymmetry determinants and causes mitotic spindle misorientation (62). However, overexpression of constitutively active Gαi^{Q205L} has essentially no effect on neuroblast ACD. Thus, in the context of *Drosophila* neuroblast ACD, the active Gα species appears not to be GTP-bound Gαi but rather Gαi·GDP in complex with Pins. It is possible that Gβγ sequestration as caused by Gαi overexpression leads to the observed aberrations to neuroblast ACD. However, significant differences are seen with respect to relative size of daughter cells and Miranda mislocalization upon Gαi overexpression versus the loss of Gβ13F (the *Drosophila* orthologue of conventional mammalian Gβ subunits β1-β4) (62). These results suggest that the effects of wild-type Gαi overexpression on ACD are not due solely to Gβ13F subunit sequestration; however, this does not necessarily exclude a role for Gβγ signaling in ACD.

In contrast to the apical localization of Gαi, Gβ13F is uniformly expressed throughout the neuroblast cortex (62). Nevertheless, Gβ13F appears to be required for neuroblast ACD because neuroblasts in Gβ13F-deficient embryos

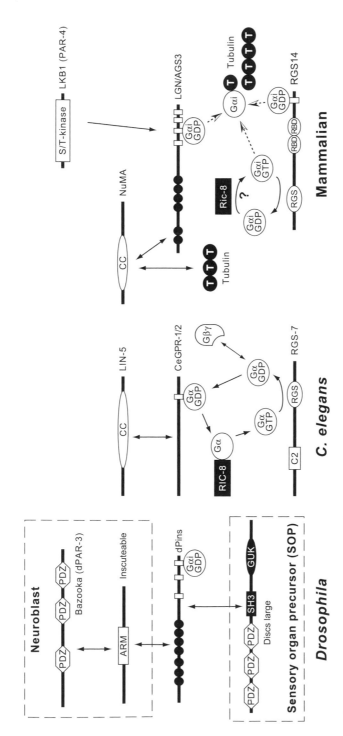

Figure 6 Comparable signal transduction complexes, centered around Pins family members, regulate *Drosophila* (*left*) and *C. elegans* (*middle*) asymmetric cell divisions and mammalian cell division (*right*). Interactions between signaling components are abstracted from genetic and direct biochemical evidence. Domain abbreviations are ARM, weakly predicted Armadillo repeats; C2, homology to conserved region 2 of protein kinase C; CC, coiled-coil region; GUK, membrane-associated guanylate kinase homology domain; PDZ, PSD-95/Discs large/ZO-1 homology domain; RBD, Ras-binding domain; RGS, regulator of G-protein signaling box; SH3, Src homology-3 domain; and S/T-kinase; serine/threonine kinase domain.

exhibit a high frequency of symmetric division (64). Fuse and colleagues (64) have found that ectopic coexpression of Gβ13F and Gγ1 reduces spindle size, in contrast to the large spindles seen in *Gβ13F*-null neuroblasts, suggesting that Gβγ signaling may act to suppress spindle formation. Confounding these analyses is the fact that one consequence of *Gβ13F* deletion is concomitant loss of Gαi expression (62). To overcome this, Yu and colleagues (65) recently generated and analyzed loss-of-function Gαi mutants, leading them to propose that Gβγ signaling is involved in ACD upstream of both Bazooka/DaPKC and Pins/Gαi complexes. Gβ13F appears to regulate the asymmetric localization and stability of both the Bazooka/DaPKC and Pins/Gαi pathway components, in a fashion not attributable to loss of Gαi expression.

In this light, recent genetic evidence from Cai et al. (66) suggests that *Drosophila* ACD may be defined by two, functionally overlapping, parallel pathways, one containing Bazooka/DaPKC and the other comprising Pins/Gαi. This is elegantly illustrated by the division of *Drosophila* sensory organ precursor (SOP) pI cells [a second model system for ACD; reviewed in (52)]. pI cells divide asymmetrically along an anterior-posterior axis to produce anterior pIIa and posterior pIIb daughter cells. The cell surface serpentine receptor Frizzled acts to reorganize an initial apical-basal polarity into an anterior-posterior polarity during interphase. In sharp contrast to neuroblasts, Pins and Gαi localize in SOP pI cells to the anterior cortex, whereas Bazooka/DaPKC localize to the posterior cortex (62, 67) and act antagonistically to Pins/Gαi (66). Division of SOP pI cells is asymmetric because the spindle is offset somewhat toward the anterior and resultant daughter cells have distinct fates as Numb is segregated solely to the pIIb cell (52).

The lack of Insc expression in pI cells can explain the phenotypic difference between SOP and neuroblast ACD. Ectopic expression of Insc in pI cells recruits the Bazooka/DaPKC complex to the anterior, and consequently, the anterior spindle becomes larger and cell division more asymmetric (66, 67). In the normal absence of Insc, the Pins/Gαi complex is recruited to the anterior by the SH3 domain of the Discs large (Dlg) protein, which can directly bind an unspecified target sequence within Pins (67) (Figure 6). Thus, weak spindle asymmetry in the SOP pI cell division, given opposing localizations of Pins/Gαi and Bazooka/DaPKC complexes, produces only a mild size difference in resultant pIIa and pIIb daughters, whereas in neuroblasts, strong asymmetry given cosegregating complexes produces a prominent size differential between GMC and neuroblast progeny.

Mechanism of G-Protein Activation in *Drosophila* ACD

In canonical G-protein signaling, the Gαβγ heterotrimer is activated by GPCR-mediated nucleotide exchange and the separation of Gβγ from Gα·GTP (1, 2). It is currently unclear whether this paradigm also holds true in ACD. In *Drosophila* neuroblasts, the Insc/Pins/Gαi·GDP complex is devoid of Gβ13F (62), consistent with the mutually exclusive nature of GoLoco- versus Gβγ-binding to GDP-loaded Gα subunits. Schaefer et al. (62) report that preincubation of neuroblast lysates with recombinant Pins, or a peptide corresponding to its final GoLoco motif, can disrupt the Gαi/Gβγ interaction, as measured by the loss of Gβ from immunoprecipitated Gαi. Coupled with the recent in vitro work of Ghosh and colleagues (47) previously described, this finding leads to a provocative hypothesis that, in ACD, Pins acts to "activate" (i.e., separate) heterotrimeric G proteins without nucleotide exchange on the Gα subunit. An alternative hypothesis is that the addition of Pins or GoLoco peptide elicits an innate Gβγ releasing factor within the neuroblast lysate.

Current opinion suggests that G-protein action in ACD is independent of any GPCR-mediated GEF activity; recent biochemical and genetic evidence for a receptor-independent GEF for Gα subunits (i.e., Ric-8; see next section) supports the lack of a strict requirement for GPCR function if exchange activity is a requirement for ACD. However, GPCR signaling has not been formally disproven to be involved in ACD. Indeed, GPCR signaling can be peripherally linked with ACD, given that SOP polarization is directed by the actions of Frizzled receptors that bear more than a passing resemblance to canonical GPCRs (68). However, as previously stated, significant differences do exist between ACD in sensory organ precursors and in neuroblasts. In particular, expression of Gαi^{Q205L} in neuroblasts is without any effect on ACD, whereas in SOP cells, ACD is perturbed (62). This finding suggests that potential G-protein effectors of ACD may differ in SOP cells versus neuroblasts and that GTP-bound Gαi has a specific role to play in SOP division.

The functional interplay within the Pins/Gαi complex has yet to be clearly delineated. It may be that the sole function of Gαi (a lipidated and, thus, membrane-associated protein) is to recruit Pins to the plasma membrane. In this vein, it has recently been shown that the GoLoco region of Pins specifies cortical recruitment, whereas the first three TPRs mediate apical localization (63). An alternative speculation, however, is that specific effectors exist, but are hitherto unrecognized, for GoLoco-complexed Gαi subunits (18).

ACD in the *C. elegans* One-Cell Embryo

Studies in the nematode worm *C. elegans* have recently provided crucial insights into the mechanisms of G-protein signaling in ACD, as reviewed by Gönczy (69). The one-cell stage *C. elegans* embryo (P$_0$) divides asymmetrically to form a large anterior blastomere (AB) and a smaller posterior blastomere (P$_1$). Anterior-posterior polarity is thought to be initiated by the sperm aster shortly after fertilization. This paternally derived cue directs the polarized distribution of an evolutionarily conserved set of proteins that regulate cell polarity and ACD. This is exemplifed by the partitioning-defective (PAR) proteins, the analysis of which has illuminated the study of cell polarity in several metazoan organisms (70). In the *C. elegans* one-cell embryo, PAR-3, PAR-6, and the atypical PKC-3 are localized to the anterior cortex, whereas PAR-1 and PAR-2 are localized to the posterior cortex. The mitotic spindle at metaphase is symmetrically positioned along the anterior-posterior axis. During anaphase, the anterior spindle position is fixed, but the posterior spindle moves toward the posterior cortex. This movement is accompanied by vigorous rocking of the spindle as it elongates. Characteristically, the spindle flattens in telophase, and because the cleavage furrow forms to bisect the spindle, the daughter cells (AB and P_1) are destined to be of unequal size, with the resultant AB blastomere being larger (60% of initial cell volume) than the P_1 cell (40%). In concert with genetic studies, spindle-severing experiments demonstrate that polarity provided by PAR proteins results in an increased pulling force on the posterior spindle (71). These "extra-spindle" pulling forces act along astral microtubules and are exerted unequally on the spindle poles. This imbalance of extra-spindle pulling forces is under the control of polarity cues (72).

Heterotrimeric G-Protein Involvement in *C. elegans* ACD

The first evidence that heterotrimeric G-protein-signaling pathways might function in the regulation of *C. elegans* cell division came when Zwaal and colleagues (73) observed that Gβ regulates centrosome migration in the *C. elegans* early embryo. Subsequently, Gotta & Ahringer (74) demonstrated that simultaneous RNA interference (RNAi)-mediated knockdown of two *C. elegans* Gα subunits, GOA-1 and GPA-16, causes a spindle positioning defect that results in a symmetric P_0 division and the production of equal sized AB and P_1 blastomeres. A genome-wide screen for cell division genes by Gönczy and colleagues (75) discovered two near-identical proteins (GPR-1 and GPR-2, for G-protein regulator; hereafter collectively called GPR-1/2) containing a single GoLoco motif that, when inactivated by RNAi, gives a spindle positioning defect indistingushible from that of *goa-1/gpa-16* (*RNAi*) embryos. This finding is reminiscent of the *Drosophila* ACD pathway with PAR-proteins translating polarity cues into spindle positioning via signal transduction through a GoLoco motif protein and heterotrimeric G proteins.

Subsequent to this, three groups have independently delineated the spatial and temporal signaling mechanisms of the GoLoco protein GPR-1/2 and its signal transduction via G proteins in the first cell division of *C. elegans* (76–78) (Figure 6). GPR-1/2 is located at the cortex and is enriched at the posterior pole. The asymmetric localization of the PAR proteins (PAR-1, -2, and -6) is not altered in *gpr-1/2* (*RNAi*) and *goa-1/gpa-16* (*RNAi*) embryos, indicating that the spindle displacement phenotype is not due to altered cell polarity. Genetic and spindle-severing studies (discussed below) have indicated that GPR-1/2 acts downstream of the PAR proteins to facilitate ACD.

It is worth contrasting *C. elegans* versus *Drosophila* ACD and their differential utilization of Gα/GoLoco signaling. In *Drosophila*, Gα signaling plays a role in establishing cell polarity and determinant segregation, whereas in *C. elegans*, Gα acts downstream of polarity cues to mediate spindle alignment. Recent studies indicate that the LIN-5 protein is another important component of the asymmetric spindle positioning pathway in *C. elegans*. LIN-5 was independently isolated, both biochemically and genetically, as a GPR-1/2-interacting protein (77, 78) (Figure 6). Although GPR-1/2 is distantly related to *Drosophila* Pins (Figure 5), LIN-5 bears no sequence resemblance to *Drosophila* Inscuteable (Figure 6). *lin-5 (RNAi)* embryos have loss-of-function phenotypes similar to *gpr-1/2 (RNAi)* and *goa-1/gpa-16 (RNAi)* embryos, indicating they may all signal in the same pathway. LIN-5 also regulates the localization of GPR-1/2. Collectively, these findings suggest that LIN-5 may act as a scaffold or cofactor for GPR-1/2 and/or function in a pathway parallel to GPR-1/2 (77, 78). [Yet another component within this pathway, the DEP domain-containing protein LET-99, appears to function antagonistically to GPR-1/2 (79). Determining the epistatic relationships between these various signaling components is paramount.]

In vivo spindle severing with ultraviolet laser microbeams provided the definitive characterization of the role of GPR-1/2 and GOA-1/GPA-16 in force generation (76). Spindle severing experiments in wild-type embryos indicate that the peak velocity of the posterior spindle is 40% greater than the peak velocity of the anterior spindle; peak velocities are presumed to reflect the extent of pulling forces. In contrast, in *gpr-1/2 (RNAi)* and *goa-1/gpa-16 (RNAi)* embryos, peak spindle velocity is equivalent between the anterior and posterior spindles but dramatically reduced in magnitude. Therefore, the Gα-GPR1/2 protein complex is required to generate extra-spindle pulling forces.

What then is responsible for the force imbalance between the anterior and posterior poles? Elegant studies by Grill and coworkers (80) suggest that increased pulling forces on the posterior spindle pole result from an increased number of force generators at the cortex rather than an increase in the magnitude of the quantal force per se. Thus, asymmetric localization of GPR-1/2 provides a likely mechanism for the assembly of a force-generating protein complex, potentially including LIN-5, GPR-1/2, and Gα. Rather than GPR-1/2 asymmetry being unequivocal, there is instead a subtle bias (50% more posterior cortical GPR1/2 versus anterior) that correlates well with the differential spindle velocity (40% increase in posterior velocity versus anterior) (76) and the observation of 50% more active cortical force generators at the posterior (80). The force-generating complex must interact with the astral microtubules to generate pulling forces, presumably via microtubule depolymerization or motor activity (72). Direct modulation of microtubule dynamics by heterotrimeric G proteins is a potential mechanism for force generation that will be discussed in a subsequent section.

A Cycle of GDI, GEF, and GAP Activities in *C. elegans* ACD?

Although a role for GoLoco/Gα interations in both *Drosophila* and *C. elegans* ACD has been elucidated, the exact nature of the G-protein nucleotide exchange and hydrolysis cycle during ACD remains to be resolved. A combination of powerful genetic studies and in vitro biochemistry has described a new component of the G-protein cycle in the context of ACD and, potentially, for G-protein signaling in general. Miller and colleagues (81, 82) isolated a novel gene, *ric-8* (resistance to inhibitors of cholinesterase; also known as synembryn), as an upstream regulator of the *C. elegans* Gαq-like protein EGL-30. *ric-8* reduction-of-function mutants have high rates of embryonic lethality (29%), which could be augmented to almost 100% in embryos heterozyous for *goa-1* loss-of-function alleles, thus indicating that RIC-8 and GOA-1 signal in the same pathway during embryogenesis (83). Spindle rocking during the P_0 cleavage was diminished or nonexistent in *ric-8* mutant embryos, and a loss of spindle and blastomere asymmetry was also observed. Compellingly, *ric-8* mutant phenotypes are reminiscent of *gpr-1/2* (*RNAi*) and *goa-1/gpa-16* (*RNAi*) phenotypes.

A recent study has provided some insight into the possible role for RIC-8 in the G-protein cycle during ACD (Figure 6). Tall and coworkers (84) independently isolated rat Ric-8 in a yeast two-hybrid screen for Gα$_s$- and Gα$_o$-interacting proteins. Biochemical analyses of rat Ric-8A indicated it is a GEF for several mammalian Gα subunits. Ric-8, in contrast to GPCRs or Ras-superfamily GEFs, appears to have a unique mechanism of nucleotide exchange. Ric-8 preferentially interacts with GDP-bound Gα, in the absence of Gβγ, and causes nucleotide release and the formation of a stable nucleotide-free complex (84). Binding of GTP to Gα then reduces its affinity for Ric-8 and the complex dissociates. Thus, it appears the G-protein cycle during ACD, certainly in *C. elegans* if not other organisms, utilizes Gα subunits, a Gα GDI (GPR-1/2), and a Gα GEF (RIC-8) (Figure 6).

The biochemical actions on isolated Gα subunits that are ascribed to GoLoco proteins and Ric-8 are opposing ones: GDI activity versus GEF activity, respectively. This then represents an enigma because these biochemically opposite activities, when individually removed by RNAi or mutation, result in the same phenotype of symmetric division. Thus, it appears from the genetics that GPR-1/2 and RIC-8 are each required to activate G-protein signaling in *C. elegans* ACD. To resolve this enigma, it has been proposed that RIC-8 may require interdiction by the GPR-1/2 GoLoco motif to function on GOA-1/GPA-16 (77, 78). Ric-8 is unable to act as a GEF when its substrate, Gα·GDP, is bound to Gβγ (84). The suggestion is, therefore, that Gβγ is somehow removed from Gα by the GPR-1/2 GoLoco motif. It is also possible that, subsequent to Gβγ displacement, the GoLoco motif of GPR-1 may present Gα·GDP to RIC-8 and, thus, act cooperatively with RIC-8 to facilitate nucleotide exchange (Figure 6).

It also appears that Gα·GTP deactivation is required for appropriate signaling during asymmetric P_0 division. Loss of RGS-7, a potential GAP for GOA-1 and GPA-16 (Figure 6), results in overly vigorous posterior spindle rocking, a more asymmetric cell division plane, and thus more exaggerated asymmetry between resultant AB and P_1 cells (M. Koelle and H. Hess, personal communication). This suggests that elevated Gα·GTP levels result in excess force generation and further implies that Gα·GTP is the active species for force generation. In this model, GPR-1/2 may be required for appropriate spatial localization and clustering of the signaling complex. However, the notion of a GoLoco/Gα·GDP complex having specific effectors within the ACD signaling pathway cannot be readily dismissed.

Functional redundancy exists between GOA-1 and GPA16. To detect cell division defects, embryos must be lacking both GOA-1 *and* GPA-16 (78). Paradoxically, though, there appears to be differential interaction between the Gα subunits and their signaling regulators. Yeast two-hybrid studies indicate that GOA-1, but not GPA-16, interacts with GPR-1/2 (76), whereas both G proteins interact with RIC-8 (P. Gönczy, K. Colombo, and K. Afshar, personal communication). The Gα selectivity of RGS-7 has not been fully determined, but it does act as a GAP for GOA-1 (M. Koelle and H. Hess, personal communication). It is also important to note that, in addition to their role as negative regulators of G-protein signaling by virtue of Gα GAP activity, RGS-box proteins can also act as effectors for GTP-bound Gα subunits (3); hence, an effector function for RGS-7 cannot be ruled out. Another intriguing possibility is that, with both GEF and GAP activities seemingly critical for proper ACD, the nucleotide cycling rate of GOA-1 and GPA-16 in this system may determine the signaling outcome for ACD (Figure 6), as is the case for the Rho-family GTPase Cdc42 (85). Studies from mammalian systems may be able to shed some light on possible mechanisms of force generation in ACD and also underscore the conserved usage of non-conventional heterotrimeric G-protein-signaling paradigms in cell division.

Heterotrimeric G-Protein Involvement in Mammalian Cell Division

The functional interplay between components of plasma membrane-delimited GPCR signaling is well defined in mammalian systems (1, 2). In contrast, investigation into the role of heterotrimeric G-protein signaling in mammalian cell division has lagged behind studies in lower metazoans. A limited number of reports exist that preface a role for Gα signaling in mammalian mitosis [for example (86, 87)], but to a large extent, the field is open for investigation. Similarly, there is evidence that asymmetric cell division is important during mammalian neurogenesis [as reviewed by Cayouette & Raff (88)]; however, no role for heterotrimeric G-protein signaling has yet been elucidated in this realm.

The mammalian Pins orthologue LGN [named after the leucine-glycine-asparagine tripeptide present in its TPR regions (89)] has recently been described as a mitotic regulator, shedding light on possible effector systems for GoLoco proteins in cell division. Ectopic expression of LGN causes severe mitotic abnormalities in several mammalian cell lines (90). More compelling, however, is that RNAi-mediated knockdown of endogenous LGN levels disrupts microtubule organization and chromosome segregation during mitosis (90). The nuclear mitotic apparatus (NuMA) protein was identified, using a yeast two-hybrid screen, as an LGN-binding partner (90). The first two TPRs of LGN specify binding to NuMA in vitro and in vivo (Figure 6). This interaction provides a direct link between GoLoco/Gα signaling and the regulation of spindle dynamics because NuMA has been shown to regulate spindle formation and organization at the level of the centrosome (91). Assays of aster formation in *Xenopus* mitotic extracts using recombinant NuMA fragments and anti-LGN antibodies indicate that LGN acts negatively on the intrinsic ability of NuMA to stabilize microtubules and form asters (90). Mechanistically, it appears that the LGN- and tubulin-binding sites on NuMA partially overlap, suggesting that LGN sterically inhibits NuMA-mediated microtubule stabilization (92) (Figure 6).

The only reported investigation into the cellular function of the mammalian LGN paralogue AGS3 has come from Pattingre and colleagues (93); this investigation found that ectopic expression of AGS3 fragments can attenuate amino acid deprivation-induced autophagy. As previously described, an interaction between the mammalian PAR-4 homologue LKB-1 and AGS3 has recently been demonstrated by in vitro and cellular cotransfection studies (43), but the functional consequences of this interaction have yet to be determined. Little evidence currently exists that AGS3 regulates either GPCR or mitotic signaling in a bona fide physiological context, but given recent findings regarding LGN function, it appears likely that AGS3 may have an analogous function in cell division processes.

Seminal studies by Rasenick and colleagues (94–98) provide a compelling potential mechanism by which G-protein signaling may regulate spindle forces during mitosis. Mammalian $G\alpha$ and $G\beta\gamma$ subunits have been shown to regulate microtubule dynamics both in vitro (95, 96) and in vivo (97, 98). $G\alpha_i$ subunits can activate the GTPase activity of tubulin and consequently accelerate microtubule dynamics. Thus, the possibility exists that GoLoco/$G\alpha$ complexes signal directly to tubulin to modulate spindle dynamics (Figure 6). Accordingly, in *C. elegans*, the cortical force-generating system could comprise GPR-1/2, $G\alpha$, and tubulin, with $G\alpha$ acting to depolymerize astral microtubules at the cortex and with the asymmetric distribution of GPR-1/2 leading to unbalanced forces applied to the spindle poles. Indeed, Labbe and colleagues (99) have recently demonstrated that microtubules at the posterior cortex are less stable during spindle displacement in the *C. elegans* embryo. In contrast, microtubules are equally stable at the anterior and posterior cortex in *goa-1/gpa-16* (*RNAi*) embryos (99), thus reinforcing a role for heterotrimeric G proteins in the control of cortical microtubule dynamics.

In line with the findings in *C. elegans* and *Drosophila*, striking evidence from studies of RGS14 now implicate heterotrimeric G-protein signaling in mammalian embryogenesis. RGS14 contains an N-terminal RGS box, tandem Ras-binding domains, and a C-terminal GoLoco motif (12, 100) (Figure 6). RGS14 protein in the mouse embryo is detectable at 28 h postfertilization, concurrent with the loss of expression of RGS12 (a mammalian paralogue of RGS14) (Figure 1). RGS14 localizes to the first mitotic spindle. Homologous inactivation of *Rgs14* in the mouse results in early embryonic lethality (101). Mouse embryos lacking RGS14 undergo preimplantation lethality at the two-cell stage due to an apparent cell cleavage failure; the one-cell stage is characterized by absence of a developed microtubular network and asymmetric distribution of the nucleus. Indications are that RGS14, via its GoLoco motif, controls microtubule dynamics at the mitotic spindle, and the loss of this function is sufficient to disrupt embryonic development (S.J.A. D'Souza and L. Martin-McCaffrey, personal communication).

CONCLUDING REMARKS

Originally identified as a putative Gα-binding sequence, trapped in a screen for GPCR modulators, and characterized biochemically as a guanine nucleotide dissociation inhibitor, the GoLoco motif was born into a presumed role as a modulator of plasma membrane-delimited G-protein–coupled receptor signal transduction. However, overwhelming recent evidence from *Drosophila, C. elegans,* and now mammalian systems suggest GoLoco motif proteins are crucial regulatory elements in animal developmental processes—serving a fundamental and evolutionarily conserved role in organism embryogenesis, apparently at the level of force generation and organization of the mitotic spindle. Future studies should therefore be targeted at the further exposition of the genetic, molecular, and cell biological bases of GoLoco protein action in the machinery of cell division.

ACKNOWLEDGMENTS

We thank Dr. Pierre Gönczy (ISREC), and University of North Carolina colleagues Drs. Miller B. Jones, Christopher A. Johnston, and Christopher R. McCudden for critical reading of this manuscript. Our work is supported by NIH grants GM062338 and GM065533. F.S.W. is an American Heart Association Postdoctoral Fellow. R.J.K. gratefully acknowledges predoctoral fellowship support from the National Institute of Mental Health (F30 MH064319). D.P.S. is a Year 2000 Neuroscience Scholar of the EJLB Foundation (Montréal, Canada) and recipient of the Burroughs-Wellcome Fund New Investigator Award in the Basic Pharmacological Sciences.

The *Annual Review of Biochemistry* is online at http://biochem.annualreviews.org

LITERATURE CITED

1. Gilman AG. 1987. *Annu. Rev. Biochem.* 56:615–49
2. Hamm HE. 1998. *J. Biol. Chem.* 273:669–72
3. Neubig RR, Siderovski DP. 2002. *Nat. Rev. Drug Discov.* 1:187–97
4. Kimple RJ, Jones MB, Shutes A, Yerxa BR, Siderovski DP, Willard FS. 2003.
    Comb. Chem. High Throughput Screen. 6:399–407
5. Siderovski DP, Strockbine B, Behe CI. 1999. *Crit. Rev. Biochem. Mol. Biol.* 34:215–51
6. Siderovski DP, Harden TK. 2003. In *Handbook of Cell Signaling*, ed. RA Bradshaw, EA Dennis, pp. 631–38. San Diego: Academic

7. Granderath S, Stollewerk A, Greig S, Goodman CS, O'Kane CJ, Klambt C. 1999. *Development* 126:1781–91

8. Siderovski DP, Diverse-Pierluissi M, De Vries L. 1999. *Trends Biochem. Sci.* 24: 340–41

9. Ponting CP. 1999. *J. Mol. Med.* 77: 695–98

10. De Vries L, Fischer T, Tronchere H, Brothers GM, Strockbine B, et al. 2000. *Proc. Natl. Acad. Sci. USA* 97:14364–69

11. Peterson YK, Bernard ML, Ma H, Hazard S 3rd, Graber SG, Lanier SM. 2000. *J. Biol. Chem.* 275:33193–96

12. Kimple RJ, De Vries L, Tronchere H, Behe CI, Morris RA, et al. 2001. *J. Biol. Chem.* 276:29275–81

13. Natochin M, Gasimov KG, Artemyev NO. 2001. *Biochemistry* 40:5322–28

14. Cismowski MJ, Takesono A, Ma CL, Lizano JS, Xie XB, et al. 1999. *Nat. Biotechnol.* 17:878–83

15. Takesono A, Cismowski MJ, Ribas C, Bernard M, Chung P, et al. 1999. *J. Biol. Chem.* 274:33202–5

16. Howard AD, McAllister G, Feighner SD, Liu Q, Nargund RP, et al. 2001. *Trends Pharmacol. Sci.* 22:132–40

17. Ross EM, Wilkie TM. 2000. *Annu. Rev. Biochem.* 69:795–827

18. Kimple RJ, Willard FS, Siderovski DP. 2002. *Mol. Interv.* 2:88–100

19. Kimple RJ, Kimple ME, Betts L, Sondek J, Siderovski DP. 2002. *Nature* 416: 878–81

20. Lambright DG, Noel JP, Hamm HE, Sigler PB. 1994. *Nature* 369:621–28

21. Lambright DG, Sondek J, Bohm A, Skiba NP, Hamm HE, Sigler PB. 1996. *Nature* 379:311–19

22. Wall MA, Posner BA, Sprang SR. 1998. *Structure* 6:1169–83

23. Bernard ML, Peterson YK, Chung P, Jourdan J, Lanier SM. 2001. *J. Biol. Chem.* 276:1585–93

24. Peterson YK, Hazard S 3rd, Graber SG, Lanier SM. 2002. *J. Biol. Chem.* 277: 6767–70

25. Scheffzek K, Ahmadian MR, Kabsch W, Wiesmuller L, Lautwein A, et al. 1997. *Science* 277:333–38

26. Rittinger K, Walker PA, Eccleston JF, Smerdon SJ, Gamblin SJ. 1997. *Nature* 389:758–62

27. Coleman DE, Berghuis AM, Lee E, Linder ME, Gilman AG, Sprang SR. 1994. *Science* 265:1405–12

28. Mixon MB, Lee E, Coleman DE, Berghuis AM, Gilman AG, Sprang SR. 1995. *Science* 270:954–60

29. Natochin M, Gasimov KG, Artemyev NO. 2002. *Biochemistry* 41:258–65

30. Kimple RJ, Willard FS, Hains MD, Jones MB, Nweke GN, Siderovski DP. 2004. *Biochem. J.* 378:801–8

31. Zhang XL, Zhang HL, Oberdick J. 2002. *Mol. Brain Res.* 105:1–10

32. Rondard P, Iiri T, Srinivasan S, Meng E, Fujita T, Bourne HR. 2001. *Proc. Natl. Acad. Sci. USA* 98:6150–55

33. Redd KJ, Oberdick J, McCoy J, Denker BM, Luo Y. 2002. *J. Neurosci. Res.* 70: 631–37

34. Luo Y, Denker BM. 1999. *J. Biol. Chem.* 274:10685–88

35. Kaushik R, Yu FW, Chia W, Yang XH, Bahri S. 2003. *Mol. Biol. Cell* 14: 3144–55

36. Jordan JD, Carey KD, Stork PJ, Iyengar R. 1999. *J. Biol. Chem.* 274:21507–10

37. Mochizuki N, Ohba Y, Kiyokawa E, Kurata T, Murakami T, et al. 1999. *Nature* 400:891–94

38. Ma H, Peterson YK, Bernard ML, Lanier SM, Graber SG. 2003. *Biochemistry* 42: 8085–93

39. Natochin M, Lester B, Peterson YK, Bernard ML, Lanier SM, Artemyev NO. 2000. *J. Biol. Chem.* 275:40981–85

40. Meng J, Glick JL, Polakis P, Casey PJ. 1999. *J. Biol. Chem.* 274:36663–69

41. Meng J, Casey PJ. 2002. *J. Biol. Chem.* 277:43417–24

42. Hollinger S, Ramineni S, Hepler JR. 2003. *Biochemistry* 42:811–19

43. Blumer JB, Bernard ML, Peterson YK,

Nezu J, Chung P, et al. 2003. *J. Biol. Chem.* 278:23217–20

44. Watts JL, Morton DG, Bestman J, Kemphues KJ. 2000. *Development* 127:1467–75
45. Martin SG, St Johnston D. 2003. *Nature* 421:379–84
46. Webb CK, Kimple RJ, Siderovski DP, Oxford GS. 2002. *Soc. Neurosci.* 542:10 (Abstr.)
47. Ghosh M, Peterson YK, Lanier SM, Smrcka AV. 2003. *J. Biol. Chem.* 278:34747–50
48. Pathirana S, Zhao D, Bownes M. 2001. *Mech. Dev.* 109:137–50
49. Schiff ML, Siderovski DP, Jordan JD, Brothers G, Snow B, et al. 2000. *Nature* 408:723–27
50. Rubinfeld B, Munemitsu S, Clark R, Conroy L, Watt K, et al. 1991. *Cell* 65:1033–42
51. Schaefer M, Shevchenko A, Knoblich JA. 2000. *Curr. Biol.* 10:353–62
52. Knoblich JA. 2001. *Nat. Rev. Mol. Cell Biol.* 2:11–20
53. Knoblich JA. 2001. *Symp. Soc. Exp. Biol.* 53:75–89
54. Chia W, Yang XH. 2002. *Curr. Opin. Genet. Dev.* 12:459–64
55. Schober M, Schaefer M, Knoblich JA. 1999. *Nature* 402:548–51
56. Wodarz A, Ramrath A, Kuchinke U, Knust E. 1999. *Nature* 402:544–47
57. Kraut R, Chia W, Jan LY, Jan YN, Knoblich JA. 1996. *Nature* 383:50–55
58. Knoblich JA, Jan LY, Jan YN. 1999. *Curr. Biol.* 9:155–58
59. Tio M, Zavortink M, Yang XH, Chia W. 1999. *J. Cell Sci.* 112:1541–51
60. Yu FW, Morin X, Cai Y, Yang XH, Chia W. 2000. *Cell* 100:399–409
61. Parmentier ML, Woods D, Greig S, Phan PG, Radovic A, et al. 2000. *J. Neurosci.* 20:RC84 (1–5)
62. Schaefer M, Petronczki M, Dorner D, Forte M, Knoblich JA. 2001. *Cell* 107:183–94

63. Yu FW, Ong CT, Chia W, Yang XH. 2002. *Mol. Cell. Biol.* 22:4230–40
64. Fuse N, Hisata K, Katzen AL, Matsuzaki F. 2003. *Curr. Biol.* 13:947–54
65. Yu FW, Cai Y, Kaushik R, Yang XH, Chia W. 2003. *J. Cell Biol.* 162:623–33
66. Cai Y, Yu FW, Lin SP, Chia W, Yang XL. 2003. *Cell* 112:51–62
67. Bellaiche Y, Radovic A, Woods DF, Hough CD, Parmentier ML, et al. 2001. *Cell* 106:355–66
68. Malbon CC, Wang H, Moon RT. 2001. *Biochem. Biophys. Res. Commun.* 287:589–93
69. Gönczy P. 2003. *Med. Sci.* 19:735–42
70. Kemphues KJ, Priess JR, Morton DG, Cheng NS. 1988. *Cell* 52:311–20
71. Grill SW, Gönczy P, Stelzer EH, Hyman AA. 2001. *Nature* 409:630–33
72. Gönczy P. 2002. *Trends Cell. Biol.* 12:332–39
73. Zwaal RR, Ahringer J, van Luenen HG, Rushforth A, Anderson P, Plasterk RH. 1996. *Cell* 86:619–29
74. Gotta M, Ahringer J. 2001. *Nat. Cell Biol.* 3:297–300
75. Gönczy P, Echeverri C, Oegema K, Coulson A, Jones SJ, et al. 2000. *Nature* 408:331–36
76. Colombo K, Grill SW, Kimple RJ, Willard FS, Siderovski DP, Gönczy P. 2003. *Science* 300:1957–61
77. Srinivasan DG, Fisk RM, Xu HH, van den Heuvel S. 2003. *Genes Dev.* 17:1225–39
78. Gotta M, Dong Y, Peterson YK, Lanier SM, Ahringer J. 2003. *Curr. Biol.* 13:1029–37
79. Tsou MF, Hayashi A, Rose LS. 2003. *Development.* 130:5717–30
80. Grill SW, Howard J, Schaffer E, Stelzer EH, Hyman AA. 2003. *Science* 301:518–21
81. Miller KG, Alfonso A, Nguyen M, Crowell JA, Johnson CD, Rand JB. 1996. *Proc. Natl. Acad. Sci. USA* 93:12593–98

82. Miller KG, Emerson MD, McManus JR, Rand JB. 2000. *Neuron* 27:289–99

83. Miller KG, Rand JB. 2000. *Genetics* 156: 1649–60

84. Tall GG, Krumins AM, Gilman AG. 2003. *J. Biol. Chem.* 278:8356–62

85. Lin R, Bagrodia S, Cerione R, Manor D. 1997. *Curr. Biol.* 7:794–97

86. Willard FS, Crouch MF. 2000. *Immunol. Cell Biol.* 78:387–94

87. Crouch MF, Osborne GW, Willard FS. 2000. *Cell. Signal.* 12:153–63

88. Cayouette M, Raff M. 2002. *Nat. Neurosci.* 5:1265–69

89. Mochizuki N, Cho G, Wen B, Insel PA. 1996. *Gene* 181:39–43

90. Du QS, Stukenberg PT, Macara IG. 2001. *Nat. Cell Biol.* 3:1069–75

91. Gaglio T, Saredi A, Compton DA. 1995. *J. Cell Biol.* 131:693–708

92. Du QS, Taylor L, Compton DA, Macara IG. 2002. *Curr. Biol.* 12:1928–33

93. Pattingre S, De Vries L, Bauvy C, Chantret I, Cluzeaud F, et al. 2003. *J. Biol. Chem.* 278:20995–1002

94. Wang N, Yan K, Rasenick MM. 1990. *J. Biol. Chem.* 265:1239–42

95. Roychowdhury S, Rasenick MM. 1997. *J. Biol. Chem.* 272:31576–81

96. Roychowdhury S, Panda D, Wilson L, Rasenick MM. 1999. *J. Biol. Chem.* 274: 13485–90

97. Sarma T, Voyno-Yasenetskaya T, Hope TJ, Rasenick MM. 2003. *FASEB J.* 17: 848–59

98. Popova JS, Rasenick MM. 2003. *J. Biol. Chem.* 278:34299–308

99. Labbe JC, Maddox PS, Salmon ED, Goldstein B. 2003. *Curr. Biol.* 13:707–14

100. Snow BE, Antonio L, Suggs S, Gutstein HB, Siderovski DP. 1997. *Biochem. Biophys. Res. Commun.* 233:770–77

101. Martin-McCaffrey L, Natale DRC, Kimple RJ, Willard FS, Oliveira-dos-Santos AJ, et al. 2002. *Critical role of regulator of G-protein signaling-14 (RGS14) in early mouse embryogenesis.* Presented at Great Lakes G-Protein–Coupled Recept. Retreat. 3rd Annu. Jt. Meet., Ann Arbor, MI

Annu. Rev. Biochem. 2004. 73:953–90
doi: 10.1146/annurev.biochem.73.011303.073940
Copyright © 2004 by Annual Reviews. All rights reserved
First published online as a Review in Advance on March 26, 2004

OPIOID RECEPTORS

Maria Waldhoer, Selena E. Bartlett, and
Jennifer L. Whistler

*Ernest Gallo Clinic and Research Center, University of California, San Francisco,
Emeryville, California 94608; e-mail: mariaw@egcrc.net, selenab@egcrc.net,
shooz2@itsa.ucsf.edu*

Key Words morphine, GPCR/7TM, RAVE, tolerance, dependence

■ **Abstract** Opioid receptors belong to the large superfamily of seven transmembrane-spanning (7TM) G protein-coupled receptors (GPCRs). As a class, GPCRs are of fundamental physiological importance mediating the actions of the majority of known neurotransmitters and hormones. Opioid receptors are particularly intriguing members of this receptor family. They are activated both by endogenously produced opioid peptides and by exogenously administered opiate compounds, some of which are not only among the most effective analgesics known but also highly addictive drugs of abuse. A fundamental question in addiction biology is why exogenous opioid drugs, such as morphine and heroin, have a high liability for inducing tolerance, dependence, and addiction. This review focuses on many aspects of opioid receptors with the aim of gaining a greater insight into mechanisms of opioid tolerance and dependence.

CONTENTS

INTRODUCTION

Opioid receptors belong to the large superfamily of seven transmembrane-spanning (7TM) G protein-coupled receptors (GPCRs). As a class, GPCRs are of fundamental physiological importance mediating the actions of the majority of known neurotransmitters and hormones. Opioid receptors are particularly intriguing members of this receptor family. They are activated both by endogenously produced opioid peptides and by exogenously administered opioid drugs, such as morphine, which are not only among the most effective analgesics known but also highly addictive drugs of abuse (1). Morphine is a prototypical opioid and has been used medicinally for centuries. The first reports of its use are by a second century Greek physician, Galen, who administered opium to patients for the relief of pain and to relax people suffering from asthma and congestive heart failure. Opium is an extract of the poppy plant, *Papaver somniferum*. In 1806, the German chemist Serturner isolated the opium alkaloids, one of them being morphine after Morpheus, the god of dreams. However, it was not until 167 years later that the pharmacology of morphine was defined at the receptor level (2). There are at least four opioid receptor subtypes; each of which has its own repertoire of ligands.

A fundamental question in addiction biology is why exogenous opioid drugs, such as morphine and heroin, have a high liability for inducing tolerance, dependence, and addiction. This review focuses on the many aspects of opioid receptors with the aim of gaining a greater insight into mechanisms of opioid tolerance and dependence. The hope is to provide new insights into the design of more effective therapeutic agents for the treatment of pain.

PHARMACOLOGY

There are several subtypes of opioid receptor defined by their unique pharmacological responses to the repertoire of opioid ligands.

Receptor Subtypes

Since opioid-binding sites were first proposed in the early 1950s (3) and 1960s (4) and discovered in mammalian brain tissue in 1973 (2, 5, 6), three decades of extensive pharmacological studies have uncovered a variety of opioid receptor types. To date, four opioid receptors have been cloned, the MOP[1] (μ = mu for morphine), the KOP (κ = kappa for ketocyclazocine), the DOP (δ = delta for deferens because it was first identified in mouse vas deferens) (7), and the NOP-R [initially called LC132 (8), ORL-1 (9), or nociceptin/orphanin FQ receptor (10)]. However, there is an obvious disparity between the existence of only four opioid receptor genes and the substantial pharmacological evidence for additional opioid receptor phenotypes. It is thus feasible to speculate that posttranslational modifications, alternative mRNA splicing (11), tissue distribution of more than one receptor gene and/or scaffolding with additional proteins, or homo- or heterodimerization (12) of the existing DOP, MOP, KOP, and NOP receptor proteins could result in these various additional pharmacological phenotypes.

MOP RECEPTOR   To date, MOP receptor (MOP-R) genes have been cloned (Figure 1) from rat (13–18), mouse (19), human (20), Rhesus and crab-eating monkey (G.M. Miller, B.K. Madras, direct GenBank deposit), pig (21), cow (22), guinea pig (S.A. Smith, J.T. Stupfel, N.A. Ilias, G.D. Olsen, direct GenBank deposit), *Catostomus commersoni* (white suckerfish) (23), zebra fish (24), and frog (*Rana pipiens*) (25). Weakly homologous sequences are also present in the genomes of *Caenorhabditis elegans* and *Drosophila melanogaster* (see also Table 1). A variety of alternative mRNA splice variants have also been cloned and/or suggested for the MOP-R gene (11, 26), and some of the resulting proteins differ in their trafficking properties, although their pharmacology does not appear to differ in ligand binding assays (27–29). Thus, it is unlikely that these splice variants explain the observation of pharmacologically distinct MOP-R subtypes.

DOP RECEPTOR   The first opioid receptor to be cloned was the mouse DOP receptor (DOP-R) in 1992 from a NG108-15 glioma hybridoma cell line (30, 31), followed by the discovery of the rat DOP-R (14), the human DOP-R in 1994 (32), and an amphibian DOP-R (25) (see Table 1). Although only one DOP-R gene has been cloned, on the basis of in vitro and in vivo pharmacological characteristics, a variety of subtypes of DOP-Rs have been proposed. Specifically, the opioid ligands [D-Pen^2, D-Pen^5]enkephalin (DPDPE), [D-Ala^2, D-Leu^5] enkephalin (DADLE), [D-Ala^2, D-Leu^5]enkephalyl-Cys (DALCE), and 7-benzylidenenaltrexone (BTNX) have been shown to bind with high affinity to DOP sites designated $\delta1$, whereas [D-Ser^2, Leu^5]enkephalyl-Thr (DSLET), [D-Ala^2]deltorphin II, naltrindole 5'-isothiocyanate (5'-NTI), and naltribene (NTB) bind to DOP sites designated $\delta2$ (33–35). Pharmacologically, the cloned

---

[1]According to the nomenclature suggested by the International Union of Pharmacology Nomenclature for Opioid receptors (http://www.iuphar.org), the opioid receptors are referred to as MOP, DOP, KOP, and NOP receptors (MOP-R, DOP-R, KOP-R, or NOP-R) for the mu (μ), delta (δ), kappa (κ), and nociceptin receptors, respectively.

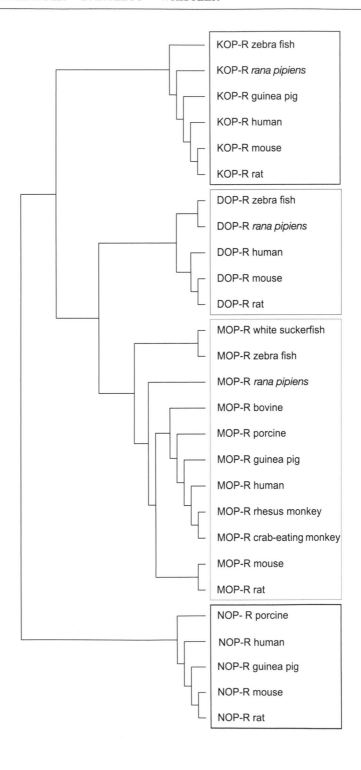

TABLE 1 Cloned opioid receptor subtypes

Species	Receptor/accession number			
	MOP-R	**DOP-Rs**	**KOP-R**	**NOP-R**
Bos taurus (cow)	P79350			
Caenorhabditis elegans	NM_063487		NM_073482	
Catostomus commersoni (white suckerfish)	O42324			
Cavia porcellus (guinea pig)	Q8CGM4		P41144	P47748
Danio rerio (zebra fish)	AAK01143	O57585	AAG60607	
Homo sapiens (human)	P35372	P41143	P41145	P41146
Macaca fascicularis (crab-eating monkey)	Q95M54			
Macaca mulatta (rhesus monkey)	Q9MYW9			
Mus musculus (mouse)	P42866	P32300	P33534	P35377
Rana pipiens (grass frog)	AF530571	AF530572	AF530573	
Rattus norvegicus (rat)	P33535	P33533	P34975	P35370
Sus scrofa (pig)	Q95247			P79292

DOP-R resemble most the δ1-subtype when expressed in heterologous systems. Whether there is a specific gene or a splice variant encoding for a DOP-R of the δ2-type remains to be seen. For an extensive review on DOP-Rs, see (36).

KOP RECEPTOR   KOP receptors (KOP-R) have been cloned from rat (37–41), mouse (42, 43), human (44–46), guinea pig (47), amphibian (25), and zebra fish (R.E. Rodriguez, direct submission to GenBank), and they may also occur in *C. elegans* (see Table 1). The cloned KOP-R has high affinity for the endogenous peptide dynorphin A (1-17) (48), and a variety of selective synthetic agonists and

---

←

Figure 1 Dendrogram of opioid receptors based on their amino acid identity. This phylogenetic tree was generated by Francois Talabot, Sereono Pharmaceuticals, Switzerland, using Clustal W 1.83 alignments of the full amino acid sequence followed by an analysis with the Jukes-Cantor Corrected Distance Program [Genetics Computer Group GrowTree (GCG), Wisconsin Package]. This correction method underestimates the true distance for more distantly related sequences. GCG was used to create a phylogenetic tree from a distance matrix created by distances using the UPGMA method (309) in order to estimate a species tree when the expected rate of gene substitution is constant and the distance measure is linear with evolutionary time (distance is measured as amino acid substitutions).

antagonists have been developed for the KOP-R (see Table 2). As for the other opioid receptor classes, a number of subtypes have also been described for the kappa type receptors pharmacologically; $\kappa 1$ sites bind to dynorphin 1-17 but not to DADLE (representing the cloned KOP-R profile), $\kappa 2$ sites bind (Arg[6], Phe[7])Met-enkephalin and DADLE, and the so-called $\kappa 3$ site is sensitive to naloxone benzoylhydrazone [for a review, see (49)].

NOP RECEPTOR   The NOP receptor (NOP-R) was first identified in 1994 and has since been identified in five species (see Table 1), human (9), rat (8, 50–54), mouse (43), pig (55), and guinea pig (G. Xie, direct submission to GenBank). These proteins share greater than 90% sequence identity and ~60% homology with the three classical opioid receptors, DOP-R, KOP-R, and MOP-R. In addition, splice variants of the NOP-R have been reported in rat (53) and mouse brain tissue (56). Because early ligand-binding studies on the cloned NOP-R found very low levels of binding to all known opioid ligands, this receptor was considered to be an orphan receptor and termed "orphanin FQ," "nociceptin," or "ORL-1" (for opioid receptor-like 1) receptor. However, soon after cloning the orphan receptor, two groups identified the endogenous peptide ligand for NOP-R from rat and porcine brain tissue. The new ligand was a hectadecapeptide closely related to the KOP-R-selective peptide dynorphin A and was termed nociceptin or orphanin FQ (OFQ) (10, 57).

OTHER OPIOID RECEPTORS   The sigma receptor (σ = sigma for SKF10047) was initially classified as an opioid receptor (58). However, since the time it was cloned in 1996 (59), it has become evident that the sigma receptor is a single transmembrane-spanning protein targeted by other drugs of abuse, for example phencyclidine and its analogues [for review see (60)], and it is no longer regarded as a member of the opioid receptor family. Moreover, a variety of other opioid receptors have been described on the basis of pharmacological profiles that did not match any of the MOP, DOP, KOP, or NOP receptors. These include a ζ (zeta) receptor, which has recently been cloned and classified as an opioid growth factor receptor (OGFr) with no homology to the classical opioid receptors (61, 62). In addition, a λ (lambda) receptor and a β-endorphin-sensitive ϵ (epsilon) opioid-binding site have been described (63). However, these receptors are poorly characterized, and proof for their existence through identification of their respective genes is still lacking.

Opioid Ligands

The endogenous opioid peptides are mainly derived from four precursors, pro-opiomelanocortin, proenkephalin, prodynorphin, and pronociceptin/orphanin FQ. Except for nociceptin/orphanin FQ, all peptides derived from the other precursors consist of a pentapeptide sequence TyrGlyGlyPheMet/Leu (YGGFM/ L). Nociceptin/orphanin FQ, however, contains a phenylalanine (F) instead of the N-terminal tyrosine, a residue necessary for high affinity binding to the classic

TABLE 2 Selective opioid receptor ligands

Receptor	Endogenous peptides	Peptide agonists	Peptide antagonists	Agonists	Antagonists
MOP-R	Endomorphin-1 Endomorphin-2 β-endorphin β-neoendorphin Dermorphin	DAMGO PL 017	CTOP Octreotide (SMS201,995)	Fentanyl Morphine Sufentanyl	β-FNA (affinity label) Naloxonazine (irreversible)
DOP-R	Leu^5-Enkephalin Met^5-Enkephalin Met^5-Enkephalin-Arg^6-Phe^7 Met^5-Enkephalin-Arg^{6}Gly^{7}Leu^8 Deltorphin Deltorphin I Deltorphin II	DADLE DPDPE DSLET	ICI 174,864 (inverse agonist) TIPP TIPP[ψ]	BW373U86 SIOM SNC 80 TAN-67	Benzylidenenaltrexone (BNTX) Naltriben (NTB) Naltrindole (NTI) NTI 5′ isothiocyanate (NTII)
KOP-R	Dynorphin A Dynorphin B	Dynorphin 1a		Bremazocine Ethylketocyclazocine (EKC) Ketocyclazocine CI-977 U-50,488 Spiradoline (U-62,066) U-69,593 ICI 199,441 ICI 197,067 BRL 52,537 BRL 52,656 6′-GNTI	DIPPA Nor-binaltorphimine (nor BNI) 5′-Guanidinonaltrindole (5′-GNTI)
NOP-R[a]	Nociceptin/orphanin FQ	[Arg^{14}, Lys^{15}]nociceptin [(pX)Phe^4]nociceptin (1-13) amide analogues NC(1-13)NH$_2$ Cyclo[Cys^{10}, Cys^{14}]NC(1-14)NH$_2$ ZP120	[N-Phe^1]NC(1-13)NH$_2$ UFP-101	Ro 64-6198	Benzimidazolinone (J-113397) JTC-801 TRK-820

[a]For references on NOP-R-selective ligands please refer to the following reviews (67, 68). For all other references, please see http://www.opioid.umn.edu

TABLE 3 Nonselective opioid receptor ligands[a]

Receptor	Mixed agonists-antagonists	Agonists	Antagonists
MOP-R/DOP-R/KOP-R	Bremazocine	Etorphine	Diprenorphine
	Buprenorphine	Levorphanol	Naloxone
	Nalbuphine	Meperidine	Naltrexone
	Nalorphine	Methadone	β-CNA (affinity label)
MOP-R/DOP-R	DIPP-NH2Ψ		

[a]For references please see http://www.opioid.umn.edu.

opioid receptors (64, 65). For the highly selective MOP-R peptides endomorphins 1 and 2, no precursor has yet been identified, and these tetrapeptides are structurally unrelated to all the other endogenous peptides (66). For a summary of the most commonly used synthetic opioid peptides and other opioid ligands, we refer the reader to Tables 2–3 (67, 68) and the Web page http://www.opioid. umn.edu (and references therein) of the Center for Opioid Research and Design (68a).

STRUCTURE-FUNCTION

Opioid receptors belong to the class A (Rhodopsin) family of G_i/G_o protein-coupled receptors with an extracellular N-terminal domain (see Figure 2), 7TM helical domains connected by three extracellular and three intracellular domains and an intracellular C-terminal tail, possibly forming a fourth intracellular loop with its putative palmitoylation site(s). On the basis of the 2.8 Å resolution 3-D structure of rhodopsin (69), it is assumed that the seven transmembrane helices of opioid receptors are arranged sequentially in a counterclockwise fashion (as viewed from the extracellular side, see Figure 2) forming a tight helical bundle. Together with the extracellular domains of the receptor, this provides a dynamic interface for the binding of various opioid ligands. The opioid receptors are about 60% identical to each other with greatest homology in the transmembrane helices and the greatest diversity in their N and C termini as well as their extracellular loops (38). In addition, all four opioid receptors possess two conserved cysteine residues in the first and second extracellular loops, which probably form a disulfide bridge. All opioid receptors are also modified with Asn-linked glycosyl chains at various sites in their N-terminal domains. All four opioid receptors contain highly conserved fingerprint residues of family A receptors, such as the Asp-Arg-Tyr (DRY) motif at the cytoplasmic end of TM3, Asn I:18 in TM1; AspII:10 in TM2; CysIII:01 in TM3; TrpIV:10 in TM4; and ProV:16, ProVI:15, and ProVII:17 in TMs 5–7 [for nomenclature, see (70)].

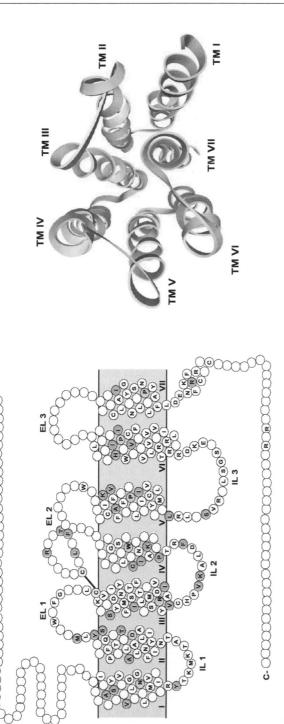

Figure 2 Structure of opioid receptors. (*left*) Serpentine model of the opioid receptor. Each transmembrane helix is labeled with a roman number. The white empty circles represent nonconserved amino acids among the MOP, DOP, KOP, and NOP receptors. White circles with a letter represent identical amino acids among all four opioid receptors. Violet circles represent further identity between the MOP-R, DOP-R, and KOP-R. Green circles highlight the highly conserved fingerprint residues of family A receptors, Asn I:18 in TM1, AspII:10 in TM2, CysIII:01 in TM3, TrpIV:10 in TM4, ProV:16 in TM5, ProVI:15 in TM6, and ProVII:17 TM7 [for nomenclature, see (70)]. Yellow circles depict the two conserved cystines in EL loops 1 and 2, likely forming a disulfide-bridge. IL = intracellular loop, and EL = extracellular loop. (*right*) Proposed arrangement of the seven transmembrane helices of opioid receptors as viewed from the top (extracellular side). The seven transmembrane helices are arranged sequentially in a counterclockwise manner. Each transmembrane helix is labeled with a roman number.

Mutations and What They Affect

Many ligands act promiscuously on MOP, DOP, and KOP receptors, yet there are selective ligands for each receptor. Extensive site-directed mutagenesis studies and studies on chimeric receptors have helped to determine specific domains involved in (*a*) ligand binding, (*b*) G protein-effector activation, (*c*) constitutive activity, and (*d*) receptor desensitization/endocytosis/downregulation.

LIGAND BINDING   On the basis of both mutational analysis and chimeric receptor studies, as well as computational modeling, it is evident that opioid receptors have some common features defining an opioid-binding pocket and more divergent aspects that delineate high affinity and selectivity for receptor subtype-specific ligands [for extensive reviews, see (36, 71–74)]. It has been suggested that all opioid receptors share a common binding cavity that is situated in an inner interhelical conserved region comprising transmembrane helices 3, 4, 5, 6, and 7. This cavity is partially covered by the extracellular loops. These highly divergent extracellular loops, together with residues from the extracellular ends of the TM segments, play a role in ligand selectivity, especially for peptides, allowing them to discriminate between the different opioid receptor types. The largest ligands [such as DPDPE or norbinaltorphimine (norBNI)] fill almost all of the available space within the binding cavity and interact with residues from both conserved and variable regions. Yet, smaller alkaloid agonists (such as morphine) interact predominantly with conserved residues in the bottom of the binding cavity. Moreover, alkaloid antagonists (such as naloxone) are thought to shift slightly deeper in the binding pocket than agonists, thereby sterically hindering a shift of TM3 and TM7, and consequently preventing an active receptor confirmation, thus leading to functional antagonism (71).

Takemori & Portoghese (75) suggested that ligand binding and selectivity may be conferred through the recognition of two distinct structural units inherent to the ligand. In this context, during binding, the common message tyramine moiety of alkaloids or Tyr1 of the cyclic peptides lies at the bottom of the binding cavity and interacts primarily with residues that are common to all opioid receptors. The chemically different address fragments of the ligands are oriented toward the extracellular surface of the transmembrane domains, interact with many residues that differ among the respective opioid receptors, and are situated along the borders of the binding cavity and/or in the extracellular loops (71). It has been further suggested that the extracellular loops might then act as a "gate" that allows for the passage of certain ligands while excluding others, rather than containing residues crucial for specific binding of the address portion of high-affinity ligands. Thus, ligands might bind equally well to the intratransmembrane cavity of all opioid receptor subtypes; however, particular elements of the loops may have unfavorable interactions with selective ligands and, in turn, prevent their binding (76).

In addition to distinct transmembrane residues, agonist ligand selectivity for MOP, DOP, and KOP receptors have been attributed to the first and third extracellular loop for the MOP-R (74, 77–79), the second extracellular loop for the KOP-R (80–84), and the third extracellular loop for the DOP-R (85–90). More specifically, over 20 residues have been shown to determine ligand binding affinity to the MOP-R, whereas selectivity seems to be highly dependent on only four amino acids, Asp128, Asn150, and Lys303 in TM6, and Trp318 in TM7 (for detailed review and references therein, see (74). For the DOP-R, key determinants for selective ligand binding have been shown to involve residues within the third extracellular loop (Val296, Val297, Arg291, as well as Trp284 on top of TM6) (87, 89, 90) and a series of residues within the TM helices, comprising residues in TMs 3, 4, 6, and 7 (Asp 128 and Tyr 129 in TM3, Trp173 in TM4, Trp 174 and Trp284 in TM6, and Tyr308 in TM7) (85, 91), and Asp95 in TM2, which is crucial for recognition of DOP receptor selective agonists but is not involved in the binding of antagonists or nonselective agonists (86). Distinct key residues in the KOP-R determine peptide versus synthetic ligand binding and comprise the negatively charged second extracellular loop, which has been suggested to form an amphiphilic helix that interacts with six positively charged residues in the endogenous peptide agonist ligand dynorphin A (68a, 80–83). On the contrary, binding of KOP-R selective agonists of the acylacetamide class (i.e., U50488H, U69593, U62066, CI-977, ICI-19,944, ICI-197,067, BRL52537, and BRL52656) have been predicted to utilize residues Asp138 in TM3 and Ile294, Leu295, and Ala298 as key anchoring points for receptor binding (92). The prototypical KOP-R antagonist norBNI (93), however, was shown to require acidic residues within the KOP-R, involving the conserved aspartate residue Asp138 in TM3 and most crucially the nonconserved Glu297 on top of TM6, very close to the start of extracellular loop 3 (94–96). Likewise, residues Asp138, Glu297, and, in addition, His291 in TM6 have been suggested to be docking sites for the even more potent and selective indolomorphinan antagonist 5′-Guanidino-naltrindole (5′-GNTI) to the KOP receptor (97, 98).

Interestingly, the NOP-R lacks some of the classical rather conserved binding sites for opioid ligands near the top of each of the TMs and displays very low affinity to opioid ligands that promiscuously bind with high affinity to the MOP, DOP, and KOP receptor regions (99, 100). In fact, reintroducing some of these conserved residues to TM5 or TM6 of the NOP-R reinstates high-affinity binding to naltrexone, naltrindole, or dynorphin A (101). This is in line with a computational analysis of the NOP-R, which predicts a binding cavity for the N-terminal part of nociceptin within TMs 3, 5, 6, and 7, a region highly conserved across the other opioid receptors (102, 103).

G PROTEIN-EFFECTOR ACTIVATION   Opioid receptors are predominantly coupled to pertussis toxin-sensitive, heterotrimeric G_i/G_o proteins, although coupling to pertussis toxin-insensitive G_s or G_z proteins has also been reported (104–106).

Upon receptor activation, both G-protein α and $\beta\gamma$ subunits interact with multiple cellular effector systems, inhibiting adenylyl cyclases and voltage-gated Ca^{2+} channels and stimulating G protein-activated inwardly rectifying K^+ channels (GIRKs) and phospholipase $C\beta$ (PLCβ) [for reviews, see (99, 107–111)].

Activation mechanisms of GPCRs/7TM receptors have been suggested to involve ligand-induced transmembrane motions, resulting in exposure of the intracellular loops and making them more readily accessible to G proteins. A plethora of studies on GPCRs/7TM receptors have established the pivotal roles of the second and the third intracellular loops plus, at least in some receptors, the proximal portion of the C-terminal tail in G-protein coupling. Intracellular loop 3 is considered a key determinant of coupling specificity among the different G-protein α subunits, whereas intracellular loop 2 is more likely involved in the efficiency of G-protein activation [for review, see (112)]. For the DOP-R, a whole-receptor mutagenesis study describing 30 activating point mutations clustering spatially into four structural groups throughout the entire receptor protein proposed a ligand-induced receptor activation scheme as follows (113): An opioid agonist binds to a hydrophobic cluster in extracellular loop 3 and possibly in N-terminal regions, thereby destabilizing the interactions of TM6 and TM7 close to the extracellular side of the receptor. The ligand then enters the binding pocket and disrupts hydrophobic and hydrophilic interactions within TM helices 3, 6, and 7. This results in movements of these helices and subsequent propagation of these movements downward within the receptor. This movement ultimately results in a break of cytoplasmic ionic locks and possibly exposure of intracellular receptor domains to G proteins and other effector proteins.

Point mutations that result in constitutive activity and/or altered agonist efficacy in the MOP-R have pinpointed the importance of certain side chains, which include residues in transmembrane helices 2 (Asp114 and Tyr106), 4 (Ser196), and 6 (His297), as well as in the conserved Asp164 of the DRY motif at the beginning of intracellular loop 2 and in intracellular loop 3 [see (74, 114–116)]. Mutation of Ser196 in TM4 is of particular interest because it has recently been shown to confer agonist properties to antagonists in vitro (117) and in vivo (118). Tissue-specific expression of such a receptor could possibly provide a new therapeutic strategy for the treatment of pain.

In the NOP-R, alanine replacement of Gln286, located on top of TM 6, yielded a mutant receptor that retained a binding profile like the wild-type receptor. However, this mutant appeared to be functionally inactive, indicating that residue Gln286 may play a pivotal role in transmitting a nociceptin signal through NOP-R (119). In addition, mutation of Asn133 in TM3 activates G proteins two- to threefold over basal levels. Eleven other point mutations in this study resulted in either wild-type or decreased G protein-coupling efficiency. However, expression levels of these mutant NOP-Rs were rather low, and internalization rates of these mutants were not assessed (120).

CONSTITUTIVE ACTIVITY   Many GPCRs display a certain level of basal signaling activity and thus can activate G proteins in the absence of agonists [for review, see (112, 121, 122)]. DOP-Rs display a certain basal level of constitutive activity when they are expressed either endogenously in neuroblastoma cells (123, 124) or heterologously in various cell lines. Furthermore, inverse agonists like ICI-174864 upregulate DOP-R expression on the cell surface, independent of G-protein activation (125–131). MOP-Rs have also been shown to exhibit basal signaling activity in SH-SY5Y cells and in transfected HEK293 cells (132, 133) and display more elevated constitutive activity following chronic exposure to morphine (132). This phenomenon has been suggested to contribute to the development of tolerance and dependence. The molecular mechanism underlying this elevated basal signaling capacity of the MOP-R following prolonged agonist treatment might involve calmodulin (CaM), which has been shown to act as a constraint at the third intracellular loop of the MOP-R. In the nonagonist exposed state of the receptor, calmodulin binds to the MOP-R and competes for G-protein coupling, thereby inhibiting constitutive activity. After prolonged morphine treatment, CaM is released from the receptor, thus resulting in an elevated ligand-independent receptor activity (132, 134).

DESENSITIZATION/ENDOCYTOSIS/DOWNREGULATION   For the purpose of this review and in the interest of clarity, receptor desensitization is defined as any process that alters the functional coupling of a receptor to its G-protein/second messenger-signaling pathway. Endocytosis is defined as the translocation of receptors from the cell surface to an intracellular compartment. Lastly, down-regulation is defined as any process that decreases the number of ligand binding sites. Overall, it is essential to keep in mind that all three of these processes can each be modulated in various and distinct ways. It is therefore absolutely necessary (and represents a real challenge) to address each of these processes individually in order to understand their concerted actions in full detail.

DESENSITIZATION   By far the most complex of these processes is that of desensitization. Following activation by alkaloid or peptide agonists, opioid receptors are regulated by multiple mechanisms, many of which have been implicated in receptor desensitization. One of these is receptor phosphorylation. As for many GPCRs, signaling from opioid receptors is rapidly regulated by a well-characterized and highly conserved cascade of events involving receptor phosphorylation by G protein-coupled receptor kinases (GRKs) and subsequent β-arrestin recruitment, which has been extensively reviewed elsewhere (135). These processes contribute directly to rapid receptor desensitization by facilitating the uncoupling of the receptor from its G protein. GRK- and β-arrestin-mediated desensitization is a rapid process occurring often within minutes of receptor activation. However, opioid receptors can also be desensitized/uncoupled from G proteins by GRK- and β-arrestin-independent mechanisms [for review, see (136)]. These are very distinct processes by which a receptor can be desensitized. However, they have not always been dissected as such, and this has led to a somewhat obscured image of the nature of desensitization.

Many kinases phosphorylate opioid receptors, a number of which have been shown to desensitize receptors as well. These include not only GRK, but also protein kinase A (PKA), protein kinase C (PKC) and calcium/calmodulin-dependent protein kinase II [for review, see (135, 137)]. Inhibitors of mitogen-activated protein kinase also prevent opioid receptor desensitization (138), possibly by blocking the recruitment of β-arrestin to the plasma membrane following agonist activation (139). Opioid receptors can also be tyrosine phosphorylated. This modification affects desensitization as well, as has been shown in a series of studies examining C-terminal residues (Tyr96, Tyr106, Tyr166, and Tyr336) in the MOP-R (140), Tyr318 in the DOP-R (141, 142), and residues in the first and second intracellular loop of the KOP-R (143). The net amount of receptor desensitization is thus controlled by many factors, which include phosphorylation, dephosphorylation, new receptor protein synthesis and secretion, as well as receptor endocytosis, recycling, and degradation (see below).

ENDOCYTOSIS   Endocytosis is one feature that distinguishes GRK/β-arrestin-mediated desensitization from other phosphorylation-dependent desensitization mechanisms. Following desensitization by GRKs and β-arrestin, opioid receptors are then rapidly endocytosed into an intracellular compartment. This process occurs following even brief agonist exposure and independently of signal transduction (131, 144). The speed and conservation of this process is ideal for modulating signaling from endogenous ligands, such as neurotransmitters, that are released in a pulsatile manner. Following their endocytosis, receptors can then be recycled back to the membrane, thereby restoring the functional complement of receptors, a process termed "resensitization." In contrast, chronic exposure of opioid receptors to agonist, for example, during exogenous drug administration, can also lead to receptor desensitization/uncoupling. This involves alternate mechanisms, including PKA- and PKC-mediated phosphorylation, mentioned above. These desensitization processes most often require prolonged agonist treatment and are dependent on signal transduction. Importantly and in contrast to GRK-phosphorylated receptors, PKA/PKC-phosphorylated receptors are not necessarily rapidly endocytosed. Therefore, receptors that have been desensitized by GRK-independent phosphorylation require a mechanism other than endocytosis to resensitize. It is thus clear that experiments that do not differentiate between the various ways in which receptors can be phosphorylated will not sufficiently describe the full phenomenon of desensitization.

DOWNREGULATION   Receptors that have been desensitized and rapidly endocytosed are uniquely poised to make an important decision with substantial impact on future signal transduction. As mentioned above, following endocytosis, receptors can be recycled, thereby restoring the functional complement of receptors. Alternatively, receptors that have been endocytosed can be targeted for degradation, thereby decreasing the functional complement of receptors ultimately resulting in receptor downregulation. Although endocytosis and subsequent degradation of receptors are not the only means of producing receptor downregulation, they can produce receptor downregulation rapidly, even following brief exposure to agonist (145). Apparent receptor downregulation can also be affected by alterations in rate of receptor synthesis and/or folding/secretion (146).

Not all GPCRs and not all opioid receptors are downregulated following their endocytosis. Although both the MOP and DOP receptors are endocytosed via clathrin-coated pits following agonist-induced activation, GRK phosphorylation, and association with cytoplasmic β-arrestins (147–149), they differ in their fate following endocytosis. Whereas MOP-Rs are recycled following their endocytosis, DOP-Rs are transported deeper into the endocytic pathway, are rapidly degraded by the lysosome (150, 151), and hence "downregulated." It has recently been proposed that the cellular fate of receptors can be controlled by a specific protein interaction between the C-terminal tails of some GPCRs and the G-protein-coupled receptor associated sorting protein (GASP) (151). In particular, GPCRs that interact with GASP, for example the DOP-R, are targeted to the degradative pathway and downregulated. The receptors that do not interact with GASP, for example, the MOP-R, recycle and hence are resensitized.

In short, the role of endocytosis in modulating signaling from the receptor comprises both endocytic events (desensitization) and post-endocytic sorting events (resensitization or downregulation). Hence, for any GPCR-ligand pair there are two distinct trafficking properties that must be assessed: Whether or not ligand causes endocytosis of the receptor, and where that receptor goes following its endocytosis.

A number of mutants have been described that affect desensitization/endocytosis/resensitization/downregulation of each of the opioid receptors. For example, a mutation that prevented desensitization of the receptor could do so in many ways: by preventing phosphorylation, by preventing endocytosis, by enhancing resensitization, or by preventing downregulation. A recent study elegantly illustrates the importance of assessing desensitization mutants for precise phenotype. Qiu et al. (152) redescribed a mutant of the MOP-R truncated at Ser363. This mutant is not phosphorylated in response to agonist; it is endocytosed more slowly than the wild-type receptor and yet is more rapidly desensitized. This paradox could lead one to conclude that phosphorylation and endocytosis are therefore not responsible for desensitization of this receptor. However, the actual reason this mutant is more rapidly desensitized is because of a decrease in the recycling rate. Likewise, the mutation of Thr394 to Ala394 in the MOP-R shows complex and paradoxical phenotypes. The Thr394Ala mutant is endocytosed more quickly than the wild-type receptor, but nevertheless it desensitizes more slowly. The reason for this phenomenon is that this mutant MOP-R recycles more quickly (29) and is downregulated more slowly (153) than the wild-type receptor.

Unfortunately, most receptor desensitization mutants have not been analyzed in this detail. Although a clearer picture of what processes in this broad category, desensitization/endocytosis/resensitization/downregulation, are affected by what receptor mutants will likely emerge over the next few years, some of the primary observations are summarized below.

Truncation of the C-terminal tails of the DOP and MOP receptors has been shown to result in agonist-independent increased internalization and recycling rates that suggest the presence of regulatory elements critical for facilitating the receptor's interaction with the endocytic machinery (154–157). More specifically, Ser344 and Ser363 in the C-terminal tail of DOP-Rs have been shown to be crucial for receptor internalization (155, 158), whereas other C-terminal tail motifs (148, 150, 154) and a recently described di-leucine motif in the third intracellular loop were shown to be involved in downregulation and/or targeting the DOP-R to lysosomes (159).

For the MOP-R, mutations in the C-terminal tail revealed a number of serine/threonine residues (Ser356, Ser363, Ser375, Thr370, and Thr394) (29, 160, 161) and acidic residues (Glu288, Glu391, and Glu393) (153) involved in receptor internalization that are likely to be involved in recycling the receptor back to the surface (152, 156, 162). Recently, the human KOP-R has been shown to interact with the ezrin-radixin-moesin-binding phosphoprotein-50 / N+/H+-exchanger regulator factor (EBP50/NHERF) via a C-terminal domain slightly distinct from that previously described in the β_2-adrenergic receptor (163) that leads to enhanced recycling of the KOP-R (164). In addition, Ser358 in the human and Ser369 in the rat KOP-R C-terminal tail may be involved in receptor internalization (165, 166). Very few studies have been conducted to investigate domains involved in trafficking of NOP-Rs; however, NOP-Rs have been shown to undergo agonist-induced internalization and to utilize a GRK/β-arrestin-dependent pathway (167).

Receptor Dimerization

Dimerization is a mechanism for regulating the signaling from several classes of plasma membrane receptors, and it has been a particularly well-studied mechanism for the receptor tyrosine kinases [reviewed in (168)]. More recently, several groups have reported dimerization of several GPCRs (169, 170), including opioid receptors (12). In fact, heterodimerization of opioid receptors has been shown to alter opioid ligand properties (12, 171) and affect receptor trafficking in cell culture model systems (172) and in vivo (173). There are also reports of heterodimerization of the opioid receptors with other classes of GPCR; these include the NK1 receptor (174) and the β2-adrenergic receptor (172, 175).

Determining whether opioid receptor heterodimers or other opioid GPCR dimers exist in vivo is a true challenge. Advances in the study and design of bivalent ligands that selectively target heterodimers may provide the necessary tools [for review, see (176, 177)]. In addition, the phenotypes of the opioid receptor knockout mice also provide hints that heterodimerization, at least among the opioid receptors, is a real phenomenon. For example some MOP-R-specific analgesia is lost in the DOP-R-deficient mice (see below) (178). One explanation for this observation could be that a MOP/DOP receptor dimer has also been eliminated in the genetically modified animals.

FUNCTION

Opioid receptor activation by endogenous and exogenous ligands results in a multitude of effects, which include analgesia, respiratory depression, euphoria, feeding, the release of hormones, inhibition of gastrointestinal transit, and effects on anxiety. In general, agonists selective for MOP or DOP receptors are analgesic and rewarding, whereas KOP-R-selective agonists are dysphoric. Morphine and other opioids remain the analgesics of choice for the treatment of chronic pain. However the major limitation to their long-term use is the development of physiological "tolerance," a profound decrease in analgesic effects observed in most patients during prolonged drug administration. In addition to tolerance, physiological "dependence," which results in the necessity for continued administration of increasing doses of drug to prevent the development of symptoms of opioid withdrawal, can ensue in some patients. For these reasons, animal models of tolerance, dependence, and addiction have been devised to determine the mechanisms that underlie these complex phenomena. The following sections will address various mechanisms by which opioids mediate their physiological actions and will try to shed light on the molecular mechanisms that underpin opioid tolerance, dependence, and addiction.

Animal Models of Opioid Receptor Function

The development of tolerance to opioids is typically measured as a change in antinociceptive or analgesic responses. The two most common behavioral assays to assess such changes are the hot plate and the tail flick tests, where a heat source is applied to either the tail or hind paw of an animal. Dependence is measured in morphine-tolerant animals by either withdrawal of the opioid agonist or the administration of an opioid antagonist, such as naloxone. The typical behaviors reflecting withdrawal/dependence symptoms in animals are increases in locomotion, jumping, and weight loss. In addition, opioids are addictive drugs. Addiction is hypothesized to result from modulation of neural brain circuits associated with stress and anxiety, reward (positive reinforcement), as well as learning and memory. In animals, reward is measured using a technique referred to as conditioned place preference, a method that monitors an animal's ability to develop a preference for a certain environment when paired with a drug. Anxiety is induced by exposing animals to a stressful situation, for example, during a forced swim test.

Opioid Receptor-Deficient Mice

Transgenic mice have been generated that are deficient for each of the opioid receptors. In addition, some reports have emerged describing double and triple opioid receptor-deficient mice. These mice have provided insights into the specific roles of the opioid receptor subtypes in responses to endogenous and exogenous opioids.

MOP RECEPTOR-DEFICIENT MICE   MOP receptor-deficient mice have been generated in a number of different laboratories (179–182). The phenotypes of these mice differ depending on whether the disruption of the MOR-1 gene was targeted in exon 1 or exon 2. Activation of MOP-Rs elicits many behavioral responses, which include analgesia, hyperlocomotion, respiratory depression, constipation, and immunosuppression. Each of these behaviors is eliminated in the exon 2-MOP-R-deficient mice (179–181, 183, 184). Exon 1-deficient MOP-R mice, however, selectively retain analgesia to the MOP-R agonists morphine-6-glucuronide (M6G) and heroin (182). Thus, it seems possible that M6G and heroin analgesia is mediated by a MOP-R that contains exon 2, but not exon 1, of the MOP-R gene, thus representing a receptor likely to be distinct from the traditional morphine receptors. Some reports describe a reduced analgesic effect by DOP-R-selective ligands in the MOP-R-deficient mice (180); this suggests that MOP and DOP receptors could be functionally linked (185). In addition to being unresponsive to MOP-R ligands, such as morphine, MOP-R-deficient mice show reduced reward in a conditioned place preference paradigm to multiple other drugs of abuse (186–189). This indicates that MOP-Rs may play a role in the neural circuitry of reward.

DOP RECEPTOR-DEFICIENT MICE   DOP-R agonists have been shown to be potent analgesics with limited side effects. Thus, there is much effort devoted to creating orally available compounds that target this receptor class (190, 191). DOP-R-deficient mice still retain sensitivity to some DOP-R-selective ligands. However, relatively high doses of DOP-R ligands are generally required to produce analgesia. At these doses, the DOP-R-selective ligands have some binding affinity at other opioid receptors, which could explain these results. DOP-Rs are localized primarily intracellularly rather than on the surface of most cells (192, 193), an observation that might explain why relatively high doses of DOP-R-selective agonists are required for analgesia. Intriguingly, the DOP-R has been implicated in the development of morphine tolerance (194, 195), and in fact, DOP-R-deficient mice do not develop morphine tolerance to the same extent as their wild-type littermates (178) (see below). An additional observation is that DOP-R-deficient mice display increased levels of anxiety in both the forced swim test paradigm and light-dark box entry tests (196, 197), suggesting that ligands targeting this receptor may represent new leads for the treatment of affective disorders such as schizophrenia, bipolar disorder, and manic depression (198).

KOP-R-DEFICIENT MICE   KOP-R-deficient animals do not show altered basal nociceptive sensitivity to thermal or mechanical pain; however, their sensitivity to peritoneal injections of acetic acid is increased (199). Therefore, KOP-R agonists might be good analgesics for the treatment of visceral pain (200). In contrast to their wild-type littermates, KOP-R-deficient mice do not show conditioned place aversion or dysphoria to KOP-R-selective agonists (199). In addition, following chronic morphine treatment, naloxone-precipitated withdrawal behavior is attenuated in these mice (198, 199). Although KOP-R-deficient mice do not show alterations in stress-induced analgesia using a single trial of the forced swim test (201), there is some suggestion that KOP-Rs may participate in stress-induced emotional responses. Specifically, stress-induced analgesia and immobility are reduced in mice with a defective prodynorphin gene. This phenomenon has also been observed in wild-type animals upon administration of the KOP-R antagonist NorBNI (202).

NOP-R-DEFICIENT MICE   There is a general agreement that NOP-R- and nociceptin-deficient mice have unaltered basal nociceptive responses and analgesic responses to morphine [for a review, see (203)]. However, they do show altered analgesia in some paradigms, for example, in measures of stress-induced analgesia (204–209).

Morphine and Opioid Receptor Trafficking

Morphine-activated MOP-Rs are relatively unique in that they are not GRK phosphorylated nor do they efficiently recruit β-arrestin, even though they are in an active receptor conformation (147, 210, 211). Additionally, morphine fails to promote endocytosis of the wild-type MOP-R in cultured cells (212, 213) and native neurons (214, 215), whereas endogenous peptide ligands, such as endorphins and the hydrolysis-resistant form of enkephalin, D-Ala^2-MePhe^4-Gly^5-ol (DAMGO) as well as several opioid drugs, such as methadone (216), readily promote receptor endocytosis (214, 215, 217). Hence morphine-activated MOP-Rs generally elude an important, highly conserved regulatory mechanism designed to rapidly modulate receptor-mediated signaling.

RAVE   The ability of an agonist to induce signaling and its ability to promote endocytosis of the MOP-R are not always inextricably linked. This indicates that MOP-R endocytosis is an independent functional property of the receptor-ligand pair (218). Agonist activity and receptor endocytosis have opposing effects on receptor-mediated signaling. Therefore the net amount of signal transmitted to the cell is a function of both processes. This "net signal" has been termed RAVE, for relative activity versus endocytosis (218). In this model, morphine would have a particularly high RAVE value as a consequence of its inability to promote receptor desensitization and endocytosis. In contrast, endorphins and opioid drugs that acutely signal with similar efficacy, yet induce receptor desensitization and endocytosis, would have lower RAVE values than morphine.

The utility of the RAVE as an index is dependent on the assumption that receptor activity and endocytosis are not always associated processes. The RA numerator part of this index is influenced by several factors, which include agonist efficacy, potency, bioavailability, affinity, and half-life, all of which affect signaling in a positive way. For example, agonists with high efficacy but poor bioavailability in vivo would have a lower RA than agonists with similar efficacy but a long half-life. Hence, long-acting ligands would have higher RAVE values than more labile ones, even if their ability to endocytose the receptor were identical. Likewise, agonists with similar efficacies but different potencies, would have different RA values when given at the same dose; however, the RA in this case could be manipulated by appropriate changes in dosing. At the other extreme, a low-efficacy agonist with a long half-life could have a higher RA in vivo than a high-efficacy agonist with a very short half-life, because the net amount of signaling transmitted to the cell could be higher over the lifetime of a stable albeit low-efficacy ligand, than that of a high-efficacy but unstable ligand. In all these cases, endocytosis would oppose the RA. Agonists with identical RA values (either intrinsically or because of appropriate dosing) but varying abilities to facilitate endocytosis would differ in their RAVE values, that is, the actual amount of signal transmitted to the cell during the lifetime of the ligand and the receptor. Furthermore, the same ligand-receptor pair could have a different RAVE value depending on the cellular environment. For example, in cell-based systems where GRKs and/or β-arrestins are overexpressed, morphine-activated MOP-Rs do indeed recruit β-arrestin, and as a consequence, MOP-Rs desensitize and endocytose (147, 219). In these overexpressing cells, morphine would have a functionally lower RAVE value than in cells expressing endogenous levels of GRKs and arrestins. It is thus possible that differing levels of GRKs and arrestins in vivo likewise impart differing RAVE values to the morphine-MOP-Rs as well.

There have been numerous studies assessing the relationship between agonist efficacy and receptor endocytosis—with somewhat mixed results. There is substantial evidence from both heterologous and ex vivo systems that there is a linear relationship between efficacy and endocytosis for some ligands (220–222). However, the efficacy of an agonist appears to be a function of (*a*) the expression system in which signaling is being measured and (*b*) the effector that is used as the readout for a positive signal. The most proximal (and hence least amplified) readout for agonist efficacy is at the level of G-protein coupling. The GTPγS assay assesses the ability of a receptor-ligand pair to stimulate nucleotide exchange on the alpha subunit of the trimeric G protein to which the receptor is coupled. However, G-protein coupling is not required for endocytosis of the opioid receptors because pretreatment of cells with pertussis toxin does not abolish endocytosis of the receptor (131, 144). Hence coupling strength/signal strength per se is likely not the decisive factor determining whether a receptor is efficiently shuttled to the endocytic pathway. In further support of these observations, for some receptor-ligand pairs, even an antagonist can promote receptor endocytosis (223).

We favor the hypothesis that a conformation of the receptor when bound to ligand determines the efficiency at which a receptor recruits the endocytic machinery. It is possible that highly efficacious agonists retain a receptor in an activated conformation longer than do lower efficacy agonists, thereby allowing more time for the recruitment of the endocytic machinery. By this model, efficacy and endocytosis should be positively correlated. Alternatively, it is possible that different receptor-ligand pairs adopt different activated conformations, some of which are more efficient substrates for endocytosis. In this model, endocytosis and efficacy would not necessarily have to be related, if the conformation for most efficient coupling were different than the conformation for most efficient endocytosis. The reality is probably somewhere in between.

Hence, given the complexity of the opioid system and pharmacology of the various opioid ligands, it has been difficult to draw strong conclusions regarding the relationship between opioid receptor endocytosis and tolerance and dependence (see RAVE and Tolerance below). Briefly, each time a different drug is used, for example, morphine versus methadone, numerous aspects of signaling are modified, including—but not limited to—endocytosis. For example, morphine has metabolites that could be contributing to both receptor signaling and endocytosis (see below). In another example, methadone has a substantially longer half-life in humans in vivo than does morphine. Furthermore, the methadone composition used clinically is a mixed enantiomer with not only MOP-R agonism but also N-methyl-D-aspartate (NMDA) glutamate receptor antagonism. There is substantial evidence that antagonism of the NMDA receptor can reduce tolerance to opioids [see below and (224, 225)]. Thus, although different opioids have different RAs and different VEs, there are no two opioid ligands that differ only in their ability to endocytose the receptor.

TOLERANCE

Even though opioids, such as morphine, remain the analgesic of choice in many cases, a major limitation to their long-term use is the development of physiological tolerance, a profound decrease in analgesic effect observed in all patients during prolonged administration of opioid drug. The development of opioid tolerance in humans varies depending on the route of administration and on the disease state for which the opioids are prescribed. For example, tolerance is usually not a problem with short-term postoperative epidural or intrathecal opioids but rather presents itself following chronic epidural or intrathecal opioid usage. In addition, patients with terminal malignancies experience an escalation in pain as their disease progresses. This situation makes it difficult to distinguish between the development of opioid tolerance and an increased requirement for opioids for efficacious treatment of their pain (226). Thus, despite considerable progress, the molecular and cellular mechanisms mediating the development of tolerance and withdrawal to morphine remain controversial.

Multiple hypotheses exist to try to explain morphine tolerance. Several manipulations of multiple systems have been observed to effect the development of morphine tolerance and will be discussed in the following sections. Many of these observations, although appearing at first to be unrelated, likely reflect a common adaptive response to nonphysiological doses of morphine.

Receptor Downregulation

In theory, downregulation of opioid receptors would lead to tolerance by reducing the number of receptors available for drug-mediated actions. In fact, in vitro, downregulation of MOP-Rs has been reported following chronic agonist treatment (227, 228). However, in vivo, results are less clear cut: Following chronic morphine treatment, up- and downregulation of receptor number has been reported as well as no changes in receptor density [for review, see (229)], making it difficult to assess the role of receptor downregulation in the development of tolerance. Furthermore, the time course of receptor downregulation observed in cultured cells fails to match the time course of the development of tolerance in vivo. As an added layer of complexity, there appear to be differences in the ability of distinct agonist ligands to promote receptor downregulation (230). Taken together, the available data suggest that it is unlikely that receptor downregulation is solely responsible for the development of morphine tolerance. This view has lead to the idea that rather than becoming downregulated, MOP-Rs may instead become desensitized or, more precisely, uncoupled from downstream signaling pathways. In this context, receptors could become desensitized without the loss of a single receptor.

Receptor Desensitization

At the cellular level, acute desensitization of receptors occurs within minutes of agonist exposure, whereas the development of tolerance in vivo takes days to weeks, depending on the paradigm (227, 229, 231, 232). Several studies have examined whether there is functional desensitization of MOP-Rs in tolerant animals. In these reports, differences in opioid receptor G-protein coupling were measured in different brain regions and in the spinal cord before and after chronic opioid exposure in rats. These detailed studies demonstrate that the extent of MOP-R desensitization is highly dependent upon the brain region examined (233, 234). Following chronic drug administration, ten regions showed a 20% to 40% decrease in MOP-R coupling, whereas no significant changes were detected in seven other brain areas (234). If indeed there was receptor reserve (see below), it is difficult to envision how partial desensitization/uncoupling of receptors in some brain regions would result in complete analgesic insensitivity to the drug. Nevertheless, it is interesting to note that the most substantial changes in desensitization/coupling were observed in brain areas that mediate nociception and where tolerance to the analgesic and respiratory depressant effects of morphine were observed. The fact that other areas of the brain—in particular those mediating the reinforcing effects of opioids—were not desensitized/uncoupled may explain why little tolerance develops to the reinforcing and discriminative stimulus effects of opioids, despite complete analgesic tolerance (235).

Assuming that at the most 40% of MOP-Rs are desensitized in the tolerant state, the remaining 60% are presumably still coupled to G proteins. However, it has not been determined whether these receptors are coupled to G_i/G_o or to other G proteins, such as G_s or G_z (104, 105). In fact, in vitro, it has been shown that excitatory effects of opioids (106) are indeed blocked by cholera toxin (236), suggesting a G_s-mediated effect. Importantly, very low concentrations of opioid antagonists also block the excitatory effects of opioids, enhance the antinociceptive potency of morphine, and attenuate the development of morphine tolerance and dependence in vivo (237). Furthermore, low doses of the antagonist naloxone can itself produce/enhance analgesia (238–240). These observations created a paradox. One explanation is that, at low doses, naloxone antagonizes the G_s-coupled opioid receptors more completely than the G_i-coupled opioid receptors, therefore eliminating the excitatory (antianalgesic) properties of the opioid agonists. This observation has been transferred successfully to the clinic and highlights the need for studies to verify and dissect the mechanisms that underpin the effectiveness of a low-antagonist/agonist combination therapy in the treatment of chronic pain.

Bohn et al. (241, 242) have shown that there is a significant increase in MOP-R-G-protein coupling, enhanced analgesia, and decreased morphine tolerance in mice lacking β-arrestin-2, although numerous studies have shown that morphine-activated MOP-Rs are not efficiently desensitized via a GRK/β-arrestin-mediated route (147, 211, 230). This finding raises yet another paradox. Perhaps β-arrestin-mediated desensitization accounts, at least in part, for the 40% decrease in G-protein coupling observed in some brain regions (see above). If these regions were particularly rich in GRK and/or β-arrestins, MOP-R would desensitize, as observed in cell culture models, and thus result in morphine-induced MOP-R desensitization and endocytosis (147, 211). Consistent with this, G-protein coupling in the β-arrestin-deficient animals was altered in the brain stem. However, in the striatum, the levels of forskolin-stimulated adenylyl cyclase activity were unchanged in these mice. Although receptor G-protein coupling was not assessed in this region, these results suggest that there were no alterations in MOP-R G-protein coupling (241). Furthermore, the β-arrestin-2-deficient mice did not show changes in naloxone-precipitated withdrawal, suggesting that MOP-Rs must still be coupled in these animals; otherwise, the antagonist would have no effect. It will be interesting to examine the pattern of MOP-R desensitization and downregulation in these knockout animals following chronic administration of other opioid ligands (for example, etorphine), which consistently utilize arrestins for receptor desensitization even in the absence of GRK/β-arrestin overexpression. This is especially important in light of the studies mentioned above that suggest that morphine- and etorphine-induced tolerance may be mediated by different mechanisms (230).

Animals lacking β-arrestin-2 will also likely fail to properly regulate desensitization of receptors other than the MOP-R, for example, α2-adrenergic receptors, which could account for the enhanced basal- and morphine-induced analgesia in these animals. Mice deficient in the norepinephrine (NE) transporter, also show potentiated morphine analgesia, presumably as a result of elevated levels of norepinephrine (243). Because norepinephrine is a ligand for α2-adrenergic receptors (244, 245), some of which are desensitized by β-arrestins (246), one would assume that NE effects would be enhanced in the the β-arrestin knockout mice, leading to enhanced morphine analgesia.

Uncoupling From cAMP Pathway—cAMP Superactivation

Following chronic morphine treatment, cellular levels of cAMP are elevated, a phenomenon termed cAMP superactivation. At first sight this observation suggests that opioid receptors are no longer coupled to inhibitory G proteins. However, such is not the case because antagonist treatment following chronic morphine exposure leads to an even further increase in cAMP levels, demonstrating that MOP-Rs are indeed still functionally coupled to G_i/G_o proteins. Instead the elevated cAMP levels reflect cellular adaptive changes, which include increased expression of certain types of adenylyl cyclase, protein kinase A (PKA), and cAMP response element binding protein (CREB) [reviewed in (110, 231)]. For example, phosphorylation of type II adenylyl cyclase isoforms can significantly increase their stimulatory responsiveness to $G_s\alpha$ and $G\beta\gamma$ (247). Importantly, any of these compensatory changes produces cells/animals that appear tolerant to morphine simply because cAMP levels are no longer as effectively regulated by morphine as they were in the naive state. This phenomenon has been reported by several groups both in vitro and in brain regions implicated in addiction; these include the ventral tegmental area (248) and locus coeruleus (231). Here, it has been shown that the elevated levels of cAMP are responsible for changes in gene expression as well as for alterations in neurotransmitter release (248–250). These changes are long term and difficult to reverse. In fact, many of the observations discussed below may be directly attributed to cAMP superactivation.

RAVE and Tolerance

Whistler and colleagues (173, 218) have previously proposed that the regulation of opioid receptors by endocytosis serves a protective role in reducing the development of tolerance and dependence to opioid drugs. This model, termed the RAVE hypothesis (see above), is outlined as follows. First, as a consequence of endocytosis, cells are rapidly desensitized to agonist. Second, following endocytosis, receptors can be recycled to the cell surface in a fully active state, thereby resensitizing cells to agonist. This highly dynamic cycle of receptor regulation may be designed to mediate the actions of endogenous opioid peptides, which are typically released in a phasic or pulsatile manner. Opioid drugs, in contrast, persist in the extracellular milieu for a prolonged period of time because of their slow clearance. The opioid receptors are thus activated in an abnormally prolonged manner. Hence, drugs with high RAVE values, such as morphine, would have an enhanced propensity to produce cAMP superactivation. This phenomenon will ultimately result in tolerance and dependence precisely because the morphine-activated opioid receptors signal for aberrantly long periods of time.

Studies in cell culture have demonstrated that mutations of the MOP-R that enhance morphine-induced endocytosis do indeed ameliorate cAMP superactivation (150). Conversely, mutations that prevent endocytosis of the MOP-R in response to ligands, such as methadone and DAMGO, enhance methadone- and DAMGO-evoked cAMP superactivation (150). Furthermore, in vivo, facilitation of MOP-R endocytosis in response to morphine by coadministration of a small dose of DAMGO prevented the development of morphine tolerance (173). In a similar manner, Ueda et al. (251) found that intraplantar injections of morphine and DAMGO in mice produced opposite effects on receptor internalization. Specifically whereas DAMGO internalized MOP-Rs, morphine failed to do so. Likewise, they found an inverse relationship between endocytosis and tolerance. Although morphine was able to produce acute analgesic tolerance, DAMGO failed to do so. However, when an inhibitor of PKC activation was applied to inhibit DAMGO-induced MOP-R internalization, DAMGO was able to induce acute tolerance. This finding again exemplifies the importance of receptor internalization in modulating the level of signaling through the MOP-R.

It has been shown that morphine fails to internalize MOP-Rs in most brain regions, as examined by binding and autoradiography studies. However, a recent study reports that morphine is capable of inducing rapid redistribution of MOP-Rs in the dendrites of nucleus accumbens neurons. This observation suggests that the trafficking of opioid receptors may be differentially modulated by morphine (252) depending on the cellular environment. Thus, a finer understanding of brain region–specific differences in receptor trafficking will be needed in the future in order to understand its contribution to the development of opioid tolerance. Furthermore, it is tantalizing to speculate that the development of analgesic tolerance in the spinal cord and in peripheral tissues may involve separate biochemical mechanisms to that in the brain.

Efficacy

The degree to which the maximal effect of a given drug can be achieved without full receptor occupancy defines its intrinsic efficacy. Each opioid ligand has its own intrinsic efficacy. The fraction of the total receptor pool not required for maximal effect is the receptor reserve (or spare receptors). In this context, high-efficacy ligands would be expected to produce maximal analgesia leaving many spare receptors. On the other hand, low-efficacy ligands would leave fewer spare receptors.

Some studies have suggested that the intrinsic efficacy of an opioid analgesic can determine its ability to induce the development of tolerance. In this model, highly efficacious ligands, such as etorphine and fentanyl, leave more spare receptors than low-efficacy ligands, such as morphine. Provided that the occupied receptors become tolerant, i.e., desensitized, high-efficacy ligands will leave more receptors in the nontolerant state. The model thus assumes that receptor occupancy leads to irreversible desensitization. However, this is primarily not the case during chronic opioid treatment (see desensitization above). The model also assumes that there is an inverse relationship between intrinsic efficacy and tolerance (253–255). However, this relationship is observed only when drugs are administered continuously in animal models. Importantly, when drugs are given intermittently, an inverse relationship between intrinsic efficacy and tolerance was not observed (256). This observation is important given that, in a clinical setting, drugs are given intermittently.

Gebhart (226) explains these discrepancies of intermittent and continual dosing as a possible result of the differences in the lipophilicity of high-efficacy ligands, such as etorphine and fentanyl, compared to lower efficacy ligands, such as morphine. In this model, the high-efficacy ligands (e.g., fentanyl) are also more lipophilic. Thus, in cases such as chronic infusion, they can more easily redistribute to sites other than the spinal cord, increasing their efficacy for the treatment of pain.

What is the evidence for spare receptors or receptor reserve? The development of MOP-R-deficient mice has provided a tool to address this issue. In MOP-R-heterozygote mice in which only half of the MOP-Rs remain, the level of naloxone-precipitated withdrawal jumping is similar to that of wild-type animals, suggesting that there may, in fact, be a substantial receptor reserve (257). In contrast, in all other behavioral measures of opioid effects, including locomotion, conditioned place preference, self-administration, reinforcement, or analgesia, the genotype did in fact linearly correlate with behavior. This apparent discrepancy might suggest that more spare receptors are needed for some opioid effects but not others. On a tissue level, in the guinea pig ileum, maximal morphine responses were demonstrated even following inactivation of ~90% of the initial MOP-R population (258), suggesting this tissue has a receptor reserve of 90%. The receptor reserve was reduced in morphine-tolerant animals. However, reduction in the spare receptor fraction does not necessarily imply reduction in the number of binding sites; it merely implies that more receptors are required to produce the same analgesic effect in tolerant animals. This loss of receptor reserve, i.e., the need for more receptors, in tolerant animals is exactly what one would expect if second messenger-signaling cascades were upregulated due to cAMP superactivation.

Interaction of MOP-R With Other Opioid Receptor Subtypes

DOP-R-deficient mice show reduced development of morphine tolerance (178). In addition, following morphine treatment for at least 48 h, DOP-Rs are redistributed from intracellular compartments, where they normally reside, to the cell surface in the dorsal horn (259, 260). Thus the question arises as to whether these two observations could in fact be related. What if the recruitment of the DOP receptor to the plasma membrane changes the signaling and/or trafficking properties of the MOP receptor? If this were the case, one would assume that other treatments that have been shown to cause redistribution of DOP-Rs to the plasma membrane, such as neurotrophins or swim stress (259–262), should enhance the development of morphine tolerance and dependence. Furthermore, because the timeline of DOP-R redistribution mirrors that of cAMP superactivation, one might also imagine that these phenomena are interrelated.

A partial loss of the development of morphine tolerance has also been reported in NOP-R-deficient mice (209). However, morphine is a more potent antinociceptive agent in NOP-R-deficient mice as well, making it difficult to distinguish between enhanced antinociception and reduced tolerance. In cell culture models, NOP/MOP receptors dimerize and have an altered pharmacological phenotype (263). This leads one to speculate that the partial loss of morphine tolerance in the NOP-R-deficient mice may result from altered signaling of the MOP receptor. However, there is little information about whether chronic morphine administration alters the distribution of the NOP-R. In short, alterations in the composition of opioid receptor subtypes (DOP/MOP, NOP/MOP?) in a given cell/brain area appear to be a prerequisite for the development of tolerance after chronic morphine treatment.

Endogenous Antiopioid System(s)

It has been proposed that an antiopioid system exists to regulate and compensate the opioid system and as such maintain the homeostasis (229, 264). One view is that antireward systems can contribute to motivational changes during chronic administration of drugs (265). Investigators have proposed that, following chronic opioid administration, endogenous antiopioid peptides are released to compensate for prolonged inhibition by morphine. The most well-studied endogenous antiopioid peptides are the Tyr-Pro-Leu-Gly-NH$_2$ (Tyr-MIF-1) family of peptides, neuropeptide FF, and orphanin FQ/nociceptin [for extensive review, see (229)]. In addition, cholecystokinin (CCK) signaling [for review, see (266)], NMDA receptors [for review, see (267)], and excitatory opioid receptors coupled through G$_s$ [for review, see (106)] are also antipioid systems.

Other factors modulating the antinociceptive effect of morphine have also been described. Two of these are the active metabolites of morphine, morphine-6-glucuronide, a potent analgesic, and morphine-3-glucuronide (M3G), an antianalgesic (268). M3G has been shown to cause excitatory behavioral effects in animals and to directly modulate neurotransmitter release (269). These effects could be inhibited by NMDA receptor antagonists (270) or low doses of naloxone (271).

The precise molecular mechanisms mediating upregulation of these antiopioid systems is unknown. However it is tempting to speculate that superactivation of the cAMP/PKA pathway, which is known to affect Cre-mediated gene expression, may also be responsible for upregulation of these systems. A brief outline of some of the compensatory antiopioids follows.

CCK   Cholecystokinins belong to the gastrin family of peptides and are some of the most potent antagonists of the opioid system (272, 273) [for review, see (266, 274)]. CCK peptides bind to two GPCRs, CCK-A (also called CCK-1) and CCK-B (also called CCK-2) [for review, see (264, 266, 275)]. CCK-B/CCK-2 receptor-deficient mice have significantly increased amounts of endogenous opioids in the brain (276), suggesting CCK-B/2 activation prevents release of endogenous opioids. In addition, CCK reduces the effectiveness of opioids in behavioral and electrophysiological tests. Moreover, CCK acting at the CCK-B receptors is required for the development of associative morphine tolerance, which develops after repeated administration of opioids in the presence of specific environmental cues (277, 278). Importantly, morphine-induced in vivo release of spinal CCK has been shown to be mediated by DOP-Rs (279). As noted above, DOP-Rs are translocated during the development of morphine tolerance, suggesting that further increases in CCK release may be directly related to morphine-induced translocation of DOP-Rs.

NMDA RECEPTORS   The glutamate or NMDA receptors are another proposed antiopioid system (267). The NMDA receptor antagonist, MK801, inhibits the development of morphine tolerance (224, 280) and dependence (224). Importantly, the NR2A (281) and the NR1 (282) subunits of the NMDA receptor are upregulated following chronic morphine treatment, implicating NMDA receptor upregulation in the behavioral manifestation of tolerance. NR1 subunit upregulation is blocked by NMDA-R antagonists. One key additional observation is that, although NMDA receptor antagonists can inhibit the induction and development of morphine tolerance, they cannot reverse these effects once they have been established [for review, see (267)]. One explanation might be that once NMDA-Rs are upregulated and likely have initiated changes in plasticity responsible for tolerance and dependence, mere antagonism of the NMDA receptors will not be able to reverse these plasticity changes. Therefore, NMDA receptor antagonists do not reverse morphine tolerance and dependence.

In summary, a theme emerges whereby, following chronic morphine treatment, several systems are upregulated. It remains to be determined if a common mechanism exists that results in the upregulation of any or all of these systems. Perhaps, for example, superactivation of the cAMP pathway, which leads to upregulation of PKA and thus CREB-induced gene expression, is responsible for these changes. If this were the case, mechanisms that modified cAMP superactivation, for example, alterations in RAVE, would be expected to influence all of these systems.

DEPENDENCE AND WITHDRAWAL

In addition to the side effects of tolerance, respiratory suppression, constipation, allodynia, myoclonic jerks, and seizures, long-term use of opioids also causes physiological dependence in some patients. Dependence is reflected by a need for continued administration of increasing doses of drug to prevent the development of opioid withdrawal. The physiological drug-dependent state is revealed following cessation of the drug, manifested with the classic withdrawal syndrome referred to as "cold turkey." Neuronal adaptations resulting from the chronic use of opioid drugs were first hypothesized to explain the development of drug dependence (283). However, drug users who are no longer physically dependent can still show psychological dependence and relapse. Therefore, it is thought that different areas of the CNS are responsible for controlling the physiological and psychological components of drug dependence (231, 284, 285).

Tolerance and dependence may share a common mechanism because the severity of withdrawal signs and the extent of the development of tolerance correlate well in vivo (286) and in vitro (287). In addition, when the development of tolerance is blocked using NMDA receptor antagonists, dependence is not observed, again suggesting the two phenomena are interrelated (224, 288, 289). In fact, cAMP superactivation has been suggested as a common mechanism contributing to both tolerance and dependence [for review, see (110)].

However, genetically modified mice have demonstrated that the development of tolerance and dependence can be dissociated. In particular, dependence, as measured by opioid antagonist-induced withdrawal, has been demonstrated in mice that do not develop morphine tolerance. For example, DOP-R-deficient, β-arrestin-2-deficient, and the 129S6 inbred strain of mice all display deficits in the development of morphine tolerance, but opioid withdrawal behaviors were still observed (241, 290). Importantly, one has to keep in mind that morphine analgesia is enhanced in each of these genetically modified mice, thus making it difficult to distinguish enhanced analgesia from reduced tolerance.

One region of the brain that has been studied extensively as a model for opioid dependence and withdrawal is the locus coeruleus (LC) [reviewed in (110, 283, 291)]. Following the chronic administration of morphine, the LC neurons show tolerance to morphine as well as cellular correlates of dependence and withdrawal [(292–294) and for review, see (110)]. Some of these cellular changes include increases in the levels of G-protein subunits, adenylyl cyclases, PKA (295–301), and CREB (302, 303). Interestingly, upregulation of the cAMP pathway has been consistently observed in the LC, nucleus accumbens, amygdala, doral raphe nucleus, and ventral tegmental area (248, 301, 304, 305). These considerations suggest that, once again, this pathway may be a cellular hallmark of the development of not only morphine tolerance (see above) but also dependence (110, 250, 251, 284, 306–308).

In conclusion, although there have been active studies of opioids and opioid receptors for decades, the field continues to be a dynamic and lively one. The cloning of the receptors over 10 years ago initiated a renewed surge of interest in the opioid field. A plethora of structure-function studies ensued, leading to clear models explaining the distinct and varied actions of opioids and opioid receptors. Likewise, improvements in the behavioral assays utilized to model the physiological and psychological effects of opioids in animals—in particular tolerance, dependence, and reward—have ushered in a new opioid era. Advances in technology have also improved our ability to dissect the molecular mechanisms underlying these complex behaviors. We hope that this review has demonstrated that many effects that occur as a consequence of prolonged opiate use, which at first glance may appear unrelated, converge on a common theme. Perhaps a few primary events, such as cAMP superactivation, may initiate a cascade of cellular and synaptic changes during chronic exposure to opioids. Together, these events underpin the manifestation of opiate tolerance, dependence, and addiction.

Nevertheless, many questions still remain to be answered, and it is precisely these unsolved mysteries that keep interest in the opioids and their receptors alive and well.

ACKNOWLEDGMENTS

We thank Christian Elling, Phil Portoghese, and David E. Ferguson for helpful discussions on this manuscript, and Francois Talabot for help in creating the phylogenetic tree. Further, the authors thank the members of the Whistler Lab, Joseph Kim, Johan Enquist, Randall Armstrong, and Lisa Daitch for critically reading this manuscript. This work was supported by funds provided by NIH/NIDA (#R01DA015232) and by the State of California for medical research on alcohol and substance abuse through the University of California, San Francisco to J.L.W. S.E.B. is a recipient of a NARSAD young investigator award.

The *Annual Review of Biochemistry* is online at http://biochem.annualreviews.org

LITERATURE CITED

1. Hughes J, Kosterlitz HW. 1983. *Br. Med. Bull.* 39:1–3
2. Pert CB, Snyder SH. 1973. *Science* 179: 1011–14
3. Beckett AH, Casy AF. 1954. *J. Pharm. Pharmacol.* 6:986–1001
4. Portoghese PS. 1965. *J. Med. Chem.* 8: 609–16
5. Terenius L. 1973. *Acta Pharmacol. Toxicol.* 32:317–20
6. Simon EJ, Hiller JM, Edelman I. 1973. *Proc. Natl. Acad. Sci. USA* 70:1947–49
7. Lord JA, Waterfield AA, Hughes J, Kosterlitz HW. 1977. *Nature* 267: 495–99
8. Bunzow JR, Saez C, Mortrud M, Bouvier C, Williams JT, et al. 1994. *FEBS Lett.* 347:284–88
9. Mollereau C, Parmentier M, Mailleux P, Butour JL, Moisand C, et al. 1994. *FEBS Lett.* 341:33–38
10. Meunier JC, Mollereau C, Toll L, Suaudeau C, Moisand C, et al. 1995. *Nature* 377:532–35
11. Pasternak GW. 2001. *Life Sci.* 68: 2213–19
12. Jordan BA, Devi LA. 1999. *Nature* 399: 697–700
13. Chen Y, Mestek A, Liu J, Hurley JA, Yu L. 1993. *Mol. Pharmacol.* 44:8–12
14. Fukuda K, Kato S, Mori K, Nishi M, Takeshima H. 1993. *FEBS Lett.* 327: 311–14
15. Wang JB, Imai Y, Eppler CM, Gregor P, Spivak CE, Uhl GR. 1993. *Proc. Natl. Acad. Sci. USA* 90:10230–34

16. Thompson RC, Mansour A, Akil H, Watson SJ. 1993. *Neuron* 11:903–13

17. Minami M, Onogi T, Toya T, Katao Y, Hosoi Y, et al. 1994. *Neurosci. Res.* 18: 315–22

18. Bunzow JR, Zhang G, Bouvier C, Saez C, Ronnekleiv OK, et al. 1995. *J. Neurochem.* 64:14–24

19. Min BH, Augustin LB, Felsheim RF, Fuchs JA, Loh HH. 1994. *Proc. Natl. Acad. Sci. USA* 91:9081–85

20. Wang JB, Johnson PS, Persico AM, Hawkins AL, Griffin CA, Uhl GR. 1994. *FEBS Lett.* 338:217–22

21. Pampusch MS, Osinski MA, Brown DR, Murtaugh MP. 1998. *J. Neuroimmunol.* 90:192–98

22. Onoprishvili I, Andria ML, Vilim FS, Hiller JM, Simon EJ. 1999. *Brain Res. Mol. Brain Res.* 73:129–37

23. Darlison MG, Greten FR, Harvey RJ, Kreienkamp HJ, Stuhmer T, et al. 1997. *Proc. Natl. Acad. Sci. USA* 94:8214–19

24. Barrallo A, Gonzalez-Sarmiento R, Alvar F, Rodriguez RE. 2000. *Brain Res. Mol. Brain Res.* 84:1–6

25. Stevens C. 2003. *Rev. Analg.* 7:69–82

26. Pasternak GW. 2001. *Trends Pharmacol. Sci.* 22:67–70

27. Koch T, Schulz S, Pfeiffer M, Klutzny M, Schroder H, et al. 2001. *J. Biol. Chem.* 276:31408–14

28. Koch T, Schulz S, Schroder H, Wolf R, Raulf E, Hollt V. 1998. *J. Biol. Chem.* 273:13652–57

29. Wolf R, Koch T, Schulz S, Klutzny M, Schroder H, et al. 1999. *Mol. Pharmacol.* 55:263–68

30. Kieffer BL, Befort K, Gaveriaux RC, Hirth CG. 1992. *Proc. Natl. Acad. Sci. USA* 89:12048–52

31. Evans CJ, Keith DE Jr, Morrison H, Magendzo K, Edwards RH. 1992. *Science* 258:1952–55

32. Knapp RJ, Malatynska E, Fang L, Li XP, Babin E, et al. 1994. *Life Sci.* 54: L463–69

33. Vanderah T, Takemori AE, Sultana M, Portoghese PS, Mosberg HI, et al. 1994. *Eur. J. Pharmacol.* 252:133–37

34. Portoghese PS, Sultana M, Nagase H, Takemori AE. 1992. *Eur. J. Pharmacol.* 218:195–96

35. Jiang Q, Takemori AE, Sultana M, Portoghese PS, Bowen WD, et al. 1991. *J. Pharmacol. Exp. Ther.* 257:1069–75

36. Quock RM, Burkey TH, Varga E, Hosohata Y, Hosohata K, et al. 1999. *Pharmacol. Rev.* 51:503–32

37. Nishi M, Takeshima H, Fukuda K, Kato S, Mori K. 1993. *FEBS Lett.* 330:77–80

38. Chen Y, Mestek A, Liu J, Yu L. 1993. *Biochem. J.* 295(Part 3):625–28

39. Meng F, Xie GX, Thompson RC, Mansour A, Goldstein A, et al. 1993. *Proc. Natl. Acad. Sci. USA* 90:9954–58

40. Li S, Zhu J, Chen C, Chen YW, Deriel JK, et al. 1993. *Biochem. J.* 295(Part 3):629–33

41. Minami M, Toya T, Katao Y, Maekawa K, Nakamura S, et al. 1993. *FEBS Lett.* 329:291–95

42. Yasuda K, Raynor K, Kong H, Breder CD, Takeda J, et al. 1993. *Proc. Natl. Acad. Sci. USA* 90:6736–40

43. Nishi M, Takeshima H, Mori M, Nakagawara K, Takeuchi T. 1994. *Biochem. Biophys. Res. Commun.* 205:1353–57

44. Mansson E, Bare L, Yang D. 1994. *Biochem. Biophys. Res. Commun.* 202: 1431–37

45. Simonin F, Gaveriaux-Ruff C, Befort K, Matthes H, Lannes B, et al. 1995. *Proc. Natl. Acad. Sci. USA* 92:7006–10

46. Zhu J, Chen C, Xue JC, Kunapuli S, DeRiel JK, Liu-Chen LY. 1995. *Life Sci.* 56:PL201–7

47. Xie GX, Meng F, Mansour A, Thompson RC, Hoversten MT, et al. 1994. *Proc. Natl. Acad. Sci. USA* 91:3779–83

48. Chavkin C, James IF, Goldstein A. 1982. *Science* 215:413–15

49. Akil H, Watson SJ. 1994. *Prog. Brain Res.* 100:81–86

50. Fukuda K, Kato S, Mori K, Nishi M,

Takeshima H, et al. 1994. *FEBS Lett.* 343: 42–46

51. Chen Y, Fan Y, Liu J, Mestek A, Tian M, et al. 1994. *FEBS Lett.* 347:279–83

52. Lachowicz JE, Shen Y, Monsma FJ Jr, Sibley DR. 1995. *J. Neurochem.* 64: 34–40

53. Wang JB, Johnson PS, Imai Y, Persico AM, Ozenberger BA, et al. 1994. *FEBS Lett.* 348:75–79

54. Wick MJ, Minnerath SR, Lin X, Elde R, Law PY, Loh HH. 1994. *Brain Res. Mol. Brain Res.* 27:37–44

55. Osinski MA, Pampusch MS, Murtaugh MP, Brown DR. 1999. *Eur. J. Pharmacol.* 365:281–89

56. Pan YX, Xu J, Wan BL, Zuckerman A, Pasternak GW. 1998. *FEBS Lett.* 435: 65–68

57. Reinscheid RK, Nothacker HP, Bourson A, Ardati A, Henningsen RA, et al. 1995. *Science* 270:792–94

58. Martin WR, Eades CG, Thompson JA, Huppler RE, Gilbert PE. 1976. *J. Pharmacol. Exp. Ther.* 197:517–32

59. Hanner M, Moebius FF, Flandorfer A, Knaus HG, Striessnig J, et al. 1996. *Proc. Natl. Acad. Sci. USA* 93:8072–77

60. Monassier L, Bousquet P. 2002. *Fundam. Clin. Pharmacol.* 16:1–8

61. Zagon IS, Gibo DM, McLaughlin PJ. 1991. *Brain Res.* 551:28–35

62. Zagon IS, Verderame MF, Allen SS, McLaughlin PJ. 2000. *Brain Res.* 856: 75–83

63. Wuster M, Schulz R, Herz A. 1979. *Neurosci. Lett.* 15:193–98

64. Lapalu S, Moisand C, Mazarguil H, Cambois G, Mollereau C, Meunier JC. 1997. *FEBS Lett.* 417:333–36

65. Lapalu S, Moisand C, Butour JL, Mollereau C, Meunier JC. 1998. *FEBS Lett.* 427:296–300

66. Zadina JE, Hackler L, Ge LJ, Kastin AJ. 1997. *Nature* 386:499–502

67. Zaveri N, Polgar WE, Olsen CM, Kelson AB, Grundt P, et al. 2001. *Eur. J. Pharmacol.* 428:29–36

68. Zaveri N. 2003. *Life Sci.* 73:663–78

68a. Portoghese PS, Ferguson DM. 2004. *CORD center for opioid research and design.* http://www.opioid.umn.edu

69. Palczewski K, Kumasaka T, Hori T, Behnke CA, Motoshima H, et al. 2000. *Science* 289:739–45

70. Schwartz TW. 1994. *Curr. Opin. Biotechnol.* 5:434–44

71. Pogozheva ID, Lomize AL, Mosberg HI. 1998. *Biophys. J.* 75:612–34

72. Minami M, Satoh M. 1995. *Neurosci. Res.* 23:121–45

73. Law P-Y, Loh HH. 1999. *J. Pharmacol. Exp. Ther.* 289:607–24

74. Chavkin C, McLaughlin JP, Celver JP. 2001. *Mol. Pharmacol.* 60:20–25

75. Takemori AE, Portoghese PS. 1992. *Annu. Rev. Pharmacol. Toxicol.* 32: 239–69

76. Metzger TG, Ferguson DM. 1995. *FEBS Lett.* 375:1–4

77. Onogi T, Minami M, Katao Y, Nakagawa T, Aoki Y, et al. 1995. *FEBS Lett.* 357:93–97

78. Mansour A, Taylor LP, Fine JL, Thompson RC, Hoversten MT, et al. 1997. *J. Neurochem.* 68:344–53

79. Ulens C, Van Boven M, Daenens P, Tytgat J. 2000. *J. Pharmacol. Exp. Ther.* 294: 1024–33

80. Xue JC, Chen C, Zhu J, Kunapuli S, DeRiel JK, et al. 1994. *J. Biol. Chem.* 269:30195–99

81. Wang JB, Johnson PS, Wu JM, Wang WF, Uhl GR. 1994. *J. Biol. Chem.* 269: 25966–99

82. Paterlini G, Portoghese PS, Ferguson DM. 1997. *J. Med. Chem.* 40:3254–62

83. Zhang L, DeHaven RN, Goodman M. 2002. *Biochemistry* 41:61–68

84. Kong H, Raynor K, Yano H, Takeda J, Bell GI, Reisine T. 1994. *Proc. Natl. Acad. Sci. USA* 91:8042–46

85. Befort K, Tabbara L, Kling D, Maigret B, Kieffer BL. 1996. *J. Biol. Chem.* 271: 10161–68

86. Kong H, Raynor K, Yasuda K, Moe ST,

Portoghese PS, et al. 1993. *J. Biol. Chem.* 268:23055–58

87. Meng F, Ueda Y, Hoversten MT, Thompson RC, Taylor L, et al. 1996. *Eur. J. Pharmacol.* 311:285–92

88. Pepin MC, Yue SY, Roberts E, Wahlestedt C, Walker P. 1997. *J. Biol. Chem.* 272:9260–67

89. Valiquette M, Vu HK, Yue SY, Wahlestedt C, Walker P. 1996. *J. Biol. Chem.* 271:18789–96

90. Varga EV, Li X, Stropova D, Zalewska T, Landsman RS, et al. 1996. *Mol. Pharmacol.* 50:1619–24

91. Befort K, Tabbara L, Bausch S, Chavkin C, Evans C, Kieffer B. 1996. *Mol. Pharmacol.* 49:216–23

92. Subramanian G, Paterlini MG, Larson DL, Portoghese PS, Ferguson DM. 1998. *J. Med. Chem.* 41:4777–89

93. Portoghese PS, Lipkowski AW, Takemori AE. 1987. *Life Sci.* 40:1287–92

94. Hjorth SA, Thirstrup K, Grandy DK, Schwartz TW. 1995. *Mol. Pharmacol.* 47:1089–94

95. Larson DL, Jones RM, Hjorth SA, Schwartz TW, Portoghese PS. 2000. *J. Med. Chem.* 43:1573–76

96. Jones RM, Hjorth SA, Schwartz TW, Portoghese PS. 1998. *J. Med. Chem.* 41:4911–14

97. Jones RM, Portoghese PS. 2000. *Eur. J. Pharmacol.* 396:49–52

98. Stevens WC Jr, Jones RM, Subramanian G, Metzger TG, Ferguson DM, Portoghese PS. 2000. *J. Med. Chem.* 43:2759–69

99. New DC, Wong YH. 2002. *Neurosignals* 11:197–212

100. Meunier J, Mouledous L, Topham CM. 2000. *Peptides* 21:893–900

101. Meng F, Taylor LP, Hoversten MT, Ueda Y, Ardati A, et al. 1996. *J. Biol. Chem.* 271:32016–20

102. Topham CM, Mouledous L, Poda G, Maigret B, Meunier JC. 1998. *Protein Eng.* 11:1163–79

103. Topham CM, Mouledous L, Meunier JC. 2000. *Protein Eng.* 13:477–90

104. Garzon J, Castro M, Sanchez-Blazquez P. 1998. *Eur. J. Neurosci.* 10:2557–64

105. Hendry IA, Kelleher KL, Bartlett SE, Leck KJ, Reynolds AJ, et al. 2000. *Brain Res.* 870:10–19

106. Crain SM, Shen KF. 1990. *Trends Pharmacol. Sci.* 11:77–81

107. Loh HH, Smith AP. 1990. *Annu. Rev. Pharmacol. Toxicol.* 30:123–47

108. Ikeda K, Kobayashi T, Kumanishi T, Yano R, Sora I, Niki H. 2002. *Neurosci. Res.* 44:121–31

109. Connor M, Christie MJ. 1999. *Clin. Exp. Pharmacol. Physiol.* 26:493–99

110. Williams JT, Christie MJ, Manzoni O. 2001. *Physiol. Rev.* 81:299–343

111. Borgland SL. 2001. *Clin. Exp. Pharmacol. Physiol.* 28:147–54

112. Gether U. 2000. *Endocr. Rev.* 21:90–113

113. Decaillot FM, Befort K, Filliol D, Yue S, Walker P, Kieffer BL. 2003. *Nat. Struct. Biol.* 10:629–36

114. Huang P, Visiers I, Weinstein H, Liu-Chen LY. 2002. *Biochemistry* 41:11972–80

115. Huang P, Li J, Chen C, Visiers I, Weinstein H, Liu-Chen LY. 2001. *Biochemistry* 40:13501–9

116. Law PY, Wong YH, Loh HH. 1999. *Biopolymers* 51:440–55

117. Claude PA, Wotta DR, Zhang XH, Prather PL, McGinn TM, et al. 1996. *Proc. Natl. Acad. Sci. USA* 93:5715–19

118. Yang W, Law PY, Guo X, Loh HH. 2003. *Proc. Natl. Acad. Sci. USA* 100:2117–21

119. Mouledous L, Topham CM, Moisand C, Mollereau C, Meunier JC. 2000. *Mol. Pharmacol.* 57:495–502

120. Kam KW, New DC, Wong YH. 2002. *J. Neurochem.* 83:1461–70

121. Lefkowitz RJ, Cotecchia S, Samama P, Costa T. 1993. *Trends Pharmacol. Sci.* 14:303–7

122. Rosenkilde MM, Waldhoer M, Luttichau

HR, Schwartz TW. 2001. *Oncogene* 20: 1582–93

123. Costa T, Herz A. 1989. *Proc. Natl. Acad. Sci. USA* 86:7321–25

124. Costa T, Lang J, Gless C, Herz A. 1990. *Mol. Pharmacol.* 37:383–94

125. Chiu TT, Yung LY, Wong YH. 1996. *Mol. Pharmacol.* 50:1651–57

126. Mullaney I, Carr IC, Milligan G. 1996. *Biochem. J.* 315(Part 1):227–34

127. Merkouris M, Mullaney I, Georgoussi Z, Milligan G. 1997. *J. Neurochem.* 69: 2115–22

128. Hosohata K, Burkey TH, Alfaro-Lopez J, Hruby VJ, Roeske WR, Yamamura HI. 1999. *Eur. J. Pharmacol.* 380:R9–10

129. Neilan CL, Akil H, Woods JH, Traynor JR. 1999. *Br. J. Pharmacol.* 128:556–62

130. Labarre M, Butterworth J, St-Onge S, Payza K, Schmidhammer H, et al. 2000. *Eur. J. Pharmacol.* 406:R1–3

131. Zaki PA, Keith DE Jr, Thomas JB, Carroll FI, Evans CJ. 2001. *J. Pharmacol. Exp. Ther.* 298:1015–20

132. Wang Z, Bilsky EJ, Porreca F, Sadee W. 1994. *Life Sci.* 54:PL339–50

133. Burford NT, Wang DX, Sadee W. 2000. *Biochem. J.* 348(Part 3):531–37

134. Wang DX, Surratt CK, Sadee W. 2000. *J. Neurochem.* 75:763–71

135. Ferguson SS. 2001. *Pharmacol. Rev.* 53: 1–24

136. Liu JG, Anand KJ. 2001. *Brain Res. Brain Res. Rev.* 38:1–19

137. Krupnick JG, Benovic JL. 1998. *Annu. Rev. Pharmacol. Toxicol.* 38:289–319

138. Polakiewicz RD, Schieferl SM, Dorner LF, Kansra V, Comb MJ. 1998. *J. Biol. Chem.* 273:12402–6

139. Eisinger DA, Ammer H, Schulz R. 2002. *J. Neurosci.* 22:10192–200

140. Pak Y, O'Dowd BF, Wang JB, George SR. 1999. *J. Biol. Chem.* 274:27610–16

141. Kramer HK, Andria ML, Kushner SA, Esposito DH, Hiller JM, Simon EJ. 2000. *Brain Res. Mol. Brain Res.* 79: 55–66

142. Kramer HK, Andria ML, Esposito DH,

Simon EJ. 2000. *Biochem. Pharmacol.* 60: 781–92

143. Appleyard SM, McLaughlin JP, Chavkin C. 2000. *J. Biol. Chem.* 275:38281–85

144. Remmers AE, Clark MJ, Liu XY, Medzihradsky F. 1998. *J. Pharmacol. Exp. Ther.* 287:625–32

145. Tsao PI, von Zastrow M. 2000. *J. Biol. Chem.* 275:11130–40

146. Petaja-Repo UE, Hogue M, Bhalla S, Laperriere A, Morello JP, Bouvier M. 2002. *EMBO J.* 21:1628–37

147. Whistler JL, von Zastrow M. 1998. *Proc. Natl. Acad. Sci. USA* 95:9914–19

148. Whistler JL, Tsao P, von Zastrow M. 2001. *J. Biol. Chem.* 276:34331–38

149. Lowe JD, Celver JP, Gurevich VV, Chavkin C. 2002. *J. Biol. Chem.* 277: 15729–35

150. Finn AK, Whistler JL. 2001. *Neuron* 32: 829–39

151. Whistler JL, Enquist J, Marley A, Fong J, Gladher F, et al. 2002. *Science* 297: 615–20

152. Qiu Y, Law PY, Loh HH. 2003. *J. Biol. Chem.* 278:36733–39

153. Pak Y, O'Dowd BF, George SR. 1997. *J. Biol. Chem.* 272:24961–65

154. Trapaidze N, Cvejic S, Nivarthi RN, Abood M, Devi LA. 2000. *DNA Cell Biol.* 19:93–101

155. Trapaidze N, Keith DE, Cvejic S, Evans CJ, Devi LA. 1996. *J. Biol. Chem.* 271: 29279–85

156. Segredo V, Burford NT, Lameh J, Sadee W. 1997. *J. Neurochem.* 68:2395–404

157. Murray SR, Evans CJ, von Zastrow M. 1998. *J. Biol. Chem.* 273:24987–91

158. Xiang B, Yu GH, Guo J, Chen L, Hu W, et al. 2001. *J. Biol. Chem.* 276:4709–16

159. Wang W, Loh HH, Law PY. 2003. *J. Biol. Chem.* 278:36848–58

160. El Kouhen R, Burd AL, Erickson-Herbrandson LJ, Chang CY, Law PY, Loh HH. 2001. *J. Biol. Chem.* 276:12774–80

161. Burd AL, El Kouhen R, Erickson LJ, Loh HH, Law PY. 1998. *J. Biol. Chem.* 273:34488–95

162. Tanowitz M, Von Zastrow M. 2003. *J. Biol. Chem.* 278:45978–86
163. Cao TT, Deacon HW, Reczek D, Bretscher A, von Zastrow M. 1999. *Nature* 401:286–90
164. Li JG, Chen C, Liu-Chen LY. 2002. *J. Biol. Chem.* 277:27545–52
165. Zhang F, Li J, Li JG, Liu-Chen LY. 2002. *J. Pharmacol. Exp. Ther.* 302:1184–92
166. McLaughlin JP, Xu M, Mackie K, Chavkin C. 2003. *J. Biol. Chem.* 278:34631–40
167. Spampinato S, Di Toro R, Qasem AR. 2001. *NeuroReport* 12:3159–63
168. Hackel PO, Zwick E, Prenzel N, Ullrich A. 1999. *Curr. Opin. Cell Biol.* 11:184–89
169. Angers S, Salahpour A, Bouvier M. 2001. *Life Sci.* 68:2243–50
170. Lee SP, Xie Z, Varghese G, Nguyen T, O'Dowd BF, George SR. 2000. *Neuropsychopharmacology* 23:S32–40
171. George SR, Fan T, Xie ZD, Tse R, Tam V, et al. 2000. *J. Biol. Chem.* 275:26128–35
172. Jordan BA, Trapaidze N, Gomes I, Nivarthi R, Devi LA. 2001. *Proc. Natl. Acad. Sci. USA* 98:343–48
173. He L, Fong J, von Zastrow M, Whistler JL. 2002. *Cell* 108:271–82
174. Pfeiffer M, Kirscht S, Stumm R, Koch T, Wu DF, et al. 2003. *J. Biol. Chem.* 278:51630–37
175. McVey M, Ramsay D, Kellett E, Rees S, Wilson S, et al. 2001. *J. Biol. Chem.* 276:14092–99
176. Portoghese PS. 2001. *J. Med. Chem.* 44:2259–69
177. Lutz RA, Pfister HP. 1992. *J. Recept. Res.* 12:267–86
178. Zhu Y, King MA, Schuller AG, Nitsche JF, Reidl M, et al. 1999. *Neuron* 24:243–52
179. Matthes HWD, Maldonado R, Simonin F, Valverde O, Slowe S, et al. 1996. *Nature* 383:819–23
180. Sora I, Funada M, Uhl GR. 1997. *Eur. J. Pharmacol.* 324:R1–2
181. Loh HH, Liu HC, Cavalli A, Yang W, Chen YF, Wei LN. 1998. *Brain Res. Mol. Brain Res.* 54:321–26
182. Schuller AG, King MA, Zhang J, Bolan E, Pan YX, et al. 1999. *Nat. Neurosci.* 2:151–56
183. Fuchs PN, Roza C, Sora I, Uhl G, Raja SN. 1999. *Brain Res.* 821:480–86
184. Qiu C, Sora I, Ren K, Uhl G, Dubner R. 2000. *Eur. J. Pharmacol.* 387:163–69
185. Matthes HWD, Smadja C, Valverde O, Vonesch JL, Foutz AS, et al. 1998. *J. Neurosci.* 18:7285–95
186. Lichtman AH, Sheikh SM, Loh HH, Martin BR. 2001. *J. Pharmacol. Exp. Ther.* 298:1007–14
187. Ghozland S, Matthes HWD, Simonin F, Filliol D, Kieffer BL, Maldonado R. 2002. *J. Neurosci.* 22:1146–54
188. Hall FS, Sora I, Uhl GR. 2001. *Psychopharmacology* 154:43–49
189. Berrendero F, Kieffer BL, Maldonado R. 2002. *J. Neurosci.* 22:10935–40
190. Wei ZY, Brown W, Takasaki B, Plobeck N, Delorme D, et al. 2000. *J. Med. Chem.* 43:3895–905
191. Dondio G, Ronzoni S, Eggleston DS, Artico M, Petrillo P, et al. 1997. *J. Med. Chem.* 40:3192–98
192. Svingos AL, Cheng PY, Clarke CL, Pickel VM. 1995. *Brain Res.* 700:25–39
193. Cheng PY, Svingos AL, Wang H, Clarke CL, Jenab S, et al. 1995. *J. Neurosci.* 15:5976–88
194. Abdelhamid EE, Sultana M, Portoghese PS, Takemori AE. 1991. *J. Pharmacol. Exp. Ther.* 258:299–303
195. Fundytus ME, Schiller PW, Shapiro M, Weltrowska G, Coderre TJ. 1995. *Eur. J. Pharmacol.* 286:105–8
196. Filliol D, Ghozland S, Chluba J, Martin M, Matthes HWD, et al. 2000. *Nat. Genet.* 25:195–200
197. Roberts AJ, Gold LH, Polis I, McDonald JS, Filliol D, et al. 2001. *Alcohol. Clin. Exp. Res.* 25:1249–56

198. Gaveriaux-Ruff C, Kieffer BL. 2002. *Neuropeptides* 36:62–71

199. Simonin F, Valverde O, Smadja C, Slowe S, Kitchen I, et al. 1998. *EMBO J.* 17:886–97

200. Gebhart GF, Su X, Joshi S, Ozaki N, Sengupta JN. 2000. *Ann. NY Acad. Sci.* 909:41–50

201. Gaveriaux-Ruff C, Simonin F, Filliol D, Kieffer BL. 2003. *J. Neuroimmunol.* 134: 72–81

202. McLaughlin JP, Marton-Popovici M, Chavkin C. 2003. *J. Neurosci.* 23: 5674–83

203. Mogil JS, Pasternak GW. 2001. *Pharmacol. Rev.* 53:381–415

204. Mamiya T, Noda Y, Ren X, Nagai T, Takeshima H, et al. 2001. *J. Neural. Transm.* 108:1349–61

205. Koster A, Montkowski A, Schulz S, Stube EM, Knaudt K, et al. 1999. *Proc. Natl. Acad. Sci. USA* 96:10444–49

206. Kest B, Hopkins E, Palmese CA, Chen ZP, Mogil JS, Pintar JE. 2001. *Neuroscience* 104:217–22

207. Bertorelli R, Bastia E, Citterio F, Corradini L, Forlani A, Ongini E. 2002. *Peptides* 23:1589–96

208. Mogil JS, Grisel JE, Reinscheid RK, Civelli O, Belknap JK, Grandy DK. 1996. *Neuroscience* 75:333–37

209. Ueda H, Yamaguchi T, Tokuyama S, Inoue M, Nishi M, Takeshima H. 1997. *Neurosci. Lett.* 237:136–38

210. Blake AD, Bot G, Freeman JC, Reisine T. 1997. *J. Biol. Chem.* 272:782–90

211. Zhang J, Ferguson SS, Barak LS, Bodduluri SR, Laporte SA, et al. 1998. *Proc. Natl. Acad. Sci. USA* 95:7157–62

212. Arden JR, Segredo V, Wang Z, Lameh J, Sadee W. 1995. *J. Neurochem.* 65: 1636–45

213. Keith DE, Murray SR, Zaki PA, Chu PC, Lissin DV, et al. 1996. *J. Biol. Chem.* 271:19021–24

214. Sternini C, Spann M, Anton B, Keith DE Jr, Bunnett NW, et al. 1996. *Proc. Natl. Acad. Sci. USA* 93:9241–46

215. Keith DE, Anton B, Murray SR, Zaki PA, Chu PC, et al. 1998. *Mol. Pharmacol.* 53:377–84

216. Garrido MJ, Troconiz IF. 1999. *J. Pharmacol. Toxicol. Methods* 42:61–66

217. Trapaidze N, Gomes I, Cvejic S, Bansinath M, Devi LA. 2000. *Brain Res. Mol. Brain Res.* 76:220–28

218. Whistler JL, Chuang HH, Chu P, Jan LY, von Zastrow M. 1999. *Neuron* 23: 737–46

219. Kovoor A, Nappey V, Kieffer BL, Chavkin C. 1997. *J. Biol. Chem.* 272: 27605–11

220. Kovoor A, Celver JP, Wu A, Chavkin C. 1998. *Mol. Pharmacol.* 54:704–11

221. Selley DE, Liu Q, Childers SR. 1998. *J. Pharmacol. Exp. Ther.* 285:496–505

222. Borgland SL, Connor M, Osborne PB, Furness JB, Christie MJ. 2003. *J. Biol. Chem.* 278:18776–84

223. Willins DL, Berry SA, Alsayegh L, Backstrom JR, Sanders-Bush E, et al. 1999. *Neuroscience* 91:599–606

224. Trujillo KA, Akil H. 1991. *Science* 251: 85–87

225. Tiseo PJ, Inturrisi CE. 1993. *J. Pharmacol. Exp. Ther.* 264:1090–96

226. Gebhart GF. 1990. *Anesthesiology* 73: 1065–66

227. Law PY, Hom DS, Loh HH. 1983. *Mol. Pharmacol.* 24:413–24

228. Puttfarcken PS, Cox BM. 1989. *Life Sci.* 45:1937–42

229. Harrison LM, Kastin AJ, Zadina JE. 1998. *Peptides* 19:1603–30

230. Patel MB, Patel CN, Rajashekara V, Yoburn BC. 2002. *Mol. Pharmacol.* 62: 1464–70

231. Nestler EJ. 1996. *Neuron* 16:897–900

232. Pak Y, Kouvelas A, Scheideler MA, Rasmussen J, O'Dowd BF, George SR. 1996. *Mol. Pharmacol.* 50:1214–22

233. Sim LJ, Selley DE, Dworkin SI, Childers SR. 1996. *J. Neurosci.* 16:2684–92

234. Sim-Selley LJ, Selley DE, Vogt LJ, Childers SR, Martin TJ. 2000. *J. Neurosci.* 20:4555–62

235. Slot LAB, Colpaert FC. 2003. *Behav. Pharmacol.* 14:167–71
236. Shen KF, Crain SM. 2001. *Brain Res.* 919:20–30
237. Crain SM, Shen KF. 1995. *Proc. Natl. Acad. Sci. USA* 92:10540–44
238. Levine JD, Gordon NC, Fields HL. 1979. *Nature* 278:740–41
239. Levine JD, Gordon NC, Taiwo YO, Coderre TJ. 1988. *J. Clin. Investig.* 82: 1574–77
240. Gan TJ, Ginsberg B, Glass PS, Fortney J, Jhaveri R, Perno R. 1997. *Anesthesiology* 87:1075–81
241. Bohn LM, Gainetdinov RR, Lin FT, Lefkowitz RJ, Caron MG. 2000. *Nature* 408:720–23
242. Bohn LM, Lefkowitz RJ, Gainetdinov RR, Peppel K, Caron MG, Lin FT. 1999. *Science* 286:2495–98
243. Bohn LM, Xu F, Gainetdinov RR, Caron MG. 2000. *J. Neurosci.* 20:9040–45
244. Fairbanks CA, Stone LS, Kitto KF, Nguyen HO, Posthumus IJ, Wilcox GL. 2002. *J. Pharmacol. Exp. Ther.* 300: 282–90
245. Bie B, Fields HL, Williams JT, Pan ZZ. 2003. *J. Neurosci.* 23:7950–57
246. Wu G, Krupnick JG, Benovic JL, Lanier SM. 1997. *J. Biol. Chem.* 272:17836–42
247. Chakrabarti S, Rivera M, Yan SZ, Tang WJ, Gintzler AR. 1998. *Mol. Pharmacol.* 54:655–62
248. Bonci A, Williams JT. 1997. *J. Neurosci.* 17:796–803
249. Vaughan CW, Ingram SL, Connor MA, Christie MJ. 1997. *Nature* 390:611–14
250. Ingram SL, Vaughan CW, Bagley EE, Connor M, Christie MJ. 1998. *J. Neurosci.* 18:10269–76
251. Ueda H, Inoue M, Matsumoto T. 2001. *J. Neurosci.* 21:2967–73
252. Haberstock-Debic H, Wein M, Barrot M, Colago EE, Rahman Z, et al. 2003. *J. Neurosci.* 23:4324–32
253. Stevens CW, Yaksh TL. 1989. *J. Pharmacol. Exp. Ther.* 251:216–23
254. Stevens CW, Yaksh TL. 1989. *J. Pharmacol. Exp. Ther.* 250:1–8
255. Mjanger E, Yaksh TL. 1991. *J. Pharmacol. Exp. Ther.* 258:544–50
256. Duttaroy A, Yoburn BC. 1995. *Anesthesiology* 82:1226–36
257. Sora I, Elmer G, Funada M, Pieper J, Li XF, et al. 2001. *Neuropsychopharmacology* 25:41–54
258. Chavkin C, Goldstein A. 1984. *Proc. Natl. Acad. Sci. USA* 81:7253–57
259. Cahill CM, Morinville A, Lee MC, Vincent JP, Collier B, Beaudet A. 2001. *J. Neurosci.* 21:7598–607
260. Morinville A, Cahill CM, Esdaile MJ, Aibak H, Collier B, et al. 2003. *J. Neurosci.* 23:4888–98
261. Commons KG. 2003. *J. Comp. Neurol.* 464:197–207
262. Kim KA, von Zastrow M. 2003. *J. Neurosci.* 23:2075–85
263. Pan YX, Bolan E, Pasternak GW. 2002. *Biochem. Biophys. Res. Commun.* 297: 659–63
264. Crawley JN, Corwin RL. 1994. *Peptides* 15:731–55
265. Koob GF, Le Moal M. 2001. *Neuropsychopharmacology* 24:97–129
266. Noble F, Roques BP. 1999. *Prog. Neurobiol.* 58:349–79
267. Trujillo KA. 2000. *Psychopharmacology* 151:121–41
268. Smith MT, Watt JA, Cramond T. 1990. *Life Sci.* 47:579–85
269. Hemstapat K, Monteith GR, Smith D, Smith MT. 2003. *Anesth. Analg.* 97: 494–505
270. Bartlett SE, Cramond T, Smith MT. 1994. *Life Sci.* 54:687–94
271. Halliday AJ, Bartlett SE, Colditz P, Smith MT. 1999. *Life Sci.* 65:225–36
272. Faris PL, Komisaruk BR, Watkins LR, Mayer DJ. 1983. *Science* 219:310–12
273. Magnuson DS, Sullivan AF, Simonnet G, Roques BP, Dickenson AH. 1990. *Neuropeptides* 16:213–18
274. Rotzinger S, Vaccarino FJ. 2003. *J. Psychiatry Neurosci.* 28:171–81

275. Wank SA. 1995. *Am. J. Physiol. Gastrointest. Liver Physiol.* 269:G628–46

276. Pommier B, Beslot F, Simon A, Pophillat M, Matsui T, et al. 2002. *J. Neurosci.* 22:2005–11

277. Kim JA, Siegel S. 2001. *Behav. Neurosci.* 115:704–9

278. Mitchell JM, Basbaum AI, Fields HL. 2000. *Nat. Neurosci.* 3:47–53

279. Gustafsson H, Afrah AW, Stiller CO. 2001. *J. Neurochem.* 78:55–63

280. Marek P, Ben-Eliyahu S, Gold M, Liebeskind JC. 1991. *Brain Res.* 547:77–81

281. Inoue M, Mishina M, Ueda H. 2003. *J. Neurosci.* 23:6529–36

282. Zhu H, Brodsky M, Gorman AL, Inturrisi CE. 2003. *Brain Res. Mol. Brain Res.* 114:154–62

283. Koob GF, Maldonado R, Stinus L. 1992. *Trends Neurosci.* 15:186–91

284. Nestler EJ. 2001. *Nat. Rev. Neurosci.* 2: 119–28

285. Nestler EJ. 2002. *Neurobiol. Learn. Mem.* 78:637–47

286. Way EL, Loh HH, Shen FH. 1969. *J. Pharmacol. Exp. Ther.* 167:1–8

287. Rezvani A, Way EL. 1983. *Life Sci.* 33(Suppl. 1): 349–52

288. Kolesnikov YA, Ferkany J, Pasternak GW. 1993. *Life Sci.* 53:1489–94

289. Gonzalez P, Cabello P, Germany A, Norris B, Contreras E. 1997. *Eur. J. Pharmacol.* 332:257–62

290. Nitsche JF, Schuller AG, King MA, Zengh M, Pasternak GW, Pintar JE. 2002. *J. Neurosci.* 22:10906–13

291. Nestler EJ, Aghajanian GK. 1997. *Science* 278:58–63

292. Aghajanian GK. 1978. *Nature* 276: 186–88

293. Christie MJ, Williams JT, North RA. 1987. *Mol. Pharmacol.* 32:633–38

294. Kogan JH, Nestler EJ, Aghajanian GK. 1992. *Eur. J. Pharmacol.* 211:47–53

295. Duman RS, Tallman JF, Nestler EJ. 1988. *J. Pharmacol. Exp. Ther.* 246: 1033–39

296. Nestler EJ, Tallman JF. 1988. *Mol. Pharmacol.* 33:127–32

297. Nestler EJ, Erdos JJ, Terwilliger R, Duman RS, Tallman JF. 1989. *Brain Res.* 476:230–39

298. Gold SJ, Han MH, Herman AE, Ni YG, Pudiak CM, et al. 2003. *Eur. J. Neurosci.* 17:971–80

299. Matsuoka I, Maldonado R, Defer N, Noel F, Hanoune J, Roques BP. 1994. *Eur. J. Pharmacol.* 268:215–21

300. Widnell KL, Self DW, Lane SB, Russell DS, Vaidya VA, et al. 1996. *J. Pharmacol. Exp. Ther.* 276:306–15

301. Shaw-Lutchman TZ, Barrot M, Wallace T, Gilden L, Zachariou V, et al. 2002. *J. Neurosci.* 22:3663–72

302. Maldonado R, Blendy JA, Tzavara E, Gass P, Roques BP, et al. 1996. *Science* 273:657–59

303. Lane-Ladd SB, Pineda J, Boundy VA, Pfeuffer T, Krupinski J, et al. 1997. *J. Neurosci.* 17:7890–901

304. Terwilliger RZ, Beitner-Johnson D, Sevarino KA, Crain SM, Nestler EJ. 1991. *Brain Res.* 548:100–10

305. Jolas T, Nestler EJ, Aghajanian GK. 2000. *Neuroscience* 95:433–43

306. Roth BL, Willins DL. 1999. *Neuron* 23: 629–31

307. Mamiya T, Noda Y, Ren X, Hamdy M, Furukawa S, et al. 2001. *Br. J. Pharmacol.* 132:1111–17

308. Kieffer BL, Evans CJ. 2002. *Cell* 108: 587–90

309. Sneath PH, Sokal RR. 1962. *Nature* 193: 855–60

Annu. Rev. Biochem. 2004. 73:991–1018
doi: 10.1146/annurev.biochem.73.011303.073711
First published online as a Review in Advance on March 30, 2004

Structural Aspects of Ligand Binding to and Electron Transfer in Bacterial and Fungal P450s

Olena Pylypenko[1] and Ilme Schlichting[2]

[1]Max Planck Institute for Molecular Physiology, Department of Physical Biochemistry, 44227 Dortmund, Germany; email: elena.pylypenko@mpi-dortmund.mpg.de
[2]Max Planck Institute for Medical Research, Department of Biomolecular Mechanisms, 69120 Heidelberg, Germany; email: ilme.schlichting@mpimf-heidelberg.mpg.de

Key Words   heme protein, cytochrome P450, crystal structure, ligand binding, induced-fit mechanism, hemoprotein

■ **Abstract**   Cytochrome P450 enzymes are heme-containing monooxygenases that are named after an absorption band at 450 nm when complexed with carbon monoxide. They catalyze a wide variety of reactions and are unique in their ability to hydroxylate nonactivated hydrocarbons. P450 enzymes are involved in numerous biological processes, which include the biosynthesis of lipids, steroids, antibiotics, and the degradation of xenobiotics. In line with the variety of reactions catalyzed, the size of their substrates varies significantly. Some P450s have open active sites (e.g., BM3), and some have shielded active sites that open only transiently (e.g., P450cam), whereas others bind the substrate only when attached to carrier proteins (e.g., Oxy proteins). Structural aspects of both organic and gaseous ligand binding and electron transfer are described.

CONTENTS

CYTOCHROME P450: FROM THE MICROSOMAL CARBON MONOXIDE-BINDING PIGMENT TO THE P450 SUPERFAMILY

The first experimental evidence relating to cytochromes P450 was discovered in 1955 (1, 2) when an enzyme system oxidizing xenobiotic compounds was identified in the endoplasmic reticulum of the liver. In 1958, a carbon monoxide-(CO) binding pigment having an absorption maximum at 450 nm was detected in liver microsomes (3, 4). Later the microsomal carbon monoxide-binding pigment was demonstrated to be a hemoprotein containing Fe-protoporphyrin IX (5, 6), which was named cytochrome P450 after the characteristic feature in its absorption spectrum on complex formation with CO. Unique spectral properties of P450 suggested that it is a low-spin ferric hemoprotein (7) with a thiolate group as the fifth axial Fe heme ligand (8–10). Raman spectroscopy provided confirmation of the presence of a Fe-S bond and identified this as a covalently bonded cysteine residue (11).

The role of cytochrome P450 in the synthesis of steroid hormones in the adrenal cortex microsomes (12, 13) and in the oxidation of various drugs by liver microsomes (14) was first discovered in 1963. Since this initial discovery, a number of P450 enzymes were found in essentially all mammalian tissues, for instance brain (15), skin (16), and intestine (17). Additionally, with the continued identification of unique P450s that catalyze biosynthetic and biodegradative reactions in mammals, as well as in archaea, bacteria, fungi and plants, it has been realized that the role of P450s is much more diverse than had been previously suspected (18). P450s often confer on prokaryotes the ability to catabolize compounds used as a carbon source or to detoxify xenobiotics. Other functions described for prokaryotic P450s include fatty acid metabolism and biosynthesis of antibiotics (19). The physiological functions of eukaryotic P450 enzymes include many aspects of the biosynthesis and catabolism of signaling molecules, steroid hormones, vitamin D_3, retinoic acid, and oxylipins (20–22). In fungi, they are involved in the synthesis of membrane sterols and mycotoxins, detoxification of phytoalexins, and metabolism of lipid carbon sources. P450s from plants are involved in the biosynthesis or catabolism of many types of hormones, in the oxygenation of fatty acids, in the synthesis of cutins, and in all the pathways of secondary metabolism, in lignification, and in the synthesis of flower pigments and defense chemicals, which are also aromas, flavors, antioxidants, phytoestrogens, anticancer drugs, and other medicines (23, 24). P450s from all organisms participate in the detoxification or sometimes the activation of xenobiotics. They have been shown to contribute to carcinogenesis and are essential determinants of drug and pesticide metabolism, tolerance, selectivity, and compatibility (19, 21, 22, 25, 26). Thus, cytochrome P450s (CYPs or P450s) comprise a whole superfamily of proteins with extremely diverse functions widely distributed in virtually all organisms. It was proposed that the P450 gene superfamily evolved from a common ancestor over a period of some 3.5 billion years (27), which would explain why P450 proteins are ubiquitous in the biosphere.

Because of the obvious importance of P450s in the metabolism of physiologically important compounds, there is a broad interest in predicting the substrate specificity of P450s and thus to a large extent the pharmacokinetics of drugs, in understanding the mechanisms of disease states resulting from mutations in P450s, and in engineering known P450 proteins. With this knowledge, P450s may be used for genetic modification of microorganisms able to biodegrade pollutant chemicals and for production of new chemicals and drugs (28).

P450 Catalytic Mechanism

Until recently the catalytic function of P450 proteins appeared to be the transfer of one oxygen atom from O_2 into various substrates

$$RH + O_2 + 2e^- + 2H^+ \rightarrow ROH + H_2O.$$

For this monooxygenase reaction, cytochrome P450s receive electrons from NADPH or NADH via electron transfer proteins (29). Other catalytic functions, different from oxygen atom transfers, have also been reported for several cytochrome P450s. Some of these atypical cytochrome P450-dependent reactions are oxidations involving C-C or C-N bond cleavage, dehydrations, isomerizations, and reductions. An overview of the diverse reactions catalyzed by cytochrome P450s is given in Reference 18.

In its resting state (ferric form), P450 generally exists as a mixture of a hexacoordinate low-spin Fe(III) with a water molecule *trans* to its endogenous cysteinate ligand and a pentacoordinate high-spin Fe(III) with the cysteinate as only axial ligand (Figure 1*a*). Binding of substrates in a protein site close to the heme generally shifts the equilibrium between the two Fe(III) states toward the pentacoordinate complex (Figure 1*b*). One-electron reduction of the complex to a ferrous state (Figure 1*c*) produces the third intermediate, P450 Fe(II), a pentacoordinate high-spin complex that is able to bind various ligands including O_2. Carbon monoxide can bind to ferrous P450 instead of dioxygen (Figure 1*e*), inducing a shift of the Soret band to 450 nm, which is characteristic for P450 enzymes (4, 6). CO binds with high affinity and prevents binding and activation of O_2, resulting in inhibition of P450 activity. Binding of molecular oxygen to P450 Fe(II) creates a superoxide complex (Figure 1*d*). A second reduction step leads to very short-lived activated oxygen species (Figure 1*f,g,h*). Initial protonation of the distal oxygen in the peroxo-iron complex (Figure 1*f*) produces a hydroperoxo-iron species (Figure 1*g*), and a subsequent protonation of the distal oxygen results in a species that loses water to make an iron-oxo complex (Figure

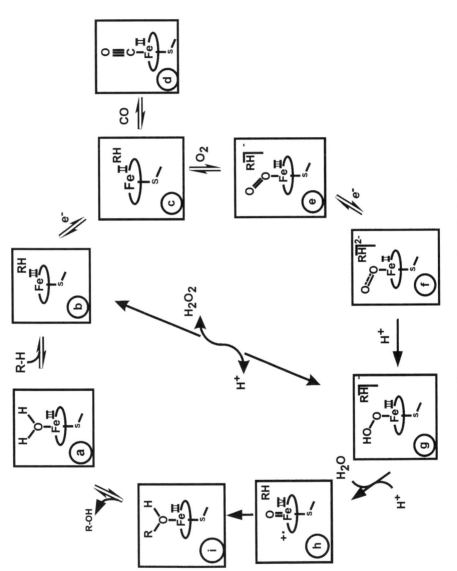

Figure 1 The reaction cycle of P450 monooxygenase catalyzed hydroxylation.

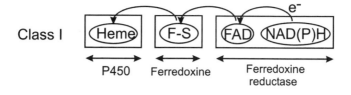

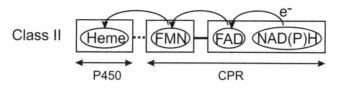

Figure 2   Domain organization of P450-containing monooxygenase systems. Independent structural domains are shown as boxes. Cofactors and heme are shown as ellipses and labeled. Curved arrows show the electron transfer pathway.

1*h*) (29). The diverse oxidations effected by these enzymes suggest that different types of oxidants are involved in some specific functions (30, 31). Oxygen atom transfer from an iron-oxo complex (Figure 1*h*) to the substrate yields the oxidized product (ROH) (Figure 1*i*) and regenerates the resting state. In the presence of external oxygenation agents, such as peracids, the complex (Figure 1*b*) may directly yield the hydroperoxo complex (Figure 1*g*) via a "shunt" pathway.

The great diversity of reactions catalyzed by P450 enzymes appears to be based on two unique properties of these hemoproteins: the ability of their iron to exist under a great variety of oxidation states with different reactivities and active sites that provide a great variety of substrates access to their iron (18). In addition to providing access to the business end of the enzyme, the active sites must be structurally diverse to provide the required specificity yet conserve a mechanism of catalysis and solvent exclusion. Water access to the active site is restricted to prevent side reactions, such as the conversion of activated dioxygen to superoxide or peroxide.

Classes of P450

P450s, like other redox active metalloproteins, separate key steps in the catalytic cycle by the input of reducing equivalents into the metal center. P450s can be divided into four classes depending on how electrons are delivered from NAD(P)H to the active site.

Class I and II P450s require the successive delivery of two electrons, which are provided by redox protein partners. All P450-containing monooxygenase systems of these two classes share a common structural and functional domain architecture (Figure 2). There are no fundamental differences between the protein domain and the individual protein components in P450-containing systems. They are composed of three functional domains: (*a*) NAD(P)H-dependent FAD-containing reductase (FAD domain), (*b*) an iron-sulfur protein (in class I) or FMN-binding domain (in class II), and (*c*) a P450 protein (heme domain) (32).

Class I P450s take electrons from a specific ferredoxin (an iron-sulfur protein) that shuttles electrons from a NAD(P)H-dependent ferredoxin reductase (33). Class II P450s receive electrons directly from a NADPH–cytochrome-P450 reductase (CPR), composed of FAD- and FMN-domains. In a microsomal CPR-P450 system, different P450s interact with a universal P450 reductase. P450s of class II can also exist as a single polypeptide chain with two functional parts, a heme and a reductase domain (34); in this one-component system, the reductase domain can interact only with its own P450 domain (e.g., P450 BM3).

Class III P450s are self-sufficient and do not require an electron donor. Class IV P450s receive electrons directly from NADH.

PROGRESS IN STRUCTURAL STUDIES ON BACTERIAL AND FUNGAL P450s

Most P450s whose structures have been determined are of prokaryotic origin, principally because these forms are soluble, whereas eukaryotic P450s are generally attached to microsomal or mitochondrial membranes by an N-terminal peptide anchor, with other regions embedded in the membrane bilayer. Soluble bacterial P450s correspond to the signal anchor-truncated form of eukaryotic P450s. The first P450 purified and crystallized was a soluble bacterial P450, P450cam of *Pseudomonas putida* (see paragraph below). It was purified by Gunsalus and coworkers in 1970 (35) and crystallized in 1974 (36). The total amino acid sequence of P450cam was determined by chemical sequencing in 1982 (37), and its tertiary camphor-bound structure was published in 1985 (38). Several other bacterial P450s were successfully crystallized in the following years, and their molecular structures were determined. At present, there are 91 P450 structures (12 unique bacterial, 1 fungal, and 2 different eukaryotic proteins) deposited in the Protein Data Bank (http://www.rcsb.org).

Despite low sequence identities—less than 20% with only three absolutely conserved amino acids across this gene superfamily—the P450 overall fold and topologies are very similar (39) (Figure 3*a*). The structures consist generally of 14 α-helixes named A to L and 5 β-sheets that form a triagonal-prism-shaped molecule with predominately α-helical (in Figure 3*a* on the right) and β-structured (on the left) parts. The conserved P450 structural core is formed by a four-helix bundle composed of three parallel helices, D, L, and I, and one antiparallel helix E (40). The prosthetic heme group is confined between the distal I-helix and proximal L-helix and attached to the adjacent Cys-heme-ligand loop (in Figure 3*a* the loop is hidden by the heme), which contains the P450 signature amino acid sequence FxxGx(H/R)xCxG. The absolutely conserved cysteine is the fifth ligand of the heme iron and the origin of the characteristic

450-nm Soret absorbance found in the CO-bound proteins. The I-helix contains the signature sequence (A/G)Gx(E/D)T located above the heme, and because of this sequence, there is a kink in the middle of the helix. The highly conserved threonine preceded by an acidic residue is positioned in the active site and believed to be involved in catalysis (41–45). The conserved core also includes the helices J and K. The C-terminal part of the K-helix contains the absolutely conserved sequence ExxR; the residues stabilize a structurally highly conserved coil termed the meander. There are two structurally conserved β-sheets: β1 containing five strands and β2 containing two strands. The presence of β5 is variable.

Substrate Recognition Sequences

Although the P450 fold is highly conserved, different P450s have enough structural diversity to accomplish their dissimilar specific roles. The most variable structural elements in P450s are the helices A, B', B, F, G, H, K', the β sheets β3 and β4, and the loops. The variable regions contain residues associated with substrate recognition/binding and include six substrate recognition sequences (SRS) (Figure 3b) (46). The B'-helix region (SRS1), parts of the F- and G-helices (SRS2 and SRS3), a part of the I-helix (SRS4), the K-helix β2 connecting region (SRS6), and the β4 hairpin (SRS5) line the P450 active site. In particular, the SRS predetermine P450 substrate specificity; point mutations within SRSs significantly affect the substrate specificities (46). The SRSs are considered flexible protein regions, which move upon substrate binding so as to favor binding and subsequently the catalytic reaction. In many cases, the resting form of an enzyme does not perfectly fit the structure of its substrate. The perfect molecular fit occurs only after substrate binding to the protein by a structural rearrangement of the protein enveloping the substrate. This induced-fit model of substrate protein interaction was proposed in 1958 by Koshland (47) who addressed the functional importance of structural flexibility in proteins. Comparison of substrate-free and different ligand-bound P450 structures gives insight into the structural changes in P450s upon ligand binding, which can be explained by an induced-fit model.

P450 BM3, Open and Closed Conformations

The *Bacillus megaterium* flavocytochrome P450 BM3 (CYP102) fatty acid hydroxylase uses a class II redox system: an FAD- and FMN-containing NADPH-cytochrome P450 reductase (34). The P450 BM3 reductase is fused to the C terminus of the P450 in a single continuous polypeptide.

P450 BM3 is the paradigm of substrate-binding induced structural changes. In the substrate-free P450 BM3 (48), the active site is open (Figure 4a). The SRSs1,2,4,5,6 with the N-terminal part of the A-helix and β1 form a wide substrate access channel. A crystal structure of the heme domain of P450 BM3

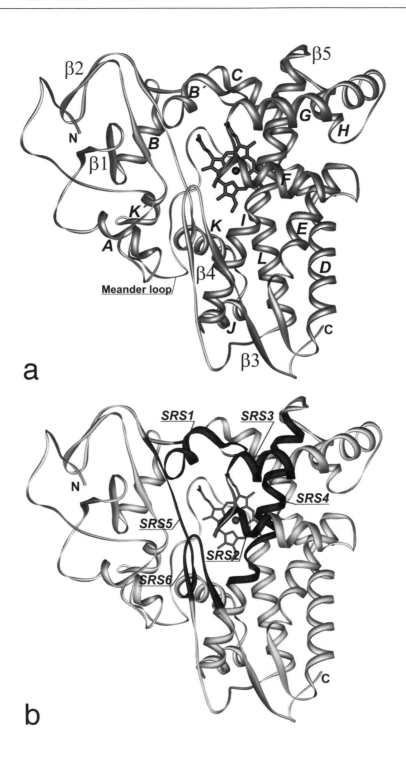

complexed to palmitoleic acid revealed a significant conformational change upon substrate binding, which involved closure of the active site channel.

In the structure of the P450 BM3-substrate complex, palmitoleic acid is bound within a long, narrow channel that leads to the heme (49). The active site is inaccessible to solvent (Figure 4a) predominantly because of changes in the orientation of the F- and G-helices such that SRS2 becomes a lid of the closed substrate-binding pocket. This changes the I-helix conformation from bent to somewhat straighter, which affects the whole protein structure and favors the closing of the active site. The sites of hydroxylation by the enzyme, the $\omega-1$ and $\omega-2$ positions of the fatty acid, are located 9 Å from the heme iron in the structure of the palmitoleate-bound P450 BM3. NMR studies confirmed the large distance between substrate and heme, and they showed that heme reduction provides the trigger for a second conformational change that drives the fatty acid into position for oxidative attack closer to the heme iron (3.0 Å) (50).

The structure of the palmitoleate-bound form of P450 BM3 provided the basis for a rational investigation of the roles of amino acids that define the substrate-binding pocket (51). The F87 side chain, located close to the heme in the substrate-binding pocket, rotates from a horizontal to a perpendicular orientation to the heme upon substrate binding. The mutation F87G increases the enzyme affinity for laurate, affecting the regioselectivity of fatty acid oxidation (52). At the opposite end of the active site are R47 and Y51 interacting with the substrate carboxylate (51). Engineering alternative carboxylate-binding residues closer to the heme in the hydrophobic active site core results in improved binding and turnover of short chain alkanoic acids, and removal of the R47–Y51 motif hinders the recognition and catalysis of carboxylic acid substrates of all chain lengths, emphasizing its importance in the initial recognition of substrates (53). Removal of the carboxylate-binding motif of P450 BM3 in the double mutant R47L Y51F increases the capacity of the enzyme to oxidize pyrene and other polycyclic aromatic hydrocarbons (54). The triple mutant F87V, L188Q, A74G, was observed to hydroxylate indole to produce indigo and indirubin (55).

The high-resolution crystal structure of P450 BM3 with the alternative, more tightly binding substrate, N-palmitoglycine, revealed conformational changes in SRS4 that have functional relevance (56). In the substrate-free form, a water molecule is inserted between I263-G267 and the A264-T268 carbonyl oxygen atoms and backbone amides, resulting in a 13° kink of the I-helix. The water is displaced upon substrate binding, resulting in a straightening of the I-helix, the formation of a hydrogen bond between I263 and E267, and the binding of a new

---

Figure 3   (a) A ribbon representation of the P450cam structure (distal face). The α-helixes are labeled with capital letters in italic, β-sheets are labeled with numbers. The N and C termini are labeled with N and C, respectively. (b) Substrate recognition sequence (SRS) regions are shown in black and labeled.

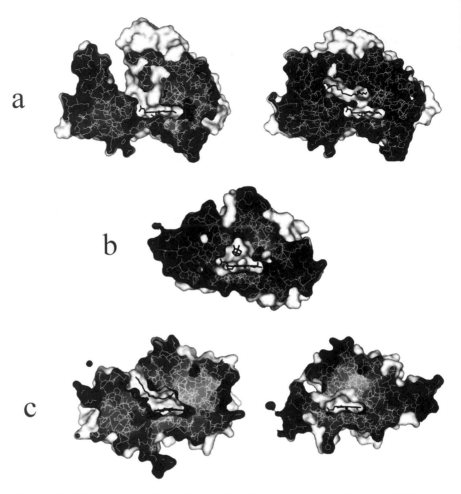

Figure 4 Solvent accessible surfaces clipped to show active site accessibility. (*a*) Substrate-free P450 BM3 (*left*) exhibits an open active site. In N-palmitoglycine–bound P450 BM3 (*right*), the active site is closed. (*b*) Camphor-bound P450cam. The active site is closed in both substrate-bound and substrate-free forms. (*c*) Fatty acid–bound P450Bsβ exhibits two channels. One is occupied by the substrate (*left*); the second one (*right*) is open.

water molecule (Wat501). This causes movement of the carbonyl of A264 (corresponding to G248 in P450cam) away from the heme iron and a concurrent 1 Å shift of the iron-bound water molecule (Wat500). A low-spin to high-spin conversion results that frees the oxygen binding site and positions a water molecule close to oxygen binding site. The conformational changes correspond to a molecular switch in which substrate binding converts the enzyme from an inactive form to a catalytically competent one (56).

CYP119, a P450 From a Thermophile

CYP119 is the first P450 identified in archaea (*Sulfolobus solfataricus*). As expected, CYP119 exhibits unusual thermal stability with a melting temperature of ~ 90°C (57) compared with ~ 55°C for bacterial P450cam. So far, the function and thus the natural substrate of CYP119 are unknown. Cocrystal structures of CYP119 with the inhibitors imidazole and 4-phenylimidazole that bind to the heme iron show dramatic changes in the structure of the F/G region and the β4 hairpin compared to the ligand-free structure (58, 59). In the uncomplexed CYP119, the F/G loop extends away from the remainder of the protein, resulting in a rather open active site. In contrast, in the inhibitor-bound structures, the F/G loop dips into the active site (58). The loop movement involves untwisting of one G-helix turn and twisting or untwisting of one F-helix turn in case of the complexes with phenylimidazole and imidazole, respectively. This results in lengthening of the F/G loop in case of the imidazole complex, and it allows a deeper F/G loop penetration in the active site to make contacts with the bound ligand. The conformational changes observed between the free and complexed CYP119 structures suggest that both movement and adoption of drastically different conformations of the F/G loop and the β4 hairpin may be responsible for the accommodation of and interactions with active site ligands of differing sizes and shapes (59).

P450cam, Access to a Shielded Active Site

P450cam from *Pseudomonas putida* catalyzes the 5-exo hydroxylation of camphor. It was the first P450 to have its crystal structure determined (38). In contrast to the P450s described above, the active site is shielded from solvent; there is no obvious substrate access channel. In the substrate-free state, the active site is filled with six ordered water molecules (60), one of which interacts with the heme iron and results in a low-spin complex (61). Another one of the water molecules forms a hydrogen bond with Y96 located in the B'-helix at the "ceiling" of the active site. In the camphor-bound complex, a similar interaction takes place with the carbonyl oxygen atom of camphor. In the crystal structures of complexes with compounds that bind with less affinity, either because they are considerably smaller than camphor (such as norcamphor), or because they lack the ability to form a hydrogen bond to Y96 (such as camphane), a water molecule occupies the position of the sixth iron ligand (62), and the Fe spin has intermediate values (63, 64). The iron spin state has been interpreted as an indicator of water accessibility of the iron and the degree of solvation of the active site (65). Structural analysis of the active site cavity and of the interactions between camphor and the protein was used to modify the substrate [resulting in novel P450cam inhibitors (66)] or the protein [aiming at engineering enzymes for the biotransformation of unnatural substrates, with the long-term goal of applications in the synthesis of fine chemicals and bioremediation of environmental contaminants (67)].

The question of how the substrate gains access to the on average closed active site of P450cam has been studied computationally and experimentally by determination of structures of complexes with large ligands. First direct clues were obtained from the crystal structure of the complex with both enantiomers of a chiral, multifunctional inhibitor (Pfizer compound UK-39,671) (68) where the aforementioned Y96 points up into the substrate channel inferred from the temperature factor distribution in the substrate-free protein (60), and Phe193 is displaced toward the enzyme surface. Despite these and other structural changes, the channel is not wide enough to permit the transit of a substrate-sized molecule. Structures of even larger ruthenium (Ru)-linker substrate complexes show the substrate moiety bound at the active site, the biphenyl linkers in a channel that likely gives natural substrates access to the buried active center, and the Ru-sensitizers near the protein surface. To accommodate the linker, a channel opens up by movement of the F- (residues 173–185) and G- (192–214) helices against the perpendicular I-helix (234–267), and the F/G loop (185–192) retracts from the β-sheet domain in the enzyme (69). The functional relevance of the artificially pried-open channel is underlined by the similarity of the conformations found in the open structures of P450cam, P450 BM3, and P450nor. The functional relevance of this channel is also supported by molecular dynamics calculations on camphor exit pathways in P450cam (70, 71). The most likely pathway, "pathway 2a," allows the ligand to exit between the F/G loop and the B'-helix (71) and is identical to the channel that is held open by the Ru-sensitizer-linked substrates.

The F/G loop movement in open and closed conformations of P450cam is coupled to structural changes in functionally important residues located in the I-helix, such as T252, D251, and G248. The G248 peptide carbonyl forms a hydrogen bond with the hydroxyl of T252 in the closed camphor-bound state but forms a 2.6 Å hydrogen bond with the iron-bound water molecule in the open state (69). The latter interaction stabilizes the water-bound, low-potential ferric heme, preventing the binding of oxygen in the absence of substrate, and thus wasteful side reactions. The situation is analogous to the one in P450 BM3 described above (56).

In the ternary oxygen complex of P450cam (72) and the structurally isomorphous cyanide complex (73), the 251–252 peptide flips, its carbonyl forms a hydrogen bond with the N255 peptide amide, and the T252 side chain rotates to interact with the bound oxygen or cyanide and a new water molecule. The peptide flip introduces binding of another new water molecule, positioned to attack the bound oxygen. So again, the ligand-induced conformational changes of the I-helix change the conformation of critical residues, which in this case position the tools (catalytic T252) and ingredients (water molecules) for catalysis to occur.

ACCESS OF GASEOUS LIGANDS TO THE ACTIVE SITE OF P450cam   An open question central to a large number of gaseous ligand-binding metallo-proteins is whether ligand access or escape occurs through distinct transiently opening channels or randomly through the protein matrix. Time-resolved spectroscopy and crystallography and molecular dynamics simulations on wild-type and mutant myoglobins (Mb) suggest that there are a limited number of pathways that open up transiently on a nanosecond timescale and that involve cavities in the protein matrix that have been shown to bind xenon (74, 75). Their identification with CO docking sites strongly supports this hypothesis (76–81). This has implications for other systems as well as exemplified in Ni-Fe hydrogenase where xenon-binding cavities have also been observed and correlated with gas access (82). In Mb, the cavities may play a role during both ligand entrance and exit from the protein; following dissociation, they may stabilize the dissociated ligand long enough to prevent geminate rebinding and allow for protein fluctuations that open ligand escape channels (77). On rebinding, the pockets may serve as a local storehouse for ligands that must compete with high-solvent concentrations to bind at the active site. This means that the cavities and putative channels must be lined up for efficient transfer. In turn, this implies that the cavities and the dynamics of the molecule have an evolutionary survival value because they affect ligand affinities (83) and thus function. It is therefore of interest that there are four xenon binding pockets in P450cam (92; pdb code 1UYU). The first one, Xenon1, is lined by L245, L246, A218, I220, A167, and C242, and the second one, Xenon2, is lined by I367, L371, M261, and L257. Xenon3 is occupied much more weakly and is lined by P278, K372, L375, and I275. Xenon1 and Xenon4 (Figure 5) are close to one of the ligand exit pathways (pathway 1) predicted by steered molecular dynamics (70). The maximum energy barrier in this pathway is correlated in time with displacement of the I-helix around residue 247 and the heme vinyl side chain. This pathway has a much higher energy barrier for camphor exit than pathway 2, which is identical to the channel observed in P450 BM3 and in the ruthenium linker complexes of P450cam, and pathway 3. Based on the xenon binding data, we suggest that pathway 1 may serve as an entry route for gaseous ligands, such as oxygen.

P450s WITH POLYKETIDE SYNTHASE OR NONRIBOSOMAL PEPTIDE SYNTHETASE–PRODUCED SUBSTRATES

Most of the currently identified bacterial biosynthetic gene clusters involved in the production of biologically active secondary metabolites include cytochrome P450 genes. The P450 gene products catalyze a broad array of site-specific tailoring reactions, which include hydroxylation and oxidative phenol cross-linking, at different stages of product formation. In polyketide- and peptide-based antibiotic biosynthetic gene clusters, the P450 genes are typically located downstream to polyketide synthase (PKS) or nonribosomal peptide synthetase (NRPS) genes. PKS and NRPS are multimodular enzymes catalyzing synthesis

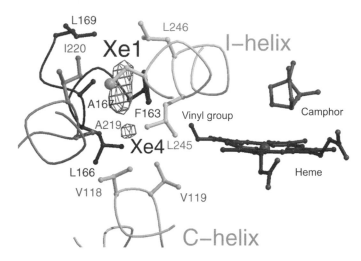

Figure 5   Xenon binding sites in P450cam. Fobs-Fcalc electron density is contoured at 4.5 σ.

of the polyketide or polypeptide core structure of the antibiotics (84). During core structure synthesis, the component amino acids and the growing peptide chain are attached to the phosphopantetheine arm of peptidyl carrier protein domains (PCP) of NRPS (Figure 6) [in case of PKS the component carboxylic acid monomers and the growing polymer chain are attached to acyl carrier protein (ACP) domains of PKS]. Chain termination is effected by hydrolysis, aminolysis, or intramolecular cyclization (85).

P450eryF, CYP154C1, and CYP121

In the macrolide antibiotic erythromycin biosynthesis pathway in *Saccharopolyspora erythraea*, the PKS produces a cyclic molecule, 6-deoxyerythronolide B (6dEB). After 6dEB release from the PKS, it becomes the substrate for P450eryF. P450eryF catalyzes hydroxylation of C-6 of the 14-membered macrolactone 6dEB ring. Crystal structures of both P450eryF-substrate and ketoconazole-inhibitor complexes show a closed active site occupied by the ligands (86). There is no substrate-free structure available for comparison. In the P450eryF/6dEB complex, part of the active site is occupied by water molecules (87). Eight of these are displaced from the active site to accommodate the larger ketoconazole compound. The large water-binding region in the active site of the P450eryF/6dEB structure appears to allow molecules that are larger than 6dEB

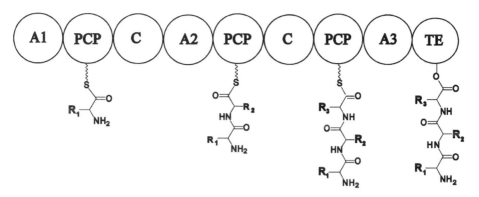

Figure 6 Domain organization of nonribosomal peptide synthetase. NRPS is a multidomain protein composed of condensation domains (C) catalyzing peptide bond formation, amino acid adenylation domains (A), peptidyl carrier domains (PCP) and a chain-terminating thioesterase domain (TE). One elongation module, consisting of condensation, adenylation, and peptidyl carrier domains adds one amino acid to the growing peptide chain. The NRPS template for a peptide of n amino acid residues is: A-PCP-(C-A-PCP)$_{n-1}$-TE. During nonribosomal synthesis, the growing peptide chain is attached to the phosphopantetheine arm of the PCP-domain.

to be accommodated in the active site. The ketoconazole- and 6dEB-bound structures also differ in the B′-helix and the I-helix kink. The latter has been proposed to provide space for dioxygen and water to bind (87), and it collapses in the ketoconazole-bound complex. Interestingly, the inhibitor can be washed out and replaced by the substrate (6dEB) in the crystalline state; the protein structure is restored and becomes identical to the original P450eryF-6dEB complex structure. This implies (*a*) that the substrate access channel in P450eryF is a flexible region capable of undergoing reversible conformational changes even in the crystalline state (86) and (*b*) that there is an equilibrium between open and closed forms that allows ligand exchange. The local ligand-dependent structural reorganization in P450eryF shows that the protein active site can adopt multiple conformations as a result of ligand binding.

CYP154C1 from *Streptomyces coelicolor* A3(2) is another structurally resolved macrolide hydroxylase (88). CYP154C1 exhibits catalytic activity toward both 12- and 14-membered ring macrolactones in vitro. CYP154C1 has 37% sequence identity and 51% sequence similarity with the macrolide monooxygenase P450eryF, which is unusually high for unrelated P450s (88). The structural similarity of both proteins suggests that the conformational differences between substrate-bound P450eryF and substrate-free CYP154C1 might reflect the conformational changes that occur when substrate binds. The crystal structure of the substrate-free CYP154C1 macrolide monooxygenase differs from the homologous P450eryF substrate/inhibitor-bound structures by significant dislocations of the F-, G- and B′-helices. Together with the shorter length of the

G-helix in CYP154C1 and the F/G loop being open compared to EryF, a significant separation between the two α and β protein domains is achieved (88). These changes result in an open active site similar to the substrate-free P450 BM3 (48). Taken together, these results suggest that CYP154C1 might rearrange its conformation from open to closed and vice versa driven by substrate binding and product release.

CYP121 from *Mycobacterium tuberculosis* shows amino acid sequence similarity to a number of polyketide monooxygenase P450s, which include P450-eryF, suggesting a potential role in polyketide metabolism in *Mtb*. Whereas the physiological role remains unclear, the enzyme binds bulky azole antifungal drugs with high affinity, and the binding constants for these drugs are perturbed by the presence of erythromycin and other large polyketides, indicating that CYP121 may metabolize polyketides or bulky polycyclics in vivo (89). The substrate-free CYP121 crystal structure is quite similar to that of substrate or inhibitor-bound P450eryF (90). The CYP121 active site is apparently inaccessible from the surface, and the remarkably rigid active site cavity is filled with water molecules similar to substrate-free P450cam, suggesting significant movements of the secondary structure elements and loops lining the active site to allow substrate access.

P450s With Carrier Protein–Linked Substrates

Some cytochrome P450s modify amino acids or growing peptides directly on the NRPS before chain termination. The Novobiocin gene cluster from *Streptomyces spheroids* (91) contains the NovH-NovI gene pair, which represents a PCP domain and a P450 enzyme, respectively. It was demonstrated that P450 NovI is responsible for the formation of a β-OH-Tyr intermediate in novobiocine biosynthesis (91) and that the tyrosine must be covalently linked to the phosphopantetheine arm of the PCP domain (NovH) in order to be recognized as a NovI substrate. The NikQ gene in nikkomycin biosynthesis has a similar function as NovI. NikQ is a P450 that catalyzes the β-hydroxylation of histidine while conjugated to a carrier protein (93).

OXY PROTEINS   Vancomycin is a glycopeptide antibiotic produced by *Amycolatopsis orientalis* (94). The backbone of this antibiotic consists of a nonribosomally synthesized heptapeptide modified by oxidative cyclization of phenolic side chains by cytochrome P450s (95) named Oxy proteins (96) (see Figure 7). Crystal structures of substrate-free OxyB (97) and OxyC (98) proteins, performing the first and the third cyclizations (99), revealed open active sites and high flexibility in F-, G-, and B'-helix regions, allowing access of the relatively large substrate to the active sites. Compared to the typical P450 fold, OxyB and C proteins contain an additional β-hairpin $\beta 0$, which is also present in P450nor, although in this case the β-strand is replaced by a loop conformation. Other P450s lack this structural element. The side chain of R13 (the number corresponds to the OxyB sequence) from the β-hairpin is turned toward the inside of

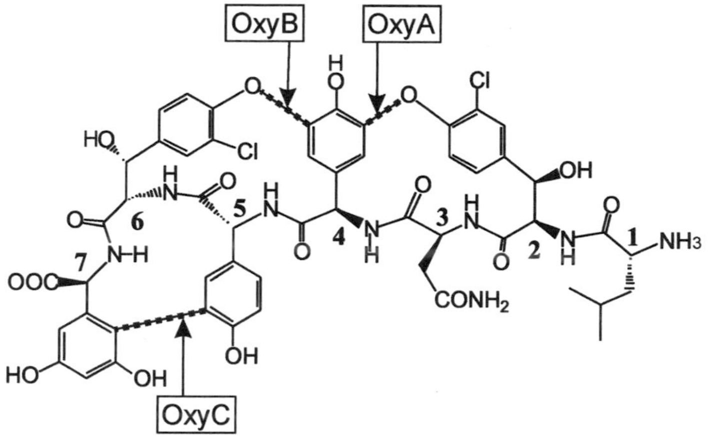

Figure 7 Scheme of the vancomycin heptapeptide backbone with OxyA, B, C catalyzed cyclizations indicated.

the molecule and contacts amino acid residues from the substrate-binding pocket (Figure 8). The hydrogen bond network in this region involves residues from the three-dimensionally highly conserved meander loop located on the surface of the protein (see Figure 3) and extends to the residues from the loop connecting the K-helix and $\beta 4$ located near the heme. In all known P450 structures, the first and second β-strands in $\beta 1$ are connected with a flexible loop, which is presumably involved in substrate recognition/binding. There OxyC has an insertion that forms two turns of an additional α-helix (A″). Both, $\beta 0$ and the A″-helix are lining the active site from the β domain side. The described structural features might be required for positioning the residues in the active site and the substrate binding/recognition. The heptapeptide-OxyB binding experiments showed a rather weak interaction of the peptide with the enzyme (97). The timing of the heptapeptide release from the PCP domain of the NRPS remains unclear, and it was suggested that the heptapeptide might be still attached through its C terminus to a PCP domain in order to act as a substrate for Oxy proteins analogously to NovI (91, 97). A cluster of residues conserved among Oxy proteins lining the active site entrance from the side of the B′-helix might be implicated in the binding of the PCP domain–linked heptapeptide (98).

ELECTRON TRANSFER

The monooxygenation reaction catalyzed by P450s of class I and II requires two distinct electron transport (ET) processes in the single turnover. In the first ET process, the P450, whose heme iron is in the ferric state, accepts one electron from the redox partner to produce the ferrous form of P450. After binding of

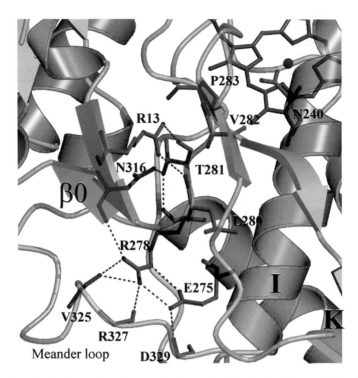

Figure 8   The hydrogen bond network (shown in *blue dashed lines*) of the OxyB β0, K-helix, and meander loop. The residues involved in the interaction are shown in ball-and-stick representation and are labeled.

molecular oxygen to the ferrous heme iron to form oxygenated P450, the second electron is transferred to activate the O-O bond and to initiate the oxygen transfer to the substrate (see Figure 1). Therefore, the ET reactions between P450 and the redox partner are essential for the catalytic cycle of P450, making the study of P450 redox-partner molecular recognition a subject of intensive research. An important question is are any structural features in P450s of different classes determinative for redox-partner recognition?

Experiments with manipulation of ionic strength show that a complementary charge interaction is involved in redox-partner docking (100). Mutation studies indicate that the P450 provides positively charged residues (101–103) and that the electron donors have conserved negative charges (104), which are critical for proper interaction of the partners leading to electron transfer. In most structurally resolved P450s, a positively charged patch is present at the proximal surface. This region is considered a P450 interaction site with redox partners, such as ferredoxin and P450 reductase (CPR) (see classes of P450 above) (39, 101–103). The distal face of the P450s is covered with a large cluster of negative residues. This asymmetric charge distribution leads to the formation of a molecular dipole

(39). It has been suggested that the interaction and orientation of a P450 with its reductase is facilitated by the dipole of the P450 molecule. The electrostatic dipole of the P450 points from this large patch of negative residues toward a positively charged one on the proximal side and may help in directing the negatively charged region of the redox partner (39).

The crystal structure of a complex between the heme- and FMN-binding domains of P450 BM3 (class II P450) is the only available structure representing a specific electron-transfer complex in a P450 monooxygenase system (105). Calculation of surface potentials demonstrates that, indeed, the complex formation between P450 BM3-heme and FMN-binding domains might be facilitated by long-range electrostatic attraction with the involvement of complementary charged surfaces on both molecules. In the flavin domain, acidic amino acids are clustered on the side of the molecule that faces and interacts with the positively charged proximal side of the heme domain (105).

There are only two direct hydrogen bonds, one salt bridge, and several water-mediated contacts between the heme- and FMN-binding domains in the P450 BM3 structure. The presence of only a few direct contacts in the 967 $Å^2$ area of interface indicates that the interaction between the heme and flavin domains is not strong. It should be noted that the residue analogous to H100 from the C-helix in P450 BM3, which forms a direct contact with E494 from the FMN-binding domain in the crystal structure of the complex, is conserved as a basic amino acid in P450s. The H100–E494 hydrogen bond may play an important role in P450–redox partner electron transfer. The E494 main chain nitrogen makes a contact with the hydroxyl group of S488 (3.4 Å), which interacts with the oxygen atom of the FMN ribityl-phosphate chain (2.4 Å) (Figure 9a). Moreover, H100 forms a hydrogen bond with a water molecule contacting the D propionic acid of the heme, and the H100 imidazole ring plane is 3.6 Å distant from the W96 indol ring plane; the indol nitrogen of the latter forms a hydrogen bond with the D propionic acid of the heme. The described interactions form a possible "communication pathway" between the heme and FMN molecules in the complex required for electron transfer.

In class I P450s, a conserved arginine residue occupies a position equivalent to H100 from P450 BM3. One of its $N\omega$ atoms forms a hydrogen bond with the D propionic group of the heme and two hydrogen bonds with the main chain oxygen of the residue that is followed by the heme iron ligating Cys (Figure 9b), while the other $N\omega$ atom and the $N\delta$ atom of the arginine are solvent accessible and may form contacts with a protein partner. The arginine position, conformation, and interactions are conserved in all known class I P450 structures, which include mammalian P450 2C5. Moreover, available evidence from site-directed mutagenesis of the arginine residue in P450cam (R112) shows that the arginine plays an essential role in the electron transfer from reduced putidaredoxin (Pdx), its physiological reductant, to P450cam (class I P450). In fact, the substitution of R112 evokes a drastic decrease in the catalytic activity of P450cam in the

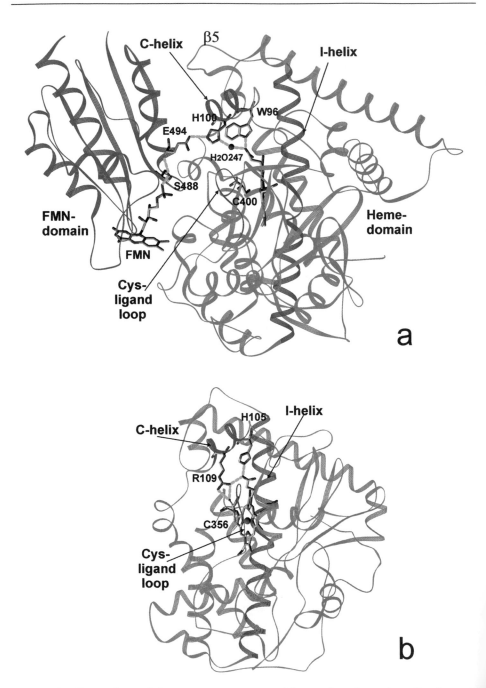

Figure 9 Comparison of the arrangement, composition, and conformation of amino acid residues implicated in interactions of the heme- and FMN-binding domains in P450 BM3 (*a*), with the corresponding residues in OxyC (*b*), P450nor (*c*), and P450Bsβ (*d*).

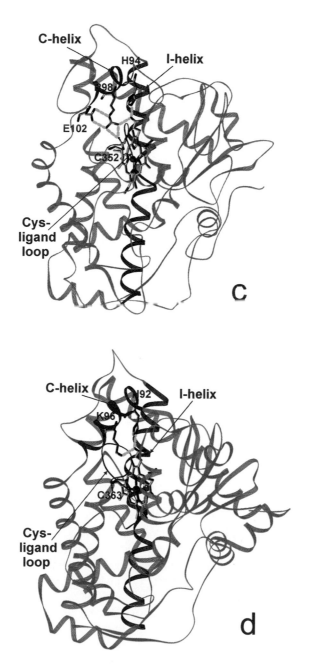

Figure 9   (*Continued*)

presence of Pdx (103); it was inferred that R112 is located at the Pdx binding site. On the basis of these mutational studies, Pochapsky et al. (106) proposed a model structure of the P450cam-Pdx complex, which assumes a salt bridge between R112 and D38 in Pdx. The theoretical analysis of the ET pathway in the proposed P450cam-Pdx complex indicates that R112 is also involved in the coupling between the 2Fe-2S cluster in Pdx and the heme iron of P450cam (107); this is consistent with the mutational study of R112. These modeling and mutational studies support the crucial roles of Arg112 in the specific binding to Pdx. (W/H)XXX(H/R), a motive in the C-helix, which interacts with the heme propionate, is considered as one of characteristic motifs that aid in sequence alignment of P450s (108). The arrangement of this region is very similar in P450s of both class I and II and may indicate the possibility of a common redox protein partner binding site for the proteins.

Mammalian nitric oxide synthase (NOS) contains FAD, FMN, and heme and was the first eukaryotic catalytically self-sufficient P450-like system described (109). The C-terminal domain of NOS is clearly homologous to CPR, whereas its N-terminal domain has low similarity with P450 sequences, but it also represents a heme thiolate protein. Reactions catalyzed by NOS, such as L-arginine $N\omega$-hydroxylase, $N\omega$-hydroxy-L-arginine monooxygenase, and NADPH:cytochrome c reductase, prove the functional identity of heme and reductase domains of NOS with P450 and of CPR, respectively (32).

NOS catalyze the formation of nitric oxide (NO) and citrulline from arginine, O_2, and NAPDH derived reducing equivalents via two monooxygenation steps, with the first one being mechanistically similar, but not identical, to that of the cytochrome P450 system. The transfer of electrons is from NADPH to FAD to FMN to the heme, reminiscent of the cytochrome P450 system. Binding of calmodulin (CaM), however, is required for the NOSs to transfer electrons from the reductase domain to the heme where catalysis occurs.

The first reaction catalyzed by NOS is probably mediated by a P450-like ferryl species, although it is generated by a distinct process in which a tetrahydrobiopterin (H4B) molecule in NOS serves as a transient electron donor (110). NOS has an unique fold with no similarities to previous P450 structures or any other proteins (111, 112, 113). In NOS there is a direct hydrogen-bonding interaction between the heme propionate and the N-3 acidic proton of H4B. The H4B-heme interaction is reminiscent of the described Arg/His-heme interaction in P450s. Thus, H4B might serve as an adaptor between the NOS C-terminal reductase domain (114), structurally similar to CPR, and the NOS heme domain, structurally diverse from P450.

Nitric oxide reductase (class IV P450) (P450nor) is involved in denitrification by the fungus *Fusarium oxysporum* (115). In contrast to other P450s, this enzyme does not possess monooxygenase activity but is able to reduce NO in the presence of NADH to form nitrous oxide (N_2O). Moreover, the reaction of NO reduction by P450nor does not require other protein components (116). This suggests that P450 should receive electrons directly from NADH.

In P450nor, the conserved arginine, R98, forms the described contacts, but its $N\omega$ and $N\delta$ atoms are involved in hydrogen bonds with E102 (Figure 9c), thereby preventing any additional contacts and "switching off" the arginine residue. Thus, the P450nor redox protein partner binding site is blocked by an intramolecular salt bridge. This is in line with the finding that P450nor does not require a redox protein partner for electron transfer.

The crystal structure (117) revealed a unique configuration of P450nor despite its overall structural similarity to other P450s. The B'-helix and the loop between the F- and G-helices (F/G loop) form an opening to the heme-distal pocket. The F/G loop rises toward the outside to form a wide entrance and a large cavity inside the distal pocket. In addition, positive charges due to Arg and Lys residues are concentrated outside and inside the heme-distal pocket. Mutant proteins in which the arginine and lysine residues were replaced with glutamine, glutamic acid, or alanine showed that these structural features are associated with a direct NAD(P)H interaction and showed that the positive charge cluster functions in attracting the negatively charged NAD(P)H (118). SRS1 that contains the B'-helix, the most variable structural element among P450 species, may also play an important role in direct interaction with NAD(P)H. Mutational analysis showed that only a few residues in the P450nor B'-helix region determine the cofactor specificity (119). These data indicate that NAD(P)H presumably binds at the proximal side of P450nor.

Recently, the first crystal structure of a class III P450, P450$_{BS\beta}$ from *Bacillus subtilis*, was solved (120). P450$_{BS\beta}$ catalyzes the hydroxylation of a long-chain fatty acid, using hydrogen peroxide (H_2O_2) in place of $O_2/2e^-/2H^+$. In P450$_{BS\beta}$, K96,which occupies the position of the conserved arginine, forms the described contacts and cannot establish any additional interactions (Figure 9d). Moreover, K96 is covered with a bulge in the Cys-ligand loop that is not present in the other known P450 structures. The substrate-bound P450$_{BS\beta}$ structure (120) revealed two channels connecting the heme active site to the protein surface. The first channel is occupied by a substrate fatty acid, whereas the second one is open and might allow access of a reducing agent or oxygen to the active site.

Almost all structurally resolved P450s of class I and II have a β5 hairpin that resembles the hairpin of P450 BM3, which uses this structural element to interact with the FMN-domain of its redox partner (105). This structural motif might be considered as a feature of P450s requiring redox protein partners for electron transport because the β5 hairpin is replaced by the tight loop between the H- and I-helices in P450nor, which does not require a redox protein partner for electron transfer. In CYP119, the β5 hairpin is tight (58), resembling that of P450nor (117). In contrast to other P450s, Cyp119 lacks a predominately negative distal face, and it has a large positive electrostatic potential at the tip region of the C terminus (59). It was reported that the ferric heme CYP119 is only very slowly reduced by putidaredoxin, ferredoxin, and P450 reductase (57). It was proposed that CYP119 redox interactions differ from class I and class II P450s (59). Later it was published that the D77R mutation improves binding of the heterologous redox-partner putidaredoxin to CYP119 and speeds up the rate of electron transfer from it to the heme group (122). The D77 residue was found to be located in a highly electronegative patch on the C-helix turn in the previously described (W/H)XXX(H/R) motive. The arginine substitution helps to form an electropositive region that preferably interacts with the negatively charged region of the redox partner.

Thus, comparison of the arrangement, composition, and conformation of amino acid residues implicated in interactions of the heme- and FMN-binding domains in P450 BM3 (class II P450) with corresponding residues in P450s of class I, II, III, and IV with known 3-D structures revealed that the organization of this region is very similar in class I and II P450s. This region may represent a possible common redox protein partner binding site in these proteins. In contrast, in P450nor (class IV) and P450$_{BS\beta}$ (class III), which do not require a redox protein partner for electron transfer, the redox protein partner binding site is blocked by intramolecular interactions.

CONCLUSIONS

Cytochrome P450s form a superfamily of cysteine-ligated hemoproteins that share a common fold despite low sequence similarity. P450s catalyze a number of monooxygenase reactions using many diverse substrates with a broad range of specificities (http://drnelson.utmem.edu/CytochromeP450.html). Promiscuity is particularly striking in human P450 3A4 that metabolizes up to half of all drugs in use (123). A common feature to all members of the family is that the active sites must be structurally diverse to provide the required specificity, yet conserve a mechanism of catalysis and prevention of unregulated solvent access. This requirement, in addition to the restrictions imposed by the conserved fold despite extremely different substrates, is met by an induced-fit mechanism that acts as a substrate-binding induced molecular switch. The enzymes have open and closed conformations, depending on the absence and presence of substrate, respectively, with the equilibrium between the conformations differing in the P450 family. The open form exists only transiently in P450cam, but it is stable in other forms such as P450 BM3 (CYP102). Nevertheless, common conformational states among different procaryotic and mammalian (124) P450 enzymes indicate that the channel allowing access of the substrate to the active site is structurally conserved and provided by movement of the F-, G-, and B'-helices. The F- and G-helices are cantilevered above the I-helix, which contains the catalytically important residues and serves as a fulcrum allowing the F- and G-helices to rise and fall above the substrate-binding cavity to accommodate substrate entry, binding, and exit. The F-, G-, and B'-helices act as a lid closing the active site upon substrate binding. Variability in the B'-helix (SRS1) and the F/G region (SRS2/3) plays a key role in the induced fit upon substrate/ligand binding. Ligand-induced movement of the I-helix (SRS4) not only positions the catalytic residues located therein, but it also changes the water structure surrounding the heme iron. This regulates the heme reduction potential and oxygen binding by either stabilizing or destabilizing the met-water molecule bound to the heme iron. Thus, structural interplay couples oxygen binding to the presence of substrate, thereby reducing side reactions, such as the production of superoxide, peroxide, and other toxic forms of reduced dioxygen.

The *Annual Review of Biochemistry* is online at http://biochem.annualreviews.org

LITERATURE CITED

1. Axelrod J. 1955. *J. Pharmacol.* 114: 430–38
2. Brodie B, Axelrod J, Cooper JR, Gaudette L, LaDu BN, et al. 1955. *Science* 121:603–4
3. Garfinkel D. 1957. *Arch. Biochem. Biophys.* 71:111–20
4. Klingenberg M. 1958. *Arch. Biochem. Biophys.* 75:376–86
5. Omura T, Sato R. 1962. *J. Biol. Chem.* 237:1375–76
6. Omura T, Sato R. 1964. *J. Biol. Chem.* 239:2370–78
7. Hashimoto Y, Yamano T, Mason HS. 1962. *J. Biol. Chem.* 237:3843–44
8. Muramaki K, Mason HS. 1967. *J. Biol. Chem.* 242:1102–10
9. Bayer E, Hill HAO, Röder A, Williams RJP. 1969. *Chem. Commun.* 1969:109
10. Hill HAO, Röder A, Williams RJP. 1970. *Struct. Bond.* 8:123–51
11. Champion PM, Gunsalus IC. 1977. *J. Am. Chem. Soc.* 99:2000–2
12. Greengard P, Psychoyos S, Tallan HH, Cooper DY, Rosenthal O, Estabrook RW. 1967. *Arch. Biochem. Biophys.* 121: 298–303
13. Omura T, Sato R, Cooper DY, Rosenthal O, Estabrook RW. 1965. *Fed. Proc.* 24: 1181–89
14. Remmer H, Schenkman J, Estabrook RW, Sasame H, Gillette J, et al. 1966. *Mol. Pharmacol.* 2:187–90
15. Foidart A, de Clerck A, Harada N, Balthazart J. 1994. *Physiol. Behav.* 55: 453–64
16. Jugert FK, Agarwal R, Kuhn A, Bickers DR, Merk HF, Mukhtar H. 1994. *J. Investig. Dermatol.* 102:970–75
17. Macica C, Balazy M, Falck JR, Mioskowski C, Carroll MA. 1993. *Am. J. Physiol. Gastrointest. Liver Physiol.* 265: G735–41
18. Mansuy D. 1998. *Comp. Biochem. Physiol. C* 121:5–14
19. Werck-Reichhart D, Feyereisen R. 2000. *Genome Biol.* 1:REVIEWS3003
20. Nelson D. 2003. *Cytochrome P450 homepage. http://drnelson.utmem.edu/CytochromeP450.html*
21. Hasler JA, Estabrook R, Murray M, Pikuleva I, Waterman M, et al. 1999. *Mol. Asp. Med.* 20:1–137
22. Feyereisen R. 1999. *Annu. Rev. Entomol.* 44:507–33
23. Chapple C. 1998. *Annu Rev. Plant Physiol. Plant Mol. Biol.* 49:311–43
24. Kahn R, Durst F. 2000. *Recent Adv. Phytochem.* 34:151–89
25. Werck-Reichhart D, Hehn A, Didierjean L. 2000. *Trends Plant Sci.* 5:116–23
26. Gonzalez FJ, Kimura S. 1999. *Cancer Lett.* 143:199–204
27. Lewis DF. 1998. *Xenobiotica* 28:617–61
28. Guengerich FP. 2002. *Nat. Rev. Drug Discov.* 1:359–66
29. Ortiz de Montellano PR, ed. 1995. *Cytochrome P450: Structure, Mechanism, and Biochemistry.* New York/London: Plenum
30. Newcomb M, Toy PH. 2000. *Acc. Chem. Res.* 33:449–55

31. Newcomb M, Shen R, Lu Y, Coon MJ, Hollenberg PF, et al. 2002. *J. Am. Chem. Soc.* 124:6879–86

32. Degtyarenko KN. 1995. *Protein Eng.* 8: 737–47

33. Munro AW, Lindsay JG. 1996. *Mol. Microbiol.* 20:1115–25

34. Narhi LO, Fulco AJ. 1987. *J. Biol. Chem.* 262:6683–90

35. Dus K, Katagiri M, Yu CA, Erbes DL, Gunsalus IC. 1970. *Biochem. Biophys. Res. Commun.* 40:1423–30

36. Yu C, Gunsalus IC, Katagiri M, Suhara K, Takemori S. 1974. *J. Biol. Chem.* 249: 94–101

37. Haniu M, Armes LG, Tanaka M, Yasunobu KT, Shastry BS, et al. 1982. *Biochem. Biophys. Res. Commun.* 105: 889–94

38. Poulos TL, Finzel BC, Gunsalus IC, Wagner GC, Kraut J. 1985. *J. Biol. Chem.* 260:16122–30

39. Hasemann CA, Kurumbail RG, Boddupalli SS, Peterson JA, Deisenhofer J. 1995. *Structure* 3:41–62

40. Presnell SR, Cohen FE. 1989. *Proc. Natl. Acad. Sci. USA* 86:6592–96

41. Imai M, Shimada H, Watanabe Y, Matsushima-Hibiya Y, Makino R, et al. 1989. *Proc. Natl. Acad. Sci. USA* 86: 7823–27

42. Martinis SA, Atkins WM, Stayton PS, Sligar SG. 1989. *J. Am. Chem. Soc.* 111: 9252–53

43. Kimata Y, Shimada H, Hirose T, Ishimura Y. 1995. *Biochem. Biophys. Res. Commun.* 208:96–102

44. Vidakovic M, Sligar SG, Li H, Poulos TL. 1998. *Biochemistry* 37:9211–19

45. Taraphder S, Hummer G. 2003. *J. Am. Chem. Soc.* 125:3931–40

46. Gotoh O. 1992. *J. Biol. Chem.* 267: 83–90

47. Koshland DE Jr. 1958. *Proc. Natl. Acad. Sci. USA* 44:98–104

48. Ravichandran KG, Boddupalli SS, Hasermann CA, Peterson JA, Deisenhofer J. 1993. *Science* 261:731–36

49. Li H, Poulos TL. 1994. *Structure* 2: 461–64

50. Modi S, Sutcliffe MJ, Primrose WU, Lian LY, Roberts GC. 1996. *Nat. Struct. Biol.* 3:414–17

51. Li H, Poulos TL. 1997. *Nat. Struct. Biol.* 4:140–46

52. Oliver CF, Modi S, Primrose WU, Lian LY, Roberts GC. 1997. *Biochem. J.* 327(Pt. 2):537–44

53. Ost TW, Miles CS, Murdoch J, Cheung Y, Reid GA, et al. 2000. *FEBS Lett.* 486: 173–77

54. Carmichael AB, Wong LL. 2001. *Eur. J. Biochem.* 268:3117–25

55. Li QS, Schwaneberg U, Fischer P, Schmid RD. 2000. *Chemistry* 6:1531–36

56. Haines DC, Tomchick DR, Machius M, Peterson JA. 2001. *Biochemistry* 40: 13456–65

57. Koo LS, Tschirret-Guth RA, Straub WE, Moenne-Loccoz P, Loehr TM, Ortiz de Montellano PR. 2000. *J. Biol. Chem.* 275: 14112–23

58. Yano JK, Koo LS, Schuller DJ, Li H, Ortiz de Montellano PR, Poulos TL. 2000. *J. Biol. Chem.* 275:31086–92

59. Park SY, Yamane K, Adachi S, Shiro Y, Weiss KE, et al. 2002. *J. Inorg. Biochem.* 91:491–501

60. Poulos TL, Finzel BC, Howard AJ. 1986. *Biochemistry* 25:5314–22

61. Tsai R, Yu CA, Gunsalus IC, Peisach J, Blumberg W, et al. 1970. *Proc. Natl. Acad. Sci. USA* 66:1157–63

62. Raag R, Poulos TL. 1991. *Biochemistry* 30:2674–84

63. Fisher MT, Sligar SG. 1985. *Biochemistry* 24:6696–701

64. Atkins WM, Sligar SG. 1988. *Biochemistry* 27:1610–16

65. Fisher MT, Scarlata SF, Sligar SG. 1985. *Arch. Biochem. Biophys.* 240:456–63

66. Helms V, Deprez E, Gill E, Barret C, Hui Hoa GHB, Wade RC. 1996. *Biochemistry* 35:1485–99

67. Chen XH, Christopher A, Jones JP, Bell

SG, Guo Q, et al. 2002. *J. Biol. Chem.* 277:37519–26

68. Raag R, Li H, Jones BC, Poulos TL. 1993. *Biochemistry* 32:4571–78

69. Dunn AR, Dmochowski IJ, Bilwes AM, Gray HB, Crane BR. 2001. *Proc. Natl. Acad. Sci. USA* 98:12420–25

70. Ludemann SK, Lounnas V, Wade RC. 2000. *J. Mol. Biol.* 303:797–811

71. Ludemann SK, Lounnas V, Wade RC. 2000. *J. Mol. Biol.* 303:813–30

72. Schlichting I, Berendzen J, Chu K, Stock AM, Maves SA, et al. 2000. *Science* 287: 1615–22

73. Fedorov R, Ghosh DK, Schlichting I. 2003. *Arch. Biochem. Biophys.* 409: 25–31

74. Tilton RF Jr, Kuntz ID Jr, Petsko GA. 1984. *Biochemistry* 23:2849–57

75. Scott EE, Gibson QH. 1997. *Biochemistry* 36:11909–17

76. Chu K, Vojtechovsky J, McMahon BH, Sweet RM, Berendzen JR, Schlichting I 2000. *Nature* 403.921–23

77. Östermann A, Waschipky R, Parak FG, Nienhaus GU. 2000. *Nature* 404:205–8

78. Brunori M, Vallone B, Cutruzzola F, Travaglini-Allocatelli C, Berendzen J, et al. 2000. *Proc. Natl. Acad. Sci. USA* 97: 2058–63

79. Srajer V, Ren Z, Teng TY, Schmidt M, Ursby T, et al. 2001. *Biochemistry* 40: 13802–15

80. Bourgeois D, Vallone B, Schotte F, Arcovito A, Miele AE, et al. 2003. *Proc. Natl. Acad. Sci. USA* 100:8704–9

81. Schotte F, Lim M, Jackson TA, Smirnov AV, Soman J, et al. 2003. *Science* 300: 1944–47

82. Montet Y, Amara P, Volbeda A, Vernede X, Hatchikian EC, et al. 1997. *Nat. Struct. Biol.* 4:523–26

83. Brunori M, Cutruzzola F, Savino C, Travaglini-Allocatelli C, Vallone B, Gibson QH. 1999. *Trends Biochem. Sci.* 24:253–55

84. Mootz HD, Schwarzer D, Marahiel MA. 2002. *ChemBioChem* 3:491–504

85. Keating TA, Ehmann DE, Kohli RM, Marshall CG, Trauger JW, Walsh CT. 2001. *ChemBioChem* 2:99–107

86. Cupp-Vickery JR, Garcia C, Hofacre A, McGee-Estrada K. 2001. *J. Mol. Biol.* 311: 101–10

87. Cupp-Vickery JR, Poulos TL. 1995. *Nat. Struct. Biol.* 2:144–53

88. Podust LM, Kim Y, Arase M, Neely BA, Beck BJ, et al. 2003. *J. Biol. Chem.* 278: 12214–21

89. Leys D, Mowat CG, McLean KJ, Richmond A, Chapman SK, et al. 2003. *J. Biol. Chem.* 278:5141–47

90. Cupp-Vickery JR, Poulos TL. 1997. *Steroids* 62:112–16

91. Chen H, Walsh CT. 2001. *Chem. Biol.* 8:301–12

92. Wade RC, Winn PJ, Schlichting I, Sudarko. 2004. *J. Inorg. Biochem.* In press

93. Chen H, Hubbard BK, O'Connor SE, Walsh CT. 2002. *Chem. Biol.* 9:103–12

94. McCormick MH, Stark WM, Pittenger GE, McGuire JM. 1955. *Antibiot. Annu.* 3:606–11

95. Hubbard BK, Walsh CT. 2003. *Angew. Chem. Int. Ed. Engl.* 42:730–65

96. Pelzer S, Sussmuth R, Heckmann D, Recktenwald J, Huber P, et al. 1999. *Antimicrob. Agents Chemother.* 43: 1565–73

97. Zerbe K, Pylypenko O, Vitali F, Zhang W, Rouset S, et al. 2002. *J. Biol. Chem.* 277:47476–85

98. Pylypenko O, Vitali F, Zerbe K, Robinson JA, Schlichting I. 2003. *J. Biol. Chem.* 278:46727–33

99. Bischoff D, Pelzer S, Bister B, Nicholson GJ, Stockert S, et al. 2001. *Angew. Chem. Int. Ed.* 40:4688–91

100. Hintz MJ, Mock DM, Peterson LL, Tuttle K, Peterson JA. 1982. *J. Biol. Chem.* 257:14324–32

101. Stayton PS, Sligar SG. 1990. *Biochemistry* 29:7381–86

102. Stayton PS, Poulos TL, Sligar SG. 1989. *Biochemistry* 28:8201–5

103. Koga H, Sagara Y, Yaoi T, Tsujimura M, Nakamura K, et al. 1993. *FEBS Lett.* 331:109–13

104. Geren L, Tuls J, O'Brien P, Millett F, Peterson JA. 1986. *J. Biol. Chem.* 261: 15491–95

105. Sevrioukova IF, Li H, Zhang H, Peterson JA, Poulos TL. 1999. *Proc. Natl. Acad. Sci. USA* 96:1863–68

106. Pochapsky TC, Lyons TA, Kazanis S, Arakaki T, Ratnaswamy G. 1996. *Biochimie* 78:723–33

107. Roitberg AE, Holden MJ, Mayhew MP, Kurnikov IV, Beratan DN, Vilker VL. 1998. *J. Am. Chem. Soc.* 120:8927–32

108. Li H, Yano JK, Poulos TL. 2002. *Methods Enzymol.* 357:79–93

109. White KA, Marletta MA. 1992. *Biochemistry* 31:6627–31

110. Nishida CR, Knudsen G, Straub W, Ortiz de Montellano PR. 2002. *Drug Metab. Rev.* 34:479–501

111. Crane BR, Arvai AS, Ghosh DK, Wu C, Getzoff ED, et al. 1998. *Science* 279: 2121–26

112. Raman CS, Li H, Martasek P, Kral V, Masters BS, Poulos TL. 1998. *Cell* 95: 939–50

113. Fischmann TO, Hruza A, Niu XD, Fossetta JD, Lunn CA, et al. 1999. *Nat. Struct. Biol.* 6:233–42

114. Ortiz de Montellano PR, Nishida C, Rodriguez-Crespo I, Gerber N. 1998. *Drug Metab. Dispos.* 26:1185–89

115. Kizawa H, Tomura D, Oda M, Fukamizu A, Hoshino T, et al. 1991. *J. Biol. Chem.* 266:10632–37

116. Nakahara K, Tanimoto T, Hatano K, Usuda K, Shoun H. 1993. *J. Biol. Chem.* 268:8350–55

117. Park SY, Shimizu H, Adachi S, Nakagawa A, Tanaka I, et al. 1997. *Nat. Struct. Biol.* 4:827–32

118. Kudo T, Takaya N, Park SY, Shiro Y, Shoun H. 2001. *J. Biol. Chem.* 276: 5020–26

119. Zhang L, Kudo T, Takaya N, Shoun H. 2002. *J. Biol. Chem.* 277:33842–47

120. Lee DS, Yamada A, Sugimoto H, Matsunaga I, Ogura H, et al. 2003. *J. Biol. Chem.* 278:9761–67

121. Deleted in proof

122. Koo LS, Immoos CE, Cohen MS, Farmer PJ, Ortiz de Montellano PR. 2002. *J. Am. Chem. Soc.* 124:5684–91

123. Guengerich FP. 1999. *Annu. Rev. Pharmacol. Toxicol.* 39:1–17

124. Wester MR, Johnson EF, Marques-Soares C, Dijols S, Dansette PM, et al. 2003. *Biochemistry* 42:9335–45

Annu. Rev. Biochem. 2004. 73:1019–49
doi: 10.1146/annurev.biochem.73.011303.073752
Copyright © 2004 by Annual Reviews. All rights reserved
First published online as a Review in Advance on March 25, 2004

ROLES OF N-LINKED GLYCANS IN THE ENDOPLASMIC RETICULUM

Ari Helenius[1] and Markus Aebi[2]

Institute of Biochemistry[1] and Institute of Microbiology,[2] Swiss Federal Institute of Technology Zurich, Zurich 8093, Switzerland; e-mail: ari.helenius@bc.biol.ethz.ch, markus.aebi@micro.biol.ethz.ch

Key Words   glycoprotein folding, quality control, ER-associated degradation, calnexin/calreticulin, glycan biosynthesis

■ **Abstract**   From a process involved in cell wall synthesis in archaea and some bacteria, N-linked glycosylation has evolved into the most common covalent protein modification in eukaryotic cells. The sugars are added to nascent proteins as a core oligosaccharide unit, which is then extensively modified by removal and addition of sugar residues in the endoplasmic reticulum (ER) and the Golgi complex. It has become evident that the modifications that take place in the ER reflect a spectrum of functions related to glycoprotein folding, quality control, sorting, degradation, and secretion. The glycans not only promote folding directly by stabilizing polypeptide structures but also indirectly by serving as recognition "tags" that allow glycoproteins to interact with a variety of lectins, glycosidases, and glycosyltranferases. Some of these (such as glucosidases I and II, calnexin, and calreticulin) have a central role in folding and retention, while others (such as α-mannosidases and EDEM) target unsalvageable glycoproteins for ER-associated degradation. Each residue in the core oligosaccharide and each step in the modification program have significance for the fate of newly synthesized glycoproteins.

CONTENTS

INTRODUCTION

The majority of proteins synthesized in the endoplasmic reticulum (ER) are glycoproteins. The N-linked oligosaccharide moieties of these proteins serve highly diverse functions. They are ligands in a multitude of recognition processes: They stabilize the proteins against denaturation and proteolysis, enhance solubility, modulate immune responses, facilitate orientation of proteins relative to a membrane, confer structural rigidity to proteins, regulate protein turnover, fine-tune the charge and isoelectric point of proteins, and mediate interactions with pathogens. No other covalent protein modification is as common and as complex chemically, and no other modification is employed for so many different purposes.

The biosynthesis of glycoproteins is a task shared by the ER and the Golgi apparatus. The division of labor is such that together with the cytosol the ER is responsible for the synthesis of the polypeptide and the core oligosaccharides, for the covalent coupling of glycan and polypeptide, and for initial modification of the glycans. Once the glycoproteins have folded and oligomerized properly, they move to the Golgi complex where the N-linked glycans are subjected to further trimming and modification. New saccharides are often added to generate the complex glycans found in the mature glycoproteins.

For a long time, it was unclear why cells have evolved such a complicated and apparently wasteful biosynthetic strategy. Why synthesize a large oligosaccharide in the ER, and then—after transferring it to a polypeptide—subject it to trimming in order to build it up again with different sugars? Why would this process be so important as to be conserved and virtually unchanged in all eukaryotes? The apparent lack of acceptable logic has not escaped the thousands of biochemistry and cell biology students who have had to memorize each step in the pathway.

The logic turns out not to be so complicated: The different configurations of the N-linked glycan are not merely intermediates in a biosynthesis pathway but have specific functions of their own. In fact, they play a role in at least three different stages during the existence of a glycoprotein. For each of them the N-linked glycans have to look different.

The first phase occurs in the ER where partially trimmed versions of the core oligosaccharide are needed for proper protein folding and quality control. Here, the glycans help to secure the fidelity of protein production. This is the phase described in this review. The second phase involves a role in intracellular transport and targeting exemplified by the role of mannose 6-phosphate in targeting of lysosomal hydrolases. This phase occurs in the ER, in the Golgi complex, and in the *trans*-Golgi network. The third phase takes place after extensive modification in the Golgi. It occurs when the mature protein has reached the extracellular space, the lysosome, the plasma membrane, or wherever the protein is targeted. The functions of the glycans in the mature proteins are as varied as the structures themselves.

In this review, we focus on the role of N-linked glycans in the ER. We describe the synthetic events and the trimming enzymes. We also describe the role of glycans in folding and in the regulation of glycoprotein degradation by ER-associated degradation (ERAD). These events involve a network of lectins, folding sensors, glycosyltransferases, and glycosidases. Finally we discuss the evolutionary origin of this machinery. For a general backround on the role of N-linked glycans and their functions, several reviews can be recommended (1, 2). We particularly recommend a recent review by Trombetta (3), which covers many of the same topics that we do with comprehensive references to the original literature.

N-LINKED GLYCANS AS SEMI-INDEPENDENT APPENDICES

After analyzing the SWISS-PROT database, Apweiler et al. (4) recently predicted that more than half of all eukaryotic protein species are glycoproteins. About 90% of these are likely to carry N-linked glycans, and there is an average of 1,9 N-linked glycans per polypeptide chain. With a molecular weight up to 3 kDa, the oligosaccharide groups in mammalian glycoproteins frequently make up a sizable portion of the mass of a glycoprotein and cover a large fraction of its surface.

The glycans are exposed on the surface. They form flexible, hydrated branches that can extend 3 nm or further into the solvent. X-ray crystal structures of glycoproteins and NMR studies indicate that, aside from interactions with the first two N-acetylglucosamines (GlcNAcs), the majority of glycans have few contacts with the surface of the protein (5).

Recently, Petrescu and colleagues (6) surveyed 506 glycoproteins listed in the Protein Data Bank crystallographic database and found the glycans tend to be located in positions where the secondary structure changes. They also observed that some glycans (10%) occur in invaginations of the protein surface, and others are bound to asparagines at the edge of indentations in the protein surface partially filled by the glycans (20%). Even in cases where there are interactions between the oligosaccharide and the protein surface, the terminal sugars are generally free. The N-linked glycans and the protein moiety are thus as a rule relatively independent of each other. Glycans can often be enzymatically removed from the surface of folded glycoproteins without appreciable effects on protein structure and function. X-ray crystallographers take advantage of this to make glycoproteins more amenable to crystallization.

That glycans behave like semi-independent appendices has several consequences. First, they can be modified without appreciable effects on the protein. Every N-linked glycan is, in fact, subject to extensive modification. This allows cells to fine-tune the biophysical and biological properties of glycoproteins and to generate the microheterogeneity so characteristic of glycoproteins (7). Importantly, the semi-independent nature of glycans also allows cell types and cells in different stages of differentiation and transformation to imprint on their glycoprotein pool their own specific biochemical characteristics, and thus give their exposed surface a "corporate identity." This makes cells recognizable to other cells in a multicellular environment. It allows self-recognition and provides a central theme in development, differentiation, physiology, and disease (8).

Both extra- and intracellularly, N-linked glycans are used as specific tags or signals recognized by a spectrum of carbohydrate-binding proteins (lectins) (9, 10). Glycans are ideally suited for signal functions. They are prominently exposed on the surface of glycoproteins, they are polar, and in each protein they can be present in multiple copies. During evolution, single point mutations can generate or eliminate N-linked glycosylation sites thus providing genetic versatility. Because the chemistry of sugars allows multiple types of linkages, cells can package and expose a large amount of information in a small space. The loss or addition of a single residue, or the alteration in a single bond, can, for example, dramatically alter binding to lectins (9).

As illustrated below, the notion of glycans as modifiable tags for lectin binding in diverse recognition processes is a central theme in glycobiology. That the affinity between carbohydrates and proteins is generally rather low (dissociation constants are typically in the micromolar range or higher) does not seem to be a major disadvantage. In the confined and crowded space of the ER, lectins are abundant and interactions are meant to be transient. In addition, most lectins are multivalent, which dramatically enhances their avidity for ligands.

BI-COMPARTMENTAL BIOSYNTHESIS OF CORE OLIGOSACCHARIDE

N-linked glycans are added to proteins en bloc in the lumen of the ER as presynthesized oligosaccharides. The core oligosaccharides have a clearly defined structure. In virtually all eukaryotes they are composed of a branched oligosaccharide unit made of three glucoses, nine mannoses, and two N-acetyl-glucosamines ($Glc_3Man_9GlcNAc_2$) (Figure 1).

The core glycan is the product of a biosynthesis pathway in which monosaccharides are added to a lipid carrier (dolichol-pyrophosphate) by monosaccharyl-transferases in the ER membrane (1, 11). Synthesis occurs on both sides of the

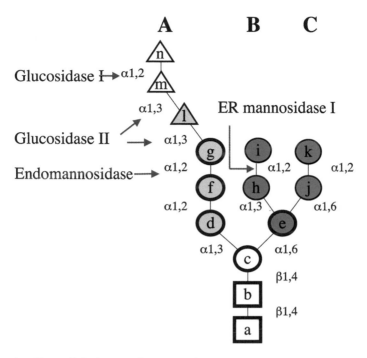

Figure 1   The N-linked core oligosaccharide. The core glycan has 14 saccharides: 3 glucoses (*triangles*), 9 mannoses (*circles*), and 2 N-acetylglucosamines. Each saccharide has a letter assigned to it, and these are used in the text to identify the saccharide (3). There are three branches named A, B, and C. The cleavage sites of some glycosidases involved in trimming are indicated. Residues a-g (shown by bold symbols) are added to the glycan on the cytosolic surface of the ER-membrane during biosynthesis; the rest are added lumenally. The blue residues (d, f, g, and l) are involved in the interaction of monoglucosylated glycans with calnexin and calreticulin. The green residues, with the exception of i, are likely to interact with EDEM.

ER membrane (Figure 2). Seven sugars are added on the cytosolic surface after which the sugar moiety is translocated, "flipped," to the lumenal side. Flipping is catalyzed by an ATP-independent, bi-directional flippase (12). In yeast, there is evidence that the flippase is the RFT1 protein, a polytopic membrane protein with about 10 *trans*-membrane spans (13). Unlike flippases in the plasma membrane that translocate phospholipids (14), RFT1 does not have ATP-binding cassettes and hence does not belong to the ABC family of translocators. Genes for homologous proteins occur in the genomes of other eukaryotes (15).

Whether facing the lumen or the cytosol, each individual glycosyltransferase displays strong preference toward a single oligosaccharide substrate (16). This leads to a linear, stepwise biosynthetic pathway of the branched oligosaccharide. The final step is the addition of a terminal α-1,2 linked glucose residue (residue

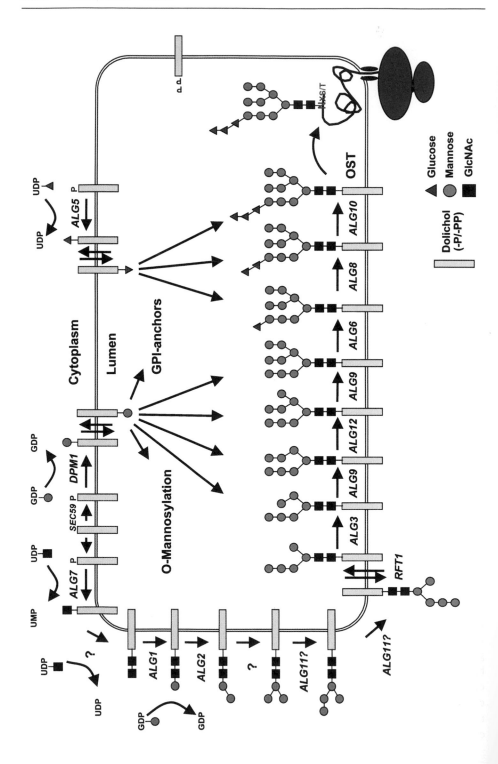

n) (Figure 1). This residue is needed for efficient recognition by the oligosac-charyltransferase, the enzyme that transfers the finished oligosaccharide from the lipid-bound precursor to the polypeptide (17, 18).

Oligosaccharyltransferase

The oligosaccharyltransferase (OST) is associated with the translocon complex. Together with the signal peptidase, BiP, calnexin, and possibly other factors, it is a member of the welcoming committee that every nascent polypeptide meets as it exits the translocon complex and enters the ER lumen. OST scans the emerging polypeptide for glycosylation sequons (Asn-X-Ser/Thr) and adds N-linked oligosaccharides to the side chain nitrogen of the Asn residue by an N-glycosidic bond. Because the oligosaccharides are added when the sequon is only 12 to 14 residues into the ER lumen, the active site of the enzyme can be no further than 5 nm away from the mouth of the translocon (19). As it passes the OST with an average rate of about 5 residues per second, the polypeptide is still unfolded (20).

The transfer of a glycan to the chemically rather inert side chain of the Asn requires formation of a loop in the polypeptide so that the hydroxyl groups of Ser or Thr can contact the Asn amide and render it more nucleophilic (21, 22). This explains why the middle residue X in the sequon cannot be a proline; proline prevents the formation of such a loop. It also explains why folded polypeptides are poor substrates.

OST is composed of multiple *trans*-membrane subunits. In *Saccharomyces cerevisiae*, there are eight, Ost1p, Ost2p, Wbp1, Swp1, Stt3p, Ost3p/Ost6p, Ost4p, and Ost5p, of which the five first are essential (23). Ost3p and Ost6p are homologues that define distinct oligosaccharyltransferase isoforms (23–25).

←——————————————————————————————

Figure 2 Synthesis of the N-linked core oligosaccharide and its transfer to a polypeptide chain. Biosynthesis occurs on both sides of the ER membrane. The yeast loci required for the individual biosynthetic steps are indicated. Synthesis starts on the cytoplasmic side where GlcNAc-1-phosphate is transferred from UDP-GlcNAc to dolichylpyrophosphate, followed by an additional GlcNAc and five mannose residues. The Man_5 $GlcNAc_2$ oligosaccharide, thus generated, is translocated (flipped) into the lumen of the ER (13). On the lumenal side, the lipid-linked Man_5 $GlcNAc_2$ is extended by the addition of four mannose and three glucose residues. The enzymes involved differ from most other glycosyltransferases in spanning the membrane several times and in being quite hydrophobic (16, 17, 34, 179–181). Unlike the cytosol-oriented glycosyltransferases that initiate core oligosaccharide assembly and Golgi-localized glycosyltransferases, these use lipids (dolichol-P-Man and dolichol-P-Glc) as saccharide donors. Little is known about the catalytic mechanism of these interesting enzymes. However, based on sequence similarity, a common origin has been suggested (182).

Except for the small Ost5p and Ost4p, all subunits present in yeast OST have homologues in the mammalian enzyme (26, 27). The homologues of the five essential subunits in yeast (ribophorin I, DAD1, OST48, ribophorin II, and STT3-A/STT3-B) are thought to form a central core complex to which N33 and IAP are associated peripherally (28). There are two isoforms of OST in mammalian cells that differ with respect to the STT3-A/STT3-B subunit. Expression of these alternative subunits is tissue- and cell-type specific (28).

As outlined above, the terminal α-1,2 linked glucose (n) (Figure 1) in the lipid-linked oligosaccharide is a central element in substrate recognition by OST. In vivo analysis of glycosylation efficiency and oligosaccharide specificity in yeast show that when the complete $Glc_3Man_9GlcNAc_2$ oligosaccharide is limiting, priority is given to the transfer of the correct oligosaccharide structure even with the risk of underglycosylation. However, if the $Glc_3Man_9GlcNAc_2$ oligosaccharide is not available at all, incompletely assembled oligosaccharides are transferred. Gilmore and coworkers (29) developed an allosteric model for yeast OST to explain this interesting ambiguity in substrate specificity.

In higher eukaryotes, in contrast to yeast, the choice between faithful addition of the complete glycan structure and occupancy of glycosylation sites is subject to regulation (28). Mammalian cells contain two OST complexes that differ in selectivity toward the lipid-linked donor substrate. Complexes containing the STT3-A subunit preferentially use the complete oligosaccharide, although those with STT3-B have a higher v_{max} but can also use incomplete core glycans. The variable expression of the two complexes observed in different cell types may reflect differences in the balance between maximal glycosylation versus faithful use of the complete oligosaccharide.

The complexity of OST has made it difficult to assign specific functions to subunits. Cross-linking experiments suggest that STT3 (30, 31), ribophorinI/Ost1 (32), and the Ost48 subunit make direct contact with the substrate, and biochemical data supports an essential role for Wbp1p (Ost48p) (33). However, the distinct influence of different STT3 subunits on the catalytic properties of the mammalian OST (28) and the recent finding that a bacterial homologue of the STT3 subunit, the PglB protein in *Campylobacter jejuni*, is sufficient for OST activity imply that the STT3 subunit is the catalytic subunit (34, 35).

The efficiency with which OST transfers a core glycan to individual sequons varies. Analysis of well-characterized glycoproteins in the SWISS-PROT database and of glycoproteins in the PDB crystallographic database indicates that sequon occupancy by oligosaccharides is about 2/3 (4, 6). Some sequons are ignored, some are glycosylated with partial efficiency, and some are glycosylated with full efficiency. Numerous factors affect efficiency, including the subunit composition of the OST complex, the amino acids in the sequon itself and immediately adjacent to it, the location of the sequon in the polypeptide chain, the availability of dolichol precursor in the ER, and the rate of protein folding.

GLYCOPROTEIN FOLDING

Protein folding in the ER begins cotranslocationally during the entry of a polypeptide chain through the translocon complex either into the lumen of the ER or, in the case of polytopic membrane proteins, into the ER membrane (36–38). It continues after the polypeptide chain has dissociated from the ribosome and the translocon complex. In contrast to proteins in the cytosol, most proteins that fold in the ER acquire disulfide bonds through an oxidation process catalyzed by thiol-disulfide oxidoreductases (39). Finally, because many proteins are oligomeric in their native form, assembly into homo- or hetero-oligomers often completes the folding process.

That many glycoproteins need their N-linked glycans for efficient secretion was realized when tunicamycin, an inhibitor of N-linked glycosylation, became available in the 1970s [see (1, 40)]. It is now clear that when glycosylation is blocked many polypeptides undergo improper or incomplete folding. Failing to reach the native conformation, they do not pass ER quality control (41). They are retained in the ER and eventually are degraded (42).

It is important to point out that all glycoproteins are not equally dependent on their glycans for folding and secretion. If devoid of glycans, many suffer only partial loss of folding and secretion efficiency, others become temperature sensitive, and many remain unaffected [reviewed in (40, 43)]. The importance of glycosylation is thus highly variable. As a rule, the folding of those glycoproteins that have a large number of glycans is more glycan dependent.

When individual sequons in glycoproteins are mutated, it was observed that only some are essential. In the hemagglutinin (HA) of influenza virus (Aichi strain), for example, only one of six glycans (N81) is absolutely essential when glycans are removed one by one (44). Without this glycan, HA fails to acquire any of the six native intrachain disulfide bonds, and it does not exit the ER. The most likely explanation is a local perturbation that prevents oxidization of a nearby disulfide bond (C67-C76). This bond is needed for the onset of further oxidative folding of the molecule. Removing the other five glycans in HA individually does not affect folding, but removing several of them together can lead to misfolding (38, 44). This example illustrates a phenomenon often observed: Although a single glycan may not be essential, the same glycan may prove important when more than one is eliminated. Again, great variability is seen among individual glycoproteins. It is evident that glycans have local effects where their precise location is important, and global effects where their presence is important—but their precise location is not.

Direct Effects on Folding

Part of the fold-promoting effect of oligosaccharides is biophysical. The addition of large, polar carbohydrates affects the properties of a polypeptide chain directly. Although systematic biophysical studies are lacking, it is known that the presence of a glycan can profoundly influence the conformational profile of short glycopeptides (45). Often, it rigidifies their conformation by limiting the conformational space accessible to the polypeptide chain. An interaction between the N-acetyl group of the first GlcNAcs and the polypeptide, moreover, promotes the formation of β-turns (46). It is likely that some glycans have a role in promoting and stabilizing local structure (6). Other effects on folding are likely to be more global, such as increased solubility of folding intermediates.

Comparison of native glycoproteins to nonglycosylated versions of the same shows that the presence of glycans increases stability, solubility, and resistance to proteases (45, 47–49). The stabilization effect is mainly entropic. For example, the two N-linked glycans stabilize the first domain of ovomucoid against thermal denaturation by about 2.5 kcal/mol (50). It is hypothesized that glycans stabilize the folded conformation by decreasing the freedom of mobility of unfolded conformations.

The Calnexin/Calreticulin Cycle

Glycans have also an indirect role in protein folding. It is based on binding of the newly synthesized glycoproteins to lectins in the ER. In this process, the glycans serve as sorting signals that the cell modifies to reflect the folding status of the protein. That such a tagging paradigm applies to glycoprotein folding and quality control emerged from four key observations in the early 1990s: (*a*) N-linked glycans in misfolded glycoproteins retain a glucose residue (residue 1 in Figure 1) in the A branch, and they undergo a cycle of re- and deglucosylation (51); (*b*) the glucosyltransferase in the ER responsible for the reglucosylation is selective for misfolded glycoproteins (52); (*c*) newly synthesized glycoproteins bind transiently and selectively to calnexin, a resident ER protein (53); and (*d*) the binding to calnexin is blocked by inhibitors of ER glucosidases (54).

The existence of a cycle (Figure 3) was first postulated in 1993 for calnexin (54, 55), and later the cycle was expanded to include the soluble, luminal ER protein calreticulin (56, 57). Although controversial in the beginning, the calnexin/calreticulin cycle is now generally accepted. Each step has been extensively analyzed and confirmed in many systems. The literature has been frequently reviewed (3, 58–62).

Before discussing the individual proteins involved in the calnexin/calreticulin cycle in detail, it may be useful to go through the cycle step-by-step (Figure 3). The process begins when a core glycan has been added to the growing, nascent polypeptide chain. The first glucose is rapidly removed by glucosidase I and is followed by removal of the second glucose by glucosidase II. The monoglucosylated core ligand, thus generated, binds the glycoprotein to calnexin or calreticulin. These sequester the nascent or newly synthesized glycopolypeptide chains and serve as molecular chaperones, preventing aggregation and export of the incompletely folded chains from the ER. In many cases, they also protect the folding intermediates against premature degradation and, at the same time, expose them to ERp57, a thiol-disulfide oxidoreductase. It is a cofactor that provides assistance for proper disulfide bond formation during the ongoing folding process.

To release the bound chains from calnexin and calreticulin, glucosidase II removes the remaining glucose residue. The glycoprotein no longer binds to the lectins and is now free to leave the ER unless recognized by a soluble enzyme, UDP-Glc:glycoprotein glucosyltransferase (GT). GT only reglucosylates incompletely folded glycoproteins, and it serves as a folding sensor in the cycle. If reglucosylated by GT, a glycoprotein rebinds to the lectins. A glycoprotein stays in the cycle until it is either properly folded and oligomerized, or degraded. Virtually all newly synthesized glycoproteins seem to undergo a phase in which they associate transiently with calnexin, calreticulin, or both. Calnexin association is observed for both membrane-bound and soluble glycoproteins. Calreticulin can also bind to both types of proteins, but it is more frequently associated with soluble substrates.

If association with the cycle is inhibited, for example, by blocking the action of the glucosidases, the folding rate of glycoproteins is often increased and folding efficiency decreased. In some cases, quality control breaks down with the consequence that misfolded glycoproteins exit the ER. In other cases, the glycoproteins associate with BiP, an important ER chaperone that cooperates and competes with the calnexin cycle for substrates and that serves as a back-up.

Calnexin and Calreticulin

Calnexin (a transmembrane protein of Type I) and calreticulin (a soluble lumenal ER protein) are related members of the legume lectin family. Both are monomeric, calcium-binding proteins, and have ER localization signals. Transgenic mice devoid of calreticulin or calnexin have embryonic lethal phenotypes or are born with strong debilitating phenotypes (63, 64).

The NMR structure of the calreticulin P domain and the X-ray structure of calnexin's ectodomain show unusual domain architectures (60, 65). There are two separate domains, a globular β-sandwich domain with homology to legume lectins and a long extended hairpin fold corresponding to a proline-rich domain (the P domain) not seen in legume lectins (Figure 4). Calnexin has, in addition, a single transmembrane sequence and a cytoplasmic domain of 91 residues that can undergo phosphorylation and allows interactions with ribosomes (66).

The globular domain contains a concave and a convex β-sheet with the sugar binding site located on the concave surface partially shielded by the P domain (60) (Figure 4). The structure is stabilized by a calcium atom, which is not part of the lectin site. That calnexin and calreticulin (in contrast to ERGIC-53 and most other legume lectins) are monomers (67) is most likely explained by the presence of a short helix behind the convex β-sheet that may prevent intermolecular interactions between these surfaces. With its β-sandwich structure, the globular domain is not only similar to legume lectins, to ERGIC-53, and to galectins, but it is also similar to the neurexin family of neuronal cell surface receptors with hundreds of members (68–70).

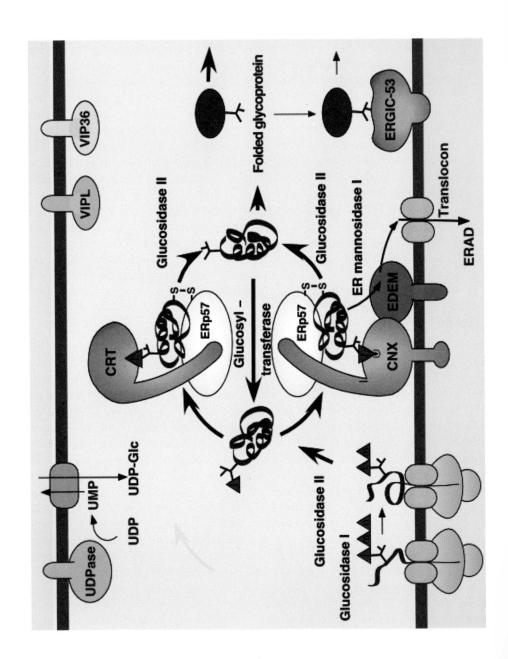

Extending from a loop that connects two β-strands, the P domain curves away from the globular domain as a narrow hairpin fold stabilized by short antiparallel β-sheets and small hydrophobic clusters (60, 71). The length of the arm is 14 nm in calnexin and 11 nm in calreticulin. The structure suggests a model in which the glycosylated substrate protein, with a glycan moiety bound to the lectin site, occupies the space between the curved P domain arm and the globular domain. In this protected location, the substrate may be shielded from contacts with external factors and other incompletely folded proteins and thus prevented from aggregation. NMR studies suggest that the arm is somewhat flexible, which may allow the chaperone to adapt the space between the arm and the globular domain to the dimensions of a substrate (65).

Calnexin and calreticulin have virtually identical carbohydrate specificities (72–74). In addition to the single $\alpha-1,3$ linked glucose (l), which is essential, at least three downstream mannoses in the A branch (d, f, and g) contribute to binding (Figure 4). IgG with a single monoglucosylated core glycan binds to calreticulin with a K_d of ~ 2 μM (75, 76).

Modeling and mutational analysis show that the presence of the $\alpha-1,3$ linked glucose is essential for calreticulin binding mainly because the equatorially oriented 2-hydroxyl forms hydrogen bonds with Asp317 and Tyr109 (75, 77). Similar interactions tie down the glucose in the calnexin lectin domain (60). The three mannoses in the A arm enhance the binding affinity by 25- to 50-fold by

---

Figure 3 The calnexin/calreticulin cycle. Immediately after addition of the core glycan to a growing polypeptide chain by OST, the outermost of the three glucose residues (n) is removed by glucosidase I. Soon thereafter, glucosidase II removes the middle glucose (m). Via the monoglucosylated core glycans thus generated, the glycoprotein binds to calnexin (CNX) and calreticulin (CRT). These sequester the nascent or newly synthesized chains and expose them to ERp57, a thiol-disulfide oxidoreductase that provides assistance during disulfide bond formation. When glucosidase II removes the remaining glucose (l), the glycoprotein dissociates from calnexin and calreticulin. The protein now encounters one of three possible fates. If properly folded, it is free to leave the ER. Exit may be assisted by mannose lectins, such as ERGIC-53, VIP36, and VIPL. If it is incompletely folded, UDP-Glc:glycoprotein glucosyltransferase uses UDP-glucose transported by a UDP-glucose/UMP exchanger from the cytosol to reglucosylate the high-mannose glycans located in improperly folded regions. Through these glycans, the glycoprotein rebinds to calnexin and calreticulin. The third fate is ER-associated degradation (ERAD) after retrotranslocation of the misfolded glycoprotein to the ER most likely through the translocon complex. ERAD of glycoproteins occurs when they have stayed in the ER lumen for some time and when they are recognized by a putative lectin (EDEM) because they have lost a mannose (i) through the action of ER mannosidase I. Red triangles are glucose residues. Abbreviations used are EDEM, ER degradation-enhancing α-mannosidase-like protein; VIP36, vesicular integral protein 36; VIPL, VIP36-like protein; ERAD, ER-associated protein degradation; ERGIC, ER-Golgi intermediate compartment; and ERp57, ER protein 57.

A

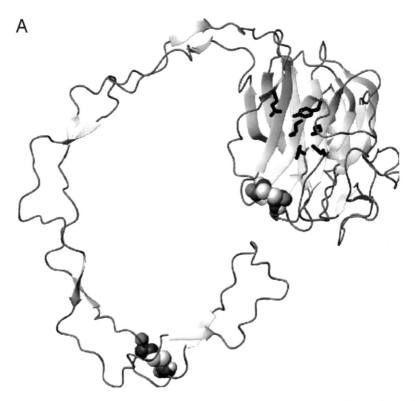

Figure 4 The structure of the calnexin ectodomain and a model of the calreticulin oligosaccharide-binding site. In calnexin (*A*) the oligosaccharide-binding site is situated in the globular domain composed mainly of a β-sandwich (60). The side chains of the residues interacting with the terminal glucose (l) are shown. The P domain is seen as a long antiparallel loop with four repeat units that have the same fold. The two intrachain disulfides are also shown. (*B*) A model of the oligosaccharide-binding site of calreticulin (77). The tetrasaccharide bound is Glcα_{1-3} Manα_{1-2} Manα_{1-2} Man. Mutational analysis shows that several of the residues that interact with the sugar in the model are essential for binding (76).

occupying additional subsites for hexapyranosyl residues and thus increasing the number of hydrogen bonds and van der Waals interactions. By involving the entire A branch of the monoglucosylated oligosaccharide, the binding site is larger than generally observed for carbohydrate-protein interactions. Hence, the thermodynamics of the interaction shows unusual properties, such as enthalpy-entropy compensation (76).

Although the lectin activity is no longer questioned as the main principle of substrate binding to calnexin and calreticulin, the significance of protein-protein interactions remains an open issue. One main argument in favor of functionally

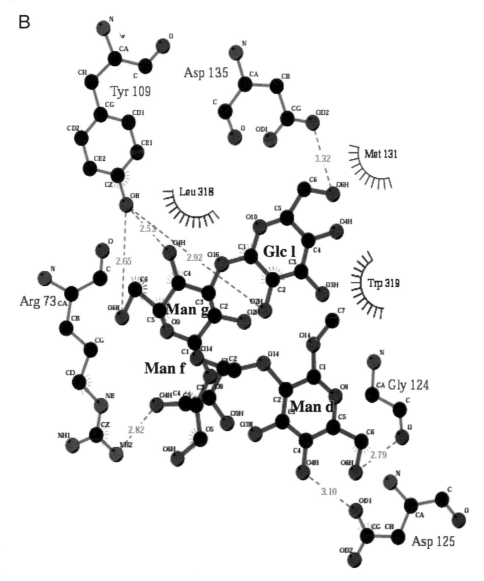

Figure 4   (*Continued*)

significant interactions between the protein moieties is that calnexin can be coimmunoprecipitated from cell lysates with some incompletely folded cargo proteins lacking glycans. Moreover, calreticulin and the soluble ectodomain of calnexin associate selectively with certain denatured nonglycosylated proteins in vitro and preserve them in a refolding competent state at elevated temperatures (78, 79). For further arguments, see Danilczyk & Williams (80).

The facts that weigh against a central role for protein-protein interactions are by no means more compelling. The vast majority of calnexin and calreticulin substrates fail to bind if glucose trimming is inhibited. Provided they have the monoglucosylated glycans, binding of model glycoproteins, such as RNase B and IgG, is independent of polypeptide conformation, suggesting that calnexin does not distinguish between conformers (75, 81, 82). In the case of IgG, the protein moiety does not contribute to the binding energy (76). Finally, the X-ray structure of the calnexin ectodomain does not reveal obvious binding sites for hydrophobic peptides or patches.

Other open issues include the significance of ATP binding and hydrolysis. The addition of ATP causes a change in the fluorescence emission and other properties of calnexin and calreticulin (78, 79, 83, 84). Also, a weak ATPase activity has been reported for both calnexin and calreticulin (78, 79). The reported presence of calreticulin in the cytosol and nucleus of cells has also generated discussion, as well as the presence of calnexin and calreticulin on the surface of cells (85).

ERp57

The majority of proteins synthesized in the ER acquire disulfide bridges. Oxidation is catalyzed by protein disulfide isomerase (PDI) and other thiol-disulfide oxidoreductases (39). Formation of correct disulfides is generally essential for proper folding. With four thioredoxin-like domains, ERp57 is a close homologue of PDI (86). The extreme N- and C-terminal domains have characteristic CXXC active site sequences. In vivo, ERp57 has been shown to form mixed disulfides with incompletely folded glycoproteins (87), and it acts as a reductase in the case of the partially folded major histocompatibility complex Class I heavy chain (88).

ERp57 differs from PDI in that it forms a complex with calnexin and calreticulin, and it specifically interacts with glycoproteins (89, 90). NMR studies and deletion mutants show that it binds to the distal end of the P domain (91, 92). Although the interaction is weak ($K_d \sim 9~\mu M$) the complex is likely to be stabilized by the formation of mixed disulfides with the substrate glycoprotein. The presence of ERp57 at the tip of the P domain is likely to further confine the space in which a glycoprotein substrate is trapped.

Glucosidase I

Glucosidase I removes the outermost glucose residue attached via an $\alpha-1,2$ linkage to the middle glucose (residue n) (Figure 1). It is a membrane glyco-protein (type II) of about 82 kDa with a short N-terminal cytosolic peptide, a single transmembrane sequence, and a large, glucosidase-active ectodomain. Together with glucosidase II, glucosidase I prevents binding of the protein-bound glycan to OST and makes possible the entry of a glycoprotein into the calnexin/

calreticulin cycle. Castanospermine and other polyhydroxylated indolizidine alkaloids inhibit both glucosidase I and II (93). They are frequently used to inhibit the entry of newly synthesized glycoproteins into the calnexin/calreticulin cycle.

A glucosidase I defect was recently described in a neonate with severe hypotonia and dysmorphic features (94). The syndrome has now been named congenital disorder of glycosylation type IIb. Disruption of the glucosidase I gene (gsc-1) in *Arabidopsis* results in defects in the accumulation of seed storage proteins and in the formation of protein bodies, as well as in deficiencies in cell differentiation during embryonal development (95). Tissue culture cell lines lacking glucosidase I activity are viable.

Glucosidase II

Glucosidase II plays a double role in the calnexin/calreticulin cycle. It prepares the substrate for entry into the cycle and allows substrate exit from the cycle. By using the same enzyme for feeding substrate into the cycle and removing the product, cells may make sure that the cycle cannot be oversaturated.

Glucosidase II is a soluble lumenal enzyme composed of two tightly associated glycopolypeptide chains, α and β, with molecular weights of 107 and 54 kDa (96–98). The sequence in the C-terminal half of the α chain contains a catalytic domain belonging to the hydrolase family 31 (99). Both α and β chains occur in several differentially spliced forms (100).

The β chain is a highly conserved glycoprotein. In addition to an N-terminal signal sequence and a C-terminal Lys-Asp-Glu-Leu (KDEL) sequence, it contains a domain homologous to the lectin domain of mannose 6-P receptors (81, 101). As the α chain lacks known ER retention sequences, it is likely that the β chain might serve as a localization subunit. It is also possible that the β chain is needed to allow the α chain to fold properly because coexpression studies have shown that both subunits are essential for enzymatic activity, solubility and/or stability, as well as ER retention of the enzyme (102–104).

Inhibition or genetic disruption of glucosidase II in tissue culture cells leads to accelerated but less efficient glycoprotein folding and secretion, partial breakdown of the quality control system with incompletely folded proteins being secreted, induction of the unfolded protein response, and premature degradation of misfolded glycoproteins (102, 105–108). Disruption of the α chain in *Schizosaccharomyces pombe* results in total loss of glucosidase II activity, induction of the unfolded protein response, and accumulation of cargo proteins in the ER (102, 109). Disruption of the β chain has similar effects except for the formation of small amounts of $Glc_1Man_{8-7}GlcNAc_2$ chains. Germline mutations in the β subunit in humans are associated with autosomal dominant polycystic liver disease (110, 111). This is an inherited condition in which multiple cysts of biliary epithelial origin occur in the liver.

UDP-Glucose:Glycoprotein Glucosyltransferase

GT, the folding sensor, is undoubtedly the most interesting of the enzymes in the calnexin/calreticulin cycle. Because the properties of GT have been recently reviewed in this series (59), only some of the most important features will be discussed here. GT is a large, soluble, lumenal protein with a C-terminal ER-retrieval sequence (112–114). In addition to its localization in the ER, it is present in ER exit sites and in the ER-Golgi intermediate compartment (ERGIC) (115). GT has a C-terminal, catalytic segment of 300 amino acids with homology to members of the glycosyltransferase family 8 (113, 114, 116). The N-terminal sequence of about 1200 residues are presumed to participate in substrate glycoprotein recognition. The N- and C-terminal domains of the enzyme are intimately connected functionally and structurally (117, 118).

GT transfers a glucose residue from UDP-glucose to protein-bound high-mannose glycans. Efficiency decreases with decreasing number of mannoses in branches B and C (52). This is probably important for the interplay between the calnexin cycle and ERAD (see below). Interestingly, glucosidase II shows a similar decrease in efficiency with decreasing mannose number (119). Taken together these observations suggest that a glycoprotein passes more slowly through the cycle as mannoses are progressively lost.

How does GT sense the folding status of a substrate glycoprotein? Currently there is only a partial answer to this question. It is clear that selectivity must be based on general features shared by misfolded proteins. The enzyme makes, for example, no distinction between glycoproteins and neoglycoproteins of bacterial or nonsecretory origin (59). There are at least two general ways in which a protein might be able to sense whether another protein is incompletely folded: through exposed hydrophobic peptides or patches or through excessive surface dynamics. Which of these apply to GT is not clear. That the data is still incomplete is in part due to unsatisfactory in vitro assays that make use of aggregated substrates with heterogeneous conformations and glycan configurations.

The enzyme does not use free glycans or small glycopeptides as substrate, hence the name glycoprotein glucosyltransferase. Studies with monomeric, nonaggregated, well-characterized glycoprotein substrates, such as RNase B, show that a random coil is not a substrate; whereas molten globule-like, partially folded proteins are good substrates (120–122). However, a recent study using $Man_9GlcNAc_2$-containing tryptic peptides showed that GT does reglucosylate glycopeptides with high efficiency provided that they have more than 12 amino acid residues and that the sequence includes hydrophobic residues on either the N-terminal or C-terminal side of the glycan (123). Such short peptides are unlikely to have secondary or tertiary structure.

Because the presence of misfolded, nonglycosylated proteins does not inhibit glucose transfer to denatured thyroglobulin (124), it was hypothesized that GT recognizes the protein and the glycan moieties together. This was supported by the observation that the presence of the innermost GlcNAc residue of the N-linked glycan was sufficient to make a denatured protein inhibitory to the enzyme. The tendency of GT to bind to hydrophobic peptides in the absence of glycans argues, on the other hand, that GT can recognize and bind to proteins in the absence of glycans (124). Recent studies in our group have shown, furthermore, that misfolded RNase A devoid of glycans interferes with GT activity, suggesting sugar-independent interactions (C. Ritter, K. Quirin, and A. Helenius, unpublished results).

In studies with RNase B dimers, it was found that GT reglucosylates glycans that are located in a misfolded domain while ignoring glycans in a nearby identical but folded domain of the same protein (121). When local misfolding was induced by mutating one of the disulfide bonds, only glycans in the misfolded region were glucosylated, suggesting highly localized glucosylation (C. Ritter, K. Quirin, and A. Helenius, unpublished results).

In summary, current data indicate that GT interacts with glycoproteins and possibly nonglycoproteins but only if they are improperly folded. If the incompletely folded regions contain high-mannose glycans, these are selectively reglucosylated. The enzyme recognizes relatively small local folding defects, and the presence of exposed hydrophobic residues improves recognition. A partially ordered, molten globule-like structure seems to be optimal for efficient recognition.

UDPase

GT uses as a substrate a nucleotide sugar, UDP-Glc, transported into the ER lumen from the cytosol. An import activity for UDP-Glc has been described in the ER of rat liver and in *S. cerevisiae*, and it has been established that import of UDP-Glc is coupled to the exit of UMP (125). To make use of this antiporter, the ER must first convert the UDP generated by GT to UMP. In the secretory pathway of mammalian cells, three nucleoside diphosphatases are known, two in the ER and one in the Golgi complex.

Two phosphatases have been reported in the ER, a soluble and a membrane-bound enzyme (126, 127). The soluble UDPase is a glycoprotein that needs Ca^{2+}, Mg^{2+}, or Mn^{2+} to work, whereas the membrane-bound UDPase only functions with Ca^{2+}. Both enzymes belong to the ecto-nucleoside triphosphate diphosphohydrolase (E-NTPDase) family, and hydrolyze UDP, GDP, and IDP but not nucleoside triphosphates or ADP. Although ubiquitously expressed, the two enzymes have somewhat different tissue distribution. The soluble UDPase is enriched in the ER, but it has no known ER retention sequence. When expressed in Cos-7 cells, it is secreted (128). The membrane-bound enzyme occurs on the ER and the intermediate compartment. A RXRXR sequence at the N terminus may serve as an ER retention sequence. The protein is anchored to the membrane via sequences close to the N terminus and has its active domain in the ER lumen.

ER-DEPENDENT DEGRADATION OF GLYCOPROTEINS

As outlined above, N-linked oligosaccharides play an essential role in the folding of glycoproteins in the ER. If glycoproteins fail to fold or to oligomerize, they are retained in the ER and eventually degraded. The degradation process, termed ER-associated degradation (ERAD), is important because it prevents accumulation of unsalvageable, misfolded proteins in the ER.

ERAD involves three functionally distinct steps: the recognition of a glycoprotein as malfolded, its retrotranslocation to the cytoplasm, and the subsequent, ubiquitin-dependent degradation by the proteasome. Several excellent articles have recently reviewed aspects of this process (129–132). Here, we will focus exclusively on the recognition step.

Targeting of substrates to ERAD is a delicate process. Misfolded and unassembled protein subunits should be degraded, but bona fide folding intermediates should not. Current data shows that such a distinction does indeed take place. It is based on the length of time that a glycoprotein has spent in the ER. In mammalian cells, degradation of a misfolded glycoprotein typically starts after a lag period of 30–90 min followed by exponential decay (133–135). This means that cells give newly synthesized proteins a fair chance to fold and assemble before degradation sets in.

The use of a timer mechanism does, however, create a problem: how to spare resident ER glycoproteins destined to stay permanently in the ER? Unless misfolded, they should not be targeted for destruction after the initial lag. It is possible that the system does not rely exclusively on a timer but also on a folding sensor (136).

The delay in degradation of glycoproteins by ERAD is clearly linked to trimming of mannoses. The most important mannosidase is an $\alpha-1,2$ exomannosidase, a membrane-bound ER enzyme that specifically removes the terminal mannose (residue i) from the B branch of the oligosaccharide to yield the $Man_8GlcNAc_2$ B-isomer (Figure 1) (137). If the mannosidase is inhibited by kifunensin (a specific inhibitor) or mutated, glycoprotein degradation is dramatically slower (138). In contrast, if ER mannosidase I is overexpressed, the onset of degradation is accelerated (139). Because the activity of this and other mannosidases in the ER is relatively low (136), the mannoses are removed over a period that varies from 10 min in yeast to more than an hour in some mammalian cells. The slow-acting mannosidase is likely to serve as the timer that protects newly synthesized proteins and nascent chains. That the lag times vary greatly for different proteins may depend on the number of glycans, their locations in the protein, and other effects.

Genetic evidence in *S. cerevisiae* suggests that the $Man_8GlcNAc_2$ B-isomer generated by ER mannosidase I is recognized by a membrane-bound ER protein called Htm1/Mnl1 (140, 141). Htm1/Mnl1 is a mannosidase homologue, but because it does not have detectable mannosidase activity, it is thought to serve as a mannose lectin and to be responsible for directing glycoproteins into the retrotranslocation and degradation pathway. Elimination of this membrane protein retards ERAD of glycoproteins but not of nonglycosylated proteins (140). The generation of $Man_8GlcNAc_2$ B-isomer is required for efficient ERAD, yet it is not sufficient because only malfolded proteins with $Man_8GlcNAc_2$ are degraded (136).

In mammalian cells, ER mannosidase I, the homologue of the yeast mannosidase, plays a similar role. Glycoprotein degradation is dramatically reduced by kifunensin, and when overexpressed, a homologue of Htm1p/Mnl1p, called EDEM [for ER degradation-enhancing α-mannosidase-like protein (142)], accelerates the degradation of malfolded glycoproteins (134, 143). The presence of a functional calnexin/calreticulin cycle makes the system more complex than in yeast. In some cases, the calnexin cycle seems to protect glycoproteins against degradation during the initial lag period. If glycoproteins are not allowed to enter the calnexin cycle, ERAD starts without lag (134, 135). It is likely that once mannose trimming has occurred, EDEM begins competing for substrates with the calnexin/calreticulin cycle. Recall that when mannoses are lost from the B and C branches, both GT and glucosidase II become less efficient (52, 119). The substrate is therefore likely to pass more slowly through the cycle and may in this way give EDEM an advantage. It has also been shown that EDEM and calnexin coimmunoprecipitate, suggesting that they associate with each other in a complex (143) It is possible that EDEM acquires substrate glycoproteins directly from calnexin.

The role of additional mannosidases is not yet fully clarified. In addition to the ER $\alpha-1,2$-mannosidase I, mammalian cells have an ER mannosidase II that removes a mannose (residue k) from the C branch (Figure 1) (137). Further $\alpha-1,2$ mannosidases of the same enzyme family occur in the Golgi of mammalian cells. Because some of the mannosidases preferentially act on either the A or the C arm (137), they could, if present in the ER, be used to extract glycoproteins from the calnexin cycle or prevent ERAD. Indeed, it has been recently reported that ERAD substrates can be trimmed to $Man_7GlcNAc_2$ or even $Man_5GlcNAc_2$, suggesting exposure to several mannosidases (144, 145). These data suggest that a degradation pathway independent of the $Man_8GlcNAc_2$ isomer B exists in mammalian cells (146, 147).

It is noteworthy that most mammalian cells have in the ERGIC and in the Golgi complex an endomannosidase that cleaves the linkage between the first glucose-substituted mannose (g) and the second mannose (f) in the A branch (Figure 1) (18). This enzyme is thought to remove the glucose from monoglucosylated chains that have somehow escaped the action of ER glucosidases and thus permit normal processing of the glycans in the Golgi complex.

OTHER LECTINS IN THE SECRETORY PATHWAY

Lectins Related to Mannose 6-Phosphate Receptors

That the secretory pathway contains functionally important lectins was first recognized during an effort to clarify the molecular causes of the hereditary disease mucolipidosis II (I-cells disease). It was realized that modified N-linked glycans are address tags in the intracellular targeting of lysosomal enzymes and that lectins serve as receptors (148). The lectins in question are the two mannose 6-phosphate binding receptors (MPRs) that transport lysosomal enzymes from the *trans*-Golgi network to endosomes. They have been thoroughly analyzed with respect to function and structure (149).

The crystal structure of the lectin domain in Ca^{2+}-dependent (CD)-MPR, the smaller of the receptors, shows a flattened β-barrel structure with three disulfide bonds (150). The carbohydrate binding site is located in a relatively deep pocket formed by the β-sheets and the loops that connect them. The site contains a bound Mn^{2+} that interacts with the phosphate group of mannose 6-phosphate. Binding of the ligand causes a conformational change in the receptor, and binding is reversed by mildly acidic pH.

Analysis of sequence databases recently allowed Munro (101) to identify additional MPR-like domains in a variety of genes and proteins. He classified these as members of a family and called them mannose 6-phosphate receptor homology domains (MRHs). The conservation of sequences in the region of the glycan binding pocket in CD-MPR suggested that they may also be lectins. Consistent with this, he observed that in addition to genes of unknown function, the proteins with MRH domains include two N-glycan modifying enzymes: the ER glucosidase II β-subunit discussed above and the lysosomal hydrolase N-acetylglucosamine-1-phosphotransferase (GlcNAc-phosphotransferase) γ-subunit (151). This is the *cis*-Golgi enzyme responsible for adding GlcNAc-phosphate to lysosomal hydrolases. Whether these enzyme-associated MRH domains act as lectins remains to be determined. Loss of the γ-subunit leads to a defect in mannose 6-phosphate signal generation, mistargeting of lysosomal hydrolases, and a variant of pseudohurler polydystrophy (mucolipidosis IIIC) (152).

Leguminous Lectin Homologues

In addition to calnexin and calreticulin, the early secretory pathway contains several other lectins that have been classified as members of the leguminous lectin family. Best characterized is the ubiquitously expressed and abundant ERGIC-53 (p58, MR60), a homo-oligomeric, type I transmembrane protein with mannose lectin activity (153). The X-ray crystal structure of the carbohydrate-binding domain shows the β-sandwich structure characteristic for this lectin family (68). No long appendices like the P domain in calnexin and calreticulin are present.

Equipped with a complex repertoire of cytoplasmic tail signals for interaction with COPI and COPII coats, ERGIC-53 cycles between the ERGIC and the *cis*-Golgi. Although it has been difficult to define physiological ligands, it is apparent that ERGIC-53 is responsible for ER-to-Golgi export of a subset of secretory and lysosomal glycoproteins. Association with cathepsin Z, a lysosomal enzyme, seems to be dependent on glucose trimming, implying that

ERGIC-53 binds to glycoproteins that have successfully emerged from the calnexin/calreticulin cycle (154). Mutations in ERGIC-53 can, in humans, result in an autosomal recessive bleeding disorder caused by a deficiency in coagulation factors V and VIII (F5F8D) (155).

Vesicular integral protein 36 (VIP36) is another membrane-bound leguminous lectin concentrated primarily in the early secretory pathway (156–158). This lectin is homologous with ERGIC-53 but lacks the coiled-coil structure in the stem. It is specific for high-mannose type N-glycans containing β-1,2 mannoses, and it is thought to mediate forward transport of glycoproteins in the secretory pathway, and possibly their targeting to the apical plasma membrane in polarized epithelial cells.

Examination of genomic and EST databases has led to the identification of additional mammalian proteins with homology to ERGIC-53 and VIP36 (159, 160). The homologues have been classified into three main phylogenetic groups: the VIP36/VIPL, the ERGIC-53/ERGL, and a group of fungal homologues that include Emp46p and Emp47p in *S. cerevisiae* (159). Of these, VIPL is a highly conserved VIP36-like protein with orthologues in many species from humans to Xenopus (160). Similar to VIP36, it lacks the coiled stem domain present in ERGIC-53 and in the fungal homologues and is therefore less likely to assemble into an oligomer. It has a cytosolic ER retention motif and is mainly located in the ER. Although the primary sequence suggests that it is a mannose-binding lectin, lectin activity has not been experimentally confirmed for any of these proteins except ERGIC-53 and VIP36. Overexpression of VIPL interferes with ERGIC-53 cycling, suggesting that it may regulate ERGIC-53 function (160). Knockdown experiments have revealed retardation in the export of a subset of proteins (87, 122, 159). Emp46p and Emp47p are homologous membrane proteins in the early secretory pathway of *S. cerevisiae* that cause a partial secretion defect when disrupted (161).

Taken together, it is apparent that proteins with lectin domains occur throughout the secretory pathway. They belong to several families, and many of them are likely to bind mannose (162). Their functions are related to folding, quality control, sorting, and forward transport of glycoprotein cargo. Because they are functionally redundant, and their affinities for their ligands are weak, their biochemical characterization is challenging. However, they are an interesting new class of proteins that have their functions in the lumen of secretory organelles promoting sorting, cargo selection, and efficiency.

N-GLYCANS: THE EVOLUTION OF A SIGNAL

The functions of N-glycans can be further illuminated by considering the evolutionary origin of ER glycosylation. As in most functions of the ER, the synthesis of N-linked glycans stems from homologous processes in the plasma membrane of bacteria or archaea (11, 163). Proteins with N-linked glycans are, in fact, present in the outermost layer of the archaeal cell wall (164) and in the cell wall of certain gram-negative bacteria, such as *Campylobacter jejuni* (35, 165). The synthesis of prokaryotic glycoproteins shows such striking similarities to the process in eukaryotic cells that there can be no doubt that the processes are homologous.

In archaea, the oligosaccharide moiety is assembled at the cytoplasmic face of the plasma membrane on a lipid carrier, dolichylphosphate or dolichylpyrophosphate. It is then translocated through the plasma membrane and enzymically transferred to a protein on the outside surface of the plasma membrane. The use of a lipid-bound oligosaccharide precursor makes sense in these organisms, because a soluble precursor would diffuse away from the cell. The acceptor sequence in the polypeptide is none other than the familiar Asn-X-Ser/Thr (166). In bacteria, the oligosaccharide precursors are probably bound to bactoprenol, an isoprenoid lipid very similar to dolichol. A homologue of the STT3 subunit of OST in *C. jejuni* is essential and sufficient for oligosaccharide transfer to protein (35; M. Feldman, M. Wacker, and M. Aebi, unpublished results), and Asn-X-Ser/Thr serves as an acceptor sequence in the bacterial system.

Although the membrane topology and oligosaccharide transfer mechanisms are similar, there are differences between eukaryotes and prokaryotes. The limited information available suggests, for example, that there is great diversity among oligosaccharides transferred to proteins in different prokaryote species (164, 167, 168). In contrast, almost all eukaryotes transfer the same structure, $Glc_3Man_9GlcNAc_2$ (Figure 1). A notable exception is within the clade of the trypanosomatids, primitive eukaryotes, in which the $Man_{6-9}GlcNAc_2$ oligosaccharides are transferred to proteins (169, 170). These organisms lack specific Dol-P-Man or Dol-P-Glc-dependent mannosyl- and glycosyltransferase required for the assembly of the complete $Glc_3Man_9GlcNAc_2$, but they do have GT.

Thus, we can view the $Glc_3Man_9GlcNAc_2$ oligosaccharide as a bipartite structure in the evolutionary sense: The $Man_5GlcNAc_2$ oligosaccharide, assembled in the cytoplasm, probably corresponds to the glycan transferred to protein in the archaeal ancestor of all eukaryotic cells, whereas the three glucose and the four mannose residues, added in the lumen of the ER, are extensions that have occurred during eukaryotic evolution. The situation in trypanosomatids might represent an evolutionary intermediate.

The $Man_5GlcNAc_2$ oligosaccharide (residues a-g) (Figure 1) is closer to the protein and functionally distinct. These sugars seem to be primarily responsible for the direct effects that N-linked glycans have in protein folding (3, 6). Five of them escape trimming in the ER and the Golgi complex when complex glycans are synthesized.

We speculate that the addition of further saccharides (mannoses h, i, j, and k, and glucoses m, n, and l) to the archaeal $Man_5GlcNAc_2$ oligosaccharide during eukaryotic evolution was driven by the internalization of glycoprotein biosynthesis from the plasma membrane to the ER and by the concomitant need to export newly synthesized proteins to the cell surface. Internalization made it possible to sequester biosynthesis, folding, and early maturation in a closed compartment in which conditions could be better controlled than in a space connected to the extracellular medium. It also opened new possibilities to increase the volume of production and to expand the product repertoire to include more complex proteins and proteins that required assistance from a variety of chaperones. The disadvantages included the need to control the fidelity of protein maturation and necessity to deal with misfolded proteins, an inevitable side effect of protein folding and oligomeric assembly. The distal mannose and glucose residues (h-n) of the $Glc_3Man_9GlcNAc_2$ oligosaccharide were introduced to serve as recognition tags in the calnexin/calreticulin cycle during quality control and during substrate selection for ERAD. These distal sugars made it possible for the cell to use the N-linked glycans as a composite signal.

THE N-LINKED GLYCAN, A COMPOSITE, MULTIFUNCTIONAL SIGNAL

The different signaling functions of the protein-bound core oligosaccharide are organized in such a way that the A branch provides information about the folding status, and the B and C arms are used for ERAD (Figure 1). The signals are interpreted by specific lectins: calnexin and calreticulin in the case of the A branch and EDEM/HTM1 in the case of the B and C arms. As discussed above, efficient binding of calnexin to the A branch requires the presence of the α-1,3-linked glucose residue (l); however, the three mannose residues g, f, and d significantly enhance the affinity. The monoglucosylated A branch acts as a positive signal for binding. However, this signal can be efficiently masked by the presence of the additional α-1,3-linked glucose (m). This residue functions as a negative signal for calnexin and calreticulin binding.

Similarly the three mannose residues (h, k, and l) of the B and C arm of the oligosaccharide may act as an optimal substrate for EDEM and as a positive signal in the ERAD pathway. Indeed, the structure of the EDEM homologue ER mannosidase I with a bound Man_8 oligosaccharide (171) is compatible with this hypothesis. Again, genetic evidence suggests that the α-1,2-linked mannose residue i of the B branch acts as a negative signal for EDEM/HTM1 binding.

On the basis of these considerations, we can now propose functions for each of the individual trimming steps in the ER. The outermost α-1,2 glucose residue (n) is needed for efficient recognition of the processor oligosaccharide by OST. OST actually sees the complete A branch, but glucose n seems to be a key residue (16). Hydrolysis of this glucose residue is likely to promote release of the glycoprotein substrate from OST and prevent rebinding of the substrate after transfer to a polypeptide chain. Removal of this glucose, furthermore, exposes the α-1,3-linked glucose residue (m), which as discussed above serves as a negative signal for binding to calnexin and calreticulin. Only after removal of m is the critical third glucose (l) exposed terminally, allowing association of calnexin and calreticulin with the A branch. Similarly, removal of mannose is likely to expose a branched oligomannose structure for binding to EDEM/HTM1. Our guess would be that EDEM binds a tetra mannose unit of h, e, j, and k with h as the key residue. Loss of mannoses from the C branch may weaken the interaction (136).

PERSPECTIVES

The ER is a highly specialized compartment. Although hard numbers are missing, it is estimated that many of the chaperones and folding enzymes are present in nearly millimolar concentrations. With such extreme crowding, it is not surprising that the resident proteins tend to form complexes and extensive networks (173–175). It is a world where affinities tend to be low (in the micro- to millimolar range), and molecular interactions transient. The spectrum of chaperones, lectins, and enzymes is delicately balanced to best serve the secretory process of each specific cell type under physiological conditions (176–178). In most cell types, the conditions are optimized for the synthesis of glycoproteins.

Among the challenges in this field are a better grasp of the complexities of the physiological context and more complete understanding of the rules that prevail in this compartment. It is important to learn to manipulate the conditions, to develop cures for various ER-storage diseases, to control the production of glycoproteins, and to explore in further detail the complex pathways by which proteins mature in the ER.

The analysis of glycoprotein folding and quality control in the ER has opened a window into the intracellular functions of oligosaccharides. One of the most important functions is clearly to secure efficient protein production. In the eukaryotic cell, the fidelity of this biosynthetic process has top priority. What might have started as a simple structural component of the extracellular matrix of an archaebacterium has evolved into a highly coded recognition signal whose idiom is now modulated and interpreted by a spectrum of enzymes and lectins in the ER and other intracellular compartments.

ACKNOWLEDGMENTS

We thank Lars Ellgaard, Matthias Gautschi, Paola Deprez, Sergio Trombetta, and Claude Jakob for advice. Work in our labs was supported by the Swiss National Science Foundation and ETH Zurich.

The *Annual Review of Biochemistry* is online at http://biochem.annualreviews.org

LITERATURE CITED

1. Kornfeld R, Kornfeld S. 1985. *Annu. Rev. Biochem.* 54:631–64
2. Varki A. 1993. *Glycobiology* 3:97–130
3. Trombetta ES. 2003. *Glycobiology* 13: R77–91
4. Apweiler R, Hermjakob H, Sharon N. 1999. *Biochim. Biophys. Acta* 1473:4–8
5. Wormald MR, Petrescu AJ, Pao YL, Glithero A, Elliott T, Dwek RA. 2002. *Chem. Rev.* 102:371–86
6. Petrescu AJ, Milac A-L, Petrescu SM, Dwek RA, Wormald MR. 2004. *Glycobiology* 14:103–14
7. Rudd PM, Wormald MR, Stanfield RL, Huang M, Mattsson N, et al. 1999. *J. Mol. Biol.* 293:351–66
8. Lowe JB, Marth JD. 2003. *Annu. Rev. Biochem.* 72:643–91
9. Weis WI, Drickamer K. 1996. *Annu. Rev. Biochem.* 65:441–73
10. Lis H, Sharon N. 1998. *Chem. Rev.* 98: 637–74
11. Burda P, Aebi M. 1999. *Biochim. Biophys. Acta* 1426:239–57
12. Hirschberg CB, Snider MD. 1987. *Annu. Rev. Biochem.* 56:63–87
13. Helenius J, Ng DTW, Marolda CL, Walter P, Valvano MA, Aebi M. 2002. *Nature* 415:447–50
14. Balasubramanian K, Schroit AJ. 2003. *Annu. Rev. Physiol.* 65:701–34
15. Helenius J, Aebi M. 2002. *Semin. Cell Dev. Biol.* 13:171–78
16. Burda P, Jakob CA, Beinhauer J, Hegemann JH, Aebi M. 1999. *Glycobiology* 9:617–25
17. Burda P, Aebi M. 1998. *Glycobiology* 8: 455–62
18. Spiro RG. 2000. *J. Biol. Chem.* 275: 35657–60
19. Nilsson IM, von Heijne G. 1993. *J. Biol. Chem.* 268:5798–801
20. Kowarik M, Kung S, Martoglio B, Helenius A. 2002. *Mol. Cell* 10:769–78
21. Bause E. 1983. *Biochem. J.* 209:331–36
22. Imperiali B, Shannon KL, Unno M, Rickert KW. 1992. *J. Am. Chem. Soc.* 114:7944–45
23. Knauer R, Lehle L. 1999. *J. Biol. Chem.* 274:17249–56
24. Karaoglu D, Kelleher DJ, Gilmore R. 1995. *J. Cell Biol.* 130:567–78
25. Karaoglu D, Kelleher DJ, Gilmore R. 1997. *J. Biol. Chem.* 272:32513–20
26. Dempski RE Jr, Imperiali B. 2002. *Curr. Opin. Chem. Biol.* 6:844–50
27. Knauer R, Lehle L. 1999. *Biochim. Biophys. Acta* 1426:259–73
28. Kelleher DJ, Karaoglu D, Mandon EC, Gilmore R. 2003. *Mol. Cell* 12:101–11
29. Karaoglu D, Kelleher D, Gilmore R. 2001. *Glycobiology* 11:54
30. Yan G, Lennarz WJ. 2002. *J. Biol. Chem.* 277:47692–700
31. Nilsson I, Kelleher DJ, Miao YW, Shao YL, Kreibich G, et al. 2003. *J. Cell Biol.* 161:715–25
32. Yan Q, Prestwich GD, Lennarz WJ. 1999. *J. Biol. Chem.* 274:5021–25
33. Pathak R, Hendrickson TL, Imperiali B. 1995. *Biochemistry* 34:4179–85
34. Burda P, te Heesen S, Brachat A, Wach A, Düsterhöft A, Aebi M. 1996. *Proc. Natl. Acad. Sci. USA* 93:7160–65
35. Wacker M, Linton D, Hitchen PG, Nita-Lazar M, Haslam SM, et al. 2002. *Science* 298:1790–93
36. Bergman LW, Kuehl WM. 1979. *J. Biol. Chem.* 254:8869–76
37. Chen W, Helenius J, Braakman I, Helenius A. 1995. *Proc. Natl. Acad. Sci. USA* 92:6229–33
38. Daniels R, Kurowski B, Johnson AE, Hebert DN. 2003. *Mol. Cell* 11:79–90
39. Fassio A, Sitia R. 2002. *Histochem. Cell Biol.* 117:151–57
40. Olden K, Parent JB, White SL. 1982. *Biochim. Biophys. Acta* 650:209–32
41. Hurtley SM, Helenius A. 1989. *Annu. Rev. Cell Biol.* 5:277–307
42. Klausner RD, Sitia R. 1990. *Cell* 62: 611–14

43. Helenius A. 1994. *Mol. Biol. Cell* 5: 253–65

44. Hebert DN, Zhang JX, Chen W, Foellmer B, Helenius A. 1997. *J. Cell Biol.* 139:613–23

45. Imperiali B, O'Connor SE. 1999. *Curr. Opin. Chem. Biol.* 3:643–49

46. O'Connor SE, Imperiali B. 1996. *Chem. Biol.* 3:803–12

47. Imberty A, Perez S. 1995. *Protein Eng.* 8:699–709

48. Wormald MR, Dwek RA. 1999. *Struct. Fold. Des.* 7:R155–60

49. Kundra R, Kornfeld S. 1999. *J. Biol. Chem.* 274:31039–46

50. DeKoster GT, Robertson AD. 1997. *Biochemistry* 36:2323–31

51. Suh K, Bergmann JE, Gabel CA. 1989. *J. Cell Biol.* 108:811–19

52. Sousa MC, Ferrero-Garcia MA, Parodi AJ. 1992. *Biochemistry* 31:97–105

53. Ou WJ, Cameron PH, Thomas DY, Bergeron JJ. 1993. *Nature* 364:771–76

54. Hammond C, Helenius A. 1994. *Science* 266:456–58

55. Hammond C, Helenius A. 1993. *Curr. Biol.* 3:884–85

56. Nauseef WM, McCormick SJ, Clark RA. 1995. *J. Biol. Chem.* 270:4741–47

57. Peterson JR, Ora A, Van PN, Helenius A. 1995. *Mol. Biol. Cell* 6:1173–84

58. Ellgaard L, Molinari M, Helenius A. 1999. *Science* 286:1882–88

59. Parodi A. 2000. *Annu. Rev. Biochem.* 69: 69–93

60. Schrag JD, Bergeron JJ, Li Y, Borisova S, Hahn M, et al. 2001. *Mol. Cell* 8: 633–44

61. Chevet E, Cameron PH, Pelletier MF, Thomas DY, Bergeron JJ. 2001. *Curr. Opin. Struct. Biol.* 11:120–24

62. Ellgaard L, Helenius A. 2003. *Nat. Rev. Mol. Cell Biol.* 4:181–91

63. Denzel A, Molinari M, Trigueros C, Martin JE, Velmurgan S, et al. 2002. *Mol. Cell. Biol.* 22:7398–404

64. Michalak M, Lynch J, Groenendyk J, Guo L, Parker JMR, Opas M. 2002. *Biochim. Biophys. Acta* 1600:32–37

65. Ellgaard L, Riek R, Braun D, Herrmann T, Helenius A, Wuthrich K. 2001. *FEBS Lett.* 488:69–73

66. Chevet E, Wong HN, Gerber D, Cochet C, Fazel A, et al. 1999. *EMBO J.* 18: 3655–66

67. Bouvier M, Stafford WF. 2000. *Biochemistry* 39:14950–59

68. Velloso LM, Svensson K, Schneider G, Pettersson RF, Lindqvist Y. 2002. *J. Biol. Chem.* 277:15979–84

69. Rudenko G, Nguyen T, Chelliah Y, Sudhof TC, Deisenhofer J. 1999. *Cell* 99: 93–101

70. Loris R. 2002. *Biochim. Biophys. Acta* 1572:198–208

71. Ellgaard L, Riek R, Herrmann T, Guntert P, Braun D, et al. 2001. *Proc. Natl. Acad. Sci. USA* 98:3133–38

72. Ware FE, Vassilakos A, Peterson PA, Jackson MR, Lehrman MA, Williams DB. 1995. *J. Biol. Chem.* 270:4697–704

73. Vassilakos A, Michalak M, Lehrman MA, Williams DB. 1998. *Biochemistry* 37:3480–90

74. Spiro RG, Zhu Q, Bhoyroo V, Soling HD. 1996. *J. Biol. Chem.* 271:11588–94

75. Patil AR, Thomas CJ, Surolia A. 2000. *J. Biol. Chem.* 275:24348–56

76. Kapoor M, Srinivas H, Kandiah E, Gemma E, Ellgaard L, et al. 2003. *J. Biol. Chem.* 278:6194–200

77. Kapoor M, Ellgaard L, Gopalakrishnapai J, Schirra C, Gemma E, et al. 2004. *Biochemistry* 43:97–106

78. Ihara Y, Cohen-Doyle MF, Saito Y, Williams DB. 1999. *Mol. Cell* 4:331–41

79. Saito Y, Ihara Y, Leach MR, Cohen-Doyle MF, Williams DB. 1999. *EMBO J.* 18:6718–29

80. Danilczyk UG, Williams DB. 2001. *J. Biol. Chem.* 276:25532–40

81. Rodan AR, Simons JF, Trombetta ES, Helenius A. 1996. *EMBO J.* 15:6921–30

82. Zapun A, Petrescu SM, Rudd PM, Dwek

RA, Thomas DY, Bergeron JJM. 1997. *Cell* 88:29–38

83. Corbett EF, Michalak KM, Oikawa K, Johnson S, Campbell ID, et al. 2000. *J. Biol. Chem.* 275:27177–85

84. Ou WJ, Bergeron JJ, Li Y, Kang CY, Thomas DY. 1995. *J. Biol. Chem.* 270: 18051–59

85. Papp S, Opas M. 2003. In *Calreticulin*, ed. P Eggleton, M Michalak, pp. 38–45. Georgetown, TX: Landes Biosci.

86. Hirano N, Shibasaki F, Sakai R, Tanaka T, Nishida J, et al. 1995. *Eur. J. Biochem.* 234:336–42

87. Molinari M, Helenius A. 1999. *Nature* 402:90–93

88. Antoniou AN, Ford S, Alphey M, Osborne A, Elliott T, Powis SJ. 2002. *EMBO J.* 21:2655–63

89. Elliott JG, Oliver JD, High S. 1997. *J. Biol. Chem.* 272:13849–55

90. Oliver JD, Roderick HL, Llewellyn DH, High S. 1999. *Mol. Biol. Cell* 10: 2573–82

91. Frickel EM, Riek R, Jelesarov I, Helenius A, Wuthrich K, Ellgaard L. 2002. *Proc. Natl. Acad. Sci. USA* 99:1954–59

92. Leach MR, Cohen-Doyle MF, Thomas DY, Williams DB. 2002. *J. Biol. Chem.* 277:29686–97

93. Elbein AD. 1983. *Methods Enzymol.* 98: 135–54

94. De Praeter CM, Gerwig GJ, Bause E, Nuytinck LK, Vliegenthart JF, et al. 2000. *Am. J. Hum. Genet.* 66:1744–56

95. Boisson M, Gomord V, Audran C, Berger N, Dubreucq B, et al. 2001. *EMBO J.* 20:1010–19

96. Brada D, Dubach UC. 1984. *Eur. J. Biochem.* 141:149–56

97. Hettkamp H, Legler G, Bause E. 1984. *Eur. J. Biochem.* 142:85–90

98. Trombetta ES, Simons JF, Helenius A. 1996. *J. Biol. Chem.* 271:27509–16

99. Henrissat B, Bairoch A. 1993. *Biochem. J.* 293(Pt. 3):781–88

100. Arendt CW, Dawicki W, Ostergaard HL. 1999. *Glycobiology* 9:277–83

101. Munro S. 2001. *Curr. Biol.* 11: R499–501

102. D'Alessio C, Fernandez F, Trombetta ES, Parodi AJ. 1999. *J. Biol. Chem.* 274: 25899–905

103. Pelletier MF, Marcil A, Sevigny G, Jakob CA, Tessier DC, et al. 2000. *Glycobiology* 10:815–27

104. Treml K, Meimaroglou D, Hentges A, Bause E. 2000. *Glycobiology* 10: 493–502

105. Hebert DN, Foellmer B, Helenius A. 1996. *EMBO J.* 15:2961–68

106. Branza-Nichita N, Petrescu AJ, Dwek RA, Wormald MR, Platt FM, Petrescu SM. 1999. *Biochem. Biophys. Res. Commun.* 261:720–25

107. Machold RP, Ploegh HL. 1996. *J. Exp. Med.* 184:2251–59

108. Zhang JX, Braakman I, Matlack KE, Helenius A. 1997. *Mol. Biol. Cell* 8: 1943–54

109. Labriola C, Cazzulo JJ, Parodi AJ. 1995. *J. Cell Biol.* 130:771–79

110. Drenth JP, te Morsche RH, Smink R, Bonifacino JS, Jansen JB. 2003. *Nat. Genet.* 33:345–47

111. Li AR, Davila S, Furu L, Qian Q, Tian X, et al. 2003. *Am. J. Hum. Genet.* 72: 691–703

112. Trombetta S, Bosch M, Parodi AJ. 1989. *Biochemistry* 28:8108–16

113. Parker CG, Fessler LI, Nelson RE, Fessler JH. 1995. *EMBO J.* 14:1294–303

114. Arnold SM, Fessler LI, Fessler JH, Kaufman RJ. 2000. *Biochemistry* 39: 2149–63

115. Roth J. 2002. *Chem. Rev.* 102:285–303

116. Tessier DC, Dignard D, Zapun A, Radominska-Pandya A, Parodi AJ, et al. 2000. *Glycobiology* 10:403–12

117. Arnold SM, Kaufman RJ. 2003. *J. Biol. Chem.* 278:43320–28

118. Guerin M, Parodi AJ. 2003. *J. Biol. Chem.* 278:20540–46

119. Grinna LS, Robbins PW. 1980. *J. Biol. Chem.* 255:2255–58

120. Trombetta ES, Helenius A. 2000. *J. Cell Biol.* 148:1123–29

121. Ritter C, Helenius A. 2000. *Nat. Struct. Biol.* 7:278–80

122. Caramelo JJ, Castro OA, Alonso LG, De Prat-Gay G, Parodi AJ. 2003. *Proc. Natl. Acad. Sci. USA* 100:86–91

123. Taylor SC, Thibault P, Tessier DC, Bergeron JJ, Thomas DY. 2003. *EMBO Rep.* 4:405–11

124. Sousa M, Parodi AJ. 1995. *EMBO J.* 14:4196–203

125. Perez M, Hirschberg CB. 1986. *J. Biol. Chem.* 261:6822–30

126. Trombetta SE, Parodi AJ. 1992. *J. Biol. Chem.* 267:9236–40

127. Failer BU, Braun N, Zimmermann H. 2002. *J. Biol. Chem.* 277:36978–86

128. Mulero JJ, Yeung G, Nelken ST, Ford JE. 1999. *J. Biol. Chem.* 274:20064–67

129. Tsai B, Ye Y, Rapoport TA. 2002. *Nat. Rev. Mol. Cell Biol.* 3:246–55

130. Kostova Z, Wolf DH. 2003. *EMBO J.* 22:2309–17

131. McCracken AA. 2003. *BioEssays* 25:868–77

132. Yoshida Y. 2003. *J. Biochem.* 134:183–90

133. Lippincott-Schwartz J, Bonifacino JS, Yuan LC, Klausner RD. 1988. *Cell* 54:209–20

134. Molinari M, Calanca V, Galli C, Lucca P, Paganetti P. 2003. *Science* 299:1397–400

135. Mancini R, Aebi M, Helenius A. 2003. *J. Biol. Chem.* 278:46895–905

136. Jakob CA, Burda P, Roth J, Aebi M. 1998. *J. Cell Biol.* 142:1223–33

137. Herscovics A. 2001. *Biochimie* 83:757–62

138. Liu Y, Choudhury P, Cabral CM, Sifers RN. 1999. *J. Biol. Chem.* 274:5861–67

139. Wu Y, Swulius MT, Moremen KW, Sifers RN. 2003. *Proc. Natl. Acad. Sci. USA* 100:8229–34

140. Jakob CA, Bodmer D, Spirig U, Battig P, Marcil A, et al. 2001. *EMBO Rep.* 2:423–30

141. Nakatsukasa K, Nishikawa S, Hosokawa N, Nagata K, Endo T. 2001. *J. Biol. Chem.* 276:8635–38

142. Hosokawa N, Wada I, Hasegawa K, Yorihuzi T, Tremblay LO, et al. 2001. *EMBO Rep.* 2:415–22

143. van Anken E, Romijn EP, Maggioni C, Mezghrani A, Sitia R, et al. 2003. *Immunity* 18:243–53

144. Frenkel Z, Gregory W, Kornfeld S, Lederkremer GZ. 2003. *J. Biol. Chem.* 278:34119–24

145. Kitzmuller C, Caprini A, Moore SE, Frenoy JP, Schwaiger E, et al. 2003. *Biochem. J.* 376:687–96

146. Ermonval M, Duvet S, Zonneveld D, Cacan R, Buttin G, Braakman I. 2000. *Glycobiology* 10:77–87

147. Cabral CM, Choudhury P, Liu Y, Sifers RN. 2000. *J. Biol. Chem.* 275:25015–22

148. Kornfeld S. 1987. *FASEB J.* 1:462–68

149. Ghosh P, Dahms NM, Kornfeld S. 2003. *Nat. Rev. Mol. Cell Biol.* 4:202–12

150. Dahms NM, Hancock MK. 2002. *Biochim. Biophys. Acta* 1572:317–40

151. Bao M, Booth JL, Elmendorf BJ, Canfield WM. 1996. *J. Biol. Chem.* 271:31437–45

152. Raas-Rothschild A, Cormier-Daire V, Bao M, Genin E, Salomon R, et al. 2000. *J. Clin. Investig.* 105:673–81

153. Hauri HP, Kappeler F, Andersson H, Appenzeller C. 2000. *J. Cell Sci.* 113:587–96

154. Appenzeller C, Andersson H, Kappeler F, Hauri HP. 1999. *Nat. Cell Biol.* 1:330–34

155. Zhang B, Cunningham MA, Nichols WC, Bernat JA, Seligsohn U, et al. 2003. *Nat. Genet.* 34:220–25

156. Fiedler K, Simons K. 1994. *Cell* 77:625–26

157. Fullekrug J, Scheiffele P, Simons K. 1999. *J. Cell Sci.* 112:2813–21

158. Hara-Kuge S, Ohkura T, Seko A, Yamashita K. 1999. *Glycobiology* 9:833–39

159. Neve EP, Svensson K, Fuxe J, Pettersson RF. 2003. *Exp. Cell Res.* 288:70–83

160. Nufer O, Mitrovic S, Hauri HP. 2003. *J. Biol. Chem.* 278:15886–96

161. Sato K, Nakano A. 2002. *Mol. Biol. Cell* 13:2518–32

162. Schrag JD, Procopio DO, Cygler M, Thomas DY, Bergeron JJ. 2003. *Trends Biochem. Sci.* 28:49–57

163. Bugg TD, Brandish PE. 1994. *FEMS Microbiol. Lett.* 119:255–62

164. Messner P, Schaffer C. 2003. *Fortschr. Chem. Org. Naturstoffe* 85:51–124

165. Young NM, Brisson JR, Kelly J, Watson DC, Tessier L, et al. 2002. *J. Biol. Chem.* 277:42530–39

166. Lechner J, Wieland F. 1989. *Annu. Rev. Biochem.* 58:173–94

167. Moens S, Vanderleyden J. 1997. *Arch. Microbiol.* 168:169–75

168. Schaffer C, Graninger M, Messner P. 2001. *Proteomics* 1:248–61

169. Parodi AJ. 1993. *Glycobiology* 3:193–99

170. McConville MJ, Mullin KA, Ilgoutz SC, Teasdale RD. 2002. *Microbiol. Mol. Biol. Rev.* 66:122–54

171. Vallee F, Lipari F, Yip P, Sleno B, Herscovics A, Howell PL. 2000. *EMBO J.* 19:581–88

172. Deleted in proof

173. Booth C, Koch LE. 1990. *Cell* 59:729–37

174. Tatu U, Helenius A. 1997. *J. Cell Biol.* 10:555–65

175. Meunier L, Usherwood YK, Chung KT, Hendershot LM. 2002. *Mol. Biol. Cell* 13:4456–69

176. Kaufman RJ. 1999. *Genes Dev.* 13:1211–33

177. Ng DT, Spear ED, Walter P. 2000. *J. Cell Biol.* 150:77–88

178. Mori K. 2003. *Traffic* 4:519–28

179. Aebi M, Gassenhuber J, Domdey H, Heesen ST. 1996. *Glycobiology* 6:439–44

180. Stagljar I, te Heesen S, Aebi M. 1994. *Proc. Natl. Acad. Sci. USA* 91:5977–81

181. Reiss G, te Heesen S, Zimmerman J, Robbins PW, Aebi M. 1996. *Glycobiology* 6:493–98

182. Oriol R, Martinez-Duncker I, Chantret I, Mollicone R, Codogno P. 2002. *Mol. Biol. Evol.* 19:1451–63

Annu. Rev. Biochem. 2004. 73:1051–87
doi: 10.1146/annurev.biochem.73.011303.073950
Copyright © 2004 by Annual Reviews. All rights reserved
First published online as a Review in Advance on March 24, 2004

Analyzing Cellular Biochemistry in Terms of Molecular Networks

Yu Xia,[1,][*] Haiyuan Yu,[1,][*] Ronald Jansen,[4,][*]
Michael Seringhaus,[1] Sarah Baxter,[1] Dov Greenbaum,[1]
Hongyu Zhao,[2] and Mark Gerstein[1,3]

[1]*Department of Molecular Biophysics and Biochemistry, [2]Department of Epidemiology
and Public Health, [3]Department of Computer Science, Yale University, New Haven,
Connecticut 06520; e-mail: yuxia@csb.yale.edu, haiyuan.yu@yale.edu,
michael.seringhaus@yale.edu, sarah.baxter@yale.edu, dov.greenbaum@yale.edu,
hongyu.zhao@yale.edu, mark.gerstein@yale.edu*
[4]*Computational Biology Center, Memorial Sloan-Kettering Cancer Center, New York,
New York 10021; e-mail: jansenr@mskcc.org*

Key Words genome wide high-throughput experiments, protein-protein
interaction networks, regulatory networks, integration and prediction, network
topology

■ **Abstract**   One way to understand cells and circumscribe the function of proteins
is through molecular networks. These networks take a variety of forms including
webs of protein-protein interactions, regulatory circuits linking transcription factors
and targets, and complex pathways of metabolic reactions. We first survey experimental techniques for mapping networks (e.g., the yeast two-hybrid screens). We
then turn our attention to computational approaches for predicting networks from
individual protein features, such as correlating gene expression levels or analyzing
sequence coevolution. All the experimental techniques and individual predictions
suffer from noise and systematic biases. These problems can be overcome to some
degree through statistical integration of different experimental datasets and predictive
features (e.g., within a Bayesian formalism). Next, we discuss approaches for
characterizing the topology of networks, such as finding hubs and analyzing subnet-
works in terms of common motifs. Finally, we close with perspectives on how
network analysis represents a preliminary step toward a systems approach for
modeling cells.

---

*These authors contributed equally to this review.

CONTENTS

INTRODUCTION

An important idea emerging in postgenomic biology is that the cell can be viewed as a complex network of interacting proteins, nucleic acids, and other biomolecules (1, 2). Similarly complex networks are also used to describe the structure of a number of wide-ranging systems, including the Internet, power grids, the ecological food web, and scientific collaborations. Despite the seemingly vast differences among these systems, they all share common features in terms of network topology (3–11). Therefore, networks may provide a framework for describing biology in a universal language understandable to a broad audience.

Many fundamental cellular processes involve interactions among proteins and other biomolecules. Comprehensively identifying these interactions is an important step toward systematically defining protein function (2, 12) because clues about the function of an unknown protein can be obtained by investigating its interaction with other proteins of known function.

A biomolecular interaction network can be viewed as a collection of nodes (representing biomolecules), some of which are connected by links (representing interactions). There are many classes of molecular networks in a cell, each with different types of nodes and links. We list a representative subset below:

- Protein-protein physical interaction networks. Here nodes represent proteins, and links represent direct physical contacts between proteins. In addition to direct interaction, two proteins can interact indirectly through other proteins when they belong to the same complex.

- Protein-protein genetic interaction networks. In general, two genes are said to interact genetically if a mutation in one gene either suppresses or enhances the phenotype of a mutation in its partner gene (13). Some researchers restrict the term "genetic interaction" to a pair of so-called synthetic lethal genes, meaning that cell death occurs when this pair of genes is deleted simultaneously, though neither deletion alone is lethal. Synthetic lethal relationships may exist between functionally redundant genes, and therefore can be used to determine the function of unknown genes.

- Expression networks. Large-scale microarray experiments probing mRNA expression levels yield vast quantities of data useful for constructing expression networks. In an expression network, genes that are coexpressed are considered connected (14–16). Genes linked in an expression network are not necessarily coregulated because unrelated genes can sometimes show correlated expression simply by coincidence. The structure of an expression network can vary greatly across different experiments, and even within the same experiment, networks produced by different clustering algorithms are often distinct.

- Regulatory networks. Protein-DNA interactions are an important and common class of interactions. Most DNA-binding proteins are transcription factors that regulate the expression of target genes. Combinatorial use of transcription factors further complicates simple interactions of target genes for a given transcription factor. A regulatory network consists of transcription factors and their targets with a specific directionality to the connection between a transcription factor and its target (17, 18). Transcription factors can either up- or downregulate expression of their target genes.

- Metabolic networks. These networks describe the biochemical reactions within different metabolic pathways in the cell. Nodes represent metabolic substrates and products, and links represent metabolic reactions (19).

- Signaling networks. These networks represent signal transduction pathways through protein-protein and protein-small molecule interactions (20). Nodes represent proteins or small molecules (21), and links represent signal transduction events.

These biomolecular networks are the focus of this review. We first discuss how networks can be reconstructed from a combined experimental and computational perspective. Later, we discuss how networks can be analyzed to yield biological insight.

SURVEY OF EXPERIMENTAL TECHNIQUES

There are several experimental methods for uncovering protein-protein and protein-DNA interactions in biological systems on a large scale. Here we review the most current, powerful, and common of these.

Yeast Two-Hybrid Screens

The yeast two-hybrid (Y2H) system (22) has been widely used in protein-protein physical interaction assays. The system uses putative interacting proteins to broker an in vivo reconstitution of the DNA binding domain (DB) and activation domain (AD) of the yeast transcription factor Gal4p. Hybrid proteins are created by fusing the two proteins or domains of interest (generally called "bait" and "prey") to the DB and AD regions of Gal4p, respectively. These two-hybrid proteins are introduced into yeast, and if transcription of Gal4p-regulated reporter genes is observed, the two proteins of interest are deemed to have formed an interaction—thereby bringing the DB and AD domains of Gal4p together and reconstituting the functional transcriptional activator.

Unlike most biochemical analyses of protein-protein interaction, such as coimmunoprecipitation, cross-linking, and chromatographic cofractionation (22), the two-hybrid system does not demand any protein purification, isolation, or manipulation. The proteins to be tested are expressed by the yeast cells, and a result is easily seen by in vivo reporter gene assays. The two-hybrid technique is therefore applicable to nearly any pair of putative interacting proteins.

There exist three main approaches for large-scale two-hybrid studies (23). The matrix approach (one versus one) systematically tests pairs of proteins for an interaction phenotype; a positive result can indicate that these particular proteins interact. Array experiments (one versus all) examine the interactions of a single DB fusion protein against a pool of AD fusions. Depending on the size of the AD pool, whole-proteome coverage can be achieved against the single DB fusion. Pooling studies (all versus all) involve yeast strains expressing different DB fusions mass-mated with strains expressing AD hybrids. With such experiments, it is conceptually possible to test every protein in the organism against every other protein.

The first large-scale, systematic search for yeast protein-protein interactions was conducted in 1997 (24). In 2000, Uetz et al. (25) published the results of two different large-scale screens on all full-length predicted open reading frames (ORFs). The first approach involved a protein array of roughly 6000 yeast transformants, which each expressed one yeast ORF-AD fusion. One hundred ninety-two yeast proteins were screened against this array. In the second screen, a library of cells was generated and pooled such that all 6000 AD fusions were present. Nearly all predicted yeast proteins, expressed as DB fusions, were screened against this library, and positives were identified by sequencing. Later, Ito et al. (26, 27) reported another systematic identification of yeast interacting protein pairs with a whole-genome-level two-hybrid screen. Their comprehensive approach involved cloning all yeast ORFs as both bait and prey, and they tested about 4×10^6 mating reactions (roughly 10% of all possible combinations). The researchers pooled constructs such that each pool expressed either 96 DB fusions or 96 AD fusions, and they screened all possible combinations of these pools. False positives were controlled by requiring a positive interaction result on at least three independent occasions. Overlap between the Ito and Uetz screens was low, indicating that both studies, though extensive, sampled only a small subset of yeast protein interactions (28, 29).

It is also possible to use large-scale two-hybrid screens to explore interactions relevant to a specific pathway or biological process. Drees et al. (30) screened 68 Gal4p DB fusions of yeast proteins associated with cell polarity against an array of yeast transformants expressing roughly 90% of predicted yeast ORFs. In addition, large-scale two-hybrid screens are not confined to yeast proteins: Working with proteins involved in vulval development, Walhout et al. (31) conducted large-scale interaction mapping in the nematode *Caenorhabditis elegans*, and Boulton et al. (32) combined protein-protein interaction mapping with phenotypic analysis in *C. elegans* to explore DNA damage response interaction networks.

Comprehensive In Vivo Pull-Down Techniques

In vivo pull-down describes a class of techniques that use either a native or modified bait protein to identify and precipitate interacting partners. Most experiments concerned with studying protein-protein interactions through pull-down techniques consist of three parts: bait presentation, affinity purification, and analysis of the recovered complex (33).

Compared with the two-hybrid system, the main advantages to in vivo pull-down techniques are the relative ease of analyzing complete complexes, and the use of native, processed, and posttranslationally modified protein as a reagent to target potential interactors in their natural environment and at normal abundance levels (34). If a suitable antibody exists to the native protein, endogenous protein can be used. However, because insufficient antibodies exist to attack most unmodified proteins with the requisite specificity and affinity, more general techniques, such as tagging, are typically used for large-scale assays. Generic tagging involves the addition of a sequence onto the gene of interest, encoding a tag recognized by a convenient antibody. HA-tagging is a common epitope-tagging approach that has been used successfully (35). A recent tagging strategy facilitating recovery of highly pure protein preparations is a tandem affinity purification (TAP) system consisting of a calmodulin-binding domain and the protein-A Ig-binding domain separated by the tobacco etch virus (TEV) protease target sequence (36). Bait protein is recovered with an immunoglobulin-bound solid support, and after washing, it is released from this support by protease cleavage. Following this initial purification, the recovered sample is passed over a calmodulin column, pending elution with ethylene glycol-bis (β-aminoethyl ether)-N,N,N',N'-tetraacetic acid or other Ca^{2+} chelators. This two-stage purification ensures low background noise and correspondingly high sample purity, but risks losing weak interacting partners or complex components due to the harsh purification procedure.

After the bait/interactor complex is purified, components of this complex can be identified by mass spectrometry (MS). The many recent advances in MS technology, such as matrix-assisted laser desorption/ionization time-of-flight (MALDI-TOF), electrospray ionization (ESI), and tandem MS/MS, among others, have enabled accuracy to increase while permitting ionization (and therefore, characterization) of larger biomolecules. In general, MS proteomics experiments comprise five stages (33): The first three involve purification (typically culminating in one-dimensional gel electrophoresis), tryptic digestion to generate short peptides, and high-pressure liquid chromatography separation of the tryptic digest; the final two stages are tandem mass spectrometry assays. The high accuracy of MS spectra, combined with knowledge of the genomic sequence of the organism in question, permits rapid and accurate identification of the proteins involved in the recovered complex.

Two large-scale projects dealing with the yeast interactome were recently completed by Gavin et al. (37) and Ho et al. (38). Gavin et al. purified 589 bait proteins from a library of 1548 tagged strains, and from these identified 1440 distinct participant proteins in 232 complexes. Ho et al. purified 725 bait proteins from which 1578 interacting proteins were identified. Both studies used extensive literature comparisons to characterize the complexes they found, and both reported significant participation by previously unknown or unannotated genes (35, 37, 38).

Protein Chips

The application of microarray technology to proteomics yielded the protein chip, an advanced in vitro technique for protein functional assays on a large scale. Protein chip technology is directly applicable to protein interaction networks because the large number of immobilized proteins can be probed with labeled substrate in a single experiment.

Arenkov et al. (39) reported the creation of a polyacrylamide-based protein microchip, containing 0.2 nanoliter (nl) spots of gel substrate in which proteins were immobilized; this platform allowed electrophoresis to be used to enhance mixing of substrate. MacBeath and Schreiber's protein chip (40) uses microarray technology and robotics to spot nanoliter volumes of protein onto aldehyde-coated glass slides. The abundance of lysine residues in most proteins, combined with a reactive N-terminal amine, permits proteins to become covalently linked to the slide surface in a number of possible orientations.

Shortly thereafter, Zhu et al. (41) described another type of protein chip, also mounted on a glass slide but comprising a system of 300 nl silicone elastomer microwells for physical separation of samples during processing. As with the MacBeath protein arrays, the target protein was covalently linked to the chip, though here the chemical cross-linker GPTS was used. The following year, the same group announced the creation of the first whole-proteome chip (42), a glass slide similar to MacBeath & Schreiber's initial protein chip, but it contained over 80% of known yeast ORF gene products attached to nickel-coated slides via 6-His tags. Zhu et al. demonstrated the effectiveness of the proteome chip for protein-protein interaction studies by probing with biotinylated calmodulin in the presence of calcium; calmodulin binding partners were visualized by probing with Cy3-labeled streptavidin. This demonstrated that biotinylated constructs of virtually any protein could be used to probe the proteome chip, thereby visualizing protein-protein interactions. In addition to uncovering several known calmodulin interactors, the researchers found a significant number of novel potential interaction partners, many of which share a motif believed to be important for calmodulin binding.

Structure Determination of Biomolecular Complexes

An atomic view of physical interactions between biomolecules can be achieved by solving three-dimensional (3D) structures of biomolecular complexes, most often accomplished with X-ray crystallography and NMR spectroscopy. In particular, X-ray crystallography is able to produce the most spatially accurate description of biomolecular interactions. Though technically challenging, significant advances have been made in recent years, and X-ray crystallography can now be applied to complexes as large as several megadaltons. For a detailed review of various structural determination methods for biomolecular complexes, see Reference 43.

Comparing In Vivo and In Vitro Techniques

The caveats associated with genomic-level datasets stem largely from the experimental techniques used to generate them, and in particular, care should be taken to note whether interaction results originate from in vivo or in vitro studies. A major advantage of in vivo pull-down techniques is that near-native interactions can be probed, provided that tagging and bait expression do not interfere with the replication of endogenous levels of protein activity—proper folding, posttranslational modification, and the accessibility of biologically relevant binding partners are generally assumed. Still, the abundance of proteins and solutes in the cell means contaminants often copurify and potentially yield misleading results. In vivo experiments generally offer little or no direct control over reaction conditions (especially in the case of large-scale studies), whereas in vitro assays permit exquisite control over ion concentration, temperature, and other factors. The assumption that in vivo assay conditions are biologically meaningful is sometimes inapplicable to interactions probed by the yeast two-hybrid technique, which must occur in the yeast nucleus. In vitro and two-hybrid approaches are unlikely to recover only significant binding partners, and they risk false-positive results if interacting proteins localize to different cell compartments, express at different times in the cell cycle, or are otherwise inaccessible to binding under normal conditions. Still, in vitro techniques, such as protein chip assays, are convenient to record because results can be visualized for individual putative interacting partners. Compare this situation to the grouped results of many pooling techniques in which over- or underrepresentation in bait/prey pools can influence results, and positives must be identified by sequencing or bar code analysis.

Methods for Determining Protein-Protein Genetic Interactions

Synthetic lethal screens are used to identify genetic interactions between proteins. Small-scale synthetic lethal screens have been used to identify genes involved in many cellular processes (44–46). Recently, Tong et al. (13) introduced a systematic method to construct large-scale double mutant arrays, termed synthetic genetic array (SGA) analysis, in which double mutants were created by crossing a query mutation to an array of roughly 4700 deletion mutants, and nonviable double-mutant meiotic progeny were identified. SGA analysis has generated a genetic network of 291 interactions among 204 genes.

Methods for Determining Protein-DNA Interactions

Protein-DNA interactions can be determined by three core methods:

- Gel shift. Compared with protein molecules, DNA molecules are much smaller and therefore have much higher mobility in a polyacrylamide gel. Under favorable conditions, unbound DNA can be distinguished from DNA associated with proteins because of their relative mobility (47, 48). Recently, several enhanced methods, such as capillary electrophoretic mobility shift assay (CEMSA) (49), have been proposed to improve the performance of this approach.

- DNA footprinting. A 5′-end-labeled, double-stranded target DNA segment is partially degraded by DNase in both the presence and absence of the putative binding protein. Degraded fragments are visualized by electrophoresis and autoradiography. The binding site on the DNA will be protected by the binding protein from DNAase degradation (48, 50). Compared with gel shift methods, DNA footprinting not only confirms the interaction between the DNA and the binding protein, but it can also elucidate the specific binding site of the protein.

- In vivo cross-linking and immunoprecipitation. The binding protein is first covalently linked to DNA in situ using any of a variety of common cross-linking reagents; among these, UV and formaldehyde have been widely used. After cross-linking, chromosomal DNA is sheared; the protein is precipitated using a specific antibody, and bound DNA fragments coprecipitate. Reversal of cross-links releases bound DNA, so fragments can be identified by polymerase chain reaction (PCR) and electrophoresis (51, 52). This method is also called chromatin immunoprecipitation (ChIP).

Recently, with the advent of microarray technology, novel methods have been introduced to rapidly determine the binding sites of transcription factors on a genome-wide scale (17, 18, 53, 54).

- Chromatin-immunoprecipitation and microarray/chip technique (ChIP-chip). This method combines the ChIP technique with DNA microarray technology. Thousands of DNA fragments purified by the ChIP method are identified simultaneously by microarray experiments (53). Using ChIP-chip, Lee et al. (17) were able to create a yeast regulatory network consisting of 106 transcription factors and 2363 target genes.

- DNA adenine methyltransferase identification (DamID). The use of cross-linking reagents can produce artifacts in ChIP-chip experiments. To overcome this problem, van Steensel and Henikoff (55, 56) introduced a new technique to map protein-DNA interactions, termed DamID. The DNA binding protein of interest is genetically fused with *Escherichia coli* DNA adenine methyltransferase (Dam). Dam methylates the N^6-position of adenine in the sequence GATC, which occurs on average every 200–300 base pairs in the fly genome. Upon In vivo binding of the protein to its target DNA sites, DNA around the target sites is preferentially methylated by the tethered Dam enzyme. Subsequently, genomic DNA is digested into small fragments by *DpnI*. DNA fragments without methylated GATCs are removed by *DpnII* digestion. The remaining methylated fragments are amplified by selective PCR and quantified by microarray analysis (54–56). Recently, Sun et al. (54) successfully mapped protein-DNA interactions at high resolution along large segments of genomic DNA from *Drosophila melanogaster* using the DamID technique and genomic DNA tiling path microarrays.

Conceivably, data generated by these different methods can be used to cross validate one another, thereby producing more comprehensive information. Although each method yields only a subset of the total interactions present, a more complete yeast regulatory network consisting of 180 transcription factors and 3474 target genes has been produced through the synthesis of all available datasets (57).

Databases for Biomolecular Interactions

Many databases have been created to store the tremendous amount of data required for and contained in these networks; some of which are summarized in Table 1 (58–67). Some databases are more comprehensive than others; for instance, MIPS contains not only protein-protein physical interaction data but genetic interaction information as well (60).

TABLE 1 Summary of the databases for biomolecular interactions

Name	URL	Type of networks	References
Database of Interacting Proteins (DIP)	http://dip.doe-mbi.ucla.edu/	Physical	(58)
Biomolecular Interaction Network Database (BIND)	http://www.bind.ca/	Physical	(59)
Human Protein Reference Database (HPRD)	http://www.hprd.org/	Physical	(67)
Munich Information Center for Protein Sequences (MIPS)	http://mips.gsf.de/	Physical & genetic	(60)
The Yeast Proteome Database (YPD)	http://www.incyte.com/sequence/proteome/index.shtml	Physical, genetic, & regulatory	(61)
The Transcription Factor Database (TRANSFAC)	http://transfac.gbf.de/TRANSFAC/	Regulatory	(62)
Regulon Data Base (RegulonDB)	http://www.cifn.unam.mx/Computational_Genomics/regulondb/	Regulatory	(63)
Kyoto Encyclopedia of Genes and Genomes (KEGG)	http://www.kegg.com/	Metabolic	(64)
Encyclopedia of Metabolic Pathways (MetaCyc)	http://metacyc.org/	Metabolic	(65)
Alliance for Cellular Signaling (AfCS)	http://www.cellularsignaling.org/	Signaling	(66)

COMPUTATIONAL APPROACHES FOR PREDICTING INTERACTIONS

In addition to experimentally determined interaction datasets, a vast amount of biological information is contained in the ever-growing datasets of protein sequences, structures, functions, and expressions, and in the literature. Here we review computational methods that extract interaction information from these datasets.

Computational Approaches for Predicting Protein-Protein Interactions

We first review computational approaches for predicting interactions between proteins. Protein-protein interactions can be inferred on the basis of comparative genomics, detailed sequence and structural analysis, correlation of protein functional genomic features, and the existence of conserved interactions in other organisms. In addition, protein-protein interactions can be extracted from literature in an automated way.

PREDICTING PROTEIN FUNCTIONAL RELATIONSHIPS ON THE BASIS OF COMPARATIVE GENOMICS   Several methods exist to predict functional relationships between pairs of proteins on the basis of their patterns of occurrence and their location across multiple genomes. The first method identifies protein pairs that are adjacent along the chromosome. Protein pairs are likely to share similar functions if such chromosomal proximity is conserved across multiple genomes (68–70). In addition, conserved gene order can also be used as an indicator for functional interaction (71). These methods are inspired by the experimental observation that functionally related proteins in bacteria tend to cluster along the chromosome to form operons; their applicability in eukaryotes is less clear.

The second method predicts protein functional interaction on the basis of patterns of domain fusion (72, 73). Sometimes two protein domains exist as separate proteins in one genome, but they are fused together into a single protein in another genome. In such a case, the domains are likely to be functionally related (74).

The third method analyzes patterns of occurrence of proteins in multiple genomes. For each protein, a phylogenetic profile is constructed that indicates whether or not the protein is present in each genome. From an evolutionary standpoint, protein pairs with similar phylogenetic profiles tend to travel together and are candidates for functional interaction (75–77).

PREDICTING PROTEIN-PROTEIN INTERACTIONS ON THE BASIS OF DETAILED SEQUENCE AND STRUCTURAL ANALYSIS   Two methods exploit the hypothesis that interacting proteins tend to coevolve. In the first method, the coevolution of interacting protein families is measured by the similarity of phylogenetic trees constructed from multiple sequence alignments of the two protein families (78, 79). When this technique is applied on a genomic scale, phylogenetic trees for all proteins can be constructed. Proteins with similar phylogenetic trees are more likely to interact with one another (80). In the second method, the coevolutionary signal in multiple sequence alignments is further analyzed in terms of correlated mutations: A protein pair is likely to interact if there is accumulation of correlated mutations between the interacting partners (81).

Certain pairs of sequence motifs and structural families preferentially interact. To identify such pairs, one first classifies known protein interactions in terms of interactions between sequence motifs and structural families (82, 83). Pairs of sequence motifs and structural families that are overrepresented in the interaction dataset can then be identified. A new protein pair is likely to interact if it can be classified into one of these overrepresented sequence motif or structural family pairs.

It is also possible to predict protein-protein interactions from sequence information using machine-learning techniques. For example, using a database of known interactions, a support vector machine learning system can be trained to predict interactions based on sequence information and associated physicochemical properties, such as charge, hydrophobicity, and surface tension (84).

With progress in structural genomic projects and structure prediction methods, structural models can be built with varying degrees of accuracy for an increasing fraction of genomic proteins. For two candidate proteins, each equipped with accurate structural models, it is possible to assess the likelihood of interaction in vitro by calculating the lowest free energy for the protein complex. This process, called docking, has proven increasingly successful in structure prediction of protein complexes, as indicated in the Critical Assessment of Prediction of Interactions (CAPRI) meetings (85). However, docking is a time-consuming procedure, and its accuracy needs further improvement. In its current form, it is not feasible to predict protein interactions on a genomic scale with this technique.

Databases of solved 3D structures for protein complexes provide additional information that can be exploited for predicting protein-protein interactions. The full set of known 3D complexes can be used to search for all complex homologues in yeast (86). In this method, called multimeric threading, sequences of every protein pair are aligned (or threaded) to a 3D complex template to optimize a compatibility scoring function compiled from known 3D complexes. Top protein pairs with the best compatibility scores are likely to interact in a way similar to the 3D complex template.

EXTRACTING PROTEIN INTERACTIONS FROM LITERATURE   A number of methods have been developed to extract protein interactions from literature. These methods can be grouped into two categories. Methods in the first category use machine learning techniques to screen the literature for articles containing information about protein interactions (87); selected articles are then curated by hand. Methods in the second category automatically extract protein interaction events from biomedical articles. Techniques used range from statistical analysis of co-occurrence of names of biomolecules (88) to natural language processing

(89). For detailed reviews of information extraction methods for molecular biology, see (90, 91).

ANNOTATION TRANSFER OF PROTEIN INTERACTIONS   Sequence homology offers an efficient way to map genome-wide interaction datasets between different organisms, using the concept of interolog. This will be discussed below in the section entitled "Cross-referencing Different Networks."

CORRELATION OF PROTEIN FUNCTIONAL GENOMIC FEATURES AS PREDICTORS FOR PROTEIN INTERACTIONS   In addition to sequence and structural information, functional genomic datasets are also available for certain organisms. Much of this functional genomic information is applicable to the study of protein interactions. Consider each class of functional genomic data as a protein feature; two proteins are therefore more likely to interact if these genomic features are correlated. A list of potential functional genomic features for proteins is given below.

- mRNA expression. Interacting proteins tend to have correlated expression profiles (16, 92). Protein abundance can be indirectly and quite crudely measured by the presence or absence of the corresponding mRNA transcripts, though large differences can exist between the mRNA and protein abundance (93). Still, several studies have reported a significant correlation of mRNA transcript levels among proteins that interact (92, 94, 95). This correlation is more prominent for proteins in permanent complexes and less noticeable for those participating in transient complexes (92).

- The phenotype of knockout mutants (96, 97) can serve as another potential indicator, suggesting whether two proteins are subunits of the same complex. The genetic deletion of different subunits of the same complex may disturb the function of a complex in the same way, thus producing a similar phenotype. Synthetic lethal interactions are generally enriched in genes that encode members of the same complex (13). More generally, if proteins function in related cellular processes, they have an increased chance of being in the same complex.

- To form an interaction, proteins must localize to the same subcellular compartment at the same time. Colocalization thus serves as a useful predictor for protein interaction. A large amount of protein subcellular localization data is available for yeast (98).

Circumstantial evidence, such as the indicators given above, is rarely strong enough to directly predict protein-protein interactions. However, when these datasets are properly combined, quite reliable predictions can result.

Integration of Protein-Protein Interaction Datasets

We have seen that protein-protein interaction datasets come from a variety of different experimental and computational sources. To gain a comprehensive understanding of the interactome, we must integrate these disparate interaction datasets. There are two key reasons for integrating multiple protein-protein interaction datasets. First, different interaction datasets cover different subsets of the proteome, so it is reasonable to consider their union. Second, the degree of confidence in a protein-protein interaction depends upon how much evidence supports it (99–102). Usually, when multiple, distinct data sources all contribute evidence for a predicted interaction, we gain increased confidence in the validity of our prediction. It is important to note that different experimental methods carry with them different systematic errors—errors that cannot be corrected by repetition.

INTEGRATION OF MULTIPLE DATASETS OF PHYSICAL PROTEIN-PROTEIN INTERAC-TIONS: RNA POLYMERASE II   The value of integrating multiple datasets of physical protein-protein interactions was demonstrated in a recent study by Edwards et al. (29), who compared the crystal structure of RNA polymerase II with protein-protein interaction experiments on the same set of proteins. The protein-protein interaction experiments, including cross-linking, pull-down and far western blotting studies, were carried out while this structure was still unknown (29, 103–107). The subsequent publication of the crystal structure allowed a retrospective assessment of the success of these experiments.

The comparison showed that the individual protein-protein interaction experiments tended to measure subsets of the potential interactions in the RNA polymerase II structure. Furthermore, individual experiments missed many interactions present in the true structure (false negatives) among the protein pairs that were tested, and they found spurious protein-protein interactions absent from the true structure (false positives). The best pull-down experiment was inconsistent with the crystal structure for 23% of the protein pairs, whereas some experiments were incorrect nearly 50% of the time.

To reduce these error rates, different datasets can be combined. The simplest rules for integration of multiple datasets are the and- and or rules. The "and rule" predicts a positive interaction only when all datasets agree (intersection); the "or rule" predicts an interaction when at least one dataset gives a positive result (union). The and rule tends to give more accurate results, but offers low coverage because few cases exist where all available datasets agree. The or rule tends to yield maximum sensitivity (that is, the discovery of the highest number of true positives), but it simultaneously produces the highest number of false positives.

An intuitive method of combining the datasets is a majority voting procedure (Figure 1) in which the different experimental results contribute an additive positive or negative vote toward the final result. If the majority of datasets detect an interaction between a protein pair, the pair is predicted to interact, whereas the pair is considered noninteracting if the majority of datasets do not measure an interaction. A major caveat of this procedure is that each dataset implicitly carries the same weight, despite the fact that some datasets contain more reliable results, and other datasets may be redundant. In fact, in the RNA polymerase II example, the prediction by the voting procedure offers virtually no improvement in accuracy compared with the results of the individual interaction experiments (Figure 1). Altogether, the voting procedure has higher coverage than the individual experiments, a trivial result of the integration.

Machine-learning methods provide more sophisticated data integration procedures that take into account data reliability and redundancy, often leading to better results in both coverage and accuracy. An effective method is the Bayesian network, in particular the naive Bayesian network in its simplest form. Bayesian networks have previously been applied successfully in computational biology research, ranging from the prediction of subcellular localization of proteins (108) to the combination of different gene prediction algorithms (109, 110).

The Bayesian network combines different interaction datasets in a probabilistic manner, assigning a probability to the prediction result rather than just a binary classification. Each individual dataset is essentially weighted by its accuracy and redundancy. The naive Bayesian network yields optimal results when the different datasets contain uncorrelated evidence; but even when this condition is not met, the results are often useful. In the RNA polymerase II example, naive Bayesian network integration leads to an increase in accuracy ranging from 5% to 26% compared to the individual experiments (Figure 1). Details on using Bayesian networks for integrating interaction datasets can be found in the Appendix.

INTEGRATION OF GENOME-SCALE PROTEIN-PROTEIN INTERACTION DATA   Similar data integration methods can be used on a genomic scale. This is important because several studies have demonstrated that a large number of false positives occurs in the results of individual interaction experiments carried out in a high-throughput manner and on a large scale, calling into question the general validity of such experiments. A fair estimate is that the number of false positives in high-throughput studies is on the same order of magnitude as the actual number of true positive interactions; this reflects the fact that the number of interacting proteins in any cell is perhaps several orders of magnitude smaller than the number of all possible combinations between the proteins in the entire proteome. Screening for protein-protein interactions in the proteome is therefore equivalent to using a diagnostic test for screening for people with a rare disease in the general population: An experiment with a small false positive rate would still yield a high absolute number of false positives simply because the pool of tested candidates is so large. Thus, a natural strategy to overcome this problem is the combination of multiple interaction data sources and other genomic data.

DE NOVO PREDICTION OF PROTEIN COMPLEXES   Jansen et al. (111) recently showed how protein complexes can be predicted de novo with high confidence when multiple genomic datasets are integrated. In their study, the MIPS complexes catalog was used as a sample of well-characterized protein complexes (determined from more reliable small-scale interaction studies), and a list of

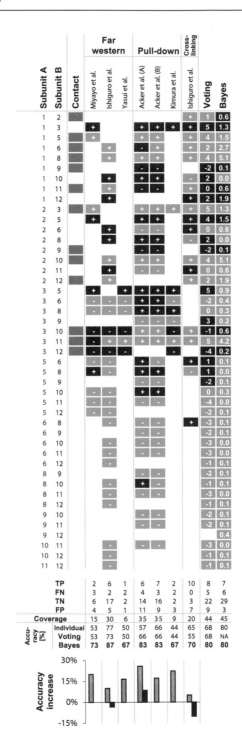

negative examples (noninteracting protein pairs) was constructed from proteins that were observed to have different subcellular localizations (60, 98). Although such a list of negatives may be imperfect, it is expected to be strongly enriched in noninteracting protein pairs when compared to randomly chosen proteins. These datasets ("gold standards") serve as a reference for observing whether the prediction results are correct (testing) and for determining the parameters of possible integration methods (training).

It is possible to quantify how the different values in the individual genomic features fare in predicting whether two proteins are members of the same complex (Table 2). More details can be found at http://www.genecensus.org/intint/. These different genomic features can then be combined using naive Bayesian networks (analogous to the method employed in the aforementioned example of RNA polymerase II). Cross validation with the reference datasets shows that the predictions are highly enriched in positive protein pairs (interact-

←—————————————————————————————————————→

Figure 1   Comparison of the crystal structure of RNA polymerase II and the protein-protein interaction experiments. The RNA polymerase II structure consists of 10 protein subunits, which allows for 45 different pairings between these proteins (shown in the first two columns). The third column shows which of the protein pairs are in physical contact (with a contact interface area ≥ 800 Å^2, shown by the *gray* squares). The following columns show the results from three far western, three pull-down and one cross-linking experiment. The experimental results are indicated as either positive $(+)$, when an interaction was found, or negative $(-)$ when no interaction was detected. The results are either true (*green*), when they agreed with the crystal structure contacts, or false (*red*), when they disagreed. Blank fields in the table correspond to protein pairs that were not tested in the experiments. The two columns on the right show the results of integrating the seven experiments into one prediction of interactions. The Voting column shows the difference between positive and negative experimental results for each protein pair. The Bayes column shows the posterior odds, calculated with a naive Bayesian network, of having a real protein-protein interaction on the basis of experimental data. The Coverage row shows how many protein pairs were measured in each experiment or how many protein pairs were covered by the voting and the Bayesian network procedure. These can be divided into true positive (TP), false negative (FN), true negative (TN), and false positive (FP). The bottom three rows show the accuracy of individual experiments, defined as (TP + TN)/Coverage. The Voting and Bayes rows show the accuracy of the voting procedure and the naive Bayesian network for the same subset of protein pairs that were measured in the individual experiments. The graph at the bottom shows the Accuracy increase, the difference between the accuracies of the integration methods and the individual experiments. Gray bars represent Bayes, and black bars represent Voting. Note that the results here were obtained without cross validation. It can be shown, however, that the naive Bayesian network or various other machine-learning methods achieve higher sensitivity and accuracy than the majority voting in a leave-one-out prediction—where the protein pair for which a prediction is made is excluded from the training of the parameters of the machine-learning method (data not shown).

TABLE 2 Combining genomic features using Bayesian networks to predict yeast protein-protein interactions

Essentiality[a]	Number of protein pairs[b]	Gold standard overlap		Sum (pos)[d]	Sum (neg)[e]	Sum (pos)/ sum (neg)[f]	P (Ess|pos)[g]	P (Ess|neg)[g]	L[g,h]
		Pos[c]	Neg[c]						
Values									
EE	384,126	1,114	81,924	1,114	81,924	0.014	5.18E-01	1.43E-01	3.6
NE	2,767,812	624	285,487	1,738	367,411	0.005	2.90E-01	4.98E-01	0.6
NN	4,978,590	412	206,313	2,150	573,724	0.004	1.92E-01	3.60E-01	0.5
Sum	8,130,528	2,150	573,724	—	—	—	1.00E+00	1.00E+00	1.0

Expression correlation	Number of protein pairs	Gold standard overlap		Sum (pos)	Sum (neg)	Sum (pos)/ sum (neg)	P (exp|pos)	P (exp|neg)	L
		Pos	Neg						
Values									
0.9	678	16	45	16	45	0.36	2.10E-03	1.68E-05	124.9
0.8	4,827	137	563	153	608	0.25	1.80E-02	2.10E-04	85.5
0.7	17,626	530	2,117	683	2,725	0.25	6.96E-02	7.91E-04	88.0
0.6	42,815	1,073	5,597	1,756	8,322	0.21	1.41E-01	2.09E-03	67.4
0.5	96,650	1,089	14,459	2,845	22,781	0.12	1.43E-01	5.40E-03	26.5
0.4	225,712	993	35,350	3,838	58,131	0.07	1.30E-01	1.32E-02	9.9
0.3	529,268	1,028	83,483	4,866	141,614	0.03	1.35E-01	3.12E-02	4.3
0.2	1,200,331	870	183,356	5,736	324,970	0.02	1.14E-01	6.85E-02	1.7
0.1	2,575,103	739	368,469	6,475	693,439	0.01	9.71E-02	1.38E-01	0.7
0	9,363,627	894	1,244,477	7,369	1,937,916	0.00	1.17E-01	4.65E-01	0.3
−0.1	2,753,735	164	408,562	7,533	2,346,478	0.00	2.15E-02	1.53E-01	0.1
−0.2	1,241,907	63	203,663	7,596	2,550,141	0.00	8.27E-03	7.61E-02	0.1
−0.3	485,524	13	84,957	7,609	2,635,098	0.00	1.71E-03	3.18E-02	0.1
−0.4	160,234	3	28,870	7,612	2,663,968	0.00	3.94E-04	1.08E-02	0.0
−0.5	48,852	2	8,091	7,614	2,672,059	0.00	2.63E-04	3.02E-03	0.1
−0.6	17,423	—	2,134	7,614	2,674,193	0.00	0.00E+00	7.98E-04	0.0
−0.7	7,602	—	807	7,614	2,675,000	0.00	0.00E+00	3.02E-04	0.0
−0.8	2,147	—	261	7,614	2,675,261	0.00	0.00E+00	9.76E-05	0.0
−0.9	67	—	12	7,614	2,675,273	0.00	0.00E+00	4.49E-06	0.0
Sum	18,773,128	7,614	2,675,273	—	—	—	1.00E+00	1.00E+00	1.0

TABLE 2 *(Continued)*

MIPS function similarity	Number of protein pairs	Gold standard overlap				Sum (pos)/ sum (neg)	P (MIPS| pos)	P (MIPS| neg)	L
		Pos	Neg	Sum (pos)	Sum (neg)				
Values									
1–9	6,584	171	1,094	171	1,094	0.16	2.12E-02	8.33E-04	25.5
10–99	25,823	584	4,229	755	5,323	0.14	7.25E-02	3.22E-03	22.5
100–1000	88,548	688	13,011	1,443	18,334	0.08	8.55E-02	9.91E-03	8.6
1000–10000	255,096	6,146	47,126	7,589	65,460	0.12	7.63E-01	3.59E-02	21.3
10000–Inf	5,785,754	462	1,248,119	8,051	1,313,579	0.01	5.74E-02	9.50E-01	0.1
Sum	6,161,805	8,051	1,313,579	—	—	—	1.00E+00	1.00E+00	1.0

GO biological process similarity	Number of protein pairs	Gold standard overlap				Sum (pos)/ sum (neg)	P (GO| pos)	P (GO| neg)	L
		Pos	Neg	Sum (pos)	Sum (neg)				
Values									
1–9	4,789	88	819	88	819	0.11	1.17E-02	1.27E-03	9.2
10–99	20,467	555	3,315	643	4,134	0.16	7.38E-02	5.14E-03	14.4
100–1000	58,738	523	10,232	1,166	14,366	0.08	6.95E-02	1.59E-02	4.4
1000–10000	152,850	1,003	28,225	2,169	42,591	0.05	1.33E-01	4.38E-02	3.0
10000–Inf	2,909,442	5,351	602,434	7,520	645,025	0.01	7.12E-01	9.34E-01	0.8
Sum	3,146,286	7,520	645,025	—	—	—	1.00E+00	1.00E+00	1.0

[a]The first column describes the genomic feature. Protein pairs in the essentiality data can take on three discrete values (EE, both essential; NN, both nonessential; and NE, one essential and one not). The values for mRNA expression correlations range on a continuous scale between -1.0 and $+1.0$. Functional similarity counts, calculated based on MIPS and gene ontology (GO) classification schemes, are integers between 1 and ~ 18 million. We binned the mRNA expression correlation values into 19 intervals and the functional similarity counts into 5 intervals.

[b]The second column gives the number of protein pairs with a particular feature value (e.g., EE) drawn from the whole yeast interactome (~ 18 million pairs).

[c]Columns Pos and Neg give the overlap of these pairs with the gold-standard positives and the gold-standard negatives.

[d]The column Sum (pos) shows how many gold-standard positives are among the protein pairs with likelihood ratio greater than or equal to L, which can be computed by summing up the values in the column Pos to the left.

[e]The column Sum (neg) shows the number of gold-standard negatives among the protein pairs with likelihood ratio greater than or equal to L.

[f]The column Sum (pos)/sum (neg) is a measure of how well each feature predicts protein-protein interactions (given a certain likelihood ratio cutoff).

[g]The last three columns on the right give the conditional probabilities of the feature values—$P(feature\ value\ |\ pos)$ and $P(feature\ value\ |\ neg)$—and the likelihood ratio L, the ratio of these two conditional probabilities.

[h]The likelihood ratios of the individual features can be combined using a naive Bayesian network, as explained in Equation 1 in the Appendix. The prior odds were set to 1/600, which corresponds to a very conservative estimate that there are at most 30,000 pairs of proteins in the same complex among the 18 million protein pairs in yeast.

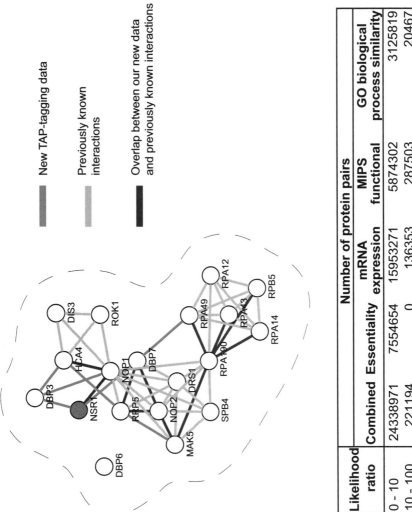

ing proteins) rather than negative protein pairs (negatives). Figure 2 shows an example of the de novo prediction results: A set of rRNA processing proteins were predicted to be present in the same complex and were subsequently validated with TAP-tagging experiments. Figure 2 also shows the value of integrating multiple datasets: The confidence with which proteins can be predicted to be in the same complex (here measured in terms of the likelihood ratio) is low in the individual datasets but high in the combined data.

To conclude, the integration of multiple interaction data sources—or data providing circumstantial evidence about protein-protein interactions—can lead to reliable predictions of protein-protein interactions, even if the individual datasets are related to these interactions only in a statistical sense and contain many false positives. If performed correctly, integration of multiple interaction datasets should yield an error rate lower than that of the component datasets. Machine-learning methods, such as Bayesian networks, have advantages over more simple-minded integration procedures.

We have seen how Bayesian networks can be used as a means to integrate multiple data sources. But in addition to integrating and correlating sets of data, Bayesian networks can also be used to model the regulatory relationships between individual proteins. In the former case, the Bayesian network is used primarily as a tool for integration and classification, whereas the latter application aims at modeling the interdependency of gene and protein activities, as we discuss below.

Reconstructing Biological Pathways and Regulatory Networks From Quantitative Measurements

A large amount of data has been produced by quantitatively monitoring the concentrations of biomolecules in a cell, such as mRNA expression levels. Many computational methods, including correlation metric construction (112), Boolean networks (113–115), and Bayesian networks (116, 117), have been developed to reconstruct biological pathways and networks from these quantitative measurements. Here we discuss Boolean networks and Bayesian networks in detail.

Figure 2 (*top*) Graphic representation of a complex around the protein Nsr1 that was predicted by combining genomic features, such as essentiality, expression, and function. All proteins shown have posterior odds of being in the same complex greater than 1 (assuming prior odds of 1/600). Some of the protein pairs were experimentally verified by TAP-tagging, whereas other protein pairs were shown to be interacting by previous proteomics studies (25, 27, 37, 38). (*bottom*) A table of the distribution of likelihood ratios among the individual experiments and the combined data shows the value of integrating multiple data sources: The combined data contain a much larger number of protein pairs with high likelihood ratios.

A Boolean network is a system of interconnected binary elements, defined by a set of nodes and a group of Boolean functions. Each node exists in one of two states; this is applicable to any binary condition, for example on/off or active/inactive. In general, these two states are assigned numerical values of 1 and 0. A Boolean operation is a function taking input from a set of binary variables and producing output to a single binary variable. Boolean networks can be used to describe the dynamics of a biological system in that all nodes are updated synchronously, moving the system into its next state. Because the number of all possible states of the system is limited and the transition rules are defined deterministically and do not depend on time, the system either reaches a cycle or converges to an attractor. The attractor can be a steady state or a limit cycle. Attractors can be regarded as the target area of the organism, for instance, cell types following differentiation and development. Although Boolean networks have been considered and developed as an approximation model for biological networks, they are inherently deterministic, and thus they do not reflect the inherent randomness that is an integral part of biology. Probabilistic Boolean networks incorporate stochastic variations (115), but the identification of models and the estimation of model parameters under these generalized Boolean networks can pose both theoretical and computational challenges. Another serious limitation of the Boolean network is that all possible variables must be assigned to binary states, whereas most biological activities exhibit continuous measurements. Most recent studies have focused more on the properties of Boolean networks, so the usefulness of Boolean networks as a general modeling and computational tool for biological pathways has yet to be demonstrated.

Recently, there has been enormous interest in modeling gene expression data with Bayesian networks [see for example (118)]. Owing to the stochastic nature of biological processes and various measurement errors, the Bayesian network has won support as a suitable technique with which to study gene expression data. Simply put, a Bayesian network is a graphical representation of a joint probability distribution. It consists of two parts: B_s and B_p, where B_s is a directed acyclic graph (DAG), meaning a directed graph where no path starts and ends at the same node, and B_p is a set of local joint probability distributions describing statistical associations. Causal inferences can be made from these associations by statistically testing the associations between variables or by using a certain measure to score all possible structures and searching for those with high scores. In general, the scoring method is better and more intuitive, and much research has focused on this issue (89, 119).

Dynamic Bayesian networks (DBN) represent a generalization of Bayesian networks. With DBN modeling, we can model the stochastic evolution of a set of random variables over time (120). Bayesian networks have been used to model gene expression data at various scales. Some studies have modeled roughly 800 yeast cell-cycle genes (116). Other groups have focused on a more limited number of genes. For example, the yeast pheromone response pathway ($\sim$32 genes) was recently studied (117). A detailed analysis of just three genes involved in the yeast galactose pathway was reported (118). Although the application of both Bayesian networks and DBN to modeling gene expression has been discussed, their usefulness remains to be shown, and analyses of more well-understood genetic pathways are needed.

There are some limitations to current Bayesian network and DBN approaches. From a statistical perspective, expression levels must be discretized, undoubtedly leading to loss of information. Although we can simplify the computation (as well as obtain a stable result) through such discretization, we need to explore alternative ways to discretize data and, more importantly, to find reliable approaches to analyze continuous data.

Two major limitations exist to using Bayesian networks to model biological pathways. First, all observations are assumed to stem from the same distribution, which clearly cannot model the dynamics of biological systems and responses to environmental perturbations. Second, there is the identifiability problem; many distinct DAGs may result in the same joint probability distributions. Although the DBN may partially address these problems, the computational and theoretical implications of extension to more general models require further investigation. It has been reported in the literature (117) that the Bayesian network methodology was able to correctly identify the true biological model from two competing hypotheses, yet it became clear that this particular analysis was driven by 2 outlying observations from a total of 55 observations (H. Zhao and B. Wu, unpublished results). The Bayesian networks also failed to detect the galactose pathway from genomics data reported in (121). Furthermore, when a DBN was applied to time-course data in *Drosophila* (122), it failed to identify the correct transcriptional regulatory network among three genes showing expression patterns clearly consistent with known biology (H. Zhao and B. Wu, unpublished results). A closer inspection of the cause of DBN failure showed that the stationarity assumption underlying this approach may be too strong and inappropriate. Our experience with Bayesian networks and DBN suggests that a considerable amount of work needs to be done to improve current methods before meaningful results can be reliably extracted from genomic data.

Clearly, better statistical methods are needed to reconstruct biological pathways from quantitative measurements. In addition, improvements along other directions are possible. First, additional quantitative measurements performed on a systematically perturbed network can help define the network architecture with increasing accuracy (121, 123). Second, cross-species comparison can help reveal the conserved core network. Evolutionarily conserved coexpression implies selective advantage and, therefore, a functional relationship (124). Third, the aforementioned analyses need to be combined with other types of information, such as shared functional classification (125), shared promoter motifs (126), protein-protein interaction data (127), and protein-DNA binding data (128). In the end, all these diverse genomic datasets need to be integrated in a proper way for an accurate reconstruction of biomolecular networks.

APPROACHES FOR ANALYZING LARGE NETWORKS OF INTERACTIONS

Once molecular networks have been reconstructed, we can then proceed to compare and contrast them in terms of global and local topology and to relate structural properties of networks to protein properties, such as function or essentiality. These topics, generally termed network analysis, are reviewed here.

Network Topology

The classical random network theory, introduced by Erdös and Rényi (10, 129), has been generally used to model complex networks. This model assumes that each node in a network is connected to another node randomly with probability p, and the degrees of the nodes follow a Poisson distribution, which has a strong peak at the average degree, K. Most random networks are highly homogeneous in that most nodes have the same number of links (degree), $k_i \approx K$, where k_i is the degree of the ith node. The chance of having nodes with k links falls off exponentially [i.e., $P(k) \approx e^{-k}$] for large k.

To explain the heterogeneous nature of complex networks, Barabási and colleagues (5) recently proposed a "scale-free" model in which the degree distribution in many large networks follows a power-law distribution [$P(k) \approx k^{-r}$]. The most remarkable point about this distribution is that most of the nodes within these networks have very few links, but a few (the hubs) are exceptionally highly connected. Concurrently, Watts & Strogatz (3) found that many networks also have a "small-world" property, meaning they are defined as being both highly clustered and containing small characteristic path lengths.

The relevance of such structures is apparent in multiple disciplines. A recent practical example is the North American power grid structure. Although haphazardly constructed, the grid has evolved into a network that is defined by a power law; most nodes in the grid are linked to few other nodes, yet some hubs are highly connected to many other nodes. The power law adds a level of robustness to the network, meaning many individual nodes can fail without destroying the whole grid. Conversely, when several hubs or too many nodes (130) fail, the entire network will collapse, and a large regional blackout will result. This example highlights two important characteristics of networks that are pertinent to the protein interaction network. First, networks often evolve naturally into power law networks—hubs evolve naturally from nodes because of the inherent characteristics of the original node, i.e., its importance, fitness, or relative age within the network (10). Second, power-law networks are robust: A network defined by a power law has an inherent design that makes it less susceptible to random destabilizing events. The small-world concept is important in social networks, connecting multiple and otherwise unassociated cliques, which cause the networks to have a higher than otherwise logically suspected degree of clustering (the so-called six degrees of separation). Similarly, protein networks are often found to adhere to the small-world property, primarily due to the interconnectivity of a select group of nodes.

Topological analysis of these networks provides quantitative insight into their basic organization. Generally, there are four topological parameters in network analysis (Figure 3) (3–5, 7–11, 19):

- Average degree (K). The degree of a node is the number of links that this node has with other nodes. The average degree of the whole network is the average of the degree of all its individual nodes.

- Clustering coefficient (C), defined as the ratio of the number of existing links between the neighbors of a node and the maximum possible number of links between them. The clustering coefficient of the network is the average of the individual coefficients. This statistic can be used to determine the completeness of the network.

- Characteristic path length (L). The graph theoretical distance between two nodes is the minimum number of edges that is necessary to traverse from one node to the other. The characteristic path length of a network is the average of these minimum distances: It gives a measure of how close nodes are connected within the network.

- The diameter (D) of a network is the longest graph theoretical distance between any two nodes in the graph.

Networks can be divided into two broad categories: directed and undirected. Physical interaction, genetic interaction, and expression networks are "undirected," meaning no directionality or causality is implied in the interactions. Stated differently, "node A is linked to node B" is equivalent to "node B is linked to node A." These undirected networks should be sharply distinguished from other biological networks, such as regulatory networks, metabolic networks, and signaling networks; these "directed" networks do imply directionality in their linkages. A node in the directed network may have an incoming degree and an outgoing degree (see Figure 3C), which are completely independent. The incoming degree of a node is the number of edges pointing toward this node, whereas its outgoing degree is the number of edges pointing out of this node. The clustering coefficient cannot be calculated for directed networks (10).

Substructures Within Networks

Complex networks, such as protein-protein physical interaction networks (herein referred to simply as interaction networks) and regulatory networks, usually contain biologically meaningful substructures. Within interaction networks, protein complexes will theoretically appear as a clique, a fully connected subgraph. However, because of the limitation of the interaction identification techniques, some of the links within the same complex may be missing. Therefore, in reality, most complexes are quasi cliques within the interaction networks (131).

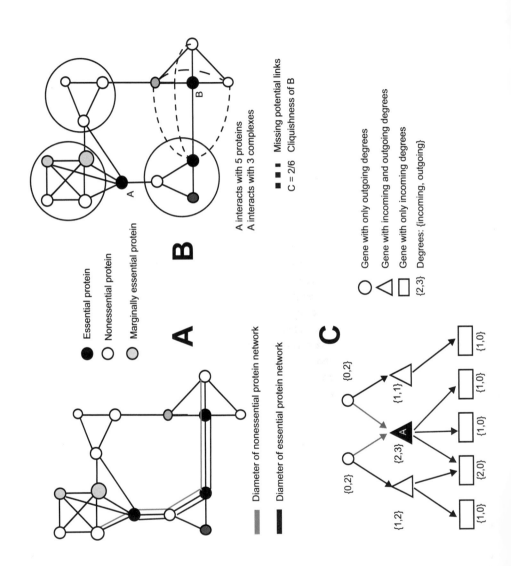

A

Diameter of nonessential protein network
Diameter of essential protein network

- ● Essential protein
- ○ Nonessential protein
- ◐ Marginally essential protein

B

A interacts with 5 proteins
A interacts with 3 complexes

■ ■ ■ Missing potential links
C = 2/6   Cliquishness of B

C

○ Gene with only outgoing degrees
△ Gene with incoming and outgoing degrees
▢ Gene with only incoming degrees
{2,3} Degrees: {incoming, outgoing}

{0,2}   {0,2}

{1,1}

{1,2}

{2,3} A

{1,0}   {1,0}   {1,0}   {2,0}   {1,0}

Compared with yeast two-hybrid methods, in vivo pull-down methods detect protein complexes, rather than binary protein-protein interactions. In order to break down the complexes into binary interaction pairs, Bader & Hogue (132) proposed two models: spoke and matrix. The "spoke" model assumes that only the bait proteins directly interact with each component of the complex. The "matrix" model assumes that each component interacts with all other components in the same complex. An important assumption in extrapolating gene function is "guilt by association," i.e., two interacting proteins should share the same function. In the paper, the authors were able to show that interacting pairs produced by the spoke model are more likely to share common functions than those produced by the matrix model. In addition, Bader & Hogue proposed a new method to determine protein complexes within interaction networks, termed "k-core." A k-core is a graph of minimal degree k. To date, high-throughput interaction identification methods, such as yeast two-hybrid methods and in vivo pull-down methods, all have high false-positive rates. However, within a k-core, links between proteins are strengthened by one another as a joint probability, which largely increases the accuracy of the predicted interactions. Using k-core method, Bader & Hogue were able to identify many well-known complexes, as well as some novel but reasonable ones, within the yeast interaction networks (132).

As discussed above, regulatory networks are different from interaction networks in that regulatory networks are directed networks. There are six basic network motifs within regulatory networks (Figure 4) (17, 133, 134): (*a*) single-input motif, in which a set of targets are regulated by only one transcription factor; (*b*) multi-input motif, where a set of targets are regulated by more than one regulator, and these are the only regulators for these targets; (*c*) feed-forward loop, where one transcription factor (TF1) regulates another transcription factor (TF2), and both factors regulate their targets together; (*d*) autoregulation, in which one transcription factor regulates itself; (*e*) multicomponent loop, where

---

Figure 3   Schematic illustration of different topological parameters within undirected and directed networks. (*A*) In an undirected network, the diameter of the essential protein network (shown as the *red line*) is the maximum distance between any two essential proteins. The path can go through nonessential proteins, but it has to start and end at essential ones. The diameter of the nonessential protein network can be similarly defined. (*B*) Protein B has four interaction partners, among which there are only two connections, whereas there could potentially be six (shown as the *dotted lines*). Therefore, the clustering coefficient for B is 1/3. Protein A interacts with five different proteins, which belong to three different complexes. Therefore, the complex degree of A is 3. In (*A*) and (*B*), there are essential proteins (*black circles*), nonessential proteins (*open circles*), and marginally essential proteins (*gray circles*). The size of the circle represents the degree of the node. (*C*) The outgoing degree of transcription factor A is 3 and its incoming degree is 2. Genes without transcription factor activities (shown as *rectangles*) only have incoming degrees.

1. Single-input motif (SIM)   **2. Multi-input motif (MIM)**   **3. Feed-forward loop (FFL)**

4. Autoregulation (Auto)   **5. Multicomponent loop (MCL)**   **6. Regulator chain (RC)**

Figure 4   Depiction of the six basic regulatory motifs. The figure shows transcription factors (*circles*) and targets (*rectangles*). Detailed descriptions are given in the section entitled "Substructures Within Networks."

one factor regulates a second factor, which in turn regulates the first one; and (*f*) regulator chain, which for several regulators, one regulates another in a chain fashion.

Application of Topological Analysis

The ultimate goal of functional genomics is to determine the function of every gene product in fully sequenced genomes. Different prediction schemes have been proposed, such as the concept that coexpressed genes share similar functions or that interacting proteins have the same functions. Given the relative ease with which large-scale protein-protein interaction datasets can now be produced, functional genomics relies on interaction data to determine the function of an unclassified protein based on its interacting partners. Traditionally, pairs of interacting proteins have been thought to share similar functions (25, 26); but because proteins normally interact with more than one partner, and the interacting partners for the same protein do not generally share the same functions, this idea is clearly problematic. A better method, known as the "majority rule" method (135, 136), assigns an unknown protein to the functional class to which the majority of its partners belong. Obviously, the method is still inefficient because only a small portion of the genes in fully sequenced genomes have functional annotations, and the functional assignment of an unknown protein will affect the assignment of its interaction partners, which will in turn affect the assignment of this protein itself. This circular reasoning serves to amplify possible errors in function prediction. Therefore, in order to efficiently predict protein functions based on interaction networks, the global topological structures of the networks have to be taken into consideration. Here, two such methods will be discussed in detail.

Bu et al. (131) introduced a method to determine quasi cliques within interaction networks using spectral analysis. These quasi cliques were proven to be biologically relevant functional groups; this is similar to the concept of protein complexes discussed above. In order to perform spectral analysis, an interaction network is represented by an $N \times N$ adjacency matrix. N is the total number of proteins within the network. The adjacency matrix is defined as $A = (a_{ij})$, where $a_{ij} = 1$ if protein i interacts with protein j, and $a_{ij} = 0$ if not. For each eigenvector of the matrix with a positive eigenvalue, the proteins corresponding to absolutely larger components tend to form a quasi clique, meaning every two of them tend to interact with each other. The quasi cliques were defined based on the following criteria: (*a*) each quasi clique must contain at least 10 proteins; (*b*) the proteins were sorted by their absolute weight value in an eigenvector, and the top 10% were selected; and (*c*) each protein in a quasi clique must interact with at least 20% of the clique members. The clustering coefficient of each quasi clique was tuned for a high degree of interconnectivity. Within yeast interaction networks, 48 quasi cliques were successfully identified. The proteins of each quasi clique indeed tend to share common functions based on MIPS functional classification (131).

Concurrently, Vazquez et al. (137) proposed a global optimization method to predict functions of unknown proteins within the interaction networks. Simply put, the global optimization method computes a score for any particular configuration of the functional assignment for the whole protein interaction network. The score is lower if fewer interacting pairs are assigned to distinct functional classes. The method aims to find the configuration of functional assignment with the lowest possible score through global optimization. Because there can be more than one optimal solution for this kind of problem, a "simulated annealing" technique was introduced to determine the optimal configurations, and the most frequent functional assignment for a certain protein in all optimal solutions was assigned as its function. In order to evaluate the success rate, a fraction of classified proteins were considered as unclassified in the input data; for proteins with more than one interacting partner, the performance of the global optimization method is much better than that of the majority rule method (137).

Cross-referencing Different Networks

So far, we have discussed many distinct types of networks. The fact that such networks exist in all species means that the total number of different networks is far larger. It is impossible to investigate every network of every organism in equal detail; however, in model organisms, particularly *Saccharomyces cerevisiae*, a vast amount of data has been accumulated for all these types of networks. Mapping the networks in model organisms to other species by homology provides insight into how to exploit the usefulness (and prevent the potential pitfalls) of annotating unknown genes in other less characterized species. To this end, Walhout et al. (31) introduced the concept of "interolog" to transfer interaction networks from one species to another. Interologs are defined as orthologous pairs of interacting proteins in different organisms. Thus if interacting proteins X and Y in one organism have interacting orthologs X' and Y' in another species, the X-Y and X'-Y' interactions are called interologs (31). Subsequently, based on 216 worm protein pairs and 72 yeast protein pairs, Matthews et al. (138) experimentally estimated the accuracy of this interolog method to be in the range of 16% to 31% for two species that are evolutionarily distant by about 900 million years.

Cross-species comparison of interaction networks tells us how these networks evolve. Similarly, comparison of different networks within the same organism often sheds light on the basic organization principles of the cell. Yu et al. analyzed the regulatory and expression networks for *S. cerevisiae* and were able to show that coregulated genes are generally coexpressed, and the correlation in expression profiles is highest for genes targeted by multiple transcription factors. Furthermore, coregulated gene pairs tend to share cellular functions, and there are subdivisions within individual network motifs that separate the regulation of genes of distinct functions. The expression profiles of transcription factors and their target genes display more complex relationships than simple correlation, with the regulatory response of target genes often being delayed (57).

INTERACTION NETWORKS AND SYSTEMS BIOLOGY

Mapping and understanding molecular interaction networks represent the first steps toward modeling how a cell actually operates in time and space. As a result of genome-wide high-throughput experiments, we are now generating a comprehensive parts list of functional elements for many genomes, and soon we will also have a comprehensive catalog of how these functional elements interact with and regulate each other. The grand challenge then will be to put all the pieces back together to create predictive models of cellular behavior. Such is the goal of systems biology, an emerging field that quantitatively measures and models the behavior of a cell from a systems perspective, which is a result of the collective spatial-temporal dynamics of its interacting components (139, 140). Here we briefly review some of the challenges and initial successes of using interaction networks to model cellular behavior.

The first challenge is to create a 3D view of molecular interaction networks in a cell. This is important because biomolecules are 3D objects; they function and interact through spatially precise atomic interactions in crowded microenvironments. The second challenge is to capture the dynamic and context-dependent nature of interaction networks. In addition to mapping out all possible interactions, it is also important to know under which conditions (cellular state, environment type, and protein modification type) each interaction is present in a cell. Finally, the third challenge is to quantitatively measure interaction networks. Interaction networks tell us whether or not two molecules interact. In order to model cellular behavior, it is also important to know how strongly and how quickly they interact. Such requirements call for continuing improvements in quantitative high-throughput methods.

Despite these considerable challenges, a number of recent modeling successes indicate a promising future for systems biology. For well-characterized interaction networks with a small number of genes and proteins, it is possible to build a detailed kinetic model of the system, and simulations are generally in good agreement with experiments. Such systems include, among others, bacteriophages (141), bacteria chemotaxis (142), circadian clocks (143), and signaling pathways (144). Available software could potentially scale up these detailed simulations to the level of an entire cell (145, 146), but the predictive power of the whole-cell simulations is limited by the fact that the vast majority of underlying kinetic parameters remain unknown.

Alternatively, it is possible to model the behavior of interaction networks in a coarse manner without knowing detailed kinetic parameters. Such modeling can be applied on a genomic scale or on less well-characterized systems because it requires only a few parameters beyond the topology and stoichiometry of the network. A prime example is the flux balance analysis of metabolic networks (147), where the steady-state behavior of the entire network can be modeled reasonably well. Similarly, much of the logic and dynamics of a bacteria cell-cycle regulatory network can be understood and possibly modeled without a full set of kinetic parameters (148).

These modeling efforts are providing us with an increasing number of insights into the design principles of biomolecular networks. For example, biomolecular networks can be grouped into modules (1). Functional elements within a module interact strongly with each other and carry out a common function in a concerted fashion. Biomolecular networks are resilient toward common external and internal perturbations (149). Furthermore, noise is an integral part of the functioning of biomolecular networks (150).

In summary, molecular interaction networks are at the core of current functional genomic research. These networks represent an appealing framework upon which different genomic data can be integrated, and analysis of these networks has yielded the first clues about their organizational and design principles. Furthermore, these networks lay the foundation for systems biology analysis of the cell. With combined experimental, computational, and theoretical efforts, a complete mapping of all interaction networks, and ultimately a rational understanding of cellular behavior, will become a reality.

ACKNOWLEDGMENTS

We thank Anuj Kumar for reading the experimental section of the draft and reference selection. R. Jansen is supported by the Seraph Foundation. H. Zhao is supported by NSF grant 0241160. This work is supported by a grant from NIH/NIGMS (GM054160-07).

APPENDIX

Details on Using Bayesian Networks for Integrating Interaction Datasets

Given multiple experimental results e_i (from N different experiments, with $i = 1 \ldots N$), the posterior odds of a protein-protein interaction can be computed as follows with a naive Bayesian network:

$$O_{post} = \prod_{i=1}^{N} L_i(e_i) O_{prior} \qquad 1.$$

Here, O_{post} is defined as

$$O_{post} = \frac{P(I = +|e_1, e_2 \ldots e_N)}{P(I = -|e_1, e_2 \ldots e_N)} = \frac{P(I = +|e_1, e_2 \ldots e_N)}{1 - P(I = +|e_1, e_2 \ldots e_N)} \qquad 2.$$

whereas O_{prior} is

$$O_{prior} = \frac{P(I = +)}{P(I = -)} = \frac{P(I = +)}{1 - P(I = +)}. \qquad 3.$$

Thus the posterior odds describe the odds of having a protein-protein interaction ($I = +$) given that we have the information from the N experiments, whereas the prior odds are related to the chance of randomly finding a protein-protein interaction when no experimental data are known. If $O_{post} > 1$, the chances of having an interaction are higher than the chances of having no interaction. For the RNA polymerase II example given in the main text, the prior odds were set to $13/(45–13) \approx 0.41$, i.e., the ratio of protein pairs observed to be in contact in the crystal structure of RNA polymerase II divided by the remaining protein pairs, but they could also be determined by counting the number of protein-protein interactions in comparable protein structures.

$L_i(e_i)$ describes the likelihood ratio of the experimental result e_i, and it can be computed from the table in Figure 1 as follows:

$$L_i(e_i = +1) = \frac{TP_i}{TP_i + FN_i} \frac{FP_i + TN_i}{FP_i} \qquad 4.$$

and

$$L_i(e_i = -1) = \frac{FN_i}{TP_i + FN_i} \frac{FP_i + TN_i}{TN_i} \qquad 5.$$

where the subscript i refers to a particular column in Figure 1. (We assume here for simplicity that an experiment either has a positive or a negative result, i.e., $e_i = \pm1$). For a perfect experiment with no errors, one would observe $FP_i \rightarrow 0$ and $FN_i \rightarrow 0$, such that $L(e_i = +1) \rightarrow \infty$ and $L(e_i = -1) \rightarrow 0$.

The naive Bayes procedure can be intuitively understood by comparing it to the voting procedure. In the voting procedure the experimental results are simply added up:

$$s = \sum_{i=1}^{N} e_i . \qquad 6.$$

Then, when $s \geq 0$, we consider the protein pair to be interacting (and noninteracting otherwise). Note that all experiments are weighted equally. In contrast, the naive Bayes procedure weights each experiment differently based on the likelihood ratio values. The analogy to a weighted voting procedure can be seen if we take the logarithm of Equation 1:

$$\log(O_{post}) = \sum_{i=1}^{N} \log(L_i(e_i)) + \log(O_{prior}) . \qquad 7.$$

Here, a protein pair is considered to be interacting if $\log(O_{post}) \geq 0$, whereas the term $\log(L_i(e_i))$ corresponds to the weight of experiment e_i. The difference with the voting procedure is the inclusion of the term $\log(O_{prior})$, which represents the chance of randomly finding a protein-protein interaction without experimental information.

The *Annual Review of Biochemistry* is online at http://biochem.annualreviews.org

LITERATURE CITED

1. Hartwell LH, Hopfield JJ, Leibler S, Murray AW. 1999. *Nature* 402:C47–52
2. Eisenberg D, Marcotte EM, Xenarios I, Yeates TO. 2000. *Nature* 405:823–26
3. Watts DJ, Strogatz SH. 1998. *Nature* 393: 440–42
4. Albert R, Jeong H, Barabási A-L. 1999. *Nature* 401:130–31
5. Barabási A-L, Albert R. 1999. *Science* 286:509–12
6. Huberman BA, Adamic LA. 1999. *Nature* 401:131
7. Albert R, Jeong H, Barabasi AL. 2000. *Nature* 406:378–82
8. Amaral LA, Scala A, Barthelemy M, Stanley HE. 2000. *Proc. Natl. Acad. Sci. USA* 97:11149–52
9. Jeong H, Mason SP, Barábasi AL, Oltvai ZN. 2001. *Nature* 411:41–42
10. Albert R, Barabási AL. 2002. *Rev. Mod. Phys.* 74:47–97
11. Girvan M, Newman ME. 2002. *Proc. Natl. Acad. Sci. USA* 99:7821–26
12. Lan N, Montelione GT, Gerstein M. 2003. *Curr. Opin. Chem. Biol.* 7:44–54
13. Tong AH, Evangelista M, Parsons AB, Xu H, Bader GD, et al. 2001. *Science* 294: 2364–68

14. Eisen MB, Spellman PT, Brown PO, Botstein D. 1998. *Proc. Natl. Acad. Sci. USA* 95:14863–68

15. Altman RB, Raychaudhuri S. 2001. *Curr. Opin. Struct. Biol.* 11:340–47

16. Qian J, Dolled-Filhart M, Lin J, Yu H, Gerstein M. 2001. *J. Mol. Biol.* 314: 1053–66

17. Lee TI, Rinaldi NJ, Robert F, Odom DT, Bar-Joseph Z, et al. 2002. *Science* 298: 799–804

18. Horak CE, Luscombe NM, Qian J, Bertone P, Piccirrillo S, et al. 2002. *Genes Dev.* 16:3017–33

19. Jeong H, Tombor B, Albert R, Oltvai ZN, Barabási AL. 2000. *Nature* 407: 651–54

20. Pawson T, Scott JD. 1997. *Science* 278: 2075–80

21. Sambrano GR, Chandy G, Choi S, Decamp D, Hsueh R, et al. 2002. *Nature* 420:708–10

22. Fields S, Song O. 1989. *Nature* 340: 245–46

23. Walhout AJ, Vidal M. 2001. *Nat. Rev. Mol. Cell. Biol.* 2:55–62

24. Fromont-Racine M, Rain JC, Legrain P. 1997. *Nat. Genet.* 16:277–82

25. Uetz P, Giot L, Cagney G, Mansfield TA, Judson RS, et al. 2000. *Nature* 403: 623–27

26. Ito T, Tashiro K, Muta S, Ozawa R, Chiba T, et al. 2000. *Proc. Natl. Acad. Sci. USA* 97:1143–47

27. Ito T, Chiba T, Ozawa R, Yoshida M, Hattori M, Sakaki Y. 2001. *Proc. Natl. Acad. Sci. USA* 98:4569–74

28. von Mering C, Krause R, Snel B, Cornell M, Oliver SG, et al. 2002. *Nature* 417: 399–403

29. Edwards AM, Kus B, Jansen R, Greenbaum D, Greenblatt J, Gerstein M. 2002. *Trends Genet.* 18:529–36

30. Drees BL, Sundin B, Brazeau E, Caviston JP, Chen GC, et al. 2001. *J. Cell Biol.* 154:549–71

31. Walhout AJ, Sordella R, Lu X, Hartley JL, Temple GF, et al. 2000. *Science* 287: 116–22

32. Boulton SJ, Gartner A, Reboul J, Vaglio P, Dyson N, et al. 2002. *Science* 295: 127–31

33. Aebersold R, Mann M. 2003. *Nature* 422: 198–207

34. Ashman K, Moran MF, Sicheri F, Pawson T, Tyers M. 2001. *Science STKE* 2001: PE33

35. Kumar A, Snyder M. 2002. *Nature* 415: 123–24

36. Rigaut G, Shevchenko A, Rutz B, Wilm M, Mann M, Seraphin B. 1999. *Nat. Biotechnol.* 17:1030–32

37. Gavin AC, Bosche M, Krause R, Grandi P, Marzioch M, et al. 2002. *Nature* 415: 141–47

38. Ho Y, Gruhler A, Heilbut A, Bader GD, Moore L, et al. 2002. *Nature* 415: 180–83

39. Arenkov P, Kukhtin A, Gemmell A, Voloshchuk S, Chupeeva V, Mirzabekov A. 2000. *Anal. Biochem.* 278:123–31

40. MacBeath G, Schreiber SL. 2000. *Science* 289:1760–63

41. Zhu H, Klemic JF, Chang S, Bertone P, Casamayor A, et al. 2000. *Nat. Genet.* 26:283–89

42. Zhu H, Bilgin M, Bangham R, Hall D, Casamayor A, et al. 2001. *Science* 293: 2101–5

43. Sali A, Glaeser R, Earnest T, Baumeister W. 2003. *Nature* 422:216–25

44. Bender A, Pringle JR. 1991. *Mol. Cell. Biol.* 11:1295–305

45. Wang T, Bretscher A. 1997. *Genetics* 147: 1595–607

46. Mullen JR, Kaliraman V, Ibrahim SS, Brill SJ. 2001. *Genetics* 157:103–18

47. Garner MM, Revzin A. 1981. *Nucleic Acids Res.* 9:3047–60

48. Seguin C, Hamer DH. 1987. *Science* 235: 1383–87

49. Fraga MF, Uriol E, Diego LB, Berdasco M, Esteller M, et al. 2002. *Electrophoresis* 23:1677–81

50. Galas DJ, Schmitz A. 1978. *Nucleic Acids Res.* 5:3157–70

51. Kuo MH, Allis CD. 1999. *Methods* 19: 425–33

52. Simpson RT. 1999. *Curr. Opin. Genet. Dev.* 9:225–29

53. Iyer VR, Horak CE, Scafe CS, Botstein D, Snyder M, Brown PO. 2001. *Nature* 409:533–38

54. Sun LV, Chen L, Greil F, Negre N, Li TR, et al. 2003. *Proc. Natl. Acad. Sci. USA* 100:9428–33

55. van Steensel B, Henikoff S. 2000. *Nat. Biotechnol.* 18:424–28

56. van Steensel B, Delrow J, Henikoff S. 2001. *Nat. Genet.* 27:304–8

57. Yu H, Luscombe NM, Qian J, Gerstein M. 2003. *Trends Genet.* 19:422–27

58. Xenarios I, Salwinski L, Duan XJ, Higney P, Kim SM, Eisenberg D. 2002. *Nucleic Acids Res.* 30:303–5

59. Bader GD, Betel D, Hogue CW. 2003. *Nucleic Acids Res.* 31:248–50

60. Mewes HW, Frishman D, Guldener U, Mannhaupt G, Mayer K, et al. 2002. *Nucleic Acids Res.* 30:31–34

61. Csank C, Costanzo MC, Hirschman J, Hodges P, Kranz JE, et al. 2002. *Methods Enzymol.* 350:347–73

62. Wingender E, Chen X, Fricke E, Geffers R, Hehl R, et al. 2001. *Nucleic Acids Res.* 29:281–83

63. Salgado H, Santos-Zavaleta A, Gama-Castro S, Millan-Zarate D, Diaz-Peredo E, et al. 2001. *Nucleic Acids Res.* 29: 72–74

64. Kanehisa M, Goto S. 2000. *Nucleic Acids Res.* 28:27–30

65. Karp PD, Riley M, Paley SM, Pellegrini-Toole A. 2002. *Nucleic Acids Res.* 30: 59–61

66. Gilman AG, Simon MI, Bourne HR, Harris BA, Long R, et al. 2002. *Nature* 420:703–6

67. Peri S, Navarro JD, Amanchy R, Kristiansen TZ, Jonnalagadda CK, et al. 2003. *Genome Res.* 13:2363–71

68. Tamames J, Casari G, Ouzounis C, Valencia A. 1997. *J. Mol. Evol.* 44:66–73

69. Overbeek R, Fonstein M, D'Souza M, Pusch GD, Maltsev N. 1999. *Proc. Natl. Acad. Sci. USA* 96:2896–901

70. Yanai I, Mellor JC, DeLisi C. 2002. *Trends Genet.* 18:176–79

71. Dandekar T, Snel B, Huynen M, Bork P. 1998. *Trends Biochem. Sci.* 23:324–28

72. Marcotte EM, Pellegrini M, Ng HL, Rice DW, Yeates TO, Eisenberg D. 1999. *Science* 285:751–53

73. Enright AJ, Iliopoulos I, Kyrpides NC, Ouzounis CA. 1999. *Nature* 402:86–90

74. Yanai I, Derti A, DeLisi C. 2001. *Proc. Natl. Acad. Sci. USA* 98:7940–45

75. Tatusov RL, Koonin EV, Lipman DJ. 1997. *Science* 278:631–37

76. Gaasterland T, Ragan MA. 1998. *Microb. Comp. Genomics* 3:199–217

77. Pellegrini M, Marcotte EM, Thompson MJ, Eisenberg D, Yeates TO. 1999. *Proc. Natl. Acad. Sci. USA* 96:4285–88

78. Goh CS, Bogan AA, Joachimiak M, Walther D, Cohen FE. 2000. *J. Mol. Biol.* 299:283–93

79. Goh CS, Cohen FE. 2002. *J. Mol. Biol.* 324:177–92

80. Pazos F, Valencia A. 2001. *Protein Eng.* 14:609–14

81. Pazos F, Valencia A. 2002. *Proteins* 47: 219–27

82. Sprinzak E, Margalit H. 2001. *J. Mol. Biol.* 311:681–92

83. Park J, Lappe M, Teichmann SA. 2001. *J. Mol. Biol.* 307:929–38

84. Bock JR, Gough DA. 2001. *Bioinformatics* 17:455–60

85. Janin J, Henrick K, Moult J, Eyck LT, Sternberg MJ, et al. 2003. *Proteins* 52:2–9

86. Lu L, Arakaki AK, Lu H, Skolnick J. 2003. *Genome Res.* 13:1146–54

87. Marcotte EM, Xenarios I, Eisenberg D. 2001. *Bioinformatics* 17:359–63

88. Stapley BJ, Benoit G. 2000. *Pac. Symp. Biocomput.,* pp. 529–40. Singapore: World Sci. Publ.

89. Friedman C, Kra P, Yu H, Krauthammer

M, Rzhetsky A. 2001. *Bioinformatics* 17(Suppl. 1): S74–82

90. Hirschman L, Park JC, Tsujii J, Wong L, Wu CH. 2002. *Bioinformatics* 18: 1553–61

91. Blaschke C, Hirschman L, Valencia A. 2002. *Brief. Bioinform.* 3:154–65

92. Jansen R, Greenbaum D, Gerstein M. 2002. *Genome Res.* 12:37–46

93. Greenbaum D, Colangelo C, Williams K, Gerstein M. 2003. *Genome Biol.* 4:117

94. Kemmeren P, van Berkum NL, Vilo J, Bijma T, Donders R, et al. 2002. *Mol. Cell* 9:1133–43

95. Ge H, Liu Z, Church GM, Vidal M. 2001. *Nat. Genet.* 29:482–86

96. Ross-Macdonald P, Coelho PS, Roemer T, Agarwal S, Kumar A, et al. 1999. *Nature* 402:413–18

97. Giaever G, Chu AM, Ni L, Connelly C, Riles L, et al. 2002. *Nature* 418:387–91

98. Kumar A, Agarwal S, Heyman JA, Matson S, Heidtman M, et al. 2002. *Genes Dev.* 16:707–19

99. Gerstein M, Lan N, Jansen R. 2002. *Science* 295:284–87

100. Tong AH, Drees B, Nardelli G, Bader GD, Brannetti B, et al. 2002. *Science* 295: 321–24

101. Marcotte EM, Pellegrini M, Thompson MJ, Yeates TO, Eisenberg D. 1999. *Nature* 402:83–86

102. Jansen R, Lan N, Qian J, Gerstein M. 2002. *J. Struct. Funct. Genomics* 2: 71–81

103. Acker J, de Graaff M, Cheynel I, Khazak V, Kedinger C, Vigneron M. 1997. *J. Biol. Chem.* 272:16815–21

104. Kimura M, Ishihama A. 2000. *Nucleic Acids Res.* 28:952–59

105. Ulmasov T, Larkin RM, Guilfoyle TJ. 1996. *J. Biol. Chem.* 271:5085–94

106. Miyao T, Yasui K, Sakurai H, Yamagishi M, Ishihama A. 1996. *Genes Cells* 1:843–54

107. Ishiguro A, Kimura M, Yasui K, Iwata A, Ueda S, Ishihama A. 1998. *J. Mol. Biol.* 279:703–12

108. Drawid A, Gerstein M. 2000. *J. Mol. Biol.* 301:1059–75

109. Pavlovic V, Garg A, Kasif S. 2002. *Bioinformatics* 18:19–27

110. Troyanskaya OG, Dolinski K, Owen AB, Altman RB, Botstein D. 2003. *Proc. Natl. Acad. Sci. USA* 100:8348–53

111. Jansen R, Yu HY, Greenbaum D, Kluger Y, Krogan NJ, et al. 2003. *Science* 302: 449–53

112. Arkin A, Shen P, Ross J. 1997. *Science* 277:1275–79

113. Liang S, Fuhrman S, Somogyi R. 1998. *Pac. Symp. Biocomput.*, pp. 18–29. Singapore: World Sci. Publ.

114. Akutsu T, Miyano S, Kuhara S. 2000. *Bioinformatics* 16:727–34

115. Shmulevich I, Dougherty ER, Kim S, Zhang W. 2002. *Bioinformatics* 18: 261–74

116. Friedman N, Linial M, Nachman I, Pe'er D. 2000. *J. Comput. Biol.* 7:601–20

117. Hartemink AJ, Gifford DK, Jaakkola TS, Young RA. 2002. *Pac. Symp. Biocomput.*, pp. 437–49 Singapore: World Sci. Publ.

118. Hartemink AJ, Gifford DK, Jaakkola TS, Young RA. 2001. *Pac. Symp. Biocomput.*, pp. 422–33 Singapore: World Sci. Publ.

119. Heckerman D, Geiger D, Chickering DM. 1995. *Mach. Learn.* 20:197–243

120. Dean T, Kanazawa K. 1988. *Probabilistic temporal reasoning. Proc. Am. Assoc. Artif. Intell.*, St. Paul, MN, pp. 524–29. Menlo Park, CA: AAAI Press

121. Ideker T, Thorsson V, Ranish JA, Christmas R, Buhler J, et al. 2001. *Science* 292: 929–34

122. Arbeitman MN, Furlong EE, Imam F, Johnson E, Null BH, et al. 2002. *Science* 297:2270–75

123. Gardner TS, di Bernardo D, Lorenz D, Collins JJ. 2003. *Science* 301:102–5

124. Stuart JM, Segal E, Koller D, Kim SK. 2003. *Science* 302:249–55

125. Ihmels J, Friedlander G, Bergmann S, Sarig O, Ziv Y, Barkai N. 2002. *Nat. Genet.* 31:370–77

126. Pilpel Y, Sudarsanam P, Church GM. 2001. *Nat. Genet.* 29:153–59

127. Segal E, Wang H, Koller D. 2003. *Bioinformatics* 19(Suppl. 1):I264–72

128. Bar-Joseph Z, Gerber GK, Lee TI, Rinaldi NJ, Yoo JY, et al. 2003. *Nat. Biotechnol.* 21:1337–42

129. Erdös P, Rényi A. 1959. *Public Math.* 6:290–97

130. Cohen R, Erez K, ben-Avraham D, Havlin S. 2000. *Phys. Rev. Lett.* 85:4626–28

131. Bu D, Zhao Y, Cai L, Xue H, Zhu X, et al. 2003. *Nucleic Acids Res.* 31:2443–50

132. Bader GD, Hogue CW. 2002. *Nat. Biotechnol.* 20:991–97

133. Shen-Orr SS, Milo R, Mangan S, Alon U. 2002. *Nat. Genet.* 31:64–68

134. Milo R, Shen-Orr S, Itzkovitz S, Kashtan N, Chklovskii D, Alon U. 2002. *Science* 298:824–27

135. Schwikowski B, Uetz P, Fields S. 2000. *Nat. Biotechnol.* 18:1257–61

136. Hishigaki H, Nakai K, Ono T, Tanigami A, Takagi T. 2001. *Yeast* 18:523–31

137. Vazquez A, Flammini A, Maritan A, Vespignani A. 2003. *Nat. Biotechnol.* 21:697–700

138. Matthews LR, Vaglio P, Reboul J, Ge H, Davis BP, et al. 2001. *Genome Res.* 11:2120–26

139. Ideker T, Galitski T, Hood L. 2001. *Annu. Rev. Genomics Hum. Genet.* 2:343–72

140. Kitano H. 2002. *Science* 295:1662–64

141. Arkin A, Ross J, McAdams HH. 1998. *Genetics* 149:1633–48

142. Barkai N, Leibler S. 1997. *Nature* 387:913–17

143. Barkai N, Leibler S. 2000. *Nature* 403:267–68

144. Bhalla US, Iyengar R. 1999. *Science* 283:381–87

145. Takahashi K, Ishikawa N, Sadamoto Y, Sasamoto H, Ohta S, et al. 2003. *Bioinformatics* 19:1727–29

146. Loew LM, Schaff JC. 2001. *Trends Biotechnol.* 19:401–6

147. Edwards JS, Ibarra RU, Palsson BO. 2001. *Nat. Biotechnol.* 19:125–30

148. McAdams HH, Shapiro L. 2003. *Science* 301:1874–77

149. Alon U, Surette MG, Barkai N, Leibler S. 1999. *Nature* 397:168–71

150. McAdams HH, Arkin A. 1999. *Trends Genet.* 15:65–69

AUTHOR INDEX

SUBJECT INDEX

A

3AB
  telomerase regulation and, 194

Abruptex mutants
  developmental glycobiology and, 501

ACC deaminase
  pyridoxal 5'-phosphate enzymes and, 385

ACC synthase
  pyridoxal 5'-phosphate enzymes and, 410

Acetylation
  FLAP endonuclease 1 and, 607–9

Acetylcholine
  palmitoylation of intracellular signaling proteins and, 578

Acetyl-CoA transferase
  Sir2 family of protein deacetylases and, 424

p-Acetylphenylalanine
  incorporation of nonnatural amino acids into proteins and, 153

O-Acetylserine sulfhydrylase
  pyridoxal 5'-phosphate enzymes and, 384, 392–93

Acidic-glutamine-arginine triad
  GoLoco motif and, 929–34

α-Actinin
  lysophospholipid receptors and, 342

Actinobacillus spp.
  TolC protein and, 469

Actins

lysophospholipid receptors and, 341–42
mechanical processes in biochemistry and, 729
polymerization dynamics spatial and temporal regulation of, 209–35

Activators
  cytochrome *c*-mediated apoptosis and, 89–92, 97–99
  opioid receptors and, 963–64
  pyridoxal 5'-phosphate enzymes and, 408–10
  Sir2 family of protein deacetylases and, 417, 422–23
  SUMO in protein modification and, 357

Active sites
  DNA polymerase *γ* and, 303–4, 306–7
  ligand binding to/electron transfer in cytochrome P450, 991, 997–1003

Active transport
  ion pumping by Ca^{2+}-ATPase of sarcoplasmic reticulum, 269–90

Actomyosin
  lysophospholipid receptors and, 333

Actuator domain
  ion pumping by Ca^{2+}-ATPase of sarcoplasmic reticulum, 287

Acute promyelocytic leukemia (APL)

SUMO in protein modification and, 368–69

Acylation
  lipid-modified secreted protein signals and, 891, 905–6, 908

Acyl carrier protein
  ligand binding to/electron transfer in cytochrome P450, 1004

Acyl-CoA binding protein
  palmitoylation of intracellular signaling proteins and, 566

Acyl-CoA synthetase
  Sir2 family of protein deacetylases and, 425

Acyl phosphatase
  two-state folding rates and, 838, 846

Acyl-protein thioesterase 1 (APT1)
  palmitoylation of intracellular signaling proteins and, 559, 577–82

Acylthioesterases
  palmitoylation of intracellular signaling proteins and, 559

Acyltransferases
  lipid-modified secreted protein signals and, 908
  palmitoylation of intracellular signaling proteins and, 559, 564, 566–70, 582

ADAR1 protein
  history of research, 32–33

Addiction biology